Copyright © 1999, 1996, 1990, 1986 by Harcourt, Inc.

All rights reserved. No part of this publication may be reproduced or transmitted in any form or by any means, electronic or mechanical, including photocopy, recording, or any information storage and retrieval system, without permission in writing from the publisher.

Requests for permission to make copies of any part of the work should be mailed to the following address: Permissions Department, Harcourt, Inc., 6277 Sea Harbor Drive, Orlando, FL 32887-6777.

Publisher: Emily Barrosse
Publisher: John Vondeling
Acquisitions/Developmental Editor: Kent Porter Hamann
Product Manager: Pauline Mula
Project Editor: Frank Messina
Production Manager: Charlene Catlett Squibb
Art Director/Text Designer: Caroline McGowan
Cover Designer: Ruth Hoover

Cover Image and Credit: Hippuric acid as viewed in polarized light. (Copyright Alfred Pasieka/Peter Arnold, Inc.)

Frontmatter Photo Credits: p. vii: Potassium hydroxide flame (Charles D. Winters); p. xiii: Crystals of alum (Charles D. Winters); p. xv: Ice and water (Charles D. Winters).

Printed in the United States of America

Principles of Modern Chemistry, Fourth Edition
ISBN 0-03-024427-7

Library of Congress Catalog Card Number: 98-84384

1  2  3  4  5  6  7  8  9    048    15  14  13  12  11  10  9  8  7  6

## Physical Constants

| | |
|---|---|
| Standard acceleration of terrestrial gravity | $g = 9.80655 \text{ m s}^{-2}$ (exactly) |
| Avogadro's number | $N_0 = 6.022137 \times 10^{23}$ |
| Bohr radius | $a_0 = 0.52917725 \text{ Å} = 5.2917725 \times 10^{-11} \text{ m}$ |
| Boltzmann's constant | $k_B = 1.38066 \times 10^{-23} \text{ J K}^{-1}$ |
| Electron charge | $e = 1.6021773 \times 10^{-19} \text{ C}$ |
| Faraday constant | $\mathscr{F} = 96{,}485.31 \text{ C mol}^{-1}$ |
| Masses of fundamental particles: | |
| Electron | $m_e = 9.109390 \times 10^{-31} \text{ kg}$ |
| Proton | $m_p = 1.672623 \times 10^{-27} \text{ kg}$ |
| Neutron | $m_n = 1.674929 \times 10^{-27} \text{ kg}$ |
| Ratio of proton mass to electron mass | $m_p/m_e = 1836.15270$ |
| Permittivity of vacuum | $\epsilon_0 = 8.8541878 \times 10^{-12} \text{ C}^2 \text{ J}^{-1} \text{ m}^{-1}$ |
| Planck's constant | $h = 6.626076 \times 10^{-34} \text{ J s}$ |
| Speed of light in a vacuum | $c = 2.99792458 \times 10^8 \text{ m s}^{-1}$ (exactly) |
| Universal gas constant | $R = 8.31451 \text{ J mol}^{-1} \text{ K}^{-1}$ |
| | $= 0.0820578 \text{ L atm mol}^{-1} \text{ K}^{-1}$ |

Values are taken from *Quantities, Units and Symbols in Physical Chemistry*, International Union of Pure and Applied Chemistry, Blackwell Scientific Publications, 1988.

## Conversion Factors

| | |
|---|---|
| Standard atmosphere | $1 \text{ atm} = 1.01325 \times 10^5 \text{ Pa}$ |
| | $= 1.01325 \times 10^5 \text{ kg m}^{-1} \text{ s}^{-2}$ (exactly) |
| Atomic mass unit | $1 \text{ u} = 1.660540 \times 10^{-27} \text{ kg}$ |
| | $1 \text{ u} = 1.492419 \times 10^{-10} \text{ J} = 931.4943 \text{ MeV}$ (energy equivalent from $E = mc^2$) |
| Calorie | $1 \text{ cal} = 4.184 \text{ J}$ (exactly) |
| Electron volt | $1 \text{ eV} = 1.6021773 \times 10^{-19} \text{ J}$ |
| | $= 96.48531 \text{ kJ mol}^{-1}$ |
| Foot | $1 \text{ ft} = 12 \text{ in} = 0.3048 \text{ m}$ (exactly) |
| Gallon (U.S.) | $1 \text{ gallon} = 4 \text{ quarts} = 3.78541 \text{ L}$ (exactly) |
| Liter-atmosphere | $1 \text{ L atm} = 101.325 \text{ J}$ (exactly) |
| Metric ton | $1 \text{ metric ton} = 1000 \text{ kg}$ (exactly) |
| Pound | $1 \text{ lb} = 16 \text{ oz} = 0.45359237 \text{ kg}$ (exactly) |

PRINCIPLES OF

# MODERN CHEMISTRY

FOURTH EDITION

To the memory of Norman Nachtrieb, our teacher, colleague, and friend, whose many contributions fill this book and whose spirit continues to inspire us.

*The search for truth is in one way hard and in another easy,*
*for it is evident that no one can master it fully or miss it completely.*
*But each adds a little to our knowledge of nature, and from all*
*the facts assembled there arises a certain grandeur.*

(Greek inscription, taken from Aristotle, on the facade of
the National Academy of Sciences building in Washington, D.C.)

# Preface

The fourth edition of *Principles of Modern Chemistry* continues in its tradition of addressing students in honors and high-mainstream general chemistry courses. The new edition represents a substantial revision, providing an early introduction to molecular structure and emphasizing structural arguments in the presentation of the remaining topics. The connection of molecular structure with function and with properties is an approach that pervades contemporary research in chemistry, from molecular biology to materials science. The new organization is directed to the students of the future, who will learn to think in terms of molecular structure and who will be introduced to the great predictive power of computation and simulation, which complement more qualitative forms of chemical reasoning.

Potassium hydroxide flame.

## CHANGES IN THIS EDITION

The chapters in the fourth edition are organized within larger units to emphasize the conceptual structure of chemistry. Each unit begins with an overview that enumerates the pedagogical goals of the chapters comprising the unit and that describes the key concepts that are developed to meet these goals. Several changes in organization have been made in this edition: eight chapters have been extensively rewritten, and eleven have been moderately revised. The major changes are as follows:

- The material in Chapter 2 of the third edition, "Chemical Periodicity and Inorganic Reactions," has been completely reorganized. In the fourth edition, Chapter 3 provides a thorough introduction to chemical bonding and molecular structure based on the classical description, addressing ionization potential, electron affinity, electronegativity, Coulomb stabilization in ionic bonding, the Lewis electron pair model of covalent bonding, and the valence shell electron-pair repulsion (VSEPR) model. The surveys of reaction types have been moved to later chapters where a more thorough and unified discussion of each reaction type is provided in appropriate context.

- The treatment of intermolecular forces (Chapter 5) is significantly expanded. The nature and strength of these forces are related to molecular structure and are used to interpret macroscopic properties of substances.

- The treatment of the second law of thermodynamics (Chapter 8) now provides greater emphasis on the conceptual and molecular basis of the law. Calculations of entropy changes and predictions of spontaneity have been extended.

- The description of chemical equilibrium (Chapter 9) is now based on thermodynamics, which in this edition is discussed in Chapters 7 and 8. After a brief empirical introduction to equilibrium, the mass action law is obtained by minimizing the Gibbs free energy. With the proper choice of reference states for products and reactants, this approach guarantees that the equilibrium constant is a dimensionless quantity. Thermodynamics provides the basis for calculating equilibrium constants, the direction of change for an initial reaction mixture, and the temperature dependence of the equilibrium constant. Each of these concepts is illustrated with gas-phase applications. The subsequent chapters on acid–base, heterogeneous, and electrochemical equilibria build on this background in thermodynamics.

- The classification of acid–base behavior according to the Brønsted–Lowry and Lewis definitions (Chapter 10) receives a unified presentation. The Arrhenius classification is introduced earlier (Chapter 6) in the context of acid–base titrations.

- The presentation of quantum mechanics and atomic structure (Chapter 15) has been revised to provide experimental illustrations of quantal phenomena and magnitudes before the discussion of quantal concepts. For this purpose, the descriptions of blackbody radiation, the photoelectric effect, and photoelectron spectroscopy have been extended. Measurements of atomic excitation thresholds by electron energy loss methods (the Franck–Hertz experiment) have been introduced to demonstrate explicitly that energy is quantized in atoms. The discussion of shielding and penetration in periodic trends in atomic properties has been extended.

- Chapter 16 presents a unified theoretical and experimental treatment of quantum mechanics and molecular structure, combining the molecular orbital descriptions of Chapter 14 in the third edition with an expanded treatment of spectroscopy from Chapter 15 of the third edition. The description of optical spectroscopy now includes the Beer–Lambert Law and extinction coefficients; the description of nuclear magnetic resonance includes an introduction to the chemical shift and to spin-spin splitting. Both theoretical and experimental concepts are illustrated by selected topics in conjugated systems and atmospheric photochemistry.

- The descriptive material (Chapters 20 to 25) has been rearranged and grouped explicitly into two areas of focus: chemical processes and materials.

## TEACHING OPTIONS

As in the previous editions, an instructor may choose to alter the order in which the chapters are presented for a chemistry course; the text is structured to permit this kind of flexibility. For example, some instructors may wish to cover the material on the quantum theory of atoms and molecules earlier in their courses. In this case, they can move directly to Chapters 15 and 16 after Chapter 3. Others may wish to present the chapters on chemical equilibria (Chapters 9 to 11) earlier in the course in a more empirical fashion, before the presentation of thermodynamics. Certain topics may be omitted without loss of continuity. For example, a principles-oriented course might cover the first 19 chapters thoroughly and then select only one or two of the last chapters for close attention. A course with a more descriptive orientation might omit the sections entitled "A Deeper Look . . . ," which are more advanced mathematically than the sections in the main part of the book, and cover the last six chapters more systematically. Additional suggestions are given in the *Instructor's Manual*.

## FEATURES

### Mathematical Level

The book presupposes a solid high school background in algebra and coordinate geometry. The concepts of slope and area are introduced in the physical and chemical contexts in which they arise, and differential and integral notation is used only when necessary. The book is designed to be fully self-contained in its use of mathematical methods. In this context, Appendix C should prove particularly useful to the student and the instructor.

Key equations in the text are indicated by boldface numbers on the right-hand side of the text column, and practice should be gained in using them for chemical calculations. Other equations, such as intermediate steps in mathematical derivations, are less central to the overall line of reasoning in the book.

## Updated Design and New Illustrations and Photographs

The fourth edition features a modern and well-organized design, as well as new illustrations, all of which have been digitally rendered in an exacting three-dimensional style. In the spirit of previous editions, great care has been taken in the selection of photographs and illustrations to ensure that each piece relates functionally and pedagogically to appropriate concepts in the narrative. The authors are indebted to Joseph J. BelBruno at Dartmouth College for his help in reviewing the photographs and illustrations for accuracy, completeness, and clarity.

## Worked Examples

This textbook includes worked examples that demonstrate the methods of reasoning applied in solving chemical problems. The examples are inserted immediately after the presentation of the corresponding principles, and cross-references are made to related problems appearing at the end of the chapter.

## A Deeper Look . . .

Sections entitled "A Deeper Look . . ." provide students with the physical origins of observed chemical behavior. The material that they present is sometimes more advanced mathematically than that in the main parts of the book. Their use affords instructors considerable flexibility in modifying the level of the course material that they wish to teach.

## Key Terms

Key terms appear in boldface where they are first introduced. In addition, definitions for most key terms are included in the Index/Glossary for ready reference.

## Concepts & Skills

Each chapter concludes with a list of concepts and skills for review by the student. Included in this list are cross-references back to the section in which the topic was covered and forward to problems that help to test mastery of the particular skill involved. This feature is helpful for self-testing and review of material.

## Cumulative Exercises

At the end of each of Chapters 2 through 19 is a Cumulative Exercise that focuses on a problem of chemical interest and draws on material from the entire chapter for its solution. Working through a chapter's Cumulative Exercise provides a useful review of material in the chapter, helps students to put principles into practice, and prepares students to solve the problems that follow.

## Problems

Problems are grouped into three categories. Answers to odd-numbered "paired problems" are collected in Appendix G; they enable the student to check the answer to the first problem in a pair before undertaking the second problem, which is paral-

lel to it. The Additional Problems, which are unpaired, provide further applications of the principles developed in the chapter. The Cumulative Problems integrate material from the chapter with topics presented earlier in the book. The more challenging problems are indicated with asterisks.

## Appendices

Appendices A, B, and C are important pedagogically. Appendix A discusses experimental error and scientific notation, while Appendix B introduces the SI system of units used throughout the book and describes the methods used for converting units. A new section in Appendix B provides a brief review of some fundamental principles in physics, which may be particularly helpful to students in understanding topics covered in Chapters 3, 4, 7, 15, and 16. Appendix C provides a review of mathematics for general chemistry. Appendices D, E, and F are compilations of thermodynamic, electrochemical, and physical data, respectively.

## Index/Glossary

The Index/Glossary at the back of the book gives brief definitions of key terms, as well as cross-references to the pages on which the terms appear.

## SUPPLEMENTS

### Study and Problem-Solving Guide

The best way to master physical and chemical concepts is to solve problems; even students with outstanding mathematical skills can encounter difficulties when working chemistry problems. Trouble usually arises in passing from the words of the problem through formal statements of concepts to the manipulation of mathematical equations. The *Study and Problem-Solving Guide*, authored by Wade A. Freeman of the University of Illinois at Chicago, helps students meet such difficulties head-on. It summarizes definitions, concepts, and equations, gives some additional insights into the material presented in the text, provides problem-solving hints, and, as a practical illustration, presents detailed solutions to all of the odd-numbered problems in this book. Each student is strongly encouraged to purchase a copy of this study guide.

### Instructor's Manual

The *Instructor's Manual* contains solutions to the even-numbered problems, as well as suggestions for ways to use this textbook in courses with different plans of organization.

### Overhead Transparencies

A set of at least 110 full-color figures and tables from the text is available.

### Saunders Interactive General Chemistry CD-ROM

Authored by John Kotz of SUNY–Oneonta and William Vining of the University of Massachusetts, this CD-ROM serves as a powerful multimedia companion for the course. The CD-ROM includes original animation and graphics, interactive tools, pop-up definitions, over 100 video clips of chemical experiments that are enhanced by sound effects and narration, over 300 molecular models and anima-

tions, and portions of the molecular modeling tools from the Oxford Molecular Group. In addition, Version 2.5 incorporates ActivChemistry$^{TM}$, which makes it possible for students to design and perform simulated laboratory experiments. By varying experimental parameters and observing results, students discover concepts on their own.

## Saunders General Chemistry Web Site

The Saunders General Chemistry Web Site at **www.saunderscollege.com/chem/ general** includes the kind of support material that cannot be provided by the text:

- On-line Problem-Solving
- Interactive Math Review for General Chemistry
- Real-World Applications Reference Base
- Chemical Research and General Chemistry
- On-line Instructor Resources (including some specifically relating to *Principles of Modern Chemistry, Fourth Edition*)
- Links to Useful Chemistry-Related Web Sites
- New Ancillary Course Material

Through these seven resource features, the Saunders Web Site (1) provides a new context for interactive learning, (2) effectively incorporates animation, three-dimensional depiction of chemical phenomena, and molecular modeling, and (3) stands as an extensive reference base.

## CalTech Chemistry Animation Project (CAP)

This package consists of a set of six video units that cover, with unmatched quality and clarity, the chemical topics of atomic orbitals, valence shell electron-pair repulsion (VSEPR) theory, crystals and units cells, molecular orbitals in diatomic molecules, periodic trends, and hybridization and resonance.

Saunders College Publishing may provide complementary instructional aids and supplements or supplement packages to those adopters qualified under our adoption policy. Please contact your sales representative for more information. If as an adopter or potential user you receive supplements you do not need, please return them to your sales representative or send them to

Attn: Returns Department
Troy Warehouse
465 South Lincoln Drive
Troy, MO 63379

## ACKNOWLEDGMENTS

In preparing this fourth edition, we have benefited greatly from the comments of students who used the first three editions over the past several years. We would also like to acknowledge the many helpful suggestions of colleagues at the University of

Chicago, the University of California—Los Angeles, and other colleges and universities who have taught from this book.

We extend special thanks to the following professors who completed user surveys for the third edition or reviewed the manuscript for the fourth edition:

Mario E. Baur, University of California—Los Angeles

Joseph J. BelBruno, Dartmouth College

Alan Campion, University of Texas at Austin

Emily A. Carter, University of California—Los Angeles

Robin Garrell, University of California—Los Angeles

Malcolm F. Nicol, University of California—Los Angeles

Philip Pechukas, Columbia University

William P. Reinhardt, University of Washington

Michael P. Rosynek, Texas A & M University

John Weare, University of California—San Diego

Peter M. Weber, Brown University

We are particularly grateful to friends and colleagues who provided original scientific illustrations for this new edition of the book. They are Drs. Gilberto Medeiros-Ribeiro and R. Stanley Williams (Hewlett-Packard Research Laboratories), Drs. Andrew J. Pounds and Mark Iken (Scientific Visualization Laboratory, Georgia Institute of Technology), Dr. Stuart Watson and Professor Emily Carter (University of California—Los Angeles), Professor Nathan Lewis (California Institute of Technology), Dr. Don Eigler (IBM Almaden Research Center), Dr. Gerard Parkinsen and Mr. William Gerace (OMICRON Vakuumphysik), Dr. Richard P. Muller and Professor W. A. Goddard III (California Institute of Technology), Dr. Jane Strouse (University of California—Los Angeles), Professor James Speck and Professor Stephen Den Baars (University of California—Santa Barbara), and Professor John Baldeschwieler (California Institute of Technology). We are also indebted to Ms. Patti Kellett, Lecture Demonstrator in the Department of Chemistry and Biochemistry at UCLA, for the development of several illustrations.

The staff members at Saunders College Publishing have been most helpful in preparing this fourth edition. In particular, we acknowledge the key role of our Acquisitions Editor, Kent Porter Hamann, who guided us toward revisions in this fourth edition, and our Project Editor, Frank Messina, who kept the schedule moving smoothly. We acknowledge also the contributions of the Senior Art Director, Caroline McGowan, who designed the text and developed the new illustrations in this edition, as well as our Photo Researcher, Jane Sanders, who assisted in obtaining the photographs. The continuing support and encouragement of Publisher John Vondeling is gratefully acknowledged.

David W. Oxtoby                         H. P. Gillis
The University of Chicago               University of California—Los Angeles

April 1998                              April 1998

# Contents Overview

Crystals of alum.

## UNIT 5

### QUANTAL DESCRIPTION OF ATOMIC AND MOLECULAR STRUCTURE    521

## UNIT 6

### CHEMICAL PROCESSES    727

## UNIT 7

### MATERIALS    807

### APPENDICES    A.1

# Contents

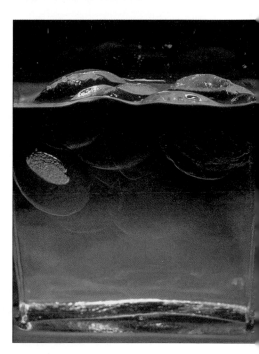

Ice and water.

## UNIT 2

### KINETIC MOLECULAR EXPLANATION OF THE STATES OF MATTER   97

## UNIT 3

# EQUILIBRIUM IN CHEMICAL REACTIONS    201

---

## UNIT 4

### RATES OF CHEMICAL AND PHYSICAL PROCESSES    445

---

## UNIT 5

### QUANTAL DESCRIPTION OF ATOMIC AND MOLECULAR STRUCTURE    521

---

**UNIT 6**

# CHEMICAL PROCESSES    727

---

**UNIT 7**

# MATERIALS    807

## Appendices   A.1

# About the Authors

## DAVID W. OXTOBY

David W. Oxtoby is a physical chemist who studies the statistical mechanics of liquids, including nucleation, phase transitions, and liquid-state reaction and relaxation. He received the B.A. (Chemistry and Physics) from Harvard University and the Ph.D. (Chemistry) from the University of California at Berkeley. After a postdoctoral position at the University of Paris, he joined the faculty of The University of Chicago, where he has taught general chemistry, thermodynamics, and statistical mechanics. He currently serves as Dean of Physical Sciences at The University of Chicago.

## H. P. GILLIS

H. P. Gillis conducts experimental research in the physical chemistry of electronic materials, with special emphasis on phenomena at solid surfaces and interfaces. Dr. Gillis received the B.S. (Chemistry and Physics) at Louisiana State University and the Ph.D. (Chemical Physics) at The University of Chicago. After postdoctoral research at the University of California—Los Angeles and ten years with the technical staff at Hughes Research Laboratories in Malibu, California, Dr. Gillis joined the academic world. Since 1988 he has taught courses in general chemistry, physical chemistry, quantum mechanics, and surface science at Georgia Institute of Technology and at the University of California—Los Angeles.

# UNIT 1 INTRODUCTION TO THE STUDY OF MODERN CHEMISTRY

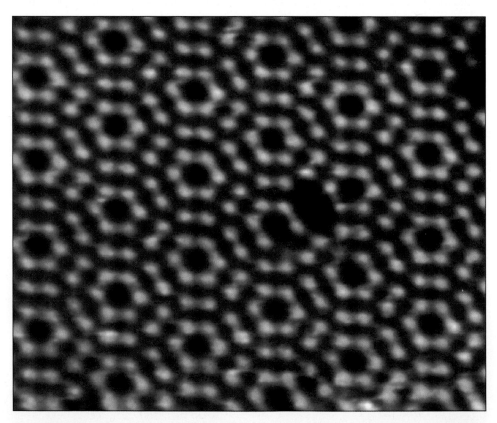

The surface of a silicon crystal imaged in the scanning tunneling microscope. The surface was cleaned and imaged in vacuum to remove all impurity atoms. Individual silicon atoms are edged in red. The open space between atoms is shown as the black background. Vacancies (or defects) where missing silicon atoms should be located are readily identified. (*Courtesy of Drs. Gilberto Medeiros-Ribeiro and R. Stanley Williams/Hewlett-Packard Research Laboratories, Palo Alto, CA*)

Atoms, molecules, and the chemical bond are the central concepts in modern chemistry. From the speculations of ancient philosophers, through the classic 18th and 19th century experiments that led to the atomic hypothesis of Dalton, and continuing with present-day experiments in which atoms and molecules are directly manipulated in beams and are detected individually by the scanning tunneling microscope (STM), the search for atoms and molecules has been a fascinating detective story.

**CHAPTER 1**
The Nature and Conceptual Basis of Modern Chemistry

**CHAPTER 2**
Chemical Equations and Reaction Yields

**CHAPTER 3**
Chemical Bonding: The Classical Description

THE GOAL OF UNIT 1 IS to tell this story of atoms, molecules, and bonds in three parts essential to the study of modern chemistry: **(1)** What is the evidence for the existence of atoms and molecules? **(2)** How do atoms and molecules control the transformation of one substance into another in chemical reactions? **(3)** What are the structures and shapes of molecules, as described by classical methods before the advent of quantum mechanics?

# The Nature and Conceptual Basis of Modern Chemistry

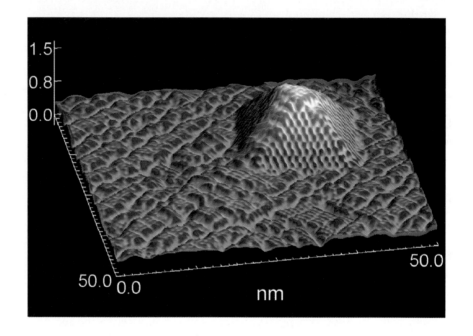

---

## 1.1

### NATURE OF MODERN CHEMISTRY

*Illustration*
Germanium atoms deposited on a clean silicon surface spontaneously form pyramidal structures. The individual germanium atoms in the pyramid are revealed in this image obtained with a scanning tunneling microscope. (*Courtesy of Drs. Gilberto Medeiros-Ribeiro and R. Stanley Williams/Hewlett-Packard Research Laboratories, Palo Alto, CA*)

Chemistry is the study of the properties of substances and, most especially, of the reactions that transform one substance into another. Chemistry provides routes for tailoring the properties of existing substances to meet some particular need or application and for creating entirely new materials designed from the beginning to have particular desired properties. Through such accomplishments chemistry has made dramatic contributions to improving agricultural production, controlling the spread of disease, increasing the efficiency of energy production, and reducing environmental pollution. Much of the excitement of modern chemistry attaches to studies of change, of the dynamics of these transformations into new substances. Dynamics of chemical changes are pervasive, governing phenomena as diverse as evolution of small carbon-containing molecules in interstellar space, changes in terrestrial atmospheric and climatic patterns induced by pollutants, and the unfolding of life processes in living organisms. A single set of fundamental scientific principles unifies the description of these diverse phenomena.

Chemistry is a relatively young science—its foundations were established in the last quarter of the 18th century—that arose from desires of the ancient and medieval alchemists to transform the properties of materials for economic gain (Fig. 1.1). For many centuries the alchemists attempted without success to transform "base" metals into gold. They operated under the erroneous assumption that the properties of a material could somehow be "extracted" from that material and transferred to another. If the essential properties—such as yellow color, softness, and ductility—could be assembled from various inexpensive sources, then gold could be created at great profit.

The alchemists persisted in their efforts for more than a thousand years. They collected many useful, empirical results that have been incorporated into modern chemistry, but they never transformed base metals into gold. Toward the middle of the 17th century, certain individuals began to challenge the validity of the basic assumptions of the alchemists. These doubts culminated in publication of *The Sceptical Chymist* by Robert Boyle in England in the 1660s, one of the pioneering steps toward modern chemistry. Another century was required to complete the conceptual foundations of modern chemistry, which then flourished throughout the 19th and 20th centuries.

To observers in the late 20th century, the mistake of the alchemists is immediately clear. They did not follow the scientific method. In the scientific method, a new idea is accepted only temporarily, in the form of a **hypothesis.** It is then subjected to rigorous testing, under tightly controlled conditions, in experiments. Only by surviving many such tests does the hypothesis earn sufficient confidence in its validity to merit the label **scientific law.** Concepts or ideas that have earned the status of scientific laws by direct and repeated testing can then be applied in new environments with confidence. Had a proper set of tests been made in separate, independent experiments, the alchemists would have recognized that the properties of a material are in fact intrinsic, inherent characteristics of that material and cannot be separated from it.

The history of the alchemists shows the origin of a certain duality in the nature of modern chemistry, which persists to the present. On the one hand, one sees the urge to apply established chemical knowledge for profit, since chemistry contributes to the foundations of numerous professions and industries. On the other hand, one sees the urge to create new chemical knowledge, driven by intellectual curiosity and by the desire to have reliable information for applications. Both aspects invest the activities of numerous scientists and engineers in addition to professional chemists. No matter what the specific context, the second aspect requires scrupulous adherence to the scientific method, in which new knowledge is subjected to rigorous scrutiny before it earns the confidence of the scientific community.

In their careers, most students of chemistry will be more concerned with applying chemistry than with generating new chemical knowledge. Nevertheless, a useful stratagem for learning the science of chemistry is to take the point of view that one personally is establishing the scientific foundations of chemistry for the very first time. Upon encountering a new topic, imagine that you are the first person ever to see the laboratory results on which it is based. Imagine that you must construct the new concepts and explanations to interpret these results, and that you will present and defend your conclusions before the scientific community. Be suspicious,

**FIGURE 1.1** Alchemists searched in vain for procedures that would turn base metals into gold. Their apparatus foreshadowed equipment in modern chemical laboratories. (*Painting* The Alchemist *by Hendrick Heerschop, 1671. Courtesy of Dr. Alfred Bader*)

cross-check everything, and demand independent confirmations. Always remain with Boyle the "skeptical chemist"; follow the scientific method in your acquisition of knowledge, even from textbooks. In this way you will make the science of chemistry your own, and you will experience the intellectual excitement and satisfaction of discovery and interpretation. But most important, you will recognize that chemistry is not merely a closed set of facts and formulas; it is a living, growing method of investigating all aspects of human experience that depend on changes in the composition of substances.

Chemical reasoning, both in applications and in basic research, proceeds somewhat in the manner of a detective story in which tangible clues lead to a mental picture of events not witnessed directly by the detective. Chemical experiments are conducted in laboratories equipped with beakers, flasks, analytical balances, pipettes, optical spectrophotometers, lasers, vacuum pumps, pressure gauges, mass spectrometers, centrifuges, and other apparatus all of which operate on the macroscopic scale perceptible to ordinary human senses. Yet the actual chemical transformation events occur in the microscopic world of atoms and molecules, tiny objects far too small to be detected directly by humans. The laboratory instruments are the bridge between these worlds, giving the experimenter not only the means to influence the actions of the atoms and molecules but also the means to measure their response. Chemists *think* in the highly visual microscopic world of atoms and molecules, but *work* in the tangible world of macroscopic laboratory apparatus. These two aspects of chemical science cannot be divorced, and their interplay is emphasized throughout this book. Students of chemistry must equally master the concepts of chemistry, which describe the microscopic world of atoms and molecules, and the macroscopic procedures of chemistry, on which the concepts are founded.

The concepts of chemistry rest on two fundamental principles: the conservation of matter and the conservation of energy. The total amount of matter involved in a chemical reaction is conserved; that is, remains constant throughout the reaction. Matter is neither created nor destroyed in chemical reactions; matter is rearranged from one substance into another. Rearrangements of matter are inevitably accompanied by changes in energy. The amounts of chemical energy stored in the molecules of two different substances are intrinsically different, and chemical energy may be converted into thermal or electrical energy during reactions. Nonetheless, energy is neither created nor destroyed during chemical reactions; the total amount of energy involved in a chemical reaction is conserved. These principles must be modified slightly for the case of nuclear reactions, which occur at energies so high that matter and energy can be converted into one another through Einstein's relation $E = mc^2$. During nuclear reactions, the sum of energy and matter is conserved.

---

## 1.2

## MACROSCOPIC METHODS AND APPROACHES OF MODERN CHEMISTRY

The main concern of chemistry is to study the transformation of one pure substance into another through chemical reactions. This involves two traditions—**analysis** (taking things apart) and **synthesis** (putting things together)—that can be traced from the earliest periods of Greek science to the present. The early Greek philosophers attempted to analyze the constituents of matter in terms of four elements: air, earth, fire, and water. Early artisans and metallurgists synthesized new materials of practical importance—including bronze, glass, and cement—both by design and by accident. Today, when a medicinally important product is discovered in a natural source, the first step is to analyze the product to find its composition and structure, and then a strategy can be designed to synthesize it from easily available starting materials. Penicillin was discovered in 1929 through the accidental contamination of a bacterial culture by mold; its laboratory synthesis was not achieved until 1957.

## Substances and Mixtures

The study of chemical reactions can be greatly complicated and often obscured by the presence of extraneous materials. The first step is therefore to learn how to analyze and classify materials to be sure one is working with pure substances before commencing with reactions (Fig. 1.2). Suppose we take a sample of a material, which

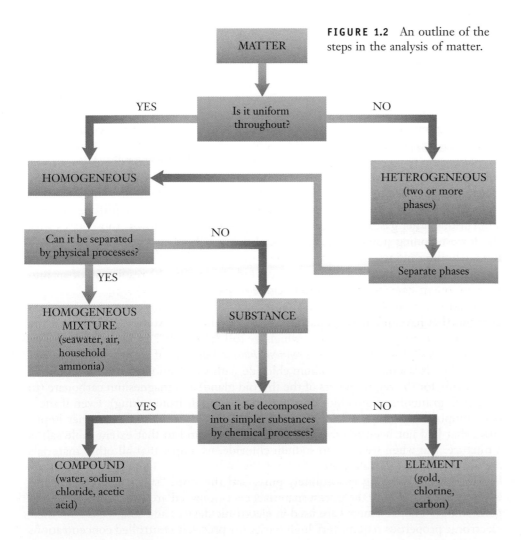

**FIGURE 1.2**  An outline of the steps in the analysis of matter.

(a)

(c)

(b)

**FIGURE 1.3** (a) A solid mixture of blue $Cu(NO_3)_2 \cdot 6H_2O$ and yellow CdS is added to water. (b) Although the $Cu(NO_3)_2 \cdot 6H_2O$ dissolves readily and passes through the filter, the CdS remains largely undissolved and is held on the filter. (c) Evaporation of the solution leaves nearly pure crystals of $Cu(NO_3)_2 \cdot 6H_2O$. *(Leon Lewandowski)*

may be gas, liquid, or solid, and examine various properties or distinguishing characteristics, such as color, odor, or density. A piece of wood, for example, has regions that possess different properties (some portions are darker than others) and is said to be **heterogeneous.** Other materials, such as air or a mixture of salt and water, are **homogeneous** because their properties do not vary through the sample. They are still **mixtures,** however, because it is possible to separate them into components by ordinary physical means such as melting, freezing, boiling, or dissolving in solvents (Fig. 1.3). These operations provide ways of separating materials from one another on the basis of their properties (freezing point, boiling point, solubility, and so on). For example, air is a mixture of several components (oxygen, nitrogen, argon, and various other gases). If it is liquefied and then warmed slowly, those gases with the lowest boiling points evaporate first, leaving behind in the liquid those with higher boiling points. Such a separation would not be perfect, but the process of liquefaction and evaporation could be repeated to improve the resolution of air into its component gases to any required degree of purity.

If all these physical procedures (and many more) fail to separate matter into portions that have different properties, the material is a **substance.** What about the common material sodium chloride, which we call "table salt"? Is it a substance? The answer is "yes" if we use the term *sodium chloride* but "no" if we use the term *table salt.* Table salt is a mixture of sodium chloride with small amounts of sodium iodide (to provide for the requirements of the thyroid gland) and magnesium carbonate (to coat each grain of sodium chloride and prevent the salt from caking). Even if these two components were not deliberately added, table salt would contain other impurities that had not been removed in its preparation, and to that extent table salt is a mixture. But when we refer to sodium chloride, we imply that all other materials are absent, and it is a substance.

In practice, nothing is absolutely pure, and the word "substance" is therefore an idealization. Among the purest materials ever prepared are silicon (Fig. 1.4) and germanium. These elements are used in electronic devices and solar cells, and their electronic properties require very high purity (or precisely controlled concentrations of deliberately added impurities). Meticulous chemical and physical methods have enabled scientists to prepare germanium and silicon that contain concentrations not exceeding one part per billion of impurities that would alter their electrical properties.

## The Elements

Literally millions of substances have so far been discovered (or synthesized) and identified. Are these the fundamental building blocks of matter? If they were, their classification alone would pose an insurmountable task. In fact, all these substances are combinations of much smaller numbers of building blocks called **elements.** Elements are substances that cannot be decomposed into two or more simpler substances by ordinary physical or chemical means. Here again, the word "ordinary" is important, because we exclude the processes of radioactive decay (either natural or artificial) and of high-energy nuclear reactions that *do* transform one element into another. The term "**compound**" refers to substances that contain two or more chemical elements. Hydrogen and oxygen are elements because no further chemical separation is possible, but water is a compound because it can be separated into hydrogen and oxygen by passing an electric current through it (Fig. 1.5). *Binary* compounds are substances (such as water) that contain two elements, *ternary* compounds contain three elements, *quaternary* four, and so on.

At present some 112 chemical elements have been identified. A few have been known since before recorded history, principally because they occur in nature as elements rather than in combination with one another in compounds. Gold, silver, lead, copper, and sulfur are chief among them. Gold is found in streams in the form of little granules (placer gold) or nuggets in loosely consolidated rock. Sulfur is associated with volcanoes, and copper can often be found in its native state in shallow mines. Iron occurs in its elemental state only rarely (in meteorites). Usually it is combined with oxygen or other elements. In the second millennium B.C., ancient metallurgists somehow learned to reduce iron oxide to iron with charcoal in forced-draft fires, and the Iron Age was inaugurated.

The names of the chemical elements and the symbols that designate them are fascinating subjects. Many have Latin roots, such as gold (*aurum*, symbol Au), copper

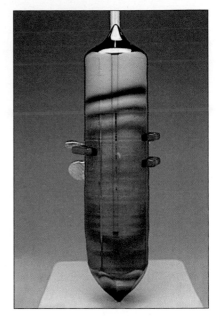

**FIGURE 1.4**   Nearly pure elemental silicon is produced by pulling a 10-inch-long solid cylinder (called a boule) out of the melt, leaving most of the impurities behind. *(Charles D. Winters)*

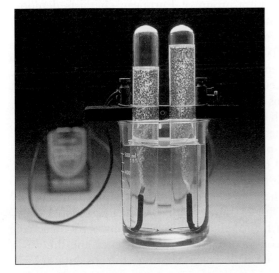

**FIGURE 1.5**   As electric current passes through water containing dissolved sulfuric acid, gaseous hydrogen and oxygen form as bubbles at the electrodes. This method of electrolysis can be used to prepare very pure gases. Note that the volume of hydrogen produced (left) is twice that of the oxygen (right). This is a consequence of the law of combining volumes, discussed in Section 1.3. *(Charles D. Winters)*

(*cuprum*, Cu), iron (*ferrum*, Fe), and mercury (*hydrargyrum*, Hg). Hydrogen (H) means "water former." Potassium (*kalium*, K) takes its common name from potash (potassium carbonate), a useful chemical obtained in early times by leaching the ashes of wood fires with water. Many elements take their names from Greek and Roman mythology: cerium (Ce) from Ceres, goddess of plenty; tantalum (Ta) from Tantalus; niobium (Nb) from Niobe, daughter of Tantalus. Some elements are named for continents: europium (Eu), americium (Am). Others are named after countries: germanium (Ge), francium (Fr), polonium (Po). Cities provide the names of others: holmium (Stockholm, Ho), ytterbium (Ytterby, Yb), berkelium (Berkeley, Bk). Some are named for the planets: uranium (U), plutonium (Pu), neptunium (Np). Others take their names from colors: praseodymium (green, Pr), rubidium (red, Rb), cesium (sky blue, Cs). Still others honor great scientists: curium (Marie Curie, Cm), mendelevium (Dmitri Mendeleev, Md), fermium (Enrico Fermi, Fm), einsteinium (Albert Einstein, Es), seaborgium (Glenn Seaborg, Sg).

---

## 1.3

# CONCEPTS AND TOOLS OF MODERN CHEMISTRY: LAWS OF CHEMICAL COMBINATION

The existence of the elements as the most fundamental substances, which can be combined chemically to form compound substances, provides no information on the microscopic structure of matter or how that microscopic structure controls, and is revealed by, chemical reactions. Ancient philosophers dealt with these fascinating questions by proposing assumptions, or *postulates*, about the structure of matter. The Greek philosopher Democritus (c. 460–c. 370 B.C.) postulated the existence of unchangeable *atoms* of the elements, which he imagined to undergo continuous random motion in the vacuum, a remarkably modern point of view. It follows from this postulate that matter is not divisible without limit; there is a lower limit to which a compound can be divided before it becomes separated into atoms of the elements from which it is made. Lacking both experimental capabilities and the essentially modern scientific view that theories must be tested and refined by experiment, the Greek philosophers were content to leave their views in the form of assertions.

More than 2000 years passed before a group of European chemists demonstrated experimentally that elements combine only in masses with very definite ratios when forming compounds and that compounds react with each other only in masses with very definite ratios. These results could only be interpreted by inferring that smallest indivisible units of the elements (atoms) combined to form smallest indivisible units of the compounds (molecules) in definite proportions. The very definite mass ratios involved in reactions were interpreted as convenient means for counting the number of atoms of each element participating in the reaction. These results, summarized as the laws of chemical combination, provided overwhelming, if indirect, evidence for the existence of atoms and molecules.

In present times, we are so used to speaking of atoms that we rarely stop to consider the experimental evidence for their existence collected in the 18th and 19th centuries. Twentieth-century science has developed a number of sophisticated techniques to measure the properties of single atoms, and powerful microscopes even allow us to observe them (see Section 1.6). Well before single atoms were detected, however, chemists could speak with confidence about their existence and the ways in which they combine to form molecules. Moreover, although the absolute masses of single atoms of oxygen and hydrogen were not measured until the early 20th century, some 50 years earlier chemists could assert (correctly) that the *ratio* of the two masses was close to 16:1. The chemical evidence for the existence of atoms and for the scale of relative atomic masses provides a fascinating and important story, both in its own right and as an illustration of the way in which science progresses. That story is presented in the present section.

## The Law of Conservation of Mass

The first key steps toward understanding the microscopic nature of matter were taken during the 18th century in the course of studies of heat and combustion. It had been observed that an organic material (such as wood) left a solid residue of ash when burned, and a metal heated in air was transformed into a "calx," which we now call an oxide. The explanation for this phenomenon popular in the early 18th century was that a property called "phlogiston" was driven out of wood or metal by the heat of a fire. From the modern perspective this seems quite absurd because the ash weighs less than the original wood whereas the calx weighs more than the metal. At the time, however, the principle of conservation of mass was not yet established, and people saw no reason why the mass of a material should not change upon heating.

Further progress could be made only through the careful use of the chemical balance to determine the changes in mass[1] that occur in chemical reactions. The balance had been known since antiquity, but had been used principally as an assayer's tool and for verifying the masses of coins or commodities in commerce. The great French chemist Antoine Lavoisier used the balance to demonstrate that the sum of the masses of the products of a chemical reaction equals the sum of the masses of the reactants. He heated mercury in a sealed flask containing air. After several days a red substance, mercury(II) oxide, was produced. The gas remaining in the flask was reduced in mass and could no longer support life or combustion; a candle was extinguished by it, and animals suffocated when forced to breathe it. We now know that this residual gas was nitrogen, and that the oxygen in the air had reacted with the mercury. Lavoisier then took a carefully weighed amount of the red oxide of mercury and heated it strongly (Fig. 1.6). He weighed both the mercury and the gas that were produced and showed that their combined mass was the same as that of the mercury(II) oxide with which he had started. After further experiments Lavoisier was able to state the **law of conservation of mass:**

> *In every chemical operation an equal quantity of matter exists before and after the operation.*

Lavoisier was the first to observe that a chemical reaction is analogous to an algebraic equation. We would write his second reaction as

$$2\ HgO \longrightarrow 2\ Hg + O_2$$

although at that time the identity of the gas (oxygen) was not known.

**FIGURE 1.6**   When the red solid mercury(II) oxide is heated, it decomposes to mercury and oxygen. Note the drops of liquid mercury condensing on the side of the test tube. *(Charles D. Winters)*

---

[1] Chemists sometimes use the term "weight" in place of "mass." Strictly speaking, weight and mass are not the same. The mass of a body is an invariant quantity, but its weight is the force exerted upon it by gravitational attraction (usually by the Earth). Newton's second law relates the two ($w = m \times g$, where $g$ is the acceleration due to gravity). As $g$ varies from place to place on the Earth's surface, so does the weight of a body. In chemistry we deal mostly with ratios, which are the same for masses and weights. We will use the term "mass" exclusively, but "weight" is still in colloquial chemical use.

## The Law of Definite Proportions

Rapid progress ensued as chemists began to make accurate determinations of the masses of reactants and products. A controversy arose between two schools of thought, led by the French chemists Claude Berthollet and Joseph Proust. Berthollet believed that the proportions (by mass) of the elements in a particular compound were not fixed but could vary over a certain range. Water, for example, rather than containing 11.1% by mass of hydrogen, might have somewhat less or more than this mass percentage. Proust disagreed, arguing that any apparent variation was due to impurities and experimental errors. He also stressed the difference between homogeneous mixtures and chemical compounds. Through his careful work, Proust demonstrated the fundamental **law of definite proportions:**

> *In a given chemical compound, the proportions by mass of the elements that compose it are fixed, independent of the origin of the compound or its mode of preparation.*

Pure sodium chloride contains 60.66% chlorine by mass, whether we obtain it from salt mines, or by crystallizing it from waters of the oceans or inland salt seas, or by synthesizing it from its elements, sodium and chlorine.[2]

The law of definite proportions was a crucial step in the development of modern chemistry, and by 1808 Proust's conclusions had become widely accepted. We now recognize that this law is not strictly true in all cases. Certain solids exist over a small range of compositions and are called **nonstoichiometric compounds.** An example is wüstite, which has the nominal chemical formula $FeO$ (with 77.73% iron by mass) but whose composition in fact ranges continuously from $Fe_{0.95}O$ (with 76.8% iron) down to $Fe_{0.85}O$ (74.8% iron), depending on the method of preparation. Such compounds, called **berthollides** in honor of Berthollet, are discussed further in Section 19.5. All gaseous compounds, however, obey the law of definite proportions. This account illustrates a common pattern of scientific progress: experimental observation of parallel behavior leads to the establishment of a law, or principle. More accurate studies may then reveal exceptions to the general principle. Explanation of the exceptions leads to deeper understanding.

## The Atomic Theory of Dalton

The English scientist John Dalton was by no means the first person to propose the existence of atoms; as we have seen, such ideas date back to Greek times (the word "atom" comes from Greek *a-* [not] plus *tomos* [cut]: "not divisible"). Dalton's major contribution was to marshal the evidence for their existence. He showed that the mass relationships found by Lavoisier and Proust could be interpreted most simply by postulating the existence of atoms of the various elements.

In 1808 Dalton published *A New System of Chemical Philosophy*, in which the following five statements comprise the **atomic theory of matter:**

1. Matter consists of indivisible atoms.
2. All of the atoms of a given chemical element are identical in mass and in all other properties.

---

[2] This statement needs some qualification. As we will see in the next section, many elements have several *isotopes*, which are species whose atoms have almost identical chemical properties but different masses. Natural variation in isotope abundances leads to small variations in the mass proportions of elements in a compound, and larger variations can be induced by artificial isotopic enrichment.

**3.** Different chemical elements have different kinds of atoms; in particular, their atoms have different masses.

**4.** Atoms are indestructible and retain their identities in chemical reactions.

**5.** A compound forms from its elements through the combination of atoms of unlike elements in small whole-number ratios.

Dalton's fourth postulate is clearly related to the law of conservation of mass, and the fifth is an attempt to explain the law of definite proportions. Suppose that one rejects the atomic theory and adopts a view in which compounds are subdivisible without limit. What, then, ensures the constancy of composition of a substance such as sodium chloride? On the other hand, if each sodium atom in sodium chloride is matched by one chlorine atom, then the constancy of composition can be understood. In this argument for the law of definite proportions, it does not matter how small the atoms of sodium and chlorine are. It is important merely that there be some lower bound to the subdivisibility of matter, because the moment we put in such a lower bound, arithmetic steps in. Matter becomes countable, and the units of counting are simply atoms. Believing in the law of definite proportions as an established experimental fact, Dalton *postulated* the atom.

## The Law of Multiple Proportions

The composition of a compound is shown by its **chemical formula.** The symbol "$H_2O$" for water indicates that the substance water contains two atoms of hydrogen for each atom of oxygen. It is now known that in water the atoms in each group of three (two H and one O) are linked together by attractive forces strong enough to keep the group together for a reasonable period of time. Such a group is called a **molecule.** In the absence of knowledge about a compound's molecules, the numerical subscripts in the chemical formula simply give the relative proportions of the elements in the compound. How do we know that these are the true proportions? The determination of chemical formulas (and the accompanying determination of relative atomic masses) was a major accomplishment of 19th-century chemistry, building on the atomic hypothesis of Dalton.

In the simplest kind of compounds, two elements combine, contributing equal numbers of atoms to the union to form **diatomic molecules,** which consist of two atoms each. Eighteenth- and nineteenth-century chemists knew, however, that two elements often combine in different proportions, forming more than one compound.

For example, carbon (C) and oxygen (O) combine under different conditions to form two different compounds. Let us call them A and B. Analysis shows that A contains 1.333 grams (g) of oxygen per 1.000 g of carbon, and that B contains 2.667 g of oxygen per 1.000 g of carbon. Although at this point we know nothing about the chemical formulas of the two oxides of carbon, we can immediately say that molecules of compound A contain half as many oxygen atoms per carbon atom as do molecules of compound B. The evidence for this is that the ratio of the masses of oxygen in A and B, for a fixed mass of carbon in each, is $1.333:2.667$, or $1:2$. If the formula of compound A were CO, then the formula of compound B would have to be $CO_2$, $C_2O_4$, $C_3O_6$, or some other multiple of $CO_2$. If compound A were $CO_2$, then compound B would be $CO_4$ or $C_2O_8$, and so on. From these data, we cannot say which of these (or an infinite number of other possibilities) are true formulas of the molecules of compounds A and B, but we do know that the number of oxygen atoms per carbon atom in the two compounds is the *quotient of integers*.

Consider another example. Arsenic (As) and sulfur (S) combine to form two sulfides, A and B, in which the masses of sulfur per 1.000 g of arsenic are respectively

0.428 g and 0.642 g. The ratio of these sulfur masses is $0.428 : 0.642 = 0.667 \approx 2 : 3$. We conclude that *if* the formula of compound A is a multiple of AsS, then the formula of compound B must be a multiple of $As_2S_3$.

These two examples illustrate the **law of multiple proportions:**

*When two elements form a series of compounds, the masses of one element that combine with a fixed mass of the other element are in the ratio of small integers to each other.*

In our first example, the ratio of the masses of oxygen in the two compounds, for a given mass of carbon, was $1 : 2$. In the second example, the ratio of the masses of sulfur in the two compounds, for a given mass of arsenic, was $2 : 3$. Today we know that the carbon oxides are CO (carbon monoxide) and $CO_2$ (carbon dioxide), and the arsenic sulfides are $As_4S_4$ and $As_2S_3$. Dalton could not have known this, however, because he had no information from which to decide how many atoms of carbon and oxygen are in one molecule of the carbon–oxygen compounds, or how many atoms of arsenic and sulfur are in the arsenic–sulfur compounds.

## EXAMPLE 1.1

Chlorine (Cl) and oxygen form four different binary compounds. Analysis gives these results:

| Compound | Mass of O Combined with 1.0000 g of Cl |
|----------|------------------------------------------|
| A | 0.22564 g |
| B | 0.90255 g |
| C | 1.3539 g |
| D | 1.5795 g |

**(a)** Show that the law of multiple proportions holds for these compounds.
**(b)** If the formula of compound A is a multiple of $Cl_2O$, then determine the formulas of compounds B, C, and D.

### Solution

**(a)** Form ratios by dividing each mass of oxygen by the smallest, which is 0.22564 g:

$$0.22564 \text{ g} : 0.22564 \text{ g} = 1.0000 \text{ for compound A}$$

$$0.90255 \text{ g} : 0.22564 \text{ g} = 4.0000 \text{ for compound B}$$

$$1.3539 \text{ g} : 0.22564 \text{ g} = 6.0003 \text{ for compound C}$$

$$1.5795 \text{ g} : 0.22564 \text{ g} = 7.0001 \text{ for compound D}$$

The ratios are whole numbers to a high degree of precision, and the law of multiple proportions is satisfied. Ratios of whole numbers would also have satisfied that law.
**(b)** If compound A has a formula that is some multiple of $Cl_2O$, then compound B is $Cl_2O_4$ (or $ClO_2$, or $Cl_3O_6$, and so forth) because it is four times richer in oxygen than compound A. Similarly, compound C, which is six times richer in oxygen than compound A, is $Cl_2O_6$ (or $ClO_3$, or $Cl_3O_9$, and so forth), and compound D, which is seven times richer in oxygen than compound A, is $Cl_2O_7$ (or a multiple thereof).

**Related Problems: 7, 8, 9, 10**

Dalton made a sixth assumption to resolve the dilemma of the absolute number of atoms present in a molecule, and he called it the "rule of greatest simplicity." He maintained that if two elements form only a single compound, its molecules will have the simplest possible formula: AB. Thus, he assumed that when hydrogen and oxygen combine to form water, the reaction is

$$H + O \longrightarrow HO$$

Dalton was wrong, as we now know, and the correct reaction is

$$2\ H_2 + O_2 \longrightarrow 2\ H_2O$$

## The Law of Combining Volumes

At this time a French chemist, Joseph Gay-Lussac, conducted some important experiments on the volumes of gases that react with one another to form new gases. He discovered the **law of combining volumes:**

> *The volumes of two reacting gases (at the same temperature and pressure) are in the ratio of simple integers. Moreover, the ratio of the volume of each product gas to the volume of either reacting gas is the ratio of simple integers.*

We consider three examples:

2 volumes of hydrogen + 1 volume of oxygen $\longrightarrow$ 2 volumes of water vapor

1 volume of nitrogen + 1 volume of oxygen $\longrightarrow$ 2 volumes of nitrogen oxide

3 volumes of hydrogen + 1 volume of nitrogen $\longrightarrow$ 2 volumes of ammonia

Gay-Lussac did not theorize on his experimental findings, but shortly after their publication an Italian chemist, Amedeo Avogadro, used them to formulate an important hypothesis.

## Avogadro's Hypothesis

In 1811 Avogadro stated a postulate that has become known as **Avogadro's hypothesis:**

> *Equal volumes of different gases (at the same temperature and pressure) contain equal numbers of particles.*

The question immediately arose: are "particles" of the elements the same as Dalton's atoms? Avogadro took the point of view that they were not; rather, elements could exist as diatomic molecules. With his hypothesis Avogadro could explain Gay-Lussac's law of combining volumes (Fig. 1.7). Thus, the reactions we wrote out in words become

$$2\ H_2 + O_2 \longrightarrow 2\ H_2O$$

$$N_2 + O_2 \longrightarrow 2\ NO$$

$$3\ H_2 + N_2 \longrightarrow 2\ NH_3$$

all of whose coefficients are proportional to the volumes of the reactant and product gases in Gay-Lussac's experiments and whose chemical formulas agree with modern results. Dalton, on the other hand, would have written

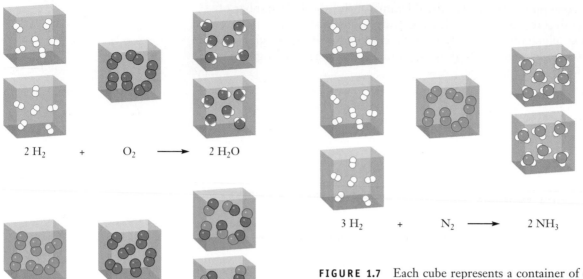

2 H₂    +    O₂    ⟶    2 H₂O

3 H₂    +    N₂    ⟶    2 NH₃

N₂    +    O₂    ⟶    2 NO

**FIGURE 1.7** Each cube represents a container of equal volume under the same conditions. If each contains the same number of molecules (Avogadro's hypothesis), and if hydrogen, oxygen, and nitrogen exist as diatomic molecules, then the combining volumes observed by Gay-Lussac in the three reactions can be understood.

$$H + O \longrightarrow OH$$
$$N + O \longrightarrow NO$$
$$H + N \longrightarrow NH$$

all of whose coefficients disagree with Gay-Lussac's observations of their relative volumes.

In addition to predicting correct molecular formulas, Avogadro's hypothesis gives correct results for the relative atomic masses of the elements. Analysis by chemists during the 18th century had revealed that 1 g of hydrogen combines fully with 8 g of oxygen to make 9 g of water. If Dalton's formula for water, HO, were correct, then an atom of oxygen would have to weigh eight times as much as an atom of hydrogen; that is, Dalton's assumption requires the **relative atomic mass** of oxygen to be 8 on a scale where the relative atomic mass of hydrogen is set at 1. From Avogadro's hypothesis, however, each water molecule contains twice as many atoms of hydrogen as oxygen, so to achieve the experimental mass relationship, each oxygen atom must have twice as large a relative atomic mass. This gives a relative atomic mass of 16 for oxygen, a result consistent with modern measurements.

One would think that Dalton would have welcomed Avogadro's brilliant hypothesis, but he did not. Dalton and others remained firm in their conviction that elements could not exist as diatomic molecules. One reason for this was the belief that molecules were held together by a force called "affinity," which expressed the attraction of opposites, just as we think of the attraction between positive and negative electric charges. If this were true, why should two *like* atoms be held together in a molecule? Moreover, if like atoms somehow did hold together in pairs, why should they not aggregate further to form molecules with three, four, or six atoms and so forth? In the face of loyalty to the affinity theory, Avogadro's reasoning did not attract the attention it deserved. Because different chemists adopted different chemical formulas for molecules, confusion reigned. A textbook published by August Kekulé in 1861 gave 19 different chemical formulas for acetic acid!

It was 50 years after Avogadro's work that an Italian chemist, Stanislao Cannizzaro, presented a paper at the First International Chemical Congress in 1860 in Karlsruhe, Germany, that convinced others to accept Avogadro's approach. Cannizzaro analyzed many gaseous compounds and showed that their chemical formulas could be established with a consistent scheme that used Avogadro's hypothesis but avoided any extra assumptions about molecular formulas. Gaseous hydrogen, oxygen, and nitrogen (and fluorine, chlorine, bromine, and iodine as well) indeed turn out to consist of diatomic molecules under ordinary conditions.

---

## 1.4

## CONCEPTS AND TOOLS OF MODERN CHEMISTRY: PHYSICAL STRUCTURE OF ATOMS

In the original Greek conception, carried over into Dalton's time and re-enforced by the laws of chemical combination, the atom was considered the ultimate and indivisible building block of matter. At the end of the 19th century this notion began to be replaced by the view that atoms were themselves composed of smaller, *elementary* particles. Scientists had carried the process of analysis (Section 1.2) down to the subatomic level.

### Electrons

One important piece of evidence for the existence of particles smaller than atoms came from studies of the effects of large electric fields on atoms and molecules. When a gas is placed in a glass tube between two conducting plates, and a large electrical potential difference is imposed across the plates, current passes through the gas; the magnitude of this current can be measured in the external circuit connecting the two metal plates. This result suggests that the electric field has broken down the atoms into new species that carry charge. (The same effect occurs when an electrical discharge from a lightning bolt passes through air.) The magnitude of the current is proportional to the amount of gas in the tube. But, when the gas is nearly all removed, the current does not go to zero; this suggests that the current originates from one of the metal plates. The mysterious invisible current carriers appeared to travel in straight lines from the cathode (the plate at negative potential) and produce a luminous spot where they impinged upon the glass tube near the anode (the plate at positive potential). These current carriers were called **cathode rays,** or **beta rays.** Several decades of research revealed that cathode rays could be deflected by both magnetic and electric fields and could heat a piece of metal foil in the tube until it glowed. Influenced by Maxwell's recent electromagnetic theory of light, one school of physicists believed the cathode rays to be a strange form of invisible light, while another group considered them to be a stream of negatively charged particles.

In 1897 the British physicist J. J. Thomson carried out a series of experiments that resolved the controversy by proving that cathode rays are negatively charged particles called **electrons.** His key experiment is depicted schematically in Figure 1.8. In a highly evacuated tube, a beam of cathode rays was produced by a cathode

**FIGURE 1.8** Thomson's apparatus to measure the electron charge-to-mass ratio, $e/m_e$. Electrons (cathode rays) stream across the tube from left to right. The electric field alone deflects the beam down, and the magnetic field alone deflects it up. By adjusting the two field strengths, Thomson could achieve a condition of zero net deflection. ($\ell$ indicates the length of the deflection plates.)

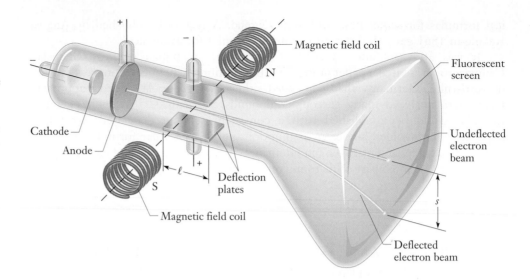

and anode in the usual way. A hole in the anode allowed some of the rays to pass between a second pair of plates that could be charged positively and negatively to establish an electric field perpendicular to the cathode-ray trajectory. For the arrangement shown in Figure 1.8, the cathode rays were deflected downward (indicating that they carried a negative charge), and the deviation could be measured accurately from the displacement of the luminous spot on a screen at the end of the tube. A magnetic field was established in the same region by passing an electric current through a pair of coils, its direction perpendicular to both that of the electric field and that of the cathode rays. The magnetic field also produced a deflection of the rays that could be made to oppose the deflection caused by the electric field. By varying the strengths of the two fields, Thomson could cause the cathode rays to pass undeviated through the tube. Under these conditions the beam experienced two equal but opposing forces. One, due to the electric field $E$, was

$$f_E = Ee$$

and the other, due to the magnetic field $H$, was

$$f_H = Hev$$

where $e$ is the electric charge on the supposed particles and $v$ is their velocity. The velocity of the cathode rays was therefore

$$v = \frac{E}{H}$$

When the magnetic field was turned off, only the electric field acted on the beam, giving it a displacement

$$s = \tfrac{1}{2}at^2$$

away from the central axis of the tube at the point where it emerged from the field between the plates as is shown in Figure 1.8. The particles acquired the acceleration $a$ toward the positive plate in the time $t$ it took for them to traverse the length of the condenser $\ell$. Because $v = \ell/t$, $t$ and then $a$ can be calculated. From Newton's second law,

$$f_E = m_e a = Ee$$

Successively substituting the above expressions for $a$, $t$, and $v$ leaves only measurable quantities:

$$\frac{e}{m_e} = \frac{a}{E} = \frac{2s}{t^2 E} = \frac{2sv^2}{\ell^2 E} = \frac{2sE}{\ell^2 H^2}$$

From this, the charge-to-mass ratio for the electron can be determined. The currently accepted value is $e/m_e = 1.7588196 \times 10^{11}$ C kg$^{-1}$, where charge is measured in coulombs and mass in kilograms. (See Appendix B for a full discussion of units.)

Thomson's apparatus is the forerunner of the modern cathode ray tube (crt) display widely used as a video monitor. Electrons emitted from the cathode are steered by rapidly varying electric and magnetic fields to "write" the video image on light-emitting materials deposited on the inside wall of the tube. The image is viewed through the glass wall at the end of the tube.

Thomson's experiment allowed determination of only the *ratio* of the charge to the mass of the electron. The actual value of the electric charge was measured in 1906 by the American physicist Robert Millikan with his student H. A. Fletcher. In their elegant experiment (Fig. 1.9), tiny drops of oil became charged by collisions with electrons or ionized gas molecules in the surrounding air. A charged oil drop (with charge $Q$ and mass $M$) situated in an electric field between two plates was subject to two forces: the force of gravity $-Mg$, causing it to fall, and a force $QE$ from the electric field, causing it to rise. By adjusting the electric field to balance the two forces and independently determining the drops' masses, $M$, from their falling speeds in the absence of an electric field, Millikan showed that the charge $Q$ was always an integral multiple of the same basic charge, $1.59 \times 10^{-19}$ C. He suggested that the different oil drops carried integral numbers of a fundamental charge, which he took to be the charge of a single electron. More accurate modern measurements led to the value $e = 1.6021773 \times 10^{-19}$ C. Together with the $e/m_e$ ratio found by Thomson, this gives $m_e = 9.109390 \times 10^{-31}$ kg for the electron mass.

## The Nucleus

Along with the electrons pulled off atoms by an electric field, positively charged particles also form. These **canal rays** were observed in cathode ray tubes but traveled in the opposite direction and were thousands of times more massive than the electron. This led Thomson to propose an atomic model in which the atom was

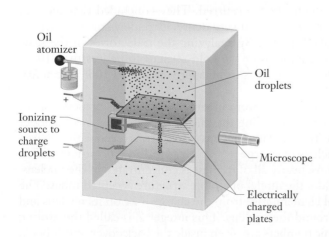

Oil atomizer

Oil droplets

Ionizing source to charge droplets

Microscope

Electrically charged plates

**FIGURE 1.9** Millikan's apparatus to measure the charge on an electron, $e$. By adjusting the electric field strength between the charged plates, Millikan could halt the fall of negatively charged oil drops and determine their net charge.

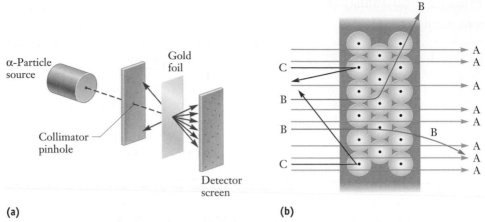

**(a)**                                                                        **(b)**

**FIGURE 1.10**   (a) Flashes of light mark the arrival of alpha particles at the detector screen. In the Rutherford experiment, the rate of hits on the screen varied from around 20 per minute at high angles to nearly 132,000 per minute at low angles. (b) The interpretation of Rutherford's experiment. Most of the alpha particles pass through the space between nuclei and undergo only small deflections (A). A few pass close to a nucleus and are more strongly deflected (B). Some are even scattered backward (C). The nucleus is far smaller proportionately than the dots shown.

regarded as consisting of electrons imbedded in a jelly-like fluid of positive charge. In this "plum pudding" model, Thomson supposed that each atom had an integral number of electrons $Z$ whose charge was exactly balanced by the jelly-like positive charge, making the atom electrically neutral.

In 1911 Ernest Rutherford made a startling discovery. He and his students had been studying the properties of alpha particles emitted from radium, an element recently isolated by Marie and Pierre Curie. Rutherford's experiments consisted in bombarding thin gold foil ($6 \times 10^{-5}$ cm thick) with these particles and observing their deflections via the scintillations they produced on a ZnS fluorescent screen (Fig. 1.10). Almost all of the alpha particles passed straight through the foil, but occasionally one was found to have been deflected through a large angle. Even less frequently, a particle was observed to be scattered backwards! Rutherford was astounded, because the alpha particles were relatively massive and fast-moving. In his words, "It was almost as incredible as if you fired a 15-inch shell at a piece of tissue paper and it came back and hit you." He and his students studied the frequency with which such large deflections occurred. They concluded that most of the mass in the gold foil was concentrated in very dense, extremely small, positively charged particles that they called **nuclei.** By analyzing the trajectories of the scattered particles, they estimated the radius of the gold nucleus to be less than $10^{-12}$ cm and the positive charge on each nucleus to be approximately $+100e$ (the actual value is $+79e$).

Rutherford proposed a model of the atom in which the nucleus possesses a net charge of $+Ze$, with $Z$ electrons surrounding the nucleus out to a distance of about $10^{-8}$ cm. In the Rutherford picture, a gold atom has 79 electrons (each with a charge of $-1e$) arranged about a nucleus of charge $+79e$. Nearly all of the volume of an atom is occupied by its electrons; nearly all of its mass is concentrated in the nucleus.

Rutherford's theory provides the model of atomic structure accepted today. The properties of a given chemical element arise from the charge $+Ze$ on its nucleus and the presence of $Z$ electrons around the nucleus. This integer $Z$ is called the **atomic number** of the element. Atomic numbers are given inside the back cover of this book.

## Protons, Neutrons, and Isotopes

The smallest and simplest nucleus is that of the hydrogen atom—the **proton.** It has a positive charge of exactly the same magnitude as the charge on the electron, but its mass is $1.67262 \times 10^{-27}$ kg, which is 1836 times greater than the electron mass. Nuclei of other elements contain $Z$ times the charge on the proton, but their masses are greater than $Z$ times the proton mass. The helium nucleus, for example, has atomic number $Z = 2$, but its mass is approximately four times the proton mass. In 1920 Rutherford suggested the existence of an uncharged particle in nuclei, the **neutron,** with a mass close to that of the proton. In this case, the nucleus consists of $Z$ protons and $N$ neutrons. The **mass number** $A$ is defined as $Z + N$.

A nuclear species (**nuclide**) is characterized by its atomic number $Z$ (that is, the nuclear charge in units of $e$, or the number of protons in the nucleus) and its mass number $A$ (the sum of protons plus neutrons in the nucleus). We denote an atom that contains such a nuclide with the symbol $^A_Z X$, where X is the chemical symbol for the element. The atomic number $Z$ is sometimes omitted because it is implied by the chemical symbol for the element. Thus, $^1_1 H$ (or $^1 H$) is a hydrogen atom, and $^{12}_6 C$ (or $^{12} C$) is a carbon atom whose nucleus contains six protons and six neutrons. **Isotopes** are nuclides of the same chemical species (that is, having the same $Z$) with different mass numbers $A$ and therefore different numbers of neutrons in the nucleus. The nuclides of hydrogen, deuterium, and tritium, represented by $^1_1 H$, $^2_1 H$, and $^3_1 H$, respectively, are all members of the family of isotopes that belong to the element hydrogen.

---

### EXAMPLE 1.2

Radon-222 ($^{222} Rn$) has recently received publicity because its presence in basements may increase the number of cancer cases in the general population, especially among smokers. State the number of electrons, protons, and neutrons that make up an atom of $^{222} Rn$.

### Solution

From the table at the back of the book, the atomic number of radon is 86, so the nucleus contains 86 protons and $222 - 86 = 136$ neutrons. The atom has 86 electrons to balance the positive charge on the nucleus.

**Related Problems: 15, 16, 17, 18**

---

### 1.5

## CONCEPTS AND TOOLS OF MODERN CHEMISTRY: THE PERIODIC TABLE

The number of known chemical compounds is already vast, and it increases rapidly with new research. The number of reactions these compounds undergo is unlimited. The resultant body of chemical knowledge, viewed as a collection of facts, is overwhelming in its size, range, and complexity. It has been made manageable by the observation that the properties of the elements naturally display certain regularities; these enable classification of the elements into families whose members have similar chemical and physical properties. When the elements are listed in order of increasing atomic number Z, these families display a remarkable fact: patterns of chemical behavior recur regularly as a function of Z. This discovery is summarized concisely by **the periodic law:**

> The chemical properties of the elements are periodic functions of the atomic number Z.

Consequently the elements listed in order of increasing Z can be arranged in a chart called the **periodic table** that displays at a glance the patterns of chemical similarity. The periodic table then permits systematic classification, interpretation, and prediction of all chemical information.

The modern periodic table, shown in Figure 1.11 and inside the front cover of this book, places elements in **groups** (arranged vertically) and **periods** (arranged horizontally). There are eight groups of **representative elements** or "main-group" elements. In addition, there are ten groups (and three periods) of **transition-metal elements,** a period of elements from atomic number 57 through 71 called the rare-earth or **lanthanide elements,** and a period of elements from atomic number 89 through 103 called the **actinides,** all of which are unstable and most of which must be produced artificially. The lanthanide and actinide elements are usually placed below the rest of the table to conserve space. The groups of representative elements are numbered (with Roman numerals) from I to VIII, with the letter A sometimes added to differentiate them from the transition-metal groups, which are labeled from IB to VIIIB. In this book we will use group numbers exclusively for the representative elements (dropping the "A"), and we will refer to transition-metal elements by the first element in the corresponding group. For example, the elements in the carbon group (C, Si, Ge, Sn, Pb) are designated as Group IV, and the elements chromium (Cr), molybdenum (Mo), and tungsten (W) as the chromium group.[3]

## Survey of Chemical Properties: The Representative Elements

Empirical classification of the elements is based on similarities in physical or chemical properties. Elements are classified as **metals** or **nonmetals,** depending on the presence (or absence) of a characteristic metallic luster, good (or poor) ability to conduct electricity and heat, and malleability (or brittleness). Certain elements (antimony, arsenic, boron, silicon, and tellurium) resemble metals in some respects and nonmetals in others and are therefore called **semimetals** or **metalloids.** Empirical formulas of the binary compounds of the elements with chlorine (their *chlorides*), with oxygen (their *oxides*), and with hydrogen (their *hydrides*), reveal chemical periodicity.

Group I, the **alkali metals** (lithium, sodium, potassium, rubidium, and cesium), are all relatively soft metals with low melting points that form 1 : 1 compounds with chlorine, having chemical formulas such as $NaCl$ and $RbCl$. The alkali metals react with water to liberate hydrogen; potassium, rubidium, and cesium liberate sufficient heat of reaction to ignite the hydrogen. Group II, the **alkaline-earth metals** (beryllium, magnesium, calcium, strontium, barium, and radium), react in a 1 : 2 atom ratio with chlorine, giving compounds such as $MgCl_2$ and $CaCl_2$.

Of the nonmetallic elements, Group VI, the **chalcogens** (oxygen, sulfur, selenium, and tellurium), form 1 : 1 compounds with alkaline-earth metals (e.g., $CaO$

---

[3] Recently, several international organizations recommended a new system of group designation in which the main-group elements make up Groups 1, 2, and 13 through 18, and the transition-metal elements fill Groups 3 through 12.

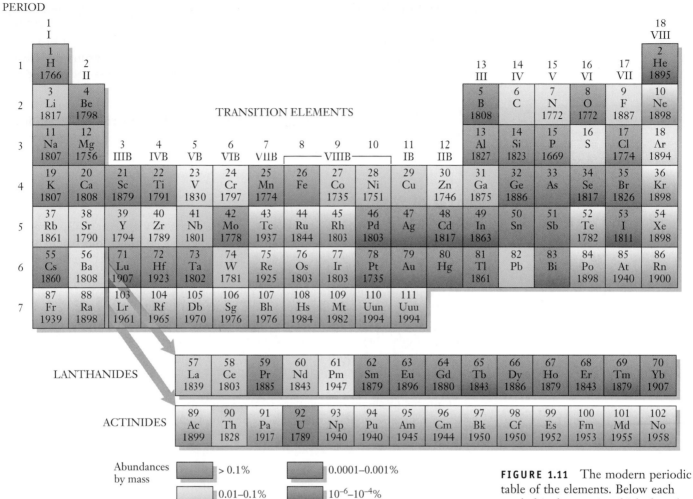

**FIGURE 1.11**   The modern periodic table of the elements. Below each symbol is the year in which that element was discovered; elements with no dates have been known since ancient times. Above each symbol is the atomic number. The color coding indicates the relative abundance by mass of the elements in the world (the atmosphere, oceans and freshwater bodies, and the Earth's crust to a depth of 40 km). Oxygen alone comprises almost 50% of this mass, and silicon comprises more than 25%.

and BaS ) but 2:1 compounds with the alkali metals (e.g., $Li_2O$ and $Na_2S$). Members of Group VII, the **halogens** (fluorine, chlorine, bromine, and iodine) differ significantly in physical properties (the first two are gases at room temperature, bromine is a liquid, and iodine a solid), but their chemical behaviors are similar. Any alkali-metal element will combine with any halogen in 1:1 proportion to form a compound such as LiF or RbI, called an **alkali halide.** Alkaline-earth metals react with halogens in 1:2 proportion, giving alkaline-earth halides such as $CaF_2$ and $MgBr_2$.

Other elements fall into three additional groups that are somewhat less clearly defined in their chemical and physical properties than those already mentioned. Group III contains a semimetal (boron) and four metals (aluminum, gallium, indium, and thallium). All form 1:3 chlorides (such as $GaCl_3$) and 2:3 oxides (such as $Al_2O_3$). Group IV consists of the elements carbon, silicon, germanium, tin, and lead. All of these elements form 1:4 chlorides (such as $SiCl_4$), 1:4 hydrides (such as $GeH_4$), and 1:2 oxides (such as $SnO_2$). Tin and lead are metals with low melting points, and silicon and germanium are semimetals. Elemental carbon exists in numerous forms: graphite, diamond, and the recently discovered fullerenes. Group V includes nitrogen, phosphorus, arsenic, antimony, and bismuth. With hydrogen and oxygen these elements form binary compounds with empirical formulas such as $PH_3$ and $N_2O_5$. The hydrides become increasingly unstable as their molar masses increase, and $BiH_3$ can only be kept below $-45°C$. A similar trend exists for the oxides, and $Bi_2O_5$ has never been obtained in pure form. The lighter members of this group are clearly nonmetals (nitrogen and phosphorus)—bismuth is clearly a metal—and arsenic and antimony are classed as semimetals.

Group VIII, the noble gases (helium, neon, argon, krypton, xenon, and radon), are sometimes called the **inert gases** because of their relative inertness toward chemical combination. These all appear in monatomic form, unlike the other permanent gases (hydrogen, oxygen, nitrogen, fluorine, chlorine) which are diatomic.

## Survey of Physical Properties

Physical as well as chemical properties of the elements vary systematically across the periodic table. Important physical properties include melting points and boiling points, thermal and electrical conductivities, densities, atomic sizes, and the energy changes involved in adding an electron to or removing an electron from a neutral atom. Appendix F presents numerical values for most of these properties. In general, the elements on the left side of the table (especially in the later periods) are metallic solids and good conductors of electricity. On the right side (especially in earlier periods) they are poor conductors of electricity and are generally found in the gaseous state at room temperature. In between, the semimetals form a zigzag line of division between metallic and nonmetallic elements (see inside the front cover of this book).

Patterns in chemical reactivity of the elements, evidenced by the arrangement of the groups and periods in the table, correlate with patterns in the physical structure of the atom, evidenced by $Z$. Reading across the table horizontally shows that each main group element (Groups I–VIII) in Period 3 contains exactly 8 more electrons than the element immediately above it. Similarly, each main group element in Periods 4 and 5 contains exactly 18 more electrons than the element immediately above it. The explanation why increasing $Z$ by 8 in period 2 and increasing $Z$ by 18 in periods 3 and 4 produces similar chemical reactivity will be provided by the quantum mechanical theory of atomic structure in Chapter 15.

## Development of the Periodic Table

The periodic table is a monumental scientific achievement. Its development is a fascinating story which illustrates the essential interplay between observation, prediction, and testing required for scientific progress. By the late 1860s more than 60 chemical elements had been identified, and much was known about their descriptive chemistry. Various proposals were put forth to arrange the elements into groups based on similarities in chemical and physical properties. The next step was to recognize a connection between group properties and atomic masses. When the elements known at the time were written in order of increasing atomic mass, repetitive sequences of the groups appeared. When the series of elements was written so as to begin a new row with each alkali metal, elements of the same groups were automatically assembled in columns in a periodic table of the elements that was the forerunner of the modern table.

When the German chemist Lothar Meyer and (independently) the Russian Dmitri Mendeleev first introduced the periodic table in 1869–70, one third of the naturally occurring chemical elements had not yet been discovered. Yet both chemists were sufficiently farsighted to leave gaps where their analyses of periodic physical and chemical properties indicated that new elements should be located. Mendeleev was bolder than Meyer and even assumed that if a measured atomic mass put an element in the wrong place in the table, the atomic mass was wrong. In some cases this was true. Indium, for example, had previously been assigned an atomic mass between those of arsenic and selenium. Because there is no space in the periodic table between these two elements, Mendeleev suggested that the atomic mass of indium be changed to a completely different value, where it would fill an empty space between cadmium and tin. Subsequent work has shown that the elements are *not* ordered strictly by atomic mass in the periodic system. For example, tellurium comes before iodine in the periodic table, even though its atomic mass is slightly greater. Such anomalies are due to the relative abundances of the isotopes of the elements involved. We now know (Section 1.4) that atomic number, not mass number, determines chemical behavior.

Mendeleev went further than Meyer in another respect: he predicted the properties of six elements yet to be discovered. For example, a gap just below aluminum suggested a new element would be found with properties analogous to those of aluminum. Mendeleev designated this element "eka-aluminum" (Sanskrit *eka*, "next") and predicted its properties. Just five years later an element with the proper atomic mass was isolated and named *gallium* by its discoverer. The close correspondence between the observed properties of gallium and Mendeleev's predictions for eka-aluminum lent strong support to the periodic law. Additional support came in 1885 when eka-silicon, which had also been described in advance by Mendeleev, was discovered and named *germanium*.

The structure of the periodic table appeared to limit the number of possible elements. It was therefore quite surprising when John William Strutt, Lord Rayleigh, discovered a gaseous element in 1894 that did not fit into the previous classification scheme. A century earlier, Henry Cavendish had noted the existence of a residual gas when oxygen and nitrogen are removed from air, but its importance had not been realized. Together with William Ramsay, Rayleigh isolated the gas and named it argon. Ramsay then studied a gas that was present in natural gas deposits and discovered that it was helium, an element whose presence in the sun had been noted earlier in the spectrum of sunlight but that had not previously been known on Earth. Rayleigh and Ramsay postulated the existence of a new group of elements, and in 1898 other members of the series (neon, krypton, and xenon) were isolated.

---

### 1.6

## THE MOLE CONCEPT: WEIGHING AND COUNTING MOLECULES

The laws of chemical combination assert that chemical reactions occur not through arbitrary amounts of material but by regrouping atoms in numbers specific to the substances involved in the reaction. How does one weigh out a sample containing exactly the number of atoms or molecules needed for a particular chemical reaction? What is the mass of an atom or a molecule? These questions must be answered indirectly, since atoms and molecules are far too small to permit direct determination of their masses.

While developing the laws of chemical combination, chemists had determined indirectly the relative masses of atoms of different elements; for example, 19th-century chemists concluded that oxygen atoms weigh 16 times more than hydrogen atoms. In the early 20th century, atomic and molecular relative masses were determined much more accurately by the work of J. J. Thomson, F. W. Aston, and others who developed the technique of **mass spectrometry** as an extension of Thomson's measurement of the charge–mass ratio of the electron. These relative masses on the atomic scale are related to absolute masses on the gram scale through a conversion factor. Finally, one weighs out a sample of the mass required to provide the desired number of atoms or molecules of a substance for a particular reaction. These concepts and methods are developed in the present section.

## Direct Physical Measurements of Relative Atomic and Molecular Masses: Mass Spectrometry

**Mass spectrometry** is the chemist's most accurate method for determining relative atomic and molecular masses. In a mass spectrometer (Fig. 1.12), one or more electrons are removed from each such atom or molecule. The resulting positively charged species, called **ions,** are accelerated by an electric field then passed through a magnetic field. The extent of curvature of the particle trajectories depends on the ratio of their charges to their masses, just as in Thomson's experiments on cathode rays (electrons) described in Section 1.4. This technique allows species of different mass to be separated and detected. Early experiments in mass spectrometry demon-

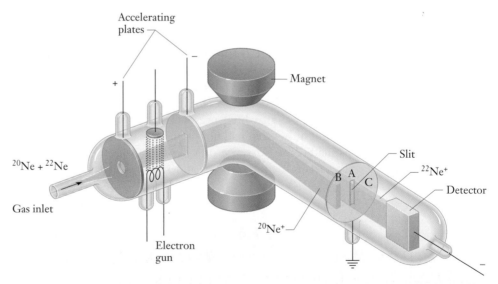

**FIGURE 1.12** A simplified representation of a modern mass spectrometer. A gas mixture containing the isotopes $^{20}$Ne and $^{22}$Ne is introduced through the gas inlet. Some of these atoms are ionized by collisions with electrons as they pass through the electron gun. The resulting ions are accelerated to a particular kinetic energy by the electric field between the accelerating plates. The ion beam passes into a magnetic field, where it is separated into components, each containing ions with a characteristic charge-to-mass ratio. Here, the spectrometer has been adjusted to detect the less strongly deflected $^{22}$Ne$^+$ ions in the inlet mixture. By changing the magnitude of the electric or magnetic field, one can move the beam of $^{20}$Ne$^+$ ions from B to A, and the beam of $^{22}$Ne$^+$ ions from B to C, so that $^{20}$Ne$^+$ ions can be detected.

strated, for example, a mass ratio of 16:1 for oxygen relative to hydrogen, confirming by *physical* techniques a relationship deduced originally on *chemical* grounds.

Because many elements have more than one naturally occurring isotope, however, the relationship between the chemist's and the physicist's relative atomic masses was not always so simple. The atomic mass of chlorine determined by chemical means was 35.45 (relative to an oxygen atomic mass of 16), but instead of showing a single peak in the mass spectrum corresponding to a relative atomic mass of 35.45, chlorine showed two peaks, with relative masses near 35 and 37. Approximately three quarters of all chlorine atoms appear to have relative atomic mass 35 (the $^{35}$Cl atoms), and one quarter have relative atomic mass 37 (the $^{37}$Cl atoms). Naturally occurring chlorine is thus a mixture of two isotopes with different masses but very nearly identical chemical properties. Elements in nature usually are mixtures, after all. Dalton's second assumption (that all atoms of a given element are identical in mass) is thus shown to be wrong in most cases. Because the *chemical* properties of different isotopes are so similar, their existence was not discovered until the development of the mass spectrometer.

Until 1900 chemists worked with a scale of relative atomic masses in which the average relative atomic mass of hydrogen was set at 1. At about that time, they changed to a scale in which the average relative atomic mass of naturally occurring oxygen (a mixture of $^{16}$O, $^{17}$O, and $^{18}$O) was set at 16. In 1961, by international agreement, the atomic mass scale was further revised, with the adoption of exactly 12 as the relative atomic mass of $^{12}$C. There are two stable isotopes of carbon: $^{12}$C and $^{13}$C ($^{14}$C and other isotopes of carbon are unstable and of very low terrestrial abundance). Natural carbon contains 98.892% $^{12}$C and 1.108% $^{13}$C by mass. The relative atomic masses of the elements as found in nature can be obtained as averages over the masses of the isotopes of each element, weighted by their observed fractional abundances. If an element consists of $n$ isotopes, of which the $i$th isotope has a mass $A_i$ and a fractional abundance $p_i$, then the average relative atomic mass of the element in nature (its chemical relative atomic mass) will be

$$A = A_1 p_1 + A_2 p_2 + \cdots + A_n p_n \equiv \sum_{i=1}^{n} A_i p_i \qquad \textbf{[1.1]}$$

The relative atomic mass of a nuclide is close to but (except for $^{12}$C) not exactly equal to its mass number.

## EXAMPLE 1.3

Calculate the chemical relative atomic mass of carbon, taking the relative atomic mass of $^{13}$C to be 13.003354 on the $^{12}$C scale.

### Solution
Set up the following table:

| Isotope | Isotopic Mass × Abundance |
|---------|---------------------------|
| $^{12}$C | $12.000000 \times 0.98892 = 11.867$ |
| $^{13}$C | $13.003354 \times 0.01108 = 0.144$ |
| | Chemical relative atomic mass $= 12.011$ |

**Related Problems: 23, 24, 25, 26**

The number of significant figures in a table of chemical or natural relative atomic masses (see inside the back cover of this book) is limited not only by the accuracy of the mass spectrometric data but also by any variability in the natural abundances of the isotopes. If lead from one mine has a relative atomic mass of 207.18 and lead from another has 207.23, there is no way a result more precise than 207.2 can be obtained. In fact, geochemists now are able to use small variations in the $^{16}O:^{18}O$ isotopic abundance ratio as a "thermometer" to deduce the temperatures at which different oxygen-containing rocks were formed in the Earth's crust over geologic time scales. They also find anomalies in the oxygen isotopic compositions of certain meteorites, implying that their origins may lie outside our solar system.

*Relative atomic masses have no units* because they are ratios of two masses measured in whatever units we choose (grams, kilograms, pounds, and so forth). The **relative molecular mass** of a compound is the sum of the relative atomic masses of the elements that constitute it, each one multiplied by the number of atoms of that element in a molecule. The formula of water is $H_2O$, for example, so its relative molecular mass is

2 (relative atomic mass of H) + 1 (relative atomic mass of O) =
$$2(1.0079) + 1(15.9994) = 18.0152$$

## Relation of Atomic and Macroscopic Masses: Avogadro's Number

What are the actual masses (in grams) of individual atoms and molecules? To answer this, it is necessary to establish a connection between the macroscopic scale of masses used in the laboratory and the microscopic scale of the masses of individual atoms and molecules. The link between the two is provided by **Avogadro's number** ($N_0$), defined as the number of atoms in exactly 12 g of $^{12}C$. We will later consider some of the many experimental methods that have been developed to determine the numerical value of $N_0$. Its currently accepted value is

$$N_0 = 6.022137 \times 10^{23}$$

The mass of a single $^{12}C$ atom is then found by dividing exactly 12 g by $N_0$:

$$\text{mass of a } ^{12}C = \frac{12.0000 \text{ g}}{6.022137 \times 10^{23}} = 1.992648 \times 10^{-23} \text{ g}$$

This is truly a very small mass, reflecting the very large number of atoms in a 12-g sample of carbon.

Avogadro's number is defined relative to the $^{12}C$ atom because that isotope is the basis for the modern scale of relative atomic masses. We can apply it to other substances as well, in a particularly simple fashion. Consider sodium, which has a relative atomic mass of 22.98977. A sodium atom is 22.98977/12 times as heavy as a $^{12}C$ atom. If $N_0$ atoms of $^{12}C$ have a mass of 12 g, then the mass of $N_0$ atoms of sodium must be

$$\frac{22.98977}{12} (12 \text{ g}) = 22.98977 \text{ g}$$

The mass, in grams, of $N_0$ atoms of *any* element is numerically equal to the relative atomic mass of that element. The same conclusion applies to molecules. From the relative molecular mass of water calculated earlier, the mass of $N_0$ molecules of water is 18.0152 g.

| EXAMPLE 1.4 |
|---|

One of the heaviest atoms found in nature is $^{238}U$. It has a relative atomic mass of 238.0508 on a scale in which 12 is the atomic mass of $^{12}C$. Calculate the mass (in grams) of one atom of $^{238}U$.

### Solution

Because $N_0$ atoms of $^{238}U$ have a mass of 238.0508 g and $N_0$ is $6.022137 \times 10^{23}$, one atom must have a mass equal to

$$\frac{238.0508 \text{ g}}{6.022137 \times 10^{23}} = 3.952929 \times 10^{-22} \text{ g}$$

**Related Problems: 27, 28**

## The Mole Concept

Atoms and molecules have very small masses, so chemical experiments with measurable amounts of substances involve many of these atoms and molecules. It is convenient to group atoms or molecules in counting units of $N_0 = 6.022137 \times 10^{23}$ to measure the **chemical amount** of a substance. One of these counting units is called a **mole** (abbreviated "mol"; from Latin *moles*, meaning "heap" or "pile"). One mole of a substance is the amount that contains Avogadro's number of atoms, molecules, or other entities. In other words, 1 mol of $^{12}C$ contains $N_0$ $^{12}C$ atoms, 1 mol of water contains $N_0$ water molecules, and so forth. We must be careful in some cases, because a phrase such as "1 mol of oxygen" is ambiguous. We should refer instead to "1 mol of $O_2$" if there are $N_0$ oxygen *molecules*, and "1 mol of O" if there are $N_0$ oxygen *atoms*.

The mass of one mole of atoms of an element—the **molar mass,** with units of grams per mole—is numerically equal to the dimensionless relative atomic mass of that element, and the same relationship holds between the molar mass of a compound and its relative molecular mass. Thus, the relative molecular mass of water is 18.0152, and its molar mass is 18.0152 g $mol^{-1}$.

To determine the chemical amount, in moles, of a given substance, we use the chemist's most powerful tool, the laboratory balance. If a sample of iron weighs 8.232 g, then

$$\text{chemical amount of iron} = \frac{\text{number of grams of iron}}{\text{molar mass of iron}}$$

$$= \frac{8.232 \text{ g Fe}}{55.847 \text{ g mol}^{-1}}$$

$$= 0.1474 \text{ mol Fe}$$

where the molar mass of iron was obtained from a table of relative atomic masses. The calculation can be turned around, as well. Suppose a certain chemical amount, 0.2000 mol, of water is needed in a chemical reaction. We have

$$(\text{chemical amount of water}) \times (\text{molar mass of water}) = \text{mass of water}$$

$$(0.2000 \text{ mol H}_2\text{O}) \times (18.015 \text{ g mol}^{-1}) = 3.603 \text{ g H}_2\text{O}$$

**FIGURE 1.13** One-mole quantities of several substances. Clockwise from top: graphite (C), potassium permanganate (KMnO₄), copper sulfate pentahydrate (CuSO₄·5H₂O), copper (Cu), sodium chloride (NaCl), and potassium dichromate (K₂Cr₂O₇). Antimony (Sb) is at the center. *(Leon Lewandowski)*

Thus, we simply measure out 3.603 g of water. In both cases, the molar mass is the conversion factor between the substance's mass and its chemical amount in moles.

Although numbers of moles are frequently determined by weighing, it is still preferable to think of a mole as a fixed number of particles (Avogadro's number) rather than as a fixed mass. The term "mole" is thus analogous to a term such as "dozen": one dozen pennies weighs 26 g, substantially less than the mass of one dozen nickels, 60 g; but each group contains 12 coins. Figure 1.13 shows mole quantities of several substances.

### EXAMPLE 1.5

Nitrogen dioxide ($NO_2$) is a major component of urban air pollution. For a sample containing 4.000 g of $NO_2$, calculate **(a)** the chemical amount (the number of moles) of $NO_2$ and **(b)** the number of molecules of $NO_2$.

### Solution

**(a)** From the tabulated molar masses of nitrogen (14.007 g mol$^{-1}$) and oxygen (15.999 g mol$^{-1}$), the molar mass of $NO_2$ is

$$14.007 \text{ g mol}^{-1} + (2 \times 15.999 \text{ g mol}^{-1}) = 46.005 \text{ g mol}^{-1}$$

The chemical amount of $NO_2$ is then

$$\text{moles NO}_2 = \frac{4.000 \text{ g NO}_2}{46.005 \text{ g mol}^{-1}} = 0.08695 \text{ mol NO}_2$$

**(b)** To convert from moles to number of molecules, multiply by Avogadro's number:

$$\text{molecules NO}_2 = (0.08695 \text{ mol NO}_2) \times 6.0221 \times 10^{23} \text{ mol}^{-1}$$

$$= 5.236 \times 10^{22} \text{ molecules NO}_2$$

**Related Problems: 33, 34**

## Density and Molecular Size

The **density** of a sample is the ratio of its mass to its volume:

$$\text{density} = \frac{\text{mass}}{\text{volume}} \qquad \textbf{[1.2]}$$

The base unit of mass in the International System of Units (discussed in Appendix B) is the kilogram (kg), but this is often inconveniently large for practical purposes in chemistry. Frequently the gram is used instead; moreover, the gram is the standard unit for expressing molar masses. Several units for volume are in frequent use. The base SI unit of the cubic meter ($m^3$) is also quite unwieldy for laboratory purposes (1 $m^3$ of water weighs 1000 kg, or 1 metric ton). For volume we will therefore use the liter (1 L = $10^{-3}$ $m^3$) and the cubic centimeter, which is identical to the milliliter (1 $cm^3$ = 1 mL = $10^{-3}$ L = $10^{-6}$ $m^3$). Table 1.1 lists the densities of some substances in units of grams per cubic centimeter.

The density of a substance is not a fixed, invariant quantity but depends on the pressure and temperature at the time of measurement. For some substances (especially gases and liquids), the volume may be more convenient to measure than the mass, and when the density is known, it provides the conversion factor between volume and mass. For example, near room temperature liquid benzene ($C_6H_6$) has a density of 0.8765 g $cm^{-3}$. Suppose that 0.2124 L of benzene is measured into a container. The mass of benzene is then the product of volume and density:

$$0.2124 \text{ L} \times (1 \times 10^3 \text{ cm}^3 \text{ L}^{-1}) \times (0.8765 \text{ g cm}^{-3}) = 186.2 \text{ g}$$

Dividing this by the molar mass of benzene (78.114 g $mol^{-1}$) gives the corresponding chemical amount, 2.384 mol.

Knowing the density and molar mass of a substance, we readily compute its **molar volume,** the volume occupied by one mole of substance:

$$V_m = \frac{\text{molar mass (g mol}^{-1})}{\text{density (g cm}^{-3})} = \text{molar volume (cm}^3 \text{ mol}^{-1})$$

Near 0°C, for example, ice has a density of 0.92 g $cm^{-3}$, so the molar volume of solid water under these conditions is

$$V_m = \frac{18.0 \text{ g mol}^{-1}}{0.92 \text{ g cm}^{-3}} = 20 \text{ cm}^3 \text{ mol}^{-1}$$

The molar volume of a gas is much larger than this. For $O_2$ at room conditions, the data in Table 1.1 give a molar volume of 24,600 $cm^3$ $mol^{-1}$ = 24.6 L $mol^{-1}$, which is more than 1000 times larger than the molar volume just computed for ice under the same conditions of temperature and pressure. How can we interpret this fact on a microscopic level? We note also that the volumes of liquids and solids do not shift very much with changes in temperature or pressure, but gas volumes are quite sensitive to these changes. A hypothesis to explain such observations is that molecules in liquids and solids are close enough to touch one another, but in a gas they are separated by large distances. If this hypothesis is correct (and it has been borne out by further study), then the sizes of the molecules themselves can be estimated from the volume per molecule in the liquid or solid state. The volume per molecule is the molar volume divided by Avogadro's number; for ice this gives

$$\text{volume per } H_2O \text{ molecule} = \frac{20 \text{ cm}^3 \text{ mol}^{-1}}{6.02 \times 10^{23} \text{ mol}^{-1}} = 3.3 \times 10^{-23} \text{ cm}^3$$

### TABLE 1.1

### *Densities of Some Substances*[†]

| Substance | Density (g cm$^{-3}$) |
|---|---|
| Hydrogen | 0.000082 |
| Oxygen | 0.00130 |
| Water | 1.00 |
| Magnesium | 1.74 |
| Sodium chloride | 2.16 |
| Quartz | 2.65 |
| Aluminum | 2.70 |
| Iron | 7.86 |
| Copper | 8.96 |
| Silver | 10.5 |
| Lead | 11.4 |
| Mercury | 13.5 |
| Gold | 19.3 |
| Platinum | 21.4 |

[†] These densities were measured at room temperature and at average atmospheric pressure near sea level.

This corresponds to a cube with edges about $3.2 \times 10^{-8}$ cm long. We conclude from this and other density measurements that the characteristic size of atoms and small molecules is on the order of $10^{-8}$ cm. Avogadro's number provides the link between laboratory-scale measurements and the masses and volumes of single atoms and molecules.

### Direct Physical Detection of the Atom: Scanning Tunneling Microscopy

The laws of chemical combination provide indirect evidence for the existence of atoms, and Avogadro's number permits inference of the size scale of atoms. The scanning tunneling microscope (STM) provides much more direct evidence for the existence and the size of atoms.

Microscopy began with the fabrication of simple magnifying glasses and evolved by the late 17th century to the first optical microscopes through which single biological cells could be observed. By the 1930s the electron microscope had been developed for detection of objects too small to be seen in optical microscopes. The electron microscope revealed single atoms, but at the cost of damage inflicted to the sample by the high-energy beam of electrons required to resolve such small objects. In the 1980s Gerd Binnig and Heinrich Rohrer in Switzerland developed the scanning tunneling microscope (STM), which images atoms using low-energy electrons. For this accomplishment they received the 1986 Nobel Prize in physics. This device uses a very fine-pointed electrically conducting probe that is passed over the surface of the sample being examined (Fig. 1.14a). When the probe is nearly in contact with atoms of the sample, a small electrical current (called the "tunneling current") can pass from the sample to the probe. The magnitude of this current is

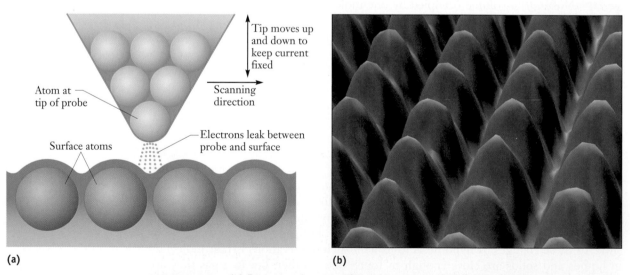

(a)                                                                 (b)

**FIGURE 1.14**    (a) In a scanning tunneling microscope (STM), an electric current passes through a single atom or a small group of atoms in the probe tip and then into the surface of the sample being examined. As the probe moves over the surface, its distance is adjusted to keep the current constant, allowing a tracing out of the shapes of the atoms or molecules on the surface. (b) The STM image of the surface of a nickel crystal illustrates the corrugated nature of the surface produced by the troughs between rows of Ni atoms. *(b courtesy of Dr. Don Eigler/IBM Almaden Research Center, San Jose, CA)*

extremely sensitive to the distance of the probe from the surface, decreasing by a factor of 1000 as the probe moves away from the surface by a distance as small as $10^{-8}$ cm. Feedback circuitry holds the current constant while the probe is swept laterally across the surface, moving up and down in the vertical direction as it passes over structural features in the surface. The vertical position is monitored and the information is stored in a computer. By sweeping the probe tip along each of a series of closely spaced parallel tracks, a three-dimensional image of the surface can be constructed and displayed (see Fig. 1.14b, as well as the Unit 1 opening figure preceding page 1 and the Chapter 1 opening figure on page 2).

STM images confirm visually many features, such as size of atoms and the distances between them, already known from other techniques. However, much new information has been obtained as well. The STM images have revealed the positions and shapes of molecules undergoing chemical reactions on surfaces, which helps to guide the search for new ways of carrying out such reactions. They have also revealed the shape of the surface of the molecules of the nucleic acid DNA, which plays a central role in genetics.

---

### 1.7

## THE CONCEPT OF ENERGY: FORMS, MEASUREMENT, AND CONSERVATION

Sections 1.3 through 1.6 emphasized the key roles of the structure and conservation of matter in forming chemical compounds. This discussion will be extended in Chapter 2 to develop methods for calculating the amounts of compounds that are formed in specific chemical reactions. Transformation of one substance into another through chemical reactions is accompanied and influenced by changes in energy. Energy changes will be studied in several different contexts throughout this book; for example, we will measure the energy required to remove an electron from an atom, the energy liberated when an electron is added to an atom, and the energy required to rupture a particular chemical bond. The present section summarizes general features of energy changes as background for these more specialized applications described later. This summary assumes familiarity with the concepts of force, velocity, acceleration, and work. Adequate background material on each is provided in Appendix B.

### Forms of Energy

The concept of **energy** originated in the science of mechanics and was defined as the capacity to perform work, that is, to move an object from one position to another. It is now understood that energy appears in many different forms, each of which can be used to cause particular kinds of physical and chemical changes. From daily experience, we recognize the kinetic energy due to the speed of an onrushing automobile. To stop the automobile, its kinetic energy must be overcome by work done by the brakes. Otherwise the automobile crashes into stationary objects and expends its kinetic energy in deforming these objects, as well as itself. We recognize the potential energy of a mass of snow on a ski slope, which becomes the kinetic energy of an avalanche. We recognize the electrical energy stored in a battery, which can move objects by driving motors or warm objects through electric heaters. We recognize the chemical energy stored in gasoline, which can move objects by powering the internal combustion engine. We recognize the thermal energy of hot steam, which can move objects by driving the steam engine.

Each of these forms of energy plays a role in chemistry and each will be described at the appropriate point later in the book. Here we concentrate on the nature of potential and kinetic energy and their interconversion.

The **kinetic energy** of a moving object is defined by

$$KE = \tfrac{1}{2} mv^2$$

where $m$ is the mass of the object and $v$ is its speed. A stationary object has no kinetic energy. In the SI systems of units (see Appendix B), energy is expressed in joules (J). Thus 0.5 joule is the kinetic energy of an object with mass 1 kg moving at a speed of 1 m s$^{-1}$.

**Potential energy** is the energy stored in an object due to its location relative to a specified reference position. Potential energy therefore depends explicitly on the position $x$ of the object, and is expressed as a mathematical function $V(x)$. The change in potential energy when an object is moved from position $x_1$ to position $x_2$ is equal to the work done in moving the object:

$$\text{change in PE} = \text{force} \cdot \text{displacement} = \text{force} \cdot (x_2 - x_1)$$

For example, to lift an object of mass $m$ from the surface of the Earth to the height $h$, some agency must perform work in the amount $mgh$ against the downward-directed force of gravity, which has constant acceleration denoted by $g$. The resulting change in potential energy is $mgh$. Potential energy is expressed in joules in the SI system of units.

## Conservation of Energy

The science of mechanics deals with idealized motions of objects in which friction does not occur. During the motions of an object, its potential and kinetic energy are interconverted subject to the restriction that their sum always remains constant. For example, consider a soccer ball rolling down the side of a steep gully at the edge of the playing field. The ball rolls down one side, across the bottom, partway up the opposite side, then reverses direction, and continues to oscillate across the bottom of the gully. On each downward leg of its journey, the ball loses potential energy and gains kinetic energy, but their sum remains constant. On each upward leg, the ball loses kinetic energy and gains potential energy, but their sum remains constant.

The description in the previous paragraph is clearly an idealization, since eventually the ball comes to rest at the bottom of the gully. On each upward leg and on each downward leg the ball loses some of its kinetic energy through friction with the surface of the gully. Both the ball and the gully surface are slightly warmed as a result. The energy lost from the purely mechanical motion is added to the internal energy of the ball and the gully surface. Internal energy, heat, and friction are discussed in detail in Chapter 7 as part of the science of thermodynamics. The total amount of energy has not been changed; energy has been neither created nor destroyed in this process. Rather, a new mode of energy storage or transfer has been identified to interpret the new effects beyond those explained by basic mechanics.

By similar arguments, the law of conservation of energy has been extended from the idealized motions of mechanics to include a broad variety of phenomena in which several different forms of energy are involved. This law is one of the securest building blocks in scientific reasoning and can be used as the starting point for interpreting and relating a great range of superficially different processes. In chemistry, this law provides the foundation for studying complex processes in which kinetic, potential, electrical, chemical, and thermal energy are interconverted without net loss or gain.

## Potential Energy Curves

Consider again the soccer ball rolling down the walls of the gully. This process is shown schematically in Figure 1.15, where the curve $V(x)$ represents a "cross-sectional" sketch of the gully. The curve $V(x)$ also represents the potential energy of the ball relative to its value at the bottom of the gully. The $x$ coordinate locates the distance of the ball from the bottom of the gully to a position along its side. Suppose the ball is held in place at the position $x_1$. It possesses only potential energy, which we represent by the value $E_1$. If the ball is released, it falls down the slope and passes across the bottom where its potential energy is zero and its kinetic energy is $E_1$. It then climbs the opposite side until it rises to position $-x_1$, where its kinetic energy is zero. The ball promptly reverses direction and retraces its path back to position $x_1$ where its total energy $E_1$ is all potential energy.

Knowledge of the potential energy curve as a function of position enables prediction of the net force on the object at each position. From either side, the force is directed toward the bottom; the force is always in that direction in which the slope of the potential energy curve is negative. The physical explanation is the following. From the definition of potential energy stated earlier, it can be shown that

$$\text{force} = -\ (\text{slope of PE})$$

The net force drives the object toward the position where the potential energy is a minimum and its slope is zero.

Similar potential energy diagrams can be used to represent the interaction between a pair of objects, such as Earth–moon, Earth–Mars, electron–nucleus, electron–electron, nucleus–nucleus, atom–atom, molecule–molecule. Such diagrams will be constructed at several points later in this book and used to interpret the relative motions of the pair of objects. These considerations are extremely important in describing the formation of chemical bonds, the states of matter, and the role of molecular collisions in chemical reactions.

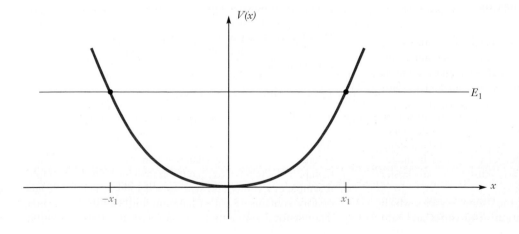

**FIGURE 1.15** Sketch of the potential energy of a soccer ball rolling down the sides of a gully. The ball is released at position $x_1$ with total energy $E_1$. In idealized motion, it continues to oscillate back and forth across the bottom of the gully. The net force on the ball is always in the direction in which the potential energy has negative slope. In real systems, there is sufficient friction between ball and gully to bring the ball to rest at the bottom of the gully.

## CONCEPTS & SKILLS

*After studying this chapter and working the problems that follow, you should be able to*

1. Describe in operational terms how to distinguish among mixtures, compounds, and elements (Section 1.2).
2. Outline Dalton's atomic theory of matter and describe its experimental basis (Section 1.3).
3. Describe the reasoning that permits chemical formulas to be determined by purely chemical means (Section 1.3, problems 7–10).
4. Describe the experiments that led to the discovery of the electron and the measurement of its mass and charge, and describe those that revealed the nature of the nucleus (Section 1.4).
5. State the numbers of protons, neutrons, and electrons in particular atoms (Section 1.4, problems 15–18).
6. Describe the structure of the periodic table, and predict chemical and physical properties of an element based on those of others in its group or period (Section 1.5, problems 19–20).
7. Given the atomic masses and natural abundances of the isotopes of an element, calculate its chemical atomic mass (Section 1.6, problems 23–24).
8. Interconvert the mass, chemical amount, number of molecules, and (using the density) volume of a substance (Section 1.6, problems 29–38).
9. Calculate the kinetic energy of moving objects and apply the law of conservation of energy to the interconversion of potential and kinetic energy (Section 1.7, problems 39–40).
10. Use a potential energy diagram to predict the net forces acting on an object at each location.

## PROBLEMS

*Answers to problems whose numbers are boldface appear in Appendix G. Problems that are more challenging are indicated with asterisks.*

### Macroscopic Methods and Approaches of Modern Chemistry

1. Classify the following materials as substances or mixtures: table salt, wood, mercury, air, water, seawater, sodium chloride, mayonnaise. If they are mixtures, subclassify them as homogeneous or heterogeneous; if they are substances, subclassify them as compounds or elements.
2. Classify the following materials as substances or mixtures: absolute (pure) alcohol, milk (as purchased in a store), copper wire, rust, barium bromide, concrete, baking soda, baking powder. If they are mixtures, subclassify them as homogeneous or heterogeneous; if they are substances, subclassify them as compounds or elements.
3. A 17th-century chemist wrote of the "simple bodies which enter originally into the composition of mixtures and into

which these mixtures resolve themselves or may be finally resolved." What is being discussed?
4. Since 1800, nearly 200 sincere but erroneous reports of the discovery of new chemical elements have been made. Why have mistaken reports of new elements been so numerous? Why is it relatively easy to prove that a material is *not* a chemical element but difficult to prove absolutely that a material *is* an element?

### The Laws of Chemical Combination

5. A sample of ascorbic acid (vitamin C) is synthesized in the laboratory. It contains 30.0 g of carbon and 40.0 g of oxygen. Another sample of ascorbic acid, isolated from lemons (an excellent source of the vitamin), contains 12.7 g of carbon. Compute the mass of oxygen (in grams) in the second sample.
6. A sample of a compound synthesized and purified in the laboratory contains 25.0 g of hafnium and 31.5 g of tellurium. The identical compound is discovered in a rock formation.

A sample from the rock formation contains 0.125 g of hafnium. Determine how much tellurium is in the sample from the rock formation.

7. Nitrogen (N) and silicon (Si) form two binary compounds with the following compositions:

| Compound | Mass % N | Mass % Si |
|----------|----------|-----------|
| 1 | 33.28 | 66.72 |
| 2 | 39.94 | 60.06 |

(a) Compute the mass of silicon that combines with 1.0000 g of nitrogen in each case.

(b) Show that these compounds satisfy the law of multiple proportions. If the second compound has the formula $Si_3N_4$, what is the formula of the first compound?

8. Iodine (I) and fluorine (F) form a series of binary compounds with the following compositions:

| Compound | Mass % I | Mass % F |
|----------|----------|----------|
| 1 | 86.979 | 13.021 |
| 2 | 69.007 | 30.993 |
| 3 | 57.191 | 42.809 |
| 4 | 48.829 | 51.171 |

(a) Compute in each case the mass of fluorine that combines with 1.0000 g of iodine.

(b) By figuring out small whole-number ratios among the four answers in part (a), show that these compounds satisfy the law of multiple proportions.

9. Vanadium and oxygen form a series of compounds with the following compositions:

| Mass % V | Mass % O |
|----------|----------|
| 76.10 | 23.90 |
| 67.98 | 32.02 |
| 61.42 | 38.58 |
| 56.02 | 43.98 |

What are the relative numbers of atoms of oxygen in the compounds for a given mass of vanadium?

10. Tungsten (W) and chlorine (Cl) form a series of compounds with the following compositions:

| Mass % W | Mass % Cl |
|----------|-----------|
| 72.17 | 27.83 |
| 56.45 | 43.55 |
| 50.91 | 49.09 |
| 46.36 | 53.64 |

If a molecule of each compound contains only one tungsten atom, what are the formulas for the four compounds?

11. A liquid compound containing only hydrogen and oxygen is placed in a flask. Two electrodes are dipped into the liquid, and an electric current is passed between them. Gaseous hydrogen forms at one electrode and gaseous oxygen at the other. After a time, 14.4 mL of hydrogen has evolved at the negative terminal and 14.4 mL of oxygen has evolved at the positive terminal.

(a) Assign a chemical formula to the compound in the cell.

(b) Explain why more than one formula is possible as the answer to part (a).

12. A sample of liquid $N_2H_4$ is decomposed to give gaseous $N_2$ and gaseous $H_2$. The two gases are separated, and the nitrogen occupies 13.7 mL at room conditions of pressure and temperature. Determine the volume of the hydrogen under the same conditions.

13. Pure nitrogen dioxide ($NO_2$) forms when dinitrogen oxide ($N_2O$) and oxygen ($O_2$) are mixed in the presence of a certain catalyst. What volumes of $N_2O$ and oxygen are needed to produce 4.0 L of $NO_2$ if all gases are held at the same conditions of temperature and pressure?

14. Gaseous methanol ($CH_3OH$) reacts with oxygen ($O_2$) to produce water vapor and carbon dioxide. What volumes of water vapor and carbon dioxide will be produced from 2.0 L of methanol if all gases are held at the same temperature and pressure conditions?

## Physical Structure of Atoms

15. The isotope of plutonium used for nuclear fission is $^{239}Pu$. Determine (a) the ratio of the number of neutrons in a $^{239}Pu$ nucleus to the number of protons and (b) the number of electrons in a single Pu atom.

16. The last "missing" element from the first six periods was promethium, which was finally discovered in 1947 among the fission products of uranium. Determine (a) the ratio of the number of neutrons in a $^{145}Pm$ nucleus to the number of protons and (b) the number of electrons in a single Pm atom.

17. The americium isotope $^{241}Am$ is used in smoke detectors. Describe the composition of a neutral atom of this isotope in terms of protons, neutrons, and electrons.

18. In 1982, the production of a single atom of $^{266}_{109}Mt$ (meitnerium-266) was reported. Describe the composition of a neutral atom of this isotope in terms of protons, neutrons, and electrons.

## The Periodic Table

19. Before the element scandium was discovered in 1879, it was known as "eka-boron." Predict the properties of scandium from averages of the corresponding properties of its neighboring elements in the periodic table (which follow). Compare your predictions with the observed values in Appendix F.

| Element | Symbol | Melting Point (°C) | Boiling Point (°C) | Density (g cm$^{-3}$) |
|---------|--------|-------------------|-------------------|----------------------|
| Calcium | Ca | 839 | 1484 | 1.55 |
| Titanium | Ti | 1660 | 3287 | 4.50 |
| Scandium | Sc | ? | ? | ? |

20. The element technetium (Tc) is not found in nature but has been produced artificially through nuclear reactions. Use the data in the following table for several neighboring elements to estimate the melting point, boiling point, and density of technetium. Compare your predictions with the observed values in Appendix F.

| Element | Symbol | Melting Point (°C) | Boiling Point (°C) | Density (g cm$^{-3}$) |
|---|---|---|---|---|
| Manganese | Mn | 1244 | 1962 | 7.2 |
| Molybdenum | Mo | 2610 | 5560 | 10.2 |
| Rhenium | Re | 3180 | 5627 | 20.5 |
| Ruthenium | Ru | 2310 | 3900 | 12.3 |

21. Use the group structure of the periodic table to predict the empirical formulas for the binary compounds that hydrogen forms with the elements antimony, bromine, tin, and selenium.

22. Use the group structure of the periodic table to predict the empirical formulas for the binary compounds that hydrogen forms with the elements germanium, fluorine, tellurium, and bismuth.

## Weighing and Counting Molecules

23. The natural abundances and isotopic masses of the element silicon (Si) relative to $^{12}C = 12.00000$ are

| Isotope | % Abundance | Isotopic Mass |
|---|---|---|
| $^{28}Si$ | 92.21 | 27.97693 |
| $^{29}Si$ | 4.70 | 28.97649 |
| $^{30}Si$ | 3.09 | 29.97376 |

Calculate the atomic mass of naturally occurring silicon.

24. The natural abundances and isotopic masses of the element neon (Ne) are

| Isotope | % Abundance | Isotopic Mass |
|---|---|---|
| $^{20}Ne$ | 90.00 | 19.99212 |
| $^{21}Ne$ | 0.27 | 20.99316 |
| $^{22}Ne$ | 9.73 | 21.99132 |

Calculate the atomic mass of naturally occurring neon.

25. Only two isotopes of boron (B) occur in nature; their atomic masses and abundances are given in the following table. Complete the table by computing the relative atomic mass of $^{11}B$ to four significant figures, taking the tabulated relative atomic mass of natural boron as 10.811.

| Isotope | % Abundance | Atomic Mass |
|---|---|---|
| $^{10}B$ | 19.61 | 10.013 |
| $^{11}B$ | 80.39 | ? |

26. Over half of all the atoms in naturally occurring zirconium are $^{90}Zr$. The other four stable isotopes of zirconium have the following relative atomic masses and abundances:

| Isotope | % Abundance | Atomic Mass |
|---|---|---|
| $^{91}Zr$ | 11.27 | 90.9056 |
| $^{92}Zr$ | 17.17 | 91.9050 |
| $^{94}Zr$ | 17.33 | 93.9063 |
| $^{96}Zr$ | 2.78 | 95.9083 |

Compute the relative atomic mass of $^{90}Zr$ to four significant figures, using the tabulated relative atomic mass 91.224 for natural zirconium.

27. Compute the mass, in grams, of a single iodine atom if the relative atomic mass of iodine is 126.90447 on the accepted scale of atomic masses (based on 12 as the relative atomic mass of $^{12}C$).

28. Determine the mass, in grams, of exactly 100 million atoms of fluorine if the relative atomic mass of fluorine is 18.998403 on a scale on which exactly 12 is the relative atomic mass of $^{12}C$.

29. Compute the relative molecular masses of the following compounds on the $^{12}C$ scale.
    (a) $P_4O_{10}$       (d) $KMnO_4$
    (b) $BrCl$            (e) $(NH_4)_2SO_4$
    (c) $Ca(NO_3)_2$

30. Compute the relative molecular masses of the following compounds on the $^{12}C$ scale.
    (a) $[Ag(NH_3)_2]Cl$         (d) $H_2SO_4$
    (b) $Ca_3[Co(CO_3)_3]_2$     (e) $Ca_3Al_2(SiO_4)_3$
    (c) $OsO_4$

31. Suppose that a person counts out gold atoms at the rate of one each second for the entire span of an 80-year life. Has the person counted enough atoms to be detected with an ordinary balance? Explain.

32. A gold atom has a diameter of $2.88 \times 10^{-10}$ m. Suppose the atoms in 1.00 mol of gold atoms are arranged just touching their neighbors in a single straight line. Determine the length of the line.

33. The vitamin A molecule has the formula $C_{20}H_{30}O$, and a molecule of vitamin $A_2$ has the formula $C_{20}H_{28}O$. Determine how many moles of vitamin $A_2$ contain the same number of atoms as 1.000 mol of vitamin A.

34. Arrange the following in order of increasing mass: 1.06 mol $SF_4$; 117 g $CH_4$; $8.7 \times 10^{23}$ molecules of $Cl_2O_7$; $417 \times 10^{23}$ atoms of argon (Ar).

35. Mercury is traded by the "flask," a unit that has a mass of 34.5 kg. Determine the volume of a flask of mercury if the density of mercury is 13.6 g cm$^{-3}$.

36. Gold costs $400 per troy ounce, and one troy ounce equals 31.1035 g. Determine the cost of 10.0 cm$^3$ of gold if the density of gold is 19.32 g cm$^{-3}$ at room conditions.

37. Aluminum oxide ($Al_2O_3$) occurs in nature as a mineral called corundum, which is noted for its hardness and its resis-

tance to attack by acids. Its density is 3.97 g cm$^{-3}$. Calculate the number of atoms of aluminum in 15.0 cm$^3$ of corundum.

38. Calculate the number of atoms of silicon (Si) in 415 cm$^3$ of the colorless gas disilane at 0°C and atmospheric pressure, where its density is 0.00278 g cm$^{-3}$. The molecular formula of disilane is $Si_2H_6$.

## Energy: Forms and Interconversions

**39.** After spiking, a volleyball travels with speed near 100 miles per hour. Calculate the kinetic energy of the volleyball.

40. The fastball of a famous pitcher in the National League has been clocked in excess of 95 mph. Calculate the work done by the pitcher in accelerating the ball to that speed.

## Additional Problems

41. Soft-wood chips weighing 17.2 kg are placed in an iron vessel and mixed with 150.1 kg of water and 22.43 kg of sodium hydroxide. A steel lid seals the vessel, which is then placed in an oven at 250°C for 6 hours. Much of the wood fiber decomposes under these conditions; the vessel and lid do not react.
    (a) Classify each of the materials mentioned as a substance or mixture. Subclassify the substances as elements or compounds.
    (b) Determine the mass of the contents of the iron vessel after the reaction.

* 42. In a reproduction of the Millikan oil-drop experiment, a student obtains the following values for the charges on nine different oil droplets.

| | | |
|---|---|---|
| 6.563 × 10$^{-19}$ C | 13.13 × 10$^{-19}$ C | 19.71 × 10$^{-19}$ C |
| 8.204 × 10$^{-19}$ C | 16.48 × 10$^{-19}$ C | 22.89 × 10$^{-19}$ C |
| 11.50 × 10$^{-19}$ C | 18.08 × 10$^{-19}$ C | 26.18 × 10$^{-19}$ C |

   (a) Based on these data alone, what is your best estimate of the number of electrons on each of the above droplets? (*Hint:* Begin by considering differences in charges between adjacent data points, and see what groups these fall into.)
   (b) Based on these data alone, what is your best estimate of the charge on the electron?
   (c) Is it conceivable that the actual charge is half the charge you calculated in (b)? What evidence would help you decide one way or the other?

43. A rough estimate of the radius of a nucleus is provided by the formula $r = kA^{1/3}$, where $k$ is approximately $1.3 \times 10^{-13}$ cm and $A$ is the mass number of the nucleus. Estimate the density of the nucleus of $^{127}$I (which has a nuclear mass of $2.1 \times 10^{-22}$ g), in grams per cubic centimeter. Compare with the density of solid iodine, 4.93 g cm$^{-3}$.

44. In a neutron star, gravity causes the electrons to combine with protons to form neutrons. A typical neutron star has a mass half that of the sun, compressed into a sphere of radius 20 km. If such a neutron star contains $6.0 \times 10^{56}$ neutrons, calculate its density in grams per cubic centimeter. Compare this with the density inside a $^{232}$Th nucleus, in which 142 neutrons and 90 protons occupy a sphere of radius $9.1 \times 10^{-13}$ cm. Take the mass of a neutron to be $1.675 \times 10^{-24}$ g and that of a proton to be $1.673 \times 10^{-24}$ g.

45. Dalton's 1808 version of the atomic theory of matter included five general statements (Section 1.3). According to modern understanding, four of those statements require amendment or extension. List the modifications that have been made to four of the five original postulates.

46. Naturally occurring rubidium (Rb) consists of two isotopes: $^{85}$Rb (atomic mass 84.9117) and $^{87}$Rb (atomic mass 86.9092). The atomic mass of the isotope mixture found in nature is 85.4678. Calculate the percent abundances of the two isotopes in rubidium.

* 47. The inhabitants of a planet in a distant galaxy measure mass in units of "margs," where 1 marg = 4.8648 g. Their scale of atomic masses is based on the isotope $^{32}$S (atomic mass on Earth = 31.972), so they define one "elom" of $^{32}$S as the amount of sulfur in exactly 32 margs of $^{32}$S. Furthermore, they define $N_{or}$, or "Ordagova's number" (after the well-known scientist Oedema Ordagova), as the number of atoms of sulfur in exactly 32 margs of $^{32}$S.
    (a) Calculate the numerical value of $N_{or}$ to four significant figures.
    (b) Calculate the atomic mass of phosphorus in margs per elom to five significant figures, assuming that the abundances of phosphorus isotopes are the same as those found on Earth.

48. A tennis ball weighs approximately two ounces on a postage scale. A student practices his serve against the wall of the chemistry building, and the ball achieves the speed of 98 miles per hour. Calculate the kinetic energy of the ball after the serve. How much work is done on the chemistry building in one collision?

# Chemical Equations and Reaction Yields

***Illustration***
An "assay balance of careful construction" of the type used by Lavoisier before 1788. This balance became the production model that served as a general, all-purpose balance for about 40 years. Users of this type of balance included Sir Humphrey Davy and his young assistant, Michael Faraday. (*Courtesy of the Chandler Museum at Columbia University. Photo by Charles D. Winters*)

I n Chapter 1, we showed how chemical and physical methods are used to establish chemical formulas and relative atomic and molecular masses. In this chapter, we take a further step and consider chemical reactions. We examine the balanced chemical equations that summarize these reactions and show how the masses of substances consumed and produced are related. This is an immensely practical and important subject. The question of how much of a substance will react with a given amount of another substance and how much product will be produced is central to all chemical processes, whether industrial, geological, or biological.

## 2.1

# EMPIRICAL AND MOLECULAR FORMULAS

According to the laws of chemical combination, each substance is described by a chemical formula that indicates the relative numbers of atoms of the elements in that substance. Now we distinguish between two types of formulas: the molecular formula and the empirical formula. The **molecular formula** of a substance specifies the number of atoms of each element in one molecule of that substance. Thus, the molecular formula of carbon dioxide is $CO_2$: each molecule of carbon dioxide contains one atom of carbon and two atoms of oxygen. The molecular formula of glucose is $C_6H_{12}O_6$: each glucose molecule contains 6 atoms of carbon, 6 of oxygen, and 12 of hydrogen. Molecular formulas can be defined for all gaseous substances and for those liquids or solids that, like glucose, possess well-defined molecular subunits.

By contrast, the **empirical formula** of a compound is the simplest formula that gives the correct relative numbers of atoms of each kind in a compound. The empirical formula of glucose, for example, is $CH_2O$, indicating that the numbers of atoms of carbon, hydrogen, and oxygen are in a ratio of $1:2:1$. When a molecular formula is known, it is clearly preferable because it conveys more information. In some solids and liquids, however, distinct small molecules do not exist and the only meaningful chemical formula is an empirical one. Solid cobalt(II) chloride, which has the empirical formula $CoCl_2$, is an example. There are strong attractive forces between a cobalt atom and two adjoining chlorine atoms in solid cobalt(II) chloride, but it is not possible to distinguish the forces *within* such a "molecule" of $CoCl_2$ from those operating *between* it and a neighbor; the latter are equally strong. The solid is, in effect, a single giant molecule. Consequently, cobalt(II) chloride is represented with an empirical formula and referred to by a **formula unit** of $CoCl_2$, rather than by "a molecule of $CoCl_2$."

In some cases, small molecules are incorporated into a solid structure, and the chemical formula is written to show this fact explicitly. Thus, cobalt and chlorine form not only the anhydrous salt $CoCl_2$ just mentioned, but also the hexahydrate $CoCl_2 \cdot 6H_2O$, in which six water molecules are incorporated per $CoCl_2$ formula unit (Fig. 2.1). The dot in this formula is used to set off a well-defined molecular component of the solid, such as water.

**FIGURE 2.1** When cobalt(II) chloride crystallizes from solution, it brings with it six water molecules per formula unit, giving a red solid with the empirical formula $CoCl_2 \cdot 6H_2O$. This solid melts at 86°C; above approximately 110°C it loses some water and forms a lavender solid with the empirical formula $CoCl_2 \cdot 2H_2O$. *(Leon Lewandowski)*

---
## 2.2

# CHEMICAL FORMULA AND PERCENTAGE COMPOSITION

The empirical formula $H_2O$ indicates that for every atom of oxygen in water, there are two atoms of hydrogen. Equivalently, one mole of $H_2O$ contains two moles of hydrogen atoms and one mole of oxygen atoms. The number of atoms and the number of moles of each element are present in the same ratio, namely $2:1$. The empirical formula for a substance is clearly related to the percentage composition by mass of that substance. This connection can be used in various ways.

## Empirical Formula and Percentage Composition

The empirical formula of a compound and the percentage composition by mass of the elements that constitute it are related simply through the mole concept. For example, the empirical formula of ethylene (molecular formula $C_2H_4$) is $CH_2$. Its composition by mass is calculated from the masses of carbon and hydrogen in 1 mol of $CH_2$ formula units:

$$\text{mass of C} = 1 \text{ mol C} \times (12.011 \text{ g mol}^{-1}) = 12.011 \text{ g}$$

$$\text{mass of H} = 2 \text{ mol H} \times (1.00794 \text{ g mol}^{-1}) = 2.0159 \text{ g}$$

Adding these gives a total mass of 14.027 g. The mass percentages of carbon and hydrogen in the compound are then found by dividing each of their masses by this total mass and multiplying by 100%, giving 85.628% C and 14.372% H.

## Determination of Empirical Formula from Measured Mass Composition

We can reverse the procedure just described and determine the empirical formula from the elemental analysis of a compound, as illustrated by the following example.

### EXAMPLE 2.1

A 60.00-g sample of a dry-cleaning fluid was analyzed and found to contain 10.80 g C, 1.36 g H, and 47.84 g Cl. Determine the empirical formula of the compound, using a table of atomic masses.

### Solution

The chemical amounts of the elements in the sample are

$$\text{For carbon:} \quad \frac{10.80 \text{ g C}}{12.011 \text{ g mol}^{-1}} = 0.8992 \text{ mol C}$$

$$\text{For hydrogen:} \quad \frac{1.36 \text{ g H}}{1.008 \text{ g mol}^{-1}} = 1.35 \text{ mol H}$$

$$\text{For chlorine:} \quad \frac{47.84 \text{ g Cl}}{35.453 \text{ g mol}^{-1}} = 1.349 \text{ mol Cl}$$

The ratio of the chemical amount of carbon to that of chlorine (or hydrogen) is $0.8992:1.349 = 0.6666$, which is very close to $2:3$. The numbers of moles are then in a ratio of 2:3:3, so the empirical formula is $C_2H_3Cl_3$. Further measurements would be necessary to find the actual molecular mass and the correct *molecular* formula from among $C_2H_3Cl_3$, $C_4H_6Cl_6$, or any higher multiples, $(C_2H_3Cl_3)_n$.

**Related Problems: 7, 8, 9, 10, 11, 12**

## Empirical Formula Determined from Elemental Analysis by Combustion

A **hydrocarbon** is a compound containing only carbon and hydrogen. Its empirical formula can be determined by using the combustion train shown in Figure 2.2. In this device, a weighed amount of the hydrocarbon is burned completely in oxygen, yielding carbon dioxide and water. The masses of water and carbon dioxide can then be determined, and from these data the empirical formula is found, as in the following example.

### EXAMPLE 2.2

A certain compound, used as a welding fuel, contains carbon and hydrogen only. Burning a small sample of it completely in oxygen gives 3.38 g of $CO_2$, 0.692 g of water, and no other products. What is the empirical formula of the compound?

### Solution

We first compute the chemical amounts of the $CO_2$ and $H_2O$. Because all of the carbon has been converted to $CO_2$, and all of the hydrogen to water, the chemical amounts of C and H in the unburned gas can be determined:

$$\text{moles of C} = \text{moles of CO}_2 = \frac{3.38 \text{ g}}{44.01 \text{ g mol}^{-1}} = 0.0768 \text{ mol}$$

$$\text{moles of H} = 2(\text{moles of H}_2\text{O}) = 2\left(\frac{0.692 \text{ g}}{18.02 \text{ g mol}^{-1}}\right) = 0.0768 \text{ mol}$$

Note that the number of moles of water is multiplied by 2 to find the number of moles of hydrogen atoms, because each water molecule contains two hydrogen atoms. Because the compound contains equal chemical amounts (numbers of moles) of carbon and hydrogen, its empirical formula is CH. Its molecular formula may be CH, or $C_2H_2$, or $C_3H_3$, and so on.

**Related Problems: 13, 14**

**FIGURE 2.2** A combustion train to determine amounts of carbon and hydrogen in hydrocarbons. A weighed sample is burned in a flow of oxygen to form water and carbon dioxide. These products of combustion pass over a desiccant such as magnesium perchlorate, $Mg(ClO_4)_2$, which absorbs the water. In a second stage, the carbon dioxide passes on and is absorbed on finely divided particles of sodium hydroxide, NaOH, mixed with calcium chloride, $CaCl_2$. The changes in mass of the two absorbers give the amounts of water and carbon dioxide produced.

Heaters

$O_2 \rightarrow$

$CO_2 + H_2O \rightarrow$

$CO_2 \rightarrow$

Sample

$Mg(ClO_4)_2$ desiccant for $H_2O$ absorption

$NaOH + CaCl_2$ for $CO_2$ absorption

## Connection between Empirical Formula and Molecular Formula

The molecular formula is some whole-number multiple of the empirical formula. To determine the molecular formula, it is necessary to know the approximate molar mass of the compound under study. From Avogadro's hypothesis, the ratio of molar masses of two gaseous compounds is the same as the ratio of their densities, provided that those densities are measured at the same temperature and pressure. (This is true because a given volume contains the same number of molecules of the two gases.) The welding gas from Example 2.2 has density $1.06$ g $L^{-1}$ at 25°C and atmospheric pressure. Under the same conditions, gaseous oxygen (which exists as diatomic $O_2$ molecules with molar mass $32.0$ g $mol^{-1}$) has a density of $1.31$ g $L^{-1}$. The approximate molar mass of the welding gas is then

$$\text{molar mass of welding gas} = \frac{1.06 \text{ g L}^{-1}}{1.31 \text{ g L}^{-1}} (32.0 \text{ g mol}^{-1}) = 25.9 \text{ g mol}^{-1}$$

The molar mass corresponding to the *empirical* formula CH is $13.0$ g $mol^{-1}$. Because $25.9$ g $mol^{-1}$ is approximately twice this, there must be two CH units per molecule, so the molecular formula is $C_2H_2$. The gas is acetylene.

---

### 2.3

# WRITING BALANCED CHEMICAL EQUATIONS

Chemical reactions combine elements into compounds, decompose compounds back into elements, and transform existing compounds into new compounds. Because atoms are indestructible in chemical reactions, the same number of atoms (or moles of atoms) of each element must be present before and after any reaction. The conservation of matter in a chemical change is represented in a balanced chemical equation for that process.

An equation can be balanced using stepwise reasoning. Consider the decomposition of ammonium nitrate ($NH_4NO_3$) upon gentle heating to dinitrogen oxide ($N_2O$) and water. The *unbalanced* equation for this process is

$$NH_4NO_3 \longrightarrow N_2O + H_2O$$

The formulas on the left side of the arrow denote **reactants,** and those on the right side denote **products.** This equation is unbalanced because there are 3 mol of O atoms on the left (and 4 of H) but only 2 mol of O atoms and 2 of H atoms on the right. To balance the equation, first assign 1 as the coefficient of one species, usually the one containing the most elements—in this case, $NH_4NO_3$. Next, seek out the elements that appear in only one other place in the equation and assign coefficients to balance the numbers of their atoms. Here nitrogen appears in only one other place ($N_2O$), and a coefficient of 1 for the $N_2O$ ensures that there are 2 mol of N atoms on each side of the equation. Hydrogen appears in $H_2O$, so its coefficient is 2 to balance the 4 mol of H atoms on the left side. This gives

$$NH_4NO_3 \longrightarrow N_2O + 2 H_2O$$

Finally, verify that the last element, oxygen, is also balanced by noting that there are 3 mol of O atoms on each side. The coefficients of 1 in front of the $NH_4NO_3$ and $N_2O$ are omitted by convention.

As a second example, consider the reaction in which butane ($C_4H_{10}$) is burned in oxygen to form carbon dioxide and water:

$$\_\_ \ C_4H_{10} + \_\_ \ O_2 \longrightarrow \_\_ \ CO_2 + \_\_ \ H_2O$$

Spaces have been left for the coefficients that tell the number of moles of each reactant and product. Start with 1 mol of butane, $C_4H_{10}$. It contains 4 mol of carbon atoms and thus must produce 4 mol of carbon dioxide molecules, because there is nowhere else the carbon can go. The coefficient of $CO_2$ is therefore 4. In the same way, the 10 mol of hydrogen *atoms* must form 5 mol of water *molecules*, because each water molecule contains two hydrogen atoms; thus, the coefficient of the $H_2O$ is 5:

$$C_4H_{10} + \_\_ \ O_2 \longrightarrow 4 \ CO_2 + 5 \ H_2O$$

Four moles of $CO_2$ contains 8 mol of oxygen atoms, and 5 mol of $H_2O$ contains 5 mol of oxygen atoms, giving a total of 13 mol of oxygen atoms. Thirteen moles of oxygen atoms are equivalent to $\frac{13}{2}$ mol of oxygen molecules, so the coefficient of $O_2$ is $\frac{13}{2}$. The balanced equation is

$$C_4H_{10} + \tfrac{13}{2} \ O_2 \longrightarrow 4 \ CO_2 + 5 \ H_2O$$

There is nothing wrong with fractions such as $\frac{13}{2}$ in a balanced equation, because fractions of moles are perfectly meaningful. However, it is often the custom to eliminate any such fractions. Here, multiplying all coefficients in the equation by 2 gives

$$2 \ C_4H_{10} + 13 \ O_2 \longrightarrow 8 \ CO_2 + 10 \ H_2O$$

Let us summarize the steps in balancing a chemical equation:

1. Assign 1 as the coefficient of one species. The best choice is the most complicated species, with the largest number of elements.
2. Identify, in sequence, elements that appear in only one chemical species whose coefficient is not yet determined. Choose that coefficient to balance the number of moles of atoms of that element. Continue until all coefficients are identified.
3. If desired, multiply the whole equation by the smallest integer that will eliminate any fractions.

This method of balancing equations "by inspection" works in many, but not all, cases. In Section 12.1, we present techniques for balancing certain more complex chemical equations.

Once the reactants and products are known, balancing chemical equations is a routine, mechanical process of accounting. The difficult part (and the part where chemistry comes in) is to know what substances will react with each other and to determine what products are formed. This is a question to which we will return many times throughout this book.

---

**EXAMPLE 2.3**

The Hargreaves process is an industrial procedure for making sodium sulfate ($Na_2SO_4$) for use in papermaking. The starting materials are sodium chloride (NaCl), sulfur dioxide ($SO_2$), water, and oxygen. Hydrogen chloride (HCl) is generated as a byproduct. Write a balanced chemical equation for this process.

**Solution**

The unbalanced equation is

$$\_ \text{ NaCl} + \_ \text{ SO}_2 + \_ \text{ H}_2\text{O} + \_ \text{ O}_2 \longrightarrow \_ \text{ Na}_2\text{SO}_4 + \_ \text{ HCl}$$

Begin by assigning a coefficient of 1 to the $\text{Na}_2\text{SO}_4$ because it is the most complex species, composed of three different elements. There are 2 mol of Na atoms on the right, giving 2 as the coefficient of the NaCl. By the same argument, the coefficient of the $\text{SO}_2$ must be 1 to balance the 1 mol of sulfur on the right. This gives

$$2 \text{ NaCl} + \text{SO}_2 + \_ \text{ H}_2\text{O} + \_ \text{ O}_2 \longrightarrow \text{Na}_2\text{SO}_4 + \_ \text{ HCl}$$

Next we note that there are 2 mol of Cl atoms on the left (reactant) side, so the coefficient of the HCl must be 2. Hydrogen is the next element to balance, with 2 mol on the right side and therefore a coefficient of 1 for the $\text{H}_2\text{O}$:

$$2 \text{ NaCl} + \text{SO}_2 + \text{H}_2\text{O} + \_ \text{ O}_2 \longrightarrow \text{Na}_2\text{SO}_4 + 2 \text{ HCl}$$

Finally, the oxygen atoms must be balanced. There are 4 mol of oxygen atoms on the right, but the left contains 2 mol from the $\text{SO}_2$ and 1 from the $\text{H}_2\text{O}$; therefore, 1 mol of oxygen *atoms* must come from the $\text{O}_2$. The coefficient of the $\text{O}_2$ is therefore $\frac{1}{2}$:

$$2 \text{ NaCl} + \text{SO}_2 + \text{H}_2\text{O} + \tfrac{1}{2} \text{ O}_2 \longrightarrow \text{Na}_2\text{SO}_4 + 2 \text{ HCl}$$

Multiply all coefficients in the equation by 2, giving

$$4 \text{ NaCl} + 2 \text{ SO}_2 + 2 \text{ H}_2\text{O} + \text{O}_2 \longrightarrow 2 \text{ Na}_2\text{SO}_4 + 4 \text{ HCl}$$

In balancing the equation, oxygen was considered last because it appears in several places on the left side of the equation.

**Related Problems: 19, 20**

---

## 2.4

## MASS RELATIONSHIPS IN CHEMICAL REACTIONS

A balanced chemical equation makes quantitative statements about the relative masses of the substances reacting. The chemical equation for the combustion of butane

$$2 \text{ C}_4\text{H}_{10} + 13 \text{ O}_2 \longrightarrow 8 \text{ CO}_2 + 10 \text{ H}_2\text{O}$$

can be interpreted as saying either that

$$2 \text{ molecules of C}_4\text{H}_{10} + 13 \text{ molecules of O}_2 \longrightarrow$$
$$8 \text{ molecules of CO}_2 + 10 \text{ molecules of H}_2\text{O}$$

or that

$$2 \text{ mol of C}_4\text{H}_{10} + 13 \text{ mol of O}_2 \longrightarrow 8 \text{ mol of CO}_2 + 10 \text{ mol of H}_2\text{O}$$

Multiplying the molar mass of each substance in the reaction by the number of moles represented in the balanced equation gives

$$116.3 \text{ g of C}_4\text{H}_{10} + 416.0 \text{ g of O}_2 \longrightarrow 352.1 \text{ g of CO}_2 + 180.2 \text{ g of H}_2\text{O}$$

The study of the relationships between masses of reactants and products is called **stoichiometry** (from the Greek *stoicheion*, "element," + *metron*, "measure"). Stoichiometry is fundamental to all of chemistry.

The coefficients in a balanced chemical equation give "chemical conversion factors" between the chemical amounts of substances consumed in or produced by a reaction. If 6.16 mol of butane reacts according to the preceding equation, the amounts of $O_2$ consumed and $CO_2$ generated are

$$\text{moles } O_2 = 6.16 \text{ mol } C_4H_{10} \times \left( \frac{13 \text{ mol } O_2}{2 \text{ mol } C_4H_{10}} \right) = 40.0 \text{ mol } O_2$$

$$\text{moles } CO_2 = 6.16 \text{ mol } C_4H_{10} \times \left( \frac{8 \text{ mol } CO_2}{2 \text{ mol } C_4H_{10}} \right) = 24.6 \text{ } CO_2$$

For most practical purposes we will be interested in the *masses* of reactants and products, because those are the quantities that are directly measured. In this case, the molar masses (calculated from a table of atomic masses) are used to convert the chemical amount of a substance (in moles) to its mass (in grams), as illustrated by the following example.

## EXAMPLE 2.4

Calcium hypochlorite, $Ca(OCl)_2$, is used as a bleaching agent. It is produced from sodium hydroxide, calcium hydroxide, and chlorine according to the overall equation

$$2 \text{ NaOH} + Ca(OH)_2 + 2 \text{ Cl}_2 \longrightarrow Ca(OCl)_2 + 2 \text{ NaCl} + 2 \text{ H}_2O$$

How many grams of chlorine and sodium hydroxide react with 1067 g of $Ca(OH)_2$, and how many grams of calcium hypochlorite are produced?

### Solution

The chemical amount of $Ca(OH)_2$ consumed is

$$\frac{1067 \text{ g } Ca(OH)_2}{74.09 \text{ g mol}^{-1}} = 14.40 \text{ mol } Ca(OH)_2$$

where the molar mass of $Ca(OH)_2$ has been obtained from the molar masses of calcium, oxygen, and hydrogen as

$$40.08 + 2(15.999) + 2(1.0079) = 74.09 \text{ g mol}^{-1}$$

According to the balanced equation, 1 mol of $Ca(OH)_2$ reacts with 2 mol of NaOH and 2 mol of $Cl_2$ to produce 1 mol of $Ca(OCl)_2$. If 14.40 mol of $Ca(OH)_2$ reacts, then

$$\text{moles NaOH} = 14.40 \text{ mol } Ca(OH)_2 \left( \frac{2 \text{ mol NaOH}}{1 \text{ mol } Ca(OH)_2} \right)$$

$$= 28.80 \text{ mol NaOH}$$

$$\text{moles Cl}_2 = 14.40 \text{ mol } Ca(OH)_2 \left( \frac{2 \text{ mol Cl}_2}{1 \text{ mol } Ca(OH)_2} \right)$$

$$= 28.80 \text{ mol Cl}_2$$

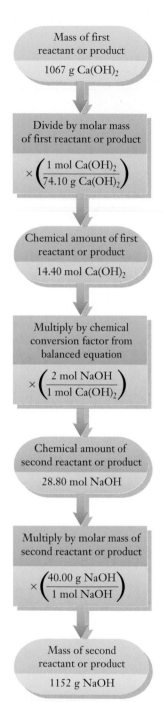

**FIGURE 2.3** The steps in a stoichiometric calculation. In a typical calculation, the mass of one reactant or product is known, and the masses of one or more other reactants or products are to be calculated using the balanced chemical equation and a table of relative atomic masses.

$$\text{moles Ca(OCl)}_2 = 14.40 \text{ mol Ca(OH)}_2 \left( \frac{1 \text{ mol Ca(OCl)}_2}{1 \text{ mol Ca(OH)}_2} \right)$$

$$= 14.40 \text{ mol Ca(OCl)}_2$$

From the chemical amounts and molar masses of reactants and products, the desired masses are found:

$$\text{mass NaOH reacting} = (28.80 \text{ mol})(40.00 \text{ g mol}^{-1}) = 1152 \text{ g}$$

$$\text{mass Cl}_2 \text{ reacting} = (28.80 \text{ mol})(70.91 \text{ g mol}^{-1}) = 2042 \text{ g}$$

$$\text{mass Ca(OCl)}_2 \text{ produced} = (14.40 \text{ mol})(142.98 \text{ g mol}^{-1}) = 2059 \text{ g}$$

**Related Problems: 21, 22, 23, 24**

In calculations such as that in Example 2.4, a known mass of one substance takes part in a reaction, and we need to calculate the masses of one or more other reactants or products. Figure 2.3 summarizes the three-step process used. With experience, it is possible to write down the answers in a shorthand form so that all three conversions are carried out at the same time. The chemical amount of NaOH reacting in the preceding example can be written as

$$\left( \frac{1067 \text{ g Ca(OH)}_2}{74.10 \text{ g mol}^{-1}} \right) \times \left( \frac{2 \text{ mol NaOH}}{1 \text{ mol Ca(OH)}_2} \right) \times 40.00 \text{ g mol}^{-1} = 1152 \text{ g NaOH}$$

It is better at first, however, to follow a stepwise procedure for such calculations.

---

## 2.5

### LIMITING REACTANT AND PERCENTAGE YIELD

In the cases considered so far, the reactants were present in the exact ratios necessary for them to be completely consumed in forming products. This is not the usual case, however. It is necessary to have methods for describing cases where one of the reactants may not be present in sufficient amount and where conversion to products is less than complete.

### Limiting Reactant

Suppose arbitrary amounts of reactants are mixed and allowed to react. The one that is used up first is the **limiting reactant,** and a portion of the other reactants remains after completion of the reaction. These other reactants are present **in excess.** An increase in the amount of the limiting reactant will lead to an increase in the amount of product formed. This is not true of the other reactants. In an industrial process, the limiting reactant is frequently the most expensive one, to ensure that none of it is wasted. For instance, the silver nitrate used in preparing silver chloride for photographic film by the reaction

$$\text{AgNO}_3 + \text{NaCl} \longrightarrow \text{AgCl} + \text{NaNO}_3$$

is far more expensive than the sodium chloride (ordinary salt). Thus, it makes sense to carry out the reaction with an excess of sodium chloride, to ensure that as much of the silver nitrate as possible reacts to form products.

There is a systematic method to find the limiting reactant and determine the maximum possible amounts of products. Take each reactant in turn, assume that it is used up completely in the reaction, and calculate the mass of one of the products that will be formed. Whichever reactant gives the *smallest* mass of this product is the limiting reactant. Once it has reacted fully, no further product can be formed.

## EXAMPLE 2.5

Sulfuric acid ($H_2SO_4$) forms in the chemical reaction

$$2 SO_2 + O_2 + 2 H_2O \longrightarrow 2 H_2SO_4$$

Suppose 400 g $SO_2$, 175 g $O_2$, and 125 g $H_2O$ are mixed and the reaction proceeds until one of the reactants is used up. Which is the limiting reactant? What mass of $H_2SO_4$ is produced, and what masses of the other reactants remain?

### Solution

The chemical amount of each reactant originally present is calculated by dividing each mass by the corresponding molar mass:

$$\frac{400 \text{ g } SO_2}{64.06 \text{ g mol}^{-1}} = 6.24 \text{ mol } SO_2$$

$$\frac{175 \text{ g } O_2}{32.00 \text{ g mol}^{-1}} = 5.47 \text{ mol } O_2$$

$$\frac{125 \text{ g } H_2O}{18.02 \text{ g mol}^{-1}} = 6.94 \text{ mol } H_2O$$

If all the $SO_2$ reacted, it would give

$$6.24 \text{ mol } SO_2 \times \left(\frac{2 \text{ mol } H_2SO_4}{2 \text{ mol } SO_2}\right) = 6.24 \text{ mol } H_2SO_4$$

If all the $O_2$ reacted, it would give

$$5.47 \text{ mol } O_2 \times \left(\frac{2 \text{ mol } H_2SO_4}{1 \text{ mol } O_2}\right) = 10.94 \text{ mol } H_2SO_4$$

Finally, if all the water reacted, it would give

$$6.94 \text{ mol } H_2O \times \left(\frac{2 \text{ mol } H_2SO_4}{2 \text{ mol } H_2O}\right) = 6.94 \text{ mol } H_2SO_4$$

In this case $SO_2$ is the limiting reactant because the computation based on its amount gives the smallest amount of product (6.24 mol $H_2SO_4$). Oxygen and water are present in excess. After the reaction the amount of each that remains is the original amount minus the amount reacting:

$$\text{moles } O_2 = 5.47 \text{ mol } O_2 - \left(6.24 \text{ mol } SO_2 \times \frac{1 \text{ mol } O_2}{2 \text{ mol } SO_2}\right)$$

$$= 5.47 - 3.12 \text{ mol } O_2 = 2.35 \text{ mol } O_2$$

$$\text{moles } H_2O = 6.94 \text{ mol } H_2O - \left(6.24 \text{ mol } SO_2 \times \frac{2 \text{ mol } H_2O}{2 \text{ mol } SO_2}\right)$$

$$= 6.94 - 6.24 \text{ mol } H_2O = 0.70 \text{ mol } H_2O$$

The masses of reactants and products after the reaction are

$$\text{mass } H_2SO_4 \text{ produced} = (6.24 \text{ mol})(98.07 \text{ g mol}^{-1}) = 612 \text{ g}$$

$$\text{mass } O_2 \text{ remaining} = (2.35 \text{ mol})(32.00 \text{ g mol}^{-1}) = 75 \text{ g}$$

$$\text{mass } H_2O \text{ remaining} = (0.70 \text{ mol})(18.02 \text{ g mol}^{-1}) = 13 \text{ g}$$

The total mass at the end is $612 + 13 + 75 = 700$ g, which is, of course, equal to the total mass originally present, $400 + 175 + 125 = 700$ g, as required by the law of conservation of mass.

**Related Problems: 35, 36**

## Percentage Yield

The amounts of products calculated so far have been **theoretical yields,** determined by assuming that the reaction goes cleanly and completely. The **actual yield** of a product (the amount present after separating it from other products and reactants and purifying it) is usually less than the theoretical yield. There are several possible reasons for this. The reaction may stop short of completion so that reactants remain unreacted. There may be competing reactions that give other products and so reduce the yield of the desired one. Finally, in the process of separation and purification some of the product is invariably lost, although that amount can be reduced by careful experimental techniques. The ratio of the actual yield to the theoretical yield (multiplied by 100%) gives the **percentage yield** for that product in the reaction.

### EXAMPLE 2.6

The sulfide ore of zinc (ZnS) is reduced to elemental zinc by "roasting" it (heating it in air) to give ZnO and then heating the ZnO with carbon monoxide. The two reactions can be written as

$$ZnS + \tfrac{3}{2} O_2 \longrightarrow ZnO + SO_2$$

$$ZnO + CO \longrightarrow Zn + CO_2$$

Suppose 5.32 kg of ZnS is treated in this way and 3.30 kg of pure Zn is obtained. Calculate the theoretical yield of Zn and its actual percentage yield.

### Solution

From the molar mass of ZnS ($97.46$ g mol$^{-1}$), the chemical amount of ZnS initially present is

$$\frac{5320 \text{ g ZnS}}{97.46 \text{ g mol}^{-1}} = 54.6 \text{ mol ZnS}$$

Because each mole of ZnS gives 1 mol of ZnO in the first chemical equation, and each mole of ZnO then gives 1 mol of Zn, the theoretical yield of Zn is 54.6 mol. In grams, this is

$$54.6 \text{ mol Zn} \times 65.39 \text{ g mol}^{-1} = 3570 \text{ g Zn}$$

The ratio of actual yield to theoretical yield, multiplied by 100%, gives the percentage yield of Zn:

$$\% \text{ yield} = \left(\frac{3.30 \text{ kg}}{3.57 \text{ kg}}\right) \times 100\% = 92.4\%$$

**Related Problems: 37, 38**

It is clearly desirable to achieve as high a percentage yield of product as possible in order to reduce consumption of raw materials. In some synthetic reactions (especially in organic chemistry), the final product is the result of many successive reactions. In such processes the yields in the individual steps must be quite high if the synthetic method is to be a practical success. Suppose, for example, that ten consecutive reactions must be carried out to reach the product, and each has a percentage yield of only 50% (a fractional yield of 0.5). The overall yield is the product of the fractional yields of the steps:

$$\underset{\text{10 terms}}{(0.5) \times (0.5) \times \cdots \times (0.5)} = (0.5)^{10} = 0.001$$

This overall percentage yield of 0.1% makes the process useless for synthetic purposes. If all the individual percentage yields could be increased to 90%, however, the overall yield would be $(0.9)^{10} = 0.35$, or a 35% overall yield. This is much more reasonable and might make the process feasible.

## CUMULATIVE EXERCISE

### Titanium

Metallic titanium and its alloys (especially those with aluminum and vanadium) combine the advantages of high strength and light weight, and so are widely used in the aerospace industry for the bodies and engines of airplanes. The major natural source for titanium is the ore rutile, which contains titanium dioxide ($TiO_2$).

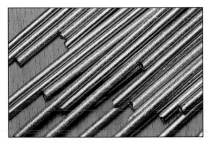

Bars of titanium. *(Copyright Klaus Guldbrandsen/Science Photo Library/ Photo Researchers, Inc.)*

(a) An intermediate in the preparation of elemental titanium from $TiO_2$ is a volatile chloride of titanium (boiling point 136°C) that contains 25.24% titanium by mass. Determine the empirical formula of this compound.

(b) At 136°C and atmospheric pressure, the density of this gaseous chloride is 5.6 g $L^{-1}$. Under the same conditions, the density of gaseous nitrogen ($N_2$, molar mass 28.0 g $mol^{-1}$) is 0.83 g $L^{-1}$. Determine the molecular formula of this compound.

(c) The titanium chloride dealt with in parts (a) and (b) is produced by the reaction of chlorine with a hot mixture of titanium dioxide and coke (carbon), with carbon dioxide generated as a byproduct. Write a balanced chemical equation for this reaction.

(d) What mass of chlorine is needed to produce 79.2 g of the titanium chloride?

(e) The titanium chloride then reacts with liquid magnesium at 900°C to give titanium and magnesium chloride ($MgCl_2$). Write a balanced chemical equation for this step in the refining of titanium.

(f) Suppose the reaction chamber for part (e) contains 351 g of the titanium chloride and 63.2 g of liquid magnesium. Which is the limiting reactant? What maximum mass of titanium could result?

(g) Isotopic analysis of the titanium from a particular ore gave these results:

| Isotope | Relative Mass | Abundance (%) |
| --- | --- | --- |
| $^{46}$Ti | 45.952633 | 7.93 |
| $^{47}$Ti | 46.95176 | 7.28 |
| $^{48}$Ti | 47.947948 | 73.94 |
| $^{49}$Ti | 48.947867 | 5.51 |
| $^{50}$Ti | 49.944789 | 5.34 |

Calculate the mass of a single $^{48}$Ti atom and the *average* mass of the titanium atoms in this ore sample.

**Answers**
(a) $TiCl_4$
(b) $TiCl_4$
(c) $TiO_2 + C + 2\ Cl_2 \rightarrow TiCl_4 + CO_2$
(d) 59.2 g
(e) $TiCl_4 + 2\ Mg \rightarrow Ti + 2\ MgCl_2$
(f) Mg; 62.3 g
(g) $7.961949 \times 10^{-23}$ g, $7.950 \times 10^{-23}$ g

## CONCEPTS & SKILLS

*After studying this chapter and working the problems that follow, you should be able to*

1. Given the percentages by mass of the elements in a compound, determine its empirical formula (Section 2.1, problems 7–12).
2. Use ratios of gas densities to estimate molecular mass and determine molecular formulas (Section 2.2, problems 15–18).
3. Balance simple chemical equations (Section 2.3, problems 19–20).
4. Given the mass of a reactant or product in a chemical reaction, use a balanced chemical equation to calculate the masses of other reactants consumed and other products formed (Section 2.4, problems 21–30).
5. Given a set of initial masses of reactants and a balanced chemical equation, determine the limiting reactant and calculate the masses of reactants and products after the reaction has gone to completion (Section 2.5, problems 35–36).
6. Determine the percentage yield of a reaction from its calculated theoretical yield and its measured actual yield (Section 2.5, problems 37–38).

## PROBLEMS

*Answers to problems whose numbers are boldface appear in Appendix G. Problems that are more challenging are indicated with asterisks.*

### Chemical Formulas and Chemical Equations

1. A newly synthesized compound has the molecular formula $ClF_2O_2PtF_6$. Compute, to four significant figures, the mass percentage of each of the four elements in this compound.
2. Acetaminophen is the generic name of the pain reliever in Tylenol and some other headache remedies. The compound has the molecular formula $C_8H_9NO_2$. Compute, to four significant figures, the mass percentage of each of the four elements in acetaminophen.
3. Arrange the following compounds from left to right in order of increasing percentage by mass of hydrogen: $H_2O$, $C_{12}H_{26}$, $N_4H_6$, LiH.
4. Arrange the following compounds from left to right in order of increasing percentage by mass of fluorine: HF, $C_6HF_5$, BrF, $UF_6$.

5. "Q-gas" is a mixture of 98.70% helium and 1.30% butane ($C_4H_{10}$) by mass. It is used as a filling for gas-flow Geiger counters. Compute the mass percentage of hydrogen in Q-gas.

6. A pharmacist prepares an antiulcer medicine by mixing 286 g of $Na_2CO_3$ with water, adding 150 g of glycine ($C_2H_5NO_2$), and stirring continuously at 40°C until a firm mass results. She heats the mass gently until all of the water has been driven away. No other chemical changes occur in this step. Compute the mass percentage of carbon in the resulting white crystalline medicine.

7. Zinc phosphate is used as a dental cement. A 50.00-mg sample is broken down into its constituent elements and gives 16.58 mg of oxygen, 8.02 mg of phosphorus, and 25.40 mg of zinc. Determine the empirical formula of zinc phosphate.

8. Bromoform is 94.85% bromine, 0.40% hydrogen, and 4.75% carbon by mass. Determine its empirical formula.

9. Fulgurites are the products of the melting that occurs when lightning strikes the earth. Microscopic examination of a sand fulgurite reveals that it is a globule with variable composition that contains some grains of the definite chemical composition Fe 46.01%, Si 53.99%. Determine the empirical formula of these grains.

10. A sample of a "suboxide" of cesium gives up 1.6907% of its mass as gaseous oxygen when gently heated, leaving pure cesium behind. Determine the empirical formula of this binary compound.

11. Barium and nitrogen form two binary compounds containing 90.745% and 93.634% barium, respectively. Determine the empirical formulas of these two compounds.

12. Carbon and oxygen form no fewer than five different binary compounds. The mass percentages of carbon in the five compounds are as follows: A, 27.29; B, 42.88; C, 50.02; D, 52.97; and E, 65.24. Determine the empirical formulas of the five compounds.

13. A sample of 1.000 g of a compound containing carbon and hydrogen reacts with oxygen at elevated temperature to yield 0.692 g of $H_2O$ and 3.381 g of $CO_2$.
    (a) Calculate the masses of C and H in the sample.
    (b) Does the compound contain any other elements?
    (c) What are the mass percentages of C and H in the compound?
    (d) What is the empirical formula of the compound?

14. Burning a compound of calcium, carbon, and nitrogen in oxygen in a combustion train generates calcium oxide (CaO), carbon dioxide ($CO_2$), nitrogen dioxide ($NO_2$), and no other substances. A small sample gives 2.389 g of CaO, 1.876 g of $CO_2$, and 3.921 g of $NO_2$. Determine the empirical formula of the compound.

15. The empirical formula of a gaseous fluorocarbon is $CF_2$. At a certain temperature and pressure, a 1-liter volume holds 8.93 g of this fluorocarbon, while under the same conditions the 1-liter volume holds only 1.70 g of gaseous fluorine ($F_2$). Determine the molecular formula of this compound.

16. At its boiling point (280°C) and at atmospheric pressure, phosphorus has a gas density of 2.7 g $L^{-1}$. Under the same

conditions nitrogen has a gas density of 0.62 g $L^{-1}$. How many atoms of phosphorus are there in one phosphorus molecule under these conditions?

17. A gaseous binary compound has a vapor density that is 1.94 times that of oxygen at the same temperature and pressure. When 1.39 g of the gas is burned in an excess of oxygen, 1.21 g of water is formed, removing all the hydrogen originally present.
    (a) Estimate the molecular mass of the gaseous compound.
    (b) How many H atoms are there in a molecule of the compound?
    (c) What is the maximum possible value of the atomic mass of the second element in the compound?
    (d) Are other values possible for the atomic mass of the second element? Use a table of atomic masses to identify the element that best fits the data.
    (e) What is the molecular formula of the compound?

18. A gaseous binary compound has a vapor density that is 2.53 times that of nitrogen at 100°C and atmospheric pressure. When 8.21 g of the gas reacts with $AlCl_3$ at 100°C, 1.62 g of gaseous nitrogen is produced, removing all the nitrogen originally present.
    (a) Estimate the molecular mass of the gaseous compound.
    (b) How many N atoms are there in a molecule of the compound?
    (c) What is the maximum possible value of the atomic mass of the second element?
    (d) Are other values possible for the atomic mass of the second element? Use a table of atomic masses to identify the element that best fits the data.
    (e) What is the molecular formula of the compound?

19. Balance the following chemical equations.
    (a) $H_2 + N_2 \rightarrow NH_3$
    (b) $K + O_2 \rightarrow K_2O_2$
    (c) $PbO_2 + Pb + H_2SO_4 \rightarrow PbSO_4 + H_2O$
    (d) $BF_3 + H_2O \rightarrow B_2O_3 + HF$
    (e) $KClO_3 \rightarrow KCl + O_2$
    (f) $CH_3COOH + O_2 \rightarrow CO_2 + H_2O$
    (g) $K_2O_2 + H_2O \rightarrow KOH + O_2$
    (h) $PCl_5 + AsF_3 \rightarrow PF_5 + AsCl_3$

20. Balance the following chemical equations.
    (a) $Al + HCl \rightarrow AlCl_3 + H_2$
    (b) $NH_3 + O_2 \rightarrow NO + H_2O$
    (c) $Fe + O_2 + H_2O \rightarrow Fe(OH)_2$
    (d) $HSbCl_4 + H_2S \rightarrow Sb_2S_3 + HCl$
    (e) $Al + Cr_2O_3 \rightarrow Al_2O_3 + Cr$
    (f) $XeF_4 + H_2O \rightarrow Xe + O_2 + HF$
    (g) $(NH_4)_2Cr_2O_7 \rightarrow N_2 + Cr_2O_3 + H_2O$
    (h) $NaBH_4 + H_2O \rightarrow NaBO_2 + H_2$

## Mass Relationships in Chemical Reactions

21. For each of the following chemical reactions, calculate the mass of the underlined reactant that is required to produce 1.000 g of the underlined product.

(a) $Mg + 2 HCl \rightarrow \underline{H_2} + MgCl_2$
(b) $2 \underline{CuSO_4} + 4 KI \rightarrow 2 CuI + \underline{I_2} + 2 K_2SO_4$
(c) $\underline{NaBH_4} + 2 H_2O \rightarrow NaBO_2 + 4 \underline{H_2}$

22. For each of the following chemical reactions, calculate the mass of the underlined product that is produced from 1.000 g of the underlined reactant.
    (a) $\underline{CaCO_3} + H_2O \rightarrow \underline{Ca(OH)_2} + CO_2$
    (b) $\underline{C_3H_8} + 5 O_2 \rightarrow 3 \underline{CO_2} + 4 H_2O$
    (c) $2 \underline{MgNH_4PO_4} \rightarrow \underline{Mg_2P_2O_7} + 2 NH_3 + H_2O$

23. An 18.6-g sample of $K_2CO_3$ was treated in such a way that all of its carbon was captured in the compound $K_2Zn_3[Fe(CN)_6]_2$. Compute the mass (in grams) of this product.

24. A chemist dissolves 1.406 g of pure platinum (Pt) in an excess of a mixture of hydrochloric and nitric acids and then, after a series of subsequent steps involving several other chemicals, isolates a compound of molecular formula $Pt_2C_{10}H_{18}N_2S_2O_6$. Determine the maximum possible yield of this compound.

25. Disilane ($Si_2H_6$) is a gas that reacts with oxygen to give silica ($SiO_2$) and water. Calculate the mass of silica that would form if 25.0 $cm^3$ of disilane (with a density of $2.78 \times 10^{-3}$ g $cm^{-3}$) reacted with excess oxygen.

26. Tetrasilane ($Si_4H_{10}$) is a liquid with a density of 0.825 g $cm^{-3}$. It reacts with oxygen to give silica ($SiO_2$) and water. Calculate the mass of silica that would form if 25.0 $cm^3$ of tetrasilane reacted completely with excess oxygen.

27. Cryolite ($Na_3AlF_6$) is used in the production of aluminum from its ores. It is made by the reaction

$$6 NaOH + Al_2O_3 + 12 HF \longrightarrow 2 Na_3AlF_6 + 9 H_2O$$

Calculate the mass of cryolite that can be prepared by the complete reaction of 287 g of $Al_2O_3$.

28. Carbon disulfide ($CS_2$) is a liquid that is used in the production of rayon and cellophane. It is manufactured from methane and elemental sulfur via the reaction

$$CH_4 + 4 S \longrightarrow CS_2 + 2 H_2S$$

Calculate the mass of $CS_2$ that can be prepared by the complete reaction of 67.2 g of sulfur.

29. Potassium nitrate ($KNO_3$) is used as a fertilizer for certain crops. It is produced through the reaction

$$4 KCl + 4 HNO_3 + O_2 \longrightarrow 4 KNO_3 + 2 Cl_2 + 2 H_2O$$

Calculate the minimum mass of KCl required to produce 567 g of $KNO_3$. What mass of $Cl_2$ will be generated as well?

30. Elemental phosphorus can be prepared from calcium phosphate via the overall reaction

$$2 Ca_3(PO_4)_2 + 6 SiO_2 + 10 C \longrightarrow$$
$$6 CaSiO_3 + P_4 + 10 CO$$

Calculate the minimum mass of $Ca_3(PO_4)_2$ required to produce 69.8 g of $P_4$. What mass of $CaSiO_3$ is generated as a byproduct?

*31. An element X has a dibromide with the empirical formula $XBr_2$ and a dichloride with the empirical formula $XCl_2$. The dibromide is completely converted to the dichloride when it is heated in a stream of chlorine according to the equation

$$XBr_2 + Cl_2 \longrightarrow XCl_2 + Br_2$$

When 1.500 g of $XBr_2$ is treated, 0.890 g of $XCl_2$ results.
(a) Calculate the atomic mass of the element X.
(b) By reference to a list of the atomic masses of the elements, identify the element X.

*32. An element A has a triiodide with the formula $AI_3$ and a trichloride with the formula $ACl_3$. The triiodide is quantitatively converted to the trichloride when it is heated in a stream of chlorine, according to the equation

$$AI_3 + \tfrac{3}{2} Cl_2 \longrightarrow ACl_3 + \tfrac{3}{2} I_2$$

If 0.8000 g of $AI_3$ is treated, 0.3776 g of $ACl_3$ is obtained.
(a) Calculate the atomic mass of the element A.
(b) Identify the element A.

*33. A mixture consisting of only sodium chloride (NaCl) and potassium chloride (KCl) weighs 1.0000 g. When the mixture is dissolved in water and an excess of silver nitrate is added, all of the chloride ions associated with the original mixture are precipitated as insoluble silver chloride (AgCl). The mass of the silver chloride is found to be 2.1476 g. Calculate the mass percentages of sodium chloride and potassium chloride in the original mixture.

*34. A mixture of aluminum and iron weighing 9.62 g reacts with hydrogen chloride in aqueous solution according to the parallel reactions

$$2 Al + 6 HCl \longrightarrow 2 AlCl_3 + 3 H_2$$
$$Fe + 2 HCl \longrightarrow FeCl_2 + H_2$$

A 0.738-g quantity of hydrogen is evolved when the metals react completely. Calculate the mass of iron in the original mixture.

35. When ammonia is mixed with hydrogen chloride (HCl), the white solid ammonium chloride ($NH_4Cl$) is produced. Suppose 10.0 g of ammonia is mixed with the same mass of hydrogen chloride. What substances will be present after the reaction has gone to completion, and what will their masses be?

36. The poisonous gas hydrogen cyanide (HCN) is produced by the high-temperature reaction of ammonia with methane ($CH_4$). Hydrogen is also produced in this reaction.
(a) Write a balanced chemical equation for the reaction that occurs.
(b) Suppose 500.0 g of methane is mixed with 200.0 g of ammonia. Calculate the masses of the substances present after the reaction is allowed to proceed to completion.

37. The iron oxide $Fe_2O_3$ reacts with carbon monoxide, CO, to give iron and carbon dioxide:

$$Fe_2O_3 + 3 CO \longrightarrow 2 Fe + 3 CO_2$$

The reaction of 433.2 g of $Fe_2O_3$ with excess CO yields 254.3 g of iron. Calculate the theoretical yield of iron (assuming complete reaction) and its percentage yield.

38. Titanium dioxide, $TiO_2$, reacts with carbon and chlorine to give gaseous $TiCl_4$:

$$TiO_2 + 2\ C + 2\ Cl_2 \longrightarrow TiCl_4 + 2\ CO$$

The reaction of 7.39 kg of titanium dioxide with excess C and $Cl_2$ gives 14.24 kg of titanium tetrachloride. Calculate the theoretical yield of $TiCl_4$ (assuming complete reaction) and its percentage yield.

### Additional Problems

39. Human parathormone has the impressive molecular formula $C_{691}H_{898}N_{125}O_{164}S_{11}$. Compute the mass percentages of all of the elements in this compound.

40. A white oxide of tungsten is 79.2976% tungsten by mass. A blue tungsten oxide also contains exclusively tungsten and oxygen, but it is 80.8473% tungsten by mass. Determine the empirical formulas of white tungsten oxide and blue tungsten oxide.

41. A dark brown binary compound contains oxygen and a metal. It is 13.38% oxygen by mass. Heating it moderately drives off some of the oxygen and gives a red binary compound that is 9.334% oxygen by mass. Strong heating drives off more oxygen and gives still another binary compound that is only 7.168% oxygen by mass.
    (a) Compute the mass of oxygen that is combined with 1.000 g of the metal in each of these three oxides.
    (b) Assume that the empirical formula of the first compound is $MO_2$ (where M stands for the metal). Give the empirical formulas of the second and third compounds.
    (c) Name the metal.

42. A binary compound of nickel and oxygen contains 78.06% nickel by mass. Is this a stoichiometric or a nonstoichiometric compound? Explain.

43. Two binary oxides of the element manganese contain, respectively, 30.40% and 36.81% oxygen by mass. Calculate the empirical formulas of the two oxides.

*44. A sample of a gaseous binary compound of boron and chlorine weighing 2.842 g occupies 0.153 L. This sample is decomposed to give 0.664 g of solid boron and enough gaseous chlorine ($Cl_2$) to occupy 0.688 L at the same temperature and pressure. Determine the molecular formula of the compound.

45. A possible practical way to eliminate oxides of nitrogen (such as $NO_2$) from automobile exhaust gases employs cyanuric acid, $C_3N_3(OH)_3$. When heated to the relatively low temperature of 625°F, cyanuric acid converts to gaseous isocyanic acid (HNCO). Isocyanic acid reacts with $NO_2$ in the exhaust to form nitrogen, carbon dioxide, and water, all of which are normal constituents of the air.
    (a) Write balanced equations for these two reactions.
    (b) If the process just described became practical, how much cyanuric acid (in kilograms) would be required to absorb the $1.7 \times 10^{10}$ kg of $NO_2$ generated annually in auto exhaust in the United States?

46. Aspartame (molecular formula $C_{14}H_{18}N_2O_5$) is a sugar substitute in soft drinks. Under certain conditions, one mole of aspartame reacts with two moles of water to give one mole of aspartic acid (molecular formula $C_4H_7NO_4$), one mole of methanol (molecular formula $CH_3OH$), and one mole of phenylalanine. Determine the molecular formula of phenylalanine.

47. 3′-Methylphthalanilic acid is used commercially as a "fruit set" to prevent premature drop of apples, pears, cherries, and peaches from the tree. It is 70.58% carbon, 5.13% hydrogen, 5.49% nitrogen, and 18.80% oxygen. If eaten, the fruit set reacts with water in the body to produce an innocuous product, which contains carbon, hydrogen, and oxygen only, and m-toluidine ($NH_2C_6H_4CH_3$), which causes anemia and kidney damage. Compute the mass of the fruit set that would produce 5.23 g of m-toluidine.

48. Aluminum carbide ($Al_4C_3$) reacts with water to produce gaseous methane ($CH_4$). Calculate the mass of methane formed from 63.2 g of $Al_4C_3$.

49. Citric acid ($C_6H_8O_7$) is made by fermentation of sugars such as sucrose ($C_{12}H_{22}O_{11}$) in air. Oxygen is consumed and water generated as a byproduct.
    (a) Write a balanced equation for the overall reaction that occurs in the manufacture of citric acid from sucrose.
    (b) What mass of citric acid is made from 15.0 kg of sucrose?

50. A sample that contains only $SrCO_3$ and $BaCO_3$ weighs 0.800 g. When it is dissolved in excess acid, 0.211 g of carbon dioxide is liberated. What percentage of $SrCO_3$ did the sample contain? Assume all the carbon originally present is converted to carbon dioxide.

51. A sample of a substance with the empirical formula $XBr_2$ weighs 0.5000 g. When it is dissolved in water and all of its bromide is converted to insoluble AgBr by addition of an excess of silver nitrate, the mass of the resulting AgBr is found to be 1.0198 g. The chemical reaction is

$$XBr_2 + 2\ AgNO_3 \longrightarrow 2\ AgBr + X(NO_3)_2$$

    (a) Calculate the molecular mass (i.e., formula mass) of $XBr_2$.
    (b) Calculate the atomic mass of X and give its name and symbol.

52. A newspaper article about the danger of global warming from the accumulation of greenhouse gases such as carbon dioxide states that "reducing driving your car by 20 miles a week would prevent release of over 1000 pounds of $CO_2$ per year into the atmosphere." Is this a reasonable statement? Assume that gasoline is octane (molecular formula $C_8H_{18}$) and that it is burned completely to $CO_2$ and $H_2O$ in the engine of your car. Facts (or reasonable guesses) about your car's gas mileage, the density of octane, and other factors will also be needed.

53. In the Solvay process for producing sodium carbonate ($Na_2CO_3$), the following reactions occur in sequence:

$$NH_3 + CO_2 + H_2O \longrightarrow NH_4HCO_3$$

$$NH_4HCO_3 + NaCl \longrightarrow NaHCO_3 + NH_4Cl$$

$$2\ NaHCO_3 \xrightarrow{heat} Na_2CO_3 + H_2O + CO_2$$

How many metric tons of sodium carbonate would be produced per metric ton of $NH_3$ if the process were 100% efficient? (1 metric ton = 1000 kg.)

54. A yield of 3.00 g of $KClO_4$ is obtained from the (unbalanced) reaction

$$KClO_3 \longrightarrow KClO_4 + KCl$$

when 4.00 g of the reactant is used. What is the percentage yield of the reaction?

55. An industrial-scale process for making acetylene consists of the following sequence of operations:

$$CaCO_3 \longrightarrow CaO \ + \ CO_2$$

limestone                 lime        carbon dioxide

$$CaO + 3\ C \longrightarrow CaC_2 \ + \ CO$$

calcium carbide   carbon monoxide

$$CaC_2 + 2\ H_2O \longrightarrow Ca(OH)_2 \ + \ C_2H_2$$

calcium        acetylene
hydroxide

What is the percentage yield of the overall process if 2.32 metric tons of $C_2H_2$ is produced from 10.0 metric tons of limestone? (1 metric ton = 1000 kg.)

56. Silicon nitride ($Si_3N_4$), a valuable ceramic, is made by the direct combination of silicon and nitrogen at high temperature. How much silicon must react with excess nitrogen to prepare 125 g of silicon nitride if the yield of the reaction is 95.0%?

# Chemical Bonding:
# The Classical Description

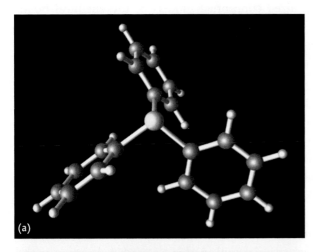

(a)

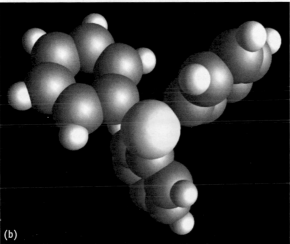

(b)

*Illustration*
(a) The shape of the molecule triphenyl-phosphine [$(C_6H_5)_3P$] is determined by locating the valence shell electron pairs in those positions that minimize the overall energy of the molecule. (b) The space-filled representation aids the analysis and understanding of the steric environment responsible for the specific molecular geometry. *(Courtesy of Drs. Andrew J. Pounds and Mark A. Iken, Scientific Visualization Laboratory, Georgia Institute of Technology, Atlanta, GA)*

C hapters 1 and 2 revealed how the law of conservation of mass, through the mole concept, determines quantitative mass relationships in chemical reactions. That discussion assumed prior knowledge of the chemical formulas of the reactants and products in each equation. The far more open-ended question of what compounds are found in nature (or can be made in the laboratory) and what types of reactions they undergo now arises. Why are some elements and compounds fiercely reactive and others inert? Why are there compounds with chemical formulas

$H_2O$ and $NaCl$, but never $H_3O$ or $NaCl_2$? Why are helium and the other noble gases monatomic, when molecules of hydrogen and chlorine are diatomic? These questions can be answered by examining the formation of chemical bonds between atoms.

When two or more atoms are in close proximity, their electrons interact and form new arrangements of electrons around the nuclei that have lower total potential energy than the isolated atoms. The reduction in energy stabilizes the new arrangements relative to the isolated atoms through formation of chemical bonds. Chemical bonds are formed by sharing or transferring electrons between atoms. These processes lead to two idealized conceptual models of the chemical bond. When electrons are shared between atoms, the bond between them is called **covalent.** When electrons are transferred from one atom to another, the resulting bond is called **ionic.** Although numerous real examples of these two extreme idealized models are known, most real bonds are neither fully ionic nor fully covalent. Real molecules exhibit a continuum from purely ionic to purely covalent bonding, and most possess a mixture of ionic and covalent character. Bonds in which there is partial transfer of charge are **polar covalent.** A quantitative description of chemical bonding depends on the detailed arrangement of electrons in each atom and requires quantum mechanics for its explanation. The quantum treatment of chemical bonding will be provided in UNIT 5.

The purpose of the present chapter is to provide an approximate, introductory description of chemical bond formation and the shapes of molecules that does not require quantum mechanics. One essential tool in this classical description is the **Lewis electron dot diagram,** which represents the transfer or sharing of electrons between atoms. The Lewis dot model of bonding rationalizes the formulas of compounds, but gives no information on the three-dimensional shapes of molecules. To explain shape, the Lewis model must be supplemented by a second tool, the Valence Shell Electron Pair Repulsion Theory (VSEPR). In addition to the Lewis model and VSEPR theory, this chapter develops a system of inorganic nomenclature. Together, these three tools provide a basis for organizing vast amounts of chemical information in a rational and systematic way.

---

## 3.1

# ELECTRONEGATIVITY: THE TENDENCY OF ATOMS TO ATTRACT ELECTRONS

The type of bond formed between a pair of atoms is determined by the ability of each of the atoms to attract electrons from the other. A positively charged **ion** (called a **cation**) forms when an atom loses one or more electrons, and a negatively charged ion (called an **anion**) forms when an atom adds electrons. For a free, isolated atom, the ability to lose an electron is measured by its **ionization energy,** while the ability to gain an electron is measured by the **electron affinity.** The average between these two properties of isolated atoms is used to define a new quantity called the **electronegativity** that measures the net tendency of one atom to attract electrons from another atom to which it is bonded. Comparing the electronegativity values for two atoms indicates whether they will form an ionic, covalent, or polar covalent bond. The present section describes ionization energy, electron affinity, and electronegativity and relates their values for each atom to its location in the periodic table. Building on this background, remaining sections in this chapter systematically describe ionic, covalent, and polar covalent bonds and introduce the properties of the structures produced when these bonds are formed.

Changing the electron arrangement around an atom by gain or loss of electrons also changes the energy of the atom, as evidenced by the fact that ionization energy and electron affinity have physical dimensions of energy. Section 1.7, which describes the energy concept, should be reviewed in preparation for studying these properties of atoms.

## Ionization Energies

The ionization energy of an atom (sometimes also referred to as the first ionization energy) is the minimum energy necessary to detach an electron from the neutral gaseous atom. It is the change in energy $\Delta E$ for the process

$$X(g) \longrightarrow X^+(g) + e^- \qquad\qquad \Delta E = IE_1$$

The Greek letter $\Delta$ is widely used to symbolize the difference in value of a property caused by a process or change carried out in the laboratory. Here $\Delta E =$ [Energy of reaction products] $-$ [Energy of reactants]. By convention, $\Delta E$ is positive when energy must be invested to drive the process, and $\Delta E$ is negative if the process liberates energy. Ionization energies are always positive; an input of energy is required to detach an electron from an atom. The ionization energy influences how an atom interacts with other atoms to form chemical bonds. It is a measure of the stability of the outer electron configuration of the free atom: the greater its magnitude, the more energy is required to remove an outer electron from the atom and therefore the more stable the electronic structure of the atom.

The *second* ionization energy, $IE_2$, is the minimum energy required to remove a second electron, or $\Delta E$ for the process

$$X^+(g) \longrightarrow X^{2+}(g) + e^- \qquad\qquad \Delta E = IE_2$$

The third, fourth, and higher ionization energies are defined in an analogous fashion. Appendix F lists measured ionization energies of the elements, which generally increase across a period to become large for each noble gas atom and then fall abruptly for the alkali atom at the beginning of the next period. Ionization energy is thus a periodic property of the atoms (Section 1.5). This periodicity correlates with the fact that each main group element (Groups I–VIII) in Period 3 (through element 20, Ca) contains exactly 8 more electrons than the element immediately above it, and each main group element in Periods 4 and 5 (from element 31, Ga, through element 56, Ba) contains exactly 18 more electrons than the element immediately above it. The connection will be made precise by the shell model of the atom to be described by quantum mechanics in Chapter 15. Here, our main concern is to use the experimentally determined values of $IE_1$ to estimate the ease with which atoms participate in ionic bonding.

## Electron Affinity

The extent to which an atom can accept an extra electron is indicated by the stability of the corresponding gaseous anion. The stability of the anion can be measured directly as the energy required to convert the anion back to the neutral atom by removing the electron. For example, the reaction

$$Cl^-(g) \longrightarrow Cl(g) + e^-$$

requires energy $\Delta E = +349$ kJ mol$^{-1}$ to be supplied. For historical reasons, the energetics of such reactions are defined for the reverse reaction, namely the attachment of an electron to a neutral atom. The electron affinity $EA$ of an atom is the energy *released* when an electron is added to it. Thus $EA = -\Delta E$ for the reaction

$$X(g) + e^- \longrightarrow X^-(g)$$

For example the reaction

$$Cl(g) + e^- \longrightarrow Cl^-(g)$$

shows $\Delta E = -349$ kJ mol$^{-1}$; thus $EA = +349$ kJ mol$^{-1}$ for Cl. The electron affinity of X is the same as the electron detachment energy of X$^-$—that is, $\Delta E$ for the process

$$X^-(g) \longrightarrow X(g) + e^-$$

An electron affinity is either positive or negative, depending on the element involved. A negative electron affinity means that the electron must be forced to "stick" and will spontaneously fly away from the negative ion if it is in free space. Appendix F lists the electron affinities of the elements.

The periodic trends in electron affinity largely parallel those in ionization energy, increasing across a period to become large and positive for the halogen atom, then falling abruptly to a negative value for the noble gas atom.

No gaseous atom has a positive electron affinity for a *second* electron, because a gaseous ion with a net charge of $-2e$ is always unstable with respect to ionization. Attaching a second electron means bringing it close to a species that is already negatively charged. The two repel each other, and the energy rises. In crystalline environments, however, doubly negative ions such as $O^{2-}$ can be stabilized by electrostatic interaction with neighboring positive ions.

## Mulliken's Electronegativity Scale

The values of $IE_1$ and $EA$ indicate whether an atom readily forms a negative ion or a positive ion. In order to summarize these results in a single property suitable for showing trends with respect to location in the periodic table, in 1934 Robert Mulliken defined the electronegativity of each atom. Electronegativity is a measure of the tendency of an atom to draw electrons to itself in a chemical bond. Two years earlier, Linus Pauling had given a different definition of electronegativity. Their separate scales produce values that are nearly proportional.[1] Mulliken observed that elements toward the lower left corner of the periodic table have low ionization energies and small electron affinities. This means that they give up electrons readily (to form positive ions) but do not readily accept electrons (to form negative ions). They tend to act as electron *donors* in interactions with other elements. In contrast, elements in the upper right corner of the periodic table have high ionization energies and also (except for the noble gases) have large electron affinities. As a result, these elements accept electrons easily but give them up only reluctantly; such atoms act as electron *acceptors*.

---

[1] Numerical values are tabulated in Appendix F on Pauling's scale.

Mulliken then defined electronegativity as proportional to the average of the ionization energy and the electron affinity.

$$\text{electronegativity (Mulliken)} \propto \tfrac{1}{2}(IE_1 + EA)$$

Electron acceptors (such as the halogens) have high ionization energies and electron affinities and are thus labeled highly **electronegative.** Electron donors (such as the alkali metals) have low ionization energies and electron affinities and therefore low electronegativities; they are labeled **electropositive.** The noble gases only rarely participate in chemical bonding. Their high ionization energies and negative electron affinities mean that they are reluctant either to give up or to accept electrons. Electronegativities are generally not assigned to the noble gases.

The essential points of this section can be summed up as follows. The tendency of an atom to donate or accept electrons in a chemical bond is indicated by the value of its electronegativity. Highly electronegative atoms, on the right side of the periodic table, readily accept electrons and form negative ions. Highly electropositive atoms, on the left side of the table, readily donate electrons to form positive ions. Ionic bonds, in which electrons are transferred, are formed between atoms with large differences in electronegativity. Covalent bonds, in which electrons are shared between atoms, are formed between atoms with small differences in electronegativity.

---

### 3.2

## IONIC BONDING: COULOMB STABILIZATION ENERGY

This section describes formation of ionic bonds between atoms with large differences in electronegativity, for example, Na and F. Formation of the positive and negative ions by electron transfer between the atoms is represented by Lewis dot diagrams. The Lewis diagrams provide a rationale for which ions can be formed by each atom and for the chemical formulas of ionic compounds. The oppositely charged ions are stabilized by the attractive Coulomb force between them; the magnitude of the stabilization energy can be estimated by calculating the Coulomb potential energy between the ions. Very large stabilization energies are achieved when large numbers of ions are placed in an ordered arrangement wherein each ion is surrounded by those of the opposite charge. Ionic bonding produces large extended crystals in the solid state; pure ionic bonding is observed in free molecules in the gaseous state only at very high temperatures.

### Lewis Dot Diagrams for Ionic Bonding

The Lewis model begins by recognizing that not all the electrons in an atom participate in chemical bonding. Electrons appear to occupy a set of **shells** surrounding the nucleus, and those in inner shells (called **core electrons**) are not significantly involved in the formation of bonds between atoms. The outermost, partially filled shell (called the **valence shell**) contains the electrons that need to be included in most descriptions of chemical bonding, the **valence electrons.** Atoms with filled shells possess great chemical stability. This observation is explained by the quantum description of atomic structure and reflects the strength of the electrostatic force that binds the electron to the nucleus. Progress through the elements in order of increasing atomic number reveals a filled shell whenever a noble-gas element such as helium, neon, or argon is reached. The additional electron in atoms of the first element past a noble-gas element (that is, the outermost electron in an alkali-metal atom) is the first occupant of a new shell, so an alkali-metal atom has one valence electron. With the exception of helium, the number of valence electrons in a neutral atom of a main-group element (those in Groups I through VIII) is equal to the element's group number in the periodic table.

The main-group elements that follow a series of transition-metal elements require some special attention. Atoms of bromine, for example, have 17 more electrons than atoms of argon, the preceding noble gas. We still say that bromine has 7 valence electrons (like chlorine), not 17. The reason is that in the fourth, fifth, and sixth row the ten electrons added in the course of the transition-metal series (although they are very important for the bonding of those elements) have become *core* electrons by the time the end of the transition-metal series is reached. The bonding properties of an element such as bromine resemble those of the lighter elements in its group.

The Lewis model represents valence electrons with dots; core electrons are not shown. The first four dots are displayed singly around the four sides of the symbol of the element. If there are more than four valence electrons, their dots are paired with those already present. The result is a **Lewis dot symbol** for that atom. The Lewis notation for the elements of the first two periods is

$$\cdot H \qquad\qquad\qquad\qquad\qquad\qquad\qquad\qquad\qquad\qquad :He$$

$$\cdot Li \quad \cdot Be\cdot \quad \cdot \overset{\cdot}{B}\cdot \quad \cdot \overset{\cdot}{\underset{\cdot}{C}}\cdot \quad :\overset{\cdot}{\underset{\cdot}{N}}\cdot \quad \cdot\overset{\cdot\cdot}{\underset{\cdot\cdot}{O}}\cdot \quad :\overset{\cdot\cdot}{\underset{\cdot\cdot}{F}}\cdot \quad :\overset{\cdot\cdot}{\underset{\cdot\cdot}{Ne}}:$$

## The Formation of Ionic Compounds

The creation of ions is indicated by removing dots from or adding them to the Lewis dot symbol and also by writing the net electric charge of the ion as a right superscript. For example:

| | | | |
|---|---|---|---|
| Na· | Na$^+$ | ·Ca· | Ca$^{2+}$ |
| Sodium atom | Sodium ion | Calcium atom | Calcium ion |
| :$\overset{\cdot\cdot}{F}$· | :$\overset{\cdot\cdot}{\underset{\cdot\cdot}{F}}$:$^-$ | ·$\overset{\cdot\cdot}{S}$· | :$\overset{\cdot\cdot}{\underset{\cdot\cdot}{S}}$:$^{2-}$ |
| Fluorine atom | Fluoride ion | Sulfur atom | Sulfide ion |

Special stability results when an atom, by either losing or gaining electrons, forms an ion whose outermost shell has the same number of electrons as the outermost shell of a noble-gas atom. For example, Na with ionization energy of 495.8 kJ mol$^{-1}$ easily gives up its valence electron to achieve the same number of valence electrons as Ne, which is extraordinarily stable, with an ionization energy of 2080.6 kJ mol$^{-1}$. Except for hydrogen and helium, whose valence shells are completed with two electrons, atoms of the first few periods of the periodic table have a maximum of eight electrons in their valence shells. We say that a chlorine ion (:$\overset{\cdot\cdot}{\underset{\cdot\cdot}{Cl}}$:$^-$) or an argon atom (:$\overset{\cdot\cdot}{\underset{\cdot\cdot}{Ar}}$:) has a **completed octet** in its valence shell.

The tendency of atoms to achieve valence octets describes much chemical reactivity. Atoms of elements in Groups I and II achieve an octet by losing electrons to form cations; atoms of elements in Groups VI and VII do so by gaining electrons to form anions. Reactions of the metallic elements on the left side of the periodic table with the nonmetallic elements on the right side always transfer just enough electrons to form ions with completed octets. The following equations, in which $e^-$ stands for an electron, use Lewis symbols to show the formation first of a cation and an anion and then of an **ionic compound.**

$$\text{Na} \cdot \longrightarrow \text{Na}^+ + e^- \qquad \text{Loss of a valence electron}$$

$$e^- + \text{:}\overset{..}{\text{Cl}}\cdot \longrightarrow \text{:}\overset{..}{\underset{..}{\text{Cl}}}\text{:}^- \qquad \text{Gain of a valence electron}$$

$$\text{Na}^+ + \text{:}\overset{..}{\underset{..}{\text{Cl}}}\text{:}^- \longrightarrow \text{Na}^+ \text{:}\overset{..}{\underset{..}{\text{Cl}}}\text{:}^- \qquad \text{Combination to form the compound NaCl}$$

Another example is the formation of $CaBr_2$:

$$\cdot \text{Ca} \cdot + 2 \text{:}\overset{..}{\underset{..}{\text{Br}}}\cdot \longrightarrow \text{Ca}^{2+} + 2 \text{:}\overset{..}{\underset{..}{\text{Br}}}\text{:}^- \longrightarrow \text{Ca}^{2+}(\text{Br}^-)_2$$

| Neutral atoms | Positive ion | Negative ions | Ionic |
| not having octets | with octet | with octets | compound |

The model predicts a 1:1 compound between Na and Cl and a 1:2 compound between Ca and Br, in agreement with experiment.

Ionic compounds, except those with $OH^-$ as the anion, are often called **salts** by analogy with NaCl, common table salt. They are solids at room conditions and generally have high melting and boiling points (for example, NaCl melts at 801°C and boils at 1413°C). Solid ionic compounds usually conduct electricity poorly, but their melts (the molten liquids) conduct well.

## Names and Formulas of Ionic Compounds

The combination of cations with anions results in ionic compounds. Each is named by listing the name of the cation followed by that of the anion. Ions can be either monatomic or polyatomic; the latter are also referred to as molecular ions.

A monatomic cation bears the name of the parent element. We have already encountered the sodium ion ($Na^+$) and the calcium ion ($Ca^{2+}$); ions of the other elements in Groups I and II are named in the same way. The transition metals and the metallic elements of Groups III, IV, and V differ from the Group I and II metals in that they often form several stable ions in compounds and in solution. Although calcium compounds never contain $Ca^{3+}$ ions (always $Ca^{2+}$), the element iron forms both the $Fe^{2+}$ and $Fe^{3+}$ ions, and thallium forms both the $Tl^+$ and $Tl^{3+}$ ions. When a metal forms ions of more than one charge, we distinguish them by placing a Roman numeral in parentheses after the name of the metal:

| | | | | | |
|---|---|---|---|---|---|
| $Cu^+$ | copper(I) ion | $Fe^{2+}$ | iron(II) ion | $Sn^{2+}$ | tin(II) ion |
| $Cu^{2+}$ | copper(II) ion | $Fe^{3+}$ | iron(III) ion | $Sn^{4+}$ | tin(IV) ion |

An earlier method for distinguishing between such pairs of ions used the suffixes -*ous* and -*ic* added to the root of the (usually Latin) name of the metal to indicate the ions of lower and higher charge, respectively. Thus, $Fe^{2+}$ was called the ferrous ion and $Fe^{3+}$ the ferric ion. This method, although still sometimes used, is not recommended for systematic nomenclature and will not appear again in this book.

A few polyatomic cations are of importance in inorganic chemistry. These include the ammonium ion, $NH_4^+$ (obtained by adding $H^+$ to ammonia); the hydronium ion, $H_3O^+$ (obtained by adding $H^+$ to water); and the particularly interesting molecular ion formed by mercury: $Hg_2^{2+}$, the mercury(I) ion. This species must be carefully distinguished from $Hg^{2+}$, the mercury(II) ion. The Roman numeral I in

**TABLE 3.1**

### Formulas and Names of Some Common Anions

| | |
|---|---|
| $F^-$ | fluoride |
| $Cl^-$ | chloride |
| $Br^-$ | bromide |
| $I^-$ | iodide |
| $H^-$ | hydride |
| $O^{2-}$ | oxide |
| $S^{2-}$ | sulfide |
| $O_2^{2-}$ | peroxide |
| $O_2^-$ | superoxide |
| $OH^-$ | hydroxide |
| $CN^-$ | cyanide |
| $CNO^-$ | cyanate |
| $SCN^-$ | thiocyanate |
| $MnO_4^-$ | permanganate |
| $CrO_4^{2-}$ | chromate |
| $Cr_2O_7^{2-}$ | dichromate |
| $CO_3^{2-}$ | carbonate |
| $HCO_3^-$ | hydrogen carbonate |
| $NO_2^-$ | nitrite |
| $NO_3^-$ | nitrate |
| $SiO_4^{4-}$ | silicate |
| $PO_4^{3-}$ | phosphate |
| $HPO_4^{2-}$ | hydrogen phosphate |
| $H_2PO_4^-$ | dihydrogen phosphate |
| $SO_3^{2-}$ | sulfite |
| $SO_4^{2-}$ | sulfate |
| $HSO_4^-$ | hydrogen sulfate |
| $ClO^-$ | hypochlorite |
| $ClO_2^-$ | chlorite |
| $ClO_3^-$ | chlorate |
| $ClO_4^-$ | perchlorate |

parentheses means in this case that the average charge on each of the two mercury atoms is +1. Compounds with the empirical formulas HgCl and HgBr correspond to $Hg_2Cl_2$ and $Hg_2Br_2$.

A monatomic anion is named by adding the suffix *-ide* to the first portion of the name of the element. Thus, chlor*ine* becomes the chlor*ide* ion, and oxy*gen* becomes the ox*ide* ion. The other monatomic anions of Groups V, VI, and VII are named similarly. Many polyatomic anions exist, and the naming of these species is more complex. The names of the oxoanions (each contains oxygen in combination with a second element) are derived by adding the ending *-ate* to the stem of the name of that second element. Some elements form two oxoanions. The *-ate* ending is then used for the oxoanion with the larger number of oxygen atoms (e.g., $NO_3^-$, nitr*ate*), and the ending *-ite* is added for the name of the anion with the smaller number (e.g., $NO_2^-$, nitr*ite*). For elements such as chlorine, which form more than two oxoanions, we use the additional prefixes *per-* (largest number of oxygen atoms) and *hypo-* (smallest number of oxygen atoms). An oxoanion containing hydrogen as a third element includes that word in its name. The $HCO_3^-$ oxoanion, for example, is called the hydrogen carbonate ion in preference to its common (nonsystematic) name, "bicarbonate ion," and $HSO_4^-$, often called "bisulfate ion," is better designated as the hydrogen sulfate ion. Table 3.1 lists some of the most important anions. It is important to be able to recognize and name the ions from that table, bearing in mind that the electric charge is an essential part of the formula.

The composition of an ionic compound is determined by overall charge neutrality: the total positive charge on the cations must exactly balance the total negative charge on the anions. The following names and formulas of ionic compounds illustrate this point.

| | | |
|---|---|---|
| Tin(II) bromide | One 2+ cation, two 1− anions | $SnBr_2$ |
| Potassium permanganate | One 1+ cation, one 1− anion | $KMnO_4$ |
| Ammonium sulfate | Two 1+ cations, one 2− anion | $(NH_4)_2SO_4$ |
| Iron(II) dihydrogen phosphate | One 2+ cation, two 1− anions | $Fe(H_2PO_4)_2$ |

**EXAMPLE 3.1**

Give the chemical formulas of **(a)** calcium cyanide and **(b)** copper(II) phosphate.

### Solution

**(a)** Calcium cyanide is composed of $Ca^{2+}$ and $CN^-$ ions. For the overall charge to be 0, there must be two $CN^-$ ions for each $Ca^{2+}$ ion. Thus, the chemical formula of calcium cyanide is $Ca(CN)_2$.

**(b)** The ions present in this compound are $Cu^{2+}$ and $PO_4^{3-}$. To ensure charge neutrality, there must be three $Cu^{2+}$ ions (total charge +6) and two $PO_4^{3-}$ ions (total charge −6) per formula unit. Thus, the chemical formula of copper(II) phosphate is $Cu_3(PO_4)_2$.

**Related Problems: 7, 8**

### Ionic Bonding: Coulomb Stabilization Energy

By invoking the octet rule, the Lewis electron dot model rationalizes the formation of ionic bonds between atoms from Groups I and II and atoms from Groups VI and VII. The next level of sophistication is to estimate the stabilization energy of the ion pair relative to the free atoms, since this quantity measures the strength of the bond formed.

Consider the energy change that accompanies the formation of an isolated ionic bond. Suppose a highly electropositive element, such as potassium, reacts with a highly electronegative element, such as fluorine. Potassium is a good electron donor; its ionization energy is

$$\text{K} \longrightarrow \text{K}^+ + e^- \qquad\qquad \Delta E = +419 \text{ kJ mol}^{-1}$$

Fluorine is a good electron acceptor; its electron affinity ($EA = -\Delta E$) is

$$\text{F} + e \longrightarrow \text{F}^- \qquad\qquad \Delta E = -328 \text{ kJ mol}^{-1}$$

Even in this favorable case, there is a *cost* in energy to transfer an electron from a potassium atom to a fluorine atom; the energy that must be invested to form $\text{K}^+$ is greater than the energy released when $\text{F}^-$ is formed. An examination of the ionization energies and electron affinities of the elements shows that this is always the case. The smallest ionization energy (for Cs, 376 kJ mol$^{-1}$) is greater in magnitude than the largest electron affinity (for Cl, 349 kJ mol$^{-1}$). For atoms separated by large distances in the gas phase, electron transfer to form ions is always energetically unfavorable.

How, then, does an ionic bond form? As the two oppositely charged ions approach each other, the Coulomb force attracts them together. According to Coulomb's law (see Appendix B), the potential energy of interaction between two charges $Q_1$ and $Q_2$ separated by distance $R$ is

$$\text{potential energy} = \frac{Q_1 Q_2}{4\pi\epsilon_0 R} \qquad\qquad \textbf{[3.1]}$$

where $\epsilon_0$, called the **permittivity of the vacuum,** is a proportionality constant with value $8.854 \times 10^{-12}$ C$^2$ J$^{-1}$ m$^{-1}$. Here $R$ is the distance between the ion centers and $Q_1$ and $Q_2$ are the charges on the two ions (for $\text{K}^+$ and $\text{F}^-$, $Q_1$ is $+e$ and $Q_2$ is $-e$). Because $Q_1$ and $Q_2$ have opposite signs in this case, this Coulomb energy is negative and the total energy is *lowered* as the ions move toward each other:

$$\Delta E_{\text{Coulomb}} = \frac{Q_1 Q_2}{4\pi\epsilon_0 R}$$

At short enough distances, the Coulomb attraction more than compensates for the energy required to transfer the electron, and an ionic bond forms. Figure 3.1 shows how the energy varies with separation $R$ for the ionic and neutral states. For large separations the neutral atoms are always more stable, but at shorter distances the ionic species may be favored. At very short distances the electrons of the two atoms or ions begin to repel each other and the energy rises steeply. The equilibrium bond length $R_e$ of an ionic bond is determined by a balance of attractive and repulsive forces.

For an ionic bond such as the one in KF, the energy of dissociation to neutral atoms can be approximated as

$$\Delta E_d \approx -\frac{Q_1 Q_2}{4\pi\epsilon_0 R_e} - \Delta E_\infty \qquad\qquad \textbf{[3.2]}$$

where $\Delta E_\infty = IE_1(\text{K}) - EA(\text{F})$ in this case. In writing this equation we have assumed the bond to be purely ionic and have neglected the relatively small contribution of short-range repulsive interactions at the bond length $R_e$. The first term in this equation is the energy cost to separate the two *ions* (a positive number because $Q_2$ is

**FIGURE 3.1**  The potential energies of the ionic and neutral states of K + F as a function of separation $R$. Note that the two curves cross, so that the ionic state has a lower energy at short separations and the neutral state has a lower energy at large separations.

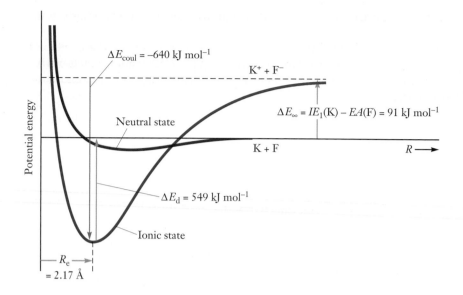

negative), and the second term is the energy released when electrons are transferred between ions to give neutral atoms (a negative number including the minus sign).

## EXAMPLE 3.2

Estimate the energy of dissociation to neutral atoms for KF, which has a bond length of 2.17 Å.

### Solution

$$\Delta E_d \approx -\frac{Q_1 Q_2}{4\pi\epsilon_0 R_e} - \Delta E_\infty$$

$$= -\frac{-(1.602 \times 10^{-19}\text{ C})^2(6.022 \times 10^{23}\text{ mol}^{-1})}{(4)(3.1416)(8.854 \times 10^{-12}\text{ C}^2\text{ J}^{-1}\text{ m}^{-1})(2.17 \times 10^{-10}\text{ m})}$$
$$- 9.1 \times 10^4\text{ J mol}^{-1}$$

$$= 64.0 \times 10^4\text{ J mol}^{-1} - 9.1 \times 10^4\text{ J mol}^{-1}$$

$$= 549\text{ kJ mol}^{-1}$$

This estimate compares fairly well with the experimentally measured dissociation energy of 498 kJ mol$^{-1}$. Note that we multiplied the Coulomb energy per molecule of KF by Avogadro's number to obtain the result in joules per mole of KF.

**Related Problems: 11, 12**

There are several reasons for the discrepancy between our simple calculations and experiment. First, our calculations omitted repulsive interactions; including them would tend to reduce $\Delta E_d$. Second, we assumed the KF bond to be purely ionic, whereas in fact the bond has a small amount of covalent (nonionic) character. Third, we assumed the charge distributions of the ions to be point charges centered on the respective nuclei, ignoring the distortion in the charge distribution of the F$^-$ ion caused by the electric field of the nearby K$^+$ ion. This distortion, termed **polar-**

**ization,** means that the charge distribution around each nucleus is not strictly symmetric.

Gaseous molecules of potassium fluoride exist only at high temperatures. At ordinary temperatures, a collection of such molecules would condense to form an ionic solid, liberating large amounts of energy. At room temperature, KF forms a three-dimensional lattice of great stability in which positive and negative ions occupy alternate sites.

---

## 3.3

# STRUCTURE OF ISOLATED MOLECULES: PROPERTIES OF THE COVALENT CHEMICAL BOND

Charge transfer followed by Coulomb stabilization accounts for formation of ionic bonds in solid compounds. When atoms with similar or equal values of electronegativity interact, they share electrons and form covalent bonds between the atoms to produce a molecule. Many substances exist as molecules in the gas phase at room temperature. Therefore it is possible to prepare isolated molecules and determine their structure by experiment. The present section surveys the structure of molecules to provide familiarity with features that must be explained by the theory of covalent bonding described in the following section.

A molecule is a discrete, recognizable aggregate of atoms whose number and identities correspond to the molecular formula for a compound, for example, $H_2$, $SO_2$, $NH_3$, $C_2H_3O_2$. A molecule is stable, in the sense that it does not spontaneously rearrange or fly apart into fragments. A molecule is said to be reactive if when brought near another molecule it regroups with atoms from the neighbor to produce new molecules different from the first two. A given molecule may be reactive with certain neighbors, but nonreactive with others.

The structure of a stable molecule is defined by the three-dimensional arrangement of the atoms within the molecule. Pictorial representations of structure—for example, the familiar "ball and stick" models and sketches—show the nuclei of the constituent atoms, but not their electrons; the electrons provide the chemical bonds that hold the atomic nuclei together as the molecule. Certain properties summarize features of the three-dimensional molecular structure (Fig. 3.2). Bond length measures the distance between the atomic nuclei in a particular bond; the set of all bond lengths together qualitatively indicate the overall size of the molecule. Bond angles, defined as the relative orientation of two adjacent bonds, measure the shape of the molecule. Molecules are not rigid structures with unchanging bond lengths and angles; their atoms can vibrate, usually with small amplitudes, about their average positions. Average bond lengths and bond angles can be measured by spectroscopic techniques (Chapter 16) and by x-ray diffraction (Section 19.1).

In preparation for describing the formation of chemical bonds, let us examine two of the crucial properties that characterize bonds: length and energy. These two properties are related, in turn, to another property: the bond order.

## Bond Lengths

For a diatomic molecule, the only relevant structural parameter is the bond length, the distance between the nuclei of the two atoms. Table 3.2 lists the bond lengths of a number of diatomic molecules, expressed in units of Ångstroms (defined as

**FIGURE 3.2** Three-dimensional molecular structures of (a) $H_2$, (b) $SO_2$, (c) $NH_3$, and (d) $C_2H_3O_2$, showing bond lengths and bond angles. *(Courtesy of Drs. Andrew J. Pounds and Mark A. Iken, Scientific Visualization Laboratory, Georgia Institute of Technology, Atlanta, GA)*

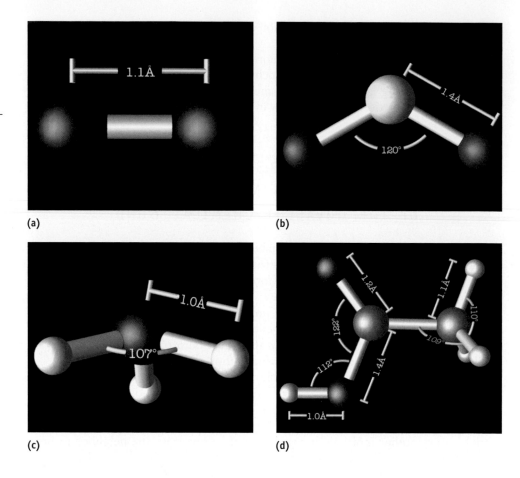

(a)

(b)

(c)

(d)

$1\ \text{Å} = 10^{-10}$ m.) Certain systematic trends are immediately obvious. Within a group in the periodic table, bond lengths usually increase with increasing atomic number Z. Thus, bond lengths increase in the series $F_2$ to $Cl_2$ to $Br_2$ to $I_2$. Similarly, ClF has a shorter bond length than ClBr. These trends are directly related to the corresponding trends in atomic sizes, which usually increase with increasing atomic number in a given group of the periodic table. (Chapter 15 discusses the quantum basis for atomic sizes.) This qualitative connection cannot be made quantitative, however, because bond lengths are not simply sums of atomic radii. Too much rearrangement of electron density occurs during bond formation to allow any such approximation.

A significant result from experimental studies is that the length of the bond between a given pair of atoms changes very little from one molecule to another. A C—H bond has almost the same length in acetylsalicylic acid (aspirin, $C_9H_8O_4$) as in methane ($CH_4$). Table 3.3 shows that the lengths of OH, CC, and CH bonds in numerous molecules are constant to within a few percent.

## Bond Energies

Energy is the crucial factor that determines the stability of molecules. The quantum theory to be presented in Chapter 16 shows that a covalent chemical bond forms because the involved atoms have lower energy when they are close to each other than when they are far apart. The quantum theory also shows that the molecular

**TABLE 3.2**

### Properties of Diatomic Molecules

| | Average Bond Length (Å) | Bond Energy (kJ mol$^{-1}$) | | Average Bond Length (Å) | Bond Energy (kJ mol$^{-1}$) |
|---|---|---|---|---|---|
| $H_2$ | 0.751 | 433 | HBr | 1.424 | 363 |
| $N_2$ | 1.100 | 942 | HI | 1.620 | 295 |
| $O_2$ | 1.211 | 495 | ClF | 1.632 | 252 |
| $F_2$ | 1.417 | 155 | BrF | 1.759 | 282 |
| $Cl_2$ | 1.991 | 240 | BrCl | 2.139 | 216 |
| $Br_2$ | 2.286 | 190 | ICl | 2.324 | 208 |
| $I_2$ | 2.669 | 148 | NO | 1.154 | 629 |
| HF | 0.926 | 565 | CO | 1.131 | 1073 |
| HCl | 1.284 | 429 | | | |

geometry observed experimentally is the one that gives the molecule the lowest energy. The energy that must be absorbed when a bond is broken reveals the *strength* of that bond. The **bond energy,** sometimes also called the **bond dissociation energy,** is the energy required to break one mole of the particular bond under discussion. The bond energy is denoted by $\Delta E_d$ ("d" stands for *dissociation* here) and is measured directly in units of kJ per mole. Table 3.2 lists bond energies for selected diatomic molecules. Again, certain systematic trends with changes in atomic number are evident. Bonds generally grow weaker with increasing atomic number, as shown by the decrease in the bond energies of the hydrogen halides in the order HF > HCl > HBr > HI. Note, however, the unusual weakness of the bond in the fluorine molecule $F_2$ (its bond energy is significantly *smaller* than that of $Cl_2$ and comparable to that of $I_2$). Bond strength decreases dramatically in the diatomic molecules from $N_2$ (945 kJ mol$^{-1}$) to $O_2$ (498 kJ mol$^{-1}$) to $F_2$ (158 kJ mol$^{-1}$). What accounts for this behavior? A successful theory of bonding must explain both the general trends and the reasons for particular exceptions.

Bond energies, like bond lengths, are fairly reproducible (within about 10%) from one compound to another. It is therefore possible to tabulate *average* bond energies from measurements on a series of compounds. The energy of any given bond in different compounds will deviate somewhat from those shown, but in most cases the deviations are small.

### Bond Order

Sometimes the length and energy of the bond between two specific kinds of atoms are *not* reproduced from one chemical compound to another but are sharply different. Table 3.4 shows the great differences in bond lengths and bond energies of carbon–carbon bonds in ethane ($H_3CCH_3$), ethylene ($H_2CCH_2$), and acetylene (HCCH). Carbon–carbon bonds from many other molecules fit into one of the three classes given in the table (that is, some carbon–carbon bond lengths are close to 1.54 Å, others are close to 1.34 Å, and still others are close to 1.20 Å). This confirms the existence of not one, but three types of carbon–carbon bonds. These are classified according to their bond order as follows. The weakest and longest (as in ethane) is a single bond represented by C—C; that of intermediate strength (as in ethylene) is a double bond, C═C; and the strongest and shortest (as in acetylene) is a triple bond, C≡C. In the next section (Section 3.4), the **bond order** will be shown to be the number of shared electron pairs for these bonds: 1, 2, and 3, respectively.

**TABLE 3.3**

### Reproducibility of Bond Lengths

| Bond | Molecule | Bond Length (Å) |
|---|---|---|
| O—H | $H_2O$ | 0.958 |
| | $H_2O_2$ | 0.960 |
| | HCOOH | 0.95 |
| | $CH_3OH$ | 0.956 |
| C—C | diamond | 1.5445 |
| | $C_2H_6$ | 1.536 |
| | $CH_3CHF_2$ | 1.540 |
| | $CH_3CHO$ | 1.50 |
| C—H | $CH_4$ | 1.091 |
| | $C_2H_6$ | 1.107 |
| | $C_2H_4$ | 1.087 |
| | $C_6H_6$ | 1.084 |
| | $CH_3Cl$ | 1.11 |
| | $CH_2O$ | 1.06 |

**TABLE 3.4**

### Three Types of Carbon–Carbon Bonds

| Bond | Molecule | Bond Length (Å) | Bond Energy (kJ mol$^{-1}$) |
|------|----------|-----------------|------------------------------|
| C—C | $C_2H_6$ (or $H_3CCH_3$) | 1.536 | 345 |
| C=C | $C_2H_4$ (or $H_2CCH_2$) | 1.337 | 612 |
| C≡C | $C_2H_2$ (or HCCH) | 1.204 | 809 |

Even these three types do not cover all the carbon–carbon bonds found in nature, however. In benzene ($C_6H_6$) the experimental carbon–carbon bond length is 1.397 Å and the bond energy is 505 kJ mol$^{-1}$. This bond is intermediate between a single bond and a double bond (it has bond order $1\frac{1}{2}$). In fact, we shall see in Chapter 17 that the bonding in compounds such as benzene differs from that in many other compounds. Although many bonds have properties that depend primarily on the two atoms forming the bond (and thus are similar from one compound to another), bonding in benzene and certain related molecules depends on the nature of the whole molecule.

Multiple bond order is seen with atoms other than carbon, and indeed occurs between unlike atoms. Lengths of representative bonds are listed in Table 3.5. All values are in units of Ångstroms.

In summary, bond properties are often quite reproducible between compounds, but we must be alert for exceptions that may signal new types of chemical bonding.

---

### 3.4

## COVALENT BONDING: ELECTRON PAIR MODEL

When atoms with comparable or identical electronegativity values interact, electron transfer does not occur. What kind of bond forms when valence electrons can be pulled from one atom toward the other, but not completely transferred? An even greater puzzle is presented by formation of homonuclear diatomic molecules. The $H_2$ molecule is quite stable, for example, with a bond dissociation energy of 432 kJ mol$^{-1}$, yet it consists of two identical atoms. There is no possibility of a net charge transfer from one to the other to form an ionic bond. The explanation of the stability of $H_2$ comes from a different type of bonding, *covalent* bonding, which arises from the sharing of electrons between atoms.

**TABLE 3.5**

### Average Bond Lengths (in Å)

| | | | | | |
|------|------|------|------|------|------|
| C—C | 1.54 | N—N | 1.45 | C—H | 1.10 |
| C=C | 1.34 | N=N | 1.25 | N—H | 1.01 |
| C≡C | 1.20 | N≡N | 1.10 | O—H | 0.96 |
| C—O | 1.43 | N—O | 1.43 | C—N | 1.47 |
| C=O | 1.20 | N=O | 1.18 | C≡N | 1.16 |

## The Chemical Bond As a Shared Electron Pair

Consider the bonding of the $H_2^+$ molecular ion. This, the simplest of all polyatomic species, is stable but rather reactive. Experiment and theoretical calculations agree that the H—H bond length $R_e$ is 1.06 Å and the dissociation energy $\Delta E_d$ is 255.5 kJ mol$^{-1}$. To explain the existence of $H_2^+$, we use a simple classical model for chemical bonding. The electron is treated as a point negative charge interacting with the two hydrogen nuclei that are separated by a distance $R$ (Fig. 3.3). The electron exerts an attractive Coulomb force on the nuclei. If the electron lies *between* the two (Fig. 3.3a), the force tends to pull the nuclei together, *strengthening* the bond. If, on the other hand, it lies *outside* the region between the nuclei (Fig. 3.3b), the repulsive force between the positively charged nuclei tends to push the two apart, *weakening* the bond. According to this simple picture, covalent bonding occurs when an electron spends most of its time in the region between nuclei so that it is shared between them. When, in contrast, an electron spends most of its time outside the region between the nuclei, the bond is weakened and the molecule tends to dissociate.

This picture of covalent bonding in the $H_2^+$ molecular ion can be applied to other molecules. The Lewis model (to be developed in the next several paragraphs) represents covalent bonds as shared valence-electron pairs positioned between two nuclei, where they experience net attractive interactions with each nucleus and contribute to the strengthening of the bond through the mechanism of Figure 3.3a.

## Lewis Diagrams for Covalent Bonding

The concept of covalent bonding is required because the elements in Groups III through V of the periodic table (especially in the first two periods) have a lesser tendency to form ions than those at the left and right sides of the table. What happens when atoms with comparable values of electronegativity form bonds? For example, K (electronegativity 0.82) and F (electronegativity 3.98) readily form KF by ionic bonding. Now consider the interaction of carbon (electronegativity 2.55) and hydrogen (electronegativity 2.20) to form methane ($CH_4$). Unlike ionic compounds, this substance is a gas at room temperature, not a solid. Cooling methane to low temperatures condenses it to a solid in which the $CH_4$ molecules retain their identities. Methane dissolves in water to a slight extent, but it does not ionize. Thus, it is not useful to think of methane as an ionic substance made up of $C^{4-}$ and $H^+$ ions (or $C^{4+}$ and $H^-$ ions). It is a nonionic compound.

The Lewis model for covalent bonding starts with the recognition that electrons are not transferred from one atom to another in a nonionic compound, but are *shared* between atoms to form covalent bonds. Hydrogen and chlorine combine, for example, to form the **covalent compound** hydrogen chloride. This can be indicated with a

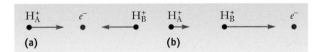

**FIGURE 3.3**  Electrons attract hydrogen nuclei because the two have opposite charge. (a) If the electron lies between two hydrogen nuclei, the attraction creates a net force that pulls the two nuclei together. (b) If the electron lies outside the internuclear region, it exerts a much greater force on the nearer nucleus, giving a net force that pulls the two nuclei apart.

**Lewis diagram** for the molecule of the product, in which the valence electrons from each atom are redistributed so that one electron from the hydrogen atom and one from the chlorine atom are now shared by the two atoms. The two dots representing this electron pair are placed between the symbols for the two elements:

$$H\cdot \; + \; \cdot \ddot{\underset{..}{Cl}}: \; \longrightarrow \; H\!:\!\ddot{\underset{..}{Cl}}:$$

The basic rule that governs the writing of Lewis diagrams is the **octet rule:** whenever possible, the electrons in a covalent compound are distributed in such a way that each main-group element (except hydrogen) is surrounded by eight electrons (an *octet* of electrons). Hydrogen has two electrons in such a structure. When the octet rule is satisfied, the atom attains the special stability of a noble-gas shell. In the structure for HCl just shown, the H nucleus, through sharing, is close to two valence electrons (like the noble gas, He), and the Cl has eight valence electrons near it (like Ar). Electrons that are shared between two atoms are counted as contributing to the filling of the valence shell of each atom.

A shared pair of electrons can also be represented by a short line (—):

$$H\!\!-\!\!\ddot{\underset{..}{Cl}}:$$

The unshared electron pairs around the chlorine atom in the Lewis diagram are called **lone pairs,** and they make no contribution to the bond between the atoms. Lewis diagrams of some simple covalent compounds are

$$NH_3 \qquad\qquad H_2O \qquad\qquad CH_4$$

$$H\!:\!\ddot{N}\!:\!H \qquad H\!:\!\ddot{\underset{..}{O}}\!:\!H \qquad \begin{array}{c} H \\ H\!:\!\ddot{C}\!:\!H \\ H \end{array}$$
$$\underset{H}{\phantom{H:N:H}}$$

$$H\!\!-\!\!\overset{..}{\underset{|}{N}}\!\!-\!\!H \qquad H\!\!-\!\!\ddot{\underset{..}{O}}\!\!-\!\!H \qquad H\!\!-\!\!\overset{\overset{\textstyle H}{|}}{\underset{\underset{\textstyle H}{|}}{C}}\!\!-\!\!H$$
$$\underset{H}{\phantom{H-N-H}}$$

Lewis diagrams indicate how bonds connect the atoms in a molecule, but they do not show the molecule's spatial geometry. The ammonia molecule is not planar but pyramidal, for example, with the nitrogen atom at the apex. The water molecule is bent rather than straight. Three-dimensional geometry can be indicated by ball-and-stick models such as those in Figure 3.4.

In some molecules we must suppose that more than one pair of electrons are shared by two atoms. In the oxygen molecule, each atom has six valence electrons, so for each to achieve an octet configuration, *two* pairs of electrons must be shared, making a **double bond** between the atoms:

$$\ddot{\underset{..}{O}}\!:\!:\!\ddot{\underset{..}{O}} \qquad or \qquad \ddot{\underset{..}{O}}\!\!=\!\!\ddot{\underset{..}{O}}$$

Similarly, the $N_2$ molecule has a **triple bond,** involving three shared electron pairs:

$$:\!N\!:\!:\!:\!N\!: \qquad or \qquad :\!N\!\equiv\!N\!:$$

while the $F_2$ molecule has a single bond. The number of shared electron pairs in a bond determines the order of the bond, which has already been connected with bond energy and bond length in Section 3.3. The decrease in bond order from 3 to 2 to 1 explains the dramatic trend in the bond energies of the sequence of diatomic

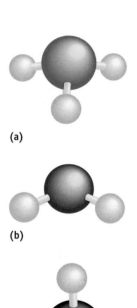

**(a)**

**(b)**

**(c)**

**FIGURE 3.4** Molecules of three familiar substances, drawn in ball-and-stick fashion. The sizes of the balls have been reduced somewhat to show the bonds more clearly, but the relative sizes of the balls are correct. (a) Ammonia, $NH_3$. (b) Water, $H_2O$. (c) Methane, $CH_4$.

molecules $N_2$, $O_2$, and $F_2$ left as a question in Section 3.3. A carbon–carbon bond can involve the sharing of one, two, or three electron pairs. A progression from single to triple bonding is found in the hydrocarbons ethane ($C_2H_6$), ethylene ($C_2H_4$), and acetylene ($C_2H_2$):

This progression corresponds to the three types of carbon–carbon bonds whose properties were related to bond order in Section 3.3. and summarized in Table 3.4 and Table 3.5. Multiple bonding to attain an octet most often involves the elements carbon, nitrogen, oxygen, and, to a lesser degree, sulfur. Double and triple bonds are shorter than a single bond between the same pair of atoms. Illustrative examples are included in Table 3.5 in Section 3.3.

## Formal Charges

In **homonuclear** diatomic molecules (in which both atoms are the same, as in $H_2$ and $Cl_2$) the electrons are equally shared between the two atoms, and for such molecules the covalency is nearly ideal.

Consider, however, a molecule of carbon monoxide (CO). Its Lewis diagram has a triple bond:

$$:C:::O:$$

This uses the ten valence electrons (four from the C and six from the O) and gives each atom an octet. If the six bonding electrons were shared equally, the carbon atom would own five valence electrons (one *more* than its group number) and the oxygen atom would own five electrons (one *less* than its group number). Equal sharing implies that formally the carbon atom must gain an electron, and the oxygen atom must lose an electron. This situation is described by assigning a formal charge to each atom, defined as the charge an atom in a molecule would have if the electrons in its Lewis diagram were divided equally between the atoms that share them. Thus in CO, C has a formal charge of $-1$, and O has a formal charge of $+1$:

$$\overset{\ominus}{:}C:::O:\overset{\oplus}{}$$

Carbon monoxide is a covalent compound, and the assignment of formal charges does not make it ionic.

We emphasize that the equal sharing of electrons in the bonds of such **heteronuclear** diatomic molecules is without experimental basis and is simply a postulate. It has a useful purpose in cases in which two or more Lewis diagrams are possible for the same molecule. When this happens, the diagram with the smallest formal charges is the preferred one, and it generally gives the best description of bonding in the molecule.

The formal charge on an atom in a Lewis diagram is simple to calculate. If the valence electrons were removed from an atom, it would have a positive charge equal to its group number in the periodic table (elements in Group VI, the chalcogens, have six valence electrons and therefore a charge of $+6$ when those electrons are removed). From this positive charge, subtract the number of lone-pair valence

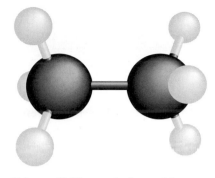

Ethane, $C_2H_6$, can be burned in oxygen as a fuel, and if strongly heated it reacts to form hydrogen and ethylene.

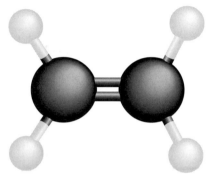

Ethylene, $C_2H_4$, is the largest-volume organic (carbon-containing) chemical produced.

Acetylene, $C_2H_2$, has a triple bond that makes it highly reactive.

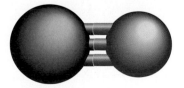

Carbon monoxide, CO, is a colorless, odorless, and toxic gas produced by the incomplete burning of hydrocarbons in air. It is used in the production of elemental metals from their oxide ores.

electrons possessed by the atom in the Lewis diagram, and then subtract one half of the number of bonding electrons shared by it:

formal charge = group number − number of electrons in lone pairs
$$- \tfrac{1}{2}(\text{number of electrons in bonding pairs})$$

## EXAMPLE 3.3

Compute the formal charges on the atoms in the following Lewis diagram, which represents the azide ion ($N_3^-$):

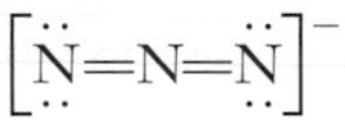

$$\left[ \ddot{N}{=}N{=}\ddot{N} \right]^-$$

### Solution

Nitrogen is in Group V. Hence, each N atom contributes five valence electrons to the bonding, and the negative charge on the ion contributes one more electron. The Lewis diagram correctly represents 16 electrons.

Each of the terminal nitrogen atoms has four electrons in lone pairs and four bonding electrons (which comprise a double bond) associated with it. Therefore,

$$\text{formal charge}_{(\text{terminal N})} = 5 - 4 - \tfrac{1}{2}(4) = -1$$

The nitrogen atom in the center of the structure has no electrons in lone pairs. Its entire octet lies in the eight bonding electrons:

$$\text{formal charge}_{(\text{central N})} = 5 - 0 - \tfrac{1}{2}(8) = +1$$

The sum of the three formal charges is −1, which is the true overall charge on this polyatomic ion. Failure of this check indicates an error in either the Lewis diagram or the arithmetic.

**Related Problems: 15, 16**

## Drawing Lewis Diagrams

In drawing Lewis diagrams, we shall assume that the molecular "skeleton" (that is, a plan of the bonding of specific atoms to other atoms) is known. In this respect, it helps to know that hydrogen and fluorine are always terminal atoms in Lewis diagrams, bonded to only one other atom. A systematic procedure for drawing Lewis diagrams can then be used, as expressed by the following rules:

1.  Count up the total number of valence electrons available by first using the group numbers to add up the valence electrons from all the atoms present. If the species is a negative ion, *add* the absolute value of the total charge; if it is a positive ion, *subtract* the value of the charge.
2.  Calculate the total number of electrons that would be needed if each atom had its *own* noble-gas shell of electrons around it (two for hydrogen, eight for carbon and heavier elements).
3.  Subtract the number in step 1 from the number in step 2. This is the number of shared (or bonding) electrons present.

4. Assign two bonding electrons (one pair) to each bond in the molecule or ion.
5. If bonding electrons remain, assign them in pairs by making some of the bonds double or triple bonds. In some cases there may be more than one way to do this. In general, double bonds form only between atoms of the elements C, N, O, and S. Triple bonds are usually restricted to C, N, or O.
6. Assign the remaining electrons as lone pairs to the atoms, giving octets to all atoms except hydrogen.
7. Determine the formal charge on each atom and write it next to that atom. Check that the formal charges add to give the correct total charge on the molecule or polyatomic ion.

The last rule identifies diagrams that are undesirable because they imply large separations of negative from positive charge, and catches inadvertent errors (e.g., the wrong number of dots).

The use of these rules is illustrated by the following example.

## EXAMPLE 3.4

Write a Lewis electron-dot diagram for phosphoryl chloride, $POCl_3$ (Fig. 3.5). Assign formal charges to all of the atoms.

### Solution

The first step is to calculate the total number of valence electrons available in the molecule. For $POCl_3$ it is

$$5 \text{ (from P)} + 6 \text{ (from O)} + [3 \times 7 \text{ (from Cl)}] = 32$$

Next, calculate how many electrons would be necessary if each atom were to have its own noble-gas shell of electrons around it. Because there are five atoms in the present case (none of them hydrogen), 40 electrons would be required. From the difference of these numbers ($40 - 32 = 8$), each atom can achieve an octet only if eight electrons are shared between pairs of atoms. Eight electrons correspond to four electron pairs, so each of the four linkages in $POCl_3$ must be a single bond. (If the number of shared electron pairs were *larger* than the number of bonds, double or triple bonds would be present.)

The other 24 valence electrons are assigned as lone pairs to the atoms in such a way that each achieves an octet configuration. The resulting Lewis diagram is

$$
\begin{array}{c}
\overset{\displaystyle \overset{..}{:}\text{O}\overset{..}{:}\;^{\ominus}}{\underset{\displaystyle :\overset{..}{\underset{..}{\text{Cl}}}}{\overset{\displaystyle |\,^{\oplus}}{:\overset{..}{\text{Cl}}-\text{P}-\overset{..}{\text{Cl}}:}}}
\end{array}
$$

Formal charges are already indicated in this diagram. Phosphorus has the group number 5, and it shares eight electrons with no lone-pair electrons, so

$$\text{formal charge on P} = 5 - 4 = +1$$

Oxygen has the group number 6 with six lone-pair electrons and two shared electrons, so

$$\text{formal charge on O} = 6 - 6 - \tfrac{1}{2}(2) = -1$$

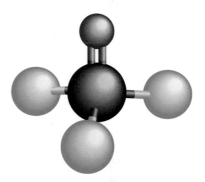

**FIGURE 3.5** Phosphoryl chloride, $POCl_3$, is a reactive compound used to introduce phosphorus into organic molecules in synthesis reactions. Experimental studies show that the P—O bond is more like a double bond than a single bond. Lewis diagrams rationalizing the P—O double bond can be constructed as an example of valence shell expansion, which is discussed on pages 75 and 76.

All three chlorine atoms have zero formal charge, computed by

$$\text{formal charge on Cl} = 7 - 6 - \tfrac{1}{2}(2) = 0$$

**Related Problems: 21, 22, 23, 24**

## Resonance Forms

For certain molecules or molecular ions, two or more equivalent Lewis diagrams can be written. An example is ozone ($O_3$), for which there are two possible Lewis diagrams:

Ozone, $O_3$, is a pale-blue gas with a pungent odor that condenses to a deep-blue liquid below $-112°C$.

These diagrams suggest that one O—O bond is a single bond and the other is double, so the molecule would be asymmetric. In fact, the two O—O bond lengths are found experimentally to be identical, with a length of 1.28 Å, intermediate between the O—O single bond length in $H_2O_2$ (1.49 Å) and the O=O double bond length in $O_2$ (1.21 Å). The Lewis diagram picture fails, but it can be patched up by saying that the actual bonding in $O_3$ is represented as a **resonance hybrid** of the two Lewis diagrams in which each of the bonds is intermediate between a single and a double bond. This is represented by connecting the diagrams with a double-headed arrow:

The term "resonance" does not mean that the molecule physically oscillates back and forth from one of these bonding structures to the other. Rather, within the limitations of the Lewis dot model of bonding, the best representation of the actual bonding is a hybrid diagram that includes features of each of the acceptable individual diagrams. This awkwardness can be avoided through a treatment of bonding using molecular orbitals, as shown in Chapter 16.

### EXAMPLE 3.5

Draw three resonance forms for the nitrate ion, $NO_3^-$ (Fig. 3.6), and estimate the bond lengths.

### Solution

In $NO_3^-$ there are 24 valence electrons. For each atom to have its own octet, $4 \times 8 = 32$ electrons would be required. Therefore, $32 - 24 = 8$ electrons must be shared between atoms, implying a total of four bonding pairs. These can be distributed in one double and two single bonds, leading to the equivalent resonance diagrams

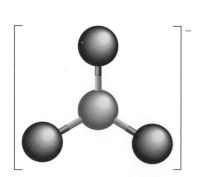

**FIGURE 3.6**   The nitrate ion, $NO_3^-$, has a symmetric planar structure.

The two singly bonded oxygen atoms carry formal charges of $-1$ and that of the nitrogen atom is $+1$. The bond lengths should all be equal and should lie between the values given in Table 3.5 for N—O (1.43 Å) and N=O (1.18 Å). The experimentally measured value is 1.24 Å.

**Related Problems: 27, 28, 29, 30, 31, 32**

## Breakdown of the Octet Rule

Lewis diagrams with the octet rule are useful in predicting the types of molecules that will be stable under ordinary conditions of temperature and pressure. For example, we can write a simple Lewis diagram for water ($H_2O$),

$$H : \overset{..}{\underset{..}{O}} : H$$

in which each atom has a noble-gas configuration. This is impossible to do for OH or for $H_3O$, which suggests that these species are either unstable or highly reactive.

There are several situations in which the octet rule is *not* satisfied.

***Case 1: Odd-Electron Molecules***   The electrons in a Lewis diagram that satisfies the octet rule must occur in pairs—bonding pairs or lone pairs. Any molecule that has an odd number of electrons cannot satisfy the octet rule. Most stable molecules have even numbers of electrons, but a few have odd numbers. An example is nitrogen oxide (NO), a stable (though reactive) molecule that is an important factor in air pollution. It has 11 electrons, and the best electron-dot diagram for it is

$$: \overset{..}{N} : \ : \overset{..}{\underset{..}{O}}$$

in which only the oxygen atom has a noble-gas configuration. The stability of NO contradicts the octet rule.

***Case 2: Octet-Deficient Molecules***   Some molecules are stable even though they have too few electrons to achieve an octet. For example, the standard rules for $BF_3$ would lead to the Lewis diagram

$$\begin{array}{c} : \overset{..}{F} : \\ | \quad \oplus \\ : \overset{..}{\underset{..}{F}} {-} B {=} \overset{..}{\underset{..}{F}} \\ \ominus \end{array}$$

but experimental evidence strongly indicates that there are no double bonds in $BF_3$ (fluorine never forms double bonds). Moreover, the placement of a positive formal charge on fluorine is never correct. A more nearly correct diagram is

$$\begin{array}{c} : \overset{..}{F} : \\ .. \quad | \quad .. \\ : \overset{..}{\underset{..}{F}} {-} B {-} \overset{..}{\underset{..}{F}} : \end{array}$$

Although this Lewis diagram denies an octet to the boron atom, it does at least assign zero formal charges to all atoms.

***Case 3: Valence Shell Expansion***   Lewis diagrams become more complex in the compounds of elements from the third and subsequent periods. Sulfur, for example, forms some compounds that are readily described by Lewis diagrams that give closed shells to all atoms. An example is hydrogen sulfide ($H_2S$), which is analogous to

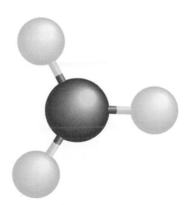

Boron trifluoride, $BF_3$, is a highly reactive gas that condenses to a liquid at $-100°C$. Its major use is in speeding up a large class of reactions involving carbon compounds.

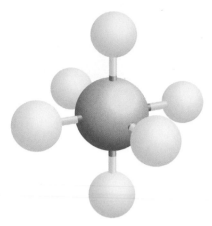

Sulfur hexafluoride, $SF_6$, is an extremely stable, dense, and unreactive gas. It is used as an insulator in high-voltage generators and switches.

water in its Lewis diagram. In sulfur hexafluoride ($SF_6$), however, the central sulfur is bonded to six fluorine atoms. This cannot be described by a Lewis diagram unless more than eight electrons are allowed around the sulfur atom, a process called **valence shell expansion.** The resulting Lewis diagram is

$$
\begin{array}{ccc}
 & :\!\ddot{F}\!: & \\
:\ddot{F} & \big| & \ddot{F}: \\
 & \diagdown S \diagup & \\
:\ddot{F} & \big| & \ddot{F}: \\
 & :\!\ddot{F}\!: &
\end{array}
$$

The fluorine atoms have octets, but the central sulfur atom shares a total of 12 electrons.

In the standard procedure for writing Lewis diagrams, the need for valence shell expansion is signaled when the number of shared electrons is not sufficient to place a bonding pair between each pair of atoms that are supposed to be bonded. In $SF_6$, for example, 48 electrons are available but 56 are needed to form separate octets on seven atoms. This means that $56 - 48 = 8$ electrons would be shared. Four electron pairs are not sufficient to make even single bonds between the central S atom and the six terminal F atoms. In this case we still follow rule 4 (assign one bonding pair to each bond in the molecule or ion) even though this will use more than eight electrons. Rule 5 becomes irrelevant, because there are no extra shared electrons. Rule 6 is now replaced with a new rule:

**6′.** Assign lone pairs to the *terminal* atoms to give them octets. If any electrons still remain, assign them to the central atoms as lone pairs.

The effect of rule 6′ is to abandon the octet rule for the central atom but preserve it for the terminal atoms.

---

**EXAMPLE 3.6**

Write a Lewis diagram for the linear $I_3^-$ (triiodide) ion.

**Solution**

There are 7 valence electrons from each iodine atom plus 1 from the overall ion charge, giving a total of 22. Because $3 \times 8 = 24$ electrons would be needed in separate octets, $24 - 22 = 2$ are shared according to the original rules. Two electrons are not sufficient to make two different bonds, however, so valence expansion is necessary.

A pair of electrons is placed in each of the two bonds, and rule 6′ is used to complete the octets of the two terminal I atoms. This leaves

$$:\ddot{I}\!-\!I\!-\!\ddot{I}:$$

At this stage, 16 valence electrons have been used. The remaining 6 are placed as lone pairs on the central I atom. A formal charge of $-1$ then resides on this atom:

$$:\ddot{I}\!-\!\overset{\ominus}{\ddot{I}}\!-\!\ddot{I}:$$

Note the valence expansion on the central atom: it shares or owns a total of ten electrons rather than the eight required by adherence to the octet rule.

**Related Problems: 33, 34**

# POLAR COVALENT BONDING:
# ELECTRONEGATIVITY AND DIPOLE MOMENTS

Laboratory measurements show that most real bonds are neither fully ionic nor fully covalent but instead possess a mixture of ionic and covalent character. Bonds in which there is a partial transfer of charge are **polar covalent.** This section provides an approximate description of the polar covalent bond based on the ability of one atom to draw toward it electrons from another atom without complete transfer. This ability is estimated by comparing the electronegativity values for the two atoms.

## Pauling's Electronegativity Scale

Mulliken's electronegativity scale, discussed in Section 3.1, is precisely defined in terms of properties of the isolated atom; consequently it does not reflect the detailed competition for electrons in a particular bond. In 1932, two years before Mulliken's treatment, Linus Pauling proposed a helpful way to quantify the extent of ionic bonding between a particular pair of atoms, A and B, in a molecule. Suppose the dissociation energy of an A—A bond is $\Delta E_{AA}$, and that of a B—B bond is $\Delta E_{BB}$; both bonds are covalent because the atoms in them are identical. Then an estimate of the *covalent* contribution to the dissociation energy of an A—B bond is the (geometric) mean of these two energies, $\sqrt{\Delta E_{AA}\Delta E_{BB}}$. However, the actual A—B bond includes ionic character because some charge transfer occurs between the atoms. The ionic character tends to strengthen the bond and increase its value of $\Delta E_{AB}$. Pauling suggested that the difference between the actual and covalent bond energies,

$$\Delta \equiv \Delta E_{AB} - \sqrt{\Delta E_{AA}\Delta E_{BB}}$$

is a measure of the **electronegativity** difference between the two atoms A and B. Specifically, he defined

$$\chi_A - \chi_B = 0.102\Delta^{1/2}$$

where $\chi_A$ and $\chi_B$ (Greek chi) are the electronegativities of A and B and $\Delta$ is measured in kJ mol$^{-1}$. The atom that more readily accepts an electron (and thus tends to carry a net negative charge) has the larger value of $\chi$. On the Pauling scale shown in Figure 3.7 and used in Appendix F, electronegativities range from 3.98 (for fluorine) to 0.79 (for cesium). These numerical values are useful for exploring periodic trends and making semiquantitative comparisons. They are not highly precise experimental results but average numbers based on measured bond energies from a large range of compounds.

The periodic trends in electronegativity, as shown in Figure 3.7, are quite interesting. Electronegativity decreases down a group at the left or right side of the periodic table; that is, cesium is less electronegative than lithium and iodine is less electronegative than fluorine. High electronegativity is favored by circumstances that stabilize additional electronic charge density drawn toward the atom. These conditions can be explained only by the quantal description of atomic structure (see Chapter 15).

The absolute value of the difference in electronegativity of two bonded atoms tells the degree of *polarity* in their bond. A large difference (greater than about 2.0) means that the bond is ionic and electrons transfer completely or nearly completely to the more electronegative atom. A small difference (less than about 0.4) means

**FIGURE 3.7** Average electronegativities of atoms, computed with the method developed by Linus Pauling. Electronegativity values have no units.

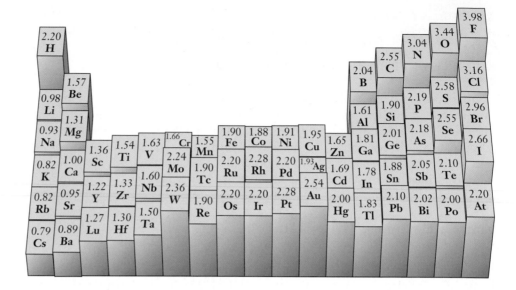

that the bond is largely covalent, with electrons in the bond fairly evenly shared. Intermediate values of the difference signify a polar covalent bond with intermediate character.

---

### EXAMPLE 3.7

Using Figure 3.7, arrange the following bonds in order of decreasing polarity: H—C, O—O, H—F, I—Cl, Cs—Au.

#### Solution

The differences in electronegativity between the five pairs of atoms (without regard to sign) are 0.35, 0.00, 1.78, 0.50, and 1.75, respectively. The order of decreasing polarity is the order of decrease in this difference: H—F, Cs—Au, I—Cl, H—C, and O—O. The last bond in this listing is nonpolar.

**Related Problems: 35, 36**

---

## Dipole Moments and Percent Ionic Character

In a bond that is almost purely ionic, such as that of KF, almost complete transfer of an electron from the electropositive to the electronegative species takes place; KF can be written fairly accurately as $K^+F^-$ with charges $+e$ and $-e$ on the two ions. The charge distribution for a molecule such as HF, with significant covalent character, is more complex. If we wish to approximate it by two point charges, then it is best described as $H^{\delta+}F^{\delta-}$, where some fraction $\delta$ of the full charge $\pm e$ is on each nucleus. A useful measure of ionic character and of electronegativity differences, especially for diatomic molecules, is the **dipole moment** of the molecule. If two charges of equal magnitude and opposite sign, $+Q$ and $-Q$, are separated by a distance $R$, the dipole moment $\mu$ (Greek mu) of that charge distribution is

$$\mu = QR \qquad [3.3]$$

In SI units, $\mu$ is measured in coulomb meters, an inconveniently large unit for discussing molecules. The unit most often used is the debye (D), which is related to SI units by [2]

$$1 \text{ debye} = 3.336 \times 10^{-30} \text{ coulomb meter}$$

The debye can also be defined as the dipole moment of two charges $\pm e$ that are 0.2082 Å apart. If $\delta$ is the fraction of a unit charge on each atom in a diatomic molecule ($Q = e\delta$), then

$$\mu \text{ (debye)} = \frac{R(\text{Å})}{0.2082 \text{ Å D}^{-1}} \delta$$

Dipole moments are measured experimentally by electrical and spectroscopic methods and provide useful information about the nature of bonding. In HF, for example, the value of $\delta$ calculated from the dipole moment ($\mu = 1.82$ D) and bond length ($R = 0.917$ Å) is 0.41, substantially less than the value of 1 for a purely ionic bond. We convert $\delta$ to a "percent ionic character" by multiplying by 100% and say that the bond in HF is 41% ionic. Deviations from 100% ionic bonding occur for two reasons: first, covalent contributions lead to electron sharing between atoms; second, the electron charge distribution around one ion is distorted by the electric field of the other ion (polarization). When polarization is extreme, regarding the ions as point charges is no longer a good approximation, and a more accurate description of the distribution of electric charge is necessary.

Table 3.6 gives a scale of ionic character for diatomic molecules, based on the definition of $\delta$. The ionic character defined from the dipole moment is reasonably

## TABLE 3.6

### Dipole Moments of Diatomic Molecules

| Molecule | Average Bond Length (Å) | Dipole Moment (D) | Percent Ionic Character (100$\delta$) |
|---|---|---|---|
| $H_2$ | 0.751 | 0 | 0 |
| CO | 1.131 | 0.112 | 2 |
| NO | 1.154 | 0.159 | 3 |
| HI | 1.620 | 0.448 | 6 |
| ClF | 1.632 | 0.888 | 11 |
| HBr | 1.424 | 0.828 | 12 |
| HCl | 1.284 | 1.109 | 18 |
| HF | 0.926 | 1.827 | 41 |
| CsF | 2.347 | 7.884 | 70 |
| LiCl | 2.027 | 7.129 | 73 |
| LiH | 1.604 | 5.882 | 76 |
| KBr | 2.824 | 10.628 | 78 |
| NaCl | 2.365 | 9.001 | 79 |
| KCl | 2.671 | 10.269 | 82 |
| KF | 2.176 | 8.593 | 82 |
| LiF | 1.570 | 6.327 | 84 |
| NaF | 1.931 | 8.156 | 88 |

[2] This definition appears unusual in SI units. If electrostatic units (esu) (which we do not use here) rather than coulombs are used to measure charge, then 1 debye $=10^{-18}$ esu cm. The debye is named after the Dutch-born Nobel laureate Peter Debye, who pioneered the measurement of dipole moments and the theory of polar molecules.

**FIGURE 3.8** Two measures of ionic character for diatomic molecules are the electronegativity difference (from Fig. 3.7) and the percent ionic character $100\delta$ (calculated from the observed dipole moment and bond length). The curve in this graph shows that the two correlate approximately but that there are many exceptions.

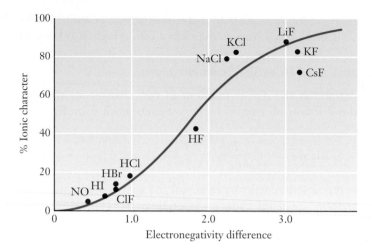

well correlated with the Pauling electronegativity difference (Fig. 3.8): high ionic character usually corresponds to large electronegativity difference, with the more electropositive atom carrying the charge $+\delta$. There are exceptions to this general trend, however. Carbon is less electronegative than oxygen, so one would predict a charge distribution of $C^{\delta+}O^{\delta-}$ in the CO molecule. In fact, the measured dipole moment is quite small in magnitude and in the other direction: $C^{\delta-}O^{\delta+}$, with $\delta = 0.02$. The discrepancy arises because of the lone-pair electron density on the carbon atom (which, interestingly, is reflected in the formal charge of $-1$ carried by that atom, as discussed in Section 3.4).

---

### 3.6

# THE SHAPES OF MOLECULES: VSEPR THEORY

When two molecules approach one another to start a chemical reaction, the probability of a successful encounter can depend critically on the three-dimensional shapes and the relative orientation of the molecules, as well as on their chemical identity. Shape is especially important in biological and biochemical reactions where molecules must fit precisely onto specific sites on membranes and templates—drug and enzyme activity are important examples. Characterization of molecular shape is therefore an essential part of the study of molecular structure.

Molecular shape or geometry is governed by energy; a molecule assumes that geometry that gives it the lowest potential energy. Sophisticated quantum mechanical calculations consider numerous possible geometrical arrangements for a molecule, calculate the total potential energy of the molecule for each arrangement, and identify the arrangement that gives the lowest potential energy. This procedure can be mimicked within the approximate classical model described in this chapter by considering numerous possible arrangements of bond angles and identifying the one that corresponds to lowest potential energy of the molecule. Because a covalent bond is formed by the sharing of a pair of electrons between two atoms (as described in the Lewis model of Section 3.4), changes in bond angles change the relative positions of the electron pairs around a given central atom. Electrons tend to repel each other through the electrostatic (Coulomb) repulsion between like charges and through quantum mechanical effects. Consequently, it is desirable in terms of energy for electrons to avoid each other. VSEPR theory provides the procedures for predicting molecular geometry by minimizing potential energy due to electron pair repulsions.

## The VSEPR Theory

The **valence shell electron-pair repulsion (VSEPR) theory** starts with the fundamental idea that electron pairs in the valence shell of an atom repel each other. This includes both lone pairs, which are localized on the atom and not involved in bonding, and bonding pairs, which are covalently shared with other atoms. The electron pairs position themselves as far apart as possible to minimize their repulsions. The molecular geometry, which is defined by the positions of the *atoms*, is then traced from the relative locations of the electron pairs.

The arrangement that minimizes repulsions naturally depends on the number of electron pairs. Figure 3.9 shows the configuration of minimum energy for two to six electron pairs around a central atom. Two electron pairs place themselves on opposite sides of the atom in a linear arrangement, three pairs form a trigonal planar structure, four arrange themselves at the corners of a tetrahedron, five define a trigonal bipyramid, and six an octahedron. To find which geometry applies, we determine the **steric number** $SN$ of the central atom, which is defined by

$$SN = \left(\begin{array}{c}\text{number of atoms}\\\text{bonded to central atom}\end{array}\right) + \left(\begin{array}{c}\text{number of lone pairs}\\\text{on central atom}\end{array}\right)$$

The steric number of an atom in a molecule can be determined by drawing the Lewis diagram of the molecule and adding the number of atoms bonded to it and its number of lone pairs.

## EXAMPLE 3.8

Calculate steric numbers for iodine in $IF_4^-$ and for bromine in $BrO_4^-$. These molecular ions have central $I^-$ or $Br^-$ surrounded by the other four atoms.

### Solution

The central $I^-$ has eight valence electrons. Each F atom has seven valence electrons of its own and needs to share one of the electrons from the $I^-$ to achieve a noble-gas configuration. Thus, four of the $I^-$ valence electrons take part in covalent bonds, leaving the remaining four to form two lone pairs. The steric number is given by

$$SN = 4 \text{ (bonded atoms)} + 2 \text{ (lone pairs)} = 6$$

In $BrO_4^-$, each oxygen atom needs to share two of the electrons from the $Br^-$ in order to achieve a noble-gas configuration. Because this accounts for all eight of the $Br^-$ valence electrons, there are no lone pairs on the central atom and

$$SN = 4 \text{ (bonded atoms)} + 0 \text{ (lone pairs)} = 4$$

Double-bonded or triple-bonded atoms count the same as single-bonded atoms in determining the steric number. In $CO_2$, for example, two double-bonded oxygen

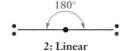

**2: Linear**

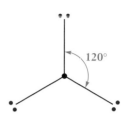

**3: Trigonal planar**

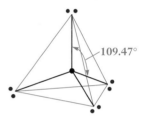

**4: Tetrahedral**

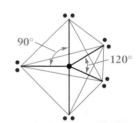

**5: Trigonal bipyramidal**

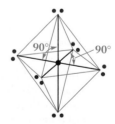

**6: Octahedral**

**FIGURE 3.9** The positions of minimum energy for electron pairs on a sphere centered on an atom's nucleus. The angles between the electron pairs are indicated. For two, three, and four electron pairs they are 180°, 120°, and 109.47°, respectively.

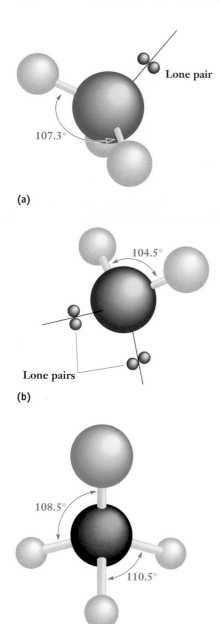

**FIGURE 3.10** (a) Ammonia (NH₃) has a pyramidal structure in which the bond angles are smaller than 109.5°. (b) Water (H₂O) has a bent structure, with a bond angle smaller than 109.5° and smaller than that of NH₃. (c) CH₃Cl has a distorted tetrahedral structure.

atoms are attached to the central carbon and there are no lone pairs on that atom, and so $SN = 2$.

The steric number is used to predict molecular geometries. In molecules $XY_n$ in which there are no lone pairs on the central atom X (the simplest case),

$$SN = \text{number of bonded atoms} = n$$

The $n$ bonding electron pairs (and therefore the outer atoms) position themselves as shown in Figure 3.9 to minimize electron-pair repulsion. Thus, $CO_2$ is predicted (and found experimentally) to be linear, $BF_3$ is trigonal planar, $CH_4$ tetrahedral, $PF_5$ trigonal bipyramidal, and $SF_6$ octahedral.

When lone pairs are present, the situation changes slightly. There can now be three different types of repulsions: (1) bonding pair against bonding pair, (2) bonding pair against lone pair, and (3) lone pair against lone pair. Consider the ammonia molecule ($NH_3$), for example, which has three bonding electron pairs and one lone pair (Fig. 3.10a). The steric number is four, and the electron pairs arrange themselves into an approximately tetrahedral structure. The lone pair is not identical to the three bonding pairs, however, and so there is no reason the electron pair structure should be *exactly* tetrahedral. It is found that lone pairs tend to occupy more space than bonding pairs (because they are held closer to the central atom), and so the angles of bonds opposite them are reduced. The geometry of the *molecule*, as distinct from that of the electron pairs, is named for the sites occupied by actual atoms. The description of the molecular geometry makes no reference to lone pairs that may be present on the central atom, even though their presence affects that geometry. The structure of the ammonia molecule is thus predicted to be a trigonal pyramid in which the H—N—H bond angle is smaller than the tetrahedral angle of 109.5°. The observed structure has an H—N—H bond angle of 107.3°. The H—O—H bond angle in water, which has two lone pairs and two bonding pairs, is still smaller at 104.5° (Fig. 3.10b).

A similar distortion takes place when two types of outer atoms are present. In $CH_3Cl$, the bonding electron pair in the C—Cl bond is not the same as those in the C—H bonds, and so the structure is a distorted tetrahedron (Fig. 3.10c). Because Cl is more electronegative than H, it tends to attract electrons away from the central atom, reducing the electron-pair repulsion. This allows the Cl—C—H bond angles to become 108.5°, smaller than tetrahedral, while the H—C—H angles become 110.5°, larger than tetrahedral. In effect, electropositive substituents repel other substituents more strongly than do electronegative substituents.

The fluorides $PF_5$, $SF_4$, $ClF_3$, and $XeF_2$ all have steric number 5 but have different numbers of lone pairs (0, 1, 2, and 3, respectively). What shapes do their molecules have? We have already mentioned that $PF_5$ is trigonal bipyramidal. Two of the fluorine atoms occupy **axial** sites (Fig. 3.11a), and the other three occupy **equatorial** sites. Because the two kinds of sites are not equivalent, there is no reason for all of the P—F bond lengths to be equal. Experiment shows that the equatorial P—F bond length is 1.534 Å, shorter than the axial P—F lengths, which are 1.577 Å.

$SF_4$ has four bonded atoms and one lone pair. Does the lone pair occupy an axial or an equatorial site? In the VSEPR theory, electron pairs repel each other much more strongly when they form a 90° angle with respect to the central atom than when the angle is larger. A single lone pair therefore finds a position that minimizes the number of 90° repulsions it has with bonding electron pairs. It occupies an equatorial position with two 90° repulsions (Fig. 3.11b) rather than an axial position with three 90° repulsions (Fig. 3.11c). The axial S—F bonds are bent slightly away from the lone pair, and so the molecular structure of $SF_4$ is a distorted seesaw. A second lone pair

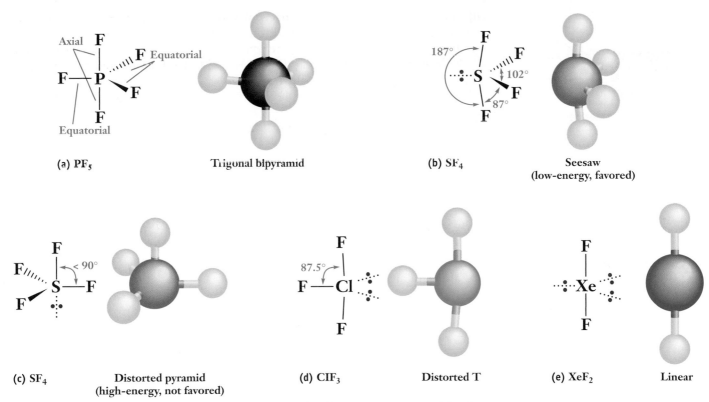

**FIGURE 3.11** Molecules with steric number 5. The molecular geometry is named for the sites occupied by the *atoms*, not the underlying trigonal–bipyramidal structure of electron pairs.

(in $ClF_3$, for example) also takes an equatorial position, leading to a distorted T-shaped molecular structure (Fig. 3.11d). A third lone pair (in $XeF_2$ or $I_3^-$, for example) occupies the third equatorial position, and the molecular geometry is linear (Fig. 3.11e).

## EXAMPLE 3.9

Predict the geometry of the following molecules and ions: **(a)** $ClO_3^+$, **(b)** $ClO_2^+$, **(c)** $SiH_4$, **(d)** $IF_5$.

### Solution

**(a)** The central Cl atom has all of its valence electrons (six, because the ion has a net positive charge) involved in bonds to the surrounding three oxygen atoms and has no lone pairs. Its steric number is 3. In the molecular ion the central Cl should be surrounded by the three O atoms in a trigonal planar structure.

**(b)** The central Cl atom in this ion also has a steric number of 3, composed of two bonded atoms and a single lone pair. The predicted molecular geometry is a bent molecule with an angle somewhat less than 120°.

**(c)** The central Si has a steric number of 4 and no lone pairs. The molecular geometry should consist of the Si atom surrounded by a regular tetrahedron of H atoms.

**FIGURE 3.12** The structure of $IF_5$. Note the distortions of F—I—F bond angles from 90° because of the lone pair at the bottom (not shown).

**(d)** Iodine has seven valence electrons of which five are shared in bonding pairs with F atoms. This leaves two electrons to form a lone pair, and so the steric number is 5 (bonded atoms) + 1 (lone pair) = 6. The structure will be based on the octahedron of electron pairs from Figure 3.9, with five F atoms and one lone pair. The lone pair can be placed on any one of the six equivalent sites and will cause the four F atoms to bend away from it toward the fifth F atom, giving the distorted structure shown in Figure 3.12.

**Related Problems: 41, 42, 43, 44**

The VSEPR theory is a simple but remarkably powerful model for predicting the geometries and approximate bond angles of molecules that have a central atom. In fact, there are many cases in which it is more successful than theories that require extensive calculation. It has its limitations, however. VSEPR theory does not account for the fact that observed bond angles in the Group V and VI hydrides $H_2S$ (92°), $H_2Se$ (91°), $PH_3$ (93°), and $AsH_3$ (92°) are so far from tetrahedral (109.5°) and so close to right angles (90°).

## Dipole Moments of Polyatomic Molecules

Polyatomic molecules, like the diatomic molecules considered in the last section, may have dipole moments. Often, a good approximation is to assign a dipole moment (which is now a vector, shown as an arrow) to each bond and then obtain the total moment by carrying out a vector sum of the bond dipoles. In $CO_2$, for example, each C—O has a bond dipole (Fig. 3.13a). Because the dipoles are equal in magnitude and point in opposite directions, however, the total molecular dipole moment vanishes. In OCS, which is also linear with a central carbon atom, the C—O and C—S bond dipole moments have different magnitudes, leaving a net nonzero dipole moment (Fig. 3.13b). In the water molecule (Fig. 3.13c) the two bond dipoles add vectorially to give a net molecular dipole moment. The more symmetric molecule $CCl_4$, on the other hand, has no net dipole moment (Fig. 3.13d). Even though each of the four C—Cl bonds (pointing from the four corners of a tetrahedron) has a dipole moment, their vector sum is 0. Molecules such as $H_2O$ and OCS, with nonzero dipole moments, are **polar;** those such as $CO_2$ and $CCl_4$, with no net dipole moment, are **nonpolar,** even though they contain polar bonds. As discussed in Chapter 5, intermolecular forces differ between polar and nonpolar molecules, a fact that significantly affects their physical and chemical properties.

---

**EXAMPLE 3.10**

Predict whether the molecules $NH_3$ and $SF_6$ will have dipole moments.

### Solution

The $NH_3$ molecule has a dipole moment, because the three N—H bond dipoles add to give a net dipole pointing downward from the N atom to the base of the pyramid that defines the $NH_3$ structure. The $SF_6$ molecule has no dipole moment, because each S—F bond dipole is balanced by an equal one pointing in the opposite direction on the other side of the molecule.

**Related Problems: 47, 48**

**FIGURE 3.13** The total dipole moment of a molecule is obtained by vector addition of its bond dipoles. This operation is performed by placing the arrows head to tail. (a) $CO_2$. (b) OCS. (c) $H_2O$. (d) $CCl_4$.

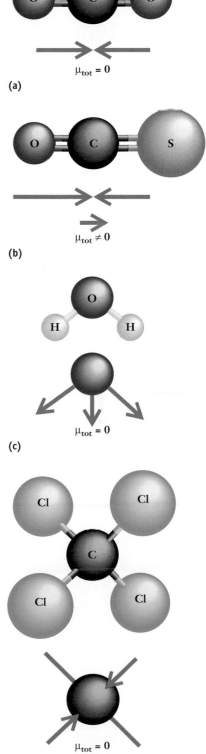

---

## 3.7

# INORGANIC NOMENCLATURE AND OXIDATION NUMBERS

In Sections 3.2 and 3.4 we used the periodic table and the Lewis model to describe the transfer and sharing of electrons in chemical compounds. We conclude this chapter by discussing the systematic nomenclature of inorganic compounds. In Section 3.2, ionic compounds were named by following the name of the positive ion with the name of the negative ion. If a metal atom can form more than one stable ion, the ionic charge is given by a Roman numeral enclosed in parentheses; for example, iron(III) represents the $Fe^{3+}$ ion. This Roman numeral notation can be generalized to apply to compounds with covalent character as well, giving rise to a quantity called *oxidation number*. Oxidation numbers are used in naming compounds, in classifying types of reactions, and in exploring the systematic chemistry of the elements.

## Oxidation Number

The terms "ionic" and "covalent" have been used up to now as though they represent the complete transfer of one or more electrons from one atom to another or the equal sharing of pairs of electrons between atoms in a molecule. In actuality, both types of bonding are idealizations that rarely apply exactly. To account for the transfer of electrons from one molecule or ion to another in oxidation–reduction reactions and to name different binary compounds of the same elements, it is not necessary to have detailed knowledge of the exact electron distributions in molecules, whether they are chiefly ionic or covalent. We can assign convenient fictitious charges to the atoms in a molecule and call them **oxidation numbers,** making certain the law of charge conservation is strictly observed. Oxidation numbers are chosen so that in ionic compounds the oxidation number coincides with the charge on the ion. The following simple conventions are useful:

1. The oxidation numbers of the atoms in a neutral molecule must add up to zero, and those in an ion must add up to the charge on the ion.
2. Alkali metal atoms have oxidation number $+1$, alkaline earth atoms $+2$, in their compounds.
3. Fluorine always has oxidation number $-1$ in its compounds. The other halogens have oxidation number $-1$ in their compounds except those with oxygen and with other halogens, where they can have positive oxidation numbers.
4. Hydrogen is assigned oxidation number $+1$ in its compounds except in metal hydrides such as LiH, where convention 2 takes precedence and hydrogen has oxidation number $-1$.
5. Oxygen is assigned oxidation number $-2$ in compounds. There are two exceptions: in compounds with fluorine, convention 3 takes precedence, and in compounds

that contain O—O bonds, conventions 2 and 4 take precedence. Thus, the oxidation number of oxygen in $OF_2$ is $+2$; in peroxides (such as $H_2O_2$ and $Na_2O_2$) it is $-1$. In superoxides (such as $KO_2$) oxygen's oxidation number is $-\frac{1}{2}$.

Convention 1 is fundamental because it guarantees charge conservation: the total number of electrons must remain constant in chemical reactions. This rule also makes the oxidation numbers of the neutral atoms of elements all zero. Conventions 2 to 5 are based on the principle that in ionic compounds the oxidation number should equal the charge on the ion. Note that fractional oxidation numbers, though not common, are allowed and in fact necessary to be consistent with this set of conventions.

With the preceding conventions in hand, chemists can assign oxidation numbers to the atoms in the vast majority of compounds. Apply conventions 2 through 5 as listed above, noting the exceptions given; then assign oxidation numbers to the other elements in such a way that convention 1 is always obeyed. Note that convention 1 applies not only to free ions but to the components that make up ionic solids. Chlorine has oxidation number $-1$ not only as a free $Cl^-$ ion but in the ionic solid $AgCl$ and in covalent $CH_3Cl$. It is important to recognize common ionic species (especially molecular ions) and to know the total charges they carry. Table 3.1 listed the names and formulas of many common anions. Inspection of the table reveals that several elements exhibit different oxidation numbers in different compounds. In $Ag_2S$ sulfur appears as the sulfide ion and has oxidation number $-2$, but in $Ag_2SO_4$ it appears as part of a sulfate ($SO_4^{2-}$) ion. In this case,

$$(\text{oxidation number of S}) + [4 \times (\text{oxidation number of O})] = \text{total charge on ion}$$

$$x + [4(-2)] = -2$$

$$\text{oxidation number of S} = x = +6$$

A convenient way to indicate the oxidation number of an atom is to write it directly above the corresponding symbol in the formula of the compound:

$$\overset{+1\ -2}{N_2O} \qquad \overset{+1\ -1}{LiH} \qquad \overset{0}{O_2} \qquad \overset{+6\ -2}{SO_4^{2-}}$$

## EXAMPLE 3.11

Assign oxidation numbers to the atoms in the following chemical compounds and ions: $NaCl$, $ClO^-$, $Fe_2(SO_4)_3$, $SO_2$, $I_2$, $KMnO_4$, $CaH_2$.

### Solution

$\overset{+1\ \ -1}{NaCl}$       From conventions 2 and 3.

$\overset{+1\ -2}{ClO^-}$       From conventions 1 and 5.

$\overset{+3\ \ +6\ -2}{Fe_2(SO_4)_3}$       From conventions 1 and 5. This is solved by recognizing the presence of sulfate ($SO_4^{2-}$) groups.

$\overset{+4\ -2}{SO_2}$       From conventions 1 and 5.

$\overset{0}{I_2}$          From convention 1. $I_2$ is an element.

$\overset{+1\ +7\ -2}{KMnO_4}$     From conventions 1, 2, and 5.

$\overset{+2\ -1}{CaH_2}$       From conventions 1 and 2 (metal hydride case).

**Related Problems: 53, 54**

**FIGURE 3.14** Several oxides of manganese. They are arranged in order of increasing oxidation number of the Mn, counterclockwise from bottom left: $MnO$, $Mn_3O_4$, $Mn_2O_3$, and $MnO_2$. A compound of still higher oxidation state, $Mn_2O_7$, is a dark red liquid that explodes easily. (*Leon Lewandowski*)

Oxidation numbers must not be confused with the formal charges on Lewis dot diagrams described in Section 3.4. They resemble formal charges to the extent that both are assigned, by arbitrary conventions, to symbols in formulas for specific purposes. The purposes differ, however. Formal charges are used solely to identify preferred Lewis diagrams. Oxidation numbers are used in nomenclature, in identifying oxidation–reduction reactions, and in exploring trends in chemical reactivity across the periodic table. Oxidation numbers are often not the same as the true charges on atoms. In $KMnO_4$ or $Mn_2O_7$, for example, Mn has the oxidation number $+7$, but the actual net charge on the atom is much less than this. A high oxidation state usually indicates significant covalent character in the bonding of that compound. The oxides of manganese with lower oxidation states (Fig. 3.14) have more ionic character.

Figure 3.15 shows the most common oxidation states of the elements of the main groups. The strong diagonal lines for these elements reflect the stability of closed electron octets. For example, the oxidation states $+6$ and $-2$ for sulfur would correspond to losing or gaining enough electrons to attain a noble-gas octet. The octet-based Lewis model is not useful for compounds of the transition elements, which are discussed further in Chapter 18.

**FIGURE 3.15** Important oxidation states of the main-group elements.

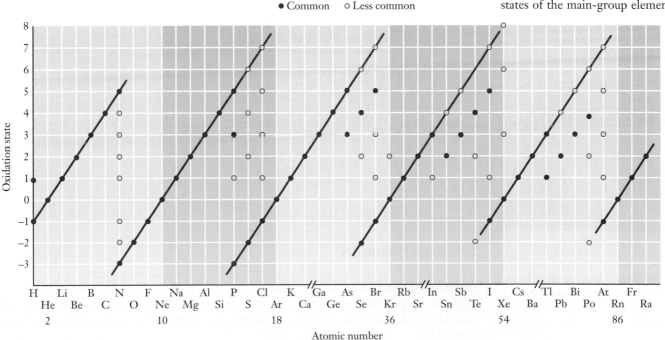

**TABLE 3.7**

*Prefixes Used for Naming Binary Covalent Compounds*

| Number | Prefix |
|--------|--------|
| 1 | *mono-* |
| 2 | *di-* |
| 3 | *tri-* |
| 4 | *tetra-* |
| 5 | *penta-* |
| 6 | *hexa-* |
| 7 | *hepta-* |
| 8 | *octa-* |
| 9 | *nona-* |
| 10 | *deca-* |

## Naming Binary Covalent Compounds

How do we name nonionic (covalent) compounds? If a pair of elements forms only one compound, begin with the name of the element appearing first in the chemical formula, followed by the second element with the suffix *-ide* added to its root. This is quite analogous to the naming of ionic compounds. Just as NaBr is sodium bromide, so the following names designate typical covalent compounds:

| | | | |
|---|---|---|---|
| HBr | hydrogen bromide | $H_2S$ | hydrogen sulfide |
| $BeCl_2$ | beryllium chloride | BN | boron nitride |

A number of well-established nonsystematic names continue to be used (even the strictest chemist does not call water "hydrogen oxide"!) They include

| | | | |
|---|---|---|---|
| $H_2O$ | water | $PH_3$ | phosphine |
| $NH_3$ | ammonia | $AsH_3$ | arsine |
| $N_2H_4$ | hydrazine | $COCl_2$ | phosgene |

If a pair of elements forms more than one compound, two methods can be used to distinguish between them.

1. Use Greek prefixes (Table 3.7) to specify the number of atoms of each element in the molecular formula of the compound (*di-* for two, *tri-* for three, and so forth). If the compound is a solid without well-defined molecules, name the empirical formula in this way. The prefix for one (*mono-*) is omitted except in the case of carbon monoxide.
2. Write the oxidation number of the first-named element in Roman numerals and place it in parentheses after the name of that element.

Applying the two methods to the oxides of nitrogen gives

| | | |
|---|---|---|
| $N_2O$ | dinitrogen oxide | nitrogen(I) oxide |
| NO | nitrogen oxide | nitrogen(II) oxide |
| $N_2O_3$ | dinitrogen trioxide | nitrogen(III) oxide |
| $NO_2$ | nitrogen dioxide | nitrogen(IV) oxide |
| $N_2O_4$ | dinitrogen tetraoxide | nitrogen(IV) oxide |
| $N_2O_5$ | dinitrogen pentaoxide | nitrogen(V) oxide |

The first method has some advantages over the second. It distinguishes between $NO_2$ (nitrogen dioxide) and $N_2O_4$ (dinitrogen tetraoxide), two distinct compounds that would both be called nitrogen(IV) oxide under the second system of nomenclature. Two of these oxides have common (nonsystematic) names that may be encountered elsewhere: $N_2O$ is often called nitrous oxide, and NO is called nitric oxide.

On the right is a sample of rutile, $TiO_2$, which is the primary ore of titanium. Behind it is a large quartz crystal through which slender rutile hairs penetrate. Crystals of this type are called Venus hairstone. *(Michael Dalton/Fundamental Photographs)*

## CUMULATIVE EXERCISE

### Oxides and Peroxides

Consider the three compounds $KO_2$, $BaO_2$, and $TiO_2$. Each contains two oxygen atoms per metal atom, but the oxygen occurs in different forms in the three compounds. (a) The oxygen in $TiO_2$ occurs as $O^{2-}$ ions. Give the Lewis dot symbol for this ion. How many valence electrons does this ion have? What is the chemical name for $TiO_2$?

(b) Recall that Group II elements form stable +2 ions. By referring to Table 3.1, identify the oxygen-containing ion in $BaO_2$ and give the name of the compound. Draw a Lewis diagram for the oxygen-containing ion, showing formal charge. Is the bond in this ion a single or a double bond?

(c) Recall that Group I elements form stable +1 ions. By referring to Table 3.1, identify the oxygen-containing ion in $KO_2$ and give the name of the compound. Show that the oxygen-containing ion is an odd-electron species. Draw the best Lewis diagram you can for it.

**Answers**

(a) The ion $:\ddot{O}:^{2-}$ has eight valence electrons (an octet). $TiO_2$ is titanium(IV) oxide.

(b) The ion in $BaO_2$ must be the peroxide ion ($O_2^{2-}$), and the compound is barium peroxide. The Lewis diagram for the peroxide ion is

$$\overset{\ominus}{\phantom{:}}\ddot{\phantom{O}}\quad\ddot{\phantom{O}}\overset{\ominus}{\phantom{:}}$$
$$:\ddot{O}:\ddot{O}:$$

and the O—O bond is a single bond.

(c) The ion in $KO_2$ must be the superoxide ion ($O_2^-$), and the compound is potassium superoxide. The superoxide ion has 13 valence electrons. The best Lewis diagram is a pair of resonance diagrams

$$:\dot{O}:\overset{\ominus}{\ddot{O}}:\quad\text{and}\quad\overset{\ominus}{\ddot{O}}:\dot{O}:$$

in which only one of the oxygen atoms attains an octet electron configuration.

# CONCEPTS & SKILLS

*After studying this chapter and working the problems that follow, you should be able to*

1. Describe the trends in ionization energy, electron affinity, and electronegativity across the periodic table (Section 3.1).
2. Use the principle of charge balance to write chemical formulas for ionic compounds (Section 3.2, problems 3–10).
3. Calculate the energy of dissociation of gaseous diatomic ionic compounds into neutral atoms and ions (Section 3.2, problems 11–12).
4. Describe the relationships between bond order, bond length, and bond energy (Section 3.3, problems 13–14).
5. Given a molecular formula, draw a Lewis diagram for the molecule (Section 3.4, problems 21–24).
6. Assign formal charges and identify resonance diagrams for a given Lewis diagram (Section 3.4, problems 27–32).
7. Estimate the percent ionic character of a bond from its dipole moment (Section 3.5, problems 37–40).
8. Predict the geometries of molecules by use of the VSEPR model (Section 3.6, problems 41–46).
9. Determine whether a polyatomic molecule is polar or nonpolar (Section 3.6, problems 47–48).
10. Assign oxidation numbers to atoms in compounds (Section 3.7, problems 53–54).
11. Name inorganic compounds, given their chemical formulas, and write chemical formulas for named inorganic compounds (Section 3.7, problems 55–58).

*Answers to problems whose numbers are boldface appear in Appendix G. Problems that are more challenging are indicated with asterisks.*

## Ionic Bonding: Coulomb Stabilization Energy

1. For each of the following atoms or ions, give the Lewis dot symbol and state the total number of electrons, the number of valence electrons, and the number of core electrons.
   (a) Rn   (c) $Se^{2-}$
   (b) $Sr^+$   (d) $Sb^-$

2. For each of the following atoms or ions, give the Lewis dot symbol and state the total number of electrons, the number of valence electrons, and the number of core electrons.
   (a) $Ra^{2+}$   (c) $Bi^{2-}$
   (b) Br   (d) $Ga^+$

3. Give the name and formula of an ionic compound involving only the elements in each pair that follows. Write Lewis symbols for the elements both before and after chemical combination.
   (a) Chlorine and cesium   (c) Aluminum and sulfur
   (b) Calcium and astatine   (d) Potassium and tellurium

4. Give the name and formula of an ionic compound involving only the elements in each pair that follows. Write Lewis symbols for the elements both before and after chemical combination.
   (a) Gallium and bromine
   (b) Strontium and polonium
   (c) Magnesium and iodine
   (d) Lithium and selenium

5. Give systematic names to the following compounds.
   (a) $Al_2O_3$   (d) $Ca(NO_3)_2$
   (b) $Rb_2Se$   (e) $Cs_2SO_4$
   (c) $(NH_4)_2S$   (f) $KHCO_3$

6. Give systematic names to the following compounds.
   (a) $KNO_2$   (d) $NaH_2PO_4$
   (b) $Sr(MnO_4)_2$   (e) $BaCl_2$
   (c) $MgCr_2O_7$   (f) $NaClO_3$

7. Write the chemical formulas for the following compounds.
   (a) Silver cyanide
   (b) Calcium hypochlorite
   (c) Potassium chromate
   (d) Gallium oxide
   (e) Potassium superoxide
   (f) Barium hydrogen carbonate

8. Write the chemical formulas for the following compounds.
   (a) Cesium sulfite
   (b) Strontium thiocyanate
   (c) Lithium hydride
   (d) Sodium peroxide
   (e) Ammonium dichromate
   (f) Rubidium hydrogen sulfate

9. Trisodium phosphate (TSP) is a heavy-duty cleaning agent. Write its chemical formula. What would be the systematic name for this ionic compound?

10. Monoammonium phosphate is a compound made up of $NH_4^+$ and $H_2PO_4^-$ ions that is employed as a flame retardant (its use for this purpose was first suggested by Gay-Lussac in 1821). Write its chemical formula. What is the systematic chemical name of this compound?

11. In a gaseous KCl molecule, the internuclear distance is 2.67 Å. Using data from Appendix F and neglecting the small short-range repulsion between the ion cores of $K^+$ and $Cl^-$, estimate the dissociation energy of gaseous KCl into K and Cl atoms (in kilojoules per mole).

12. In a gaseous RbF molecule, the bond length is 2.274Å. Using data from Appendix F and neglecting the small short-range repulsion between the ion cores of $Rb^+$ and $F^-$, estimate the dissociation energy of gaseous RbF into Rb and F atoms (in kilojoules per mole).

## Structure of Isolated Molecules: Properties of the Covalent Chemical Bond

13. The bond lengths of the X—H bonds in $NH_3$, $PH_3$, and $SbH_3$ are 1.02, 1.42, and 1.71 Å, respectively. Estimate the length of the As—H bond in $AsH_3$, the gaseous compound that decomposes on a heated glass surface in Marsh's test for arsenic. Which of these four hydrides has the weakest X—H bond?

14. Arrange the following covalent diatomic molecules in order of the lengths of the bonds: BrCl, ClF, IBr. Which of the three has the weakest bond (the smallest bond energy)?

## Covalent Bonding: Electron Pair Model

15. Assign formal charges to all atoms in the following Lewis diagrams.
   (a) $SO_4^{2-}$   (b) $S_2O_3^{2-}$

   (c) $SbF_3$   (d) $SCN^-$

16. Assign formal charges to all atoms in the following Lewis diagrams.
   (a) $ClO_4^-$   (b) $SO_2$

   (c) $BrO_2^-$   (d) $NO_3^-$

17. Determine the formal charges on all the atoms in the following Lewis diagrams:

$$H—\overset{..}{N}=\overset{..}{O} \quad \text{and} \quad H—\overset{..}{\underset{..}{O}}=N$$

Which one would best represent bonding in the molecule HNO?

18. Determine the formal charges on all the atoms in the following Lewis diagrams:

$$:\overset{..}{\underset{..}{Cl}}—\overset{..}{\underset{..}{Cl}}—\overset{..}{\underset{..}{O}}: \quad \text{and} \quad :\overset{..}{\underset{..}{Cl}}—\overset{..}{\underset{..}{O}}—\overset{..}{\underset{..}{Cl}}:$$

Which one would best represent bonding in the molecule $Cl_2O$?

19. In each of the following Lewis diagrams, Z represents a main-group element. Name the group to which Z belongs in each case, and give an example of such a compound or ion that actually exists.

(a)

$$\overset{..}{\underset{..}{O}}=Z=\overset{..}{\underset{..}{O}}$$

(b)

$$:\overset{..}{\underset{..}{O}}: \quad :\overset{..}{\underset{..}{O}}:$$
$$:\overset{..}{\underset{..}{O}}—Z—\overset{..}{\underset{..}{O}}—Z—\overset{..}{\underset{..}{O}}:$$
$$:\overset{..}{\underset{..}{O}}: \quad :\overset{..}{\underset{..}{O}}:$$

(c)

$$\left[\overset{..}{\underset{..}{O}}=\overset{..}{Z}—\overset{..}{\underset{..}{O}}:\right]^{-}$$

(d)

$$\left[\begin{array}{c} :\overset{..}{\underset{..}{O}}: \\ | \\ H—\overset{..}{\underset{..}{O}}—Z—\overset{..}{\underset{..}{O}}: \\ | \\ :\overset{..}{\underset{..}{O}}: \end{array}\right]^{-}$$

20. In each of the following Lewis diagrams, Z represents a main-group element. Name the group to which Z belongs in each case, and give an example of such a compound or ion that actually exists.

(a)

$$\left[:C\equiv Z:\right]^{-}$$

(b)

$$\left[\begin{array}{c} :\overset{..}{\underset{..}{O}}: \\ | \\ :\overset{..}{\underset{..}{O}}—Z—\overset{..}{\underset{..}{O}}: \\ | \\ :\overset{..}{\underset{..}{O}}: \end{array}\right]^{-}$$

(c)

$$\left[\begin{array}{c} :\overset{..}{\underset{..}{O}}: \\ | \\ :\overset{..}{\underset{..}{O}}—Z—\overset{..}{\underset{..}{O}}: \end{array}\right]^{2-}$$

(d)

$$H—\overset{..}{Z}—\overset{..}{Z}—H$$
$$\qquad | \quad | $$
$$\qquad H \quad H$$

21. Draw Lewis electron-dot diagrams for the following species.
    (a) $AsH_3$      (c) $KrF^-$
    (b) HOCl         (d) $PO_2Cl_2^-$ (central P atom)

22. Draw Lewis electron-dot diagrams for the following species.
    (a) Methane
    (b) Carbon dioxide
    (c) Phosphorus trichloride
    (d) Perchlorate ion

23. Urea is an important chemical fertilizer with the chemical formula $(H_2N)CO(NH_2)$. The carbon atom is bonded to both nitrogen atoms and to the oxygen atom. Draw urea's Lewis diagram and use Table 3.5 to estimate its bond lengths.

24. Acetic acid is the active ingredient of vinegar. Its chemical formula is $CH_3COOH$, and the second carbon atom is bonded to the first carbon atom and to both oxygen atoms. Draw acetic acid's Lewis diagram and use Table 3.5 to estimate its bond lengths.

25. Under certain conditions the stable form of sulfur consists of rings of eight sulfur atoms. Draw the Lewis diagram for such a ring.

26. White phosphorus ($P_4$) consists of four phosphorus atoms arranged at the corners of a tetrahedron. Draw the valence electrons on this structure to give a Lewis diagram that satisfies the octet rule.

27. Draw Lewis electron-dot diagrams for the following species, indicating formal charges and resonance diagrams where applicable.
    (a) $H_3NBF_3$
    (b) $CH_3COO^-$ (acetate ion)
    (c) $HCO_3^-$ (hydrogen carbonate ion)

28. Draw Lewis electron-dot diagrams for the following species, indicating formal charges and resonance diagrams where applicable.
    (a) HNC (central N atom)
    (b) $SCN^-$ (thiocyanate ion)
    (c) $H_2CNN$ (the first N atom is bonded to the carbon and the second N)

29. Draw Lewis diagrams for the two resonance forms of the nitrite ion, $NO_2^-$. In what range do you expect the nitrogen–oxygen bond length to fall? (*Hint:* Use Table 3.5.)

30. Draw Lewis diagrams for the three resonance forms of the carbonate ion, $CO_3^{2-}$. In what range do you expect the carbon–oxygen bond length to fall? (*Hint:* Use Table 3.5.)

31. Methyl isocyanate, which was involved in the disaster in Bhopal, India, in 1984, has the chemical formula $CH_3NCO$. Draw its Lewis diagram, including resonance forms. (*Note:* The N atom is bonded to the two C atoms.)

32. Peroxyacetyl nitrate (PAN) is one of the prime irritants in photochemical smog. It has the formula $CH_3COOONO_2$, with the structure

$$CH_3—C\overset{\displaystyle O}{\underset{\displaystyle O—O—NO_2}{\diagdown\diagup}}$$

Draw its Lewis diagram, including resonance forms.

33. Draw Lewis diagrams for the following compounds. In the formula, the symbol of the central atom is given first. (*Hint:* The valence octet may be expanded for the central atom.)
    (a) $PF_5$      (b) $SF_4$      (c) $XeO_2F_2$

34. Draw Lewis diagrams for the following ions. In the formula, the symbol of the central atom is given first. (*Hint:* The valence octet may be expanded for the central atom.)
    (a) $BrO_4^-$      (b) $PCl_6^-$      (c) $XeF_3^+$

## Polar Covalent Bonding: Electronegativity and Dipole Moments

**35.** Ionic compounds tend to have higher melting and boiling points and to be less volatile (that is, have lower vapor pressures) than covalent compounds. For each of the following pairs, use electronegativity differences to predict which compound has the higher vapor pressure at room temperature.
(a) $CI_4$ or $KI$
(b) $BaF_2$ or $OF_2$
(c) $SiH_4$ or $NaH$

**36.** For each of the following pairs, use electronegativity differences to predict which compound has the higher boiling point.
(a) $MgBr_2$ or $PBr_3$
(b) $OsO_4$ or $SrO$
(c) $Cl_2O$ or $Al_2O_3$

**37.** Estimate the percent ionic character of the bond in each of the following diatomic molecules, based upon the dipole moment.

| | Average Bond Length (Å) | Dipole Moment (D) |
|---|---|---|
| ClO | 1.573 | 1.239 |
| KI | 3.051 | 10.82 |
| TlCl | 2.488 | 4.543 |
| InCl | 2.404 | 3.79 |

**38.** Estimate the percent ionic character of the bond in each of the following species. All the species are unstable or reactive under ordinary laboratory conditions, but they can be observed in interstellar space.

| | Average Bond Length (Å) | Dipole Moment (D) |
|---|---|---|
| OH | 0.980 | 1.66 |
| CH | 1.131 | 1.46 |
| CN | 1.175 | 1.45 |
| $C_2$ | 1.246 | 0 |

**39.** The percent ionic character of a bond can be approximated by the formula $16\Delta + 3.5\Delta^2$, where $\Delta$ is the magnitude of the difference in the electronegativities of the atoms (see Fig. 3.7). Calculate the percent ionic character of HF, HCl, HBr, HI, and CsF, and compare the results with those in Table 3.6.

**40.** The percent ionic character of the bonds in several interhalogen molecules (as estimated from their measured dipole moments and bond lengths) are ClF (11%), BrF (15%), BrCl (5.6%), ICl (5.8%), and IBr (10%). Estimate the percent ionic characters for each of these molecules, using the equation in the preceding problem, and compare them with the given values.

## The Shapes of Molecules: VSEPR Theory

**41.** For each of the following molecules, give the steric number and sketch and name the approximate molecular geometry.

In each case, the central atom is listed first and the other atoms are all bonded directly to it.
(a) $CBr_4$     (d) $SOCl_2$
(b) $SO_3$     (e) $ICl_3$
(c) $SeF_6$

**42.** For each of the following molecules or molecular ions, give the steric number and sketch and name the approximate molecular geometry. In each case, the central atom is listed first and the other atoms are all bonded directly to it.
(a) $PF_3$     (d) $ClO_2^-$
(b) $SO_2Cl_2$     (e) $GeH_4$
(c) $PF_6^-$

**43.** For each of the following molecules or molecular ions, give the steric number, sketch and name the approximate molecular geometry, and describe the directions of any *distortions* from the approximate geometry due to lone pairs. In each case, the central atom is listed first and the other atoms are all bonded directly to it.
(a) $ICl_4^-$     (c) $BrO_3^-$
(b) $OF_2$     (d) $CS_2$

**44.** For each of the following molecules or molecular ions, give the steric number, sketch and name the approximate molecular geometry, and describe the direction of any *distortions* from the approximate geometry due to lone pairs. In each case, the central atom is listed first and the other atoms are all bonded directly to it.
(a) $TeH_2$     (c) $PCl_4^+$
(b) $AsF_3$     (d) $XeF_5^+$

**45.** Give an example of a molecule or ion having a formula of each of the following types and structures.
(a) $AB_3$ (planar)     (c) $AB_2^-$ (bent)
(b) $AB_3$ (pyramidal)     (d) $AB_3^{2-}$ (planar)

**46.** Give an example of a molecule or ion having a formula of each of the following types and structures.
(a) $AB_4^-$ (tetrahedral)     (c) $AB_6^-$ (octahedral)
(b) $AB_2$ (linear)     (d) $AB_3^-$ (pyramidal)

**47.** For each of the answers in problem 41, state whether the species is polar or nonpolar.

**48.** For each of the answers in problem 42, state whether the species is polar or nonpolar.

**49.** The molecules of a certain compound contain one atom each of nitrogen, fluorine, and oxygen. Two possible structures are NOF (O as central atom) and ONF (N as central atom). Does the information that the molecule is bent limit the choice to one of these two possibilities? Explain.

**50.** Mixing $SbCl_3$ and $GaCl_3$ in a one-to-one molar ratio (using liquid sulfur dioxide as a solvent) gives a solid ionic compound of empirical formula $GaSbCl_6$. A controversy arises over whether this compound is $(SbCl_2^+)(GaCl_4^-)$ or $(GaCl_2^+)(SbCl_4^-)$.
(a) Predict the molecular structures of the two anions.
(b) It is learned that the cation in the compound has a bent structure. Based on this fact, which formulation is more likely to be correct?

**51.** (a) Use the VSEPR theory to predict the structure of the NNO molecule.

(b) The substance NNO has a small dipole moment. Which end of the molecule is more likely to be the positive end, based only on electronegativity?

52. Ozone ($O_3$) has a nonzero dipole moment. In the molecule of $O_3$, one of the oxygen atoms is directly bonded to the other two, which are not bonded to each other.

(a) Based on this information, state which of the following structures are possible for the ozone molecule: symmetric linear, unsymmetric linear (for example, different O—O bond lengths), and bent. (*Note:* Even an O—O bond can have a bond dipole if the two oxygen atoms are bonded to different atoms or if only one of the oxygen atoms is bonded to a third atom.)

(b) Use the VSEPR theory to predict which of the structures of part (a) is observed.

## Inorganic Nomenclature and Oxidation Numbers

53. Assign oxidation numbers to the atoms in each of the following species: $SrBr_2$, $Zn(OH)_4^{2-}$, $SiH_4$, $CaSiO_3$, $Cr_2O_7^{2-}$, $Ca_5(PO_4)_3F$, $KO_2$, CsH.

54. Assign oxidation numbers to the atoms in each of the following species: $NH_4NO_3$, $CaMgSiO_4$, $Fe(CN)_6^{4-}$, $B_2H_6$, $BaH_2$, $PbCl_2$, $Cu_2O(SO_4)$, $S_4O_6^{2-}$.

55. Write the chemical formula for each of the following compounds.

(a) Silicon dioxide
(b) Ammonium carbonate
(c) Lead(IV) oxide
(d) Diphosphorus pentaoxide
(e) Calcium iodide
(f) Iron(III) nitrate

56. Write the chemical formula for each of the following compounds.

(a) Lanthanum(III) sulfide
(b) Cesium sulfate
(c) Dinitrogen trioxide
(d) Iodine pentafluoride
(e) Chromium(III) sulfate
(f) Potassium permanganate

57. Give the systematic name for each of the following compounds.

(a) $Cu_2S$ and CuS     (d) $ZrCl_4$
(b) $Na_2SO_4$     (e) $Cl_2O_7$
(c) $As_4O_6$     (f) $Ga_2O$

58. Give the systematic name for each of the following compounds.

(a) $Mg_2SiO_4$     (d) $(NH_4)_2HPO_4$
(b) $Fe(OH)_2$ and $Fe(OH)_3$     (e) $SeF_6$
(c) $As_2O_5$     (f) $Hg_2SO_4$

## Additional Problems

59. Many important fertilizers are ionic compounds containing the elements nitrogen, phosphorus, and potassium because these are frequently the limiting plant-growth nutrients in soil.

(a) Write the chemical formulas for the following chemical fertilizers: ammonium phosphate, potassium nitrate, ammonium sulfate.

(b) Calculate the mass percentage of nitrogen, phosphorus, and potassium for each of the compounds in part (a).

60. Two possible Lewis diagrams for sulfine ($H_2CSO$) are

(a) Compute the formal charges on all atoms.
(b) Draw a Lewis diagram for which all the atoms in sulfine have formal charges of zero.

61. There is persuasive evidence for the brief existence of the unstable molecule OPCl.

(a) Draw a Lewis diagram for this molecule in which the octet rule is satisfied on all atoms and the formal charges on all atoms are zero.

(b) The compound OPCl reacts with oxygen to give $O_2PCl$. Draw a Lewis diagram of $O_2PCl$ for which all formal charges are equal to zero. Draw a Lewis diagram in which the octet rule is satisfied on all atoms.

62. The compound $SF_3N$ has been synthesized.

(a) Draw the Lewis diagram of this molecule, supposing that the three fluoride atoms and the nitrogen atom surround the sulfur atom. Indicate the formal charges. Repeat, but assume that the three fluorine atoms and the sulfur atom surround the nitrogen atom.

(b) On the basis of the results in part (a), speculate about which arrangement is more likely to correspond to the actual molecular structure.

63. In nitryl chloride ($NO_2Cl$), the chlorine atom and the two oxygen atoms are bonded to a central nitrogen atom, and all the atoms lie in a plane. Draw the two electron-dot resonance forms that satisfy the octet rule and that together are consistent with the fact that the two nitrogen–oxygen bonds are equivalent.

64. The molecular ion $S_3N_3^-$ has the cyclic structure

$$\begin{bmatrix} & & S & & \\ N & & & & N \\ | & & & & | \\ S & & & & S \\ & & N & & \end{bmatrix}^-$$

All S—N bonds are equivalent.

(a) Give six equivalent resonance hybrid Lewis diagrams for this molecular ion.

(b) Compute the formal charges on all atoms in the molecular ion in each of the six Lewis diagrams.

(c) Determine the charge on each atom in the polyatomic ion, assuming that the true distribution of electrons is the *average* of the six Lewis diagrams arrived at in parts (a) and (b).

(d) An advanced calculation suggests that the actual charge resident on each N atom is $-0.375$, and that on each S atom is $+0.041$. Show that this result is consistent with the overall $-1$ charge on the molecular ion.

*65. The two compounds nitrogen dioxide and dinitrogen tetraoxide are mentioned in Section 3.7.

(a) $NO_2$ is an odd-electron compound. Draw the best Lewis diagrams you can for it, recognizing that one atom cannot achieve an octet configuration. Use formal charges to decide whether that should be the (central) nitrogen atom or one of the oxygen atoms.

(b) Draw resonance forms for $N_2O_4$ that obey the octet rule. The two N atoms are bonded in this molecule.

66. Although magnesium and the alkaline-earth metals situated below it in the periodic table form ionic chlorides, beryllium chloride ($BeCl_2$) is a covalent compound.

(a) Follow the usual rules to write a Lewis diagram for $BeCl_2$, in which each atom attains an octet configuration. Indicate formal charges.

(b) The Lewis diagram that results from part (a) is an extremely unlikely one because of the double bonds and formal charges it shows. By relaxing the requirement of placing an octet on the beryllium atom, show how a Lewis diagram without formal charges can be written.

67. (a) The first noble-gas compound, prepared by Neil Bartlett in 1962, was an orange-yellow ionic solid consisting of $XeF^+$ and $Pt_2F_{11}^-$ ions. Draw a Lewis diagram for the $XeF^+$ ion.

(b) Shortly after the preparation of the ionic compound discussed in part (a), it was found that the irradiation of mixtures of xenon and fluorine with sunlight produced white crystalline $XeF_2$. Draw a Lewis diagram for this molecule, allowing valence expansion on the central xenon atom.

* 68. Represent the bonding in $SF_2$ (F—S—F) with Lewis diagrams. Include the formal charges on all atoms. The dimer of this compound has the formula $S_2F_4$. It was isolated in 1980 and shown to have the structure $F_3S$—SF. Draw a possible Lewis diagram to represent the bonding in the dimer, indicating the formal charges on all atoms. Is it possible to draw a Lewis diagram for $S_2F_4$ in which all atoms have valence octets? Explain why or why not.

69. Consult Figure 3.7 and compute the difference in electronegativity between the atoms in LiCl and those in HF. Based on their physical properties (see below), are the two similar or different in terms of bonding?

|  | LiCl | HF |
|---|---|---|
| Melting point | 605°C | −83.1°C |
| Boiling point | 1350°C | 19.5°C |

70. Use the data in Appendix F to compute the energy changes ($\Delta E$) of the following pairs of reactions.

(a) $Na(g) + I(g) \rightarrow Na^+(g) + I^-(g)$ and
$$Na(g) + I(g) \rightarrow Na^-(g) + I^+(g)$$

(b) $K(g) + Cl(g) \rightarrow K^+(g) + Cl^-(g)$ and
$$K(g) + Cl(g) \rightarrow K^-(g) + Cl^+(g)$$

Explain why $Na^+I^-$ and $K^+Cl^-$ form in preference to $Na^-I^+$ and $K^-Cl^+$.

* 71. At large interatomic separations, an alkali halide molecule MX has a lower energy as two neutral atoms, M + X; at short separations, the ionic form $(M^+)(X^-)$ has a lower energy. At a certain distance $R_c$ the energies of the two forms become equal, and it is near this distance that the electron will jump from the metal to the halogen atom during a collision. Because the forces between neutral atoms are weak at large distances, a reasonably good approximation can be made by ignoring any variation in potential $V(R)$ for the neutral atoms between $R_c$ and $R = \infty$. For the ions in this distance range, $V(R)$ is dominated by their Coulomb attraction.

(a) Express $R_c$ in terms of the first ionization energy of the metal M and the electron affinity of the halogen X.

(b) Calculate $R_c$ for LiF, KBr, and NaCl, using data from Appendix F.

72. The gaseous potassium chloride molecule has a measured dipole moment of 10.3 D, which indicates that it is a very polar molecule. The separation between the nuclei in this molecule is 2.67 Å. What would the dipole moment of a KCl molecule be if there were opposite charges of one fundamental unit ($1.60 \times 10^{-19}$ C) at the nuclei?

73. (a) Predict the geometry of the $SbCl_5^{2-}$ ion, using the VSEPR method.

(b) The ion $SbCl_6^{3-}$ is prepared from $SbCl_5^{2-}$ by treatment with $Cl^-$ ion. Determine the steric number of the central antimony atom in this ion, and discuss the extension of the VSEPR theory that would be needed for the prediction of its molecular geometry.

74. The element xenon (Xe) is by no means chemically inert; it forms a number of chemical compounds with electronegative elements such as fluorine and oxygen. The reaction of xenon with varying amounts of fluorine produces $XeF_2$ and $XeF_4$. Subsequent reaction of one or the other of these compounds with water produces (depending on conditions) $XeO_3$, $XeO_4$, and $H_4XeO_6$ as well as mixed compounds such as $XeOF_4$. Predict the structures of these six xenon compounds, using the VSEPR theory.

75. Predict the arrangement of the atoms about the sulfur atom in $F_4S{=}O$, assuming that double-bonded atoms require more space than single-bonded atoms.

76. Draw Lewis diagrams and predict the geometries of the following molecules. State which are polar and which are nonpolar.

(a) ONCl     (d) $SCl_4$
(b) $O_2NCl$     (e) $CHF_3$
(c) $XeF_2$

* 77. Suppose that any given kind of bond, such as an O—H bond, has a characteristic electric dipole. In other words, suppose that electric dipole moments can be assigned to bonds just as bond energies can be. Both are usefully accurate approximations. Consider the water molecule

Show that if $\mu_{OH}$ is the dipole moment of the OH bond, then the dipole moment of water is $\mu(H_2O) = 2\mu_{OH} \cos(\theta/2)$. What is the dipole moment $\mu_{OH}$ if $\mu(H_2O)$ is 1.86 D?

78. The carbon–carbon bond length in $C_2H_2$ is 1.20 Å; that in $C_2H_4$ is 1.34 Å, and that in $C_2H_6$ is 1.53 Å. Near which of these values would you predict the bond length of $C_2$ to lie? Is the experimentally observed value, 1.31 Å, consistent with your prediction?

79. Bismuth forms an ion with the formula $Bi_5^{3+}$. Arsenic and fluorine form a complex ion $[AsF_6]^-$, with fluorine atoms arranged around a central arsenic atom. Assign oxidation numbers to each of the atoms in the bright yellow crystalline solid with the formula $Bi_5(AsF_6)_3 \cdot 2\ SO_2$.

80. A good method of preparing very pure oxygen on a small scale is the decomposition of $KMnO_4$ in a vacuum above 215°C:

$$2\ KMnO_4(s) \longrightarrow K_2MnO_4(s) + MnO_2(s) + O_2(g)$$

Assign an oxidation number to each atom and verify that the total number of electrons lost is equal to the total number gained.

* 81. (a) Determine the oxidation number of lead in each of the following oxides: $PbO$, $PbO_2$, $Pb_2O_3$, $Pb_3O_4$.
   (b) The only known lead ions are $Pb^{2+}$ and $Pb^{4+}$. How can you reconcile this statement with your answer to part (a)?

82. In some forms of the periodic table, hydrogen is placed in Group I; in others, it is placed in Group VII. Give arguments in favor of each location.

---

## CUMULATIVE PROBLEMS

83. A certain element, M, is a main-group metal that reacts with chlorine to give a compound with the chemical formula $MCl_2$, and with oxygen to give the compound $MO$.
   (a) To which group in the periodic table does element M belong?
   (b) The chloride contains 44.7% chlorine by mass. Name the element M.

* 84. An ionic compound used as a chemical fertilizer has the composition (by mass) 48.46% O, 23.45% P, 21.21% N, 6.87% H. Give the name and chemical formula of the compound, and draw Lewis diagrams for the two types of ions that make it up.

85. A compound is being tested for use as a rocket propellant. Analysis reveals that it contains 18.54% F, 34.61% Cl, and 46.85% O.
   (a) Determine the empirical formula for this compound.
   (b) Assuming that the molecular formula is the same as the empirical formula, draw a Lewis diagram for this mole-cule. Review examples elsewhere in the chapter to decide which atom is most likely to lie at the center.
   (c) Use the VSEPR theory to predict the structure of the molecule from part (b).

* 86. A stable triatomic molecule can be formed that contains one atom each of nitrogen, sulfur, and fluorine. Three bonding structures are possible, depending on which is the central atom: NSF, SNF, and SFN.
   (a) Write a Lewis diagram for each of these molecules, indicating the formal charge on each atom.
   (b) Frequently, the structure with the least separation of formal charge is the most stable. Is this statement consistent with the observed structure for this molecule—namely NSF, which has a central sulfur atom?
   (c) Does consideration of the electronegativities of N, S, and F from Figure 3.7 help to rationalize this observed structure? Explain.

# U N I T  2

# KINETIC MOLECULAR EXPLANATION OF THE STATES OF MATTER

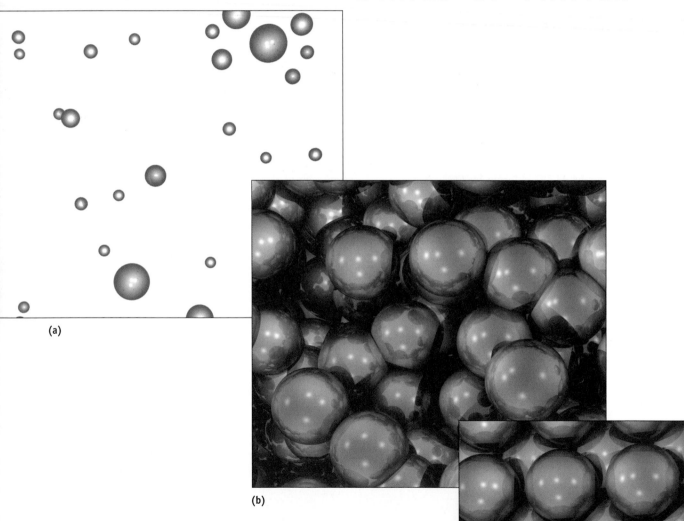

(a)

(b)

(c)

Spatial arrangement of molecules in a gas (a), a liquid (b), and a solid (c). Because the molecules in a solid or a liquid are in close contact with one another, the volume per molecule of the substance is nearly equal to the volume of the molecule itself. The molecules in a gas are much farther apart, with considerable open space between them. *(Courtesy of Dr. Stuart C. Watson and Professor Emily A. Carter/UCLA)*

We recognize gases, liquids, and solids quite easily because we are surrounded by important examples of each: air, water, and earth. Gases and liquids are fluid, but solids are rigid. A gas expands to fill any container it occupies; a liquid has a fixed volume but flows to conform to the shape of its container; a solid has a fixed volume *and* a fixed shape that resists deformation. These characteristics originate from the arrangements and motions of molecules, which are in turn determined by the forces between molecules. In a solid or liquid, molecules (shown in the sketches on the previous page as spheres) are in very close contact. At these distances, molecules experience strong mutual forces of attraction. In a liquid, molecules can slide past one another while in a solid molecules remain attached to their home positions. In a gas, the distances between molecules are much greater, and forces between them are essentially negligible. Molecules in a gas can roam freely throughout the container.

**CHAPTER 4**
The Gaseous State

**CHAPTER 5**
Solids, Liquids, and Phase Transitions

**CHAPTER 6**
Solutions

THE GOALS OF UNIT 2 ARE  **(1)** To define the essential properties of solids, liquids, and gases; and **(2)** To relate their magnitudes to motions of molecules and to the forces between molecules.

# The Gaseous State

***Illustration***
Gaseous nitrogen dioxide is generated by the reaction of copper with nitric acid. *(Charles D. Winters)*

Gases provide the simplest opportunity for relating macroscopic properties of a substance to the structure and interactions of its molecules. On the macroscopic level, gases are distinguished from liquids and solids by their much smaller values of mass density (conveniently measured in grams per $cm^3$). On the microscopic level, the number density (number of molecules in one $cm^3$ of the sample) is smaller—and the distances between molecules are much greater—than in liquids and solids. Molecules with no net electrical charge exert significant forces on each other only when they are close. Consequently, when studying gases it is a legitimate simplification to ignore interactions between molecules or to consider collisions between at most two molecules at the time.

The central goal of this chapter is to provide a deeper understanding of the behavior of gases as temperature, pressure, and volume are changed. The main theme of the chapter is to develop and apply the ideal gas law, which has been shown by experiment to represent accurately the bulk properties of gases at low density. Corrections to the ideal gas law will be obtained by explicit consideration of collisions between pairs of molecules. The results and insights obtained provide essential background for later study of the rate and extent of chemical reactions that occur through molecular collisions.

---

## 4.1

## THE CHEMISTRY OF GASES

The Greeks considered air to be one of the four fundamental elements in nature, and as early as the 17th century its physical properties (such as resistance to compression) were being studied. The chemical composition of air (Table 4.1) was not appreciated until late in the 18th century when Lavoisier, Priestley, and others showed that air consists primarily of two different substances: oxygen and nitrogen. Oxygen was characterized by its ability to support life. It was recognized that once a volume of oxygen was used up (by burning a candle in a closed container, for example), the nitrogen that remained could no longer keep animals alive. More than 100

---

### TABLE 4.1

### *Composition of Air*[†]

| Constituent | Formula | Fraction by Volume |
| --- | --- | --- |
| Nitrogen | $N_2$ | 0.78110 |
| Oxygen | $O_2$ | 0.20953 |
| Argon | Ar | 0.00934 |
| Carbon dioxide | $CO_2$ | 0.00034 |
| Neon | Ne | $1.82 \times 10^{-5}$ |
| Helium | He | $5.2 \times 10^{-6}$ |
| Methane | $CH_4$ | $1.5 \times 10^{-6}$ |
| Krypton | Kr | $1.1 \times 10^{-6}$ |
| Hydrogen | $H_2$ | $5 \times 10^{-7}$ |
| Dinitrogen oxide | $N_2O$ | $3 \times 10^{-7}$ |
| Xenon | Xe | $8.7 \times 10^{-8}$ |

[†] Air also contains other constituents, the abundances of which are quite variable in the atmosphere. Examples are water ($H_2O$), 0–0.07; ozone ($O_3$), $0–7 \times 10^{-8}$; carbon monoxide (CO), $0–2 \times 10^{-8}$; nitrogen dioxide ($NO_2$), $0–2 \times 10^{-8}$; sulfur dioxide ($SO_2$), $0–1 \times 10^{-6}$.

years elapsed before a careful reanalysis of air showed that oxygen and nitrogen account for only about 99% of the total volume, most of the remaining 1% being a new gas, given the name "argon." The other noble gases (helium, neon, krypton, and xenon) are present in air to lesser extents.

Additional gases are found on the earth's surface. Methane ($CH_4$) is produced by bacterial processes, especially in swampy areas (hence its older name, "marsh gas"). It is a major constituent of natural-gas deposits formed over many millennia by decaying plant matter under the earth's surface. The recovery of methane from municipal landfills for use as a fuel gas is now a commercially feasible process. Gases also form when liquids evaporate. Water vapor in the air from the evaporation of liquid water is the best known example; it causes the humidity of the air.

Gases are also formed by chemical reactions. When heated, some solids decompose to give gaseous products. The decomposition of mercury(II) oxide to mercury and oxygen (Fig. 1.6) has already been mentioned:

$$2\ HgO(s) \xrightarrow{\text{Heat}} 2\ Hg(\ell) + O_2(g)$$

Carbon dioxide, $CO_2$, is dissolved in aqueous solution to carbonate beverages. Carbon dioxide also reacts with water to produce $H_2CO_3(aq)$, which provides some of the acidity in soft drinks. Solid carbon dioxide (dry ice) is used for refrigeration.

This reaction was used by Joseph Priestley in his discovery of oxygen. Even earlier (1756) Joseph Black had shown that marble, which consists primarily of calcium carbonate ($CaCO_3$), dissociates on heating to give quicklime (CaO) and carbon dioxide:

$$CaCO_3(s) \xrightarrow{\text{Heat}} CaO(s) + CO_2(g)$$

Ammonium chloride ($NH_4Cl$) can be decomposed by heat to produce two gases, ammonia and hydrogen chloride:

$$NH_4Cl(s) \xrightarrow{\text{Heat}} NH_3(g) + HCl(g)$$

Some gas-forming reactions proceed explosively. The decomposition of nitroglycerin is a detonation in which all the products are gaseous:

$$4\ C_3H_5(NO_3)_3(\ell) \longrightarrow 6\ N_2(g) + 12\ CO_2(g) + O_2(g) + 10\ H_2O(g)$$

Several elements react with oxygen to form gaseous oxides. Carbon dioxide, for example, forms during animal respiration but is also produced by burning coal, oil, and other materials containing compounds of carbon. Oxides of sulfur are produced by burning elemental sulfur, and those of nitrogen from elemental nitrogen:

$$S(s) + O_2(g) \longrightarrow SO_2(g)$$

$$2\ SO_2(g) + O_2(g) \longrightarrow 2\ SO_3(g)$$

$$N_2(g) + O_2(g) \longrightarrow 2\ NO(g)$$

$$2\ NO(g) + O_2(g) \longrightarrow 2\ NO_2(g)$$

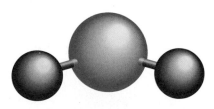

Sulfur dioxide, $SO_2$, is produced by burning sulfur in air, as the first step in the production of sulfuric acid.

The role of these compounds in air pollution will be considered further in Section 16.9.

The actions of acids on certain ionic solids are another important class of reactions that form gases. Carbon dioxide is produced from the reaction of acids with carbonates (Fig. 4.1):

$$CaCO_3(s) + 2\ HCl(g) \longrightarrow CaCl_2(s) + CO_2(g) + H_2O(\ell)$$

Other examples of this type of reaction are

$$Na_2S(s) + 2\ HCl(g) \longrightarrow 2\ NaCl(s) + H_2S(g)$$

$$K_2SO_3(s) + 2\ HCl(g) \longrightarrow 2\ KCl(s) + SO_2(g) + H_2O(\ell)$$

$$NaCl(s) + H_2SO_4(\ell) \longrightarrow NaHSO_4(s) + HCl(g)$$

In these reactions, the metal ions (sodium, calcium, and potassium) play no direct role, and parallel reactions can be written for other metal ions.

The gases just mentioned show extremely varied chemical behavior. Some, such as HCl and SO₃, are very reactive, acidic, and corrosive, whereas others, such as N₂O and nitrogen, are much less reactive. Although the chemical properties of gases vary significantly, their physical properties are much simpler to understand. At sufficiently low densities the physical properties of all gases behave in the same way and approach those of an "ideal" gas, the subject of the following sections.

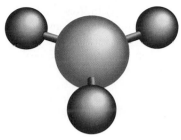

Sulfur trioxide, SO₃, is a corrosive gas produced by the further oxidation of sulfur dioxide.

---

**4.2**

## PRESSURE AND TEMPERATURE OF GASES

Three useful properties for describing gases are volume ($V$), pressure ($P$), and temperature ($T$). Volume is self-evident and needs no comment; as stated in Section 1.6, we shall use the liter as a unit of volume. Pressure and temperature require a little more care to define.

### Pressure

The force exerted by a gas on a unit area of the walls of its container is called the **pressure** of that gas. We do not often stop to think that the air around us exerts a pressure on us and on everything else at the earth's surface. This fact was demonstrated in an ingenious experiment by Evangelista Torricelli (1608–1647), an Italian scientist who had been an assistant to Galileo. He sealed a long glass tube at one end and filled it with mercury. He then covered the open end with his thumb, turned the tube upside down and immersed it in a dish of liquid mercury (Fig. 4.2a), taking care that no air leaked in. The mercury in the tube fell, leaving a nearly perfect vacuum at the closed end, but it did not all run out of the tube. It stopped when its top was about 76 cm above the level of the mercury in the dish. Torricelli showed that the exact height varied somewhat from day to day and from place to place.

This simple device, called a **barometer,** works like a balance, one arm of which is loaded with the mass of mercury in the tube and the other with a column of air of the same cross-sectional area that extends to the top of the earth's atmosphere, approximately 150 km up (Fig. 4.2b). The height of the mercury column adjusts itself so that its mass and that of the air column (and thus the two forces on the surface of the mercury in the dish) become equal. Day-to-day changes in the height of the column occur as the force exerted by the atmosphere varies with the weather. A barometer at a high elevation has a portion of the atmosphere below it and gives a lower reading than one at sea level, if the effect of the weather is excluded.

The relationship between Torricelli's invention and the pressure of the atmosphere can be understood from Newton's second law of motion, which states that

**FIGURE 4.1** The calcium carbonate in a piece of chalk reacts with an aqueous solution of hydrochloric acid to produce bubbles of carbon dioxide. (*Charles D. Winters*)

**FIGURE 4.2** (a) In Torricelli's barometer, the top of the mercury in the tube is approximately 76 cm higher than that in the open beaker. (b) The mass of mercury in the column of height $h$ exactly balances that of a column of air of the same diameter extending to the top of the atmosphere.

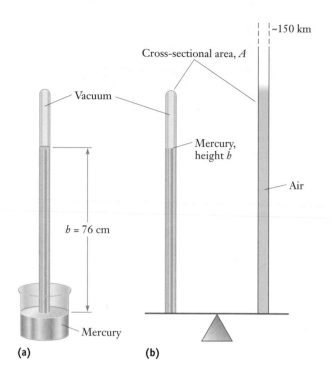

$$\text{force} = \text{mass} \times \text{acceleration}$$

$$F = ma$$

where the acceleration of a body ($a$) is the rate at which its velocity changes. All bodies attract one another because of gravity, and the force of attraction imparts an acceleration to each body. The standard acceleration due to the earth's gravitational field (usually denoted by $g$ instead of $a$) is $g = 9.80665$ m s$^{-2}$. Pressure is the force per unit area, or the total force $F$ divided by the area $A$:

$$P = \frac{F}{A} = \frac{mg}{A}$$

Because the volume of mercury in the barometer is $V = Ah$,

$$P = \frac{mg}{A} = \frac{mg}{V/h} = \frac{mgh}{V}$$

If the density is stated as $\rho = m/V$, this equation can be rewritten:

$$P = \rho g h \qquad\qquad\qquad \textbf{[4.1]}$$

With this equation we can calculate the pressure exerted by the atmosphere. The density of mercury at 0°C, in SI units (see Appendix B), is

$$\rho = 13.5951 \text{ g cm}^{-3} = 1.35951 \times 10^4 \text{ kg m}^{-3}$$

and the height of the mercury column under ordinary atmospheric conditions near sea level is close to 0.76 m (760 mm). Let us use exactly this value for the height in our computation:

$$P = \rho g h = (1.35951 \times 10^4 \text{ kg m}^{-3})(9.80665 \text{ m s}^{-2})(0.760000 \text{ m})$$
$$= 1.01325 \times 10^5 \text{ kg m}^{-1} \text{ s}^{-2}$$

| TABLE 4.2 | |
|---|---|
| **Units for Pressure** | |
| **Unit** | **Definition or Relationship** |
| pascal (Pa) | $1 \text{ kg m}^{-1} \text{ s}^{-2}$ |
| bar | $1 \times 10^5$ Pa |
| atmosphere (atm) | 101,325 Pa |
| torr | 1/760 atm |
| 760 mm Hg (at 0°C) | 1 atm |
| 14.6960 pounds per square inch (psi, lb in$^{-2}$) | 1 atm |

Pressure can be expressed in various units. The natural SI unit is the $\text{kg m}^{-1} \text{ s}^{-2}$, which arose in the preceding calculation and is called the *pascal* (Pa). One **standard atmosphere** (1 atm) is defined as exactly $1.01325 \times 10^5$ Pa. The standard atmosphere is a useful unit because the pascal is inconveniently small and because "atmospheric pressure" is important as a standard of reference. Pressures must be expressed in pascals if calculations are to be carried out entirely in SI units.

Unfortunately, pressure units in common use have proliferated. Although we shall work primarily with the standard atmosphere, it is important to be able to recognize other units and convert among them. Weather reports express atmospheric pressure in terms of the height (in millimeters or inches) of the column of mercury it can support. A pressure of 1 atm supports a 760-mm column of mercury at 0°C, so we often speak of 1 atm pressure as 760 mm (of Hg). Because the density of mercury depends slightly on temperature, for accurate work it is necessary to specify the temperature and make the proper corrections to the density. A more precise term is the *torr*, defined as 1 torr = 1/760 atm (or 760 torr = 1 atm) at *any* temperature. Only at 0°C do the torr and the millimeters of Hg coincide. These and other units of pressure are summarized in Table 4.2.

## Boyle's Law

In the 17th century, Robert Boyle, an English natural philosopher and theologian, studied the properties of confined gases. He noted that a gas tends to spring back to its original volume after being compressed or expanded. These properties are much like those of metal springs, which were being investigated by his collaborator Robert Hooke. Boyle's experiments on the compression and expansion of air were published in 1662 in a monograph, "The Spring of the Air and Its Effects."

Boyle worked with a simple piece of apparatus: a J-tube in which air was trapped at the closed end by a column of mercury (Fig. 4.3). If the difference in height $h$ between the two mercury levels in such a tube is 0, then the pressure of the air in the closed part exactly balances that of the atmosphere, so the pressure $P$ is 1 atm. Adding mercury to the open end of the tube increases the pressure of the confined air by 1 mm Hg for every millimeter by which the level of mercury on the open side of the tube exceeds the level on the closed side. Expressed in atmospheres, the pressure is

$$P = 1 \text{ atm} + \frac{h \text{ (in mm)}}{760 \text{ mm atm}^{-1}}$$

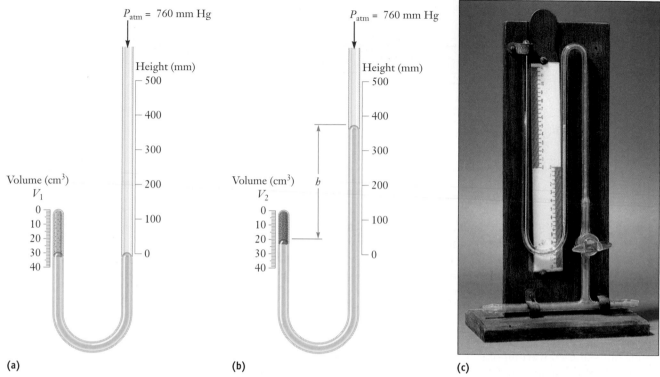

**FIGURE 4.3** (a) Boyle's J-tube. When the heights of mercury on the two sides of the tube are the same, the pressure of the confined gas must equal that of the atmosphere— 1 atm or 760 mm Hg. (b) After mercury has been added, the pressure of the gas is increased by the number of millimeters of mercury in the height difference $h$. The compression of the gas causes it to occupy a smaller volume. (c) The same principle applies in a manometer, used to measure the pressure of a gas. The glass tube at the bottom is connected to the gas sample, and the difference in height between the levels of the mercury in the two arms is read off on a calibrated scale. (*c, Leon Lewandowski*)

The volume of the confined air can be read off from the previously calibrated tube. Boyle discovered that the product of pressure and volume, $PV$, is a constant $C$ at a given temperature for a sample of gas:

$$PV = C \qquad \text{(fixed temperature and fixed amount of gas)} \qquad \textbf{[4.2]}$$

This result is known as **Boyle's law.**

The equation $P = C/V$ is the equation of a hyperbola, shown in Figure 4.4a. Many curves that look like hyperbolas in fact are not. How are we to know whether Boyle's experiments are best described by $PV = C$ and not some more complicated law? There are two ways of plotting the data to answer this question. The first is to rewrite Boyle's law in the form

$$P = \frac{C}{V} = C\left(\frac{1}{V}\right)$$

and to note that $P$ is proportional to $1/V$. A graph of $P$ against $1/V$ (rather than $V$) should be a straight line passing through the origin with slope $C$ (see Fig. 4.4b).

Alternatively, because $PV = C$, a plot of $PV$ against $P$ should be independent of $P$, as shown in Figure 4.4c.

It is important to remember that the constant $C$ depends on temperature and on the amount (number of moles) of gas used. At 0°C and for 1 mol of gas (e.g., 31.999 g $O_2$, 28.013 g $N_2$, or 2.0159 g $H_2$), an extrapolation of $PV$ to zero pressure (as shown in Fig. 4.4c) gives a limiting value of 22.414 L atm for $C$, so that

$$PV = 22.414 \text{ L atm} \qquad \text{(for one mole of gas at 0°C)}$$

If the pressure is 1.00 atm, the volume should be 22.4 L; if $P$ is 4.00 atm, $V$ should be 22.414/4.00 = 5.60 L. Boyle's law is an idealization that is satisfied exactly by all gases at *very* low pressures; for real gases near 1 atm pressure, small corrections may be necessary for highly accurate work. For high-pressure gases (beyond 50 to 100 atm), substantial corrections are needed.

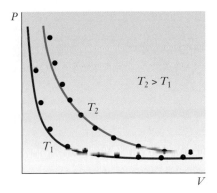

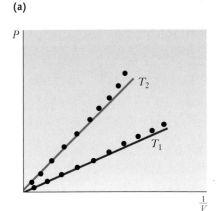

## EXAMPLE 4.1

The long cylinder of a bicycle pump has a volume of 1131 cm$^3$ and is filled with air at a pressure of 1.02 atm. The outlet valve is sealed shut and the pump handle is pushed down until the volume of the air is 517 cm$^3$. Compute the pressure inside the pump, and express it in atmospheres and pounds per square inch.

### Solution

Note that the temperature and amount of gas are not stated in this problem, so the value of 22.414 L atm cannot be used for the constant $C$. All that is necessary, however, is to assume that the temperature does not change as the pump handle is pushed down. If $P_1$ and $P_2$ are the initial and final pressures, and $V_1$ and $V_2$ the initial and final volumes, then

$$P_1V_1 = P_2V_2$$

because the temperature and amount of air in the pump do not change. Substitution gives

$$(1.02 \text{ atm})(1131 \text{ cm}^3) = P_2(517 \text{ cm}^3)$$

which can be solved for $P_2$:

$$P_2 = 2.23 \text{ atm}$$

In pounds per square inch (Table 4.2), this pressure is

$$P_2 = 2.23 \text{ atm} \times 14.696 \text{ psi atm}^{-1} = 32.8 \text{ psi}$$

**Related Problems: 11, 12**

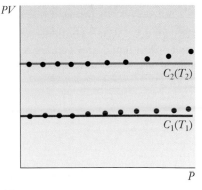

**FIGURE 4.4** Three ways of depicting Boyle's law. Small deviations from the law (especially apparent in parts b and c) arise for real gases at higher pressures or smaller volumes.

## Temperature and Charles's Law

**Temperature** is one of those elusive properties that all of us think we understand but that are very difficult to pin down in a quantitative fashion. We have an instinctive feeling (through the sense of touch) of *hot* and *cold*; for example, water at its freezing point is colder than at its boiling point and thus should have a lower temperature. On the Celsius scale of temperature, the freezing point is assigned a temperature of 0°C and the boiling point (at 1 atm pressure) is 100°C. On the Fahrenheit scale the same two temperatures are 32°F and 212°F.

**FIGURE 4.5** When a balloon is cooled with liquid nitrogen, its volume shrinks drastically as the gas inside cools. *(Leon Lewandowski)*

Assigning two fixed points in this way does not resolve the problem of defining a temperature scale. Ether, for example, boils at atmospheric pressure somewhere between 0 and 100°C, but what temperature should be assigned to its boiling point? Further arbitrary choices are certainly not the answer. The problem is that temperature is not a mechanical quantity like pressure, and thus is more difficult to define.

There are mechanical properties that *depend* on temperature, however. For example, as liquid mercury is heated from 0°C to 100°C its volume increases by 1.82%. This change in volume could be used as the basis for the definition of a temperature scale. Temperature could be *assumed* to be linearly related to the volume of mercury so that temperatures could be read off from the volume of mercury in a tube (a mercury thermometer). The problem with this definition, of course, is that it is tied to the properties of a single substance, mercury. If the volume of some other substance is measured on the mercury scale, it will *not* be a linear function of temperature. Why should mercury be chosen rather than another substance? How can temperature be defined below the freezing point or above the boiling point of mercury? To answer this question, let us examine the behavior of gases upon heating or cooling.

Boyle observed that the product of the pressure and volume of a confined gas changes upon heating and thus depends on temperature (Fig. 4.5), but the first quantitative experiments were carried out by the French scientist Jacques Charles more than a century later. The crucial physical content of Charles's observations is that, at sufficiently low pressures, *all* gases expand by the same relative amount if they have equal initial and final temperatures. For example, heating a sample of $N_2$ from the freezing point of water to the boiling point causes the gas to expand to 1.366 times its original volume; the same 36.6% increase in volume is found for $O_2$, $CO_2$, and other gases. (In contrast, liquids and solids vary widely in thermal expansion.) This universal behavior suggests that temperature can be expressed as a linear function of gas volume. We therefore write

$$t = c\left(\frac{V}{V_0} - 1\right)$$

where $V$ is the volume of the gas at temperature $t$, $V_0$ is its volume at the freezing point of water, and $c$ is a constant, the same for all gases. We have written the linear equation in this form to ensure that the freezing point of water (when $V = V_0$) will be at $t = 0$, corresponding to the zero of the Celsius scale of temperature. From the measured fractional increase in $V$ to the boiling point (taken to be 100°C), the value of $c$ can be determined. In 1802, Gay-Lussac reported a value for $c$ of 267°C. Subsequent experiments have refined this result to give $c = 273.15°C$. The definition of temperature (in degrees Celsius) is then

$$t = 273.15°C\left(\frac{V}{V_0} - 1\right)$$

The temperature of a gas can be measured by taking a gas sample at low pressure and comparing its volume with that at the freezing point of water. For many gases atmospheric pressure is sufficiently low, but for highly accurate temperature determinations it is necessary to use lower pressures or to apply small corrections.

Once temperature has been defined in this way, we can return to mercury and measure its *actual* changes in volume with temperature. The result found is almost, but not quite, linear. For example, if a mercury thermometer is calibrated to match the gas thermometer at 0°C and 100°C and if the scale in between is divided evenly into 100 parts to mark off degrees, a small error will result from the use of this thermometer. A temperature of 40.00°C on the gas thermometer will be read as 40.11°C on the mercury thermometer, which indicates that the volume of liquid mercury does not vary exactly linearly with temperature.

The preceding equation can be rewritten to express the gas volume in terms of the temperature. The result is

$$V = V_0\left(1 + \frac{t}{273.15°C}\right) \qquad \text{[4.3]}$$

In words, the volume of a gas varies linearly with its temperature. This is the most common statement of **Charles's law,** but it is misleading in a way because the linearity is built in through the definition of temperature. The key observation is the universal nature of the constant 273.15°C, which is the same for all gases at low pressures. Writing Charles's law in this form suggests a very interesting lower limit to the temperatures that can be obtained. Negative temperatures on the Celsius scale correspond to temperatures below the freezing point of water and of course are meaningful. But what would happen as $t$ approached −273.15°C? The volume would then approach 0 (Fig. 4.6), and if $t$ could go below this value the volume would become negative, clearly an impossible result. We therefore surmise that $t = -273.15°C$ is a fundamental limit below which the temperature cannot be lowered. All real gases condense to a liquid or solid before they reach this **absolute zero** of temperature, but more rigorous arguments show that *no* substance (gas, liquid, or solid) can be cooled below −273.15°C. In fact, it becomes increasingly difficult to cool a substance as absolute zero is approached. The coldest temperatures reached to date are on the order of several microdegrees (1 microdegree = $10^{-6}$ °C) above absolute zero.

The absolute zero of temperature is a compellingly logical choice as the zero point of a temperature scale. The easiest way to create such a new scale is to add 273.15 to the Celsius temperature. This gives the **Kelvin temperature scale:**

$$T \text{ (Kelvin)} = 273.15 + t \text{ (Celsius)} \qquad \text{[4.4]}$$

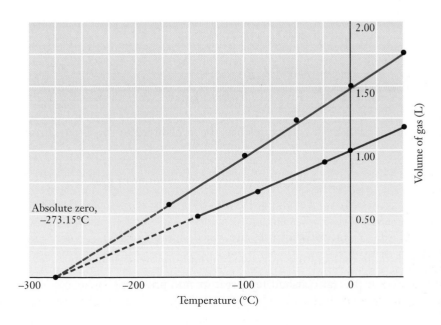

**FIGURE 4.6** The volume of a sample of a gas is a function of its temperature under conditions of constant pressure. The observed straight-line response of volume to temperature illustrates Charles's law. A particular sample of a gas (red line) has a volume of 1.0 L at a temperature of 0°C. Another sample of a gas (blue line) held at the same pressure takes up more volume at 0°C but shrinks faster as cooled. Its *percentage* change in volume is the same as that of the first sample for every degree of temperature change. Extrapolation of the trends (dashed lines) predicts that the volumes of the samples go to zero at a temperature of −273.15°C. Similar observations are made irrespective of the chemical identity of the gas.

The capital $T$ signifies that this is an absolute scale, whose unit is the *kelvin* (K). Thus, a temperature of 25.00°C corresponds to $273.15 + 25.00 = 298.15$ K. On this scale, Charles's law takes the form

$$V \propto T \qquad \text{(fixed pressure and fixed amount of gas)}$$

where the proportionality constant is determined by the pressure and the amount of gas present. The ratios of volumes occupied at two different temperatures by a fixed amount of gas at fixed pressure follow:

$$\frac{V_1}{V_2} = \frac{T_1}{T_2}$$

---

**EXAMPLE 4.2**

A scientist studying the properties of hydrogen at low temperature takes a volume of 2.50 L of hydrogen at atmospheric pressure and a temperature of 25.00°C and cools the gas at constant pressure to −200.00°C. Predict the volume of the hydrogen.

**Solution**

The first step is always to convert temperatures to kelvins:

$$t_1 = \quad 25.00°\text{C} \implies T_1 = 273.15 + 25.00 = 298.15 \text{ K}$$

$$t_2 = -200.00°\text{C} \implies T_2 = 273.15 - 200.00 = 73.15 \text{ K}$$

The ratio in Charles's law is

$$\frac{V_1}{T_1} = \frac{2.50 \text{ L}}{298.15 \text{ K}} = \frac{V_2}{T_2} = \frac{V_2}{73.15 \text{ K}}$$

$$V_2 = \frac{(73.15 \text{ K})(2.50 \text{ L})}{298.15 \text{ K}} = 0.613 \text{ L}$$

**Related Problems: 13, 14, 15, 16, 17, 18**

---

**4.3**

# THE IDEAL GAS LAW

Several properties of gases at low pressures have been deduced to this point. From Boyle's law,

$$V \propto \frac{1}{P} \qquad \text{(at constant temperature, fixed amount of gas)}$$

and from Charles's law,

$$V \propto T \qquad \text{(at constant pressure, fixed amount of gas)}$$

where $T$ is the absolute temperature in kelvins. In addition, it is clear that

$$V \propto n \qquad \text{(at constant temperature and pressure)}$$

where $n$ is the chemical amount of substance (in moles), because at a given temperature and pressure, doubling the amount of *any* substance (gas, liquid, or solid) must double the volume.

These three statements may be combined in the form

$$V \propto \frac{nT}{P}$$

A proportionality constant called $R$ converts this proportionality to an equation:

$$V = R\frac{nT}{P} \qquad \text{or} \qquad \boxed{PV = nRT} \qquad \text{[4.5]}$$

This is the **ideal gas law,** an empirical law that holds approximately for all gases near atmospheric pressure and that becomes increasingly accurate as the pressure is decreased.

Situations frequently arise in which a gas undergoes a change that takes it from some initial condition (described by $P_1$, $V_1$, $T_1$, and $n_1$) to a final condition (described by $P_2$, $V_2$, $T_2$, and $n_2$). Because $R$ is a constant,

$$\frac{P_1 V_1}{n_1 T_1} = \frac{P_2 V_2}{n_2 T_2} \qquad \text{[4.6]}$$

This is a useful alternative form of the ideal gas law.

## EXAMPLE 4.3

A weather balloon filled with helium has a volume of $1.0 \times 10^4$ L at 1.00 atm and 30°C. It rises to an altitude at which the pressure has dropped to 0.60 atm and the temperature is $-20$°C. What is the volume of the balloon then? Assume that the balloon stretches in such a way that the pressure inside stays close to the pressure outside.

### Solution

Because the amount of helium does not change, we can set $n_1$ equal to $n_2$ and cancel it out of Equation 4.6, giving

$$\frac{P_1 V_1}{T_1} = \frac{P_2 V_2}{T_2}$$

Solving this for the only unknown quantity, $V_2$, gives

$$V_2 = V_1\left(\frac{P_1}{P_2}\right)\left(\frac{T_2}{T_1}\right)$$
$$= 1.0 \times 10^4 \text{ L}\left(\frac{1.00 \text{ atm}}{0.60 \text{ atm}}\right)\left(\frac{253 \text{ K}}{303 \text{ K}}\right)$$
$$= 1.4 \times 10^4 \text{ L} = 14,000 \text{ L}$$

Note that temperatures must be converted to kelvins before the ideal gas law is used.

**Related Problems: 19, 20, 21, 22**

The approach outlined in Example 4.3 can be applied when other combinations of the four variables ($P$, $V$, $T$, and $n$) remain constant (Fig. 4.7). For example, when $n$

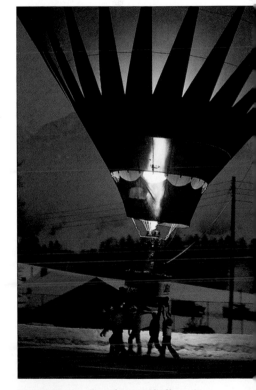

**FIGURE 4.7** In a hot-air balloon, the volume remains nearly constant (the balloon is rigid) and the pressure is nearly constant as well (unless the balloon rises very high). Thus, $n$ is inversely proportional to $T$; as the air inside the balloon is heated, its amount decreases and its density falls. This reduced density gives the balloon its lift. (*Copyright Bernard Asset/Photo Researchers, Inc.*)

and $T$ remain constant, they can be canceled from the equation and Boyle's law, which was used in solving Example 4.1, results. When $n$ and $P$ remain constant, this relationship reduces to Charles's law, used in Example 4.2.

We are not quite finished. For a given gas, $PV/nT$ is a constant, called $R$, but it has not yet been established whether $R$ is a *universal* constant or whether it changes from one gas to another (e.g., from oxygen to nitrogen to hydrogen). To deal with this issue, we invoke Avogadro's hypothesis, whose validity is established by the success of the laws of chemical combination considered in Chapter 1. According to this hypothesis, fixed volumes of different gases at the same temperature and pressure contain the same number of molecules (and therefore of moles). In other words, for fixed $V$, $P$, and $T$, the value of $n$ is fixed as well and is independent of the particular gas studied. Therefore, $R$ is a **universal gas constant** that is the same for all gaseous substances.

The numerical value of $R$ depends on the units chosen for $P$ and $V$. At the freezing point of water ($T = 273.15$ K) the product $PV$ for 1 mol of any gas approaches the value 22.414 L atm at low pressures; hence, $R$ has the value

$$R = \frac{PV}{nT} = \frac{22.414 \text{ L atm}}{(1.0000 \text{ mol})(273.15 \text{ K})} = 0.082058 \text{ L atm mol}^{-1} \text{ K}^{-1}$$

If $P$ is measured in SI units of pascals (kg m$^{-1}$ s$^{-2}$) and $V$ in cubic meters, then $R$ has the value

$$R = \frac{(1.01325 \times 10^5 \text{ kg m}^{-1} \text{ s}^{-2})(22.414 \times 10^{-3} \text{ m}^3)}{(1.0000 \text{ mol})(273.15 \text{ K})}$$

$$= 8.3145 \text{ kg m}^2 \text{ s}^{-2} \text{ mol}^{-1} \text{ K}^{-1}$$

Because 1 kg m$^2$ s$^{-2}$ is defined to be one *joule* (the SI unit of energy, abbreviated J), the gas constant is

$$R = 8.3145 \text{ J mol}^{-1} \text{ K}^{-1}$$

## EXAMPLE 4.4

What mass of helium is needed to fill the weather balloon from Example 4.3?

### Solution
First, solve the ideal gas law for $n$:

$$n = \frac{PV}{RT}$$

If $P$ is expressed in atmospheres and $V$ is expressed in liters, then the value $R = 0.08206$ L atm mol$^{-1}$ K$^{-1}$ must be used.

$$n = \frac{(1.00 \text{ atm})(1.00 \times 10^4 \text{ L})}{(0.08206 \text{ L atm mol}^{-1} \text{ K}^{-1})(303.15 \text{ K})} = 402 \text{ mol}$$

Because the molar mass of helium is 4.00 g mol$^{-1}$, the mass of helium required is

$$(402 \text{ mol})(4.00 \text{ g mol}^{-1}) = 1610 \text{ g} = 1.61 \text{ kg}$$

**Related Problems: 23, 24**

## Chemical Calculations for Gases

One of the most important applications of the gas laws in chemistry is in the calculation of volumes of gases consumed or produced in chemical reactions. If the conditions of pressure and temperature are known, then the ideal gas law can be used to convert between chemical amount and gas volume. Instead of working with the mass of each gas taking part in the reaction, we can then use its volume, which is easier to measure. This is illustrated by the following example.

---

**EXAMPLE 4.5**

Concentrated nitric acid acts on copper to give nitrogen dioxide and dissolved copper ions (Fig. 4.8) according to the balanced chemical equation

$$Cu(s) + 4\ H^+(aq) + 2\ NO_3^-(aq) \longrightarrow 2\ NO_2(g) + Cu^{2+}(aq) + 2\ H_2O(\ell)$$

Suppose that 6.80 g of copper is consumed in this reaction and that the $NO_2$ is collected at a pressure of 0.970 atm and a temperature of 45°C. Calculate the volume of $NO_2$ produced.

### Solution

The first step (as in Fig. 2.3) is to convert from the mass of the known reactant or product (in this case, 6.80 g of Cu) to its chemical amount by using the molar mass of copper, 63.55 g mol$^{-1}$:

$$\frac{6.80\ \text{g Cu}}{63.55\ \text{g mol}^{-1}} = 0.107\ \text{mol Cu}$$

Next, the chemical amount of $NO_2$ generated in the reaction is found by forming the "chemical conversion factor" 2 mol of $NO_2$ per 1 mol of Cu from the balanced equation and multiplying the chemical amount of Cu by it:

$$0.107\ \text{mol Cu} \times \left(\frac{2\ \text{mol } NO_2}{1\ \text{mol Cu}}\right) = 0.214\ \text{mol } NO_2$$

Finally, the ideal gas law is used to find the volume from the chemical amount (recall that the temperature must first be expressed in kelvins by adding 273.15):

$$V = \frac{nRT}{P} = \frac{(0.214\ \text{mol})(0.08206\ \text{L atm mol}^{-1}\ \text{K}^{-1})(273 + 45\ \text{K})}{0.970\ \text{atm}} = 5.76\ \text{L}$$

Therefore, 5.76 L of $NO_2$ is produced under these conditions.

**Related Problems: 25, 26, 27, 28, 29, 30, 31, 32**

---

**FIGURE 4.8**  When concentrated nitric acid is added to a copper penny, the copper is oxidized and an aqueous solution of blue copper(II) nitrate forms. In addition, some of the nitrate ion is reduced to brown gaseous nitrogen dioxide, which bubbles off. *(Leon Lewandowski)*

---

**4.4**

## MIXTURES OF GASES

Suppose a mixture of gases occupies a container at a certain temperature. We define the **partial pressure** of each gas as the pressure that gas would exert if it alone were present in the container. **Dalton's law** (named after its discoverer, John Dalton) then states that the total pressure measured, $P_{\text{tot}}$, is the sum of the partial pressures of

**FIGURE 4.9** According to Dalton's law, the total pressure of a gas mixture is the sum of the pressures exerted by the individual gases. Note that the total volume is the same.

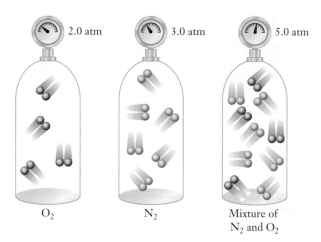

the individual gases (Fig. 4.9). This law holds under the same conditions as the ideal gas law itself: it is approximate at moderate pressures, but becomes increasingly more accurate as the pressure is lowered.

For a gas mixture at low pressure, the partial pressure of one component, A, is

$$P_A = n_A \frac{RT}{V}$$

The total pressure is the sum of the partial pressures:

$$P_{tot} = P_A + P_B + P_C + \cdots = (n_A + n_B + n_C + \cdots)\frac{RT}{V} = n_{tot}\frac{RT}{V} \qquad [4.7]$$

where $n_{tot}$ is the total number of moles in the gas mixture. Dividing the first equation by the second gives

$$\frac{P_A}{P_{tot}} = \frac{n_A}{n_{tot}} \qquad \text{or} \qquad P_A = \frac{n_A}{n_{tot}} P_{tot}$$

We define

$$X_A = \frac{n_A}{n_{tot}}$$

as the **mole fraction** of A in the mixture—the number of moles of A divided by the total number of moles present. Then

$$P_A = X_A P_{tot} \qquad [4.8]$$

The partial pressure of any component in a mixture of ideal gases is the total pressure multiplied by the mole fraction of that component. Note that the volume fractions in Table 4.1 are the same as mole fractions.

---

### E X A M P L E  4.6

When $NO_2$ is cooled to room temperature, some of it reacts to form a *dimer*, $N_2O_4$, through the reaction

$$2\,NO_2 \longrightarrow N_2O_4$$

Suppose 15.2 g of $NO_2$ is placed in a 10.0-L flask at high temperature and the flask is cooled to 25°C. The total pressure is measured to be 0.500 atm. What partial pressures and mole fractions of $NO_2$ and $N_2O_4$ are present?

## Solution

Initially there is 15.2 g/46.01 g mol$^{-1}$ = 0.330 mol $NO_2$, and therefore the same number of moles of nitrogen atoms. If at 25°C there are $n_{NO_2}$ moles of $NO_2$ and $n_{N_2O_4}$ moles of $N_2O_4$, then, because the total number of moles of nitrogen atoms is unchanged,

**(a)**
$$n_{NO_2} + 2\,n_{N_2O_4} = 0.330 \text{ mol}$$

To find a second relation between $n_{NO_2}$ and $n_{N_2O_4}$, use Dalton's law:

$$P_{NO_2} + P_{N_2O_4} = 0.500 \text{ atm}$$

$$\frac{RT}{V}\,n_{NO_2} + \frac{RT}{V}\,n_{N_2O_4} = 0.500 \text{ atm}$$

$$n_{NO_2} + n_{N_2O_4} = 0.500 \text{ atm}\,\frac{V}{RT}$$

$$= \frac{(0.500 \text{ atm})(10.0 \text{ L})}{(0.08206 \text{ L atm mol}^{-1}\text{ K}^{-1})(298 \text{ K})}$$

**(b)**
$$n_{NO_2} + n_{N_2O_4} = 0.204 \text{ mol}$$

Subtracting Equation b from Equation a gives

$$n_{N_2O_4} = 0.126 \text{ mol}$$

$$n_{NO_2} = 0.078 \text{ mol}$$

From these results, $NO_2$ has a mole fraction of 0.38 and a partial pressure of (0.38)(0.500 atm) = 0.19 atm, and $N_2O_4$ has a mole fraction of 0.62 and a partial pressure of (0.62)(0.500 atm) = 0.31 atm.

**Related Problems: 33, 34, 35, 36, 37, 38**

---

## 4.5

## THE KINETIC THEORY OF GASES

The ideal gas law summarizes certain physical properties of gases at low pressures. It is an empirical law, the consequence of experimental observations, but its simplicity and generality prompt us to ask whether it has some underlying microscopic explanation that involves the properties of atoms and molecules in a gas. Such an explanation must allow other properties of gases at low pressures to be predicted and must explain why real gases deviate from the ideal gas law to small but measurable extents. Such a theory was developed in the 19th century, notably by the physicists Rudolf Clausius, James Clerk Maxwell, and Ludwig Boltzmann. The **kinetic theory of gases** is one of the great milestones of science, and its success provides strong evidence for the atomic theory of matter discussed in Chapter 1.

This section involves a type of reasoning different from that used so far. Instead of proceeding from experimental observations to empirical laws, we shall begin with a model and, using basic laws of physics with mathematical reasoning, show how this model helps to explain the measured properties of gases. Boyle's law is identified with the result of such reasoning, and a connection is established between temperature and the distribution of speeds in a group of gas molecules.

The underlying assumptions of the kinetic theory of gases are very simple:

**1.** A pure gas consists of a large number of identical molecules separated by distances that are great compared with their size.
**2.** The gas molecules are constantly moving in random directions with a distribution of speeds.
**3.** The molecules exert no forces on one another between collisions, so between collisions they move in straight lines with constant velocities.
**4.** The collisions of molecules with the walls of the container are elastic; no energy is lost during a collision.

## The Meaning of Temperature

We first use the kinetic theory of gases to find a relationship among pressure, volume, and the motions of molecules in an ideal gas. Comparison with the ideal gas law ($PV = nRT$) gives a deeper understanding of the meaning of temperature.

Suppose a container has the shape of a rectangular box of length $\ell$, with end faces each of which has area $A$ (see Fig. 4.10). A single molecule that is moving with speed $u$ in some direction is placed in the box. It is important to distinguish between **speed** and **velocity.** The velocity of a molecule specifies both the rate at which it is moving (its speed, in meters per second) and the direction of motion. As shown in Figure 4.11, the velocity can be indicated by an arrow (vector) $\vec{v}$, which has a length equal to the speed $u$ and which points in the direction of motion of the molecule. The velocity can also be represented by its components along three coordinate axes: $v_x$, $v_y$, and $v_z$. These are related to the speed $u$ by the Pythagorean theorem:

$$u^2 = v_x^2 + v_y^2 + v_z^2$$

The **momentum** of a molecule, $\vec{p}$, is its velocity multiplied by its mass. When the molecule collides elastically with a wall of the box, such as one of the end faces of area $A$, the $y$ and $z$ components of the velocity, $v_y$ and $v_z$, are unchanged, but the $x$ component (perpendicular to $A$) reverses sign (Fig. 4.12). The change in the $x$ component of the momentum of the molecule, $\Delta p_{x,\text{mol}}$, is

$$\Delta p_{x,\text{mol}} = \text{final momentum} - \text{initial momentum} = m(-v_x) - mv_x = -2\,mv_x$$

However, the total momentum of the system (molecule plus box) must be conserved, so this momentum change is balanced by an equal and opposite momentum change given to the wall:

$$\Delta p_{x,\text{wall}} = 2\,mv_x$$

After this collision with the wall, the molecule reverses direction, strikes the opposite face of the box, and then again approaches the original face. In between, it may strike the top, bottom, and sides. These collisions do not change $v_x$, however, so they do not affect the time between collisions with the original end face (see Fig. 4.10). The distance traveled in the $x$ direction is $2\ell$ and the magnitude of the velocity component in this direction is $v_x$, so the time elapsed between collisions with this end face is

$$\Delta t = \frac{2\ell}{v_x}$$

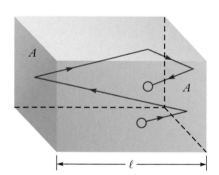

**FIGURE 4.10** The path of a molecule in a box.

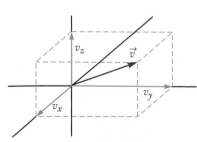

**FIGURE 4.11** The velocity is shown by an arrow of length $v$. It can be separated into three components, $v_x$, $v_y$, and $v_z$, along the three Cartesian coordinate axes.

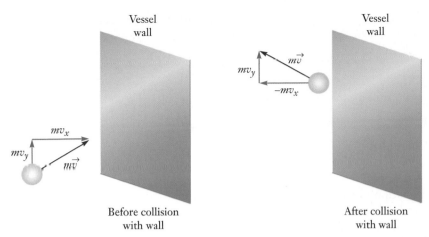

Vessel wall

Vessel wall

Before collision with wall

After collision with wall

**FIGURE 4.12**   An elastic collision of a molecule with a wall. The component of the molecule's momentum perpendicular to the wall reverses sign, from $mv_x$ to $-mv_x$. The component parallel to the wall, $mv_y$, is unchanged. The total momentum is shown by the red arrow. Although its direction is changed by the collision, its length is not, reflecting the fact that molecular speeds are unaffected by elastic wall collisions.

The momentum transferred to the wall per second is the momentum change per collision divided by $\Delta t$:

$$\frac{\Delta p_{x,\text{wall}}}{\Delta t} = \frac{2mv_x}{2\ell/v_x} = \frac{mv_x^2}{\ell}$$

From Newton's second law, this is the force exerted on the original face by the repeatedly colliding molecule:

$$f = ma = m\frac{\Delta v}{\Delta t} = \frac{\Delta p}{\Delta t} = \frac{mv_x^2}{\ell}$$

Suppose now that a large number, $N$, of molecules of mass $m$ are moving independently in the box with $x$ components of velocity $v_{x1}$, $v_{x2}$, $v_{x3}$, and so forth. Then the total force exerted on the face by the $N$ molecules is the sum of the forces exerted by the individual molecules:

$$F = \frac{mv_{x1}^2}{\ell} + \frac{mv_{x2}^2}{\ell} + \cdots + \frac{mv_{xN}^2}{\ell} = \frac{Nm}{\ell}\,\overline{v_x^2}$$

where

$$\overline{v_x^2} = \frac{1}{N}(v_{x1}^2 + v_{x2}^2 + \cdots + v_{xN}^2)$$

Here $\overline{v_x^2}$ is the average of the square of the $x$ component of the velocity of the $N$ molecules, obtained by summing $v_x^2$ for the $N$ molecules and dividing by $N$. The pressure is the total force on the wall divided by the area $A$, so

$$P = \frac{F}{A} = \frac{Nm}{A\ell}\,\overline{v_x^2}$$

Because $A\ell$ is the volume $V$ of the box, we conclude that

$$PV = Nm\,\overline{v_x^2}$$

The gas molecules have no preferred direction of motion, so $\overline{v_x^2}$, $\overline{v_y^2}$, and $\overline{v_z^2}$ should be equal to each other. We conclude that

$$\overline{u^2} = \overline{v_x^2} + \overline{v_y^2} + \overline{v_z^2} = 3\overline{v_x^2}$$

so

$$PV = \tfrac{1}{3}Nm\overline{u^2} \qquad\qquad [4.9]$$

where $\overline{u^2}$ is the **mean-square speed** of the gas molecules. From the ideal gas law,

$$PV = nRT$$

so we conclude that

$$\tfrac{1}{3}Nm\overline{u^2} = nRT$$

With this equation our major goal has been achieved: a relation between speeds of molecules and temperature. It can be simplified, however. The equation has the number of molecules, $N$, on the left and the number of moles, $n$, on the right. Because $N$ is just $n$ multiplied by Avogadro's number $N_0$, we can divide both sides by $n$ to find

$$\tfrac{1}{3}N_0 m\overline{u^2} = RT \qquad\qquad [4.10]$$

Let us examine this equation in two ways. First, we note that the kinetic energy of a molecule of mass $m$ moving at speed $u$ is equal to $\tfrac{1}{2}mu^2$, so the *average* kinetic energy of $N_0$ molecules (1 mol) is $\tfrac{1}{2}N_0 m\overline{u^2}$. This is exactly the left side of the equation, with the factor $\tfrac{1}{2}$ instead of $\tfrac{1}{3}$:

$$\text{kinetic energy per mole} = \tfrac{1}{2}N_0 m\overline{u^2} = \tfrac{3}{2} \times (\tfrac{1}{3}N_0 m\overline{u^2}) = \tfrac{3}{2}RT$$

The average kinetic energy of the molecules of a gas depends only on the temperature and is independent of the mass of the molecules or their density.

A second way to look at the equation is to recall that if $m$ is the mass of a single molecule, then $N_0 m$ is the mass of 1 mol of molecules—the molar mass, abbreviated $\mathcal{M}$. Solving the equation for the mean-square speed we find:

$$\overline{u^2} = \frac{3RT}{\mathcal{M}} \qquad\qquad [4.11]$$

The higher the temperature and the lighter the molecules, the greater the mean-square speed.

## Distribution of Molecular Speeds

We define the **root-mean-square speed** $u_{\text{rms}}$ as the square root of the mean-square speed $3RT/\mathcal{M}$:

$$u_{\text{rms}} = \sqrt{\overline{u^2}} = \sqrt{\frac{3RT}{\mathcal{M}}}$$

This equation should be used only when all the terms in it are expressed in self-consistent units such as SI units. The value used for $R$ will then be

$$R = 8.3145 \text{ J mol}^{-1}\text{ K}^{-1} = 8.3145 \text{ kg m}^2 \text{ s}^{-2}\text{ mol}^{-1}\text{ K}^{-1}$$

Molar masses $\mathcal{M}$ must be converted to *kilograms* per mole for use in the equation. The final result will then be expressed in the expected SI unit of speed, meters per second.

## EXAMPLE 4.7

Calculate $u_{rms}$ for (a) a helium atom, (b) an oxygen molecule, and (c) a xenon atom at 298 K.

### Solution

Because the factor $3RT$ will appear in all three expressions for $u_{rms}$, let us calculate it first:

$$3RT = (3)(8.3145 \text{ kg m}^2 \text{ s}^{-2} \text{ mol}^{-1} \text{ K}^{-1})(298 \text{ K}) = 7.43 \times 10^3 \text{ kg m}^2 \text{ s}^{-2} \text{ mol}^{-1}$$

The molar masses of He, $O_2$, and Xe are 4.00 g mol$^{-1}$, 32.00 g mol$^{-1}$, and 131.3 g mol$^{-1}$, respectively. Convert them to $4.00 \times 10^{-3}$ kg mol$^{-1}$, $32.00 \times 10^{-3}$ kg mol$^{-1}$, and $131.3 \times 10^{-3}$ kg mol$^{-1}$, and insert them together with the value for $3RT$ into the formula for $u_{rms}$:

$$u_{rms}(\text{He}) = \sqrt{\frac{7.43 \times 10^3 \text{ kg m}^2 \text{ s}^{-2} \text{ mol}^{-1}}{4.00 \times 10^{-3} \text{ kg mol}^{-1}}} = 1360 \text{ m s}^{-1}$$

$$u_{rms}(\text{O}_2) = \sqrt{\frac{7.43 \times 10^3 \text{ kg m}^2 \text{ s}^{-2} \text{ mol}^{-1}}{32.00 \times 10^{-3} \text{ kg mol}^{-1}}} = 482 \text{ m s}^{-1}$$

$$u_{rms}(\text{Xe}) = \sqrt{\frac{7.43 \times 10^3 \text{ kg m}^2 \text{ s}^{-2} \text{ mol}^{-1}}{131.3 \times 10^{-3} \text{ kg mol}^{-1}}} = 238 \text{ m s}^{-1}$$

At the same temperature the He, $O_2$, and Xe molecules all have the same average kinetic energy; lighter molecules move faster to compensate for their lesser masses. The average molecule moves along quite rapidly at room temperature. These rms speeds convert to 3050, 1080, and 532 mph, respectively.

**Related Problems: 41, 42, 43, 44**

The root-mean-square speed gives some idea of the typical speed of molecules in a gas, but it would be useful to have a more complete picture of the entire distribution of molecular speeds. In particular, we should like to know the fraction of molecules, $\Delta N/N$, that have speeds between $u$ and $u + \Delta u$. This fraction is given by the speed distribution function $f(u)$:

$$\frac{\Delta N}{N} = f(u) \, \Delta u$$

The function $f(u)$ was determined independently by Maxwell and Boltzmann for a gas of molecules of mass $m$ at temperature $T$. It is called the **Maxwell–Boltzmann speed distribution** and has the form

$$f(u) = 4\pi \left(\frac{m}{2\pi k_B T}\right)^{3/2} u^2 \exp\left(-mu^2/2k_B T\right) \qquad \text{[4.12]}$$

where **Boltzmann's constant** $k_B$ is equal to $R/N_0$. This distribution is plotted in Figure 4.13 for several temperatures. As the temperature is raised, the entire distribution of molecular speeds shifts toward higher values. Very few molecules have either very low or very high speeds, so $f(u)$ is small in these limits and it has a maximum at some intermediate speed. The speed distribution of the molecules in a gas

**FIGURE 4.13**  The Maxwell–Boltzmann distribution of molecular speeds in nitrogen at three temperatures. The peak in each curve gives the most probable speed $u_{mp}$, which is slightly smaller than the root-mean-square speed $u_{rms}$. The average speed $u_{av}$ (obtained simply by adding the speeds and dividing by the number of molecules in the sample) lies in between. All three measures give comparable estimates of typical molecular speeds and show how these speeds increase with temperature.

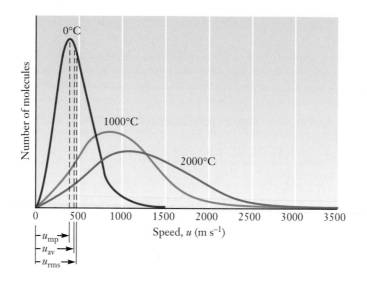

has been measured experimentally by an apparatus such as the one in Figure 4.14, giving results consistent with the predictions of Maxwell and Boltzmann.

Another way to think of $f(u)$ is as a *probability distribution*: $f(u) \, \Delta u$ is the probability that any given molecule will have a speed between $u$ and $u + \Delta u$. The **most probable speed** $u_{mp}$ is the speed at which $f(u)$ has its maximum. For the Maxwell–Boltzmann distribution function this is

$$u_{mp} = \sqrt{\frac{2k_B T}{m}} = \sqrt{\frac{2RT}{\mathcal{M}}}$$

Another important quantity is the **average speed** $\overline{u}$ which is

$$\overline{u} = \sqrt{\frac{8k_B T}{\pi m}} = \sqrt{\frac{8RT}{\pi \mathcal{M}}}$$

**FIGURE 4.14**  This device can be used to measure the distribution of molecular speeds experimentally. Only those molecules with the correct speed to pass through *both* rotating sectors will reach the detector, where they can be counted. By changing the rate of rotation of the sectors, the speed distribution can be determined.

We have already seen that

$$u_{rms} = \sqrt{\frac{3k_B T}{m}} = \sqrt{\frac{3RT}{\mathcal{M}}}$$

These three speeds are close to each other, but not equal. They stand in a ratio:

$$u_{mp} : \overline{u} : u_{rms} = 1.000 : 1.128 : 1.225$$

For each temperature there is a unique distribution curve that defines temperature in the kinetic theory of gases. Unless the molecules in a gas have a speed dis-

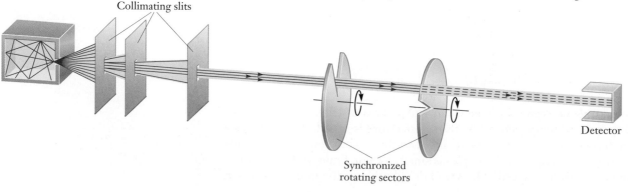

tribution that conforms to such a curve, temperature has no meaning for the gas. Temperature describes a system of gaseous molecules only when their speed distribution is represented by the Maxwell–Boltzmann function. Consider a closed container filled with molecules whose speed distribution is not "Maxwellian." Such a situation is possible (for example, just after an explosion) but it cannot persist for very long. Any distribution of molecular speeds other than a Maxwell–Boltzmann distribution quickly becomes Maxwellian through collisions of molecules, which exchange energy. Once attained, the Maxwell–Boltzmann distribution persists indefinitely. The gas molecules have come to **thermal equilibrium** with one another, and we can speak of a system as having a temperature only if the condition of thermal equilibrium exists.

# A DEEPER LOOK...

## 4.6 APPLICATIONS OF THE KINETIC THEORY

The kinetic theory we have developed can be applied to several important properties of gases. A study of the rates at which atoms and molecules collide with a wall and with one another helps to explain phenomena ranging from isotope separation based on gaseous diffusion to gas-phase chemical kinetics.

### Wall Collisions, Effusion, and Diffusion

Let us call $Z_w$ the rate of collisions of gas molecules with a section of wall of area $A$. A full mathematical calculation of $Z_w$ requires the use of integral calculus and solid geometry. We present instead some simple physical arguments to demonstrate the form of the dependence of this rate on the properties of the gas.

First of all, $Z_w$ should be proportional to the area $A$, because doubling the area will double the number of collisions with the wall. Second, $Z_w$ should be proportional to the average molecular speed $\bar{u}$, because if molecules move twice as fast they will collide twice as often with a given wall area. Finally, the wall collision rate should be proportional to the number density $N/V$, because if there are twice as many molecules in a given volume they will collide twice as often with the wall. All of these arguments are consistent with the kinetic theory of gases and are confirmed by the full mathematical analysis. We conclude that

$$Z_w \propto \frac{N}{V} \bar{u} A$$

Note that the units of both sides are inverse seconds, as required for a rate. The proportionality constant, calculated from a complete analysis of the directions from which molecules impinge on the wall, turns out to have the value $\frac{1}{4}$, so the wall collision rate is

$$Z_w = \frac{1}{4}\frac{N}{V}\bar{u}A = \frac{1}{4}\frac{N}{V}\sqrt{\frac{8RT}{\pi\mathcal{M}}}A \qquad \text{[4.13]}$$

using the result given earlier for $\bar{u}$. This simple equation has many uses. It sets an upper limit upon the rate at which a gas may react with a solid, for example. It also permits calculation of the rate at which molecules effuse through a small hole in the wall of a vessel.

---

**EXAMPLE 4.8**

Calculate the number of collisions that oxygen molecules make per second on 1.00 cm$^2$ of the surface of the vessel containing them if the pressure is $1.00 \times 10^{-6}$ atm and the temperature is 25°C (298 K).

**Solution**

First compute the quantities that appear in the equation for $Z_w$:

$$\frac{N}{V} = \frac{N_0 n}{V} = \frac{N_0 P}{RT}$$

$$= \frac{(6.022 \times 10^{23} \text{ mol}^{-1})(1.00 \times 10^{-6} \text{ atm})}{(0.08206 \text{ L atm mol}^{-1} \text{ K}^{-1})(298 \text{ K})}$$

$$= 2.46 \times 10^{16} \text{ L}^{-1} = 2.46 \times 10^{19} \text{ m}^{-3}$$

$$A = 1.00 \text{ cm}^2 = 1.00 \times 10^{-4} \text{ m}^2$$

$$\bar{u} = \sqrt{\frac{8RT}{\pi\mathcal{M}}}$$

$$= \sqrt{\frac{8(8.3145 \text{ J mol}^{-1} \text{ K}^{-1})(298 \text{ K})}{\pi(32.00 \times 10^{-3} \text{ kg mol}^{-1})}}$$

$$= 444 \text{ m s}^{-1}$$

Then the collision rate is

$$Z_w = \frac{1}{4} \frac{N}{V} \bar{u} A$$

$$= \tfrac{1}{4}(2.46 \times 10^{19} \text{ m}^{-3})(444 \text{ m s}^{-1})(1.00 \times 10^{-4} \text{ m}^2)$$

$$= 2.73 \times 10^{17} \text{ s}^{-1}$$

**Related Problems: 47, 48**

The formula just given for the rate of collisions with a wall can be used to explain **Graham's law of effusion.** In 1846 Thomas Graham showed experimentally that the rate of effusion of a gas through a small hole into a vacuum (Fig. 4.15) is inversely proportional to the square root of its molar mass. Assuming different gases are studied at the same temperature and pressure, the number density $N/V$ does not change and the rate of effusion of each gas depends only on the $1/\sqrt{M}$ factor in Equation 4.13, exactly as observed by Graham. This agreement provides additional support for the kinetic theory of gases.

Graham's law can be applied to the effusion of a mixture of two gases through a small hole. The ratio of the rates of effusion of the two species, A and B, is

$$\frac{\text{rate of effusion of A}}{\text{rate of effusion of B}} = \frac{\dfrac{1}{4}\dfrac{N_A}{V}\sqrt{\dfrac{8RT}{\pi M_A}}A}{\dfrac{1}{4}\dfrac{N_B}{V}\sqrt{\dfrac{8RT}{\pi M_B}}A}$$

$$= \frac{N_A}{N_B}\sqrt{\frac{M_B}{M_A}} \qquad \textbf{[4.14]}$$

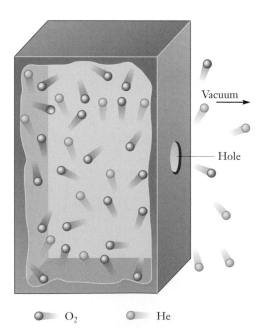

O₂          He

**FIGURE 4.15** A small hole in the box permits molecules to effuse out into a vacuum. The lighter particles (here, helium atoms) effuse at greater rates than the heavier oxygen molecules because their speeds are greater on the average.

This ratio is equal to the ratio of the numbers of molecules of the two species coming out through the hole in a given short time interval. The gas that emerges is enriched in the lighter component because lighter molecules effuse more rapidly than heavier ones. If B is heavier than A, the **enrichment factor** is $\sqrt{M_B/M_A}$ for the lighter species, A.

### EXAMPLE 4.9

Calculate the enrichment factors from effusion for a mixture of $^{235}\text{UF}_6$ and $^{238}\text{UF}_6$, uranium hexafluoride with two different uranium isotopes. The atomic mass of $^{235}\text{U}$ is 235.04, and that of $^{238}\text{U}$ is 238.05. The atomic mass of fluorine is 19.00.

### Solution

The two molar masses are

$$M(^{238}\text{UF}_6) = 238.05 + 6(19.00) = 352.05 \text{ g mol}^{-1}$$

$$M(^{235}\text{UF}_6) = 235.04 + 6(19.00) = 349.04 \text{ g mol}^{-1}$$

$$\text{enrichment factor} = \sqrt{\frac{M(^{238}\text{UF}_6)}{M(^{235}\text{UF}_6)}} = \sqrt{\frac{352.05 \text{ g mol}^{-1}}{349.04 \text{ g mol}^{-1}}}$$

$$= 1.0043$$

**Related Problems: 49, 50**

Graham's law of effusion holds true only if the opening in the vessel is small enough and the pressure low enough so that most molecules follow straight-line trajectories through the opening without undergoing collisions with one another. A related, but more complex, phenomenon is **gaseous diffusion** through a porous barrier. This differs from effusion in that molecules undergo many collisions with one another and with the barrier during their passage through it. Interestingly, the diffusion rate is also inversely proportional to the square root of the molar mass of the gas, although the reasons for this dependence are not those outlined for the effusion process. If a mixture of gases is placed in contact with a porous barrier, the gas passing through is enriched in the lighter component, A, by a factor of $\sqrt{M_B/M_A}$, and the gas remaining behind is enriched in the heavier component.

During World War II, it became necessary to separate the more easily fissionable isotope $^{235}\text{U}$ from $^{238}\text{U}$ in order to develop the atomic bomb. Because $^{235}\text{U}$ has a natural abundance of only 0.7%, its isolation in nearly pure form appeared a daunting task. The procedure adopted allowed uranium to react with fluorine to form the relatively volatile compound $\text{UF}_6$ (boiling point 56°C), which can be enriched in $^{235}\text{U}$ by passing it through a porous barrier. As shown in Example 4.9, however, each passage gives an enrichment factor of only 1.0043, so many such barriers in succession were necessary to provide sufficient enrichment of the $^{235}\text{U}$ component. Such a gaseous diffusion chamber was constructed in a very short time at the Oak Ridge National Laboratories, and the enriched $^{235}\text{U}$ was used in the Hiroshima atomic bomb. Similar methods are still used today to enrich uranium for use as fuel in nuclear power plants (Fig. 4.16).

## EXAMPLE 4.10

How many diffusion stages are required if $^{235}$U is to be enriched from 0.70% to 7.0% by means of the gaseous $UF_6$ diffusion process?

### Solution

From Example 4.9, the enrichment factor per stage is 1.0043, so that for the first stage there is the relation

$$\left(\frac{N^{235}UF_6}{N^{238}UF_6}\right)_1 = \left(\frac{N^{235}UF_6}{N^{238}UF_6}\right)_0 (1.0043)$$

where the subscripts 0 and 1 denote the concentrations initially and after the first stage. For the ratio of the numbers after $n$ stages, therefore,

$$\left(\frac{N^{235}UF_6}{N^{238}UF_6}\right)_n = \left(\frac{N^{235}UF_6}{N^{238}UF_6}\right)_0 (1.0043)^n$$

$$\frac{7.0}{93.0} = \frac{0.70}{99.30} (1.0043)^n$$

$$10.677 = (1.0043)^n$$

This equation can be solved by taking logarithms of both sides (see Appendix C):

$$\log_{10} (10.677) = n \log_{10} (1.0043)$$

$$1.02845 = 0.0018635 \, n$$

$$n = 552 \text{ stages}$$

**Related Problems: 51, 52**

## Frequency of Molecular Collisions

Collisions lie at the very heart of chemistry, because chemical reactions can occur only when molecules collide with one another. The kinetic theory of gases lets us estimate the frequency of molecular collisions and the average distance traveled by a molecule between collisions.

Molecules are assumed to be approximately spherical, with diameters, $d$, on the order of $10^{-10}$ m (Section 1.6). We initially suppose that a given molecule moves through a gas of stationary molecules of the same diameter. Such a molecule "sweeps out" a cylinder with cross-sectional area $A = \pi d^2$ (see Fig. 4.17). It collides with any molecule whose center lies inside the cylinder, within a distance $d$ of the center of the moving molecule. In one second, the length of such a cylinder is $\bar{u}\Delta t = \bar{u} \times 1$ s, where $\bar{u}$ is the average molecular speed. It does not matter that the moving molecule is scattered in another direction when it collides with other molecules; these little cylinders are merely joined together, end to end, to effect a cylinder whose length is still $\bar{u} \times 1$ s.

**FIGURE 4.16**    A gaseous diffusion plant for separation of uranium isotopes in uranium hexafluoride. If the uranium is to be used as fuel in nuclear power plants, a total enrichment is necessary over many stages, to take the gas from 0.7% to 3% $^{235}$U. *(Courtesy of Paducah Laboratory, Martin Marietta Energy Systems)*

In one second, the moving molecule therefore sweeps out a volume

$$V_{cyl} = \pi d^2 \, \bar{u}$$

If $N/V$ is the number of molecules per unit volume in the gas (the number density of the gas), then the number of collisions per second is

$$\text{collision rate} = Z_1 = \frac{N}{V} V_{cyl} = \frac{N}{V} \pi d^2 \bar{u} \qquad \text{(approximately)}$$

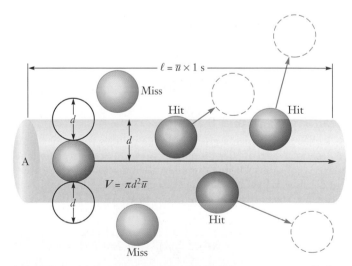

**FIGURE 4.17**    An average molecule (red) sweeps out a cylinder of volume $\pi d^2 \bar{u}$ in one second. It will collide with any molecules whose centers lie within the cylinder. From this, the rate of its collisions with other molecules can be calculated.

Actually, this equation is only approximate, because the other molecules are not standing still but are moving. A full calculation gives an extra factor of $\sqrt{2}$:

$$Z_1 = \sqrt{2}\frac{N}{V}\,\pi d^2 \bar{u}$$

Inserting the result for $\bar{u}$ from the kinetic theory gives

$$Z_1 = 4\frac{N}{V}d^2\sqrt{\frac{\pi RT}{\mathcal{M}}} \qquad \text{[4.15]}$$

**FIGURE 4.18** A gas molecule follows a straight-line path for only a short time before undergoing a collision, so its overall path is a zigzag one. The displacement $\Delta r$ of a particular molecule in time $t$ is shown. The mean-square displacement, $\overline{\Delta r^2}$, is equal to $6Dt$, where $D$ is the diffusion constant of the molecules in the gas.

## EXAMPLE 4.11

Calculate the collision frequency for **(a)** a molecule in a sample of oxygen at 1.00 atm pressure and 25°C and **(b)** a molecule of hydrogen in a region of interstellar space where the number density is $1.0 \times 10^{10}$ molecules per cubic meter and the temperature is 30 K. Take the diameter of $O_2$ to be $2.92 \times 10^{-10}$ m and that of $H_2$ to be $2.34 \times 10^{-10}$ m.

### Solution

**(a)** 
$$\frac{N}{V} = \frac{N_0 P}{RT} = \frac{(6.022 \times 10^{23} \text{ mol}^{-1})(1.00 \text{ atm})}{(0.08206 \text{ L atm mol}^{-1} \text{ K}^{-1})(298 \text{ K})}$$
$$= 2.46 \times 10^{22} \text{ L}^{-1} = 2.46 \times 10^{25} \text{ m}^{-3}$$

In Example 4.8 we found $\bar{u}$ to be 444 m s$^{-1}$. Thus, the collision frequency is

$$Z_1 = \sqrt{2}\,\pi\,(2.46 \times 10^{25} \text{ m}^{-3}) \times (2.92 \times 10^{-10} \text{ m})^2(444 \text{ m s}^{-1})$$
$$= 4.14 \times 10^9 \text{ s}^{-1}$$

This is the average number of collisions undergone by each molecule per second.

**(b)** An analogous calculation gives

$$Z_1 = 1.4 \times 10^{-6} \text{ s}^{-1}$$

In other words, the average molecule under these conditions waits $7.3 \times 10^5$ s, or 8.5 days, between collisions.

## Mean Free Path and Diffusion

$Z_1$ is the rate at which a given molecule collides with other molecules. Its inverse, $Z_1^{-1}$, is therefore a measure of the average time between collisions. During this interval a molecule travels an average distance $uZ_1^{-1}$. This average distance traveled by a molecule between collisions is called the **mean free path** $\lambda$.

$$\lambda = \bar{u}\,Z_1^{-1} = \frac{\bar{u}}{\sqrt{2}\dfrac{N}{V}\,\pi d^2 \bar{u}} = \frac{1}{\sqrt{2}\,\pi d^2 N/V} \qquad \text{[4.16]}$$

The mean free path, unlike the collision frequency, is independent of the molar mass. It must be large compared to the molecular diameter to result in ideal gas behavior. For a molecule of

diameter $3 \times 10^{-10}$ m, the mean free path at 25°C and atmospheric pressure is $1 \times 10^{-7}$ m, or 300 times larger.

Molecules in a gas move in straight lines only over rather short distances before being deflected by collisions and changing direction (Fig. 4.18). Because each one follows a zigzag course, molecules take a longer time to achieve a given displacement from their starting points than if there were no collisions. This helps to explain the slowness of diffusion in gases. Recall that at room temperature the speed of a molecule is on the order of $1 \times 10^3$ m s$^{-1}$. If an odorous gas were released in one part of a room and the molecules followed straight-line trajectories, the odor would reach other points in the room almost instantaneously. Instead, there is a time lapse due to the irregular paths ("random walks") followed by the molecules.

We can describe diffusion in a gas by means of averaged quantities such as the mean-square displacement $\overline{\Delta r^2} = \overline{\Delta x^2} + \overline{\Delta y^2} + \overline{\Delta z^2}$ (analogous to the mean-square speed considered earlier). If there are no gas currents to perturb the motion of the gas molecules (a rather strong condition, requiring strict isolation from the surroundings), then $\overline{\Delta r^2}$ is found to be proportional to the time elapsed, $t$:

$$\overline{\Delta r^2} = 6Dt \qquad \text{[4.17]}$$

The proportionality constant is $6D$, where $D$ is the **diffusion constant** of the molecules. The root-mean-square displacement $\sqrt{\overline{\Delta r^2}}$, is therefore equal to $\sqrt{6Dt}$.

The diffusion constant has units of m$^2$ s$^{-1}$. It is proportional to the mean free path $\lambda$ and to the mean molecular speed $\bar{u}$, but the proportionality constant is difficult to work out. For the simplest case, a single-component gas, it has the value $3\pi/16$, so

$$D = \frac{3\pi}{16}\lambda\bar{u} = \frac{3\pi}{16}\sqrt{\frac{8RT}{\pi\mathcal{M}}}\,\frac{1}{\sqrt{2}\,\pi d^2 N/V}$$
$$= \frac{3}{8}\sqrt{\frac{RT}{\pi\mathcal{M}}}\,\frac{1}{d^2 N/V} \qquad \text{[4.18]}$$

## EXAMPLE 4.12

Calculate the mean free path and the diffusion constant for the molecules in Example 4.11.

**Solution**

**(a)** The mean free path is

$$\lambda = \frac{1}{\sqrt{2}\pi(2.92 \times 10^{-10} \text{ m})^2(2.46 \times 10^{25} \text{ m}^{-3})}$$

$$= 1.07 \times 10^{-7} \text{ m}$$

The diffusion constant is

$$D = \frac{3\pi}{16}\lambda\bar{u} = \frac{3\pi}{16}(1.07 \times 10^{-7} \text{ m})(444 \text{ m s}^{-1})$$

$$= 2.80 \times 10^{-5} \text{ m}^2 \text{ s}^{-1}$$

**(b)** An analogous calculation for an average molecule of $H_2$ in interstellar space gives

$$\lambda - 4.1 \times 10^8 \text{ m}$$

For comparison, the distance from earth to moon is $3.8 \times 10^8$ m. The diffusion constant is

$$D = 1.4 \times 10^{11} \text{ m}^2 \text{ s}^{-1}$$

**Related Problems: 53, 54**

---

## 4.7

# REAL GASES: INTERMOLECULAR FORCES

The ideal gas law, $PV = nRT$, is a particularly simple example of an **equation of state**—an equation relating the pressure, temperature, chemical amount, and volume to one another. Such equations of state can be obtained from either theory or experiment. They are useful not only for ideal gases but for real gases, liquids, and solids.

Real gases follow the ideal gas equation of state only at sufficiently low densities. Deviations appear in a variety of forms. Boyle's law, $PV = C$, is no longer satisfied at high pressures, and Charles's law, $V \propto T$, begins to break down when the temperature becomes low. Avogadro's hypothesis also does not hold strictly for real gases at moderate pressures. At atmospheric pressure the ideal gas law is quite well satisfied for most gases, but for some (e.g., water vapor and ammonia) there are deviations of 1 to 2%. These deviations are revealed in the **compressibility factor** $z$:

$$z = \frac{PV}{nRT} \qquad \qquad \textbf{[4.19]}$$

When $z$ differs from 1 (Fig. 4.19), the ideal gas law is inadequate. A more accurate equation of state is needed.

### The van der Waals Equation of State

One of the earliest and most important improvements on the ideal gas equation of state was proposed in 1873 by the Dutch physicist Johannes van der Waals. The **van der Waals equation of state** has the form

$$\left(P + a\frac{n^2}{V^2}\right)(V - nb) = nRT \qquad \qquad \textbf{[4.20]}$$

$$P = \frac{nRT}{V - nb} - a\frac{n^2}{V^2}$$

Comparison with the ideal gas law shows two modifications arising from the forces between molecules, which are repulsive at short distances and attractive at large distances. Because of repulsive forces, molecules cannot overlap. They exclude other

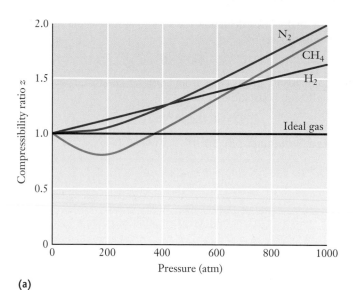

**(a)**

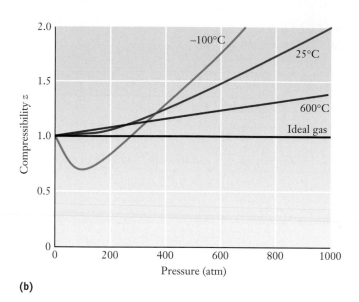

**(b)**

**FIGURE 4.19**  A plot of $z = PV/nRT$ against pressure shows deviations from the ideal gas law quite clearly. For an ideal gas, the plot is the straight horizontal line. (a) The behavior of several real gases at 25°C and (b) of nitrogen at several temperatures deviate as shown.

molecules from the volumes they occupy; in this way the effective volume available to a given molecule is not $V$ but $V - nb$, where $b$ is a constant describing the volume excluded per mole of molecules. Attractive forces hold pairs or groups of molecules together. Any tendency to cluster together reduces the effective number of independent molecules in the gas and therefore reduces the rate of collisions made with the walls of the container. Fewer wall collisions means a lower pressure. Van der Waals argued that this effect, because it depends on attractions between *pairs* of molecules, should be proportional to the *square* of the number of molecules per unit volume ($N^2/V^2$) or, equivalently, proportional to $n^2/V^2$. This effect reduces the pressure by an amount $a(n/V)^2$, where $a$ is a positive constant that depends on the strength of the attractive forces. Putting these two corrections into the ideal gas law produces the van der Waals equation just presented.

The constants $a$ and $b$ are obtained by fitting experimental $PVT$ data for real gases (see Table 4.3). They have the units

## TABLE 4.3

### Van der Waals Constants of Several Gases

| Name | Formula | $a$ (atm L$^2$ mol$^{-2}$) | $b$ (L mol$^{-1}$) |
|---|---|---|---|
| Ammonia | $NH_3$ | 4.170 | 0.03707 |
| Argon | Ar | 1.345 | 0.03219 |
| Carbon dioxide | $CO_2$ | 3.592 | 0.04267 |
| Hydrogen | $H_2$ | 0.2444 | 0.02661 |
| Hydrogen chloride | HCl | 3.667 | 0.04081 |
| Methane | $CH_4$ | 2.253 | 0.04278 |
| Nitrogen | $N_2$ | 1.390 | 0.03913 |
| Nitrogen dioxide | $NO_2$ | 5.284 | 0.04424 |
| Oxygen | $O_2$ | 1.360 | 0.03183 |
| Sulfur dioxide | $SO_2$ | 6.714 | 0.05636 |
| Water | $H_2O$ | 5.464 | 0.03049 |

$$a: \quad \text{atm L}^2 \text{ mol}^{-2}$$

$$b: \quad \text{L mol}^{-1}$$

when $R$ has the units L atm mol$^{-1}$ K$^{-1}$.

## EXAMPLE 4.13

A sample consists of 8.00 kg of gaseous nitrogen and fills a 100-L flask at 300°C. What is the pressure of the gas, using the van der Waals equation of state? What pressure would be predicted by the ideal gas equation?

### Solution

Because the molar mass of N$_2$ is 28 g mol$^{-1}$,

$$n = \frac{8.00 \times 10^3 \text{ g}}{28.0 \text{ g mol}^{-1}} = 286 \text{ mol}$$

The temperature (in kelvins) is $T = 300 + 273 = 573$ K, and the volume $V$ is 100 L. Using $R = 0.08206$ L atm mol$^{-1}$ K$^{-1}$ and the van der Waals constants for nitrogen given in Table 4.3, we calculate $P = 151 - 11 = 140$ atm. If the ideal gas law is used instead, a pressure of 134 atm is calculated. This illustrates the magnitude of deviations from the ideal gas law at higher pressures.

**Related Problems: 55, 56, 57, 58**

The effects of the two van der Waals parameters can be seen clearly by writing the compressibility factor for this equation of state:

$$z = \frac{PV}{nRT} = \frac{V}{V - nb} - \frac{a}{RT}\frac{n}{V} = \frac{1}{1 - bn/V} - \frac{a}{RT}\frac{n}{V}$$

Repulsive forces (through $b$) increase $z$ above 1, while attractive forces (through $a$) reduce $z$. At low particle densities, $n/V$ is small. After the approximation

$$\frac{1}{1 - bn/V} \cong 1 + b\frac{n}{V} + \cdots$$

is made, the compressibility factor becomes

$$z \cong 1 + \left(b - \frac{a}{RT}\right)\frac{n}{V} + \cdots = 1 + B(T)\frac{n}{V} + \cdots$$

The temperature at which the **second virial coefficient** $B(T)$ multiplying the density $n/V$ vanishes is called the Boyle temperature, $T_B$. In the van der Waals model it has the value

$$T_B = \frac{a}{Rb}$$

At temperatures above $T_B$, repulsive forces dominate and $z > 1$; at temperatures below $T_B$, attractive forces dominate and $z < 1$ (at low enough density). A gas at the Boyle temperature is still not ideal, but deviations from $z = 1$ do not appear until higher densities.

The constant $b$ is the volume excluded by 1 mol of molecules and should therefore be close to the volume per mole in the liquid state, where molecules are essentially in contact with each other. For example, the density of liquid nitrogen is $0.808 \text{ g cm}^{-3}$. One mole of $N_2$ weighs 28.0 g, so

$$\text{molar volume of } N_2(\ell) = \frac{28.0 \text{ g mol}^{-1}}{0.808 \text{ g cm}^{-3}} = 34.7 \text{ cm}^3 \text{ mol}^{-1}$$

$$= 0.0347 \text{ L mol}^{-1}$$

This is reasonably close to the van der Waals $b$ parameter of $0.03913 \text{ L mol}^{-1}$, obtained by fitting the equation of state to $PVT$ data for nitrogen.

## Intermolecular Forces

The attractive force parameter $a$ and the excluded volume per mole, $b$, arise from the forces between the atoms or molecules in a gas. Real atoms are not impenetrable, hard spheres. As a pair of molecules approach one another, the forces between them generate potential energy, which must be considered in addition to the kinetic energy associated with their speeds. The simplest case is illustrated by atoms of the noble gases. As two atoms are brought together, they attract each other until the distance between their centers, $R$, becomes short enough. If the atoms are pushed still closer together, they repel each other with increasing force as the distance between them is reduced. This can be described by a **potential energy curve** $V(R)$ (see Section 1.7 and Appendix B) such as that shown for argon in Figure 4.20. If two molecules can lower their energy by moving closer together, then a net attractive force exists between them; but if they can lower their energy only by moving apart, there is a net repulsive force. The force changes from attractive to repulsive at the minimum of $V(R)$, where the net force between atoms is zero. Potential energy curves for atoms are obtained by fitting equations for $V(R)$ to measured properties of real gases. More accurate information is provided by experiments in which beams of atoms collide with one another.

For many purposes the detailed shape of the potential is less important than aspects such as the depth and location of the potential minimum. A simple expres-

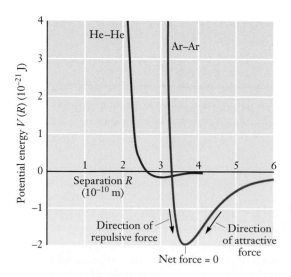

**FIGURE 4.20** Potential energy curves $V(R)$ for pairs of helium atoms (red) and pairs of argon atoms (blue). Where $V$ decreases with increasing $R$, the atoms repel each other; where $V$ increases with $R$, the atoms attract.

**TABLE 4.4**

*Lennard–Jones Parameters for Atoms and Molecules*

| Substance | $\sigma$(m) | $\epsilon$(J) |
|---|---|---|
| He | $2.56 \times 10^{-10}$ | $1.41 \times 10^{-22}$ |
| Ne | $2.75 \times 10^{-10}$ | $4.92 \times 10^{-22}$ |
| Ar | $3.40 \times 10^{-10}$ | $1.654 \times 10^{-21}$ |
| Kr | $3.60 \times 10^{-10}$ | $2.36 \times 10^{-21}$ |
| Xe | $4.10 \times 10^{-10}$ | $3.06 \times 10^{-21}$ |
| $H_2$ | $2.93 \times 10^{-10}$ | $5.11 \times 10^{-22}$ |
| $O_2$ | $3.58 \times 10^{-10}$ | $1.622 \times 10^{-21}$ |
| CO | $3.76 \times 10^{-10}$ | $1.383 \times 10^{-21}$ |
| $N_2$ | $3.70 \times 10^{-10}$ | $1.312 \times 10^{-21}$ |
| $CH_4$ | $3.82 \times 10^{-10}$ | $2.045 \times 10^{-21}$ |

sion that is frequently used to model these interactions between atoms is the **Lennard–Jones potential:**

$$V_{LJ}(R) = 4\epsilon\left[\left(\frac{\sigma}{R}\right)^{12} - \left(\frac{\sigma}{R}\right)^{6}\right]$$   **[4.21]**

where $\epsilon$ is the depth and $\sigma$ is the distance at which $V(R)$ passes through 0. This potential has an attractive part, proportional to $R^{-6}$, and a repulsive part, proportional to $R^{-12}$. The minimum in this potential is located at $2^{1/6}\sigma$, or $1.22\sigma$. Table 4.4 lists Lennard–Jones parameters for a number of atoms. Note that the depth and range of the potential increase for the heavier noble-gas atoms. Molecules such as $N_2$ that are nearly spherical can also be treated approximately with Lennard–Jones potentials. The two parameters $\epsilon$ and $\sigma$ in the Lennard–Jones potential, like the van der Waals parameters $a$ and $b$, are simple ways of characterizing the interactions between molecules in real gases.

## CUMULATIVE EXERCISE

### Ammonium Perchlorate

Ammonium perchlorate ($NH_4ClO_4$) is a solid rocket fuel used in space shuttles. When heated above 200°C, it decomposes to a variety of gaseous products, of which the most important are nitrogen, chlorine, oxygen, and water vapor.
(a) Write a balanced chemical equation for the decomposition of $NH_4ClO_4$, assuming the products just listed are the only ones generated.
(b) The sudden appearance of hot gaseous products in a small initial volume leads to rapid increases in pressure and temperature, which give the rocket its thrust. What total pressure of gas would be produced at 800°C by igniting $7.00 \times 10^5$ kg of $NH_4ClO_4$ (a typical charge of the booster rockets in the space shuttle) and allowing it to expand to fill a volume of 6400 m³ ($6.40 \times 10^6$ L)? Use the ideal gas law.
(c) Calculate the mole fraction of chlorine and its partial pressure in the mixture of gases produced.

A space shuttle taking off. *(NASA)*

(d) The van der Waals equation applies strictly to pure real gases, not to mixtures. For a mixture like the one resulting from the reaction of part (a), it may still be possible to define effective $a$ and $b$ parameters to relate total pressure, volume, temperature, and total chemical amount. Suppose the gas mixture has $a = 4.00$ atm L$^2$ mol$^{-2}$ and $b = 0.0330$ L mol$^{-1}$. Recalculate the pressure of the gas mixture in part (b), using the van der Waals equation.

(e) Calculate and compare the root-mean-square speeds of water and chlorine molecules under the conditions of part (b).

(f) The gas mixture from part (b) cools and expands until it reaches a temperature of 200°C and a pressure of 3.20 atm. Calculate the volume occupied by the gas mixture after this expansion has occurred. Assume ideal gas behavior.

**Answers**

(a) $2 \, NH_4ClO_4(s) \rightarrow N_2(g) + Cl_2(g) + 2 \, O_2(g) + 4 \, H_2O(g)$

(b) 328 atm

(c) $X_{Cl_2} = \frac{1}{8} = 0.125$ (exactly); $P_{Cl_2} = 41.0$ atm

(d) 318 atm

(e) $u_{rms}(H_2O) = 1220$ m s$^{-1}$; $u_{rms}(Cl_2) = 614$ m s$^{-1}$

(f) $2.89 \times 10^5$ m$^3$

# Concepts & Skills

*After studying this chapter and working the problems that follow, you should be able to*

1. Write chemical equations for several reactions that lead to gas formation (Section 4.1, problems 1–4).

2. Describe how pressure and temperature are defined and measured (Section 4.2, problems 5–10).

3. Use the ideal gas law to relate pressure, volume, temperature, and chemical amount of an ideal gas, and to do stoichiometric calculations involving gases (Section 4.3, problems 19–32).

4. Use Dalton's law to calculate partial pressures in gas mixtures (Section 4.4, problems 33–38).

5. Use the Maxwell–Boltzmann distribution of molecular speeds to calculate root-mean-square, most probable, and average speeds of molecules in a gas (Section 4.5, problems 41–44).

6. Describe the connection between temperature and the speeds or kinetic energies of the molecules in a gas (Section 4.5, problems 45–46).

7. Calculate the rate of collisions of molecules with a wall, and from that determine the effusion rate of a gas through a small hole of known area (Section 4.6, problems 47–48).

8. Calculate the enrichment factor for lighter molecules when a gas consisting of a mixture of light and heavy molecules effuses through a small aperture in a vessel wall (Section 4.6, problems 49–52).

9. Calculate the collision frequency, the mean free path, and the diffusion constant for gases (Section 4.6, problems 53–54).

10. Use the van der Waals equation to relate the pressure, volume, temperature, and chemical amount of a nonideal gas (Section 4.7, problems 55–58).

11. Discuss how forces between atoms and molecules vary with distance (Section 4.7).

## PROBLEMS

*Answers to problems whose numbers are boldface appear in Appendix G. Problems that are more challenging are indicated with asterisks.*

### The Chemistry of Gases

**1.** Solid ammonium hydrosulfide ($NH_4HS$) decomposes entirely to gases when it is heated. Write a chemical equation representing this change.

2. Solid ammonium carbamate ($NH_4CO_2NH_2$) decomposes entirely to gases when it is heated. Write a chemical equation representing this change.

**3.** Ammonia ($NH_3$) is an important and useful gas. Suggest a way to generate it from ammonium bromide ($NH_4Br$). Include a balanced chemical equation.

4. Hydrogen cyanide (HCN) is a poisonous gas. Explain why solutions of potassium cyanide (KCN) should never be acidified. Include a balanced chemical equation.

### Pressure and Temperature of Gases

**5.** Suppose a barometer were designed using water (with a density of $1.00 \text{ g cm}^{-3}$) rather than mercury as its fluid. What would be the height of the column of water balancing 1.00 atm pressure?

6. A vessel that contains a gas has two pressure gauges attached to it. One contains liquid mercury, the other an oil such as dibutylphthalate. The difference in levels of mercury in the two arms of the mercury gauge is observed to be 9.50 cm. Given

$$\text{density of mercury} = 13.60 \text{ g cm}^{-3}$$
$$\text{density of oil} = 1.045 \text{ g cm}^{-3}$$
$$\text{acceleration due to gravity} = 9.806 \text{ m s}^{-2}$$

(a) What is the pressure of the gas?
(b) What is the difference in heights of the oil in the two arms of the oil pressure gauge?

**7.** Calcium dissolved in the ocean is used by marine organisms to form $CaCO_3(s)$ in skeletons and shells. When the organisms die, their remains fall to the bottom. The amount of calcium carbonate that can be dissolved in seawater depends on the pressure. At great depths, where the pressure exceeds about 414 atm, the shells slowly redissolve. This prevents all the world's calcium from being tied up as insoluble $CaCO_3(s)$ at the bottom of the sea. Estimate the depth (in feet) of water that exerts a pressure great enough to dissolve seashells.

8. Suppose that the atmosphere were perfectly uniform, with a density throughout equal to that of air at 0°C, $1.3 \text{ g L}^{-1}$. Calculate the thickness of such an atmosphere that would cause a pressure of exactly one standard atmosphere at the earth's surface.

**9.** The "critical pressure" of mercury is 172.00 MPa. Above this pressure mercury cannot be liquefied, no matter what the temperature. Express this pressure in atmospheres and in bars (1 bar = $10^5$ Pa).

10. Experimental studies of solid surfaces and the chemical reactions that occur on them require very low gas pressures to avoid surface contamination. High-vacuum apparatus for such experiments can routinely reach pressures of $5 \times 10^{-10}$ torr. Express this pressure in atmospheres and in pascals.

**11.** Some nitrogen is held in a 2.00-L tank at a pressure of 3.00 atm. The tank is connected to a 5.00-L tank that is completely empty (evacuated), and a valve is opened to connect the two. No temperature change occurs in the process. Determine the total pressure in this two-tank system after the nitrogen stops flowing.

12. The Stirling engine, a heat engine invented by a Scottish minister, has been considered for use in automobile engines because of its efficiency. In such an engine, a gas goes through a four-step cycle of (1) expansion at constant $T$, (2) cooling at constant $V$, (3) compression at constant $T$ to its original volume, and (4) heating at constant $V$ to its original temperature. Suppose the gas starts at a pressure of 1.23 atm, and the volume of the gas changes from 0.350 L to 1.31 L during its expansion at constant $T$. Calculate the pressure of the gas at the end of this step in the cycle.

**13.** The absolute temperature of a 4.00-L sample of gas is doubled at constant pressure. Determine the volume of the gas after this change.

14. The Celsius temperature of a 4.00-L sample of gas is doubled from 20.0°C to 40.0°C at constant pressure. Determine the volume of the gas after this change.

**15.** The gill is an obscure unit of volume. If some $H_2(g)$ has a volume of 17.4 gills at 100°F, what volume would it have if the temperature were reduced to 0°F, assuming that its pressure stayed constant?

16. A gas originally at a temperature of 26.5°C is cooled at constant pressure. Its volume decreases from 5.40 L to 5.26 L. Determine its new temperature in degrees Celsius.

**17.** Calcium carbide ($CaC_2$) reacts with water to produce acetylene ($C_2H_2$) according to the equation

$$CaC_2(s) + 2 \text{ } H_2O(\ell) \longrightarrow Ca(OH)_2(s) + C_2H_2(g)$$

A certain mass of $CaC_2$ reacts completely with water to give 64.5 L of $C_2H_2$ at 50°C and $P = 1.00$ atm. If the same mass of $CaC_2$ reacts completely at 400°C and $P = 1.00$ atm, what volume of $C_2H_2$ will be collected at the higher temperature?

18. A convenient laboratory source for oxygen of very high purity is the decomposition of potassium permanganate at 230°C:

$$2 \text{ KMnO}_4(s) \longrightarrow K_2MnO_4(s) + MnO_2(s) + O_2(g)$$

Suppose 3.41 L of oxygen is needed at atmospheric pressure and a temperature of 20°C. What volume of oxygen should be collected at 230°C and the same pressure in order to give this volume when cooled?

## The Ideal Gas Law

19. A bicycle tire is inflated to a gauge pressure of 30.0 psi at a temperature of $t = 0$°C. What will its gauge pressure be at 32°C if the tire is considered nonexpandable? (*Note:* The gauge pressure is the *difference* between the tire pressure and atmospheric pressure, 14.7 psi.)

20. The pressure of a poisonous gas inside a sealed container is 1.47 atm at 20°C. If the barometric pressure is 0.96 atm, to what temperature (in degrees Celsius) must the container and its contents be cooled so that the container can be opened with no risk of gas spurting out?

21. A 20.6-L sample of "pure" air is collected in Greenland at a temperature of $-20.0$°C and a pressure of 1.01 atm and is forced into a 1.05-L bottle for shipment to Europe for analysis.
    (a) Compute the pressure inside the bottle just after it is filled.
    (b) Compute the pressure inside the bottle as it is opened in the 21.0°C comfort of the European laboratory.

22. Iodine heptafluoride ($IF_7$) can be made at elevated temperatures by the reaction

$$I_2(g) + 7\ F_2(g) \longrightarrow 2\ IF_7(g)$$

Suppose 63.6 L of gaseous $IF_7$ is made by this reaction at 300°C and a pressure of 0.459 atm. Calculate the volume this gas will occupy if heated to 400°C at a pressure of 0.980 atm.

23. According to a reference handbook, "The weight of one liter [of $H_2Te(g)$ is] 6.234 g." Why is this information nearly valueless? Assume that $H_2Te(g)$ is an ideal gas, and calculate the temperature (in degrees Celsius) at which this statement is true if the pressure is 1.00 atm.

24. A scuba diver's tank contains 0.30 kg of oxygen ($O_2$) compressed into a volume of 2.32 L.
    (a) Use the ideal gas law to estimate the gas pressure inside the tank at 5°C, and express it in atmospheres and in pounds per square inch.
    (b) What volume would this oxygen occupy at 30°C and a pressure of 0.98 atm?

25. Hydrogen is produced by the complete reaction of 6.24 g of sodium with an excess of gaseous hydrogen chloride.
    (a) Write a balanced chemical equation for the reaction that occurs.
    (b) How many liters of hydrogen will be produced at a temperature of 50.0°C and a pressure of 0.850 atm?

26. Aluminum reacts with excess aqueous hydrochloric acid to produce hydrogen.
    (a) Write a balanced chemical equation for the reaction. (*Hint:* Water-soluble $AlCl_3$ is the stable chloride of aluminum.)
    (b) Calculate the mass of pure aluminum that will furnish 10.0 L of hydrogen at a pressure of 0.750 atm and a temperature of 30.0°C.

27. The classic method for manufacturing hydrogen chloride—which is still in use today to a small extent—involves the reaction of sodium chloride with excess sulfuric acid at elevated temperatures. The overall equation for this process is

$$NaCl(s) + H_2SO_4(\ell) \longrightarrow NaHSO_4(s) + HCl(g)$$

What volume of hydrogen chloride is produced at 550°C and a pressure of 0.97 atm from the reaction of 2500 kg of sodium chloride?

28. In 1783 the French physicist Jacques Charles supervised and took part in the first human flight in a hydrogen balloon. Such balloons rely on the low density of hydrogen relative to air for their buoyancy. In Charles's balloon ascent, the hydrogen was produced (together with iron(II) sulfate) from the action of aqueous sulfuric acid on iron filings.
    (a) Write a balanced chemical equation for this reaction.
    (b) What volume of hydrogen is produced at 300 K and a pressure of 1.0 atmosphere when 300 kg of sulfuric acid is consumed in this reaction?
    (c) What would be the radius of a spherical balloon filled by the gas in part (b)?

29. Potassium chlorate decomposes when heated, giving oxygen and potassium chloride:

$$2\ KClO_3(s) \longrightarrow 2\ KCl(s) + 3\ O_2(g)$$

A test tube holding 87.6 g of $KClO_3$ is heated and the reaction goes to completion. What volume of $O_2$ will be evolved if it is collected at a pressure of 1.04 atm and a temperature of 13.2°C?

30. Elemental chlorine was first produced by Carl Wilhelm Scheele in 1774 using the reaction of pyrolusite ($MnO_2$) with sulfuric acid and sodium chloride:

$$4\ NaCl(s) + 2\ H_2SO_4(\ell) + MnO_2(s) \longrightarrow$$
$$2\ Na_2SO_4(s) + MnCl_2(s) + 2\ H_2O(\ell) + Cl_2(g)$$

Calculate the minimum mass of $MnO_2$ required to generate 5.32 L of gaseous chlorine, measured at a pressure of 0.953 atm and a temperature of 33°C.

31. Elemental sulfur can be recovered from gaseous hydrogen sulfide ($H_2S$) through the reaction

$$2\ H_2S(g) + SO_2(g) \longrightarrow 3\ S(s) + 2\ H_2O(\ell)$$

    (a) What volume of $H_2S$ (in liters at 0°C and 1.00 atm) is required to produce 2.00 kg (2000 g) of sulfur by this process?
    (b) What minimum mass and volume (at 0°C and 1.00 atm) of $SO_2$ are required to produce 2.00 kg of sulfur by this reaction?

32. When ozone ($O_3$) is placed in contact with dry, powdered KOH at $-15$°C, the red-brown solid potassium ozonide ($KO_3$) forms, according to the balanced equation

$$5\ O_3(g) + 2\ KOH(s) \longrightarrow 2\ KO_3(s) + 5\ O_2(g) + H_2O(s)$$

Calculate the volume of ozone needed (at a pressure of 0.134 atm and $-15$°C) to produce 4.69 g of $KO_3$.

## Mixtures of Gases

33. Sulfur dioxide reacts with oxygen in the presence of platinum to give sulfur trioxide:

$$2\ SO_2(g) + O_2(g) \longrightarrow 2\ SO_3(g)$$

Suppose that at one stage in the reaction 26.0 mol $SO_2$, 83.0 mol $O_2$, and 17.0 mol $SO_3$ are present in the reaction vessel at a total pressure of 0.950 atm. Calculate the mole fraction of $SO_3$ and its partial pressure.

34. The synthesis of ammonia from the elements is carried out at high pressures and temperatures.

$$N_2(g) + 3\ H_2(g) \longrightarrow 2\ NH_3(g)$$

Suppose that at one stage in the reaction 13 mol $NH_3$, 31 mol $N_2$, and 93 mol $H_2$ are present in the reaction vessel at a total pressure of 210 atm. Calculate the mole fraction of $NH_3$ and its partial pressure.

35. The atmospheric pressure at the surface of Mars is $5.92 \times 10^{-3}$ atm. The Martian atmosphere is 95.3% $CO_2$ and 2.7% $N_2$ by volume, with small amounts of other gases also present. Compute the mole fraction and partial pressure of $N_2$ in the atmosphere of Mars.

36. The atmospheric pressure at the surface of Venus is 90.8 atm. The Venusian atmosphere is 96.5% $CO_2$ and 3.5% $N_2$ by volume, with small amounts of other gases also present. Compute the mole fraction and partial pressure of $N_2$ in the atmosphere of Venus.

37. A gas mixture at room temperature contains 10.0 mol of CO and 12.5 mol of $O_2$.
    (a) Compute the mole fraction of CO in the mixture.
    (b) The mixture is then heated, and the CO starts to react with the $O_2$ to give $CO_2$:

$$CO(g) + \tfrac{1}{2} O_2(g) \longrightarrow CO_2(g)$$

At a certain point in the heating, 3.0 mol of $CO_2$ is present. Determine the mole fraction of CO in the new mixture.

38. A gas mixture contains 4.5 mol $Br_2$ and 33.1 mol $F_2$.
    (a) Compute the mole fraction of $Br_2$ in the mixture.
    (b) The mixture is heated above 150°C and starts to react to give $BrF_5$:

$$Br_2(g) + 5\ F_2(g) \longrightarrow 2\ BrF_5(g)$$

At a certain point in the reaction, 2.2 mol of $BrF_5$ is present. Determine the mole fraction of $Br_2$ in the mixture at that point.

39. The partial pressure of water vapor in saturated air at 20°C is 0.0230 atm.
    (a) How many molecules of water are in 1.00 $cm^3$ of saturated air at 20°?
    (b) What volume of saturated air at 20°C contains 0.500 mol of water?

40. The partial pressure of oxygen in a mixture of oxygen and hydrogen is 0.200 atm, and that of hydrogen is 0.800 atm.
    (a) How many molecules of oxygen are in a 1.500-L container of this mixture at 40°C?

(b) If a spark is introduced into the container, how many grams of water will be produced?

## The Kinetic Theory of Gases

41. (a) Compute the root-mean-square speed of $H_2$ molecules in hydrogen at a temperature of 300 K.
    (b) Repeat for $SF_6$ molecules in gaseous sulfur hexafluoride at 300 K.

42. Researchers recently reported the first optical atomic trap. In this device, beams of laser light replace the physical walls of conventional containers. The laser beams are tightly focused. They briefly (for 0.5 s) exert enough pressure to confine 500 sodium atoms in a volume of $1.0 \times 10^{-15}$ $m^3$. The temperature of this gas is 0.00024 K, the lowest temperature ever reached for a gas. Compute the root-mean-square speed of the atoms in this confinement.

43. Compare the root-mean-square speed of helium atoms near the surface of the sun, where the temperature is approximately 6000 K, with that of helium atoms in an interstellar cloud, where the temperature is 100 K.

44. The "escape velocity" necessary for objects to leave the gravitational field of the earth is 11.2 km s$^{-1}$. Calculate the ratio of the escape velocity to the root-mean-square speed of helium, argon, and xenon atoms at 2000 K. Does your result help to explain the low abundance of the light gas helium in the atmosphere? Explain.

45. Chlorine dioxide ($ClO_2$) is used for bleaching wood pulp. In a gaseous sample held at thermal equilibrium at a particular temperature, 35.0% of the molecules have speeds exceeding 400 m s$^{-1}$. If the sample is heated slightly, will the percentage of molecules with speeds in excess of 400 m s$^{-1}$ then be greater than or less than 35%? Explain.

46. The $ClO_2$ from problem 45 is heated further until it explodes, yielding $Cl_2$, $O_2$, and other gaseous products. The mixture is then cooled until the original temperature is reached. Is the percentage of *chlorine* molecules with speeds in excess of 400 m s$^{-1}$ greater than or less than 35%? Explain.

## Applications of the Kinetic Theory

47. A spherical bulb with a volume of 500 $cm^3$ is evacuated to a negligibly small residual gas pressure and then closed off. One hour later the pressure in the vessel is found to be $1.00 \times 10^{-7}$ atm due to the fact that the bulb has a tiny hole in it. Assume that the surroundings are at atmospheric pressure, $T = 300$ K, and the average molar mass of molecules in the atmosphere is 28.8 g mol$^{-1}$. Calculate the radius of the hole in the vessel wall, assuming it to be circular.

48. A 200-$cm^3$ vessel contains hydrogen gas at a temperature of 25°C and a pressure of 0.990 atm. Unfortunately, the vessel has a tiny hole in its wall, and over a period of 1.0 hour the pressure drops to 0.989 atm. What is the radius of the hole (assumed to be circular)?

49. Methane ($CH_4$) effuses through a small opening in the side of a container at the rate of $1.30 \times 10^{-8}$ mol s$^{-1}$. An unknown

gas effuses through the same opening at the rate of $5.42 \times 10^{-9}$ mol s$^{-1}$ when maintained at the same temperature and pressure as the methane. Determine the molar mass of the unknown gas.

50. Equal chemical amounts of two gases, fluorine and bromine pentafluoride, are mixed. Determine the ratio of the rates of effusion of the two gases through a very small opening in their container.

51. Calculate the theoretical number of stages that would be needed to enrich $^{235}U$ to 95% purity by means of the barrier diffusion process, using $^{235}UF_6$ and $^{238}UF_6$ as the gaseous compounds. The natural abundance of $^{238}U$ is 99.27%, and that of $^{235}U$ is 0.72%. Take the relative atomic masses of $^{235}U$ and $^{238}U$ to be 235.04 and 238.05, respectively.

52. A mixture of $H_2$ and He at 300 K effuses from a very tiny hole in the vessel that contains it. What is the mole fraction of $H_2$ in the original gas mixture if 3.00 times as many He atoms as $H_2$ molecules escape from the orifice in unit time? If the same mixture is to be separated by a barrier-diffusion process, how many stages are necessary to achieve $H_2$ of 99.9% purity?

53. At what pressure does the mean free path of Kr atoms ($d = 3.16 \times 10^{-10}$ m) become comparable with the diameter of the 1-L spherical vessel that contains them at 300 K? Calculate the diffusion constant at this pressure.

54. At what pressure does the mean free path of Kr atoms ($d = 3.16 \times 10^{-10}$ m) become comparable with the diameter of a Kr atom if $T = 300$ K? Calculate the diffusion constant at this pressure. Assume that krypton obeys the ideal gas law even at these high pressures.

## Real Gases

55. Oxygen is supplied to hospitals and chemical laboratories under pressure in large steel cylinders. Typically, such a cylinder has an internal volume of 28.0 L and contains 6.80 kg of oxygen. Use the van der Waals equation to estimate the pressure inside such a cylinder at 20°C. Express it in atmospheres and in pounds per square inch.

56. Steam at high pressures and temperatures is used to generate electrical power in utility plants. A large utility boiler has a volume of 2500 m$^3$ and contains 140 metric tons (1 metric ton = $10^3$ kg) of steam at a temperature of 540°C. Use the van der Waals equation to estimate the pressure of the steam under these conditions, in atmospheres and in pounds per square inch.

57. Using (a) the ideal gas law and (b) the van der Waals equation, calculate the pressure exerted by 50.0 g of carbon dioxide in a 1.00-L vessel at 25°C. Do attractive or repulsive forces dominate?

58. When 60.0 g of methane ($CH_4$) is placed in a 1.00-L vessel, the pressure is measured to be 130 atm. Calculate the temperature of the gas using (a) the ideal gas law and (b) the van der Waals equation. Do attractive or repulsive forces dominate?

## Additional Problems

59. The earth is approximately a sphere of radius 6370 km. Taking the average barometric pressure on the earth's surface to be 730 mm (Hg), estimate the total mass of the earth's atmosphere.

\* 60. After a flood fills a basement to a depth of 9.0 ft and completely saturates the surrounding earth, the owner buys an electric pump and quickly pumps the water out of the basement. Suddenly a basement wall collapses, the structure is severely damaged, and mud oozes in. Explain this event by estimating the difference between the outside pressure at the base of the basement walls and the pressure inside the drained basement. Assume that the mud has a density of 4.9 g cm$^{-3}$. Report the answer both in atmospheres and in pounds per square inch.

61. The density of mercury is 13.5955 g cm$^{-3}$ at 0.0°C, but only 13.5094 g cm$^{-3}$ at 35°C. Suppose that a mercury barometer is read on a hot summer day when the temperature is 35°C. The column of mercury is 760.0 mm long. Correct for the expansion of the mercury and compute the true pressure in atmospheres.

62. When a gas is cooled at constant pressure, it is found that the volume decreases linearly:

$$V = 209.4 \text{ L} + \left(0.456\frac{\text{L}}{°\text{F}}\right) \times t_F$$

where $t_F$ is the temperature in degrees Fahrenheit. From this relationship, estimate the absolute zero of temperature in degrees Fahrenheit.

63. Amontons's law relates pressure to absolute temperature. Consider the ideal gas law and then write a statement of Amontons's law in a form that is analogous to the statements of Charles's law and Boyle's law in the text.

64. A gas has a density of 2.94 g L$^{-1}$ at 50°C and $P = 1.00$ atm. What will be its density at 150°C? Calculate the molar mass of the gas, assuming it obeys the ideal gas law.

65. A lighter-than-air balloon contains 1005 mol of helium at 1.00 atm and 25.0°C.
    (a) Compute the difference between the mass of the helium it contains and the mass of the air it displaces, assuming the molar mass of air to be 29.0 g mol$^{-1}$.
    (b) The balloon now ascends to an altitude of 10 miles, where the temperature is −80.0°C. The walls of the balloon are elastic enough that the pressure inside it equals the pressure outside. Repeat the calculation of part (a).

66. Baseball reporters say that long fly balls that would have carried for home runs in July "die" in the cool air of October and are caught. The idea behind this observation is that a baseball carries better when the air is less dense. Dry air is a mixture of gases with an effective molar mass of 29.0 g mol$^{-1}$.
    (a) Compute the density of dry air on a July day when the temperature is 95.0°F and the pressure is 1.00 atm.
    (b) Compute the density of dry air on an October evening when the temperature is 50.0°F and the pressure is 1.00 atm.
    (c) Suppose that the humidity on the July day is 100%, so that the air is saturated with water vapor. Is the density

of this hot, moist air less than, equal to, or greater than the density of the hot, dry air computed in part (a)? In other terms, does high humidity favor the home run?

67. Sulfuric acid reacts with sodium chloride to produce gaseous hydrogen chloride according to the equation

$$NaCl(s) + H_2SO_4(\ell) \longrightarrow NaHSO_4(s) + HCl(g)$$

A 10.0-kg mass of NaCl reacts completely with sulfuric acid to give a certain volume of HCl(g) at 50°C and $P = 1.00$ atm. If the same volume of hydrogen chloride is collected at 500°C and $P = 1.00$ atm, what mass of NaCl has reacted?

68. Exactly 1.0 lb of Hydrone, an alloy of sodium with lead, yields (at 0.0°C and 1.00 atm) 2.6 ft³ of hydrogen when it is treated with water. All of the sodium reacts according to the equation

$$2\ Na_{in\ alloy} + 2\ H_2O(\ell) \longrightarrow 2\ NaOH(aq) + H_2(g)$$

and the lead does not react with water. Compute the percentage by mass of sodium in the alloy.

69. A sample of limestone (calcium carbonate, CaCO₃) is heated at 950 K until it is completely converted to calcium oxide (CaO) and CO₂. The CaO is then all converted to calcium hydroxide by addition of water, yielding 8.47 kg of solid Ca(OH)₂. Calculate the volume of CO₂ produced in the first step, assuming it to be an ideal gas at 950 K and a pressure of 0.976 atm.

70. A gas exerts a pressure of 0.740 atm in a certain container. Suddenly a chemical change occurs that consumes half of the molecules originally present and forms two new molecules for every three consumed. Determine the new pressure in the container if the volume of the container and the temperature are unchanged.

71. The following arrangement of flasks is set up. Assuming no temperature change, determine the final pressure inside the system after all stopcocks are opened. The connecting tube has zero volume.

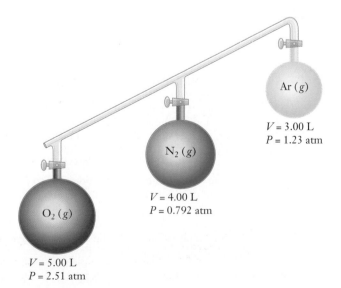

Ar (g)

V = 3.00 L
P = 1.23 atm

N₂ (g)

V = 4.00 L
P = 0.792 atm

O₂ (g)

V = 5.00 L
P = 2.51 atm

*72. A mixture of CS₂(g) and excess O₂(g) in a 10.0-L reaction vessel at 100.0°C is under a pressure of 3.00 atm. When the mixture is ignited by a spark, it explodes. The vessel successfully contains the explosion, in which all of the CS₂(g) reacts to give CO₂(g) and SO₂(g). The vessel is cooled back to its original temperature of 100.0°C, and the pressure of the mixture of the two product gases and the unreacted O₂(g) is found to be 2.40 atm. Calculate the mass (in grams) of CS₂(g) originally present.

73. Acetylene reacts with hydrogen in the presence of a catalyst to form ethane according to the equation

$$C_2H_2(g) + 2\ H_2(g) \longrightarrow C_2H_6(g)$$

The pressure of a mixture of acetylene and an excess of hydrogen decreases from 0.100 atm to 0.042 atm in a vessel of a given volume after the catalyst is introduced, and the temperature is restored to its initial value after the reaction reaches completion. What was the mole fraction of acetylene in the original mixture?

74. Refer to the atomic trap described in problem 42.
    (a) Assume ideal gas behavior to compute the pressure exerted on the "walls" of the optical bottle in this experiment.
    (b) In this gas, the mean free path (the average distance traveled by the sodium atoms between collisions) is 3.9 m. Compare this to the mean free path of the atoms in gaseous sodium at room conditions.

75. Deuterium (²H), when heated to sufficiently high temperature, undergoes a nuclear fusion reaction that results in the production of helium. The reaction proceeds rapidly at a temperature $T$, at which the average kinetic energy of the deuterium atoms is $8 \times 10^{-16}$ J. (At this temperature, deuterium molecules dissociate completely into deuterium atoms.)
    (a) Calculate $T$ in kelvins. (Atomic mass of ²H = 2.015.)
    (b) For the fusion reaction to occur with ordinary hydrogen atoms, the average energy of the atoms must be about $32 \times 10^{-16}$ J. By what factor does the average speed of the ¹H atoms differ from that of the ²H atoms of part (a)?

76. Molecules of oxygen of the following isotopic composition are separated in an oxygen enrichment plant: ¹⁶O¹⁶O, ¹⁶O¹⁷O, ¹⁶O¹⁸O, ¹⁷O¹⁷O, ¹⁷O¹⁸O, ¹⁸O¹⁸O.
    (a) Compare the average translational kinetic energy of the lightest and heaviest molecular oxygen species at 200°C and at 400°C.
    (b) Compare the average speeds of the lightest and heaviest molecular oxygen species at the same two temperatures.

*77. What is the probability that an O₂ molecule in a sample of oxygen at 300 K has a speed between $5.00 \times 10^2$ m s⁻¹ and $5.10 \times 10^2$ m s⁻¹? (*Hint:* Try approximating the area under the Maxwell–Boltzmann distribution by small rectangles.)

*78. Molecules in a spherical container make wall collisions in a great circle plane of the sphere. All paths traveled between collisions are equal in length.
    (a) Relate the path length $\Delta\ell$ to the angle $\theta$ and the sphere radius $r$.

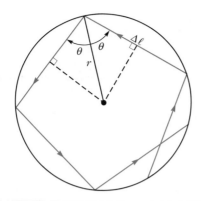

(b) Express the component of the molecular momentum transferred to the wall in a collision in terms of the molecular mass $m$, the speed $u$, and the angle $\theta$.

(c) Using parts (a) and (b), calculate the average force exerted on the wall by one molecule.

(d) Derive the relation $PV = \frac{1}{3} Nm\overline{u^2}$ for a spherical container. This is the same relation found for a rectangular box in Section 4.5.

79. A vessel with a small hole in its wall is filled with oxygen to a pressure of 1.00 atm at 25°C. In a one-minute period, 3.25 g of oxygen effuses out through the hole into a vacuum. The vessel is evacuated and filled with an unknown gas at the same pressure and volume. In this case, 5.39 g of the unknown gas effuses out in one minute. Calculate the molar mass of the unknown gas.

80. A cylindrical storage tank for natural gas (mostly methane, $CH_4$) with a 20-ft radius and a 50-ft height is filled to a pressure of 2000 psi at 20°C. A small leak of 1.0-mm$^2$ area develops at one of the welds. Calculate the mass of $CH_4$, in grams, that leaks out of the tank in one day. What fraction of the total gas escapes per day?

81. A thermos bottle (Dewar vessel) has an evacuated space between its inner and outer walls to diminish the rate of transfer of thermal energy to or from the bottle's contents. For good insulation the mean free path of the residual gas (air; average molecular mass = 29) should be at least ten times the distance between the inner and outer walls, which is about 1.0 cm. What should be the maximum residual gas pressure in the evacuated space if $T = 300$ K? Take an average diameter of $d = 3.1 \times 10^{-10}$ m for the molecules in the air.

82. A tanker truck carrying liquid ammonia overturns, releasing ammonia vapor into the air.

(a) Approximating ammonia, oxygen, and nitrogen as spheres of equal diameter ($3 \times 10^{-10}$ m), estimate the diffusion constant of ammonia in air at atmospheric pressure and 20°C.

(b) Calculate the time required for a 100-m root-mean-square displacement of ammonia from the truck, and

express this time in everyday units (seconds, minutes, hours, days, or years). The actual time for the ammonia to travel this distance is far shorter because of the existence of air currents (even when there is no wind).

83. Molecules of $UF_6$ are approximately 175 times more massive than $H_2$ molecules; however, Avogadro's number of $H_2$ molecules confined at a set temperature exert the same pressure on the walls of the container as the same number of $UF_6$ molecules. Explain how this is possible.

84. The number density of atoms (chiefly hydrogen) in interstellar space is about 10 per cubic centimeter, and the temperature is about 100 K.

(a) Calculate the pressure of the gas in interstellar space, and express it in atmospheres.

(b) Under these conditions, an atom of hydrogen collides with another atom once every $1 \times 10^9$ seconds (that is, once every 30 years). By using the root-mean-square speed, estimate the distance traveled by a hydrogen atom between collisions. Compare this distance with the distance from the earth to the sun (150 million km).

85. A sample of 2.00 mol of argon is confined at low pressure in a volume at a temperature of 50°C. Describe quantitatively the effects of each of the following changes on the pressure, the average energy per atom in the gas, the root-mean-square speed, the rate of collisions with a given area of wall, the frequency of Ar–Ar collisions, and the mean free path:

(a) The temperature is decreased to −50°C.

(b) The volume is doubled.

(c) The amount of argon is increased to 3.00 mol.

86. By assuming that the collision diameter of a $CH_4$ molecule is given by its Lennard–Jones $\sigma$ parameter (Table 4.4), estimate the rate at which methane molecules collide with one another at 25°C and a pressure of (a) 1.00 atm; (b) $1.0 \times 10^{-7}$ atm.

*87. The van der Waals constant $b$ is related to the volume excluded per mole of molecules, so it should be proportional to $N_0\sigma^3$, where $\sigma$ is the distance parameter in the Lennard–Jones potential.

(a) Make a plot of $b$ against $N_0\sigma^3$ for Ar, $H_2$, $CH_4$, $N_2$, and $O_2$, using data from Tables 4.3 and 4.4. Do you see an overall correlation between the two?

(b) The van der Waals constant $a$ has dimensions of pressure times the square of the molar volume. Rewrite the units of $a$ in terms of energy, length, and number of moles, and suggest a relationship between $a$ and some combination of the constants $\epsilon$, $\sigma$, and $N_0$. Make a plot of $a$ against this combination of constants for the gases of part (a). Do you see an overall correlation in this case?

*88. Take the derivative of the Lennard–Jones potential to express the force exerted on one atom by another in terms of the distance $R$ between them. Calculate the forces (in joules per meter) on a pair of interacting argon atoms at distances of 3.0, 3.4, 3.8, and $4.2 \times 10^{-10}$ m. Is the force attractive or repulsive at each of these distances?

## CUMULATIVE PROBLEMS

89. A gaseous hydrocarbon, in a volume of 25.4 L at 400 K and a pressure of 3.40 atm, reacts in an excess of oxygen to give 47.4 g of $H_2O$ and 231.6 g of $CO_2$. Determine the molecular formula of the hydrocarbon.

90. A sample of a gaseous binary compound of boron and chlorine weighing 2.842 g occupies 0.153 L at 0°C and 1.00 atm pressure. This sample is decomposed to give solid boron and gaseous chlorine ($Cl_2$). The chlorine occupies 0.688 L at the same temperature and pressure. Determine the molecular formula of the compound.

91. A mixture of calcium carbonate, $CaCO_3$, and barium carbonate, $BaCO_3$, weighing 5.40 g reacts fully with hydrochloric acid, $HCl(aq)$, to generate 1.39 L of $CO_2(g)$, measured at 50°C and 0.904 atm pressure. Calculate the percentages by mass of $CaCO_3$ and $BaCO_3$ in the original mixture.

92. A solid sample of $Rb_2SO_3$ weighing 6.24 g reacts with 1.38 L of gaseous HBr, measured at 75°C and 0.953 atm pressure. The solid RbBr, extracted from the reaction mixture and purified, has a mass of 7.32 g.
    (a) What is the limiting reactant?
    (b) What is the theoretical yield of RbBr, assuming complete reaction?
    (c) What is the actual percentage yield of product?

# CHAPTER 5

# Solids, Liquids, and Phase Transitions

*Illustration*
Solid iodine is converted directly to a vapor (sublimes) when warmed. Here, purple iodine is redeposited as solid on the cooler upper surfaces of the tube. *(Charles D. Winters)*

Solids, liquids, and gases are differentiated by the magnitudes of their bulk properties. Bulk properties are associated with an assembly of molecules as a whole, not with the members of the assembly. Bulk properties measure the response of the assembly to an externally applied disturbance (e.g., increased pressure or temperature) and reflect the extent to which molecules can move in response to the disturbance. Molecular responses depend on the average distance between molecules and on the magnitude of the forces between them. The interplay between number density of molecules and the magnitude of intermolecular forces establishes a characteristic local structure, or arrangement of molecules, in each state of matter. This local structure determines the bulk properties of matter.

We begin with a brief survey of bulk properties and their molecular interpretation, followed by a description of several types of forces between molecules and of the origin of these forces in the structure of the molecules. Then we survey the states of matter and changes between them as consequences of the forces between molecules. In the previous chapter, we explained the ideal gas law from kinetic molecular arguments by neglecting all intermolecular forces in the gas. In the present chapter, we apply similar kinetic molecular arguments to solids and liquids but cannot neglect the intermolecular forces. Consequently, the discussion of solids and liquids is more qualitative than that for gases.

---

## 5.1

# BULK PROPERTIES OF GASES, LIQUIDS, AND SOLIDS: MOLECULAR INTERPRETATION

Each of the following measurements provides clear distinctions between gases, liquids, and solids by probing the strength of intermolecular forces, albeit indirectly. Each demonstrates that in gases at low density molecules are on average far apart and interact weakly; in condensed phases, molecules are closely packed together and interact quite strongly. These general conclusions apply almost always to substances consisting of small, nearly rigid molecules such as $N_2$, $CO_2$, $CH_4$, and $C_2H_3O_2$. By contrast, biological materials and synthetic plastic or polymeric materials contain complex chainlike molecules that can become strongly entangled and classification as solids or liquids becomes difficult.

Qualitative molecular interpretation of bulk properties requires only the recognition of long-range attractive forces and short-range repulsive forces between molecules, without detailed considerations of molecular shapes. These forces are conveniently represented by the Lennard–Jones 6–12 potential, already used in Section 4.7 to obtain corrections to the ideal gas law.

### Molar Volume

One mole of a typical solid or liquid occupies a volume of 10 to 100 $cm^3$ at room conditions, but the **molar volume** of a gas under the same conditions is about 24,000 $cm^3$ $mol^{-1}$. This large difference explains why solids and liquids are called the condensed states of matter. Since a mole of any substance contains Avogadro's number of molecules, molar volume is inversely related to the **number density** (number of molecules $cm^{-3}$) of materials; liquids and solids have high number densities, and gases have very low number densities. Upon melting, most solids change volume by only 2 to 10%, showing that the solid and liquid states of a given substance are condensed, relative to the gaseous state, by roughly the same amount.

The similar molar volumes of solid and liquid forms of the same substance suggest that the distances between neighboring molecules in the two states must be approximately the same. Density measurements (Section 1.6) show that the intermolecular contacts, the distances between the nuclei of atoms at the far edge of one molecule and the near edge of a neighbor, usually range from $3 \times 10^{-10}$ m to $5 \times 10^{-10}$ m in solids and liquids. At these distances, longer range attractive forces and shorter range repulsive forces just balance one another, giving a minimum in the potential energy (Section 4.7). Although these intermolecular separations are significantly longer than chemical bond distances (which range from 0.5 to $2.5 \times 10^{-10}$ m), they are much shorter than the intermolecular separations in gases, which average around $30 \times 10^{-10}$ m under room conditions. The distinction is shown schematically in Figure 5.1.

## Compressibility

The **compressibility** of a substance is defined as the fractional decrease in volume as applied pressure is increased. Both solids and liquids are nearly incompressible, but gases have large compressibilities. According to Boyle's law, doubling the pressure on a gas from 1 to 2 atm reduces it to one half of its original volume (at constant temperature). Doubling the pressure on water or steel scarcely changes the volume. The high compressibility of gases and the low compressibilities of solids and liquids suggest that in gases there is space between the molecules, but in the condensed states the particles of a substance are in contact or nearly in contact.

The fact that gases are highly compressible while liquids and solids are almost totally incompressible is consistent with strong intermolecular forces in the condensed states and their absence in the gaseous state. Only modest energy is required to bring the widely separated gas molecules closer together, since forces between them are negligible. Attempts to compress liquids and solids, however, require great expenditure of energy against the strongly repulsive forces operating once molecules are in contact (Section 4.7).

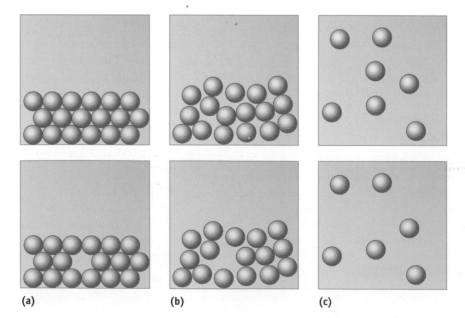

**FIGURE 5.1** Intermolecular forces create structure in liquids and solids. If a single atom is removed from a snapshot of the atomic arrangement in a solid (a), it is easy to figure out exactly where to put it back in. For a liquid (b), the choices are limited. If a single atom is removed from a gas (c), no clue remains to tell where it came from.

(a)                    (b)                    (c)

## Thermal Expansion

The **coefficient of thermal expansion** is defined as the fractional increase in the volume of a substance per degree increase in temperature. Charles's law shows that this coefficient is the same for all gases and takes the value $1/273.15(°C)^{-1}$ at 0°C. Increasing the temperature by 1°C thus causes a gas to expand by 1/273.15, or 0.366% of its original volume at 0°C, as long as the pressure is constant. The thermal expansion coefficients of liquids and solids are much smaller. Heating water from 20 to 21°C increases its volume by only 0.0212%, and the volume of mercury goes up by only 0.0177% over the same temperature interval. The coefficients of thermal expansion of solids are mostly less than 0.02% per degree Celsius.

The difference in thermal expansion between condensed states and gases is consistent with strong intermolecular forces in the condensed states and their absence in the gaseous state. An increase in volume in a solid or liquid requires that attractive forces between each molecule and its neighbors be partially overcome. Because the intermolecular distances in a solid or liquid fall in the region where forces from intermolecular attractions are strongest, relatively small expansion is produced by increasing the temperature. By contrast, molecules in a gas are so far apart that forces from attractions are essentially negligible; the same temperature increase produces much greater expansion in a gas than in condensed phases.

## Fluidity and Rigidity

The most characteristic property of gases and liquids is their **fluidity,** which contrasts with the **rigidity** of solids. A rigid material retains its shape under stress (externally applied mechanical force); it manifests structural strength by resisting flow when stress is applied. Liquids possess definite volumes but keep no definite shapes of their own; they flow easily under stress. The resistance of a material to macroscopic flow is measured by its **shear viscosity;** on the microscopic level, shear viscosity measures the resistance when one thin layer of molecules is "dragged across" another thin layer. The shear viscosities of most liquids are about 16 orders of magnitude smaller than those of most solids, and those of gases are smaller yet. The properties of **hardness** (resistance to indentation) and **elasticity** (capacity to recover shape when a deforming stress is removed) are closely related to rigidity, or high shear viscosity. Solids possess these properties in good measure; gases and liquids do not.

## Diffusion

If two different substances are placed in contact (e.g., a drop of red ink into a beaker of water) they start to mix; molecules of one migrate, or **diffuse,** into the other. Molecules in gases at room conditions diffuse at rates on the order of centimeters per second. Molecules in liquids and solids diffuse far more slowly. The **diffusion constant** of a substance measures the rate of diffusive mixing. At room temperature and pressure, diffusion constants for the diffusion of liquids into liquids are about four orders of magnitude smaller than those for gases into gases; diffusion constants of solids into solids are many orders of magnitude smaller yet. Diffusion in solids is really quite slow.

Figure 5.1 showed a "snapshot" of a liquid and a solid, fixing the positions of the atoms at a particular instant in time. The paths followed by the molecules in these two states can also be examined over a short time interval (Fig. 5.2). In liquids, molecules are free to travel through the sample, changing neighbors constantly

**FIGURE 5.2**  In this computer-simulated picture of the motions of atoms in a tiny melting crystal, the atoms at the center (in the solid) move erratically about particular sites. The atoms at the surface (in the liquid) move over much greater distances.

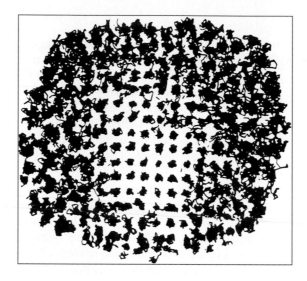

in the course of their diffusive motion. In a solid, the molecules vibrate about their home positions, and mostly they stay close to those positions. The low shear viscosity of a liquid implies that its molecules can quickly change neighbors, finding new interactions as the liquid flows in response to an external stress. The rigidity of solids suggests, in contrast, a durable arrangement of neighbors about any given molecule. The durable arrangement of molecules in a solid, as opposed to the freedom of molecules to diffuse in a liquid at comparable packing density, is the crucial difference between the solid and liquid states.

In a liquid, individual molecules experience interactions with neighbors that lead, at any instant, to a local environment closely resembling that in a solid, but they quickly move on. Their trajectories consist of "rattling" motions in a temporary cage formed by neighbors and superimposed upon erratic displacements over larger distances. In this respect, a liquid is intermediate between a gas and a solid. A gas (see Fig. 4.18) provides no temporary cages, so each molecule of a gas travels a longer distance before colliding with a second molecule. Consequently, the diffusion constant of a gas is larger than that of a liquid. In a solid, the cages are nearly permanent, so diffusion is very slow. Melting occurs as thermal energy increases the amplitude of vibration of the molecules about their home positions in a solid to such a degree that they are set free to make major excursions. These explanations also apply to the observation that liquids can dissolve substances much more rapidly than do solids. Individual molecules of a liquid quickly wander into contact with molecules of an added, second substance, and new attractions between the unlike molecules have an early chance to replace those existing originally in the pure liquid.

Liquids and gases mix through **convection,** as well as by diffusion. In convection, the net flow of a whole region of fluid with respect to another leads to mixing at far greater rates than occurs through simple diffusion; convection is not available to solids. Convection is the primary mechanism through which mixing takes place in the oceans and in the atmosphere.

## Surface Tension

Boundaries between phases have special importance in chemistry and biology. Each type of boundary has unique characteristics. The surface of water (or any liquid) in contact with air (or any gas) resists attempts to increase its area (Fig. 5.3a). This

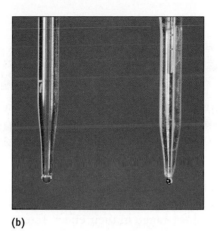

(a)                                                    (b)

**FIGURE 5.3**    (a) Surface tension causes the spherical shape of the water droplet in this photograph, which was taken an instant after a drop of water hit the surface of a pool and bounced up, pulling with it a column of water. *(Hermann Eisenbeiss/Photo Researchers, Inc.)* (b) The mercury drop at the dropper tip on the right is a nearly perfect sphere, whereas the water drop on the left sags slightly. This is evidence of the higher surface tension of the mercury, the drops of which resist the deforming pull of gravity more effectively than those of water. *(Leon Lewandowski)*

**surface tension** causes the surface to behave like a weak, elastic skin. Effects of surface tension are particularly apparent under zero gravity, where liquids float around as spherical drops because spheres contain the largest volume for the smallest surface area of any geometrical shape. If two small drops encounter each other, they tend to coalesce into a larger drop because one large drop has a smaller surface area than two small drops. The surface tension of water is larger than that of most other liquids at room temperature, but is about six times smaller than that for the liquid metal mercury, which has one of the highest values known. (Fig. 5.3b).

Surface tension results from the intermolecular attractions among the molecules in a liquid (Fig. 5.4). Increasing the surface area of a liquid requires redistributing some of the molecules that were originally buried in the interior to positions at the enlarged boundaries. Molecules at the edges have no neighbors on one side and experience attractions only from molecules in the bulk of the liquid. Their potential energy is higher than if they were in the interior. Increasing the surface area of a liquid therefore requires an input of energy. Liquids (such as water and mercury) that have large surface tensions have particularly strong intermolecular attractions, as confirmed by measurement of other properties.

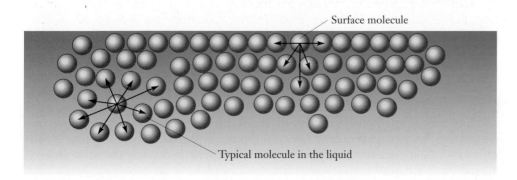

Surface molecule

Typical molecule in the liquid

**FIGURE 5.4**    The intermolecular attractions acting on a molecule at the surface of a liquid pull it downward and to the sides, but not upward. In the interior, a molecule is pulled more or less equally in all directions.

---
**5.2**
---

# Intermolecular Forces: Origins in Molecular Structure

In order to provide a more quantitative explanation of the magnitudes of the properties of different materials, it is necessary to consider several types of intermolecular forces in addition to simple attraction and repulsion. These are described in the following paragraphs, and related to the structure of the molecules. Intermolecular forces are distinguished from intramolecular forces, which are the covalent chemical bonds discussed in Chapter 3. Covalent bonds among atoms establish and maintain the structure of discrete molecules and are strong, directional, and comparatively short-range in their effect. Intermolecular forces differ in several important ways:

1. Intermolecular forces are generally weaker than covalent chemical bonds; for example, it takes 239 kJ to break a mole of Cl–Cl covalent bonds but only 1.2 kJ to overcome a mole of Ar–Ar attractions.
2. Intermolecular forces are much less directional than covalent chemical bonds.
3. Intermolecular forces operate at longer range than covalent chemical bonds.

All intermolecular and intramolecular forces arise because matter is composed of electrically charged particles. Although they all spring ultimately from the same source, it is very useful to distinguish types of forces based on their strength, directionality, and range. Often the physical and chemical properties of liquids and solids can be interpreted or even predicted according to the types of intermolecular force that predominate in their internal organization.

## Ion–Ion Forces

The units of organization in ionic solids and liquids are electrically charged entities, sometimes monatomic, like $Na^+$, $Cl^-$, and $Ca^{2+}$, and sometimes polyatomic, like $NH^{4+}$ and $SO_4^{2-}$. The main interaction among these ions is the **Coulomb force** of electrostatic attraction or repulsion, which leads to the Coulomb potential described in Section 3.2. Ions of like charge repel one another, and ions of unlike charge attract one another. These **ion–ion forces** can be as strong as the covalent bond and are relatively long-range; their potential energy is proportional to $R^{-1}$, and decreases less rapidly with distance than do the strengths of other types of interactions. Ion–ion forces are not directional; each ion interacts equally with neighboring ions on all sides. Ion–ion forces lead to the formation of ionic bonds through the Coulomb stabilization energy (Section 3.2).

## Dipole–Dipole Forces

Two polar molecules interact through **dipole–dipole forces**. As shown in Figure 5.5, these forces depend on the orientations of the two molecules, and consequently can be either attractive or repulsive. Random motions of polar molecules in gases and liquids lead to a variety of energetically favorable temporary dipole–dipole orientations. The potential energy between dipoles separated by the distance $R$ falls off as $R^{-3}$. Consequently, this potential decreases with increasing distance between

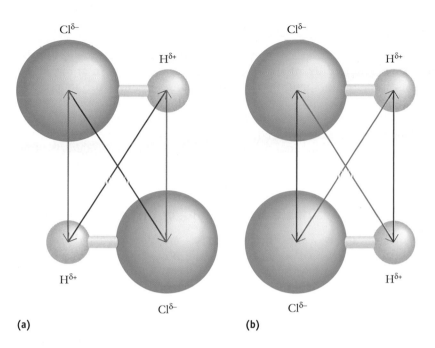

$Cl^{\delta-}$ $H^{\delta+}$

$H^{\delta+}$

$Cl^{\delta-}$

(a)

$Cl^{\delta-}$ $H^{\delta+}$

$H^{\delta+}$

$Cl^{\delta-}$

(b)

**FIGURE 5.5** A molecule of HCl can be represented as having a small net negative charge on the Cl end, balanced by a small net positive charge on the H end. The forces between two HCl molecules depend on their orientations. In (a), the oppositely charged ends (blue arrows) are closer than the ends with the same charge (red arrows). This gives a net attractive force. In (b), the opposite is true, and the net force is repulsive.

dipoles much more rapidly than does the Coulomb potential between ions. In liquids, thermal energy can overcome dipole–dipole attractions and disrupt favorable orientations; dipole–dipole interactions are too weak to hold molecules in a liquid together in a nearly rigid arrangement. Nonetheless, they are sufficiently strong to influence numerous physical properties, including boiling points, melting points, and molecular orientations in solids.

*dipole dipole weaker than both!*
*— ionic*
*— covalent bonds.*

## Ion–Dipole Forces

A polar molecule interacts with both positive and negative ions: positive ions are attracted by the negative end of the dipole and repelled by the positive end, and the reverse is true for negative ions. The interaction between a polar solvent molecule, such as water, and a dissolved ion is the most common case of ion–dipole interaction. Figure 5.6 shows dissolved $Na^+$ and $Cl^-$ ions surrounded by water dipoles. This case will be discussed more thoroughly in Chapter 6.

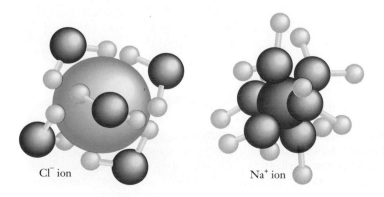

$Cl^-$ ion

$Na^+$ ion

**FIGURE 5.6** Both positive and negative ions are attracted to neighboring water molecules in aqueous solution by ion–dipole forces. The orientations of the water molecules are reversed in the two cases, however. Oxygen atoms in molecules of water bear small negative charges; hydrogen atoms bear small positive charges.

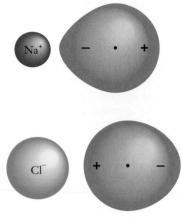

**FIGURE 5.7** As an ion approaches an atom or molecule, its electrostatic field distorts the distribution of the outer electrons. Here the average electron densities are shown by shading. The effect of this distortion is to create a dipole moment that exerts an attractive force back on the ion.

*If solving for these forces make sure they fit this range.*

## Induced Dipole Forces

The electrons in a nonpolar molecule or atom are distributed symmetrically, but the distribution can be distorted by an approaching electrical charge. An argon atom, for example, has no dipole moment, but an approaching Na$^+$ ion (with its positive charge) attracts the electrons on its near side more strongly than the ones on its far side. By tugging the nearby electrons harder, the Na$^+$ ion induces polarity in the argon atom (Fig. 5.7). Once the induced dipole is present, the situation is like the ion–dipole case just described. **Induced dipole forces** also can be caused by a negative ion or by another dipole. These are weak and are effective only at short range.

## Induced Dipole–Induced Dipole Forces: London Dispersion Forces

Helium atoms, like the atoms of the other noble gases, are uncharged and nonpolar, so none of the forces mentioned so far explains the observed fact that there are attractions between helium atoms. Such attractions derive from **dispersion forces,** which exist between all atoms and molecules. They arise from fluctuations over time in the distribution of the electrons on two neighboring atoms or molecules. A fluctuation in one molecule (making a temporary dipole) will induce a second temporary dipole in the other. The interaction of these two dipoles will then cause an attractive force between the two molecules. Figure 5.8 provides a simple picture of the source of this interaction. The ability to distort the electron cloud on an atom or molecule is determined by its polarizability, which increases with the number of electrons in the atom or molecule. Therefore heavier atoms or molecules interact more strongly by dispersion forces than do lighter ones. Dispersion forces are always attractive and depend on intermolecular separation as $R^{-6}$. Consequently they decrease even more rapidly with increasing distance than do dipole–dipole interactions. The energy of dispersion forces ranges from zero to around 1 kJ mol$^{-1}$. Dispersion forces provide the attractive term in the Lennard–Jones potential described previously in Section 4.7.

## Repulsive Forces

As atoms or molecules approach each other closely, **repulsive forces** come into play and soon dominate the attractive forces considered so far. The source of these forces

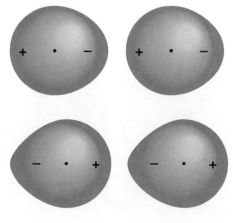

**FIGURE 5.8** A fluctuation of the electron distribution on one atom induces a corresponding temporary dipole moment on a neighboring atom. The two dipole moments interact to give a net attractive force, called a "dispersion force."

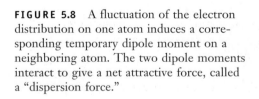

is the strong repulsion between the core (nonvalence) electron clouds when neighboring atoms are forced close to each other. Because the potential energy due to repulsive forces varies as $R^{-n}$ where $n$ is quite large ($n = 12$ in the Lennard–Jones potential), this contribution is negligible until the distance between centers becomes small, at which point the repulsive energy increases rapidly as distance is reduced further. This steep, repulsive interaction at extremely small distances justifies modeling atoms as hard, nearly incompressible spheres with characteristic dimensions called the **van der Waals radii.** This label honors the early contributions of Johannes van der Waals to the study of nonbonded interactions between molecules and their influence on the properties of materials. The minimum distance between molecules in a condensed phase is determined by the sum of the van der Waals radii of their atoms. Space-filling models and drawings are usually designed to approximate the van der Waals surface of molecules, which represents the distance of closest approach by neighboring molecules.

## EXAMPLE 5.1

State which attractive intermolecular forces are likely to predominate in the associations among molecules in the following substances:
**(a)** $F_2(s)$
**(b)** $HBr(\ell)$
**(c)** $NH_4Cl(s)$

### Solution
**(a)** Molecules of $F_2$ are nonpolar, so the predominant attractive forces between molecules in $F_2(s)$ come from dispersion.
**(b)** The HBr molecule has a permanent dipole moment. The predominant forces between molecules are dipole–dipole. Dispersion forces will also contribute to associations, especially because Br is a rather heavy atom.
**(c)** The ammonium ions are attracted to the chloride ions primarily by ion–ion forces.

**Related Problems: 15, 16, 17, 18, 19, 20**

## Comparison of Potential Energy Curves

The relative strengths and effective ranges of several intermolecular forces are illustrated in Figure 5.9, which shows how the potential energy depends on the center-to-center distance for several pairs of ions, atoms, and molecules. The forces illustrated here include the Coulomb ($R^{-1}$), dipole–dipole ($R^{-3}$), dispersion ($R^{-6}$), and repulsive ($R^{-12}$). The species shown in Figure 5.9 were chosen so that the interacting atoms, ions, or molecules have the same number of electrons (Ar, $Cl^-$, $K^+$, HCl). For comparison, the covalent bond (*intra*molecular force) for $Cl_2$ is also shown. The ion–ion interaction of $K^+$ with $Cl^-$ is the strongest (stronger even than the covalent interaction in $Cl_2$), followed by the interaction between two HCl molecules (dipole–dipole and dispersion), and the Ar–Ar interaction (dispersion only).

**FIGURE 5.9** The potential energy of a pair of atoms, ions, or molecules depends on the distance between the members of a pair. Here, the potential energy at large separations (to the right side of the graph) is arbitrarily set to zero. As pairs of particles approach each other, the potential energy becomes negative because attractive forces come into effect. The lowest point in each curve occurs at the distance where attractive and repulsive forces exactly balance. The relative potential energy values at these minima measure the relative strength of the attractive forces in the various cases illustrated. Note the very shallow potential energy minimum for HCl and Ar. The inset shows these same two curves with the vertical scale expanded by a factor of 100. [The HCl–HCl curve was computed for the relative orientations of Figure 5.5(a).]

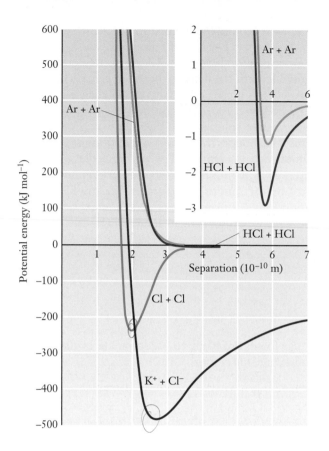

---

### 5.3

# INTERMOLECULAR FORCES IN LIQUIDS

The same intermolecular forces that make gases nonideal (Section 4.7 and Section 5.2) are responsible for the existence of solids and liquids. At very high temperatures, these forces are negligible because the very high kinetic energy of the molecules disrupts all possible attractions; all materials are gaseous at sufficiently high temperature. At lower temperatures where materials are in the liquid state, molecules are close together, and details of the intermolecular potential energy determine the properties of materials. Section 5.2 described the influence of molecular structure on the intermolecular potential energy. The present section surveys the resulting correlation between properties of liquids and the structure of their constituent molecules. Special emphasis is given to the unusual properties of water.

The stronger the attractive forces are, the greater the stability of the liquid relative to the gas phase and the higher the normal boiling point $T_b$. Ionic liquids generally have the strongest attractions, due to the Coulomb interaction of charged ions, and have very high boiling points. Molten NaCl, for example, boils at 1686 K under atmospheric pressure. At the opposite extreme, the normal boiling point of helium is only 4.2 K. Within a series of related compounds, those of higher molar mass tend to have higher normal boiling points. Thus, from helium to xenon, normal boiling points increase (see Fig. 5.10) along with the strengths of the attractive forces among the noble gases (represented by the well depth $\epsilon$ in Table 4.4).

Between the noble-gas and ionic liquids falls a class of liquids called **polar liquids.** In liquid HCl, the molecules arrange themselves to the extent possible with

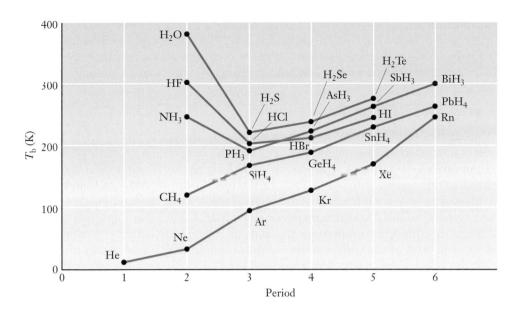

neighboring dipoles oriented to minimize the dipole–dipole potential energy. As described in Section 5.2, the dipole–dipole intermolecular forces in such polar liquids are weaker than the ion–ion coulombic forces in ionic liquids but stronger than the dispersion forces in **nonpolar liquids** such as $N_2$. These three forces operate respectively between molecules in which the bonding is polar covalent, fully ionic, and fully covalent. As Figure 5.10 shows, HCl has a higher boiling point than argon (a nonpolar fluid of atoms with nearly the same molar mass) because of its polar nature. The magnitudes of the intermolecular forces in HCl and argon were compared explicitly in Figure 5.9.

## Hydrogen Bonds

Figure 5.10 showed the normal boiling points of several series of hydrides. As just stated, the boiling point usually increases with increasing molar mass in a series of related compounds. The dramatic deviations from these systematic trends shown by HF, $NH_3$, and especially $H_2O$ indicate the strength and importance of the special type of bond that is common to these cases, a **hydrogen bond.** Such a bond forms when a hydrogen atom bonded to an O, N, or F atom (highly electronegative atoms) also interacts with the lone electron pair of another such atom nearby. Figure 5.11 shows the interaction of a pair of water molecules to form a dimer in the gas phase. The hydrogen bond that forms is weaker than an ordinary O—H covalent bond but significantly stronger than most intermolecular interactions. Like most hydrogen bonds, that in water is linear but asymmetric, with the hydrogen atom closer to and more strongly bound to one of the oxygen atoms. It is indicated as O—H$\cdots$O.

Like HCl or $H_2S$ molecules, the water molecule is polar. The water molecule is bent, from VSEPR theory, and the orientation of its dipole moment (positive end toward the hydrogen atoms and negative end toward the oxygen atom) has been related to its structure in Section 3.6. In the liquid, these molecules orient themselves in directions that minimize the potential energy between them; consequently, hydrogen atoms on one molecule are close to oxygen atoms on neighboring molecules. The hydrogen atom in a bond such as O—H is surrounded by a relatively low density of negative charge since, unlike all other elements, it has no electrons other than valence electrons. As a result, it can approach very close to the lone-pair

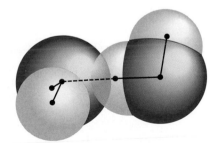

**FIGURE 5.11**  A single hydrogen bond between water molecules forms a dimer. This bond is far weaker than a covalent bond but still strong enough to resist dissociation at room temperature. The shared hydrogen atom at the center approaches the neighboring oxygen atom quite closely.

electrons on a neighboring oxygen atom, causing a strong electrostatic (Coulomb) interaction between the two. In addition, a small amount of covalent bonding arises from the sharing of electrons between the two oxygen atoms and the intervening hydrogen atom. These effects combine to make the interaction unusually strong.

## Special Properties of Water

Water makes up about 0.023% of the total mass of the earth. About $1.4 \times 10^{21}$ kg of it is distributed above, on, and below the earth's surface. The volume of this vast amount of water is about 1.4 billion $km^3$. Most of the earth's water (97.7%) is contained in the oceans, with about 1.9% in the form of ice or snow and most of the remainder (a very small fraction of the total) available as fresh water in lakes, rivers, underground sources, and atmospheric water vapor. A small but important fraction is bound to cations in certain minerals, such as clays and hydrated crystalline salts. More than 80% of the surface of the earth is covered with water—as ice and snow near the poles, as relatively pure water in lakes and rivers, and as a salt solution in the world's oceans.

The unusual properties of water, which come from its network of hydrogen bonds, have profound effects on life on earth. Figure 5.10 compared the boiling points of water and hydrides that lack hydrogen bonds. An extrapolation of the trends from the latter compounds would give a boiling point for "water without hydrogen bonds" near 150 K ($-123°C$). Life as we know it would not be possible under these circumstances.

If all possible hydrogen bonds form in a mole ($N_0$ molecules) of pure water, then every oxygen atom is surrounded tetrahedrally by four hydrogen atoms: its own two and two from neighboring molecules. The hydrogen bonds number $2N_0$ and form a three-dimensional network. In ice, this full complement of hydrogen bonds does form. The result is an array of interlocking six-membered rings of water molecules (Fig. 5.12), echoed macroscopically in the characteristic six-fold symmetry of snowflakes.

**FIGURE 5.12** The structure of ice is quite open. Each water molecule has only four nearest neighbors with which it interacts by means of hydrogen bonds (blue dashed lines).

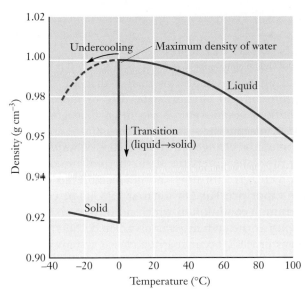

**FIGURE 5.13**    The density of water rises to a maximum as it is cooled to 3.98°C, then starts to decrease slowly. Undercooled water (water chilled below its freezing point but not yet converted to ice) continues the smooth decrease in density. When liquid water freezes, the density drops abruptly.

Water has its maximum density at 4°C (Fig. 5.13), and it expands upon freezing. This unusual behavior, which is seen in very few other liquids, also is due to hydrogen bonds. When ice melts, some of the hydrogen bonds that maintain the open structure of Figure 5.12 break and the structure partially collapses, producing a liquid with a smaller volume (higher density). The reverse process, a sudden expansion of water on freezing, can cause bursting of water pipes and freeze–thaw cracking of concrete. Such expansion has many beneficial effects as well, however. If ice were denser than water, the winter ice that forms at the surface of a lake would sink to the bottom and the lake would freeze from the bottom up. Instead, the ice remains at the surface and the water near the bottom achieves a stable wintertime temperature near 4°C, which allows the lake's fish to survive.

## EXAMPLE 5.2

Predict the order of increase in the normal boiling points of the following substances: $F_2$, HBr, $NH_4Cl$, and HF.

### Solution

As an ionic substance, $NH_4Cl$ should have the highest boiling point of the four (measured value: 520°C). HF should have a higher boiling point than HBr, because its molecules form hydrogen bonds (Fig. 5.10) that are stronger than the dipolar interactions in HBr (measured: 20°C for HF, −67°C for HBr). Fluorine, $F_2$, is non-polar and contains light atoms and so should have the lowest boiling point of the four substances (measured: −188°C).

**Related Problems: 23, 24**

---

### 5.4

## PHASE EQUILIBRIUM

Liquids and solids, like gases, are **phases**—samples of matter that are uniform throughout in both chemical constitution and physical state. Two or more phases can coexist. Suppose a small quantity of liquid water is put in an evacuated flask, with the temperature held at 25°C by placing the system in a constant-temperature bath. A pressure gauge is used to monitor changes in the pressure of water vapor inside the flask. Immediately after the water enters the flask, the pressure of water vapor begins to rise from zero. It increases with time and gradually levels off at a value of 0.03126 atm, which is the **vapor pressure** of water at 25°C. The contents of the flask have reached **equilibrium,** a condition in which no further changes in macroscopic properties occur as long as the system remains isolated. This passage toward equilibrium is a spontaneous process, occurring in a closed system without any external influence. If some of the water vapor that has formed is removed, additional water evaporates from the liquid to reestablish the same vapor pressure, $P_{vap}(H_2O) = 0.03126$ atm.

What is happening on a microscopic scale to cause this spontaneous movement of the system toward equilibrium? According to the kinetic theory, the molecules of water in the liquid are in a constant state of thermal motion. Some of those near the surface are moving fast enough to escape the attractive forces holding them in the liquid; this process of **evaporation** causes the pressure of the water vapor to increase. As the number of molecules in the vapor phase increases, the reverse process begins to occur: molecules in the vapor strike the surface of the liquid and some are captured, leading to **condensation.** As the pressure of the gas increases, the rate of condensation increases until it balances the rate of evaporation from the surface (Fig. 5.14). Once this occurs, there is no further net flow of matter from one phase to the other; the system has reached **phase equilibrium,** characterized by a particular value of the water vapor pressure. Water molecules continue to evaporate from the surface of the liquid, but other water molecules return to the liquid from the vapor at an equal rate. A similar phase equilibrium is established between an ice cube and liquid water at the freezing point.

The vapor pressure of the water is independent of the size and shape of the container. If the experiment is duplicated in a larger flask, then a greater *amount* of

**FIGURE 5.14** The approach to equilibrium in evaporation and condensation. Initially the pressure above the liquid is very low, and many more molecules leave the liquid surface than return to it. As time passes, more molecules fill the gas phase until the equilibrium vapor pressure $P_{vap}$ is approached; the rates of evaporation and condensation then become equal.

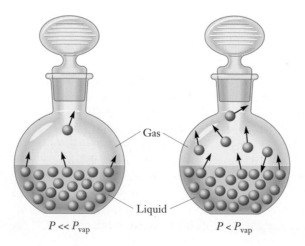
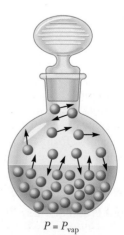

Gas

Liquid

$P \ll P_{vap}$          $P < P_{vap}$          $P = P_{vap}$

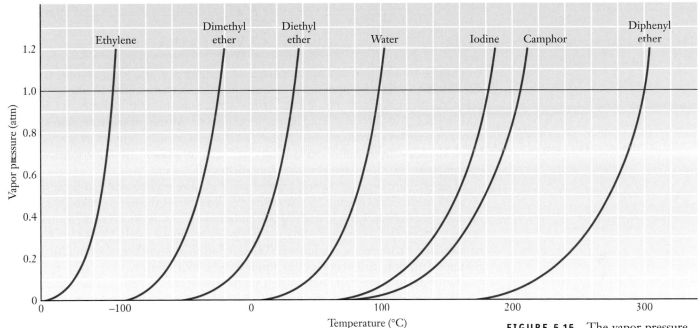

**FIGURE 5.15** The vapor pressure of a solid or liquid depends strongly on temperature. The temperature at which the vapor pressure becomes 1 atm defines the normal boiling point of a liquid and the normal sublimation point of a solid.

water evaporates on the way to equilibrium, but the final pressure in the flask at 25°C is still 0.03126 atm as long as some liquid water is present. If the experiment is repeated at a temperature of 30.0°C, everything happens as just described except that the pressure in the space above the water reaches 0.04187 atm. A higher temperature corresponds to a larger average kinetic energy for the water molecules. A new balance between the rates of evaporation and condensation is struck, but at a higher vapor pressure. The vapor pressure of water, and of all other substances, increases with increasing temperature (Fig. 5.15 and Table 5.1).

Phase equilibrium is a *dynamic* process that is quite different from the static equilibrium achieved as a marble rolls to a stop after being spun into a bowl. In the equilibrium between liquid water and water vapor, the partial pressure levels off, not because evaporation and condensation stop, but because at equilibrium their rates become equal. One further feature of equilibrium evidenced by the liquid–vapor system is that the properties of the system at equilibrium are independent of the direction from which equilibrium is approached. If we inject enough water vapor into the empty flask so that initially the pressure of the vapor is *above* the vapor pressure of liquid water, $P_{vap}(H_2O)$, then liquid water will condense until the same equilibrium vapor pressure is achieved (0.03126 atm at 25°C). Of course, if we do not use enough water vapor to exceed a pressure of 0.03126 atm, all the water will remain in the vapor phase and two-phase equilibrium will not occur.

The presence of water vapor above an aqueous solution has an important practical consequence. If a reaction in aqueous solution generates gases, these gases are "wet," containing water vapor at a partial pressure given by the equilibrium vapor pressure of water at the temperature of the experiment. The amount of gas generated is determined not by the total pressure but by the partial pressure of the gas. Dalton's law (Section 4.4) must be used to subtract the partial pressure of water as listed in Table 5.1. This correction is significant in quantitative work.

**TABLE 5.1**

### Vapor Pressure of Water at Various Temperatures

| Temperature (°C) | Vapor Pressure (atm) |
|---|---|
| 15.0 | 0.01683 |
| 17.0 | 0.01912 |
| 19.0 | 0.02168 |
| 21.0 | 0.02454 |
| 23.0 | 0.02772 |
| 25.0 | 0.03126 |
| 30.0 | 0.04187 |
| 50.0 | 0.1217 |

---

**5.5**

---

# PHASE TRANSITIONS

Suppose 1 mol of gaseous sulfur dioxide is compressed at a temperature fixed at 30.0°C. The volume is measured at each pressure, and a graph of volume against pressure is constructed (Fig. 5.16). At low pressures the graph shows the inverse dependence ($V \propto 1/P$) predicted by the ideal gas law. As the pressure increases, deviations appear because the gas is not ideal. At this temperature, attractive forces dominate; therefore, the volume falls below its ideal gas value and approaches 4.74 L (rather than 5.50 L) as the pressure approaches 4.52 atm.

For pressures up to 4.52 atm this behavior is quite regular and can be described by the van der Waals equation. At 4.52 atm, something surprising takes place: the volume drops abruptly by a factor of 100 and remains low as the pressure is increased further. What has happened? The gas has been liquefied solely by the application of pressure. If the compression of $SO_2$ is continued, another abrupt (but small) change in volume will occur as the liquid freezes to a solid.

Condensed phases also arise when the temperature of a gas is lowered at constant pressure. If steam (water vapor) is cooled at 1 atm pressure, it condenses to liquid water at 100°C and freezes to solid ice at 0°C. Liquids and solids form at low temperatures once the attractive forces between molecules become strong enough to dominate the random effects of kinetic energy.

Six **phase transitions** occur among the three states of matter (Fig. 5.17). Typically, solids melt to give liquids when they are heated, and liquids boil to give gases. **Boiling** is an extension of evaporation (in which the vapor escapes from the surface only). In boiling, bubbles of gaseous substance form actively throughout the body of a liquid and rise up to escape at the surface. Only when the vapor pressure of a liquid exceeds the external pressure can the liquid start to boil. The **boiling point** is the temperature at which the vapor pressure of a liquid equals the external pressure. The external pressure influences boiling points quite strongly; for example, water boils at 25°C if the external pressure is reduced below 0.03126 atm (recall that the vapor pressure of water at 25°C is just this number) but requires a tem-

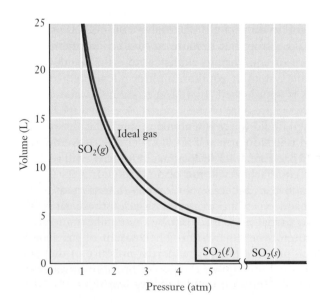

**FIGURE 5.16** As one mole of $SO_2$ is compressed at a constant temperature of 30°C, the volume at first falls somewhat below its ideal gas value. Then, at 4.52 atm, the volume falls abruptly as the gas condenses to a liquid. At a much higher pressure, a further transition to the solid takes place.

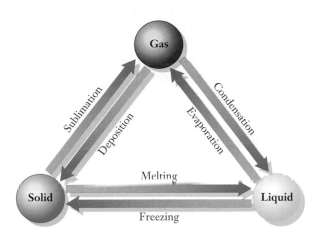

**FIGURE 5.17** Direct transitions among all three states of matter not only are possible but are observed in everyday life.

perature of 121°C to boil under an external pressure of 2.0 atm. At high elevations the pressure of the atmosphere is lower than 1 atm, so water boils at a temperature less than 100°C and food cooks more slowly in boiling water than it would at a lower elevation. By contrast, the use of a pressure cooker increases the boiling temperature and speeds the cooking of food. The **normal boiling point** is defined as the temperature at which the vapor pressure of the liquid equals 1 atm. Figure 5.15 shows that in general a lower normal boiling point implies a higher vapor pressure at any fixed temperature (and therefore a more volatile liquid).

**Melting** is the conversion of a solid to the liquid state. The **normal melting point** of a solid is the temperature at which solid and liquid are in equilibrium under a pressure of 1 atm. The normal melting point of ice is 0.00°C, so liquid water and ice coexist indefinitely (are in equilibrium) at this temperature at a pressure of 1 atm. If the temperature is lowered by even a small amount, then all of the water eventually freezes; if the temperature is infinitesimally raised, all of the ice eventually melts. The qualifying term "normal" is often omitted in talking about melting points because they depend only weakly on pressure.

It is sometimes possible to overshoot a phase boundary, with the new phase appearing only after some delay. An example is the superheating of a liquid. Liquid water, for example, can reach a temperature somewhat above 100°C if heated rapidly. When vaporization of a superheated liquid does occur, it can be quite violent, with liquid thrown out of the container. To avoid this superheating in the laboratory, boiling chips (pieces of porous fired clay) may be added to the liquid. They help to initiate boiling as soon as the normal boiling point is reached, by providing sites where gas bubbles can form. Undercooling of liquids below their freezing points is also possible. In careful experiments, undercooled liquid water has been studied at temperatures below −30°C (at atmospheric pressure).

Many materials react chemically when heated, before they have a chance to melt or boil. Substances that change their chemical identities before changing state do not have normal melting or boiling points. For example, sucrose (table sugar) melts, but quickly begins to darken and eventually chars (Figure 5.18). A temperature hot enough to overcome the intermolecular attractions in sugar is also sufficient to break apart the sugar molecules themselves.

Intermolecular forces exert strong influence on phase transitions. Data already presented in Section 5.3 illustrate the trend that the normal boiling point in a series of liquids increases as the strength of intermolecular forces in the liquids increases. The stronger the intermolecular attractions in a liquid, the lower its vapor pressure

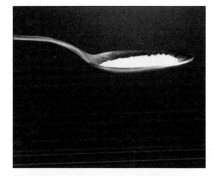

(a)

(b)

**FIGURE 5.18** When sugar (a) is heated, it melts and simultaneously decomposes to a dark-colored caramelized mixture (b). *(Leon Lewandowski)*

at any temperature, and the higher its temperature must be raised to produce vapor pressure equal to 1 atm. Melting points depend more sensitively than do boiling points on molecular shapes and on details of the molecular interactions, and so vary less systematically with the strength of the attractive forces.

---

## 5.6

## PHASE DIAGRAMS

If the temperature of a substance is held constant and its pressure is changed, a phase transition will be observed at some particular pressure between two of its three phases: solid, liquid, and gas. Repeating the process at many different temperatures gives the data necessary to draw the **phase diagram** of that substance—a plot of pressure against temperature that shows the stable state for every pressure–temperature combination. Figure 5.19 shows a sketch of the phase diagram for water. A great deal of information can be read from such diagrams. For each substance there is a unique combination of pressure and temperature, called the **triple point** (marked "T"), at which the gas, liquid, and solid phases coexist in equilibrium. Extending from the triple point are three lines, each denoting the conditions for the equilibrium coexistence of two phases. Along line TA, solid and gas are in equilibrium; along TB, solid and liquid; and along TC, liquid and gas. The areas between lines encompass combinations of pressure and temperature where only one phase exists.

The gas–liquid coexistence curve extends upward in temperature and pressure from the triple point. This line, stretching from T to C in the phase diagrams, is the vapor pressure curve of the liquid substance, portions of which were shown in Figure 5.15. The gas–liquid coexistence curve does not continue indefinitely, but instead terminates at the **critical point** (point C in Fig. 5.19). Along this coexistence curve there is an abrupt, discontinuous change in the density and other properties from one side to the other. The differences between the properties of the

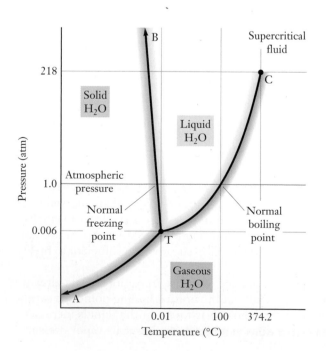

**FIGURE 5.19** Phase diagram for water. The pressures and temperatures are not drawn to scale.

liquid and the gas become smaller as the critical point is approached, and disappear altogether at that point. If the substance is placed in a closed container and is gradually heated, a **meniscus** is observed at the boundary between liquid and gas (Fig. 5.20); at the critical point this meniscus disappears. For pressures above the critical pressure (218 atm for water) it is no longer possible to identify a particular state as gas or liquid. A substance beyond its critical point is called a **supercritical fluid** because the term "fluid" includes both gases and liquids.

The liquid–solid coexistence curve does not terminate, as the gas–liquid curve does at the critical point, but continues to indefinitely high pressures. In practice, such a curve is almost vertical because very large changes in pressure are necessary to change the freezing temperature of a liquid. For most substances, this curve inclines slightly to the right (Fig. 5.21a and b): an increase in pressure increases the freezing point of the liquid. Another way of saying this is that at constant temperature an increase in pressure leads to the formation of a phase with higher density (smaller volume), and for most substances the solid is denser than the liquid. Water and a few other substances are anomalous (Fig. 5.21c); for them, the liquid–solid coexistence curve slopes up initially to the *left*, showing that an increase in pressure causes the solid to melt. This anomaly is related to the densities of the liquid and solid phases: ice is less dense than water (that is why ice cubes float on water), so when ice is compressed at 0°C, it melts.

**FIGURE 5.20**   When a gas and liquid coexist, the interface between them is clearly visible as a meniscus. The meniscus is useful for reading the volume of a liquid in a buret. It disappears as the critical point is reached. (*Leon Lewandowski*)

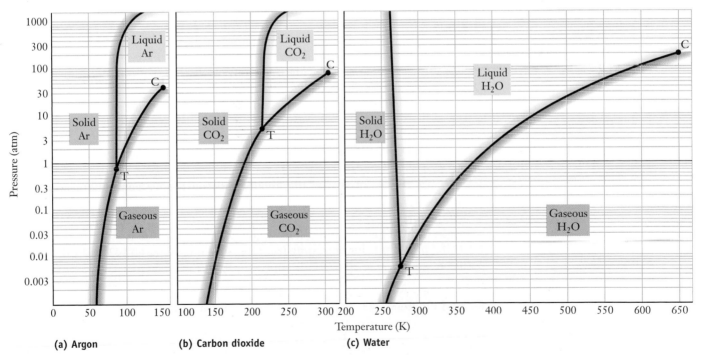

(a) Argon          (b) Carbon dioxide          (c) Water

**FIGURE 5.21**   In these phase diagrams, the pressure increases by a factor of ten at regular intervals along the vertical axis. This method of graphing allows large ranges of pressure to be plotted. The light horizontal and vertical lines mark a pressure of 1 atm and a temperature of 298.15K, or 25°C. Their intersection identifies room conditions. Argon and carbon dioxide are gases at room conditions, but water is a liquid. The letter "T" marks the triple points of the substances, and the letter "C" marks their critical points. The region of stability of liquid water is larger than that of either carbon dioxide or argon.

Sublimation of solid carbon dioxide (dry ice). The white clouds are drops of water vapor (moisture in the air) that condense at the low temperatures near the solid surface. Gaseous carbon dioxide itself is transparent. (*Charles D. Winters*)

For most substances, including water (Fig. 5.21c), atmospheric pressure occurs somewhere between the triple-point pressure and the critical pressure, so in our ordinary experience all three phases—gas, liquid, and solid—are observed. For a few substances the triple-point pressure lies *above* $P = 1$ atm, however, and under atmospheric conditions there is a direct transition called **sublimation** from solid to gas, without an intermediate liquid state. Carbon dioxide is such a substance (see Fig. 5.21b); its triple-point pressure is 5.117 atm (the triple-point temperature is $-56.57°C$). Solid $CO_2$ (dry ice) sublimes directly to gaseous $CO_2$ at atmospheric pressure. In this respect it differs from ordinary ice, which melts before it evaporates and sublimes only at pressures below its triple-point pressure, 0.0060 atm. This fact is used in freeze-drying, a process in which foods are frozen and then put in a vacuum chamber at a pressure of less than 0.0060 atm. The ice crystals that formed on freezing then sublime, leaving a dried food that can be reconstituted by adding water.

Many substances exhibit more than one solid phase as the temperature and pressure are varied. At ordinary pressures the stable state of carbon is graphite, a rather soft, black solid, but at high enough pressures the hard, transparent diamond form is stable. Below 13.2°C (and at atmospheric pressure) elemental tin undergoes a slow transformation from the metallic white form to a powdery gray form, a process referred to as "tin disease." This has caused the destruction of tin organ pipes in unheated buildings. No fewer than nine solid forms of ice are known, some of which exist only over a very limited range of temperatures and pressures.

### EXAMPLE 5.3

Consider a sample of argon held at $P = 600$ atm, $T = 100$ K in Figure 5.21a.
**(a)** What phase(s) is (are) present at equilibrium?
**(b)** Suppose the argon originally held at $P = 600$ atm, $T = 100$ K is heated at constant pressure. What happens?
**(c)** Describe a procedure to convert all the argon to gas without changing the temperature.

### Solution

**(a)** Because this point lies on the liquid–solid coexistence curve, both liquid and solid phases are present.
**(b)** An increase in temperature at constant pressure corresponds to a movement to the right in the figure, and the solid argon melts to a pure liquid.
**(c)** If the pressure is lowered sufficiently at constant temperature (below the triple-point pressure of 0.75 atm, for example) the argon will be converted completely to gas.

**Related Problems: 43, 44, 45, 46, 47, 48, 49, 50, 51, 52**

### CUMULATIVE EXERCISE

#### Bismuth

Bismuth is a rather rare element in the earth's crust, but its oxides and sulfides appear at sufficient concentrations as impurities in lead and copper ores to make its recovery from these sources practical. Annual production of bismuth amounts to several million kilograms worldwide. Although elemental bismuth is a metal, its electrical conductivity is quite poor and it is relatively brittle. The major uses of bismuth arise from its low melting point (271.3°C) and the even lower melting points of its alloys, which range down to 47°C. These alloys are used as temperature sensors in fire detectors and automatic sprinkler systems because, in case of fire, they melt at a well-defined temperature, breaking an electrical connection and triggering an alarm or deluge.

Elemental bismuth    (*Charles D. Winters*)

(a) At its normal melting point, the density of solid bismuth is 9.73 g cm$^{-3}$, and that of liquid bismuth is 10.05 g cm$^{-3}$. Does the volume of a sample of bismuth increase or decrease on melting? Does bismuth more closely resemble water or argon (see Fig. 5.21) in this regard?

(b) Since 1450 (ten years after Gutenberg), bismuth alloys have been used to cast metal type for printing. Explain why those alloys that share the melting behavior of bismuth discussed in part (a) would be especially useful for this application.

(c) A sample of solid bismuth is held at a temperature of 271.0°C and compressed. What will be observed?

(d) The vapor pressure of liquid bismuth has been measured to be 5.0 atm at a temperature of 1850°C. Does its normal boiling point lie above or below this temperature?

(e) At 1060°C, the vapor pressure of liquid bismuth is 0.013 atm. Calculate the number of bismuth atoms per cubic centimeter at equilibrium in the vapor above liquid bismuth at this temperature.

(f) The normal boiling point of liquid tin is 2270°C. Do you predict that liquid tin will be more volatile or less volatile than liquid bismuth at 1060°C?

(g) Bismuth forms two fluorides: $BiF_3$ and $BiF_5$. As is usually the case, the compound with the metal in the lower oxidation state has more ionic character, whereas that with the metal in the higher oxidation state has more covalent (molecular) character. Predict which bismuth fluoride will have the higher boiling point.

(h) Will $AsF_5$ have a higher or a lower normal boiling point than $BiF_5$?

**Answers**

(a) Bismuth resembles water in that its volume decreases on melting.

(b) The volume of bismuth increases on freezing, so that as the liquid alloy is cast, it fits tightly into its mold rather than shrinking away from the mold as most other metals do. This gives more sharply defined metal type.

(c) The bismuth will melt.

(d) Its normal boiling point lies below this temperature.

(e) $7.2 \times 10^{16}$ atoms cm$^{-3}$.

(f) Liquid tin will be less volatile.

(g) The $BiF_3$; in fact, $BiF_3$ boils at 900°C and $BiF_5$ boils at 230°C.

(h) It will have a lower normal boiling point.

## CONCEPTS & SKILLS

*After studying this chapter and working the problems that follow, you should be able to*

1. Relate trends in values of bulk properties to the strength and range of intermolecular forces (Section 5.1, problems 1–12).

2. Relate magnitudes and distance dependence of intermolecular forces to the structure of molecules (Section 5.2, problems 13–20).

3. Describe the effects of different kinds of intermolecular forces on properties of liquids (Section 5.3, problems 21–30).
4. Discuss the evidence that phase equilibrium is a dynamic process at the molecular level (Section 5.4, problems 31–38).
5. Describe the effects of different kinds of intermolecular forces on phase transitions (Section 5.5, problems 39–42).
6. Sketch the pressure–temperature phase diagram for a typical substance, and identify the lines, areas, and singular points (Section 5.6, problems 43–52).

## PROBLEMS

*Answers to problems whose numbers are boldface appear in Appendix G. Problems that are more challenging are indicated with asterisks.*

### Bulk Properties of Gases, Liquids, and Solids: Molecular Interpretation

1. A substance is nearly nonviscous and quite compressible, and it has a large coefficient of thermal expansion. Is it most likely to be a solid, a liquid, or a gas?
2. A substance is viscous, nearly incompressible, and not elastic. Is it most likely to be a solid, a liquid, or a gas?
3. A sample of volume 258 cm$^3$ has a mass of 2.71 kg.
   (a) Is the material gaseous or condensed?
   (b) If the molar mass of the material is 108 g mol$^{-1}$, calculate its molar volume.
4. A sample of volume 18.3 L has a mass of 57.9 g.
   (a) Is the material gaseous or condensed?
   (b) If the molar mass of the material is 123 g mol$^{-1}$, calculate its molar volume.
5. Heating a sample of matter from 20°C to 40°C at constant pressure causes its volume to increase from 546.0 cm$^3$ to 547.6 cm$^3$. Classify the material as (a) a nearly ideal gas, (b) a nonideal gas, or (c) condensed.
6. Cooling a sample of matter from 70°C to 10°C at constant pressure causes its volume to decrease from 873.6 cm$^3$ to 712.6 cm$^3$. Classify the material as (a) a nearly ideal gas, (b) a nonideal gas, or (c) condensed.
7. At 1.00 atm pressure and a temperature of 25°C, the volume of 1.0 g of water is 1.0 mL. At the same pressure and a temperature of 101°C, the volume of 1.0 g of water is nearly 1700 times larger. Give the reason for this large change in volume.
8. Doubling the absolute temperature of a gas essentially doubles its volume at constant pressure. Doubling the temperature of many metals, however, often increases their volumes by only a few percent. Explain.
9. Will solid sodium chloride be harder (that is, more resistant to indentation) or softer than solid carbon tetrachloride? Explain.
10. Will the surface tension of molten sodium chloride be higher than or lower than that of carbon tetrachloride? Explain.
11. Do you expect that the diffusion constant will increase or decrease as the density of a liquid is increased (by compressing it) at constant temperature? Explain. What will happen

to the diffusion constant of a gas and a solid as the density increases?
12. Do you anticipate that the diffusion constant will increase as the temperature of a liquid increases at constant pressure? Why or why not? Will the diffusion constant increase with temperature for a gas and a solid? Explain.

### Intermolecular Forces: Origins in Molecular Structure

13. Compare ion-dipole forces with induced dipole forces. In what ways are they similar and different? Give an example of each.
14. Compare dipole-dipole forces with dispersion forces. In what ways are they similar and different? Give an example of each.
15. Name the types of attractive forces that will contribute to the interactions between atoms, molecules, or ions in the following substances. Indicate the one(s) you expect to predominate.
    (a) KF    (b) HI    (c) Rn    (d) N$_2$
16. Name the types of attractive forces that will contribute to the interactions between atoms, molecules, or ions in the following substances. Indicate the one(s) you expect to predominate.
    (a) Ne    (b) ClF    (c) F$_2$    (d) BaCl$_2$
17. Predict whether a sodium ion will be most strongly attracted to a bromide ion, a molecule of hydrogen bromide, or an atom of krypton.
18. Predict whether an atom of argon will be most strongly attracted to another atom of argon, to an atom of neon, or to an atom of krypton.
19. (a) Use Figure 5.9 to estimate the length of the covalent bond in Cl$_2$ and the length of the ionic bond in K$^+$Cl$^-$. Note: the latter corresponds to the distance between the atoms in a isolated single molecule of K$^+$Cl$^-$, not in KCl(s) (solid potassium chloride).
    (b) A book states: "the shorter the bond, the stronger the bond." What features of Figure 5.9 show that this is not always true?
20. True or false: Any two atoms held together by nonbonded attractions must be farther apart than any two atoms held together by a chemical bond. Explain.

## Intermolecular Forces in Liquids

**21.** Under room conditions, fluorine and chlorine are gases, bromine is a liquid, and iodine is a solid. Explain the origin of this trend in the physical state of the halogens.

**22.** The later halogens form pentafluorides: $ClF_5$, $BrF_5$, and $IF_5$. At 0°C, one of these is a solid, one a liquid, and one a gas. Specify which is which, and explain your reasoning.

**23.** List the following substances in order of increasing normal boiling points $T_b$, and explain your reasoning: NO, $NH_3$, Ne, RbCl.

**24.** List the following substances in order of increasing normal boiling points $T_b$, and explain your reasoning: $SO_2$, He, HF, $CaF_2$, Ar.

**25.** As a vapor, methanol exists to an extent as a tetramer, $(CH_3OH)_4$, in which four $CH_3OH$ molecules are held together by hydrogen bonds. Propose a reasonable structure for this tetramer.

**26.** Hypofluorous acid (HOF) is the simplest possible compound that allows comparison between fluorine and oxygen in their abilities to form hydrogen bonds. Although F attracts electrons more strongly than O, solid HOF unexpectedly contains no H···F hydrogen bonds! Draw a proposed structure for chains of HOF molecules in the crystalline state. The bond angle in HOF is 101°.

**27.** Hydrazine ($N_2H_4$) is used as a reducing agent and in the manufacture of rocket fuels. How do you expect its boiling point to compare with that of ethylene ($C_2H_4$)?

**28.** Hydrogen peroxide ($H_2O_2$) is a major industrial chemical that is produced on a scale approaching $10^9$ kg per year. It is widely used as a bleach and in chemical manufacturing processes. How do you expect its boiling point to compare with those of fluorine ($F_2$) and hydrogen sulfide ($H_2S$), two substances with molar masses comparable to that of hydrogen peroxide?

**29.** A flask contains 1.0 L (1.0 kg) of room-temperature water. Calculate the number of possible hydrogen bonds among the water molecules present in this sample. Each water molecule can accept two hydrogen bonds and also furnish the hydrogen atoms for two hydrogen bonds.

**30.** What is the maximum number of hydrogen bonds that can be formed in a sample containing 1.0 mol ($N_0$ molecules) of liquid HF? Compare with the maximum number that can form in 1.0 mol of liquid water.

## Phase Equilibrium

**31.** Hydrogen at a pressure of 1 atm condenses to a liquid at 20.3 K and solidifies at 14.0 K. The vapor pressure of liquid hydrogen is 0.213 atm at 16.0 K. Calculate the volume of 1.00 mol of $H_2$ vapor under these conditions and compare it with the volume of 1.00 mol of $H_2$ at STP.

**32.** Helium condenses to a liquid at 4.224 K under atmospheric pressure and remains a liquid down to the absolute zero of temperature. (It is used as a coolant to reach very low temperatures.) The vapor pressure of liquid helium at 2.20 K is 0.05256 atm. Calculate the volume occupied by 1.000 mol of helium vapor under these conditions, and compare it with the volume of the same amount of helium at STP.

**33.** The vapor pressure of liquid mercury at 27°C is $2.87 \times 10^{-6}$ atm. Calculate the number of mercury atoms per cubic centimeter in the "empty" space above the top of the column of mercury in a barometer at 27°C.

**34.** The tungsten filament in an incandescent light bulb ordinarily operates at a temperature of about 2500°C. At this temperature, the vapor pressure of solid tungsten is $7.0 \times 10^{-9}$ atm. Estimate the number of gaseous tungsten atoms per cubic centimeter under these conditions.

**35.** Calcium carbide reacts with water to produce acetylene ($C_2H_2$) and calcium hydroxide. The acetylene is collected over water at 40.0°C under a total pressure of 0.9950 atm. The vapor pressure of water at this temperature is 0.0728 atm. Calculate the mass of acetylene per liter of "wet" acetylene collected in this way, assuming ideal gas behavior.

**36.** A metal reacts with aqueous hydrochloric acid to produce hydrogen. The hydrogen ($H_2$) is collected over water at 25°C under a total pressure of 0.9900 atm. The vapor pressure of water at this temperature is 0.0313 atm. Calculate the mass of hydrogen per liter of "wet" hydrogen above the water, assuming ideal gas behavior.

**37.** Carbon dioxide is liberated by the reaction of aqueous hydrochloric acid with calcium carbonate:

$$CaCO_3(s) + 2\ H^+(aq) \longrightarrow Ca^{2+}(aq) + CO_2(g) + H_2O(\ell)$$

A volume of 722 mL of $CO_2(g)$ is collected over water at 20°C and a total pressure of 0.9963 atm. At this temperature, water has a vapor pressure of 0.0231 atm. Calculate the mass of calcium carbonate that has reacted, assuming no losses of carbon dioxide.

**38.** When an excess of sodium hydroxide is added to an aqueous solution of ammonium chloride, gaseous ammonia is produced:

$$NaOH(aq) + NH_4Cl(aq) \longrightarrow$$
$$NaCl(aq) + NH_3(g) + H_2O(\ell)$$

Suppose 3.68 g of ammonium chloride reacts in this way at 30°C and a total pressure of 0.9884 atm. At this temperature, the vapor pressure of water is 0.0419 atm. Calculate the volume of ammonia saturated with water vapor that will be produced under these conditions, assuming no leaks or other losses of gas.

## Phase Transitions

**39.** High in the Andes, an explorer notes that the water for tea is boiling vigorously at a temperature of 90°C. Use Figure 5.15 to estimate the atmospheric pressure at the altitude of the camp. What fraction of the earth's atmosphere lies below the level of the explorer's camp?

**40.** The total pressure in a pressure cooker filled with water rises to 4.0 atm when it is heated, and this pressure is maintained by the periodic operation of a relief valve. Use Figure 5.21c to estimate the temperature of the water in the pressure cooker.

41. Iridium melts at a temperature of 2410°C and boils at 4130°C, whereas sodium melts at a temperature of 97.8°C and boils at 904°C. Predict which of the two molten metals has the larger surface tension at its melting point. Explain your prediction.

42. Aluminum melts at a temperature of 660°C and boils at 2470°C, whereas thallium melts at a temperature of 304°C and boils at 1460°C. Which metal will be more volatile at room temperature?

## Phase Diagrams

43. At its melting point (624°C), the density of solid plutonium is 16.24 g cm$^{-3}$. The density of liquid plutonium is 16.66 g cm$^{-3}$. A small sample of liquid plutonium at 625°C is strongly compressed. Predict what phase changes, if any, will occur.

44. Phase changes occur between different solid forms, as well as from solid to liquid, liquid to gas, and solid to gas. When white tin at 1.00 atm is cooled below 13.2°C, it spontaneously changes (over a period of weeks) to gray tin. The density of gray tin is *less* than the density of white tin (5.75 g cm$^{-3}$ versus 7.31 g cm$^{-3}$). Some white tin is compressed to a pressure of 2.00 atm. At this pressure, should the temperature be higher or lower than 13.2°C for the conversion to gray tin to take place? Explain your reasoning.

45. The following table gives several important points on the pressure-temperature diagram of ammonia:

|  | *P* (atm) | *T* (K) |
|---|---|---|
| Triple point | 0.05997 | 195.42 |
| Critical point | 111.5 | 405.38 |
| Normal boiling point | 1.0 | 239.8 |
| Normal melting point | 1.0 | 195.45 |

Use this information to sketch the phase diagram of ammonia.

46. The following table gives several important points on the pressure-temperature diagram of nitrogen:

|  | *P* (atm) | *T* (K) |
|---|---|---|
| Triple point | 0.123 | 63.15 |
| Critical point | 33.3978 | 126.19 |
| Normal boiling point | 1.0 | 77.35 |
| Normal melting point | 1.0 | 63.29 |

Use this information to sketch the phase diagram of nitrogen. The density of $N_2(s)$ is 1.03 g cm$^{-3}$, and that of $N_2(\ell)$ is 0.808 g cm$^{-3}$.

47. Determine whether argon is a solid, a liquid, or a gas at each of the following combinations of temperature and pressure (use Fig. 5.21).
    (a) 50 atm and 100 K  (c) 1.5 atm and 25 K
    (b) 8 atm and 150 K  (d) 0.25 atm and 120 K

48. Some water starts out at a temperature of 298 K and a pressure of 1 atm. It is compressed to 500 atm at constant temperature and then heated to 750 K at constant pressure. Next

it is decompressed at 750 K back to 1 atm and finally cooled to 400 K at constant pressure.
    (a) What was the state (solid, liquid, or gas) of the water at the start of the experiment?
    (b) What is the state (solid, liquid, or gas) of the water at the end of the experiment?
    (c) Did any phase transitions occur during the four steps described? If so, at what temperature and pressure did they occur? (*Hint*: Trace out the various changes on the phase diagram of water [see Fig. 5.21].)

49. The vapor pressure of solid acetylene at −84.0°C is 760 torr.
    (a) Does the triple-point temperature lie above or below −84.0°C? Explain.
    (b) Suppose a sample of solid acetylene is held under an external pressure of 0.80 atm and heated from 10 K to 300 K. What phase change(s), if any, will occur?

50. The triple point of hydrogen occurs at a temperature of 13.8 K and a pressure of 0.069 atm.
    (a) What is the vapor pressure of solid hydrogen at 13.8 K?
    (b) Suppose a sample of solid hydrogen is held under an external pressure of 0.030 atm and heated from 5 K to 300 K. What phase change(s), if any, will occur?

51. The density of nitrogen at its critical point is 0.3131 g cm$^{-3}$. At a very low temperature, 0.3131 g of solid nitrogen is sealed into a thick-walled glass tube with a volume of 1.000 cm$^3$. Describe what happens inside the tube as the tube is warmed past the critical temperature, 126.19 K.

52. At its critical point, ammonia has a density of 0.235 g cm$^{-3}$. You have a special thick-walled glass tube that has a 10.0-mm outside diameter, a wall thickness of 4.20 mm, and a length of 155 mm. How much ammonia must you seal into the tube if you wish to observe the disappearance of the meniscus as you heat the tube and its contents past 132.23°C, the critical temperature?

## Additional Problems

53. Would you classify candle wax as a solid or a liquid? What about rubber? Discuss.

* 54. When a particle diffuses, its *mean-square displacement* in a time interval $\Delta t$ is $6D\Delta t$, where $D$ is the diffusion constant. Its *root-mean-square displacement* is the square root of this (recall the analogous root-mean-square speed from Section 4.6). Calculate the root-mean-square displacement at 25°C after 1.00 hour of (a) an oxygen molecule in air ($D = 2.1 \times 10^{-5}$ m$^2$ s$^{-1}$), (b) a molecule in liquid water ($D = 2.26 \times 10^{-9}$ m$^2$ s$^{-1}$), and (c) an atom in solid sodium ($D = 5.8 \times 10^{-13}$ m$^2$ s$^{-1}$). Note that for solids with melting points higher than sodium, the diffusion constant can be many orders of magnitude smaller.

55. Liquid hydrogen chloride will dissolve many ionic compounds. Diagram how molecules of hydrogen chloride will tend to distribute themselves about a negative ion and about a positive ion in such solutions.

* 56. We saw in Section 4.7 that the van der Waals constant $b$ (with units of L mol$^{-1}$) is related to the volume per molecule in the liquid and thus to the sizes of the molecules. The combination of van der Waals constants $a/b$ has units of L atm

$mol^{-1}$. Because the liter atmosphere is a unit of energy (1 L atm = 101.325 J), $a/b$ is proportional to the energy per mole for interacting molecules and thus to the strength of the attractive forces between molecules, as shown in Figure 5.9. By using the van der Waals constants in Table 4.3, rank the attractive forces in the following from strongest to weakest: $N_2$, $H_2$, $SO_2$, and HCl.

57. Describe how the average kinetic and potential energies per mole change as a sample of water is heated from 10 K to 1000 K at a constant pressure of 1 atm.

58. As a sample of water is heated from 0.0°C to 4.0°C, its density increases from 0.99987 to 1.00000 g cm$^{-3}$. What can you conclude about the coefficient of thermal expansion of water in this temperature range? Is water unusual in its behavior? Explain.

59. At 25°C, the equilibrium vapor pressure of water is 0.03126 atm. A humidifier is placed in a room of volume 110 m$^3$ and is operated until the air becomes saturated with water vapor. Assuming that no water vapor leaks out of the room and that initially there was no water vapor in the air, calculate the number of grams of water that have passed into the air.

60. The text states that at 1000°C, the vapor pressure of tungsten is $2 \times 10^{-25}$ atm. Calculate the volume occupied per tungsten *atom* in the vapor under these conditions.

61. You boil a small quantity of water inside a 5.0-L metal can. When the can is filled with water vapor (and all air has been expelled), you quickly seal the can with a screw-on cap and remove it from the source of heat. It cools to 60°C, and most of the steam condenses to liquid water. Determine the pressure inside the can. (*Hint*: Refer to Figure 5.15.)

* 62. The air over an unknown liquid is saturated with the vapor of that liquid at 25°C and a total pressure of 0.980 atm. Suppose that a sample of 6.00 L of the saturated air is collected, and the vapor of the unknown liquid is removed from that sample by cooling and condensation. The pure air remaining occupies a volume of 3.75 L at $-50$°C and 1.000 atm. Calculate the vapor pressure of the unknown liquid at 25°C.

63. If it is true that all solids and liquids have vapor pressures, then at low enough external pressures, every substance should start to boil. In space, there is effectively zero external pressure. Explain why spacecraft do not just boil away as vapors when placed in orbit.

* 64. Butane-fueled cigarette lighters, which give hundreds of lights each, typically contain 4 to 5 g of butane ($C_4H_{10}$) that is confined in a 10-mL plastic container and exerts a pressure of 3.0 atm at room temperature (25°C). The butane boils at $-0.5$°C under normal pressure. Butane lighters have been known to explode in use, inflicting serious injury. A person hoping to end such accidents suggests that there be less butane placed in the lighters, so that the pressure inside them does not exceed 1.0 atm. Estimate how many grams of butane would be contained in such a lighter.

65. A cooling bath is prepared in a laboratory by mixing chunks of solid $CO_2$ with ethanol. $CO_2(s)$ sublimes at $-78.5$°C to $CO_2(g)$. Ethanol freezes at $-114.5$°C and boils at $+78.4$°C. State the temperature of this cooling bath, and describe what will be seen when it is prepared under ordinary laboratory conditions.

66. Oxygen melts at 54.8 K and boils at 90.2 K at atmospheric pressure. At the normal boiling point, the density of the liquid is 1.14 g cm$^{-3}$ and the vapor can be approximated as an ideal gas. The critical point is defined by $T_c = 154.6$ K, $P_c = 49.8$ atm, and (density) $d_c = 0.436$ g cm$^{-3}$. The triple point is defined by $T_t = 54.4$ K, $P_t = 0.0015$ atm, a liquid density equal to 1.31 g cm$^{-3}$, and a solid density of 1.36 g cm$^{-3}$. At 130 K, the vapor pressure of the liquid is 17.25 atm. Use this information to construct a phase diagram showing $P$ versus $T$ for oxygen. You need not draw the diagram to scale, but you should give numerical labels to as many points as possible on both axes.

67. It can be shown that if a gas obeys the van der Waals equation, its critical temperature, its critical pressure, and its molar volume at the critical point are given by the equations

$$T_c = \frac{8a}{27Rb} \qquad P_c = \frac{a}{27b^2} \qquad \left(\frac{V}{n}\right)_c = 3b$$

where $a$ and $b$ are the van der Waals constants of the gas. Use the van der Waals constants for oxygen, carbon dioxide, and water (from Table 4.3) to estimate the critical point properties of these substances. Compare with the observed values given in Figure 5.21 and in problem 66.

* 68. Each increase in pressure of 100 atm decreases the melting point of ice by about 1.0°C.
    (a) Estimate the temperature at which liquid water freezes under a pressure of 400 atm.
    (b) One possible explanation of why a skate moves smoothly over ice is that the pressure exerted by the skater on the ice lowers its freezing point and causes it to melt. The pressure exerted by an object is the force (its mass times the acceleration of gravity, 9.8 m s$^{-2}$) divided by the area of contact. Calculate the change in freezing point of ice when a skater with a mass of 75 kg stands on a blade of area $8.0 \times 10^{-5}$ m$^2$ in contact with the ice. Is this sufficient to explain the ease of skating at a temperature of, for example, $-5$°C (23°F)?

* 69. (a) Sketch the phase diagram of *temperature* versus *molar volume* for carbon dioxide, indicating the region of each of the phases (gas, liquid, solid) and the coexistence regions for two phases.
    (b) Liquid water has a maximum at 4°C in its curve of density against temperature at $P = 1$ atm, and the solid is less dense than the liquid. What happens if you try to draw a phase diagram of $T$ versus molar volume for water?

70. The critical temperature of HCl is 51°C, lower than those of both HF, 188°C, and HBr, 90°C. Explain this by analyzing the nature of the intermolecular forces in each case.

71. The normal boiling points of the fluorides of the second-period elements are as follows: LiF, 1676°C; BeF$_2$, 1175°C; BF$_3$, $-100$°C; CF$_4$, $-128$°C; NF$_3$, $-129$°C; OF$_2$, $-145$°C; F$_2$, $-188$°C. Describe the nature of the intermolecular forces in this series of liquids, and account for the trends in boiling point.

## CUMULATIVE PROBLEMS

72. The tungsten filament in an incandescent light bulb ordinarily operates at a temperature of about 2500°C. At this temperature, the vapor pressure of solid tungsten is $7.0 \times 10^{-9}$ atm. Estimate the number of gaseous tungsten atoms per cubic centimeter under these conditions.

73. Other things being equal, ionic character in compounds of metals decreases with increasing oxidation number. Rank the following compounds from lowest to highest normal boiling points: $AsF_5$, $SbF_3$, $SbF_5$, $F_2$.

74. The air over an unknown liquid is saturated with the vapor of that liquid at 25°C and a total pressure of 0.980 atm. Suppose that a 6.00-L sample of the saturated air is collected and that the vapor of the unknown liquid is removed from the sample by cooling and condensation. The pure air remaining occupies a volume of 3.75 L at −50°C and 1.000 atm. Calculate the vapor pressure of the unknown liquid at 25°C.

# Solutions

**H**omogeneous systems that contain two or more substances are called **solutions.** Usually a solution is thought of as a liquid that contains some dissolved substance, such as a solid or gas. This is the sense in which the term is used throughout most of this chapter, but solutions of one solid in another are also common (an example is an alloy of gold and silver). In fact, any homogeneous system of two or more substances (liquid, solid, or gas) is a solution. The major component

***Illustration***
Dissolution of sugar in water. *(Copyright Richard Megna/Fundamental Photographs)*

**163**

is usually called the **solvent,** and minor components are the **solutes.** The solvent is regarded as a "carrier" or medium for the solute, which can participate in chemical reactions in the solution or leave the solution through precipitation or evaporation. Description of these phenomena requires quantitative specifications of the amount of solute in the solution, or the *composition* of the solution. Solutions are formed by mixing two or more pure substances whose molecules interact directly in the mixed state. The change in intermolecular forces experienced by molecules in moving from pure solute or solvent into the mixed state influences both the ease of formation and the stability of a solution. Solutions can be in *phase equilibrium* with gases, solids, or other liquids; these equilibria frequently show interesting effects determined by the molecular weight of the solute. All these aspects of solutions will be described in the present chapter.

---

## 6.1

### COMPOSITION OF SOLUTIONS

Several measures can be used to specify the composition of a solution. **Mass percentage** (colloquially called weight percentage) is frequently employed in everyday life and is defined as the percentage by mass of a given substance in the solution. In chemistry, the most useful measures of composition are *mole fraction*, *molarity*, and *molality*.

The mole fraction of a substance in a mixture is the number of moles of that substance divided by the total number of moles present. This term was introduced in the discussion of gas mixtures and Dalton's law (Section 4.4). In a binary mixture containing $n_1$ mol of species 1 and $n_2$ mol of species 2, the mole fractions $X_1$ and $X_2$ are

$$X_1 = \frac{n_1}{n_1 + n_2}$$

$$X_2 = \frac{n_2}{n_1 + n_2} = 1 - X_1$$

The mole fractions of all the species present must add up to 1. When it is possible to make a clear distinction between solvent and solutes, the label 1 denotes the solvent, and higher numbers are given to the solutes. If comparable amounts of two liquids such as water and alcohol are mixed, however, the assignment of the labels 1 and 2 is arbitrary.

The **concentration** of a substance is the number of moles per unit volume. The SI units moles per cubic meter are of inconvenient size for chemical work, so the **molarity,** defined as the number of moles of solute per liter of solution, is used instead:

$$\text{molarity} = \frac{\text{moles solute}}{\text{liters solution}} = \text{mol L}^{-1}$$

"M" is the abbreviation for "moles per liter." A 0.1 M (read "0.1 molar") solution of HCl has 0.1 mol of HCl (dissociated into ions, of course) per liter of solution. Molarity is the most common way of specifying the compositions of dilute solutions. For accurate measurements it has the disadvantage that it depends slightly on temperature. If a solution is heated or cooled, its volume changes, so the number of moles of solute per liter of solution also changes.

The **molality,** in contrast, involves a ratio of masses and is independent of temperature. It is defined as the number of moles of solute per kilogram of *solvent*:

$$\text{molality} = \frac{\text{moles solute}}{\text{kilograms solvent}} = \text{mol kg}^{-1}$$

Because water has a density of 1.00 g cm$^{-3}$ at 20°C, 1.00 liter of water weighs 1.00 × 10$^3$ g, or 1.00 kg. It follows that in a dilute aqueous solution the number of moles of solute per liter is approximately the same as the number of moles per kilogram of water, so that molarity and molality have nearly equal values. For nonaqueous solutions and concentrated aqueous solutions, this approximate equality fails.

## EXAMPLE 6.1

A solution is prepared by dissolving 22.4 g of MgCl$_2$ in 0.200 L of water. Taking the density of pure water to be 1.00 g cm$^{-3}$ and the density of the resulting solution to be 1.089 g cm$^{-3}$, calculate the mole fraction, molarity, and molality of MgCl$_2$ in this solution.

### Solution

We are given the *mass* of the MgCl$_2$ and the *volume* of the water. The chemical amounts are

$$\text{moles MgCl}_2 = \frac{22.4 \text{ g}}{95.22 \text{ g mol}^{-1}} = 0.235 \text{ mol}$$

$$\text{moles water} = \frac{(0.200 \text{ L})(1000 \text{ cm}^3 \text{ L}^{-1})(1.00 \text{ g cm}^{-3})}{18.02 \text{ g mol}^{-1}}$$

$$= 11.1 \text{ mol}$$

$$\text{mole fraction MgCl}_2 = \frac{0.235 \text{ mol}}{(11.1 + 0.235) \text{ mol}} = 0.0207$$

To calculate the molarity, we must first determine the volume of solution. Its mass is 200 g water + 22.4 g MgCl = 222.4 g, and its density is 1.089 g cm$^{-3}$, so the volume is

$$\text{volume solution} = \frac{222.4 \text{ g}}{1.089 \text{ g cm}^{-3}} = 204 \text{ cm}^3 = 0.204 \text{ L}$$

$$\text{molarity MgCl}_2 = \frac{0.235 \text{ mol MgCl}_2}{0.204 \text{ L}} = 1.15 \text{ M}$$

$$\text{molality MgCl}_2 = \frac{0.235 \text{ mol MgCl}_2}{0.200 \text{ kg H}_2\text{O}} = 1.18 \text{ mol kg}^{-1}$$

**Related Problems: 3, 4**

**EXAMPLE 6.2**

A 9.386 M aqueous solution of sulfuric acid has a density of 1.5091 g cm$^{-3}$. Calculate the molality, the percentage by mass, and the mole fraction of sulfuric acid in this solution.

**Solution**

It is convenient to choose one liter of the solution, whose mass is

$$(1000 \text{ cm}^3)(1.5091 \text{ g cm}^{-3}) = 1509.1 \text{ g} = 1.5091 \text{ kg}$$

One liter contains 9.386 mol $H_2SO_4$, or

$$9.386 \text{ mol } H_2SO_4 \times 98.08 \text{ g mol}^{-1} = 920.6 \text{ g } H_2SO_4$$

The mass of water in this liter of solution is then obtained by subtraction:

$$\text{mass of water in 1 L of solution} = 1.5091 - 0.9206 \text{ kg} = 0.5885 \text{ kg}$$

The molality is now directly obtained as

$$\text{molality } H_2SO_4 = \frac{9.386 \text{ mol } H_2SO_4}{0.5885 \text{ kg } H_2O} = 15.95 \text{ mol kg}^{-1}$$

and the mass percentage is

$$\text{mass percentage } H_2SO_4 = \frac{0.9206 \text{ kg}}{1.5091 \text{ kg}} \times 100\% = 61.00\%$$

The chemical amount of water is

$$\text{moles } H_2O = \frac{588.5 \text{ g}}{18.02 \text{ g mol}^{-1}} = 32.66 \text{ mol}$$

so that the mole fraction of $H_2SO_4$ is

$$\text{mole fraction } H_2SO_4 = X_2 = \frac{9.386 \text{ mol}}{9.386 + 32.66 \text{ mol}} = 0.2232$$

**Related Problems: 5, 6**

## Preparation of Solutions

Examples 6.1 and 6.2 show that if a known mass of solute is added to a known volume of solvent, the molarity can be calculated only if the density of the resulting solution is known. If one liter of solvent is used, for example, the volume of the resulting solution is less than one liter in some cases and greater in others. If a solution is to have a given molarity, it is clearly inconvenient to need to know the solution density. This is avoided in practice by dissolving the measured amount of solute in a smaller amount of solvent, then adding solvent continuously until the desired total volume is reached. For accurate work, a carefully calibrated **volumetric flask** is used (Fig. 6.1).

On occasion it may be necessary to prepare a dilute solution of specified concentration from a more concentrated solution of known concentration by adding pure solvent to it. Suppose that the initial concentration (molarity) is $c_i$ and the initial solution volume is $V_i$. The chemical amount of solute is $(c_i \text{ mol L}^{-1})(V_i \text{ L}) =$

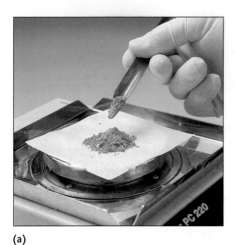

(a)

(b)

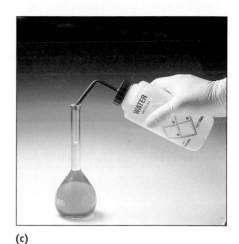

(c)

$c_iV_i$ mol. This does not change upon dilution to a final solution volume $V_f$ because only solvent, and not solute, is being added. Thus, $c_iV_i = c_fV_f$, and the final molarity is

$$c_f = \frac{\text{moles solute}}{\text{final solution volume}} = \frac{c_iV_i}{V_f}$$    **[6.1]**

This equation can be used either to calculate the final concentration after dilution to a given final volume or to determine what final volume should be used to obtain a given concentration.

**FIGURE 6.1**   To prepare a solution of nickel chloride, $NiCl_2$, with accurately known concentration, weigh out an amount of the solid (a), transfer it to a volumetric flask (b), dissolve it in somewhat less than the required amount of water, and dilute to the total volume marked on the neck of the flask (c). *(Charles D. Winters)*

## EXAMPLE 6.3

**(a)** Describe how to prepare 0.500 L of a 0.100 M aqueous solution of potassium hydrogen carbonate ($KHCO_3$).
**(b)** Describe how to dilute this solution to a final concentration of 0.0400 M $KHCO_3$.

### Solution
**(a)**         moles solute = $(0.500 \text{ L})(0.100 \text{ mol L}^{-1})$ = 0.0500 mol

grams solute = $(0.0500 \text{ mol})(100.12 \text{ g mol}^{-1})$ = 5.01 g

because 100.12 is the molar mass of $KHCO_3$. We would therefore dissolve 5.01 g of $KHCO_3$ in a small amount of water and dilute the solution to 0.500 L.
**(b)** Rearranging Equation 6.1 gives

$$V_f = \frac{c_i}{c_f}V_i$$

$$= \left(\frac{0.100 \text{ mol L}^{-1}}{0.0400 \text{ mol L}^{-1}}\right)(0.500 \text{ L}) = 1.25 \text{ L}$$

To achieve this, the solution from part **(a)** is diluted to a total volume of 1.25 L by adding water.

**Related Problems: 9, 10**

---
6.2
---

# NATURE OF DISSOLVED SPECIES

During dissolution, the attractions among the particles in the original phases (solvent-to-solvent and solute-to-solute attractions) are broken up and replaced, at least in part, by new solvent-to-solute attractions. Unlike a compound, a solution has its components present in *indefinite* proportions and cannot be represented by a chemical formula. Equations for dissolution reactions do not include the solvent as a reactant. They indicate the original state of the solute in parentheses on the left-hand side of the equation and identify the solvent in parentheses on the right-hand side. For example, solid *(s)* sucrose dissolves in water to give an **aqueous** *(aq)* solution of sucrose

$$C_{12}H_{22}O_{11}(s) \longrightarrow C_{12}H_{22}O_{11}(aq)$$

Although the solute and solvent can be any combination of solid, liquid, and gas phases, liquid water is indisputably the best known and most important solvent. Consequently, we emphasize aqueous solutions with the understanding that dissolution also occurs in many other solvents. We describe formation of aqueous solutions by considering the intermolecular forces between the solute and water molecules.

## Aqueous Solutions of Molecular Species

Molecular substances that have polar molecules are readily dissolved by water. Examples are the sugars, which have the general formula $C_m(H_2O)_n$. Specific cases are sucrose $C_{12}H_{22}O_{11}$ (table sugar), fructose $C_6H_{12}O_6$ (fruit sugar), and ribose $C_5H_{10}O_5$, a subunit in the biomolecules known as ribonucleic acids. Despite their general formula, the sugars do not contain water molecules, but they do include polar O—H (hydroxyl) groups bonded to carbon atoms, which provide sites for dipole–dipole interactions with water molecules. These attractions replace the solute–solute interactions, and the individual aquated sugar molecules move off into the solution (Fig. 6.2). Many other molecular substances follow the same pattern, provided they are sufficiently polar. Nonpolar substances such as carbon tetrachloride, octane, and the common oils and waxes are not dissolved in water.

## Aqueous Solutions of Ionic Species (Electrolytes)

Potassium sulfate is an ionic solid that dissolves in water up to 120 g $L^{-1}$ at 25°C; this maximum mass that can be dissolved in 1 L at 25°C is called the **solubility** in water. The chemical equation for this **dissolution reaction** is written as

$$K_2SO_4(s) \longrightarrow 2\ K^+(aq) + SO_4^{2-}(aq)$$

The dissolution of ionic species, illustrated in Figure 6.3, occurs through the ion–dipole forces described in Section 5.2. Each negative ion in solution is surrounded by water molecules oriented with the positive end of their dipole moments toward the negative ion. Each positive ion is surrounded by water molecules of the opposite orientation.

Each ion dissolved in water and its surrounding **solvation shell** of water molecules constitute an entity held together by ion–dipole forces. These solvated ions can move as intact entities when an electric field is applied (Fig. 6.4). Because the

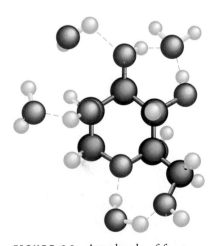

**FIGURE 6.2** A molecule of fructose in aqueous solution. Note the attractions between the hydroxyl (O—H) groups of the fructose and molecules of water. The fructose molecule is aquated; the exact number and arrangement of the attached water molecules fluctuate.

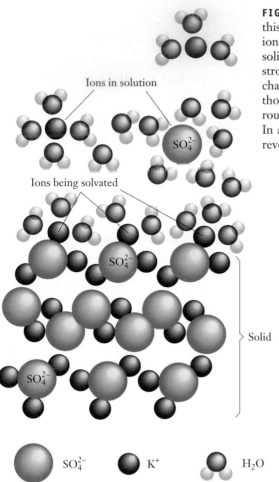

**FIGURE 6.3** When an ionic solid (in this case $K_2SO_4$) dissolves in water, the ions move away from their sites in the solid, where they were attracted strongly by ions of opposite electrical charge. New strong attractions replace those lost as each ion becomes surrounded by a group of water molecules. In a precipitation reaction the process is reversed.

Ions in solution

$SO_4^{2-}$

Ions being solvated

$SO_4^{2-}$

$SO_4^{2-}$

Solid

$SO_4^{2-}$

$SO_4^{2-}$    $K^+$    $H_2O$

resisting solution is a good conductor of electricity, $K_2SO_4$ is called a **strong electrolyte.**

In Example 6.1, it is important to note that, although the molarity of $MgCl_2$ is 1.15 M, the molarity of $Cl^-$ ions in the solution is twice as large, or 2.30 M, because each formula unit of $MgCl_2$ dissociates to give two $Cl^-$ ions.

Different compounds dissolve to different extents in water. Only a small amount (0.0025 g) of solid barium sulfate dissolves per liter of water at 25°C according to the reaction $BaSO_4(s) \longrightarrow Ba^{2+}(aq) + SO_4^{2-}(aq)$. The near-total insolubility of barium sulfate suggests that if a large enough amount of aqueous barium ion were mixed with aqueous sulfate ion, the reverse reaction would occur and solid barium sulfate would appear. Of course, it is not possible to prepare a beaker containing ions of one charge only. Ions of opposite charge must be present to maintain overall charge neutrality. It is possible, however, to prepare one beaker that contains a soluble barium compound in water (such as barium chloride) and a second beaker that contains a soluble sulfate compound in water (such as potassium sulfate). Mixing the two solutions (Fig. 6.5) then produces solid barium sulfate through the reaction

$$Ba^{2+}(aq) + SO_4^{2-}(aq) \longrightarrow BaSO_4(s)$$

which is called a **precipitation reaction.**

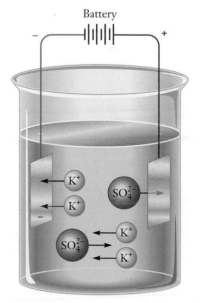

Battery

$K^+$

$SO_4^{2-}$

$K^+$

$K^+$

$SO_4^{2-}$

$K^+$

**FIGURE 6.4**  An aqueous solution of potassium sulfate conducts electricity. When metallic plates (electrodes) charged by a battery are put in the solution, positive ions ($K^+$) migrate toward the negative plate, and negative ions ($SO_4^{2-}$) toward the positive plate.

**FIGURE 6.5** A solution of potassium sulfate is being added to one of barium chloride. A cloud of white solid barium sulfate is formed; the potassium chloride remains in solution. (*Leon Lewandowski*)

Such precipitation reactions are sometimes written as

$$BaCl_2(aq) + K_2SO_4(aq) \longrightarrow BaSO_4(s) + 2\ KCl(aq)$$

which suggests ionic exchange: the two anions exchange places. This is somewhat misleading, however, because $BaCl_2$, $K_2SO_4$, and $KCl$ are all dissociated into ions in aqueous solution. It is more accurate to write

$$Ba^{2+}(aq) + 2\ Cl^-(aq) + 2\ K^+(aq) + SO_4^{2-}(aq) \longrightarrow BaSO_4(s) + 2\ K^+(aq) + 2\ Cl^-(aq)$$

The potassium and chloride ions appear on both sides of the equation. They are **spectator ions,** which ensure charge neutrality but do not take part directly in the chemical reaction. Omitting such spectator ions from the balanced chemical equation leads to the **net ionic equation**

$$Ba^{2+}(aq) + SO_4^{2-}(aq) \longrightarrow BaSO_4(s)$$

A net ionic equation includes only the ions (and molecules) that actually take part in the reaction.

In dissolution and precipitation reactions, ions retain their identities and, in particular, oxidation states do not change. The ions simply exchange the positions they had in a solid (surrounded by other ions) for new positions in solution (surrounded by solvent molecules), or they undergo the reverse process.

---

### EXAMPLE 6.4

An aqueous solution of sodium carbonate is mixed with an aqueous solution of calcium chloride, and a white precipitate immediately forms. Write a net ionic equation to account for this precipitate.

#### Solution

Aqueous sodium carbonate contains $Na^+(aq)$ and $CO_3^{2-}(aq)$ ions, and aqueous calcium chloride contains $Ca^{2+}(aq)$ and $Cl^-(aq)$ ions. Mixing the two solutions places $Na^+(aq)$ and $Cl^-(aq)$ ions and also $Ca^{2+}(aq)$ and $CO_3^{2-}(aq)$ ions in contact for the first time. The precipitate forms by the reaction

$$Ca^{2+}(aq) + CO_3^{2-}(aq) \longrightarrow CaCO_3(s)$$

because the other combination of ions leads to sodium chloride, a compound that is known to be soluble in water.

**Related Problems: 13, 14**

---

### 6.3

## STOICHIOMETRY OF REACTIONS IN SOLUTION: INTRODUCTION TO TITRATIONS

The great majority of all chemical reactions that occur on the earth's surface, whether in living organisms or among inorganic substances, take place in aqueous solution. When a chemical reaction is carried out between substances in solution, the same requirements of stoichiometry apply that were discussed in Chapter 2, in the sense that the conservation laws (embodied in balanced chemical equations) are always in force. It is now necessary to modify our use of them, however. Instead of a conversion between masses and chemical amounts (using the molar mass as a conversion factor), the conversion is now between *solution volumes* and chemical amounts, with the concentration as the conversion factor.

For instance, consider the reaction that is used commercially to prepare elemental bromine from its salts in solution:

$$2 \ Br^-(aq) + Cl_2(aq) \longrightarrow 2 \ Cl^-(aq) + Br_2(aq)$$

Suppose there is 50.0 mL of a 0.0600 M solution of NaBr. What volume of a 0.0500 M solution of $Cl_2$ is needed to react completely with the $Br^-$? To answer this, find the chemical amount of bromide ion present:

$$0.0500 \ L \times (0.0600 \ mol \ L^{-1}) = 3.00 \times 10^{-3} \ mol \ Br^-$$

Next, use the chemical conversion factor 1 mol of $Cl_2$ per 2 mol of $Br^-$ to find

$$\text{moles } Cl_2 \text{ reacting} = 3.00 \times 10^{-3} \ mol \ Br^- \left( \frac{1 \ mol \ Cl_2}{2 \ mol \ Br^-} \right) = 1.50 \times 10^{-3} \ mol \ Cl_2$$

Finally, find the necessary volume of aqueous chlorine:

$$\frac{1.50 \times 10^{-3} \ mol}{0.0500 \ mol \ L^{-1}} = 3.00 \times 10^{-2} \ L \ solution$$

The reaction requires $3.00 \times 10^{-2}$ L, or 30.0 mL, of the $Cl_2$ solution. In practice, an excess of $Cl_2$ solution would be used to ensure more nearly complete conversion of the bromide ion to bromine.

The chloride ion concentration after completion of the reaction might also be of interest. Because each mole of bromide ion that reacts gives one mole of chloride ion in the products, the chemical amount of $Cl^-$ produced is $3.00 \times 10^{-3}$ mol. The final volume of the solution is 0.0800 L, so the final concentration of $Cl^-$ is

$$[Cl^-] = \frac{3.00 \times 10^{-3} \ mol}{0.0800 \ L} = 0.0375 \ \text{M}$$

where square brackets around a chemical symbol are introduced to refer to molarity.

## EXAMPLE 6.5

When potassium dichromate is added to concentrated hydrochloric acid, it reacts according to the chemical equation

$$K_2Cr_2O_7(s) + 14 \ HCl(aq) \longrightarrow$$
$$2 \ K^+(aq) + 2 \ Cr^{3+}(aq) + 8 \ Cl^-(aq) + 7 \ H_2O(\ell) + 3 \ Cl_2(g)$$

producing a mixed solution of chromium(III) chloride and potassium chloride and evolving gaseous chlorine. Suppose that 6.20 g of $K_2Cr_2O_7$ reacts with concentrated HCl and that the final volume of the solution is 100.0 mL. Calculate the final concentration of $Cr^{3+}(aq)$ and the chemical amount of chlorine produced.

### Solution
The first step is to convert the mass of $K_2Cr_2O_7$ to its chemical amount:

$$\frac{6.20 \ g \ K_2Cr_2O_7}{294.19 \ g \ mol^{-1}} = 0.0211 \ mol \ K_2Cr_2O_7$$

The balanced chemical equation states that 1 mol of $K_2Cr_2O_7$ reacts to give 2 mol of $Cr^{3+}$ and 3 mol of $Cl_2$. Use of these two chemical conversion factors gives

$$\text{moles } Cr^{3+} = 0.0211 \text{ mol } K_2Cr_2O_7\left(\frac{2 \text{ mol } Cr^{3+}}{1 \text{ mol } K_2Cr_2O_7}\right)$$

$$= 0.0422 \text{ mol } Cr^{3+}$$

$$\text{moles } Cl_2 = 0.0211 \text{ mol } K_2Cr_2O_7\left(\frac{3 \text{ mol } Cl^{3+}}{1 \text{ mol } K_2Cr_2O_7}\right)$$

$$= 0.0633 \text{ mol } Cl_2$$

Because the final volume of the solution is 0.100 L, the concentration of $Cr^{3+}(aq)$ is

$$[Cr^{3+}] = \frac{0.0422 \text{ mol}}{0.100 \text{ L}} = 0.422 \text{ M}$$

**Related Problems: 15, 16**

## Titration

One of the most important techniques in analytical chemistry is **titration**—the addition of a carefully measured volume of one solution, containing substance A in known concentration, to a second solution containing substance B in unknown concentration, with which it will react quantitatively. Completion of the reaction, the **end point,** is signaled by a change in some physical property, such as the color of the reacting mixture. End points can be detected in colorless reaction mixtures by adding a substance called an **indicator** that changes color at the end point. At the end point, the known chemical amount of substance A that has been added is uniquely related to the unknown chemical amount of B initially present by the balanced equation for the titration reaction. Titration enables chemists to determine the unknown amount of a substance present in a sample. The two most common applications of titrations involve acid–base neutralization reactions and oxidation–reduction (or redox) reactions. Both are described briefly here to illustrate the fundamental importance of solution stoichiometry calculations in titrations. Detailed discussion of equilibrium aspects of acid–base and redox reactions will be presented later in the book.

## Brief Background on Acid–Base Reactions

Table 6.1 lists the names and formulas of a number of important acids. Acids and bases have been known and characterized since ancient times. Chemical description and explanation of their properties and behavior has progressed through several stages of sophistication and generality. A broadly applicable modern treatment will be presented in Chapter 10. Meanwhile an introduction to titrations is based on the work of the Swedish chemist Svante Arrhenius, who defined acids and bases in terms of their behavior when dissolved in water.

In pure water, small but equal amounts of hydrogen ions ($H^+$) and hydroxide ions ($OH^-$) are present.[1] These arise from the partial ionization of water:

Acetic acid, $CH_3COOH$, is a common weak organic acid that is found in vinegar as a 3 to 5% solution by mass.

---

[1] As we will see in Chapter 10, a hydrogen ion in aqueous solution is closely held by a water molecule and is better represented as a hydronium ion, $H_3O^+(aq)$. Here, we employ the simpler notation $H^+(aq)$.

## TABLE 6.1

### *Names of Common Acids*

| Binary Acids | Oxoacids | Organic Acids |
|---|---|---|
| HF, hydrofluoric acid | $H_2CO_3$, carbonic acid | HCOOH, formic acid |
| HCl, hydrochloric acid | $H_3PO_4$, phosphoric acid | $CH_3COOH$, acetic acid |
| HCN, hydrocyanic acid[†] | $HNO_2$, nitrous acid | $C_6H_5COOH$, benzoic acid |
| $H_2S$, hydrosulfuric acid | $HNO_3$, nitric acid | HOOC—COOH, oxalic acid |
| | $H_2SO_3$, sulfurous acid | |
| | $H_2SO_4$, sulfuric acid | |
| | HClO, hypochlorous acid | |
| | $HClO_2$, chlorous acid | |
| | $HClO_3$, chloric acid | |
| | $HClO_4$, perchloric acid | |

[†] Contains three elements but is named as a binary acid.

$$H_2O(\ell) \longrightarrow H^+(aq) + OH^-(aq)$$

Following Arrhenius, we define an **acid** as a substance that, when dissolved in water, increases the amount of hydrogen ion over that present in pure water. Gaseous hydrogen chloride reacts with water to give hydrochloric acid:

$$HCl(g) \longrightarrow H^+(aq) + Cl^-(aq)$$

A **base** is defined as a substance that, when dissolved, leads to an increase in the amount of hydroxide ion over that present in pure water. Sodium hydroxide dissolves extensively in water according to the equation

$$NaOH(s) \longrightarrow Na^+(aq) + OH^-(aq)$$

and is a strong base. Ammonia is another base, as shown by the products of its reaction with water:

$$NH_3(aq) + H_2O(\ell) \longrightarrow NH_4^+(aq) + OH^-(aq)$$

When an acidic solution is mixed with a basic solution, a **neutralization reaction** occurs:

$$H^+(aq) + OH^-(aq) \longrightarrow H_2O(\ell)$$

This is the reverse of the water ionization reaction shown earlier. If the spectator ions are put back into the equation it reads, for example,

$$\underset{\text{Acid}}{HCl} + \underset{\text{Base}}{NaOH} \longrightarrow \underset{\text{Water}}{H_2O} + \underset{\text{Salt}}{NaCl}$$

showing that a salt can be defined as the product (other than water) of the reaction of an acid with a base. It is usually preferable, however, to omit the spectator ions and to indicate explicitly only the reacting ions.

## Acid–Base Titration

In most acid–base reactions, there is no sharp color change at the endpoint. In such cases it is necessary to add a small amount of an indicator, a dye that changes color when the reaction is complete (indicators will be discussed in detail in Section 10.3).

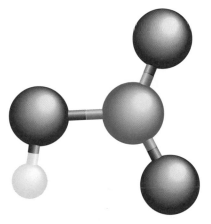

Nitric acid, $HNO_3$, is manufactured from ammonia.

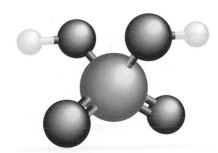

Sulfuric acid, $H_2SO_4$, is the industrial chemical produced on the largest scale in the world, in amounts exceeding 100 million tons per year.

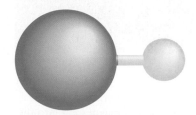

Hydrochloric acid, HCl, is used in the pickling of steel and other metals to remove oxide layers on the surface.

Phenolphthalein is such an indicator, changing from colorless to pink when a solution changes from acidic to basic. The concentration of acetic acid in an aqueous solution can be determined by adding a few drops of a phenolphthalein solution and then titrating it with a solution of sodium hydroxide of accurately known concentration. At the first permanent appearance of a pink color, the stopcock of the buret is closed. At this point the reaction

$$CH_3COOH(aq) + OH^-(aq) \longrightarrow CH_3COO^-(aq) + H_2O(\ell)$$

has gone stoichiometrically to completion.

---

### EXAMPLE 6.6

A sample of vinegar is to be analyzed for its acetic acid content. A volume of 50.0 mL is measured out and titrated with a solution of 1.306 M NaOH; 31.66 mL of that titrant is required to reach the phenolphthalein endpoint. Calculate the concentration of acetic acid in the vinegar (in moles per liter).

### Solution

The chemical amount of NaOH reacting is found by multiplying the volume of NaOH solution (31.66 mL = 0.03166 L) by its concentration (1.306 M):

$$0.03166 \text{ L} \times 1.306 \text{ mol L}^{-1} = 4.135 \times 10^{-2} \text{ mol NaOH}$$

Because 1 mol of acetic acid reacts with 1 mol of $OH^-(aq)$, the chemical amount of acetic acid originally present must also have been $4.135 \times 10^{-2}$ mol. Its concentration was then

$$[CH_3COOH] = \frac{4.135 \times 10^{-2} \text{ mol}}{0.0500 \text{ L}} = 0.827 \text{ M}$$

**Related Problems: 25, 26**

---

**FIGURE 6.6** Magnesium burning in air gives off an extremely bright light. This characteristic led to the incorporation of magnesium into the flash powder used in early photography. Magnesium powder is still used in fireworks for the same reason.
*(Leon Lewandowski)*

## Brief Background on Oxidation–Reduction (Redox) Reactions

In **oxidation–reduction** (or **redox**) **reactions** electrons are transferred between reacting species as they combine to form products. This exchange is described as a change in the oxidation number of the reactants: the oxidation number of the species giving up electrons increases, while that for the species accepting electrons decreases. Oxidation numbers were defined and methods for their calculation were presented in Section 3.7. A prototype redox reaction is that of magnesium with oxygen (Fig. 6.6). When this reaction is carried to completion, the product is magnesium oxide:

$$2 \text{ Mg}(s) + O_2(g) \longrightarrow 2 \text{ MgO}(s)$$

Magnesium is **oxidized** in this process; it *gives up* electrons as its oxidation number *increases* from 0 (in elemental Mg) to +2 (in MgO). Oxygen, which accepts these electrons is said to be **reduced;** its oxidation number *decreases* from 0 to −2. The transfer of electrons ($e^-$) can be indicated with arrows:

$$2 \overset{0}{\text{Mg}} + \overset{0}{O_2} \longrightarrow 2 \overset{+2 \; -2}{\text{MgO}}$$
$$\downarrow \qquad \uparrow$$
$$2 \times 2e^- \; = 2 \times 2e^-$$

The arrows point *away from* the species being oxidized (giving up electrons) and *toward* the species being reduced (accepting electrons). The electron "bookkeeping" beneath the equation ensures that the same number of electrons are taken up by oxygen as are given up by magnesium:

2 Mg atoms $\times$ 2 electrons per Mg atom =
$$2 \text{ oxygen atoms per formula unit} \times 2 \text{ electrons per oxygen atom}$$

Originally, the term "oxidation" referred only to reactions with oxygen. It now is used to describe any process in which the oxidation number of a species increases, even if oxygen is not involved in the reaction. For example, when calcium combines with chlorine to form calcium chloride,

$$\overset{0}{Ca}(s) + \overset{0}{Cl_2}(g) \longrightarrow \overset{+2\ -1}{CaCl_2}$$
$$\underset{2e^-}{\downarrow} \qquad \underset{= 2 \times 1e^-}{\uparrow}$$

the calcium has been oxidized and the chlorine has been reduced.

Oxidation–reduction reactions are among the most important in chemistry, biochemistry, and industry. Combustion of coal, natural gas, and gasoline (for heat and power) are redox reactions, as are the recovery of metals such as iron and aluminum from their oxide ores and the production of chemicals such as sulfuric acid from sulfur, air, and water. The human body metabolizes sugars through redox reactions to obtain energy; the reaction products are liquid water and gaseous carbon dioxide.

## EXAMPLE 6.7

Determine whether the following equations represent oxidation–reduction reactions.
**(a)** $SnCl_2(s) + Cl_2(g) \rightarrow SnCl_4(\ell)$
**(b)** $CaCO_3(s) \rightarrow CaO(s) + CO_2(g)$
**(c)** $2 H_2O_2(\ell) \rightarrow 2 H_2O(\ell) + O_2(g)$

### Solution
**(a)** Tin increases its oxidation number from +2 to +4, and the oxidation number of chlorine in $Cl_2$ is reduced from 0 to $-1$. Some other chlorine atoms are unchanged in oxidation number, but this is still a redox reaction.
**(b)** The oxidation state of Ca remains at +2, that of O at $-2$, and that of C at +4. Thus, this is not a redox reaction.
**(c)** The oxidation number of oxygen in $H_2O_2$ is $-1$. This changes to $-2$ in the product $H_2O$ and 0 in $O_2$. This is a redox reaction in which the same element is both oxidized and reduced.

**Related Problems: 27, 28, 29, 30**

Numerous redox reactions occur in aqueous solution. For example, potassium permanganate readily oxidizes $Fe^{2+}$ in acidic solution, while the Mn is reduced:

$$MnO_4^-(aq) + 5 Fe^{2+}(aq) + 8 H^+(aq) \longrightarrow Mn^{2+}(aq) + 5 Fe^{3+}(aq) + 4 H_2O(\ell)$$

### Redox Titration

Redox titrations have the special advantage that many involve intensely colored species that show a striking color change at the endpoint. For example, $MnO_4^-$ is deep

(a)

(b)

**FIGURE 6.7** (a) Addition of a small amount of potassium permanganate from a buret gives a dash of purple color to the $Fe^{2+}$ solution, which disappears quickly as the permanganate ion reacts to give $Fe^{3+}$ and $Mn^{2+}$. (b) When all the $Fe^{2+}$ ions have been consumed, additional drops of permanganate solution give a pale purple color to the solution. *(Charles D. Winters)*

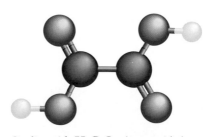

Oxalic acid, $H_2C_2O_4$, is an acid that can give up two hydrogen ions in aqueous solution to form the oxalate ion, $C_2O_4^{2-}$.

purple in color, while $Mn^{2+}$ is colorless. Thus when $MnO_4^-$ has been added to $Fe^{2+}$ in very slight excess, the color of the solution changes permanently to purple.

The titration is begun by opening the stopcock of the buret and letting a small volume of permanganate solution run into the flask that contains the $Fe^{2+}$ solution. A dash of purple colors the solution (Fig. 6.7a) but rapidly disappears as the permanganate ion reacts with $Fe^{2+}$ ions to give the nearly colorless $Mn^{2+}$ and $Fe^{3+}$ products. The addition of incremental volumes of permanganate solution is continued until the $Fe^{2+}$ is almost completely converted to $Fe^{3+}$. At this stage, the addition of just one drop of $KMnO_4$ imparts a pale purple color to the reaction mixture (Fig. 6.7b) and signals the completion of the reaction. The volume of the titrant $KMnO_4$ solution is calculated by subtracting the initial reading of the solution meniscus in the buret from the final volume reading.

Suppose that the permanganate solution has a concentration of 0.09625 M, and 26.34 mL (0.02634 L) of it is added to reach the endpoint. The chemical amount of $Fe^{2+}$ in the original solution is calculated in two steps:

$$\text{amount of } MnO_4^- \text{ reacting} = 0.02634 \text{ L} \times 0.09625 \text{ mol L}^{-1}$$
$$= 2.535 \times 10^{-3} \text{ mol } MnO_4^-$$

From the balanced chemical equation, each mole of $MnO_4^-$ used causes the oxidation of 5 mol of $Fe^{2+}$, so

$$\text{amount of } Fe^{2+} \text{ reacting} = 2.535 \times 10^{-3} \text{ mol } MnO_4^- \times \left( \frac{5 \text{ mol } Fe^{2+}}{1 \text{ mol } MnO_4^-} \right)$$
$$= 1.268 \times 10^{-2} \text{ mol } Fe^{2+}$$

These direct titrations form the basis of more complicated analytical procedures. Many analytical procedures are indirect and involve additional preliminary reactions of the sample before the titration can be carried out. For example, a soluble calcium salt will not take part in a redox reaction with potassium permanganate. However, adding ammonium oxalate to the solution containing $Ca^{2+}$ causes the quantitative precipitation of calcium oxalate:

$$Ca^{2+}(aq) + C_2O_4^{2-}(aq) \longrightarrow CaC_2O_4(s)$$

After the precipitate is filtered and washed, it is dissolved in sulfuric acid to form oxalic acid:

$$CaC_2O_4(s) + 2 \text{ H}^+(aq) \longrightarrow Ca^{2+}(aq) + H_2C_2O_4(aq)$$

Finally, the oxalic acid is titrated with permanganate solution of accurately known concentration, taking advantage of the redox reaction

$$2 \text{ } MnO_4^-(aq) + 5 \text{ } H_2C_2O_4(aq) + 6 \text{ H}^+(aq) \longrightarrow$$
$$2 \text{ } Mn^{2+}(aq) + 10 \text{ } CO_2(g) + 8 \text{ } H_2O(\ell)$$

In this way, the quantity of calcium can be determined indirectly by reactions that involve precipitation, acid–base, and redox steps.

---

### 6.4

## Cᴏʟʟᴉɢᴀᴛᴉᴠᴇ Pʀᴏᴘᴇʀᴛᴉᴇs ᴏꜰ Sᴏʟᴜᴛᴉᴏɴs

We turn now from the chemical reactions of solutions to such physical properties as their vapor pressures and phase diagrams. Consider first a solution made by dissolving a nonvolatile solute in a solvent. By "nonvolatile" we mean that the vapor pressure of the solute above the solution is negligible. An example is a solution of sucrose (cane sugar) in water, in which the vapor pressure of sucrose above the solution is zero. The case of a volatile solute is taken up in Section 6.5.

The *solvent* vapor pressure is not zero and changes with the composition of the solution at a fixed temperature. If the mole fraction of solvent ($X_1$) is 1, then the vapor pressure is $P_1^\circ$, the vapor pressure of pure solvent at the temperature of the experiment. When $X_1$ approaches zero (giving pure solute), the vapor pressure $P_1$ of the solvent must go to zero also, because solvent is no longer present. As the mole fraction $X_1$ changes from 1 to 0, $P_1$ drops from $P_1^\circ$ to 0. What is the shape of the curve?

The French chemist François-Marie Raoult found that for some solutions a plot of solvent vapor pressure against solvent mole fraction can be fitted closely by a straight line (Fig. 6.8). Solutions that conform to this straight-line relationship obey the simple equation

$$P_1 = X_1 P_1^\circ \qquad [6.2]$$

which is known as **Raoult's law.** Such solutions are called **ideal solutions.** Other solutions deviate from straight-line behavior and are called **nonideal solutions.** They may show positive deviations (with vapor pressures higher than those predicted by Raoult's law) or negative deviations (with lower vapor pressures). On a molecular level, negative deviations arise when the solute attracts solvent molecules especially strongly, reducing their tendency to escape into the vapor phase. Positive deviations arise in the opposite case, when solvent and solute molecules are not strongly attracted to each other. Even nonideal solutions with nondissociating solutes approach Raoult's law as $X_1$ approaches 1, however, just as all real gases obey the ideal gas law at sufficiently low densities.

Raoult's law forms the basis for four properties of dilute solutions that are called **colligative properties** (from Latin *colligare*, "to collect together") because they depend on the collective effect of the *number* of dissolved particles rather than on the *nature* of the particular particles involved. These four properties are

1. The lowering of the vapor pressure of a solution relative to pure solvent
2. The elevation of the boiling point
3. The depression of the freezing point
4. The phenomenon of osmotic pressure

## Vapor-Pressure Lowering

Because $X_1 = 1 - X_2$ for a two-component solution, Raoult's law can be rewritten as

$$\Delta P_1 = P_1 - P_1^\circ = X_1 P_1^\circ - P_1^\circ = -X_2 P_1^\circ \qquad [6.3]$$

so that the *change* in vapor pressure of the solvent is proportional to the mole fraction of solute. The negative sign implies **vapor-pressure lowering;** the vapor pressure is always lower above a dilute solution than it is above the pure solvent.

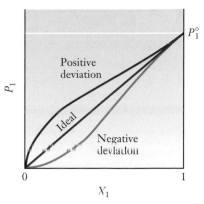

**FIGURE 6.8**  In an ideal solution, a graph of solvent vapor pressure $P_1$ versus mole fraction of solvent, $X_1$, is a straight line. Nonideal solutions behave differently; examples of positive and negative deviations from the ideal solution are shown. The vapor pressure of pure solvent is $P_1^\circ$.

---

## EXAMPLE 6.8

At 25°C the vapor pressure of pure benzene is $P_1^\circ = 0.1252$ atm. Suppose 6.40 g of naphthalene, $C_{10}H_8$ (molar mass 128.17 g mol$^{-1}$), is dissolved in 78.0 g of benzene (molar mass 78.0 g mol$^{-1}$). Calculate the vapor pressure of benzene over the solution, assuming ideal behavior.

*X = mole fraction*

## Solution

The chemical amount of solvent, $n_1$, is 1.00 mol in this case (because 78.0 g = 1.00 mol benzene has been used). The chemical amount of solute is $n_2 = 6.40$ g/128.17 g mol$^{-1}$ = 0.0499 mol $C_{10}H_8$, so the mole fraction $X_1$ is

$$X_1 = \frac{n_1}{n_1 + n_2} = \frac{1.00 \text{ mol}}{1.00 + 0.0499 \text{ mol}} = 0.9525$$

From Raoult's law, the vapor pressure of benzene above the solution is

$$P_1 = X_1 P_1^\circ = 0.9525 \times 0.1252 \text{ atm} = 0.119 \text{ atm}$$

**Related Problems: 33, 34**

## Boiling-Point Elevation

*Boiling Point*

*pressure = 1 atm*

The normal boiling point of a pure liquid or a solution is the temperature at which the vapor pressure reaches 1 atm. Because a dissolved solute reduces the vapor pressure, the temperature of the solution must be increased to make it boil. That is, the boiling point of a solution is higher than that of the pure solvent. This phenomenon, referred to as **boiling-point elevation,** provides an alternative method for determining molar masses.

The vapor-pressure curve of a dilute solution lies slightly below that for the pure solvent. In Figure 6.9, $\Delta P_1$ is the decrease of vapor pressure at $T_b$, and $\Delta T_b$ is the change in temperature necessary to hold the vapor pressure at 1 atm (in other words, $\Delta T_b = T_b' - T_b$ is the increase in boiling point caused by addition of solute to the pure solvent). For small concentrations of nondissociating solutes, the two curves are parallel, so

$$-\frac{\Delta P_1}{\Delta T_b} = \text{slope of curve} = S$$

$$\Delta T_b = -\frac{\Delta P_1}{S} = \frac{X_2 P_1^\circ}{S}$$

$$= \frac{1}{S}\left(\frac{n_2}{n_1 + n_2}\right) \qquad \text{(from Raoult's law, with } P_1^\circ = 1 \text{ atm)}$$

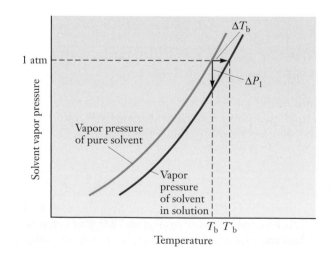

**FIGURE 6.9**  The vapor pressure of the solvent above a dilute solution is lower than that of the pure solvent at all temperatures. As a result, in order for the solution to boil (i.e., for the vapor pressure to reach 1 atm), a higher temperature is required for the solution than for the pure solvent. This amounts to an elevation of the boiling point.

## TABLE 6.2

### Boiling-Point Elevation and Freezing-Point Depression Constants

| Solvent | Formula | $T_b(°C)$ | $K_b$ (K kg mol$^{-1}$) | $T_f(°C)$ | $K_f$ (K kg mol$^{-1}$) |
|---|---|---|---|---|---|
| Acetic acid | $CH_3COOH$ | 118.1 | 3.07 | 17 | 3.9 |
| Benzene | $C_6H_6$ | 80.1 | 2.53 | 5.5 | 4.9 |
| Carbon tetrachloride | $CCl_4$ | 76.7 | 5.03 | −22.9 | 32 |
| Diethyl ether | $C_4H_{10}O$ | 34.7 | 2.02 | −116.2 | 1.8 |
| Ethanol | $C_2H_5OH$ | 78.4 | 1.22 | −114.7 | 1.9 |
| Naphthalene | $C_{10}H_0$ | — | — | 80.5 | 6.8 |
| Water | $H_2O$ | 100.0 | 0.512 | 0.0 | 1.86 |

The constant $S$ is a property of the pure solvent only, because it is the slope of the vapor pressure curve $-\Delta P_1/\Delta T_b$ near 1 atm pressure. In other words, $S$ is independent of the solute species involved.

 *S can be universally used.*

For very dilute solutions, $n_1 \gg n_2$, and this may be simplified to

$$\Delta T_b = \frac{1}{S}\frac{n_2}{n_1} = \frac{1}{S}\left(\frac{m_2/M_2}{m_1/M_1}\right)$$

where $m_1$ and $m_2$ are the masses of solvent and solute (in grams) and $M_1$ and $M_2$ are their molar masses in grams per mole. Because $M_1$, like $S$, is a property of the solvent only, it is convenient to combine the two and define a new constant $K_b$ through

$$K_b = \frac{M_1}{(1000 \text{ g kg}^{-1})S}$$

Then

$$\Delta T_b = K_b\left(\frac{m_2/M_2}{m_1/(1000 \text{ g kg}^{-1})}\right)$$

Because $m_1$ is measured in grams, $m_1/(1000 \text{ g kg}^{-1})$ is the number of kilograms of solvent. Also, $m_2/M_2$ is the number of moles of solute. The expression in parentheses is therefore the molality ($m$) of the solution.

$$\Delta T_b = K_b m \qquad\qquad \textbf{[6.4]}$$

For a given solvent, $K_b$ is obtained by measuring the boiling-point elevations for dilute solutions of known molality (i.e., containing a known amount of solute with known molar mass). Table 6.2 gives values of $K_b$ for a number of solvents. Once $K_b$ has been found, it can be used either to predict boiling-point elevations for solutes of known molar mass or to determine molar masses from measured boiling-point elevations, as illustrated in part **(b)** of the following example.

## EXAMPLE 6.9

**(a)** When 5.50 g of biphenyl ($C_{12}H_{10}$) is dissolved in 100.0 g of benzene, the boiling point increases by 0.903°C. Calculate $K_b$ for benzene.

(b) When 6.30 g of an unknown hydrocarbon is dissolved in 150.0 g of benzene, the boiling point of the solution increases by 0.597°C. What is the molar mass of the unknown substance?

### Solution

(a) Because the molar mass of biphenyl is 154.2 g mol$^{-1}$, 5.50 g contains 5.50 g/ 154.2 g mol$^{-1}$ = 0.0357 mol. The molality $m$ is

$$m = \frac{\text{mol solvent}}{\text{kg solvent}} = \frac{0.0357 \text{ mol}}{0.1000 \text{ kg}} = 0.357 \text{ mol kg}^{-1}$$

$$K_b = \frac{\Delta T_b}{m} = \frac{0.903 \text{ K}}{0.357 \text{ mol kg}^{-1}} = 2.53 \text{ K kg mol}^{-1} \qquad \text{for benzene}$$

(b) Solving $\Delta T_b = K_b m$ for $m$ gives

$$m = \frac{\Delta T_b}{K_b} = \frac{0.597 \text{ K}}{2.53 \text{ K kg mol}^{-1}} = 0.236 \text{ mol kg}^{-1}$$

The chemical amount of solute is the product of the molality of the solution and the mass of the solvent, $m_1$:

$$n_2 = (0.236 \text{ mol kg}^{-1}) \times (0.1500 \text{ kg}) = 0.0354 \text{ mol}$$

Finally, the molar mass of the solute is its mass divided by its chemical amount:

$$\text{molar mass of solute} = M_2 = \frac{m_2}{n_2} = \frac{6.30 \text{ g}}{0.0354 \text{ mol}} = 178 \text{ g mol}^{-1}$$

The unknown hydrocarbon might be anthracene ($C_{14}H_{10}$), which has a molar mass of 178.24 g mol$^{-1}$.

**Related Problems: 35, 36, 37, 38**

---

So far, only *nondissociating* solutes have been considered. Colligative properties depend on the total number of moles per liter of dissolved species present. If a solute dissociates (as sodium chloride dissolves to furnish $Na^+$ and $Cl^-$ ions in aqueous solution), then the molality $m$ to be used is the *total* molality. One mole of NaCl dissolves to furnish 2 mol of ions, so the total molality and the boiling-point elevation are twice as large as they would be if NaCl molecules were present in solution. One mole of $Ca(NO_3)_2$ dissolves to furnish 3 mol of ions (1 mol of $Ca^{2+}$ ions and 2 mol of $NO_3^-$ ions), giving three times the boiling-point elevation. The corresponding vapor-pressure lowering is greater, as well. Ions are not exactly like neutral molecules in solution, however, and nonideal behavior appears at lower concentrations in solutions that contain ions.

### EXAMPLE 6.10

Lanthanum (III) chloride ($LaCl_3$) is a salt that is completely dissociated into ions in dilute aqueous solution,

$$LaCl_3(s) \longrightarrow La^{3+}(aq) + 3 \, Cl^-(aq)$$

yielding 4 mol of ions per mole of $LaCl_3$. Suppose 0.2453 g of $LaCl_3$ is dissolved in 10.00 g of $H_2O$. What is the boiling point of the solution at atmospheric pressure, assuming ideal solution behavior?

## Solution

The molar mass of $LaCl_3$ is 245.3 g mol$^{-1}$.

$$\text{moles of LaCl}_3 = \frac{0.2453 \text{ g}}{245.3 \text{ g mol}^{-1}}$$

$$= 1.000 \times 10^{-3} \text{ mol}$$

$$\text{total molality} = m = \frac{(4)(1.000 \times 10^{-3}) \text{ mol of ions}}{0.0100 \text{ kg solvent}}$$

$$= 0.400 \text{ mol kg}^{-1}$$

This is inserted into the equation for the boiling-point elevation:

$$\Delta T_b = K_b m = (0.512 \text{ K kg mol}^{-1})(0.400 \text{ mol kg}^{-1}) = 0.205 \text{ K}$$

$$T_b = 100.205°C$$

The actual boiling point is slightly lower than this because the solution is nonideal.

## Freezing-Point Depression

The phenomenon of **freezing-point depression** is analogous to that of boiling-point elevation. Here we consider only cases in which the first solid that crystallizes from solution is the pure solvent. If solute crystallizes out with solvent, the situation is more complicated.

Pure solid solvent coexists at equilibrium with a certain vapor pressure of solvent, determined by the temperature. Solvent in solution likewise coexists with a certain vapor pressure of solvent. If solid solvent and the solvent in solution are to coexist, they must have the *same* vapor pressure. This means that the freezing temperature of a solution can be identified as the temperature at which the vapor-pressure curve of the pure solid solvent intersects that of the solution (Fig. 6.10). As solute is added to the solution, the vapor pressure of the solvent falls and the

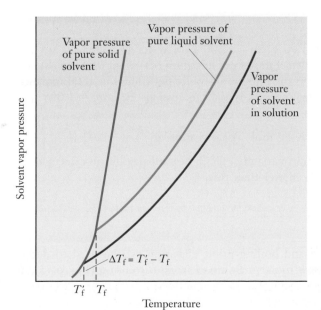

**FIGURE 6.10** The vapor pressure of solvent above a dilute solution, compared with that above pure liquid and solid solvent. The depression of the freezing point from $T_f$ to $T_f'$ is shown.

freezing point, the temperature at which the first crystals of pure solvent begin to appear, drops. The difference $\Delta T_f = T'_f - T_f$ is therefore negative, and a freezing-point depression is observed.

The change in temperature $\Delta T_f$ is once again proportional to the change in vapor pressure $\Delta P_1$. For small enough concentrations of solute, the freezing-point depression is related to the total molality $m$ (by analogy with the case of boiling-point elevation) through

$$\Delta T_f = T'_f - T_f = -K_f m \qquad \text{[6.5]}$$

where $K_f$ is a positive constant that depends only on the properties of the solvent (Table 6.2). The phenomenon of freezing-point depression is responsible for the fact that seawater, containing dissolved salts, has a slightly lower freezing point than fresh water. Concentrated salt solutions have still lower freezing points. Salt spread on an icy road reduces the freezing point of the ice, and the ice melts.

Measurements of the drop in the freezing point, like those of elevation of the boiling point, can be used to determine molar masses of unknown substances. If a substance dissociates in solution, the *total* molality of all species present (ionic or neutral) must be used in the calculation.

## EXAMPLE 6.11

The chemical amounts of the major dissolved species in a 1.000-L sample of seawater follow. Estimate the freezing point of the seawater, assuming $K_f = 1.86$ K kg mol$^{-1}$ for water.

| | | | |
|---|---|---|---|
| $Na^+$ | 0.458 mol | $Cl^-$ | 0.533 mol |
| $Mg^{2+}$ | 0.052 mol | $SO_4^{2-}$ | 0.028 mol |
| $Ca^{2+}$ | 0.010 mol | $HCO_3^-$ | 0.002 mol |
| $K^+$ | 0.010 mol | $Br^-$ | 0.001 mol |
| Neutral species | 0.001 mol | | |

### Solution

Because water has a density of 1.00 g cm$^{-3}$, 1.00 L of water weighs 1.00 kg. For dilute *aqueous* solutions, the number of moles per kilogram of solvent (the molality $m$) is therefore approximately equal to the number of moles per liter. The total molality, obtained by adding the individual species molalities just given, is $m = 1.095$ mol kg$^{-1}$. Then

$$\Delta T = -K_f m = -(1.86 \text{ K kg mol}^{-1})(1.095 \text{ mol kg}^{-1}) = -2.04 \text{ K}$$

The seawater should freeze at approximately $-2°C$. Nonideal solution effects make the actual freezing point slightly higher than this.

**Related Problems: 39, 40**

Both freezing-point depression and boiling-point elevation can be used to determine whether a species of known molar mass dissociates in solution (Fig. 6.11), as the following example shows.

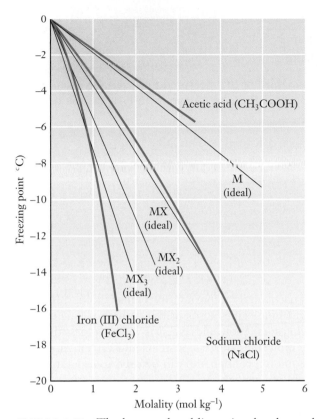

**FIGURE 6.11**    The heavy colored lines give the observed depression of the freezing point of water by acetic acid, NaCl, and FeCl$_3$ as the molality of the solutions increases. Straight black lines sketch the predicted ideal behavior for one through four moles of particles per mole in solution. The experimental curve for NaCl (which gives *two* moles of dissolved particles) stays close to the ideal straight line for MX; the experimental curve for FeCl$_3$ (which gives *four* moles of dissolved particles) stays fairly close to the ideal straight line for MX$_3$. The pattern suggests that acetic acid dissolves to give one mole of particles per mole of solute. As the molalities of the solutions increase, the observed freezing-point depressions deviate in varying ways from the straight lines.

## EXAMPLE 6.12

When 0.494 g of K$_3$Fe(CN)$_6$ is dissolved in 100.0 g of water, the freezing point is found to be −0.093°C. How many ions are present for each formula unit of K$_3$Fe(CN)$_6$ dissolved?

### Solution

The total molality of all species in solution is

$$m = \frac{-\Delta T_f}{K_f} = \frac{0.093\ \text{K}}{1.86\ \text{K kg mol}^{-1}} = 0.050\ \text{mol kg}^{-1}$$

Because the molar mass of K$_3$Fe(CN)$_6$ is 329.25 g mol$^{-1}$, the total molality if *no* dissociation had taken place would be

$$\frac{\left(\dfrac{0.494 \text{ g}}{329.25 \text{ g mol}^{-1}}\right)}{0.100 \text{ kg}} = 0.0150 \text{ mol kg}^{-1}$$

This is between one fourth and one third of the measured total molality in solution, so each $K_3Fe(CN)_6$ must dissociate into three to four ions. In fact, the dissociation that takes place is

$$K_3Fe(CN)_6(s) \longrightarrow 3 \text{ K}^+(aq) + [Fe(CN)_6]^{3-}(aq)$$

Deviations from ideal solution behavior have reduced the effective total molality from 0.060 to 0.050 mol kg$^{-1}$.

**Related Problems: 43, 44**

## Osmotic Pressure

The fourth colligative property is particularly important in cellular biology, because it plays a vital role in the transport of molecules across cell membranes. Such membranes are **semipermeable,** allowing small molecules (such as water) to pass through but blocking the passage of large molecules such as proteins and carbohydrates. A semipermeable membrane (e.g., common cellophane) can be used to separate small solvent molecules from large solute molecules.

Suppose a solution is contained in an inverted tube, the lower end of which is covered by a semipermeable membrane. This solution has a solute concentration of $c$ moles per liter. When the end of the tube is inserted in a beaker of pure solvent (Fig. 6.12), solvent flows from the beaker into the tube. The volume of the solution increases, and the solvent rises in the tube until, at equilibrium, it reaches a height $h$ above the solvent in the beaker. The pressure on the solution side of the membrane is greater than the atmospheric pressure on the surface of the pure solvent by an amount given by the **osmotic pressure, $\pi$:**

$$\pi = \rho g h \qquad \text{[6.6]}$$

where $\rho$ is the density of the solution (1.00 g cm$^{-3}$ for a dilute aqueous solution) and $g$ the acceleration due to gravity (9.807 m s$^{-2}$).

For example, a height $h$ of 0.17 m corresponds to an osmotic pressure for a dilute aqueous solution of

$$\pi = [(1.00 \text{ g cm}^{-3})(10^{-3} \text{ kg g}^{-1})(10^6 \text{ cm}^3 \text{ m}^{-3})](9.807 \text{ m s}^{-2})(0.17 \text{ m})$$

$$= 1.7 \times 10^3 \text{ kg m}^{-1} \text{ s}^{-2} = 1.7 \times 10^3 \text{ Pa}$$

$$\pi \text{ (atm)} = \frac{1.7 \times 10^3 \text{ Pa}}{1.013 \times 10^5 \text{ Pa atm}^{-1}} = 0.016 \text{ atm}$$

This illustrates how accurately very small osmotic pressures can be measured.

In 1887 Jacobus van't Hoff discovered an important relationship among osmotic pressure $\pi$, concentration $c$, and absolute temperature $T$:

$$\pi = cRT \qquad \text{[6.7]}$$

**Start**

**Equilibrium**

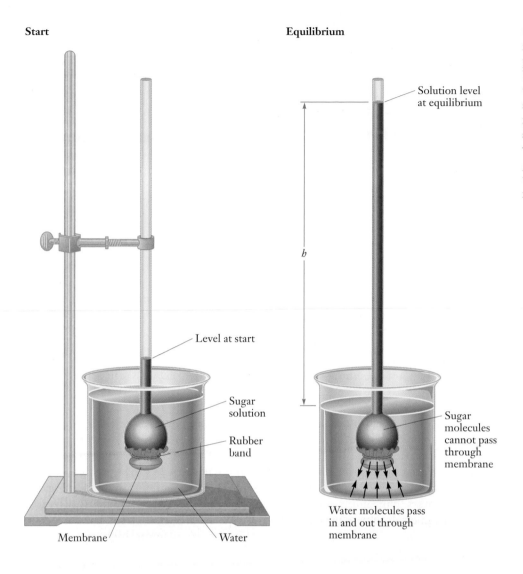

Solution level
at equilibrium

$h$

Level at start

Sugar
solution

Rubber
band

Membrane

Water

Sugar
molecules
cannot pass
through
membrane

Water molecules pass
in and out through
membrane

**FIGURE 6.12**  In this device to measure osmotic pressure, the semi-permeable membrane allows solvent, but not solute, molecules to pass through. This results in a net flow of solvent into the tube until equilibrium is achieved, with the level of solution at a height $h$ above the solvent in the beaker. Once this happens, the solvent molecules pass through the membrane at the same rate in both directions.

$R$ is the gas constant, equal to 0.08206 L atm mol$^{-1}$ K$^{-1}$ if $\pi$ is expressed in atmospheres and $c$ in moles per liter. Because $c = n/V$, where $n$ is the chemical amount of solute and $V$ is the volume of the solution, van't Hoff's equation can be rewritten as

$$\pi V = nRT$$

which bears a striking similarity to the ideal gas law. With this relationship the molar mass of a dissolved substance can be determined from the osmotic pressure of its solution.

**EXAMPLE 6.13**

A chemist dissolves 2.00 g of a protein in 0.100 L water. The osmotic pressure is 0.021 atm at 25°C. What is the approximate molar mass of the protein?

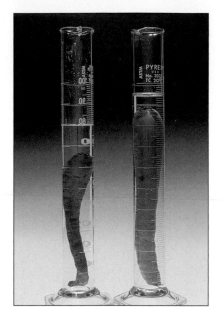

When a carrot is immersed in salt water (left), the osmotic pressure outside the cells of the vegetable is higher than that inside. Water flows out into the solution, causing the carrot to shrink. A carrot left in pure water (right) does not shrivel. (*Charles D. Winters*)

### Solution

The concentration in moles per liter is

$$c = \frac{\pi}{RT} = \frac{0.021 \text{ atm}}{(0.08206 \text{ L atm mol}^{-1} \text{ K}^{-1})(298 \text{ K})}$$

$$= 8.6 \times 10^{-4} \text{ mol L}^{-1}$$

Now 2.00 g dissolved in 0.100 L gives the same concentration as 20.0 g in 1.00 L. Therefore, $8.6 \times 10^{-4}$ mol of protein must weigh 20.0 g, and the molar mass is

$$\mathcal{M} = \frac{20.0 \text{ g}}{8.6 \times 10^{-4} \text{ mol}} = 23,000 \text{ g mol}^{-1}$$

**Related Problems: 45, 46**

---

Osmotic pressure is particularly useful for measuring molar masses of large molecules such as proteins, whose solubilities may be low. In the case given in Example 6.13, for instance, the height difference $h$ is 22 cm, an easily measured quantity. By contrast, the other three colligative properties in this example would show very small effects:

$$\text{vapor-pressure lowering} = 4.8 \times 10^{-7} \text{ atm}$$

$$\text{boiling-point elevation} = 0.00044 \text{ K}$$

$$\text{freezing-point depression} = 0.0016 \text{ K}$$

All of these changes are too small for accurate measurement. As with the other techniques, the *total* number of moles of solute species determines the osmotic pressure if dissociation occurs.

Osmosis has other important uses. In some parts of the world potable water is a precious commodity. It can be obtained much more economically by desalinizing brackish waters, through a process called **reverse osmosis,** than by distillation. When an ionic solution in contact with a semipermeable membrane has a pressure applied to it that exceeds its osmotic pressure, water of quite high purity passes through. Reverse osmosis is also used to control water pollution.

---

## 6.5

## Mɪxᴛᴜʀᴇs ᴀɴᴅ Dɪsᴛɪʟʟᴀᴛɪᴏɴ

The preceding section described the properties of solutions of nonvolatile solutes in liquid solvents. The concept of an ideal solution can be extended to mixtures of two or more components, each of which is volatile. An ideal solution is one in which the vapor pressure of *each* species present is proportional to its mole fraction in solution over the whole range of mole fraction:

$$P_i = X_i P_i^\circ$$

where $P_i^\circ$ is the vapor pressure (at a given temperature) of pure substance $i$, $X_i$ is its mole fraction in solution, and $P_i$ is its partial vapor pressure over the solution. This is a generalization of Raoult's law to each component of a solution.

For an ideal mixture of two volatile substances, the vapor pressure of component 1 is

$$P_1 = X_1 P_1^\circ$$

and that of component 2 is

$$P_2 = X_2 P_2^\circ = (1-X_1)P_2^\circ$$

The vapor pressures for such an ideal solution are shown in Figure 6.13, together with typical vapor pressures for a nonideal solution.

## Henry's Law

At sufficiently low mole fraction $X_2$, the vapor pressure of component 2 (even in a nonideal solution) is proportional to $X_2$:

$$P_2 = k_2 X_2 \qquad\qquad \textbf{[6.8]}$$

where $k_2$ is a constant. For $X_1$ small ($X_2$ near 1),

$$P_1 = k_1 X_1 = k_1(1-X_2)$$

This linear vapor pressure of what is called the solute (because it is present at small mole fraction) is known as **Henry's law:** the vapor pressure of a volatile dissolved substance is proportional to the mole fraction of that substance in solution. Whenever Raoult's law is valid for a solvent, Henry's law is valid for the solute (Fig. 6.13).

*if raoult's law is valid, so to is Henry's.*

One familiar application of Henry's law is in the carbonation of beverages. If the partial pressure of $CO_2$ above a solution is increased, the amount dissolved in the solution increases proportionately. When the beverage can is opened, dissolved gas bubbles out of solution in response to the lower $CO_2$ pressure outside. Henry's law is important in biology (gases such as oxygen dissolve in blood and other bodily fluids) and in environmental chemistry (volatile pollutants can move between bodies of water and the atmosphere).

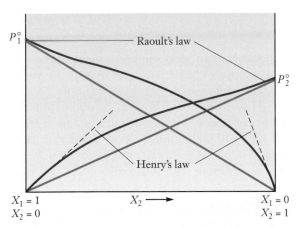

**FIGURE 6.13** Vapor pressures above a mixture of two volatile liquids. Both ideal behavior (blue lines) and nonideal behavior (red curves) are shown. Positive deviations from ideal solution behavior are illustrated here, although negative deviations are observed for other nonideal solutions. Raoult's and Henry's laws are shown as dilute solution limits for the nonideal mixture; the markers explicitly identify regions where Raoult's law represents actual behavior.

## EXAMPLE 6.14

The Henry's law constant for oxygen dissolved in water is $4.34 \times 10^4$ atm at 25°C. If the partial pressure of oxygen in air is 0.20 atm under ordinary atmospheric conditions, calculate the concentration (in moles per liter) of dissolved oxygen in water that is in equilibrium with air at 25°C.

### Solution

Henry's law is used to calculate the mole fraction of oxygen in water:

$$X_{O_2} = \frac{P_{O_2}}{k_{O_2}} = \frac{0.20 \text{ atm}}{4.34 \times 10^4 \text{ atm}} = 4.6 \times 10^{-6}$$

Next, the mole fraction is converted to molarity. One liter of water weighs 1000 g, so it contains

$$\frac{1000 \text{ g } H_2O}{18.02 \text{ g mol}^{-1}} = 55.5 \text{ mol water}$$

Because $X_{O_2}$ is so small, $n_{H_2O} + n_{O_2}$ is very close to $n_{H_2O}$, and it can be written

$$X_{O_2} = \frac{n_{O_2}}{n_{H_2O} + n_{O_2}} \approx \frac{n_{O_2}}{n_{H_2O}}$$

$$4.6 \times 10^{-6} = \frac{n_{O_2}}{55.5 \text{ mol}}$$

Thus, the chemical amount of oxygen in one liter of water is

$$n_{O_2} = (4.6 \times 10^{-6})(55.5 \text{ mol}) = 2.6 \times 10^{-4} \text{ mol}$$

and the concentration of dissolved $O_2$ is $2.6 \times 10^{-4}$ M.

**Related Problems: 49, 50**

## Distillation

The vapor pressures of the pure components of an ideal solution usually differ, and for this reason such a solution has a composition different from that of the vapor phase with which it is in equilibrium. This can best be seen in an example.

Hexane ($C_6H_{14}$) and heptane ($C_7H_{16}$) form a nearly ideal solution over the whole range of mole fractions. At 25°C, the vapor pressure of pure hexane is $P_1^\circ = 0.198$ atm, and that of pure heptane is $P_2^\circ = 0.0600$ atm. Suppose a solution contains 4.00 mol of hexane and 6.00 mol of heptane, so that its mole fractions are $X_1 = 0.400$ and $X_2 = 0.600$. The vapor in equilibrium with this ideal solution has partial pressures

$$P_{\text{hexane}} = P_1 = X_1 P_1^\circ = (0.400)(0.198 \text{ atm}) = 0.0792 \text{ atm}$$

$$P_{\text{heptane}} = P_2 = X_2 P_2^\circ = (0.600)(0.0600 \text{ atm}) = 0.0360 \text{ atm}$$

From Dalton's law, the total pressure is the sum of these partial pressures:

$$P_{\text{total}} = P_1 + P_2 = 0.1152 \text{ atm}$$

If $X_1'$ and $X_2'$ are the mole fractions in the vapor, then

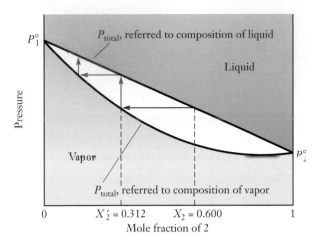

**FIGURE 6.14** The composition of the vapor above a solution differs from the composition of the liquid with which it is in equilibrium. Here the upper (straight) line is the total pressure of the vapor in equilibrium with an ideal solution having mole fraction $X_2$ of component 2. By moving horizontally from that line to a point of intersection with the lower curve, we can locate the mole fraction $X_2'$ of component 2 in the vapor (red arrow). Subsequent condensations and vaporizations are shown by blue arrows.

$$X_1' = \frac{0.0792 \text{ atm}}{0.1152 \text{ atm}} = 0.688$$

$$X_2' = 1 - X_1' = \frac{0.0360 \text{ atm}}{0.1152 \text{ atm}} = 0.312$$

The liquid and the vapor with which it is in equilibrium have different compositions (Fig. 6.14), and the vapor is enriched in the more volatile component.

Suppose some of this vapor were removed and condensed to a liquid. The vapor in equilibrium with this new solution would be still richer in the more volatile component, and the process could be continued further (Fig. 6.14). This progression underlies the technique of separating a mixture into its pure components by **fractional distillation**, a process in which the components are successively evaporated and recondensed. What has been described so far corresponds to a constant-temperature process, but actual distillation is carried out at constant total pressure. The vapor pressure–mole fraction plot is transformed into a boiling temperature–mole fraction plot, shown in Figure 6.15. Note that the component with the lower vapor pressure (2) has the higher boiling point, $T_b^2$. If the temperature of a solution of a certain composition is raised until it touches the liquid line in the plot, the vapor in equilibrium with the solution is richer in the more volatile component 1. Its composition lies at the intersection of the horizontal constant-temperature line and the equilibrium vapor curve.

A liquid can be vaporized in different ways. It can simply be boiled until it is entirely vaporized and the final composition of the vapor is the same as that of the original liquid. It is clear that such a mixture boils over a range of temperatures, rather than at a single $T_b$ like a pure liquid. Alternatively, if the boiling is stopped midway, the vapor fraction that has boiled off can be collected and recondensed. The resulting liquid (i.e., the condensate) will be richer in component 1 than the original solution was. By repeating the process again and again, mixtures successively richer in 1 will be obtained. This is the principle behind the **distillation**

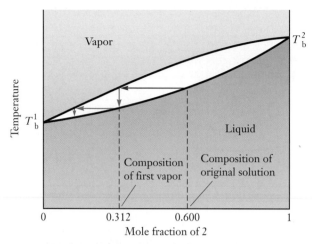

**FIGURE 6.15**    The boiling point of an ideal solution varies with the composition of the solution. The upper curve is the boiling temperature referred to the vapor composition, and the lower curve is the boiling temperature referred to the liquid composition. The vapors boiling off of a solution that has a 0.600 mole fraction of component 2 are enriched in the more volatile component 1 to the extent that their mole fraction of component 2 is only 0.312 (red arrow). The subsequent blue arrows show the further steps used in obtaining nearly pure component 1 by fractional distillation.

**column** (Fig. 6.16). Throughout the length of the tube, such evaporations and recondensations take place, and this allows mixtures to be separated into their constituent substances. Such a process is used to separate nitrogen and oxygen in air; the air is liquefied and then distilled, with the lower-boiling nitrogen ($T_b = -196°C$) vaporizing before the oxygen ($T_b = -183°C$).

Nonideal solutions may have more complicated behavior. A mixture showing large *negative* deviations from Raoult's law (one in which solute–solvent forces are

**FIGURE 6.16**    In a distillation column, temperature decreases with height in the column. The less volatile components condense and fall back to the flask, but the more volatile ones continue up the column into the water-cooled condenser, where they condense and are recovered in the receiver.

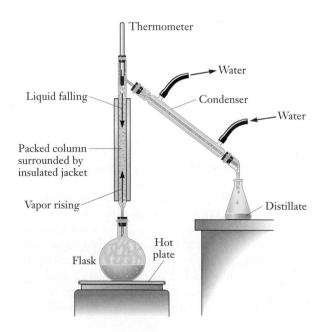

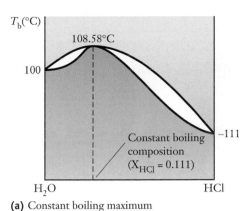

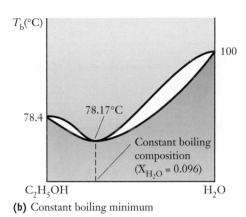

**(a)** Constant boiling maximum

**(b)** Constant boiling minimum

**FIGURE 6.17** Dependence of boiling temperature on mole fraction for maximum- and minimum-boiling azeotropes. The coordinates are not to scale.

strongly attractive) will show a boiling-point *maximum* (Fig. 6.17a). A solution at the maximum is called a **maximum-boiling azeotrope;** an example is that formed by the $H_2O$/HCl system. The boiling-point maximum occurs in this case at 108.58°C and 1 atm pressure for a composition of 20.22% HCl by mass. A mixture showing large *positive* deviations from ideal behavior may show a boiling-point *minimum* (Fig. 6.17b) and a corresponding **minimum-boiling azeotrope.** Ethyl alcohol and water form such an azeotrope with a normal boiling point of 78.17°C and a composition of 4% water by mass. In this case, attractive forces between ethyl alcohol molecules and between water molecules are stronger than those between ethyl alcohol and water, so the solution boils at a lower temperature than either pure component. An azeotrope behaves like a single-component fluid in that it boils at a well-defined temperature and the solution and vapor have the same composition. A mixture of two substances that has an azeotrope cannot be separated by fractional distillation into two pure substances, but only into one pure substance and a mixture with the azeotropic composition. A mixture of 50% ethanol and water, for example, can be distilled to obtain pure water and an azeotropic mixture containing 4% water and 96% ethanol. The last 4% of water cannot be removed by distillation to obtain pure ethanol.

---

## 6.6

## COLLOIDAL SUSPENSIONS

A **colloid** is a mixture of two or more substances in which one phase is suspended as a large number of very small particles in a second phase. The dispersed substance and the background medium may be any combination of gas, liquid, or solid. Examples of colloids include aerosol sprays (liquid suspended in gas), smoke (solid particles in air), milk (fat droplets and solids in water), mayonnaise (water droplets in oil), and paint (solid pigment particles in oil for oil-based paints, or pigment and oil dispersed in water for latex paints). Colloidal particles are larger than single molecules but are too small to be seen by the eye; their dimensions typically range from $10^{-9}$ to $10^{-6}$ m in diameter. Their presence can be seen most dramatically in the way in which they scatter light; a familiar example is the passage of light from a movie projector through a suspension of small dust particles in air. The gemstone opal has remarkable optical properties that arise from colloidal water suspended in solid silicon dioxide (Fig. 6.18).

**FIGURE 6.18** A natural opal. (*Alfred Pasieka/Science Photo Library/Photo Researchers, Inc.*)

**FIGURE 6.19** When a salt is added to a colloidal dispersion (a), the repulsive forces between the colloidal particles are reduced and aggregation occurs (b). Eventually the aggregated particles fall to the bottom of the container as a low-density sediment (c).

**(a)** Dispersion         **(b)** Aggregation         **(c)** Sedimentation

Although some colloids settle out into two separate phases if left standing long enough, others persist indefinitely; a suspension of gold particles prepared by Michael Faraday in 1857 shows no apparent settling to date. In many colloids, the particles have net positive or negative charges on their surfaces, balanced by an opposite charge of ions in solution. The settling out of such colloids is speeded by dissolving salts in the solution, a process called **flocculation.** The salts reduce the repulsive electrostatic forces between the suspended particles, causing aggregation and sedimentation (Fig. 6.19). Flocculation occurs in river deltas; when river water containing suspended clay particles meets the salt water of the ocean, the clay settles out as open, low-density sediments. Flocculating agents are deliberately added to paints so that the pigment will settle in a loosely packed sediment. When the paint is stirred, the pigment is redispersed through the medium. In the absence of such agents, the suspended particles tend to settle in compact sediments that are difficult to resuspend.

In some cases the formation of a colloid is not desirable, as in the precipitation of a solid from solution (Section 6.2). Especially with metal sulfides, the solid precipitate may appear as a colloidal suspension with particles small enough to pass through ordinary filter paper (Fig. 6.20). If this happens, precipitated solid can be separated out only by flocculation or centrifugation or by forcing the suspension through a membrane such as cellophane that permits passage of only the small solvent molecules.

**FIGURE 6.20** (a) This colloidal suspension of $PbCrO_4$ appears cloudy. (b) After flocculation, the precipitate settles to the bottom. *(Leon Lewandowski)*

**(a)**                **(b)**

Suspended particles are in a constant state of motion called **Brownian motion** after Robert Brown, a Scottish botanist who used a microscope to observe the motion of pollen particles in water. Brownian motion results from the constant random buffeting of the particles by solvent molecules. In 1905 Albert Einstein showed how the motion of Brownian particles could be described on a microscopic level; his work provided one of the most striking and convincing verifications of the molecular hypothesis and of the kinetic theory of matter and led to a fairly accurate determination of Avogadro's number.

## CUMULATIVE EXERCISE

### Maple Syrup

The sap in a maple tree can be described approximately as a 3.0% (by mass) solution of sucrose ($C_{12}H_{22}O_{11}$) in water. Sucrose does not dissociate to any significant extent in aqueous solution.

(a) At 20°C, the density of sap is 1.010 g cm$^{-3}$. Calculate the molarity of sucrose in sap.

(b) A typical maple tree yields about 12 gal of sap per year. Calculate how many grams of sucrose are contained in this volume of sap (1 gal = 3.785 L).

(c) The rising of sap in trees is caused largely by osmosis; the concentration of dissolved sucrose in sap is higher than that of the groundwater outside the tree. Calculate the osmotic pressure of a sap solution and the height to which the sap should rise above the ground on a day when the temperature is 20°C. Approximate the groundwater as pure (although in fact it typically contains 0.01 to 0.03 M dissolved species). Express the answer in meters and in feet (1 m = 3.28 ft).

(d) To produce maple syrup from sap, one boils the sap to reduce its water content. Calculate the normal boiling point of a sap solution.

(e) Maple syrup is the concentrated sap solution that results when most of the water is boiled off. The syrup has a composition of approximately 64% (by mass) sucrose and 36% water, with flavoring components present in small concentrations. If the density of the maple syrup is 1.31 g cm$^{-3}$, calculate the mole fraction, molarity, and molality of sucrose in maple syrup.

(f) What volume (in gallons) of maple syrup can be obtained from the sap in one typical tree?

(g) In the presence of vanadium(V) oxide, dinitrogen tetraoxide oxidizes sucrose to oxalic acid ($H_2C_2O_4$) according to the equation

$$C_{12}H_{22}O_{11}(aq) + 9\ N_2O_4(s) \longrightarrow 6\ H_2C_2O_4(aq) + 18\ NO(g) + 5\ H_2O(\ell)$$

Calculate the mass of $N_2O_4$ that will react completely with 7.00 L of the sap solution from part (a), and give the concentration of oxalic acid that results.

Tapping syrup from sugar maple trees. *(Copyright 1989 Blair Seitz/Photo Researchers, Inc.)*

### Answers

(a) 0.089 M

(b) $1.4 \times 10^3$ g sucrose

(c) $\pi = 2.1$ atm; height = 22 m = 72 ft

(d) 100.046°C

(e) Mole fraction = 0.086; molarity = 2.4 M; molality = 5.2 mol kg$^{-1}$

(f) 0.56 gal

(g) $5.2 \times 10^2$ g $N_2O_4$; 0.53 M

## CONCEPTS & SKILLS

*After studying this chapter and working the problems that follow, you should be able to*

1. Express the concentration of a solute in solution in units of mass percentage, molarity, molality, and mole fraction (Section 6.1, problems 3–8).
2. Describe how a solution of a given molarity is prepared and the effect on molarity of dilution (Section 6.1, problems 9–12).
3. Describe the formation of a solution in molecular terms by comparing intermolecular forces in the pure phases and in the solution (Section 6.2).
4. Calculate the chemical amounts of substances reacting during a solution-phase reaction such as titration (Section 6.3, problems 15–32).
5. Calculate the molar mass of a nonvolatile solute from the changes it causes in the colligative properties (vapor-pressure lowering, boiling-point elevation, freezing-point lowering, or osmotic pressure) of its dilute solution (Section 6.4, problems 33–48).
6. Discuss the meaning of Henry's law and use it to calculate the solubilities of gases in liquids (Section 6.5, problems 49–52).
7. Relate the total pressure and composition of the vapor in equilibrium with an ideal two-component solution to the composition of the solution and the vapor pressures of its pure components (Section 6.5, problems 53–56).
8. Explain how distillation is used to separate the volatile components of a binary liquid solution (Section 6.5).
9. Describe the physical properties of a colloidal suspension (Section 6.6).

## PROBLEMS

*Answers to problems whose numbers are boldface appear in Appendix G. Problems that are more challenging are indicated with asterisks.*

### Composition of Solutions

1. A patient has a "cholesterol count" of 214. Like many blood-chemistry measurements, this result has the units mg $dL^{-1}$.
   (a) Determine the molar concentration of cholesterol in this patient's blood, taking the molar mass of cholesterol to be 386.64 g $mol^{-1}$.
   (b) Estimate the molality of cholesterol in the patient's blood.
   (c) If 214 is a typical cholesterol reading among men in the United States, determine the volume of such blood required to furnish 8.10 g of cholesterol.

2. In many states, a person is legally intoxicated if his or her blood has a concentration of 0.1 g (or more) of ethyl alcohol ($C_2H_5OH$) per deciliter. Express this "threshold concentration" in mol $L^{-1}$.

3. A solution of hydrochloric acid in water is 38.00% hydrochloric acid by mass. Its density is 1.1886 g $cm^{-3}$ at 20°C. Compute its molarity, mole fraction, and molality at this temperature.

4. A solution of acetic acid and water contains 205.0 g $L^{-1}$ of acetic acid and 820.0 g $L^{-1}$ of water.
   (a) Compute the density of the solution.

   (b) Compute the molarity, molality, mole fraction, and mass percentage of acetic acid in this solution.
   (c) Take the acetic acid as the solvent, and do the same for water as the solute.

5. A 6.0835 M aqueous solution of acetic acid ($C_2H_4O_2$) has a density of 1.0438 g $cm^{-3}$. Compute its molality.

6. A 1.241 M aqueous solution of $AgNO_3$ (used to prepare silver chloride photographic emulsions) has a density of 1.171 g $cm^{-3}$. Compute its molality.

7. Water is slightly soluble in liquid nitrogen. At −196°C (the boiling point of liquid nitrogen), the mole fraction of water in a saturated solution is $1.00 \times 10^{-5}$. Compute the mass of water that can dissolve in 1.00 kg of boiling liquid nitrogen.

8. Some water dissolves in liquid methane at −161°C to give a solution in which the mole fraction of water is $6.0 \times 10^{-5}$. Determine the mass of water dissolved in 1.00 L of this solution if the density of the solution is 0.78 g $cm^{-3}$.

9. Concentrated phosphoric acid as sold for use in the laboratory is usually 90% $H_3PO_4$ by mass (the rest is water). Such a solution contains 12.2 mol of $H_3PO_4$ per liter of solution at 25°C.
   (a) Compute the density of this solution.
   (b) What volume of this solution should be used in mixing 2.00 L of a 1.00 M phosphoric acid solution?

10. A perchloric acid solution is 60.0% $HClO_4$ by mass. It is simultaneously 9.20 M at 25°C.
    (a) Compute the density of this solution.
    (b) What volume of this solution should be used in mixing 1.00 L of a 1.00 M perchloric acid solution?
11. Suppose 25.0 g of solid NaOH is added to 1.50 L of an aqueous solution that is already 2.40 M in NaOH. Then water is added until the final volume is 4.00 L. Determine the concentration of the NaOH in the resulting solution.
12. Suppose 0.400 L of a solution of 0.0700 M nitric acid is added to 0.800 L of a solution of 0.0300 M nitric acid, giving a total volume of 1.200 L. Calculate the concentration (molarity) of nitric acid in the resulting solution.

## Nature of Dissolved Species

13. Rewrite the following balanced equations as net ionic equations.
    (a) $NaCl(aq) + AgNO_3(aq) \rightarrow AgCl(s) + NaNO_3(aq)$
    (b) $K_2CO_3(s) + 2\ HCl(aq) \rightarrow 2\ KCl(aq) + CO_2(g) + H_2O(\ell)$
    (c) $2\ Cs(s) + 2\ H_2O(\ell) \rightarrow 2\ CsOH(aq) + H_2(g)$
    (d) $2\ KMnO_4(aq) + 16\ HCl(aq) \rightarrow$
        $\qquad 5\ Cl_2(g) + 2\ MnCl_2(aq) + 2\ KCl(aq) + 8\ H_2O(\ell)$
14. Rewrite the following balanced equations as net ionic equations.
    (a) $Na_2SO_4(aq) + BaCl_2(aq) \rightarrow BaSO_4(s) + 2\ NaCl(aq)$
    (b) $6\ NaOH(aq) + 3\ Cl_2(g) \rightarrow$
        $\qquad NaClO_3(aq) + 5\ NaCl(aq) + 3\ H_2O(\ell)$
    (c) $Hg_2(NO_3)_2(aq) + 2\ KI(aq) \rightarrow Hg_2I_2(s) + 2\ KNO_3(aq)$
    (d) $3\ NaOCl(aq) + KI(aq) \rightarrow$
        $\qquad NaIO_3(aq) + 2\ NaCl(aq) + KCl(aq)$

## Stoichiometry of Reactions in Solution: Introduction to Titrations

15. When treated with acid, lead(IV) oxide is reduced to a lead(II) salt, with the liberation of oxygen:

    $2\ PbO_2(s) + 4\ HNO_3(aq) \longrightarrow$
    $\qquad\qquad 2\ Pb(NO_3)_2(aq) + 2\ H_2O(\ell) + O_2(g)$

    What volume of a 7.91 M solution of nitric acid is just sufficient to react with 15.9 g of lead(IV) oxide according to this equation?
16. Phosphoric acid is made industrially by the reaction of fluorapatite, $Ca_5(PO_4)_3F$, in phosphate rock with sulfuric acid:

    $Ca_5(PO_4)_3F(s) + 5\ H_2SO_4(aq) + 10\ H_2O(\ell) \longrightarrow$
    $\qquad 3\ H_3PO_4(aq) + 5(CaSO_4 \cdot 2H_2O)(s) + HF(aq)$

    What volume of 6.3 M phosphoric acid is generated by the reaction of 2.2 metric tons (2200 kg) of fluorapatite?
17. The carbon dioxide produced (together with hydrogen) from the industrial-scale oxidation of methane in the presence of nickel is removed from the gas mixture in a scrubber containing an aqueous solution of potassium carbonate:

    $CO_2(g) + H_2O(\ell) + K_2CO_3(aq) \longrightarrow 2\ KHCO_3(aq)$

    Calculate the volume of carbon dioxide (at 50°C and 1.00 atm pressure) that will react with 187 L of a 1.36 M potassium carbonate solution.
18. Nitrogen oxide can be generated on a laboratory scale by the reaction of dilute sulfuric acid with aqueous sodium nitrite:

    $6\ NaNO_2(aq) + 3\ H_2SO_4(aq) \longrightarrow$
    $\qquad 4\ NO(g) + 2\ HNO_3(aq) + 2\ H_2O(\ell) + 3\ Na_2SO_4(aq)$

    What volume of 0.646 M aqueous $NaNO_2$ should be used in this reaction to generate 5.00 L of nitrogen oxide at a temperature of 20°C and a pressure of 0.970 atm?
19. Write a balanced equation for the acid–base reaction that leads to the production of each of the following salts. Name the acid, base, and salt.
    (a) $CaF_2$          (c) $Zn(NO_3)_2$
    (b) $Rb_2SO_4$      (d) $KCH_3COO$
20. Write a balanced equation for the acid–base reaction that leads to the production of each of the following salts. Name the acid, base, and salt.
    (a) $Na_2SO_3$              (c) $PbSO_4$
    (b) $Ca(C_6H_5COO)_2$     (d) $CuCl_2$
21. Hydrogen sulfide can be removed from natural gas by reaction with excess sodium hydroxide. Name the salt that is produced in this reaction. (*Note:* Hydrogen sulfide loses both of its hydrogen atoms in the course of this reaction.)
22. During the preparation of viscose rayon, cellulose is dissolved in a bath containing sodium hydroxide and later reprecipitated as rayon using a solution of sulfuric acid. Name the salt that is a byproduct of this process. Rayon production is in fact a significant commercial source for this salt.
23. Phosphorus trifluoride is a highly toxic gas that reacts slowly with water to give a mixture of phosphorous acid and hydrofluoric acid.
    (a) Write a balanced chemical equation for this reaction.
    (b) Determine the concentration (in moles per liter) of each of the acids that result from the reaction of 1.94 L of phosphorus trifluoride (measured at 25°C and 0.970 atm pressure) with water to give a solution volume of 872 mL.
24. Phosphorus pentachloride reacts violently with water to give a mixture of phosphoric acid and hydrochloric acid.
    (a) Write a balanced chemical equation for this reaction.
    (b) Determine the concentration (in moles per liter) of each of the acids that result from the complete reaction of 1.22 L of phosphorus pentachloride (measured at 215°C and 0.962 atm pressure) with enough water to give a solution volume of 697 mL.
25. To determine the concentration of a solution of nitric acid, a 100.0-mL sample is placed in a flask and titrated with a 0.1279 M solution of potassium hydroxide. A volume of 37.85 mL is required to reach the phenolphthalein endpoint. Calculate the concentration of nitric acid in the original sample.
26. The concentration of aqueous ammonia in a cleaning solution is determined by titration with hydrochloric acid. A volume of 23.18 mL of 0.8381 M HCl is needed to titrate a 50.0-mL sample of the ammonia solution to a methyl red

endpoint. Calculate the concentration of ammonia in the cleaning solution.

27. For each of the following balanced equations, write the oxidation number above the symbol of each atom that changes oxidation state in the course of the reactions.
    (a) $2 \, PF_2I(\ell) + 2 \, Hg(\ell) \rightarrow P_2F_4(g) + Hg_2I_2(s)$
    (b) $2 \, KClO_3(s) \rightarrow 2 \, KCl(s) + 3 \, O_2(g)$
    (c) $4 \, NH_3(g) + 5 \, O_2(g) \rightarrow 4 \, NO(g) + 6 \, H_2O(g)$
    (d) $2 \, As(s) + 6 \, NaOH(\ell) \rightarrow 2 \, Na_3AsO_3(s) + 3 \, H_2(g)$

28. For each of the following balanced equations, write the oxidation number above the symbol of each atom that changes oxidation state in the course of the reaction.
    (a) $N_2O_4(g) + KCl(s) \rightarrow NOCl(g) + KNO_3(s)$
    (b) $H_2S(g) + 4 \, O_2F_2(g) \rightarrow SF_6(g) + 2 \, HF(g) + 4 \, O_2(g)$
    (c) $2 \, POBr_3(s) + 3 \, Mg(s) \rightarrow 2 \, PO(s) + 3 \, MgBr_2(s)$
    (d) $4 \, BCl_3(g) + 3 \, SF_4(g) \rightarrow 4 \, BF_3(g) + 3 \, SCl_2(\ell) + 3 \, Cl_2(g)$

29. Selenic acid($H_2SeO_4$) is a powerful oxidizing acid that dissolves not only silver (as does the related acid $H_2SO_4$) but gold, through the reaction

    $$2 \, Au(s) + 6 \, H_2SeO_4(aq) \longrightarrow$$
    $$Au_2(SeO_4)_3(aq) + 3 \, H_2SeO_3(aq) + 3 \, H_2O(\ell)$$

    Determine the oxidation numbers of the atoms in this equation. Which species is oxidized and which is reduced?

30. Diiodine pentaoxide oxidizes carbon monoxide to carbon dioxide under room conditions, yielding iodine as the second product:

    $$I_2O_5(s) + 5 \, CO(g) \longrightarrow I_2(s) + 5 \, CO_2(g)$$

    This can be used in an analytical method to measure the amount of carbon monoxide in a sample of air. Determine the oxidation numbers of the atoms in this equation. Which species is oxidized and which is reduced?

31. Potassium dichromate in acidic solution is used to titrate a solution of iron(II) ions, with which it reacts according to

    $$Cr_2O_7^{2-}(aq) + 6 \, Fe^{2+}(aq) + 14 \, H^+(aq) \longrightarrow$$
    $$2 \, Cr^{3+}(aq) + 6 \, Fe^{3+}(aq) + 7 \, H_2O(\ell)$$

    A potassium dichromate solution is prepared by dissolving 5.134 g of $K_2Cr_2O_7$ in water and diluting to a total volume of 1.000 L. A total of 34.26 mL of this solution is required to reach the endpoint in a titration of a 500.0-mL sample containing $Fe^{2+}(aq)$. Determine the concentration of $Fe^{2+}$ in the original solution.

32. Cerium(IV) ions are strong oxidizing agents in acidic solution, oxidizing arsenious acid to arsenic acid according to the equation

    $$2 \, Ce^{4+}(aq) + H_3AsO_3(aq) + H_2O(\ell) \longrightarrow$$
    $$2 \, Ce^{3+}(aq) + H_3AsO_4(aq) + 2 \, H^+(aq)$$

    A sample of $As_2O_3$ weighing 0.217 g is dissolved in basic solution and then acidified to make $H_3AsO_3$. Its titration with a solution of acidic cerium(IV) sulfate requires 21.47 mL. Determine the original concentration of $Ce^{4+}(aq)$ in the titrating solution.

## Colligative Properties of Solutions

33. The vapor pressure of pure acetone ($CH_3COCH_3$) at 30°C is 0.3270 atm. Suppose 15.0 g of benzophenone, $C_{13}H_{10}O$, is dissolved in 50.0 g of acetone. Calculate the vapor pressure of acetone above the resulting solution.

34. The vapor pressure of diethyl ether (molar mass, 74.12 g mol$^{-1}$) at 30°C is 0.8517 atm. Suppose 1.800 g of maleic acid, $C_4H_4O_4$, is dissolved in 100.0 g of diethyl ether at 30°C. Calculate the vapor pressure of diethyl ether above the resulting solution.

35. Pure toluene ($C_7H_8$) has a normal boiling point of 110.60°C. A solution of 7.80 g of anthracene ($C_{14}H_{10}$) in 100.0 g of toluene has a boiling point of 112.06°C. Calculate $K_b$ for toluene.

36. When 2.62 g of the nonvolatile solid anthracene, $C_{14}H_{10}$, is dissolved in 100.0 g of cyclohexane, $C_6H_{12}$, the boiling point of the cyclohexane is raised by 0.41°C. Calculate $K_b$ for cyclohexane.

37. When 39.8 g of a nondissociating, nonvolatile sugar is dissolved in 200.0 g of water, the boiling point of the water is raised by 0.30°C. Estimate the molar mass of the sugar.

38. When 2.60 g of a substance that contains only indium and chlorine is dissolved in 50.0 g of tin(IV) chloride, the normal boiling point of the tin(IV) chloride is raised from 114.1°C to 116.3°C. If $K_b = 9.43$ K kg mol$^{-1}$ for $SnCl_4$, what are the approximate molar mass and the probable molecular formula of the solute?

39. The Rast method for determining molar masses employs camphor as the solvent. Camphor melts at 178.4°C, and its large $K_f$(37.7 K kg mol$^{-1}$) makes it especially useful for accurate work. A sample of an unknown substance that weighs 0.840 g lowers the freezing point of 25.0 g of camphor to 170.8°C. What is its molar mass?

40. Barium chloride has a freezing point of 962°C and a freezing-point depression constant of 108 K kg mol$^{-1}$. If 12 g of an unknown substance dissolved in 562 g of barium chloride gives a solution with a freezing point of 937°C, compute the molar mass of the unknown, assuming no dissociation takes place.

41. Ice cream is made by freezing a liquid mixture that, as a first approximation, can be considered a solution of sucrose ($C_{12}H_{22}O_{11}$) in water. Estimate the temperature at which the first ice crystals begin to appear in a mix that consists of 34% (by mass) sucrose in water. As ice crystallizes out, the remaining solution becomes more concentrated. What happens to its freezing point?

42. The solution to problem 41 shows that to make homemade ice cream, temperatures ranging downward from −3°C are needed. Ice cubes from a freezer have a temperature of about −12°C (+10°F), which is cold enough, but contact with the warmer ice cream mixture causes them to melt to liquid at 0°C, which is too warm. To obtain a liquid that is cold enough, salt (NaCl) is dissolved in water, and ice is added to the salt water. The salt lowers the freezing point of the water enough so that it can freeze the liquid inside the ice cream

maker. The instructions for an ice cream maker say to add one part salt to eight parts water (by mass). What is the freezing point of this solution (in degrees Celsius and degrees Fahrenheit)? Assume that the NaCl dissociates fully into ions and that the solution is ideal.

43. An aqueous solution is 0.8402 molal in $Na_2SO_4$. It has a freezing point of $-4.218°C$. Determine the effective number of particles arising from each $Na_2SO_4$ formula unit in this solution.

44. The freezing-point depression constant of pure $H_2SO_4$ is $6.12$ K kg $mol^{-1}$. When 2.3 g of ethanol ($C_2H_5OH$) is dissolved in 1.00 kg of pure sulfuric acid, the freezing point of the solution is 0.92 K lower than the freezing point of pure sulfuric acid. Determine how many particles are formed as one molecule of ethanol goes into solution in sulfuric acid.

45. A 200-mg sample of a purified compound of unknown molar mass is dissolved in benzene and diluted with that solvent to a volume of 25.0 $cm^3$. The resulting solution is found to have an osmotic pressure of 0.0105 atm at 300 K. What is the molar mass of the unknown compound?

46. Suppose 2.37 g of a protein is dissolved in water and diluted to a total volume of 100.0 mL. The osmotic pressure of the resulting solution is 0.0319 atm at 20°C. What is the molar mass of the protein?

47. A polymer of large molar mass is dissolved in water at 15°C, and the resulting solution rises to a final height of 15.2 cm above the level of the pure water, as water molecules pass through a semipermeable membrane into the solution. If the solution contains 4.64 g polymer per liter, calculate the molar mass of the polymer.

48. Suppose 0.125 g of a protein is dissolved in 10.0 $cm^3$ of ethyl alcohol ($C_2H_5OH$), whose density at 20°C is 0.789 g $cm^{-3}$. The solution rises to a height of 26.3 cm in an osmometer (an apparatus for measuring osmotic pressure). What is the approximate molar mass of the protein?

## Mixtures and Distillation

49. The Henry's law constant at 25°C for carbon dioxide dissolved in water is $1.65 \times 10^3$ atm. If a carbonated beverage is bottled under a $CO_2$ pressure of 5.0 atm:
    (a) Calculate the chemical amount of carbon dioxide dissolved per liter of water under these conditions, using 1.00 g $cm^{-3}$ as the density of water.
    (b) Explain what happens on a microscopic level after the bottle cap is removed.

50. The Henry's law constant at 25°C for nitrogen dissolved in water is $8.57 \times 10^4$ atm, that for oxygen is $4.34 \times 10^4$ atm, and that for helium is $1.7 \times 10^5$ atm.
    (a) Calculate the number of moles of nitrogen and oxygen dissolved per liter of water in equilibrium with air at 25°C. Use Table 4.1.
    (b) Air is dissolved in blood and other bodily fluids. As a deep-sea diver descends, the pressure increases and the concentration of dissolved air in the blood increases. If the diver returns to the surface too quickly, gas bubbles

out of solution within the body so rapidly that it can cause a dangerous condition called the bends. Use Henry's law to show why divers sometimes use a combination of helium and oxygen in their breathing tanks in place of compressed air.

51. At 25°C, some water is added to a sample of gaseous methane ($CH_4$) at 1.00 atm pressure in a closed vessel and the vessel is shaken until as much methane as possible dissolves. Then 1.00 kg of the solution is removed and boiled to expel the methane, yielding a volume of 3.01 L of $CH_4(g)$ at 0°C and 1.00 atm. Determine the Henry's law constant for methane in water.

52. When exactly the procedure of problem 51 is carried out using benzene ($C_6H_6$) in place of water, the volume of methane that results is 0.510 L at 0°C and 1.00 atm. Determine the Henry's law constant for methane in benzene.

53. At 20°C, the vapor pressure of toluene is 0.0289 atm and the vapor pressure of benzene is 0.0987 atm. Equal chemical amounts (equal numbers of moles) of toluene and benzene are mixed and form an ideal solution. Compute the mole fraction of benzene in the vapor in equilibrium with this solution.

54. At 90°C, the vapor pressure of toluene is 0.534 atm and the vapor pressure of benzene is 1.34 atm. Benzene (0.400 mol) is mixed with toluene (0.900 mol) to form an ideal solution. Compute the mole fraction of benzene in the vapor in equilibrium with this solution.

55. At 40°C, the vapor pressure of pure carbon tetrachloride ($CCl_4$) is 0.293 atm and the vapor pressure of pure dichloroethane ($C_2H_4Cl_2$) is 0.209 atm. A nearly ideal solution is prepared by mixing 30.0 g of carbon tetrachloride with 20.0 g of dichloroethane.
    (a) Calculate the mole fraction of $CCl_4$ in the solution.
    (b) Calculate the total vapor pressure of the solution at 40°C.
    (c) Calculate the mole fraction of $CCl_4$ in the vapor in equilibrium with the solution.

56. At 300 K, the vapor pressure of pure benzene ($C_6H_6$) is 0.1355 atm and the vapor pressure of pure n-hexane ($C_6H_{14}$) is 0.2128 atm. Mixing 50.0 g of benzene with 50.0 g of n-hexane gives a solution that is nearly ideal.
    (a) Calculate the mole fraction of benzene in the solution.
    (b) Calculate the total vapor pressure of the solution at 300 K.
    (c) Calculate the mole fraction of benzene in the vapor in equilibrium with the solution.

## Additional Problems

57. Veterinarians use Donovan's solution to treat skin diseases in animals. The solution is prepared by mixing 1.00 g of $AsI_3(s)$, 1.00 g of $HgI_2(s)$, and 0.900 g of $NaHCO_3(s)$ in enough water to make a total volume of 100.0 mL.
    (a) Compute the total mass of iodine per liter of Donovan's solution, in grams per liter.
    (b) You need a lot of Donovan's solution to treat an outbreak of rash in an elephant herd. You have plenty of mercury(II) iodide and sodium hydrogen carbonate, but the only arsenic(III) iodide you can find is 1.50 L of a 0.100 M

aqueous solution. Explain how to prepare 3.50 L of Donovan's solution starting with these materials.

58. Relative solubilities of salts in liquid ammonia can differ significantly from those in water. Thus, silver bromide is soluble in ammonia, but barium bromide is not (the reverse of the situation in water).

   (a) Write a balanced equation for the reaction of an ammonia solution of barium nitrate with an ammonia solution of silver bromide. Silver nitrate is soluble in liquid ammonia.

   (b) What volume of a 0.50 M solution of silver bromide will react completely with 0.215 L of a 0.076 M solution of barium nitrate in ammonia?

   (c) What mass of barium bromide will precipitate from the reaction in part (b)?

* 59. Suppose 150 mL of a 10.00% by mass solution of sodium chloride (density = 1.0726 g cm$^{-3}$) is acidified with sulfuric acid and then treated with an excess of $MnO_2(s)$. Under these conditions, all of the chlorine is liberated as $Cl_2(g)$. The chlorine is collected without loss and reacts with excess $H_2(g)$ to form $HCl(g)$. The $HCl(g)$ is dissolved in enough water to make 250 mL of solution. Compute the molarity of this solution.

* 60. The amount of ozone in a mixture of gases can be determined by passing the mixture through an acidic aqueous solution of potassium iodide, where the ozone reacts according to

$$O_3(g) + 3\ I^-(aq) + H_2O(\ell) \longrightarrow$$
$$O_2(g) + I_3^-(aq) + 2\ OH^-(aq)$$

to form the triiodide ion $I_3^-$. The amount of triiodide produced is then determined by titrating with thiosulfate solution:

$$I_3^-(aq) + 2\ S_2O_3^{2-}(aq) \longrightarrow 3\ I^-(aq) + S_4O_6^{2-}(aq)$$

A small amount of starch solution is added as an indicator because it forms a deep-blue complex with the triiodide solution. Disappearance of the blue color thus signals the completion of the titration. Suppose 53.2 L of a gas mixture at a temperature of 18°C and a total pressure of 0.993 atm is passed through a solution of potassium iodide until the ozone in the mixture has reacted completely. The solution requires 26.2 mL of a 0.1359 M solution of thiosulfate ion to titrate to the endpoint. Calculate the mole fraction of ozone in the original gas sample.

61. The vapor pressure of pure liquid $CS_2$ is 0.3914 atm at 20°C. When 40.0 g of rhombic sulfur is dissolved in 1.00 kg of $CS_2$, the vapor pressure of $CS_2$ falls to 0.3868 atm. Determine the molecular formula for the sulfur molecules dissolved in $CS_2$.

62. The expressions for boiling-point elevation and freezing-point depression apply accurately to *dilute* solutions only. A saturated aqueous solution of NaI (sodium iodide) in water has a boiling point of 144°C. The mole fraction of NaI in the solution is 0.390. Compute the molality of this solution. Compare the boiling-point elevation predicted by the expression in this chapter to the elevation actually observed.

63. You take a bottle of soft drink out of your refrigerator. The contents are liquid and stay liquid, even when you shake them.

Presently you remove the cap, and the liquid freezes solid. Offer a possible explanation for this observation.

64. Mercury(II) chloride ($HgCl_2$) freezes at 276.1°C and has a freezing-point depression constant $K_f$ of 34.3 K kg mol$^{-1}$. When 1.36 g of solid mercury(I) chloride (empirical formula HgCl) is dissolved in 100 g of $HgCl_2$, the freezing point is lowered by 0.99°C. Calculate the molar mass of the dissolved solute species, and give its molecular formula.

* 65. The vapor pressure of an aqueous solution of $CaCl_2$ at 25°C is 0.02970 atm. The vapor pressure of pure water at the same temperature is 0.03126 atm. Estimate the freezing point of the solution.

66. Ethylene glycol ($CH_2OHCH_2OH$) is used in antifreeze because, when mixed with water, it lowers the freezing point below 0°C. What mass percentage of ethylene glycol in water must be used to reduce the freezing point of the mixture to −5.0°C, assuming ideal solution behavior?

67. A new compound has the empirical formula $GaCl_2$. This surprises some chemists who, based on the position of gallium in the periodic table, expect a chloride of gallium to have the formula $GaCl_3$ or possibly GaCl. They suggest that the "$GaCl_2$" is really $Ga[GaCl_4]$, in which the bracketed group behaves as a unit with a −1 charge. Suggest experiments to test this hypothesis.

* 68. Suppose two beakers are placed in a small closed container at 25°C. One contains 400 mL of a 0.100 M aqueous solution of NaCl; the second contains 200 mL of a 0.250 M aqueous solution of NaCl. Small amounts of water evaporate from both solutions. As time passes, the volume of solution in the second beaker gradually increases, and that in the first gradually decreases. Why? If we wait long enough, what will the final volumes and concentrations be?

* 69. The walls of erythrocytes (red blood cells) are permeable to water. In a salt solution, they shrivel (lose water) when the outside salt concentration is high, and swell (take up water) when the outside salt concentration is low. In an experiment at 25°C, an aqueous solution of NaCl that has a freezing point of 0.406°C causes erythrocytes neither to swell nor to shrink, indicating that the osmotic pressure of their contents is equal to that of the NaCl solution. Calculate the osmotic pressure of the solution inside the erythrocytes under these conditions, assuming that its molarity and molality are equal.

70. Silver dissolves in molten lead. Compute the osmotic pressure of a 0.010 M solution of silver in lead at 423°C. Compute the height of a column of molten lead ($\rho = 11.4$ g cm$^{-3}$) to which this pressure corresponds.

71. Henry's law is important in environmental chemistry, where it predicts the distribution of pollutants between water and the atmosphere. Benzene ($C_6H_6$) emitted in wastewater streams, for example, can pass into the air, where it is degraded by processes induced by light from the sun. The Henry's law constant for benzene in water at 25°C is 301 atm. Calculate the partial pressure of benzene vapor in equilibrium with a solution of 2.0 g of benzene per 1000 L of water. How many benzene molecules are present in each cubic centimeter?

* 72. Refer to the data of problem 54. Calculate the mole fraction of toluene in a mixture of benzene and toluene that boils at 90°C under atmospheric pressure.

73. What is the difference between a solution and a colloidal suspension? Give examples of each, and show how, in some cases, it may be difficult to classify a mixture as one or the other.

---

## CUMULATIVE PROBLEMS

74. A student prepares a solution by dissolving 1.000 mol of $Na_2SO_4$ in water. She accidentally leaves the container uncovered, and comes back the next week to find only a white, solid residue. The mass of the residue is 322.2 g. Determine the chemical formula of this residue.

75. Complete combustion of 2.40 g of a compound of carbon, hydrogen, and oxygen yielded 5.46 g $CO_2$ and 2.23 g $H_2O$. When 8.69 g of the compound was dissolved in 281 g of water, the freezing point of the solution was found to be $-0.97$°C. What is the molecular formula of the compound?

# U N I T 3 EQUILIBRIUM IN CHEMICAL REACTIONS

Stalactites (top) and stalagmites (bottom) consist of calcium carbonate. They form when a water solution containing $Ca^{2+}$ and $HCO_3^-$ ions enters a cave. Carbon dioxide is released, and calcium carbonate precipitates: $Ca^{2+}(aq) + 2\ HCO_3^-(aq) \rightarrow CaCO_3(s) + H_2O + CO_2(g)$. *(Arthur N. Palmer)*

How far do chemical reactions proceed toward completion and what determines the extent of their progress? Experience shows that many reactions do not proceed to completion but approach instead an *equilibrium state* in which both products and unconsumed reactants are present in definite relative amounts. Once equilibrium has been achieved, no further changes in composition take place. The equilibrium state is described quantitatively by the equilibrium constant for the reaction, which depends on the temperature at which the reaction is carried out. The science of thermodynamics, which quantifies heat and temperature, predicts the equilibrium constant from simple physical properties of the reactants and products.

THE GOALS OF UNIT 3 ARE **(1)** To relate composition in the equilibrium state to the equilibrium constant, **(2)** To describe the influence of temperature on the equilibrium constant, and **(3)** To apply thermodynamics to describe these connections.

# Thermodynamic Processes and Thermochemistry

E xperience shows that heat plays a key role in determining the extent of chemical reactions. Heat drives some reactions forward and retards others. It is appropriate to begin the study of chemical equilibrium with an introduction to thermodynamics, which describes the nature of heat and gives procedures for measuring heat transfer quantitatively. Then the principles of thermodynamics will be applied to study chemical equilibrium in a very large variety of reactions.

Thermodynamics, in which a few apparently simple laws summarize a rich variety of observed behavior, is one of the surest and most powerful conceptual tools we possess in science. The central feature of thermodynamics is the universality of

*Illustration*
The reaction of sodium with water is highly exothermic. *(Charles D. Winters)*

its basic laws, and the beauty of the subject is that so many physical conclusions can be deduced from those few laws. The laws of thermodynamics cannot themselves be derived or proved. Instead, they are generalizations that have been drawn from a vast number of observations of the behavior of matter. The history of thermodynamics, like that of many fields of science, has been fraught with misconceptions. As we look back upon the beginnings of the discipline in the 19th century, it appears to have been agonizingly slow to develop. But it *has* developed, and its laws are the pillars upon which much of modern science rests. Most of thermodynamics is in finished form today, but it is still being applied to research in the forefront of science, on systems ranging from black holes in distant parts of the universe to the growth and development of the biological cell.

Thermodynamics is an *operational* science, concerned with macroscopic properties that, in principle, can be measured. Its goal is to predict what types of chemical and physical processes are possible, and under what conditions, and to calculate quantitatively the properties of the equilibrium state that ensues when a process is carried out. For example, with thermodynamics the following kinds of questions can be raised.

1. If hydrogen and nitrogen are mixed, will they tend to react? If so, in what percentage yield will they produce ammonia?
2. What effect will a particular change in temperature or pressure have on the extent of their reaction?
3. How can the conditions for a process be optimized so as to maximize its efficiency?

Thermodynamics is an immensely practical subject. The knowledge that a particular chemical process is impossible under certain conditions can save a great deal of time spent vainly trying to make it occur. Thermodynamics can also suggest ways in which conditions can be changed to make a process possible.

The power of thermodynamics lies in its generality: it rests upon no particular model of the structure of matter. In fact, if the entire atomic theory of matter were to be overthrown (a *very* unlikely event!), the foundations of thermodynamics would be unshaken. At the same time, thermodynamics has some important limitations. It is a tool that enables the prediction of certain macroscopic properties, but it cannot explain why a substance has a particular property. It can determine whether a process can occur, but it cannot say how rapidly it will occur. Thermodynamic arguments show that diamond is an unstable substance at atmospheric pressure and will eventually revert to graphite, but they cannot state how long the process will take.

## 7.1

## SYSTEMS, STATES, AND PROCESSES

Thermodynamics uses abstract models to represent real-world laboratory systems and processes. This requires precise definitions of thermodynamic terms. Many apparent contradictions can be traced to loose or inaccurate use of language. Some terms have everyday meanings that are different from their thermodynamic usage. This section contains a brief introduction to the language of thermodynamics.

A **system** is that part of the universe of immediate interest in a particular experiment. The system always contains a certain amount of matter and is described by certain parameters that are controlled in the experiment. For example, the gas confined in a closed box may be a system, characterized by the number of moles of the gas and the fixed volume of the box. But the gas molecules in a particular cubic centimeter of space in the middle of a room may also be considered a system. In the first case the boundaries are physical ones; in the second they are conceptual or mathematical, and the system is characterized by its volume, which is definite, and by the number of moles of gas within it, which may fluctuate as the system exchanges molecules with the surrounding regions. A **closed system** is one whose boundaries prevent the flow of matter into and out of it (they are impermeable), whereas in an **open system** the boundaries permit such flow. The amount of matter in an open system can change with time. The portion of the remainder of the universe that can exchange energy with the system during the process of interest is called the **surroundings.** The system and the surroundings together constitute the **thermodynamic universe** for that process.

Thermodynamics is concerned with macroscopic properties of systems and the ways in which they change. Such properties are of two kinds, *extensive* and *intensive*. To distinguish between them, consider the following "thought experiment." Place a thin wall through the middle of a system, dividing it into two subsystems each of which, like the system itself, is characterized by certain properties. An **extensive property** of the system is one that can be written as the *sum* of the corresponding properties of the subsystems. Volume, mass, and energy are typical extensive properties; the volume of a system, for example, is the sum of the volumes of its subsystems. An **intensive property** of the system is one that is the *same* as the corresponding properties of each of the subsystems. Temperature and pressure are typical intensive properties; if a system at 298 K is divided in half, the temperature of each half will still be 298 K.

A **thermodynamic state** is a macroscopic condition of a system whose properties are uniquely determined by the laboratory apparatus that holds them at selected fixed values and independent of time. For example, a system comprising 2 moles of He gas can be held in a piston-cylinder apparatus that maintains the system pressure at 1.5 atm, and the apparatus may be immersed in a heat bath that maintains the system temperature at 298 K. The properties $P$ and $T$ are then said to be **constrained** to the values 1 atm and 298 K respectively. The piston-cylinder and the heat bath are the **constraints** that maintain the selected values of the properties $T$ and $P$.

After the system has been prepared subject to the constraints in the laboratory and when all disturbances due to the preparation have ceased and none of its properties change with time, the system is said to have reached **equilibrium.** The same equilibrium state can be reached from different directions. The thermodynamic state of a system composed of a given quantity of a liquid or gaseous pure substance is fixed when any two of its independent properties are given. Thus, the specification of $P$ and $T$ for 1 mol of a pure gas fixes not merely the volume $V$ but all other properties of the material, such as the internal energy $E$, which will be defined later in Section 7.2. These relationships can be indicated by three-dimensional plots of any property as a function of two independent properties. For 1 mol of a gas, a plot can be made of pressure against volume and temperature, as in Fig. 7.1. The resulting surface can be represented by an equation giving $P$ as a function of $V$ and $T$; this is

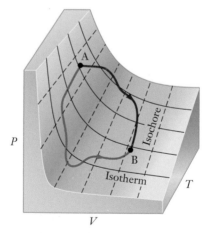

**FIGURE 7.1** The $PVT$ surface of 1 mol of ideal gas. Each point on the surface represents a combination of pressure, volume, and temperature allowed by the equation of state of the gas. Along an isotherm ($T$ constant), the pressure varies inversely with volume; along an isochore ($V$ constant), it varies linearly with temperature. Two processes are shown connecting states A and B along paths that satisfy the equation of state at every point.

called the *equation of state* of the substance (see Section 4.7.) If the experimenter avoids regions of small $V$ and low $T$ where gas nonideality becomes important, the experimentally determined surface is that shown in Figure 7.1, and the equation of state is the ideal gas law $PV = nRT$. The points on this surface ($A$, $B$, …) represent experimentally determined equilibrium thermodynamic states of the system characterized by particular values of $P$, $V$, and $T$. Experience shows that the values of all other macroscopic properties take on definite values at each of these points. For example, one can visualize the internal energy $E$ as a similar three-dimensional plot versus $T$ and $V$.

A **thermodynamic process** leads to a change in the thermodynamic state of a system. Such a process may be a *physical* process (e.g., a change in the pressure of a gaseous system or the boiling of a liquid) or a *chemical* process, in which there is a change in the distribution of matter among different chemical species (e.g., the reaction of solid $CaCO_3$, at 900 K and 1 atm pressure, to give solid $CaO$ and gaseous $CO_2$ at the same temperature and pressure).

Since a process changes the state of a system, the process must start with the system in a particular equilibrium state and must also end with the system in a particular equilibrium state. Two such states A and B are indicated in Figure 7.1. It is intuitively appealing to sketch a *path* on the surface of equilibrium thermodynamic states to summarize the progress of the system during a process. However, only very special processes of the type called *reversible* can be represented in this way, as explained in the next paragraphs.

Many conditions of a system do not correspond to any equilibrium thermodynamic state. For example, suppose a gas is confined by a piston in a cylinder with volume $V_1$ (thermodynamic state A). If the piston is abruptly pulled out to increase the volume to $V_2$ (Fig. 7.2) chaotic gas currents arise as the molecules begin to move into the larger volume. These intermediate stages are not thermodynamic states, because such properties as density and temperature are changing rapidly both through space and in time. Eventually the currents cease and the system approaches a new equilibrium thermodynamic state, B. States A and B are thermodynamic states, but the conditions in between cannot be described by only a few macroscopic variables and so are not thermodynamic states. Such a process is called **irreversible;** it cannot be represented as a path on a thermodynamic surface (as in Fig. 7.1) because the intermediate stages do not correspond to points on the equation-of-state surface.

By contrast, a **reversible process** proceeds through a continuous series of thermodynamic states and thus can be indicated as a path on the equation-of-state surface. Such a process is an idealization, because true equilibrium is reached only after an infinite length of time and therefore such a process could never occur in a finite time. However, if a real process is conducted slowly enough and in small enough steps, the real (irreversible) process can be considered an approximation to the idealized limiting reversible process. The term "reversible" is used because an infinitesimal change in external conditions suffices to reverse the direction of motion of the system. For example, if a gas is expanded by pulling out a piston very slowly, only a tiny change in the forces exerted from the outside is required to change the

*[handwritten margin note: MUST END AND FINISH AT EQ'BM.]*

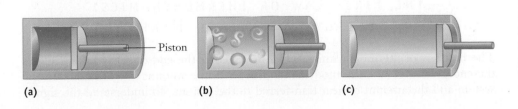

(a)        (b)        (c)      — Piston

**FIGURE 7.2** Stages in an irreversible expansion of a gas from an initial state (a) of volume $V_1$ to a final state (c) of volume $V_2$. In the intermediate stage shown (b), the gas is not in equilibrium; because of turbulence, pressure and temperature cannot be defined.

**FIGURE 7.3** Differences in state properties (such as the difference in altitude between two points) are independent of the path followed. Other properties (such as the total distance traveled) depend on the particular path.

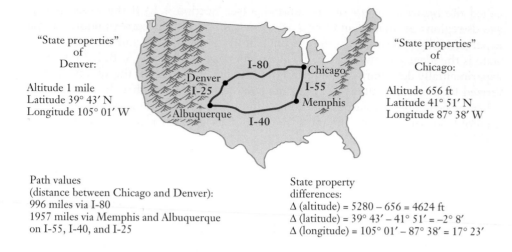

"State properties"
of
Denver:

Altitude 1 mile
Latitude 39° 43′ N
Longitude 105° 01′ W

"State properties"
of
Chicago:

Altitude 656 ft
Latitude 41° 51′ N
Longitude 87° 38′ W

Path values
(distance between Chicago and Denver):
996 miles via I-80
1957 miles via Memphis and Albuquerque
on I-55, I-40, and I-25

State property
differences:
$\Delta$ (altitude) = 5280 − 656 = 4624 ft
$\Delta$ (latitude) = 39° 43′ − 41° 51′ = −2° 8′
$\Delta$ (longitude) = 105° 01′ − 87° 38′ = 17° 23′

direction of motion of the piston and begin to compress the gas. A concrete illustration clarifies the terminology. A gas confined inside a piston–cylinder arrangement will experience an irreversible compression when a kilogram of sand is suddenly dropped onto the piston. The same compression can be achieved (almost) reversibly by transferring the same kilogram of sand onto the piston one grain at the time.

An infinite number of reversible paths can be identified between any two thermodynamic states A and B. Two of them shown in Figure 7.1 could be realized by very slowly changing the values of $T$ and $V$ in the proper sequence in the laboratory apparatus. Such reversible paths are frequently invoked throughout this book as a tool for calculating changes produced by processes.

Certain properties of a system, called **state functions,** are uniquely determined by the thermodynamic state of the system. Volume, temperature, pressure, and the internal energy $E$ are examples of state functions. The Greek letter $\Delta$ (delta) is used to indicate *changes* in state functions in a thermodynamic process. Thus, $\Delta V = V_{final} - V_{initial}$ (or $V_f - V_i$) is the change in volume between initial and final states, and $\Delta E = E_{final} - E_{initial}$ is the corresponding change in energy (note that these changes are taken to be the final-state property minus the initial-state property). Because $\Delta E$ (or $\Delta V$ or $\Delta T$) depends only on the initial and final states, the same change, $\Delta E$, will be measured no matter which path (reversible or irreversible) is followed between any given pair of thermodynamic states; *the change in any state function is independent of path.* The converse statement is also true: if the change in a property of a system is independent of path, the property is a state function. Fig. 7.3 illustrates two different paths that connect a given initial state and a common final state.

---

**7.2**

## THE FIRST LAW OF THERMODYNAMICS:
## ENERGY, WORK, AND HEAT

The first law of thermodynamics, which is stated at the end of this section, relates the energy change in a thermodynamic process to the amount of work done on the system and the amount of heat transferred to the system. To understand the significance of this law, it is first necessary to examine the ways in which amounts of heat and work are measured. You will see that heat and work are simply different means by which energy is transferred into or out of a system.

## Work

The mechanical definition of **work** is the product of the external force on a body times the distance through which the force acts. If a body moves in a straight line from point $r_i$ to $r_f$ with a constant force $F$ applied along the direction of the path, the work done on the body is

$$w = F(r_f - r_i) \qquad \text{(force along direction of path)}$$

What is the relationship between work and energy in mechanical systems? Suppose a block of mass $M$ is moving with initial velocity $v_i$ along a frictionless surface. If a constant force $F$ is exerted on it in the direction of its motion, it will experience a constant acceleration $a = F/M$. After a time $t$, the block's velocity will have increased from $v_i$ to $v_f$ and its position will have changed from $r_i$ to $r_f$. The work done on the block is

$$w = F(r_f - r_i) = Ma(r_f - r_i)$$

The distance traveled, $r_f - r_i$, is given by the average velocity, in this case $(v_i + v_f)/2$, multiplied by the elapsed time $t$:

$$r_f - r_i = \left(\frac{v_i + v_f}{2}\right)t$$

When the acceleration is constant, it is equal to the change in velocity, $v_f - v_i$, divided by the elapsed time:

$$a = (v_f - v_i)/t$$

Substituting both of these results into the expression for the work done gives

$$w = M\left(\frac{v_f - v_i}{t}\right)\left(\frac{v_i + v_f}{2}\right)t$$

$$= \frac{M}{2}(v_f - v_i)(v_f + v_i)$$

$$= \frac{M}{2}v_f^2 - \frac{M}{2}v_i^2$$

The expression on the right side is the change in kinetic energy, $\frac{1}{2}Mv^2$, of the block. For this idealized example with a frictionless surface, the work done is equal to the change in energy (in this case, kinetic) of the block.

As a second example, consider the work done in lifting an object in a gravitational field. To raise a mass $M$ from an initial height $h_i$ to a final height $h_f$, an upward force sufficient to counteract the downward force of gravity $Mg$ must be exerted. The work done on the object in this case is

$$w = Mg(h_f - h_i) = Mg\Delta h$$

This is the change in *potential* energy $Mgh$ of the object, showing once again a connection between mechanical work done and a change in energy.

One important kind of mechanical work in chemistry is **pressure–volume work,** which results when a gas is compressed or expanded under the influence of

**FIGURE 7.4** As the gas inside this cylinder is heated, it expands, pushing the piston against the pressure $P_{ext}$ exerted by the gas outside. As the piston is displaced over a distance $h_f - h_i = \Delta h$, the volume of the cylinder increases by an amount $A\,\Delta h$, where $A$ is the surface area of the piston.

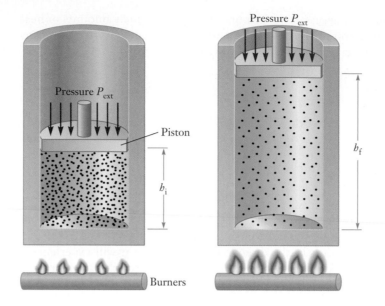

an outside pressure, such as occurs in the internal combustion engine of an automobile. Imagine that a gas has pressure $P_i$ and is confined in a cylinder by a frictionless piston of cross-sectional area $A$ and negligible mass (Fig. 7.4). The force exerted on the piston face by the gas is $F_i = P_i A$, because pressure is force divided by area. If there is a gas on the other side of the piston with pressure $P_{ext}$ ("ext" for "external"), then if $P_{ext} = P_i$ the piston will experience no net force; if $P_{ext}$ is increased, the gas will be compressed, and if it is decreased, the gas will expand. Consider first the case in which the external force is less than the initial force exerted by the gas, $P_i A$. Then the gas will expand and lift the piston from $h_i$ to $h_f$. The work in this case is

$$w = -F_{ext}(h_f - h_i)$$

The negative sign is inserted because the force from the gas outside opposes the expansion of the gas inside the cylinder. This is rewritten as

$$w = -P_{ext}A\Delta h$$

The product $A\Delta h$ is the volume change of the system, $\Delta V$, so the work is

$$w = -P_{ext}\Delta V \qquad \text{[7.1]}$$

For an expansion, $\Delta V > 0$, so $w < 0$ and the system does work; it pushes back the surroundings. If the gas had been compressed (by making $P_{ext}$ larger than $P_i$), work would have been done *on* the system. As before, $w = -P_{ext}\Delta V$, but now $\Delta V < 0$, so $w > 0$. If there is no volume change, $\Delta V = 0$ and no pressure–volume work is done. Finally, if there is no mechanical link to the surroundings (that is, if $P_{ext} = 0$), then once again no pressure–volume work can be performed.

If the pressure $P_{ext}$ is expressed in pascals and the volume in cubic meters, their product is in joules (J). These are the SI units. For many purposes, however, it is more convenient to express pressures in atmospheres and volumes in liters, resulting in liter-atmospheres (L atm) as a unit of work. The two work units are related by

$$1 \text{ L atm} = (10^{-3} \text{ m}^3)(1.01325 \times 10^5 \text{ kg m}^{-1}\text{ s}^{-2}) = 101.325 \text{ J}$$

## EXAMPLE 7.1

A cylinder confines 2.00 L of gas under a pressure of 1.00 atm. The external pressure is also 1.00 atm. The gas is heated slowly, with the piston sliding freely to maintain the pressure of the gas close to 1.00 atm. Suppose the heating continues until a final volume of 3.50 L is reached. Calculate the work done on the gas and express it in joules.

*[handwritten: $w = -P_{ext} + \Delta V$ $= 1 atm \times 1.50 L$]*

### Solution

This is an expansion of a system from 2.00 L to 3.50 L against a constant external pressure of 1.00 atm. The work done on the system is then

$$w = -P_{ext}\Delta V = -(1.00 \text{ atm})(3.50 \text{ L} - 2.00 \text{ L}) = -1.50 \text{ L atm}$$

Conversion to joules gives

$$w = (-1.50 \text{ L atm})(101.325 \text{ J L}^{-1} \text{ atm}^{-1}) = -152 \text{ J}$$

Because $w$ is negative, we can say either that $-152$ J of work is done *on* the gas or that $+152$ J of work is done *by* the gas.

**Related Problems: 1, 2**

## Heat

So far, two types of energy have been considered: the kinetic energy of a moving object and the potential energy of an object in a gravitational field. A third type of energy, less visible but equally important, is the **internal energy,** defined as the total energy content of a system due to potential energy between molecules, due to the kinetic energy of molecular motions, and due to chemical energy stored in chemical bonds. Potential energy between molecules appears as the lattice energy of solids and the attractive and repulsive interactions between molecules in gases and liquids. Kinetic energy appears in the translation and the internal motions of individual molecules (Fig. 7.5).

*[handwritten margin notes: Kinetic, Potential, Internal, E between bonds]*

Gas molecules are in a constant state of motion (Section 4.5) even when no overall gas flow is taking place in the container; the same is true of molecules in liquids and solids. If a piece of hot metal is plunged into a container of water, the temperature of the water rises as its molecules begin to move faster, corresponding to an increase in the internal energy of the water. The amount of energy transferred between two objects initially at different temperatures is called **heat,** or **thermal energy.** When a hot body is brought into contact with a colder body, the two temperatures change until they become equal. This process is sometimes described as the "flow" of heat from the hotter to the colder body. Although this picture is useful, it is somewhat misleading because it implies that heat is a substance that is contained in matter. Instead, heat (like work) is a way in which energy is exchanged between a system and its surroundings.

How can amounts of heat be measured? One simple way is to use an **ice calorimeter** (Fig. 7.6), which consists of a bath containing ice and water, well insulated to prevent heat transfer to the surroundings. If heat is transferred to the bath from the system, some of the ice melts. Because a given mass of water has a smaller volume than the same mass of ice, the total volume of the ice–water mixture decreases as heat enters. If twice as much heat is transferred, twice as much ice will melt and

**FIGURE 7.5** A ball being dropped from a height exhibits four kinds of energy. (a) It begins with a high *potential energy* associated with its attraction to the earth's surface through the force of gravity. (b) As it falls, this potential energy is converted to *kinetic energy* of motion of the ball. (c) After impact with the ground, the molecules near the surface of the ball are pushed against one another, increasing the *internal potential energy* of the ball. (d) As the ball stops bouncing, the molecules readjust their positions, but they move a little faster. The ball has a higher *internal kinetic energy* and a higher temperature now than just before impact with the ground.

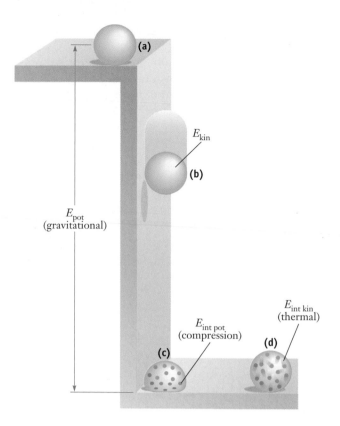

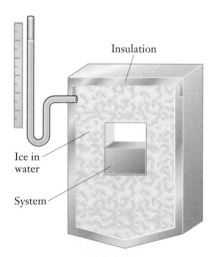

**FIGURE 7.6** An ice calorimeter. As the ice melts, the volume of the ice–water mixture decreases, an effect that can be read off the scale on the left.

the volume change will be twice as great. The amount of heat transferred is determined from the change in volume of the contents of the calorimeter. Heat transferred from the ice bath *into* the system causes water to freeze and *increases* the total volume of the ice bath.

An alternative way to measure heat is to take advantage of the fact that if heat is transferred to or removed from a substance in a single phase at constant pressure, the temperature changes in a reproducible way. The **specific heat capacity** of a material is the amount of heat required to raise the temperature of a one-gram mass by 1°C. If twice as much heat is transferred, twice as great a temperature change will be seen (provided the specific heat capacity itself does not change appreciably with temperature). The temperature change of a fixed amount of a given substance can thus be used as a measure of the quantity of heat transferred to or from it. This is described by

$$q = Mc_{\mathrm{s}}\Delta T \qquad \textbf{[7.2]}$$

where $q$ is the heat transferred to a body of mass $M$ with specific heat capacity $c_{\mathrm{s}}$ in order to cause a temperature change of $\Delta T$.

Because heat, like work, is a form in which energy is transferred, the appropriate unit for it is also the joule. Historically, however, the connections among work, heat, and energy were not appreciated until the middle of the 19th century, by which time a separate unit for heat, the calorie, was already well established. One calorie was defined as the amount of heat required to raise the temperature of one gram of water from 14.5°C to 15.5°C (or, in other words, the specific heat capacity of water, $c_{\mathrm{s}}$, at 15°C was *defined* as 1.00 cal K$^{-1}$ g$^{-1}$).

The qualitative equivalence of heat and work as means of energy transfer was suggested in 1798 by Benjamin Thompson, Count Rumford. In the course of his work as military advisor to the King of Bavaria, Thompson observed that the quantity of heat produced in the boring of cannons was proportional to the amount of work done in the process. Moreover, the operation could be continued indefinitely, demonstrating that heat was not a substance contained in the metal of the cannon. More quantitative measurements were carried out by the German physician Julius Mayer and by the English physicist James Joule. In the 1840s these scientists showed that the temperature of a substance could be raised by doing work on the substance as well as by adding heat to it. Figure 7.7 shows an apparatus in which a paddle, driven by a falling weight, churns the water in a tank. Work is performed on the water, and the temperature increases. The work done is $-Mg\,\Delta h$, where $\Delta h$ is the (negative) change in the height of the weight and $M$ is its mass. If all of this work goes to increase the water temperature (that is, if no heat enters the container or leaks out to the surroundings), then the specific heat capacity of the water in *joules* per gram per degree is equal to the quantity of work done divided by the product of the mass of water and its temperature rise.

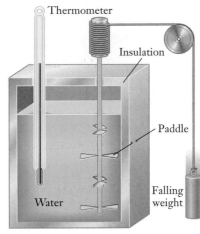

**FIGURE 7.7**   The falling weight turns a paddle that does work on the system (the water), causing an increase in its temperature.

---

## EXAMPLE 7.2

Suppose a 10.00-kg mass drops through a height difference of 3.00 m, and the resulting work is used to turn a paddle in 200.0 g of water, initially at 15.00°C. The final water temperature is found to be 15.35°C. Assuming that the work done is used entirely to raise the water temperature, calculate the conversion factor between joules and calories.

### Solution

The total work done is

$$w = -Mg\,\Delta h = -(10.00 \text{ kg})(9.806 \text{ m s}^{-2})(-3.00 \text{ m}) = 294 \text{ J}$$

The heat (in calories) required to raise the water temperature by the same amount is

$$q = Mc_s\Delta T = (200.0 \text{ g})(1.000 \text{ cal K}^{-1} \text{ g}^{-1})(0.35 \text{ K}) = 70 \text{ cal}$$

Because the work done has the same effect on the water as direct transfer of heat, these two expressions can be set equal to each other, giving

$$70 \text{ calories} = 294 \text{ joules}$$

$$1 \text{ calorie} \approx 4.2 \text{ joules}$$

**Related Problems: 3, 4**

---

These and other experiments eliminated the need for the calorie as an independent unit, and the calorie is now *defined* as

$$1 \text{ cal} = 4.184 \text{ J} \qquad \text{(exactly)}$$

In this book the joule is used as the primary unit for heat and energy. However, much of the chemical literature continues to use the calorie as the unit of heat, so it is important to be familiar with both units.

## The First Law of Thermodynamics

Both heat and work are forms in which energy is transferred into and out of a system; they can be thought of as energy in transit. If the energy change is caused by *mechanical* contact of the system with its surroundings, work is done; if it is caused by *thermal* contact (leading to equalization of temperatures), heat is transferred. In many processes both heat and work cross the boundary of a system, and that system's internal energy change is the sum of the two contributions. This statement, called the **first law of thermodynamics,** takes the mathematical form[1]

$$\Delta E = q + w \tag{7.3}$$

A system cannot be thought to "contain" work or heat, because both work and heat refer not to states of the system but to *processes* that transform one state into another. In the Joule experiment of Figure 7.7, the work done on the water (the system) by the falling weight increased the water's temperature. Work was performed on the system without heat being transferred, and from the first law, $\Delta E = w$. The same change in state of the system can be brought about by transferring heat to the system without work being done, so that $\Delta E = q$. Because $q$ and $w$ depend on the particular process (or path) connecting the states, they are not state functions. However, their sum, $\Delta E = q + w$, is independent of path, and so internal energy is a function of state. The fundamental physical content of the first law of thermodynamics is the following observation:

> Although $q$ and $w$ depend individually on the path followed between a given pair of states, their sum does not.

As was already asserted, the laws of thermodynamics cannot be derived or proved; they are the results of countless experiments on a tremendous variety of substances. It is not possible even to "check" the first law by independently measuring $\Delta E$, $w$, and $q$, because no "energy gauges" exist to determine energy changes $\Delta E$. But what *can* be done is to measure $w$ and $q$ for a series of different processes connecting the same initial and final states and to find that their sum $q + w$ is always the same. This reliance upon the first law is what defines the energy change $\Delta E$.

In any process, the heat *added to* the system is *removed from* the surroundings, and so

$$q_{sys} = -q_{surr}$$

In the same way, the work done *on* the system is done *by* the surroundings, and so

$$w_{sys} = -w_{surr}$$

---

[1] Some books, especially older ones, define work as positive when it is done *by* the system. The reason is that many engineering applications focus on the work done by a particular heat engine, and so it is helpful to define that quantity as positive. Thus, the work is given by

$$w = P_{ext}\Delta V \qquad \text{(older convention)}$$

and the first law reads

$$\Delta E = q - w \qquad \text{(older convention)}$$

Although this book does not use the older convention, you should check which convention is employed when consulting other books.

Adding these two and invoking the first law give

$$\Delta E_{sys} = -\Delta E_{surr}$$

Thus, the energy changes of system and surroundings have the same magnitude but opposite signs. The total energy change of the thermodynamic universe for a given process (system plus surroundings) is then

$$\Delta E_{univ} = \Delta E_{sys} + \Delta E_{surr} = 0 \qquad \text{[7.4]}$$

Our conclusion is that in any process, the total energy of the thermodynamic universe remains unchanged; energy is conserved. This is another statement of the first law of thermodynamics.

---

### 7.3

## HEAT CAPACITY, ENTHALPY, AND CALORIMETRY

In the previous section, specific heat capacity was defined as the amount of heat required to raise the temperature of 1 g of material by 1 K. That definition is somewhat imprecise, because in fact the amount of heat required depends on whether the process is carried out at constant volume or at constant pressure. In this section, precise methods are described for measuring the amount of energy transferred as heat during a process and for relating this amount to the thermodynamic properties of the system under investigation.

### Heat Capacity and Specific Heat Capacity

The **heat capacity** $C$ is defined as the amount of energy that must be added to the system to raise its temperature by 1 K. The heat capacity is a property of the system as a whole and has units of $J\ K^{-1}$.

$$q = C\Delta T$$

Now consider an experiment in which two systems containing identical masses of the same substance are heated to produce identical changes in temperature; System 1 is held at constant volume, System 2 at constant pressure. Which system has absorbed more heat in these identical temperature changes? All the energy gained by System 1 has contributed to increasing the temperature of the gas, and therefore the speed of the molecules, subject to the fixed volume. But in System 2, some of the energy gained has been promptly lost as the system performed work of expansion against the surroundings at constant pressure. Consequently, System 2 must absorb more thermal energy from the surroundings than does System 1 in order to take identical gas samples through identical temperature changes. Two independent heat capacity functions must be defined: $C_P$, the heat capacity at constant pressure and $C_V$, the heat capacity at constant volume. For a given system, $C_P$ is greater than $C_V$. This difference can be quite large for gases, but is usually negligible for solids and liquids where the only volume change at constant pressure is the very small expansion or contraction upon heating and cooling, respectively.

In thermodynamics the *molar* heat capacities $c_V$ and $c_P$ (the system heat capacities $C_V$ and $C_P$ divided by the number of moles of substance in the system) are particularly useful: $c_V$ is the amount of heat required to raise the temperature of 1 mol of substance by 1 K at constant volume, and $c_P$ is the corresponding amount required

**TABLE 7.1**

**Specific Heat Capacities at Constant Pressure (at 25°C)**

| Substance | Specific Heat Capacity (J K$^{-1}$ g$^{-1}$) |
|---|---|
| Hg($\ell$) | 0.140 |
| Cu($s$) | 0.385 |
| Fe($s$) | 0.449 |
| SiO$_2$($s$) | 0.739 |
| CaCO$_3$($s$) | 0.818 |
| O$_2$($g$) | 0.917 |
| H$_2$O($\ell$) | 4.18 |

at constant pressure. If the total heat transferred to $n$ mol at constant volume is $q_V$, then

$$q_V = nc_V(T_2 - T_1) = nc_V\Delta T \qquad [7.5]$$

If an amount $q_P$ is transferred at constant pressure, then

$$q_P = nc_P\Delta T \qquad [7.6]$$

provided that $c_V$ and $c_P$ do not change significantly between the initial and final temperatures. The **specific heat capacity** at constant $V$ or constant $P$ is the system heat capacity reported per gram of substance. Extensive tabulations of the molar and specific heat capacities are available. Representative values are illustrated in Table 7.1. The molecular interpretation of heat capacity for ideal gases will be introduced in Section 7.4.

When two objects at different temperatures are brought into contact with each other, energy in the form of heat is exchanged between them until they reach a common temperature. If they are insulated from their surroundings, the amount of heat $q_2$ taken up by the cooler object is equal to $-q_1$, the amount of heat given up by the hotter object. As always, the convention followed is that energy transferred to an object has a positive sign, so $q_2$ is positive when $q_1$ is negative. This analysis is broadly applicable; a typical example follows.

### EXAMPLE 7.3

A piece of iron weighing 72.4 g is heated to 100.0°C and plunged into 100.0 g of water that is initially at 10.0°C in a Styrofoam cup calorimeter (Fig. 7.8). Calculate the final temperature that is reached, assuming no heat is lost to the surroundings.

### Solution

The coffee cup calorimeter operates at constant pressure determined by the atmosphere; specific heat data at constant pressure are therefore appropriate. Because the data involve masses, it is easier to work with specific heat capacities (Table 7.1) than with molar heat capacities. If $t_f$ is the final temperature (in degrees Celsius), then the equation for heat balance gives

$$M_1(c_{s1})\Delta T_1 = -M_2(c_{s2})\Delta T_2$$

$$(100.0 \text{ g H}_2\text{O})(4.18 \text{ J °C}^{-1} \text{ g}^{-1})(t_f - 10.0°C) =$$
$$-(72.4 \text{ g Fe})(0.449 \text{ J °C}^{-1} \text{ g}^{-1})(t_f - 100.0°C)$$

This is a linear equation for the unknown temperature $t_f$, and its solution is

$$418t_f - 4180 = -32.51t_f + 3251$$

$$t_f = 16.5°C$$

Note that specific heat capacities are numerically the same whether expressed in J K$^{-1}$ g$^{-1}$ or J (°C)$^{-1}$ g$^{-1}$ (because the degree Celsius and the kelvin have the same size). Converting 10.0°C and 100.0°C to kelvins and using specific heat capacities in units of J K$^{-1}$ g$^{-1}$ gives $t_f = 289.7$ K, an equivalent answer.

**Related Problems: 11, 12**

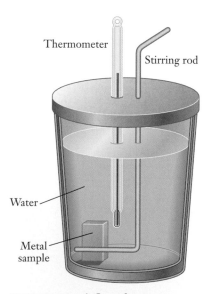

Thermometer

Stirring rod

Water

Metal sample

**FIGURE 7.8** A Styrofoam cup calorimeter. As the piece of metal cools, it releases heat to the water. The amount of heat released can be determined from the temperature change of the water.

## Heat Transfer at Constant Volume: Bomb Calorimeters

Suppose some reacting species are sealed in a small closed container (called a bomb) and the container is placed in a calorimeter like the one in Figure 7.9. As the molecules react chemically, heat is given off or taken up and the change in temperature of the calorimetric fluid is measured. Because the container is sealed tightly, its volume is constant and no pressure–volume work is done. Therefore, the change in internal energy is equal to the measured heat absorbed in the chemical reaction at constant volume:

$$\Delta E = q_V$$

Such experiments at constant volume are often inconvenient or difficult to carry out. They require the use of a well-constructed reaction vessel that can resist the large pressure changes that occur in many chemical reactions.

## Heat Transfer at Constant Pressure: Enthalpy

Most experiments in chemistry are done under constant (atmospheric) *pressure* conditions rather than at constant volume, however, and so it would be desirable to relate the heat transferred at constant pressure, $q_P$, to some state property analogous to the internal energy $E$.

   If the work done is entirely pressure–volume work, and if the external pressure is held constant, then

$$\Delta E = q_P + w = q_P - P_{\text{ext}} \Delta V$$

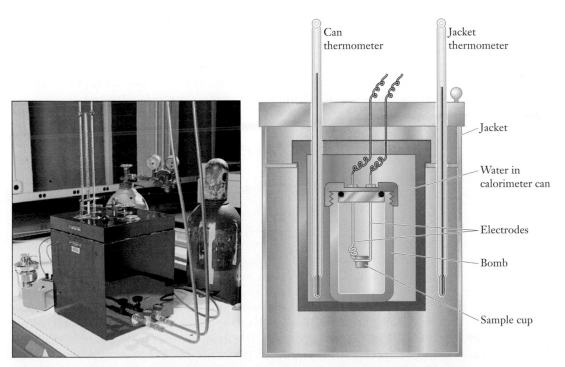

**FIGURE 7.9**   The combustion calorimeter is also called a "bomb calorimeter"; the combustion reaction in it is carried out at a fixed volume. *(Leon Lewandowski)*

If the external pressure is now assumed to be equal to the internal pressure of the system $P$, then

$$\Delta E = q_P - P\,\Delta V$$

$$q_P = \Delta E + P\,\Delta V$$

Because $P$ is constant, $P\,\Delta V = \Delta(PV)$, and this equation becomes

$$q_P = \Delta(E + PV)$$

The combination $E + PV$ appearing on the right side is now defined as the **enthalpy** $H$:

$$H = E + PV \tag{7.7}$$

and so

$$q_P = \Delta(E + PV) = \Delta H$$

Because $E$, $P$, and $V$ are state functions, $H$ must be one as well. Heat transfer at constant pressure has therefore been related to a state function.

It is important to remember that

$$\Delta H = q_P = \Delta E + P\,\Delta V \qquad \text{(constant pressure)}$$

holds only at constant pressure. If the pressure changes, the more general relationship

$$\Delta H = \Delta E + \Delta(PV)$$

must be used. Like the energy, the enthalpy change is determined by the initial and final states and is independent of the particular path along which the process is carried out. This is always true for a state function.

Physical interpretation of the enthalpy function follows immediately from the equation $\Delta H = \Delta E + P\,\Delta V$ at constant pressure. Clearly $H$ has physical dimensions of energy and is in effect a "corrected" internal energy that reflects the consequences of changing $V$ while thermal energy is being absorbed at constant pressure. The "correction term" $P\,\Delta V$ accounts precisely for the energy used in expansion work, rather than for increasing the temperature of the system. Thus $\Delta H$ is physically the appropriate state function to measure $q$ in constant pressure processes.

*[handwritten margin note: ΔH used ~~when~~ to measure q when P = constant]*

---

### 7.4

# ILLUSTRATIONS OF THE FIRST LAW OF THERMODYNAMICS IN IDEAL GAS PROCESSES

The first law of thermodynamics has been stated in a very general form, applicable to any process that begins and ends in equilibrium states. The heat and work terms have been analyzed separately and methods presented for calculating, measuring, and interpreting each. All the concepts are now in place for applying thermodynamics to the discussion of particular laboratory processes. Applications require data on certain properties of the substance being studied, however, such as its equation of state and its heat capacities. Thermodynamic arguments alone cannot provide the actual values of such properties; instead, thermodynamics establishes universal relations among such properties. The actual values must be obtained by methods other than thermodynamics, such as experimental measurements or theoretical calculations in statistical thermodynamics. To illustrate these points, data on the heat capacities of ideal gases are obtained by methods outside thermodynamics in the next few paragraphs. Thermodynamics is then applied to analyze particular processes carried out on ideal gases.

## Heat Capacities of Ideal Gases

The pair of molar heat capacities $c_V$ and $c_P$ for an ideal monatomic gas can be calculated from the results of the kinetic theory of gases and the ideal gas equation of state. From Section 4.5, the average translational kinetic energy of $n$ mol of an ideal gas is

$$E_k = \tfrac{3}{2}nRT$$

As will be seen in Chapter 16, the rotations and vibrations of diatomic or polyatomic molecules make additional contributions to the energy. In a *monatomic* gas, however, these other contributions are not present, and changes in the total energy $\Delta E$ measured in thermodynamics can be equated to changes in the translational kinetic energy. If $n$ mol of a monatomic gas is taken from a temperature $T_1$ to a temperature $T_2$, the energy change is

$$\Delta E = \tfrac{3}{2}nR(T_2 - T_1) = \tfrac{3}{2}nR\,\Delta T$$

Note that the pressure and volume do not affect $E$ explicitly (except through temperature changes), and so this result is independent of the change in pressure or volume of the gas.

Now consider changing the temperature of an ideal gas at constant volume from the point of view of thermodynamics. Because the volume is constant (the gas is confined in a vessel with rigid, diathermal walls), the pressure–volume work $w$ must be zero, and therefore

$$\Delta E = q_V = nc_V\,\Delta T \qquad \text{(ideal gas)}$$

where $q_V$ is the heat transferred at constant volume. Equating this thermodynamic relation with the previous expression for $\Delta E$ (from kinetic theory) shows that

$$c_V = \tfrac{3}{2}R \qquad \text{(monatomic ideal gas)} \qquad\qquad \textbf{[7.8]}$$

Similarly, the molar heat capacity at constant pressure, $c_P$, is calculated by examining the heating of a monatomic ideal gas at constant pressure from temperature $T_1$ to $T_2$. Experimentally, such a process can be carried out by placing the gas in a cylinder with a piston that moves out as the gas is heated, keeping the gas pressure equal to the outside pressure. In this case,

$$\Delta E = \tfrac{3}{2}nR\,\Delta T = nc_V\,\Delta T$$

still holds (because the energy change depends only on the temperatures for an ideal gas), but we now have

$$\Delta E = q_P + w$$

because the work is no longer zero. The work for a constant-pressure process is easily calculated from

$$w = -P\,\Delta V = -P(V_2 - V_1)$$

and the heat transferred is

$$q_P = nc_P\,\Delta T$$

Because $w$ is negative, $q_P$ is larger than $q_V$ by the amount of work done by the gas as it expands. This gives

$$\Delta E = q + w$$

$$nc_V \Delta T = nc_P \Delta T - P(V_2 - V_1)$$

From the ideal gas law, $PV_1 = nRT_1$ and $PV_2 = nRT_2$, so that

$$nc_V \Delta T = nc_P \Delta T - nR \Delta T$$

$$c_V = c_P - R$$

$$c_P = c_V + R$$

For a monatomic ideal gas, this shows that $c_P = \frac{5}{2}R$. It is important to use the proper units for $R$ in these expressions for $c_V$ and $c_P$. If heat is to be measured in joules, $R$ must be expressed as

$$R = 8.315 \text{ J K}^{-1} \text{ mol}^{-1}$$

For a diatomic or polyatomic ideal gas, $c_V$ is greater than $\frac{3}{2}R$, because energy can be stored in rotational and vibrational motions of the molecules; a greater amount of heat must be transferred to achieve a given temperature change. However, it is still true that

$$c_P = c_V + R \qquad \text{(any ideal gas)} \qquad \text{[7.9]}$$

and it is also still true that energy changes depend only on the temperature change; so, for a small temperature change $\Delta T$,

$$\Delta E = nc_V \Delta T \qquad \text{(any ideal gas)} \qquad \text{[7.10]}$$

For an ideal gas process,

$$\Delta H = \Delta E + \Delta(PV) = nc_V \Delta T + nR \Delta T$$

$$\Delta H = nc_P \Delta T \qquad \text{(ideal gas)} \qquad \text{[7.11]}$$

because $c_P = c_V + R$. This result holds for any ideal gas process and shows that enthalpy changes, like energy changes, depend only on the temperature difference between initial and final states.

## EXAMPLE 7.4

Suppose that 1.00 kJ of heat is transferred to 2.00 mol of argon (at 298 K, 1 atm). What will the final temperature $T_f$ be if the heat is transferred **(a)** at constant volume, or **(b)** at constant pressure? Calculate the energy change $\Delta E$ in each case.

### Solution

Because argon is a monatomic, approximately ideal gas,

$$c_V = \frac{3}{2}R = 12.47 \text{ J K}^{-1} \text{ mol}^{-1}$$

$$c_P = \frac{5}{2}R = 20.79 \text{ J K}^{-1} \text{ mol}^{-1}$$

At constant volume,

$$q_V = nc_V \Delta T$$

$$1000 \text{ J} = (2.00 \text{ mol})(12.47 \text{ J K}^{-1} \text{ mol}^{-1}) \Delta T$$

$$\Delta T = 40.1 \text{ K}; \qquad T_f = 298 + 40.1 = 338 \text{ K}$$

$$\Delta E = nc_V \Delta T = q_V = 1000 \text{ J}$$

At constant pressure,

$$q_P = nc_P \Delta T$$

$$1000 \text{ J} = (2.00 \text{ mol})(20.79 \text{ J K}^{-1} \text{ mol}^{-1}) \Delta T$$

$$\Delta T = 24.0 \text{ K}; \qquad T_f = 298 + 24.0 = 322 \text{ K}$$

$$\Delta E = nc_V \Delta T = (2.00 \text{ mol})(12.47 \text{ J K}^{-1} \text{ mol}^{-1})(24 \text{ K}) = 600 \text{ J}$$

Note that the expression for $\Delta E$ involves $c_V$ even though the process is carried out at constant pressure. The difference of 400 J between the input $q_P$ and $\Delta E$ is the work done by the gas as it expands.

**Related Problems: 17, 18, 19, 20, 21, 22**

## Heat and Work for Ideal Gases

Calculations of heat and work can now be carried out for a variety of processes involving an ideal gas. To illustrate the fact that $q$ and $w$ depend individually on the path followed but their sum does not, consider the expansion of 1.00 mol of an ideal monatomic gas following two different paths. The system begins at state A ($P_A =$ 2.00 atm, $V_A = 10.0$ L) and reaches a final state, B ($P_B = 1.00$ atm, $V_B = 30.0$ L), via either of two paths shown in Figure 7.10. Along path ACB (red arrows) the system is first heated at constant pressure ($P_{ext} = P_A = 2$ atm) until the volume has tripled; then it is cooled at constant volume until the pressure is halved. Along path ADB (blue arrows) the system is cooled at constant volume until the pressure is halved and then heated at constant pressure ($P_{ext} = P_B = 1$ atm) until the volume has tripled.

The calculations of heat and work for each step are of the type already performed and are straightforward. Thus,

$$w_{AC} = -P_{ext} \Delta V = -P_A(V_B - V_A) \qquad w_{CB} = 0 \text{ because } V_C = V_B$$

$$q_{AC} = q_P = nc_P \Delta T = \tfrac{5}{2}nR(T_C - T_A)$$

$$q_{CB} = q_V = nc_V \Delta T = \tfrac{3}{2}nR(T_B - T_C)$$

From the ideal gas law, $nRT_A = P_AV_A$, $nRT_B = P_BV_B$, and $nRT_C = P_CV_C = P_AV_B$ (because $P_A = P_C$ and $V_B = V_C$). Using these relations and summing over the two steps give

$$w_{ACB} = w_{AC} + w_{CB} = -P_A(V_B - V_A) = -40.0 \text{ L atm} = -4050 \text{ J}$$

$$q_{ACB} = q_{AC} + q_{CB} = \tfrac{5}{2}nR(T_C - T_A) + \tfrac{3}{2}nR(T_B - T_C)$$

$$= \tfrac{5}{2}P_A(V_B - V_A) + \tfrac{3}{2}V_B(P_B - P_A) = (100.0 - 45.0) \text{ L atm} = 5570 \text{ J}$$

The sum of these is $\Delta E = w_{ACB} + q_{ACB} = 1520$ J. (This could also have been obtained by using the ideal gas law to calculate the initial and final temperatures $T_A$ and $T_B$.)

The corresponding calculation for path ADB gives

$$w_{ADB} = -2030 \text{ J} \qquad \text{and} \qquad q_{ADB} = 3550 \text{ J}$$

Even though both the work and the heat have changed, their sum is still 1520 J, illustrating that $E$ is a state function, while $q$ and $w$ are not.

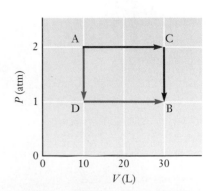

**FIGURE 7.10** States A and B of a system are connected by two different ideal-gas processes, one passing through state C and the other through state D.

---

### 7.5

## Thermochemistry

The energy changes considered so far have arisen from direct mechanical work on a system (as in the paddle wheel driven by a falling weight) or from the establishment of thermal contact between two systems at different temperatures. In chemistry, an important additional source of energy change in a system comes from the heat given off or taken up in the course of a chemical reaction. The study of these heat effects is referred to as **thermochemistry.** Because chemical reactions are usually studied at constant pressure, heats of reaction are measured at constant pressure. The tabulated values are therefore in the form of **reaction enthalpies.**

### Enthalpies of Reaction

When carbon monoxide is burned in oxygen to carbon dioxide,

$$CO(g) + \tfrac{1}{2} O_2(g) \longrightarrow CO_2(g)$$

heat is given off. Because this heat is transferred out of the reaction vessel (the system) and into the calorimeter, it has a *negative* sign. Careful calorimetric measurements show that 1.000 mol of CO reacted completely with 0.500 mol of $O_2$, at 25°C and a constant pressure of 1 atm, leads to an enthalpy change of

$$\Delta H = q_P = -2.830 \times 10^5 \text{ J} = -283.0 \text{ kJ}$$

The kilojoule (kJ), equal to $10^3$ J, is used because most enthalpy changes for chemical reactions lie in the range of thousands of joules per mole.

When heat is given off by a reaction ($\Delta H$ is negative), the reaction is said to be **exothermic** (Fig. 7.11). Reactions in which heat is taken up ($\Delta H$ positive) are called

**FIGURE 7.11** The thermite reaction, 2 Al(*s*) + Fe₂O₃(*s*) → 2 Fe(*s*) + Al₂O₃(*s*), is among the most exothermic of all reactions, liberating 16 kJ of heat for every gram of Al that reacts. The iron(III) oxide and aluminum are mixed in finely divided form. Like the combustion of paper, the reaction requires a source of ignition, but then continues on its own. *(Charles D. Winters)*

**(a)**                                                                          **(b)**

**FIGURE 7.12**    (a) When mixed in a flask, the two solids $Ba(OH)_2 \cdot 8H_2O(s)$ and $NH_4NO_3(s)$ undergo an acid–base reaction: $Ba(OH)_2 \cdot 8\ H_2O(s) + 2\ NH_4NO_3(s) \longrightarrow Ba(NO_3)_2\ (aq) + 2\ NH_3(aq) + 10\ H_2O(\ell)$. (b) The water produced dissolves excess ammonium nitrate in an endothermic reaction. The dissolution absorbs so much heat that the water on the surface of the wet wooden block freezes to the bottom of the flask, and the block can be lifted up with the flask.    *(Charles D. Winters)*

**endothermic** (Fig. 7.12). An obvious example of an endothermic reaction is the preceding reaction written in the opposite direction:

$$CO_2(g) \longrightarrow CO(g) + \tfrac{1}{2}\ O_2(g) \qquad\qquad \Delta H = +283.0 \text{ kJ}$$

If the direction of a chemical reaction is reversed, the enthalpy change reverses sign. Heat is required to convert $CO_2$ to CO and $O_2$ at constant pressure. The decomposition of $CO_2$ into CO and $O_2$ is very difficult to carry out in the laboratory, but the reverse reaction presents no practical obstacle. Thermodynamics allows prediction of the $\Delta H$ of the decomposition reaction with complete confidence, even if a calorimetric experiment is never actually carried out.

The convention is used that a reaction enthalpy written after a balanced chemical equation refers to the enthalpy change that accompanies the complete conversion of stoichiometric amounts of reactants to products; the numbers of moles of reactants and products are given by the coefficients in the equation. The preceding equation shows the enthalpy change when 1 mol of $CO_2$ is converted to 1 mol of CO and $\tfrac{1}{2}$ mol of $O_2$. If this equation is multiplied by a factor of 2, the enthalpy change must also be doubled because twice as many moles are then involved (enthalpy, like energy, is an *extensive* property).

$$2\ CO_2(g) \longrightarrow 2\ CO(g) + O_2(g) \qquad\qquad \Delta H = +566.0 \text{ kJ}$$

The molar amounts need not be integers, as the following example illustrates.

## EXAMPLE 7.5

Red phosphorus reacts with liquid bromine in an exothermic reaction (Fig. 7.13):

$$2\ P(s) + 3\ Br_2(\ell) \longrightarrow 2\ PBr_3(g) \qquad\qquad \Delta H = -243 \text{ kJ}$$

Calculate the enthalpy change when 2.63 g of phosphorus reacts with an excess of bromine in this way.

**FIGURE 7.13**    Red phosphorus reacts exothermically in liquid bromine. The rising gases are a mixture of the product $PBr_3$ and unreacted bromine that has boiled off. *(Charles D. Winters)*

## Solution

First, convert from grams of phosphorus to moles, using the molar mass of phosphorus, 30.97 g mol$^{-1}$:

$$\text{moles P} = \frac{2.63 \text{ g P}}{30.97 \text{ g mol}^{-1}} = 0.0849 \text{ mol}$$

Given that an enthalpy change of $-243$ kJ is associated with 2 mol of P, it is readily seen that the enthalpy change associated with 0.0849 mol is

$$\Delta H = 0.0849 \text{ mol P} \times \left( \frac{-243 \text{ kJ}}{2 \text{ mol P}} \right) = -10.3 \text{ kJ}$$

**Related Problems: 23, 24, 25, 26**

---

The enthalpy change for the reaction of 1 mol of carbon monoxide with oxygen was stated to be $-283.0$ kJ. In a second experiment, the heat evolved when 1 mol of carbon (graphite) is burned in oxygen to carbon dioxide at 25°C is readily measured to be

$$C(s,gr) + O_2(g) \longrightarrow CO_2(g) \qquad\qquad \Delta H = -393.5 \text{ kJ}$$

Suppose the enthalpy change for the reaction

$$C(s,gr) + \tfrac{1}{2} O_2(g) \longrightarrow CO(g) \qquad\qquad \Delta H = ?$$

is wanted. This reaction cannot be carried out simply in the laboratory. If 1 mol of graphite is heated with $\frac{1}{2}$ mol of oxygen, almost half of the carbon burns to $CO_2(g)$ and the remainder is left as carbon at equilibrium. But the heat that would be evolved if it were possible to perform the experiment can be calculated from the fact that $\Delta H$ is independent of the path followed from reactants to products (Fig. 7.14). A path is chosen in which 1 mol of C is burned with $O_2$ to $CO_2$ (with $\Delta H = -393.5$ kJ), and to this is added the calculated enthalpy change for the hypothetical process in which $CO_2$ is converted to CO and $O_2$ ($\Delta H = +283.0$ kJ). The total $\Delta H$ is the algebraic sum of the two known enthalpy changes, $-393.5$ kJ $+ 283.0$ kJ $= -110.5$ kJ. To see this more clearly, we write out the reactions:

$$C(s,gr) + O_2(g) \longrightarrow CO_2(g) \qquad\qquad \Delta H_1 = -393.5 \text{ kJ}$$

$$\underline{CO_2(g) \longrightarrow CO(g) + \tfrac{1}{2} O_2(g) \qquad\qquad \Delta H_2 = +283.0 \text{ kJ}}$$

$$C(s,gr) + \tfrac{1}{2} O_2(g) \longrightarrow CO(g) \qquad\qquad \Delta H = \Delta H_1 + \Delta H_2 = -110.5 \text{ kJ}$$

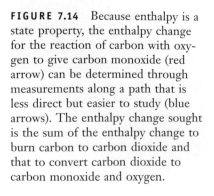

**FIGURE 7.14** Because enthalpy is a state property, the enthalpy change for the reaction of carbon with oxygen to give carbon monoxide (red arrow) can be determined through measurements along a path that is less direct but easier to study (blue arrows). The enthalpy change sought is the sum of the enthalpy change to burn carbon to carbon dioxide and that to convert carbon dioxide to carbon monoxide and oxygen.

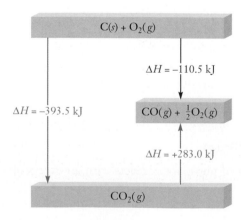

If two or more chemical equations are added to give another chemical equation, the corresponding enthalpies of reaction must be added.

This is known as **Hess's law** and derives from the fact that enthalpy is a state function.

The corresponding energy change $\Delta E$ might be desired for this reaction. The quantity $\Delta E$ is simple to calculate because of the relation

$$\Delta H = \Delta E + \Delta(PV)$$

or

$$\Delta E = \Delta H - \Delta(PV)$$

The gases can be assumed to obey the ideal gas law, so that

$$\Delta(PV) = \Delta(nRT) = RT\,\Delta n_g$$

because the temperature is constant at 25°C. Here $\Delta n_g$ is the change in the number of moles of *gas* in the reaction as written:

$$\Delta n_g = \text{total moles of product gases} - \text{total moles of reactant gases}$$

$$= 1 \text{ mol} - \tfrac{1}{2} \text{ mol} = \tfrac{1}{2} \text{ mol}$$

(Graphite is a solid, and its volume is negligible compared with the volumes of the gases.) Hence,

$$\Delta(PV) = RT\,\Delta n_g = (8.315 \text{ J K}^{-1} \text{ mol}^{-1})(298 \text{ K})(\tfrac{1}{2} \text{ mol}) = 1.24 \times 10^3 \text{ J} = 1.24 \text{ kJ}$$

Note that $R$ must be expressed in J K$^{-1}$ mol$^{-1}$ if a result in joules is desired. Therefore,

$$\Delta E = -110.5 \text{ kJ} - 1.24 \text{ kJ} = -111.7 \text{ kJ}$$

For reactions in which only liquids and solids are involved, or those in which the number of moles of gas does not change, the enthalpy and energy changes are almost equal and their difference can be neglected.

Phase changes are not chemical reactions, but they can be considered in the same context. Heat must be transferred to ice in order to transform it to water, and so the phase change is endothermic, with $\Delta H$ positive:

$$H_2O(s) \longrightarrow H_2O(\ell) \qquad\qquad \Delta H_{fus} = +6.007 \text{ kJ mol}^{-1}$$

Here $\Delta H_{fus}$ is the **molar enthalpy of fusion,** the heat that must be transferred at constant pressure to melt one mole of substance. When a liquid freezes, the reaction is reversed, and an equal amount of heat is given off to the surroundings; that is, $\Delta H_{freez} = -\Delta H_{fus}$. The vaporization of one mole of liquid at constant pressure and temperature requires an amount of heat called the **molar enthalpy of vaporization** $\Delta H_{vap}$,

$$H_2O(\ell) \longrightarrow H_2O(g) \qquad\qquad \Delta H_{vap} = +40.7 \text{ kJ mol}^{-1}$$

whereas the condensation of a liquid from a vapor is an exothermic process, with $\Delta H_{cond} = -\Delta H_{vap}$. Table 7.2 lists enthalpies of fusion and vaporization.

---

**EXAMPLE 7.6**

---

To vaporize 100.0 g of carbon tetrachloride at its normal boiling point, 349.9 K, and $P = 1$ atm, 19.5 kJ of heat is required. Calculate $\Delta H_{vap}$ for CCl$_4$ and compare it with $\Delta E$ for the same process.

**TABLE 7.2**

*Enthalpy Changes of Fusion and Vaporization*[†]

| Substance | $\Delta H_{fus}$ (kJ mol$^{-1}$) | $\Delta H_{vap}$ (kJ mol$^{-1}$) |
|-----------|----------------------------------|----------------------------------|
| $NH_3$    | 5.65  | 23.35 |
| HCl       | 1.992 | 16.15 |
| CO        | 0.836 | 6.04  |
| $CCl_4$   | 2.5   | 30.0  |
| $H_2O$    | 6.007 | 40.66 |
| NaCl      | 28.8  | 170   |

[†] The enthalpy changes are measured at the normal melting point and the normal boiling point, respectively.

**Solution**

The molar mass of $CCl_4$ is 153.8 g mol$^{-1}$, and so the chemical amount in 100.0 g is

$$\frac{100.0 \text{ g } CCl_4}{153.8 \text{ g mol}^{-1}} = 0.6502 \text{ mol } CCl_4$$

The enthalpy change for one mole of $CCl_4$ is then

$$\left(\frac{19.5 \text{ kJ}}{0.6502 \text{ mol } CCl_4}\right) \times 1.00 \text{ mol } CCl_4 = +30.0 \text{ kJ} = \Delta H_{vap}$$

The energy change is then

$$\Delta E = \Delta H_{vap} - \Delta(PV) = \Delta H_{vap} - RT \, \Delta n_g$$

Inserting $T = 349.9$ K and $\Delta n_g = 1$ (because there is an increase of one mole of gaseous products for each mole of liquid that is vaporized) gives

$$\Delta E = 30.0 \text{ kJ} - (8.315 \text{ J K}^{-1} \text{ mol}^{-1})(349.9 \text{ K})(1.00 \text{ mol})(10^{-3} \text{ kJ J}^{-1})$$

$$= (30.0 - 2.9) \text{ kJ} = +27.1 \text{ kJ mol}^{-1}$$

Thus, of the 30.0 kJ of energy transferred from the surroundings in the form of heat, 27.1 kJ is used to increase the internal energy of the molecules ($\Delta E$), and 2.9 kJ is used to expand the resulting vapor, $\Delta(PV)$.

## Standard-State Enthalpies

Absolute enthalpies of substances, like absolute energies, cannot be measured or calculated. Only *changes* in enthalpy can be measured. Just as altitudes are measured relative to a standard altitude (sea level), it is necessary to adopt a reference state for the enthalpies of substances. To cope with this problem, chemists define **standard states** for chemical substances as follows:

For solids and liquids, the standard state is the thermodynamically stable state at pressure of 1 atm and at a specified temperature.

For gases, the standard state is the gaseous phase at pressure of 1 atm, at a specified temperature, and exhibiting ideal gas behavior.

For dissolved species, the standard state is a 1 M solution at pressure of 1 atm, at a specified temperature, and exhibiting ideal solution behavior.

Standard-state values of enthalpy and other quantities are designated by attaching a superscript ° (pronounced "naught") to the symbol for the quantity and writing the specified temperature as a subscript. Any temperature may be chosen as the "specified temperature." The most common choice is 298.15 K (25°C exactly); if the temperature of a standard state is not explicitly indicated, 298.15 K should be assumed to be the value.

Once standard states have been defined, the zero of the enthalpy scale is defined by arbitrarily setting the enthalpies of selected reference substances to zero in their standard states. This is completely analogous to assigning zero as the altitude at sea level. Chemists have agreed to the following *the chemical elements in standard states at 298.15 K have zero enthalpy.* A complication immediately arises because some elements exist in various allotropic forms that differ in structure and all physical properties, including enthalpy. For example, oxygen can be prepared as $O_2(g)$ or $O_3(g)$ (ozone), and carbon exists in numerous allotropic forms, including graphite, diamond, and the fullerenes (see Section 1.5). The convention agreed upon is to assign zero enthalpy to the form that is most stable at 1 atm and 298.15 K. Thus $O_2(g)$ is assigned zero enthalpy in its standard state at 298.15 K, while $O_3(g)$ has non-zero enthalpy in its standard state at 298.15 K. The most stable form of carbon at 1 atm and 298.15 K is graphite, which is assigned zero enthalpy; diamond and all the fullerenes are assigned non-zero enthalpy.[2]

The enthalpy change for a chemical reaction in which all reactants and products are in their standard states and at a specified temperature is called the **standard enthalpy** (written $\Delta H°$) for that reaction. The standard enthalpy is the central tool in thermochemistry because it provides a systematic means for comparing the energy changes due to bond rearrangements in different reactions. Standard enthalpies can be calculated from tables of reference data. For this purpose one additional concept must be introduced. By definition, the **standard enthalpy of formation** $\Delta H_f°$ of a compound is the enthalpy change for the reaction that produces one mole of the compound from its elements in their stable states, all at 25°C and 1 atm pressure. For example, the standard enthalpy of formation of liquid water is the enthalpy change for the reaction

$$H_2(g) + \tfrac{1}{2} O_2(g) \longrightarrow H_2O(\ell) \qquad\qquad \Delta H° = -285.83 \text{ kJ}$$

$$\Delta H_f°(H_2O(\ell)) = -285.83 \text{ kJ mol}^{-1}$$

Here the superscript ° indicates standard-state conditions, and the subscript "f" stands for *formation.*

The $\Delta H_f°$ for an *element* that is already in its standard state is clearly zero, because no further change is needed to bring it to standard-state conditions. The standard enthalpy of formation of a mole of *atoms* of an element, however, is frequently a large positive quantity. That is, the reaction is endothermic:

$$\tfrac{1}{2}H_2(g) \longrightarrow H(g) \qquad\qquad \Delta H° = +217.96 \text{ kJ}$$

$$\Delta H_f°(H(g)) = 217.96 \text{ kJ mol}^{-1}$$

For dissolved species, the standard state is defined as an ideal solution with a concentration of 1 M (this is obtained in practice by extrapolating the dilute solution behavior up to this concentration). A special comment is in order on the standard

---

[2] There is one exception to this choice of standard state. The standard state of phosphorus is taken to be white phosphorus, rather than the more stable red or black form, because the latter are less well characterized and less reproducible in their properties.

enthalpies of formation of ions. When a strong electrolyte dissolves in water, both positive and negative ions form; it is impossible to produce one without the other. It is therefore also impossible to measure the enthalpy change of formation of ions of only one charge. Only the sum of the enthalpies of formation of the positive and negative ions is accessible to calorimetric experiments. A further arbitrary choice is therefore made, namely that $\Delta H_f^\circ$ of $H^+(aq)$ is set to zero.

The reason for introducing standard enthalpies of formation is that it is then necessary only to tabulate $\Delta H_f^\circ$ for compounds (a long list at that), rather than $\Delta H^\circ$ for all the conceivable reactions those compounds may undergo with one another (a vastly longer list!). Appendix D gives a short table of standard enthalpies of formation at 25°C. The following example shows how they can be used to determine enthalpy changes for reactions carried out at 25°C and 1 atm pressure.

## EXAMPLE 7.7

Using Appendix D, calculate $\Delta H^\circ$ for the following reaction at 25°C and 1 atm pressure:

$$2\ NO(g) + O_2(g) \longrightarrow 2\ NO_2(g)$$

### Solution

Because enthalpy is a function of state, $\Delta H$ can be calculated along any convenient path. In particular, two steps can be chosen for which $\Delta H$ is found easily. In step 1, the reactants are decomposed into the elements in their standard states:

$$2\ NO(g) \longrightarrow N_2(g) + O_2(g) \qquad \Delta H_1 = -2\ \Delta H_f^\circ(NO)$$

The minus sign appears because the process chosen is the reverse of the formation of NO; the factor of 2 is present because 2 mol of NO is involved. Because oxygen is already an element in its standard state, it does not need to be changed [equivalently, $\Delta H_f^\circ(O_2(g))$ is 0].

In step 2, the elements are combined to form products:

$$N_2(g) + 2\ O_2(g) \longrightarrow 2\ NO_2(g) \qquad \Delta H_2 = 2\ \Delta H_f^\circ(NO_2)$$

The enthalpy change of the overall reaction is then the sum of these two enthalpies:

$$\Delta H^\circ = \Delta H_1 + \Delta H_2 = -2\ \Delta H_f^\circ(NO) + 2\ \Delta H_f^\circ(NO_2)$$
$$= -(2\ mol)(90.25\ kJ\ mol^{-1}) + (2\ mol)(33.18\ kJ\ mol^{-1}) = -114.14\ kJ$$

**Related Problems: 35, 36, 37, 38**

The general pattern should be clear from this example. The change $\Delta H^\circ$ for a reaction at atmospheric pressure and 25°C is the sum of the $\Delta H_f^\circ$ for the *products* (multiplied by their coefficients in the balanced chemical equation) minus the sum of the $\Delta H_f^\circ$ for the *reactants* (also multiplied by their coefficients). For a general reaction of the form

$$a A + b B \longrightarrow c C + d D$$

the standard enthalpy change is

$$\Delta H^\circ = c\ \Delta H_f^\circ(C) + d\ \Delta H_f^\circ(D) - a\ \Delta H_f^\circ(A) - b\ \Delta H_f^\circ(B)$$

## Bond Enthalpies

Chemical reactions between molecules require existing bonds to break and new ones to form with the atoms arranged differently. Chemists have developed methods to study highly reactive intermediates, species in which bonds have been broken and not yet re-formed. For example, a hydrogen atom can be removed from a methane molecule,

$$CH_4(g) \longrightarrow CH_3(g) + H(g)$$

leaving two fragments, neither of which has a stable valence electron structure in the Lewis electron-dot picture. Both will go on to react rapidly with other molecules or fragments and eventually form the stable products of the reaction. During the brief presence of these reactive species, however, many of their properties can be measured.

An important quantity that has been measured is the enthalpy change when a bond is broken in the gas phase, called the **bond enthalpy.** This is invariably positive, because heat must be added to a collection of stable molecules to break their bonds. For example, the bond enthalpy of a C—H bond in methane is 438 kJ mol$^{-1}$, the measured standard enthalpy change for the reaction

$$CH_4(g) \longrightarrow CH_3(g) + H(g) \qquad \Delta H° = +438 \text{ kJ}$$

in which 1 mol of C—H bonds is broken, one for each molecule of methane. Bond enthalpies are fairly reproducible from one compound to another. Each of the following gas-phase reactions involves the breaking of a C—H bond:

$$C_2H_6(g) \longrightarrow C_2H_5(g) + H(g) \qquad \Delta H° = +410 \text{ kJ}$$

$$CHF_3(g) \longrightarrow CF_3(g) + H(g) \qquad \Delta H° = +429 \text{ kJ}$$

$$CHCl_3(g) \longrightarrow CCl_3(g) + H(g) \qquad \Delta H° = +380 \text{ kJ}$$

$$CHBr_3(g) \longrightarrow CBr_3(g) + H(g) \qquad \Delta H° = +377 \text{ kJ}$$

The approximate constancy of the measured enthalpy changes (all lie within 8% of their average value) suggests that the C—H bonds in all five molecules are rather similar. Because such bond enthalpies are reproducible from one molecule to another, it is useful to tabulate *average* bond enthalpies from measurements on a series of compounds (Table 7.3). Any given bond enthalpy will differ somewhat from those shown, but in most cases the deviations are small. The reproducibility of bond energies in a series of molecules was introduced in Section 3.5, and representative values were listed in Table 3.4. Bond energy values are related to bond enthalpy values by the relation $\Delta H = \Delta E + \Delta(PV)$, illustrated in the paragraphs following Example 7.5.

The bond enthalpies of Table 7.3 can be used, together with enthalpies of atomization of the elements from the same table, to estimate standard enthalpies of formation $\Delta H_f°$ for molecules in the gas phase, and enthalpy changes $\Delta H°$ for gas-phase reactions. This is illustrated by the following example.

---

**EXAMPLE 7.8**

Estimate the standard enthalpy of formation of dichlorodifluoromethane, $CCl_2F_2(g)$ (Fig. 7.15). This compound is also known as Freon-12 and has been used as a refrigerant because of its low reactivity and high volatility. It and other, related

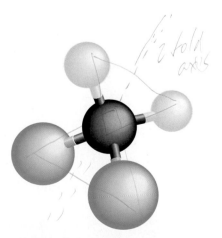

**FIGURE 7.15** Dichlorodifluoromethane, $CCl_2F_2$, also known as Freon-12.

chlorofluorocarbons (CFCs) are being phased out because of their role in depleting the ozone layer in the outer atmosphere, as discussed in Section 16.9.

### Solution

The standard enthalpy of formation of $CCl_2F_2(g)$ is the enthalpy change for the process in which it is formed from the elements in their standard states at 25°C:

$$C(s,gr) + Cl_2(g) + F_2(g) \longrightarrow CCl_2F_2(g)$$

This reaction can be replaced by a hypothetical two-step process: all the species appearing on the left are atomized and then the atoms are combined to make $CCl_2F_2$.

$$C(s,gr) + Cl_2(g) + F_2(g) \longrightarrow C(g) + 2\ Cl(g) + 2\ F(g)$$

$$C(g) + 2\ Cl(g) + 2\ F(g) \longrightarrow CCl_2F_2(g)$$

The enthalpy change $\Delta H_1$ for the first step is the sum of the atomization energies:

$$\Delta H_1 = \Delta H_f^\circ(C(g)) + 2\ \Delta H_f^\circ(Cl(g)) + 2\ \Delta H_f^\circ(F(g))$$

$$= 716.7 + 2(121.7) + 2(79.0) = 1118\ kJ$$

The enthalpy change $\Delta H_2$ for the second step can be estimated from the bond enthalpies of Table 7.3. This step involves the formation (with release of heat and therefore negative enthalpy change) of two C—Cl and two C—F bonds per molecule. The net $\Delta H$ for this step is then

$$\Delta H_2 \approx -[2(328) + 2(441)] = -1538\ kJ$$

$$\Delta H_1 + \Delta H_2 = -1538 + 1118 = -420\ kJ$$

This $\Delta H_f^\circ$, $-420$ kJ mol$^{-1}$, compares fairly well with the experimental value, $-477$ kJ mol$^{-1}$. In general, much better agreement than this is not to be expected, because tabulated bond enthalpies are only average values.

**Related Problems: 45, 46, 47, 48**

---

### TABLE 7.3

*Average Bond Enthalpies*

| | Molar Enthalpy of Atomization (kJ mol$^{-1}$)[†] | Bond Enthalpy (kJ mol$^{-1}$)[‡] | | | | | | | | |
|---|---|---|---|---|---|---|---|---|---|---|
| | | H— | C— | C= | C≡ | N— | N= | N≡ | O— | O= |
| H | 218.0 | 436 | 413 | | | 391 | | | 463 | |
| C | 716.7 | 413 | 348 | 615 | 812 | 292 | 615 | 891 | 351 | 728 |
| N | 472.7 | 391 | 292 | 615 | 891 | 161 | 418 | 945 | | |
| O | 249.2 | 463 | 351 | 728 | | | | | 139 | 498 |
| S | 278.8 | 339 | 259 | 477 | | | | | | |
| F | 79.0 | 563 | 441 | | | 270 | | | 185 | |
| Cl | 121.7 | 432 | 328 | | | 200 | | | 203 | |
| Br | 111.9 | 366 | 276 | | | | | | | |
| I | 106.8 | 299 | 240 | | | | | | | |

[†] From Appendix D.

[‡] From L. Pauling, *The Nature of the Chemical Bond*, 3rd ed. (Cornell University Press, Ithaca, New York, 1960).

| **7.6** |

## REVERSIBLE PROCESSES IN IDEAL GASES

Most thermodynamic processes carried out in laboratory work are irreversible, in the sense of Section 7.1. Except in the initial and final states, the system is not at equilibrium, and the relationship between observable properties is not defined. Consequently, changes in thermodynamic quantities during an irreversible process cannot in general be calculated. Nonetheless, the changes in those quantities that are state functions are well defined, so long as the initial and final equilibrium states are known. Since these changes are independent of the detailed path of the process, they can be evaluated for any known process that connects these initial and final states. Changes can be directly calculated along *reversible* paths, during which the system proceeds through a sequence of equilibrium states in which observable properties are related by the equation of state.

In the present section, calculations are illustrated for changes in macroscopic properties caused during several specific reversible processes in ideal gases. These will serve as auxiliary calculation pathways for evaluating changes in state functions during irreversible processes. This procedure will be used extensively in Chapter 8 on spontaneous processes and the second law.

Recall from Section 7.1 that a true reversible process is an idealization; it is a process in which the system proceeds with infinitesimal speed through a series of equilibrium states. The external pressure $P_{ext}$ therefore can never differ by more than an infinitesimal amount from the pressure $P$ of the gas itself. The heat, work, energy, and enthalpy changes for ideal gases at constant volume (called **isochoric processes**) and at constant pressure **(isobaric processes)** have already been considered. In this section, *isothermal* (constant temperature) and *adiabatic* ($q = 0$) processes are examined.

### Isothermal Processes

An **isothermal process** is one carried out at constant temperature. This is accomplished by placing the system in a large reservoir (bath) at fixed temperature and allowing heat to be transferred as required between system and reservoir. The reservoir is large enough that its temperature is almost unchanged by this heat transfer. In Section 7.4, $E$ for an ideal gas was shown to depend only on temperature, and therefore $\Delta E = 0$ for any isothermal gas process. From the first law it follows that

$$w = -q \qquad \text{(isothermal process, ideal gas)}$$

In a *reversible* process, $P_{ext} = P_{gas} \equiv P$. However, the relation $w = -P_{ext}\Delta V$ from Section 7.2 cannot be used to calculate the work, because that expression applies only if the external pressure remains constant as the volume changes. In the reversible isothermal expansion of an ideal gas,

$$P_{ext} = P = \frac{nRT}{V}$$

By Boyle's law the pressure falls as the volume is increased from $V_1$ to $V_2$, as shown by the solid line in Figure 7.16. In this case, the work is calculated by approximating the process as a *series* of expansions by small amounts $\delta V$, during each of which

**FIGURE 7.16**  The area under the graph of external pressure against volume can be approximated as the sum of the areas of the rectangles shown.

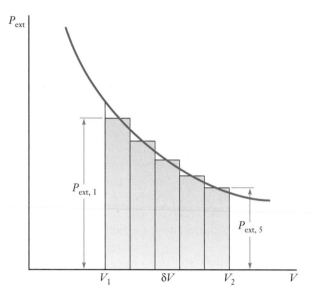

$P_{ext}$ is held constant at $P_i$, with $i = 1, 2, 3, \ldots$ labeling the step. The work done in step $i$ is $-P_i\delta V$, and so the total work done is the sum of the work in all steps:

$$w = -P_1\delta V - P_2\delta V - \cdots$$

The summation is the sum of the areas of the rectangles in the figure. As the step size $\delta V$ is made smaller, it approaches the infinitesimal volume change that is written $dV$, and the corresponding work becomes

$$dw = -P_{ext}\, dV$$

In this limit the sum of the areas of the rectangles approaches the area under the graph of $P$ versus $V$. This limiting sum is the *integral* of $P_{ext}$ from $V_1$ to $V_2$ and is written symbolically using an integral sign:

$$w = -\int_{V_1}^{V_2} P\, dV \qquad \text{(reversible process)}$$

This is a compact way of saying that the work $w$ for a reversible expansion process is (with a minus sign) the area under the graph of $P$ plotted against $V$ from $V_1$ to $V_2$.

Mathematical methods of defining and calculating integrals are beyond the scope of this book, but Appendix C gives some supplementary information about several integrals that are important in chemistry, and it should be read by students unfamiliar with calculus. The main idea is the relationship between a thermodynamic quantity (in this case, the work) and the area under a curve.

In the present case the ideal gas law is used to write this area as

$$w = -nRT\int_{V_1}^{V_2} \frac{1}{V}dV$$

This integral is the area under a graph of $1/V$ against $V$ (a hyperbola) from $V_1$ to $V_2$. It defines the *natural logarithm* function, symbolized "ln" (see Appendix C). In particular,

$$w = -nRT \ln \frac{V_2}{V_1}$$

$$q = -w = nRT \ln \frac{V_2}{V_1}$$

$$\Delta E = 0 \qquad \text{because} \qquad \Delta T = 0$$

$$\Delta H = \Delta E + \Delta(PV) = \Delta E + \Delta(nRT) = 0 \qquad \qquad \text{[7.12]}$$

The enthalpy change, like the energy change, depends only on temperature for an ideal gas.

The fact that $V_2$ is greater than $V_1$ implies that $w < 0$ and $q > 0$; in an isothermal expansion, the system does work against the surroundings and heat must be transferred into it in order to maintain $T$ constant. In an isothermal compression, the reverse is true: the surroundings do work on the system, and the system must then lose heat to the bath in order to maintain $T$ constant.

## EXAMPLE 7.9

Calculate the heat and the work associated with a process in which 5.00 mol of gas expands reversibly at constant temperature $T = 298$ K from a pressure of 10.0 atm to 1.00 atm.

### Solution

At constant $T$ and $n$,

$$\frac{V_2}{V_1} = \frac{P_1}{P_2} = \frac{10.0 \text{ atm}}{1.00 \text{ atm}} - 10.0$$

Thus,

$$w = -(5.00 \text{ mol})(8.315 \text{ J K}^{-1} \text{ mol}^{-1})(298 \text{ K}) \ln 10.0$$
$$= -2.85 \times 10^4 \text{ J} = -28.5 \text{ kJ}$$

$$q = -w = 28.5 \text{ kJ}$$

**Related Problems: 51, 52**

## Adiabatic Processes

An **adiabatic process** is one in which there is no transfer of heat into or out of the system. This is accomplished by placing insulation around the system to prevent heat flow.

$$q = 0$$

$$\Delta E = w$$

Consider a very small adiabatic change. The volume changes by an amount $dV$ and the temperature by an amount $dT$. Now $E$ depends only on temperature for an ideal gas, and so

$$dE = nc_V \, dT$$

As always, the work is given by $-P_{ext}\,dV$. Setting these equal gives

$$nc_V\,dT = -P_{ext}\,dV$$

In other words, the temperature change $dT$ is related to the volume change $dV$ in such a process.

If the process is *reversible* as well as adiabatic, so that $P_{ext} \approx P$, the ideal gas law can be used to write

$$nc_V\,dT = -P\,dV = -\frac{nRT}{V}\,dV$$

The equation is simplified by dividing both sides through by $nT$, making the left side depend only on $T$ and the right side only on $V$:

$$\frac{c_V}{T}\,dT = -\frac{R}{V}\,dV$$

Suppose now that the change is not infinitesimal but large. How are temperature and volume related in this case? If a series of such infinitesimal changes is added together, the result is an integral of both sides of the equation from the initial state (specified by $T_1$ and $V_1$) to the final state (specified by $T_2$ and $V_2$):

$$c_V\int_{T_1}^{T_2}\frac{1}{T}\,dT = -R\int_{V_1}^{V_2}\frac{1}{V}\,dV$$

Here $c_V$ has been assumed to be approximately independent of $T$ over the range from $T_1$ to $T_2$. Evaluating the integrals gives

$$c_V\ln\frac{T_2}{T_1} = -R\ln\frac{V_2}{V_1} = R\ln\frac{V_1}{V_2}$$

A more useful form results from taking antilogarithms of both sides:

$$\left(\frac{T_2}{T_1}\right)^{c_V} = \left(\frac{V_1}{V_2}\right)^{R} = \left(\frac{V_1}{V_2}\right)^{c_P-c_V}$$

The last step used the fact that $R = c_P - c_V$. Thus,

$$\left(\frac{T_2}{T_1}\right) = \left(\frac{V_1}{V_2}\right)^{c_P/c_V-1} = \left(\frac{V_1}{V_2}\right)^{\gamma-1}$$

where $\gamma = c_P/c_V$ is the ratio of specific heats. This can be rearranged to give

$$T_1V_1^{\gamma-1} = T_2V_2^{\gamma-1} \qquad \textbf{[7.13]}$$

In many situations, the initial thermodynamic state and therefore $T_1$ and $V_1$ are known. If the final volume $V_2$ is known, $T_2$ can be calculated; if $T_2$ is known, $V_2$ can be calculated.

In some cases, only the final pressure, $P_2$, of an adiabatic process is known. In this case the ideal gas law gives

$$\frac{P_1V_1}{T_1} = \frac{P_2V_2}{T_2}$$

Multiplying this by Equation 7.12 gives

$$P_1V_1^{\gamma} = P_2V_2^{\gamma} \qquad \textbf{[7.14]}$$

This can be used to calculate $V_2$ from a known $P_2$.

Once the pressure, temperature, and volume of the final state are known, the energy and enthalpy changes and the work done are straightforward to calculate:

$$\Delta E = nc_V(T_2 - T_1) = w \qquad \text{(reversible adiabatic process for ideal gas)}$$

$$\Delta H = \Delta E + \Delta(PV) = \Delta E + (P_2V_2 - P_1V_1)$$

or, more simply,

$$\Delta H = nc_P\, \Delta T$$

## EXAMPLE 7.10

Suppose 5.00 mol of an ideal monatomic gas at an initial temperature of 298 K and pressure of 10.0 atm is expanded adiabatically and reversibly until the pressure has dropped to 1.00 atm. Calculate the final volume and temperature, the energy and enthalpy changes, and the work done.

### Solution

The initial volume is

$$V_1 = \frac{nRT_1}{P_1} = \frac{(5.00 \text{ mol})(0.08206 \text{ L atm K}^{-1} \text{ mol}^{-1})(298 \text{ K})}{(10.0 \text{ atm})} = 12.2 \text{ L}$$

and the heat capacity ratio for a monatomic gas is

$$\gamma = \frac{c_P}{c_V} = \frac{\frac{5}{2}R}{\frac{3}{2}R} = \frac{5}{3}$$

For a reversible adiabatic process,

$$\frac{P_1}{P_2}V_1^{\gamma} = V_2^{\gamma}$$

$$(10.0)(12.2 \text{ L})^{5/3} = V_2^{5/3}$$

$$V_2 = (12.2 \text{ L})(10.0)^{3/5} = 48.7 \text{ L}$$

The final temperature can now be calculated from the ideal gas law:

$$T_2 = \frac{P_2V_2}{nR} = 119 \text{ K}$$

From this, the work done and the energy change can be found,

$$w = \Delta E = nc_V\, \Delta T = (5.00 \text{ mol})(\tfrac{3}{2} \times 8.315 \text{ J K}^{-1} \text{ mol}^{-1})(119 \text{ K} - 298 \text{ K})$$

$$= -11{,}200 \text{ J}$$

as well as the enthalpy change:

$$\Delta H = nc_P\, \Delta T = (5.00 \text{ mol})(\tfrac{5}{2} \times 8.315 \text{ J K}^{-1} \text{ mol}^{-1})(119 \text{ K} - 298 \text{ K}) = -18{,}600 \text{ J}$$

**Related Problems: 53, 54**

Figure 7.17 compares the adiabatic expansion of this example to the isothermal expansion of Example 7.9. Note that the initial states were the same in the two cases, as were the final pressures. However, the final volume is larger by more than a

**FIGURE 7.17** A comparison of reversible isothermal and adiabatic expansions. Using the technique shown in Figure 7.16, we see that the adiabatic work is 40% of the isothermal work.

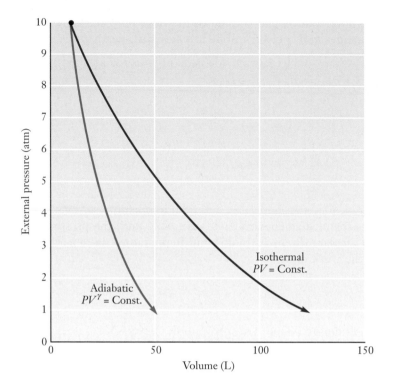

factor of 2 for the isothermal expansion, and the work output for the adiabatic case is only 40% of the output from the isothermal expansion. Because $\gamma > 1$, the adiabatic line falls off more rapidly with increasing volume than the isothermal line. Because no heat is transferred in an adiabatic expansion, the work comes from the internal energy of the gas, and thus the temperature drops.

A methanol-powered bus. *(Copyright Vanessa Vick/Photo Researchers, Inc.)*

## CUMULATIVE EXERCISE

### Methanol as a Gasoline Substitute

Methanol ($CH_3OH$) is employed as a substitute for gasoline in certain high-performance vehicles. To design engines that will run on it, its thermochemistry must be understood.

(a) The methanol in an automobile engine must be in the gas phase before it can react. Calculate the heat (in kilojoules) that must be added to 1.00 kg of liquid methanol to raise its temperature from 25.0°C to its normal boiling point, 65.0°C. The molar heat capacity of liquid methanol is 81.6 J $K^{-1}$ $mol^{-1}$.

(b) Once the methanol has reached its boiling point, it must be vaporized. The molar enthalpy of vaporization for methanol is 38 kJ $mol^{-1}$. How much heat must be added to vaporize 1.00 kg of methanol?

(c) Once it is in the vapor phase, the methanol can react with oxygen in the air according to

$$CH_3OH(g) + \tfrac{3}{2}\,O_2(g) \longrightarrow CO_2(g) + 2\,H_2O(g)$$

Use average bond enthalpies to estimate the enthalpy change in this reaction, for one mole of methanol reacting.

(d) Use data from Appendix D to calculate the actual enthalpy change in this reaction, assuming it to be the same at 65°C as at 25°C.

(e) Calculate the heat released when 1.00 kg of gaseous methanol is burned in air at constant pressure. Use the more accurate result of part (d) rather than that of part (c).

(f) Calculate the *difference* between the change in enthalpy and the change in internal energy when 1.00 kg of gaseous methanol is oxidized to gaseous $CO_2$ and $H_2O$ at 65°C.

(g) Suppose now that the methanol is burned inside the cylinder of an automobile. Taking the radius of the cylinder to be 4.0 cm and the distance moved by the piston during one stroke to be 12 cm, calculate the work done on the gas per stroke as it expands against an external pressure of 1.00 atm. Express your answer in liter-atmospheres and in joules.

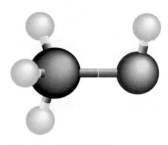

Methanol, $CH_3OH$.

**Answers**

(a) 102 kJ

(b) $1.2 \times 10^3$ kJ

(c) −508 kJ

(d) −676.49 kJ

(e) $2.11 \times 10^4$ kJ

(f) $\Delta H - \Delta E = 43.9$ kJ

(g) −0.60 L atm = −61 J

## CONCEPTS & SKILLS

*After studying this chapter and working the problems that follow, you should be able to*

1. Give precise definitions for the terms thermodynamic system, open system, closed system, thermodynamic state, reversible and irreversible process (Section 7.1).

2. Define and give examples of properties that are state functions of a system (Section 7.1).

3. Calculate the work done on an ideal gas when it is compressed reversibly (Section 7.2, problems 1–2).

4. Give a physical interpretation to the concept of heat, and calculate the change in temperature of a given quantity of a substance from its heat capacity and the amount of heat transferred to it (Section 7.2, problems 5–8).

5. Calculate the final temperature reached when two substances of different mass, heat capacity, and temperature are placed in thermal contact (Section 7.3, problems 11–16).

6. Calculate the amounts of heat and work and the change in energy of an ideal gas during expansions and compressions (Section 7.4, problems 17–22).

7. Calculate the energy and enthalpy changes for chemical reactions from the standard molar enthalpies of formation of reactants and products (Section 7.5, problems 35–40).

8. Use bond enthalpies to estimate enthalpies of formation of gaseous compounds (Section 7.5, problems 45–48).

9. Calculate the heat absorbed and work done by an ideal gas when it expands reversibly and either isothermally or adiabatically (Section 7.6, problems 51–54).

## PROBLEMS

*Answers to problems whose numbers are boldface appear in Appendix G. Problems that are more challenging are indicated with asterisks.*

### The First Law of Thermodynamics: Energy, Work, and Heat

1. Some nitrogen for use in synthesizing ammonia is heated slowly, maintaining the external pressure close to the internal pressure of 50.0 atm, until its volume has increased from 542 L to 974 L. Calculate the work done on the nitrogen as it is heated, and express it in joules.

2. The gas mixture inside one of the cylinders of an automobile engine expands against a constant external pressure of 0.98 atm, from an initial volume of 150 mL (at the end of the compression stroke) to a final volume of 800 mL. Calculate the work done on the gas mixture during this process, and express it in joules.

3. When a ball of mass $m$ is dropped through a height difference $\Delta h$, its potential energy changes by the amount $mg \, \Delta h$ where $g$ is the acceleration of gravity, equal to $9.81 \text{ m s}^{-2}$. Suppose that when the ball hits the ground, all that energy is converted to heat, increasing the temperature of the ball. If the specific heat capacity of the material in the ball is $0.850 \text{ J K}^{-1} \text{ g}^{-1}$, calculate the height from which the ball must be dropped to increase the temperature of the ball by $1.00°C$.

4. During his honeymoon in Switzerland, James Joule is said to have used a thermometer to measure the temperature difference between the water at the top and at the bottom of a waterfall. Take the height of the waterfall to be $\Delta h$ and the acceleration of gravity $g$ to be $9.81 \text{ m s}^{-2}$. Assuming that all the potential energy change $mg \, \Delta h$ of a mass $m$ of water is used to heat that water by the time it reaches the bottom, calculate the temperature difference between the top and the bottom of a waterfall 100 m high. Take the specific heat capacity of water to be $4.18 \text{ J K}^{-1} \text{ g}^{-1}$.

5. The specific heat capacities of Li($s$), Na($s$), K($s$), Rb($s$), and Cs($s$) at 25°C are 3.57, 1.23, 0.756, 0.363, and 0.242 J K$^{-1}$ g$^{-1}$, respectively. Compute the molar heat capacities of these elements and identify any periodic trend. If there is a trend, use it to predict the molar heat capacity of francium, Fr($s$).

6. The specific heat capacities of $F_2(g)$, $Cl_2(g)$, $Br_2(g)$, and $I_2(g)$ are 0.824, 0.478, 0.225, and 0.145 J K$^{-1}$ g$^{-1}$, respectively. Compute the molar heat capacities of these elements and identify any periodic trend. If there is a trend, use it to predict the molar heat capacity of astatine, $At_2(g)$.

7. The specific heat capacities of the metals nickel, zinc, rhodium, tungsten, gold, and uranium at 25°C are 0.444, 0.388, 0.243, 0.132, 0.129, and 0.116 J K$^{-1}$ g$^{-1}$, respectively. Calculate the molar heat capacities of these six metals. Note how closely the molar heat capacities for these metals, which were selected at random, cluster about a value of 25 J K$^{-1}$ mol$^{-1}$. The rule of Dulong and Petit states that the molar heat capacities of the metallic elements are approximately 25 J K$^{-1}$ mol$^{-1}$.

8. Use the empirical rule of Dulong and Petit stated in problem 7 to estimate the specific heat capacities of vanadium, gallium, and silver.

9. A chemical system is sealed in a strong, rigid container at room temperature and then heated vigorously.
    (a) State whether $\Delta E$, $q$, and $w$ of the system are positive, negative, or zero during the heating process.
    (b) Next, the container is cooled to its original temperature. Determine the signs of $\Delta E$, $q$, and $w$ for the cooling process.
    (c) Designate heating as step 1 and cooling as step 2. Determine the signs of $(\Delta E_1 + \Delta E_2)$, $(q_1 + q_2)$, and $(w_1 + w_2)$ if possible.

10. A battery harnesses a chemical reaction to extract energy in the form of useful electrical work.
    (a) A certain battery runs a toy truck and becomes partially discharged. In the process, it performs a total of 117.0 J of work on its immediate surroundings. It also gives off 3.0 J of heat, which the surroundings absorb. No other work or heat is exchanged with the surroundings. Compute $q$, $w$, and $\Delta E$ of the battery, making sure each quantity has the proper sign.
    (b) The same battery is now recharged exactly to its original condition. This requires 210.0 J of electrical work from an outside generator. Determine $q$ for the battery in this process. Explain why $q$ has the sign that it does.

### Heat Capacity, Enthalpy, and Calorimetry

11. Suppose 61.0 g of hot metal, which is initially at 120.0°C, is plunged into 100.0 g of water that is initially at 20.00°C. The metal cools down and the water heats up until they reach a common temperature of 26.39°C. Calculate the specific heat capacity of the metal, using 4.18 J K$^{-1}$ g$^{-1}$ as the specific heat capacity of the water.

12. A piece of zinc at 20.0°C that weighs 60.0 g is dropped into 200.0 g of water at 100.0°C. The specific heat capacity of zinc is 0.389 J K$^{-1}$ g$^{-1}$, and that of water near 100°C is 4.22 J K$^{-1}$ g$^{-1}$. Calculate the final temperature reached by the zinc and the water.

13. Very early in the study of the nature of heat it was observed that if two bodies of equal mass but different temperatures are placed in thermal contact, their specific heat capacities depend inversely on the change in temperature each undergoes on reaching its final temperature. Write a mathematical equation in modern notation to express this fact.

14. Iron pellets with total mass 17.0 g at a temperature of 92.0°C are mixed in an insulated container with 17.0 g of water at a temperature of 20.0°C. The specific heat capacity of water is ten times greater than that of iron. What is the final temperature inside the container?

15. In their *Memoir on Heat*, published in 1783, Lavoisier and Laplace reported, "The heat necessary to melt ice is equal to three quarters of the heat that can raise the same mass of water from the temperature of the melting ice to that of boiling water" (English translation). Use this 18th-century observation to compute the amount of heat (in joules) needed to melt 1.00 g of ice. Assume that heating 1.00 g of water requires 4.18 J of heat for each 1.00°C throughout the range from 0°C to 100°C.

16. Galen, the great physician of antiquity, suggested scaling temperature from a reference point defined by mixing equal masses of ice and boiling water in an insulated container. Imagine that this is done with the ice at 0.00°C and the water at 100.0°C. Assume that the heat capacity of the container is negligible and that it takes 333.4 J of heat to melt 1.000 g of ice at 0.00°C to water at 0.00°C. Compute Galen's reference temperature in degrees Celsius.

## Illustrations of the First Law of Thermodynamics in Ideal Gas Processes

17. If 0.500 mol of neon at 1.00 atm and 273 K expands against a constant external pressure of 0.100 atm until the gas pressure reaches 0.200 atm and the temperature reaches 210 K, calculate the work done on the gas, the internal energy change, and the heat absorbed by the gas.

18. Hydrogen behaves as an ideal gas at temperatures above 200 K and at pressures below 50 atm. Suppose 6.00 mol of hydrogen is initially contained in a 100-L vessel at a pressure of 2.00 atm. The average molar heat capacity of hydrogen at constant pressure, $c_P$, is 29.3 J K$^{-1}$ mol$^{-1}$ in the temperature range of this problem. The gas is cooled reversibly at constant pressure from its initial state to a volume of 50 L. Calculate the following quantities for this process.
    (a) Temperature of the gas in the final state, $T_2$
    (b) Work done on the gas, $w$, in joules
    (c) Internal energy change of the gas, $\Delta E$, in joules
    (d) Heat absorbed by the gas, $q$, in joules

19. Suppose 2.00 mol of an ideal, monatomic gas is initially at a pressure of 3.00 atm and a temperature $T = 350$ K. It is expanded *irreversibly* and *adiabatically* ($q = 0$) against a constant external pressure of 1.00 atm until the volume has doubled.
    (a) Calculate the final volume.
    (b) Calculate $w$, $q$, and $\Delta E$ for this process, in joules.
    (c) Calculate the final temperature of the gas.

20. Consider the free, isothermal (constant $T$) expansion of an ideal gas. "Free" means that the external force is zero, perhaps because a stopcock has been opened and the gas is allowed to expand into a vacuum. Calculate $\Delta E$ for this irreversible process. Show that $q = 0$ so that the expansion is also adiabatic ($q = 0$) for an ideal gas. This is analogous to a classic experiment first carried out by Joule.

21. If 6.00 mol of argon in a 100-L vessel initially at 300 K is compressed adiabatically ($q = 0$) and irreversibly until a final temperature of 450 K is reached, calculate the energy change of the gas, the heat added to the gas, and the work done on the gas.

22. A gas expands against a constant external pressure of 2.00 atm until its volume has increased from 6.00 L to 10.00 L. During this process it absorbs 500 J of heat from the surroundings.
    (a) Calculate the energy change of the gas $\Delta E$.
    (b) Calculate the work $w$ done on the gas in an irreversible adiabatic ($q = 0$) process connecting the same initial and final states.

## Thermochemistry

23. For each of the following reactions, the enthalpy change written is that measured when the numbers of moles of reactants and products taking part in the reaction are as given by their coefficients in the equation. Calculate the enthalpy change when 1.00 *gram* of the underlined substance is consumed or produced.
    (a) $4 \text{ Na}(s) + \text{O}_2(g) \rightarrow 2 \underline{\text{Na}_2\text{O}}(s)$    $\Delta H = -828$ kJ
    (b) $\text{CaMg(CO}_3)_2(s) \rightarrow \text{CaO}(s) + \underline{\text{MgO}}(s) + 2 \text{ CO}_2(g)$
    $\Delta H = +302$ kJ
    (c) $\text{H}_2(g) + 2 \underline{\text{CO}}(g) \rightarrow \text{H}_2\text{O}_2(\ell) + 2 \text{ C}(s)$    $\Delta H = +33.3$ kJ

24. For each of the following reactions, the enthalpy change given is that measured when the numbers of moles of reactants and products taking part in the reaction are as given by their coefficients in the equation. Calculate the enthalpy change when 1.00 *gram* of the underlined substance is consumed or produced.
    (a) $\text{Ca}(s) + \underline{\text{Br}_2}(\ell) \rightarrow \text{CaBr}_2(s)$    $\Delta H = -683$ kJ
    (b) $6 \text{ Fe}_2\text{O}_3(s) \rightarrow 4 \underline{\text{Fe}_3\text{O}_4}(s) + \text{O}_2(g)$    $\Delta H = +472$ kJ
    (c) $2 \underline{\text{NaHSO}_4}(s) \rightarrow 2 \text{ NaOH}(s) + 2 \text{ SO}_2(g) + \text{O}_2(g)$
    $\Delta H +806$ kJ

25. Liquid bromine dissolves readily in aqueous NaOH:

    $\text{Br}_2(\ell) + 2 \text{ NaOH}(aq) \longrightarrow$
    $\text{NaBr}(aq) + \text{NaOBr}(aq) + \text{H}_2\text{O}(\ell)$

    Suppose $2.88 \times 10^{-3}$ mol of $\text{Br}_2(\ell)$ is sealed in a glass capsule that is then immersed in a solution containing excess $\text{NaOH}(aq)$. The capsule is broken, the mixture is stirred, and a measured 121.3 J of heat evolves. In a separate experiment, simply breaking an empty capsule and stirring the solution in the same way evolves 2.34 J of heat. Compute the heat evolved as 1.00 mol of $\text{Br}_2(\ell)$ dissolves in excess $\text{NaOH}(aq)$.

26. A chemist mixes 1.00 g of $\text{CuCl}_2$ with an excess of $(\text{NH}_4)_2\text{HPO}_4$ in dilute aqueous solution. He measures the evolution of 670 J of heat as the two substances react to give $\text{Cu}_3(\text{PO}_4)_2(s)$. Compute the $\Delta H$ that would result from the reaction of 1.00 mol of $\text{CuCl}_2$ with an excess of $(\text{NH}_4)_2\text{HPO}_4$.

27. Calculate the enthalpy change when 2.38 g of carbon monoxide (CO) vaporizes at its normal boiling point. Use data from Table 7.2.

28. Molten sodium chloride is used for making elemental sodium and chlorine. Suppose the electrical power to a vat containing 56.2 kg of molten sodium chloride is cut off, and the salt crystallizes (without changing its temperature). Calculate the enthalpy change, using data from Table 7.2.

29. Suppose an ice cube weighing 36.0 g at a temperature of $-10°C$ is placed in 360 g of water at a temperature of 20°C. Calculate the temperature after thermal equilibrium is reached, assuming no heat loss to the surroundings. The enthalpy of fusion of ice is $\Delta H_{fus} = 6.007$ kJ mol$^{-1}$, and the molar heat capacities $c_P$ of ice and water are 38 and 75 J K$^{-1}$ mol$^{-1}$, respectively.

30. You have a supply of ice at 0.0°C and a glass containing 150 g of water at 25°C. The enthalpy of fusion for ice is $\Delta H_{fus} = 333$ J g$^{-1}$, and the specific heat capacity of water is 4.18 J K$^{-1}$ g$^{-1}$. How many grams of ice must be added to the glass (and melted) to reduce the temperature of the water to 0°C?

31. The measured enthalpy change for burning ketene ($CH_2CO$)

$$CH_2CO(g) + 2 O_2(g) \longrightarrow 2 CO_2(g) + H_2O(g)$$

is $\Delta H_1 = -981.1$ kJ at 25°C. The enthalpy change for burning methane

$$CH_4(g) + 2 O_2(g) \longrightarrow CO_2(g) + 2 H_2O(g)$$

is $\Delta H_2 = -802.3$ kJ at 25°C. Calculate the enthalpy change at 25°C for the reaction

$$2 CH_4(g) + 2 O_2(g) \longrightarrow CH_2CO(g) + 3 H_2O(g)$$

32. Given the following two reactions and corresponding enthalpy changes,

$$CO(g) + SiO_2(s) \longrightarrow SiO(g) + CO_2(g) \quad \Delta H = +520.9 \text{ kJ}$$

$$8 CO_2(g) + Si_3N_4(s) \longrightarrow 3 SiO_2(s) + 2 N_2O(g) + 8 CO(g)$$
$$\Delta H = +461.05 \text{ kJ}$$

compute the $\Delta H$ of the reaction

$$5 CO_2(g) + Si_3N_4(s) \longrightarrow 3 SiO(g) + 2 N_2O(g) + 5 CO(g)$$

33. The enthalpy change to make diamond from graphite is 1.88 kJ mol$^{-1}$. Which gives off more heat when burned—a pound of diamonds or a pound of graphite? Explain.

34. The enthalpy change of combustion of monoclinic sulfur to $SO_2(g)$ is $-9.376$ kJ g$^{-1}$. Under the same conditions, the rhombic form of sulfur has an enthalpy change of combustion to $SO_2(g)$ of $-9.293$ kJ g$^{-1}$. Compute the $\Delta H$ of the reaction

$$S(\text{monoclinic}) \longrightarrow S(\text{rhombic})$$

per gram of sulfur reacting.

35. Calculate the standard enthalpy change $\Delta H°$ at 25°C for the reaction

$$N_2H_4(\ell) + 3 O_2(g) \longrightarrow 2 NO_2(g) + 2 H_2O(\ell)$$

using the standard enthalpies of formation ($\Delta H_f°$) of reactants and products at 25°C from Appendix D.

36. Using the data in Appendix D, calculate $\Delta H°$ for each of the following processes.
    (a) $2 NO(g) + O_2(g) \rightarrow 2 NO_2(g)$
    (b) $C(s) + CO_2(g) \rightarrow 2 CO(g)$
    (c) $2 NH_3(g) + \frac{7}{2} O_2(g) \rightarrow 2 NO_2(g) + 3 H_2O(g)$
    (d) $C(s) + H_2O(g) \rightarrow CO(g) + H_2(g)$

37. Zinc is commonly found in nature in the form of the mineral sphalerite (ZnS). A step in the smelting of zinc is the roasting of sphalerite with oxygen to produce zinc oxide:

$$2 ZnS(s) + 3 O_2(g) \longrightarrow 2 ZnO(s) + 2 SO_2(g)$$

    (a) Calculate the standard enthalpy change $\Delta H°$ for this reaction, using data from Appendix D.
    (b) Calculate the heat absorbed when 3.00 metric tons (1 metric ton = $10^3$ kg) of sphalerite is roasted under constant-pressure conditions.

38. The thermite process (see Fig. 7.11) is used for welding railway track together. In this reaction, aluminum reduces iron(III) oxide to metallic iron:

$$2 Al(s) + Fe_2O_3(s) \longrightarrow 2 Fe(s) + Al_2O_3(s)$$

Igniting a small charge of barium peroxide mixed with aluminum triggers the reaction of a mixture of aluminum powder and iron(III) oxide; the molten iron produced flows into the space between the steel rails that are to be joined.
    (a) Calculate the standard enthalpy change $\Delta H°$ for this reaction, using data from Appendix D.
    (b) Calculate the heat given off when 3.21 g of iron(III) oxide is reduced by aluminum at constant pressure.

39. The dissolution of calcium chloride in water

$$CaCl_2(s) \longrightarrow Ca^{2+}(aq) + 2 Cl^-(aq)$$

is used in first-aid hot packs. In these packs, an inner pouch containing the salt is broken, allowing the salt to dissolve in the surrounding water.
    (a) Calculate the standard enthalpy change $\Delta H°$ for this reaction, using data from Appendix D.
    (b) Suppose 20.0 g of $CaCl_2$ is dissolved in 0.100 L of water at 20.0°C. Calculate the temperature reached by the solution, assuming it to be an ideal solution with a heat capacity close to that of 100 g of pure water (418 J K$^{-1}$).

40. Ammonium nitrate dissolves in water according to the reaction

$$NH_4NO_3(s) \longrightarrow NH_4^+(aq) + NO_3^-(aq)$$

    (a) Calculate the standard enthalpy change $\Delta H°$ for this reaction, using data from Appendix D.
    (b) Suppose 15.0 g of $NH_4NO_3$ is dissolved in 0.100 L of water at 20.0°C. Calculate the temperature reached by the solution, assuming it to be an ideal solution with a heat capacity close to that of 100 g of pure water (418 J K$^{-1}$).
    (c) From a comparison with the results of problem 39, can you suggest a practical application of this dissolution reaction?

41. The standard enthalpy change of combustion [to $CO_2(g)$ and $H_2O(\ell)$] at 25°C of the organic liquid cyclohexane, $C_6H_{12}(\ell)$ is $-3923.7$ kJ mol$^{-1}$. Determine the $\Delta H_f°$ of $C_6H_{12}(\ell)$. Use data from Appendix D.

42. The standard enthalpy change of combustion [to $CO_2(g)$ and $H_2O(\ell)$] at 25°C of the organic liquid cyclohexene, $C_6H_{10}(\ell)$, is $-3731.7$ kJ mol$^{-1}$. Determine the $\Delta H_f°$ of $C_6H_{10}(\ell)$.

43. A sample of pure solid naphthalene ($C_{10}H_8$) weighing 0.6410 g is burned completely with oxygen to $CO_2(g)$ and

$H_2O(\ell)$ in a constant-volume calorimeter at 25°C. The amount of heat evolved is observed to be 25.79 kJ.
(a) Write and balance the chemical equation for the combustion reaction.
(b) Calculate the standard change in internal energy ($\Delta E°$) for the combustion of 1.000 mol of naphthalene to $CO_2(g)$ and $H_2O(\ell)$.
(c) Calculate the standard enthalpy change ($\Delta H°$) for the same reaction as in part (b).
(d) Calculate the standard enthalpy of formation per mole of naphthalene, using data for the standard enthalpies of formation of $CO_2(g)$ and $H_2O(\ell)$ from Appendix D.

44. A sample of solid benzoic acid ($C_6H_5COOH$) that weighs 0.000 g is burned in an excess of oxygen to $CO_2(g)$ and $H_2O(\ell)$ in a constant-volume calorimeter at 25°C. The temperature rise is observed to be 2.15°C. The heat capacity of the calorimeter and its contents is known to be $9382 \text{ J K}^{-1}$.
(a) Write and balance the equation for the combustion of benzoic acid.
(b) Calculate the standard change in internal energy ($\Delta E°$) for the combustion of 1.000 mol of benzoic acid to $CO_2(g)$ and $H_2O(\ell)$ at 25°C.
(c) Calculate the standard enthalpy change ($\Delta H°$) for the same reaction as in part (b).
(d) Calculate the standard enthalpy of formation per mole of benzoic acid, using data for the standard enthalpies of formation of $CO_2(g)$ and $H_2O(\ell)$ from Appendix D.

45. A second chlorofluorocarbon used as a refrigerant and in aerosols (besides that discussed in Example 7.8) is $CCl_3F$. Use the atomization enthalpies and average bond enthalpies from Table 7.3 to estimate the standard enthalpy of formation ($\Delta H_f°$) of this compound in the gas phase.

46. The compound $CF_3CHCl_2$ (with a C—C bond) has been proposed as a substitute for $CCl_3F$ and $CCl_2F_2$ because it decomposes more quickly in the atmosphere and is much less liable to reduce the concentration of ozone in the stratosphere. Use the atomization enthalpies and average bond enthalpies from Table 7.3 to estimate the standard enthalpy of formation ($\Delta H_f°$) of $CF_3CHCl_2$ in the gas phase.

47. Propane has the structure $H_3C—CH_2—CH_3$. Use average bond enthalpies from Table 7.3 to estimate the change in enthalpy $\Delta H°$ for the reaction

$$C_3H_8(g) + 5\ O_2(g) \longrightarrow 3\ CO_2(g) + 4\ H_2O(g)$$

48. Use average bond enthalpies from Table 7.3 to estimate the change in enthalpy $\Delta H°$ for the reaction

$$C_2H_4(g) + H_2(g) \longrightarrow C_2H_6(g)$$

Refer to the molecular structures on page 71.

49. The reaction

$$BBr_3(g) + BCl_3(g) \longrightarrow BBr_2Cl(g) + BCl_2Br(g)$$

has a $\Delta H$ very close to 0. Sketch the Lewis structures of the four compounds, and explain why $\Delta H$ is so small.

50. At 381 K, the following reaction takes place:

$$Hg_2Cl_4(g) + Al_2Cl_6(g) \longrightarrow 2\ HgAlCl_5(g) \quad \Delta H = +10 \text{ kJ}$$

(a) Offer an explanation for the very small $\Delta H$ for this reaction in terms of the known structures of the compounds

(b) Explain why the small $\Delta H$ in this reaction is evidence against

as the structure of $Hg_2Cl_4(g)$.

## Reversible Processes in Ideal Gases

51. If 2.00 mol of an ideal gas at 25°C expands isothermally and reversibly from 9.00 to 36.00 L, calculate the work done on the gas and the heat absorbed by the gas in the process. What are the changes in energy ($\Delta E$) and in enthalpy ($\Delta H$) of the gas in the process?

52. If 54.0 g of argon at 400 K is compressed isothermally and reversibly from a pressure of 1.50 atm to 4.00 atm, calculate the work done on the gas and the heat absorbed by the gas in the process. What are the changes in energy ($\Delta E$) and in enthalpy ($\Delta H$) of the gas?

53. Suppose 2.00 mol of a monatomic ideal gas ($c_V = \frac{3}{2}R$) is expanded adiabatically and reversibly from a temperature $T = 300$ K, where the volume of the system is 20.0 L, to a volume of 60.0 L. Calculate the final temperature of the gas, the work done on the gas, and the energy and enthalpy changes.

54. Suppose 2.00 mol of an ideal gas is contained in a heat-insulated cylinder with a moveable frictionless piston. Initially the gas is at 1.00 atm and 0°C. The gas is compressed reversibly to 2.00 atm. The molar heat capacity at constant pressure, $c_P$, equals $29.3 \text{ J K}^{-1} \text{ mol}^{-1}$. Calculate the final temperature of the gas, the change in its internal energy $\Delta E$, and the work done on the gas.

## Additional Problems

55. At one time, it was thought that the molar mass of indium was near $76 \text{ g mol}^{-1}$. By referring to the law of Dulong and Petit (problem 7), show how the measured specific heat of metallic indium, $0.233 \text{ J K}^{-1} \text{ g}^{-1}$, makes this value unlikely.

56. The following table shows how the specific heat at constant pressure of liquid helium changes with temperature. Note the sharp increase over this temperature range:

Temperature (K):

| 1.80 | 1.85 | 1.90 | 1.95 | 2.00 | 2.05 | 2.10 | 2.15 |

$c_s$ ($\text{J K}^{-1} \text{ g}^{-1}$):

| 2.81 | 3.26 | 3.79 | 4.42 | 5.18 | 6.16 | 7.51 | 9.35 |

Estimate how much heat it takes at constant pressure to raise the temperature of 1.00 g of He($\ell$) from 1.8 K to 2.15 K. (*Hint:* For each temperature interval of 0.05 K, take the average, $c_s$, as the sum of the values at the ends of the interval divided by 2.)

57. Imagine that 2.00 mol of argon, confined by a moveable, frictionless piston in a cylinder at a pressure of 1.00 atm and a temperature of 398 K, is cooled to 298 K. Argon gas may be considered ideal, and its molar heat capacity at constant pressure is $c_P = (5/2)R$ where $R = 8.315$ J K$^{-1}$ mol$^{-1}$. Calculate:
    (a) The work done on the system, $w$
    (b) The heat absorbed by the system, $q$
    (c) The energy change of the system, $\Delta E$
    (d) The enthalpy change of the system, $\Delta H$

58. Suppose 1.00 mol of ice at $-30°$C is heated at atmospheric pressure until it is converted to steam at 140°C. Calculate $q$, $w$, $\Delta H$, and $\Delta E$ for this process. For ice, water, and steam, $c_P$ is 38, 75, and 36 J K$^{-1}$ mol$^{-1}$, respectively, and can be taken to be approximately independent of temperature. $\Delta H_{fus}$ for ice is 6.007 kJ mol$^{-1}$, and $\Delta H_{vap}$ for water is 40.66 kJ mol$^{-1}$. Use the ideal gas law for steam, and assume that the volume of 1 mol of ice or water is negligible relative to that of 1 mol of steam.

59. The gas inside a cylinder expands against a constant external pressure of 1.00 atm from a volume of 5.00 L to a volume of 13.00 L. In doing so, it turns a paddle immersed in 1.00 L of water. Calculate the temperature rise of the water, assuming no loss of heat to the surroundings or frictional losses in the mechanism. Take the density of water to be 1.00 g cm$^{-3}$ and its specific heat to be 4.18 J K$^{-1}$ g$^{-1}$.

60. Suppose 1.000 mol of argon (assumed to be an ideal gas) is confined in a strong, rigid container of volume 22.41 L at 273.15 K. The system is heated until 3.000 kJ (3000 J) of heat has been added. The molar heat capacity of the gas does not change during the heating and equals 12.47 J K$^{-1}$ mol$^{-1}$.
    (a) Calculate the original pressure inside the vessel (in atmospheres).
    (b) Determine $q$ for the system during the heating process.
    (c) Determine $w$ for the system during the heating process.
    (d) Compute the temperature of the gas after the heating, in degrees Celsius. Assume the container has zero heat capacity.
    (e) Compute the pressure (in atmospheres) inside the vessel after the heating.
    (f) Compute $\Delta E$ of the gas during the heating process.
    (g) Compute $\Delta H$ of the gas during the heating process.
    (h) The correct answer to part (g) exceeds 3.000 kJ. The increase in enthalpy (which at one time was misleadingly called the "heat content") in this system exceeds the amount of heat actually added. Why is this not a violation of the law of conservation of energy?

61. When glucose, a sugar, reacts fully with oxygen, carbon dioxide and water are produced:

$$C_6H_{12}O_6(s) + 6 \ O_2(g) \longrightarrow 6 \ CO_2(g) + 6 \ H_2O(\ell)$$
$$\Delta H° = -2820 \ kJ$$

Suppose a person weighing 50 kg (mostly water, with specific heat capacity 4.18 J K$^{-1}$ g$^{-1}$) eats a candy bar containing 14.3 g of glucose. If all the glucose reacted with oxygen and the heat produced were used entirely to increase the person's body temperature, what temperature increase would result? (In fact, most of the heat produced is lost to the surroundings before such a temperature increase occurs.)

62. In walking a kilometer, you use about 100 kJ of energy. This energy comes from the oxidation of foods, which is about 30% efficient. How much energy do you save by walking a kilometer instead of driving a car that gets 8.0 km L$^{-1}$ of gasoline (19 miles per gallon)? The density of gasoline is 0.68 g cm$^{-3}$ and its enthalpy of combustion is $-48$ kJ g$^{-1}$.

63. Liquid helium and liquid nitrogen are both used as coolants; He($\ell$) boils at 4.21 K, and N$_2$($\ell$) boils at 77.35 K. The specific heat of liquid helium near its boiling point is 4.25 J K$^{-1}$ g$^{-1}$, and the specific heat of liquid nitrogen near *its* boiling point is 1.95 J K$^{-1}$ g$^{-1}$. The enthalpy of vaporization of He($\ell$) is 25.1 J g$^{-1}$, and the enthalpy of vaporization of N$_2$($\ell$) is 200.3 J g$^{-1}$ (these data are calculated from the values in Appendix F). Discuss which liquid is the better coolant (on a per-gram basis) *near* its boiling point and which is better *at* its boiling point.

64. When 1.00 g of potassium chlorate (KClO$_3$) is dissolved in 50.0 g of water in a Styrofoam calorimeter of negligible heat capacity, the temperature drops from 25.00 to 23.36°C. Calculate $q$ for the water and $\Delta H°$ for the process.

$$KClO_3(s) \longrightarrow K^+(aq) + ClO_3^-(aq)$$

The specific heat of water is 4.184 J K$^{-1}$ g$^{-1}$.

65. The enthalpy of combustion and the standard enthalpy of formation of a fuel can be determined by measuring the temperature change in a calorimeter when a weighed amount of the fuel is burned in oxygen.
    (a) Write a balanced chemical equation for the combustion of isooctane, C$_8$H$_{18}$($\ell$), to CO$_2$($g$) and H$_2$O($\ell$). Isooctane is a component of gasoline and is used as a reference standard in determining the "octane rating" of a fuel mixture.
    (b) Suppose 0.542 g of isooctane is placed in a fixed-volume (bomb) calorimeter, which contains 750 g of water, initially at 20.450°C, surrounding the reaction compartment. The heat capacity of the calorimeter itself (excluding the water) has been measured to be 48 J K$^{-1}$ in a separate calibration. After the combustion of the isooctane is complete, the water temperature is measured to be 28.670°C. Taking the specific heat of water to be 4.184 J K$^{-1}$ g$^{-1}$, calculate $\Delta E$ for the combustion of 0.542 g of isooctane.
    (c) Calculate $\Delta E$ for the combustion of 1 mol of isooctane.
    (d) Calculate $\Delta H$ for the combustion of 1 mol of isooctane.
    (e) Calculate $\Delta H_f°$ for the isooctane.

66. The enthalpy change to form 1 mol of Hg$_2$Br$_2$($s$) from the elements at 25°C is $-206.77$ kJ mol$^{-1}$, and that of HgBr($g$) is 96.23 kJ mol$^{-1}$. Compute the enthalpy change for the decomposition of 1 mol of Hg$_2$Br$_2$($s$) to 2 mol of HgBr($g$):

$$Hg_2Br_2(s) \rightarrow 2 \ HgBr(g)$$

* 67. The gas most commonly used in welding is acetylene [$C_2H_2(g)$]. When acetylene is burned in oxygen, the reaction is

$$C_2H_2(g) + \tfrac{5}{2} O_2(g) \longrightarrow 2 CO_2(g) + H_2O(g)$$

(a) Using data from Appendix D, calculate $\Delta H°$ for this reaction.

(b) Calculate the total heat capacity of 2.00 mol of $CO_2(g)$ and 1.00 mol of $H_2O(g)$, using $c_P(CO_2(g)) = 37$ J $K^{-1}$ $mol^{-1}$ and $c_P(H_2O(g)) = 36$ J $K^{-1}$ $mol^{-1}$.

(c) When this reaction is carried out in an open flame, almost all the heat produced in part (a) goes to raise the temperature of the products. Calculate the maximum flame temperature that is attainable in an open flame burning acetylene in oxygen. The actual flame temperature would be lower than this because heat is lost to the surroundings.

* 68. The enthalpy of reaction changes somewhat with temperature. Suppose we wish to calculate $\Delta H$ for a reaction at a temperature $T$ that is different from 298 K. To do this, we can replace the direct reaction at $T$ with a three-step process. In the first step, the temperature of the reactants is changed from $T$ to 298 K. $\Delta H$ for this step can be calculated from the molar heat capacities of the reactants, which are assumed to be independent of temperature. In the second step, the reaction is carried out at 298 K with an enthalpy change $\Delta H°$. In the third step, the temperature of the products is changed from 298 K to $T$. The sum of these three enthalpy changes is $\Delta H$ for the reaction at temperature $T$.

An important process contributing to air pollution is the chemical reaction

$$SO_2(g) + \tfrac{1}{2} O_2(g) \longrightarrow SO_3(g)$$

For $SO_2(g)$ the heat capacity $c_P$ is 39.9, for $O_2(g)$ it is 29.4, and for $SO_3(g)$ it is 50.7 J $K^{-1}$ $mol^{-1}$. Calculate $\Delta H$ for the preceding reaction at 500 K, using the enthalpies of formation at 298.15 K from Appendix D.

* 69. At the top of the compression stroke in one of the cylinders of an automobile engine (that is, at the minimum gas volume), the volume of the gas–air mixture is 150 mL, the temperature is 600 K, and the pressure is 12.0 atm. The ratio of the number of moles of octane vapor to the number of moles of air in the combustion mixture is 1.00:80.0. What is the maximum temperature attained in the gas if octane burns explosively before the power stroke of the piston (gas expansion) begins? The gases may be considered to be ideal, and their heat capacities at constant pressure (assumed to be temperature-independent) are

$$c_P(C_8H_{18}(g)) = 327 \text{ J } K^{-1} \text{ mol}^{-1}$$
$$c_P(O_2(g)) = 35.2 \text{ J } K^{-1} \text{ mol}^{-1}$$
$$c_P(N_2(g)) = 29.8 \text{ J } K^{-1} \text{ mol}^{-1}$$
$$c_P(CO_2(g)) = 45.5 \text{ J } K^{-1} \text{ mol}^{-1}$$
$$c_P(H_2O(g)) = 38.9 \text{ J } K^{-1} \text{ mol}^{-1}$$

The enthalpy of formation of $C_8H_{18}(g)$ at 600 K is $-57.4$ kJ $mol^{-1}$.

70. Initially, 46.0 g of oxygen is at a pressure of 1.00 atm and a temperature of 400 K. It expands adiabatically and reversibly until the pressure is reduced to 0.60 atm, and then it is compressed isothermally and reversibly until the volume returns to its original value. Calculate the final pressure and temperature of the oxygen, the work done and heat added to the oxygen in this process, and the energy change $\Delta E$. Take $c_P(O_2) = 29.4$ J $K^{-1}$ $mol^{-1}$.

71. A young chemist buys a "one-lung" motorcycle but, before learning how to drive it, wants to understand the processes that occur in its engine. The manual says the cylinder has a radius of 5.00 cm, a piston stroke of 12.00 cm, and a (volume) compression ratio of 8:1. If a mixture of gasoline vapor (taken to be $C_8H_{18}$) and air in mole ratio 1:62.5 is drawn into the cylinder at 80°C and 1.00 atm, calculate:

(a) The temperature of the compressed gases just before the spark plug ignites them. (Assume the gases are ideal, the compression is adiabatic, and the average heat capacity of the mixture of gasoline vapor and air is $c_P = 35$ J $K^{-1}$ $mol^{-1}$.)

(b) The volume of the compressed gases just before ignition.

(c) The pressure of the compressed gases just before ignition.

(d) The maximum temperature of the combustion products, assuming combustion is completed before the piston begins its downstroke. Take $\Delta H_f°(C_8H_{18}) = -57.4$ kJ $mol^{-1}$.

(e) The temperature of the exhaust gases, assuming the expansion stroke to be adiabatic.

## CUMULATIVE PROBLEMS

72. Suppose 32.1 g of $ClF_3(g)$ and 17.3 g of $Li(s)$ are mixed and allowed to react at atmospheric pressure and 25°C until one of the reactants is used up, producing $LiCl(s)$ and $LiF(s)$. Calculate the amount of heat evolved.

73. Calculate $\Delta H_f°$ and $\Delta E_f°$ for the formation of silane, $SiH_4(g)$, from its elements at 298 K, if 250 cm$^3$ of the gaseous compound at $T = 298$ K and $P = 0.658$ atm is burned in a constant-volume gas calorimeter in an excess of oxygen and causes the evolution of 9.757 kJ of heat. The combustion reaction is

$$SiH_4(g) + 2 O_2(g) \longrightarrow SiO_2(s, \text{quartz}) + 2 H_2O(\ell)$$

and the formation of silane from its elements is

$$Si(s) + 2 H_2(g) \longrightarrow SiH_4(g)$$

74. (a) Draw Lewis diagrams for $O_2$, $CO_2$, $H_2O$, $CH_4$, $C_8H_{18}$, and $C_2H_5OH$. In $C_8H_{18}$, the eight carbon atoms form a

chain with single bonds; in $C_2H_5OH$ the two carbon atoms are bonded to one another. Using average bond enthalpies from Table 7.3, compute the enthalpy change in each of the following reactions, if 1 mol of each carbon compound is burned and all reactants and products are in the gas phase.

(b) $CH_4 + 2\ O_2 \rightarrow CO_2 + 2\ H_2O$

(burning methane, or natural gas)

(c) $C_8H_{18} + \frac{25}{2}\ O_2 \rightarrow 8\ CO_2 + 9\ H_2O$

(burning octane, in gasoline)

(d) $C_2H_5OH + 3\ O_2 \rightarrow 2\ CO_2 + 3\ H_2O$

(burning ethanol, in gasohol)

75. By considering the nature of the intermolecular forces in each case, rank the following substances from smallest to largest enthalpy of vaporization: KBr, Ar, $NH_3$, and He. Explain your reasoning.

* 76. A supersonic nozzle is a cone-shaped object with a small hole in the end through which a gas is forced. As it moves through the nozzle opening, the gas expands in a manner that can be approximated as reversible and adiabatic. Such nozzles are used in molecular beams (Section 13.6) and in supersonic air-craft engines to provide thrust, because as the gas cools, its random thermal energy is converted into directed motion of the molecules with average velocity $v$. Very little thermodynamic work is done because the external pressure is low, so the net effect is to convert thermal energy to net translational motion of the gas molecules. Suppose the gas in the nozzle is helium; its pressure is 50 atm and its temperature is 400 K before it begins its expansion.

(a) What are the average speed and the average velocity of the molecules *before* the expansion?

(b) What will be the temperature of the gas after its pressure has dropped to 1.0 atm in the expansion?

(c) What is the average velocity of the molecules at this point in the expansion?

77. (a) Draw a Lewis diagram for carbonic acid, $H_2CO_3$, with a central carbon atom bonded to the three oxygen atoms.

(b) Carbonic acid is unstable in aqueous solution and converts to dissolved carbon dioxide. Use bond enthalpies to estimate the enthalpy change for the reaction

$$H_2CO_3 \longrightarrow H_2O + CO_2$$

# Spontaneous Processes and Thermodynamic Equilibrium

The introduction to Unit 3 identified predicting the equilibrium composition for a chemical reaction as the central goal of this Unit. An experimenter mixes the reactants, regulates external conditions such as temperature and pressure, and waits to see whether the reaction proceeds under the selected conditions. Two questions arise immediately: (1) Does the reaction proceed *spontaneously* (i.e., without further intervention from the experimenter) at the selected conditions? (2) If the

*Illustration*
The reaction between sodium and chlorine proceeds imperceptibly if at all until the addition of a drop of water sets it off. *(Charles D. Winters)*

reaction is spontaneous, what determines the ratio of products and reactants at equilibrium?

The second law, which is a brilliant abstraction and generalization from the observed facts of directionality in processes, provides a route for predicting spontaneity and directionality in physical and chemical processes involving heat transfer. This is accomplished by introducing a new state function called *entropy* and denoted by $S$ defined so that its total value in the system and the surroundings increases in the direction of a spontaneous process. The present chapter develops the second law and demonstrates methods for calculating $\Delta S$ and predicting spontaneity for simple physical processes in which no chemical reactions occur.

When spontaneous processes are carried out at constant $T$ and at constant $P$, the conditions for spontaneity and for equilibrium are stated more conveniently in terms of another state function called the *Gibbs Free Energy* (denoted by $G$) which is derived from $S$. Because chemical reactions are usually conducted at constant $T$ and constant $P$, their thermodynamic description is based on $\Delta G$ rather than $\Delta S$. The present chapter ends by identifying conditions for spontaneity of chemical reactions in terms of $\Delta G$. The next chapter will develop methods for characterizing the state of chemical equilibrium in terms of $\Delta G$.

---

## 8.1

# THE NATURE OF SPONTANEOUS PROCESSES

A spontaneous change is one that, given enough time, occurs by itself without outside intervention. One of the most striking features of spontaneous change in nature is that it has a direction. A particularly dramatic example is shown in Figure 8.1. One of the goals of thermodynamics is to account for this *directionality* of spontaneous change. The first law of thermodynamics does not help in this regard. Energy is conserved both in a forward process and in its reverse; nothing in the first law indicates a preference for one direction of change over the other.

Spontaneous processes more familiar in chemical laboratories also exhibit intrinsic directionality. (1) We measure *heat flowing* from a hot body to a cold one when they are brought into thermal contact, but we never detect heat flowing spontaneously in the opposite direction. (2) We observe a gas *expanding* into a region of lower pressure but we never see the reverse process, a gas compressing itself spontaneously into a small part of its container. (3) We place a drop of red ink into a beaker of water and observe spreading of the color by *diffusion* of the ink particles until the water is uniformly pink, but we never see the ink spontaneously reappear as a small red drop in a volume of otherwise colorless water. (4) We place 10 g of sucrose (ordinary table sugar) in a beaker and add 100 mL of water at 80°C. The sucrose *dissolves* to form a uniform solution. We never observe the spontaneous reappearance of a mound of sucrose at the bottom of a beaker of water.[1] (5) We open a container of acetone on the laboratory bench. We detect the aroma of acetone

---

[1] By intentional evaporation of the water, we can recover the sucrose. This constitutes a second process, beyond the spontaneous process of dissolution that is our immediate concern.

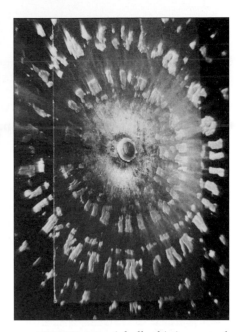

**FIGURE 8.1** A bullet hitting a steel plate at a speed of 1600 feet per second melts as its kinetic energy is converted to heat and sprays in all directions. These three photographs make sense only in the order shown; the reverse process is unmistakably unlikely. *(Copyright Dr. Harold Edgerton/Palm Press, Inc.)*

because some of the molecules have *evaporated* from the liquid and then diffused through the atmosphere to our position. (6) We mix solid red phosphorus with liquid bromine and observe a violent, exothermic chemical *reaction* that produces gaseous $PBr_3$ (see Example 7.5 and Fig. 7.13). We never observe the spontaneous reappearance of solid P at the bottom of a pool of liquid bromine in the reaction vessel.

The direction of spontaneity of each of these six processes is readily apparent from observation of the initial and final states, regardless of their paths. This suggests the existence of a new *state function* that indicates the directionality of spontaneous processes. That state function will turn out to be **entropy,** and it will be defined to increase in the direction of spontaneity.

These six spontaneous processes all share certain common features. Initially, two systems or objects in different thermodynamic states are brought into close proximity. They are placed in contact and allowed to interact: barriers (constraints) previously in place between them are removed. A spontaneous process may occur in which the systems exchange energy and matter and in which the volume of both systems may change. Since the two systems together constitute a *thermodynamic universe*, the total amount of energy, of volume, and of matter shared between them is fixed; during the process, these quantities are redistributed between the two systems.

What determines whether a process under consideration will be spontaneous? Where does a spontaneous process end? How are energy, volume, and matter partitioned between the two systems at equilibrium? What is the nature of the final equilibrium state? These questions cannot be answered by the first law. Their answers require the second law and properties of the entropy. Several developments are necessary before questions of this type can be examined. Entropy will be defined in terms of molecular motions in Section 8.2, and in terms of macroscopic process variables in Section 8.3. Methods for calculating entropy changes and for predicting whether a process is spontaneous will be presented in Section 8.5.

8.2

# ENTROPY AND SPONTANEITY: A MOLECULAR STATISTICAL INTERPRETATION

What is entropy and why should it be related to the spontaneity of processes in nature? These are deep questions that we can only begin to answer here. To do so, we must step outside the confines of classical thermodynamics, which is concerned only with macroscopic properties, and examine the microscopic molecular basis for the second law. Such an approach, called **statistical thermodynamics,** shows that spontaneous change in nature can be understood by using probability theory to predict and explain the behavior of the many atoms and molecules that make up a macroscopic sample of matter. Statistical thermodynamics also provides theoretical procedures for calculating the entropy of a system from molecular properties.

## Spontaneity

Consider a particularly simple spontaneous process: the free *adiabatic* expansion of 1 mol of an ideal gas into a vacuum (Fig. 8.2). The gas is initially held in the left bulb in volume $V/2$, while the right bulb is evacuated. After the stopcock is opened, the gas expands to fill the entire volume $V$.

Now examine the same free expansion from a microscopic point of view. Imagine that the path of one particular tagged molecule can be followed during the expansion and for some period of time after final equilibrium has been established, perhaps through a series of time–lapse snapshots showing the locations of all molecules in the gas. From its starting position on the left side at the beginning of the experiment, the tagged molecule will cross to the right, then back to the left, and so forth. If enough time elapses, the molecule will eventually spend equal amounts of time on the two sides, because there is no physical reason for it to prefer one side to the other. Recall from Section 4.5 that the molecules in an ideal gas do not collide with each other, and their kinetic energy does not change in collisions with the walls. The energy of the molecules remains the same at all locations, and the two sides of the container are identical. Put in other terms, the *probability* that the molecule is on the left side at any given instant will be $\frac{1}{2}$, the same as the probability that it is on the right side. This concept of probability provides the key to understanding spontaneous change because it enables comparing the likelihood of finding all the molecules on the left side—after the constraint has been removed—with the likelihood of finding the molecules uniformly distributed over the combined volume of both sides.

Exactly how improbable is such a spontaneous compression of 1 mol of gas from the combined volume back to the left side? The probability that one particular molecule is on the left side at a given time is $\frac{1}{2}$. A second specific molecule may be on either the left or the right, and so the probability that both are on the left is $\frac{1}{2} \times \frac{1}{2} = \frac{1}{4}$. In a gas containing a total of four molecules, all four molecules will be on the left only $\frac{1}{2} \times \frac{1}{2} \times \frac{1}{2} \times \frac{1}{2} = \frac{1}{16}$ of the time (Fig. 8.3). Continuing this argument for all $N_0 = 6.0 \times 10^{23}$ molecules leads to the probability that all $N_0$ molecules are on the left:

$$\frac{1}{2} \times \frac{1}{2} \times \cdots \times \frac{1}{2} = \left(\frac{1}{2}\right)^{6.0 \times 10^{23}}$$

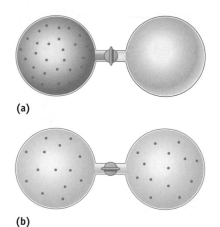

**(a)**

**(b)**

**FIGURE 8.2** The free expansion of a gas into a vacuum. (a) The stopcock is closed with all gas in the left bulb. (b) With the stopcock open, half the gas is in each bulb.

This can be rewritten in scientific notation as 1 divided by 10 raised to the power $a$ (or, equivalently, as $10^{-a}$):

$$\frac{1}{10^a} = \left(\frac{1}{2}\right)^{6.0 \times 10^{23}} = \frac{1}{2^{6.0 \times 10^{23}}}$$

A calculator can handle numbers this large (or this small) only if logarithms are used. Taking the base-10 logarithms of both denominators gives

$$a = \log 2^{(6.0 \times 10^{23})} = 6.0 \times 10^{23} \log 2$$
$$= (6.0 \times 10^{23})(0.30)$$
$$= 1.8 \times 10^{23}$$

The probability that all the molecules will be on the left is 1 in $10^{1.8 \times 10^{23}}$.

This is a vanishingly small probability, because $10^{1.8 \times 10^{23}}$ is an unimaginably large number. It is vastly larger than the number $1.8 \times 10^{23}$ (which is a large number already). To realize this, consider how such numbers would be written. The number $1.8 \times 10^{23}$ is fairly straightforward to write out. It contains 22 zeros:

$$180{,}000{,}000{,}000{,}000{,}000{,}000{,}000$$

The number $10^{1.8 \times 10^{23}}$ is 1 followed by $1.8 \times 10^{23}$ zeros. Written out, such a number would more than fill all the books in the world. Put in other terms, $1.8 \times 10^{23}$ corresponds to the number of molecules in about 5 cm³ of water, but $10^{1.8 \times 10^{23}}$ is far larger than the number of molecules in the entire universe!

The statistical molecular picture explains that when the constraint is removed the gas expands to fill the whole available volume because this more uniform configuration of the molecules is overwhelmingly more probable. In the same way, the gas is never observed to compress spontaneously into a smaller volume, because this nonuniform configuration of the molecules is overwhelmingly improbable in the absence of a constraint. Nothing in the laws of mechanics prevents a gas from compressing spontaneously, but the event is so improbable, it is not seen. *The directionality of spontaneous change is a consequence of the large numbers of molecules in the macroscopic systems treated by thermodynamics.*

Spontaneity in nature results from the random, statistical behavior of large numbers of molecules. In systems containing fewer molecules the situation can be quite different, because the uniform configuration of molecules, while still the most probable, is no longer so overwhelmingly the most probable configuration. For example, if there were only 6 molecules, instead of $6 \times 10^{23}$, it would not be surprising to find them all on the left side at a given time. In fact, the probability that 6 molecules are on the left side is $(\frac{1}{2})^6 = \frac{1}{64}$, so there is 1 chance in 64 that this will occur. Such small systems exhibit *statistical fluctuations* of the molecular configuration. Some of the fluctuations correspond to the initial nonuniform configuration maintained by the constraint, despite the absence of the constraint.

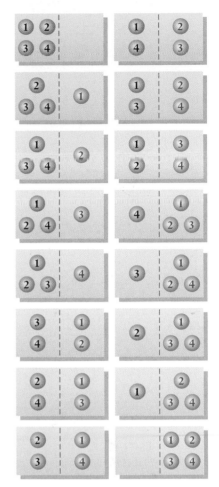

**FIGURE 8.3** The 16 possible states of a system of 4 molecules that may occupy either side of a container. In only one of these are all four molecules on the left side.

## EXAMPLE 8.1

Calculate the probability of a spontaneous compression of 1.00 mol of gas by 0.01%—that is, the probability that all the molecules will be found in a volume $V' = 0.9999V$ at a certain time.

### Solution

In this case, the probability that a given molecule is in $V'$ is not $\frac{1}{2}$ but 0.9999 (the probability that it is in the remainder of $V$ is 0.0001). The probability that all $N_0$ molecules are in $V'$ is

$$(0.9999)^{N_0} = 1/10^a$$

$$a = -N_0 \log (0.9999) = (6.0 \times 10^{23})(4.3 \times 10^{-5})$$

$$= 2.6 \times 10^{19}$$

The chance of such a compression occurring spontaneously is 1 in $10^{2.6 \times 10^{19}}$, which is still vanishingly small. Thus, a spontaneous compression of even a fraction of a percent will not be seen.

**Related Problems: 7, 8**

## Relation of Entropy to the Number of Microstates

Spontaneous processes occur when constraints are removed from a system; the molecules respond by moving to explore the suddenly increased range of motions now available to them. In the free expansion of a gas, the molecules are initially confined in one part of the container. After the constraint is removed (the stopcock is opened), they are free to stay where they were, but they also are free to move throughout the larger combined volume of the two regions. Qualitatively, the numerical value of the entropy of a macroscopic system held in a particular thermodynamic state is a measure of the *range of possible motions* (i.e., the range of possible positions and momenta) available to the molecules while the system is held in that particular thermodynamic state. Any change in the macroscopic properties that enables the molecules to move through larger regions of space or that increases molecular speeds increases the entropy of the system.

The precise connection between entropy and molecular motions is made through the number of microscopic, mechanical states, or **microstates,** available to molecules of the system. This number, denoted by $\Omega$, counts all the possible combinations of positions and momenta available to the $N$ molecules in the system when the system has energy $E$ and volume $V$. $\Omega$ is thus a function of $E$, $V$, and $N$ denoted as $\Omega(E, V, N)$. If the system is a monatomic ideal gas, the atoms do not interact. The position of each atom ranges freely throughout the entire volume $V$. The internal energy consists of the total kinetic energy of the atoms given as

$$E = \sum_{i=1}^{N} \epsilon_i = \sum_{i=1}^{N} \frac{[p_{xi}^2 + p_{yi}^2 + p_{zi}^2]}{2m}$$

so the momenta of the atoms range through all values that satisfy the condition

$$2mE = \sum_{i=1}^{N} [p_{xi}^2 + p_{yi}^2 + p_{zi}^2]$$

For a monatomic ideal gas, the value of $\Omega$ is given by

$$\Omega(E, V, N) = g \, V^N E^{(3N/2)}$$

where $g$ is a collection of constants. Now imagine the walls defining the system are manipulated in a laboratory process to change the values of $E$ and $V$. The range of positions and momenta available to the molecules will increase or decrease accordingly. Since $N$ is a very large number—of order $10^{23}$—the value of $\Omega$ will increase or decrease dramatically when the volume of the system is increased or decreased in an expansion or compression, respectively. It will increase or decrease dramatically when the internal energy of the system is increased or decreased by heating or cooling, respectively.

The equation connecting entropy $S$ and the number of available states $\Omega$ is

$$S = k_B \ln \Omega \qquad\qquad [8.1]$$

which was originally discovered by the Austrian physicist Ludwig Boltzmann in the late 19th century (Fig. 8.4). **Boltzmann's constant** $k_B$ is identified as $R/N_0$, the ratio of the universal gas constant $R$ to Avogadro's number $N_0$. Thus entropy has the physical dimensions J K$^{-1}$. It is impossible to overstate the importance of Boltzmann's relation, for it provides the link between the microscopic world of atoms and molecules and the macroscopic world of bulk matter and thermodynamics. Although this equation holds quite generally, it is difficult to use because calculation of $\Omega$ is a daunting theoretical task except for the simplest ideal systems. Other equations are used for practical applications in statistical thermodynamics. For present purposes, the equation provides qualitative insight into the physical meaning of entropy and qualitative interpretation of the magnitude and sign of entropy changes caused by specific thermodynamic processes. The following example illustrates these insights in a simple case in which only the volume of the system changes in the process.

**FIGURE 8.4** The fundamental relationship between entropy ($S$) and the number of microstates ($W$) was derived by Ludwig Boltzmann in 1868. On his tombstone in Vienna is carved the equation he obtained, $S = k \log W$. We would write "ln" instead of "log" for the natural logarithm. *(David Oxtoby)*

## EXAMPLE 8.2

Consider the free expansion of 1 mol of gas from $V/2$ to $V$, illustrated in Figure 8.2. Use Boltzmann's relation to estimate the change in entropy for this process.

### Solution

Consider the entire apparatus consisting of the filled and the evacuated bulbs to be a thermodynamic universe, so any exchange of thermal energy occurs solely between them. Then examine the effects of doubling $V$ on $\Omega$. If the volume is doubled, the number of positions available to a given molecule is doubled also. Therefore, the number of states available to the molecule should be proportional to the volume $V$:

$$\text{number of states available per molecule} = cV$$

where $c$ is a proportionality constant. The state of a *two*-molecule system is given jointly by the states of the molecules in it, so the number of microscopic states available is just the product of the number of states for each molecule, $(cV) \times (cV) = (cV)^2$. For an $N$-molecule system,

$$\text{microscopic states available} = \Omega = (cV)^N$$

Inserting this expression into Boltzmann's relation gives the entropy change for the free expansion of 1 mol ($N_0$ atoms) of gas from a volume $V/2$ to $V$:

$$\Delta S \text{ (microscopic)} = N_0 k_B \ln (cV) - N_0 k_B \ln (cV/2)$$

$$= N_0 k_B \ln \left( \frac{cV}{cV/2} \right) = N_0 k_B \ln 2$$

Note that the constant $c$ has dropped out. The calculated change in entropy upon expansion of the gas is clearly positive, which is consistent with the increase in the number of available microstates upon expansion.

In the next section, the entropy change for this process from the macroscopic view will be calculated to be

$$\Delta S \text{ (thermodynamic)} = R \ln 2$$

illustrating in a particular case that Boltzmann's constant, $k_B$, is identified as $R/N_0$.

This example illustrates why Boltzmann's relation must involve the logarithmic function. The entropy, because it is an extensive variable, is proportional to $N$, but $\Omega$ depends on $N$ through the *power* to which $cV$ is raised. Doubling $N$ doubles $S$, but leads to $\Omega$ being squared. The only mathematical function that can connect two such quantities is the logarithm.

## Entropy and Disorder

Return again to the spontaneous expansion of a gas into a vacuum and again consider the entire apparatus to form a thermodynamic universe. The previous discussion has shown that the entropy of this isolated composite system increases with the number of microscopic states available to the molecules. Another common way to describe this phenomenon is to relate the entropy to disorder: an *ordered* system has low entropy because its molecules are constrained to occupy only certain positions in space. As constraints are removed and the molecules are free to occupy more locations, the *disorder* increases and the entropy rises. An isolated thermodynamic system is not observed to become ordered spontaneously (i.e, to have the molecules reappear in the same locations in which they were originally held by the constraint) because the total number of states accessible after the constraint has been removed, each equally probable, is overwhelmingly greater than the number of states corresponding to the ordered configuration.

The melting of a solid illustrates these points. In a solid, the atoms or molecules are constrained to stay near their equilibrium positions in the crystal lattice, whereas in the liquid they can move far away from these fixed (ordered) positions. More states are available to the molecules of the liquid, corresponding to an increase in entropy: upon melting of the substance, $\Delta S_{sys}$ is positive. In the same way, when a liquid evaporates, the disorder and the entropy increase because the number of microstates available to the molecules increases enormously. Although the molecules in a liquid remain at the bottom of their container, those in a gas move throughout the container. They are in a greater state of disorder, so the entropy change $\Delta S_{sys}$ on vaporization is also positive.

Although the concepts of order and disorder are useful and are related qualitatively to entropy, the more fundamental connection is to the number of microstates available. In some cases (such as in certain phase transformations between solid phases), it is hard to say which phase is more ordered. A calculation of the relative number of microstates available, however, correctly predicts the sign of the entropy change in a transformation from one such phase to the other.

---

8.3

---

# ENTROPY AND HEAT: EXPERIMENTAL BASIS OF THE SECOND LAW OF THERMODYNAMICS

The purpose of this section is to define the entropy function in terms of measurable macroscopic quantities, so that changes in entropy can be calculated for specific laboratory processes. The definition is provided by the second law, which is stated as an abstraction and generalization of engineering observations on the efficiency of heat engines. We present a nonmathematical qualitative summary of the arguments on efficiency and then state the second law. The definition will be used in Section 8.5 to calculate entropy changes in processes and to predict spontaneity of processes.

Section 8.4 presents a mathematical version of the same arguments set forth here. Either Section 8.3 or Section 8.4 provides adequate background for the calculations in Section 8.5, which are the heart of the second law.

## Background of the Second Law

The second law originated in practical concerns over the efficiency of steam engines at the dawn of the Industrial Age late in the 18th century and required about a century for its complete development as an engineering tool. The central issues in that development are summarized as follows. In each stroke of a steam engine, a quantity of hot steam at high pressure is injected into a piston–cylinder arrangement, where it immediately expands and pushes the piston outward against an external load, doing useful work by moving the load. The external mechanical arrangement to which the piston is attached includes a reciprocating mechanism that returns the piston to its original position at the end of the stroke, so that a new quantity of steam can be injected to start the next stroke. The engine operates as a cyclic process, returning to the same state at the beginning of each stroke. In each stroke, the process of expansion is highly irreversible, since the steam is cooled and exhausted from the cylinder at the end of the stroke. In essence, the engine extracts thermal energy from a hot reservoir, uses some of this energy to accomplish useful work, then discards the remainder into a cooler reservoir (the environment). The energy lost to the environment cannot be recovered by the engine. The internal combustion engine operates in a similar manner, with injection of hot steam replaced by *in situ* ignition of combustible fuel that burns in highly exothermic reactions. Its expansion process is similarly irreversible, since some of the energy expended by the hot gases is not recoverable.

In both engines, the efficiency (i.e., the ratio of work accomplished *by* the engine in a cycle to the heat invested to drive that cycle) can be improved by reducing the unrecoverable losses to the environment in each cycle. Seeking to maximize efficiency Sadi Carnot, an officer in Napoleon's French Army Corps of Engineers, modeled operation of the engine with an idealized cyclic process now known as the *Carnot cycle*. He concluded that unrecoverable losses of energy to the environment cannot be completely eliminated, no matter how carefully the engine is designed. Even if the engine is operated as a reversible process (in which case displacement of the external load is too slow to be of practical interest), its efficiency cannot exceed a fundamental limit known as the *thermodynamic efficiency*. Thus an engine with 100 percent efficiency cannot be constructed.

Carnot's conclusion has been restated in more general terms by Rudolf Clausius in the form

There is no device that can transfer heat from a colder to a warmer reservoir without net expenditure of work.

and by Lord Kelvin in the form

There is no device that can transform heat withdrawn from a reservoir completely into work with no other effect.

These statements are consistent with ordinary experience that (1) heat always flows spontaneously from a hotter body to a colder body, and (2) work is always required to refrigerate a body. With confidence based on experience Clausius, Kelvin, and later scientists and engineers have assumed these statements to be valid for *all* heat transfer processes and labeled them as equivalent formulations of the second law of thermodynamics.

## Definition of Entropy

How do we apply these general statements to laboratory processes such as the six examples described in Section 8.1, which at first glance bear no resemblance to heat engines? We use the same reasoning that led to these statements to define entropy and obtain an equation for the entropy change during a process. Highlights of the argument are summarized here, and a more detailed development is presented in the following section.

Carnot's analysis of efficiency for a heat engine operating *reversibly* showed that in each cycle $q/T$ at the high temperature reservoir and $q/T$ at the low temperature environment summed to zero:

$$\frac{q_h}{T_h} + \frac{q_l}{T_l} = 0$$

This result suggests that $q/T$ is a state function, since its value does not change in a cyclic process. Clausius extended this result to show that the quantity $\int (1/T)\, dq_{rev}$ is independent of path in *any* reversible process and is therefore a state function. Clausius then *defined* the entropy change $\Delta S = S_f - S_i$ of a system in a process starting in state i and ending in state f by the equation

$$\Delta S = \int_i^f \frac{dq_{rev}}{T} \qquad\qquad \textbf{[8.2]}$$

Entropy is therefore a state function and has physical dimensions $J\,K^{-1}$. Calculation of $\Delta S = S_f - S_i$ is accomplished by evaluating this integral along any reversible path between states i and f for which $dq_{rev}$ and $T$ are known. Section 8.5 gives detailed procedures for calculating $\Delta S$ for numerous types of systems and processes. The calculated values of $\Delta S$ are used to predict whether a particular contemplated process will be spontaneous.

The preceding discussion emphasized the fact that the second law, like all other laws of science, is a bold extrapolation of the results of a great deal of direct experimental observation under controlled conditions. The second law is not proven to be true by a single definitive experiment, it is not derived as a consequence of some more general theory, and it is not handed down by some higher authority. Rather, it is invoked as one of the "starting points" or postulates of thermodynamics, and conclusions drawn from its application to a great variety of irreversible processes (not necessarily involving heat engines) are compared with the results obtained in laboratory studies. To date, no disagreements have been found between predictions properly made from the second law and the results of properly designed experiments.

# A DEEPER LOOK...

## 8.4 CARNOT CYCLES, EFFICIENCY, AND ENTROPY

This section provides a mathematical development of the relation between entropy and heat already presented qualitatively in Section 8.3. No additional results are obtained, but considerably greater insight is provided.

### The Carnot Cycle

In a Carnot cycle (Fig. 8.5), a system traverses two isothermal and two adiabatic paths to return to its original state. Each path is carried out reversibly (that is, in thermal equilibrium with internal and external forces nearly balanced at every step). As the system proceeds from state A to state C through state B, the system performs work (Fig. 8.5a). Of course, work performed *on* a system in a reversible process was defined to be

$$w_{ABC} = -\int_{V_A}^{V_C} P\, dV$$

(Path ABC)

and it is clearly negative for this (expansion) process. If the system then returns from state C to A through state D,

$$w_{CDA} = -\int_{V_C}^{V_A} P\, dV$$

(Path CDA)

which is now a positive quantity for this (compression) process; work is performed upon the system (Fig. 8.5b). In the course of the cycle, the work performed by the system is the area under curve ABC, whereas that performed on the system is the (smaller) area under curve ADC. The net result of the whole cycle is that work is performed by the system, and the amount of this work is the difference between the two areas, which is the area enclosed by the cycle (Fig. 8.5c). As in any cyclic process, the overall energy change $\Delta E$ is zero, and so the net work equals the negative of the total heat added to the system.

We have not yet specified the material contained in the system. Assume initially that it is an ideal gas, for which the results from Section 7.6 apply directly, with $T_h$ defined to be the higher temperature in the cycle and $T_l$ the lower.

### Path AB: Isothermal Expansion (temperature $T_h$)

$$w_{AB} = -q_{AB} = -nRT_h \ln\left(\frac{V_B}{V_A}\right)$$

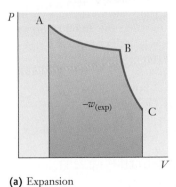 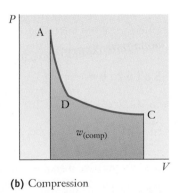 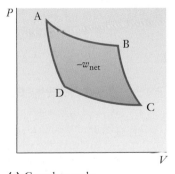

**(a)** Expansion  **(b)** Compression  **(c)** Complete cycle

**FIGURE 8.5** Stages of the Carnot cycle. The work done by the system in expansion and on the system by compression is shown in a and b respectively, by the shaded areas. The net work done per cycle is the area enclosed by the curve ABCDA.

### Path BC: Adiabatic Expansion

$$q_{BC} = 0$$

$$w_{BC} = nc_V(T_1 - T_h) = -nc_V(T_h - T_1)$$

### Path CD: Isothermal Compression (temperature $T_1$)

$$w_{CD} = -q_{CD} = -nRT_1 \ln\left(\frac{V_D}{V_C}\right)$$

$$= nRT_1 \ln\left(\frac{V_C}{V_D}\right)$$

### Path DA: Adiabatic Compression

$$q_{DA} = 0$$

$$w_{DA} = nc_V(T_h - T_1)$$

The net work done on the system is

$$w_{net} = w_{AB} + w_{BC} + w_{CD} + w_{DA}$$

$$= -nRT_h \ln\left(\frac{V_B}{V_A}\right) - nc_V(T_h - T_1)$$

$$+ nRT_1 \ln\left(\frac{V_C}{V_D}\right) + nc_V(T_h - T_1)$$

$$= -nRT_h \ln\left(\frac{V_B}{V_A}\right) + nRT_1 \ln\left(\frac{V_C}{V_D}\right)$$

This can be simplified by noting that $V_B$ and $V_C$ lie on one adiabatic path, and $V_A$ and $V_D$ lie on another. In Section 7.6, the relation for a reversible adiabatic process was found:

$$\frac{T_2}{T_1} = \left(\frac{V_1}{V_2}\right)^{\gamma-1}$$

Hence,

$$\frac{T_h}{T_1} = \left(\frac{V_C}{V_B}\right)^{\gamma-1} \qquad \text{for path BC}$$

and

$$\frac{T_h}{T_1} = \left(\frac{V_D}{V_A}\right)^{\gamma-1} \qquad \text{for path DA}$$

Equating these expressions gives

$$\left(\frac{V_C}{V_B}\right)^{\gamma-1} = \left(\frac{V_D}{V_A}\right)^{\gamma-1}$$

or

$$\frac{V_C}{V_B} = \frac{V_D}{V_A} \qquad \text{and} \qquad \frac{V_B}{V_A} = \frac{V_C}{V_D}$$

Hence, the net work done in one passage around the Carnot cycle is

$$w_{net} = -nR(T_h - T_1) \ln\frac{V_B}{V_A} \qquad \textbf{[8.3]}$$

### Heat Engines

The Carnot cycle is an idealized model for a heat engine. When a certain amount of heat, $q_{AB}$, is added to the system at the higher temperature $T_h$, a net amount of work $-w_{net}$ is obtained from the system. In addition, some heat $q_{CD}$ is discharged at the lower temperature, but this energy is "degraded" and is no longer available for use in the engine. The **efficiency** $\epsilon$ of such an engine is the ratio of the negative of the net work done on the system, $-w_{net}$, to the heat added along the high-temperature isothermal path:

$$\epsilon = \frac{-w_{net}}{q_{AB}}$$

It is the net work that is available for the performance of useful mechanical tasks such as turning electrical generators or dynamos, but it is the heat $q_{AB}$ absorbed at the higher temperature $T_h$ that must be "paid for" in terms of coal or oil consumed to supply it. The efficiency $\epsilon$ must be maximized to get out the most work possible for the lowest cost.

For the ideal gas Carnot cycle, the efficiency is easily calculated:

$$\epsilon = \frac{-w_{net}}{q_{AB}} = \frac{nR(T_h - T_1) \ln(V_B/V_A)}{nRT_h \ln(V_B/V_A)}$$

$$= \frac{T_h - T_1}{T_h} = 1 - \frac{T_1}{T_h} \qquad \textbf{[8.4]}$$

This result places a fundamental limit on the efficiency with which heat can be converted to mechanical work. Only if the high temperature $T_h$ were infinite or the low temperature $T_1$ were 0 would it be possible to have a heat engine that operates with 100% efficiency. To maximize efficiency, the greatest possible temperature difference should be used. Although this result has been derived only for the ideal gas, it will be shown later in this section that it applies to *any* reversible engine operating between two temperatures. For a *real* engine (which must operate irreversibly) the actual efficiency must be lower than this ideal Carnot efficiency, or **thermodynamic efficiency.**

### EXAMPLE 8.3

Suppose a heat engine absorbs 10.0 kJ of heat from a high-temperature source at $T_h = 450$ K and discards heat to a low-temperature reservoir at $T_1 = 350$ K. Calculate the thermodynamic efficiency $\epsilon$ of conversion of heat to work; the amount of work performed, $-w_{net}$; and the amount of heat discharged at $T_1$, $q_{CD}$.

### Solution

$$\epsilon = \frac{T_h - T_1}{T_h} = \frac{450 \text{ K} - 350 \text{ K}}{450 \text{ K}} = 0.222$$

so that the engine can be, at most, 22.2% efficient. Because $\epsilon = -w_{net}/q_{AB}$,

$$w_{net} = -\epsilon q_{AB} = -(0.222)(10.0 \text{ kJ}) = -2.22 \text{ kJ}$$

Because $\Delta E$ for the whole cycle is 0,

$$\Delta E = 0 = q_{AB} + q_{CD} + w_{net}$$

$$q_{CD} = -q_{AB} - w_{net} = -10.0 + 2.22 = -7.8 \text{ kJ}$$

Therefore, 7.8 kJ is discharged at 350 K. This heat must be removed from the vicinity of the engine by a cooling system, as otherwise it will cause $T_l$ to increase and reduce the efficiency of the engine.

**Related Problems: 11, 12**

## Efficiency of General Carnot Engines

In this subsection, all Carnot engines operating reversibly between two temperatures $T_h$ and $T_l$ are shown to have the same efficiency:

$$\epsilon = \frac{-w_{net}}{q_h} = \frac{T_h - T_l}{T_h} \qquad \text{[8.5]}$$

In other words, the efficiency that was calculated for an ideal gas applies equally to any other working fluid. To demonstrate this, the *contrary* is assumed, and that assumption is shown to lead to a contradiction with experience; the assumption is therefore deemed false.

Assume that there *are* two reversible machines operating between the same two temperatures $T_h$ and $T_l$, one of which has an efficiency $\epsilon_1$ that is greater than the efficiency $\epsilon_2$ of the other. The two machines are adjusted so that the total work output is the same for both. The more efficient machine is run as a heat engine so that it produces mechanical work $-w_1$. This work is used to operate the other machine in the opposite sense (as a heat pump), so that $w_2 = -w_1$. The net work input to the *combined* machines is then 0, because the work produced by the first machine is used to run the second.

Now examine what happens to heat in this situation. Because engine 1 is more efficient than engine 2, and because the work is the same for both, engine 1 must withdraw less heat from the hot reservoir at $T_h$ than is discharged into the same reservoir by engine 2; there is a net transfer of heat *into* the high-temperature reservoir. For the combined engines, $\Delta E_{tot} = w_{tot} = 0$; so $q_{tot}$ must also be 0 and a net transfer of heat *out of* the low-temperature reservoir must therefore occur. What has been devised is an apparatus that can transfer heat from a low-temperature reservoir to a high-temperature reservoir with no net expenditure of work.

*But this is impossible.* All of our experience shows that one cannot make a device that permits transfer of heat from a cold body to a hot body without doing work. In fact, exactly the opposite is seen: heat flows spontaneously from hotter to colder bodies. Our experience can be summarized in the statement

It is impossible to construct a device that will transfer heat from a cold reservoir to a hot reservoir in a continuous cycle with no net expenditure of work.

This is one form of the second law of thermodynamics, as stated by Rudolf Clausius.

## Further Discussion of Efficiency

An assumption was made that led to a conclusion that contradicted experience. Therefore, the original assumption must have been wrong, and there *cannot* be two reversible engines operating between the same two temperatures with different efficiencies. In other words, all Carnot engines must have the same efficiency, which has already been calculated to be

$$\epsilon = \frac{T_h - T_l}{T_h}$$

for the ideal gas. For any substance, whether ideal or not, undergoing a Carnot cycle,

$$\epsilon = \frac{-w_{net}}{q_h} = \frac{q_h + q_l}{q_h} = \frac{T_h - T_l}{T_h}$$

The Carnot cycle forms the basis for a *thermodynamic* scale of temperature. Because $\epsilon = 1 - (T_l/T_h)$, the Carnot efficiencies determine temperature ratios and thereby establish a temperature scale. In practice, the difficulty of operating real engines close to the reversible limit makes this procedure rather impractical; instead, real gases at low pressures are used to define and determine temperatures.

The last two terms of the preceding equation can be rewritten as

$$1 + \frac{q_l}{q_h} = 1 - \frac{T_l}{T_h}$$

and simplified to

$$\frac{q_h}{T_h} + \frac{q_l}{T_l} = 0 \qquad \text{[8.6]}$$

This simple equation has profound importance, because it contains the essence of the second law of thermodynamics, namely that $q/T$ is a state function. To see this, consider a *general* reversible cyclic process, and draw a series of closely spaced adiabats, as shown in Figure 8.6. (An adiabat is a representation on the phase diagram of thermodynamic states connected by a reversible adiabatic process.) Now replace each segment along the given cycle (ABCDA) with a series of alternating isothermal and adiabatic segments. It is clear that by taking more and more closely spaced adiabats, a path can be obtained that is arbitrarily close to the desired curve.

Now follow the evolution of $\Sigma_i(q_i/T_i)$ along this curve. The key observation is that the contributions $q_i/T_i$ appear in pairs: for each $q_{h,i}/T_{h,i}$ from an isothermal segment along the ABC path, there is a $q_{l,i}/T_{l,i}$ along the CDA path. Any given pair forms two sides of a Carnot cycle, with the other two sides determined by adiabats, and so

$$\frac{q_{h,i}}{T_{h,i}} + \frac{q_{l,i}}{T_{l,i}} = 0$$

Summing over $i$ shows that

$$\sum_i \left( \frac{q_{h,i}}{T_{h,i}} + \frac{q_{l,i}}{T_{l,i}} \right) = 0$$

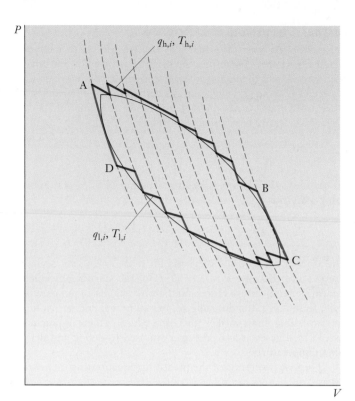

**FIGURE 8.6** A general cyclic process (ABCDA) can be approximated to arbitrary accuracy by the sum of a series of Carnot cycles.

Because this summation follows the original reversible path to an arbitrary accuracy (if the lengths of the segments are made short enough), it follows that

$$\int \frac{1}{T} \, dq_{rev} = 0$$

for *any* closed, reversible path. Although $q$ is not a state function, $q_{rev}/T$ *is* a state function because, like energy and enthalpy, its total change is 0 for any process that begins and ends in the same state. Based on this result, Clausius *defined* the entropy change $\Delta S = S_f - S_i$ of a system in a process starting in state i and ending in state f by the equation

$$\Delta S = \int_i^f \frac{dq_{rev}}{T} \qquad [8.7]$$

As we have stated, the principles of thermodynamics are based on observations of nature and are not subject to rigorous proof. We have accomplished something quite significant, however. From the premise (based on physical observation) that heat cannot be transferred from a low-temperature body to a high-temperature body without expenditure of work, we have derived the result that $\int (1/T) \, dq_{rev}$ is independent of path and is therefore a state function. Some might say, "If we have to make some assumption anyway, why don't we just assume that $\int (1/T) \, dq_{rev}$ is path-independent to start with?" This is certainly possible and is practiced in many presentations of thermodynamics. The approach followed here has been different; we have preferred to base our assumptions directly on physical observation rather than on abstract mathematical axioms.

---

**8.5**

## ENTROPY CHANGES AND SPONTANEITY

The present section outlines procedures for calculating entropy changes that occur in the system and in the surroundings during a variety of processes. The final subsection shows how entropy changes calculated for the thermodynamic universe (system plus surroundings) predict whether a particular contemplated process can occur spontaneously when attempted in the laboratory.

### $\Delta S_{sys}$ for Isothermal Processes

If the reversible process selected as the pathway for calculation is isothermal, the integral simplifies immediately:

$$\Delta S = \int_i^f \frac{dq_{rev}}{T} = \frac{1}{T} \int_i^f dq_{rev} = \frac{q_{rev}}{T} \qquad [8.8]$$

Here $q_{rev}$ is the *finite* amount of heat absorbed by the system during the entire reversible isothermal process.

*Compression/Expansion of an Ideal Gas*    Consider an ideal gas enclosed in a piston–cylinder arrangement that is maintained at constant temperature in a heat bath. The gas can be compressed (or expanded) reversibly by changing the position of the piston to accomplish a specified change in volume. In Section 7.6, the heat transferred between system and bath when the gas is expanded (or compressed) isothermally and reversibly from volume $V_1$ to $V_2$ was shown to be

$$q_{rev} = nRT \ln \left( \frac{V_2}{V_1} \right)$$

The resulting change in entropy is therefore

$$\Delta S = nR \ln \left( \frac{V_2}{V_1} \right) \qquad (\text{constant } T) \qquad \textbf{[0.9]}$$

The entropy of a gas increases during an isothermal expansion and decreases during a compression. Boltzmann's relation provides the molecular interpretation. $\Omega$, the number of microstates available to the system, increases as the volume of the system increases and decreases as volume decreases; the entropy of the system increases or decreases accordingly.

*Phase Transitions*    Another type of constant-temperature process is a phase transition such as the melting of a solid at constant pressure. This occurs reversibly at the freezing temperature $T_f$ because an infinitesimal change in external conditions (e.g., a lowering of the temperature) serves to reverse the process. The reversible heat when 1 mol of substance melts is $q_{rev} = \Delta H_{fus}$, and

$$\Delta S_{fus} = \frac{q_{rev}}{T_f} = \frac{\Delta H_{fus}}{T_f} \qquad \textbf{[8.10]}$$

The entropy increases when a solid melts or a liquid vaporizes, and it decreases when the phase transition occurs in the opposite direction. Again Boltzmann's relation provides the molecular interpretation. When a solid melts or a liquid vaporizes, the system becomes more disordered (or more precisely, the number of accessible microstates $\Omega$ increases), and the entropy increases.

---

### EXAMPLE 8.4

Calculate the entropy change when 3.00 mol of benzene vaporizes reversibly at its normal boiling point of 80.1°C. The molar enthalpy of vaporization of benzene at this temperature is 30.8 kJ mol$^{-1}$.

### Solution

The entropy change when 1 mol of benzene is vaporized at 80.1°C (= 353.25 K) is

$$\Delta S_{vap} = \frac{\Delta H_{vap}}{T_b} = \frac{30{,}800 \text{ J mol}^{-1}}{353.25 \text{ K}} = +87.2 \text{ J K}^{-1} \text{ mol}^{-1}$$

When 3.00 mol is vaporized, the entropy change is three times as great:

$$\Delta S = (3.00 \text{ mol})(+87.2 \text{ J K}^{-1} \text{ mol}^{-1}) = +262 \text{ J K}^{-1}$$

**Related Problems: 13, 14**

Remarkably, most liquids have about the same molar entropy of vaporization. **Trouton's rule** states the magnitude of this entropy change:

$$\Delta S_{\text{vap}} = 88 \pm 5 \text{ J K}^{-1} \text{ mol}^{-1} \qquad \text{[8.11]}$$

Note that the $\Delta S_{\text{vap}}$ of benzene, calculated in Example 8.4, is within this range. The constancy of $\Delta S_{\text{vap}}$ means that $\Delta H_{\text{vap}}$ and $T_{\text{b}}$, which vary widely from substance to substance, must do so in the same proportion. The rule allows enthalpies of vaporization to be estimated from boiling temperatures. Exceptions exist, however; water has a molar entropy of vaporization of 109 J K$^{-1}$ mol$^{-1}$. The value for water is unusually high because hydrogen bonding in liquid water enforces much greater order (lower entropy) than is present in other liquids.

## $\Delta S_{\text{sys}}$ for Processes with Changing Temperature

Now consider a reversible process in which the temperature changes. In this case, Equation 8.7 must be used

$$\Delta S = \int_{A}^{B} \frac{1}{T} \, dq_{\text{rev}}$$

In the integral, A and B represent respectively the initial and final equilibrium states for the process. The calculation must be carried out along a reversible path connecting A and B. For a reversible *adiabatic* process, $q = 0$ and therefore $\Delta S = 0$. Such a process is also called **isentropic** (that is, the entropy is constant).

In a reversible *isochoric* process, the volume is held constant and the system is heated or cooled by contact with a reservoir that differs from it in temperature by an infinitesimal amount $dT$. The heat transferred in this case is

$$dq_{\text{rev}} = nc_{\text{V}} \, dT$$

and the entropy change of the system as it is heated from $T_1$ to $T_2$ is

$$\Delta S = \int_{T_1}^{T_2} \frac{1}{T} \, dq_{\text{rev}} = \int_{T_1}^{T_2} \frac{nc_{\text{V}}}{T} \, dT$$

If $c_{\text{V}}$ is independent of $T$ over the temperature range of interest, it can be removed from the integral, giving the result

$$\Delta S = nc_{\text{V}} \int_{T_1}^{T_2} \frac{1}{T} \, dT = nc_{\text{V}} \ln\left(\frac{T_2}{T_1}\right) \qquad \text{(constant } V\text{)} \qquad \text{[8.12]}$$

The analogous result for the entropy change of the system in a reversible *isobaric* process (constant *pressure*) is

$$\Delta S = \int_{T_1}^{T_2} \frac{nc_{\text{P}}}{T} \, dT = nc_{\text{P}} \ln\left(\frac{T_2}{T_1}\right) \qquad \text{(constant } P\text{)} \qquad \text{[8.13]}$$

Entropy always increases with increasing temperature. From the kinetic theory of ideal gases in Chapter 4, it is clear that increasing the temperature of the gas increases the magnitude of the average kinetic energy per molecule and therefore the range of momenta available to molecules. This in turn increases $\Omega$ for the gas, and by Boltzmann's relation, the entropy of the gas.

Now consider an experiment in which identical samples of a gas are taken through identical temperature increases, one at constant $V$ and the other at constant $P$, and compare the entropy changes in the two processes. From the previous discussion it follows that $\Delta S_P > \Delta S_V$ since $c_P > c_V$. The molecular interpretation is based on the discussion of $c_P$ and $c_V$ in Section 7.3. The gas heated at constant $P$ experiences an increase in volume as well as in temperature; its molecules therefore gain access to a greater range of positions as well as a greater range of momenta. Consequently, the gas heated at constant $P$ experiences a *greater* increase in $\Omega$ than does the gas heated at constant $V$ and therefore a greater increase in $S$.

The following example illustrates the fact that the entropy is a state function, for which changes are independent of the path followed.

*[handwritten margin notes:]*
*$\Omega$ – disorder states*
*$\Omega_p > \Omega_v$*
*where P is constant pressure and V is constant volume.*
*When P increases, the molecules have access to a greater number of states.*

## EXAMPLE 8.5

(a) Calculate the entropy change for the process described in Example 7.9: 5.00 mol of argon expands reversibly at a constant temperature of 298 K from a pressure of 10.0 atm to 1.00 atm.
(b) Calculate the entropy change for the same initial and final states as in part (a) but along a different path. First the 5.00 mol of argon expands reversibly and *adiabatically* between the same two pressures. This is the path followed in Example 7.10; it causes the temperature to fall to 118.6 K. Then the gas is heated at constant pressure back to 298 K.

### Solution
(a) At constant temperature, the entropy change is

$$\Delta S = nR \ln \left(\frac{V_2}{V_1}\right) = nR \ln \left(\frac{P_1}{P_2}\right)$$

$$= (5.00 \text{ mol})(8.315 \text{ J K}^{-1} \text{ mol}^{-1}) \ln 10.0$$

$$= +95.7 \text{ J K}^{-1}$$

(b) For the adiabatic part of this path, the entropy change is 0. When the gas is then heated reversibly at constant pressure from 118.6 K to 298 K, the entropy change is

$$\Delta S = nc_P \ln \left(\frac{T_2}{T_1}\right)$$

$$= (5.00 \text{ mol})(\tfrac{5}{2} \times 8.315 \text{ J K}^{-1} \text{ mol}^{-1}) \ln \frac{298 \text{ K}}{118.6 \text{ K}}$$

$$= +95.7 \text{ J K}^{-1}$$

This is the same as the result from part (a), an illustration of the fact that the entropy is a state function. By contrast, the amounts of heat for the two paths are different: 28.5 kJ and 18.6 kJ.

**Related Problems: 17, 18**

## $\Delta S$ for Surroundings

Usually the surroundings can be treated as a large *heat bath* that interacts with the system by transferring heat at the fixed temperature of the bath. In such cases, the heat capacity of the surroundings (heat bath) must be sufficiently large that the heat

$\text{at constant } P$

$q_{surr} = -\Delta H_{sys}$

$\therefore \Delta S_{surr} = -\dfrac{\Delta H_{sys}}{T_{surr}}$

transferred during the process does not change the temperature of the bath. The heat gained by the surroundings during the process is the heat lost from the system. If the process occurs at constant $P$, then

$$q_{surr} = -\Delta H_{sys}$$

and the entropy change of the surroundings is

$$\Delta S_{surr} = \frac{-\Delta H_{sys}}{T_{surr}} \qquad \textbf{[8.14]}$$

If the process occurring in the system is exothermic, the surroundings gain heat and the entropy change of the surroundings is positive. Similarly, an endothermic process in the system is accompanied by a negative entropy change in the surroundings, which give up heat during the process.

If the surroundings lack sufficient heat capacity to maintain constant temperature during the process, then entropy changes for the surroundings must be calculated by the methods just demonstrated for the system, taking explicit account of the temperature change of the surroundings. Examples of both cases are included in the problems at the end of this chapter.

## $\Delta S_{tot}$ for System plus Surroundings

The tools are now in place for applying the second law to predict whether specific processes will be spontaneous. The procedure is illustrated first for spontaneous cooling of a hot body and then for irreversible expansion of an ideal gas.

***Spontaneous Cooling of a Hot Body***   Consider a spontaneous process in which a sample of hot metal is cooled by sudden immersion in a cold bath. Heat flows from the metal into the bath until they arrive at the same temperature. This spontaneous process is accompanied by an increase in the total entropy for the thermodynamic universe of the process, as illustrated by the following example.

### EXAMPLE 8.6

A well-insulated ice-water bath at 0.0°C contains 20 g ice. Throughout this experiment, the bath is maintained at the constant pressure of 1 atm. When a piece of nickel at 100°C is dropped into the bath, 10.0 g of the ice melts. Calculate the total entropy change for the thermodynamic universe of this process. (Specific heats at constant $P$: nickel, 0.46 J K$^{-1}$ g$^{-1}$; water, 4.18 J K$^{-1}$ g$^{-1}$; ice, 2.09 J K$^{-1}$ g$^{-1}$. Enthalpy of fusion of ice, 334 J K$^{-1}$ g$^{-1}$.)

### Solution

Consider the nickel to be the *system* and the ice-water bath to be the *surroundings* in this experiment. Heat flows from the Ni into the bath and melts some of the ice. Consequently the entropy of the Ni decreases and the entropy of the bath increases. The final equilibrium temperature of both system and bath is 0.0°C, as indicated by the presence of some ice in the bath at equilibrium.

Before calculating $\Delta S_{Ni}$ it is necessary to calculate the mass of the Ni from the calorimetry equation as follows:

heat lost by Ni = heat gained by ice bath = heat used in melting ice

$$-M (0.46 \text{ J K}^{-1} \text{ g}^{-1})(273.15 \text{ K} - 373.15 \text{ K}) = (10.0\text{g})(334 \text{ J g}^{-1})$$

$$M = 73 \text{ g}$$

Since the Ni is cooled at constant $P$, the entropy change for the Ni is calculated as

$$\Delta S_{\text{Ni}} = (73 \text{ g})(0.46 \text{ J g}^{-1} \text{ K}^{-1}) \ln (0.73) = -10 \text{ J K}^{-1}$$

Since the ice bath has remained at constant T throughout the experiment, it can be treated as a "large heat bath," and its entropy change is calculated as

$$\Delta S_{\text{bath}} = \frac{-\Delta H_{\text{sys}}}{T_{\text{bath}}} = \frac{-(-334 \text{ J g}^{-1})(10.0 \text{ g})}{273.15 \text{ K}} = 12 \text{ J K}^{-1}$$

Now $\Delta S_{\text{tot}} = \Delta S_{\text{Ni}} + \Delta S_{\text{bath}} = -10 + 12 = +2 \text{ J K}^{-1}$

Thus the process is spontaneous, driven by the fact that the entropy gain of the melting ice exceeds the entropy loss of the cooling metal.

**Related Problems: 20, 21, 22**

*Irreversible Expansion of an Ideal Gas*    Consider a gas confined within a piston–cylinder arrangement and held at constant temperature in a heat bath. Suppose the external pressure is abruptly reduced and held constant at the new lower value. The gas immediately expands against the piston until its internal pressure falls to match the new external pressure. The total entropy of system plus surroundings will increase during this expansion. In preparation for a quantitative example, a general comparison of irreversible and reversible processes connecting the same initial and final states provides insight into why the total entropy increases in a spontaneous process.

The work performed *by* a system $(-w)$ as it undergoes an irreversible isothermal expansion is always less than when the expansion is carried out reversibly. To see this, return to the definition of work done *on* a system:

$$w = -\int P_{\text{ext}} dV$$

During an expansion, $P_{\text{ext}}$ must be less than $P$, the pressure of the gas. For a reversible expansion $P_{\text{ext}}$ is only infinitesimally smaller (so the system is always very close to equilibrium), but for an irreversible expansion $P_{\text{ext}}$ is measurably smaller. Therefore, the area under a graph of $P_{\text{ext}}$ plotted against $V$ is less than that of a graph of $P$ against $V$ (Fig. 8.7), so that

$$-w_{\text{irrev}} = \int P_{\text{ext}} \, dV < \int P \, dV = -w_{\text{rev}}$$

and the work performed by the system, $-w_{\text{irrev}}$, is algebraically less than $-w_{\text{rev}}$. If the system is viewed as an "engine" for performing useful work on the surroundings, a reversible process is always more efficient than an irreversible one.

If the reversible and irreversible processes have the same initial and final states, then $\Delta E$ is the same for both.

$$\Delta E = w_{\text{irrev}} + q_{\text{irrev}} = w_{\text{rev}} + q_{\text{rev}}$$

But because

$$-w_{\text{irrev}} < -w_{\text{rev}}$$

we must have

$$w_{\text{irrev}} > w_{\text{rev}}$$

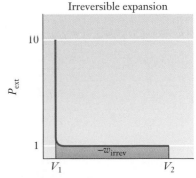

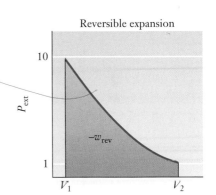

**FIGURE 8.7**  Work done by a system in reversible and irreversible expansions between the same initial and final states. The work performed is greater for the reversible process.

and

$$q_{\text{irrev}} < q_{\text{rev}}$$

The heat absorbed is a *maximum* when the process is carried out reversibly. An example illustrates these inequalities.

## EXAMPLE 8.7

Calculate the heat absorbed and the work done by a system of 5.00 mol of an ideal gas as it expands irreversibly at constant temperature $T = 298$ K from a pressure of 10.0 atm to 1.00 atm. The external pressure is held constant at 1.00 atm.

### Solution

The initial volume $V_1$ is

$$V_1 = \frac{nRT}{P_1} = \frac{(5.00 \text{ mol})(0.08206 \text{ L atm K}^{-1} \text{ mol}^{-1})(298 \text{ K})}{10.0 \text{ atm}} = 12.2 \text{ L}$$

The final volume is 10 times this, or

$$V_2 = 122 \text{ L}$$

For a constant external pressure,

$$w_{\text{irrev}} = -P_{\text{ext}}\Delta V = -(1.00 \text{ atm})(122 \text{ L} - 12.2 \text{ L}) = -110 \text{ L atm}$$

$$= -11.1 \text{ kJ}$$

At constant $T$, $\Delta E = 0$, however, so that

$$q_{\text{irrev}} = -w_{\text{irrev}} = 11.1 \text{ kJ}$$

In Example 7.9, a reversible expansion between the same two states was carried out, with the result that

$$w_{\text{rev}} = -28.5 \text{ kJ}$$

This demonstrates that

$$-w_{\text{irrev}} < -w_{\text{rev}}$$

and

$$q_{\text{irrev}} < q_{\text{rev}}$$

For reversible and irreversible processes connecting the same pair of initial and final states, it is always true that

$$q_{\text{rev}} > q_{\text{irrev}}$$

Dividing this expression by $T$, the temperature at which the heat is transferred, gives

$$\frac{q_{\text{rev}}}{T} > \frac{q_{\text{irrev}}}{T}$$

The left-hand side is the entropy change $\Delta S$,

$$\Delta S = \frac{q_{\text{rev}}}{T}$$

so

$$\Delta S > \frac{q_{irrev}}{T}$$

The last two equations can be combined as

$$\Delta S \geq \frac{q}{T}$$

where the equality applies only to a reversible process. This expression, called the **inequality of Clausius,** states that in any spontaneous process the heat absorbed by the system from a surroundings at the same temperature is always less than $T\Delta S$. In a reversible process, the heat absorbed is equal to $T\Delta S$.

Now apply Clausius's inequality to processes occurring within an *isolated* system. In this case there is no transfer of heat into or out of the system, and $q - 0$. For spontaneous processes within an isolated system, then, $\Delta S > 0$.

The thermodynamic universe of a process (i.e., a system plus its surroundings) is clearly an isolated system to which Clausius's inequality can be applied. It follows, therefore, that

1. In a *reversible* process, the total entropy of a system plus its surroundings is unchanged.
2. In an *irreversible* process, the total entropy of a system plus its surroundings must increase.
3. A process for which $\Delta S_{total} < 0$ is impossible.

These statements constitute the heart of the second law, for they provide its predictive power.

*[handwritten margin notes: reversible processes have ΔS=0. irreversible, ΔS > 0. ΔStotal < 0 is strictly impossible. Nothing can move to a state of greater disorder.]*

## EXAMPLE 8.8

Calculate $\Delta S_{tot} = \Delta S_{sys} + \Delta S_{surr}$ for the reversible and irreversible isothermal expansions of Examples 7.9 and 8.7.

### Solution

For the reversible expansion,

$$q_{rev} = 28.5 \text{ kJ}$$

$$\Delta S_{sys} = \frac{q_{rev}}{T} = \frac{28,500 \text{ J}}{298 \text{ K}} = +95.7 \text{ J K}^{-1}$$

The surroundings give up the same amount of heat at the same temperature. Hence,

$$\Delta S_{surr} = \frac{-28,500 \text{ J}}{298 \text{ K}} = -95.7 \text{ J K}^{-1}$$

$\Delta S_{tot} = 95.7 - 95.7 = 0$ for the reversible process.

For the irreversible expansion, it is still true that

$$\Delta S_{sys} = +95.7 \text{ J K}^{-1}$$

because $S$ is a function of state and the initial and final states are the same as for the reversible expansion. From Example 8.7, however, only 11.1 kJ of heat is given up by the surroundings in this case.[2] Hence,

$$\Delta S_{surr} = \frac{-11,100 \text{ J}}{298 \text{ K}} = -37.2 \text{ J K}^{-1}$$

and $\Delta S_{tot} = 95.7 - 37.2 = 58.5 \text{ J K}^{-1} > 0$ for the irreversible process.

---

## 8.6

# THE THIRD LAW OF THERMODYNAMICS

In thermodynamic processes, only *changes* in entropy, $\Delta S$, are measured, just as only changes in energy, $\Delta E$, or enthalpy, $\Delta H$, arise. It is nevertheless useful to define *absolute* entropies relative to some reference state. An important experimental observation that simplifies the choice of reference state is the following:

> In any thermodynamic process involving only pure phases in their equilibrium states, the entropy change $\Delta S$ approaches zero as $T$ approaches 0 K.

This observation is the **Nernst heat theorem,** named after its discoverer, Walther Nernst. It immediately suggests a choice of reference state: the entropy of any pure element in its equilibrium state is defined to approach zero as $T$ approaches 0 K. From the Nernst result, the entropy change for any chemical reaction, including one in which elements react to give a pure compound, approaches zero at 0 K. This has as a consequence the **third law of thermodynamics:**

> The entropy of any pure substance (element or compound) in its equilibrium state approaches zero at the absolute zero of temperature.

Absolute zero can never actually be reached, so a small extrapolation is needed to make use of this result.

The third law, like the two laws that precede it, is a macroscopic law based on experimental measurements. It is, however, consistent with the microscopic interpretation of the entropy presented in Section 8.2. In particular, from quantum mechanics and statistical thermodynamics, as the temperature approaches absolute zero the number of microstates available to a substance at equilibrium falls rapidly toward 1, and so the absolute entropy defined as $k_B \ln \Omega$ should approach 0. We have stated that the entropy of a substance *in its equilibrium state* approaches zero at 0 K. In practice, equilibrium may be very difficult to achieve at low temperatures as particle motion slows. In solid CO, for example, molecules remain randomly oriented (CO or OC) as the crystal is cooled, even though the equilibrium state at low temperatures would correspond to a definite orientation of each molecule. Because

---

[2] How can the heat from Example 8.7, which is irreversible from the perspective of the system, be reversible from the perspective of the surroundings? This can be accomplished by enclosing the gas in a material (such as a metal) that can efficiently transfer heat to and from the surroundings and thus remain close to equilibrium, at the same time that the gas itself is far from equilibrium due to the gas currents that occur during the irreversible expansion.

a molecule is so slow to reorient at low temperatures, such a crystal may not reach its equilibrium state in a measurable period of time. A nonzero entropy measured at low temperatures indicates that the system is not in equilibrium.

## Standard-State Entropies

Because the entropy of any substance in its equilibrium state is zero at absolute zero, its entropy at any other temperature $T$ is given by the entropy change as it is heated from 0 K to $T$. If heat is added at constant pressure,

$$\Delta S = n \int_{T_1}^{T_2} \frac{c_P}{T} \, dT$$

and so $S_T$, the absolute entropy of 1 mol of substance at temperature $T$, is

$$S_T = \int_0^T \frac{c_P}{T} \, dT$$

It is necessary merely to measure $c_P$ as a function of temperature and determine the area under a plot of $c_P/T$ versus $T$ from 0 K to any desired temperature. If a substance melts, boils, or undergoes some other phase change before reaching the temperature $T$, the entropy change for that process must be added to $\int (c_P/T) \, dT$.

In particular, the absolute molar entropy $S°$ at 298.15 K and 1 atm pressure (Fig. 8.8) is

$$S° = \int_0^{298.15} \frac{c_P}{T} \, dT + \Delta S \text{ (phase changes between 0 and 298.15 K)} \qquad \textbf{[8.15]}$$

Standard molar entropies $S°$ are tabulated for a number of elements and compounds in Appendix D. If $c_P$ is measured in J K$^{-1}$ mol$^{-1}$, then the entropy $S°$ will have the same units. For dissolved ions, the arbitrary convention $S°(H^+(aq)) = 0$ is applied (just as for the standard enthalpy of formation of $H^+$ discussed in Section 7.3). For this reason, some $S°$ values are negative for aqueous ions—an impossibility for substances.

Tabulated standard molar entropies are used to calculate entropy changes in chemical reactions at 25°C and 1 atm, just as standard enthalpies of formation are combined to obtain enthalpies of reaction according to Hess's law (Section 7.3).

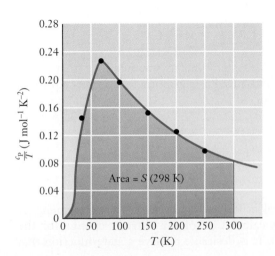

**FIGURE 8.8** A graph of $c_P/T$ versus $T$ for platinum. The black dots represent experimental measurements. The area up to any temperature (here, 298 K) is the molar entropy at that temperature.

## EXAMPLE 8.9

Using the table of standard molar entropies in Appendix D, calculate $\Delta S°$ for the chemical reaction

$$N_2(g) + 2\ O_2(g) \longrightarrow 2\ NO_2(g)$$

with reactants and products at a temperature of 25°C and a pressure of 1 atm.

### Solution
From the table,

$$S°(N_2(g)) = 191.50\ \text{J K}^{-1}\ \text{mol}^{-1}$$

$$S°(O_2(g)) = 205.03\ \text{J K}^{-1}\ \text{mol}^{-1}$$

$$S°(NO_2(g)) = 239.95\ \text{J K}^{-1}\ \text{mol}^{-1}$$

The entropy change for the reaction is the sum of the entropies of the products, minus the sum of entropies of the reactants, each multiplied by its coefficient in the balanced chemical equation:

$$\Delta S° = 2S°(NO_2(g)) - S°(N_2(g)) - 2S°(O_2(g))$$

$$= (2\ \text{mol})(239.95\ \text{J K}^{-1}\ \text{mol}^{-1}) - (1\ \text{mol})(191.50\ \text{J K}^{-1}\ \text{mol}^{-1})$$
$$- (2\ \text{mol})(205.03\ \text{J K}^{-1}\ \text{mol}^{-1})$$

$$= -121.66\ \text{J K}^{-1}$$

The factors of 2 multiply $S°$ for $NO_2$ and $O_2$ because 2 mol of each appears in the chemical equation. Note that standard molar entropies, unlike standard molar enthalpies of formation $\Delta H_f°$, are not 0 for elements at 25°C. The negative $\Delta S°$ results because this is the entropy change of the system only. The surroundings must undergo a positive entropy change in such a way that $\Delta S_{tot} \geq 0$.

**Related Problems: 23, 24, 25, 26**

---

## 8.7

# THE GIBBS FREE ENERGY

In Section 8.5, the change in entropy of a system plus its surroundings (that is, the total change of entropy $\Delta S_{tot}$) was shown to provide a criterion for whether a process is spontaneous, reversible, or impossible:

| | |
|---|---|
| $\Delta S_{tot} > 0$ | spontaneous |
| $\Delta S_{tot} = 0$ | reversible |
| $\Delta S_{tot} < 0$ | impossible |

Although $\Delta S_{tot}$ is a completely general criterion for assessing the spontaneity or impossibility of a process, it requires calculation of the entropy change for the surroundings as well as for the system. It is desirable to have a state function that predicts the feasibility of a process in the system without explicit calculations for the surroundings.

For the special case of processes at constant temperature and pressure (the most important in chemistry), such a state function, called the *Gibbs Free Energy* and denoted by *G*, exists. Development of *G* is facilitated by examination of representative laboratory processes.

## Nature of Spontaneous Processes at Fixed *T* and *P*

Consider a system enclosed in a piston–cylinder arrangement, which constrains pressure at the value *P*, and immersed in a heat bath, which constrains temperature at the value *T*. Experience shows that spontaneous processes under these conditions involve spontaneous flow of molecules across a boundary completely internal to the system, separating different regions (called *phases*) of the system (Fig. 8.9).

Here spontaneous processes are visualized as occurring by bringing the two phases A and B in Figure 8.9 into contact, both already prepared at *T* and *P*, but separated by an impermeable membrane (a constraint) that prevents exchange of matter between the phases. Removing the constraint allows spontaneous flow of molecules across the interface between phases. The system does not exchange matter with the surroundings, and the distribution of energy and volume between system and surroundings is not accounted for explicitly. The surroundings serve only to maintain *T* and *P* constant throughout the experiment. Consequently, as shown in the next subsection spontaneity of the process is determined by the change in Gibbs free energy of the *system only*, while *T* and *P* remain constant.

These processes lead to changes in the structure or composition of the phases (e.g., solutes are redistributed among immiscible solvents; phase transitions occur between the solid, liquid, and gaseous states; reactants become products). Heat may be gained or lost by the system from the large heat reservoir, while *T* is held constant. For example, the latent heat of fusion released during freezing of a liquid in the system is absorbed by the bath; endothermic reactions in the system will take heat from the bath, and exothermic reactions will give heat to the bath. *P–V* work may be done on or by the system at constant *P*, depending on whether its density increases or decreases during the transfer of matter.

Whether molecules flow spontaneously from phase A to phase B or *vice versa* is determined by the associated change in Gibbs free energy, as shown in the next subsection.

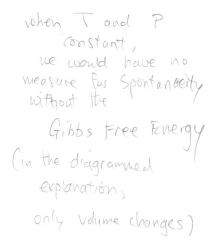

*[handwritten note: when T and P constant, we would have no measure for spontaneity without the Gibbs Free Energy. (in the diagrammed explanation, only volume changes)]*

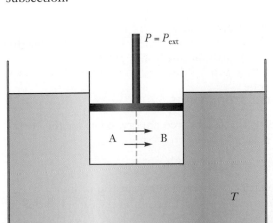

**FIGURE 8.9** After the constraint between phase A and phase B is removed, matter can flow spontaneously between phases inside a system held at constant *T* and *P* by its surroundings.

## Development of the Gibbs Free Energy

During any process carried out at constant $T$ and $P$ as described previously, note first that at constant pressure, $\Delta H_{sys} = q_P$, and so the heat transferred to the surroundings is $-q_P = -\Delta H_{sys}$. If the surroundings remain at constant temperature during the process, the transfer of heat must have the same effect on them as would a reversible transfer of the same amount of heat. Their entropy change is then

*(margin note: If surroundings at constant T, otherwise null and void.)*

$$\Delta S_{surr} = \frac{-\Delta H_{sys}}{T}$$

The total entropy change is

$$\Delta S_{tot} = \Delta S_{sys} + \Delta S_{surr} = \Delta S_{sys} - \frac{\Delta H_{sys}}{T}$$

$$= \frac{-(\Delta H_{sys} - T\,\Delta S_{sys})}{T}$$

Because the temperature $T$ is constant, we can rewrite this as

$$\Delta S_{tot} = \frac{-\Delta(H_{sys} - TS_{sys})}{T}$$

We define the **Gibbs free energy** $G$ as

$$G = H - TS \qquad\qquad \textbf{[8.16]}$$

so that

$$\Delta S_{tot} = \frac{-\Delta G_{sys}}{T}$$

*(margin note: You must be careful with ΔG.)*

Because the absolute temperature $T$ is always positive, $\Delta S_{tot}$ and $\Delta G_{sys}$ must have the opposite sign for processes occurring at constant $T$ and $P$. It follows that

*(margin note: ΔG < 0 spontaneous, however, ΔS > 0 spontaneous)*

| | |
|---|---|
| $\Delta G_{sys} < 0$ | spontaneous processes |
| $\Delta G_{sys} = 0$ | reversible processes |
| $\Delta G_{sys} > 0$ | nonspontaneous processes |

*(margin note: opposite nature due to formation of equations.)*

for processes carried out at constant temperature and pressure. If $\Delta G_{sys} > 0$ for a process, then $\Delta G_{sys} < 0$ for the *reverse* process, and that reverse process can occur spontaneously. Experimentally, one prepares the system initially at chosen values of $T$ and $P$ and then releases the appropriate constraint to allow a process. The resulting "flow" of matter goes in that direction required to reduce the value of $G$ for the system, that is, make $\Delta G < 0$, at the selected values of $T$ and $P$.

## The Gibbs Free Energy and Phase Transitions

As a simple example of the usefulness of the Gibbs free energy, consider the freezing of 1 mol of liquid water to form ice:

$$H_2O(\ell) \longrightarrow H_2O(s)$$

Examine first what happens when this process is carried out at the ordinary freezing point of water under atmospheric pressure, 273.15 K. The measured enthalpy change (the heat absorbed at constant pressure) is

$$\Delta H_{273} = q_P = -6007 \text{ J}$$

At $T_f = 273.15$ K, the freezing of water takes place reversibly (i.e., the system remains close to equilibrium as it freezes). Therefore, the entropy change is

$$\Delta S_{273} = \frac{q_{rev}}{T_f} = \frac{-6007 \text{ J}}{273.15 \text{ K}} = -21.99 \text{ J K}^{-1}$$

The Gibbs free energy change is

$$\Delta G_{273} = \Delta H_{273} - T \Delta S_{273} = -6007 \text{ J} - (273.15 \text{ K})(-21.99 \text{ J K}^{-1}) = 0$$

At the normal freezing point the Gibbs free energy change is 0, because the freezing of water under these conditions is an equilibrium process.

Now consider what happens as the water is cooled below 273.15 K to 263.15 K ($-10.00°$C). The Gibbs free energy change can now be calculated as water freezes at this lower temperature.

Assume that $\Delta H$ and $\Delta S$ for the freezing process do not depend on temperature. If that is the case,

$$\Delta G_{263} = -6007 \text{ J} - (263.15 \text{ K})(-21.99 \text{ J K}^{-1}) = -220 \text{ J}$$

An exact calculation that takes into account the fact that $\Delta H$ and $\Delta S$ *do* depend slightly on temperature leads to

$$\Delta G_{263} = -213 \text{ J}$$

for the process. Because $\Delta G < 0$, the undercooled water will freeze spontaneously at 263.15 K. At a temperature *greater* than $T_f$, $\Delta G$ is greater than 0, showing that freezing of the liquid would be impossible. This is in accord with our experience of nature: water does not freeze at atmospheric pressure if the temperature is held above 273.15 K; rather, the reverse process occurs, and ice melts.

Writing the Gibbs free energy change as $\Delta G = \Delta H - T \Delta S$ shows that a negative value of $\Delta G$ (and therefore a spontaneous process) is favored by a negative value of $\Delta H$ and a positive value of $\Delta S$. In the freezing of a liquid, $\Delta H$ is negative. However, $\Delta S$ for freezing is also negative, rather than positive (in the language of Section 8.2, the solid is more ordered than the liquid and so has a lower entropy). Whether or not a liquid freezes depends on the competition between two factors: an enthalpy effect that favors freezing and an entropy effect that disfavors it (Fig. 8.10). At low temperatures (below $T_f$) the former dominates and the liquid freezes

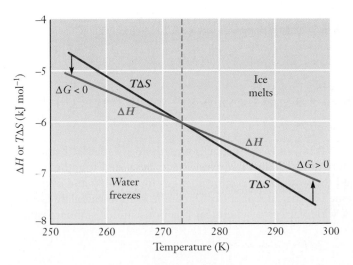

**FIGURE 8.10** Plots of $\Delta H$ and $T \Delta S$ versus temperature for the freezing of water. At 273.15 K the two curves cross, so that at this temperature $\Delta G = 0$ and ice and water coexist. Below this temperature, the freezing of water to ice is spontaneous; above it, the reverse process, the melting of ice to water, is spontaneous.

spontaneously, but at higher temperatures (above $T_f$) the latter dominates and freezing does not occur. At $T_f$ the Gibbs free energies of the two phases are equal ($\Delta G = 0$) and the phases coexist at equilibrium. Similar types of analysis can be carried out for other phase transitions, such as the condensation of a gas to a liquid.

## The Gibbs Free Energy and Chemical Reactions

The change in the Gibbs free energy provides a criterion for the spontaneity of any process occurring at constant temperature and pressure (Fig. 8.11). To predict whether a chemical reaction is spontaneous at given values of $T$ and $P$, it is necessary only to determine the sign of $\Delta G$ at these same conditions. Calculation of $\Delta G$ for a wide range of chemical reactions is accomplished systematically be defining standard state free energies, just as has been done already for enthalpies.

***Standard-State Free Energies*** The change in the Gibbs free energy for a chemical reaction carried out at constant temperature is

$$\Delta G = \Delta H - T\,\Delta S \qquad \text{[8.17]}$$

where $\Delta H$ is the enthalpy change in the reaction (considered in Section 7.5) and $\Delta S$ is the entropy change of the system (Section 8.5). Although we cannot know the *absolute* value of the Gibbs free energy of a substance (just as we cannot know the absolute value of its energy $E$) it is convenient to define a **standard molar Gibbs free energy of formation** $\Delta G_f^\circ$ that is analogous to the standard molar enthalpy of formation $\Delta H_f^\circ$ introduced in Section 7.5. The $\Delta G_f^\circ$ of a compound is the change in Gibbs free energy for the reaction in which 1 mol of the compound in its standard state is formed from its elements in their standard states. For example, $\Delta G_f^\circ$ for $CO_2(g)$ is given by the Gibbs free energy change for the reaction

$$C(s,gr) + O_2(g) \longrightarrow CO_2(g) \qquad \Delta G_f^\circ = ?$$

The $\Delta G_f^\circ$ can be calculated from $\Delta H$ and $\Delta S$ for this reaction. Here $\Delta H$ is simply $\Delta H_f^\circ$ for $CO_2(g)$ because graphite and oxygen are elements in their standard states,

$$\Delta H = \Delta H_f^\circ(CO_2) = -393.51 \text{ kJ}$$

and $\Delta S$ can be obtained from the absolute entropies of the substances involved at 25°C and 1 atm pressure (both elements and compounds, because the absolute entropy $S^\circ$ of an element is not zero in its standard state).

$$\Delta S^\circ = S^\circ(CO_2) - S^\circ(C) - S^\circ(O_2) = 213.63 - 5.74 - 205.03 \text{ J K}^{-1}$$
$$= +2.86 \text{ J K}^{-1}$$

The $\Delta G_f^\circ$ for $CO_2$ is then

$$\Delta G_f^\circ = \Delta H_f^\circ - T\,\Delta S^\circ = -393.51 \text{ kJ} - (298.15 \text{ K})(2.86 \text{ J K}^{-1})(10^{-3} \text{ kJ J}^{-1})$$
$$= -394.36 \text{ kJ}$$

From our definition, $\Delta G_f^\circ = 0$ for an *element* that is already in its standard state.

Because $G$ is a state function, chemical equations can be added together, with their $\Delta G_f^\circ$ values combined the way they were for changes in enthalpy and entropy. A table of values of $\Delta G_f^\circ$ (such as that in Appendix D) can be used to calculate Gibbs free energy changes for a wide variety of chemical reactions under standard-state conditions.

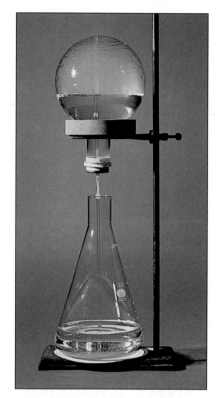

**FIGURE 8.11** The dissolution of hydrogen chloride in water

$$HCl(g) \longrightarrow HCl(aq)$$

is a spontaneous process, with $\Delta G^\circ = -35.9$ kJ. In this demonstration, the upper flask is filled with gaseous hydrogen chloride and a small amount of water is injected into it. As the hydrogen chloride dissolves spontaneously, its pressure drops. The resulting partial vacuum in the upper flask draws water up the tube from the lower flask, allowing more hydrogen chloride to dissolve. The change is so fast that a vigorous fountain of water plays into the upper flask. The free energy change of the process appears as work, raising the water. *(Leon Lewandowski)*

**EXAMPLE 8.10**

Calculate $\Delta G°$ for the following reaction, using tabulated values for $\Delta G_f°$ from Appendix D.

$$3 \text{ NO}(g) \longrightarrow \text{N}_2\text{O}(g) + \text{NO}_2(g)$$

**Solution**

$$\Delta G° = \Delta G_f°(\text{N}_2\text{O}) + \Delta G_f°(\text{NO}_2) - 3 \, \Delta G_f°(\text{NO})$$

$$= (1 \text{ mol})(104.18 \text{ kJ mol}^{-1}) + (1 \text{ mol})(51.29 \text{ kJ mol}^{-1})$$
$$- (3 \text{ mol})(86.55 \text{ kJ mol}^{-1})$$

$$= -104.18 \text{ kJ}$$

***Effects of Temperature on $\Delta G$***    Values of $\Delta G°$ calculated from the data in Appendix D are accurate only at $T = 298.15$ K. Values of $\Delta G°$ can be estimated for reactions at other temperatures and at $P = 1$ atm using the equation

$$\Delta G° = \Delta H° - T \, \Delta S°$$

and tables of standard entropies and standard enthalpies of formation. The estimates will be close to the true value if $\Delta H°$ and $\Delta S°$ are not strongly dependent on $T$, which is usually the case (Fig. 8.12).

The value of $\Delta G°$ can depend strongly on $T$, even when the values of $\Delta H°$ and $\Delta S°$ do not, because of the competition between $\Delta H°$ and $T \, \Delta S°$ in the previous equation. If $\Delta H°$ is negative and $\Delta S°$ is positive, then the reaction is spontaneous at all temperatures. If $\Delta H°$ is positive and $\Delta S°$ is negative, the reaction is never spontaneous. For the other possible combinations, there exists a special temperature $T^*$ defined by

$$T^* = \frac{\Delta H°}{\Delta S°}$$

at which $\Delta G°$ equals zero. If both $\Delta H°$ and $\Delta S°$ are positive, the reaction will be spontaneous at temperatures higher than $T^*$. If both are negative, the reaction will be spontaneous at temperatures below $T^*$. Thus, with knowledge of $\Delta H°$ and $\Delta S°$, the experimenter can manipulate conditions to make a reaction spontaneous (Fig. 8.13).

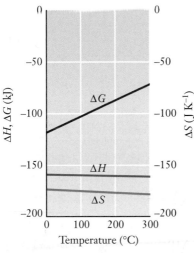

**FIGURE 8.12**    The entropy change of the reaction

$$3 \text{ NO}(g) \longrightarrow \text{N}_2\text{O}(g) + \text{NO}_2(g)$$

varies less than 5% between 0°C and 300°C; the enthalpy of reaction is even closer to constancy. The free energy change in the reaction shifts greatly over the temperature range, however, as the magnitude of $T \, \Delta S$ increases.

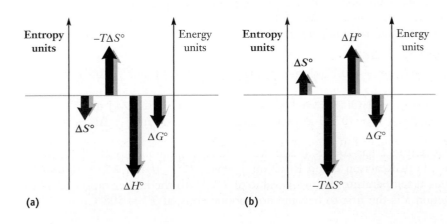

**FIGURE 8.13**    The competition between $\Delta H°$ and $\Delta S°$ determines the temperature range in which a reaction is spontaneous. (a) If both $\Delta H°$ and $\Delta S°$ are negative, the reaction is spontaneous at temperatures below $T^* = \Delta H°/\Delta S°$. (b) If both $\Delta H°$ and $\Delta S°$ are positive, the reaction is spontaneous at temperatures above $T^* = \Delta H°/\Delta S°$.

This chapter opened with the quest for criteria for predicting whether a chemical reaction could occur spontaneously under a given set of conditions. The second law provides the answer: any process can occur spontaneously under conditions where the total entropy of the system and its surroundings can increase during the process. For the particular case of constant temperature and pressure (the conditions most widely used in chemical reactions) the second law asserts that any process can occur spontaneously under conditions where the Gibbs free energy of the system can decrease during the process. Examination of the temperature dependence of the Gibbs free energy shows that knowledge of $\Delta H°$ and $\Delta S°$ enables identification of the temperature range in which a given reaction is spontaneous. From the point of view of chemistry, predicting this temperature range is the most important result from the second law. All other material in this chapter can be regarded as preliminary background for arriving at this one crucial result.

The Gibbs free energy is the thermodynamic state function most naturally suited for describing the progress of chemical reactions at constant $T$ and $P$. It will provide the basis for predicting the equilibrium composition of the reaction mixture in the next chapter.

A stockpile of sulfur near chemical plants in Los Angeles. *(Copyright David M. Doody/Uniphoto Pictures)*

## CUMULATIVE EXERCISE

### Sulfuric Acid

Sulfuric acid is produced in larger volume than any other chemical and has a tremendous number of applications ranging from fertilizer manufacture to metal treatment and chemical synthesis. Its production, properties, and uses are discussed in detail in Chapter 21.

The modern industrial production of sulfuric acid involves three steps, for which the balanced chemical equations are

**1.** $S(s) + O_2(g) \rightleftharpoons SO_2(g)$
**2.** $SO_2(g) + \frac{1}{2} O_2(g) \rightleftharpoons SO_3(g)$
**3.** $SO_3(g) + H_2O(\ell) \rightleftharpoons H_2SO_4(\ell)$

(a) Calculate $\Delta G°$ for each reaction at 25°C.
(b) Write a balanced equation for the overall reaction and calculate its value of $\Delta G°$ at 25°C.
(c) Part (a) shows all three reactions are spontaneous at 25°C. Nonetheless, reactions 1 and 2 occur too slowly at 25°C to be practical; they must be carried out at higher temperatures. Calculate $\Delta G°$ for 1, 2, and 3 at 700°C.
(d) Determine the highest temperature at which all three reactions are spontaneous.

**Answers**
(a) At 25°C: $\Delta G°_{(1)} = -300.19$ kJ mol$^{-1}$; $\Delta G°_{(2)} = -70.86$ kJ mol$^{-1}$; $\Delta G°_{(3)} = -81.64$ kJ mol$^{-1}$
(b) $S(s) + \frac{3}{2} O_2(g) + H_2O(\ell) \rightleftharpoons H_2SO_4(\ell)$; $\Delta G°_{(net)} = -452.89$ kJ mol$^{-1}$
(c) At 700°C: $\Delta G°_{(1)} = -307.81$ kJ mol$^{-1}$; $\Delta G°_{(2)} = -7.47$ kJ mol$^{-1}$; $\Delta G°_{(3)} = +32.64$ kJ mol$^{-1}$
(d) Since Reaction 1 has $\Delta H° < 0$ and $\Delta S° > 0$, it is spontaneous at all temperatures. Since both Reaction 2 and Reaction 3 have $\Delta H° < 0$ and $\Delta S° < 0$, each is spontaneous at temperatures below its value of $T^*$. With the higher ratio of $\Delta S°$ to $\Delta H°$, Reaction 3 is the first to become nonspontaneous, at $T^* = 508°C$.

## CONCEPTS & SKILLS

*After studying this chapter and working the problems that follow, you should be able to*

1. Identify the system and surroundings involved in a spontaneous process and identify the constraint that was removed to enable the process to occur (Section 8.1 and problems 1–2).
2. Provide a statistical interpretation of the change in entropy that occurs when a gas undergoes a volume change (Section 8.2, problems 3–10).
3. Calculate the entropy change for gaseous systems undergoing reversible isothermal, isochoric, and isobaric processes (Section 8.5, problems 13–18).
4. Calculate the entropy change for the system and the surroundings when irreversible processes are carried out (Section 8.5, problems 19–22).
5. Describe how the absolute entropy of a substance can be obtained (Section 8.6).
6. Calculate standard-state entropy changes for chemical reactions (Section 8.6, problems 23–30).
7. Define the Gibbs free energy function and state the criterion it provides for the spontaneity of a process (Section 8.7).
8. Calculate the change in Gibbs free energy for reversible and spontaneous phase transformations (Section 8.7, problems 31–34).
9. Calculate the change in Gibbs free energy for chemical reactions and identify temperature ranges in which a particular reaction is spontaneous (Section 8.7, problems 35–40).

---

## PROBLEMS

*Answers to problems whose numbers are boldface appear in Appendix G. Problems that are more challenging are indicated with asterisks.*

### The Nature of Spontaneous Processes

1. For each of the following processes, identify the system and the surroundings. Identify those processes that are spontaneous. For each spontaneous process identify the constraint that has been removed to enable the process to occur:
   (a) Ammonium nitrate dissolves in water.
   (b) Hydrogen and oxygen explode in a closed bomb.
   (c) A rubber band is rapidly extended by a hanging weight.
   (d) The gas in a chamber is slowly compressed by a weighted piston.
   (e) A glass shatters on the floor.
2. For each of the following processes, identify the system and the surroundings. Identify those processes that are spontaneous. For each spontaneous process identify the constraint that has been removed to enable the process to occur:
   (a) A solution of hydrochloric acid is titrated with a solution of sodium hydroxide.
   (b) Zinc pellets dissolve in aqueous hydrochloric acid.
   (c) A rubber band is slowly extended by a hanging weight.
   (d) The gas in a chamber is rapidly compressed by a weighted piston.

(e) A tray of water freezes in the freezing compartment of an electric refrigerator.

### Entropy and Spontaneity: A Molecular Statistical Interpretation

3. (a) How many "microstates" are there for the numbers that come up on a pair of dice?
   (b) What is the probability that a roll of a pair of dice will show two sixes?
4. (a) Suppose a volume is divided into three equal parts. How many microstates can be written for all possible ways of distributing four molecules among the three parts?
   (b) What is the probability that all four molecules are in the leftmost third of the volume at the same time?
5. When $H_2O(\ell)$ and $D_2O(\ell)$ are mixed, the following reaction occurs spontaneously:

$$H_2O(\ell) + D_2O(\ell) \longrightarrow 2\,HOD(\ell)$$

There is little difference between the enthalpy of an O—H bond and that of an O—D bond. What is the main driving force for this reaction?

6. The two gases $BF_3(g)$ and $BCl_3(g)$ are mixed in equal molar amounts. All B—F bonds have about the same bond enthalpy,

as do all B—Cl bonds. Explain why the mixture tends to react to form $BF_2Cl(g)$ and $BCl_2F(g)$.

7. Two large glass bulbs of identical volume are connected by means of a stopcock. One bulb initially contains 1.00 mol of $H_2$; the other, 1.00 mol of He. The stopcock is opened and the gases are allowed to mix and reach equilibrium. What is the probability that all of the $H_2$ in the first bulb will diffuse into the second bulb and all of the He gas in the second bulb will diffuse into the first bulb?

8. A mixture of 2.00 mol of nitrogen and 1.00 mol of oxygen is in thermal equilibrium in a 100-L container at 25°C. Calculate the probability that at a given time all the nitrogen will be found in the left half of the container and all the oxygen in the right half.

9. Predict the sign of the system's entropy change in each of the following processes.
   (a) Sodium chloride melts.
   (b) A building is demolished.
   (c) A volume of air is divided into three separate volumes of nitrogen, oxygen, and argon, each at the same pressure and temperature as the original air.

10. Predict the sign of the system's entropy change in each of the following processes.
   (a) A computer is constructed from iron, copper, carbon, silicon, gallium, and arsenic.
   (b) A container holding a compressed gas develops a leak and the gas enters the atmosphere.
   (c) Solid carbon dioxide (Dry Ice) sublimes to gaseous carbon dioxide.

## A Deeper Look ... Carnot Cycles, Efficiency, and Entropy

11. A thermodynamic engine operates cyclically and reversibly between two temperature reservoirs, absorbing heat from the high-temperature bath at 450 K and discharging heat to the low-temperature bath at 300 K.
   (a) What is the thermodynamic efficiency of the engine?
   (b) How much heat is discarded to the low-temperature bath if 1500 J of heat is absorbed from the high-temperature bath during each cycle?
   (c) How much work does the engine perform in one cycle of operation?

12. In each cycle of its operation, a thermal engine absorbs 1000 J of heat from a large heat reservoir at 400 K and discharges heat to another large heat sink at 300 K. Calculate:
   (a) The thermodynamic efficiency of the heat engine, operated reversibly.
   (b) The quantity of heat discharged to the low-temperature sink each cycle.
   (c) The maximum amount of work the engine can perform each cycle.

## Entropy Changes and Spontaneity

13. Tungsten melts at 3410°C and has an enthalpy change of fusion of 35.4 kJ mol$^{-1}$. Calculate the entropy of fusion of tungsten.

14. Tetraphenylgermane, $(C_6H_5)_4Ge$, has a melting point of 232.5°C, and its enthalpy increases by 106.7 J g$^{-1}$ during fusion. Calculate the molar enthalpy of fusion and molar entropy of fusion of tetraphenylgermane.

15. The normal boiling point of acetone is 56.2°C. Use Trouton's rule to estimate its molar enthalpy of vaporization.

16. The molar enthalpy of vaporization of liquid hydrogen chloride is 16.15 kJ mol$^{-1}$. Use Trouton's rule to estimate its normal boiling point.

17. If 4.00 mol of hydrogen ($c_P = 28.8$ J K$^{-1}$ mol$^{-1}$) is expanded reversibly and isothermally at 400 K from an initial volume of 12.0 L to a final volume of 30.0 L, calculate $\Delta E$, $q$, $w$, $\Delta H$, and $\Delta S$ for the gas.

18. Suppose 60.0 g of hydrogen bromide, HBr($g$), is heated reversibly from 300 K to 500 K at a constant volume of 50.0 L and then allowed to expand isothermally and reversibly until the original pressure is reached. Using $c_P(HBr(g)) =$ 29.1 J K$^{-1}$ mol$^{-1}$, calculate $\Delta E$, $q$, $w$, $\Delta H$, and $\Delta S$ for this process. Assume that HBr is an ideal gas under these conditions.

19. Exactly 1 mol of ice is heated reversibly at atmospheric pressure from −20 to 0°C, melted reversibly at 0°C, and then heated reversibly at atmospheric pressure to 20°C. $\Delta H_{fus} =$ 6007 J mol$^{-1}$, $c_P(ice) = 38$ J K$^{-1}$ mol$^{-1}$, and $c_P(water) =$ 75 J K$^{-1}$ mol$^{-1}$. Calculate $\Delta S$ for the system, the surroundings, and the universe for this process.

20. Suppose 1.00 mol of water at 25°C is flash-evaporated by allowing it to fall into an iron crucible maintained at 150°C. Calculate $\Delta S$ for the water, $\Delta S$ for the iron crucible, and $\Delta S_{tot}$, if $c_P(H_2O(\ell)) = 75.4$ J K$^{-1}$ mol$^{-1}$ and $c_P(H_2O(g)) =$ 36.0 J K$^{-1}$ mol$^{-1}$. Take $\Delta H_{vap} = 40.68$ kJ mol$^{-1}$ for water at its boiling point of 100°C.

21. In Example 7.3, a process was considered in which 72.4 g of iron initially at 100.0°C was added to 100.0 g of water initially at 10.0°C, and an equilibrium temperature of 16.5°C was reached. Take $c_P(Fe)$ to be 25.1 J K$^{-1}$ mol$^{-1}$ and $c_P(H_2O)$ to be 75.3 J K$^{-1}$ mol$^{-1}$, independent of temperature. Calculate $\Delta S$ for the iron, $\Delta S$ for the water, and $\Delta S_{tot}$ in this process.

22. Iron has a heat capacity of 25.1 J K$^{-1}$ mol$^{-1}$, approximately independent of temperature between 0 and 100°C.
   (a) Calculate the enthalpy and entropy change of 1.00 mol of iron as it is cooled at atmospheric pressure from 100°C to 0°C.
   (b) A piece of iron weighing 55.85 g and at 100°C is placed in a large reservoir of water held at 0°C. It cools irreversibly until its temperature equals that of the water. Assuming the water reservoir is large enough that its temperature remains very close to 0°C, calculate the entropy changes for the iron and the water, and the total entropy change in this process.

## The Third Law of Thermodynamics

23. (a) Use data from Appendix D to calculate the standard entropy change at 25°C for the reaction

$$N_2H_4(\ell) + 3\ O_2(g) \longrightarrow 2\ NO_2(g) + 2\ H_2O(\ell)$$

(b) Suppose the hydrazine ($N_2H_4$) is in the gaseous, rather than liquid, state. Will the entropy change for its reaction with oxygen be higher or lower than that calculated in part (a)? (*Hint:* Entropies of reaction can be added when chemical equations are added, in the same way that Hess's law allows enthalpies to be added.)

24. (a) Use data from Appendix D to calculate the standard entropy change at 25°C for the reaction

$$CH_3COOH(g) + NH_3(g) \longrightarrow$$
$$CH_3NH_2(g) + CO_2(g) + H_2(g)$$

(b) Suppose that 1.00 mol each of solid acetamide, $CH_3CONH_2(s)$, and of water, $H_2O(\ell)$, react to give the same products. Will the standard entropy change be larger or smaller than that calculated for the reaction in part (a)?

25. The alkali metals react with chlorine to give salts:

$$2 \, Li(s) + Cl_2(g) \longrightarrow 2 \, LiCl(s)$$
$$2 \, Na(s) + Cl_2(g) \longrightarrow 2 \, NaCl(s)$$
$$2 \, K(s) + Cl_2(g) \longrightarrow 2 \, KCl(s)$$
$$2 \, Rb(s) + Cl_2(s) \longrightarrow 2 \, RbCl(s)$$
$$2 \, Cs(s) + Cl_2(g) \longrightarrow 2 \, CsCl(s)$$

Using the data in Appendix D, compute $\Delta S°$ of each reaction and identify a periodic trend, if any.

26. All of the halogens react directly with $H_2(g)$ to give binary compounds. The reactions are

$$F_2(g) + H_2(g) \longrightarrow 2 \, HF(g)$$
$$Cl_2(g) + H_2(g) \longrightarrow 2 \, HCl(g)$$
$$Br_2(g) + H_2(g) \longrightarrow 2 \, HBr(g)$$
$$I_2(g) + H_2(g) \longrightarrow 2 \, HI(g)$$

Using the data in Appendix D, compute $\Delta S°$ of each reaction and identify a periodic trend, if any.

27. The dissolution of calcium chloride

$$CaCl_2(s) \longrightarrow Ca^{2+}(aq) + 2 \, Cl^-(aq)$$

is a spontaneous process at 25°C, even though the standard entropy change of the preceding reaction is negative ($\Delta S° = -44.7 \, J \, K^{-1}$). What conclusion can you draw about the change in entropy of the surroundings in this process?

28. Quartz, $SiO_2(s)$, does not spontaneously decompose to silicon and oxygen at 25°C in the reaction

$$SiO_2(s) \longrightarrow Si(s) + O_2(g)$$

even though the standard entropy change of the reaction is large and positive ($\Delta S° = +182.02 \, J \, K^{-1}$). Explain.

29. Use the microscopic interpretation of entropy from Section 8.2 to explain why the entropy change of the system in problem 28 is positive.

30. (a) Why is the entropy change of the system negative for the reaction in problem 27, when the ions become dispersed through a large volume of solution? (*Hint:* Think about the role of the solvent, water.)

(b) Use Appendix D to calculate $\Delta S°$ for the corresponding dissolution of $CaF_2(s)$. Explain why this value is even more negative than that given in problem 27.

## The Gibbs Free Energy

31. The molar enthalpy of fusion of solid ammonia is 5.65 kJ $mol^{-1}$, and the molar entropy of fusion is 28.9 J $K^{-1}$ $mol^{-1}$.
(a) Calculate the Gibbs free energy change for the melting of 1.00 mol of ammonia at 170 K.
(b) Calculate the Gibbs free energy change for the conversion of 3.60 mol of solid ammonia to liquid ammonia at 170 K.
(c) Will ammonia melt spontaneously at 170 K?
(d) At what temperature are solid and liquid ammonia in equilibrium at a pressure of 1 atm?

32. Solid tin exists in two forms: white and gray. For the transformation

$$Sn(s, \text{white}) \longrightarrow Sn(s, \text{gray})$$

the enthalpy change is $-2.1$ kJ and the entropy change is $-7.4 \, J \, K^{-1}$.
(a) Calculate the Gibbs free energy change for the conversion of 1.00 mol of white tin to gray tin at $-30$°C.
(b) Calculate the Gibbs free energy change for the conversion of 2.50 mol of white tin to gray tin at $-30$°C.
(c) Will white tin convert spontaneously to gray tin at $-30$°C?
(d) At what temperature are white and gray tin in equilibrium at a pressure of 1 atm?

33. Ethanol's enthalpy of vaporization is 38.7 kJ $mol^{-1}$ at its normal boiling point, 78°C. Calculate $q$, $w$, $\Delta E$, $\Delta S_{sys}$, and $\Delta G$ when 1.00 mol of ethanol is vaporized reversibly at 78°C and 1 atm. Assume that the vapor is an ideal gas and neglect the volume of liquid ethanol relative to that of its vapor.

34. Suppose 1.00 mol of superheated ice melts to liquid water at 25°C. Assume the specific heats of ice and liquid water have the same value and are independent of temperature. The enthalpy change for the melting of ice at 0°C is 6007 J $mol^{-1}$. Calculate $\Delta H$, $\Delta S_{sys}$, and $\Delta G$ for this process.

35. At 1200°C, the reduction of iron oxide to elemental iron and oxygen is not spontaneous:

$$2 \, Fe_2O_3(s) \longrightarrow 4 \, Fe(s) + 3 \, O_2(g) \qquad \Delta G = +840 \text{ kJ}$$

Show how this process can be made to proceed if all the oxygen generated reacts with carbon:

$$C(s) + O_2(g) \longrightarrow CO_2(g) \qquad \Delta G = -400 \text{ kJ}$$

This observation is the basis for the smelting of iron ore with coke to extract metallic iron.

36. The primary medium for free-energy storage in living cells is adenosine triphosphate (ATP). Its formation from adenosine diphosphate (ADP) is not spontaneous:

$$ADP^{3-}(aq) + HPO_4^{2-}(aq) + H^+(aq) \longrightarrow$$
$$ATP^{4-}(aq) + H_2O(\ell) \qquad \Delta G = +34.5 \text{ kJ}$$

Cells couple ATP production with the metabolism of glucose (a sugar):

$$C_6H_{12}O_6(aq) + 6\ O_2(g) \longrightarrow 6\ CO_2(g) + 6\ H_2O(\ell)$$
$$\Delta G = -2872\ kJ$$

The reaction of 1 molecule of glucose leads to the formation of 38 molecules of ATP from ADP. Show how the coupling makes this reaction spontaneous. What fraction of the free energy released in the oxidation of glucose is stored in the ATP?

37. A process at constant $T$ and $P$ can be described as spontaneous if $\Delta G < 0$ and nonspontaneous if $\Delta G > 0$. Over what range of temperatures is each of the following processes spontaneous? Assume that all gases are at a pressure of 1 atm. (*Hint:* Use Appendix D to calculate $\Delta H$ and $\Delta S$ [assumed independent of temperature and equal to $\Delta H°$ and $\Delta S°$, respectively] and then use the definition of $\Delta G$.)
    (a) The rusting of iron, a complex reaction that can be approximated as

    $$4\ Fe(s) + 3\ O_2(g) \longrightarrow 2\ Fe_2O_3(s)$$

    (b) The preparation of $SO_3(g)$ from $SO_2(g)$, a step in the manufacture of sulfuric acid:

    $$SO_2(g) + \tfrac{1}{2} O_2(g) \longrightarrow SO_3(g)$$

    (c) The production of the anesthetic dinitrogen oxide through the decomposition of ammonium nitrate:

    $$NH_4NO_3(s) \longrightarrow N_2O(g) + 2\ H_2O(g)$$

38. Follow the same procedure used in problem 37 to determine the range of temperatures over which each of the following processes is spontaneous.
    (a) The preparation of the poisonous gas phosgene:

    $$CO(g) + Cl_2(g) \longrightarrow COCl_2(g)$$

    (b) The laboratory-scale production of oxygen from the decomposition of potassium chlorate:

    $$2\ KClO_3(s) \longrightarrow 2\ KCl(s) + 3\ O_2(g)$$

    (c) The reduction of iron(II) oxide (wüstite) by coke (carbon), a step in the production of iron in a blast furnace:

    $$FeO(s) + C(s,\ gr) \longrightarrow Fe(s) + CO(g)$$

39. Explain how it is possible to reduce tungsten(VI) oxide ($WO_3$) to metal with hydrogen at an elevated temperature. Over what temperature range is this reaction spontaneous? Use the data of Appendix D.

40. Tungsten(VI) oxide can also be reduced to tungsten by heating it with carbon in an electric furnace:

    $$2\ WO_3(s) + 3\ C(s) \longrightarrow 2\ W(s) + 3\ CO_2(g)$$

    (a) Calculate the standard free energy change ($\Delta G°$) for this reaction, and comment on the feasibility of the process at room conditions.
    (b) What must be done to make the process thermodynamically feasible, assuming $\Delta H$ and $\Delta S$ do not vary much with temperature?

**Additional Problems**

41. Ethanol ($CH_3CH_2OH$) has a normal boiling point of 78.4°C and a molar enthalpy of vaporization of 38.74 kJ mol$^{-1}$. Calculate the molar entropy of vaporization of ethanol and compare it with the prediction of Trouton's rule.

42. A quantity of ice is mixed with a quantity of hot water in a sealed, rigid, insulated container. The insulation prevents heat exchange between the ice–water mixture and the surroundings. The contents of the container soon reach equilibrium. State whether the total *energy* of the contents decreases, remains the same, or increases in this process. Make a similar statement about the total *entropy* of the contents. Explain your answers.

43. (a) If 2.60 mol of $O_2(g)$ ($c_P = 29.4\ J\ K^{-1}\ mol^{-1}$) is compressed reversibly and adiabatically from an initial pressure of 1.00 atm and 300 K to a final pressure of 8.00 atm, calculate $\Delta S$ for the gas.
    (b) Suppose a different path from that in part (a) is used. The gas is first heated at constant pressure to the same final temperature and then compressed reversibly and isothermally to the same final pressure. Calculate $\Delta S$ for this path and show that it is equal to that found in part (a).

44. One mole of a monatomic ideal gas begins in a state with $P = 1.00$ atm and $T = 300$ K. It is expanded reversibly and adiabatically until the volume has doubled; then it is expanded irreversibly and isothermally into a vacuum until the volume has doubled again; then it is heated reversibly at constant volume to 400 K. Finally it is compressed reversibly and isothermally until a final state with $P = 1.00$ atm and $T = 400$ K is reached. Calculate $\Delta S_{sys}$ for this process. (*Hint:* There are two ways to solve this problem—an easy way and a hard way.)

* 45. The motion of air masses through the atmosphere can be approximated as adiabatic (because air is a poor conductor of heat) and reversible (because pressure differences in the atmosphere are small). To a good approximation, air can be treated as an ideal gas with average molar mass 29 g mol$^{-1}$ and average heat capacity 29 J K$^{-1}$ mol$^{-1}$.
    (a) Show that the displacement of the air masses occurs at constant entropy ($\Delta S = 0$).
    (b) Suppose the average atmospheric pressure near the earth's surface is $P_0$ and the temperature is $T_0$. The air is displaced upward until its temperature is $T$ and its pressure is $P$. Determine the relationship between $P$ and $T$. (*Hint:* Consider the process as occurring in two steps: first a cooling from $T_0$ to $T$ at constant pressure, and then an expansion from $P_0$ to $P$ at constant temperature. Equate the sum of the two entropy changes to $\Delta S_{tot} = 0$.)
    (c) In the lower atmosphere, the dependence of pressure on height $h$ above the earth's surface can be approximated as

    $$\ln (P/P_0) = -\mathcal{M}gh/RT$$

    where $\mathcal{M}$ is the molar mass (kg mol$^{-1}$), $g$ the acceleration due to gravity (9.8 m s$^{-2}$), and $R$ the gas constant. If the air temperature at sea level near the equator is 38°C (~100°F), calculate the air temperature at the summit of

Mount Kilimanjaro, 5.9 km above sea level. For further discussion of this problem, see L. K. Nash, *J. Chem. Educ.* 61:23, 1984.

46. Calculate the entropy change that results from mixing 54.0 g of water at 273 K with 27.0 g of water at 373 K in a vessel whose walls are perfectly insulated from the surroundings. Consider the specific heat of water to be constant over the temperature range from 273 K to 373 K and to have the value 4.18 J K$^{-1}$ g$^{-1}$.

47. Problem 22 asked for the entropy change when a piece of iron is cooled by immersion in a reservoir of water at 0°C.
    (a) Repeat problem 22(b), supposing that the iron is cooled to 50°C in a large water reservoir held at that temperature before being placed in the 0°C reservoir.
    (b) Repeat the calculation supposing that four water reservoirs at 75°C, 50°C, 25°C, and 0°C are used.
    (c) As more reservoirs are used, what happens to $\Delta S$ for the iron, for the water, and for the universe? How would you attempt to carry out a reversible cooling of the iron?

48. Problem 42 in Chapter 4 described an optical atomic trap. In one experiment, a gas of 500 sodium atoms is confined in a volume of 1000 $\mu m^3$. The temperature of the system is 0.00024 K. Compute the probability that, by chance, these 500 slowly moving sodium atoms will all congregate in the left half of the available volume. Express your answer in scientific notation.

* 49. Suppose we have several different ideal gases, $i = 1, 2, 3, \ldots, N$, each occupying its own volume $V_i$, all at the same pressure and temperature. The boundaries between the volumes are removed so that the gases mix at constant temperature in the total volume $V = \Sigma_i V_i$.
    (a) Using the microscopic interpretation of entropy, show that

    $$\Delta S = -nR \sum_i X_i \ln X_i$$

    for this process, where $n$ is the total number of moles of gas and $X_i$ is the mole fraction of gas $i$.
    (b) Calculate the entropy change when 50 g each of $O_2(g)$, $N_2(g)$, and $Ar(g)$ are mixed at 1 atm and 0°C.
    (c) Using Table 4.1, calculate the entropy change when 100 L of air (assumed to be a mixture of ideal gases) at 1 atm and 25°C is separated into its component gases at the same pressure and temperature.

* 50. The $N_2O$ molecule has the structure N—N—O. In an ordered crystal of $N_2O$, the molecules are lined up in a regular fashion, with the orientation of each determined by its position in the crystal. In a random crystal (formed on rapid freezing), each molecule has two equally likely orientations.
    (a) Calculate the number of microstates available to a random crystal of $N_0$ (Avogadro's number) of molecules.
    (b) Calculate the entropy change when 1.00 mol of a random crystal is converted to an ordered crystal.

51. By examining the following graphs, predict which element—copper or gold—has the higher absolute entropy at a temperature of 200 K.

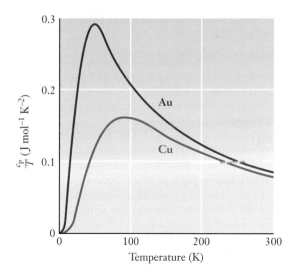

* 52. Consider the process described in problem 19 in Chapter 7. Use the results from that problem to do the following.
    (a) Calculate $\Delta S$ for the system, the surroundings, and the universe.
    (b) If the absolute entropy per mole of the gas *before* the expansion is 158.2 J K$^{-1}$ mol$^{-1}$, calculate $\Delta G_{sys}$ for the process.

53. Two different crystalline forms of sulfur are the rhombic form and the monoclinic form. At atmospheric pressure, rhombic sulfur undergoes a transition to monoclinic when it is heated above 368.5 K:

    $$S(s, \text{rhombic}) \longrightarrow S(s, \text{monoclinic})$$

    (a) What is the sign of the entropy change ($\Delta S$) for this transition?
    (b) $|\Delta H|$ for this transition is 400 J mol$^{-1}$. Calculate $\Delta S$ for the transition.

* 54. Use data from Appendix D to estimate the temperature at which $I_2(g)$ and $I_2(s)$ are in equilibrium at a pressure of 1 atm. Can this equilibrium actually be achieved? Refer to Appendix F for data on iodine.

55. The $\Delta G_f^\circ$ of $Si_3N_4(s)$ is $-642.6$ kJ mol$^{-1}$. Use this fact and the data in Appendix D to compute $\Delta G^\circ$ of the reaction

    $$3\ CO_2(g) + Si_3N_4(s) \longrightarrow$$
    $$3\ SiO_2(\text{quartz}) + 2\ N_2(g) + 3\ C(s, \text{gr})$$

56. The compound $Pt(NH_3)_2I_2$ comes in two forms, the *cis* and the *trans*, which differ in their molecular structure. The following data are available:

| | $\Delta H_f^\circ$ (kJ mol$^{-1}$) | $\Delta G_f^\circ$ (kJ mol$^{-1}$) |
|---|---|---|
| *cis* | $-286.56$ | $-130.25$ |
| *trans* | $-316.94$ | $-161.50$ |

Combine these data with data from Appendix D to compute the standard entropies ($S^\circ$) of both of these compounds at 25°C.

57. (a) Use data from Appendix D to calculate $\Delta H°$ and $\Delta S°$ at 25°C for the reaction

$$2 \, CuCl_2(s) \rightleftharpoons 2 \, CuCl(s) + Cl_2(g)$$

(b) Calculate $\Delta G$ at 590 K, assuming $\Delta H°$ and $\Delta S°$ are independent of temperature.

(c) Careful high-temperature measurements show that when this reaction is carried out at 590 K, $\Delta H_{590}$ is 158.36 kJ and $\Delta S_{590}$ is 177.74 J K$^{-1}$. Use these facts to compute an improved value of $\Delta G_{590}$ for this reaction. Determine the percentage error in $\Delta G_{590}$ that comes from using the 298-K values in place of 590-K values in this case.

58. (a) The normal boiling point of carbon tetrachloride (CCl$_4$) is 76.5°C. A student looks up the standard enthalpies of formation of CCl$_4(\ell)$ and of CCl$_4(g)$ in Appendix D. They are listed as $-135.44$ and $-102.9$ kJ mol$^{-1}$, respectively. By subtracting the first from the second, she computes $\Delta H°$ for the vaporization of CCl$_4$ to be 32.5 kJ mol$^{-1}$. But Table 7.2 states that the $\Delta H_{vap}$ of CCl$_4$ is 30.0 kJ mol$^{-1}$. Explain the discrepancy.

(b) Calculate the molar entropy change of vaporization ($\Delta S_{vap}$) of CCl$_4$ at 76.5°C.

59. The typical potassium ion concentration in the fluid outside a cell is 0.0050 M, whereas that inside a muscle cell is 0.15 M.

(a) What is the spontaneous direction of motion of ions through the cell wall?

(b) In *active transport*, cells use free energy stored in ATP (see problem 36) to move ions in the direction opposite their spontaneous direction of flow. Calculate the cost in free energy to move 1.00 mol of K$^+$ ions through the cell wall by active transport. Assume no change in K$^+$ concentrations during this process.

---

## CUMULATIVE PROBLEM

60. When a gas undergoes a reversible adiabatic expansion, its entropy remains constant even though the volume increases. Explain how this can be consistent with the microscopic interpretation of entropy developed in Section 8.2. (*Hint:* Consider what happens to the distribution of velocities in the gas.)

# Chemical Equilibrium: Basic Principles and Applications to Gas-Phase Reactions

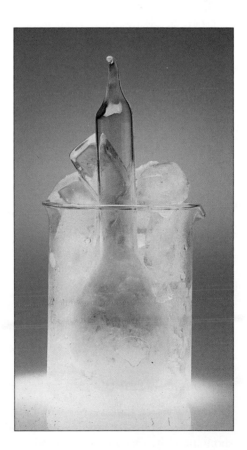

***Illustration***
In a closed vessel, colorless $N_2O_4$ spontaneously decomposes to produce reddish-brown $NO_2$:

$$N_2O_4(g) \rightleftharpoons 2\ NO_2(g)$$

The two species co-exist in the vessel, and the color of the mixture depends on the extent of decomposition. Heating the mixture favors decomposition and intensifies the color. Cooling the mixture reverses the decomposition and reduces the color. Thermodynamics explains not only the extent of decomposition at a particular temperature but also the direction and extent of change caused by heating and cooling. *(Charles D. Winters)*

From fundamental research studies to practical industrial applications of chemistry, the *yield* of chemical reactions is of great concern. Chapter 2 of this book showed how to calculate the amount of product that appears when a particular reaction is carried out by starting with particular amounts of the reactants. This calculation was based on the assumption that the reaction goes to completion—that is, all of the limiting reagent is consumed. The resulting number, called the theoretical yield, represents the maximum amount of product that could be obtained from that reaction.

In practice, many reactions do not go to completion but rather approach a state or position of **equilibrium.** This equilibrium position, at which the reaction apparently comes to an end, is a mixture of products and unconsumed reactants present in fixed relative amounts. Once equilibrium has been achieved, there is no further net conversion of reactants to products unless the experimental conditions of the reaction (temperature and pressure) are changed. The equilibrium state is characterized by the **equilibrium constant** for the reaction; from it, the composition of the equilibrium reaction mixture can be calculated. Knowing the equilibrium constant for a reaction, and its dependence on experimental conditions, the experimenter can manipulate conditions in order to maximize the practical yield of that reaction. Calculating the equilibrium composition for a particular reaction, and its dependence on experimental conditions, is therefore a matter of great importance in chemistry.

This chapter presents a general discussion of the equilibrium constant, its dependence on conditions, and its role in manipulating the yield of reactions, emphasizing those aspects broadly applicable to all chemical reactions. These general principles are illustrated with applications to reactions in the gas phase. Detailed applications to aqueous, heterogeneous, and electrochemical reactions are presented in the three following chapters.

This chapter presents the equilibrium constant, and its dependence on experimental conditions, as *consequences of the thermodynamic equilibrium* of the reaction mixture. Background material from the two previous chapters is therefore indispensable. Thermodynamics demonstrates not only the existence and the mathematical form of the equilibrium constant, but also gives procedures for calculating its value from purely thermal properties of the pure reactants and products, as well as procedures for predicting its dependence on experimental conditions.

---

## 9.1

# THE NATURE OF CHEMICAL EQUILIBRIUM

### Approach to Equilibrium

Most chemical reactions are carried out by mixing the selected reactants in a proper reaction vessel held at fixed $T$ and $P$ and waiting for the desired products to appear spontaneously. Once conditions for spontaneity have been identified, the greatest concern is the "efficiency" of the reaction. How does the experimenter obtain the greatest amount of the desired product? How close to completion does the reaction arrive?

The effects to be explained are illustrated by the reactions of cobalt(II) ions in aqueous solutions where chloride ion is also present. The cobalt(II) ions can form various different complex ions, depending on the amount of chloride present. For example, if $CoCl_2 \cdot 6H_2O$ is dissolved in pure water to the concentration 0.08 M, the resulting solution is pale pink in color due to the hexaaquacobalt(II) complex ion $[Co(H_2O)_6]^{2+}$ (Fig. 9.1a). If $CoCl_2 \cdot 6H_2O$ is dissolved in 10 M HCl to the concen-

(a)                    (b)                    (c)                    (d)

**FIGURE 9.1**   Chemical equilibrium in the cobalt chloride-HCl system. (a) The pink color is due to the hexaaqua complex ion $[Co(H_2O)_6]^{2+}$. (b) The blue color is due to the tetrachloro complex ion $[CoCl_4]^{2-}$. (c) Adding HCl to the pink solution in (a) converts some of the Co(II) to the tetrachloro complex. The lavender color is produced by the combination of pink hexaaqua species and blue tetrachloro species. (d) Adding water to the blue solution in (b) converts some of the Co(II) to the hexaaqua species. The combination of the two gives the lavender color. The same equilibrium state is reached by running the reaction from the left (c) and from the right (d). *(Charles D. Winters)*

tration 0.08 M, the solution is deep blue in color due to the tetrachlorocobalt(II) complex ion $[CoCl_4]^{2-}$ (see Fig. 9.1b). Solutions containing a mixture of both Co(II) species are purple in color.

The hexaaqua complex can be converted into the tetrachloro complex by addition of chloride ion through the reaction

$$[Co(H_2O)_6]^{2+} + 4\ Cl^- \rightleftharpoons [CoCl_4]^{2-} + 6\ H_2O$$

If the experimenter adds concentrated HCl to the pink solution in Figure 9.1a until the Co(II) concentration is 0.044 M and the HCl concentration is 5.4 M, the result is the lavender-colored solution shown in Figure 9.1c. Optical absorption spectroscopy measurements to be described in Chapter 16 confirm the presence of both Co(II) species in the solution in Figure 9.1c. Ninety-eight percent of the Co is found in the pink hexaaqua complex, and the remaining two percent is in the blue tetrachloro complex.[1] If the lavender solution is allowed to stand at constant temperature for several hours and the measurements repeated, the results are the same.

The reaction can also be carried out from "right to left" as written. If the experimenter adds pure water to the blue solution in Figure 9.1b until the cobalt concentration is 0.044 M and the HCl concentration is 5.4 M, the result is the lavender-colored solution shown in Figure 9.1d. Again, optical absorption spectroscopy confirms that 98% of the Co(II) is present in the pink hexaaqua complex and 2% is present in the blue tetrachloro complex. Again, no further change is observed after a lengthy wait.

---

[1] It may appear surprising to find such imbalance in the concentration of the Co(II) species when the color intensities of the pink and lavender solutions in Figure 9.1a and Figure 9.1c appear quite similar. This difference is explained by the fact that the blue tetrachloro complex absorbs light much more efficiently than the pink hexaaqua complex, as evidenced by the much larger molecular extinction coefficient of the blue species. Extinction coefficients will be described in Chapter 16.

**FIGURE 9.2** Sketch of the change with time of the concentrations of products and reactants in the spontaneous reactions illustrated in Figure 9.1. For ease of display, concentrations are expressed as percent of the total Co(II) present contained in each species. (a) Partial conversion of pink hexaaqua complex into blue tetrachloro complex. (b) Partial conversion of blue tetrachloro complex into pink hexaaqua complex. After changes in the slope of each species concentration become imperceptibly small, we say the reaction has arrived at chemical equilibrium.

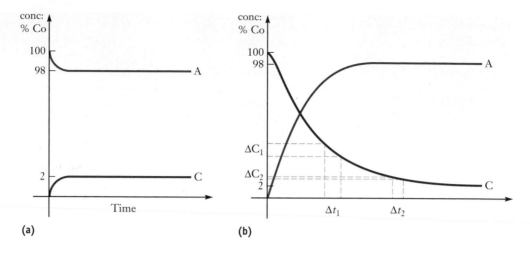

These data show that the reaction has not gone to completion, but has apparently halted at an intermediate state containing both products and unconsumed reactants. Moreover, the same final state can be achieved from either direction. This result is typical of most chemical reactions, even though it is not generally as readily apparent as here.

The questions raised in the first paragraph require quantitative investigations, which are carried out as follows. In the preceding reaction let A represent the pink hexaaqua complex, B the chloride ion, and C the blue tetrachloro complex. In the first experiment, the experimenter starts the reaction by mixing initial concentrations of A and B, denoted as $[A]_0$ and $[B]_0$. As the reaction proceeds, the experimenter periodically samples the reaction mixture, measures the concentration of A, B, and C, and plots concentration of species versus time. The results of the first experiment are represented schematically in Figure 9.2a, which shows the consumption of A and the production of C. Similarly, the second experiment is started with the initial concentration $[C]_0$, and water is added. Its results are represented schematically in Figure 9.2b, which shows the consumption of C and the production of A.

As the reaction proceeds, the concentration of each species changes progressively more slowly, as indicated by the decreasing values of the slope $[m = \Delta(\text{conc.})/\Delta t]$ sketched on the concentration curves during progressively later time intervals. In due course, the slopes become sufficiently close to zero to indicate that the concentration of each species is effectively constant in time. When this condition has been achieved, we say that the reaction has arrived at *chemical equilibrium*, and the reaction mixture has arrived at the *equilibrium composition*.[2] At later times, there is no further change in composition of the reaction mixture; the concentration of each of the species X = A, B, C remains at its equilibrium value denoted by $[X]_{eq}$.

## Characteristics of the Equilibrium State

Section 5.4 described the phase equilibrium between liquid water and water vapor, which can be represented as a chemical equation:

$$H_2O(\ell) \rightleftharpoons H_2O(g)$$

---

[2] The experimenter must exercise judgment in deciding when the slope is "sufficiently close to zero" and the concentrations are "effectively constant." There is no one instant at which equilibrium is achieved.

The double arrows ($\rightleftharpoons$) emphasize the dynamic nature of phase equilibrium: liquid water evaporates to form water vapor and at the same time vapor condenses to give liquid. An analogous dynamic description applies to a chemical equilibrium, in which bonds are broken or formed as atoms move back and forth between reactant and product molecules. When the initial concentrations of the reactants are high, collisions between their molecules cause product molecules to form. Once the concentrations of the products have increased sufficiently, the reverse reaction (forming "reactants" from "products") begins to occur. As the equilibrium state is approached, the forward and backward rates of reaction become equal and there is no further net change in reactant or product concentrations. Just as the equilibrium between liquid water and water vapor is a dynamic process on the molecular scale, with evaporation and condensation taking place simultaneously, so the chemical equilibrium between reactants and products occurs through the continuous formation of molecules of product from reactant molecules and their reaction back into reactant molecules with equal rates. Chemical equilibrium is not a static condition, although macroscopic properties such as concentrations do stop changing when equilibrium is attained. Rather, it is the consequence of a dynamic balance between forward and backward reactions.

The experimental results just presented demonstrate that the same equilibrium state is reached whether one starts with the reactants or with the products. This fact can be used to test whether a system is truly in equilibrium or whether the reaction is just so slow that changes in concentration are unmeasurably small, even though the system is far from equilibrium. If the same state is reached from either reactants or products, it may be concluded that it is a true equilibrium state.

There are four fundamental aspects of equilibrium states:

1. They display no macroscopic evidence of change.
2. They are reached through spontaneous processes.
3. They show a dynamic balance of forward and reverse processes.
4. They are the same regardless of direction of approach.

One also encounters so-called *steady states* in which the macroscopic concentrations of species are not changing with time, even though the system is not at equilibrium. Steady states are maintained not by a dynamic balance between forward and reverse processes but rather by the competition between a process that supplies the species to the system and a process that removes the species from the system. Many chemical reactions occur in living systems in steady states and do not represent an equilibrium between reactants and products. One must be certain that a reaction is at equilibrium and not in steady state before applying the methods of this chapter to explain the relative concentrations of reactants and products.

## The Empirical Law of Mass Action

Let us examine again the approach to equilibrium, as represented in Figure 9.2 and the related discussion. Extensive studies of this type for very broad classes of reactions that may be represented generally as

$$aA + bB \rightleftharpoons cC + dD$$

have demonstrated a most remarkable result. No matter what initial concentrations of reactants are selected at the beginning of the experiment, the value of the ratio

$$\frac{[C]_{eq}^c [D]_{eq}^d}{[A]_{eq}^a [B]_{eq}^b}$$

of the gas at any pressure $P_2$ to its value at 1 atm. This result can be expressed compactly as follows. If we call 1 atm the *reference state* for the gas, then the change in Gibbs free energy in taking the gas from the reference state to any pressure $P$ is given by

$$\Delta G = nRT \ln \left( \frac{P}{P_{ref}} \right) = nRT \ln P \qquad \textbf{[9.3]}$$

The last form of the equation is a short-hand version that can be used *only* when the pressure $P$ is expressed in atm. The presence of $P_{ref}$ in the denominator makes the argument of the ln function dimensionless. Choosing $P_{ref} = 1$ atm gives $P_{ref}$ the numerical value 1, which for convenience is not written explicitly. Nonetheless, this (invisible) $P_{ref}$ must always be remembered, since it is required to make the equation dimensionally correct when the general pressure $P$ in the equation is expressed in atm. If some unit of pressure other than atm is selected, $P_{ref}$ no longer has value 1, and the $P_{ref}$ selected must be carried explicitly in the equations.

***The Equilibrium Expression for Reactions in the Gas Phase*** Consider now a mixture of gases that react chemically, such as the NO, $N_2O$, and $NO_2$ given in Example 8.10:

$$3 \text{ NO}(g) \rightleftharpoons N_2O(g) + NO_2(g)$$

If all of the partial pressures are 1 atm, then $\Delta G$ for this reaction is just $\Delta G°$ at 25°C, but if the pressures differ from 1 atm, $\Delta G$ must be calculated from a three-step process (Fig. 9.3). In step 1, the partial pressure of the reactant (in this case, 3 mol of NO) is changed from its initial value, $P_{NO}$, to the reference pressure $P_{ref} = 1$ atm:

$$\Delta G_1 = 3RT \ln \left( \frac{P_{ref}}{P_{NO}} \right) = RT \ln \left( \frac{P_{ref}}{P_{NO}} \right)^3$$

In step 2, the reaction is carried out with all reactants and products at partial pressures of $P_{ref} = 1$ atm:

$$\Delta G_2 = \Delta G°$$

In step 3, the partial pressures of the products (in this case, 1 mol of $N_2O$ and 1 mol of $NO_2$) are changed from $P_{ref} = 1$ atm to $P_{N_2O}$ and $P_{NO_2}$:

$$\Delta G_3 = RT \ln \left( \frac{P_{N_2O}}{P_{ref}} \right) + RT \ln \left( \frac{P_{NO_2}}{P_{ref}} \right) = RT \ln \left[ \left( \frac{P_{N_2O}}{P_{ref}} \right) \left( \frac{P_{NO_2}}{P_{ref}} \right) \right]$$

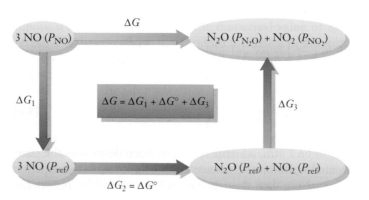

**FIGURE 9.3** A three-step process (red arrows) to calculate $\Delta G$ of a reaction (blue arrow) for which reactants and products are not in their standard states of 1 atm.

The overall Gibbs free energy change $\Delta G$ is the sum of the free energy changes for the three steps in the path:

$$\Delta G = \Delta G_1 + \Delta G_2 + \Delta G_3$$

$$= \Delta G° + RT \ln \left[ \frac{(P_{N_2O}/P_{ref})(P_{NO_2}/P_{ref})}{(P_{NO}/P_{ref})^3} \right]$$

When a chemical reaction has reached equilibrium, $\Delta G = 0$. Under these conditions the preceding equation becomes

$$-\Delta G° = RT \ln \left[ \frac{(P_{N_2O}/P_{ref})(P_{NO_2}/P_{ref})}{(P_{NO}/P_{ref})^3} \right]$$

Since $\Delta G°$ depends only on temperature, the quantity $\Delta G°/RT$ must be a constant at each value of $T$. Therefore, in the last equation, the ratio of partial pressures inside the ln function must also be constant *at equilibrium* at each value of $T$. Consequently, this ratio of partial pressures is denoted by $K(T)$ and is called the *thermodynamic equilibrium constant* for the reaction. Finally, we have

$$-\Delta G° = RT \ln K(T) \qquad\qquad \textbf{[9.4]}$$

*[handwritten margin note:] $\Delta G°$ depends only on temperature.*

For the general reaction

$$a\text{A} + b\text{B} \longrightarrow c\text{C} + d\text{D}$$

the result, obtained in the same way, is

$$-\Delta G° = RT \ln \left[ \frac{(P_C/P_{ref})^c (P_D/P_{ref})^d}{(P_A/P_{ref})^a (P_B/P_{ref})^b} \right] = RT \ln K \qquad \text{(at equilibrium)}$$

The expression for the thermodynamic equilibrium constant for reactions involving ideal gases,

$$K = \frac{(P_C/P_{ref})^c (P_D/P_{ref})^d}{(P_A/P_{ref})^a (P_B/P_{ref})^b}$$

has the same general form as the empirical law of mass action with $K_P$ for gaseous reactions introduced briefly in Section 9.1. The thermodynamic expression provides deeper understanding of that empirical result in three important ways. First, the law of mass action, previously introduced on purely empirical grounds to describe chemical equilibrium, is seen to be a consequence of the reaction system being in thermodynamic equilibrium. Second, the thermodynamic equilibrium constant can be calculated from $\Delta G°$ (that is, from $\Delta H°$ and $\Delta S°$) so that the *extent* of any equilibrium chemical reaction can be deduced from calorimetric data alone. Third, unlike $K_P$, the thermodynamic equilibrium constant $K$ is always a dimensionless quantity because the pressure of a reactant or product always appears as a ratio to the reference pressure $P_{ref}$. Collecting the terms involving $P_{ref}$ on the right side of the equation gives

$$\frac{P_C^c P_D^d}{P_A^a P_B^b} = K(P_{ref})^{(c+d-a-b)}$$

If pressures are expressed in atmospheres, then $P_{ref} = 1$ atm and the right-hand side has the same *numerical value* as $K$, with $P_{ref}$ factors serving only to make the equation dimensionally correct. If some other unit is chosen for pressures, the $P_{ref}$ factors no longer have a numerical value of unity, and must be inserted explicitly into the equilibrium expression.

The convention followed in this book is to describe chemical equilibrium in terms of the thermodynamic equilibrium constant $K$ rather than the empirical equilibrium constant $K_P$, because of the deeper understanding provided by $K$. Consequently, values of $K$ will be stated without dimensions, and all pressures will be expressed in atmospheres. The $P_{ref}$ factors will not be explicitly included because their value is unity with these choices of pressure unit and reference pressure. Following this convention, the mass action law for a general reaction involving ideal gases is written as

for $A a + B b \rightleftharpoons C c + D d$

$$K = \frac{(P_C)^c (P_D)^d}{(P_A)^a (P_B)^b}$$

with $K$ dimensionless. The following example illustrates these practices.

### EXAMPLE 9.1

Write equilibrium expressions for the following gas-phase chemical equilibria.
**(a)** $2\ NOCl(g) \rightleftharpoons 2\ NO(g) + Cl_2(g)$
**(b)** $CO(g) + \frac{1}{2} O_2(g) \rightleftharpoons CO_2(g)$

### Solution

**(a)**
$$\frac{(P_{NO})^2 (P_{Cl_2})}{(P_{NOCl})^2} = K$$

The powers of 2 come from the factors of 2 in the balanced equation.

**(b)**
$$\frac{P_{CO_2}}{(P_{CO})(P_{O_2})^{1/2}} = K$$

Fractional powers appear in the equilibrium expression whenever they are present in the balanced equation.

**Related Problems: 1, 2, 3, 4**

* Equation *

$\Delta G^\circ = -RT \ln K$

*Calculation of Equilibrium Constants from Calorimetric Data*   The thermodynamic approach provides a straightforward method for evaluating equilibrium constants through Equation 9.4:

$$\Delta G^\circ = -RT \ln K$$

wherein $\Delta G^\circ$ is evaluated from tables of data in Appendix D. The following example illustrates the procedure.

### EXAMPLE 9.2

The $\Delta G^\circ$ of the chemical reaction

$$3\ NO(g) \longrightarrow N_2O(g) + NO_2(g)$$

was calculated in Example 8.10. Now calculate the equilibrium constant of this reaction at 25°C.

## Solution

The standard free energy change for the conversion of 3 mol of NO to 1 mol of $N_2O$ and 1 mol of $NO_2$ was found to be $-104.18$ kJ. To keep the units of the calculation correct, the $\Delta G°$ is rewritten as $-104.18$ kJ $mol^{-1}$, where "per mole" signifies "per mole of the reaction as it is written," that is, per 3 mol of NO, 1 mol of $N_2O$, and 1 mol of $NO_2$, the chemical amounts in the balanced equation. Substitution gives

$$\ln K = \frac{-\Delta G°}{RT} = \frac{-(-104{,}180\,\text{J mol}^{-1})}{(8.315\,\text{J K}^{-1}\,\text{mol}^{-1})(298.15\,\text{K})} = 42.03$$

$$K = e^{42.03} = 1.8 \times 10^{18}$$

**Related Problems: 7, 8, 9, 10**

---

The conversion of $NO(g)$ to $N_2O(g)$ plus $NO_2(g)$ is spontaneous under standard conditions. The forward reaction under these conditions is scarcely observed because it is so slow, but its equilibrium constant can nonetheless be calculated! Such calculations often have enormous practical impact. For example, this calculation shows that a proposal to exploit this reaction to reduce the amount of NO in cooled automobile exhaust could succeed. The fundamental reaction tendency is there; success of the plan would hinge on increasing the reaction rate at standard conditions. Had the equilibrium constant calculated from thermodynamics been small, the plan would have been doomed at the outset, and efforts expended on it would not have been justified.

## Reactions in Ideal Solutions

Thermodynamics also applies to processes occurring in solution. Although the result is not derived here, the Gibbs free energy change for $n$ mol of a solute, as an ideal (dilute) solution changes in concentration from $c_1$ to $c_2$ mol $L^{-1}$, is

$$\Delta G = nRT \ln\left(\frac{c_2}{c_1}\right)$$

If the reference state for the solute is defined to be an ideal solution with a concentration $c_{ref} = 1$ M, then the same arguments used before for gas-phase reactions lead to expressions of the forms

$$\Delta G = \Delta G° + RT \ln\left[\frac{([C]/c_{ref})^c\,([D]/c_{ref})^d}{([A]/c_{ref})^a\,([B]/c_{ref})^b}\right]$$

and

$$\Delta G° = -RT \ln K$$

for reactions involving dissolved species. The square bracket [X] represents the concentration of species X in units of mol $L^{-1}$. Tables of standard free energies for solutes in aqueous solution at 25°C allow solution-phase equilibrium constants to be calculated (see Appendix D).

Just as with gaseous reactions, the convention followed in this book is to describe solution equilibria in terms of the thermodynamic equilibrium constant $K$ rather than the empirical $K_C$ introduced briefly in Section 9.1. Thus solution concentrations are given in units of mol $L^{-1}$ with the reference state as $c_{ref} = 1$ M, and values of $K$ are stated as dimensionless quantities. For these conditions the mass action law for solution reactions becomes

$$K = \frac{[C]^c[D]^d}{[A]^a[B]^b}$$

When working with this expression, the role of the (invisible) $c_{ref}$ factors in making the equation dimensionally correct must be kept in mind. This expression is the foundation for the discussions of acid-base equilibria in aqueous solutions in the next chapter.

---

## 9.3

# EQUILIBRIUM CALCULATIONS FOR GAS-PHASE REACTIONS

Equilibrium calculations, used throughout the science of chemistry in both basic and applied work, involve rather specific procedures. The present section presents these problem-solving techniques in the context of gas-phase reactions, but they are applicable in *all* equilibrium calculations. The strategy comprises three basic steps.

1. Always start by writing the balanced equation for the reaction under study.
2. Visualize the reaction as proceeding through three steps: (a) the reactants are brought together at their *initial partial pressures*, but not yet allowed to react; (b) reaction is initiated and produces *changes in partial pressures* of all products and reactants; (c) when the reaction reaches equilibrium, all products and reactants are present at their *equilibrium partial pressures*. (Note the similarity to the procedures described in the previous chapter for conducting spontaneous processes by manipulating constraints.)
3. Finally, develop approximation schemes when possible by neglecting a very small quantity that is added to or subtracted from a much larger quantity.

## Evaluating Equilibrium Constants from Reaction Data

The previous section presented techniques for evaluating $K$ from calorimetric data on the pure reactants and products. Occasionally, these thermodynamic data may not be available for a specific reaction, or a quick estimate of the value of $K$ may suffice. In these cases, the equilibrium constant can be evaluated from measurements made directly on the reaction mixture. If the equilibrium partial pressures of all the reactants and products can be measured, the equilibrium constant can be calculated by writing the equilibrium expression and substituting the experimental values (in atmospheres) into it. In many cases it is not practical to measure directly the equilibrium partial pressure of each separate reactant and product. Nonetheless, the equilibrium constant can usually be derived from other available data, although the determination is less direct. The method is shown in the following example.

---

**EXAMPLE 9.3**

Phosgene, $COCl_2$, forms from CO and $Cl_2$ according to the equilibrium

$$CO(g) + Cl_2(g) \rightleftharpoons COCl_2(g)$$

Adding th

Removing

and, by ir
    If a s
stant is th
same as a
of additic
librium e

**EXAM**

The con
At 25°C

is

and tha

is

Find th

**Solu**
Adding

The e
$K_1K_2$.
so $K'_3$

**Relat**

**Calc**

The
stant
dict

At 600°C, a gas mixture of CO and $Cl_2$ is prepared that has initial partial pressures (before reaction) of 0.60 atm for CO and 1.10 atm for $Cl_2$. After the reaction mixture has reached equilibrium, the partial pressure of $COCl_2(g)$ at this temperature is measured to be 0.10 atm. Calculate the equilibrium constant for this reaction. The reaction is carried out in a vessel of fixed volume.

### Solution

Only the equilibrium partial pressure of $COCl_2(g)$ and the *initial* partial pressures of the other two gases are given. To find the equilibrium constant, it is necessary to determine the equilibrium partial pressures of CO and $Cl_2$. To do this, we set up a simple table:

|  | $CO(g)$ | $+$ | $Cl_2(g)$ | $\rightleftharpoons$ | $COCl_2(g)$ |  |
|---|---|---|---|---|---|---|
| Initial partial pressure (atm) | 0.60 | | 1.10 | | 0 | I |
| Change in partial pressure (atm) | ? | | ? | | +0.10 | C |
| Equilibrium partial pressure (atm) | ? | | ? | | 0.10 | E |

Phosgene, $COCl_2$, is a chemical intermediate used in making polyurethanes for foams and surface coatings.

Note that the first two lines must add to give the third. We use the relationships built into the balanced chemical equation to fill in the blanks in the table. Because this is the only reaction taking place, every mole of $COCl_2$ produced consumes exactly 1 mol of CO and exactly 1 mol of $Cl_2$. According to the ideal gas equation, the partial pressures of gases are proportional to their chemical amounts (in moles) as long as the volume and temperature are held fixed. Therefore, the change in partial pressure of each gas must be proportional to the change in its chemical amount as the mixture goes toward equilibrium. If the partial pressure of $COCl_2$ *increases* by 0.10 atm through reaction, the partial pressures of CO and $Cl_2$ must each *decrease* by 0.10 atm. Inserting these values into the table gives

"ICE" system.

|  | $CO(g)$ | $+$ | $Cl_2(g)$ | $\rightleftharpoons$ | $COCl_2(g)$ |  |
|---|---|---|---|---|---|---|
| Initial partial pressure (atm) | 0.60 | | 1.10 | | 0 | I |
| Change in partial pressure (atm) | −0.10 | | −0.10 | | +0.10 | C |
| Equilibrium partial pressure (atm) | 0.50 | | 1.00 | | 0.10 | E |

where the last line was obtained by adding the first two.
    We now insert the equilibrium partial pressures into the equilibrium expression to calculate the equilibrium constant:

$$K = \frac{P_{COCl_2}}{(P_{CO})(P_{Cl_2})} = \frac{(0.10)}{(0.50)(1.00)} = 0.20$$

**Related Problems: 13, 14, 15, 16**

"requires care!"

For a reaction in which some of the coefficients in the balanced equation are not equal to 1, deriving expressions for the changes in the partial pressures of the products and reactants requires care. Consider the combustion of ethane at constant volume:

$$2\ C_2H_6(g) + 7\ O_2(g) \rightleftharpoons 4\ CO_2(g) + 6\ H_2O(g)$$

If the initial partial pressures are all 1.00 atm and there is a net reaction from left to right in this equation, the table is

the laboratory, and Figure 9.4, which represents the thermodynamic driving force (i.e., the Gibbs free energy) governing these events.

## EXAMPLE 9.8

The reaction between nitrogen and hydrogen to produce ammonia

$$N_2(g) + 3\ H_2(g) \rightleftharpoons 2\ NH_3(g)$$

is essential in making nitrogen-containing fertilizers. This reaction has an equilibrium constant equal to $1.9 \times 10^{-4}$ at 400°C. Suppose that 0.10 mol of $N_2$, 0.040 mol of $H_2$, and 0.020 mol of $NH_3$ are sealed in a 1.00-L vessel at 400°C. In which direction will the reaction proceed?

### Solution

The initial pressures $P_i = n_i RT/V$ are readily calculated to be

$$P_{N_2} = 5.5\ \text{atm} \qquad P_{H_2} = 2.2\ \text{atm} \qquad P_{NH_3} = 1.1\ \text{atm}$$

The initial numerical value of $Q$ is therefore

$$Q_0 = \frac{(P^\circ_{NH_3})^2}{P^\circ_{N_2}(P^\circ_{H_2})^3} = \frac{(1.1)^2}{(5.5)(2.2)^3} = 2.1 \times 10^{-2}$$

Because $Q_0 > K$, the reaction will proceed from right to left and ammonia will dissociate until equilibrium is reached.

**Related Problems: 33, 34, 35, 36**

## External Effects on $K$: Principle of Le Châtelier

Suppose a system at equilibrium is perturbed by some external stress such as a change in volume or temperature or a change in the partial pressure or concentration of one of the reactants or products. How will the system respond? The qualitative answer is embodied in a principle stated by Henri Le Châtelier in 1884:

A system in equilibrium that is subjected to a stress will react in a way that tends to counteract the stress.

**Le Châtelier's principle** provides a way to predict qualitatively the direction of change of a system under an external perturbation. It relies heavily on $Q$ as a predictive tool.

*Effects of Changing the Concentration of a Reactant or Product*   As a simple example, consider what happens when a small quantity of a reactant is added to an equilibrium mixture. The addition of reactant lowers the reaction quotient $Q$ below $K$ and a net reaction takes place in the forward direction, partially converting reactants to products, until $Q$ again equals $K$. The system partially counteracts the stress (the increase in the quantity of one of the reactants) and attains a new equilibrium state. If one of the *products* is added to an equilibrium mixture, $Q$ temporarily becomes *greater* than $K$ and a net *back* reaction occurs, partially counteracting the imposed stress by reducing the concentration of products (Fig. 9.6).

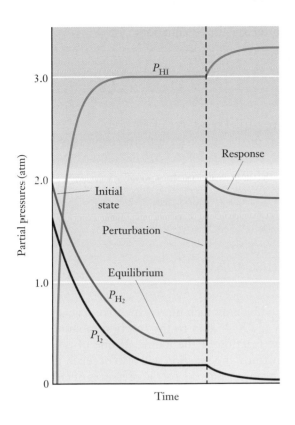

**FIGURE 9.6** Partial pressures versus time for the equilibrium

$$H_2(g) + I_2(g) \rightleftharpoons 2\,HI(g)$$

The left part of the figure shows the attainment of equilibrium starting from the initial conditions of Example 9.5. Then the equilibrium state is abruptly perturbed by an increase in the partial pressure of $H_2$ to 2.000 atm. In accordance with Le Châtelier's principle, the system responds (Example 9.9) in such a way as to decrease the partial pressure of $H_2$—that is, to counteract the perturbation that moved it away from equilibrium in the first place.

---

## EXAMPLE 9.9

An equilibrium gas mixture of $H_2(g)$, $I_2(g)$, and $HI(g)$ at 600 K has

$$P_{H_2} = 0.4756 \text{ atm} \qquad P_{I_2} = 0.2056 \text{ atm} \qquad P_{HI} = 3.009 \text{ atm}$$

This is essentially the final equilibrium state of Example 9.5. Enough $H_2$ is added to increase its partial pressure to 2.000 atm at 600 K before any reaction takes place. The mixture then once again reaches equilibrium at 600 K. What are the final partial pressures of the three gases?

### Solution

Set up the usual table, in which "initial" now means the moment after the addition of the new $H_2$ but before it reacts further.

| | $H_2(g)$ | + | $I_2(g)$ | $\rightleftharpoons$ | $2\,HI(g)$ |
|---|---|---|---|---|---|
| Initial partial pressure (atm) | 2.000 | | 0.2056 | | 3.009 |
| Change in partial pressure (atm) | $-x$ | | $-x$ | | $+2x$ |
| Equilibrium partial pressure (atm) | $2.000 - x$ | | $0.2056 - x$ | | $3.009 + 2x$ |

"Initial" – before reaction.

From Le Châtelier's principle it follows that net reaction to consume $H_2$ will follow addition of $H_2$, and this fact has been used in assigning a negative sign to the change in the partial pressure of $H_2$ in the table.

   Substitution of the equilibrium partial pressures into the equilibrium law gives

$$\frac{(3.009 + 2x)^2}{(2.000 - x)(0.256 - x)} = 92.6$$

Expansion of this expression results in the quadratic equation

$$88.60x^2 - 216.275x + 29.023 = 0$$

which can be solved to give

$$x = 0.1425 \quad \text{or} \quad 2.299$$

The second root would lead to negative partial pressures of $H_2$ and $I_2$ and is therefore physically impossible. Substitution of the first root into the expressions from the table gives

$$P_{H_2} = 2.000 - 0.1425 = 1.86 \text{ atm}$$

$$P_{I_2} = 0.2056 - 0.1425 = 0.063 \text{ atm}$$

$$P_{HI} = 3.009 + 2(0.1425) = 3.29 \text{ atm}$$

$$\text{Check:} \quad \frac{(3.29)^2}{(1.86)(0.063)} = 92.4$$

When HI is made from the elements, iodine is a much more expensive reactant than hydrogen. It therefore makes sense to add hydrogen to the reaction mixture (as in Example 9.9) to ensure more complete reaction of the iodine. If one of the products is removed from an equilibrium mixture, the reaction will also occur in the forward direction to compensate partially by increasing the partial pressures of products. Most industrial operations are designed in such a way that products can be removed continuously to achieve high overall yields, even for reactions with small equilibrium constants.

***Effects of Changing the Volume***   Le Châtelier's principle also predicts the effect of a change in volume on gas-phase equilibrium. A decrease in the volume of a gaseous system causes an increase in the total pressure, and the system responds, if possible, in such a way as to reduce the total pressure. For example, in the equilibrium

$$2 \text{ P}_2(g) \rightleftharpoons \text{P}_4(g)$$

the reaction shifts in the forward direction when the volume is decreased. This occurs because every two molecules of $P_2$ consumed produce only one molecule of $P_4$, thus reducing the total pressure and partially compensating for the external stress arising from the change in volume. In contrast, an *increase* in volume favors reactants over products in this system, and some $P_4$ dissociates to form $P_2$ (Fig. 9.7). If there is no difference in the total numbers of gas-phase molecules on the two sides of the equation, then a change in volume has no effect on the equilibrium.

**FIGURE 9.7**   An equilibrium mixture of $P_2$ and $P_4$ (center) is compressed (left). Some $P_2$ molecules combine to give $P_4$ molecules, to reduce the total number of molecules and thus the total pressure. If the volume is increased (right), some $P_4$ molecules dissociate to pairs of $P_2$ molecules to increase the total number of molecules and the pressure exerted by those molecules.

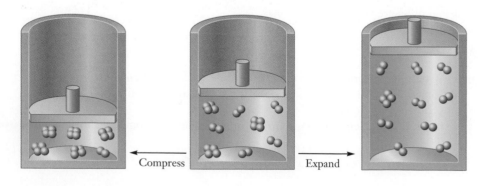

Compress          Expand

This effect of changing the volume of an equilibrium reacting mixture can also be understood by using the reaction quotient. For the phosphorus equilibrium just described, the reaction quotient is

$$Q = \frac{P_{P_4}}{(P_{P_2})^2}$$

Initially, $Q$ equals $K$. Suppose the volume is then decreased by a factor of 2; because the temperature is unchanged, this initially increases each partial pressure by a factor of 2. Because there are two powers of the pressure in the denominator and only one in the numerator, this decreases $Q$ by a factor of 2, making it lower than $K$. Reaction must then occur in the forward direction until $Q$ again equals $K$.

When the volume of a system is decreased, its total pressure increases. Another way to increase the total pressure is to add an inert gas such as argon to the reaction mixture without changing the total volume. In this case, the effect on the equilibrium is entirely different. Because the partial pressures of the reactant and product gases are unchanged by an inert gas, adding argon at constant volume has no effect on the position of the equilibrium.

*Effects of Changing the Temperature*    Chemical reactions are either **endothermic** (taking up heat from the surroundings) or **exothermic** (giving off heat). Raising the temperature of an equilibrium mixture by adding heat causes reaction to occur in such a way as to absorb some of the added heat. The equilibrium in an endothermic reaction shifts from left to right, while that in an exothermic reaction shifts from right to left, with "products" reacting to give "reactants." Equivalently, we can describe the shifts in terms of the effect of temperature on equilibrium constants. The equilibrium constant for an endothermic reaction increases with increasing temperature, while that for an exothermic reaction decreases with increasing temperature.

This effect is illustrated by the equilibrium between nitrogen dioxide ($NO_2$) and its dimer, dinitrogen tetraoxide ($N_2O_4$) (briefly considered in Example 4.6) expressed by the chemical equation

$$2\ NO_2(g) \rightleftharpoons N_2O_4(g)$$

Because $NO_2$ is a brown gas but $N_2O_4$ is colorless, the equilibrium between them can be studied by observing the color of a tube containing the two gases. At high temperatures, $NO_2$ predominates and a brown color results; as the temperature is lowered, the partial pressure of $N_2O_4$ increases and the color fades (Fig. 9.8).

The equilibrium expression for the $N_2O_4$–$NO_2$ equilibrium is

$$\frac{P_{N_2O_4}}{(P_{NO_2})^2} = K$$

$K$ has the numerical value 8.8 at $T = 25°C$, provided the partial pressures of $N_2O_4$ and $NO_2$ are expressed in atmospheres. This reaction is exothermic ($\Delta H = -58.02$ kJ mol$^{-1}$ at 298 K) because energy must be liberated when dimers are formed. Consequently, $K$ decreases as the temperature $T$ increases, so the amount of $N_2O_4$ present for a given partial pressure of $NO_2$ falls with increasing temperature as the dimer dissociates at elevated temperatures.

*Maximizing the Yield of a Reaction*    As an application of Le Châtelier's principle, consider the reaction

$$N_2(g) + 3\ H_2(g) \rightleftharpoons 2\ NH_3(g)$$

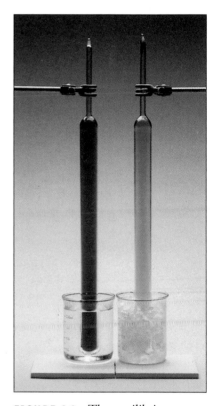

**FIGURE 9.8**    The equilibrium between $N_2O_4$ and $NO_2$ depends on temperature. The tube on the right, held in an ice bath at 0°C, contains mostly $N_2O_4$. Its color is pale because only $NO_2$ is colored. The deeper color in the tube on the left, which is held at 50°C, reflects the increased $NO_2$ present in equilibrium at the higher temperature. The tubes contain the same masses of substance, distributed in different ways between $NO_2$ and $N_2O_4$. (*Charles D. Winters*)

**FIGURE 9.9** (a) The equilibrium mole percentage of ammonia in a 1:3 mixture of $N_2$ and $H_2$ varies with temperature; low temperatures favor high yields of $NH_3$. The data shown correspond to a fixed total pressure of 300 atm. (b) At a fixed temperature (here, 500°C), the yield of $NH_3$ increases with increasing total pressure.

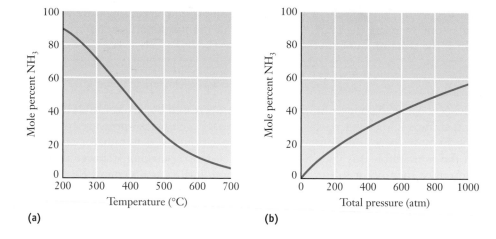

(a)

(b)

which is the basis of the industrial synthesis of ammonia. Because this reaction is exothermic, the yield of ammonia is increased by working at as low a temperature as possible (Fig. 9.9a). At too low a temperature, however, the reaction is very slow, so a compromise temperature near 500°C is usually employed. Because the number of moles of gas decreases as the reaction occurs, the yield of product is enhanced by decreasing the volume of the reaction vessel. Typically, total pressures of 150 to 300 atm are used (Fig. 9.9b), although some plants work at up to 900 atm of pressure. Even at high pressures, however, the yield of ammonia is usually only 15% to 20% because the equilibrium constant is so small. To overcome this, ammonia plants employ a cyclic process in which the gas mixture is cooled so that the ammonia liquefies (its boiling point is much higher than those of nitrogen and hydrogen) and is removed. Continuous removal of products helps to drive the reaction to completion. We will discuss the industrial production of ammonia in Chapter 21.

---

## 9.5

## TEMPERATURE DEPENDENCE OF EQUILIBRIUM CONSTANTS: THERMODYNAMIC EXPLANATION

Le Châtelier's principle is a qualitative way of describing the *stability* of equilibrium states against sudden perturbations in concentration, pressure, and temperature. The responses of the system to all three effects can be described quantitatively by thermodynamics. The effect of temperature, which is the most useful of these quantitative descriptions, is studied in this section.

The temperature dependence of the equilibrium constant is determined by the equation

$$-RT \ln K = \Delta G° = \Delta H° - T\Delta S°$$

If $\Delta H°$ and $\Delta S°$ are independent of temperature, then all the temperature dependence of $K$ lies in the factor of $T$, and the equation can be used to relate the values of $K$ at two different temperatures, as follows. At least over a limited temperature range, $\Delta H$ and $\Delta S$ do not vary much with temperature; for an example, see Fig.

8.12. To the extent that their temperature dependence may be neglected, it is evident that $\ln K$ is a linear function of $1/T$:

$$\ln K = -\frac{\Delta G°}{RT} = -\frac{\Delta H°}{RT} + \frac{\Delta S°}{R} \qquad \text{[9.6]}$$

A graph of $\ln K$ against $1/T$ is approximately a straight line with slope $-\Delta H°/R$ and intercept $\Delta S°/R$ (Fig. 9.10). If the value of $K$ is known for one temperature and $\Delta H°$ is also known, $K$ can be calculated for other temperatures.

In addition to the graphical method, an equation can be obtained to connect the values of $K$ at two different temperatures. Let $K_1$ and $K_2$ be the equilibrium constants for a reaction at temperatures $T_1$ and $T_2$, respectively. Then

$$\ln K_2 = -\frac{\Delta H°}{RT_2} + \frac{\Delta S°}{R}$$

$$\ln K_1 = -\frac{\Delta H°}{RT_1} + \frac{\Delta S°}{R}$$

Subtracting the second equation from the first gives

$$\ln\left(\frac{K_2}{K_1}\right) = -\frac{\Delta H°}{R}\left[\frac{1}{T_2} - \frac{1}{T_1}\right] \qquad \text{[9.7]}$$

This is known as the **van't Hoff equation** after the Dutch chemist Jacobus van't Hoff. Given $\Delta H°$ and $K$ at one temperature, the equation can be used to calculate $K$ at another temperature, to the approximation that $\Delta H°$ and $\Delta S°$ are constant. Alternatively, it can be used to determine $\Delta H°$ without the use of a calorimeter if the values of $K$ are known at several temperatures.

The effect of a temperature change on the equilibrium constant depends on the sign of $\Delta H°$. If $\Delta H°$ is negative (the reaction is exothermic, giving off energy as heat), then increasing the temperature *reduces K*. If $\Delta H°$ is positive (the reaction is endothermic, taking up energy as heat), then increasing the temperature *increases K*. These observations obtained from thermodynamics provide the quantitative basis for Le Châtelier's principle (Fig. 9.11).

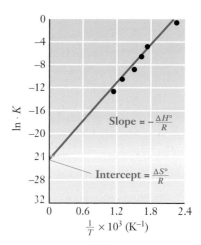

**FIGURE 9.10**   The temperature dependence of the equilibrium constant for the reaction

$$N_2(g) + 3\,H_2(g) \rightleftharpoons 2\,NH_3(g)$$

Experimental data are shown by points.

van't Hoff equation.
(given $\Delta H + \Delta S$ are constant)

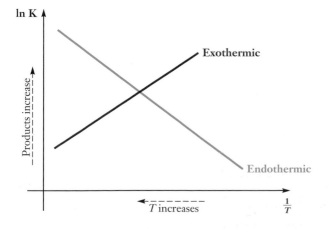

**FIGURE 9.11**   Sketch of $\ln K$ against $1/T$ for an exothermic reaction and for an endothermic reaction, as predicted by thermodynamics. Temperature increases to the left on this diagram. As $T$ increases, $K$ for the endothermic reaction increases, and $K$ for the exothermic reaction decreases in accordance with the principle of Le Châtelier.

**EXAMPLE 9.10**

Calculate $K$ for the equilibrium of Example 9.2 at $T = 400$ K, assuming $\Delta H°$ to be approximately independent of temperature over the range from 298 to 400 K.

**Solution**

The first step is to calculate $\Delta H°$ for the reaction. Appendix D provides data to calculate

$$\Delta H° = -155.52 \text{ kJ}$$

From the van't Hoff equation,

$$\ln\left(\frac{K_{400}}{K_{298}}\right) = -\frac{\Delta H°}{R}\left[\frac{1}{400 \text{ K}} - \frac{1}{298 \text{ K}}\right]$$

$$= \frac{155{,}520 \text{ J mol}^{-1}}{8.315 \text{ J K}^{-1} \text{ mol}^{-1}}\left[\frac{1}{400 \text{ K}} - \frac{1}{298 \text{ K}}\right] = -16.01$$

$$\frac{K_{400}}{K_{298}} = e^{-16.01} = 1.1 \times 10^{-7}$$

Taking $K_{298}$ to be $1.8 \times 10^{18}$ (from Example 9.2) gives

$$K_{400} = (1.8 \times 10^{18})(1.1 \times 10^{-7}) = 2.0 \times 10^{11}$$

Because the reaction is exothermic, an increase in temperature reduces the equilibrium constant.

An alternative way to do this calculation would be to determine both $\Delta H°$ and $\Delta S°$ and from them to calculate $\Delta G$ at 400 K.

**Related Problems: 51, 52**

## Temperature Dependence of Vapor Pressure

Suppose pure liquid water is in equilibrium with its vapor at temperature $T$:

$$H_2O(\ell) \rightleftharpoons H_2O(g) \qquad\qquad P_{H_2O(g)} = K$$

The temperature dependence of $K$ (and therefore of the vapor pressure $P_{H_2O(g)}$) is a special case of the van't Hoff equation. If $\Delta H_{vap}$ and $\Delta S_{vap}$ are approximately independent of temperature, then from the van't Hoff equation,

$$\ln\left(\frac{K_2}{K_1}\right) = \ln\left(\frac{P_2}{P_1}\right) = -\frac{\Delta H_{vap}}{R}\left[\frac{1}{T_2} - \frac{1}{T_1}\right]$$

where $P_2$ and $P_1$ are the vapor pressures at temperatures $T_2$ and $T_1$. From this equation, the vapor pressure at any given temperature can be estimated if its value at some other temperature is known and if the enthalpy of vaporization is also known.

At the normal boiling point of a substance, $T_b$, the vapor pressure is 1 atm. If $T_1$ is taken to correspond to $T_b$ and $T_2$ to some other temperature $T$, the van't Hoff equation is

$$\ln P = -\frac{\Delta H_{vap}}{R}\left[\frac{1}{T} - \frac{1}{T_b}\right] \tag{9.8}$$

where $P$ is the vapor pressure at temperature $T$, expressed in atmospheres.

**EXAMPLE 9.11**

The $\Delta H_{vap}$ for water is 40.66 kJ mol$^{-1}$ at the normal boiling point, $T_b = 373$ K. Assuming $\Delta H_{vap}$ and $\Delta S_{vap}$ are approximately independent of temperature from 50 to 100°C, estimate the vapor pressure of water at 50°C (323 K).

**Solution**

$$\ln P_{323} = \frac{-40660 \text{ J mol}^{-1}}{8.315 \text{ J K}^{-1} \text{ mol}^{-1}} \left[ \frac{1}{323 \text{ K}} - \frac{1}{373 \text{ K}} \right] = -2.03$$

$$P_{323} = 0.13 \text{ atm}$$

This differs slightly from the experimental value, 0.1217 atm, because $\Delta H_{vap}$ does change with temperature, an effect that was neglected in the approximate calculation.

**Related Problems: 53, 54**

## CUMULATIVE EXERCISE

### Purifying Nickel

Impure nickel, obtained from the smelting of its sulfide ores in a blast furnace, can be converted to metal of 99.9% to 99.99% purity by the Mond process, which relies on the equilibrium

$$Ni(s) + 4 CO(g) \rightleftharpoons Ni(CO)_4(g)$$

The standard enthalpy of formation of nickel tetracarbonyl, $Ni(CO)_4(g)$, is $-602.9$ kJ mol$^{-1}$, and its absolute entropy $S°$ is 410.6 J K$^{-1}$ mol$^{-1}$.

(a) Predict (without referring to a table) whether the entropy change of the system (the reacting atoms and molecules) is positive or negative in this process.

(b) At a temperature where this reaction is spontaneous, predict whether the entropy change of the surroundings is positive or negative.

(c) Use the data in Appendix D to calculate $\Delta H°$ and $\Delta S°$ for this reaction.

(d) At what temperature is $\Delta G = 0$ (and $K = 1$) for this reaction?

(e) The first step in the Mond process is the equilibration of impure nickel with CO and $Ni(CO)_4$ at about 50°C. In this step, the equilibrium constant should be as large as possible to draw as much nickel as possible into the vapor-phase complex. Calculate the equilibrium constant for the preceding reaction at 50°C.

(f) In the second step of the Mond process, the gases are removed from the reaction chamber and heated to about 230°C. At high enough temperatures, the sign of $\Delta G$ is reversed and the reaction occurs in the opposite direction, depositing pure nickel. In this step, the equilibrium constant should be as small as possible. Calculate the equilibrium constant for the preceding reaction at 230°C.

(g) The Mond process relies on the volatility of $Ni(CO)_4$ for its success. Under room conditions this compound is a liquid, but it boils at 42.2°C with an enthalpy of vaporization of 29.0 kJ mol$^{-1}$. Calculate the entropy of vaporization of $Ni(CO)_4$ and compare it with that predicted by Trouton's rule.

(h) A recently developed variation of the Mond process carries out the first step at higher pressures and at a temperature of 150°C. Estimate the maximum pressure of $Ni(CO)_4(g)$ that can be attained before the gas will liquefy at this temperature [that is, calculate the vapor pressure of $Ni(CO)_4(\ell)$ at 150°C].

*good exam question To TRY.*

**Answers**

(a) Negative

(b) Positive

(c) $\Delta H° = -160.8$ kJ; $\Delta S° = -409.5$ J K$^{-1}$

(d) 392.7 K = 119.5°C

(e) $4.0 \times 10^4$

(f) $2.0 \times 10^{-5}$

(g) 92.0 J K$^{-1}$ mol$^{-1}$, close to the Trouton's rule value of 88 J K$^{-1}$ mol$^{-1}$

(h) 16.7 atm

## CONCEPTS & SKILLS

*After studying this chapter and working the problems that follow, you should be able to*

*be able to do these.*

1. Set up an expression relating the partial pressures of the reactants and products that are in equilibrium in a chemical reaction (Section 9.1, problems 1–4).

2. Relate the equilibrium constant of a reaction to its standard Gibbs free energy change (Section 9.2, problems 7–10).

3. Calculate equilibrium constants from experimental measurements of partial and total pressures (Section 9.3, problems 11–16).

4. Combine equilibrium constants for individual reactions to obtain net equilibrium constants for combined reactions (Section 9.3, problems 17–20).

5. Calculate the equilibrium partial pressures of all species involved in a gas-phase chemical reaction from the initial pressure(s) of the reactants (Section 9.3, problems 21–30).

6. Relate concentrations to partial pressures in equilibrium calculations (Section 9.3, problems 31–32).

7. Determine the direction in which a chemical reaction will proceed spontaneously by calculating its reaction quotient (Section 9.4, problems 33–34).

8. State Le Châtelier's principle and give several applications (Section 9.4, problems 37–46).

9. Relate the change in the equilibrium constant of a reaction with temperature to its standard enthalpy change (Section 9.5, problems 47–54).

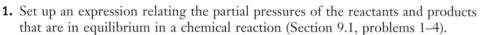

## PROBLEMS

*Answers to problems whose numbers are boldface appear in Appendix G. Problems that are more challenging are indicated with asterisks.*

### The Nature of Chemical Equilibrium

1. Write equilibrium expressions for the following gas-phase reactions.

(a) $2 H_2(g) + O_2(g) \rightleftarrows 2 H_2O(g)$

(b) $Xe(g) + 3 F_2(g) \rightleftarrows XeF_6(g)$

(c) $2 C_6H_6(g) + 15 O_2(g) \rightleftarrows 12 CO_2(g) + 6 H_2O(g)$

2. Write equilibrium expressions for the following gas-phase reactions.

(a) $2 Cl_2(g) + O_2(g) \rightleftarrows 2 Cl_2O(g)$

(b) $N_2(g) + O_2(g) + Br_2(g) \rightleftarrows 2 NOBr(g)$

(c) $C_3H_8(g) + 5 O_2(g) \rightleftarrows 3 CO_2(g) + 4 H_2O(g)$

3. At a moderately elevated temperature, phosphoryl chloride ($POCl_3$) can be produced in the vapor phase from the gaseous elements. Write a balanced chemical equation and an equilibrium expression for this system. Note that gaseous phosphorus consists of $P_4$ molecules at moderate temperatures.

4. If confined at high temperature, ammonia and oxygen quickly react and come to equilibrium with their products, water vapor and nitrogen oxide. Write a balanced chemical equation and an equilibrium expression for this system.

5. An important step in the industrial production of hydrogen is the reaction of carbon monoxide with water:

$$CO(g) + H_2O(g) \rightleftharpoons CO_2(g) + H_2(g)$$

(a) Use the law of mass action to write the equilibrium expression for this reaction.

(b) At 500°C, the equilibrium constant for this reaction is 3.9. Suppose that the equilibrium partial pressures of CO and $H_2O$ are both 0.10 atm and that of $CO_2$ is 0.70 atm. Calculate the equilibrium partial pressure of $H_2(g)$.

6. Phosgene ($COCl_2$) is an important intermediate in the manufacture of certain plastics. It is produced by the reaction

$$CO(g) + Cl_2(g) \rightleftharpoons COCl_2(g)$$

(a) Use the law of mass action to write the equilibrium expression for this reaction.

(b) At 600°C, the equilibrium constant for this reaction is 0.20. Calculate the partial pressure of phosgene in equilibrium with a mixture of CO (at 0.0020 atm) and $Cl_2$ (at 0.00030 atm).

## Thermodynamic Description of the Equilibrium State

7. Calculate $\Delta G°$ and the equilibrium constant $K$ at 25°C for the reaction

$$2\ NH_3(g) + \tfrac{7}{2}\ O_2(g) \rightleftharpoons 2\ NO_2(g) + 3\ H_2O(g)$$

using data in Appendix D.

8. Write a reaction for the dehydrogenation of gaseous ethane ($C_2H_6$) to acetylene ($C_2H_2$). Calculate $\Delta G°$ and the equilibrium constant for this reaction at 25°C, using data from Appendix D.

9. Use the thermodynamic data from Appendix D to calculate the equilibrium constant at 25°C for the following reaction:

$$SO_2(g) + \tfrac{1}{2}\ O_2(g) \rightleftharpoons SO_3(g)$$

Write an equilibrium expression for this reaction.

10. Use the thermodynamic data from Appendix D to calculate equilibrium constants at 25°C for the following reactions.
(a) $H_2(g) + N_2(g) + 2\ O_2(g) \rightleftharpoons 2\ HNO_2(g)$
(b) $Zn^{2+}(aq) + 4\ NH_3(aq) \rightleftharpoons Zn(NH_3)_4^{2+}(aq)$
Write an equilibrium expression in each case.

## Equilibrium Calculations for Gas-Phase Reactions

11. At 454 K, $Al_2Cl_6(g)$ reacts to form $Al_3Cl_9(g)$ according to the equation

$$3\ Al_2Cl_6(g) \rightleftharpoons 2\ Al_3Cl_9(g)$$

In an experiment at this temperature, the equilibrium partial pressure of $Al_2Cl_6(g)$ is 1.00 atm and the equilibrium partial pressure of $Al_3Cl_9(g)$ is $1.02 \times 10^{-2}$ atm. Compute the equilibrium constant of the preceding reaction at 454 K.

12. At 298 K, $F_3SSF(g)$ decomposes partially to $SF_2(g)$. At equilibrium, the partial pressure of $SF_2(g)$ is $1.1 \times 10^{-4}$ atm and the partial pressure of $F_3SSF$ is 0.0484 atm.
(a) Write a balanced equilibrium equation to represent this reaction.
(b) Compute the equilibrium constant corresponding to the equation you wrote.

13. The compound 1,3-di-$t$-butylcyclohexane exists in two forms that are known as the chair and boat conformations because their molecular structures resemble those objects. Equilibrium exists between the two forms, represented by the equation

$$\text{chair} \rightleftharpoons \text{boat}$$

At 580 K, 6.42% of the molecules are in the chair form. Calculate the equilibrium constant for the preceding reaction as written.

14. At 248°C and a total pressure of 1.000 atm, the fractional dissociation of $SbCl_5$ is 0.718 for the reaction

$$SbCl_5(g) \rightleftharpoons SbCl_3(g) + Cl_2(g)$$

This means that 718 out of every 1000 molecules of $SbCl_5$ originally present have dissociated. Calculate the equilibrium constant.

15. Sulfuryl chloride ($SO_2Cl_2$) is a colorless liquid that boils at 69°C. Above this temperature, the vapors dissociate into sulfur dioxide and chlorine:

$$SO_2Cl_2(g) \rightleftharpoons SO_2(g) + Cl_2(g)$$

This reaction is slow at 100°C, but it is accelerated by the presence of some $FeCl_3$ (which does not affect the final position of the equilibrium). In an experiment, 3.174 g of $SO_2Cl_2(\ell)$ and a small amount of solid $FeCl_3$ are put into an evacuated 1.000-L flask, which is then sealed and heated to 100°C. The total pressure in the flask at that temperature is found to be 1.30 atm.
(a) Calculate the partial pressure of each of the three gases present.
(b) Calculate the equilibrium constant at this temperature.

16. A certain amount of $NOBr(g)$ is sealed in a flask, and the temperature is raised to 350 K. The following equilibrium is established:

$$NOBr(g) \rightleftharpoons NO(g) + \tfrac{1}{2}\ Br_2(g)$$

The total pressure in the flask when equilibrium is reached at this temperature is 0.675 atm, and the vapor density is 2.219 g $L^{-1}$.
(a) Calculate the partial pressure of each species.
(b) Calculate the equilibrium constant at this temperature.

17. At a certain temperature, the value of the equilibrium constant for the reaction

$$CS_2(g) + 3\ O_2(g) \rightleftharpoons CO_2(g) + 2\ SO_2(g)$$

is $K_1$. How is $K_1$ related to the equilibrium constant $K_2$ for the related equilibrium

$$\tfrac{1}{3}\ CS_2(g) + O_2(g) \rightleftharpoons \tfrac{1}{3}\ CO_2(g) + \tfrac{2}{3}\ SO_2(g)$$

at the same temperature?

18. At 25°C, the equilibrium constant for the reaction

$$6\ ClO_3F(g) \rightleftharpoons 2\ ClF(g) + 4\ ClO(g) + 7\ O_2(g) + 2\ F_2(g)$$

is 32.6. Calculate the equilibrium constant at 25°C for the reaction

$$\tfrac{1}{3}\ ClF(g) + \tfrac{2}{3}\ ClO(g) + \tfrac{7}{6}\ O_2(g) + \tfrac{1}{3}\ F_2(g) \rightleftharpoons ClO_3F(g)$$

19. Suppose that $K_1$ and $K_2$ are the respective equilibrium constants for the two reactions

$$XeF_6(g) + H_2O(g) \rightleftharpoons XeOF_4(g) + 2\ HF(g)$$

$$XeO_4(g) + XeF_6(g) \rightleftharpoons XeOF_4(g) + XeO_3F_2(g)$$

Give the equilibrium constant for the reaction

$$XeO_4(g) + 2\,HF(g) \rightleftharpoons XeO_3F_2(g) + H_2O(g)$$

in terms of $K_1$ and $K_2$.

20. At 1330 K, germanium(II) oxide (GeO) and tungsten(VI) oxide ($W_2O_6$) are both gases. The following two equilibria are established simultaneously:

$$2\,GeO(g) + W_2O_6(g) \rightleftharpoons 2\,GeWO_4(g)$$

$$GeO(g) + W_2O_6(g) \rightleftharpoons GeW_2O_7(g)$$

The equilibrium constants for the two are respectively $7.0 \times 10^3$ and $38 \times 10^3$. Compute $K$ for the reaction

$$GeO(g) + GeW_2O_7(g) \rightleftharpoons 2\,GeWO_4(g)$$

21. The dehydrogenation of benzyl alcohol to make the flavoring agent benzaldehyde is an equilibrium process described by the equation

$$C_6H_5CH_2OH(g) \rightleftharpoons C_6H_5CHO(g) + H_2(g)$$

At 523 K, the value of its equilibrium constant is $K = 0.558$.
(a) Suppose 1.20 g of benzyl alcohol is placed in a 2.00-L vessel and heated to 523 K. What is the partial pressure of benzaldehyde when equilibrium is attained?
(b) What fraction of benzyl alcohol is dissociated into products at equilibrium?

22. Isopropyl alcohol can dissociate into acetone and hydrogen:

$$(CH_3)_2CHOH(g) \rightleftharpoons (CH_3)_2CO(g) + H_2(g)$$

At 179°C, the equilibrium constant for this dehydrogenation reaction is 0.444.
(a) If 10.00 g of isopropyl alcohol is placed in a 10.00-L vessel and heated to 179°C, what is the partial pressure of acetone when equilibrium is attained?
(b) What fraction of isopropyl alcohol is dissociated at equilibrium?

23. A weighed quantity of $PCl_5(s)$ is sealed in a 100.0-cm$^3$ glass bulb to which a pressure gauge is attached. The bulb is heated to 250°C, and the gauge shows that the pressure in the bulb rises to 0.895 atm. At this temperature, the solid $PCl_5$ is all vaporized and also partially dissociated into $Cl_2(g)$ and $PCl_3(g)$ according to the equation

$$PCl_5(g) \rightleftharpoons PCl_3(g) + Cl_2(g)$$

At 250°C, $K = 2.15$ for this reaction. Assume that the contents of the bulb are at equilibrium and calculate the partial pressure of the three different chemical species in the vessel.

24. Suppose 93.0 g of HI(g) is placed in a glass vessel and heated to 1107 K. At this temperature, equilibrium is quickly established between HI(g) and its decomposition products, $H_2(g)$ and $I_2(g)$:

$$2\,HI(g) \rightleftharpoons H_2(g) + I_2(g)$$

The equilibrium constant at 1107 K is 0.0259, and the total pressure at equilibrium is observed to equal 6.45 atm. Calculate the equilibrium partial pressures of HI(g), $H_2(g)$, and $I_2(g)$.

25. The equilibrium constant at 350 K for the reaction

$$Br_2(g) + I_2(g) \rightleftharpoons 2\,IBr(g)$$

has a value of 322. Bromine at an initial partial pressure of 0.0500 atm is mixed with iodine at an initial partial pressure of 0.0400 atm and held at 350 K until equilibrium is reached. Calculate the equilibrium partial pressure of each of the gases.

26. The equilibrium constant for the reaction of fluorine and oxygen to form oxygen difluoride ($OF_2$) is 40.1 at 298 K:

$$F_2(g) + \tfrac{1}{2}\,O_2(g) \rightleftharpoons OF_2(g)$$

Suppose some $OF_2$ is introduced into an evacuated container at 298 K and allowed to dissociate until its partial pressure reaches an equilibrium value of 1.00 atm. Calculate the equilibrium partial pressures of $F_2$ and $O_2$ in the container.

27. At 25°C, the equilibrium constant for the reaction

$$N_2(g) + O_2(g) \rightleftharpoons 2\,NO(g)$$

is $4.2 \times 10^{-31}$. Suppose a container is filled with nitrogen (at an initial partial pressure of 0.41 atm), oxygen (at an initial partial pressure of 0.59 atm), and nitrogen oxide (at an initial partial pressure of 0.22 atm). Calculate the partial pressures of all three gases after equilibrium is reached at this temperature.

28. At 25°C, the equilibrium constant for the reaction

$$2\,NO_2(g) \rightleftharpoons 2\,NO(g) + O_2(g)$$

is $5.9 \times 10^{-13}$. Suppose a container is filled with nitrogen dioxide at an initial partial pressure of 0.89 atm. Calculate the partial pressures of all three gases after equilibrium is reached at this temperature.

\*29. The equilibrium constant for the synthesis of ammonia

$$N_2(g) + 3\,H_2(g) \rightleftharpoons 2\,NH_3(g)$$

has the value $K = 6.78 \times 10^5$ at 25°C. Calculate the equilibrium partial pressures of $N_2(g)$, $H_2(g)$, and $NH_3(g)$ at 25°C if the total pressure is 1.00 atm and the H:N atom ratio in the system is 3:1. (*Hint:* Try the approximation that $P_{N_2}$ and $P_{H_2} \ll P_{NH_3}$ and see if the resulting equations are simplified.)

\*30. At 400°C, $K = 3.19 \times 10^{-4}$ for the reaction in problem 29. Repeat the calculation for $P_{N_2}$, $P_{H_2}$, and $P_{NH_3}$, assuming the same total pressure and composition. (*Hint:* Try the approximation that $P_{NH_3} \ll P_{N_2}$ and $P_{H_2}$ and see if the resulting equations are simplified.)

31. Calculate the concentration of phosgene ($COCl_2$) that will be present at 600°C in equilibrium with carbon monoxide (at a concentration of $2.3 \times 10^{-4}$ mol L$^{-1}$) and chlorine (at a concentration of $1.7 \times 10^{-2}$ mol L$^{-1}$). (Use the data of problem 6.)

32. The reaction

$$SO_2Cl_2(g) \rightleftharpoons SO_2(g) + Cl_2(g)$$

has an equilibrium constant at 100°C of 2.40. Calculate the concentration of $SO_2$ that will be present at 100°C in equilibrium with $SO_2Cl_2$ (at a concentration of $3.6 \times 10^{-4}$ mol L$^{-1}$) and chlorine (at a concentration of $6.9 \times 10^{-3}$ mol L$^{-1}$).

## Magnitude of $K$ and the Direction of Change

**33.** Some $Al_2Cl_6$ (at a partial pressure of 0.473 atm) is placed in a closed container at 454 K with some $Al_3Cl_9$ (at a partial pressure of $1.02 \times 10^{-2}$ atm). Enough argon is added to raise the total pressure to 1.00 atm.

(a) Calculate the initial reaction quotient for the reaction

$$3 \, Al_2Cl_6(g) \rightleftharpoons 2 \, Al_3Cl_9(g)$$

(b) As the gas mixture reaches equilibrium, will there be net production or consumption of $Al_3Cl_9$? (Use the data given in problem 11.)

**34.** Some $SF_2$ (at a partial pressure of $2.3 \times 10^{-4}$ atm) is placed in a closed container at 298 K with some $F_3SSF$ (at a partial pressure of 0.0484 atm). Enough argon is added to raise the total pressure to 1.000 atm.

(a) Calculate the initial reaction quotient for the decomposition of $F_3SSF$ to $SF_2$.

(b) As the gas mixture reaches equilibrium, will there be net formation or dissociation of $F_3SSF$? (Use the data given in problem 12.)

**35.** The progress of the reaction

$$H_2(g) + Br_2(g) \rightleftharpoons 2 \, HBr(g)$$

can be monitored visually by following changes in the color of the reaction mixture ($Br_2$ is reddish brown, and $H_2$ and HBr are colorless). A gas mixture is prepared at 700 K, in which 0.40 atm is the initial partial pressure of both $H_2$ and $Br_2$ and 0.90 atm is the initial partial pressure of HBr. The color of this mixture then fades as the reaction progresses toward equilibrium. Give a condition that must be satisfied by the equilibrium constant $K$ (for example, it must be greater than or smaller than a given number).

**36.** Recall from our discussion of the $NO_2$–$N_2O_4$ equilibrium that $NO_2$ has a brownish color. At elevated temperatures, $NO_2$ reacts with CO according to

$$NO_2(g) + CO(g) \rightleftharpoons NO(g) + CO_2(g)$$

The other three gases taking part in this reaction are colorless. When a gas mixture is prepared at 500 K, in which 3.4 atm is the initial partial pressure of both $NO_2$ and CO, and 1.4 atm is the partial pressure of both NO and $CO_2$, the brown color of the mixture is observed to fade as the reaction progresses toward equilibrium. Give a condition that must be satisfied by the equilibrium constant $K$ (for example, it must be greater than or smaller than a given number).

**37.** At $T = 1200°C$, the reaction

$$P_4(g) \rightleftharpoons 2 \, P_2(g)$$

has an equilibrium constant $K = 0.612$.

(a) Suppose the initial partial pressure of $P_4$ is 5.00 atm and that of $P_2$ is 2.00 atm. Calculate the reaction quotient $Q$ and state whether the reaction proceeds to the right or to the left as equilibrium is approached.

(b) Calculate the partial pressures at equilibrium.

(c) If the volume of the system is then increased, will there be net formation or net dissociation of $P_4$?

**38.** At $T = 100°C$, the reaction

$$SO_2Cl_2(g) \rightleftharpoons SO_2(g) + Cl_2(g)$$

has an equilibrium constant $K = 2.4$.

(a) Suppose the initial partial pressure of $SO_2Cl_2$ is 1.20 atm, and $P_{SO_2} = P_{Cl_2} = 0$. Calculate the reaction quotient $Q$ and state whether the reaction proceeds to the right or to the left as equilibrium is approached.

(b) Calculate the partial pressures at equilibrium.

(c) If the volume of the system is then decreased, will there be net formation or net dissociation of $SO_2Cl_2$?

**39.** Explain the effect of each of the following stresses on the position of the following equilibrium:

$$3 \, NO(g) \rightleftharpoons N_2O(g) + NO_2(g)$$

The reaction as written is exothermic.

(a) $N_2O(g)$ is added to the equilibrium mixture without change of volume or temperature.

(b) The volume of the equilibrium mixture is reduced at constant temperature.

(c) The equilibrium mixture is cooled.

(d) Gaseous argon (which does not react) is added to the equilibrium mixture while both the total gas pressure and the temperature are kept constant.

(e) Gaseous argon is added to the equilibrium mixture without changing the volume.

**40.** Explain the effect of each of the following stresses on the position of the equilibrium

$$SO_3(g) \rightleftharpoons SO_2(g) + \tfrac{1}{2} O_2(g)$$

The reaction as written is endothermic.

(a) $O_2(g)$ is added to the equilibrium mixture without changing volume or temperature.

(b) The mixture is compressed at constant temperature.

(c) The equilibrium mixture is cooled.

(d) An inert gas is pumped into the equilibrium mixture while the total gas pressure and the temperature are kept constant.

(e) An inert gas is added to the equilibrium mixture without changing the volume.

**41.** In a gas-phase reaction, it is observed that the equilibrium yield of products is increased by lowering the temperature and by reducing the volume.

(a) Is the reaction exothermic or endothermic?

(b) Is there a net increase or a net decrease in the number of gas molecules in the reaction?

**42.** The equilibrium constant of a gas-phase reaction increases as temperature is increased. When the nonreacting gas neon is admitted to a mixture of reacting gases (holding the temperature and the total pressure fixed and increasing the volume of the reaction vessel), the product yield is observed to decrease.

(a) Is the reaction exothermic or endothermic?

(b) Is there a net increase or a net decrease in the number of gas molecules in the reaction?

**43.** The most extensively used organic compound in the chemical industry is ethylene ($C_2H_4$). The two equations

$$C_2H_4(g) + Cl_2(g) \rightleftharpoons C_2H_4Cl_2(g)$$

$$C_2H_4Cl_2(g) \rightleftharpoons C_2H_3Cl(g) + HCl(g)$$

represent the way in which vinyl chloride ($C_2H_3Cl$) is synthesized for eventual use in polymeric plastics (polyvinyl chloride, PVC). The byproduct of the reaction, HCl, is now most cheaply made by this and similar reactions, rather than by direct combination of $H_2$ and $Cl_2$. Heat is given off in the first reaction and taken up in the second. Describe how you would design an industrial process to maximize the yield of vinyl chloride.

**44.** Methanol is made via the exothermic reaction

$$CO(g) + 2 H_2(g) \longrightarrow CH_3OH(g)$$

Describe how you would control the temperature and pressure to maximize the yield of methanol.

**45.** One way to manufacture ethanol is by the reaction

$$C_2H_4(g) + H_2O(g) \rightleftharpoons C_2H_5OH(g)$$

The $\Delta H_f^\circ$ of $C_2H_4(g)$ is 52.3 kJ mol$^{-1}$; of $H_2O(g)$, $-241.8$ kJ mol$^{-1}$; and of $C_2H_5OH(g)$, $-235.3$ kJ mol$^{-1}$. Without doing detailed calculations, suggest the conditions of pressure and temperature that will maximize the yield of ethanol at equilibrium.

**46.** Dimethyl ether ($CH_3OCH_3$) is a good substitute for environmentally harmful propellants in aerosol spray cans. It is produced by the dehydration of methanol:

$$2 CH_3OH(g) \rightleftharpoons CH_3OCH_3(g) + H_2O(g)$$

Describe reaction conditions that favor the equilibrium production of this valuable chemical. As a basis for your answer, compute $\Delta H^\circ$ and $\Delta S^\circ$ of the reaction from the data in Appendix D.

## Temperature Dependence of Equilibrium Constants: Thermodynamic Explanation

**47.** The equilibrium constant at 25°C for the reaction

$$2 NO_2(g) \rightleftharpoons N_2O_4(g)$$

is 6.8. At 200°C, the equilibrium constant is $1.21 \times 10^{-3}$. Calculate the enthalpy change ($\Delta H$) for this reaction, assuming that $\Delta H$ and $\Delta S$ of the reaction are constant over the temperature range from 25 to 200°C.

**48.** Stearic acid dimerizes when dissolved in hexane:

$$2 C_{17}H_{35}COOH(hexane) \longleftrightarrow (C_{17}H_{35}COOH)_2(hexane)$$

The equilibrium constant for this reaction is 2900 at 28°C, but it drops to 40 at 48°C. Estimate $\Delta H^\circ$ and $\Delta S^\circ$ for the reaction.

**49.** The equilibrium constant for the reaction

$$\tfrac{1}{2} Cl_2(g) + \tfrac{1}{2} F_2(g) \rightleftharpoons ClF(g)$$

is measured to be $9.3 \times 10^9$ at 298 K and $3.3 \times 10^7$ at 398 K.

(a) Calculate $\Delta G^\circ$ at 298 K for the reaction.

(b) Calculate $\Delta H^\circ$ and $\Delta S^\circ$, assuming the enthalpy and entropy changes to be independent of temperature between 298 and 398 K.

**50.** Stearic acid also dimerizes when dissolved in carbon tetrachloride:

$$2 C_{17}H_{35}COOH(CCl_4) \rightleftharpoons (C_{17}H_{35}COOH)_2(CCl_4)$$

The equilibrium constant for this reaction is 2780 at 22°C, but it drops to 93.1 at 42°C. Estimate $\Delta H^\circ$ and $\Delta S^\circ$ for the reaction.

**51.** For the synthesis of ammonia from its elements,

$$3 H_2(g) + N_2(g) \rightleftharpoons 2 NH_3(g)$$

the equilibrium constant $K = 5.9 \times 10^5$ at 298 K, and $\Delta H^\circ = -92.2$ kJ. Calculate the equilibrium constant for the reaction at 600 K, assuming no change in $\Delta H^\circ$ and $\Delta S^\circ$ between 298 K and 600 K.

**52.** The practice exercise at the end of Chapter 8 explored reaction steps in the manufacture of sulfuric acid, including the oxidation of sulfur dioxide to sulfur trioxide:

$$SO_2(g) + \tfrac{1}{2} O_2(g) \rightleftharpoons SO_3(g)$$

At 25°C the equilibrium constant for this reaction is $2.6 \times 10^{12}$, but the reaction occurs very slowly. Calculate $K$ for this reaction at 550°C, assuming $\Delta H^\circ$ and $\Delta S^\circ$ are independent of temperature in the range from 25 to 550°C.

**53.** The vapor pressure of ammonia at $-50$°C is 0.4034 atm; at 0°C, it is 4.2380 atm.

(a) Calculate the molar enthalpy of vaporization ($\Delta H_{vap}$) of ammonia.

(b) Calculate the normal boiling temperature of $NH_3(\ell)$.

**54.** The vapor pressure of butyl alcohol ($C_4H_9OH$) at 70.1°C is 0.1316 atm; at 100.8°C, it is 0.5263 atm.

(a) Calculate the molar enthalpy of vaporization ($\Delta H_{vap}$) of butyl alcohol.

(b) Calculate the normal boiling point of butyl alcohol.

## Additional Problems

**55.** At 298 K, unequal amounts of $BCl_3(g)$ and $BF_3(g)$ were mixed in a container. The gases reacted to form $BFCl_2(g)$ and $BClF_2(g)$. When equilibrium was finally reached, the four gases were present in these relative chemical amounts: $BCl_3$(90), $BF_3$(470), $BClF_2$(200), $BFCl_2$(45).

(a) Determine the equilibrium constants at 298 K of the two reactions

$$2 BCl_3(g) + BF_3(g) \rightleftharpoons 3 BFCl_2(g)$$

$$BCl_3(g) + 2 BF_3(g) \rightleftharpoons 3 BClF_2(g)$$

(b) Determine the equilibrium constant of the reaction

$$BCl_3(g) + BF_3(g) \rightleftharpoons BFCl_2 + BClF_2(g)$$

and explain why knowing this equilibrium constant really adds nothing to what you knew in part (a).

56. Methanol can be synthesized by means of the equilibrium reaction

$$CO(g) + 2 H_2(g) \rightleftharpoons CH_3OH(g)$$

for which the equilibrium constant at 225°C is $6.08 \times 10^{-3}$. Assume that the ratio of the pressures of $CO(g)$ and $H_2(g)$ is $1:2$. What values should they have if the partial pressure of methanol is to be 0.500 atm?

57. At equilibrium at 425.6°C, a sample of *cis*-1-methyl-2-ethyl-cyclopropane is 73.6% converted into the *trans* form:

$$cis \rightleftharpoons trans$$

   (a) Compute the equilibrium constant $K$ for this reaction.
   (b) Suppose that 0.525 mol of the *cis* compound is placed in a 15.00-L vessel and heated to 425.6°C. Compute the equilibrium partial pressure of the *trans* compound.

58. The equilibrium constant for the reaction

$$(CH_3)_3COH(g) \rightleftharpoons (CH_3)_2CCH_2(g) + H_2O(g)$$

   is 2.42 at 450 K.
   (a) A pure sample of the reactant, which is named "*t*-butanol," is confined in a container of fixed volume at a temperature of 450 K and at an original pressure of 0.100 atm. Calculate the fraction of this starting material that is converted to products at equilibrium.
   (b) A second sample of the reactant is confined, this time at an original pressure of 5.00 atm. Again, calculate the fraction of the starting material that is converted to products at equilibrium.

59. The hydrogenation of pyridine to piperidine

$$C_5H_5N(g) + 3 H_2(g) \rightleftharpoons C_5H_{11}N(g)$$

   is an equilibrium process whose equilibrium constant is given by the equation

$$\log_{10} K = -20.281 + \frac{10.560\ K}{T}$$

   (a) Calculate the value of $K$ at $T = 500$ K.

   (b) If the partial pressure of hydrogen is 1.00 atm, what fraction of the nitrogen is in the form of pyridine molecules at $T = 500$ K?

60. Acetic acid in the vapor phase consists of both monomeric and dimeric forms in equilibrium:

$$2 CH_3COOH(g) \rightleftharpoons (CH_3COOH)_2(g)$$

   At 110°C, the equilibrium constant for this reaction is 3.72.
   (a) Calculate the partial pressure of the dimer when the total pressure is 0.725 atm at equilibrium.
   (b) What percentage of the acetic acid is dimerized under these conditions?

*61. Repeat the calculation of problem 30 with a total pressure of 100 atm. (*Note:* Here it is necessary to solve the equation by successive approximations. See Appendix C.)

*62. At 900 K, the equilibrium constant for the reaction

$$\tfrac{1}{2} O_2(g) + SO_2(g) \rightleftharpoons SO_3(g)$$

   has the value $K = 0.587$. What will be the equilibrium partial pressure of $O_2(g)$ if a sample of $SO_3$ weighing 0.800 g is heated to 900 K in a quartz-glass vessel whose volume is 100.0 cm³? (*Note:* Here it is necessary to solve the equation by successive approximations. See Appendix C.)

63. At 300°C, the equilibrium constant for the reaction

$$PCl_5(g) \rightleftharpoons PCl_3(g) + Cl_2(g)$$

   is $K = 11.5$.
   (a) Calculate the reaction quotient $Q$ if initially $P_{PCl_3} = 2.0$ atm, $P_{Cl_2} = 6.0$ atm, and $P_{PCl_5} = 0.10$ atm. State whether the reaction proceeds to the right or to the left as equilibrium is approached.
   (b) Calculate $P_{PCl_3}$, $P_{Cl_2}$, and $P_{PCl_5}$ at equilibrium.
   (c) If the volume of the system is then increased, will the amount of $PCl_5$ present increase or decrease?

64. Although the process of dissolution of helium gas in water is favored in terms of energy, helium is only very slightly soluble in water. What keeps this gas from dissolving in great quantities in water?

---

## CUMULATIVE PROBLEMS

65. The reaction

$$P_4(g) \rightleftharpoons 2 P_2(g)$$

   is endothermic and begins to occur at moderate temperatures.
   (a) In which direction do you expect deviations to occur from Boyle's law, $P \propto 1/V$ (constant $T$), for gaseous $P_4$?
   (b) In which direction do you expect deviations to occur from Charles's law, $V \propto T$ (constant $P$), for gaseous $P_4$?

66. A 4.72-g mass of methanol, $CH_3OH$, is placed in an evacuated 1.00-L flask and heated to 250°C. It vaporizes and then reaches the following equilibrium:

$$CH_3OH(g) \rightleftharpoons CO(g) + 2 H_2(g)$$

   A tiny hole forms in the side of the container, allowing a small amount of gas to effuse out. Analysis of the escaping gases shows that the rate of escape of the hydrogen is 33 times the rate of escape of the methanol. Calculate the equilibrium constant for the preceding reaction at 250°C.

67. The triple bond in the $N_2$ molecule is very strong, but at high enough temperatures even it breaks down. At 5000 K, when the total pressure exerted by a sample of nitrogen is 1.00 atm, $N_2(g)$ is 0.65% dissociated at equilibrium:

$$N_2(g) \rightleftharpoons 2 N(g)$$

   At 6000 K with the same total pressure, the proportion of $N_2(g)$ dissociated at equilibrium rises to 11.6%. Use the van't Hoff equation to estimate the $\Delta H$ of this reaction.

# Acid–Base Equilibria

*Illustration*
Many naturally occurring dyes change color as the acidity of their surroundings changes. The compound cyanidin is blue in the basic sap of the cornflower and red in the acidic sap of the poppy. Such dyes can be used as indicators of the degree of acidity in a medium. *(Copyright Hans Reinhard/Okapia/Photo Researchers, Inc.)*

Section 6.3 introduced acids and bases, neutralization reactions, and acid–base titrations. The present chapter will apply the ideas about chemical equilibrium developed in Chapter 9 to an especially important aspect of acid–base chemistry: calculating how far such reactions proceed before reaching equilibrium.

Although many acid–base reactions take place in the gaseous and solid states and in nonaqueous solutions, in this chapter the focus is on acid–base reactions in aqueous solutions. Such solutions play roles in everyday life. Vinegar, orange juice, and battery fluid are familiar acidic aqueous solutions; basic solutions are produced when such common products as borax, baking soda, and antacids are dissolved in water. Acid–base chemistry will appear many times throughout the rest of this book. We will consider the effects of acidity on the dissolution of solids in Chapter 11, on redox reactions in electrochemical cells in Chapter 12, and on the rates of reaction in Chapter 13. Chapter 21 discusses the manufacture and uses of the workhorse industrial acids (nitric, phosphoric, and sulfuric). Chapter 25 points out the central biochemical roles of amino acids and nucleic acids. The reactions of acids and bases lie at the heart of all chemistry.

---

## 10.1

### CLASSIFICATIONS OF ACIDS AND BASES

Section 6.3 introduced acids and bases, following the definitions of Arrhenius. An **acid** is a substance that when dissolved in water increases the concentration of hydrogen ion ($H^+$) above the value it takes in pure water. A **base** increases the concentration of hydroxide ion ($OH^-$). These definitions, introduced in the 19th century, have been generalized to accommodate broader classes of compounds that were chemically similar to the familiar acids and bases. The present section introduces the modern definitions due to Brønsted and Lowry and to Lewis.

### Brønsted–Lowry Acids and Bases

A broader definition of acids and bases, which will be useful in quantitative calculations in this chapter, was proposed independently by Johannes Brønsted and Thomas Lowry in 1923. A **Brønsted–Lowry acid** is defined as a substance that can donate a hydrogen ion, and a **Brønsted–Lowry base** is a substance that can accept a hydrogen ion. In a Brønsted–Lowry acid–base reaction, hydrogen ions are transferred from the acid to the base. For example, when acetic acid is dissolved in water,

$$CH_3COOH(aq) + H_2O(\ell) \rightleftharpoons H_3O^+(aq) + CH_3COO^-(aq)$$
$$\text{Acid}_1 \qquad \text{Base}_2 \qquad\quad \text{Acid}_2 \qquad \text{Base}_1$$

hydrogen ions are transferred from acetic acid to water. Throughout this chapter, the hydronium ion, $H_3O^+(aq)$, rather than $H^+(aq)$, will be used to represent the true nature of hydrogen ions in water (Fig. 10.1). Acids and bases occur as **conjugate acid–base pairs.** $CH_3COOH$ and $CH_3COO^-$ form such a pair, where $CH_3COO^-$ is the conjugate base of $CH_3COOH$ (equivalently, $CH_3COOH$ is the conjugate acid of $CH_3COO^-$). In the same way, $H_3O^+$ and $H_2O$ form a conjugate acid–base pair. The equilibrium that is established may be pictured as the competition between two bases for hydrogen ions. For example, when ammonia is dissolved in water,

$$H_2O(\ell) + NH_3(aq) \rightleftharpoons NH_4^+(aq) + OH^-(aq)$$
$$\text{Acid}_1 \qquad \text{Base}_2 \qquad\quad \text{Acid}_2 \qquad \text{Base}_1$$

the two bases $NH_3$ and $OH^-$ compete for hydrogen ions.

*[handwritten margin notes:]*
*acid – donate*
*base – accept*
*bases compete for hydrogen ions*
*$H_3O^+ \approx H^+$ (when lazy)*

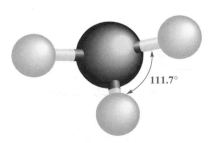

**FIGURE 10.1** The structure of the hydronium ion ($H_3O^+$).

One advantage of the Brønsted–Lowry approach is that it is not limited to aqueous solutions. An example with liquid ammonia as the solvent is

$$HCl(\text{in } NH_3) + NH_3(\ell) \rightleftharpoons NH_4^+(\text{in } NH_3) + Cl^-(\text{in } NH_3)$$

Acid$_1$          Base$_2$                          Acid$_2$          Base$_1$

The $NH_3$ acts as a base even though hydroxide ion, $OH^-$, is not present. The Arrhenius scheme that was introduced in Chapter 6 has been extended and freed from its limitation to aqueous solution.

Some molecules and ions function either as acids or bases depending on reaction conditions and are called **amphoteric.** The most common example is water itself. Water acts as an acid in donating a hydrogen ion to $NH_3$ (its conjugate base here is $OH^-$) and as a base in accepting a hydrogen ion from $CH_3COOH$ (its conjugate acid here is $H_3O^+$). In the same way, the hydrogen carbonate ion can act as an acid

$$HCO_3^-(aq) + H_2O(\ell) \rightleftharpoons H_3O^+(aq) + CO_3^{2-}(aq)$$

or as a base:

$$HCO_3^-(aq) + H_2O(\ell) \rightleftharpoons H_2CO_3(aq) + OH^-(aq)$$

## Lewis Acids and Bases

The Lewis structure model can be used to describe a more general kind of acid–base behavior of which the Arrhenius and Brønsted–Lowry definitions are special cases. A **Lewis base** is any species that donates lone-pair electrons, and a **Lewis acid** is any species that accepts such electron pairs. The Arrhenius acids and bases considered so far fit this description (with the Lewis acid, $H^+$, acting as an acceptor toward various Lewis bases such as $NH_3$ and $OH^-$, the electron pair donors). Other reactions that do *not* involve hydrogen ions can still be considered Lewis acid–base reactions. An example is the reaction between electron-deficient $BF_3$ and electron-rich $NH_3$:

$$:\overset{\displaystyle ..}{\underset{\displaystyle ..}{F}}-B \ + \ :N-H \ \longrightarrow \ :\overset{\displaystyle ..}{\underset{\displaystyle ..}{F}}-B-N-H$$

Here ammonia, the Lewis base, donates lone-pair electrons to $BF_3$, the Lewis acid or electron acceptor. The bond that forms is a **coordinate covalent bond,** in which both electrons are supplied by a lone pair on the Lewis base.

Octet-deficient compounds involving elements of Group III such as boron and aluminum are often strong Lewis acids, because Group III atoms can achieve octet configurations by forming coordinate covalent bonds. Atoms and ions from Groups V through VII have the necessary lone pairs to act as Lewis bases. Compounds of main-group elements from the later periods can also act as Lewis acids through valence expansion. In such reactions, the central atom accepts a share in additional lone pairs beyond the eight electrons needed to satisfy the octet rule. For example, $SnCl_4$ is a Lewis acid that accepts electrons from chloride ion lone pairs:

$$SnCl_4(\ell) + 2\,Cl^-(aq) \longrightarrow [SnCl_6]^{2-}(aq)$$

After the reaction, each tin atom is surrounded by 12 rather than 8 valence electrons.

The Lewis definition systematizes the chemistry of a great many binary oxides, which can be considered as anhydrides of acids or bases. An **acid anhydride** is obtained by removing water from an oxoacid (see Section 6.3) until only the oxide remains; thus, $CO_2$ is the anhydride of carbonic acid ($H_2CO_3$).

---

### EXAMPLE 10.1

What is the acid anhydride of phosphoric acid ($H_3PO_4$)?

**Solution**

If the formula unit $H_3PO_4$ (which contains an odd number of hydrogen atoms) is doubled, $H_6P_2O_8$ is obtained. Subtraction of 3 $H_2O$ from this gives $P_2O_5$, which is the empirical formula of tetraphosphorus decaoxide ($P_4O_{10}$). This compound is the acid anhydride of phosphoric acid.

**Related Problems: 9, 10**

---

The oxides of most of the nonmetals are acid anhydrides, which react with an excess of water to form acidic solutions. Examples are

$$N_2O_5(s) + H_2O(\ell) \longrightarrow 2\ HNO_3(aq) \longrightarrow 2\ H^+(aq) + 2\ NO_3^-(aq)$$

$$SO_3(g) + H_2O(\ell) \longrightarrow H_2SO_4(aq) \longrightarrow H^+(aq) + HSO_4^-(aq)$$

Silica ($SiO_2$) is the acid anhydride of the very weak silicic acid $H_2SiO_3$, a gelatinous material that is insoluble in water but readily dissolves in strongly basic aqueous solutions according to

$$H_2SiO_3(s) + 2\ OH^-(aq) \longrightarrow SiO_3^{2-}(aq) + 2\ H_2O(\ell)$$

Oxides of Group I and II metals are **base anhydrides,** obtained by removing water from the corresponding hydroxides. Calcium oxide, CaO, is the base anhydride of calcium hydroxide, $Ca(OH)_2$. The removal of water from $Ca(OH)_2$ is the reverse of the addition of water to the oxide:

$$CaO(s) + H_2O(\ell) \longrightarrow Ca(OH)_2(s)$$

$$Ca(OH)_2(s) \longrightarrow CaO(s) + H_2O(\ell)$$

The base anhydride of NaOH is $Na_2O$. Oxides of metals in the middle groups of the periodic table (III through V) lie on the border between ionic and covalent behavior and are frequently amphoteric. An example is aluminum oxide ($Al_2O_3$), which dissolves to only a limited extent in water but much more readily in either acids or bases:

Acting as a base:    $Al_2O_3(s) + 6\ H^+(aq) \longrightarrow 2\ Al^{3+}(aq) + 3\ H_2O(\ell)$

Acting as an acid:    $Al_2O_3(s) + 2\ OH^-(aq) + 3\ H_2O(\ell) \longrightarrow 2\ Al(OH)_4^-(aq)$

Figure 10.2 summarizes the acid–base character of the main-group oxides in the first periods.

Although oxoacids and hydroxides are Arrhenius acids and bases [releasing $H^+(aq)$ or $OH^-(aq)$ into aqueous solution], acid and base anhydrides do not fall into

Increasing acidity ⟶

| I | II | III | IV | V | VI | VII |
|---|---|---|---|---|---|---|
| $Li_2O$ | BeO | $B_2O_3$ | $CO_2$ | $N_2O_5$ | $(O_2)$ | $OF_2$ |
| $Na_2O$ | MgO | $Al_2O_3$ | $SiO_2$ | $P_4O_{10}$ | $SO_3$ | $Cl_2O_7$ |
| $K_2O$ | CaO | $Ga_2O_3$ | $GeO_2$ | $As_2O_5$ | $SeO_3$ | $Br_2O_7$ |
| $Rb_2O$ | SrO | $In_2O_3$ | $SnO_2$ | $Sb_2O_5$ | $TeO_3$ | $I_2O_7$ |
| $Cs_2O$ | BaO | $Tl_2O_3$ | $PbO_2$ | $Bi_2O_5$ | $PoO_3$ | $At_2O_7$ |

Increasing basicity ↓     Increasing acidity ↑

⟵ Increasing basicity

**FIGURE 10.2** Among the oxides of the main-group elements, acidity tends to increase from left to right and from bottom to top in the periodic table. Oxygen difluoride, however, has only weakly acidic properties.

*€ trend , MCQ*

this classification; they contain neither $H^+$ nor $OH^-$. Acid anhydrides are still acids in the Lewis sense however (they accept electron pairs), and base anhydrides are bases in the Lewis sense (their $O^{2-}$ ions donate electron pairs). The reaction between an acid anhydride and a base anhydride is then a Lewis acid–base reaction. An example of such a reaction is

$$CaO(s) + CO_2(g) \longrightarrow CaCO_3(s)$$

Here, the Lewis base CaO donates an electron pair (one of the lone pairs of the oxygen atom) to the Lewis acid ($CO_2$) to form a coordinate covalent bond in the $CO_3^{2-}$ ion. Similar Lewis acid–base reactions can be written for other acid–base anhydride pairs. Sulfur trioxide, for example, reacts with metal oxides to form sulfates:

$$MgO(s) + SO_3(g) \longrightarrow MgSO_4(s)$$

Note that these reactions are not redox reactions (oxidation numbers do not change). They are clearly not dissolution or precipitation reactions, nor are they acid–base reactions in the Arrhenius sense. They can be usefully classified as acid–base reactions in the Lewis sense, however.

## Comparison of Arrhenius, Brønsted–Lowry, and Lewis Definitions

The neutralization reaction between HCl and NaOH

$$\underset{\text{Acid}}{HCl} + \underset{\text{Base}}{NaOH} \longrightarrow \underset{\text{Water}}{H_2O} + \underset{\text{Salt}}{NaCl}$$

introduced in Section 6.3 shows the progressive generality in these definitions. By the Arrhenius definition, HCl is the acid and NaOH is the base. By the Brønsted–Lowry definition, $H_3O^+$ is the acid and $OH^-$ is the base. According to Lewis, $H^+$ is the acid and $OH^-$ the base, since the proton accepts the lone pair donated by $OH^-$ in the reaction

$$H^+(aq) + OH^-(aq) \longrightarrow H_2O(\ell)$$

## 10.2

# PROPERTIES OF ACIDS AND BASES IN AQUEOUS SOLUTION: THE BRØNSTED–LOWRY SCHEME

Chapters 5 and 6 described the special properties of liquid water. Because of its substantial dipole moment, water is especially effective as a solvent, stabilizing both polar and ionic solutes. Water participates in acid–base reactions as reactant as well as solvent. Biochemical reactions essential to the survival of living organisms occur in water and involve acids, bases, and ionic species. In view of the wide-ranging importance of these reactions, the remainder of this chapter is devoted to acid–base behavior and related ionic reactions in aqueous solution, for which the Brønsted–Lowry definition of acids and bases is especially well suited.

## Autoionization of Water

What happens when water acts as both acid and base in the same reaction? The resulting equilibrium is

$$H_2O(\ell) + H_2O(\ell) \rightleftharpoons H_3O^+(aq) + OH^-(aq)$$

$$\text{Acid}_1 \qquad \text{Base}_2 \qquad \text{Acid}_2 \qquad \text{Base}_1$$

or

$$2\,H_2O(\ell) \rightleftharpoons H_3O^+(aq) + OH^-(aq)$$

This reaction is responsible for the small but measurable **autoionization** of water. The equilibrium expression that results is

$$[H_3O^+][OH^-] = K_w$$

where $K_w$, the ion product constant for water, has the numerical value $1.0 \times 10^{-14}$ at 25°C. Because the concentration of the solvent, water, does not vary significantly in a dilute solution, it does not enter the equilibrium expression. The reasons for this will be discussed more fully in Section 11.1. The temperature dependence of $K_w$ is given in Table 10.1; all problems in this chapter are assumed to refer to 25°C unless otherwise stated.

Pure water contains no ions other than $H_3O^+$ and $OH^-$, and because there is overall electrical neutrality, there must be an equal number of ions of each type. Using the equilibrium expression, this gives

$$[H_3O^+] = [OH^-] = y$$

$$y^2 = 1.0 \times 10^{-14}$$

$$y = 1.0 \times 10^{-7}$$

so that in pure water at 25°C the concentrations of both $H_3O^+$ and $OH^-$ are $1.0 \times 10^{-7}$ M.

$$\rightarrow \; pH = 7$$

## Strong Acids and Bases

An aqueous acidic solution contains an excess of $H_3O^+$ over $OH^-$ ions. A **strong acid** is one that ionizes essentially completely in aqueous solution. When the strong acid HCl (hydrochloric acid) is put in water, the reaction that occurs is

$$HCl(aq) + H_2O(\ell) \longrightarrow H_3O^+(aq) + Cl^-(aq)$$

**TABLE 10.1**

***Temperature Dependence of $K_w$***

| $t$ (°C) | $K_w$ | pH of Water |
|---|---|---|
| 0 | $0.114 \times 10^{-14}$ | 7.47 |
| 10 | $0.292 \times 10^{-14}$ | 7.27 |
| 20 | $0.681 \times 10^{-14}$ | 7.08 |
| 25 | $1.01 \times 10^{-14}$ | 7.00 |
| 30 | $1.47 \times 10^{-14}$ | 6.92 |
| 40 | $2.92 \times 10^{-14}$ | 6.77 |
| 50 | $5.47 \times 10^{-14}$ | 6.63 |
| 60 | $9.61 \times 10^{-14}$ | 6.51 |

*No EQ'BM*

A single rather than a double arrow is used here to indicate that the reaction is essentially complete. Another strong acid is perchloric acid ($HClO_4$). If 0.10 mol of either of these acids is dissolved in enough water to make 1.0 L of solution, a 0.10 M concentration of $H_3O^+(aq)$ results. The acid–base properties of solutions are determined by their concentrations of $H_3O^+(aq)$, so these two strong acids have the same apparent strength in water despite differences in their intrinsic abilities to donate hydrogen ions. Water is said to have a **leveling effect** on a certain group of acids ($HCl$, $HBr$, $HI$, $H_2SO_4$, $HNO_3$, and $HClO_4$) in that they all behave as strong acids when water is the solvent; the reactions of these acids with water all lie so far to the right at equilibrium that the differences between the acids are undetectably small. The concentration of $H_3O^+$ in a 0.10 M solution of *any* strong acid that donates one hydrogen ion per molecule is simply 0.10 M; the $OH^-$ concentration is

$$[OH^-] = \frac{K_w}{[H_3O^+]} = \frac{1.0 \times 10^{-14}}{0.10} = 1.0 \times 10^{-13} \text{ M}$$

A **strong base** is defined in an analogous fashion, as a base that reacts essentially completely to give $OH^-(aq)$ ion when put in water. The amide ion ($NH_2^-$) and the hydride ion ($H^-$) are both strong bases. For every mole per liter of either of these ions that is added to water, one mole per liter of $OH^-(aq)$ forms. The other products are $NH_3$ and $H_2$, respectively:

$$H_2O(\ell) + NH_2^-(aq) \longrightarrow NH_3(aq) + OH^-(aq)$$

$$H_2O(\ell) + H^-(aq) \longrightarrow H_2(aq) + OH^-(aq)$$

The important base sodium hydroxide, a solid, increases the $OH^-$ concentration in water when it dissolves:

$$NaOH(s) \longrightarrow Na^+(aq) + OH^-(aq)$$

For every mole of $NaOH$ that dissolves in water, one mole of $OH^-(aq)$ forms, making $NaOH$ a strong base. Strong bases are leveled in aqueous solution in the same way that strong acids are leveled. If 0.10 mol of $NaOH$ or $NH_2^-$ or $H^-$ is put into enough water to make 1.0 L of solution, then in every case

$$[OH^-] = 0.10 \text{ M}$$

$$[H_3O^+] = \frac{1.0 \times 10^{-14}}{0.10} = 1.0 \times 10^{-13} \text{ M}$$

The $OH^-$ contribution from the autoionization of water is negligible in these cases, as was the contribution of $H_3O^+$ from that source in the case of the 0.10 M HCl or $HClO_4$ solution. Of course, if only a very small amount of strong acid or base were added to pure water (for example, $10^{-7}$ mol $L^{-1}$), the autoionization of water would have to be taken into account.

## The pH Function

In aqueous solution, the concentration of hydronium ion can range from 10 M to $10^{-15}$ M. It is convenient to compress this enormous range by introducing a logarithmic acidity scale, called **pH** and defined by

$$pH = -\log_{10} [H_3O^+]$$

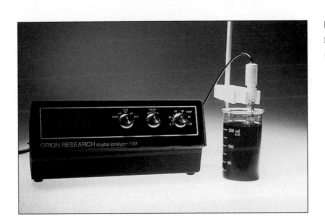

**FIGURE 10.3**  A simple pH meter with a digital read-out. *(Charles D. Winters)*

Pure water at 25°C has $[H_3O^+] = 1.0 \times 10^{-7}$ M, so

$$pH = -\log_{10}(1.0 \times 10^{-7}) = -(-7.00) = 7.00$$

A 0.10 M solution of HCl has $[H_3O^+] = 0.10$ M, so

$$pH = -\log_{10}(0.10) = -\log_{10}(1.0 \times 10^{-1}) = -(-1.00) = 1.00$$

and at 25°C a 0.10 M solution of NaOH has

$$pH = -\log_{10}\left(\frac{1.0 \times 10^{-14}}{0.10}\right) = -\log_{10}(1.0 \times 10^{-13}) = -(-13.00) = 13.00$$

As these examples show, calculating the pH is especially easy when the concentration of $H_3O^+$ is exactly a power of 10, because the logarithm is then just the power to which 10 is raised. In other cases, a calculator is needed. If the pH is given, the concentration of $H_3O^+$ can be calculated by raising 10 to the power $(-pH)$.

The most commonly encountered $H_3O^+$ concentrations are less than 1 M, so the pH function is defined with a negative sign in order to produce a number that is positive in most cases. A *high* pH means a *low* concentration of $H_3O^+$ and vice versa. At 25°C,

pH < 7        Acidic solution

pH = 7        Neutral solution

pH > 7        Basic solution

At other temperatures the pH of water differs from 7.00 (see Table 10.1). A change of one pH unit implies a change of a factor of 10 (that is, one order of magnitude) in the concentrations of $H_3O^+$ and $OH^-$. The pH is most directly measured with a **pH meter** (Fig. 10.3). The mechanism by which pH meters operate will be considered in Chapter 12. Figure 10.4 shows the pH values for several common fluids.

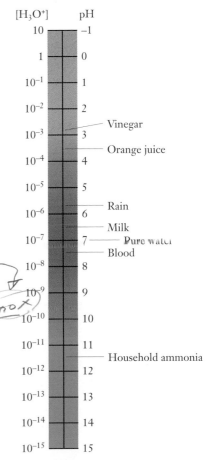

**FIGURE 10.4**  Many everyday materials are acidic or basic aqueous solutions with a wide range of pH values.

**EXAMPLE  10.2**

**(a)** A solution is prepared by dissolving 0.23 mol of NaH(*s*) in enough water to form 2.8 L of solution. Calculate its pH.
**(b)** The pH of some orange juice at 25°C is 2.85. Calculate $[H_3O^+]$ and $[OH^-]$.

**Solution**

**(a)** Because $H^-$ is a strong base, it reacts to give 0.23 mol of $OH^-$:

$$NaH(s) + H_2O(\ell) \longrightarrow Na^+(aq) + OH^-(aq) + H_2(g)$$

The concentration is

$$[OH^-] = \frac{0.23 \text{ mol}}{2.8 \text{ L}} = 8.2 \times 10^{-2} \text{ M}$$

$$[H_3O^+] = \frac{1.0 \times 10^{-14}}{8.2 \times 10^{-2}} = 1.22 \times 10^{-13} \text{ M}$$

The pH is then

$$pH = -\log_{10}(1.22 \times 10^{-13}) = 12.91$$

**(b)**
$$pH = 2.85 = -\log_{10}[H_3O^+]$$

$$[H_3O^+] = 10^{-2.85}$$

This can be evaluated by using a calculator to give

$$[H_3O^+] = 1.4 \times 10^{-3} \text{ M}$$

$$[OH^-] = \frac{1.0 \times 10^{-14}}{1.4 \times 10^{-3}} = 7.1 \times 10^{-12}$$

**Related Problems: 13, 14, 15, 16**

---

## 10.3

# ACID AND BASE STRENGTH

Acids are classified as strong or weak, depending on whether they are fully or partially ionized in solution. In a **weak acid,** the transfer of hydrogen ions to water does not proceed to completion. A weak acid such as acetic acid is thus also a **weak electrolyte;** its aqueous solutions conduct electricity less well than a strong acid of the same concentration because fewer ions are present. A weak acid produces smaller colligative properties than a strong acid (recall the effect of dissolved acetic acid on the freezing point of water in Fig. 6.11).

The Brønsted–Lowry theory helps to establish a quantitative scale for acid strength. The ionization of an acid (symbolized by "HA") in aqueous solution can be written as

$$HA(aq) + H_2O(\ell) \rightleftharpoons H_3O^+(aq) + A^-(aq)$$

where $A^-$ is the conjugate base of HA. The equilibrium expression for this chemical reaction (see pages 289 to 290) is

$$\frac{[H_3O^+][A^-]}{[HA]} = K_a$$

where the subscript "a" stands for "acid."[1] For example, if the symbol $A^-$ refers to the cyanide ion ($CN^-$), we write

$$\frac{[H_3O^+][CN^-]}{[HCN]} = K_a$$

---

[1] $H_2O(\ell)$ is in its reference state, so $c/c_{ref} = 1$ (see Section 11.1).

## TABLE 10.2

### Ionization Constants of Acids at 25°C

| Acid | HA | $A^-$ | $K_a$ | $pK_a$ |
|---|---|---|---|---|
| Hydriodic | HI | $I^-$ | $\sim 10^{11}$ | $\sim -11$ |
| Hydrobromic | HBr | $Br^-$ | $\sim 10^9$ | $\sim -9$ |
| Perchloric | $HClO_4$ | $ClO_4^-$ | $\sim 10^7$ | $\sim -7$ |
| Hydrochloric | HCl | $Cl^-$ | $\sim 10^7$ | $\sim -7$ |
| Chloric | $HClO_3$ | $ClO_3^-$ | $10^3$ | 3 |
| Sulfuric (1) | $H_2SO_4$ | $HSO_4^-$ | $\sim 10^2$ | $\sim -2$ |
| Nitric | $HNO_3$ | $NO_3^-$ | $\sim 20$ | $\sim -1.3$ |
| Hydronium ion | $H_3O^+$ | $H_2O$ | 1 | 0.0 |
| Iodic | $HIO_3$ | $IO_3^-$ | $1.6 \times 10^{-1}$ | 0.80 |
| Oxalic (1) | $H_2C_2O_4$ | $HC_2O_4^-$ | $5.9 \times 10^{-2}$ | 1.23 |
| Sulfurous (1) | $H_2SO_3$ | $HSO_3^-$ | $1.54 \times 10^{-2}$ | 1.81 |
| Sulfuric (2) | $HSO_4^-$ | $SO_4^{2-}$ | $1.2 \times 10^{-2}$ | 1.92 |
| Chlorous | $HClO_2$ | $ClO_2^-$ | $1.1 \times 10^{-2}$ | 1.96 |
| Phosphoric (1) | $H_3PO_4$ | $H_2PO_4^-$ | $7.52 \times 10^{-3}$ | 2.12 |
| Arsenic (1) | $H_3AsO_4$ | $H_2AsO_4^-$ | $5.0 \times 10^{-3}$ | 2.30 |
| Chloroacetic | $CH_2ClCOOH$ | $CH_2ClCOO^-$ | $1.4 \times 10^{-3}$ | 2.85 |
| Hydrofluoric | HF | $F^-$ | $6.6 \times 10^{-4}$ | 3.18 |
| Nitrous | $HNO_2$ | $NO_2^-$ | $4.6 \times 10^{-4}$ | 3.34 |
| Formic | HCOOH | $HCOO^-$ | $1.77 \times 10^{-4}$ | 3.75 |
| Benzoic | $C_6H_5COOH$ | $C_6H_5COO^-$ | $6.46 \times 10^{-5}$ | 4.19 |
| Oxalic (2) | $HC_2O_4^-$ | $C_2O_4^{2-}$ | $6.4 \times 10^{-5}$ | 4.19 |
| Hydrazoic | $HN_3$ | $N_3^-$ | $1.9 \times 10^{-5}$ | 4.72 |
| Acetic | $CH_3COOH$ | $CH_3COO^-$ | $1.76 \times 10^{-5}$ | 4.75 |
| Propionic | $CH_3CH_2COOH$ | $CH_3CH_2COO^-$ | $1.34 \times 10^{-5}$ | 4.87 |
| Pyridinium ion | $HC_5H_5N^+$ | $C_5H_5N$ | $5.6 \times 10^{-6}$ | 5.25 |
| Carbonic (1) | $H_2CO_3$ | $HCO_3^-$ | $4.3 \times 10^{-7}$ | 6.37 |
| Sulfurous (2) | $HSO_3^-$ | $SO_3^{2-}$ | $1.02 \times 10^{-7}$ | 6.91 |
| Arsenic (2) | $H_2AsO_4^-$ | $HAsO_4^{2-}$ | $9.3 \times 10^{-8}$ | 7.03 |
| Hydrosulfuric | $H_2S$ | $HS^-$ | $9.1 \times 10^{-8}$ | 7.04 |
| Phosphoric (2) | $H_2PO_4^-$ | $HPO_4^{2-}$ | $6.23 \times 10^{-8}$ | 7.21 |
| Hypochlorous | HClO | $ClO^-$ | $3.0 \times 10^{-8}$ | 7.53 |
| Hydrocyanic | HCN | $CN^-$ | $6.17 \times 10^{-10}$ | 9.21 |
| Ammonium ion | $NH_4^+$ | $NH_3$ | $5.6 \times 10^{-10}$ | 9.25 |
| Carbonic (2) | $HCO_3^-$ | $CO_3^{2-}$ | $4.8 \times 10^{-11}$ | 10.32 |
| Arsenic (3) | $HAsO_4^{2-}$ | $AsO_4^{3-}$ | $3.0 \times 10^{-12}$ | 11.53 |
| Hydrogen peroxide | $H_2O_2$ | $HO_2^-$ | $2.4 \times 10^{-17}$ | 11.62 |
| Phosphoric (3) | $HPO_4^{2-}$ | $PO_4^{3-}$ | $2.2 \times 10^{-13}$ | 12.67 |
| Water | $H_2O$ | $OH^-$ | $1.0 \times 10^{-14}$ | 14.00 |

where $K_a$ is the **acid ionization constant** for hydrogen cyanide in water and has a numerical value of $6.17 \times 10^{-10}$ at 25°C. Table 10.2 gives values of $K_a$ for a number of important acids. The useful quantity $pK_a = -\log_{10} K_a$ is also given. The acid ionization constant is a quantitative measure of the strength of the acid in a given solvent (here, water). A strong acid has $K_a$ greater than 1, so [HA] in the denominator is small and the acid ionizes almost completely. In a weak acid, $K_a$ is smaller than 1 and the ionized species have low concentrations; reaction with water proceeds to a limited extent before equilibrium is reached.

Hydrogen cyanide, HCN, is a highly toxic gas that dissolves in water to form equally toxic solutions of hydrocyanic acid.

The strength of a base is inversely related to the strength of its conjugate acid; the weaker the acid, the stronger its conjugate base and vice versa. To see this, note that the equation representing the ionization of a base such as ammonia in water can be written as

$$\underset{\text{Acid}_1}{H_2O(\ell)} + \underset{\text{Base}_2}{NH_3(aq)} \rightleftharpoons \underset{\text{Acid}_2}{NH_4^+(aq)} + \underset{\text{Base}_1}{OH^-(aq)}$$

which gives an equilibrium expression of the form

$$\frac{[NH_4^+][OH^-]}{[NH_3]} = K_b$$

where the subscript "b" on $K_b$ stands for "base." Because $[OH^-]$ and $[H_3O^+]$ are related through the water autoionization equilibrium expression

$$[OH^-][H_3O^+] = K_w$$

the $K_b$ expression can be written as

$$K_b = \frac{[NH_4^+]K_w}{[NH_3][H_3O^+]} = \frac{K_w}{K_a}$$

where $K_a$ is the acid ionization constant for $NH_4^+$, the conjugate acid of the base $NH_3$. This general relationship between the $K_b$ of a base and the $K_a$ of its conjugate acid shows that $K_b$ need not be tabulated separately from $K_a$, because the two are related through

$$K_w = K_a K_b$$

It is also clear that if $K_a$ is large (so that the acid is strong), then $K_b$ is small (the conjugate base is weak). Figure 10.5 summarizes the strengths of conjugate acid–base pairs.

If two bases compete for hydrogen ions, the stronger base is favored in the equilibrium that is reached. The stronger acid donates hydrogen ions to the stronger base, producing a weaker acid and a weaker base. To see this, consider the equilibrium

$$\underset{\text{Acid}_1}{HF(aq)} + \underset{\text{Base}_2}{CN^-(aq)} \rightleftharpoons \underset{\text{Acid}_2}{HCN(aq)} + \underset{\text{Base}_1}{F^-(aq)}$$

with equilibrium constant

$$\frac{[HCN][F^-]}{[HF][CN^-]} = K$$

The two bases $F^-$ and $CN^-$ compete for hydrogen ions. The overall reaction is obtained by starting with one acid ionization reaction

$$HF(aq) + H_2O(\ell) \rightleftharpoons H_3O^+(aq) + F^-(aq)$$

$$\frac{[H_3O^+][F^-]}{[HF]} = K_a = 6.6 \times 10^{-4}$$

and *subtracting* from it a second acid ionization reaction:

$$HCN(aq) + H_2O(\ell) \rightleftharpoons H_3O^+(aq) + CN^-(aq)$$

$$\frac{[H_3O^+][CN^-]}{[HCN]} = K_a' = 6.17 \times 10^{-10}$$

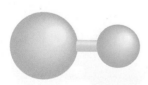

Hydrogen fluoride, HF, is a colorless liquid that boils at 19.5°C. It dissolves in water to give solutions of hydrofluoric acid, a weak acid.

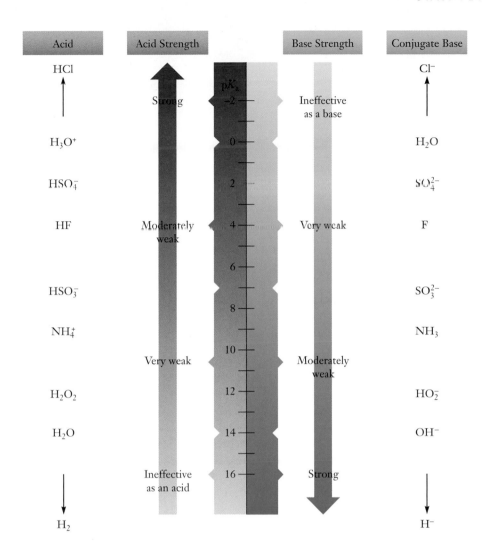

**FIGURE 10.5** The relative strengths of some acids and their conjugate bases.

*[handwritten note: HCl: given a conjugate acid state a possible conjugate base]*

When the second equation is subtracted from the first, the corresponding equilibrium constants must be divided. The numerical value of $K$ is

$$K = \frac{K_a}{K_a'} = \frac{6.6 \times 10^{-4}}{6.17 \times 10^{-10}} = 1.1 \times 10^6$$

*[handwritten note: heavily shifted to... (products side since ratio > 1)]*

Because HCN is a weaker acid than HF, $K_a'$ is smaller than $K_a$, and $K$ is larger than 1. The equilibrium described by $K$ lies strongly to the right, and there is a net donation of $H^+$ from the stronger acid (HF) to the stronger base ($CN^-$), producing the weaker acid (HCN) and the weaker base ($F^-$). The magnitudes of acid ionization constants can be used to predict the direction of net hydrogen ion transfer in reactions between acids and bases in aqueous solution.

## Electronegativity and Oxoacid Strength

Trends in the relative strength of oxoacids are explained by effects of electronegativity and bond polarity on the ease of donating a proton. **Oxoacids** donate protons in aqueous solution that previously were bonded to oxygen atoms. Examples include

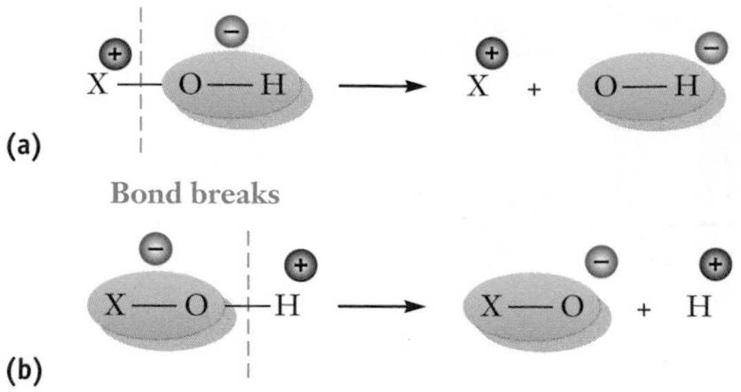

**FIGURE 10.6** In part a, the atom X is electropositive, and so extra electron density (blue areas) accumulates on the OH group. The X—O bond then breaks easily, making the compound a base. In part b, X is electronegative, and so electron density is drawn from the H atom to the X—O bond. Now it is the O—H bond that breaks easily, and the compound is an acid.

sulfuric acid ($H_2SO_4$), nitric acid ($HNO_3$), and phosphoric acid ($H_3PO_4$). If the central atom is designated X, then oxoacids have the structure

$$-X-O-H$$

where X can be bonded to additional —OH groups, to oxygen atoms, or to hydrogen atoms. How does the strength of the oxoacid change with the electronegativity of X? Consider first the extreme case in which X is a highly electropositive element, such as an alkali metal. Of course, NaOH is not an acid at all, but a base. The sodium atom in Na—O—H gives up a full electron to make $Na^+$ and $OH^-$ ions. Because the X—O bond here is almost completely ionic, the $OH^-$ group has a net negative charge that holds the $H^+$ tightly to the oxygen and prevents formation of $H^+$ ions. The less electropositive alkaline-earth elements behave similarly. They form hydroxides such as $Mg(OH)_2$ that are somewhat weaker bases than NaOH but do not in any way act as acids.

Now suppose the central atom X becomes more electronegative, reaching values between 2 and 3, as in the oxoacids of the elements B, C, P, As, S, Se, Br, and I. As X becomes more effective at withdrawing electron density from the oxygen atom, the X—O bond becomes more covalent. This leaves less negative charge on the oxygen atom, and consequently the oxoacid releases $H^+$ more readily (Fig. 10.6). Other things being equal, acid strength should increase with increasing electronegativity of the central atom. This trend is observed among the oxoacids listed in Table 10.2.

## Indicators

An **indicator** is a soluble dye that changes color noticeably over a fairly narrow range of pH. The typical indicator is a weak organic acid that has a different color from its conjugate base (Fig. 10.7). Litmus changes from red to blue as its acid form is converted to base. Good indicators have such intense colors that only a few drops of a dilute indicator solution must be added to the solution being studied. The very low concentration of indicator molecules has almost no effect on the pH of the solution. The color changes of the indicator reflect the effects of the *other* acids and bases present in the solution.

If the acid form of a given indicator is represented as HIn and the conjugate base form as $In^-$, their acid–base equilibrium is

**FIGURE 10.7**  Color differences in four indicators: bromophenol red, thymolphthalein, phenolphthalein, and bromocresol green. In each case, the acidic form is on the left and the basic form is on the right. *(Leon Lewandowski)*

$$HIn(aq) + H_2O(\ell) \rightleftharpoons H_3O^+(aq) + In^-(aq) \qquad \frac{[H_3O^+][In^-]}{[HIn]} = K_a$$

where $K_a$ is the acid ionization constant for the indicator. This expression can be rearranged to give

$$\frac{[H_3O^+]}{K_a} = \frac{[HIn]}{[In^-]}$$

If the concentration of hydronium ion $[H_3O^+]$ is large relative to $K_a$, this ratio is large, and $[HIn]$ is large compared with $[In^-]$. The solution has the color of the acid form of the indicator because most of the indicator molecules are in the acid form. Litmus, for example, has a $K_a$ near $10^{-7}$. If the pH is 5, then

$$\frac{[H_3O^+]}{K_a} = \frac{10^{-5}}{10^{-7}} = 100$$

Thus, approximately 100 times as many indicator molecules are in the acid form as in the base form, and the solution is red.

As the concentration of hydronium ion is reduced, more molecules of acid indicator ionize to give the base form. When $[H_3O^+]$ is near $K_a$, almost equal amounts of the two forms are present, and the color is a mixture of the colors of the two indicator states (violet for litmus). A further decrease in $[H_3O^+]$ to a value much smaller than $K_a$ then leads to a predominance of the base form, with the corresponding color being observed.

Different indicators have different values for $K_a$ and thus show color changes at different pH values (Fig. 10.8). The weaker an indicator is as an acid, the higher is the pH at which the color change takes place. Such color changes occur over a range of 1 to 2 pH units. Methyl red, for example, is red when the pH is below 4.8 and yellow above 6.0; shades of orange are seen for intermediate pH values. This limits the accuracy to which the pH can be determined through the use of indicators. However, in Section 10.6 it will be seen that this does not affect the analytical determination of acid or base concentrations through titration, provided that an appropriate indicator is employed.

Many natural dyes found in fruits, vegetables, and flowers act as pH indicators by changing color with changes in acidity (Fig. 10.9). A particularly striking example is cyanidin, which is responsible both for the red color of poppies and the blue color of cornflowers. The sap of the poppy is sufficiently acidic to turn cyanidin red, but the sap of the cornflower is basic and makes the dye blue. Related natural dyes called anthocyanins contribute to the colors of raspberries, strawberries, and blackberries.

**FIGURE 10.8** Indicators change their colors at very different pH values, so the best choice of indicator depends on the particular experimental conditions.

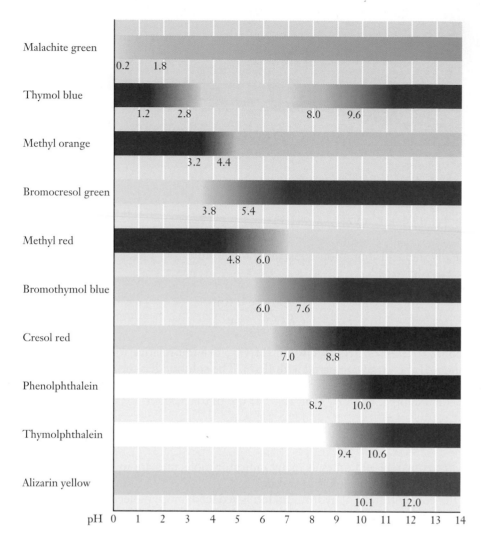

| | pH | 0 | 1 | 2 | 3 | 4 | 5 | 6 | 7 | 8 | 9 | 10 | 11 | 12 | 13 | 14 |

Malachite green — 0.2  1.8

Thymol blue — 1.2  2.8  8.0  9.6

Methyl orange — 3.2  4.4

Bromocresol green — 3.8  5.4

Methyl red — 4.8  6.0

Bromothymol blue — 6.0  7.6

Cresol red — 7.0  8.8

Phenolphthalein — 8.2  10.0

Thymolphthalein — 9.4  10.6

Alizarin yellow — 10.1  12.0

**FIGURE 10.9** Red cabbage extract is a natural pH indicator. When the solution is highly acidic, the extract gives the solution a red color. As the solution becomes less and less acidic (more basic), the color changes from red to violet to yellow. (*Charles D. Winters*)

---

| **10.4** |
| --- |

# EQUILIBRIA INVOLVING WEAK ACIDS AND BASES

Weak acids and bases react only partially with water, so to calculate the pH of their solutions we use $K_a$ or $K_b$ and the laws of chemical equilibrium. The calculations that result follow the pattern of Example 9.5 for gas equilibria. In that case, the initial gas-phase pressures $P°$ were known, and the pressures of products resulting from partial reaction were calculated. Now the initial concentration of acid or base is known, and the concentrations of products resulting from its partial reaction with water are calculated.

## Weak Acids

A weak acid has a $K_a$ smaller than 1. Values of the $pK_a$ start at zero for the strongest weak acid and range upward. (If the $pK_a$ is greater than 14, the compound is ineffective as an acid in aqueous solution.) When a weak acid is dissolved in water, the original concentration is almost always known, but partial reaction with water consumes some HA and generates $A^-$ and $H_3O^+$:

$$HA(aq) + H_2O(\ell) \rightleftharpoons H_3O^+(aq) + A^-(aq)$$

To calculate the amounts of $H_3O^+$, $A^-$, and HA at equilibrium, methods analogous to those of Chapter 9 are used, with partial pressures replaced by concentrations. A new feature here is that one of the products ($H_3O^+$) has a second source, the autoionization of the solvent, water. In the cases of usual interest this second effect is small and can be neglected; however, it is a good idea to verify at the end of each calculation that the $[H_3O^+]$ from the acid ionization alone exceeds $10^{-7}$ M by at least one order of magnitude. If this is not true, the more complete method of analysis given in Section 10.8 must be used.

---

| **EXAMPLE 10.3** |
| --- |

Acetic acid ($CH_3COOH$) has a $K_a$ of $1.76 \times 10^{-5}$ at 25°C. Suppose 1.000 mol is dissolved in enough water to give 1.000 L of solution. Calculate the pH and the fraction of acetic acid ionized at equilibrium.

### Solution

The initial concentration of acetic acid is 1.000 M. If $y$ mol $L^{-1}$ ionizes, then

$$CH_3COOH(aq) + H_2O(\ell) \rightleftharpoons H_3O^+(aq) + CH_3COO^-(aq)$$

| | $CH_3COOH$ | $H_3O^+$ | $CH_3COO^-$ |
| --- | --- | --- | --- |
| Initial concentration (M) | 1.000 | $\approx 0$ | 0 |
| Change in concentration (M) | $-y$ | $+y$ | $+y$ |
| Equilibrium concentration (M) | $1.000 - y$ | $y$ | $y$ |

Note that the $H_3O^+$ initially present from the ionization of water has been ignored, because it is small compared to $y$ for all but the weakest acids or the most dilute solutions. The equilibrium expression states that

$$\frac{[H_3O^+][CH_3COO^-]}{[CH_3COOH]} = \frac{y^2}{1.000 - y} = K_a = 1.76 \times 10^{-5}$$

This equation could be solved using the quadratic formula, as was done in Chapter 9. A quicker way is based on the fact that acetic acid is a weak acid, so that only a small fraction is ionized at equilibrium. The approximation that $y$ is small relative to 1.000 (that is, the final equilibrium concentration of $CH_3COOH$ is close to its initial concentration) is reasonable. This gives

$$\frac{y^2}{1.000 - y} \approx \frac{y^2}{1.000} = 1.76 \times 10^{-5}$$

$$y = 4.20 \times 10^{-3}$$

so that

$$[H_3O^+] = y = 4.20 \times 10^{-3} \text{ M}$$

The approximations should now be checked. First, $y$ is indeed much smaller than the original concentration of 1.000 M (by a factor of 200), so the neglect of it in the denominator is justified. Second, the concentration of $H_3O^+$ from acetic acid ($4.2 \times 10^{-3}$ M) is large compared with $10^{-7}$ M, so the neglect of water ionization is also justified. Therefore,

$$pH = -\log_{10} (4.2 \times 10^{-3}) = 2.38$$

The fraction ionized is the ratio of the concentration of $CH_3COO^-$ present at equilibrium to the concentration of $CH_3COOH$ present in the first place:

$$\frac{y}{1.000} = y = 4.2 \times 10^{-3}$$

The percentage of the acetic acid that is ionized is 0.42%. Fewer than one in a hundred of the molecules of this typical weak acid dissociate in this solution.

**Related Problems: 27, 28, 29, 30**

---

Approximation methods of the type used in this example were also employed in Example 9.6 and are discussed more extensively in Appendix C. As the solution of a weak acid becomes more dilute, an increasing fraction of it ionizes, as shown by the following example.

## EXAMPLE 10.4

Suppose that 0.00100 mol of acetic acid is used instead of the 1.000 mol in the preceding example. Calculate the pH and the percentage of acetic acid ionized.

### Solution
Once again the assumption to be made (and checked later) is that the contribution of the ionization of water to $[H_3O^+]$ is negligible. Hence,

$$[CH_3COOH] = 0.0100 - y$$

$$[CH_3COO^-] = [H_3O^+] = y$$

$$\frac{y^2}{0.00100 - y} = 1.76 \times 10^{-5}$$

In this case, $y$ is *not* small relative to 0.00100 because a substantial fraction of the acetic acid molecules ionizes. There are two alternatives for solving the problem in this case.

The first is to solve the quadratic equation that results from multiplying out the equilibrium expression:

$$y^2 + 1.76 \times 10^{-5}y - 1.76 \times 10^{-8} = 0$$

Use of the quadratic formula gives

$$y = 1.24 \times 10^{-4} \text{ M} = [\text{H}_3\text{O}^+] = [\text{CH}_3\text{COO}^-]$$

Because $y \gg 10^{-7}$ M, the neglect of water ionization is justified.

$$\text{pH} = -\log_{10}(1.24 \times 10^{-4}) = 3.91$$

$$\text{percentage ionized} = \frac{1.24 \times 10^{-4} \text{ M}}{0.00100 \text{ M}} \times 100\% = 12.4\%$$

The other alternative is to use successive approximations (Appendix C). Begin by neglecting $y$ in the denominator (relative to 0.00100), which gives

$$\frac{y^2}{0.00100} \approx 1.76 \times 10^{-5}$$

$$y \approx 1.33 \times 10^{-4} \text{ M}$$

This is just what was done in Example 10.3. Now, however, continue by reinserting this *approximate* value of $y$ into the *denominator* and recalculating $y$:

$$\frac{y^2}{0.00100 - 0.000133} = 1.76 \times 10^{-5}$$

$$y^2 = 1.53 \times 10^{-8}$$

$$y = 1.23 \times 10^{-4} \text{ M}$$

A final iteration (inserting this new value of $y$ into the denominator) gives $1.24 \times 10^{-4}$ M, the same value produced by the quadratic formula. The iteration process is terminated when two successive results for $y$ agree to the desired number of significant figures.

**Related Problems: 31, 32**

---

The method of successive approximations is often faster to apply than the quadratic formula. It is important to remember that the accuracy of a result is limited both by the accuracy of the input data (values of $K_a$ and initial concentrations) and by the fact that solutions are not ideal. It is therefore pointless to calculate equilibrium concentrations to any higher degree of accuracy than 1% to 3%.

## Weak Bases

The definition and description of weak acids, their $K_a$ values, and the production of $\text{H}_3\text{O}^+(aq)$ ion when they are dissolved in water can be translated to apply to weak bases, their $K_b$ values, and the production of $\text{OH}^-(aq)$ ion. A **weak base** such as ammonia reacts only partially with water to produce $\text{OH}^-(aq)$:

$$H_2O(\ell) + NH_3(aq) \rightleftharpoons NH_4^+(aq) + OH^-(aq)$$

$$\frac{[NH_4^+][OH^-]}{[NH_3]} = K_b = 1.8 \times 10^{-5}$$

The $K_b$ of a weak base is smaller than 1, and the weaker the base, the smaller the $K_b$. If the $K_b$ of a compound is smaller than $1 \times 10^{-14}$, that compound is ineffective as a base in aqueous solution.

The analogy to weak acids continues in the calculation of the aqueous equilibria of weak bases, as the following example shows.

## EXAMPLE 10.5

Calculate the pH of a solution made by dissolving 0.0100 mol of $NH_3$ in enough water to give 1.000 L of solution at 25°C. The $K_b$ for ammonia is $1.8 \times 10^{-5}$.

### Solution

Set up the table of the changes in concentrations that occur as the reaction goes to equilibrium, neglecting the small contribution to $[OH^-]$ from the autoionization of water:

|  | $H_2O(\ell)$ + $NH_3(aq)$ $\rightleftharpoons$ $NH_4^+(aq)$ + $OH^-(aq)$ |  |  |
|---|---|---|---|
| Initial concentration (M) | 0.0100 | 0 | ≈0 |
| Change in concentration (M) | $-y$ | $+y$ | $+y$ |
| Equilibrium concentration (M) | $0.100 - y$ | $y$ | $y$ |

Substitution into the equilibrium expression gives

$$\frac{y^2}{0.0100 - y} = K_b = 1.8 \times 10^{-5}$$

which can be solved for $y$ by either the quadratic formula or the method of successive approximations:

$$y = 4.15 \times 10^{-4} \text{ M} = [OH^-]$$

The product of $[H_3O^+]$ and $[OH^-]$ is always $K_w$. Hence,

$$[H_3O^+] = \frac{K_w}{[OH^-]} = \frac{1.0 \times 10^{-14}}{4.15 \times 10^{-4}} = 2.4 \times 10^{-11}$$

$$pH = -\log_{10}(2.4 \times 10^{-11}) = 10.62$$

The pH is greater than 7, as expected for a solution of a base.

**Related Problems: 35, 36**

## Hydrolysis

Most of the acids considered up to now have been uncharged species with the general formula HA. In the Brønsted–Lowry picture, however, there is no reason why the acid should be an electrically neutral molecule. When $NH_4Cl$, a salt, dissolves in water, $NH_4^+$ ions are present. These ionize partially by transferring hydrogen ions to water, a straightforward Brønsted–Lowry acid–base reaction:

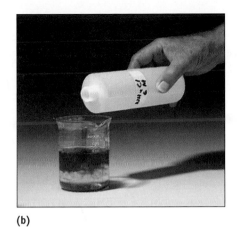

(a)                              (b)                              (c)

**FIGURE 10.10**    Proof that ammonium chloride is an acid. (a) A solution of sodium hydroxide, with the pink color of phenolphthalein indicating its basic character. (b and c) As aqueous ammonium chloride is added, a neutralization reaction takes place and the solution turns colorless from the bottom up. *(Leon Lewandowski)*

$$NH_4^+(aq) + H_2O(\ell) \rightleftharpoons H_3O^+(aq) + NH_3(aq)$$

Acid$_1$      Base$_2$         Acid$_2$        Base$_1$

$$\frac{[H_3O^+][NH_3]}{[NH_4^+]} = K_a = 5.6 \times 10^{-10}$$

$NH_4^+$ acts as an acid here, just as acetic acid did in Examples 10.3 and 10.4. Because $K_a$ is much smaller than 1, only a small amount of $H_3O^+$ is generated. $NH_4^+$ is a weak acid, but it is nonetheless an acid, and a solution of ammonium chloride has a pH below 7 (Fig. 10.10).

**Hydrolysis** is the general term given to the reaction of a substance with water, and it is applied in particular to a reaction in which the pH changes from 7 upon dissolving a salt (in this case, $NH_4Cl$) in water. There is no reason to treat hydrolysis in any special way. The hydrolysis that takes place when $NH_4Cl$ dissolves in water can be completely described as a Brønsted–Lowry reaction in which water acts as a base and $NH_4^+$ acts as an acid to give a pH below 7. In parallel fashion, a solution becomes basic when a salt whose anion is a weak base is dissolved. This, too, is a case of hydrolysis, and once again it is simply another Brønsted–Lowry acid–base reaction, with water acting now as an acid (a hydrogen ion donor).

Two salts that give basic solutions are sodium acetate and sodium fluoride. When these salts dissolve in water, they furnish acetate ($CH_3COO^-$) and fluoride ($F^-$) ions, respectively, both of which act as Brønsted–Lowry bases,

$$H_2O(\ell) + CH_3COO^-(aq) \rightleftharpoons CH_3COOH(aq) + OH^-(aq)$$

$$H_2O(\ell) + F^-(aq) \rightleftharpoons HF(aq) + OH^-(aq)$$

causing the $OH^-$ concentration to increase and giving a pH above 7.

**EXAMPLE 10.6**

Suppose 0.100 mol of $NaCH_3COO$ is dissolved in enough water to make 1.00 L of solution. What is the pH of the solution?

**Solution**

The $K_a$ for $CH_3COOH$, the conjugate acid of $CH_3COO^-$, is $1.76 \times 10^{-5}$. Therefore, $K_b$ for the acetate ion is

$$\frac{[CH_3COOH][OH^-]}{[CH_3COO^-]} = K_b = \frac{K_w}{K_a} = 5.7 \times 10^{-10}$$

The equilibrium here is

| | $H_2O(\ell) + CH_3COO^-(aq) \rightleftharpoons CH_3COOH(aq) + OH^-(aq)$ | | |
|---|---|---|---|
| Initial concentration (M) | 0.100 | 0 | 0 |
| Change in concentration (M) | $-y$ | $+y$ | $+y$ |
| Equilibrium concentration (M) | $0.100 - y$ | $y$ | $y$ |

Substitution into the equilibrium expression gives

$$\frac{y^2}{0.100 - y} = K_b = 5.7 \times 10^{-10}$$

$$y = 7.5 \times 10^{-6}\,M = [OH^-]$$

Note that $y$ is small relative to 0.100 and fairly large relative to $10^{-7}$.

$$[H_3O^+] = \frac{K_w}{[OH^-]} = \frac{1.0 \times 10^{-14}}{7.5 \times 10^{-6}} = 1.3 \times 10^{-9}\,M$$

$$pH = 8.89$$

**Related Problems: 41, 42**

Hydrolysis does not occur with all ions, only with those that are conjugate acids of weak bases or conjugate bases of weak acids. Chloride ion is the conjugate base of the strong acid HCl and consequently is ineffective as a base, unlike $F^-(aq)$ and $CH_3COO^-(aq)$. Its interaction with water would therefore scarcely change the $OH^-(aq)$ concentration. For this reason, a solution of NaCl is neutral, while one of NaF is slightly basic.

## 10.5

# BUFFER SOLUTIONS

A **buffer solution** is any solution that maintains an approximately constant pH despite small additions of acid or base. Typically, a buffer solution contains a weak acid and its conjugate weak base in approximately equal concentrations. Buffer solutions play important roles in controlling the solubility of ions in solution and in maintaining the pH in biochemical and physiological processes. Many life processes are sensitive to pH and require regulation within a small range of $H_3O^+$ and $OH^-$ concentrations. Organisms have built-in buffers to protect them against large changes in pH. Human blood, for example, has a pH near 7.4 that is maintained by a combination of carbonate, phosphate, and protein buffer systems. A blood pH below 7.0 or above 7.8 leads quickly to death.

## Calculations of Buffering Action

Consider a typical weak acid, formic acid (HCOOH), and its conjugate base, formate ion (HCOO$^-$). The latter can be obtained by dissolving a salt such as sodium formate (NaHCOO) in water. The acid–base equilibrium established between these species is given by

$$HCOOH(aq) + H_2O(\ell) \rightleftharpoons H_3O^+(aq) + HCOO^-(aq)$$

with an acid ionization constant

$$\frac{[H_3O^+][HCOO^-]}{[HCOOH]} = K_a = 1.77 \times 10^{-4}$$

In Section 10.2 the pH of a solution containing only a weak acid (such as HCOOH) or only a weak base (such as HCOO$^-$) was considered. Suppose now that the weak acid and its conjugate base are *both* present initially. The resulting calculations resemble closely the calculation of Example 9.9 in which equilibrium was established starting from a mixture in which both reactants and products were present.

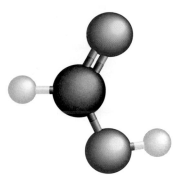

Formic acid, HCOOH, is the simplest carboxylic acid, with only a hydrogen atom attached to the —COOH group.

### EXAMPLE 10.7

Suppose 1.00 mol of HCOOH and 0.500 mol of NaHCOO are added to water and diluted to 1.00 L. Calculate the pH of the solution.

### Solution

|  | HCOOH(aq) + H$_2$O($\ell$) $\rightleftharpoons$ H$_3$O$^+$(aq) + HCOO$^-$(aq) | | |
|---|:---:|:---:|:---:|
| Initial concentration (M) | 1.00 | $\approx 0$ | 0.500 |
| Change in concentration (M) | $-y$ | $+y$ | $+y$ |
| Equilibrium concentration (M) | $1.00 - y$ | $y$ | $0.500 + y$ |

The equilibrium expression is

$$\frac{y(0.500 + y)}{1.00 - y} = K_a = 1.77 \times 10^{-4}$$

Because $y$ is likely to be small relative to 1.00 and to 0.500, the expression is approximated as

$$\frac{y(0.500)}{1.00} \approx 1.77 \times 10^{-4}$$

$$y = 3.54 \times 10^{-4} \, \text{M} = [H_3O^+]$$

A glance verifies that $y$ is indeed small relative to 1.00 and 0.500. Then

$$pH = -\log_{10}(3.54 \times 10^{-4}) = 3.45$$

**Related Problems: 43, 44**

To see how buffer solutions work, let us write the equilibrium expression for the ionization of a weak acid HA in the form

$$[H_3O^+] = K_a \frac{[HA]}{[A^-]}$$

The concentration of hydronium ion depends on the ratio of the concentration of the weak acid to the concentration of its conjugate base. The key to effective buffer action is to keep both of these concentrations nearly equal and fairly large. Adding a small amount of base to an effective buffer takes away only a few percent of the HA molecules by converting them into $A^-$ ions and adds only a few percent to the amount of $A^-$ that was originally present. The ratio $[HA]/[A^-]$ decreases, but only very slightly. Added acid consumes a small fraction of the base $A^-$ to generate a bit more HA. The ratio $[HA]/[A^-]$ now increases, but again the change is only slight. Because the concentration of $H_3O^+$ is tied directly to this ratio, it also changes only slightly. The following example illustrates buffer action quantitatively.

## EXAMPLE 10.8

Suppose 0.10 mol of a strong acid such as HCl is added to the solution in Example 10.7. Calculate the pH of the resulting solution.

### Solution

The strong acid HCl ionizes essentially completely in dilute aqueous solution. Initially, assume that *all* of the hydrogen ions from HCl are taken up by formate ions, giving formic acid, after which some formic acid ionizes back to formate ion and $H_3O^+$. This is simply a way of looking at the route by which equilibrium is approached and not a statement of the sequence of reactions that actually occurs. The position of the final equilibrium does not depend on the route by which it is attained.

Because 0.10 mol of HCl reacts with an equal number of moles of $HCOO^-$, the concentrations of $HCOO^-$ and HCOOH *before* ionization are

$$[HCOO^-]_0 = 0.50 - 0.10 = 0.40 \text{ M}$$

$$[HCOOH]_0 = 1.00 + 0.10 = 1.10 \text{ M}$$

The chart to calculate the concentrations at equilibrium is

| | $HCOOH(aq) + H_2O(\ell) \rightleftharpoons H_3O^+(aq) + HCOO^-(aq)$ | | |
|---|---|---|---|
| Initial concentration (M) | 1.10 | $\approx 0$ | 0.40 |
| Change in concentration (M) | $-y$ | $+y$ | $+y$ |
| Equilibrium concentration (M) | $1.10 - y$ | $y$ | $0.40 + y$ |

The equilibrium expression then becomes

$$\frac{y(0.40 + y)}{1.10 - y} = 1.77 \times 10^{-4}$$

Because $y$ is again likely to be small relative to both 0.40 and 1.10,

$$y \approx \left(\frac{1.10}{0.40}\right)(1.77 \times 10^{-4}) = 4.9 \times 10^{-4}$$

$$pH = 3.31$$

Even though 0.10 mol of a strong acid was added, the pH changed only slightly, from 3.45 to 3.31. By contrast, the same amount of acid added to a liter of pure water would change the pH from 7 to 1.

**Related Problems: 45, 46**

Note that this problem was solved by performing first a stoichiometric (limiting reactant) calculation and then an equilibrium calculation. A similar calculation can be performed if a strong base such as $OH^-$ is added instead of a strong acid. The base reacts with formic acid to produce formate ions. Adding 0.10 mol of $OH^-$ to the $HCOOH/HCOO^-$ buffer of Example 10.7 increases the pH only to 3.58. In the absence of the buffer system, the same base would raise the pH to 13.00.

In any buffer there is competition between the tendency of the acid to donate hydrogen ions to water (increasing the acidity) and the tendency of the base to accept hydrogen ions from water (increasing the basicity). The resulting pH depends on the magnitude of $K_a$. If $K_a$ is large relative to $10^{-7}$, the acid ionization will win out and acidity will increase, as in the $HCOOH/HCOO^-$ buffer. A basic buffer (with pH > 7) can be made by working with an acid–base pair with $K_a$ smaller than $10^{-7}$. In this case, the net reaction leads to production of $OH^-$, and $K_b$ must be used to determine the equilibrium state. A typical example is the $NH_4^+/NH_3$ buffer made by mixing ammonium chloride with ammonia.

## EXAMPLE 10.9

Calculate the pH of a solution made by adding 0.100 mol of $NH_4Cl$ and 0.200 mol of $NH_3$ to water and diluting to 1.000 L. $K_a$ for $NH_4^+$ is $5.6 \times 10^{-10}$.

### Solution

Because $K_a \ll 10^{-7}$ for $NH_4^+$ (equivalently, $K_b \gg 10^{-7}$ for $NH_3$), the net reaction is production of $OH^-$ ions. Therefore, the equilibrium is written to show the net transfer of hydrogen ions from water to $NH_3$:

$$H_2O(\ell) + NH_3(aq) \rightleftharpoons NH_4^+(aq) + OH^-(aq)$$

|  | | | |
|---|---|---|---|
| Initial concentration (M) | 0.200 | 0.100 | 0 |
| Change in concentration (M) | $-y$ | $+y$ | $+y$ |
| Equilibrium concentration (M) | $0.200 - y$ | $0.100 + y$ | $y$ |

The equilibrium expression is

$$\frac{y(0.100 + y)}{0.200 - y} = K_b = \frac{K_w}{K_a} = 1.8 \times 10^{-5}$$

If $y$ is small relative to the original concentrations of both $NH_3$ and $NH_4^+$, then

$$\frac{y(0.100)}{0.200} \approx 1.8 \times 10^{-5}$$

$$y \approx \left(\frac{0.200}{0.100}\right)(1.8 \times 10^{-5}) = 3.6 \times 10^{-5} \ll 0.100, 0.200$$

$$[H_3O^+] = \frac{1.0 \times 10^{-14}}{[OH^-]} = \frac{1.0 \times 10^{-14}}{3.6 \times 10^{-5}} = 2.8 \times 10^{-10} \text{ M}$$

$$pH = -\log_{10}(2.8 \times 10^{-10}) = 9.55$$

## Designing Buffers

Control of pH is vital in synthetic and analytical chemistry, just as it is in living organisms. Procedures that work well at a pH of 5 may fail when the concentration of hydronium ion in the solution is raised tenfold to make the pH 4. Fortunately, by the proper choice of a weak acid and the ratio in which it is mixed with its conjugate base, it is possible to prepare buffer solutions that maintain the pH close to any desired value. Consider now how to choose the best conjugate acid–base system and how to calculate the required acid–base ratio.

In Examples 10.7, 10.8, and 10.9, the equilibrium concentrations of acid and base in the buffer systems were close to the initial concentrations. When this is the case, the calculation of pH is simplified greatly, because for either an acidic *or* a basic buffer,

$$K_a = \frac{[H_3O^+][A^-]}{[HA]} \approx \frac{[H_3O^+][A^-]_0}{[HA]_0}$$

$$[H_3O^+] = \frac{[HA]_0}{[A^-]_0} K_a$$

$$pH \approx pK_a - \log_{10} \frac{[HA]_0}{[A^-]_0}$$

where the last equation was obtained by taking the logarithm and changing sign. It can be verified that this simple equation gives the correct result for the three preceding examples. However, some care must be used in employing it because it is only approximate. It will be valid only if *both* $[H_3O^+]$ and $[OH^-]$ (and therefore the extent of ionization) are small relative to $[HA]_0$ and $[A^-]_0$.

This expression relating pH to $pK_a$ can be used to design buffers with a given pH. An optimal buffer is one in which the acid and its conjugate base are as nearly equal in concentration as possible; if the differences are too great, the buffer becomes less resistant to the effects of additional acid or base. To select a buffer system, an acid with a $pK_a$ as close as possible to the desired pH should be chosen. The concentrations of acid and conjugate base can then be adjusted to give exactly the desired pH.

### EXAMPLE 10.10

Design a buffer system with pH 4.60.

### Solution

From Table 10.2, the $pK_a$ for acetic acid is 4.75, so the $CH_3COOH/CH_3COO^-$ buffer is a suitable one. The concentrations required to give the desired pH are related by

$$pH = 4.60 = pK_a - \log_{10} \frac{[CH_3COOH]_0}{[CH_3COO^-]_0}$$

$$\log_{10} \frac{[CH_3COOH]_0}{[CH_3COO^-]_0} = pK_a - pH = 4.75 - 4.60 = 0.15$$

$$\frac{[CH_3COOH]_0}{[CH_3COO^-]_0} = 10^{0.15} = 1.4$$

Such a ratio could be established by dissolving 0.100 mol of sodium acetate and 0.140 mol of acetic acid in water and diluting to 1.00 L, or 0.200 mol of $NaCH_3COO$ and 0.280 mol of $CH_3COOH$ in the same volume, and so on. As long as the ratio of the concentrations is 1.4, the solution will be buffered at approximately pH 4.60.

**Related Problems: 47, 48, 49, 50**

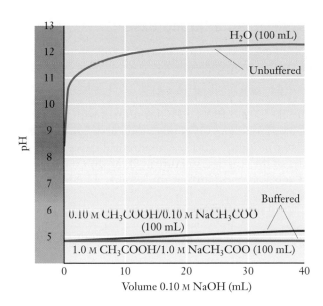

**FIGURE 10.11** Addition of a given volume of base to buffered and unbuffered solutions causes a much greater change in the pH of the unbuffered solution. Of the two buffered solutions, the one with higher buffer concentration resists pH changes more effectively.

As the preceding example shows, in the determination of the pH the *absolute* concentrations of acid and conjugate base in a buffer are much less important than is their ratio. However, the absolute concentrations do affect the capacity of the solution to resist changes in pH induced by added acid or base. The higher the concentrations of buffering species, the smaller the change in pH when a fixed amount of a strong acid or base is added. In Example 10.7, for instance, a change in buffer concentrations from 1.00 M and 0.500 M to 0.500 M and 0.250 M does not alter the original pH of 3.45, because the ratio of acid to base concentrations is unchanged. The pH after 0.100 mol of HCl is added does change to the value 3.15 rather than 3.31. The buffer at lower concentration is less resistant to pH change (Fig. 10.11). The buffering capacity of *any* buffer is exhausted if enough strong acid (or strong base) is added to use up, through chemical reaction, the original amount of weak base (or weak acid).

---

## 10.6

# ACID–BASE TITRATION CURVES

Section 6.3 described an acid–base titration as the addition of carefully metered volumes of a basic solution of known concentration to an acidic solution of unknown concentration (or the addition of acid to base) to reach an endpoint. The endpoint is signaled by the color change of an indicator or by a sudden rise or fall in pH, although the pH of the reaction mixture changes continuously over the course of an acid–base titration. A graph of the pH versus the volume $V$ of titrating solution is a **titration curve.** Its shape depends on the value of $K_a$ and on the concentrations of the acid and base reacting. The concepts of acid–base equilibria provide tools to predict the exact shapes of titration curves when these quantities are all known. The same concepts allow $K_a$ and the unknown concentration to be calculated from an experimental titration curve. Here we examine three categories of titration: strong acid reacting with strong base, weak acid reacting with strong base, and strong acid reacting with weak base. Titrations of a weak acid with a weak base (and the reverse) are not useful for analytical purposes.

### Titrations Involving a Strong Acid and a Strong Base

The addition of a strong base to a strong acid (or the reverse) is the simplest type of titration. The chemical reaction is the neutralization:

$$H_3O^+(aq) + OH^-(aq) \longrightarrow 2\ H_2O(\ell)$$

Suppose a solution of 100.0 mL (0.1000 L) of 0.1000 M HCl is titrated with 0.1000 M NaOH. What does the titration curve look like? Of course, the curve can be measured experimentally, giving the result in Figure 10.12. A theoretical curve can also be constructed by calculating the pH of the reaction mixture at many different points during the addition of the NaOH solution and plotting the results. The following example describes the latter procedure.

**1. $V = 0$ mL NaOH added**

Initially, $[H_3O^+] = 0.1000$ M, so the pH is 1.000. The chemical amount of $H_3O^+$ present initially is

$$n_{H_3O^+} = [H_3O^+](\text{volume}) = (0.1000 \text{ mol L}^{-1})(0.1000 \text{ L})$$
$$= 1.000 \times 10^{-2} \text{ mol}$$

**2. $V = 30.00$ mL NaOH added**

30.00 mL of 0.1000 M NaOH solution contains

$$(0.1000 \text{ mol L}^{-1})(0.03000 \text{ L}) = 3.000 \times 10^{-3} \text{ mol OH}^-$$

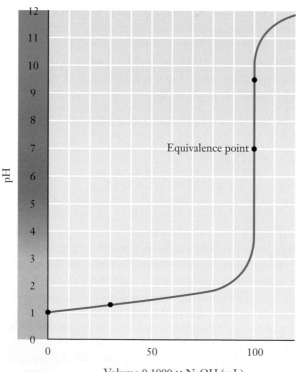

FIGURE 10.12   A titration curve for the titration of a strong acid by a strong base. The curve shown is for 100.0 mL of 0.1000 M HCl titrated with NaOH.

This reacts with (and neutralizes) an equal chemical amount of the $H_3O^+$ ion present initially and reduces $n_{H_3O^+}$ to

$$n_{H_3O^+} = (1.000 \times 10^{-2} - 3.000 \times 10^{-3})\text{mol} = 7.00 \times 10^{-3} \text{ mol}$$

In addition (and this is very important to remember), the volume of the titration mixture has increased from 100.0 to 130.0 mL (that is, from 0.1000 to 0.1300 L). The concentration of $H_3O^+$ at this point in the titration is

$$[H_3O^+] = \frac{n_{H_3O^+}}{V_{\text{tot}}} = \frac{7.00 \times 10^{-3} \text{ mol}}{0.1300 \text{ L}} = 0.0538 \text{ M}$$

$$pH = 1.27$$

### 3. $V = 100.00$ mL NaOH added

This is called the **equivalence point,** that point in the titration at which the chemical amount of base added equals the chemical amount of acid originally present. The volume of base added up to the equivalence point is the **equivalent volume,** $V_e$. At the equivalence point in the titration of a strong acid with a strong base, the concentrations of $OH^-$ and $H_3O^+$ must be equal and the pH 7.0 due to the autoionization of water. At this point the solution is simply a nonhydrolyzing salt (in this case, NaCl) in water. The pH is 7 at the equivalence point only in the titration of a strong acid with a strong base (or vice versa). The pH at the equivalence point differs from 7 if a weak acid or weak base takes part in the titration.

### 4. $V = 100.05$ mL NaOH added

Beyond the equivalence point, $OH^-$ is added to a neutral unbuffered solution. The $OH^-$ concentration can be found from the chemical amount of $OH^-$ added after the equivalence point has been reached. The volume beyond the equivalence point at this stage is 0.05 mL, or $5 \times 10^{-5}$ L. (This is the volume of approximately one drop of solution added from the buret.) The chemical amount of $OH^-$ in this volume is

$$(0.1000 \text{ mol L}^{-1})(5 \times 10^{-5} \text{ L}) = 5 \times 10^{-6} \text{ mol}$$

Meanwhile, the total volume of the titration mixture is

0.1000 L HCl solution + 0.10005 L NaOH solution = 0.20005 L solution

so that the concentration of $OH^-$ is

$$[OH^-] = \frac{\text{moles } OH^-}{\text{total volume}} = \frac{5 \times 10^{-5} \text{ mol}}{0.20005 \text{ L}} = 2.5 \times 10^{-5} \text{ M}$$

$$[H_3O^+] = 4 \times 10^{-10} \text{ M}; \qquad pH = 9.4$$

As the titration curve shows (Fig. 10.12), the pH increases dramatically in the immediate vicinity of the equivalence point: $[H_3O^+]$ changes by four orders of magnitude between 99.98 mL and 100.02 mL NaOH! Any indicator whose color changes between pH = 5.0 and pH = 9.0 therefore signals the endpoint of the titration to an accuracy of ±0.02 mL in 100.0 mL, or ±0.02%. The titration **endpoint** (the experimentally measured volume at which the indicator changes color) is then almost identical to the equivalence point (the theoretical volume at which the chemical amount of added base equals that of acid present originally).

The titration of a strong base by a strong acid is entirely parallel. In this case the pH starts at a higher value and *drops* through a pH of 7 at the equivalence point. The roles of acid and base, and of $H_3O^+$ and $OH^-$, are reversed in the equations already given.

## Titration of Weak Acids and Bases

We turn now to the titration of a weak acid with a strong base (the titration of a weak base with a strong acid is analogous). The equivalence point has the same meaning as for a strong acid titration. At the equivalence point, the chemical amount of base added (in volume $V_e$) is equal to the chemical amount of acid originally present (in volume $V_0$), so that once again

$$c_0 V_0 = c_t V_e$$

where $c_0$ is the original weak acid concentration and $c_t$ is the $OH^-$ concentration in the titrating solution.

The calculation of the titration curve differs from the strong acid–strong base case in that now equilibrium (reflected in the $K_a$ of the weak acid) enters the picture. As an example, consider the titration of 100.0 mL of a 0.1000 M solution of acetic acid ($CH_3COOH$) with 0.1000 M NaOH. For this titration,

$$V_e = \frac{c_0}{c_t} V_0 = \left( \frac{0.1000 \text{ M}}{0.1000 \text{ M}} \right)(100.0 \text{ mL})$$

$$= 100.0 \text{ mL}$$

There are four distinct ranges in the titration; each corresponds to a type of calculation that has been considered already. Figure 10.13 shows pH versus volume for this titration, and the following paragraphs outline the calculation of four typical points on the curve.

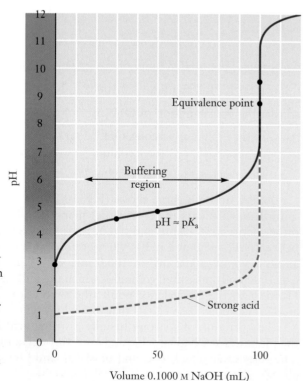

**FIGURE 10.13** A titration curve for the titration of a weak acid by a strong base. The curve shown in red is for 100.0 mL of 0.1000 M $CH_3COOH$ titrated with NaOH. For comparison, the dashed blue line shows the titration curve for a strong acid of the same amount and concentration as presented in Figure 10.12.

## 1. $V = 0$ mL NaOH added

Before any NaOH is added (that is, at the beginning of the titration), the problem is simply the ionization of a weak acid, which was considered in Section 10.4. A calculation analogous to those of Examples 10.3 and 10.4 gives a pH of 2.88.

## 2. $0 < V < V_e$

In this range the acid has been partially neutralized by added NaOH solution. Because $OH^-$ is a stronger base than acetate ion, it reacts almost completely with the acid originally present:

$$CH_3COOH(aq) + OH^-(aq) \rightleftharpoons H_2O(\ell) + CH_3COO^-(aq)$$

$$K = \frac{1}{K_b} = \frac{K_a}{K_w} = 2 \times 10^9 \gg 1$$

As a specific example, suppose that 30.00 mL of 0.1000 M NaOH has been added. The 30.00 mL NaOH contains

$$(0.1000 \text{ mol L}^{-1})(0.03000 \text{ L}) = 3.000 \times 10^{-3} \text{ mol OH}^-$$

and the original solution contained

$$(0.1000 \text{ mol L}^{-1})(0.1000 \text{ L}) = 1.000 \times 10^{-2} \text{ mol CH}_3COOH$$

The neutralization reaction generates one $CH_3COO^-$ ion for every $OH^-$ ion added. Hence, $3.000 \times 10^{-3}$ mol of $CH_3COO^-$ ion is generated. The amount of acetic acid that remains unreacted is

$$1.000 \times 10^{-2} - 3.000 \times 10^{-3} = 7.00 \times 10^{-3} \text{ mol}$$

Because the total volume is now 130.0 mL (or 0.1300 L), the nominal concentrations after reaction are

$$[CH_3COOH] \approx \frac{7.00 \times 10^{-3} \text{ mol}}{0.1300 \text{ L}} = 5.38 \times 10^{-2} \text{ M}$$

$$[CH_3COO^-] \approx \frac{3.00 \times 10^{-3} \text{ mol}}{0.1300 \text{ L}} = 2.31 \times 10^{-2} \text{ M}$$

This is nothing other than a buffer solution containing acetic acid at a concentration of $5.38 \times 10^{-2}$ M and sodium acetate at a concentration of $2.31 \times 10^{-2}$ M. Because the $K_a$ for acetic acid is larger than $10^{-7}$, hydronium ion (not hydroxide ion) predominates, and this is an acidic buffer. The pH can be found from the procedure used in Example 10.7 or, more approximately,

$$pH \approx pK_a - \log_{10} \frac{[CH_3COOH]_0}{[CH_3COO^-]_0} = 4.75 - \log_{10} \frac{5.38 \times 10^{-2}}{2.31 \times 10^{-2}} = 4.38$$

This region of the titration shows clearly the buffering action of a mixture of a weak acid with its conjugate base. At the half-equivalence point $V = V_e/2$, $[CH_3COOH]_0 = [CH_3COO^-]_0$, which corresponds to an equimolar buffer; at this point pH $\approx$ p$K_a$. On either side of this point the pH rises relatively slowly as the NaOH solution is added.

## 3. $V = V_e$

At the equivalence point, $c_t V_e$ mol of $OH^-$ has been added to an equal chemical amount of $CH_3COOH$. An identical solution could have been prepared simply by adding $c_t V_e = 1.000 \times 10^{-2}$ mol of the base $CH_3COO^-$ (in the form of

NaCH$_3$COO) to 0.2000 L of water. The pH at the equivalence point corresponds to the hydrolysis of CH$_3$COO$^-$:

$$H_2O(\ell) + CH_3COO^-(aq) \rightleftharpoons CH_3COOH(aq) + OH^-(aq)$$

as considered in Example 10.6. At the equivalence point the pH is 8.73.

Note that in the titration of a weak acid by a strong base the equivalence point comes not at pH 7 but at a higher (more basic) value. By the same token, the equivalence point in the titration of a weak base by a strong acid occurs at a pH lower than 7.

**4. $V > V_e$**

Beyond the equivalence point, OH$^-$(aq) is added to a solution of the base CH$_3$COO$^-$. The [OH$^-$] comes almost entirely from the hydroxide ion added beyond the equivalence point; very little comes from the reaction of the CH$_3$COO$^-$ with water. Beyond $V_e$, the pH for the titration of a weak acid by a strong base is very close to that for a strong acid by a strong base.

The equivalent volume $V_e$ is readily determined in the laboratory by using an indicator that changes color near pH 8.7, the pH at the equivalence point of the acetic acid titration. A suitable choice would be phenolphthalein, which changes from colorless to red over a pH range from 8.2 to 10.0. The slope of pH versus volume of strong base is less steep near the equivalence point for a weak acid than it is for a strong acid, making determinations of the equivalent volume (and of the original weak acid concentration) somewhat less accurate.

The use of titrations to determine concentrations and ionization constants for unknown acids and bases is illustrated by the following example.

---

### EXAMPLE 10.11

A volume of 50.00 mL of a weak acid of unknown concentration is titrated with a 0.1000 M solution of NaOH. The equivalence point is reached after 39.30 mL of NaOH solution has been added. At the half-equivalence point (19.65 mL) the pH is 4.85. Calculate the original concentration of the acid and its ionization constant $K_a$.

### Solution

The chemical amount of acid originally present, $c_0V_0$, is equal to the chemical amount of base added at the equivalence point, $c_tV_e$, so that

$$c_0 = \frac{V_e}{V_0}c_t = \frac{39.50 \text{ mL}}{50.00 \text{ mL}} \times 0.100 \text{ M} = 0.0786 \text{ M}$$

This is the original concentration of acid.

At the half-equivalence point,

$$pH = 4.85 \approx pK_a$$

$$K_a = 10^{-4.85} = 1.4 \times 10^{-5}$$

The unknown acid could be propionic acid, CH$_3$CH$_2$COOH (see Table 10.2).

**Related Problems: 61, 62**

---

## 10.7

# POLYPROTIC ACIDS

So far, only **monoprotic acids** have been considered. They are acids whose molecules are capable of donating only a single hydrogen ion to acceptor molecules. **Polyprotic acids** can donate two or more hydrogen ions to acceptors. Sulfuric acid is a familiar and important example. It reacts in two stages—first

$$H_2SO_4(aq) + H_2O(\ell) \longrightarrow H_3O^+(aq) + HSO_4^-(aq)$$

to give the hydrogen sulfate ion, and then

$$HSO_4^-(aq) + H_2O(\ell) \longrightarrow H_3O^+(aq) + SO_4^{2-}(aq)$$

The hydrogen sulfate ion is amphoteric, meaning that it is a base in the first reaction (with conjugate acid $H_2SO_4$) and an acid in the second (with conjugate base $SO_4^{2-}$). In its first ionization $H_2SO_4$ is a strong acid, but the product of that ionization ($HSO_4^-$) is itself only a weak acid. The $H_3O^+$ produced in a solution of $H_2SO_4$ comes primarily from the first ionization, and the solution has a pH close to that of a monoprotic strong acid of the same concentration. When this solution reacts with a strong base, however, its neutralizing power is twice that of a monoprotic acid of the same concentration, because each mole of sulfuric acid can react with and neutralize two moles of hydroxide ion.

## Weak Polyprotic Acids

Weak polyprotic acids ionize in two or more stages. Examples are carbonic acid ($H_2CO_3$), formed from solvated $CO_2$ (carbonated water; Fig. 10.14), and phosphoric acid ($H_3PO_4$). Carbonic acid can give up one hydrogen ion to form $HCO_3^-$ (hydrogen carbonate ion) or two hydrogen ions to form $CO_3^{2-}$ (carbonate ion). Phosphoric acid ionizes in three stages, giving successively $H_2PO_4^-$, $HPO_4^{2-}$, and $PO_4^{3-}$.

Two simultaneous equilibria are involved in the ionization of a diprotic acid such as $H_2CO_3$:[1]

$$H_2CO_3(aq) + H_2O(\ell) \rightleftharpoons H_3O^+(aq) + HCO_3^-(aq)$$

$$\frac{[H_3O^+][HCO_3^-]}{[H_2CO_3]} = K_{a1} = 4.3 \times 10^{-7}$$

and

$$HCO_3^-(aq) + H_2O(\ell) \rightleftharpoons H_3O^+(aq) + CO_3^{2-}(aq)$$

$$\frac{[H_3O^+][CO_3^{2-}]}{[HCO_3^-]} = K_{a2} = 4.8 \times 10^{-11}$$

---

[1] An accurate description of carbonic acid and solvated $CO_2$ is somewhat more complicated than indicated here. In fact, most of the dissolved $CO_2$ remains as $CO_2(aq)$, and only a small fraction actually reacts with water to give $H_2CO_3(aq)$. However, we will indicate by [$H_2CO_3$] the total concentration of both of these species. Approximately 0.034 mol of $CO_2$ dissolves per liter of water at 25°C and atmospheric pressure.

**FIGURE 10.14** The indicator bromothymol blue is blue in basic solution (left). When Dry Ice (solid carbon dioxide) is placed in the bottom of the cylinder (right), it dissolves to form carbonic acid. This causes the indicator to change to its colorless form. (*Charles D. Winters*)

Two important observations can be made at the outset:

**1.** The $[H_3O^+]$ in the two ionization equilibria are one and the same.
**2.** $K_{a2}$ is invariably smaller than $K_{a1}$ because the negative charge left behind by the loss of a hydrogen ion in the first ionization causes the second hydrogen ion to be more tightly bound.

Exact calculations of simultaneous equilibria can be complex. They simplify considerably when the original acid concentration is not too small and the ionization constants $K_{a1}$ and $K_{a2}$ differ substantially in magnitude (by a factor of 100 or more). The latter condition is almost always satisfied. Under such conditions, the two equilibria can be treated sequentially, as in the following example.

### E X A M P L E 10.12

Calculate the concentrations at equilibrium of $H_2CO_3$, $HCO_3^-$, $CO_3^{2-}$, and $H_3O^+$ in a saturated aqueous solution of $CO_2$, in which the original concentration of $H_2CO_3$ is 0.034 M.

### Solution

The $H_3O^+$ arises both from the ionization of $H_2CO_3$ and from the subsequent ionization of $HCO_3^-$, but because $K_{a2} \ll K_{a1}$ it is reasonable to ignore the contribution of $[H_3O^+]$ from the second ionization (as well as from the autoionization of water). These approximations will be checked later in the calculation.

If $y$ mol $L^{-1}$ of $H_2CO_3$ ionizes, the following approximations apply:

| | $H_2CO_3(aq)$ + $H_2O(\ell)$ $\rightleftharpoons$ $H_3O^+(aq)$ + $HCO_3^-(aq)$ | | |
|---|:---:|:---:|:---:|
| Initial concentration (M) | 0.034 | $\approx 0$ | 0 |
| Change in concentration (M) | $-y$ | $+y$ | $+y$ |
| Equilibrium concentration (M) | $0.034 - y$ | $y$ | $y$ |

where equating both $[HCO_3^-]$ and $[H_3O^+]$ to $y$ involves the assumption that the subsequent ionization of $HCO_3^-$ has only a small effect on its concentration. Then the first ionization equilibrium can be written as

$$\frac{[H_3O^+][HCO_3^-]}{[H_2CO_3]} = K_{a1}$$

$$\frac{y^2}{0.034 - y} = 4.3 \times 10^{-7}$$

Solving this equation for $y$ gives

$$y = 1.2 \times 10^{-4} \text{ M} = [H_3O^+] = [HCO_3^-]$$

$$[H_2CO_3] = 0.034 - y = 0.034 \text{ M}$$

The second equilibrium can be written as

$$\frac{[H_3O^+][CO_3^{2-}]}{[HCO_3^-]} = K_{a2} = \frac{(1.2 \times 10^{-4})[CO_3^{2-}]}{1.2 \times 10^{-4}} = 4.8 \times 10^{-11}$$

$$[CO_3^{2-}] = 4.8 \times 10^{-11} \text{ M}$$

The concentration of the base produced in the second ionization, $[CO_3^{2-}]$, is numerically equal to $K_{a2}$.

The assumptions must now be checked. Because

$$[CO_3^{2-}] = 4.8 \times 10^{-11} \text{ M} \ll 1.2 \times 10^{-4} \text{ M} = [HCO_3^-]$$

we were justified in ignoring the effect of the second ionization on the concentrations of $HCO_3^-$ and $H_3O^+$. The additional concentration of $H_3O^+$ furnished by $HCO_3^-$ is only $4.8 \times 10^{-11}$ M. Finally, $[H_3O^+]$ is much larger than $10^{-7}$ M, so the neglect of water autoionization was also justified.

**Related Problems: 63, 64**

---

For a triprotic acid such as $H_3PO_4$, the concentration of the base ($PO_4^{3-}$) resulting from the third ionization could have been calculated in a similar manner.

An analogous procedure applies to the reactions of a base that can accept two or more hydrogen ions. In a solution of sodium carbonate ($Na_2CO_3$), for example, the carbonate ion reacts with water to form first $HCO_3^-$ and then $H_2CO_3$:

$$H_2O(\ell) + CO_3^{2-}(aq) \rightleftharpoons HCO_3^-(aq) + OH^-(aq)$$

$$\frac{[OH^-][HCO_3^-]}{[CO_3^{2-}]} = K_{b1} = \frac{K_w}{K_{a2}} = 2.1 \times 10^{-4}$$

$$H_2O(\ell) + HCO_3^-(aq) \rightleftharpoons H_2CO_3(aq) + OH^-(aq)$$

$$\frac{[OH^-][H_2CO_3]}{[HCO_3^-]} = K_{b2} = \frac{K_w}{K_{a1}} = 2.3 \times 10^{-8}$$

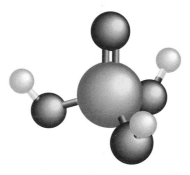

Phosphoric acid, $H_3PO_4$, is a triprotic acid that is used in the manufacture of phosphate fertilizers and in the food industry.

In this case $K_{b1} \gg K_{b2}$, so essentially all the $OH^-$ arises from the first reaction. The ensuing calculation of the concentrations of species present is just like that in Example 10.12.

## Effect of pH on Solution Composition

Changing the pH of a solution shifts the positions of all acid–base equilibria, including those involving polyprotic acids. Acid–base equilibrium expressions and equilibrium constants are used to calculate the amount of the change. For example, the two equilibria that apply for $H_2CO_3$–$HCO_3^-$–$CO_3^{2-}$ solutions can be written as

$$\frac{[HCO_3^-]}{[H_2CO_3]} = \frac{K_{a1}}{[H_3O^+]} \qquad \frac{[CO_3^{2-}]}{[HCO_3^-]} = \frac{K_{a2}}{[H_3O^+]}$$

At a given pH, the right-hand sides are known and the relative amounts of the three carbonate species can be calculated. This is illustrated by the following example.

### EXAMPLE 10.13

---

Calculate the fractions of carbonate present as $H_2CO_3$, $HCO_3^-$, and $CO_3^{2-}$ at pH 10.00.

**Solution**

At this pH, $[H_3O^+] = 1.0 \times 10^{-10}$ M, and the values of $K_{a1}$ and $K_{a2}$ for the preceding equations are used to find

$$\frac{[HCO_3^-]}{[H_2CO_3]} = \frac{4.3 \times 10^{-7}}{1.0 \times 10^{-10}} = 4.3 \times 10^3$$

$$\frac{[CO_3^{2-}]}{[HCO_3^-]} = \frac{4.8 \times 10^{-11}}{1.0 \times 10^{-10}} = 0.48$$

It is most convenient to rewrite these ratios with the same species (say, $HCO_3^-$) in the denominator. The first equation becomes

$$\frac{[H_2CO_3]}{[HCO_3^-]} = \frac{1}{4.3 \times 10^3} = 2.3 \times 10^{-4}$$

The fraction of each species present is obtained by dividing the species' concentration by the sum of the three concentrations. For $H_2CO_3$ this gives

$$\text{fraction } H_2CO_3 = \frac{[H_2CO_3]}{[H_2CO_3] + [HCO_3^-] + [CO_3^{2-}]}$$

This can be simplified by dividing numerator and denominator by $[HCO_3^-]$ and substituting in the ratios that have already been calculated:

$$\text{fraction } H_2CO_3 = \frac{[H_2CO_3]/[HCO_3^-]}{([H_2CO_3]/[HCO_3^-]) + 1 + ([CO_3^{2-}]/[HCO_3^-])}$$

$$= \frac{2.3 \times 10^{-4}}{2.3 \times 10^{-4} + 1 + 0.48} = 1.6 \times 10^{-4}$$

In the same way the fractions of $HCO_3^-$ and $CO_3^{2-}$ present are calculated to be 0.68 and 0.32, respectively.

**Related Problems: 67, 68**

---

If the calculation of Example 10.13 is repeated at a series of different pH values, the graph shown in Figure 10.15 is obtained. At high pH, $CO_3^{2-}$ predominates, and at

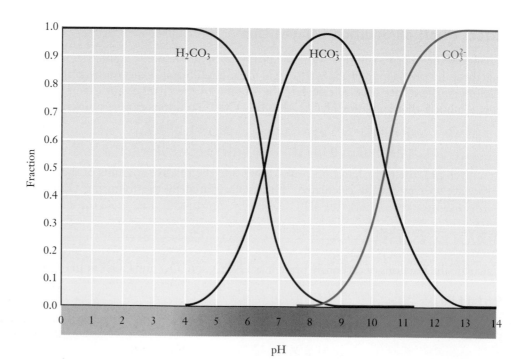

**FIGURE 10.15** The equilibrium fractions of $H_2CO_3$, $HCO_3^-$, and $CO_3^{2-}$ that are present in aqueous solution at different values of the pH.

low pH, $H_2CO_3$ is the major species. At intermediate pH (near pH 8, the approximate pH of seawater), the hydrogen carbonate ion $HCO_3^-$ is most prevalent. The variation in composition of sedimentary rocks from different locations can be traced back to the effect of pH on solution composition. Sediments containing carbonates were formed from alkaline (high-pH) lakes and oceans in which $CO_2$ was present mainly as $CO_3^{2-}$ ion. Sediments deposited from waters with intermediate pH are hydrogen carbonates or mixtures between carbonates and hydrogen carbonates. An example of the latter is trona ($2Na_2CO_3 \cdot NaHCO_3 \cdot 2H_2O$), an ore from the western United States that is an important source of both carbonates of sodium. Acidic waters did not deposit carbonates, but instead released $CO_2(g)$ to the atmosphere.

# A DEEPER LOOK...

## 10.8 EXACT TREATMENT OF ACID–BASE EQUILIBRIA

Sections 10.4 and 10.5 outlined methods for calculating equilibria involving weak acids, bases, and buffer solutions. There it was assumed that the amount of hydronium ion (or hydroxide ion) resulting from the ionization of water could be neglected in comparison with that produced by the ionization of dissolved acids or bases. In this section, that approximation is replaced by an exact treatment of acid–base equilibria (exact, that is, in the context of the mass-action law). This approach leads to somewhat more complicated equations, but it serves several purposes. It is of practical importance in cases (such as very weak acids or bases, or very dilute solutions) in which the previous approximations no longer hold. It includes as special cases the various aspects of acid–base equilibrium considered earlier. And finally, it provides a foundation for treating amphoteric equilibrium later in this section.

Consider a general case in which the initial concentration of a weak acid HA is called $c_a$, and the initial concentration of its conjugate base (which for simplicity is assumed to come from the salt NaA) is $c_b$. In solution there will be five dissolved species: HA, $A^-$, $Na^+$, $H_3O^+$, and $OH^-$. It is necessary to write down and solve five independent equations that relate the equilibrium concentrations of these species to the initial concentrations $c_a$ and $c_b$, and to $K_a$, the acid ionization constant of HA. The first equation is simply

$$[Na^+] = c_b \tag{a}$$

reflecting the fact that the $Na^+(aq)$ from the dissolved salt is a spectator ion that does not take part in the acid–base equilibrium. Next there are two equilibrium relations:

$$[H_3O^+][OH^-] = K_w \tag{b}$$

$$\frac{[H_3O^+][A^-]}{[HA]} = K_a \tag{c}$$

The fourth relation is one of stoichiometry, or conservation of "A-material":

$$c_a + c_b = [HA] + [A^-] \tag{d}$$

The original A-material was introduced either as acid or as base, with a total concentration of $c_a + c_b$. When equilibrium is reached, some redistribution has doubtless occurred, but the *total* concentration, $[HA] + [A^-]$, must be the same. The fifth and final relation results from charge balance; the solution must be electrically neutral, so the total amount of positive charge must be equal to the total amount of negative charge:

$$[Na^+] + [H_3O^+] = [A^-] + [OH^-] \tag{e}$$

These five independent equations completely determine the five unknown concentrations. To solve them, begin by substituting (a) into (e) and solving for $[A^-]$:

$$[A^-] = c_b + [H_3O^+] - [OH^-] \tag{e'}$$

Next, insert (e') into (d) and solve for $[HA]$:

$$[HA] = c_a + c_b - [A^-] = c_a - [H_3O^+] + [OH^-] \tag{d'}$$

Next, substitute both (d') and (e') into (c) to find

$$\frac{[H_3O^+](c_b + [H_3O^+] - [OH^-])}{(c_a - [H_3O^+] + [OH^-])} = K_a \tag{c'}$$

There are now two things that can be done with the general equation (c'). First, the exact solution for $[H_3O^+]$ can be set up, a procedure that is useful for very weak acids or bases or for very dilute solutions. Alternatively, (c') can be reduced in various limits to cases already considered.

For the exact solution, eliminate $[OH^-]$ in (c') by using (b). This gives

$$\frac{[H_3O^+]\left(c_b + [H_3O^+] - \dfrac{K_w}{[H_3O^+]}\right)}{\left(c_a - [H_3O^+] + \dfrac{K_w}{[H_3O^+]}\right)} = K_a$$

The numerator and denominator are multiplied by $[H_3O^+]$ and the fraction is cleared by moving the denominator to the right side:

$$[H_3O^+](c_b[H_3O^+] + [H_3O^+]^2 - K_w)$$
$$= K_a(c_a[H_3O^+] - [H_3O^+]^2 + K_w)$$

This can be rewritten as

$$[H_3O^+]^3 + (c_b + K_a)[H_3O^+]^2$$
$$- (K_w + c_aK_a)[H_3O^+] - K_aK_w = 0$$

This is a cubic equation for $[H_3O^+]$, which, although more complicated than the quadratic equations encountered so far, can nonetheless be solved with a calculator. The concentrations of $OH^-$, $A^-$, and $HA$ can then be found by successive substitutions into (b), (e′), and (d′).

Alternatively, the general equation (c′) can be examined in various limits. In an acidic buffer (as in Examples 10.7 and 10.8), if it can be assumed that $[H_3O^+] \gg [OH^-]$, then (c′) simplifies to

$$\frac{[H_3O^+](c_b + [H_3O^+])}{(c_a - [H_3O^+])} = K_a$$

which is exactly the equation used in those examples. If, additionally, $c_b = 0$, the weak-acid ionization limit of Examples 10.3 and 10.4 is reached. In a basic buffer, on the other hand, if it can be assumed that $[OH^-] \gg [H_3O^+]$, then (c′) simplifies to

$$\frac{(K_w/[OH^-])(c_b - [OH^-])}{(c_a + [OH^-])} = K_a$$

Here (b) was used to substitute for $[H_3O^+]$ where it multiplies the whole expression. This can be rewritten as

$$\frac{[OH^-](c_a + [OH^-])}{(c_b - [OH^-])} = \frac{K_w}{K_a} = K_b$$

which is exactly the equation used in Example 10.9. If no acid is present initially ($c_a = 0$), the weak-base ionization limits of Examples 10.5 and 10.6 are reached. The general approach therefore includes all of the previous calculations as special cases.

Unless conditions require the use of the exact solution, approximate equations are preferable because they are easier to apply and provide greater physical insight. If a calculation (ignoring water autoionization) of the ionization of a weak acid gives a concentration of $H_3O^+$ smaller than $10^{-6}$ M, or if a calculation of base ionization gives a concentration of $OH^-$ smaller than $10^{-6}$ M, the more exact treatment is needed. For buffer solutions, a pH near 7 does not necessarily mean that water ionization is important, unless the acid or base concentration becomes very small.

## E X A M P L E  10.14

Calculate the pH of a $1.00 \times 10^{-5}$ M solution of $HCN(aq)$. The $K_a$ of $HCN(aq)$ is $6.17 \times 10^{-10}$.

## Solution

Suppose the autoionization of water is ignored and the method of Examples 10.3 and 10.4 is used. This gives $[H_3O^+] = 7.9 \times 10^{-8}$ M, which of course makes no sense, because it is *lower* than the concentration of hydronium ion in pure water. HCN is a very weak acid, but it is nonetheless an acid, not a base.

It is necessary, therefore, to use the exact cubic equation for $[H_3O^+]$, inserting into it the proper coefficients and taking $c_a = 1.00 \times 10^{-5}$ and $c_b = 0$. This gives

$$[H_3O^+]^3 + 6.17 \times 10^{-10}[H_3O^+]^2$$
$$- 1.617 \times 10^{-14}[H_3O^+] - 6.17 \times 10^{-24} = 0$$

Unfortunately, there is no method as simple as the quadratic formula to solve a cubic equation. The easiest way to solve this equation is to try a series of values for $[H_3O^+]$ on the left side, varying them to obtain a result as close as possible to 0 (see Appendix C). It is safe to assume that the final answer will be slightly larger than $1 \times 10^{-7}$, so the initial guesses should be of that magnitude. Carrying out the procedure gives

$$[H_3O^+] = 1.27 \times 10^{-7} \text{ M} \qquad\qquad pH = 6.90$$

**Related Problems: 69, 70**

## Amphoteric Equilibria

A second situation in which an exact analysis of acid–base equilibrium is useful occurs when an amphoteric species is dissolved in water. The hydrogen carbonate ion ($HCO_3^-$) is amphoteric because it can act as an acid in the equilibrium

$$HCO_3^-(aq) + H_2O(\ell) \rightleftharpoons H_3O^+(aq) + CO_3^{2-}(aq)$$

$$\frac{[H_3O^+][CO_3^{2-}]}{[HCO_3^-]} = K_{a2} = 4.8 \times 10^{-11}$$

or as a base in the equilibrium

$$H_2O(\ell) + HCO_3^-(aq) \rightleftharpoons H_2CO_3(aq) + OH^-(aq)$$

$$\frac{[OH^-][H_2CO_3]}{[HCO_3^-]} = K_{b2} = 2.3 \times 10^{-8}$$

If sodium hydrogen carbonate ($NaHCO_3$) is dissolved in water, a competition occurs between the tendencies of $HCO_3^-$ to accept hydrogen ions and to donate them. Because $K_{b2} > K_{a2}$, there should be more production of $OH^-$ than of $H_3O^+$, so the solution should be basic.

In an exact treatment of this equilibrium, there are six unknown concentrations—those of $Na^+$, $H_2CO_3$, $HCO_3^-$, $CO_3^{2-}$, $OH^-$, and $H_3O^+$. Two equilibrium equations were already presented, and a third relates $[OH^-]$ and $[H_3O^+]$ to $K_w$. If $[HCO_3^-]_0$ is the original concentration of $NaHCO_3$, then from stoichiometry

$$[HCO_3^-]_0 = [HCO_3^-] + [H_2CO_3] + [CO_3^{2-}]$$

because the total amount of carbonate material is conserved; any reduction in $[HCO_3^-]$ must be compensated by a corresponding increase in either $[H_2CO_3]$ or $[CO_3^{2-}]$. Next the principle of conservation of charge is used. The positively charged species present are $Na^+$ and $H_3O^+$, and the negatively charged species are $HCO_3^-$, $CO_3^{2-}$, and $OH^-$. Because there is overall charge neutrality,

$$[Na^+] + [H_3O^+] = [HCO_3^-] + 2[CO_3^{2-}] + [OH^-]$$

where the coefficient 2 for $[CO_3^{2-}]$ arises because each carbonate ion is doubly charged. In addition, the $Na^+$ concentration is unchanged, so

$$[Na^+] = [HCO_3^-]_0$$

In principle, these six equations could be solved simultaneously to calculate the exact $[H_3O^+]$ for an arbitrary initial concentration of $HCO_3^-$. The result is complex and gives little physical insight. Instead, we give only a simpler, approximate solution, which is sufficient in the cases considered here. Subtracting the carbonate balance equation from the charge balance equation gives

$$[H_3O^+] = [CO_3^{2-}] - [H_2CO_3] + [OH^-]$$

The three equilibrium expressions are used to rewrite this as

$$[H_3O^+] = K_{a2}\frac{[HCO_3^-]}{[H_3O^+]} - \frac{[H_3O^+][HCO_3^-]}{K_{a1}} + \frac{K_w}{[H_3O^+]}$$

where $[CO_3^{2-}]$ and $[H_2CO_3]$ have been eliminated in favor of $[HCO_3^-]$.

Multiplying by $K_{a1}[H_3O^+]$ gives

$$K_{a1}[H_3O^+]^2 + [HCO_3^-][H_3O^+]^2 = K_{a1}K_{a2}[HCO_3^-] + K_{a1}K_w$$

$$[H_3O^+]^2 = \frac{K_{a1}K_{a2}[HCO_3^-] + K_{a1}K_w}{K_{a1} + [HCO_3^-]}$$

This equation still contains two unknown quantities, $[H_3O^+]$ and $[HCO_3^-]$. Because both $K_{a2}$ and $K_{b2}$ are small, $[HCO_3^-]$ should be close to its original value, $[HCO_3^-]_0$. If $[HCO_3^-]$ is set equal to $[HCO_3^-]_0$, this becomes

$$[H_3O^+]^2 \approx \frac{K_{a1}K_{a2}[HCO_3^-]_0 + K_{a1}K_w}{K_{a1} + [HCO_3^-]_0}$$

which can be solved for $[H_3O^+]$. In many cases of interest, $[HCO_3^-]_0 \gg K_{a1}$, and $K_{a2}[HCO_3^-]_0 \gg K_w$. When this is so, the expression simplifies to

$$[H_3O^+]^2 \approx K_{a1}K_{a2}$$

$$[H_3O^+] \approx \sqrt{K_{a1}K_{a2}}$$

$$pH \approx \frac{1}{2}(pK_{a1} + pK_{a2})$$

and so the pH of such a solution is the average of the $pK_a$ values for the two ionizations.

## EXAMPLE 10.15

What is the pH of a solution that is 0.100 M in $NaHCO_3$?

### Solution

First, the two assumptions are checked:

$$[HCO_3^-]_0 = 0.100 \gg 4.3 \times 10^{-7} = K_{a1}$$

$$[HCO_3^-]_0 K_{a2} = 4.8 \times 10^{-12} \gg 1.0 \times 10^{-14} = K_w$$

so both are satisfied. Therefore,

$$[H_3O^+] = \sqrt{K_{a1}K_{a2}} = 4.5 \times 10^{-9} \text{ M}$$

$$pH = 8.34$$

and the solution is basic, as expected.

## Titration of a Polyprotic Acid

A polyprotic acid shows more than one equivalence point. The first equivalence point occurs when the volume $V_{e1}$ of base added is sufficient to remove one hydrogen ion from each acid molecule; $V_{e2}$ is the volume sufficient to remove two hydrogen ions from each, and so forth. A diprotic acid shows two equivalence points, and a triprotic acid, three. The equivalent volumes are related to each other by

$$V_{e1} = \frac{1}{2}V_{e2} = \frac{1}{3}V_{e3}$$

Figure 10.16 shows a titration curve for triprotic phosphoric acid. The three equivalence points are at 100.0, 200.0, and 300.0 mL.

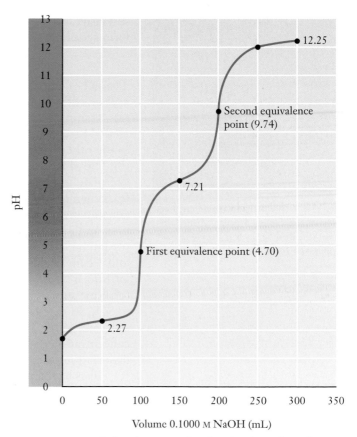

FIGURE 10.16 A titration curve for the titration of a polyprotic acid (phosphoric acid) by a strong base. The curve shown is for 100.0 mL of 0.1000 M $H_3PO_4$ titrated with NaOH. No clear third equivalence point is seen at 300 mL because $K_a$ for $HPO_4^{2-}$ is not much greater than $K_a$ for $H_2O$ in aqueous solution.

The calculation of pH as a function of the volume of added base presents no new complications beyond those already considered. The initial pH is given by a calculation analogous to that of Example 10.12, and the pH in the flat regions between equivalence points is obtained by a buffer calculation like that for a diprotic acid. For example, the pH after addition of 50.0 mL of base is that of an equimolar $H_3PO_4/H_2PO_4^-$ buffer (subsequent ioniza-

tion of $H_2PO_4^-$ can be ignored). Finally, the pH at the first equivalence point is that for a solution of $NaH_2PO_4$ and uses the amphoteric equilibria equations presented earlier in this section ($PO_4^{3-}$ can be ignored in this case). The pH at the second equivalence point is an amphoteric equilibrium in which $HPO_4^{2-}$ is in equilibrium with $H_2PO_4^-$ and with $PO_4^{3-}$.

## CUMULATIVE EXERCISE

### Acid Precipitation

Acid precipitation is a major environmental problem throughout the industrialized world. One major source is the burning of fossil fuels containing sulfur (coal, oil, and natural gas). The sulfur dioxide released into the air dissolves in water or, more seriously, may be oxidized further to sulfur trioxide. The $SO_3$ dissolves in water to form sulfuric acid:

$$SO_3(g) + H_2O(\ell) \longrightarrow H_2SO_4(aq)$$

The net effect is to increase the acidity of the rain, which damages trees, kills fish in lakes, dissolves stone, and corrodes metal.

(a) A sample of rainwater is tested for acidity by using two indicators. Addition of methyl orange to half of the sample gives a yellow color, and addition of methyl red to the other half gives a red color. Estimate the pH of the sample.

(b) The pH in acid rain can range down to 3 or even lower in heavily polluted areas. Calculate the concentrations of $H_3O^+$ and $OH^-$ in a raindrop at pH 3.30 at 25°C.

(c) When $SO_2$ dissolves in water to form sulfurous acid, $H_2SO_3(aq)$, that acid can donate a hydrogen ion to water. Write a balanced chemical equation for this reaction, and identify the stronger Brønsted–Lowry acid and base in the equation.

(d) Ignore the further ionization of $HSO_3^-$, and calculate the pH of a solution whose initial concentration of $H_2SO_3$ is $4.0 \times 10^{-4}$ M. (*Hint:* Use the quadratic equation in this case.)

(e) Now suppose that all the dissolved $SO_2$ from part (d) has been oxidized further to $SO_3$, so that $4.0 \times 10^{-4}$ mol of $H_2SO_4$ is dissolved per liter. Calculate the pH in this case. [*Hint:* Because the first ionization of $H_2SO_4$ is that of a strong acid, the concentration of $H_3O^+$ can be written as $4.0 \times 10^{-4}$ plus the unknown amount of dissociation from $HSO_4^-(aq)$.]

(f) Lakes have a natural buffering capacity, especially in regions where there is limestone that gives rise to dissolved calcium carbonate. Write an equation for the effect of a small amount of acid rain containing sulfuric acid if it falls into a lake containing carbonate ($CO_3^{2-}$) ions. Discuss how the lake will resist further pH changes. What happens if a large excess of acid rain is deposited?

(g) A sample of 1.00 L of rainwater known to contain only sulfurous (and not sulfuric) acid is titrated with 0.0100 M NaOH. The equivalence point of the $H_2SO_3/HSO_3^-$ titration is reached after 31.6 mL has been added. Calculate the orig-

The effect of acid rain on a stand of trees in the Great Smoky Mountains of the United States. *(Copyright Dembinsky Photo Associates)*

inal concentration of sulfurous acid in the sample, again ignoring any effect of $SO_3^{2-}$ on the equilibria.

(h) Calculate the pH at the half-equivalence point, after 15.8 mL has been added. (*Hint:* Use the quadratic equation.)

**Answers**

(a) 4.4 to 4.8

(b) $[H_3O^+] = 5.0 \times 10^{-4}$ M; $[OH^-] = 2.0 \times 10^{-11}$ M

(c) $H_2SO_3(aq) + H_2O(\ell) \rightleftharpoons HSO_3^-(aq) + H_3O^+(aq)$. The stronger acid is $H_3O^+$, and the stronger base is $HSO_3^-$.

(d) The pH is 3.41.

(e) The pH is 3.11.

(f) The $H_3O^+$ in the sulfuric acid solution reacts according to $H_3O^+(aq) + CO_3^{2-}(aq) \rightleftharpoons HCO_3^-(aq) + H_2O(\ell)$. The $HSO_4^-$ in the sulfuric acid reacts according to $HSO_4^-(aq) + CO_3^{2-}(aq) \rightleftharpoons SO_4^{2-}(aq) + HCO_3^{2-}(aq)$. This gives rise to a $HCO_3^-/CO_3^{2-}$ buffer that can resist further changes in pH. An excess of acid rain overwhelms the buffer and leads to the formation of $H_2CO_3$.

(g) $3.16 \times 10^{-4}$ M

(h) The pH is 3.81.

## Concepts & Skills

*After studying this chapter and working the problems that follow, you should be able to*

1. Define an acid and a base in the Brønsted–Lowry and Lewis systems and give several examples of their reaction with a solvent (Section 10.1, problems 1–12).

2. Define the pH function and convert between pH and $[H_3O^+]$ (Section 10.2, problems 13–16).

3. State the relationship between the ionization constant for an acid and that for its conjugate base (Section 10.3, problems 21–22).

4. Explain how indicators allow the pH of a solution to be estimated (Section 10.3, problems 25–26).

5. Formulate the equilibrium expression for the ionization of a weak acid or base, and use it to determine the pH and fraction ionized (Section 10.4, problems 27–36).

6. Explain the behavior of a buffer solution. Calculate its pH from the concentrations of its conjugate acid–base pair (Section 10.5, problems 43–46).

7. Design a buffer system that will regulate the pH to a particular value (Section 10.5, problems 47–50).

8. Calculate the pH at any stage in the titration of a strong acid or base by a strong base or acid (Section 10.6, problems 51–52).

9. Calculate the pH at any stage in the titration of a weak acid or base by a strong base or acid (Section 10.6, problems 53–56).

10. Calculate the concentrations of all the species present in a solution of a weak polyprotic acid (Section 10.7, problems 63–66).

11. Outline the procedure for the exact treatment of acid–base equilibrium and use it to find the pH of a very dilute solution of a weak acid or base (Section 10.8, problems 69–70).

12. Calculate the pH at selected points in the titration of a polyprotic acid (Section 10.8, problems 71–72).

## PROBLEMS

*Answers to problems whose numbers are boldface appear in Appendix G. Problems that are more challenging are indicated with asterisks.*

### Classifications of Acids and Bases

1. Which of the following can act as Brønsted–Lowry acids? Give the formula of the conjugate Brønsted–Lowry base for each of them.
   (a) $Cl^-$      (d) $NH_3$
   (b) $HSO_4^-$    (e) $H_2O$
   (c) $NH_4^+$

2. Which of the following can act as Brønsted–Lowry bases? Give the formula of the conjugate Brønsted–Lowry acid for each of them.
   (a) $F^-$      (d) $OH^-$
   (b) $SO_4^{2-}$    (e) $H_2O$
   (c) $O^{2-}$

3. Lemon juice contains citric acid ($C_6H_8O_7$). What species serves as a base when lemon juice is mixed with baking soda (sodium hydrogen carbonate) during the preparation of some lemon cookies?

4. A treatment recommended in case of accidental swallowing of ammonia-containing cleanser is to drink large amounts of diluted vinegar. Write an equation for the chemical reaction on which this procedure depends.

5. An important step in many industrial processes is the slaking of lime, in which water is added to calcium oxide to make calcium hydroxide.
   (a) Write the balanced equation for this process.
   (b) Can this be considered a Lewis acid–base reaction? If so, what is the Lewis acid and what is the Lewis base?

6. Silica ($SiO_2$) is an impurity that must be removed from a metal oxide or sulfide ore when the ore is being reduced to elemental metal. To do this, lime ($CaO$) is added. It reacts with the silica to form a slag of calcium silicate ($CaSiO_3$), which can be separated and removed from the ore.
   (a) Write the balanced equation for this process.
   (b) Can this be considered a Lewis acid–base reaction? If so, what is the Lewis acid and what is the Lewis base?

7. Chemists working with fluorine and its compounds sometimes find it helpful to think in terms of acid–base reactions in which the fluoride ion ($F^-$) is donated and accepted.
   (a) Would the acid in this system be the fluoride donor or fluoride acceptor?
   (b) Identify the acid and base in each of these reactions:

$$ClF_3O_2 + BF_3 \longrightarrow ClF_2O_2 \cdot BF_4$$

$$TiF_4 + 2\ KF \longrightarrow K_2[TiF_6]$$

8. Researchers working with glasses often think of acid–base reactions in terms of oxide donors and oxide acceptors. The oxide ion is $O^{2-}$.
   (a) In this system, is the base the oxide donor or the oxide acceptor?
   (b) Identify the acid and base in each of these reactions:

$$2\ CaO + SiO_2 \longrightarrow Ca_2SiO_4$$

$$Ca_2SiO_4 + SiO_2 \longrightarrow 2\ CaSiO_3$$

$$Ca_2SiO_4 + CaO \longrightarrow Ca_3SiO_5$$

9. Identify each of the following oxides as an acid or base anhydride. Write the chemical formula and give the name of the acid or base formed upon reaction with water.
   (a) $MgO$      (c) $SO_3$
   (b) $Cl_2O$     (d) $Cs_2O$

10. Write the chemical formula and give the name of the anhydride corresponding to each of the following acids or bases, and identify it as an acid or base anhydride.
    (a) $H_3AsO_4$    (c) $RbOH$
    (b) $H_2MoO_4$   (d) $H_2SO_3$

11. Tin(II) oxide is amphoteric. Write balanced chemical equations for its reactions with an aqueous solution of hydrochloric acid and with an aqueous solution of sodium hydroxide. (*Note:* The hydroxide complex ion of tin(II) is $[Sn(OH)_3]^-$.)

12. Zinc oxide is amphoteric. Write balanced chemical equations for its reactions with an aqueous solution of hydrochloric acid and with an aqueous solution of sodium hydroxide. (*Note:* The hydroxide complex ion of zinc is $[Zn(OH)_4]^{2-}$.)

### Properties of Acids and Bases in Aqueous Solutions: The Brønsted–Lowry Scheme

13. The concentration of $H_3O^+$ in a sample of wine is $2.0 \times 10^{-4}$ M. Calculate the pH of the wine.

14. The concentration of $OH^-$ in a solution of household bleach is $3.6 \times 10^{-2}$ M. Calculate the pH of the bleach.

15. The pH of normal human urine is in the range of 5.5 to 6.5. Compute the range of the $H_3O^+$ concentration and the range of the $OH^-$ concentration in normal urine.

16. The pH of normal human blood is in the range of 7.35 to 7.45. Compute the range of the concentration of $H_3O^+$ and the range of the $OH^-$ concentration in normal blood.

17. The $pK_w$ of seawater at 25°C is 13.776. This differs from the usual $pK_w$ of 14.00 at this temperature because dissolved salts make seawater a nonideal solution. If the pH in seawater is 8.00, what are the concentrations of $H_3O^+$ and $OH^-$ in seawater at 25°C?

18. At body temperature (98.6°F = 37.0°C), $K_w$ has the value $2.4 \times 10^{-14}$. If the pH of blood is 7.4 under these conditions, what are the concentrations of $H_3O^+$ and $OH^-$?

19. When placed in water, potassium starts to react instantly and continues to react with great vigor. On the basis of this information, select the better of the following two equations to represent the reaction.

$$2\ K(s) + 2\ H_2O(\ell) \longrightarrow 2\ KOH(aq) + H_2(g)$$

$$2\ K(s) + 2\ H_3O^+(aq) \longrightarrow 2\ K^+(aq) + H_2(g) + 2\ H_2O(\ell)$$

State the reason for your choice.

20. Molecules of *t*-butyl chloride, $(CH_3)_3CCl$, react very slowly when mixed with water at low pH to give *t*-butyl alcohol, $(CH_3)_3COH$. When the pH is raised, the reaction takes place rapidly. Write an equation or equations to explain these facts.

## Acid and Base Strength

21. Ephedrine $(C_{10}H_{15}ON)$ is a base that is used in nasal sprays as a decongestant.
    (a) Write an equation for its equilibrium reaction with water.
    (b) The $K_b$ for ephedrine is $1.4 \times 10^{-4}$. Calculate the $K_a$ for its conjugate acid.
    (c) Is ephedrine a weaker or a stronger base than ammonia?

22. Niacin $(C_5H_4NCOOH)$, one of the B vitamins, is an acid.
    (a) Write an equation for its equilibrium reaction with water.
    (b) The $K_a$ for niacin is $1.5 \times 10^{-5}$. Calculate the $K_b$ for its conjugate base.
    (c) Is the conjugate base of niacin a stronger or a weaker base than pyridine, $C_5H_5N$?

23. Use the data in Table 10.2 to determine the equilibrium constant for the reaction

$$HClO_2(aq) + NO_2^-(aq) \rightleftharpoons HNO_2(aq) + ClO_2^-(aq)$$

    Identify the stronger Brønsted–Lowry acid and the stronger Brønsted–Lowry base.

24. Use the data in Table 10.2 to determine the equilibrium constant for the reaction

$$HPO_4^{2-} + HCO_3^- \rightleftharpoons PO_4^{3-} + H_2CO_3$$

    Identify the stronger Brønsted–Lowry acid and the stronger Brønsted–Lowry base.

25. (a) Which is the stronger acid—the acidic form of the indicator bromocresol green or the acidic form of methyl orange?
    (b) A solution is prepared in which bromocresol green is green and methyl orange is orange. Estimate the pH of this solution.

26. (a) Which is the stronger base—the basic form of the indicator cresol red or the basic form of thymolphthalein?
    (b) A solution is prepared in which cresol red is red and thymolphthalein is colorless. Estimate the pH of this solution.

## Equilibria Involving Weak Acids and Bases

27. Aspirin is acetylsalicylic acid, $HC_9H_7O_4$, which has a $K_a$ of $3.0 \times 10^{-4}$. Calculate the pH of a solution made by dissolving 0.65 g of acetylsalicylic acid in water and diluting to 50.0 mL.

28. Vitamin C is ascorbic acid $(HC_6H_7O_6)$, for which $K_a$ is $8.0 \times 10^{-5}$. Calculate the pH of a solution made by dissolving a 500-mg tablet of pure vitamin C in water and diluting to 100 mL.

29. (a) Calculate the pH of a 0.20 M solution of benzoic acid at 25°C.

(b) How many moles of acetic acid must be dissolved per liter of water to obtain the same pH as that from part (a)?

30. (a) Calculate the pH of a 0.35 M solution of propionic acid at 25°C.
    (b) How many moles of formic acid must be dissolved per liter of water to obtain the same pH as that from part (a)?

31. Iodic acid $(HIO_3)$ is fairly strong for a weak acid, having a $K_a$ equal to 0.16 at 25°C. Compute the pH of a 0.100 M solution of $HIO_3$.

32. At 25°C, the $K_a$ of pentafluorobenzoic acid $(C_6F_5COOH)$ is 0.033. Suppose 0.100 mol of pentafluorobenzoic acid is dissolved in 1.00 L of water. What is the pH of this solution?

33. Papaverine hydrochloride $(papH^+Cl^-)$ is a drug used as a muscle relaxant. It is a weak acid. At 25°C, a 0.205 M solution of $papH^+Cl^-$ has a pH of 3.31. Compute the $K_a$ of the $papH^+$ ion.

34. The unstable weak acid 2-germaacetic acid $(GeH_3COOH)$ is derived structurally from acetic acid $(CH_3COOH)$ by having a germanium atom replace one of the carbon atoms. At 25°C, a 0.050 M solution of 2-germaacetic acid has a pH of 2.42. Compute the $K_a$ of 2-germaacetic acid and compare it with that of acetic acid.

35. Morphine is a weak base for which $K_b$ is $8 \times 10^{-7}$. Calculate the pH of a solution made by dissolving 0.0400 mol of morphine in water and diluting to 600 mL.

36. Methylamine is a weak base for which $K_b$ is $4.4 \times 10^{-4}$. Calculate the pH of a solution made by dissolving 0.070 mol of methylamine in water and diluting to 800.0 mL.

37. The pH at 25°C of an aqueous solution of hydrofluoric acid, HF $(K_a = 6.6 \times 10^{-4})$, is 2.13. Calculate the concentration of HF in this solution, in moles per liter.

38. The pH at 25°C of an aqueous solution of sodium cyanide (NaCN) is 11.50. ($K_a$ of HCN is $4.93 \times 10^{-10}$.) Calculate the concentration of $CN^-$ in this solution, in moles per liter.

39. You have 50.00 mL of a solution that is 0.100 M in acetic acid, and you neutralize it by adding 50.00 mL of a solution that is 0.100 M in sodium hydroxide. The pH of the resulting solution is not 7.00. Explain why. Is the pH of the solution greater than or less than 7?

40. A 75.00-mL portion of a solution that is 0.0460 M in $HClO_4$ is treated with 150.00 mL of 0.0230 M KOH(aq). Is the pH of the resulting mixture greater than, less than, or equal to 7.0? Explain.

41. Suppose a 0.100 M solution of each of the following substances is prepared. Rank the pH of the resulting solutions from lowest to highest: $NH_4Br$, NaOH, KI, $NaCH_3COO$, HCl.

42. Suppose a 0.100 M solution of each of the following substances is prepared. Rank the pH of the resulting solutions from lowest to highest: KF, $NH_4I$, HBr, NaCl, LiOH.

## Buffer Solutions

43. "Tris" is short for tris(hydroxymethyl)aminomethane. This weak base is widely used in biochemical research for the preparation of buffers. It offers low toxicity and a $pK_b$ (5.92

at 25°C) that is convenient for the control of pH in clinical applications. A buffer is prepared by mixing 0.050 mol of tris with 0.025 mol of HCl in a volume of 2.00 L. Compute the pH of the solution.

44. "Bis" is short for bis(hydroxymethyl)aminomethane. It is a weak base that is closely related to tris (see problem 43) and has similar properties and uses. Its $pK_b$ is 8.8 at 25°C. A buffer is prepared by mixing 0.050 mol of bis with 0.025 mol of HCl in a volume of 2.00 L (the same proportions as in the preceding problem). Compute the pH of the solution.

45. (a) Calculate the pH in a solution prepared by dissolving 0.050 mol of acetic acid and 0.020 mol of sodium acetate in water and adjusting the volume to 500 mL.
   (b) Suppose 0.010 mol of NaOH is added to the buffer from part (a). Calculate the pH of the solution that results.

46. Sulfanilic acid ($NH_2C_6H_4SO_3H$) is used in manufacturing dyes. It ionizes in water according to the equilibrium equation

$$NH_2C_6H_4SO_3H(aq) + H_2O(\ell) \rightleftharpoons$$
$$NH_2C_6H_4SO_3^-(aq) + H_3O^+(aq)$$
$$K_a = 5.9 \times 10^{-4}$$

A buffer is prepared by dissolving 0.20 mol of sulfanilic acid and 0.13 mol of sodium sulfanilate ($NaNH_2C_6H_4SO_3$) in water and diluting to 1.00 L.
   (a) Compute the pH of the solution.
   (b) Suppose 0.040 mol of HCl is added to the buffer. Calculate the pH of the solution that results.

47. A physician wishes to prepare a buffer solution at pH = 3.82 that efficiently resists changes in pH yet contains only small concentrations of the buffering agents. Determine which one of the following weak acids, together with its sodium salt, would probably be best to use: m-chlorobenzoic acid, $K_a = 1.04 \times 10^{-4}$; p-chlorocinnamic acid, $K_a = 3.89 \times 10^{-5}$; 2,5-dihydroxybenzoic acid, $K_a = 1.08 \times 10^{-3}$; or acetoacetic acid, $K_a = 2.62 \times 10^{-4}$. Explain.

48. Suppose you were designing a buffer system for imitation blood and wanted the buffer to maintain the blood at the realistic pH of 7.40. All other things being equal, which buffer system would be preferable—$H_2CO_3/HCO_3^-$ or $H_2PO_4^-/HPO_4^{2-}$? Explain.

49. You have at your disposal an ample quantity of a solution of 0.0500 M NaOH and 500 mL of a solution of 0.100 M formic acid (HCOOH). How much of the NaOH solution should be added to the acid solution to produce a buffer of pH 4.00?

50. You have at your disposal an ample quantity of a solution of 0.100 M HCl and 400 mL of a solution of 0.0800 M NaCN. How much of the HCl solution should be added to the NaCN solution to produce a buffer of pH 9.60?

## Acid–Base Titration Curves

51. Suppose 100.0 mL of a 0.3750 M solution of the strong base $Ba(OH)_2$ is titrated with a 0.4540 M solution of the strong acid $HClO_4$. The neutralization reaction is

$$Ba(OH)_2(aq) + 2\ HClO_4(aq) \longrightarrow$$
$$Ba(ClO_4)_2(aq) + 2\ H_2O(\ell)$$

Compute the pH of the titration solution before any acid is added, when the titration is 1.00 mL short of the equivalence point, when the titration is at the equivalence point, and when the titration is 1.00 mL past the equivalence point. (*Note:* Each mole of $Ba(OH)_2$ gives *two* moles of $OH^-$ in solution.)

52. A sample containing 26.38 mL of 0.1439 M HBr is titrated with a solution of NaOH having a molarity of 0.1219 M. Compute the pH of the titration solution before any base is added, when the titration is 1.00 mL short of the equivalence point, when the titration is at the equivalence point, and when the titration is 1.00 mL past the equivalence point.

53. A sample containing 50.00 mL of 0.1000 M hydrazoic acid ($HN_3$) is being titrated with 0.1000 M sodium hydroxide. Compute the pH before any base is added, after the addition of 25.00 mL of the base, after the addition of 50.00 mL of the base, and after the addition of 51.00 mL of the base.

54. A sample of 50.00 mL of 0.1000 M aqueous solution of chloroacetic acid, $CH_2ClCOOH$ ($K_a = 1.4 \times 10^{-3}$), is titrated with a 0.1000 M NaOH solution. Calculate the pH at the following stages in the titration, and plot the titration curve: 0, 5.00, 25.00, 49.00, 49.90, 50.00, 50.10, and 55.00 mL NaOH.

55. The base ionization constant of ethylamine ($C_2H_5NH_2$) in aqueous solution is $K_b = 6.41 \times 10^{-4}$ at 25°C. Calculate the pH for the titration of 40.00 mL of a 0.1000 M solution of ethylamine with 0.1000 M HCl at the following volumes of added HCl: 0, 5.00, 20.00, 39.90, 40.00, 40.10, and 50.00 mL.

56. Ammonia is a weak base with a $K_b$ of $1.8 \times 10^{-5}$. A 140.0-mL sample of a 0.175 M solution of aqueous ammonia is titrated with a 0.106 M solution of the strong acid HCl. The reaction is

$$NH_3(aq) + HCl(aq) \longrightarrow NH_4^+(aq) + Cl^-(aq)$$

Compute the pH of the titration solution before any acid is added, when the titration is at the half-equivalence point, when the titration is at the equivalence point, and when the titration is 1.00 mL past the equivalence point.

57. Sodium benzoate, the sodium salt of benzoic acid, is used as a food preservative. A sample containing solid sodium benzoate mixed with sodium chloride is dissolved in 50.0 mL of 0.500 M HCl, giving an acidic solution (benzoic acid mixed with HCl). This mixture is then titrated with 0.393 M NaOH. After the addition of 46.50 mL of the NaOH solution, the pH is found to be 8.2. At this point, the addition of one more drop (0.02 mL) of NaOH raises the pH to 9.3. Calculate the mass of sodium benzoate ($NaC_6H_5COO$) in the original sample. (*Hint:* At the equivalence point, the *total* number of moles of acid [here HCl] equals the *total* number of moles of base [here, both NaOH and $NaC_6H_5COO$]).

58. An antacid tablet (such as Tums or Rolaids) weighs 1.3259 g. The only acid-neutralizing ingredient in this brand of

antacid is $CaCO_3$. When placed in 12.07 mL of 1.070 M HCl, the tablet fizzes merrily as $CO_2(g)$ is given off. After all of the $CO_2$ has left the solution, an indicator is added, followed by 11.74 mL of 0.5310 M NaOH. The indicator shows that, at this point, the solution is definitely basic. Addition of 5.12 mL of 1.070 M HCl makes the solution acidic again. Then, 3.17 mL of the 0.5310 M NaOH brings the titration exactly to an endpoint, as signaled by the indicator. Compute the percentage by mass of $CaCO_3$ in the tablet.

59. What is the mass of diethylamine, $(C_2H_5)_2NH$, in 100.0 mL of an aqueous solution if it requires 15.90 mL of 0.0750 M HCl to titrate it to the equivalence point? What is the pH at the equivalence point if $K_b = 3.09 \times 10^{-4}$? What would be a suitable indicator for the titration?

60. A chemist who works in the process laboratory of the Athabasca Alkali Company makes frequent analyses for ammonia recovered from the Solvay process for making sodium carbonate. What is the pH at the equivalence point if she titrates the aqueous ammonia solution (approximately 0.10 M) with a strong acid of comparable concentration? Select an indicator that would be suitable for the titration.

61. If 50.00 mL of a 0.200 M solution of the weak base N-ethyl-morpholine ($C_6H_{13}NO$) is mixed with 8.00 mL of 1.00 M HCl and then diluted to a final volume of 100.0 mL with water, the result is a buffer with a pH of 7.0. Compute the $K_b$ of N-ethylmorpholine.

62. The sodium salt of cacodylic acid, a weak acid, has the formula $NaO_2As(CH_3)_2 \cdot 3H_2O$. Its molar mass is 214.02 g $mol^{-1}$. A solution is prepared by mixing 21.40 g of this substance with enough water to make 1.000 L of solution. Then 50.00 mL of the sodium cacodylate solution is mixed with 29.55 mL of 0.100 M HCl and enough water to bring the volume to a total of 100.00 mL. The pH of the solution is 6.00. Determine the $K_a$ of cacodylic acid.

## Polyprotic Acids

63. Arsenic acid ($H_3AsO_4$) is a weak triprotic acid. Given the three acid ionization constants from Table 10.2 and an initial concentration of arsenic acid (before ionization) of 0.1000 M, calculate the equilibrium concentrations of $H_3AsO_4$, $H_2AsO_4^-$, $HAsO_4^{2-}$, $AsO_4^{3-}$, and $H_3O^+$.

64. Phthalic acid ($H_2C_8H_4O_4$, abbreviated $H_2Ph$) is a diprotic acid. Its ionization in water at 25°C takes place in two steps:

$$H_2Ph(aq) + H_2O(\ell) \rightleftharpoons H_3O^+(aq) + HPh^-(aq)$$
$$K_{a1} = 1.26 \times 10^{-3}$$

$$HPh^-(aq) + H_2O(\ell) \rightleftharpoons H_3O^+(aq) + Ph^{2-}(aq)$$
$$K_{a2} = 3.10 \times 10^{-6}$$

If 0.0100 mol of phthalic acid is dissolved per liter of water, calculate the equilibrium concentrations of $H_2Ph$, $HPh^-$, $Ph^{2-}$, and $H_3O^+$.

65. A solution as initially prepared contains 0.050 mol $L^{-1}$ of phosphate ion ($PO_4^{3-}$) at 25°C. Given the three acid ionization constants from Table 10.2, calculate the equilibrium concentrations of $PO_4^{3-}$, $HPO_4^{2-}$, $H_2PO_4^-$, $H_3PO_4$, and $OH^-$.

66. Oxalic acid ionizes in two stages in aqueous solution:

$$H_2C_2O_4(aq) + H_2O(\ell) \rightleftharpoons H_3O^+(aq) + HC_2O_4^-(aq)$$
$$K_{a1} = 5.9 \times 10^{-2}$$

$$HC_2O_4^-(aq) + H_2O(\ell) \rightleftharpoons H_3O^+(aq) + C_2O_4^{2-}(aq)$$
$$K_{a2} = 6.4 \times 10^{-5}$$

Calculate the equilibrium concentrations of $C_2O_4^{2-}$, $HC_2O_4^-$, $H_2C_2O_4$, and $OH^-$ in a 0.10 M solution of sodium oxalate ($Na_2C_2O_4$).

67. The pH of a normal raindrop is 5.60. Compute the concentrations of $H_2CO_3(aq)$, $HCO_3^-(aq)$, and $CO_3^{2-}(aq)$ in this raindrop, if the total concentration of dissolved carbonates is $1.0 \times 10^{-5}$ mol $L^{-1}$.

68. The pH of a drop of acid rain is 4.00. Compute the concentrations of $H_2CO_3(aq)$, $HCO_3^-(aq)$, and $CO_3^{2-}(aq)$ in the acid raindrop, if the total concentration of dissolved carbonates is $3.6 \times 10^{-5}$ mol $L^{-1}$.

## A Deeper Look . . . Exact Treatment of Acid–Base Equilibria

69. Thiamine hydrochloride (vitamin $B_1$ hydrochloride, $HC_{12}H_{17}ON_4SCl_2$) is a weak acid with $K_a = 3.4 \times 10^{-7}$. Suppose $3.0 \times 10^{-5}$ g of thiamine hydrochloride is dissolved in 1.00 L of water. Calculate the pH of the resulting solution. (*Hint:* This is a sufficiently dilute solution that the autoionization of water cannot be neglected.)

70. A sample of vinegar contains 40.0 g of acetic acid ($CH_3COOH$) per liter of solution. Suppose 1.00 mL is removed and diluted to 1.00 L, and 1.00 mL of *that* solution is removed and diluted to 1.00 L. Calculate the pH of the resulting solution. (*Hint:* This is a sufficiently dilute solution that the autoionization of water cannot be neglected.)

71. At 25°C, 50.00 mL of a 0.1000 M solution of maleic acid, a diprotic acid whose ionization constants are $K_{a1} = 1.42 \times 10^{-2}$ and $K_{a2} = 8.57 \times 10^{-7}$, is titrated with a 0.1000 M NaOH solution. Calculate the pH at the following volumes of added base: 0, 5.00, 25.00, 50.00, 75.00, 99.90, 100.00, and 105.00 mL.

72. Quinine ($C_{20}H_{24}O_2N_2$) is a water-soluble base that ionizes in two stages, with $K_{b1} = 3.31 \times 10^{-6}$ and $K_{b2} = 1.35 \times 10^{-10}$, at 25°C. Calculate the pH during the titration of an aqueous solution of 1.622 g of quinine in 100.00 mL of water as a function of the volume of added 0.1000 M HCl solution at the following volumes: 0, 25.00, 50.00, 75.00, 99.90, 100.00, and 105.00 mL.

## Additional Problems

73. The ionization constant of chloroacetic acid ($ClCH_2COOH$) in water is $1.528 \times 10^{-3}$ at 0°C and $1.230 \times 10^{-3}$ at 40°C. Calculate the enthalpy of ionization of the acid in water, assuming that $\Delta H$ and $\Delta S$ are constant over this range of temperature.

74. The autoionization constant of water ($K_w$) is $1.139 \times 10^{-15}$ at 0°C and $9.614 \times 10^{-14}$ at 60°C.
    (a) Calculate the enthalpy of autoionization of water.

$$2 H_2O(\ell) \rightleftharpoons H_3O^+(aq) + OH^-(aq)$$

    (b) Calculate the entropy of autoionization of water.
    (c) At what temperature will the pH of pure water be 7.00, from these data?

75. Calculate the concentrations of $H_3O^+$ and $OH^-$ at 25°C in the following:
    (a) Orange juice (pH 2.8)
    (b) Tomato juice (pH 3.9)
    (c) Milk (pH 4.1)
    (d) Borax solution (pH 8.5)
    (e) Household ammonia (pH 11.9)

76. Try to choose which of the following is the pH of a $6.44 \times 10^{-10}$ M $Ca(OH)_2(aq)$ solution, without doing any written calculations.
    (a) 4.81     (d) 8.89
    (b) 5.11     (e) 9.19
    (c) 7.00

77. $Cl_2(aq)$ reacts with $H_2O(\ell)$ as follows:

$$Cl_2(aq) + 2 H_2O(\ell) \rightleftharpoons$$
$$H_3O^+(aq) + Cl^-(aq) + HOCl(aq)$$

    For an experiment to succeed, $Cl_2(aq)$ must be present, but the amount of $Cl^-(aq)$ in the solution must be minimized. For this purpose, should the pH of the solution be high, low, or neutral? Explain.

78. Use the data in Table 10.2 to determine the equilibrium constant for the reaction

$$H_2PO_4^-(aq) + 2 CO_3^{2-}(aq) \rightleftharpoons PO_4^{3-}(aq) + 2 HCO_3^-(aq)$$

79. Urea ($NH_2CONH_2$) is a component of urine. It is a very weak base, having an estimated $pK_b$ of 13.8 at room temperature.
    (a) Write the formula of the conjugate acid of urea.
    (b) Compute the equilibrium concentration of urea in a solution that contains no urea but starts out containing 0.15 mol L$^{-1}$ of the conjugate acid of urea.

80. Exactly 1.0 L of solution of acetic acid gives the same color with methyl red as 1.0 L of a solution of hydrochloric acid. Which solution will neutralize the greater amount of 0.10 M $NaOH(aq)$? Explain.

*81. The $K_a$ for acetic acid drops from $1.76 \times 10^{-5}$ at 25°C to $1.63 \times 10^{-5}$ at 50°C. Between the same two temperatures, $K_w$ increases from $1.00 \times 10^{-14}$ to $5.47 \times 10^{-14}$. At 50°C, the density of a 0.10 M solution of acetic acid is 98.81% of its density at 25°C. Will the pH of a 0.10 M solution of acetic acid in water increase, decrease, or remain the same when it is heated from 25°C to 50°C? Explain.

82. Calculate the pH of a solution that is prepared by dissolving 0.23 mol of hydrofluoric acid (HF) and 0.57 mol of hypochlorous acid (HClO) in water and diluting to 3.60 L. Also, calculate the equilibrium concentrations of HF, F$^-$,

HClO, and ClO$^-$. (*Hint:* The pH will be determined by the stronger acid of this pair.)

83. For each of the following compounds, indicate whether a 0.100 M aqueous solution is acidic (pH < 7), basic (pH > 7), or neutral (pH = 7): HCl, $NH_4Cl$, $KNO_3$, $Na_3PO_4$, $NaCH_3COO$.

*84. Calculate [$H_3O^+$] in a solution that contains 0.100 mol of $NH_4CN$ per liter.

$$NH_4^+(aq) + H_2O(\ell) \rightleftharpoons H_3O^+(aq) + NH_3(aq)$$
$$K_a = 5.6 \times 10^{-10}$$

$$HCN(aq) + H_2O(\ell) \rightleftharpoons H_3O^+(aq) + CN^-(aq)$$
$$K_a = 6.17 \times 10^{-10}$$

85. A buffer solution is prepared by mixing 1.00 L of 0.050 M pentafluorobenzoic acid ($C_6F_5COOH$) and 1.00 L of 0.060 M sodium pentafluorobenzoate ($NaC_6F_5COO$). The $K_a$ of this weak acid is 0.033. Determine the pH of the buffer solution.

86. A chemist needs to prepare a buffer solution with pH = 10.00, and has both $Na_2CO_3$ and $NaHCO_3$ in pure crystalline form. What mass of each should be dissolved in 1.00 L of solution if the combined mass of the two salts is to be 10.0 g?

87. Which of these procedures would *not* make a pH = 4.75 buffer?
    (a) Mix 50.0 mL of 0.10 M acetic acid and 50.0 mL of 0.10 M sodium acetate.
    (b) Mix 50.0 mL of 0.20 M acetic acid and 50.0 mL of 0.10 M NaOH.
    (c) Start with 50.0 mL of 0.20 M acetic acid and add a solution of strong base until the pH equals 4.75.
    (d) Start with 50.0 mL of 0.20 M HCl and add a solution of strong base until the pH equals 4.75.
    (e) Start with 100.0 mL of 0.20 M sodium acetate and add 50.0 mL of 0.20 M HCl.

88. It takes 4.71 mL of 0.0410 M NaOH to titrate a 50.00-mL sample of flat (no $CO_2$) GG's Cola to a pH of 4.9. At this point, the addition of one more drop (0.02 mL) of NaOH raises the pH to 6.1. The only acid in GG's Cola is phosphoric acid. Compute the concentration of phosphoric acid in this cola. Assume that the 4.71 mL of base removes only the first hydrogen from the $H_3PO_4$; that is, assume that the reaction is

$$H_3PO_4(aq) + OH^-(aq) \longrightarrow H_2O(\ell) + H_2PO_4^-(aq)$$

89. A 0.1000 M solution of a weak acid, HA, requires 50.00 mL of 0.1000 M NaOH to titrate it to its equivalence point. The pH of the solution is 4.50 when only 40.00 mL of the base has been added.
    (a) Calculate the ionization constant $K_a$ of the acid.
    (b) Calculate the pH of the solution at the equivalence point.

90. The chief chemist of Victory Vinegar Works, Ltd., interviews two chemists for employment. He states, "Quality control requires that our high-grade vinegar contain 5.00 ± 0.01%

acetic acid by mass. How would you analyze our product to ensure that it meets this specification?"

Anne Dalton says, "I would titrate a 50.00-mL sample of the vinegar with 1.000 M NaOH, using phenolphthalein to detect the equivalence point to within $\pm 0.02$ mL of base."

Charlie Cannizzaro says, "I would use a pH meter to determine the pH to $\pm 0.01$ pH units and interface it with a computer to print out the mass percentage of acetic acid."

Which candidate did the chief chemist hire? Why?

91. Phosphonocarboxylic acid

$$\begin{array}{c}
\ddot{\text{:}}\text{O} \quad \text{H} \quad \ddot{\text{O}}\text{:} \\
\| \quad | \quad \| \\
\text{H}-\ddot{\text{O}}-\text{P}-\text{C}-\text{C}-\ddot{\text{O}}-\text{H} \\
| \quad | \\
\text{H}-\ddot{\text{O}} \cdot \text{H}
\end{array}$$

effectively inhibits the replication of the herpes virus. Structurally, it is a combination of phosphoric acid and acetic acid. It can donate three protons. The equilibrium constant values are $K_{a1} = 1.0 \times 10^{-2}$, $K_{a2} = 7.8 \times 10^{-6}$, and $K_{a3} = 2.0 \times 10^{-9}$. Enough phosphonocarboxylic acid is added to blood (pH 7.40) to make its total concentration $1.0 \times 10^{-5}$ M. The pH of the blood does not change. Determine the concentrations of all four forms of the acid in this mixture.

92. Egg whites contain dissolved carbon dioxide and water, which react together to give carbonic acid ($H_2CO_3$). In the few days after an egg is laid, it loses carbon dioxide through its shell. Does the pH of the egg white increase or decrease during this period?

93. If you breathe too rapidly (hyperventilate), the concentration of dissolved $CO_2$ in your blood drops. What effect does this have on the pH of the blood?

94. A reference book states that a saturated aqueous solution of potassium hydrogen tartrate is a buffer with a pH of 3.56. Write two chemical equations that show the buffer action of this solution. (Tartaric acid is a diprotic acid with the formula $H_2C_4H_4O_6$. Potassium hydrogen tartrate is $KHC_4H_4O_6$.)

* 95. Glycine, the simplest amino acid, has both an acid group and a basic group in its structure ($H_2N-CH_2-COOH$). In aqueous solution it exists predominantly as a self-neutralized species called a zwitterion ($H_3\overset{+}{N}-CH_2-COO^-$). The zwitterion therefore behaves both as an acid and as a base, according to the equilibria at 25°C:

$$H_3\overset{+}{N}-CH_2-COO^-(aq) + H_2O(\ell) \rightleftharpoons$$
$$H_2N-CH_2-COO^-(aq) + H_3O^+(aq)$$
$$K_a = 1.7 \times 10^{-10}$$

$$H_3\overset{+}{N}-CH_2-COO^-(aq) + H_2O(\ell) \rightleftharpoons$$
$$H_3\overset{+}{N}-CH_2-COOH(aq) + OH^-(aq)$$
$$K_b = 2.2 \times 10^{-12}$$

Calculate the pH of a 0.10 M aqueous solution of glycine at 25°C. (*Hint:* You may need to take account of the autoionization of water.)

## CUMULATIVE PROBLEMS

96. Use the data from Table 10.1, together with Le Châtelier's principle, to decide whether the autoionization of water is exothermic or endothermic.

97. Baking soda (sodium hydrogen carbonate, $NaHCO_3$) is used in baking because it reacts with acids in foods to form carbonic acid ($H_2CO_3$), which in turn decomposes to water and carbon dioxide. In a batter, the carbon dioxide appears as gas bubbles that cause the bread or cake to rise.
    (a) A rule of thumb in cooking is that $\frac{1}{2}$ teaspoon baking soda is neutralized by 1 cup of sour milk. The acid component of sour milk is lactic acid ($HC_3H_5O_3$). Write an equation for the neutralization reaction.
    (b) If the density of baking soda is 2.16 g cm$^{-3}$, calculate the concentration of lactic acid in the sour milk, in moles per liter. Take 1 cup = 236.6 mL = 48 teaspoons.
    (c) Calculate the volume of carbon dioxide that is produced at 1 atm pressure and 350°F (177°C) from the reaction of $\frac{1}{2}$ teaspoon of baking soda.

98. Boric acid, $B(OH)_3$, is an acid that acts differently from the usual Brønsted–Lowry acids. It reacts with water according to

$$B(OH)_3(aq) + 2 H_2O(\ell) \rightleftharpoons B(OH)_4^-(aq) + H_3O^+(aq)$$
$$K_a = 5.8 \times 10^{-10}$$

    (a) Draw Lewis structures for $B(OH)_3$ and $B(OH)_4^-$. Can these be described as Lewis acids or Lewis bases?
    (b) Calculate the pH of a 0.20 M solution of $B(OH)_3(aq)$.

99. At 40°C and 1.00 atm pressure, a gaseous monoprotic acid has a density of 1.05 g L$^{-1}$. After 1.85 g of this gas is dissolved in water and diluted to 450 mL, the pH is measured to be 5.01. Determine the $K_a$ of this acid and use Table 10.2 to identify it.

100. At 25°C, the Henry's law constant for carbon dioxide dissolved in water is $1.8 \times 10^3$ atm. Calculate the pH of water saturated with $CO_2(g)$ at 25°C in Denver, where the barometric pressure is 0.833 atm.

# Heterogeneous Equilibria

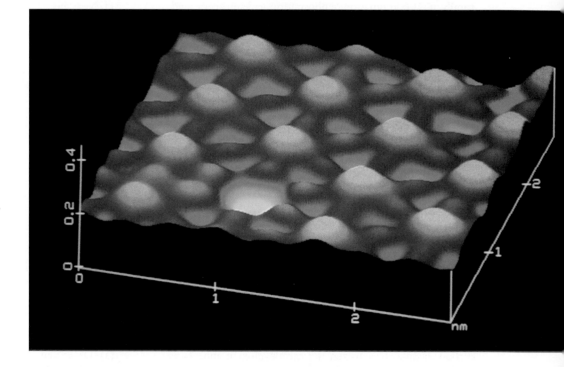

*Illustration*
Iodine atoms adsorbed on a platinum surface. *(Phototake)*

A wide variety of equilibria occur in heterogeneous systems that involve solids and liquids as well as gases and dissolved species. Several examples illustrate the variety of phenomena to be described as well as the new features introduced by the participation of solids or liquids.

1. Recall the phase equilibrium between liquid water and water vapor from Section 5.4:

$$H_2O(\ell) \rightleftharpoons H_2O(g)$$

As long as *some* liquid water is in the container, the pressure of water vapor at 25°C will be 0.03126 atm. The position of this equilibrium is not affected by the amount of liquid water present.

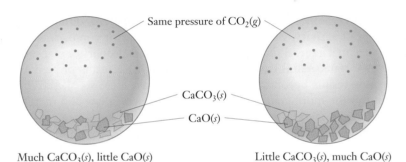

**FIGURE 11.1**   As long as both $CaO(s)$ and $CaCO_3(s)$ are present at equilibrium in a closed container, the partial pressure of $CO_2(g)$ at a fixed temperature does not depend on the amounts of the two solids present.

Same pressure of $CO_2(g)$

$CaCO_3(s)$

$CaO(s)$

Much $CaCO_3(s)$, little $CaO(s)$          Little $CaCO_3(s)$, much $CaO(s)$

2. An analogous situation occurs in the equilibrium between solid iodine and iodine dissolved in aqueous solution:

$$I_2(s) \rightleftharpoons I_2(aq)$$

The position of the equilibrium (given by the concentration of $I_2$ dissolved at a given temperature) is independent of the amount of solid present, as long as there is some.

3. Another example involves a chemical reaction, the decomposition of calcium carbonate:

$$CaCO_3(s) \rightleftharpoons CaO(s) + CO_2(g)$$

If calcium carbonate is heated, it decomposes into calcium oxide (lime) and carbon dioxide; the reverse reaction is favored at sufficiently high pressures of carbon dioxide. The equilibrium can be studied experimentally, and it is found that at any given temperature the pressure of $CO_2(g)$ is constant, independent of the amounts of $CaCO_3(s)$ and $CaO(s)$, as long as some of each is present (Fig. 11.1).

How are these pure solids and liquids, whose presence is clearly essential for the equilibria but whose exact amounts do not influence the position of the equilibria, to be described within the law of mass action?

The purpose of this chapter is to extend the law of mass action to accommodate these more complex cases. The following section develops the more general form of the law of mass action by introducing the activity concept. Subsequent sections apply the more general form to chemical separation methods and to dissolution–precipitation equilibria.

---

### 11.1

## MASS ACTION LAW FOR HETEROGENEOUS REACTIONS: THE CONCEPT OF ACTIVITY

The mass action law as presented in Chapters 9 and 10 for homogeneous reactions in ideal gases and ideal solutions can be written by straightforward inspection of the balanced equation for the reaction under study. If one or more of the reactants or products is a solid or liquid in its pure state, the procedure is less obvious, since "concentration" has no meaning for a pure species. This apparent difficulty is resolved by the concept of **activity,** which is a convenient means for comparing the properties of a substance in a general thermodynamic state to its properties in a specially selected reference state. A full treatment of equilibrium in terms of activity requires thermodynamic results beyond the scope of this book. Here we sketch the essential ideas leading to the more general form of the mass action law and merely state the range of validity of the idealized expressions presented.

The activity concept is introduced by considering the dependence of the Gibbs free energy of a system on the pressure (if a pure substance) or on the composition (if a solution), regardless of the phase of the system. Recall the discussion leading up to Equation 9.3, which showed that the change in Gibbs free energy when a gas is taken from a reference state $P_{ref}$ to any pressure $P$ is given by

$$\Delta G = nRT \ln \left( \frac{P}{P_{ref}} \right) = nRT \ln P$$

The last form is used only when pressure $P$ is expressed in atmospheres and $P_{ref} = 1$ atm. This argument can be extended to more complex systems—and the resulting equation can retain the simple form of the case for ideal gases—if we define the activity of the system by the equation

$$\Delta G = nRT \ln \left( \frac{a}{a_{ref}} \right) = nRT \ln a \qquad \textbf{[11.1]}$$

where $a_{ref}$ is the activity in the selected reference state and $a$ is the activity in a general thermodynamic state. The activity in the reference state is always assigned the value 1. It follows that the change in Gibbs free energy in taking a system from the reference state to any general thermodynamic state is determined by the activity $a$ in the general state.

The activity is related to pressure or to concentration by the **activity coefficient,** which is defined as follows. The *activity coefficient* $\gamma_i$ of a non-ideal gaseous species at pressure $P_i$ is defined by the equation

$$a_i = \frac{\gamma_i P_i}{P_{ref}} \qquad \textbf{[11.2a]}$$

The reference state is chosen to be an ideal gas at one atmosphere. Comparing Equation 11.1 and Equation 11.2a with Equation 9.3 shows that $\gamma_i$ takes the value 1 for ideal gases. It follows that the activity of an ideal gas is the ratio of its pressure to the selected reference pressure. If pressures are reported in units of atmospheres, then the activity of an ideal gas is numerically equal to its pressure. Similarly, the activity coefficient $\gamma_i$ for solute i in a solution at concentration $c_i$ is defined by the equation

$$a_i = \frac{\gamma_i c_i}{c_{ref}} \qquad \textbf{[11.2b]}$$

The activity coefficient $\gamma_i$ equals 1 in the reference state, selected as an ideal solution with convenient concentration. For solute species in the dilute solutions considered in this book, concentration is most conveniently expressed in molarity, and the reference state is selected to be an ideal solution at concentration $c_{ref} = 1$ M.

The reference states for pure solids and liquids are chosen to be those forms stable at 1 atm, just as in the definition of standard states for enthalpy of formation (see Chapter 7) and Gibbs free energy of formation (see Chapter 8). Pure substances in their reference states are assigned activity of value 1.

Once reference states have been defined, the activity coefficients $\gamma_i$ can be determined from experimental *P-V-T* and calorimetric data by procedures not described here. Consequently Equation 11.1 can be written explicitly in terms of pressure, temperature, and concentrations when needed for specific calculations. In its present version, Equation 11.1 is especially well suited for general discussions, since it elegantly summarizes much complicated information about non-ideal systems in a simple and compact form.

With Equation 11.1 available, arguments similar to those preceding Equation 9.4 then lead to the following expression for the equilibrium constant $K$ regardless of the phase of each product or reactant:

$$\frac{a_C^c \cdot a_D^d}{a_A^a \cdot a_B^b} = K \qquad \qquad [11.3]$$

Substituting the appropriate ideal expression for the activity of gaseous or dissolved species from Equation 11.1 or Equation 11.2 immediately recovers the forms of the mass action law and the equilibrium constant $K$ already derived in Chapter 9 for reactions in ideal gases or in ideal solutions. Equation 11.3 is used to write the mass action law for more general cases by substituting unity for the activity of pure liquids or solids and the appropriate ideal expression for the activity of each gaseous or dissolved species. Once a proper reference state and concentration units are identified for each reactant and product, then tabulated free energies based on these reference states can be used to calculate the equilibrium constant.

The following example illustrates the essential points.

## EXAMPLE 11.1

The compound urea, important in biochemistry, can be prepared in aqueous solution by the following reaction:

$$CO_2(g) + 2\,NH_3(g) \rightleftharpoons CO(NH_2)_2(aq) + H_2O(\ell)$$

Write the mass action law for this reaction.

### Solution
In terms of activities for each species, the equilibrium expression is

$$\frac{a_{urea} \cdot a_{H_2O}}{a_{CO_2} \cdot a_{NH_3}^2} = K$$

Choose the reference state for the gases to be $P_{ref} = 1$ atm, for the urea to be $c_{ref} = 1$ M, and for water to be the pure liquid with unit activity. Then at low pressures and concentrations, the limiting idealized mass action law is

$$\frac{[urea]}{P_{CO_2} \cdot P_{NH_3}^2} = K$$

with [urea] in units of mol $L^{-1}$ and the partial pressures in atm. The (invisible) $P_{ref}$ and $c_{ref}$ factors must always be kept in mind, since they make $K$ dimensionless in this expression. This expression can be extended to higher pressures and concentrations, where non-ideality becomes important, by inserting the appropriate activity coefficients.

**Related Problems: 1, 2, 3, 4**

The general form of the law of mass action can now be stated:

1. Gases enter equilibrium expressions as partial pressures, in atmospheres.
2. Dissolved species enter as concentrations, in moles per liter.
3. Pure solids and pure liquids do not appear in equilibrium expressions; neither does a solvent taking part in a chemical reaction, provided the solution is dilute.
4. Partial pressures and concentrations of products appear in the numerator, and those of reactants in the denominator, each raised to a power equal to its coefficient in the balanced chemical equation.

Equilibrium expressions written in this way are accurate to about 5% when the pressures of gases do not exceed several atmospheres and the concentrations of solutes do not exceed 0.1 M. These cases cover all the applications discussed in this book, and correction factors will not be applied. For concentrated ionic solutions, especially for ionic solutes with high molecular weights, the corrections can become very large. In extremely accurate studies of solution equilibria, especially in biochemical applications, activities must be used in place of partial pressures or concentrations. More advanced books should be consulted for the techniques required.

### EXAMPLE 11.2

Graphite is added to a vessel that contains only $CO_2(g)$ at a certain high temperature. The pressure rises as a reaction occurs that produces $CO(g)$:

$$C(s) + CO_2(g) \rightleftharpoons 2\ CO(g) \qquad \frac{(P_{CO})^2}{P_{CO_2}} = K = 4.2$$

and the total pressure reaches an equilibrium value of 1.40 atm. Calculate the equilibrium partial pressure of the CO.

#### Solution

There is no purpose in setting up a table of partial pressures, because there is no information on the initial partial pressure of $CO_2$. Instead, call the equilibrium partial pressure of CO $y$, so $1.40 - y$ is the equilibrium partial pressure of $CO_2$. The equilibrium expression then becomes

$$\frac{y^2}{1.40 - y} = 4.2$$

which is solved to give $y = 1.1$. The partial pressure of CO is 1.1 atm at equilibrium.

**Related Problems: 9, 10**

---

### 11.2

## DISTRIBUTION OF A SINGLE SPECIES BETWEEN IMMISCIBLE PHASES: EXTRACTION AND SEPARATION PROCESSES

An important type of heterogeneous equilibrium involves partitioning a solute species between two immiscible solvent phases. Such equilibria are used in many separation processes in chemical research and in industry.

Suppose two immiscible liquids, such as water and carbon tetrachloride, are put in a container. "Immiscible" means mutually insoluble; these liquids separate into two phases with the less dense liquid, in this case water, lying on top of the other liquid. A visible boundary, the meniscus, separates the two phases. If a solute such as iodine is added to the mixture and the vessel is shaken to distribute the iodine through the container (Fig. 11.2), the iodine is partitioned between the two phases at equilibrium with a characteristic concentration ratio, the **partition coefficient** $K$. This is the equilibrium constant for the process

$$I_2(aq) \rightleftharpoons I_2(CCl_4)$$

and can be written as

$$\frac{[I_2]_{CCl_4}}{[I_2]_{aq}} = K$$

in which $[I_2]_{CCl_4}$ and $[I_2]_{aq}$ are the concentrations (in moles per liter) of $I_2$ in the $CCl_4$ and aqueous phases, respectively. At 25°C, $K$ has the value 85 for this equilibrium. The fact that $K$ is greater than 1 shows that iodine is more soluble in $CCl_4$ than in water. If iodide ion ($I^-$) is dissolved in the water, then it can react with iodine to form the triiodide ion $I_3^-$:

$$I_2(aq) + I^-(aq) \longrightarrow I_3^-(aq)$$

This consumes $I_2(aq)$ and, by Le Châtelier's principle, causes more $I_2$ in the first equilibrium to move from the $CCl_4$ phase to the aqueous phase.

## Extraction Processes

**Extraction** takes advantage of the partitioning of a solute between two immiscible solvents to remove that solute from one solvent into another. Suppose iodine is present as a contaminant in water that also contains other solutes that are insoluble in carbon tetrachloride. In such a case, most of the iodine could be removed by shaking the aqueous solution with $CCl_4$, allowing the two phases to separate, and then pouring off the water layer from the heavier layer of carbon tetrachloride. The greater the equilibrium constant for the partition of a solute from the original solvent into the extracting solvent, the more complete such a separation will be.

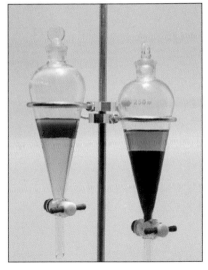

**FIGURE 11.2** Iodine is dissolved in water and poured on top of carbon tetrachloride in a separatory funnel (left). After the funnel is shaken (right), the iodine reaches a partition equilibrium between the upper (aqueous) phase and the lower ($CCl_4$) phase. The deeper color in the lower phase indicates that iodine dissolves preferentially in the denser $CCl_4$ phase. (*Leon Lewandowski*)

### EXAMPLE 11.3

An aqueous solution has an iodine concentration of $2.00 \times 10^{-3}$ M. Calculate the percentage of iodine remaining in the aqueous phase after extraction of 0.100 L of this aqueous solution with 0.050 L of $CCl_4$ at 25°C.

#### Solution

The chemical amount of $I_2$ present is

$$(2.00 \times 10^{-3} \text{ mol L}^{-1})(0.100 \text{ L}) = 2.00 \times 10^{-4} \text{ mol}$$

Suppose that $y$ mol remains in the aqueous phase and $2.00 \times 10^{-4} - y$ mol passes into the $CCl_4$ phase. Then

$$\frac{[I_2]_{CCl_4}}{[I_2]_{aq}} = K = 85$$

$$= \frac{(2.00 \times 10^{-4} - y)/0.050}{y/0.100}$$

$$= \frac{2(2.00 \times 10^{-4} - y)}{y}$$

Note that the volumes of the two solvents used are unchanged because they are immiscible. Solving for $y$ gives

$$y = 4.6 \times 10^{-6} \text{ mol}$$

The fraction remaining in the aqueous phase is $4.6 \times 10^{-6}/2.0 \times 10^{-4} = 0.023$, or 2.3%. Additional extractions could be carried out to remove more of the $I_2$ from the aqueous phase.

**Related Problems: 15, 16, 17, 18**

One extraction process used industrially on a large scale is the purification of sodium hydroxide for use in the manufacture of rayon. The sodium hydroxide produced by electrolysis (see Section 22.1) typically contains 1% sodium chloride and 0.1% sodium chlorate as impurities. If a concentrated aqueous solution of sodium hydroxide is extracted with liquid ammonia, the NaCl and $NaClO_3$ are partitioned into the ammonia phase in preference over the aqueous phase. The heavier aqueous phase is added to the top of an extraction vessel filled with ammonia, and equilibrium is reached as droplets of it settle through the ammonia phase to the bottom. This procedure reduces impurity concentrations in the sodium hydroxide solution to about 0.08% NaCl and 0.0002% $NaClO_3$.

## Chromatographic Separations

Partition equilibria are the basis of an important class of separation techniques called **chromatography.** This word comes from the Greek root *chroma*, meaning "color," and was chosen because the original chromatographic separations involved colored substances. The technique can be applied to a wide variety of mixtures of substances.

Chromatography is a continuous extraction process in which solute species are exchanged between two phases. One, the mobile phase, moves with respect to the other, stationary phase. The partition ratio $K$ of a solute A between the stationary and mobile phases is

$$\frac{[A]_{\text{stationary}}}{[A]_{\text{mobile}}} = K$$

Paper chromatography separates a mixture of watercolor paints into their component colors. The solvent rises through the paper, drawing different colors along at different rates. *(Copyright Richard Megna/Fundamental Photographs)*

### TABLE 11.1

*Chromatographic Separation Techniques*[†]

| Name | Mobile Phase | Stationary Phase |
|------|--------------|------------------|
| Gas–liquid | Gas | Liquid adsorbed on a porous solid in a tube |
| Gas–solid | Gas | Porous solid in a tube |
| Column | Liquid | Liquid adsorbed on a porous solid in a tubular column |
| Paper | Liquid | Liquid held in the pores of a thick paper |
| Thin layer | Liquid | Liquid or solid; solid is held on a glass plate and liquid may be adsorbed on it |
| Ion exchange | Liquid | Solid (finely divided ion-exchange resin) in a tubular column |

[†] Adapted from D. A. Skoog and D. M. West, *Analytical Chemistry* (Saunders College Publishing, Philadelphia, 1980), Table 18-1.

As the mobile phase containing solute passes over the stationary phase, the solute molecules move between the two phases. True equilibrium is never fully established because the motion of the fluid phase continually brings fresh solvent into contact with the stationary phase. Nevertheless, the partition coefficient $K$ provides a guide to the behavior of a particular solute. The greater $K$ is, the more time the solute spends in the stationary phase and therefore the slower its progress through the separation system. Solutes with different values of $K$ are separated by their different rates of travel.

Different types of chromatography use different mobile and stationary phases; Table 11.1 lists some of the most important. **Column chromatography** (Fig. 11.3) uses a tube packed with a porous material, frequently a silica gel on which water has been adsorbed. Water is therefore the stationary liquid phase in this case. Other solvents such as pyridine or benzene are used in the mobile phase; in some cases it is most efficient to use different solvents in succession to separate the components of a solute mixture. As the solute fractions reach the bottom of the column, they are separated for analysis or use. Column chromatography is important in industry because it is easily increased from laboratory to production scale.

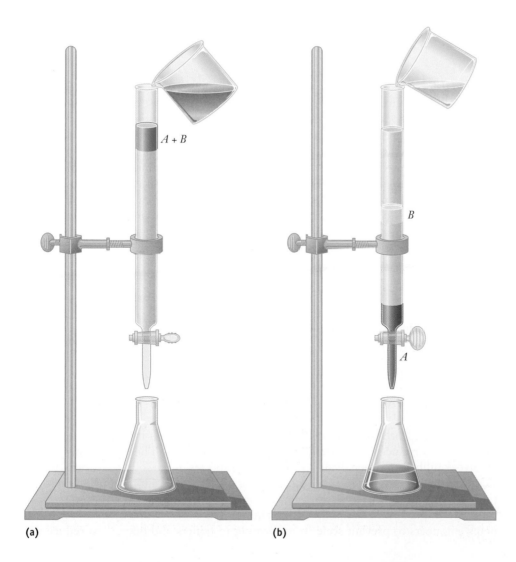

(a)                    (b)

**FIGURE 11.3** (a) In a column chromatograph, the top of the column is loaded with a mixture of solutes to be separated (green). (b) With addition of solvent, the different solutes travel at different rates, giving rise to bands. The separate fractions can be collected in different flasks for use or analysis.

**FIGURE 11.4** In a gas–liquid chromatograph, the sample is vaporized and passes through a column, carried in a stream of an inert gas such as helium or nitrogen. The residence time of any substance on the column depends on its partition coefficient from the vapor to the liquid in the column. A species leaving the column at a given time can be detected by a variety of techniques. The result is a gas chromatogram, with a peak corresponding to each substance in the mixture.

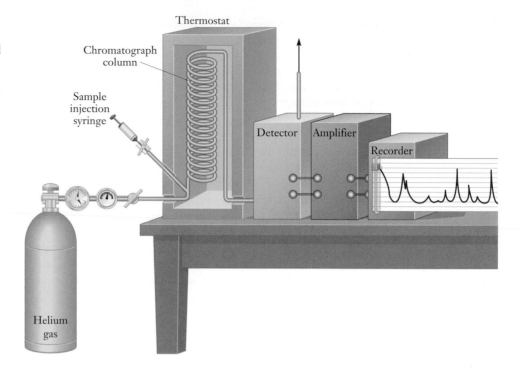

**Gas–liquid chromatography** (Fig. 11.4) is one of the most important separation techniques for modern chemical research. The stationary phase is again a liquid adsorbed on a porous solid, but the mobile phase is now gaseous. The sample is vaporized and passes through the column, carried in a stream of an inert gas such as helium or nitrogen. The residence time in the column depends on the partition coefficient of the solute species, allowing an efficient separation of mixtures. The solute leaving the column at a given time can be detected by a variety of techniques that produce a **gas chromatogram** with a peak corresponding to each solute species in the mixture. Gas–liquid chromatography is widely employed for separating the products of organic reactions. It can also be used to determine the purity of substances, because even very small amounts of impurities appear clearly as separate peaks in the chromatogram. The technique is important in the separation and identification of trace amounts of possibly toxic substances in environmental and biological samples. Amounts on the order of parts per trillion ($10^{-12}$ g in a 1-g sample) can be detected and identified.

---

## 11.3

## THE NATURE OF SOLUBILITY EQUILIBRIA

Dissolution and precipitation reactions, by which solids pass into and out of solution, were introduced and briefly discussed in Chapter 6. These reactions, which involve equilibria between dissolved species and species in the solid state, rank alongside acid–base reactions in practical importance. The dissolution and reprecipitation of solids permit chemists to isolate single products from reaction mixtures or to purify impure solid samples. Understanding the mechanisms of these reactions helps engineers to prevent scale from forming in boilers and doctors to reduce the incidence of painful kidney stones. Dissolution and precipitation control the formation of mineral deposits and profoundly affect the ecologies of rivers, lakes, and oceans.

Our theme throughout the remainder of this chapter is the manipulation of equilibria to control the solubilities of ionic solids. In this section we describe general features of the equilibria that govern dissolution and precipitation. In the remaining sections of this chapter we explore quantitative aspects of these equilibria, including the effects of additional solutes, of acids and bases, and of ligands that can bind to metal ions to form complex ions.

## General Features of Solubility Equilibria

Solubility equilibria resemble the equilibria between volatile liquids (or solids) and their vapors in a closed container. In both cases, particles from a condensed phase tend to escape and spread through a larger, but limited, volume. In both cases, equilibrium is a dynamic compromise in which the rate of escape of particles from the condensed phase is equal to their rate of return. In a vaporization–condensation equilibrium, we assumed that the vapor above the condensed phase was an ideal gas. The analogous starting assumption for a dissolution–precipitation reaction is that the solution above the undissolved solid is an ideal solution. A solution in which sufficient solute has been dissolved to establish a dissolution–precipitation equilibrium between the solid substance and its dissolved form is called a **saturated solution.**

Le Châtelier's principle applies to these equilibria, as it does to all equilibria. One way to exert a stress on a solubility equilibrium is to change the amount of solvent. Adding solvent reduces the concentration of dissolved substance; more solid then tends to dissolve to restore the concentration of the dissolved substance to its equilibrium value. If an excess of solvent is added so that all of the solid dissolves, then obviously the solubility equilibrium ceases to exist, and the solution is **unsaturated.** In a vaporization–condensation equilibrium this corresponds to the complete evaporation of the condensed phase. Removing solvent from an already saturated solution forces additional solid to precipitate in order to maintain a constant concentration. A volatile solvent is often removed by simply letting a solution stand uncovered until the solvent evaporates. When conditions are right, the solid forms as crystals on the bottom and sides of the container (Fig. 11.5).

Controlled precipitation by manipulating solubility is a widely used technique for purifying reaction products in synthetic chemistry. Side reactions can generate significant amounts of impurities; other impurities enter with the starting materials or are put in deliberately to increase the reaction rate. Running a reaction may take only hours, but the *work-up* (separation of crude product) and subsequent purification may require weeks. **Recrystallization,** one of the most powerful methods for purifying solids, relies on differences between the solubilities of the desired substance and its contaminants. An impure product is dissolved and reprecipitated, repeatedly if necessary, with careful control of the factors that influence solubility. Manipulating solubility requires an understanding of the equilibria that exist between an undissolved substance and its solution.

In recrystallization, a solution begins to deposit a compound when it is brought to the point of saturation with respect to that compound. In dissolution, the solvent attacks the solid and **solvates** it at the level of individual particles. In precipitation, the reverse occurs: solute-to-solute attractions are reestablished as the solute leaves the solution. Often, solute-to-solvent attractions persist right through the process

**FIGURE 11.5** The beaker contains an aqueous solution of $K_2PtCl_4$, a substance used as a starting material in the synthesis of anticancer drugs. It is loosely covered to keep out dust and allowed to stand. As the water evaporates, the solution becomes saturated and deposits solid $K_2PtCl_4$ in the form of long, needle-like red crystals. *(Leon Lewandowski)*

of precipitation, and solvent incorporates itself into the solid. When lithium sulfate ($Li_2SO_4$) precipitates from water, it brings with it into the solid one molecule of water per formula unit:

$$2\ Li^+(aq) + SO_4^{2-}(aq) + H_2O(\ell) \longrightarrow Li_2SO_4 \cdot H_2O(s)$$

Such loosely bound solvent is known as solvent of crystallization (Fig. 2.1). Dissolving and then reprecipitating a compound may thus furnish material that has a different chemical formula and a different mass. Consequently, recrystallization processes for purification of reaction products must be planned carefully.

Dissolution–precipitation reactions frequently come to equilibrium slowly. Days or even weeks of shaking a solid in contact with a solvent may be required before the solution becomes saturated. Moreover, solutions sometimes become **supersaturated,** a condition in which the concentration of dissolved solid *exceeds* its equilibrium value. The delay, then, is in forming, not dissolving, the solid. A super-saturated solution may persist for months or years and require extraordinary measures to be brought to equilibrium, although the tendency toward equilibrium is always there. The generally sluggish approach to equilibrium in dissolution–precipitation reactions is quite the opposite of the rapid rates at which acid–base reactions reach equilibrium.

## The Solubility of Ionic Solids

The **solubility** of a substance in a solvent is defined as the greatest amount (expressed either in grams or in moles) that will dissolve in equilibrium in a specified volume of solvent at a particular temperature. Although solvents other than water are used in many applications, aqueous solutions are the most important and are the exclusive concern here. Salts show a wide range of solubilities in water (Fig. 11.6). Silver perchlorate ($AgClO_4$) dissolves to the remarkable extent of about 5570 g (or almost 27 mol) per liter of water at 25°C, but at the same temperature only about 0.0018 g (or $1.3 \times 10^{-5}$ mol) of silver chloride ($AgCl$) dissolves per liter of water. Many salts with even lower solubilities are known. Solubilities often depend strongly on temperature (Fig. 11.7). Most dissolution reactions for ionic solids are endother-

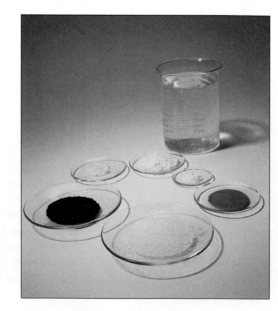

**FIGURE 11.6** Vastly different quantities of different compounds will dissolve in 1 L of water at 20°C. Clockwise from the front are borax, potassium permanganate, lead(II) chloride, sodium phosphate decahydrate, calcium oxide, and potassium dichromate. *(Leon Lewandowski)*

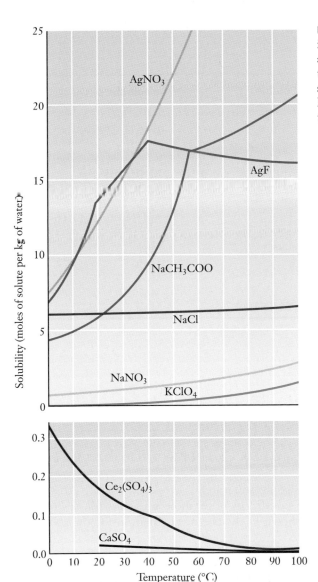

**FIGURE 11.7** Most solubilities increase with increasing temperature, but some decrease. Note that the changes are not always smooth because different solid hydrates form over different temperature ranges.

mic (heat is absorbed), so by Le Châtelier's principle the solubility increases with increasing temperature. Those dissolution reactions that are exothermic (such as for $CaSO_4$) show the opposite behavior.

Although all ionic compounds dissolve to some extent in water, those having solubilities (at 25°C) of less than 0.1 g $L^{-1}$ are called *insoluble*. Those having solubilities of more than 10 g $L^{-1}$ are *soluble*, and the intermediate cases (0.1 to 10 g $L^{-1}$) are said to be *slightly soluble*. Fortunately, it is not necessary to memorize long lists of solubility data. Table 11.2 lists some generalizations concerning groups of salts and gives enough factual data to support good predictions about precipitation or dissolution in thousands of situations of practical importance. Knowing the solubilities of ionic substances, even in these qualitative terms, provides a way to predict the courses of numerous reactions. For example, when a solution of KI is added to one of $Pb(NO_3)_2$, $K^+$ and $NO_3^-$ ions are brought into contact, as are $Pb^{2+}$ and $I^-$ ions. From the table, $KNO_3$ is a soluble salt but $PbI_2$ is insoluble, and so a precipitate of $PbI_2$ will appear (Fig. 11.8).

**FIGURE 11.8** A yellow precipitate of lead iodide forms when a drop of KI solution is added to a solution of $Pb(NO_3)_2$. (*Charles D. Winters*)

**TABLE 11.2**

## Solubilities of Ionic Compounds in Water

| Anion | Soluble[†] | Slightly Soluble | Insoluble |
|---|---|---|---|
| $NO_3^-$ (nitrate) | All | — | — |
| $CH_3COO^-$ (acetate) | Most | — | $Be(CH_3COO)_2$ |
| $ClO_3^-$ (chlorate) | All | — | — |
| $ClO_4^-$ (perchlorate) | Most | $KClO_4$ | — |
| $F^-$ (fluoride) | Group I, $AgF$, $BeF_2$ | $SrF_2$, $BaF_2$, $PbF_2$ | $MgF_2$, $CaF_2$ |
| $Cl^-$ (chloride) | Most | $PbCl_2$ | $AgCl$, $Hg_2Cl_2$ |
| $Br^-$ (bromide) | Most | $PbBr_2$, $HgBr_2$ | $AgBr$, $Hg_2Br_2$ |
| $I^-$ (iodide) | Most | — | $AgI$, $Hg_2I_2$, $PbI_2$, $HgI_2$ |
| $SO_4^{2-}$ (sulfate) | Most | $CaSO_4$, $Ag_2SO_4$, $Hg_2SO_4$ | $SrSO_4$, $BaSO_4$, $PbSO_4$ |
| $S^{2-}$ (sulfide) | Groups I and II, $(NH_4)_2S$ | — | Most |
| $CO_3^{2-}$ (carbonate) | Group I, $(NH_4)_2CO_3$ | — | Most |
| $SO_3^{2-}$ (sulfite) | Group I, $(NH_4)_2SO_3$ | — | Most |
| $PO_4^{3-}$ (phosphate) | Group I, $(NH_4)_3PO_4$ | — | Most |
| $OH^-$ (hydroxide) | Group I, $Ba(OH)_2$ | $Sr(OH)_2$, $Ca(OH)_2$ | Most |

[†] Soluble compounds are defined as those that dissolve to the extent of 10 or more grams per liter; slightly soluble compounds, 0.1 to 10 grams per liter, and insoluble compounds, less than 0.1 gram per liter at room temperature.

---

## 11.4

## IONIC EQUILIBRIA BETWEEN SOLIDS AND SOLUTIONS

When an ionic solid such as CsCl dissolves in water, it breaks up into ions that move apart from each other and become solvated by water molecules (*aquated*, or *hydrated*; Fig. 11.9). The aquated ions are shown in the chemical equation for the solubility equilibrium. Thus,

$$CsCl(s) \rightleftharpoons Cs^+(aq) + Cl^-(aq)$$

shows that the dissolved particles are ions. For a highly soluble salt (such as CsCl), the concentrations of the ions in a saturated aqueous solution are so large that the solution is very nonideal. There is much association among the ions in solution, resulting in temporary pairs of oppositely charged ions and in larger clusters, as well. In such cases, the simple type of equilibrium expression that was developed in Section 11.1 does not apply; therefore, we restrict our attention to sparingly soluble and "insoluble" salts, for which the concentrations in a saturated solution are 0.1 mol L$^{-1}$ or less—low enough that the interactions among the solvated ions are relatively small.

The sparingly soluble salt silver chloride, for example, establishes the following equilibrium when placed in water:

$$AgCl(s) \rightleftharpoons Ag^+(aq) + Cl^-(aq)$$

The equilibrium law for this reaction (written by following the rules for heterogeneous equilibria from Section 11.1) is

$$[Ag^+][Cl^-] = K_{sp}$$

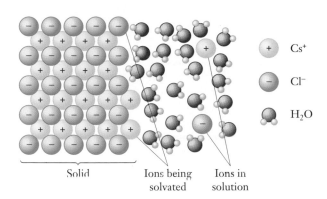

**FIGURE 11.9** The dissolution of the ionic solid CsCl in water. Note the different solvations of positive and negative ions. Water molecules arrange themselves to point the positively charged hydrogen atoms toward the negatively charged $Cl^-$ ions, and the negatively charged oxygen atoms point toward the positively charged $Cs^+$ ions.

where the subscript "sp," standing for **solubility product,** distinguishes the $K$ as referring to the dissolution of a slightly soluble ionic solid in water. At 25°C, $K_{sp}$ has the numerical value $1.6 \times 10^{-10}$ for silver chloride. In the AgCl solubility product expression, the concentrations of the two ions produced are raised to the first power because their coefficients are 1 in the chemical equation. Solid AgCl does not appear in the equilibrium expression; the amount of pure solid AgCl does not affect the equilibrium as long as *some* is present. If *no* solid is present, then the product of the two ion concentrations is no longer constrained by the solubility product expression.

## Solubility and $K_{sp}$

The molar solubility of a salt in water is not the same as its solubility product constant, but a simple relation often exists between them. For example, let us define $S$ as the molar solubility of AgCl(*s*) in water at 25°C. Then, from stoichiometry, $[Ag^+] = [Cl^-] = S$ is the molarity of either ion at equilibrium. Hence,

$$[Ag^+][Cl^-] = S^2 = K_{sp} = 1.6 \times 10^{-10}$$

Taking the square roots of both sides of the equation gives

$$S - 1.26 \times 10^{-5} \text{ M}$$

which rounds off to $1.3 \times 10^{-5}$ M. This is the molar solubility of AgCl in water. It is converted to a gram solubility by multiplying by the molar mass of AgCl:

$$(1.26 \times 10^{-5} \text{ mol L}^{-1})(143.3 \text{ g mol}^{-1}) = 1.8 \times 10^{-3} \text{ g L}^{-1}$$

Therefore, $1.8 \times 10^{-3}$ g of AgCl dissolves per liter of water at 25°C.

Solubility product constants (like solubilities) can be sensitive to temperature. At 100°C, the $K_{sp}$ for silver chloride is $2.2 \times 10^{-8}$; hot water dissolves about 12 times as much silver chloride as does water at 25°C. Refer to Table 11.3 for the solubility product constants at 25°C of a number of important sparingly soluble salts.

### EXAMPLE 11.4

The $K_{sp}$ of calcium fluoride is $3.9 \times 10^{-11}$. Calculate the concentrations of calcium and fluoride ions in a saturated solution of $CaF_2$ at 25°C, and determine the solubility of $CaF_2$ in grams per liter.

## TABLE 11.3

### Solubility Product Constants $K_{sp}$ at 25°C

**Iodates**

| | |
|---|---|
| $AgIO_3$ | $[Ag^+][IO_3^-] = 3.1 \times 10^{-8}$ |
| $CuIO_3$ | $[Cu^+][IO_3^-] = 1.4 \times 10^{-7}$ |
| $Pb(IO_3)_2$ | $[Pb^{2+}][IO_3^-]^2 = 2.6 \times 10^{-13}$ |

**Carbonates**

| | |
|---|---|
| $Ag_2CO_3$ | $[Ag^+]^2[CO_3^{2-}] = 6.2 \times 10^{-12}$ |
| $BaCO_3$ | $[Ba^{2+}][CO_3^{2-}] = 8.1 \times 10^{-9}$ |
| $CaCO_3$ | $[Ca^{2+}][CO_3^{2-}] = 8.7 \times 10^{-9}$ |
| $PbCO_3$ | $[Pb^{2+}][CO_3^{2-}] = 3.3 \times 10^{-14}$ |
| $MgCO_3$ | $[Mg^{2+}][CO_3^{2-}] = 4.0 \times 10^{-5}$ |
| $SrCO_3$ | $[Sr^{2+}][CO_3^{2-}] = 1.6 \times 10^{-9}$ |

**Chromates**

| | |
|---|---|
| $Ag_2CrO_4$ | $[Ag^+]^2[CrO_4^{2-}] = 1.9 \times 10^{-12}$ |
| $BaCrO_4$ | $[Ba^{2+}][CrO_4^{2-}] = 2.1 \times 10^{-10}$ |
| $PbCrO_4$ | $[Pb^{2+}][CrO_4^{2-}] = 1.8 \times 10^{-14}$ |

**Oxalates**

| | |
|---|---|
| $CuC_2O_4$ | $[Cu^{2+}][C_2O_4^{2-}] = 2.9 \times 10^{-8}$ |
| $FeC_2O_4$ | $[Fe^{2+}][C_2O_4^{2-}] = 2.1 \times 10^{-7}$ |
| $MgC_2O_4$ | $[Mg^{2+}][C_2O_4^{2-}] = 8.6 \times 10^{-5}$ |
| $PbC_2O_4$ | $[Pb^{2+}][C_2O_4^{2-}] = 2.7 \times 10^{-11}$ |
| $SrC_2O_4$ | $[Sr^{2+}][C_2O_4^{2-}] = 5.6 \times 10^{-8}$ |

**Sulfates**

| | |
|---|---|
| $BaSO_4$ | $[Ba^{2+}][SO_4^{2-}] = 1.1 \times 10^{-10}$ |
| $CaSO_4$ | $[Ca^{2+}][SO_4^{2-}] = 2.4 \times 10^{-5}$ |
| $PbSO_4$ | $[Pb^{2+}][SO_4^{2-}] = 1.1 \times 10^{-8}$ |

**Fluorides**

| | |
|---|---|
| $BaF_2$ | $[Ba^{2+}][F^-]^2 = 1.7 \times 10^{-6}$ |
| $CaF_2$ | $[Ca^{2+}][F^-]^2 = 3.9 \times 10^{-11}$ |
| $MgF_2$ | $[Mg^{2+}][F^-]^2 = 6.6 \times 10^{-9}$ |
| $PbF_2$ | $[Pb^{2+}][F^-]^2 = 3.6 \times 10^{-8}$ |
| $SrF_2$ | $[Sr^{2+}][F^-]^2 = 2.8 \times 10^{-9}$ |

**Chlorides**

| | |
|---|---|
| $AgCl$ | $[Ag^+][Cl^-] = 1.6 \times 10^{-10}$ |
| $CuCl$ | $[Cu^+][Cl^-] = 1.0 \times 10^{-6}$ |
| $Hg_2Cl_2$ | $[Hg_2^{2+}][Cl^-]^2 = 2 \times 10^{-18}$ |

**Bromides**

| | |
|---|---|
| $AgBr$ | $[Ag^+][Br^-] = 7.7 \times 10^{-13}$ |
| $CuBr$ | $[Cu^+][Br^-] = 4.2 \times 10^{-8}$ |
| $Hg_2Br_2$ | $[Hg_2^{2+}][Br^-]^2 = 1.3 \times 10^{-21}$ |

**Iodides**

| | |
|---|---|
| $AgI$ | $[Ag^+][I^-] = 1.5 \times 10^{-16}$ |
| $CuI$ | $[Cu^+][I^-] = 5.1 \times 10^{-12}$ |
| $PbI_2$ | $[Pb^{2+}][I^-]^2 = 1.4 \times 10^{-8}$ |
| $Hg_2I_2$ | $[Hg_2^{2+}][I^-]^2 = 1.2 \times 10^{-28}$ |

**Hydroxides**

| | |
|---|---|
| $AgOH$ | $[Ag^+][OH^-] = 1.5 \times 10^{-8}$ |
| $Al(OH)_3$ | $[Al^{3+}][OH^-]^3 = 3.7 \times 10^{-15}$ |
| $Fe(OH)_3$ | $[Fe^{3+}][OH^-]^3 = 1.1 \times 10^{-36}$ |
| $Fe(OH)_2$ | $[Fe^{2+}][OH^-]^2 = 1.6 \times 10^{-14}$ |
| $Mg(OH)_2$ | $[Mg^{2+}][OH^-]^2 = 1.2 \times 10^{-11}$ |
| $Mn(OH)_2$ | $[Mn^{2+}][OH^-]^2 = 2 \times 10^{-13}$ |
| $Zn(OH)_2$ | $[Zn^{2+}][OH^-]^2 = 4.5 \times 10^{-17}$ |

### Solution

The solubility equilibrium is

$$CaF_2(s) \rightleftharpoons Ca^{2+}(aq) + 2\,F^-(aq)$$

and the expression for the solubility product is

$$[Ca^{2+}][F^-]^2 = K_{sp}$$

The concentration of fluoride ion is squared because it has a coefficient of 2 in the chemical equation.

If $S$ mol of $CaF_2$ dissolves in 1 L, the equilibrium concentration of $Ca^{2+}$ will be $[Ca^{2+}] = S$. The concentration of $F^-$ will be $[F^-] = 2S$, because each mole of $CaF_2$ produces *two* moles of fluoride ions. Therefore,

$$[Ca^{2+}][F^-]^2 = S \times (2S)^2 = 4S^3 = K_{sp} = 3.9 \times 10^{-11}$$

Solving for $S^3$ gives

$$S^3 = \frac{1}{4}\,(3.9 \times 10^{-11})$$

Taking the cube roots of both sides of this equation gives

$$S = 2.1 \times 10^{-4}$$

The equilibrium concentrations are therefore

$$[Ca^{2+}] = S = 2.1 \times 10^{-4} \, \text{M}$$

$$[F^-] = 2S = 4.3 \times 10^{-4} \, \text{M}$$

Because the molar mass of $CaF_2$ is 78.1 g mol$^{-1}$, the gram solubility is

$$\text{gram solubility} = (2.1 \times 10^{-4} \, \text{mol L}^{-1})(78.1 \, \text{g mol}^{-1}) = 0.017 \, \text{g L}^{-1}$$

**Related Problems: 25, 26, 27, 28, 29, 30**

---

It is possible to reverse the procedure just outlined, of course, and determine the value of $K_{sp}$ from measured solubilities, as the following example illustrates.

## EXAMPLE 11.5

Silver chromate ($Ag_2CrO_4$) is a red solid that dissolves in water to the extent of 0.029 g L$^{-1}$ at 25°C. Estimate its $K_{sp}$ and compare your estimate with the value in Table 11.3.

### Solution

The molar solubility is calculated from the gram solubility and the molar mass of silver chromate (331.73 g mol$^{-1}$):

$$\text{molar solubility} = \frac{0.029 \, \text{g L}^{-1}}{331.73 \, \text{g mol}^{-1}} = 8.74 \times 10^{-5} \, \text{mol L}^{-1}$$

An extra significant digit is carried in this intermediate result to avoid round-off errors. Because each mole that dissolves gives *two* moles of silver ion, the concentration of $Ag^+(aq)$ is

$$[Ag^+] = 2 \times 8.74 \times 10^{-5} \, \text{M} = 1.75 \times 10^{-4} \, \text{M}$$

and that of $CrO_4^{2-}$ is simply $8.74 \times 10^{-5}$ M. The solubility product constant is then

$$K_{sp} = [Ag^+]^2[CrO_4^{2-}] = (1.75 \times 10^{-4})^2 \times (8.74 \times 10^{-5})$$

$$= 2.7 \times 10^{-12}$$

This is fairly close to the tabulated value, $1.9 \times 10^{-12}$.

**Related Problems: 31, 32, 33, 34**

---

Computing $K_{sp}$ from a solubility and solubility from $K_{sp}$ is valid if the solution is ideal and if there are no side reactions that reduce the concentrations of the ions after they enter solution. If such reactions are present, they cause higher solubilities than are predicted from the $K_{sp}$ expression. For example, the solubility computed for $PbSO_4(s)$ at 25°C from its $K_{sp}$ is 0.032 g L$^{-1}$, whereas that measured experimentally is 0.0425 g L$^{-1}$. The difference is caused by the presence of species other than $Pb^{2+}(aq)$ in solution, such as $PbOH^+(aq)$. Further discussion of these side reactions is deferred to Section 11.8.

## 11.5

# PRECIPITATION AND THE SOLUBILITY PRODUCT

So far, we have considered only cases in which a single slightly soluble salt attains equilibrium with its component ions in water. The relative concentrations of the cations and anions in such solutions echo their relative numbers of moles in the original salt. Thus, when AgCl is dissolved, equal numbers of moles of $Ag^+(aq)$ and $Cl^-(aq)$ ions result, and when $Ag_2SO_4$ is dissolved, twice as many moles of $Ag^+(aq)$ ions as $SO_4^{2-}(aq)$ are produced. A solubility product relationship such as

$$[Ag^+][Cl^-] = K_{sp}$$

is more general than this, however, and continues in force even if the relative chemical amounts of the two ions in solution differ from those in the pure solid compound. Such a situation often results when two solutions are mixed to give a precipitate or when another salt is present that contains an ion common to the salt under consideration.

## Precipitation from Solution

Suppose a solution is prepared by mixing one soluble salt, such as $AgNO_3$, with a solution of a second, such as NaCl. Will a precipitate of very slightly soluble silver chloride form? To reach an answer, the reaction quotient $Q$ that was defined in connection with gaseous equilibria (Section 9.4) is used. The initial reaction quotient $Q_0$, when the mixing of the solutions is complete but before any reaction occurs, is

$$Q_0 = [Ag^+]_0[Cl^-]_0$$

If $Q_0 < K_{sp}$, no solid silver chloride can appear. On the other hand, if $Q_0 > K_{sp}$, solid silver chloride precipitates until the reaction quotient $Q$ reaches $K_{sp}$ (Fig. 11.10).

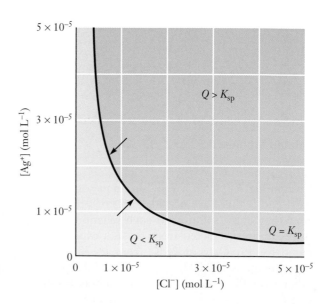

**FIGURE 11.10** Some solid silver chloride is in contact with a solution containing $Ag^+(aq)$ and $Cl^-(aq)$ ions. If a solubility equilibrium exists, then the product $Q$ of the concentrations of the ions $[Ag^+] \times [Cl^-]$ is a constant, $K_{sp}$ (curved line). When $Q$ exceeds $K_{sp}$, solid silver chloride tends to precipitate until equilibrium is attained. When $Q$ is less than $K_{sp}$, additional solid tends to dissolve. If no solid is present, $Q$ remains less than $K_{sp}$.

## EXAMPLE 11.6

An emulsion of silver chloride for photographic film is prepared by adding a soluble chloride salt to a solution of silver nitrate. Suppose 500 mL of a solution of $CaCl_2$ with a chloride ion concentration of $8.0 \times 10^{-6}$ M is added to 300 mL of a 0.0040 M solution of $AgNO_3$. Will a precipitate of $AgCl(s)$ form when equilibrium is reached?

### Solution

The "initial concentrations" to be used in calculating $Q_0$ are those *before* reaction but *after* dilution through mixing the two solutions. The initial concentration of $Ag^+(aq)$ after dilution from 300 mL to 800 mL of solution is

$$[Ag^+]_0 = 0.00400 \text{ M} \times \left(\frac{300 \text{ mL}}{800 \text{ mL}}\right) = 0.0015 \text{ M}$$

and that of $Cl^-(aq)$ is

$$[Cl^-]_0 = 8.0 \times 10^{-6} \text{ M} \times \left(\frac{500 \text{ mL}}{800 \text{ mL}}\right) = 5.0 \times 10^{-6} \text{ M}$$

The initial reaction quotient is

$$Q_0 = [Ag^+]_0[Cl^-]_0 = (0.0015)(5.0 \times 10^{-6}) = 7.5 \times 10^{-9}$$

Because $Q_0 > K_{sp}$, a precipitate of silver chloride appears at equilibrium, although there may be too little to detect visually. Another possible precipitate, calcium nitrate, is far too soluble to form in this experiment (see Table 11.2).

**Related Problems: 35, 36, 37, 38**

The equilibrium concentrations of ions after the mixing of two solutions to give a precipitate are most easily calculated by supposing that the reaction first goes to completion (consuming one type of ion) and that subsequent dissolution of the solid restores some of that ionic species to solution—just the approach used in Example 10.8 for the addition of a strong acid to a buffer solution.

## EXAMPLE 11.7

Calculate the equilibrium concentrations of silver and chloride ions resulting from the precipitation reaction of Example 11.6.

### Solution

In this case, the silver ion is clearly in excess, and so the chloride ion is the limiting reactant. If all of it were used up to make solid AgCl, the concentration of the remaining silver ion would be

$$[Ag^+] = 0.0015 - 5.0 \times 10^{-6} = 0.0015 \text{ M}$$

Set up the equilibrium as

| | $AgCl(s) \rightleftharpoons$ | $Ag^+(aq)$ + | $Cl^-(aq)$ |
|---|---|---|---|
| Initial concentration (M) | | 0.0015 | 0 |
| Change in concentration (M) | | $+y$ | $+y$ |
| Equilibrium concentration (M) | | $0.0015 + y$ | $y$ |

so that the equilibrium expression is

$$(0.0015 + y)y = K_{sp} = 1.6 \times 10^{-10}$$

This quadratic equation can be solved by use of the quadratic formula, provided the calculator carries ten significant figures. It can be solved more easily by making the approximation that $y$ is much smaller than 0.0015, so that the equation simplifies to

$$0.0015y \approx 1.6 \times 10^{-10}$$

$$y \approx 1.1 \times 10^{-7} \text{ M} = [Cl^-]$$

The assumption about the size of $y$ was justified. The concentration of silver ion is

$$[Ag^+] = 0.0015 \text{ M}$$

**Related Problems: 39, 40, 41, 42**

## The Common-Ion Effect

Suppose a small amount of NaCl($s$) is added to a saturated solution of AgCl. What happens? Sodium chloride is quite soluble in water and dissolves to form $Na^+(aq)$ and $Cl^-(aq)$ ions, raising the concentration of chloride ion. The quantity $Q_0 = [Ag^+][Cl^-]$ then exceeds the $K_{sp}$ of silver chloride, and silver chloride precipitates until the concentrations of $Ag^+(aq)$ and $Cl^-(aq)$ are sufficiently reduced that the solubility product expression once again is satisfied.

The same equilibrium may be approached from the other direction. The amount of AgCl($s$) that can dissolve in a solution of sodium chloride is less than the amount that could dissolve in the same volume of pure water. Because $[Ag^+][Cl^-] = K_{sp}$, a graph of the equilibrium concentration of silver ion against chloride concentration has the form of a hyperbola (Fig. 11.11). The presence of excess $Cl^-(aq)$ reduces the concentration of $Ag^+(aq)$ permitted, and the solubility of AgCl($s$) is reduced. In the same way, the prior presence of $Ag^+(aq)$ in the solvent (for example, when an

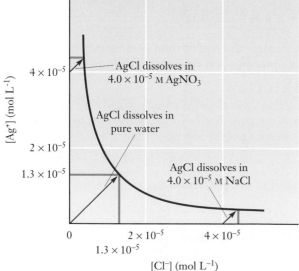

**FIGURE 11.11**   The presence of a dissolved common ion reduces the solubility of a salt in solution. As the AgCl dissolves, the concentrations of the ions follow the paths shown by the red arrows until they reach the red equilibrium curve. The molar solubilities are proportional to the lengths of the green lines: $1.3 \times 10^{-5}$ mol $L^{-1}$ for AgCl in pure water, but only $0.37 \times 10^{-5}$ mol $L^{-1}$ in either $4.0 \times 10^{-5}$ M AgNO$_3$ or $4.0 \times 10^{-5}$ M NaCl.

attempt is made to dissolve AgCl in water that already contains $AgNO_3$) reduces the amount of $Cl^-(aq)$ permitted at equilibrium and also reduces the solubility of AgCl. This is referred to as the **common-ion effect:** if the solution and the solid salt to be dissolved in it have an ion in common, then the solubility of the salt is depressed.

Let us examine the quantitative consequences of the common-ion effect. Suppose an excess of AgCl(s) is added to 1.00 L of a 0.100 M NaCl solution and the solubility is again determined. If $S$ mol of AgCl dissolves per liter, the concentration of $Ag^+(aq)$ will be $S$ mol $L^{-1}$, and that of $Cl^-(aq)$ will be

$$[Cl^-] = 0.100 + S$$

because the chloride ion comes from two sources: the 0.100 M NaCl and the dissolution of AgCl. The expression for the solubility product is written as

$$[Ag^+][Cl^-] = S(0.100 + S) = K_{sp} = 1.6 \times 10^{-10}$$

The solubility of AgCl(s) in this solution must be smaller than it is in pure water, which is much smaller than 0.100. That is,

$$S < 1.3 \times 10^{-5} \ll 0.100$$

Thus, $(0.100 + S)$ can be approximated by 0.100 (as in Example 11.7), giving

$$(0.100)S \approx 1.6 \times 10^{-10}$$

$$S \approx 1.6 \times 10^{-9}$$

This is indeed much smaller than 0.100, so the approximation was a very good one. Therefore, at equilibrium,

$$[Ag^+] = S = 1.6 \times 10^{-9} \text{ M}$$

$$[Cl^-] = 0.100 \text{ M}$$

The gram solubility of AgCl here is

$$(1.6 \times 10^{-9} \text{ mol L}^{-1})(143.3 \text{ g mol}^{-1}) = 2.3 \times 10^{-7} \text{ g L}^{-1}$$

The solubility of AgCl in 0.100 M NaCl is lower than that in pure water by a factor of about 8000.

## EXAMPLE 11.8

What is the gram solubility of $CaF_2(s)$ in a 0.100 M solution of NaF?

### Solution

Again, the molar solubility is denoted by $S$. The only source of the $Ca^{2+}(aq)$ in the solution at equilibrium is the dissolution of $CaF_2$, whereas the $F^-(aq)$ has two sources, the $CaF_2$ and the NaF. Hence,

| | $CaF_2(s) \rightleftharpoons$ | $Ca^{2+}(aq)$ + | $2\ F^-(aq)$ |
|---|---|---|---|
| Initial concentration (M) | | 0 | 0.100 |
| Change in concentration (M) | | $+S$ | $+2S$ |
| Equilibrium concentration (M) | | $S$ | $0.100 + 2S$ |

If $0.100 + 2S$ is approximated as $0.100$, then

$$[Ca^{2+}][F^-]^2 = K_{sp}$$
$$S(0.100)^2 = 3.9 \times 10^{-11}$$
$$S = 3.9 \times 10^{-9}$$

Clearly,

$$2S = 7.8 \times 10^{-9} \ll 0.100$$

so the assumption was justified. The gram solubility of $CaF_2$ is

$$(3.9 \times 10^{-9} \text{ mol L}^{-1})(78.1 \text{ g mol}^{-1}) = 3.0 \times 10^{-7} \text{ g L}^{-1}$$

and the solubility in this case is reduced by a factor of 50,000.

**Related Problems: 43, 44, 45, 46**

---

## 11.6

# THE EFFECTS OF pH ON SOLUBILITY

Some solids are only weakly soluble in water but dissolve readily in acidic solutions. Copper and nickel sulfides from ores, for example, can be brought into solution with strong acids, a fact that aids greatly in the separation and recovery of these valuable metals in their elemental forms. The effect of pH on solubility is shown dramatically in the damage done to buildings and monuments by acid precipitation (Fig. 11.12). Both marble and limestone are made up of small crystals of calcite ($CaCO_3$), which dissolves to only a limited extent in "natural" rain (with a pH of about 5.6) but dissolves much more extensively as the rainwater becomes more acidic. The reaction

$$CaCO_3(s) + H_3O^+(aq) \longrightarrow Ca^{2+}(aq) + HCO_3^-(aq) + H_2O(\ell)$$

causes this increase. This section examines the role of pH in solubility.

## Solubility of Hydroxides

One direct effect of pH on solubility occurs with the metal hydroxides. The concentration of $OH^-$ appears explicitly in the expression for the solubility product of such compounds. Thus, for the dissolution of $Zn(OH)_2(s)$,

$$Zn(OH)_2(s) \rightleftharpoons Zn^{2+}(aq) + 2\ OH^-(aq)$$

the solubility product expression is

$$[Zn^{2+}][OH^-]^2 = K_{sp} = 4.5 \times 10^{-17}$$

As the solution is made more acidic, the concentration of hydroxide ion decreases, causing an increase in the concentration of $Zn^{2+}(aq)$ ion. Zinc hydroxide is thus more soluble in acidic solution than in pure water.

**FIGURE 11.12**  The calcium carbonate in marble and limestone is very slightly soluble in neutral water. Its solubility is much greater in acidic water. Objects carved of these materials dissolve relatively rapidly in areas where rain, snow, or fog is acidified from air pollution. Shown here is the damage to a gargoyle at Lincoln Cathedral in England between 1910 (top) and 1984 (bottom). *(Courtesy of Dean of Lincoln)*

### EXAMPLE 11.9

Compare the solubility of $Zn(OH)_2$ in pure water with that in a solution buffered at pH 6.00.

## Solution

In pure water the usual solubility product calculation applies:

$$[Zn^{2+}] = S; \qquad [OH^-] = 2S$$

$$S(2S)^2 = 4S^3 = K_{sp} = 4.5 \times 10^{-17}$$

$$S = 2.2 \times 10^{-6} \, M = [Zn^{2+}]$$

and so the solubility is $2.2 \times 10^{-6}$ mol $L^{-1}$, or $2.2 \times 10^{-4}$ g $L^{-1}$. Using

$$[OH^-] = 2S = 4.5 \times 10^{-6} \, M$$

the resulting solution is found to have pH 8.65.

In the second case, it is assumed that the solution is buffered sufficiently that the pH remains 6.00 after dissolution of the zinc hydroxide. Then

$$[OH^-] = 1.0 \times 10^{-8} \, M$$

$$[Zn^{2+}] = \frac{K_{sp}}{[OH^-]^2} = \frac{4.5 \times 10^{-17}}{(1.0 \times 10^{-8})^2} = 0.45 \, M$$

so that 0.45 mol $L^{-1}$, or 45 g $L^{-1}$, should dissolve in this case. When ionic concentrations are this high, the simple form of the solubility expression will likely break down, but the qualitative conclusion is still valid: $Zn(OH)_2$ is far more soluble at pH 6.00 than in pure water.

**Related Problems: 49, 50**

---

## Solubility of Salts of Bases

Metal hydroxides can be described as salts of a strong base, the hydroxide ion. The solubility of salts in which the anion is a different weak or strong base are also affected by pH. For example, consider a solution of a slightly soluble fluoride, such as calcium fluoride. The solubility equilibrium is

$$CaF_2(s) \rightleftharpoons Ca^{2+}(aq) + 2\,F^-(aq) \qquad K_{sp} = 3.9 \times 10^{-11}$$

As the solution is made more acidic, some of the fluoride ion reacts with hydronium ion through

$$H_3O^+(aq) + F^-(aq) \rightleftharpoons HF(aq) + H_2O(\ell)$$

Because this is just the reverse of the acid ionization of HF, its equilibrium constant is the reciprocal of $K_a$ for HF, or $1/(3.5 \times 10^{-4}) = 2.9 \times 10^3$. As acid is added, the concentration of fluoride ion is reduced, so the calcium ion concentration must increase to maintain the solubility product equilibrium for $CaF_2$. As a result, the solubility of fluoride salts increases in acidic solution. The same applies to other ionic substances in which the anion is a weak or a strong base. By contrast, the solubility of a salt such as AgCl is only very slightly affected by a decrease in pH. The reason is that HCl is a strong acid, and so $Cl^-$ is ineffective as a base. The reaction

$$Cl^-(aq) + H_3O^+(aq) \longrightarrow HCl(aq) + H_2O(\ell)$$

occurs to a negligible extent in acidic solution.

# A   D E E P E R   L O O K . . .

## 11.7   Selective Precipitation of Ions

One way to analyze a mixture of ions in solution is to separate the mixture into its components by exploiting the differences in the solubilities of compounds containing the ions. To separate silver ion from lead ion, for example, a search is made for compounds of these elements that (1) have a common anion and (2) have widely different solubilities. The chlorides AgCl and $PbCl_2$ are two such compounds for which the solubility equilibria are

$$AgCl(s) \rightleftharpoons Ag^+(aq) + Cl^-(aq) \qquad K_{sp} = 1.6 \times 10^{-10}$$

$$PbCl_2(s) \rightleftharpoons Pb^{2+}(aq) + 2\,Cl^-(aq) \qquad K_{sp} = 2.4 \times 10^{-4}$$

Lead chloride is far more soluble in water than is silver chloride. Consider a solution that is 0.10 M in both $Ag^+$ and $Pb^{2+}$. Is it possible to add enough $Cl^-$ to precipitate almost all the $Ag^+$ ions but leave all the $Pb^{2+}$ ions in solution? If so, a quantitative separation of the two species can be achieved.

In order for $Pb^{2+}$ to remain in solution, its reaction quotient must remain smaller than $K_{sp}$: $Q = [Pb^{2+}][Cl^-]^2 < K_{sp}$. Inserting $K_{sp}$ and the concentration of $Pb^{2+}$ gives

$$[Cl^-]^2 < \frac{K_{sp}}{[Pb^{2+}]} = \frac{2.4 \times 10^{-4}}{0.10} = 2.4 \times 10^{-3}$$

The square root of this is

$$[Cl^-] < 4.9 \times 10^{-2} \text{ M}$$

Thus, as long as the chloride ion concentration remains smaller than 0.049 M, no $PbCl_2$ should precipitate. To reduce the *silver* ion concentration in solution as far as possible (that is, to precipitate out as much silver chloride as possible), the chloride ion concentration should be kept as high as possible without exceeding 0.049 M. If exactly this concentration of $Cl^-(aq)$ is chosen, then at equilibrium,

$$[Ag^+] = \frac{K_{sp}}{[Cl^-]} = \frac{1.6 \times 10^{-10}}{0.049} = 3.3 \times 10^{-9}$$

At that concentration of $Cl^-$, the concentration of $Ag^+$ has been reduced to $3.3 \times 10^{-9}$ from the original concentration of 0.10 M. In other words, only about three $Ag^+$ ions in $10^8$ remain in solution, but all the $Pb^{2+}$ ions are left in solution. A nearly perfect separation of the two ionic species has been achieved.

This calculation gave the optimal theoretical separation factor for the two ions. In practice, it is necessary to keep the chloride concentration lower. If $[Cl^-]$ is ten times smaller, or 0.0049 M, about three $Ag^+$ ions in $10^7$ remain in solution with the $Pb^{2+}$. This is ten times more $Ag^+$ than if $[Cl^-] = 0.049$ M, but the separation of $Ag^+$ from $Pb^{2+}$ is still very good.

Figure 11.13 shows graphically how ions can be separated based on solubility. The relation between the concentration of $Pb^{2+}$ and $Cl^-$ ions in contact with solid $PbCl_2$

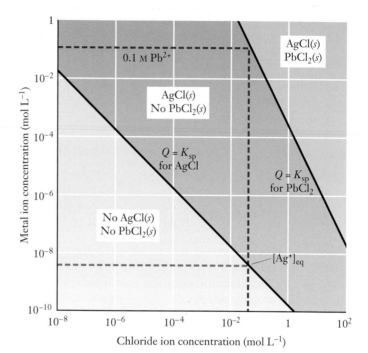

**FIGURE 11.13**  To separate a mixture of $Ag^+$ and $Pb^{2+}$ ions, a chloride ion concentration is selected that gives a $Pb^{2+}$ concentration below the equilibrium curve for $PbCl_2$ (so all $Pb^{2+}$ remains in solution) but well above the equilibrium curve for $Ag^+$. As a result, nearly all the $Ag^+$ precipitates as AgCl. If $[Pb^{2+}]_0 = [Ag^+]_0 = 0.1$ M, then the maximum $[Cl^-]$ is found by tracing the red line until it intersects the $PbCl_2$ equilibrium curve; this occurs at $[Cl^-] = 0.049$ M. The concentration of $Ag^+$ still in solution is found by tracing the blue line from the intersection of the $[Cl^-]$ line and the AgCl equilibrium curve back to the vertical axis, which indicates metal ion concentration.

$$[Pb^{2+}] = \frac{K_{sp}}{[Cl^-]^2}$$

can be rewritten in a useful form by taking the base-10 logarithms of both sides:

$$\log_{10}[Pb^{2+}] = -2\log_{10}[Cl^-] + \log_{10} K_{sp}$$

That is, a graph of $\log_{10}[Pb^{2+}]$ against $\log_{10}[Cl^-]$ is a straight line with slope $-2$. The corresponding graph for the AgCl solubility equilibrium has slope $-1$.

$$\log_{10}[Ag^+] = -\log_{10}[Cl^-] + \log_{10} K_{sp}$$

If the concentration of $Cl^-$ and the initial concentrations of the metal ions correspond to a point that lies between the two lines in Figure 11.13, then AgCl precipitates but $PbCl_2$ does not.

## Metal Sulfides

Controlling the solubility of metal sulfides has important applications. According to Table 11.2, most metal sulfides are very slightly soluble in water; only a very small amount of a compound such as $ZnS(s)$ will dissolve in water. Although it is tempting to write the resulting equilibrium as

$$ZnS(s) \rightleftharpoons Zn^{2+}(aq) + S^{2-}(aq) \qquad K_{sp} = ?$$

in analogy with the equilibria for other weakly soluble salts, this is misleading, because $S^{2-}$, like $O^{2-}$, is a very strong base (stronger than $OH^-$) and reacts almost quantitatively with water:

$$S^{2-}(aq) + H_2O(\ell) \longrightarrow HS^-(aq) + OH^-(aq)$$

Recent research has shown that $K_b$ for this reaction is on the order of $10^5$, and so essentially no $S^{2-}$ is present in aqueous solution. The net dissolution reaction is found by adding the two preceding equations,

$$ZnS(s) + H_2O(\ell) \rightleftharpoons Zn^{2+}(aq) + OH^-(aq) + HS^-(aq)$$

for which the equilibrium constant is

$$[Zn^{2+}][OH^-][HS^-] = K \approx 2 \times 10^{-25}$$

Table 11.4 gives values of the equilibrium constants for comparable reactions of other metal sulfides. As the pH decreases, the concentration of $OH^-$ decreases. At the same time, the concentration of $HS^-$ also decreases as the equilibrium

$$HS^-(aq) + H_3O^+(aq) \rightleftharpoons H_2S(aq) + H_2O(\ell)$$

shifts to the right upon addition of $H_3O^+$. If both $[OH^-]$ and $[HS^-]$ decrease, then $[Zn^{2+}]$ must increase in order to maintain

a constant value for the product of the three concentrations. As a result, the solubility of $ZnS(s)$ increases as the pH of the solution decreases. Other metal sulfides behave the same way, becoming more soluble in acidic solution.

A quantitative calculation of metal sulfide solubility requires treating several simultaneous equilibria, as the following example illustrates.

In a solution that is saturated with $H_2S$, $[H_2S]$ is fixed at 0.1 M. Calculate the molar solubility of FeS(s) in such a solution, if it is buffered at pH 3.0.

### Solution

If the pH is 3.0, then

$$[OH^-] = 1 \times 10^{-11} \text{ M}$$

In addition, the acid ionization of $H_2S$ must be considered:

$$H_2S(aq) + H_2O(\ell) \rightleftharpoons H_3O^+(aq) + HS^-(aq)$$

Substitution of $[H_2S] = 0.1$ M for a saturated solution and $[H_3O^+] = 1 \times 10^{-3}$ M (at pH 3.0) into the equilibrium expression for this reaction gives

$$\frac{[H_3O^+][HS^-]}{[H_2S]} = \frac{(1 \times 10^{-3})[HS^-]}{0.1} = K_a = 9.1 \times 10^{-8}$$

$$[HS^-] = 9 \times 10^{-6} \text{ M}$$

where $K_a$ came from Table 10.2. For the reaction

$$FeS(s) + H_2O(\ell) \rightleftharpoons Fe^{2+}(aq) + HS^-(aq) + OH^-(aq)$$

the equilibrium constant from Table 11.4 is

$$[Fe^{2+}][HS^-][OH^-] = 5 \times 10^{-19}$$

Substituting the values of $[HS^-]$ and $[OH^-]$ and solving for $[Fe^{2+}]$ give

### Equilibrium Constants for Metal Sulfide Dissolution at 25°C

| Metal Sulfide | $K^\dagger$ |
| --- | --- |
| CuS | $5 \times 10^{-37}$ |
| PbS | $3 \times 10^{-28}$ |
| CdS | $7 \times 10^{-28}$ |
| SnS | $9 \times 10^{-27}$ |
| ZnS | $2 \times 10^{-25}$ |
| FeS | $5 \times 10^{-19}$ |
| MnS | $3 \times 10^{-14}$ |

$^\dagger K$ is the equilibrium constant for the reaction $MS(s) + H_2O(\ell) \rightleftharpoons M^{2+}(aq) + OH^-(aq) + HS^-(aq)$.

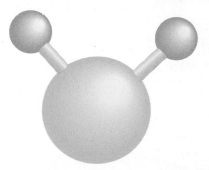

Hydrogen sulfide, $H_2S$, is a poisonous, foul-smelling gas. When dissolved in water, it gives a weak acid, hydrosulfuric acid.

$$[Fe^{2+}](9 \times 10^{-6})(1 \times 10^{-11}) = 5 \times 10^{-19}$$

$$[Fe^{2+}] = 6 \times 10^{-3} \text{ M}$$

Hence, $6 \times 10^{-3}$ mol of FeS dissolves per liter under these conditions.

**Related Problems: 59, 60**

By adjusting the pH through appropriate choice of buffers, as in Example 11.10, conditions can be selected so that metal ions of one element remain entirely in solution, whereas those of a second element in the mixture precipitate almost entirely as solid metal sulfide (Fig. 11.14). Such a procedure is important for separating metal ions in qualitative analysis.

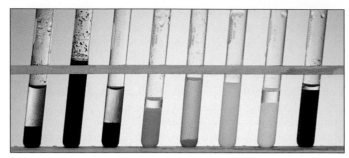

**FIGURE 11.14** These sulfides are insoluble at pH 1 and so can be separated out of a mixture containing other, more soluble sulfides. From left to right, they are PbS, $Bi_2S_3$, CuS, CdS, $Sb_2S_3$, $SnS_2$, $As_2S_3$, and HgS. *(Leon Lewandowski)*

---

## 11.8

# COMPLEX IONS AND SOLUBILITY

Many transition-metal ions form **coordination complexes** in solution or in the solid state; these consist of a metal ion surrounded by a group of anions or neutral molecules called **ligands.** The interaction involves the sharing by the metal ion of a lone pair on each ligand molecule, giving a partially covalent bond with that ligand. Such complexes frequently have strikingly deep colors. When exposed to gaseous ammonia, greenish white crystals of copper sulfate ($CuSO_4$) give a deep-blue crystalline solid with the chemical formula $Cu(NH_3)_4SO_4$ (Fig. 11.15a). The anions in the solid are still sulfate ions ($SO_4^{2-}$), but the cations are now **complex ions,** or coordination complexes of the central $Cu^{2+}$ ion with four ammonia molecules, $Cu(NH_3)_4^{2+}$. The ammonia molecules coordinate to the copper ion through their lone-pair electrons (Fig. 11.16), acting as Lewis bases toward the metal ion, the Lewis acid. When the solid is dissolved in water, the deep blue color remains. This is evidence that the complex persists in water, because when ordinary $CuSO_4$ (without ammonia ligands) is dissolved in water, a much paler blue color results (Fig. 11.15b).

Let us explore the effects of the formation of complex ions on equilibria in aqueous solutions, deferring to Chapter 18 a discussion of the microscopic structure and bonding in these complexes.

(a)

(b)

**FIGURE 11.15** (a) From left to right, crystals of $Cu(NH_3)_4SO_4$, $CuSO_4 \cdot 5H_2O$, and $CuSO_4$. (b) Aqueous solutions of copper sulfate containing (left) and not containing (right) ammonia. *(Leon Lewandowski)*

### Complex Ion Equilibria

When silver ions are dissolved in an aqueous ammonia solution, doubly coordinated silver–ammonia complexes, shown in Figure 11.17, form in two stepwise reactions:

$$Ag^+(aq) + NH_3(aq) \rightleftharpoons Ag(NH_3)^+(aq)$$

$$\frac{[Ag(NH_3)^+]}{[Ag^+][NH_3]} = K_1 = 2.1 \times 10^3$$

$$Ag(NH_3)^+(aq) + NH_3(aq) \rightleftharpoons Ag(NH_3)_2^+(aq)$$

$$\frac{[Ag(NH_3)_2^+]}{[Ag(NH_3)^+][NH_3]} = K_2 = 8 \times 10^3$$

If these two chemical equations are added (and their corresponding equilibrium laws are multiplied), the result is

$$Ag^+(aq) + 2\,NH_3(aq) \rightleftharpoons Ag(NH_3)_2^+(aq)$$

$$\frac{[Ag(NH_3)_2^+]}{[Ag^+][NH_3]^2} = K_f = K_1 K_2 = 1.7 \times 10^7$$

where $K_f$ is the **formation constant** of the full complex ion $Ag(NH_3)_2^+$. Table 11.5 gives formation constants for a representative selection of complex ions. The larger the formation constant $K_f$, the more stable the corresponding complex ion, for ions with the same number of ligands.

Because $K_1$ and $K_2$ of the $Ag(NH_3)_2^+$ complex ion are both large, a silver salt dissolved in water that contains a high concentration of ammonia will be primarily in the form of the complex ion $[Ag(NH_3)_2]^+$ at equilibrium.

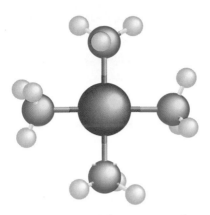

**FIGURE 11.16**   The structure of $Cu(NH_3)_4^{2+}$.

**FIGURE 11.17**   The structure of $Ag(NH_3)_2^+$.

## EXAMPLE 11.11

Suppose 0.100 mol of $AgNO_3$ is dissolved in 1.00 L of a 1.00 M solution of $NH_3$. Calculate the concentrations of the $Ag^+$ and $Ag(NH_3)^+$ ions present at equilibrium.

## TABLE 11.5

### Formation Constants of Coordination Complexes in Aqueous Solution

| | $K_f$ | $K_1$ | $K_2$ | $K_3$ | $K_4$ | $K_5$ | $K_6$ |
|---|---|---|---|---|---|---|---|
| **Ammines** | | | | | | | |
| $Ag(NH_3)_2^+$ | $1.7 \times 10^7$ | $2.1 \times 10^3$ | $8.2 \times 10^3$ | | | | |
| $Co(NH_3)_6^{2+}$ | $2.5 \times 10^4$ | $1.0 \times 10^2$ | 32 | 8.5 | 4.4 | 1.1 | 0.18 |
| $Cu(NH_3)_4^{2+}$ | $1.1 \times 10^{12}$ | $1.0 \times 10^4$ | $2 \times 10^3$ | $5 \times 10^2$ | 90 | | |
| $Ni(NH_3)_6^{2+}$ | $8 \times 10^{-7}$ | $5 \times 10^2$ | $1.3 \times 10^2$ | 40 | 12 | 3.3 | 0.8 |
| $Zn(NH_3)_4^{2+}$ | $5 \times 10^8$ | $1.5 \times 10^2$ | $1.8 \times 10^2$ | $2 \times 10^2$ | 90 | | |
| **Chlorides** | | | | | | | |
| $AgCl_2^-$ | $1.8 \times 10^5$ | $1.7 \times 10^3$ | $1.0 \times 10^2$ | | | | |
| $FeCl_4^-$ | 0.14 | 28 | 4.5 | 0.1 | $1.1 \times 10^{-2}$ | | |
| $HgCl_4^{2-}$ | $1.2 \times 10^{15}$ | $5.5 \times 10^6$ | $3 \times 10^6$ | 7 | 10 | | |
| $PbCl_4^{2-}$ | 25 | 40 | 1.5 | 0.8 | 0.5 | | |
| $SnCl_4^{2-}$ | 30 | 32 | 5.4 | 0.6 | 0.3 | | |
| **Hydroxides** | | | | | | | |
| $Co(OH)_3^-$ | $3 \times 10^{10}$ | $4 \times 10^4$ | 1 | $8 \times 10^5$ | | | |
| $Cu(OH)_4^{2-}$ | $3 \times 10^{18}$ | $1 \times 10^7$ | $5 \times 10^6$ | $2 \times 10^3$ | 30 | | |
| $Ni(OH)_3^-$ | $2 \times 10^{11}$ | $9 \times 10^4$ | $4 \times 10^3$ | $6 \times 10^2$ | | | |
| $Pb(OH)_3^-$ | $4 \times 10^{14}$ | $7 \times 10^7$ | $1.1 \times 10^3$ | $5 \times 10^3$ | | | |
| $Zn(OH)_4^{2-}$ | $5 \times 10^{14}$ | $2.5 \times 10^4$ | $8 \times 10^6$ | 70 | 33 | | |

## Solution

Suppose that most of the $Ag^+$ is present as $Ag(NH_3)_2^+$ (this will be checked later). Then

$$[Ag(NH_3)_2^+]_0 = 0.100 \text{ M}$$

$$[NH_3]_0 = 1.00 \text{ M} - (2 \times 0.100) \text{ M} = 0.80 \text{ M}$$

after each silver ion has become complexed with two ammonia molecules. The two stages of the dissociation of the $Ag(NH_3)_2^+$ ion are the reverse reaction of the complexation, and so their equilibrium constants are the reciprocals of $K_2$ and $K_1$, respectively:

$$Ag(NH_3)_2^+(aq) \rightleftharpoons Ag(NH_3)^+(aq) + NH_3(aq)$$

$$\frac{[Ag(NH_3)^+][NH_3]}{[Ag(NH_3)_2^+]} = \frac{1}{K_2} = \frac{1}{8.2 \times 10^3}$$

$$Ag(NH_3)^+(aq) \rightleftharpoons Ag^+(aq) + NH_3(aq)$$

$$\frac{[Ag^+][NH_3]}{[Ag(NH_3)^+]} = \frac{1}{K_1} = \frac{1}{2.1 \times 10^3}$$

If $y$ mol $L^{-1}$ of $Ag(NH_3)_2^+$ dissociates at equilibrium according to the first equation,

| $Ag(NH_3)_2^+(aq)$ | $\rightleftharpoons$ | $Ag(NH_3)^+(aq)$ | $+ NH_3(aq)$ |
|---|---|---|---|
| Initial concentration (M) | 0.100 | 0 | 0.80 |
| Change in concentration (M) | $-y$ | $+y$ | $+y$ |
| Equilibrium concentration(M) | $0.100 - y$ | $y$ | $0.80 + y$ |

then the first equilibrium expression becomes

$$\frac{y(0.80 + y)}{0.10 - y} = \frac{1}{K_2} = \frac{1}{8.2 \times 10^3}$$

$$y = 1.5 \times 10^{-5} \text{ M} = [Ag(NH_3)^+]$$

We can then calculate the concentration of free $Ag^+$ ions from the equilibrium law for the second step of the dissociation of the complex ion:

$$\frac{[Ag^+][NH_3]}{[Ag(NH_3)^+]} = \frac{1}{K_1}$$

$$\frac{[Ag^+](0.80)}{1.5 \times 10^{-5}} = \frac{1}{2.1 \times 10^3}$$

$$[Ag^+] = 9 \times 10^{-9} \text{ M}$$

It is clear that the original assumption was correct, and most of the silver present is tied up in the $Ag(NH_3)_2^+$ complex.

**Related Problems: 61, 62**

---

There is a close similarity between the working of Example 11.11 and an acid–base calculation. The first step (the assumption that the reaction goes to completion and is followed by a small amount of back dissociation) is analogous to the procedure for dealing with the addition of a small amount of a strong acid to a solution of weak base. The subsequent calculation of the successive dissociation steps resembles the calculation of polyprotic acid equilibria in Example 10.12. The only difference is that in complex ion equilibria it is conventional to work with formation constants, which are the inverse of the dissociation constants used in acid–base equilibria.

The formation of coordination complexes can have a large effect on the solubility of a compound in water. Silver bromide is only very weakly soluble in water,

$$AgBr(s) \rightleftharpoons Ag^+(aq) + Br^-(aq) \qquad K_{sp} = 7.7 \times 10^{-13}$$

but addition of thiosulfate ion $(S_2O_3^{2-})$ to the solution allows the complex ion $Ag(S_2O_3)_2^{3-}$ to form:

$$AgBr(s) + 2\ S_2O_3^{2-}\ (aq) \rightleftharpoons Ag(S_2O_3)_2^{3-}(aq) + Br^-(aq)$$

This greatly increases the solubility of the silver bromide (Fig. 11.18). The formation of this complex ion is an important step in the development of photographic images; thiosulfate ion is a component of the fixer that brings silver bromide into solution from the unexposed portion of the film.

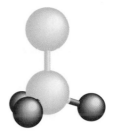

The thiosulfate ion (top) is related to the sulfate ion, $SO_4^{2-}$ (bottom), by the replacement of one oxygen atom with one sulfur atom. It is prepared, however, by the reaction of elemental sulfur with the sulfite ion $(SO_3^{2-})$.

## EXAMPLE 11.12

Calculate the solubility of AgBr in a 1.00 M aqueous solution of ammonia.

### Solution

We tentatively assume that almost all the silver that dissolves is complexed as $Ag(NH_3)_2^+$ (this will be checked later). The overall reaction is then

$$AgBr(s) + 2\ NH_3(aq) \rightleftharpoons Ag(NH_3)_2^+(aq) + Br^-(aq)$$

Note that this is the sum of the two reactions

$$AgBr(s) \rightleftharpoons Ag^+(aq) + Br^-(aq) \qquad K_{sp} = 7.7 \times 10^{-13}$$

$$Ag^+(aq) + 2\ NH_3(aq) \rightleftharpoons Ag(NH_3)_2^+(aq) \qquad K_f = 1.7 \times 10^7$$

and so its equilibrium constant is the product $K_{sp}K_f = 1.3 \times 10^{-5}$.

If $S$ mol $L^{-1}$ of AgBr dissolves, then

$$S = [Br^-] \approx [Ag(NH_3)_2^+]$$

$$[NH_3] = 1.00 = 2S$$

because 2 mol of $NH_3$ is used up for each mole of complex formed. The equilibrium expression is

$$\frac{S^2}{(1.00 - 2S)^2} = K_{sp}K_f = 1.3 \times 10^{-5}$$

$$S = 3.6 \times 10^{-3}\ M = [Ag(NH_3)_2^+] = [Br^-]$$

To check the original assumption, calculate the concentration of free silver ion:

$$[Ag^+] = \frac{K_{sp}}{[Br^-]} = \frac{K_{sp}}{S} = 2.1 \times 10^{-10} \ll [Ag(NH_3)_2^+]$$

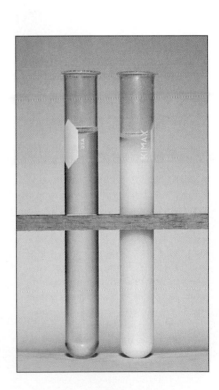

**FIGURE 11.18** An illustration of the effect of complex ion formation on solubility. Each test tube contains 2.0 g AgBr, but the one on the left also contains dissolved thiosulfate ion, which forms a complex ion with $Ag^+$. Almost none of the white solid AgBr has dissolved in pure water, but all of it has dissolved in the solution containing thiosulfate. (*Leon Lewandowski*)

verifying that almost all of the silver is complexed. The solubility is therefore $3.6 \times 10^{-3}$ mol $L^{-1}$, significantly greater than the solubility in pure water:

$$\text{solubility in pure water} = \sqrt{K_{sp}} = 8.8 \times 10^{-7} \text{ mol } L^{-1}$$

**Related Problems: 65, 66**

---

Another interesting effect of complex ions on solubilities is illustrated by the addition of iodide ion to a solution containing mercury(II) ion. After a moderate amount of iodide ion has been added, an orange precipitate forms (Fig. 11.19) through the reaction

$$\text{Hg}^{2+}(aq) + 2\,\text{I}^-(aq) \rightleftharpoons \text{HgI}_2(s)$$

With further addition of iodide ion, however, the orange solid redissolves because complex ions form:

$$\text{HgI}_2(s) + \text{I}^-(aq) \rightleftharpoons \text{HgI}_3^-(aq)$$

$$\text{HgI}_3^-(aq) + \text{I}^-(aq) \rightleftharpoons \text{HgI}_4^{2-}(aq)$$

In the same way, silver chloride will dissolve in a concentrated solution of sodium chloride by forming soluble $\text{AgCl}_2^-$ complex ions. Complex ion formation affects solubility in the opposite direction from the common-ion effect of Section 11.5.

## Acidity and Amphoterism of Complex Ions

When dissolved in water, many metal ions increase the acidity of the solution. The iron(III) ion is an example: each dissolved $\text{Fe}^{3+}$ ion is strongly solvated by six water molecules, leading to a complex ion $\text{Fe}(\text{H}_2\text{O})_6^{3+}$. This complex ion can act as a Brønsted–Lowry acid, donating hydrogen ions to the solvent, water:

$$\underset{\text{Acid}_1}{\text{Fe}(\text{H}_2\text{O})_6^{3+}(aq)} + \underset{\text{Base}_2}{\text{H}_2\text{O}(\ell)} \rightleftharpoons \underset{\text{Acid}_2}{\text{H}_3\text{O}^+(aq)} + \underset{\text{Base}_1}{\text{Fe}(\text{H}_2\text{O})_5\text{OH}^{2+}(aq)}$$

$$\frac{[\text{H}_3\text{O}^+][\text{Fe}(\text{H}_2\text{O})_5\text{OH}^{2+}]}{[\text{Fe}(\text{H}_2\text{O})_6^{3+}]} = K_a = 7.7 \times 10^{-3}$$

Metal ion hydrolysis fits into the general scheme of the Brønsted–Lowry acid–base reaction.

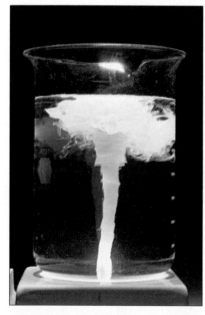

**FIGURE 11.19** The "orange tornado" is a striking demonstration of the effects of complex ions on solubility. A solution is prepared with an excess of $\text{I}^-(aq)$ over $\text{Hg}^{2+}(aq)$ so that the $\text{Hg}^{2+}$ is complexed as $\text{HgI}_3^-$ and $\text{HgI}_4^{2-}$. A magnetic stirrer is used to create a vortex in the solution. Addition of a solution containing $\text{Hg}^{2+}$ down the center of the vortex then causes the orange solid $\text{HgI}_2$ to form in a layer at the edges of the vortex, giving the tornado effect. *(Leon Lewandowski)*

### EXAMPLE 11.13

Calculate the pH of a solution that is 0.100 M in $\text{Fe}(\text{NO}_3)_3$.

### Solution

The iron(III) is present as $\text{Fe}(\text{H}_2\text{O})_6^{3+}$, which reacts as a weak acid:

$$\text{Fe}(\text{H}_2\text{O})_6^{3+} + \text{H}_2\text{O}(\ell) \rightleftharpoons \text{Fe}(\text{H}_2\text{O})_5\text{OH}^{2+} + \text{H}_3\text{O}^+(aq)$$

with $K_a$ equal to $7.7 \times 10^{-3}$. If $y$ mol $L^{-1}$ of $[\text{Fe}(\text{H}_2\text{O})_6]^{3+}$ reacts, then (neglecting the ionization of water itself)

$$[\text{H}_3\text{O}^+] = [\text{Fe}(\text{H}_2\text{O})_5\text{OH}^{2+}] = y$$

$$[\text{Fe}(\text{H}_2\text{O})_6^{3+}] = 0.100 - y$$

The equilibrium expression has the form

$$\frac{y^2}{0.100 - y} = 7.7 \times 10^{-3}$$

$$y = 2.4 \times 10^{-2} \text{ M} = [H_3O^+]$$

and so the pH is 1.62. Solutions of iron(III) salts are strongly acidic.

**Related Problems: 69, 70**

Another acceptable way to write the reaction that makes iron(III) solutions acidic is

$$Fe^{3+}(aq) + 2 H_2O(\ell) \rightleftharpoons H_3O^+(aq) + FeOH^{2+}(aq)$$

in which the specific mention of the six waters of hydration is now omitted. The $FeOH^{2+}$ complex ions are brown, but $Fe^{3+}$ ions are almost colorless. This reaction occurs to a sufficient extent to make a solution of $Fe(NO_3)_3$ in water pale brown. When strong acid is added, the equilibrium is driven back to the left, and the color fades. Table 11.6 gives values of the pH for 0.1 M solutions of several metal ions. Those that form strong complexes with hydroxide ion have low pH (Fig. 11.20), whereas those that do not form such complexes give neutral solutions (pH 7).

Different cations behave differently as water ligands are replaced by hydroxide ions in an increasingly basic solution. A particularly interesting example is $Zn^{2+}$. It forms a series of hydroxo complex ions:

$$Zn^{2+}(aq) + OH^-(aq) \rightleftharpoons ZnOH^+(aq)$$

$$ZnOH^+(aq) + OH^-(aq) \rightleftharpoons Zn(OH)_2(s)$$

$$Zn(OH)_2(s) + OH^-(aq) \rightleftharpoons Zn(OH)_3^-(aq)$$

$$Zn(OH)_3^-(aq) + OH^-(aq) \rightleftharpoons Zn(OH)_4^{2-}(aq)$$

**TABLE 11.6**

*pH of 0.1 M Aqueous Metal Nitrate Solutions at 25°C*

| Metal Nitrate | pH |
|---|---|
| $Fe(NO_3)_3$ | 1.6 |
| $Pb(NO_3)_2$ | 3.6 |
| $Cu(NO_3)_2$ | 4.0 |
| $Zn(NO_3)_2$ | 5.3 |
| $Ca(NO_3)_2$ | 6.7 |
| $NaNO_3$ | 7.0 |

**FIGURE 11.20** The acidity of metal ions is illustrated by the vigorous reaction of anhydrous $AlCl_3$ with water to generate hydrated aluminum oxides, $HCl(aq)$, and heat. The HCl turns the indicator to its red acid form. *(Leon Lewandowski)*

**FIGURE 11.21** Zinc hydroxide is insoluble in water (center) but dissolves readily in acid (left) and base (right). The indicator used is bromocresol red, which turns from red to yellow in acidic solution. *(Leon Lewandowski)*

In the Brønsted–Lowry theory, a polyprotic acid, $Zn(H_2O)_4^{2+}(aq)$, donates hydrogen ions in succession to make all the product ions. The second product, $Zn(OH)_2$, is amphoteric; it can react as either acid or base. It is only slightly soluble in pure water (its $K_{sp}$ is only $1.9 \times 10^{-17}$). If enough acid is added to solid $Zn(OH)_2$, the $OH^-$ ligands are removed, forming the soluble $Zn^{2+}$ ion; if enough base is added, $OH^-$ ligands attach to form the soluble $Zn(OH)_4^{2-}$ (zincate) ion. Thus, $Zn(OH)_2$ is soluble in strongly acidic *or* strongly basic solutions, but is only slightly soluble at intermediate pH values (Fig. 11.21). This amphoterism can be used to separate $Zn^{2+}$ from other cations that do not share the property. For example, $Mg^{2+}$ adds a maximum of two $OH^-$ ions to form $Mg(OH)_2$, a sparingly soluble hydroxide. Further addition of $OH^-$ does not lead to the formation of new complex ions. If a mixture of $Mg^{2+}$ and $Zn^{2+}$ ions is made sufficiently basic, the $Mg^{2+}$ precipitates as $Mg(OH)_2$, but the zinc remains in solution as $Zn(OH)_4^{2-}$, allowing the two to be separated. In the same way, aluminum is separated from iron industrially by dissolving solid $Al(OH)_3$ in strong base as $Al(OH)_4^-(aq)$ while $Fe(OH)_3$ remains as a precipitate.

## CUMULATIVE EXERCISE

### Carbonate Minerals

The carbonates are among the most abundant and important minerals in the earth's crust. When these minerals come into contact with fresh water or seawater, solubility equilibria are established that greatly affect the chemistry of the natural waters. Calcium carbonate ($CaCO_3$), the most important natural carbonate, makes up limestone and other forms of rock such as marble. Other carbonate minerals include dolomite, $CaMg(CO_3)_2$, and magnesite, $MgCO_3$. These compounds are sufficiently soluble that their solutions are nonideal, and so calculations based on solubility product expressions are only approximate.

(a) The rare mineral nesquehonite contains $MgCO_3$ together with water of hydration. A sample containing 21.7 g of nesquehonite is acidified and heated, and the volume of $CO_2(g)$ produced is measured to be 3.51 L at 0°C and $P = 1$ atm. Assuming all the carbonate has reacted to form $CO_2$, give the chemical formula for nesquehonite.

(b) Write a chemical equation and a solubility product expression for the dissolution of dolomite in water.

Carbonate minerals. Calcite (left) and aragonite (middle) are both $CaCO_3$, and smithsonite (right) is $ZnCO_3$. *(left, copyright Mark A. Schneider/Dembinsky Photo Associates; middle, copyright Paul Silverman/Fundamental Photographs; right, copyright Tom McHugh/Photo Researchers, Inc.)*

(c) In a sufficiently basic solution, the carbonate ion does not react significantly with water to form hydrogen carbonate ion. Calculate the solubility (in grams per liter) of limestone (calcium carbonate) in a 0.10 M solution of sodium hydroxide. Use the $K_{sp}$ from Table 11.3.

(d) In a strongly basic 0.10 M solution of $Na_2CO_3$, the concentration of $CO_3^{2-}$ is 0.10 M. What is the gram solubility of limestone in this solution? Compare your answer with that for part (c).

(e) In a mountain lake having a pH of 8.1, the total concentration of carbonate species, $[CO_3^{2-}] + [HCO_3^-]$, is measured to be $9.6 \times 10^{-4}$ M, whereas the concentration of $Ca^{2+}$ is $3.8 \times 10^{-4}$ M. Calculate the concentration of $CO_3^{2-}$ in this lake, using $K_a = 4.8 \times 10^{-11}$ for the acid ionization of $HCO_3^-$ to $CO_3^{2-}$. Is the lake unsaturated, saturated, or supersaturated with respect to $CaCO_3$?

(f) Will acid rainfall into the lake increase or decrease the solubility of limestone rocks in the lake's bed?

(g) Seawater contains a high concentration of $Cl^-$ ions, which form weak complexes such as $CaCl^+$ with calcium ions. Does the presence of such complexes increase or decrease the equilibrium solubility of $CaCO_3$ in seawater?

**Answers**

(a) $MgCO_3 \cdot 3H_2O$

(b) $CaMg(CO_3)_2(s) \rightleftharpoons Ca^{2+}(aq) + Mg^{2+}(aq) + 2\ CO_3^{2-}(aq)$
$[Ca^{2+}][Mg^{2+}][CO_3^{2-}]^2 = K_{sp}$

(c) $9.3 \times 10^{-3}$ g L$^{-1}$

(d) $8.7 \times 10^{-6}$ g L$^{-1}$, smaller than in part (c) because of the common-ion effect

(e) $5.8 \times 10^{-6}$ M. $Q = 2.2 \times 10^{-9} < K_{sp} = 8.7 \times 10^{-9}$, and so the lake is slightly less than saturated.

(f) Increase

(g) Increase

# Concepts & Skills

*After studying this chapter and working the problems that follow, you should be able to*

1. Formulate and use the mass action law expression for heterogeneous reactions (Section 11.1, problems 1–14).
2. Use the law of mass action to explain the distribution of a solute between two immiscible solvents (Section 11.2, problems 15–18).
3. Outline the basis for the separation of compounds by partition chromatography (Section 11.2).
4. Discuss the dynamical processes that lead to solubility equilibria (Section 11.3).
5. Relate the solubilities of sparingly soluble salts in water to their solubility product constants (Section 11.4, problems 25–34).
6. Use the reaction quotient to predict whether a precipitate will form when two solutions are mixed, and then calculate the equilibrium concentrations that will result (Section 11.5, problems 35–42).
7. Calculate the solubility of a sparingly soluble salt in a solution that contains a given concentration of a common ion (Section 11.5, problems 43–48).
8. Determine the dependence on pH of the solubility of a salt of a weak base (Section 11.6, problems 51–52).

9. Specify the optimum conditions for the separation of two elements on the basis of the differing solubilities of their ionic compounds (Section 11.7, problems 53–60).
10. Calculate the concentrations of molecular and ionic species in equilibrium with complex ions (Section 11.8, problems 61–64).
11. Determine the effect of complex ion formation on the solubility of sparingly soluble salts involving a common cation (Section 11.8, problems 65–66).
12. Calculate the pH of aqueous solutions containing metal cations (Section 11.8, problems 67–72).

---

## PROBLEMS

*Answers to problems whose numbers are boldface appear in Appendix G. Problems that are more challenging are indicated with asterisks.*

### Mass Action Law for Heterogeneous Reactions: The Concept of Activity

1. Using the Law of Mass Action, write the equilibrium expression for each of the following reactions.
   (a) $8 H_2(g) + S_8(s) \rightleftharpoons 8 H_2S(g)$
   (b) $C(s) + H_2O(\ell) + Cl_2(g) \rightleftharpoons COCl_2(g) + H_2(g)$
   (c) $CaCO_3(s) \rightleftharpoons CaO(s) + CO_2(g)$
   (d) $3 C_2H_2(g) \rightleftharpoons C_6H_6(\ell)$
2. Using the Law of Mass Action, write the equilibrium expression for each of the following reactions.
   (a) $3 C_2H_2(g) + 3 H_2(g) \rightleftharpoons C_6H_{12}(\ell)$
   (b) $CO_2(g) + C(s) \rightleftharpoons 2 CO(g)$
   (c) $CF_4(g) + 2 H_2O(\ell) \rightleftharpoons CO_2(g) + 4 HF(g)$
   (d) $K_2NiF_6(s) + TiF_4(s) \rightleftharpoons K_2TiF_6(s) + NiF_2(s) + F_2(g)$
3. Using the Law of Mass Action, write the equilibrium expression for each of the following reactions.
   (a) $Zn(s) + 2 Ag^+(aq) \rightleftharpoons Zn^{2+}(aq) + 2 Ag(s)$
   (b) $VO_4^{3-}(aq) + H_2O(\ell) \rightleftharpoons VO_3(OH)^{2-}(aq) + OH^-(aq)$
   (c) $2 As(OH)_6^{3-}(aq) + 6 CO_2(g) \rightleftharpoons$
   $\qquad As_2O_3(s) + 6 HCO_3^-(aq) + 3 H_2O(\ell)$
4. Using the Law of Mass Action, write the equilibrium expression for each of the following reactions.
   (a) $6 I^-(aq) + 2 MnO_4^-(aq) + 4 H_2O(\ell) \rightleftharpoons$
   $\qquad 3 I_2(aq) + 2 MnO_2(s) + 8 OH^-(aq)$
   (b) $2 Cu^{2+}(aq) + 4 I^-(aq) \rightleftharpoons 2 CuI(s) + I_2(aq)$
   (c) $\frac{1}{2} O_2(g) + Sn^{2+}(aq) + 3 H_2O(\ell) \rightleftharpoons$
   $\qquad SnO_2(s) + 2 H_3O^+(aq)$
5. $N_2O_4$ is soluble in the solvent cyclohexane; however, dissolution does not prevent $N_2O_4$ from breaking down to give $NO_2$ according to the equation

   $$N_2O_{4(cyclohexane)} \rightleftharpoons 2 NO_{2(cyclohexane)}$$

   An effort to compare this solution equilibrium to the similar equilibrium in the gas gave the following actual experimental data at 20°C:

| $[N_2O_4]$ (mol $L^{-1}$) | $[NO_2]$ (mol $L^{-1}$) |
|---|---|
| $0.190 \times 10^{-3}$ | $2.80 \times 10^{-3}$ |
| $0.686 \times 10^{-3}$ | $5.20 \times 10^{-3}$ |
| $1.54 \times 10^{-3}$ | $7.26 \times 10^{-3}$ |
| $2.55 \times 10^{-3}$ | $10.4 \times 10^{-3}$ |
| $3.75 \times 10^{-3}$ | $11.7 \times 10^{-3}$ |
| $7.86 \times 10^{-3}$ | $17.3 \times 10^{-3}$ |
| $11.9 \times 10^{-3}$ | $21.0 \times 10^{-3}$ |

   (a) Graph the *square* of the concentration of $NO_2$ versus the concentration of $N_2O_4$.
   (b) Compute the average equilibrium constant of this reaction.
6. $NO_2$ is soluble in carbon tetrachloride ($CCl_4$). As it dissolves, it dimerizes to give $N_2O_4$ according to the equation

   $$2 NO_{2(CCl_4)} \rightleftharpoons N_2O_{4(CCl_4)}$$

   A study of this equilibrium gave the following experimental data at 20°C:

| $[N_2O_4]$ (mol $L^{-1}$) | $[NO_2]$ (mol $L^{-1}$) |
|---|---|
| $0.192 \times 10^{-3}$ | $2.68 \times 10^{-3}$ |
| $0.721 \times 10^{-3}$ | $4.96 \times 10^{-3}$ |
| $1.61 \times 10^{-3}$ | $7.39 \times 10^{-3}$ |
| $2.67 \times 10^{-3}$ | $10.2 \times 10^{-3}$ |
| $3.95 \times 10^{-3}$ | $11.0 \times 10^{-3}$ |
| $7.90 \times 10^{-3}$ | $16.6 \times 10^{-3}$ |
| $11.9 \times 10^{-3}$ | $21.4 \times 10^{-3}$ |

   (a) Graph the concentration of $N_2O_4$ versus the *square* of the concentration of $NO_2$.
   (b) Compute the average equilibrium constant of this reaction.
7. At 298 K, the equilibrium constant for the reaction

   $$Fe_2O_3(s) + 3 H_2(g) \rightleftharpoons 2 Fe(s) + 3 H_2O(\ell)$$

   is $4.0 \times 10^{-6}$, and that for

   $$CO_2(g) + H_2(g) \rightleftharpoons CO(g) + H_2O(\ell)$$

is $3.2 \times 10^{-4}$. Suppose some solid $Fe_2O_3$, solid Fe, and liquid $H_2O$ are brought into equilibrium with $CO(g)$ and $CO_2(g)$ in a closed container at 298 K. Calculate the ratio of the partial pressure of $CO(g)$ to that of $CO_2(g)$ at equilibrium.

8. A sample of ammonium carbamate placed in a glass vessel at 25°C undergoes the reaction

$$NH_4OCONH_2(s) \rightleftharpoons 2\ NH_3(g) + CO_2(g)$$

The total pressure of gases in equilibrium with the solid is found to be 0.115 atm.
(a) Calculate the partial pressures of $NH_3$ and $CO_2$.
(b) Calculate the equilibrium constant at 25°C.

9. The equilibrium constant for the reaction

$$NH_3(g) + HCl(g) \rightleftharpoons NH_4Cl(s)$$

at 340°C is $K = 4.0$.
(a) If the partial pressure of ammonia is $P_{NH_3} = 0.80$ atm and solid ammonium chloride is present, what is the equilibrium partial pressure of hydrogen chloride at 340°C?
(b) An excess of solid $NH_4Cl$ is added to a container filled with ammonia at 340°C and a pressure of 1.50 atm. Calculate the pressures of $NH_3(g)$ and $HCl(g)$ reached at equilibrium.

10. The equilibrium constant for the reaction

$$H_2(g) + I_2(s) \rightleftharpoons 2\ HI(g)$$

at 25°C is $K = 0.345$.
(a) If the partial pressure of hydrogen is $P_{H_2} = 1.00$ atm and solid iodine is present, what is the equilibrium partial pressure of hydrogen iodide, $P_{HI}$, at 25°C?
(b) An excess of solid $I_2$ is added to a container filled with hydrogen at 25°C and a pressure of 4.00 atm. Calculate the pressures of $H_2(g)$ and $HI(g)$ reached at equilibrium.

11. The equilibrium constant for the "water gas" reaction

$$C(s) + H_2O(g) \rightleftharpoons CO(g) + H_2(g)$$

is $K = 2.6$ at a temperature of 1000 K. Calculate the reaction quotient $Q$ for each of the following conditions, and state which direction the reaction shifts in coming to equilibrium.
(a) $P_{H_2O} = 0.600$ atm; $P_{CO} = 1.525$ atm; $P_{H_2} = 0.805$ atm
(b) $P_{H_2O} = 0.724$ atm; $P_{CO} = 1.714$ atm; $P_{H_2} = 1.383$ atm

12. The equilibrium constant for the reaction

$$H_2S(g) + I_2(g) \rightleftharpoons 2\ HI(g) + S(s)$$

at 110°C is equal to 0.0023. Calculate the reaction quotient $Q$ for each of the following conditions and determine whether solid sulfur is consumed or produced as the reaction comes to equilibrium.
(a) $P_{I_2} = 0.461$ atm; $P_{H_2S} = 0.05$ atm; $P_{HI} = 0.0$ atm
(b) $P_{I_2} = 0.461$ atm; $P_{H_2S} = 0.05$ atm; $P_{HI} = 9.0$ atm

13. Pure solid $NH_4HSe$ is placed in an evacuated container at 24.8°C. Eventually, the pressure above the solid reaches the equilibrium pressure 0.0184 atm due to the reaction

$$NH_4HSe(s) \rightleftharpoons NH_3(g) + H_2Se(g)$$

(a) Calculate the equilibrium constant of this reaction at 24.8°C.
(b) In a different container, the partial pressure of $NH_3(g)$ in equilibrium with $NH_4HSe(s)$ at 24.8°C is 0.0252 atm. What is the partial pressure of $H_2Se(g)$?

14. The total pressure of the gases in equilibrium with solid sodium hydrogen carbonate at 110°C is 1.648 atm, corresponding to the reaction

$$2\ NaHCO_3(s) \rightleftharpoons Na_2CO_3(s) + H_2O(g) + CO_2(g)$$

($NaHCO_3$ is used in dry chemical fire extinguishers because the products of this decomposition reaction smother the fire.)
(a) Calculate the equilibrium constant at 110°C.
(b) What is the partial pressure of water vapor in equilibrium with $NaHCO_3(s)$ at 110°C if the partial pressure of $CO_2(g)$ is 0.800 atm?

## Distribution of a Single Species between Immiscible Phases: Extraction and Separation Processes

15. An aqueous solution, initially $1.00 \times 10^{-2}$ M in iodine ($I_2$), is shaken with an equal volume of an immiscible organic solvent, $CCl_4$. The iodine distributes itself between the aqueous and $CCl_4$ layers, and when equilibrium is reached at 27°C, the concentration of $I_2$ in the aqueous layer is $1.30 \times 10^{-4}$ M. Calculate the partition coefficient $K$ at 27°C for the reaction

$$I_2(aq) \rightleftharpoons I_2(CCl_4)$$

16. An aqueous solution, initially $2.50 \times 10^{-2}$ M in iodine ($I_2$), is shaken with an equal volume of an immiscible organic solvent, $CS_2$. The iodine distributes itself between the aqueous and $CS_2$ layers, and when equilibrium is reached at 25°C, the concentration of $I_2$ in the aqueous layer is $4.16 \times 10^{-5}$ M. Calculate the partition coefficient $K$ at 25°C for the reaction

$$I_2(aq) \rightleftharpoons I_2(CS_2)$$

17. Benzoic acid ($C_6H_5COOH$) dissolves in water to the extent of 2.00 g $L^{-1}$ at 15°C and in diethyl ether to the extent of $6.6 \times 10^2$ g $L^{-1}$ at the same temperature.
(a) Calculate the equilibrium constants at 15°C for the two reactions

$$C_6H_5COOH(s) \rightleftharpoons C_6H_5COOH(aq)$$

and

$$C_6H_5COOH(s) \rightleftharpoons C_6H_5COOH(ether)$$

(b) From your answers to part (a), calculate the partition coefficient $K$ for the reaction

$$C_6H_5COOH(aq) \rightleftharpoons C_6H_5COOH(ether)$$

18. Citric acid ($C_6H_8O_7$) dissolves in water to the extent of 1300 g $L^{-1}$ at 15°C and in diethyl ether to the extent of 22 g $L^{-1}$ at the same temperature.

(a) Calculate the equilibrium constants at 15°C for the two reactions

$$C_6H_8O_7(s) \rightleftharpoons C_6H_8O_7(aq)$$

and

$$C_6H_8O_7(s) \rightleftharpoons C_6H_8O_7(ether)$$

(b) From your answers to part (a), calculate the partition coefficient $K$ for the reaction

$$C_6H_8O_7(aq) \rightleftharpoons C_6H_8O_7(ether)$$

## The Nature of Solubility Equilibria

19. Gypsum has the formula $CaSO_4 \cdot 2H_2O$. Plaster of paris has the chemical formula $CaSO_4 \cdot \frac{1}{2}H_2O$. In making wall plaster, water is added to plaster of paris, and the mixture then hardens into solid gypsum. How much water (in liters, at a density of $1.00$ kg $L^{-1}$) should be added to 25.0 kg of plaster of paris to turn it into gypsum, assuming no loss from evaporation?

20. A 1.00-g sample of magnesium sulfate is dissolved in water, and the water is then evaporated away until the residue is bone dry. If the temperature of the water is kept between 48°C and 69°C, the solid that remains weighs 1.898 g. If the experiment is repeated with the temperature held between 69°C and 100°C, however, the solid has a mass of 1.150 g. Determine how many waters of crystallization per $MgSO_4$ there are in each of these two solids.

21. The following graph shows the solubility of KBr in water in units of grams of KBr per 100 g of $H_2O$. If 80 g of KBr is added to 100 g of water at 10°C and the mixture is heated slowly, at what temperature will the last KBr dissolve?

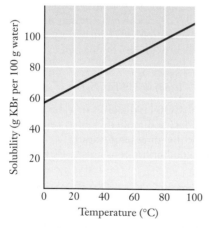

22. Figure 11.7 shows the solubility of $AgNO_3$ in water in units of moles of $AgNO_3$ per kilogram of $H_2O$. If 255 g of $AgNO_3$ is added to 100 g of water at 95°C and cooled slowly, at what temperature will the solution become saturated?

## Ionic Equilibria between Solids and Solutions

23. Iron(III) sulfate, $Fe_2(SO_4)_3$, is a yellow compound that is used as a coagulant in water treatment. Write a balanced chemical equation and a solubility product expression for its dissolution in water.

24. Lead antimonate, $Pb_3(SbO_4)_2$, is used as an orange pigment in oil-base paints and in glazes. Write a balanced chemical equation and a solubility product expression for its dissolution in water.

25. Thallium(I) iodate ($TlIO_3$) is only slightly soluble in water. Its $K_{sp}$ at 25°C is $3.07 \times 10^{-6}$. Estimate the solubility of thallium(I) iodate in water in units of grams per 100.0 mL of water.

26. Thallium thiocyanate (TlSCN) is only slightly soluble in water. Its $K_{sp}$ at 25°C is $1.82 \times 10^{-4}$. Estimate the solubility of thallium thiocyanate in units of grams per 100.0 mL of water.

27. Potassium perchlorate, $KClO_4$, has a $K_{sp}$ at 25°C of $1.07 \times 10^{-2}$. Compute its solubility in grams per liter of solution.

28. Ammonium hexachloroplatinate(IV), $(NH_4)_2(PtCl_6)$, is one of the few sparingly soluble ammonium salts. Its $K_{sp}$ at 20°C is $5.6 \times 10^{-6}$. Compute its solubility in grams per liter of solution.

29. The solubility product constant of mercury(I) iodide is $1.2 \times 10^{-28}$ at 25°C. Estimate the concentration of $Hg_2^{2+}$ and $I^-$ in equilibrium with solid $Hg_2I_2$.

30. The solubility product constant of $Hg_2Cl_2$ is $2 \times 10^{-18}$ at 25°C. Estimate the concentration of $Hg_2^{2+}$ and $Cl^-$ in equilibrium with solid $Hg_2Cl_2$ at 25°C.

31. The solubility of silver chromate ($Ag_2CrO_4$) in 500 mL of water at 25°C is 0.0129 g. Calculate its solubility product constant.

32. At 25°C, 400 mL of water can dissolve 0.00896 g of lead iodate, $Pb(IO_3)_2$. Calculate $K_{sp}$ for lead iodate.

33. At 100°C, water dissolves $1.8 \times 10^{-2}$ g of AgCl per liter. Compute the $K_{sp}$ of AgCl at this temperature.

34. A mass of 0.017 g of silver dichromate ($Ag_2Cr_2O_7$) will dissolve in 300 mL of water at 25°C. Calculate the solubility product constant $K_{sp}$ of silver dichromate.

## Precipitation and the Solubility Product

35. A solution of barium chromate ($BaCrO_4$) is prepared by dissolving $6.3 \times 10^{-3}$ g of this yellow solid in 1.00 L of hot water. Will solid barium chromate precipitate upon cooling to 25°C, according to the solubility product expression? Explain.

36. A solution is prepared by dissolving 0.090 g of $PbI_2$ in 1.00 L of hot water and cooling the solution to 25°C. Will solid precipitate result from this process, according to the solubility product expression? Explain.

37. A solution is prepared by mixing 250.0 mL of $2.0 \times 10^{-3}$ M $Ce(NO_3)_3$ and 150.0 mL of $10 \times 10^{-2}$ M $KIO_3$ at 25°C. Determine whether $Ce(IO_3)_3(s)$ ($K_{sp} = 1.9 \times 10^{-10}$) tends to precipitate from this mixture.

38. Suppose 100.0 mL of a 0.0010 M $CaCl_2$ solution is added to 50.0 mL of a $6.0 \times 10^{-5}$ M NaF solution at 25°C. Determine whether $CaF_2(s)$ ($K_{sp} = 3.9 \times 10^{-11}$) tends to precipitate from this mixture.

39. Suppose 50.0 mL of a 0.0500 M solution of $Pb(NO_3)_2$ is mixed with 40.0 mL of a 0.200 M solution of $NaIO_3$ at 25°C. Calculate the $[Pb^{2+}]$ and $[IO_3^-]$ when the mixture comes to equilibrium. At this temperature, $K_{sp}$ for $Pb(IO_3)_2$ is $2.6 \times 10^{-13}$.

40. Silver iodide (AgI) is used in place of silver chloride for the fastest photographic film because it is more sensitive to light and so can form an image in a very short exposure time. A silver iodide emulsion is prepared by adding 6.60 L of 0.10 M NaI solution to 1.50 L of 0.080 M $AgNO_3$ solution at 25°C. Calculate the concentration of silver ion remaining in solution when the mixture comes to equilibrium and its chemical amount relative to the amount present initially.

41. When 50.0 mL of 0.100 M $AgNO_3$ and 30.0 mL of 0.0600 M $Na_2CrO_4$ are mixed, a precipitate of silver chromate ($Ag_2CrO_4$) is formed. The solubility product $K_{sp}$ of silver chromate in water at 25°C is $1.9 \times 10^{-12}$. Calculate the $[Ag^+]$ and $[CrO_4^{2-}]$ remaining in solution at equilibrium.

42. When 40.0 mL of 0.0800 M $Sr(NO_3)_2$ and 80.0 mL of 0.0500 M KF are mixed, a precipitate of strontium fluoride ($SrF_2$) is formed. The solubility product $K_{sp}$ of strontium fluoride in water at 25°C is $2.8 \times 10^{-9}$. Calculate the $[Sr^{2+}]$ and $[F^-]$ remaining in solution at equilibrium.

43. Calculate the solubility (in mol $L^{-1}$) of $CaF_2(s)$ at 25°C in a 0.040 M aqueous solution of NaF.

44. Calculate the mass of AgCl that can dissolve in 100 mL of 0.150 M NaCl solution.

45. The solubility product of nickel(II) hydroxide, $Ni(OH)_2$, at 25°C is $K_{sp} = 1.6 \times 10^{-16}$.
    (a) Calculate the molar solubility of $Ni(OH)_2$ in pure water at 25°C.
    (b) Calculate the molar solubility of $Ni(OH)_2$ in 0.100 M NaOH.

46. Silver arsenate ($Ag_3AsO_4$) is a slightly soluble salt having a solubility product of $K_{sp} = 1.0 \times 10^{-22}$ at 25°C for the equilibrium
$$Ag_3AsO_4(s) \rightleftharpoons 3\,Ag^+(aq) + AsO_4^{3-}(aq)$$
    (a) Calculate the molar solubility of silver arsenate in pure water at 25°C.
    (b) Calculate the molar solubility of silver arsenate in 0.10 M $AgNO_3$.

47. A saturated solution of $Mg(OH)_2$ at 25°C is prepared by equilibrating solid $Mg(OH)_2$ with water. Concentrated NaOH is then added until the solubility of $Mg(OH)_2$ is 0.0010 times that in $H_2O$ alone. (Ignore the change in volume resulting from the addition of NaOH.) The solubility product $K_{sp}$ of $Mg(OH)_2$ is $1.2 \times 10^{-11}$ at 25°C. Calculate the concentration of hydroxide ion in the solution after the addition of the NaOH.

48. A saturated solution of $BaF_2$ at 25°C is prepared by equilibrating solid $BaF_2$ with water. Powdered NaF is then dissolved in the solution until the solubility of $BaF_2$ is 1.0% of that in $H_2O$ alone. The solubility product $K_{sp}$ of $BaF_2$ is $1.7 \times 10^{-6}$ at 25°C. Calculate the concentration of fluoride ion in the solution after addition of the powdered NaF.

## The Effects of pH on Solubility

49. Compare the molar solubility of AgOH in pure water with that in a solution buffered at pH 7.00. Note the difference between the two: when AgOH is dissolved in pure water, the pH does not remain at 7.

50. Compare the molar solubility of $Mg(OH)_2$ in pure water with that in a solution buffered at pH 9.00.

51. For each of the following ionic compounds, state whether the solubility will increase, decrease, or remain unchanged as a solution at pH 7 is made acidic,
    (a) $PbI_2$      (b) AgOH      (c) $Ca_3(PO_4)_2$

52. For each of the following ionic compounds, state whether the solubility will increase, decrease, or remain unchanged as a solution at pH 7 is made acidic.
    (a) $SrCO_3$      (b) $Hg_2Br_2$      (c) MnS

## A Deeper Look. . . Selective Precipitation of Ions

53. An aqueous solution at 25°C is 0.10 M in both $Mg^{2+}$ and $Pb^{2+}$ ions. We wish to separate the two kinds of metal ions by taking advantage of the different solubilities of their oxalates, $MgC_2O_4$ and $PbC_2O_4$.
    (a) What is the highest possible oxalate ion concentration that allows only one solid oxalate salt to be present at equilibrium? Which ion is present in the solid—$Mg^{2+}$ or $Pb^{2+}$?
    (b) What fraction of the less soluble ion still remains in solution under the conditions of part (a)?

54. An aqueous solution at 25°C is 0.10 M in $Ba^{2+}$ and 0.50 M in $Ca^{2+}$ ions. We wish to separate the two by taking advantage of the different solubilities of their fluorides, $BaF_2$ and $CaF_2$.
    (a) What is the highest possible fluoride ion concentration that allows only one solid fluoride salt to be present at equilibrium? Which ion is present in the solid—$Ba^{2+}$ or $Ca^{2+}$?
    (b) What fraction of the less soluble ion still remains in solution under the conditions of part (a)?

55. The cations in an aqueous solution that contains 0.100 M $Hg_2(NO_3)_2$ and 0.0500 M $Pb(NO_3)_2$ are to be separated by taking advantage of the difference in the solubilities of their iodides. $K_{sp}(PbI_2) = 1.4 \times 10^{-8}$ and $K_{sp}(Hg_2I_2) = 1.2 \times 10^{-28}$. What should be the concentration of iodide ion for the best separation? In the "best" separation, one of the cations should remain entirely in solution and the other should precipitate as fully as possible.

56. The cations in an aqueous solution that contains 0.150 M $Ba(NO_3)_2$ and 0.0800 M $Ca(NO_3)_2$ are to be separated by taking advantage of the difference in the solubilities of their sulfates. $K_{sp}(BaSO_4) = 1.1 \times 10^{-10}$ and $K_{sp}(CaSO_4) = 2.4 \times 10^{-5}$. What should be the concentration of sulfate ion for the best separation?

57. Calculate the $[Zn^{2+}]$ in a solution that is in equilibrium with ZnS(s) and in which $[H_3O^+] = 1.0 \times 10^{-5}$ M and $[H_2S] = 0.10$ M.

58. Calculate the $[Cd^{2+}]$ in a solution that is in equilibrium with $CdS(s)$ and in which $[H_3O^+] = 1.0 \times 10^{-3}$ M and $[H_2S] = 0.10$ M.

59. What is the highest pH at which 0.10 M $Fe^{2+}$ will remain entirely in a solution that is saturated with $H_2S$ at a concentration of $[H_2S] = 0.10$ M? At this pH, what would be the concentration of $Pb^{2+}$ in equilibrium with solid PbS in this solution?

60. What is the highest pH at which 0.050 M $Mn^{2+}$ will remain entirely in a solution that is saturated with $H_2S$ at a concentration of $[H_2S] = 0.10$ M? At this pH, what would be the concentration of $Cd^{2+}$ in equilibrium with solid CdS in this solution?

## Complex Ions and Solubility

61. Suppose 0.10 mol of $Cu(NO_3)_2$ and 1.50 mol of $NH_3$ are dissolved in water and diluted to a total volume of 1.00 L. Calculate the concentrations of $Cu(NH_3)_4^{2+}$ and of $Cu^{2+}$ at equilibrium.

62. The formation constant of the $TlCl_4^-$ complex ion is $1 \times 10^{18}$. Suppose 0.15 mol of $Tl(NO_3)_3$ is dissolved in 1.00 L of a 0.50 M solution of NaCl. Calculate the concentration at equilibrium of $TlCl_4^-$ and of $Tl^{3+}$.

63. The organic compound "18-crown-6" binds alkali metals in aqueous solution by wrapping around and enfolding the ion. It presents a niche that nicely accommodates the $K^+$ ion but is too small for the $Rb^+$ ion and too large for the $Na^+$ ion. The values of the equilibrium constants show this:

$$Na^+(aq) + \text{18-crown-6}(aq) \rightleftharpoons Na\text{-crown}^+(aq)$$
$$K = 6.6$$

$$K^+(aq) + \text{18-crown-6}(aq) \rightleftharpoons K\text{-crown}^+(aq)$$
$$K = 111.6$$

$$Rb^+(aq) + \text{18-crown-6}(aq) \rightleftharpoons Rb\text{-crown}^+(aq)$$
$$K = 36$$

An aqueous solution is initially 0.0080 M in 18-crown-6($aq$) and also 0.0080 M in $K^+(aq)$. Compute the equilibrium concentration of free $K^+$. ("Free" means not tied up with the 18-crown-6.) Compute the concentration of free $Na^+$ if the solution contains 0.0080 M $Na^+(aq)$ instead of $K^+(aq)$.

64. The organic compound 18-crown-6 (see preceding problem) also binds strongly with the alkali metal ions in methanol.

$$K^+ + \text{18-crown-6} \rightleftharpoons [\text{complex}]^+$$

In methanol solution, the equilibrium constant is $1.41 \times 10^6$. A similar reaction with $Cs^+$ has an equilibrium constant of only $2.75 \times 10^4$. A solution is made (in methanol) containing 0.020 mol $L^{-1}$ each of $K^+$ and $Cs^+$. It also contains 0.30 mol $L^{-1}$ of 18-crown-6. Compute the equilibrium concentrations of both the uncomplexed $K^+$ and the uncomplexed $Cs^+$.

65. Will silver chloride dissolve to a significantly greater extent in a 1.00 M NaCl solution than in pure water, due to the possible formation of $AgCl_2^-$ ions? Use data from Tables 11.3 and 11.5 to provide a quantitative answer to this question. What will happen in a 0.100 M NaCl solution?

66. Calculate how many grams of silver chloride will dissolve in 1.0 L of a 1.0 M $NH_3$ solution, through formation of the complex ion $Ag(NH_3)_2^+$.

67. The pH of a 0.2 M solution of $CuSO_4$ is 4.0. Write chemical equations to explain why a solution of this salt is neither basic [from the reaction of $SO_4^{2-}(aq)$ with water] nor neutral, but acidic.

68. Will a 0.05 M solution of $FeCl_3$ be acidic, basic, or neutral? Explain your answer by writing chemical equations to describe any reactions taking place.

69. The acid ionization constant for $Co(H_2O)_6^{2+}(aq)$ is $3 \times 10^{-10}$. Calculate the pH of a 0.10 M solution of $Co(NO_3)_2$.

70. The acid ionization constant for $Fe(H_2O)_6^{2+}(aq)$ is $3 \times 10^{-6}$. Calculate the pH of a 0.10 M solution of $Fe(NO_3)_2$, and compare it with the pH of the corresponding iron(III) nitrate solution from Example 11.13.

71. A 0.15 M aqueous solution of the chloride salt of the complex ion $Pt(NH_3)_4^{2+}$ is found to be weakly acidic with a pH of 4.92. This is initially puzzling because the $Cl^-$ ion in water is not acidic and $NH_3$ in water is *basic*, not acidic. Finally, it is suggested that the $Pt(NH_3)_4^{2+}$ ion as a group donates hydrogen ions. Compute the $K_a$ of this acid, assuming that just one hydrogen ion is donated.

72. The pH of a 0.10 M solution of $Ni(NO_3)_2$ is 5.0. Calculate the acid ionization constant of $Ni(H_2O)_6^{2+}(aq)$.

*73. $K_{sp}$ for $Pb(OH)_2$ is $4.2 \times 10^{-15}$, and $K_f$ for $Pb(OH)_3^-$ is $4 \times 10^{14}$. Suppose a solution whose initial concentration of $Pb^{2+}(aq)$ is 1.00 M is brought to pH 13.0 by addition of solid NaOH. Will solid $Pb(OH)_2$ precipitate, or will the lead be dissolved as $Pb(OH)_3^-(aq)$? What will be $[Pb^{2+}]$ and $[Pb(OH)_3^-]$ at equilibrium? Repeat the calculation for an initial $Pb^{2+}$ concentration of 0.050 M. (*Hint:* One way to solve this problem is to *assume* that $Pb(OH)_2(s)$ is present and calculate $[Pb^{2+}]$ and $[Pb(OH)_3^-]$ that would be in equilibrium with the solid. If the sum of these is less than the original $[Pb^{2+}]$, the remainder can be assumed to have precipitated. If not, there is a contradiction and we must assume that *no* $Pb(OH)_2(s)$ is present. In this case we can calculate $[Pb^{2+}]$ and $[Pb(OH)_3^-]$ directly from $K_f$.)

*74. $K_{sp}$ for $Zn(OH)_2$ is $4.5 \times 10^{-17}$, and $K_f$ for $Zn(OH)_4^{2-}$ is $5 \times 10^{14}$. Suppose a solution whose initial concentration of $Zn^{2+}(aq)$ is 0.010 M is brought to pH 14.0 by addition of solid NaOH. Will solid $Zn(OH)_2$ precipitate, or will the zinc be dissolved as $Zn(OH)_4^{2-}(aq)$? What will be $[Zn^{2+}]$ and $[Zn(OH)_4^{2-}]$ at equilibrium? Repeat the calculation at pH 13 for an initial $Zn^{2+}$ concentration of 0.10 M. See the hint in problem 73.

## Additional Problems

*75. At 298 K, chlorine is only slightly soluble in water. Thus, under a pressure of 1.00 atm of $Cl_2(g)$, 1.00 L of water at equilibrium dissolves just 0.091 mol of $Cl_2$.

$$Cl_2(g) \rightleftharpoons Cl_2(aq)$$

In such solutions, the $Cl_2(aq)$ concentration is 0.061 M and the concentrations of $Cl^-(aq)$ and $HOCl(aq)$ are both 0.030 M. These two additional species are formed by the equilibrium

$$Cl_2(aq) + H_2O(\ell) \rightleftharpoons H^+(aq) + Cl^-(aq) + HOCl(aq)$$

There are no other Cl-containing species. Compute the equilibrium constants $K_1$ and $K_2$ for the two reactions.

76. At 400°C, the reaction

$$BaO_2(s) + 4\ HCl(g) \rightleftharpoons BaCl_2(s) + 2\ H_2O(g) + Cl_2(g)$$

has an equilibrium constant equal to $K_1$. How is $K_1$ related to the equilibrium constant $K_2$ of the reaction

$$2\ Cl_2(g) + 4\ H_2O(g) + 2\ BaCl_2(s) \rightleftharpoons$$
$$8\ HCl(g) + 2\ BaO_2(s)$$

at 400°C?

77. Ammonium hydrogen sulfide, a solid, decomposes to give $NH_3(g)$ and $H_2S(g)$. At 25°C, some $NH_4HS(s)$ is placed in an evacuated container. A portion of it decomposes, and the total pressure at equilibrium is 0.659 atm. Extra $NH_3(g)$ is then injected into the container and, when equilibrium is reestablished, the partial pressure of $NH_3(g)$ is 0.750 atm.
    (a) Compute the equilibrium constant for the decomposition of ammonium hydrogen sulfide.
    (b) Determine the final partial pressure of $H_2S(g)$ in the container.

78. The equilibrium constant for the reaction

$$KOH(s) + CO_2(g) \rightleftharpoons KHCO_3(s)$$

is $6 \times 10^{15}$ at 25°C. Suppose 7.32 g of KOH and 9.41 g of $KHCO_3$ are placed in a closed evacuated container and allowed to reach equilibrium. Calculate the pressure of $CO_2(g)$ at equilibrium.

79. The equilibrium constant for the reduction of nickel(II) oxide to nickel at 754°C is 255.4, corresponding to the reaction

$$NiO(s) + CO(g) \rightleftharpoons Ni(s) + CO_2(g)$$

If the total pressure of the system at 754°C is 2.50 atm, calculate the partial pressures of $CO(g)$ and $CO_2(g)$.

80. Both glucose (corn sugar) and fructose (fruit sugar) taste sweet, but fructose tastes sweeter. Each year in the United States, tons of corn syrup destined to sweeten food are treated to convert glucose as fully as possible to the sweeter fructose. The reaction is an equilibrium:

$$glucose \rightleftharpoons fructose$$

   (a) A 0.2564 M solution of pure glucose is treated at 25°C with an enzyme (catalyst) that causes the preceding equilibrium to be reached quickly. The final concentration of fructose is 0.1175 M. In another experiment at the same temperature, a 0.2666 M solution of pure fructose is treated with the same enzyme, and the final concentration of glucose is 0.1415 M. Compute an average equilibrium constant for the preceding reaction.

   (b) At equilibrium under these conditions, what percentage of glucose is converted to fructose?

81. Polychlorinated biphenyls (PCBs) are a major environmental problem. These oily substances have many uses, but they resist breakdown by bacterial action when spilled in the environment and, being fat-soluble, can accumulate to dangerous concentrations in the fatty tissues of fish and animals. One little-appreciated complication in controlling the problem is that there are *209* different PCBs, all now in the environment. They are generally similar, but their solubilities in fats differ considerably. The best measure of this is $K_{ow}$, the equilibrium constant for the partition of a PCB between the fat-like solvent octanol and water.

$$PCB(aq) \rightleftharpoons PCB(octanol)$$

An equimolar mixture of PCB-2 and PCB-11 in water is treated with an equal volume of octanol. Determine the ratio between the amounts of PCB-2 and PCB-11 in the water at equilibrium. At room temperature, $K_{ow}$ is $3.98 \times 10^4$ for PCB-2 and $1.26 \times 10^5$ for PCB-11.

82. Refer to the data in problems 17 and 18. Suppose 2.00 g of a solid consisting of 50.0% benzoic acid and 50.0% citric acid by mass is added to 100.0 mL of water and 100.0 mL of diethyl ether, and the whole assemblage is shaken. When the immiscible layers are separated and the solvents are removed by evaporation, two solids result. Calculate the percentage (by mass) of the major component in each solid.

83. At 25°C, the partition coefficient for the equilibrium

$$I_2(aq) \rightleftharpoons I_2(CCl_4)$$

has the value $K = 85$. To 0.100 L of an aqueous solution, which is initially $2 \times 10^{-3}$ M in $I_2$, we add 0.025 L of $CCl_4$. The mixture is shaken in a separatory funnel and allowed to separate into two phases, and the $CCl_4$ phase is withdrawn.
   (a) Calculate the fraction of the $I_2$ remaining in the aqueous phase.
   (b) Suppose the remaining aqueous phase is shaken with another 0.025 L of $CCl_4$ and again separated. What fraction of the $I_2$ from the original aqueous solution is *now* in the aqueous phase?
   (c) Compare your answer with that of Example 11.3, in which the same total amount of $CCl_4$ (0.050 L) was used in a *single* extraction. For a given total amount of extracting solvent, which is the more efficient way to remove iodine from water?

84. Barium nitride vaporizes slightly at high temperature as it undergoes the dissociation

$$Ba_3N_2(s) \rightleftharpoons 3\ Ba(g) + N_2(g)$$

At 1000 K, the equilibrium constant is $4.5 \times 10^{-19}$. At 1200 K, the equilibrium constant is $6.2 \times 10^{-12}$.
   (a) Estimate $\Delta H°$ for this reaction.
   (b) The equation is rewritten as

$$2\ Ba_3N_2(s) \rightleftharpoons 6\ Ba(g) + 2\ N_2(g)$$

Now the equilibrium constant is $2.0 \times 10^{-37}$ at 1000 K and $3.8 \times 10^{-23}$ at 1200 K. Estimate $\Delta H°$ of *this* reaction.

85. Although iodine is not very soluble in pure water, it dissolves readily in water that contains $I^-(aq)$ ion, thanks to the reaction

$$I_2(aq) + I^-(aq) \rightleftharpoons I_3^-(aq)$$

The equilibrium constant of this reaction was measured as a function of temperature with these results:

| $t$: | 3.8°C | 15.3°C | 25.0°C | 35.0°C | 50.2°C |
|------|-------|--------|--------|--------|--------|
| $K$: | 1160  | 841    | 689    | 533    | 409    |

(a) Plot $\ln K$ on the $y$ axis as a function of $1/T$, the reciprocal of the absolute temperature.

(b) Estimate the $\Delta H°$ of this reaction.

86. The breaking of the O—O bond in peroxydisulfuryl difluoride ($FO_2SOOSO_2F$) gives $FO_2SO$:

$$(FO_2SO)_2 \rightleftharpoons 2\ FO_2SO$$

The compound on the left of this equation is a colorless liquid that boils at 67.1°C. Its vapor, when heated to about 100°C, turns brown as the product of the reaction forms. Suppose that, in a sample of the vapor, the intensity of the brown color doubles between 100°C and 110°C and that the total pressure increases only by the 2.7% predicted for an ideal gas. Estimate $\Delta H°$ for the preceding reaction.

87. Write a chemical equation for the dissolution of mercury(I) chloride in water, and give its solubility product expression.

*88. Magnesium ammonium phosphate has the formula $MgNH_4PO_4 \cdot 6H_2O$. It is only slightly soluble in water (its $K_{sp}$ is $2.3 \times 10^{-13}$). Write a chemical equation and the corresponding equilibrium law for the dissolution of this compound in water.

89. Soluble barium compounds are poisonous, but barium sulfate is routinely ingested as a suspended solid in a "barium cocktail" to improve the contrast in x-ray images. Calculate the concentration of dissolved barium per liter of water in equilibrium with solid barium sulfate.

90. A saturated aqueous solution of silver perchlorate ($AgClO_4$) contains 84.8% by mass $AgClO_4$, but a saturated solution of $AgClO_4$ in 60% aqueous perchloric acid contains only 5.63% by mass $AgClO_4$. Explain this large difference, using chemical equations.

91. Suppose 140 mL of 0.0010 M $Sr(NO_3)_2$ is mixed with enough 0.0050 M NaF to make 1.00 L of solution. Will $SrF_2(s)$ ($K_{sp} = 2.8 \times 10^{-9}$) precipitate at equilibrium? Explain.

92. The concentration of calcium ion in a town's supply of drinking water is 0.0020 M. (This water is referred to as hard water because it contains such a large concentration of $Ca^{2+}$.) Suppose the water is to be fluoridated by the addition of NaF for the purpose of reducing tooth decay. What is the maximum concentration of fluoride ion that can be achieved in the water before precipitation of $CaF_2$ begins? Will the water supply attain the level of fluoride ion recommended by the U.S. Public Health Service, about $5 \times 10^{-5}$ M (1 mg fluorine per liter)?

93. The two solids CuBr(s) and AgBr(s) are only very slightly soluble in water: $K_{sp}(CuBr) = 4.2 \times 10^{-8}$ and $K_{sp}(AgBr) = 7.7 \times 10^{-13}$. Some CuBr(s) and AgBr(s) are both mixed into a quantity of water that is then stirred until it is saturated with respect to both solutes. Next, a small amount of KBr is added and dissolves completely. Compute the ratio of $[Cu^+]$ to $[Ag^+]$ after the system reestablishes equilibrium.

*94. The two salts $BaCl_2$ and $Ag_2SO_4$ are both far more soluble in water than either $BaSO_4$ ($K_{sp} = 1.1 \times 10^{-10}$) or AgCl ($K_{sp} = 1.6 \times 10^{-10}$) at 25°C. Suppose 50.0 mL of 0.040 M $BaCl_2(aq)$ is added to 50.0 mL of 0.020 M $Ag_2SO_4(aq)$. Calculate the concentrations of $SO_4^{2-}(aq)$, $Cl^-(aq)$, $Ba^{2+}(aq)$, and $Ag^+(aq)$ that remain in solution at equilibrium.

95. The Mohr method is a technique for determining the amount of chloride ion in an unknown sample. It is based on the difference in solubility between silver chloride (AgCl; $K_{sp} = 1.6 \times 10^{-10}$) and silver chromate ($Ag_2CrO_4$; $K_{sp} = 1.9 \times 10^{-12}$). In using this method, one adds a small amount of chromate ion to a solution with unknown chloride concentration. By measuring the volume of $AgNO_3$ added before the appearance of the red silver chromate, one can determine the amount of $Cl^-$ originally present. Suppose we have a solution that is 0.100 M in $Cl^-$ and 0.00250 M in $CrO_4^{2-}$. If we add 0.100 M $AgNO_3$ solution drop by drop, will AgCl or $Ag_2CrO_4$ precipitate first? When $Ag_2CrO_4(s)$ first appears, what fraction of the $Cl^-$ originally present remains in solution?

96. Oxide ion, like sulfide ion, is a strong base. Write an equation for the dissolution of CaO in water and give its equilibrium constant expression. Write the corresponding equation for the dissolution of CaO in an aqueous solution of a strong acid, and relate its equilibrium constant to the previous one.

97. Water that has been saturated with magnesia (MgO) at 25°C has a pH of 10.16. Write a balanced chemical equation for the equilibrium between MgO(s) and the ions it furnishes in aqueous solution, and calculate the equilibrium constant at 25°C. What is the solubility, in moles per liter, of MgO in water?

98. To 1.00 L of a 0.100 M $AgNO_3$ solution is added an excess of sodium chloride. Then 1.00 L of 0.500 M $NH_3(aq)$ is added. Finally, sufficient nitric acid is added until the pH of the resulting solution is 1.0. Write balanced equations for the reactions that take place (if any) at each of the three steps in this process.

99. Only about 0.16 mg of AgBr(s) will dissolve in 1.0 L of water (this volume of solid is smaller than the head of a pin). In a solution of ammonia that contains 0.10 mol ammonia per liter of water, there are about 555 water molecules for every molecule of ammonia. However, more than 400 times as much AgBr (68 mg) will dissolve in this solution as in plain water. Explain how such a tiny change in the composition of the solution can have such a large effect on the solubility of AgBr.

*100. (a) Calculate the solubility of calcium oxalate ($CaC_2O_4$) in 1.0 M oxalic acid ($H_2C_2O_4$) at 25°C, using the two acid ionization constants for oxalic acid from Table 10.2 and the solubility product $K_{sp} = 2.6 \times 10^{-9}$ for $CaC_2O_4$.

(b) Calculate the solubility of calcium oxalate in pure water at 25°C.

(c) Account for the difference between the results of (a) and (b).

*101. When 6 M HCl is added to solid CdS, some of the solid dissolves to give the complex ion $CdCl_4^{2-}(aq)$.

(a) Write a balanced equation for the reaction that occurs.

(b) Use data from Tables 10.2 and 11.4 and the formation constant of $CdCl_4^{2-}$ ($K_f = 8 \times 10^2$) to calculate the equilibrium constant for the reaction of part (a).

(c) What is the molar solubility of CdS per liter of 6 M HCl?

*102. Using data from Table 11.5, calculate the concentrations of $Hg^{2+}(aq)$, $HgCl^{+}(aq)$, and $HgCl_2(aq)$ that result when 1.00 L of a 0.100 M $Hg(NO_3)_2$ solution is mixed with an equal volume of a 0.100 M $HgCl_2$ solution. (*Hint:* Use the analogy with amphoteric equilibria discussed in Section 10.8.)

*103. Calculate the concentration of $Cu^{2+}(aq)$ in a solution that contains 0.020 mol of $CuCl_2$ and 0.100 mol of NaCN in 1.0 L.

$$Cu^{2+}(aq) + 4\ CN^{-}(aq) \rightleftharpoons Cu(CN)_4^{2-}(aq)$$
$$K = 2.0 \times 10^{30}$$

(*Hint:* Do not overlook the reaction of $CN^{-}$ with water to give HCN.)

104. An aqueous solution of $K_2[Pt(OH)_6]$ has a pH greater than 7. Explain this fact by writing an equation showing the $Pt(OH)_6^{2-}$ ion acting as a Brønsted–Lowry base and accepting a hydrogen ion from water.

105. In Example 11.13 we included only the first acid dissociation $K_{a1}$ of a 0.100 M aqueous solution of $Fe(H_2O)_6^{3+}$. Subsequent dissociation can also occur, with $K_{a2} = 2.0 \times 10^{-5}$, to give $Fe(H_2O)_4(OH)_2^{+}$.

(a) Calculate the concentration of $Fe(H_2O)_4(OH)_2^{+}$ at equilibrium. Does the pH change significantly when this second dissociation is taken into account?

(b) We can describe the same reaction as the dissociation of a complex ion $Fe(OH)_2^{+}$ to $Fe^{3+}$ and two $OH^{-}$ ions. Calculate $K_f$, the formation constant for $Fe(OH)_2^{+}$.

---

## CUMULATIVE PROBLEMS

106. At 25°C the equilibrium constant for the reaction

$$CaSO_4(s) + 2\ H_2O(g) \rightleftharpoons CaSO_4\cdot2H_2O(s)$$

is $1.6 \times 10^3$. Over what range of relative humidities do you expect $CaSO_4(s)$ to be converted to $CaSO_4\cdot2H_2O$? (*Note:* The relative humidity is the partial pressure of water vapor divided by its equilibrium vapor pressure and multiplied by 100%. Use Table 5.1.)

107. In an extraction process, a solute species is partitioned between two immiscible solvents. Suppose the two solvents are water and carbon tetrachloride. State which phase will have the higher concentration of each of the following solutes: (a) $CH_3OH$, (b) $C_2Cl_6$, (c) $Br_2$, (d) NaCl. Explain your reasoning.

108. $\Delta G_f^{\circ}$ of $SrSO_4(s)$ is $-1340.9$ kJ mol$^{-1}$. Use this, together with information from Appendix D, to estimate the molar solubility of $SrSO_4$ in water at 25°C.

109. Calculate the equilibrium pressure (in atmospheres) of $O_2(g)$ over a sample of pure NiO(s) in contact with pure Ni(s) at 25°C. The NiO(s) decomposes according to the equation

$$NiO(s) \longrightarrow Ni(s) + \tfrac{1}{2}\ O_2(g)$$

Use data from Appendix D.

110. The sublimation pressure of solid $NbI_5$ is the pressure of gaseous $NbI_5$ present in equilibrium with the solid. It is given by the empirical equation

$$\log P = -6762/T + 8.566$$

The vapor pressure of liquid $NbI_5$, on the other hand, is given by

$$\log P = -4653/T + 5.43$$

In these two equations, $T$ is the absolute temperature in kelvins and $P$ is the pressure in atmospheres.

(a) Determine the enthalpy and entropy of sublimation of $NbI_5(s)$.

(b) Determine the enthalpy and entropy of vaporization of $NbI_5(\ell)$.

(c) Calculate the normal boiling point of $NbI_5(\ell)$.

(d) Calculate the triple-point temperature and pressure of $NbI_5$. (*Hint:* At the triple point of a substance, the liquid and solid are in equilibrium and must have the same vapor pressure. If they did not, vapor would continually escape from the phase with the higher vapor pressure and collect in the phase with the lower vapor pressure.)

111. Snow and ice sublime spontaneously when the partial pressure of water vapor is below the equilibrium vapor pressure of ice. At 0°C, the vapor pressure of ice is 0.0060 atm (the triple-point pressure of water). Taking the enthalpy of sublimation of ice to be 50.0 kJ mol$^{-1}$, calculate the partial pressure of water vapor below which ice will sublime spontaneously at −15°C.

112. The volume of a certain saturated solution is greater than the sum of the volumes of the water and salt from which it is made. Predict the effect of increased pressure on the solubility of this salt.

113. Codeine has the molecular formula $C_{18}H_{21}NO_3$. It is soluble in water to the extent of 1.00 g per 120 mL of water at room temperature and 1.00 g per 60 mL of water at 80°C. Compute the molal solubility (in mol kg$^{-1}$) at both temperatures, taking the density of water to be fixed at 1.00 g cm$^{-3}$. Is the dissolution of codeine in water endothermic or exothermic?

114. Suppose 1.44 L of a saturated solution of strontium carbonate ($SrCO_3$) in boiling water at 100°C is prepared. The solution is then strongly acidified and shaken to drive off all the gaseous $CO_2$ that forms. The volume of this gas (at a temperature of 100°C and a partial pressure of 0.972 atm) is measured to be 0.20 L (200 mL).

    (a) Calculate the molar solubility of $SrCO_3$ in water at 100°C.
    (b) Estimate the solubility product constant $K_{sp}$ of $SrCO_3$ at this temperature.
    (c) Explain why the actual $K_{sp}$ at this temperature may be lower than you predicted in part (b).

115. A buffer is prepared by adding 50.0 mL of 0.15 M $HNO_3(aq)$ to 100.0 mL of 0.12 M $NaHCOO(aq)$ (sodium formate). Calculate the solubility of $CaF_2(s)$ in this solution.

# Electrochemistry

Electrochemistry links the chemistry of oxidation–reduction reactions to the physics of charge flow. It provides an important way of utilizing the free energy available in spontaneous chemical reactions to perform useful work and of using energy to carry out reactions that would otherwise not be possible. Electrochemistry is concerned with such practical problems as the storage of energy in batteries and the efficient conversion of energy from readily available sources (such as solar or chemical energy) to forms that are useful for technological applications. In this chapter we discuss redox reactions in aqueous solution and the way in which

*Illustration*

A photoelectrochemical cell, consisting of a light-sensitive electrode immersed in an electrolyte, generating hydrogen gas from sunlight. *(Courtesy of Nathan S. Lewis, California Institute of Technology)*

electrochemical cells operate and examine the relationship between the current produced and reaction stoichiometry. We then apply thermodynamic principles to the operation of such cells and show how these principles can help to guide the search for new electrochemical technologies.

---

### 12.1

## BALANCING OXIDATION–REDUCTION EQUATIONS

Recall from Section 6.3 that in oxidation–reduction reactions the oxidation states of some of the participants change. Consider the reaction between copper and an aqueous solution of silver nitrate:

$$Cu(s) + 2\,Ag^+(aq) \longrightarrow Cu^{2+}(aq) + 2\,Ag(s)$$

The nitrate ions are spectators that do not take part in the reaction, and so they are omitted in the net equation. Two electrons are transferred from each reacting copper atom to a pair of silver ions. Copper (the electron donor) is oxidized, and silver ion (the electron acceptor) is reduced. This oxidation–reduction reaction occurs when a piece of copper is placed in an aqueous solution of silver nitrate or any other soluble silver salt (Fig. 12.1). Metallic silver immediately begins to plate out on the copper, the concentration of silver ion decreases, and blue $Cu^{2+}(aq)$ appears in solution and increases in concentration as time passes.

It is useful to consider this chemical equation as representing the sum of oxidation and reduction **half-reactions,** in which electrons ($e^-$) appear explicitly. The oxidation of copper is written as

$$Cu(s) \longrightarrow Cu^{2+}(aq) + 2\,e^-$$

and the reduction of silver ion as

$$Ag^+(aq) + e^- \longrightarrow Ag(s)$$

In the net equation, electrons must not appear explicitly. Thus, the second equation must be multiplied by 2 before being added to the first, so that the electrons cancel out on both sides. As before, this gives

$$Cu(s) + 2\,Ag^+(aq) \longrightarrow Cu^{2+}(aq) + 2\,Ag(s)$$

In Section 2.3, the chemical equation for the burning of butane in oxygen (a redox reaction) was balanced through logical reasoning, taking into account the fact that the number of moles of atoms of each element is the same before and after the reaction. Many redox reactions are more difficult to balance than this, so it is useful to have a systematic procedure. In this section we outline a procedure based on half-reactions, applying it to reactions that occur in acidic or basic aqueous solution. In these reactions, water and $H_3O^+$ (acidic solution) or $OH^-$ (basic solution) may take part either as reactants or as products, so it is necessary to *complete* the corresponding equations as well as to balance them.

As an example, let us complete and balance the chemical equation for the dissolution of copper(II) sulfide in aqueous nitric acid (Fig. 12.2):

$$CuS(s) + NO_3^-(aq) \longrightarrow Cu^{2+}(aq) + SO_4^{2-}(aq) + NO(g)$$

**FIGURE 12.1** When a piece of copper screen is inserted into a solution of silver nitrate, silver forms in a tree-like structure, and the solution turns blue as $Cu^{2+}$ ions form. *(Charles D. Winters)*

**Step 1**    *Write two unbalanced half-equations, one for the species that is oxidized and its product and one for the species that is reduced and its product.*

Here the unbalanced half-reaction involving CuS is

$$CuS \longrightarrow Cu^{2+} + SO_4^{2-}$$

The unbalanced half-reaction involving $NO_3^-$ is

$$NO_3^- \longrightarrow NO$$

**Step 2**    *Insert coefficients to make the numbers of atoms of all elements except oxygen and hydrogen equal on the two sides of each equation.*

In the present case copper, sulfur, and nitrogen are already balanced in the two half-equations, so this step is already completed.

**Step 3**    *Balance oxygen by adding $H_2O$ to one side of each half-equation.*

$$CuS + 4 H_2O \longrightarrow Cu^{2+} + SO_4^{2-}$$
$$NO_3^- \longrightarrow NO + 2 H_2O$$

**Step 4**    *Balance hydrogen. For an acidic solution, add $H_3O^+$ to the side of each half-equation that is "deficient" in hydrogen and an equal amount of $H_2O$ to the other side. For a basic solution, add $H_2O$ to the side of each half-equation that is "deficient" in hydrogen and an equal amount of $OH^-$ to the other side.*

Note that this step does not disrupt the oxygen balance achieved in Step 3. In the present case (acidic solution) the result is

$$CuS + 12 H_2O \longrightarrow Cu^{2+} + SO_4^{2-} + 8 H_3O^+$$
$$NO_3^- + 4 H_3O^+ \longrightarrow NO + 6 H_2O$$

**Step 5**    *Balance charge by inserting $e^-$ (electrons) as a reactant or product in each half-equation.*

$$CuS + 12 H_2O \longrightarrow Cu^{2+} + SO_4^{2-} + 8 H_3O^+ + 8 e^- \qquad \text{(oxidation)}$$
$$NO_3^- + 4 H_3O^+ + 3 e^- \longrightarrow NO + 6 H_2O \qquad \text{(reduction)}$$

**Step 6**    *Multiply the two half-equations by numbers chosen to make the number of electrons given off by the oxidation equal the number taken up by the reduction. Then add the two half-equations, canceling electrons. If $H_3O^+$, $OH^-$, or $H_2O$ appears on both sides of the final equation, cancel out the duplications.*

Here the oxidation half-equation must be multiplied by 3 (so that 24 electrons are produced) and the reduction half-equation by 8 (so that the same 24 electrons are consumed):

$$3 CuS + 36 H_2O \longrightarrow 3 Cu^{2+} + 3 SO_4^{2-} + 24 H_3O^+ + 24 e^-$$
$$\underline{8 NO_3^- + 32 H_3O^+ + 24 e^- \longrightarrow 8 NO + 48 H_2O}$$
$$3 CuS + 8 NO_3^- + 8 H_3O^+ \longrightarrow 3 Cu^{2+} + 3 SO_4^{2-} + 8 NO + 12 H_2O$$

This procedure balances equations that are too difficult to balance by inspection. For basic solutions, remember to add $H_2O$ and $OH^-$, rather than $H_3O^+$ and $H_2O$, at Step 4.

**FIGURE 12.2**    Copper(II) sulfide reacts with concentrated nitric acid to liberate nitrogen oxide and produces a solution of copper(II) sulfate, which displays the characteristic blue color of copper(II) ions in water. *(Leon Lewandowski)*

## EXAMPLE 12.1

Balance the following equation, which represents a reaction that takes place in basic aqueous solution:

$$Ag(s) + HS^-(aq) + CrO_4^{2-}(aq) \longrightarrow Ag_2S(s) + Cr(OH)_3(s)$$

### Solution

**Step 1**
$$Ag + HS^- \longrightarrow Ag_2S$$
$$CrO_4^{2-} \longrightarrow Cr(OH)_3$$

**Step 2**
$$2\,Ag + HS^- \longrightarrow Ag_2S$$

The other half-reaction is unchanged.

**Step 3**  $H_2O$ is now added to the second half-reaction to balance oxygen:

$$CrO_4^{2-} \longrightarrow Cr(OH)_3 + H_2O$$

**Step 4**  The right side of the silver half-reaction is deficient by one hydrogen. Add one $H_2O$ to the right and one $OH^-$ to the left:

$$2\,Ag + HS^- + OH^- \longrightarrow Ag_2S + H_2O$$

In the chromium half-reaction, the left side is deficient by five hydrogens, and so 5 $H_2O$ is added to that side and 5 $OH^-$ to the right side:

$$CrO_4^{2-} + 4\,H_2O \longrightarrow Cr(OH)_3 + 5\,OH^-$$

(Notice that the $H_2O$ on the right canceled out one of the five on the left, leaving four.)

**Step 5**  Electrons are added to the right side of the silver half-reaction and to the left side of the chromium half-reaction to balance charge:

$$2\,Ag + HS^- + OH^- \longrightarrow Ag_2S + H_2O + 2\,e^- \qquad \text{(oxidation)}$$
$$CrO_4^{2-} + 4\,H_2O + 3\,e^- \longrightarrow Cr(OH)_3 + 5\,OH^- \qquad \text{(reduction)}$$

**Step 6**  The gain and loss of electrons are equalized. The first equation is multiplied by 3 so that it absorbs six electrons, and the second is multiplied by 2 so that it liberates six electrons.

$$6\,Ag + 3\,HS^- + 3\,OH^- \longrightarrow 3\,Ag_2S + 3\,H_2O + 6\,e^- \qquad \text{(oxidation)}$$
$$2\,CrO_4^{2-} + 8\,H_2O + 6\,e^- \longrightarrow 2\,Cr(OH)_3 + 10\,OH^- \qquad \text{(reduction)}$$
$$\overline{6\,Ag + 3\,HS^- + 2\,CrO_4^{2-} + 5\,H_2O \longrightarrow 3\,Ag_2S + 2\,Cr(OH)_3 + 7\,OH^-}$$

**Related Problems: 1, 2, 3, 4**

## EXAMPLE 12.2

Balance the following equation for the reaction of arsenic(III) sulfide with aqueous chloric acid:

$$As_2S_3(s) + ClO_3^-(aq) \longrightarrow H_3AsO_4(aq) + SO_4^{2-}(aq) + Cl^-(aq)$$

## Solution

**Step 1**
$$As_2S_3 \longrightarrow H_3AsO_4 + SO_4^{2-}$$
$$ClO_3^- \longrightarrow Cl^-$$

**Step 2**
$$As_2S_3 \longrightarrow 2\,H_3AsO_4 + 3\,SO_4^{2-}$$

**Step 3**
$$As_2S_3 + 20\,H_2O \longrightarrow 2\,H_3AsO_4 + 3\,SO_4^{2-}$$
$$ClO_3^- \longrightarrow Cl^- + 3\,H_2O$$

**Step 4** The chloric acid makes this an acidic solution, and so $H_3O^+$ and $H_2O$ are used:
$$As_2S_3 + 54\,H_2O \longrightarrow 2\,H_3AsO_4 + 3\,SO_4^{2-} + 34\,H_3O^+$$
$$ClO_3^- + 6\,H_3O^+ \longrightarrow Cl^- + 9\,H_2O$$

**Step 5**
$$As_2S_3 + 54\,H_2O \longrightarrow 2\,H_3AsO_4 + 3\,SO_4^{2-} + 34\,H_3O^+ + 28\,e^-$$
$$ClO_3^- + 6\,H_3O^+ + 6\,e^- \longrightarrow Cl^- + 9\,H_2O$$

**Step 6** We must find the least common multiple of 28 and 6. This is 84, and we therefore need to multiply the first equation by $84/28 = 3$ and the second equation by $84/6 = 14$:

$$3\,As_2S_3 + 162\,H_2O \longrightarrow 6\,H_3AsO_4 + 9\,SO_4^{2-} + 102\,H_3O^+ + 84\,e^-$$
$$\underline{14\,ClO_3^- + 84\,H_3O^+ + 84\,e^- \longrightarrow 14\,Cl^- + 126\,H_2O}$$
$$3\,As_2S_3 + 14\,ClO_3^- + 36\,H_2O \longrightarrow 6\,H_3AsO_4 + 9\,SO_4^{2-} + 14\,Cl^- + 18\,H_3O^+$$

## Disproportionation

An important type of redox reaction, **disproportionation,** occurs when a single substance is both oxidized and reduced. Example 6.7(c) provided such a reaction:

$$2\,\overset{-1}{H_2O_2}(\ell) \longrightarrow 2\,\overset{-2}{H_2O}(\ell) + \overset{0}{O_2}(g)$$

The oxygen in hydrogen peroxide ($H_2O_2$) is in an intermediate oxidation state of $-1$; some of it is oxidized to $O_2$ and some is reduced to $H_2O$. The balancing of equations for disproportionation reactions is described in the following example.

### EXAMPLE 12.3

Balance the following equation for the reaction that occurs when chlorine is dissolved in basic solution:

$$Cl_2(g) \longrightarrow ClO_3^-(aq) + Cl^-(aq)$$

## Solution

**Step 1** We solve this by writing the $Cl_2$ on the left sides of *two* half-equations:

$$Cl_2 \longrightarrow ClO_3^-$$
$$Cl_2 \longrightarrow Cl^-$$

**Step 2**
$$Cl_2 \longrightarrow 2 ClO_3^-$$
$$Cl_2 \longrightarrow 2 Cl^-$$

**Step 3**   The first half-equation becomes
$$Cl_2 + 6 H_2O \longrightarrow 2 ClO_3^-$$

**Step 4**   Now the first half-equation becomes
$$Cl_2 + 12 OH^- \longrightarrow 2 ClO_3^- + 6 H_2O$$

**Step 5**   $\quad Cl_2 + 12 OH^- \longrightarrow 2 ClO_3^- + 6 H_2O + 10\, e^-$   (oxidation)

$\qquad\qquad\quad Cl_2 + 2\, e^- \longrightarrow 2 Cl^-$   (reduction)

**Step 6**   Multiply the second equation by 5 and add:

$$Cl_2 + 12 OH^- \longrightarrow 2 ClO_3^- + 6 H_2O + 10\, e^-$$
$$\underline{5 Cl_2 + 10\, e^- \longrightarrow 10 Cl^-}$$
$$6 Cl_2 + 12 OH^- \longrightarrow 2 ClO_3^- + 10 Cl^- + 6 H_2O$$

Dividing this equation by 2 gives

$$3 Cl_2 + 6 OH^- \longrightarrow ClO_3^- + 5 Cl^- + 3 H_2O$$

**Related Problems: 7, 8**

---

**12.2**

## ELECTROCHEMICAL CELLS

The reaction of copper with aqueous silver ions shown in Figure 12.1 is clearly spontaneous and therefore irreversible. Hence $\Delta G < 0$, although at this point its magnitude is unknown. Since no work is performed, the first law of thermodynamics requires that the entire energy change appear as the evolution of heat.

This same reaction can be carried out quite differently without ever bringing the two reactants into *direct* contact with each other if a galvanic cell (a battery) is constructed from them. A copper strip is partially immersed in a solution of $Cu(NO_3)_2$ and a silver strip in a solution of $AgNO_3$, as Figures 12.3 and 12.4 illustrate. The two solutions are connected by a **salt bridge,** which is an inverted U-shaped tube containing a solution of a salt such as $NaNO_3$. The ends of the bridge are stuffed with porous plugs that prevent the two solutions from mixing but allow ions to pass through. The two metal strips are connected to an **ammeter,** an instrument that measures the direction and magnitude of electric current through it.

As copper is oxidized on the left side, $Cu^{2+}$ ions enter the solution. The electrons released in the reaction pass through the external circuit from left to right, as shown by the deflection of the ammeter needle. The electrons enter the silver strip, and at the metal–solution interface they are picked up by $Ag^+$ ions, which plate out as atoms on the surface of the silver. This process would lead to an increase of positive charge in the left-hand beaker and to a decrease in the right-hand one were it not for the salt bridge; the bridge permits a net flow of positive ions through it into the right-hand beaker and of negative ions into the left-hand beaker, preserving charge neutrality in each.

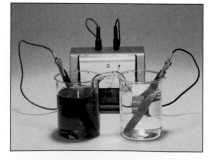

**FIGURE 12.3** A metallic copper anode reacts to give a blue solution containing copper(II) ions as silver ions plate out on a silver cathode in a galvanic cell. *(Leon Lewandowski)*

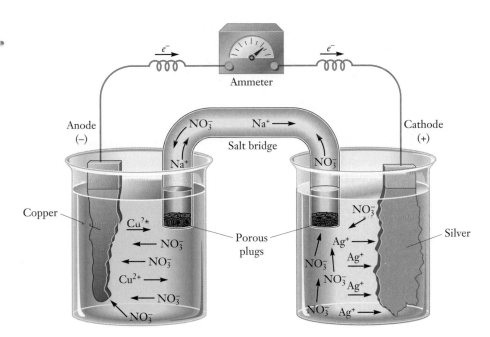

**FIGURE 12.4** In the galvanic cell of Figure 12.3, charged particles move when the circuit is completed. Electrons flow from the copper to the silver electrode through the wire. In solution, anions migrate toward the copper electrode and cations move toward the silver electrode. Sodium and nitrate ions migrate through the salt bridge to maintain electrical neutrality.

As discussed in Section 12.1, this oxidation–reduction reaction is composed of two separate half-reactions. The oxidation half-reaction in the left-hand beaker is

$$Cu(s) \longrightarrow Cu^{2+}(aq) + 2\ e^-$$

and the reduction half-reaction in the right-hand beaker is

$$Ag^+(aq) + e^- \longrightarrow Ag(s)$$

Following Michael Faraday, chemists call the site at which oxidation occurs in an electrochemical cell the **anode** and the site at which reduction occurs the **cathode.** In the galvanic cell just discussed, copper is the anode (because it is oxidized) and silver is the cathode (because $Ag^+$ is reduced). Electrons flow in the external circuit from anode to cathode. In solution both positive and negative ions are free to move. In an electrochemical cell, negative ions (anions) move toward the anode, and positive ions (cations) move toward the cathode. We will adopt a convention for galvanic cells in which the anode is shown on the left and the cathode on the right, so that electrons flow through the external circuit from left to right. Schematically the copper–silver galvanic cell is represented as

$$Cu|Cu^{2+} \parallel Ag^+|Ag$$

with the anode on the left and the cathode on the right and with the metal–solution interface represented by | and the salt bridge by ‖.

The net chemical reaction in this simple galvanic cell is the same one that takes place when a copper strip is placed in an aqueous solution of silver nitrate, but there is an essential difference in the process. Because the reaction components are separated into two compartments while electrical continuity is preserved, the direct transfer of electrons from copper atoms to silver ions is avoided, and they are forced to travel through an external circuit (wire) before they ultimately accomplish the same net effect. The current of electrons through the wire can be used for a variety of purposes. For example, if a light bulb were placed in the electric circuit, the current passing through it would cause the bulb to light. The electrochemical cell would have converted chemical energy to heat and radiant energy. Alternatively, the light bulb could be replaced by a small electric motor and the energy change of the chemical reaction used to perform mechanical work.

## Galvanic and Electrolytic Cells

What causes an electric current to flow in a galvanic cell? There must be an electrical **potential difference**, $\Delta\mathcal{E}$, between two points in the circuit to cause electrons to flow, just as a difference in gravitational potential between two points on the earth's surface causes water to flow downhill. This electrical potential difference, or **cell voltage,** can be measured with an instrument called a **voltmeter** inserted in the external circuit. The voltage measured in a galvanic cell depends on the magnitude of the current passing through the cell, and the voltage falls if the current becomes too large. The intrinsic cell voltage (the value at zero current) can be measured by placing a variable voltage source in the external circuit in such a way that its potential difference $\Delta\mathcal{E}_{ext}$ *opposes* the intrinsic potential difference $\Delta\mathcal{E}$ of the electrochemical cell. The net potential difference is then

$$\Delta\mathcal{E}_{net} = \Delta\mathcal{E} - \Delta\mathcal{E}_{ext}$$

$\Delta\mathcal{E}$ can be measured by adjusting $\Delta\mathcal{E}_{ext}$ until $\Delta\mathcal{E}_{net}$ becomes 0, at which point the current through the circuit falls to 0 as well. If $\Delta\mathcal{E}_{ext}$ is held just below $\Delta\mathcal{E}$, the net potential difference becomes small and the cell operation is close to reversible, with only a small current and a slow rate of reaction at the electrodes.

If the opposing external voltage is increased *above* the natural potential difference of the cell, the electrons reverse direction and move toward the copper electrode. Copper ions in solution accept electrons and deposit as copper metal, and silver metal dissolves and furnishes additional $Ag^+$ ions. The net reaction occurring is then the reverse of the spontaneous reaction, namely,

$$2\ Ag(s) + Cu^{2+}(aq) \longrightarrow 2\ Ag^+ + Cu(s)$$

An electrochemical cell that operates spontaneously is called a **galvanic cell** (or voltaic cell). Such a cell converts chemical energy to electrical energy, which can be used to perform work. A cell in which an opposing external potential causes the reaction to occur in the direction opposite the spontaneous direction is called an **electrolytic cell;** such a cell uses electrical energy provided by the external circuit to carry out chemical reactions that would otherwise not occur. When a cell is changed into an electrolytic cell by the addition of an external potential source that reverses the direction of electron flow, there is also a reversal in the sites of the anode and the cathode. In the electrolytic cell, oxidation takes place at the silver electrode, which therefore becomes the anode, and the copper electrode becomes the cathode.

### EXAMPLE 12.4

The final step in the production of magnesium from seawater is the electrolysis of molten magnesium chloride, in which the overall reaction is

$$Mg^{2+} + 2\ Cl^- \longrightarrow Mg(\ell) + Cl_2(g)$$

Write equations for the half-reactions occurring at the anode and at the cathode, and indicate the direction in which electrons flow through the external circuit.

## Solution

The anode is the site at which oxidation takes place—that is, where electrons are given up. The anode half-reaction must be

$$2\ Cl^- \longrightarrow Cl_2(g) + 2\ e^-$$

At the cathode, reduction takes place and the electrons are taken up. The cathode half-reaction is

$$Mg^{2+} + 2\ e^- \longrightarrow Mg(\ell)$$

Electrons move from the anode, where chlorine is liberated, through the external circuit to the cathode, where molten magnesium is produced.

**Related Problems: 9, 10**

## Faraday's Laws

The dual aspect of the electrochemical cell (galvanic and electrolytic) was realized shortly after the cell's discovery in 1800 by Alessandro Volta. Volta constructed a "battery" consisting of a number of platelets of silver and zinc that were separated from one another by porous strips of paper saturated with a salt solution. By 1807, Sir Humphry Davy had prepared elemental sodium and potassium by using a battery to electrolyze their respective hydroxides. The underlying scientific basis of the electrochemical cell was unclear, however. Michael Faraday's research showed a direct quantitative relationship between the amounts of substances that react at the cathode and the anode and the total electric charge that passes through the cell. This observation is the substance of **Faraday's laws,** which we state as

1. The mass of a given substance that is produced or consumed at an electrode is proportional to the quantity of electric charge passed through the cell.
2. Equivalent masses[1] of different substances are produced or consumed at an electrode by the passage of a given quantity of electric charge through the cell.

These laws, which summarize the stoichiometry of electrochemical processes, were discovered by Michael Faraday in 1833, more than half a century before the electron was discovered and the atomic basis of electricity was understood.

The charge $e$ on a single electron (expressed in coulombs) has been very accurately determined to be

$$e = 1.6021773 \times 10^{-19}\ C$$

so that the quantity of charge represented by 1 mol of electrons is

$$Q = (6.022137 \times 10^{23}\ mol^{-1})(1.602177 \times 10^{-19}\ C) = 96{,}485.31\ C\ mol^{-1}$$

This quantity of charge is called the **Faraday constant** (symbol $\mathscr{F}$):

$$\mathscr{F} = 96{,}485.31\ C\ mol^{-1}$$

---

[1] The equivalent mass of an element in a redox reaction, or of a compound containing that element, is its molar mass divided by the number of moles of electrons transferred per mole of substance in the corresponding half-reaction.

**Electric current** is the amount of charge flowing through a circuit per unit time. If $Q$ is the magnitude of the charge in coulombs and $t$ is the time in seconds that it takes to pass a point in the circuit, then the current $I$ is

$$I = \frac{Q}{t} \qquad\qquad \textbf{[12.1]}$$

where the units for $I$ are amperes (A) or coulombs per second. A current of $I$ amperes flowing for $t$ seconds causes $It$ coulombs of charge to pass through the circuit. The chemical amount, in moles, of electrons is

$$\frac{It}{96{,}485 \text{ C mol}^{-1}} = \text{moles of electrons}$$

From the number of moles of electrons that pass through a circuit, the number of moles (and therefore the number of grams) of substances reacting at the electrodes in the electrochemical cell can be calculated. Suppose a zinc–silver galvanic cell is constructed in which the anode half-reaction is

$$Zn(s) \longrightarrow Zn^{2+}(aq) + 2\ e^-$$

and the cathode half-reaction is

$$Ag^+(aq) + e^- \longrightarrow Ag(s)$$

Each mole of electrons that passes through the cell arises from the oxidation of $\frac{1}{2}$ mol of $Zn(s)$ (because each Zn atom gives up two electrons) and reduces 1 mol of silver ions. From the molar masses of silver and zinc, we calculate that $65.38/2 = 32.69$ g of zinc is dissolved at the anode and $107.87$ g of silver is deposited at the cathode. The same relationships hold if the cell is operated as an electrolytic cell, but in that case silver is dissolved and zinc is deposited.

### EXAMPLE 12.5

An electrolytic cell is constructed in which the silver ions in silver chloride are reduced to silver at the cathode and copper is oxidized to $Cu^{2+}(aq)$ at the anode. A current of 0.500 A is passed through the cell for 101 minutes. Calculate the mass of copper dissolved and the mass of silver deposited.

### Solution

$$t = (101 \text{ min})(60 \text{ s min}^{-1}) = 6.06 \times 10^3 \text{ s}$$

The chemical amount of electrons passed through the circuit during this time is

$$\frac{(0.500 \text{ C s}^{-1})(6.06 \times 10^3 \text{ s})}{96{,}485 \text{ C mol}^{-1}} = 3.14 \times 10^{-2} \text{ mol } e^-$$

The half-cell reactions are

$$AgCl(s) + e^- \longrightarrow Ag(s) + Cl^-(aq) \qquad \text{(cathode)}$$

$$Cu(s) \longrightarrow Cu^{2+}(aq) + 2\ e^- \qquad \text{(anode)}$$

and so the masses of silver deposited and copper dissolved are

$$(3.14 \times 10^{-2} \text{ mol } e^-) \times \left(\frac{1 \text{ mol Ag}}{1 \text{ mol } e^-}\right) \times (107.87 \text{ g mol}^{-1}) = 3.39 \text{ g Ag deposited}$$

$$(3.14 \times 10^{-2} \text{ mol } e^-) \times \left(\frac{1 \text{ mol Cu}}{2 \text{ mol } e^-}\right) \times (63.55 \text{ g mol}^{-1}) = 0.998 \text{ g Cu dissolved}$$

**Related Problems: 11, 12, 13, 14**

---

### 12.3

## THE GIBBS FREE ENERGY AND CELL VOLTAGE

Chapters 7 and 8 discussed the pressure–volume work associated with expanding and compressing gases. In electrochemistry a different kind of work, **electrical work,** is fundamental. If an amount of charge, $Q$, moves through a potential difference $\Delta\mathscr{E}$, the electrical work is

$$w_{elec} = -Q\,\Delta\mathscr{E} \qquad \textbf{[12.2]}$$

The minus sign appears in this equation because the same thermodynamic convention is followed as in Chapters 7 and 8—namely, that work done by the system (here, an electrochemical cell) has a negative sign. Because work is measured in joules and charge in coulombs, $\Delta\mathscr{E}$ has units of joules per coulomb. One volt (abbreviated V) is 1 joule per coulomb. Because the total charge $Q$ is the current $I$ multiplied by the time $t$ in seconds during which the current is flowing, this equation for the electrical work can also be written

$$w_{elec} = -It\,\Delta\mathscr{E}$$

The potential difference $\Delta\mathscr{E}$ is positive for a galvanic cell, so $w_{elec}$ is negative in this case, and net electrical work is performed by the galvanic cell. In an electrolytic cell, in contrast, $\Delta\mathscr{E}$ is negative and $w_{elec}$ is positive, corresponding to net electrical work done on the system by an external source such as an electric generator.

### EXAMPLE 12.6

A 6.00-V battery delivers a steady current of 1.25 A for a period of 1.50 hours. Calculate the total charge $Q$, in coulombs, that passes through the circuit and the electrical work done *by* the battery.

**Solution**

The total charge is

$$Q = It = (1.25 \text{ C s}^{-1})(1.50 \text{ hr})(3600 \text{ s hr}^{-1}) = 6750 \text{ C}$$

The electrical work is

$$w_{elec} = -Q\,\Delta\mathscr{E} = -(6750 \text{ C})(6.00 \text{ J C}^{-1}) = -4.05 \times 10^4 \text{ J}$$

This is the work done *on* the battery, and so the work done *by* the battery is the negative of this, or +40.5 kJ.

Thermodynamics demonstrates a fundamental relationship between the change in free energy, $\Delta G$, of a spontaneous chemical reaction at constant temperature and pressure, and the maximum electrical work that such a reaction is capable of producing:

$$-w_{\text{elec,max}} = |\Delta G| \qquad \text{(at constant } T \text{ and } P) \qquad \qquad [12.3]$$

To show this, recall the definition of the Gibbs free energy function $G$:

$$G = H - TS = E + PV - TS$$

For processes at constant pressure $P$ and constant temperature $T$ (the usual case in electrochemical cells),

$$\Delta G = \Delta E + P\,\Delta V - T\,\Delta S$$

From the first law of thermodynamics,

$$\Delta E = q + w$$

or

$$\Delta E = q + w_{\text{elec}} - P\,\Delta V$$

because there are now two kinds of work, electrical work $w_{\text{elec}}$ and pressure–volume work $-P_{\text{ext}}\,\Delta V = -P\,\Delta V$. Combining this with the equation for the change in free energy gives

$$\Delta G = q + w_{\text{elec}} - P\,\Delta V + P\,\Delta V - T\,\Delta S = q + w_{\text{elec}} - T\,\Delta S$$

If the condition of reversibility is imposed upon the galvanic cell, then

$$q = q_{\text{rev}} = T\,\Delta S$$

and the free energy change is

$$\Delta G = w_{\text{elec,rev}}$$

If the same cell were to be operated irreversibly (i.e., if a large current were permitted to flow), less electrical work would be produced. The maximum electrical work is produced by the galvanic cell when it is operated reversibly.

If $n$ mol of electrons (or $n\mathscr{F}$ coulombs of charge) passes through the external circuit of the galvanic cell when it is operated reversibly, and if $\Delta\mathscr{E}$ is the reversible cell voltage, then

$$\Delta G = w_{\text{elec}} = -Q\,\Delta\mathscr{E} = -n\mathscr{F}\,\Delta\mathscr{E} \qquad \text{(reversible)}$$

Electrical work is produced by an electrochemical cell only if $\Delta G < 0$ (or when $\Delta\mathscr{E} > 0$, which amounts to the same thing). This relationship provides a direct way to determine free energy changes from measurements of cell voltage.

## Standard States and Cell Voltages

Recall from Chapters 7, 9, and 11 that the *standard state* of a substance means a pressure of 1 atm and a specified temperature. In addition, the standard state of a solute is that for which its concentration in ideal solution is 1 M.[2] The standard free energy

---

[2] Recall that the standard state of a solute has unit *activity*, which can differ significantly from 1 M concentration for a real electrolyte solution. In what follows, we will nevertheless refer to 1 M concentration as the standard state.

change $\Delta G°$ for a reaction in which all reactants and products are in their standard states can be calculated from a table of standard free energies of formation $\Delta G_f°$ of the substances taking part in the reaction (see Appendix D). For those reactions that can be carried out in electrochemical cells, the standard free energy change $\Delta G°$ is related to a **standard cell voltage** $\Delta\mathscr{E}°$ by

$$\Delta G° = -n\mathscr{F}\,\Delta\mathscr{E}° \qquad\qquad \textbf{[12.4]}$$

Here $\Delta\mathscr{E}°$ is the cell voltage (potential difference) of a galvanic cell in which reactants and products are in their standard states (gases at 1 atm pressure, solutes at 1 M concentration, metals in their pure stable states, and a specified temperature). This standard cell voltage is an intrinsic electrical property of the cell, which can be calculated from the standard free energy change $\Delta G°$ by means of Equation 12.4. Conversely, $\Delta G°$ (and, hence, equilibrium constants for cell reactions) can be determined by measuring the standard cell voltage $\Delta\mathscr{E}°$ of a reaction in which $n$ mol of electrons passes through the external circuit.

---

### EXAMPLE 12.7

A $Zn^{2+}|Zn$ half-cell is connected to a $Cu^{2+}|Cu$ half-cell to make a galvanic cell, in which $[Zn^{2+}] = [Cu^{2+}] = 1.00$ M. The cell voltage at 25°C is measured to be $\Delta\mathscr{E}° = 1.10$ V, and Cu is observed to plate out as the reaction proceeds. Calculate $\Delta G°$ for the chemical reaction that takes place in the cell, for 1.00 mol of zinc dissolved.

### Solution
The reaction is

$$Zn(s) + Cu^{2+}(aq) \longrightarrow Zn^{2+}(aq) + Cu(s)$$

because Cu is a product. For the reaction as written, in which 1 mol of $Zn(s)$ and 1 mol of $Cu^{2+}(aq)$ react, 2 mol of electrons passes through the external circuit, so $n = 2$. Therefore,

$$\Delta G° = -n\mathscr{F}\,\Delta\mathscr{E}° = -(2.00\text{ mol})(96{,}485\text{ C mol}^{-1})(1.10\text{ V})$$

$$= -2.12 \times 10^5\text{ J} = -212\text{ kJ}$$

**Related Problems: 19, 20**

---

## Half-Cell Voltages

We could tabulate all the conceivable galvanic cells and their standard voltages, but the list would be very long. To avoid this, the half-cell potentials $\mathscr{E}°$ are tabulated; they can be combined to obtain the standard cell voltage $\Delta\mathscr{E}°$ for any complete cell.

   To see this, return to the standard cell just considered, made up of $Zn^{2+}|Zn$ and $Cu^{2+}|Cu$ half-cells. Each half-cell reaction is written as a reduction:

$$Zn^{2+}(aq) + 2\ e^- \longrightarrow Zn(s)$$

$$Cu^{2+}(aq) + 2\ e^- \longrightarrow Cu(s)$$

Each half-reaction has a certain tendency to occur. If **reduction potentials** $\mathscr{E}°(Zn^{2+}|Zn)$ and $\mathscr{E}°(Cu^{2+}|Cu)$ are associated with the two half-cells, the magnitude and sign of each of these potentials is related to the tendency of the reaction to occur as written. The more positive the half-cell reaction's potential, the greater its tendency to take place as a reduction. Of course, a single half-cell reduction reaction cannot take place in isolation, because a source of electrons is required. This can be achieved only by combining two half-cells. The reaction with the more positive (or algebraically greater) reduction potential occurs as a reduction, and that with the less positive potential is forced to run in reverse (as an oxidation) to supply electrons. The net cell potential (voltage) is then the *difference* between the individual half-cell reduction potentials. In the preceding example, copper is observed to plate out on the cathode, and so its potential must be more positive than that of zinc:

$$\Delta\mathscr{E}° = \mathscr{E}°(Cu^{2+}|Cu) - \mathscr{E}°(Zn^{2+}|Zn) = 1.10 \text{ V}$$

When two half-cells are combined to make a galvanic cell, reduction occurs in the half-cell with the algebraically greater potential (making it the cathode), and oxidation occurs in the other half-cell (making it the anode). We therefore write

$$\Delta\mathscr{E}° = \mathscr{E}°(\text{cathode}) - \mathscr{E}°(\text{anode}) \qquad \textbf{[12.5]}$$

for a galvanic cell. Cell voltages and half-cell reduction potentials are *intensive* quantities, independent of the amount of matter in a cell or the amount reacting. Half-cell reduction potentials are therefore *not* multiplied by numbers of moles reacting (as Gibbs free energies would be) to obtain the overall cell potential.

In thermodynamics, only energy differences $\Delta E$ are measurable; absolute energies are not. Therefore, energies (or enthalpies or free energies) are defined relative to a reference state for which these quantities were arbitrarily set at 0. The same reasoning applies to half-cells: because only differences are measured, we are free to define a reference potential for a particular half-cell and measure other half-cell potentials relative to it. The convention used is to define $\mathscr{E}°$ for the half-cell reaction of $H_2(g)$ and $H_3O^+(aq)$ to be 0 at all temperatures, when the gas pressure at the electrode is 1 atm and the $H_3O^+(aq)$ concentration in solution is 1 M (Fig. 12.5).

$$2 \text{ H}_3O^+(aq) + 2 \text{ } e^- \longrightarrow H_2(g) + 2 \text{ H}_2O(\ell) \qquad \mathscr{E}° = 0 \text{ V (by definition)}$$

All other half-cell potentials are then determined from this reference potential by combining their standard half-cells with the standard $H_3O^+|H_2$ half-cell in a galvanic cell and measuring the cell voltage. The magnitude and sign of that cell voltage give the standard potential of the half-cell in question, its reaction being written as an oxidation or a reduction according to whether the electrode behaves as the anode or the cathode of the galvanic cell.

For example, if a $Cu^{2+}$(1 M)$|Cu$ half-cell is connected to the standard $H_3O^+$(1 M)$|H_2$ half-cell, copper is observed to plate out, and so the copper half-cell is the cathode and the hydrogen half-cell is the anode. The observed cell voltage is 0.34 V; thus,

$$\Delta\mathscr{E}° = \mathscr{E}°(\text{cathode}) - \mathscr{E}°(\text{anode})$$
$$= \mathscr{E}°(Cu^{2+}|Cu) - \mathscr{E}°(H_3O^+|H_2) = \mathscr{E}°(Cu^{2+}|Cu) - 0$$
$$= 0.34 \text{ V}$$

The standard $Cu^{2+}|Cu$ half-cell potential is therefore 0.34 V on a scale in which the standard $H_3O^+|H_2$ half-cell potential is 0:

$$Cu^{2+}(1 \text{ M}) + 2 \text{ } e^- \longrightarrow Cu(s) \qquad\qquad \mathscr{E}° = 0.34 \text{ V}$$

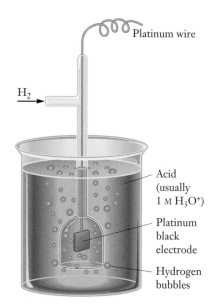

Platinum wire

$H_2$

Acid (usually 1 M $H_3O^+$)

Platinum black electrode

Hydrogen bubbles

**FIGURE 12.5** Hydrogen is a gas at room conditions, and electrodes cannot be constructed from it directly. In the hydrogen half-cell shown here, a piece of platinum covered with a fine coating of platinum black is dipped into the solution, and a stream of hydrogen is passed over the surface. The platinum itself does not react but provides a support surface for the reaction of the $H_2$ and water to give $H_3O^+$, or the reverse. It also provides the necessary electrical connection to carry away or supply electrons.

When a $Zn^{2+}(1 \text{ M})|Zn$ half-cell is connected to the standard hydrogen half-cell, zinc dissolves; its half-cell is the anode because oxidation occurs in it. The measured cell voltage is 0.76 V, and so

$$\Delta\mathscr{E}° = 0.76 \text{ V} = \mathscr{E}°(\text{cathode}) - \mathscr{E}°(\text{anode}) = 0 - \mathscr{E}°(Zn^{2+}|Zn)$$

The $Zn^{2+}(1 \text{ M})|Zn$ half-cell reduction potential is therefore $-0.76$ V:

$$Zn^{2+}(1 \text{ M}) + 2 \ e^- \longrightarrow Zn(s) \qquad \mathscr{E}° = -0.76 \text{ V}$$

If the zinc and copper half-cells are combined, the copper half-cell will be the cathode because it has the more positive half-cell reduction potential. The galvanic cell voltage under standard-state conditions (Fig. 12.6) will be

$$\Delta\mathscr{E}° = \mathscr{E}°(\text{cathode}) - \mathscr{E}°(\text{anode}) = 0.34 \text{ V} - (-0.76 \text{ V}) = 1.10 \text{ V}$$

in agreement with the measured value.

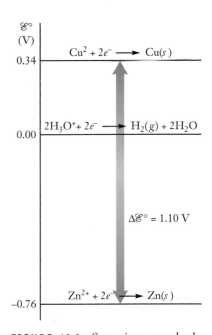

**FIGURE 12.6** Sometimes standard reduction potentials are represented to scale. In this diagram, the distance between a pair of standard half-cell potentials is proportional to the $\Delta\mathscr{E}°$ generated by the electrochemical cell that combines the two. The half-reaction with the higher reduction potential forces the other half-reaction to occur in the reverse direction.

## EXAMPLE 12.8

An aqueous solution of potassium permanganate ($KMnO_4$) has a deep purple color. In aqueous acidic solution, the permanganate ion can be reduced to the pale-pink manganese(II) ion ($Mn^{2+}$). Under standard conditions, the reduction potential of an $MnO_4^-|Mn^{2+}$ half-cell is $\mathscr{E}° = 1.49$ V. Suppose this half-cell is combined with a $Zn^{2+}|Zn$ half-cell in a galvanic cell, with $[Zn^{2+}] = [MnO_4^-] = [Mn^{2+}] = [H_3O^+] = 1$ M. **(a)** Write equations for the reactions at the anode and the cathode. **(b)** Write a balanced equation for the overall cell reaction. **(c)** Calculate the standard cell potential difference, $\Delta\mathscr{E}°$.

**Solution**

**(a)** Because $\mathscr{E}°(MnO_4^-|Mn^{2+}) = 1.49$ V is algebraically greater than $\mathscr{E}°(Zn^{2+}|Zn) = -0.76$ V, permanganate ions will be reduced at the cathode. The balanced half-cell reaction requires the presence of $H_3O^+$ ions and water, giving

$$MnO_4^-(aq) + 8 \ H_3O^+(aq) + 5 \ e^- \longrightarrow Mn^{2+}(aq) + 12 \ H_2O(\ell)$$

The half-equation for the oxidation of Zn at the anode is

$$Zn(s) \longrightarrow Zn^{2+}(aq) + 2 \ e^-$$

**(b)** In the overall reaction, the number of electrons taken up at the cathode must equal the number released at the anode, and so the first equation must be multiplied by 2 and the second by 5. Adding the two gives the overall reaction:

$$2 \ MnO_4^-(aq) + 16 \ H_3O^+(aq) + 5 \ Zn(s) \longrightarrow 2 \ Mn^{2+}(aq) + 24 \ H_2O(\ell) + 5 \ Zn^{2+}(aq)$$

**(c)** The galvanic cell potential is the difference between the standard reduction potential for permanganate (at the cathode) and that for zinc (at the anode):

$$\Delta\mathscr{E}° = \mathscr{E}°(MnO_4^-|Mn^{2+}) - \mathscr{E}°(Zn^{2+}|Zn) = 1.49 + 0.76 = 2.25 \text{ V}$$

Note that the half-cell potentials are not multiplied by their coefficients (2 and 5) before subtraction. Half-cell potentials are *intensive* properties of a galvanic cell and so are independent of the amount of the reacting species.

**Related Problems: 21, 22**

Appendix E summarizes the standard reduction potentials for a large number of half-reactions. The reactions in that table are arranged in order of decreasing reduction potentials—that is, with the most positive at the top and the most negative at the bottom. In any galvanic cell, the half-cell that is listed higher in the table will act as the cathode (if both half-cells are in the standard state).

A strong **oxidizing agent** is a chemical species that is itself easily reduced. Such species are marked by large positive reduction potentials and appear at the top left in Appendix E. Fluorine has the largest reduction potential listed, and fluorine molecules are extremely eager to accept electrons to become fluoride ions. Other strong oxidizing agents include hydrogen peroxide ($H_2O_2$) and solutions of permanganate ion ($MnO_4^-$). A strong **reducing agent,** on the other hand, is very easily oxidized, so its corresponding reduction potential is large and negative. Such reducing agents appear at the lower right in Appendix E. The alkali and alkaline earth metals are especially good reducing agents.

Oxygen itself is a good oxidizing agent in acidic solution at pH 0 because it has a fairly high reduction potential:

$$O_2 + 4\,H_3O^+ + 4\,e^- \longrightarrow 6\,H_2O \qquad\qquad \mathscr{E}° = 1.229\ V$$

Ozone ($O_3$) is still stronger, because its high free energy relative to oxygen provides an additional driving force for oxidation reactions, as shown by its high half-cell reduction potential in acidic aqueous solution:

$$O_3 + 2\,H_3O^+ + 2\,e^- \longrightarrow O_2 + 3\,H_2O \qquad\qquad \mathscr{E}° = 2.07\ V$$

Because of its great oxidizing power, ozone is used commercially as a bleach for wood pulp and as a disinfectant and sterilizing agent for water, where it oxidizes algae and organic impurities but leaves no undesirable residue. In basic aqueous solution, the half-reactions of oxygen-containing species involve hydroxide ions instead of hydronium ions, and the reduction potentials are correspondingly changed. At pH 14 (standard basic conditions),

$$O_2 + 2\,H_2O + 4\,e^- \longrightarrow 4\,OH^- \qquad\qquad \mathscr{E}° = 0.401\ V$$

$$O_3 + H_2O + 2\,e^- \longrightarrow O_2 + 2\,OH^- \qquad\qquad \mathscr{E}° = 1.24\ V$$

Both oxygen and ozone are less effective oxidizing agents in basic than in acidic solution.

## Adding and Subtracting Half-Cell Reactions

When two half-cells are combined to form a galvanic cell, their half-cell potentials are *not* multiplied by the coefficients that appear in the overall balanced chemical equation. It often happens, however, that two half-cells are to be combined to obtain *another half-cell*. For example, suppose we wish to know the standard half-cell potential corresponding to the reaction

$$Cu^{2+}(aq) + e^- \longrightarrow Cu^+(aq)$$

when the standard half-cell potentials are known for the reactions

$$Cu^{2+}(aq) + 2\,e^- \longrightarrow Cu(s) \qquad \mathscr{E}_1^° = \mathscr{E}°(Cu^{2+}|Cu) = 0.340\ V$$

$$Cu^+(aq) + e^- \longrightarrow Cu(s) \qquad \mathscr{E}_2^° = \mathscr{E}°(Cu^+|Cu) = 0.522\ V$$

Could we not simply subtract the second half-reaction from the first and, correspondingly, the second half-cell potential from the first half-cell potential, to obtain

$$Cu^{2+}(aq) + e^- \longrightarrow Cu^+(aq) \qquad \mathscr{E}_3^° = \mathscr{E}°(Cu^{2+}|Cu^+) = -0.182\ V?$$

The answer is no. Instead, the *free energy change* for each half-reaction must be calculated from

$$\Delta G^{\circ}_{hc} = -n_{hc}\mathcal{F}\mathcal{E}^{\circ}$$

and combined before calculating $\mathcal{E}^{\circ}$ for their resultant. Thus, for the free energy change for the difference of the two half-reactions,

$$\Delta G^{\circ}_3 = \Delta G^{\circ}_1 - \Delta G^{\circ}_2 = -n_1\mathcal{F}\mathcal{E}^{\circ}_1 + n_2\mathcal{F}\mathcal{E}^{\circ}_2$$

Setting this equation equal to $-n_3\mathcal{F}\mathcal{E}^{\circ}_3$ gives

$$-n_3\mathcal{F}\mathcal{E}^{\circ}_3 = -n_1\mathcal{F}\mathcal{E}^{\circ}_1 + n_2\mathcal{F}\mathcal{E}^{\circ}_2$$

$$\mathcal{E}^{\circ}_3 = \frac{n_1\mathcal{E}^{\circ}_1 - n_2\mathcal{E}^{\circ}_2}{n_3}$$

The correct standard half-cell potential for the half-cell reaction

$$Cu^{2+}(aq) + e^- \longrightarrow Cu^+(aq)$$

is

$$\mathcal{E}^{\circ}(Cu^{2+}|Cu^+) = \frac{(2 \text{ mol})(0.340 \text{ V}) - (1 \text{ mol})(0.522 \text{ V})}{1 \text{ mol}} = 0.158 \text{ V}$$

This same procedure using free energies can be used to calculate overall cell voltages from half-cell potentials. Here, however, the number of moles of electrons $n_1$ released at the anode is equal to the number $n_2$ taken up at the cathode, and is the number $n_3$ implied by the overall chemical equation. The electrons then cancel out and the result reduces to the simple form found before:

$$\Delta\mathcal{E}^{\circ} = \mathcal{E}^{\circ}(\text{cathode}) - \mathcal{E}^{\circ}(\text{anode})$$

## Reduction Potential Diagrams and Disproportionation

We can summarize the half-reactions of copper in a **reduction potential diagram** of the form

$$Cu^{2+} \xrightarrow{0.158 \text{ V}} Cu^+ \xrightarrow{0.522 \text{ V}} Cu$$
$$\underset{0.340 \text{ V}}{\underline{\hspace{6cm}}}$$

The line connecting each pair of species stands for the full half-reaction, written as a reduction and balanced by the addition of electrons (and, where necessary, water and $H_3O^+$ or $OH^-$). The number over each line is the corresponding reduction potential in volts.

Recall from Example 12.3 that *disproportionation* is the process in which a single substance is both oxidized and reduced. Reduction potential diagrams enable us to determine which species are stable with respect to disproportionation. A species can disproportionate if and only if a reduction potential that lies immediately to its right is larger than one that appears immediately to its left. To demonstrate this, note that a disproportionation reaction is the sum of a reduction half-reaction and an oxidation half-reaction. The driving force is the *difference* between the two reduction potentials, and if this difference is positive the disproportionation occurs spontaneously. In the case of $Cu^+$, the disproportionation reaction is

$$2 \text{ Cu}^+ \longrightarrow Cu^{2+} + Cu(s) \qquad\qquad \Delta\mathcal{E}^{\circ} = 0.522 - 0.158 = 0.364 \text{ V}$$

Because $\Delta\mathcal{E}^{\circ} > 0$, $\Delta G^{\circ} < 0$ and the reaction occurs spontaneously, although the rate may be slow.

**EXAMPLE 12.9**

Hydrogen peroxide, $H_2O_2$, is a possible product of the reduction of oxygen in acidic solution:

$$O_2 + 2\,H_3O^+ + 2\,e^- \longrightarrow H_2O_2 + 2\,H_2O \qquad \mathscr{E}_3^\circ = ?$$

It can then be further reduced to water:

$$H_2O_2 + 2\,H_3O^+ + 2\,e^- \longrightarrow 4\,H_2O \qquad \mathscr{E}_2^\circ = 1.77\ \text{V}$$

**(a)** Use the half-cell potential just given for the reduction of $H_2O_2$, together with that given earlier,

$$O_2 + 4\,H_3O^+ + 4\,e^- \longrightarrow 6\,H_2O \qquad \mathscr{E}_1^\circ = 1.229\ \text{V}$$

to calculate the standard half-cell potential for the reduction of $O_2$ to $H_2O_2$ in acidic solution.
**(b)** Write a reduction potential diagram for $O_2$, $H_2O_2$, and $H_2O$.
**(c)** Is $H_2O_2$ stable with respect to disproportionation in acidic solution?

**Solution**
**(a)** The desired half-cell reaction is obtained by *subtracting* the reaction with potential $\mathscr{E}_2^\circ$ from that with potential $\mathscr{E}_1^\circ$. The half-cell potentials are not subtracted, however, but combined as described earlier in this section. Taking $n_1 = 4$, $n_2 = 2$, and $n_3 = 2$ gives

$$\mathscr{E}_3^\circ = \frac{n_1\mathscr{E}_1^\circ - n_2\mathscr{E}_2^\circ}{n_3}$$

$$= \frac{(4\ \text{mol})(1.229\ \text{V}) - (2\ \text{mol})(1.77\ \text{V})}{2\ \text{mol}} = 0.69\ \text{V}$$

**(b)** The reduction potential diagram is obtained by omitting the electrons, water, and $H_3O^+$ from the corresponding half-equations:

$$O_2 \xrightarrow{\ 0.69\ \text{V}\ } H_2O_2 \xrightarrow{\ 1.77\ \text{V}\ } H_2O$$
$$\underset{1.229\ \text{V}}{\rule{5cm}{0.4pt}}$$

**(c)** $H_2O_2$ is thermodynamically unstable to disproportionation in acidic solution, because the half-cell potential to its right (1.77 V) is higher than that to its left (0.69 V). The disproportionation of $H_2O_2$ is also spontaneous in neutral solution, but is slow enough that aqueous solutions of hydrogen peroxide can be stored for a long time without deteriorating, as long as they are kept out of the light.

**Related Problems: 31, 32, 33, 34**

---

**12.4**

## CONCENTRATION EFFECTS AND THE NERNST EQUATION

Cells in which concentrations or pressures differ from their standard-state values are often of interest. It is straightforward to use the thermodynamic principles of Chapter 9 to determine the effects of concentration and pressure on the cell voltage. There it was shown that the free energy change is related to the reaction quotient $Q$ through

$$\Delta G = \Delta G° + RT \ln Q$$

Combining this equation with

$$\Delta G = -n\mathscr{F} \Delta\mathscr{E}$$

and

$$\Delta G° = -n\mathscr{F} \Delta\mathscr{E}°$$

gives

$$-n\mathscr{F} \Delta\mathscr{E} = -n\mathscr{F} \Delta\mathscr{E}° + RT \ln Q$$

and

$$\Delta\mathscr{E} = \Delta\mathscr{E}° - \frac{RT}{n\mathscr{F}} \ln Q \qquad\qquad \text{[12.6]}$$

a relationship known as the **Nernst equation.**

The Nernst equation can be rewritten in terms of common (base-10) logarithms by using the fact that

$$\ln Q \approx 2.303 \log_{10} Q$$

At 25°C (298.15 K), the combination of constants $2.303 \, RT/\mathscr{F}$ becomes

$$2.303 \, \frac{RT}{\mathscr{F}} = (2.303) \, \frac{(8.315 \text{ J K}^{-1} \text{ mol}^{-1})(298.15 \text{ K})}{96485 \text{ C mol}^{-1}}$$

$$= 0.0592 \text{ J C}^{-1} = 0.0592 \text{ V}$$

because 1 joule per coulomb is 1 volt. The Nernst equation then becomes

$$\Delta\mathscr{E} = \Delta\mathscr{E}° - \frac{0.0592 \text{ V}}{n} \log_{10} Q \qquad (\text{at } 25°C) \qquad \text{[12.7]}$$

which is its most familiar form. Here $n$ is the number of moles of electrons transferred in the overall chemical reaction as written. In a galvanic cell made from zinc, aluminum, and their ions, for example,

$$3 \text{ Zn}^{2+}(aq) + 6 \, e^- \longrightarrow 3 \text{ Zn}(s)$$

$$\underline{\phantom{3 \text{ Zn}^{2+}(aq)} 2 \text{ Al}(s) \longrightarrow 2 \text{ Al}^{3+}(aq) + 6 \, e^-}$$

$$3 \text{ Zn}^{2+}(aq) + 2 \text{ Al}(s) \longrightarrow 2 \text{ Al}^{3+}(aq) + 3 \text{ Zn}(s)$$

there is a net transfer of $n = 6$ mol of electrons through the external circuit for the equation as written.

The Nernst equation applies to half-cells in exactly the same way as it does to complete electrochemical cells. For any half-cell potential at 25°C,

$$\mathscr{E} = \mathscr{E}° - \frac{0.0592 \text{ V}}{n_{hc}} \log_{10} Q_{hc}$$

where $n_{hc}$ is the number of electrons appearing in the half-reaction and $Q_{hc}$ is the reaction quotient for the half-cell reaction written as a reduction. The electrons themselves do not appear in $Q_{hc}$. In the $\text{Zn}^{2+}|\text{Zn}$ case, the half-reaction written as a reduction is

$$\text{Zn}^{2+}(aq) + 2 \, e^- \longrightarrow \text{Zn}(s)$$

so that $n_{hc} = 2$ and $Q_{hc} = 1/[Zn^{2+}]$. If two half-cells in which reactants and products are not in their standard states are combined in a galvanic cell, the one with the more positive value of $\mathscr{E}$ will be the cathode, where reduction takes place, and the cell voltage will be

$$\Delta\mathscr{E} = \mathscr{E}(\text{cathode}) - \mathscr{E}(\text{anode})$$

## EXAMPLE 12.10

Suppose the $Zn|Zn^{2+} \parallel MnO_4^-|Mn^{2+}$ cell from Example 12.8 is operated at pH 2.00 with $[MnO_4^-] = 0.12$ M, $[Mn^{2+}] = 0.0010$ M, and $[Zn^{2+}] = 0.015$ M. Calculate the cell voltage $\Delta\mathscr{E}$ at 25°C.

### Solution

Recall that the overall equation for this cell is

$$2\,MnO_4^-(aq) + 5\,Zn(s) + 16\,H_3O^+(aq) \longrightarrow 2\,Mn^{2+}(aq) + 5\,Zn^{2+}(aq) + 24\,H_2O(\ell)$$

From Example 12.8, for every 5 mol of Zn oxidized (or 2 mol of $MnO_4^-$ reduced), 10 mol of electrons passes through the external circuit, so $n = 10$. A pH of 2.00 corresponds to a hydronium ion concentration of 0.010 M. Putting this and the other concentrations into the expression for the reaction quotient and substituting both $Q$ and the value of $\Delta\mathscr{E}°$ (from Example 12.8) into the Nernst equation give

$$\Delta\mathscr{E} = 2.25\ \text{V} - \frac{0.0592\ \text{V}}{10}\ \log_{10} \frac{[Mn^{2+}]^2[Zn^{2+}]^5}{[MnO_4^-]^2[H_3O^+]^{16}}$$

$$= 2.25\ \text{V} - \frac{0.0592\ \text{V}}{10}\ \log_{10} \frac{(0.0010)^2(0.015)^5}{(0.12)^2(0.010)^{16}}$$

$$= 2.25\ \text{V} - \frac{0.0592\ \text{V}}{10}\ \log_{10}(5.3 \times 10^{18}) = 2.14\ \text{V}$$

**Related Problems: 35, 36**

## Measuring Equilibrium Constants

Electrochemistry provides a convenient and accurate way to measure equilibrium constants for many solution-phase reactions. For an overall cell reaction,

$$\Delta G° = -n\mathscr{F}\,\Delta\mathscr{E}°$$

In addition, $\Delta G°$ is related to the equilibrium constant $K$ through

$$\Delta G° = -RT \ln K$$

so

$$RT \ln K = n\mathscr{F}\,\Delta\mathscr{E}°$$

$$\ln K = \frac{n\mathscr{F}}{RT}\,\Delta\mathscr{E}°$$

$$\log_{10} K = \frac{n}{0.0592\ \text{V}}\,\Delta\mathscr{E}° \qquad \text{(at 25°C)} \qquad\qquad \text{[12.8]}$$

The same result can be obtained in a slightly different way. Return to the Nernst equation, which reads (at 25°C)

$$\Delta\mathscr{E} = \Delta\mathscr{E}° - \frac{0.0592 \text{ V}}{n} \log_{10} Q$$

Suppose all species are present under standard-state conditions; then $Q = 1$ initially and the cell voltage $\Delta\mathscr{E}$ is the standard cell voltage $\Delta\mathscr{E}°$. As the reaction takes place, $Q$ increases and $\Delta\mathscr{E}$ decreases. As reactants are used up and products are formed, the cell voltage approaches 0. Eventually, equilibrium is reached and $\Delta\mathscr{E}$ becomes 0. At that point,

$$\Delta\mathscr{E}° = \frac{0.0592 \text{ V}}{n} \log_{10} K$$

which is the same as the relationship just obtained. This relation lets us calculate equilibrium constants from standard cell voltages, as the following example illustrates.

## EXAMPLE 12.11

Calculate the equilibrium constant for the redox reaction

$$2 \text{ MnO}_4^-(aq) + 5 \text{ Zn}(s) + 16 \text{ H}_3\text{O}^+(aq) \longrightarrow 2 \text{ Mn}^{2+}(aq) + 5 \text{ Zn}^{2+}(aq) + 24 \text{ H}_2\text{O}(\ell)$$

at 25°C using the cell voltage calculated in Example 12.8.

### Solution

From Example 12.8, $\Delta\mathscr{E}° = 2.25$ V, and from the analysis there and in Example 12.10, $n = 10$ for the preceding reaction. Therefore,

$$\log_{10} K = \frac{n}{0.0592 \text{ V}} \Delta\mathscr{E}° = \frac{10}{0.0592 \text{ V}} (225 \text{ V}) = 380$$

$$K = 10^{380}$$

This overwhelmingly large equilibrium constant reflects the strength of permanganate ion as an oxidizing agent and of zinc as a reducing agent. It means that for all practical purposes no $\text{MnO}_4^-$ ions are present at equilibrium.

**Related Problems: 43, 44**

The foregoing example illustrates how equilibrium constants for overall cell reactions can be determined electrochemically. Although the example dealt with redox equilibrium, related procedures can be used to measure the solubility product constants of sparingly soluble ionic compounds or the ionization constants of weak acids and bases. Suppose, for example, that the solubility product constant of AgCl is to be determined by means of an electrochemical cell. One half-cell contains solid AgCl and Ag metal in equilibrium with a known concentration of $\text{Cl}^-(aq)$ (established with 0.00100 M NaCl, for example) so that an unknown but definite concentration of $\text{Ag}^+(aq)$ is present. A silver electrode is used so that the half-cell reaction involved is either the reduction of $\text{Ag}^+(aq)$ or the oxidation of Ag. This is, in effect, an $\text{Ag}^+|\text{Ag}$ half-cell whose potential is to be determined. The second half-cell can be any whose potential is accurately known, and its choice is a matter of convenience. In the following example, the second half-cell is a standard $\text{H}_3\text{O}^+|\text{H}_2$ half-cell.

### EXAMPLE 12.12

A galvanic cell is constructed using a standard hydrogen half-cell (with platinum electrode) and a half-cell containing silver and silver chloride:

$$Pt|H_2(1\ atm)|H_3O^+(1\ \text{M}) \parallel Cl^-(1.00 \times 10^{-3}\ \text{M}) + Ag^+(?\ \text{M})|AgCl|Ag$$

The $H_2|H_3O^+$ half-cell is observed to be the anode, and the measured cell voltage is $\Delta\mathscr{E} = 0.397$ V. Calculate the silver ion concentration in the cell and the $K_{sp}$ of AgCl at 25°C.

### Solution

The half-cell reactions are

$$H_2(g) + 2\ H_2O(\ell) \longrightarrow 2\ H_3O^+(aq) + 2\ e^- \qquad \text{(anode)}$$

$$2\ Ag^+(aq) + 2\ e^- \longrightarrow 2\ Ag(s) \qquad\qquad \text{(cathode)}$$

$$\Delta\mathscr{E}° = \mathscr{E}°(\text{cathode}) - \mathscr{E}°(\text{anode}) = 0.800 - 0.000\ \text{V} = 0.800\ \text{V}$$

Note that $n = 2$ for the overall cell reaction, and the reaction quotient simplifies to

$$Q = \frac{[H_3O^+]^2}{[Ag^+]^2 P_{H_2}} = \frac{1}{[Ag^+]^2}$$

because $[H_3O^+] = 1$ M and $P_{H_2} = 1$ atm. The Nernst equation is

$$\Delta\mathscr{E} = \Delta\mathscr{E}° - \frac{0.0592\ \text{V}}{n} \log_{10} Q$$

$$\log_{10} Q = \frac{n}{0.0592\ \text{V}}(\Delta\mathscr{E}° - \Delta\mathscr{E}) = \frac{2}{0.0592\ \text{V}}(0.800 - 0.397\ \text{V}) = 13.6$$

$$Q = 10^{13.6} = 4 \times 10^{13} = 1/[Ag^+]^2$$

This can be solved for the silver ion concentration $[Ag^+]$ to give

$$[Ag^+] = 1.6 \times 10^{-7}\ \text{M}$$

so that

$$[Ag^+][Cl^-] = (1.6 \times 10^{-7})(1.00 \times 10^{-3})$$

$$K_{sp} = 1.6 \times 10^{-10}$$

**Related Problems: 49, 50**

## pH Meters

The voltage of a galvanic cell is sensitive to the pH if one of its electrodes is a $Pt|H_2$ electrode that dips into a solution of variable pH. A simple cell to measure pH is

$$Pt|H_2(1\ atm)|H_3O^+(\text{variable}) \parallel H_3O^+(1\ \text{M})|H_2(1\ atm)|Pt$$

If the half-cell reactions are written as

$$H_2(1\ atm) + 2\ H_2O(\ell) \longrightarrow 2\ H_3O^+(\text{var}) + 2\ e^- \qquad \text{(anode)}$$

$$2\ H_3O^+(1\ \text{M}) + 2\ e^- \longrightarrow H_2(1\ atm) + 2\ H_2O(\ell) \qquad \text{(cathode)}$$

then $n = 2$ and $Q = [H_3O^+(\text{variable})]^2$ because the other concentrations and gas pressures are unity. From the Nernst equation,

$$\Delta\mathscr{E} = \Delta\mathscr{E}° - \frac{0.0592 \text{ V}}{n} \log_{10} Q$$

and because $\Delta\mathscr{E}° = 0$,

$$\Delta\mathscr{E} = -\frac{0.0592 \text{ V}}{2} \log_{10} [H_3O^+]^2$$

$$= -0.0592 \text{ V} \log_{10} [H_3O^+] = (0.0592 \text{ V}) \text{ pH}$$

The measured cell voltage is proportional to the pH.

What we have described is a simple **pH meter.** It is inconvenient, though, to have to bubble hydrogen gas through the unknown and reference half-cells, and so a more portable and miniaturized pair of electrodes is employed to replace the hydrogen half-cells. In a typical commercial pH meter, two electrodes are dipped into the solution of unknown pH. One of these, the **glass electrode,** usually consists of an AgCl-coated silver electrode in contact with an HCl solution of known (e.g., 1.0 M) concentration in a thin-walled glass bulb. A pH-dependent potential develops across this thin glass membrane when the glass electrode is immersed in a solution of different, unknown $[H_3O^+]$. The second half-cell is frequently a **saturated calomel electrode,** consisting of a platinum wire in electrical contact with a paste of liquid mercury, calomel ($Hg_2Cl_2(s)$), and a saturated solution of KCl. The overall cell (Fig. 12.7) can be represented as

$$Ag|AgCl|Cl^- + H_3O^+(1.0 \text{ M})|\text{glass}|H_3O^+(\text{var}) \, \| \, Cl^-(\text{sat})|Hg_2Cl_2(s)|Hg|Pt$$

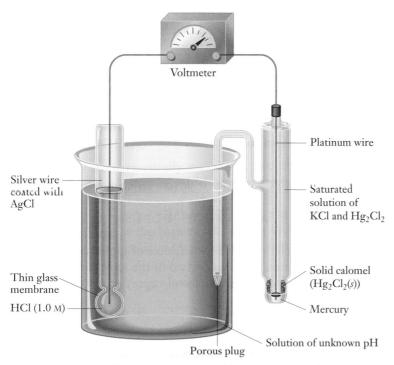

Voltmeter

Silver wire
coated with
AgCl

Thin glass
membrane

HCl (1.0 M)

Porous plug

Platinum wire

Saturated
solution of
KCl and $Hg_2Cl_2$

Solid calomel
($Hg_2Cl_2(s)$)

Mercury

Solution of unknown pH

**FIGURE 12.7**   A pH meter consists of a glass electrode (left) and a calomel electrode (right), both of which dip into a solution of unknown hydronium ion concentration.

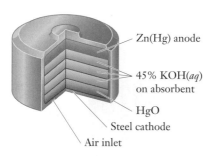

**FIGURE 12.9** A zinc–mercuric oxide dry cell, used in electric watches and cameras.

A third primary dry cell is the **zinc–mercuric oxide cell** depicted in Figure 12.9. It is frequently given the shape of a small button and is used in automatic cameras, hearing aids, digital calculators, and quartz–electric watches. This battery has an anode that is a mixture of mercury and zinc and a steel cathode in contact with solid mercury(II) oxide (HgO). The electrolyte is a 45% KOH solution that saturates an absorbent material. The anode half-reaction is the same as that in an alkaline dry cell,

$$Zn(s) + 2\ OH^-(aq) \longrightarrow Zn(OH)_2(s) + 2\ e^- \qquad \text{(anode)}$$

but the cathode half-reaction is now

$$HgO(s) + H_2O(\ell) + 2\ e^- \longrightarrow Hg(\ell) + 2\ OH^-(aq) \qquad \text{(cathode)}$$

The overall reaction is

$$Zn(s) + HgO(s) + H_2O(\ell) \longrightarrow Zn(OH)_2(s) + Hg(\ell)$$

This cell has a very stable output of 1.34 V, a fact that makes it especially valuable for use in communication equipment and scientific instruments.

## Rechargeable Batteries

In some batteries the electrodes can be regenerated after depletion by imposing an external potential across them that reverses the direction of current flow through the cell. These are called **secondary batteries,** and the process of reconstituting them to their original state is called "recharging." To recharge a run-down secondary battery, the voltage of the external source must be larger than that of the battery in its original state and, of course, opposite in polarity.

The **nickel–cadmium cell** (or *nicad battery;* Fig. 12.10) is used in hand-held electronic calculators and other cordless electric implements such as portable shavers. Its half-cell reactions during discharge are

$$Cd(s) + 2\ OH^-(aq) \longrightarrow Cd(OH)_2(s) + 2\ e^- \qquad \text{(anode)}$$

$$2\ NiO(OH)(s) + 2\ H_2O(\ell) + 2\ e^- \longrightarrow 2\ Ni(OH)_2(s) + 2\ OH^-(aq) \qquad \text{(cathode)}$$

$$Cd(s) + 2\ NiO(OH)(s) + 2\ H_2O(\ell) \longrightarrow Cd(OH)_2(s) + 2\ Ni(OH)_2(s)$$

This battery gives a fairly constant voltage of 1.4 V. When it is connected to an external voltage source, the preceding reactions are reversed as the battery is recharged.

One technically important secondary battery is the **lead–acid storage battery,** used in automobiles. A 12-V lead storage battery consists of six 2.0-V cells (Fig. 12.11) connected in series (cathode to anode) by an internal lead linkage and housed in a hard rubber or plastic case. In each cell the anode consists of metallic lead in porous form to maximize its contact area with the electrolyte. The cathode is of similar design, but its lead has been converted to lead dioxide. A sulfuric acid solution (37% by mass) serves as the electrolyte.

When the external circuit is completed, electrons are released from the anode to the external circuit and the resulting $Pb^{2+}$ ions precipitate on the electrode as insoluble lead sulfate. At the cathode, electrons from the external circuit reduce $PbO_2$ to water and $Pb^{2+}$ ions, which also precipitate as $PbSO_4$ on that electrode. The half-cell reactions are

**FIGURE 12.10** Rechargeable nickel–cadmium batteries. *(Eveready Battery Company)*

$$Pb(s) + SO_4^{2-}(aq) \longrightarrow PbSO_4(s) + 2\ e^- \qquad \text{(anode)}$$

$$PbO_2(s) + SO_4^{2-}(aq) + 4\ H_3O^+(aq) + 2\ e^- \longrightarrow PbSO_4(s) + 6\ H_2O(\ell)$$
$$\text{(cathode)}$$

$$Pb(s) + PbO_2(s) + 2\ SO_4^{2-}(aq) + 4\ H_3O^+(aq) \longrightarrow 2\ PbSO_4(s) + 6\ H_2O(\ell)$$

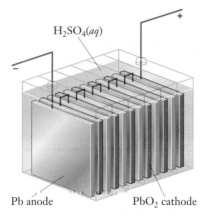

**FIGURE 12.11** In a lead–acid storage battery, anodes made of Pb alternate with cathodes of Pb coated with PbO₂. The electrolyte is sulfuric acid.

The anode and cathode are both largely converted to $PbSO_4(s)$ when the storage battery is fully discharged, and because sulfuric acid is a reactant, its concentration falls. Measuring the density of the electrolyte provides a quick way to estimate the state of charge of the battery.

When a voltage in excess of 12 V is applied across the terminals of the battery in the opposite direction, the half-cell reactions are reversed. This part of the cycle restores the battery to its initial state, ready to be used in another work-producing discharge half-cycle. Lead–acid storage batteries endure many thousands of cycles of discharge and charge before they ultimately fail because of the flaking off of $PbSO_4$ from the electrodes or the development of internal short circuits. In automobiles, they are usually not designed to undergo complete discharge–recharge cycles. Instead, a generator converts some of the kinetic energy of the vehicle to electrical energy for continuous or intermittent charging. About $1.8 \times 10^7$ J can be obtained in the discharge of an average automobile battery, and currents as large as 100 A are drawn for the short time needed to start the engine.

A drawback of the lead–acid storage battery is its low energy density, the amount of energy obtainable per kilogram of battery mass. This is not important when a battery is used to start a gasoline-powered automobile, but it precludes the battery's use in a vehicle driven by an electric motor. This difficulty of a low energy-to-mass ratio, which limits the range of vehicle operation before recharge is necessary, has spurred research by electrochemists to develop secondary batteries that have much higher energy densities.

One promising line of approach has been the development of rechargeable batteries that employ an alkali metal (lithium or sodium) as the anode and sulfur as the electron acceptor. Sulfur is a nonconductor of electricity, and so graphite is used as the cathode that conducts electrons to it. The elements must be in their liquid states, and so these batteries are high-temperature cells (sulfur melts at 112°C, lithium at 186°C, and sodium at 98°C). The sodium–sulfur cell (Fig. 12.12), for example, has an optimum operating temperature of 250°C. The half-cell reactions are

$$2\ Na \longrightarrow 2\ Na^+ + 2\ e^- \qquad \text{(anode)}$$
$$S + 2\ e^- \longrightarrow S^{2-} \qquad \text{(cathode)}$$

$$2\ Na + S \longrightarrow 2\ Na^+ + S^{2-}$$

This is actually an oversimplification of the cathodic process, because sulfide ion forms polysulfides with sulfur:

$$S^{2-} + nS \longrightarrow S_{n+1}^{2-}$$

but the fundamental principles of the cell operation are the same.

What makes the sodium–sulfur cell possible is a remarkable property of a compound called beta-alumina, which has the composition $NaAl_{11}O_{17}$. It allows sodium ions to migrate through its structure very easily, but blocks the passage of polysulfide ions. Therefore, it can function as a semipermeable medium like the membranes used in osmosis (Section 6.4). Such an ion-conducting solid electrolyte is essential to prevent direct chemical reaction between sulfur and sodium. The lithium–sulfur

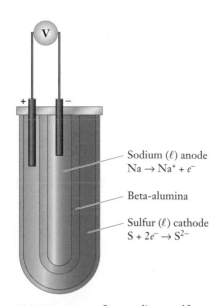

**FIGURE 12.12** In a sodium–sulfur battery, Na⁺ ions migrate through beta-alumina to the cathode to equalize the charge as electrons flow spontaneously from the anode to the cathode through the external circuit.

battery operates on similar principles, and other solid electrolytes such as calcium fluoride, which permits ionic transport of fluoride ion, may find use in cells that use those elements.

The considerable promise of the alkali metal–sulfur cell lies in its high energy density, which makes it possible to construct lightweight batteries capable of generating large currents. Their use in electric cars is an attractive possibility. When the cost of petroleum-based fuels increases and their supply becomes less reliable, cells such as the alkali metal–sulfur battery may come into their own.

## Fuel Cells

A battery is a closed system that delivers electrical energy by electrochemical reactions. Once the chemicals originally present are consumed, the battery must be either recharged or discarded. In contrast, a **fuel cell** is designed for continuous operation, with reactants (fuel) being supplied and products removed continuously. It is an energy converter, transforming chemical energy into electrical energy. Fuel cells based on the reaction

$$2 H_2(g) + O_2(g) \longrightarrow 2 H_2O(\ell)$$

were used on the Gemini and Apollo space vehicles, for example, to help meet the electrical requirements of the missions (Fig. 12.13). After the product was purified by ion exchange, it was drunk by the crew.

Figure 12.14 represents the hydrogen–oxygen fuel cell schematically. The electrodes can be any nonreactive conductor (graphite, for example), whose function is to conduct electrons into and out of the cell and to facilitate electron exchange between the gases and the ions in solution. The electrolyte transports charge through the cell, and the ions dissolved in it participate in the half-reactions at each electrode. Acidic solutions present corrosion problems, and so an alkaline solution is preferable (1 M NaOH, for instance). The anode half-reaction is then

$$H_2(g) + 2 OH^-(aq) \longrightarrow 2 H_2O(\ell) + 2 e^-$$

and the cathode reaction is

$$\tfrac{1}{2} O_2(g) + H_2O(\ell) + 2 e^- \longrightarrow 2 OH^-(aq)$$

The standard reduction potentials at 25°C (1 M concentrations, 1 atm pressure) are

$$\mathscr{E}°(H_2|H_2O) = -0.828 \text{ V} \qquad \text{and} \qquad \mathscr{E}°(O_2|OH^-) = 0.401 \text{ V}$$

The overall cell reaction is the production of water,

$$H_2(g) + \tfrac{1}{2} O_2(g) \longrightarrow H_2O(\ell)$$

**FIGURE 12.13** This hydrogen–oxygen fuel cell was used on U.S. space missions. (*United Technologies*)

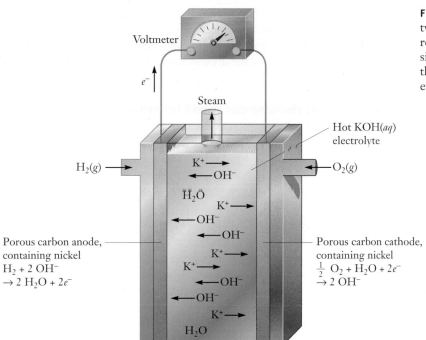

**Voltmeter**

$e^-$

Steam

H$_2$(g)

K$^+$ →
← OH$^-$

H$_2$O

K$^+$ →
← OH$^-$
← OH$^-$

K$^+$ →
K$^+$ →
← OH$^-$
← OH$^-$
K$^+$ →

H$_2$O

Hot KOH($aq$)
electrolyte

O$_2$(g)

Porous carbon anode,
containing nickel
H$_2$ + 2 OH$^-$
→ 2 H$_2$O + 2$e^-$

Porous carbon cathode,
containing nickel
$\frac{1}{2}$ O$_2$ + H$_2$O + 2$e^-$
→ 2 OH$^-$

**FIGURE 12.14**  In a hydrogen–oxygen fuel cell, the two gases are fed in separately and are oxidized or reduced on the electrodes. A hot solution of potassium hydroxide between the electrodes completes the circuit, and the steam produced in the reaction evaporates from the cell continuously.

and the cell voltage is

$$\Delta\mathscr{E}° = \mathscr{E}°(\text{cathode}) - \mathscr{E}°(\text{anode}) = 0.401 - (-0.828) - 1.229 \text{ V}$$

The overall cell voltage does not depend on pH because the OH$^-$ has dropped out; the same voltage would be obtained in an acidic fuel cell.

Another practical fuel cell accomplishes the overall reaction

$$CO(g) + \tfrac{1}{2} O_2(g) \longrightarrow CO_2(g)$$

for which the half-reactions are

$$CO(g) + 3 H_2O(\ell) \longrightarrow CO_2(g) + 2 H_3O^+(aq) + 2\ e^- \qquad \text{(anode)}$$

$$\tfrac{1}{2} O_2(g) + 2 H_3O^+(aq) + 2\ e^- \longrightarrow 3 H_2O(\ell) \qquad\qquad \text{(cathode)}$$

The electrolyte is usually concentrated phosphoric acid, and the operating temperature is between 100° and 200°C. Platinum is the electrode material of choice because it facilitates the electron transfer reactions.

Natural gas (largely CH$_4$) and even fuel oil can be "burned" electrochemically to provide electrical energy by either of two approaches. They can be converted to CO and H$_2$ or to CO$_2$ and H$_2$ prior to use in a fuel cell by reaction with steam,

$$CH_4(g) + H_2O(g) \longrightarrow CO(g) + 3 H_2(g)$$

$$CO(g) + H_2O(g) \longrightarrow CO_2(g) + H_2(g)$$

at temperatures of about 500°C. Alternatively, they can be used directly in a fuel cell at higher temperatures (up to 750°C) with a molten alkali metal carbonate as the electrolyte. Either type of fuel cell is attractive as an electrochemical energy converter in regions where hydrocarbon fuels are readily available but large-scale power plants (fossil or nuclear) are remote.

The theoretical advantage of electrochemical fuel cells over more traditional fuel technology can be seen from a thermodynamic analysis. If a chemical reaction such as the oxidation of a fuel can be carried out electrochemically, the maximum (reversible) work obtainable is equal to the free energy change $\Delta G$ (Section 12.3):

$$-w_{max}(\text{fuel cell}) = |\Delta G|$$

In traditional fuel technology, the same fuel would be burned in air, producing an amount of heat $q_P = \Delta H$, the enthalpy of combustion. The heat would then be used to run a heat engine-generator system to produce electrical power. The efficiency of conversion of heat to work is limited by the laws of thermodynamics. If the heat is supplied at temperature $T_h$ and if the lower operating temperature is $T_l$, the maximum work obtainable (Section 8.4) is

$$-w_{max}(\text{heat engine}) = \epsilon|q_P| = \frac{T_h - T_l}{T_h}|\Delta H|$$

Because the magnitude of $\Delta H$ is generally comparable to that of $\Delta G$ for fuel oxidation reactions, the fuel cell will be more efficient because the factor $(T_h - T_l)/T_h$ for a thermal engine is much less than 1. In practice, fuel cells and heat engines must be operated irreversibly (to increase the rate of energy production), and the work obtained with both is less than $w_{max}$. The overall efficiency of practical heat engines rarely exceeds 30 – 35%, whereas that of fuel cells can be in the 60 – 70% range. This advantage of fuel cell technology is partially offset by the greater expense of constructing and maintaining fuel cells, however.

---

## 12.6

## CORROSION AND ITS PREVENTION

The **corrosion** of metals is one of the most significant problems faced by advanced industrial societies (Fig. 12.15). It has been estimated that in the United States alone the annual cost of corrosion amounts to tens of *billions* of dollars. Effects of corrosion are both visible (the formation of rust on exposed iron surfaces) and invisible (the cracking and resulting loss of strength of metal beneath the surface). The mechanism of corrosion must be understood before processes can be developed for its prevention.

Although corrosion is a serious problem for many metals, we will focus on the spontaneous electrochemical reactions of iron. Corrosion can be pictured as a "short-circuited" galvanic cell, in which some regions of the metal surface act as cathodes and others as anodes, and the electric "circuit" is completed by electron flow through the iron itself. These electrochemical cells form in parts of the metal where there are impurities or in regions that are subject to stress. The anode reaction is

$$Fe(s) \longrightarrow Fe^{2+}(aq) + 2\ e^-$$

Various cathode reactions are possible. In the absence of oxygen (for example, at the bottom of a lake), the corrosion reactions are

$$
\begin{array}{ll}
Fe(s) \longrightarrow Fe^{2+}(aq) + 2\ e^- & \text{(anode)} \\
2\ H_2O(\ell) + 2\ e^- \longrightarrow 2\ OH^-(aq) + H_2(g) & \text{(cathode)} \\
\hline
Fe(s) + 2\ H_2O(\ell) \longrightarrow Fe^{2+}(aq) + 2\ OH^-(aq) + H_2(g) &
\end{array}
$$

**FIGURE 12.15** Rust and peeling paint on an iron hinge and handle. *(Copyright Eric Kamp/Phototake NYC)*

These reactions are generally slow, however, and do not cause serious amounts of corrosion. Far more extensive corrosion takes place when the iron is in contact with both oxygen and water. In this case the cathode reaction is

$$\tfrac{1}{2} O_2(g) + 2 H_3O^+(aq) + 2 e^- \longrightarrow 3 H_2O(\ell)$$

The $Fe^{2+}$ ions formed simultaneously at the anode migrate to the cathode, where they are further oxidized by $O_2$ to the $+3$ oxidation state to form rust ($Fe_2O_3 \cdot xH_2O$), a hydrated form of iron(III) oxide:

$$2 Fe^{2+}(aq) + \tfrac{1}{2} O_2(g) + (6 + x)H_2O(\ell) \longrightarrow Fe_2O_3 \cdot xH_2O + 4 H_3O^+(aq)$$

The hydronium ions produced in this reaction allow the corrosion cycle to continue.

When a portion of the paint that protects a piece of iron or steel is chipped off (Fig. 12.16), the exposed area acts as the cathode because it is open to the atmosphere (air and water) and is therefore rich in oxygen, whereas oxygen-poor areas *under* the paint act as anodes. Rust forms on the cathode (the visible, exposed region), and pitting (loss of metal through oxidation of iron and flow of metal ions to the cathode) occurs at the anode. This pitting can lead to loss of structural strength in girders and other supports. The most serious harm done by corrosion is not the visible rusting but the damage done beneath the painted surface.

A number of factors speed corrosion. Dissolved salt provides an electrolyte that improves the flow of charge through solution; a well-known example is the rapid rusting of cars in areas where salt is spread on icy roads. Higher acidity also increases corrosion, as seen by the role of $H_3O^+$ as a reactant in the reduction process at the cathode. Acidity is enhanced by the presence of dissolved $CO_2$ (which produces $H_3O^+$ and $HCO_3^-$ ions) and by air pollution from oxides of sulfur, which leads to the formation of dissolved sulfuric acid in acid precipitation.

Corrosion of iron can be inhibited in a number of ways. Coatings of paint or plastics obviously protect the metal, but they can crack or suffer other damage, thereby localizing and accentuating the process. A very important method of protecting metals arises from the phenomenon of **passivation,** in which a thin metal oxide layer forms on the surface and prevents further electrochemical reactions.

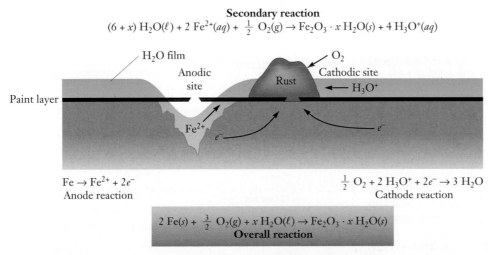

**Secondary reaction**
$$(6 + x) H_2O(\ell) + 2 Fe^{2+}(aq) + \tfrac{1}{2} O_2(g) \to Fe_2O_3 \cdot x H_2O(s) + 4 H_3O^+(aq)$$

Anode reaction
$$Fe \to Fe^{2+} + 2e^-$$

Cathode reaction
$$\tfrac{1}{2} O_2 + 2 H_3O^+ + 2e^- \to 3 H_2O$$

**Overall reaction**
$$2 Fe(s) + \tfrac{3}{2} O_2(g) + x H_2O(\ell) \to Fe_2O_3 \cdot x H_2O(s)$$

**FIGURE 12.16** The corrosion of iron. Note that pitting occurs in the anodic region, and rust appears in the cathodic region.

Some metals become passivated spontaneously upon exposure to air; aluminum, for example, reacts with oxygen to form a thin protective layer of $Al_2O_3$. Special paints designed for rust prevention contain potassium dichromate ($K_2Cr_2O_7$) and lead oxide ($Pb_3O_4$), which cause the superficial oxidation and passivation of iron. Stainless steel is an alloy of iron with chromium in which the chromium leads to passivation and prevents rusting.

A different way of preventing iron corrosion is to use a **sacrificial anode**. A comparison of the standard reduction potentials of iron and magnesium

$$Fe^{2+} + 2\ e^- \longrightarrow Fe(s) \qquad\qquad \mathscr{E}° = -0.41\ V$$

$$Mg^{2+} + 2\ e^- \longrightarrow Mg(s) \qquad\qquad \mathscr{E}° = -2.39\ V$$

shows that $Mg^{2+}$ is much harder to reduce than $Fe^{2+}$ or, conversely, that $Mg(s)$ is more easily oxidized than $Fe(s)$. A piece of magnesium in electrical contact with iron is oxidized in preference to the iron, and the iron is therefore protected. The magnesium is the sacrificial anode, and once it is consumed by oxidation it must be replaced. This method is used to protect ship hulls, bridges, and iron water pipes from corrosion. Magnesium plates are attached at regular intervals along a piece of buried pipe, and it is far easier to replace them periodically than to replace the entire pipe.

# A   D E E P E R   L O O K . . .

## 12.7   Electrolysis of Water and Aqueous Solutions

Until now we have considered primarily galvanic cells, in which chemical reactions are harnessed to produce electricity. In the reverse process, electrolysis, a source of electrical energy is used to drive a nonspontaneous reaction in the opposite direction to give products whose free energies are high enough that they may not ordinarily be present in nature. The recharging of a run-down secondary battery is an example. Electrolysis is used in the industrial production of metals, a subject discussed in Section 20.5. Here we examine only the electrolysis of water and aqueous solutions. Consider first the electrolysis of water between inert electrodes such as platinum, for which the half-cell reactions are

$$2\ H_3O^+(aq) + 2\ e^- \longrightarrow H_2(g) + 2\ H_2O(\ell) \qquad \text{(cathode)}$$

$$3\ H_2O(\ell) \longrightarrow \tfrac{1}{2}\ O_2(g) + 2\ H_3O^+(aq) + 2\ e^-$$
$$\text{(anode)}$$

$$\overline{\phantom{xxxxxxxxxx}}$$

$$H_2O(\ell) \longrightarrow H_2(g) + \tfrac{1}{2}\ O_2(g)$$

A practical problem immediately arises. The concentration of $H_3O^+(aq)$ and $OH^-(aq)$ ions in pure water at 25°C is only $1.0 \times 10^{-7}$ M, so the *rate* of electrolysis will be exceedingly small. This practical consideration is put aside for the moment, however, because it does not alter the *thermodynamic* analysis.

The potential $\mathscr{E}°$ for the cathode reaction is by definition 0 V, but because the $H_3O^+(aq)$ concentration in pure water is not 1 M but $1.0 \times 10^{-7}$ M, $\mathscr{E}$ differs from $\mathscr{E}°$ and equals

$$\mathscr{E}(\text{cathode}) = \mathscr{E}°(\text{cathode}) - \frac{0.0592\ V}{n_{hc}} \log Q_{hc}$$

$$= 0.00 - \frac{0.0592\ V}{2} \log \frac{P_{H_2}}{[H_3O^+]^2}$$

If $H_2(g)$ is produced at atmospheric pressure, this simplifies to

$$\mathscr{E}(\text{cathode}) = 0.00 - \frac{0.0592\ V}{2} \log \frac{1}{(10^{-7})^2}$$

$$= -0.414\ V$$

The anode half-reaction written as a reduction (i.e., in reverse) is

$$\tfrac{1}{2}\ O_2(g) + 2\ H_3O^+(aq) + 2\ e^- \longrightarrow 3\ H_2O(\ell)$$

A table of standard reduction potentials (Appendix E) gives $\mathscr{E}° = 1.229$ V. In the present case, the $H_3O^+(aq)$ concentration is $1.0 \times 10^{-7}$ M rather than 1 M, and so

$$\mathscr{E}(\text{anode}) = \mathscr{E}° - \frac{0.0592\ V}{2} \log \frac{1}{(P_{O_2})^{1/2}[H_3O^+]^2}$$

$$= 1.229\ V - \frac{0.0592\ V}{2} \log \frac{1}{(10^{-7})^2}$$

$$= 0.815\ V$$

if $P_{O_2} = 1$ atm. The overall cell voltage is

$$\Delta\mathscr{E} = \mathscr{E}(\text{cathode}) - \mathscr{E}(\text{anode}) = -0.414 - 0.815 \text{ V} = -1.229 \text{ V}$$

A negative $\Delta\mathscr{E}$ means that the process does not occur spontaneously; it can be made to occur only by applying an external voltage sufficient to overcome the intrinsic negative voltage of the cell. In the electrolysis of water, this minimum external voltage, which is called the **decomposition potential** of water, is 1.229 V. When this potential difference has been imposed, an overall cell reaction will occur, given by the sum of the reduction and oxidation half-reactions:

$$H_2O(\ell) \longrightarrow \tfrac{1}{2} O_2(g) + H_2(g)$$

Hydrogen will bubble off at the cathode and oxygen at the anode.

Suppose an external voltage is now used to electrolyze an electrolyte solution instead of pure water. The resultant products depend upon the concentrations of the ions present and their half-cell potentials. In a 0.10 M NaCl solution, we could conceive of the following processes taking place:

Cathode:

$$Na^+(0.1 \text{ M}) + e^- \longrightarrow Na(s)$$

$$2 H_3O^+(10^{-7} \text{ M}) + 2 e^- \longrightarrow H_2(g) + 2 H_2O(\ell)$$

Anode:

$$Cl^-(0.1 \text{ M}) \longrightarrow \tfrac{1}{2} Cl_2(g) + e^-$$

$$3 H_2O(\ell) \longrightarrow \tfrac{1}{2} O_2(g) + 2 H_3O^+ (10^{-7} \text{ M}) + 2 e^-$$

In each of these pairs of possible processes, which one will actually occur?

For the first half-reaction the reduction potential is

$$\mathscr{E}(Na^+|Na) = \mathscr{E}°(Na^+|Na) - \frac{0.0592 \text{ V}}{1} \log \frac{1}{[Na^+]}$$

$$= -2.71 - 0.06 = -2.77 \text{ V}$$

Because this result is algebraically less than the half-cell voltage $\mathscr{E}(H_3O^+(10^{-7} \text{ M})|H_2) = -0.414$ V for pure water, the reduction of $Na^+(aq)$ is impossible, and $H_2(g)$ is the cathode product.

For the third half-reaction, the reduction potential is

$$\mathscr{E}(Cl_2|Cl^-) = \mathscr{E}°(Cl_2|Cl^-) - \frac{0.0592 \text{ V}}{1} \log \frac{[Cl^-]}{P_{Cl_2}^{1/2}}$$

$$= 1.36 + 0.06 = 1.42 \text{ V}$$

Because this is more positive than the half-cell voltage $\mathscr{E}(O_2,H_3O^+(10^{-7} \text{ M})|H_2O) = 0.815$ V for pure water, $Cl_2$ has a *greater* tendency to be reduced than $O_2$, and so $Cl^-$ has a *lesser* tendency to be oxidized than $H_2O$, and the anode product is $O_2(g)$. If we try to increase the external potential above 1.229 V, all that will happen is that water will be electrolyzed at a greater rate to produce hydrogen and oxygen. Sodium and chlorine will not appear as long as sufficient water is present.

Suppose now that 0.10 M NaI is substituted for the 0.10 M NaCl solution. Sodium ions still will not be reduced; however, $\mathscr{E}(I_2|I^-)$ is

$$\mathscr{E}(I_2|I^-) = \mathscr{E}°(I_2|I^-) - \frac{0.0592 \text{ V}}{1} \log [I^-]$$

$$= 0.535 + 0.059 = 0.594 \text{ V}$$

The half-cell potential for the reduction of iodine is *less* positive than the reduction potential of $O_2(g)$ in water at pH = 7 (0.815 V), so the oxidation of 0.10 M $I^-$ occurs in preference to the oxidation of water. The anode reaction is therefore

$$I^-(0.10 \text{ M}) \longrightarrow \tfrac{1}{2} I_2(s) + e^-$$

and the overall cell reaction is

$$H_3O^+(10^{-7} \text{ M}) + I^-(0.10 \text{ M}) \longrightarrow \tfrac{1}{2} H_2(g) + \tfrac{1}{2} I_2(s) + H_2O(\ell)$$

The intrinsic cell voltage is

$$\Delta\mathscr{E} = \mathscr{E}(\text{cathode}) - \mathscr{E}(\text{anode}) = -0.414 - 0.594 = -1.008 \text{ V} < 0$$

When the decomposition potential of the solution (1.008 V) is exceeded, $H_2(g)$ and $I_2(s)$ begin to form. Of course, this causes $[I^-]$ to decrease and $\mathscr{E}(I_2|I^-)$ to increase. When the iodide ion concentration reaches about $2 \times 10^{-5}$ M, $\mathscr{E}(I_2|I^-)$ will have increased to 0.815 V and the external voltage required to maintain electrolysis will have increased to 1.229 V. At this point, water will start to be electrolyzed and oxygen will be produced at the anode.

Our results for the electrolysis of *neutral* aqueous solutions are summarized as follows:

1. A species can be reduced only if its reduction potential is algebraically greater than $-0.414$ V.
2. A species can be oxidized only if its reduction potential is algebraically smaller than 0.815 V.

In solutions with pH different from 7, these results must be modified, as shown by the following example.

## EXAMPLE 12.13

An aqueous 0.10 M solution of $NiCl_2$ is electrolyzed under 1 atm pressure. Determine the products formed at the anode and the cathode and the decomposition potential, if the pH is **(a)** 7.0; **(b)** 0.0.

### Solution

The reduction of $Ni^{2+}$ at the cathode has the half-cell potential

$$Ni^{2+}(aq) + 2 e^- \longrightarrow Ni(s)$$

$$\mathscr{E}(Ni^{2+}|Ni) = \mathscr{E}°(Ni^{2+}|Ni) - \frac{0.0592 \text{ V}}{2} \log \frac{1}{[Ni^{2+}]}$$

$$= -0.23 - 0.03 = -0.26 \text{ V}$$

and the oxidation of $Cl^-$ at the anode has the *reduction* potential

$$\tfrac{1}{2} Cl_2(g) + e^- \longrightarrow Cl^-(aq)$$

$$\mathscr{E}(Cl_2|Cl^-) = \mathscr{E}°(Cl_2|Cl^-) - \frac{0.0592 \text{ V}}{1} \log \frac{[Cl^-]}{P_{Cl_2}^{1/2}}$$

$$= 1.36 \text{ V} - (0.0592 \text{ V}) \log 0.2 = 1.40 \text{ V}$$

**(a)** In neutral solution the cathode half-reaction is

$$Ni^{2+}(aq) + 2\,e^- \longrightarrow Ni(s)$$

because $\mathscr{E}(Ni^{2+}|Ni) = -0.26$ V $> -0.414$ V. The anode half-reaction is

$$3\,H_2O(\ell) \longrightarrow 2\,H_3O^+(10^{-7}\,\text{M}) + \tfrac{1}{2}\,O_2(g) + 2\,e^-$$

because $\mathscr{E}(Cl_2|Cl^-) = 1.40$ V $> 0.815$ V (the reduction potential for this half-reaction). The cell voltage is

$$\Delta\mathscr{E} = \mathscr{E}(\text{cathode}) - \mathscr{E}(\text{anode}) = -0.26 - 0.815\ \text{V} = -1.08\ \text{V}$$

and so the decomposition potential is 1.08 V.

**(b)** In 1.0 M acid solution (pH = 0.0), the anode half-reaction is still

$$3\,H_2O(\ell) \longrightarrow 2\,H_3O^+(1\,\text{M}) + \tfrac{1}{2}\,O_2(g) + 2\,e^-$$

The reduction potential for this reaction is now $\mathscr{E}_{hc} = \mathscr{E}^{\circ}_{hc} = 1.229$ V, which is still less than 1.40 V, the reduction potential for the competing reaction involving chlorine.

The cathode half-reaction now becomes

$$2\,H_3O^+(1\,\text{M}) + 2\,e^- \longrightarrow H_2(g) + 2\,H_2O(\ell)$$

because $\mathscr{E}(H_3O^+|H_2) = 0.0$ V $> -0.26$ V for $\mathscr{E}(Ni^{2+}|Ni)$. We now have

$$\Delta\mathscr{E} = \mathscr{E}(\text{cathode}) - \mathscr{E}(\text{anode}) = 0.00 - 1.229 = -1.229\ \text{V}$$

and so the decomposition potential of the solution is now 1.229 V, just as it is for pure water.

**Related Problems: 63, 64**

---

# CUMULATIVE EXERCISE

## *Manganese*

Manganese is the 12th most abundant element on the earth's surface. Its most important ore source is pyrolusite ($MnO_2$). The preparation and uses of manganese and its compounds (which range up to +7 in oxidation state) are intimately bound up with electrochemistry.

**(a)** Elemental manganese in a state of high purity can be prepared by electrolyzing aqueous solutions of $Mn^{2+}$. At which electrode (anode or cathode) does the Mn appear? Electrolysis is also used to make $MnO_2$ in high purity from $Mn^{2+}$ solutions. At which electrode does the $MnO_2$ appear?

**(b)** The Winkler method is an analytical procedure for determining the amount of oxygen dissolved in water. In the first step, $Mn(OH)_2(s)$ is oxidized by gaseous oxygen to $Mn(OH)_3(s)$ in basic aqueous solution. Write the oxidation and reduction half-equations for this step, and write the balanced overall equation. Then use Appendix E to calculate the standard voltage that would be measured if this reaction were carried out in an electrochemical cell.

**(c)** Calculate the equilibrium constant at 25°C for the reaction in part (b).

**(d)** In the second step of the Winkler method, the $Mn(OH)_3$ is acidified to give $Mn^{3+}$, and iodide ion is added. Will $Mn^{3+}$ spontaneously oxidize $I^-$? Write a balanced equation for its reaction with $I^-$, and use data from Appendix E to calculate its equilibrium constant. Titration of the $I_2$ produced completes the use of the Winkler method.

**(e)** Manganese(IV) is an even stronger oxidizing agent than manganese(III). It oxidizes zinc to $Zn^{2+}$ in the dry cell. Such a battery has a cell voltage of 1.5 V. Calculate the electrical work done by this battery in 1.00 hour if it produces a steady current of 0.70 A.

**(f)** Calculate the mass of zinc reacting in the process described in part (e).

**(g)** The reduction potential of permanganate ion (+7 oxidation state) in acidic aqueous solution is given by

$$MnO_4^-(aq) + 8\,H_3O^+(aq) + 5\,e^- \longrightarrow Mn^{2+}(aq) + 12\,H_2O(\ell)$$
$$\mathscr{E}^{\circ} = 1.491\ \text{V}$$

Black crystals of manganite, MnOOH, an ore of manganese. *(Copyright Biophoto Associates/Photo Researchers, Inc.)*

whereas that of the analogous fifth-period species, pertechnetate ion, is

$$TcO_4^-(aq) + 8\ H_3O^+(aq) + 5\ e^- \longrightarrow Tc^{2+}(aq) + 12\ H_2O(\ell)$$

$$\mathscr{E}° = 0.500\ V$$

Which is the stronger oxidizing agent, permanganate ion or pertechnetate ion? (h) A galvanic cell is made from two half-cells. In the first, a platinum electrode is immersed in a solution at pH 2.00 that is 0.100 M in both $MnO_4^-$ and $Mn^{2+}$. In the second, a zinc electrode is immersed in a 0.0100 M solution of $Zn(NO_3)_2$. Calculate the cell voltage that will be measured.

### Answers

(a) Mn appears at the cathode and $MnO_2$ at the anode.

(b)

$$Mn(OH)_2(s) + OH^-(aq) \longrightarrow Mn(OH)_3(s) + e^- \qquad \text{(oxidation)}$$

$$\underline{O_2(g) + 2\ H_2O(\ell) + 4\ e^- \longrightarrow 4\ OH^-(aq) \qquad \text{(reduction)}}$$

$$4\ Mn(OH)_2(s) + O_2(g) + 2\ H_2O(\ell) \longrightarrow 4\ Mn(OH)_3(s)$$

$$\Delta\mathscr{E}° = 0.401 - (-0.40) = 0.80\ V$$

(c) $K = 1 \times 10^{54}$

(d) $Mn^{3+}$ will spontaneously oxidize $I^-$.

$$2\ Mn^{3+}(aq) + 2\ I^-(aq) \longrightarrow 2\ Mn^{2+}(aq) + I_2(s) \qquad K = 9 \times 10^{32}$$

(e) $3.8 \times 10^3$ J

(f) 0.85 g Zn is oxidized.

(g) Permanganate ion

(h) 2.12 V

## CONCEPTS & SKILLS

*After studying this chapter and working the problems that follow, you should be able to*

1. Balance equations for redox reactions in aqueous solution, using the half-reaction method (Section 12.1, problems 1–4).
2. Define the terms "anode" and "cathode" and give the convention that is used to represent a galvanic cell (Section 12.2, problems 9–10).
3. Use Faraday's laws to calculate the quantities of substances produced or consumed at the electrodes of electrochemical cells in relation to the total charge passing through the circuit (Section 12.2, problems 11–18).
4. Explain the relationship between the free energy change in a galvanic cell and the amount of electrical work (Section 12.3, problems 19–20).
5. Combine half-cell reactions and their standard reduction potentials to obtain the overall reactions and standard voltages of electrochemical cells (Section 12.3, problems 21–24).
6. Combine half-reactions and their standard reduction potentials to form other half-reactions and their standard reduction potentials (Section 12.3, problems 31–32).
7. Use reduction potential diagrams to determine strengths of oxidizing and reducing agents and stability toward disproportionation (Section 12.3, problems 33–34).
8. Apply the Nernst equation to calculate the voltage of a cell in which reactants and products are not in their standard states (Section 12.4, problems 35–42).

**25.** Would you expect powdered solid aluminum to act as an oxidizing agent or as a reducing agent?

**26.** Would you expect potassium perchlorate, $KClO_4(aq)$, in a concentrated acidic solution to act as an oxidizing agent or as a reducing agent?

**27.** Bromine is sometimes used in place of chlorine as a disinfectant in swimming pools. If the effectiveness of a chemical as a disinfectant depends solely on its strength as an oxidizing agent, do you expect bromine to be better or worse than chlorine as a disinfectant, at a given concentration?

**28.** Many bleaches, including chlorine and its oxides, oxidize dye compounds in cloth. Predict which of the following will be the strongest bleach at a given concentration and pH 0: $NaClO_3(aq)$, $NaClO(aq)$, $Cl_2(aq)$. How does the strongest chlorine-containing bleach compare in strength with ozone, $O_3(g)$?

**29.** Suppose you have the following reagents available at pH 0, atmospheric pressure, and 1 M concentration:

$$Co(s), Ag^+(aq), Cl^-(aq), Cr(s), BrO_3^-(aq), I_2(s)$$

(a) Which is the strongest oxidizing agent?
(b) Which is the strongest reducing agent?
(c) Which reagent will reduce $Pb^{2+}(aq)$ while leaving $Cd^{2+}(aq)$ unreacted?

**30.** Suppose you have the following reagents available at pH 0, atmospheric pressure, and 1 M concentration:

$$Sc(s), Hg_2^{2+}(aq), Cr_2O_7^{2-}(aq), H_2O_2(aq), Sn^{2+}(aq), Ni(s)$$

(a) Which is the strongest oxidizing agent?
(b) Which is the strongest reducing agent?
(c) Which reagent will oxidize $Fe(s)$ while leaving $Cu(s)$ unreacted?

**31.** (a) Use the data from Appendix E to calculate the half-cell potential $\mathscr{E}°$ for the half-reaction

$$Mn^{3+}(aq) + 3\ e^- \longrightarrow Mn(s)$$

(b) Consider the disproportionation reaction

$$3\ Mn^{2+}(aq) \rightleftharpoons Mn(s) + 2\ Mn^{3+}(aq)$$

Will $Mn^{2+}$ disproportionate in aqueous solution?

**32.** The following standard reduction potentials have been measured in aqueous solution at 25°C:

$$Tl^{3+} + e^- \longrightarrow Tl^{2+} \qquad \mathscr{E}° = -0.37\ V$$

$$Tl^{3+} + 2\ e^- \longrightarrow Tl^+ \qquad \mathscr{E}° = 1.25\ V$$

(a) Calculate the half-cell potential for the half-reaction

$$Tl^{2+} + e^- \longrightarrow Tl^+$$

(b) Consider the disproportionation reaction

$$2\ Tl^{2+}(aq) \rightleftharpoons Tl^{3+}(aq) + Tl^+(aq)$$

Will $Tl^{2+}$ disproportionate in aqueous solution?

**33.** The following reduction potentials are measured at pH 0:

$$BrO_3^- + 6\ H_3O^+ + 5\ e^- \longrightarrow \tfrac{1}{2}Br_2(\ell) + 9\ H_2O$$
$$\mathscr{E}° = 1.52\ V$$

$$Br_2(\ell) + 2\ e^- \longrightarrow 2\ Br^- \qquad \mathscr{E}° = 1.065\ V$$

(a) Will bromine disproportionate spontaneously in acidic solution?
(b) Which is the stronger reducing agent at pH 0: $Br_2(\ell)$ or $Br^-$?

**34.** The following reduction potentials are measured at pH 14:

$$ClO^- + H_2O + 2\ e^- \longrightarrow Cl^- + 2\ OH^- \qquad \mathscr{E}° = 0.90\ V$$

$$ClO_2^- + H_2O + 2\ e^- \longrightarrow ClO^- + 2\ OH^- \qquad \mathscr{E}° = 0.59\ V$$

(a) Will $ClO^-$ disproportionate spontaneously in basic solution?
(b) Which is the stronger reducing agent at pH 14: $ClO^-$ or $Cl^-$?

## Concentration Effects and the Nernst Equation

**35.** A galvanic cell is constructed that carries out the reaction

$$Pb^{2+}(aq) + 2\ Cr^{2+}(aq) \longrightarrow Pb(s) + 2\ Cr^{3+}(aq)$$

If the initial concentration of $Pb^{2+}(aq)$ is 0.15 M, that of $Cr^{2+}(aq)$ is 0.20 M, and that of $Cr^{3+}(aq)$ is 0.0030 M, calculate the initial voltage generated by the cell at 25°C.

**36.** A galvanic cell is constructed that carries out the reaction

$$2\ Ag(s) + Cl_2(g) \longrightarrow 2\ Ag^+(aq) + 2\ Cl^-(aq)$$

If the partial pressure of $Cl_2(g)$ is 1.00 atm, the initial concentration of $Ag^+(aq)$ is 0.25 M, and that of $Cl^-(aq)$ is 0.016 M, calculate the initial voltage generated by the cell at 25°C.

**37.** Calculate the reduction potential for a $Pt|Cr^{3+},Cr^{2+}$ half-cell in which $[Cr^{3+}]$ is 0.15 M and $[Cr^{2+}]$ is 0.0019 M.

**38.** Calculate the reduction potential for an $I_2(s)|I^-$ half-cell in which $[I^-]$ is $1.5 \times 10^{-6}$ M.

**39.** An $I_2(s)|I^-$ (1.00 M) half-cell is connected to an $H_3O^+|H_2$(1 atm) half-cell in which the concentration of the hydronium ion is unknown. The measured cell voltage is 0.841 V and the $I_2|I^-$ half-cell is the cathode. What is the pH in the $H_3O^+|H_2$ half-cell?

**40.** A $Cu^{2+}$(1.00 M)$|Cu$ half-cell is connected to a $Br_2(\ell)|Br^-$ half-cell in which the concentration of bromide ion is unknown. The measured cell voltage is 0.963 V, and the $Cu^{2+}|Cu$ half-cell is the anode. What is the bromide ion concentration in the $Br_2(\ell)|Br^-$ half-cell?

**41.** The following reaction occurs in an electrochemical cell:

$$3\ HClO_2(aq) + 2\ Cr^{3+}(aq) + 12\ H_2O(\ell) \longrightarrow$$
$$3\ HClO(aq) + Cr_2O_7^{2-}(aq) + 8\ H_3O^+(aq)$$

(a) Calculate $\Delta\mathscr{E}°$ for this cell.
(b) At pH 0, with $[Cr_2O_7^{2-}] = 0.80$ M, $[HClO_2] = 0.15$ M, and $[HClO] = 0.20$ M, the cell voltage is found to be 0.15 V. Calculate the concentration of $Cr^{3+}(aq)$ in the cell.

42. A galvanic cell is constructed in which the overall reaction is

$$Cr_2O_7^{2-}(aq) + 14\ H_3O^+(aq) + 6\ I^-(aq) \longrightarrow$$
$$2\ Cr^{3+}(aq) + 3\ I_2(s) + 21\ H_2O(\ell)$$

    (a) Calculate $\Delta\mathscr{E}°$ for this cell.
    (b) At pH 0, with $[Cr_2O_7^{2-}] = 1.5$ M and $[I^-] = 0.40$ M, the cell voltage is found to equal 0.87 V. Calculate the concentration of $Cr^{3+}(aq)$ in the cell.

43. By using the half-cell potentials in Appendix E, calculate the equilibrium constant at 25°C for the reaction in problem 41. Dichromate ion ($Cr_2O_7^{2-}$) is orange, and $Cr^{3+}$ is light green in aqueous solution. If 2.00 L of 1.00 M $HClO_2$ solution is added to 2.00 L of 0.50 M $Cr(NO_3)_3$ solution, what color will the resulting solution have?

44. By using the half-cell potentials in Appendix E, calculate the equilibrium constant at 25°C for the reaction

$$6\ Hg^{2+}(aq) + 2\ Au(s) \rightleftharpoons 3\ Hg_2^{2+}(aq) + 2\ Au^{3+}(aq)$$

    If 1.00 L of a 1.00 M $Au(NO_3)_3$ solution is added to 1.00 L of a 1.00 M $Hg_2(NO_3)_2$ solution, calculate the concentrations of $Hg^{2+}$, $Hg_2^{2+}$, and $Au^{3+}$ at equilibrium.

45. The following standard reduction potentials have been determined for the aqueous chemistry of indium:

$$In^{3+}(aq) + 2\ e^- \longrightarrow In^+(aq) \qquad \mathscr{E}° = -0.40\ V$$
$$In^+(aq) + e^- \longrightarrow In(s) \qquad \mathscr{E}° = -0.21\ V$$

    Calculate the equilibrium constant ($K$) for the disproportionation of $In^+(aq)$ at 25°C.

$$3\ In^+(aq) \rightleftharpoons 2\ In(s) + In^{3+}(aq)$$

46. Use data from Appendix E to compute the equilibrium constant for the reaction

$$Hg^{2+}(aq) + Hg(\ell) \rightleftharpoons Hg_2^{2+}(aq)$$

47. A galvanic cell consists of a $Pt|H_3O^+(1.00\ M)|H_2(g)$ cathode connected to a $Pt|H_3O^+(aq)|H_2(g)$ anode in which the concentration of $H_3O^+$ is unknown but is kept constant by the action of a buffer consisting of a weak acid, HA(0.10 M), mixed with its conjugate base, $A^-$(0.10 M). The measured cell voltage is $\Delta\mathscr{E} = 0.150$ V at 25°C, with a hydrogen pressure of 1.00 atm at both electrodes. Calculate the pH in the buffer solution, and from it determine the $K_a$ of the weak acid.

48. In a galvanic cell, the cathode consists of a $Ag^+(1.00\ M)|Ag$ half-cell. The anode is a platinum wire, with hydrogen bubbling over it at 1.00-atm pressure, that is immersed in a buffer solution containing benzoic acid and sodium benzoate. The concentration of benzoic acid ($C_6H_5COOH$) is 0.10 M, and that of benzoate ion ($C_6H_5COO^-$) is 0.050 M. The overall cell reaction is then

$$Ag^+(aq) + \tfrac{1}{2}\ H_2(g) + H_2O(\ell) \longrightarrow Ag(s) + H_3O^+(aq)$$

    and the measured cell voltage is 1.030 V. Calculate the pH in the buffer solution and determine the $K_a$ of benzoic acid.

49. A galvanic cell is constructed in which the overall reaction is

$$Br_2(\ell) + H_2(g) + 2\ H_2O(\ell) \longrightarrow 2\ Br^-(aq) + 2\ H_3O^+(aq)$$

    (a) Calculate $\Delta\mathscr{E}°$ for this cell.
    (b) Silver ions are added until AgBr precipitates at the cathode and $[Ag^+]$ reaches 0.060 M. The cell voltage is then measured to be 1.710 V at pH = 0 and $P_{H_2} = 1.0$ atm. Calculate $[Br^-]$ under these conditions.
    (c) Calculate the solubility product constant $K_{sp}$ for AgBr.

50. A galvanic cell is constructed in which the overall reaction is

$$Pb(s) + 2\ H_3O^+(aq) \longrightarrow Pb^{2+}(aq) + H_2(g) + 2\ H_2O(\ell)$$

    (a) Calculate $\Delta\mathscr{E}°$ for this cell.
    (b) Chloride ions are added until $PbCl_2$ precipitates at the anode and $[Cl^-]$ reaches 0.15 M. The cell voltage is then measured to be 0.22 V at pH = 0 and $P_{H_2} = 1.0$ atm. Calculate $[Pb^{2+}]$ under these conditions.
    (c) Calculate the solubility product constant $K_{sp}$ of $PbCl_2$.

## Batteries and Fuel Cells

51. Calculate the voltage $\Delta\mathscr{E}°$ of a lead–acid cell if all reactants and products are in their standard states. What will be the voltage if six such cells are connected in series?

52. Calculate the standard voltage of the zinc–mercuric oxide cell shown in Figure 12.9. (*Hint:* The easiest way to proceed is to calculate $\Delta G°$ for the corresponding overall reaction, and then find $\Delta\mathscr{E}°$ from it.) Take $\Delta G_f°(Zn(OH)_2(s)) = -553.5$ kJ $mol^{-1}$.

53. (a) What quantity of charge (in coulombs) is a fully charged 12-V lead–acid storage battery theoretically capable of furnishing if the spongy lead available for reaction at the anodes weighs 10 kg and there is excess $PbO_2$?
    (b) What is the theoretical maximum amount of work (in joules) that can be obtained from this battery?

54. (a) What quantity of charge (in coulombs) is a fully charged 1.34-V zinc–mercuric oxide watch battery theoretically capable of furnishing if the mass of HgO in the battery is 0.50 g?
    (b) What is the theoretical maximum amount of work (in joules) that can be obtained from this battery?

55. The concentration of the electrolyte, sulfuric acid, in a lead–acid storage battery diminishes as the battery is discharged. Is a discharged battery recharged by replacing the dilute $H_2SO_4$ with fresh, concentrated $H_2SO_4$? Explain.

56. One cold winter morning the temperature is well below 0°F. In trying to start your car, you run the battery down completely. Several hours later, you return to replace your fouled spark plugs and find that the liquid in the battery has now frozen even though the air temperature is actually a bit higher than it was in the morning. Explain how this can happen.

57. Consider the fuel cell that accomplishes the overall reaction

$$H_2(g) + \tfrac{1}{2}\ O_2(g) \longrightarrow H_2O(\ell)$$

If the fuel cell operates with 60% efficiency, calculate the amount of electrical work generated per gram of water produced. The gas pressures are constant at 1 atm, and the temperature is 25°C.

58. Consider the fuel cell that accomplishes the overall reaction

$$CO(g) + \frac{1}{2} O_2(g) \longrightarrow CO_2(g)$$

Calculate the maximum electrical work that could be obtained from the conversion of 1.00 mol of $CO(g)$ to $CO_2(g)$ in such a fuel cell operated with 100% efficiency at 25°C and with the pressure of each gas equal to 1 atm.

## Corrosion and Its Prevention

59. Two half-reactions proposed for the corrosion of iron in the absence of oxygen are

$$Fe(s) \longrightarrow Fe^{2+}(aq) + 2\ e^-$$

$$2\ H_2O(\ell) + 2\ e^- \longrightarrow 2\ OH^-(aq) + H_2(g)$$

Calculate the standard voltage generated by a galvanic cell running this pair of half-reactions. Is the overall reaction spontaneous under standard conditions? As the pH falls from 14, will the reaction become spontaneous?

60. In the presence of oxygen, the cathode half-reaction written in the preceding problem is replaced by

$$\frac{1}{2} O_2(g) + 2\ H_3O^+(aq) + 2\ e^- \longrightarrow 3\ H_2O(\ell)$$

but the anode half-reaction is unchanged. Calculate the standard cell voltage for *this* pair of reactions operating as a galvanic cell. Is the overall reaction spontaneous under standard conditions? As the water becomes more acidic, does the driving force for the rusting of iron increase or decrease?

61. Could sodium be used as a sacrificial anode to protect the iron hull of a ship?

62. If it is shown that titanium can be used as a sacrificial anode to protect iron, what conclusion can be drawn about the standard reduction potential of its half-reaction?

$$Ti^{3+}(aq) + 3\ e^- \longrightarrow Ti(s)$$

## A Deeper Look... Electrolysis of Water and Aqueous Solutions

63. An electrolytic cell consists of a pair of inert metallic electrodes in a solution buffered to pH = 5.0 and containing nickel sulfate ($NiSO_4$) at a concentration of 1.00 M. A current of 2.00 A is passed through the cell for 10.0 hours.
    (a) What product is formed at the cathode?
    (b) What is the mass of this product?
    (c) If the pH is changed to pH = 1.0, what product will form at the cathode?

64. A 0.100 M neutral aqueous $CaCl_2$ solution is electrolyzed using platinum electrodes. A current of 1.50 A passes through the solution for 50.0 hours.
    (a) Write the half-reactions occurring at the anode and at the cathode.

(b) What is the decomposition potential?
(c) Calculate the mass, in grams, of the product formed at the cathode.

## Additional Problems

65. The drain cleaner Drano consists of aluminum turnings mixed with sodium hydroxide. When it is added to water, the sodium hydroxide dissolves and releases heat. The aluminum reacts with water to generate bubbles of hydrogen and aqueous $Al(OH)_4^-$ ions. Write a balanced net ionic equation for this reaction.

66. Sulfur-containing compounds in the air tarnish silver, giving black $Ag_2S$. A practical method of cleaning tarnished silverware is to place the tarnished item in electrical contact with a piece of zinc and dip both into water containing a small amount of salt. Write balanced half-equations to represent what takes place.

67. A current passed through inert electrodes immersed in an aqueous solution of sodium chloride produces chlorate ion, $ClO_3^-(aq)$, at the anode and gaseous hydrogen at the cathode. Given this fact, write a balanced equation for the chemical reaction if gaseous hydrogen and aqueous sodium chlorate are mixed and allowed to react spontaneously until they reach equilibrium.

68. A galvanic cell is constructed by linking a $Co^{2+}|Co(s)$ half-cell to an $Ag^+|Ag(s)$ half-cell through a salt bridge and then connecting the cobalt and silver electrodes through an external circuit. When the circuit is closed, the cell voltage is measured to be 1.08 V and silver is seen to plate out while cobalt dissolves.
    (a) Write the half-reactions that occur at the anode and at the cathode, and the balanced overall cell reaction.
    (b) The cobalt electrode is weighed after 150 min of operation and is found to have decreased in mass by 0.36 g. By what amount has the silver electrode increased in mass?
    (c) What is the average current drawn from the cell during this period?

69. The galvanic cell $Zn(s)|Zn^{2+}(aq) \parallel Ni^{2+}(aq)|Ni(s)$ is constructed using a completely immersed zinc electrode that weighs 32.68 g and a nickel electrode immersed in 575 mL of 1.00 M $Ni^{2+}(aq)$ solution. A steady current of 0.0715 A is drawn from the cell as the electrons move from the zinc electrode to the nickel electrode.
    (a) Which reactant is the limiting reactant in this cell?
    (b) How long does it take for the cell to be completely discharged?
    (c) How much mass has the nickel electrode gained when the cell is completely discharged?
    (d) What is the concentration of the $Ni^{2+}(aq)$ when the cell is completely discharged?

70. A newly discovered bacterium can reduce selenate ion, $SeO_4^{2-}(aq)$, to elemental selenium, $Se(s)$, in reservoirs. This is significant because the soluble selenate ion is potentially toxic, but elemental selenium is insoluble and harmless.

Assume that water is oxidized to oxygen as the selenate ion is reduced. Compute the mass of oxygen produced if all the selenate in a $10^{12}$-L reservoir contaminated with 100 mg L$^{-1}$ of selenate ion is reduced to selenium.

71. Thomas Edison invented an electric meter that was nothing more than a simple coulometer, a device to measure the amount of electricity passing through a circuit. In this meter, a small, fixed fraction of the total current supplied to a household was passed through an electrolytic cell, plating out zinc at the cathode. Each month, the cathode could then be removed and weighed to determine the amount of electricity used. If 0.25% of a household's electricity passed through such a coulometer, and the cathode increased in mass by 1.83 g in a month, how many coulombs of electricity were used during that month?

*72. The chief chemist of the Brite-Metal Electroplating Co. is required to certify that the rinse solutions that are discharged from the company's tin-plating process into the municipal sewer system contain no more than 10 ppm (parts per million) by mass of $Sn^{2+}$. The chemist devises the following analytical procedure to determine the concentration. At regular intervals, a 100-mL (100 g) sample is withdrawn from the waste stream and acidified to pH = 1.0. A starch solution and 10 mL of 0.10 M potassium iodide are added, and a 25.0-mA current is passed through the solution between platinum electrodes. Iodine appears as a product of electrolysis at the anode when the oxidation of $Sn^{2+}$ to $Sn^{4+}$ is practically complete and signals its presence with the deep blue color of a complex formed with starch. What is the maximum duration of electrolysis to the appearance of the blue color that ensures that the concentration of $Sn^{2+}$ does not exceed 10 ppm?

73. Estimate the cost of the electrical energy needed to produce $1.5 \times 10^{10}$ kg (a year's supply for the world) of aluminum from $Al_2O_3(s)$ if electrical energy costs 10 cents per kilowatt-hour (1 kWh = 3.6 MJ = $3.6 \times 10^6$ J) and if the cell voltage is 5 V.

74. Titanium can be produced by electrolytic reduction from an anhydrous molten salt electrolyte that contains titanium(IV) chloride and a spectator salt that furnishes ions to make the electrolyte conduct electricity. The standard enthalpy of formation of $TiCl_4(\ell)$ is $-750$ kJ mol$^{-1}$ and the standard entropies of $TiCl_4(\ell)$, $Ti(s)$, and $Cl_2(g)$ are 253, 30, and 223 J K$^{-1}$ mol$^{-1}$. What minimum applied voltage will be necessary at 100°C?

75. A half-cell has a graphite electrode immersed in an acidic solution (pH 0) of $Mn^{2+}$ (concentration 1.00 M) in contact with solid $MnO_2$. A second half-cell has an acidic solution (pH 0) of $H_2O_2$ (concentration 1.00 M) in contact with a platinum electrode past which gaseous oxygen at a pressure of 1.00 atm is bubbled. The two half-cells are connected to form a galvanic cell.
    (a) By referring to Appendix E, write balanced chemical equations for the half-reactions at the anode and the cathode and for the overall cell reaction.
    (b) Calculate the cell voltage.

76. By considering these half-reactions and their standard reduction potentials,

$$Pt^{2+} + 2\,e^- \longrightarrow Pt \qquad \mathscr{E}° = 1.2 \text{ V}$$

$$NO_3^- + 4\,H_3O^+ + 3\,e^- \longrightarrow NO + 6\,H_2O \qquad \mathscr{E}° = 0.96 \text{ V}$$

$$PtCl_4^{2-} + 2\,e^- \longrightarrow Pt + 4\,Cl^- \qquad \mathscr{E}° = 0.73 \text{ V}$$

account for the fact that platinum will dissolve in a mixture of hydrochloric acid and nitric acid (*aqua regia*) but will not dissolve in either acid alone.

77. (a) One method to reduce the concentration of unwanted $Fe^{3+}(aq)$ in a solution of $Fe^{2+}(aq)$ is to drop a piece of metallic iron into the storage container. Write the reaction that removes the $Fe^{3+}$, and compute its standard cell potential.
    (b) By referring to problem 31, suggest a way to remove unwanted $Mn^{3+}(aq)$ from solutions of $Mn^{2+}(aq)$.

78. (a) Based only on the standard reduction potentials for the $Cu^{2+}|Cu^+$ and the $I_2(s)|I^-$ half-reactions, would you expect $Cu^{2+}(aq)$ to be reduced to $Cu^+(aq)$ by $I^-(aq)$?
    (b) The formation of solid CuI plays a role in the interaction between $Cu^{2+}(aq)$ and $I^-(aq)$.

$$Cu^{2+}(aq) + I^-(aq) + e^- \rightleftharpoons CuI(s) \qquad \mathscr{E}° = 0.86 \text{ V}$$

Taking into account this added information, do you expect $Cu^{2+}$ to be reduced by iodide ion?

79. In some old European churches, the stained-glass windows have so darkened from corrosion and age that hardly any light comes through. Microprobe analysis showed that tiny cracks and defects on the glass surface were enriched in insoluble Mn(III) and Mn(IV) compounds. From Appendix E, suggest a reducing agent and conditions that might successfully convert these compounds to soluble Mn(II) without simultaneously reducing Fe(III) (which gives the glass its colors) to Fe(II). Take $MnO_2$ as representative of the insoluble Mn(III) and Mn(IV) compounds.

80. (a) Calculate the half-cell potential for the reaction

$$O_2(g) + 4\,H_3O^+(aq) + 4\,e^- \longrightarrow 6\,H_2O(\ell)$$

at pH 7 with the oxygen pressure at 1 atm.
    (b) Explain why aeration of solutions of $I^-$ leads to their decomposition. Write a balanced equation for the redox reaction that occurs.
    (c) Will the same problem arise with solutions containing $Br^-$ or $Cl^-$? Explain.
    (d) Will decomposition be favored or opposed by increasing acidity?

81. An engineer needs to prepare a galvanic cell that uses the reaction

$$2\,Ag^+(aq) + Zn(s) \longrightarrow Zn^{2+}(aq) + 2\,Ag(s)$$

and generates an initial voltage of 1.50 V. She has 0.010 M $AgNO_3(aq)$ and 0.100 M $Zn(NO_3)_2(aq)$ solutions as well as electrodes of metallic copper and silver, wires, containers, water, and a $KNO_3$ salt bridge. Sketch the cell. Clearly indicate the concentrations of all solutions.

82. Consider a galvanic cell for which the anode reaction is

$$Pb(s) \longrightarrow Pb^{2+}(1.0 \times 10^{-2} \text{ M}) + 2 \text{ } e^-$$

and the cathode reaction is

$$VO^{2+}(0.10 \text{ M}) + 2 \text{ } H_3O^+(0.10 \text{ M}) + e^- \longrightarrow$$
$$V^{3+}(1.0 \times 10^{-5} \text{ M}) + 3 \text{ } H_2O(\ell)$$

The measured cell voltage is 0.640 V.
(a) Calculate $\mathscr{E}°$ for the $VO^{2+}|V^{3+}$ half-reaction, using $\mathscr{E}°(Pb^{2+}|Pb)$ from Appendix E.
(b) Calculate the equilibrium constant ($K$) at 25°C for the reaction

$$Pb(s) + 2 \text{ } VO^{2+}(aq) + 4 \text{ } H_3O^+(aq) \rightleftharpoons$$
$$Pb^{2+}(aq) + 2 \text{ } V^{3+}(aq) + 6 \text{ } H_2O(\ell)$$

83. Suppose we construct a pressure cell in which the gas pressures differ in the two half-cells. Suppose such a cell consists of a $Cl_2(0.010 \text{ atm})|Cl^-(1 \text{ M})$ half-cell connected to a $Cl_2(0.50 \text{ atm})|Cl^-(1 \text{ M})$ half-cell. Determine which half-cell will be the anode, write the overall equation for the reaction, and calculate the cell voltage.

84. A student decides to measure the solubility of lead sulfate in water and sets up the electrochemical cell

$$Pb|PbSO_4|SO_4^{2-}(aq, 0.0500 \text{ M}) \parallel Cl^-(aq, 1.00 \text{ M})|AgCl|Ag$$

At 25°C he finds the cell voltage to be 0.546 V, and from Appendix E he finds

$$AgCl(s) + e^- \rightleftharpoons Ag(s) + Cl^-(aq) \qquad \mathscr{E}° = 0.222 \text{ V}$$

What does he find for the $K_{sp}$ of $PbSO_4$?

85. A wire is fastened across the terminals of the Leclanchè cell in Figure 12.8. Indicate the direction of electron flow in the wire.

86. Overcharging a lead–acid storage battery can generate hydrogen. Write a balanced equation to represent the reaction taking place.

87. An ambitious chemist discovers an alloy electrode that is capable of catalytically converting ethanol reversibly to carbon dioxide at 25°C according to the half-reaction

$$C_2H_5OH(\ell) + 15 \text{ } H_2O(\ell) \longrightarrow$$
$$2 \text{ } CO_2(g) + 12 \text{ } H_3O^+(aq) + 12 \text{ } e^-$$

Believing that this discovery is financially important, the chemist patents its composition and designs a fuel cell that may be represented as

$$Alloy|C_2H_5OH(\ell)|CO_2(g) + H_3O^+(1 \text{ M}) \parallel H_3O^+(1 \text{ M})|O_2|Ni$$

(a) Write the half-reaction occurring at the cathode.
(b) Using data from Appendix E, calculate $\Delta\mathscr{E}°$ for the cell at 25°C.
(c) What is the $\mathscr{E}°$ value for the ethanol half-cell?

88. Iron or steel is often covered by a thin layer of a second metal to prevent rusting: tin cans consist of steel covered with tin, and galvanized iron is made by coating iron with a layer of zinc. If the protective layer is broken, however, iron will rust more readily in a tin can than in galvanized iron. Explain this observation by comparing the half-cell potentials of iron, tin, and zinc.

89. An electrolysis cell contains a solution of 0.10 M $NiSO_4$. The anode and cathode are both strips of Pt foil. Another electrolysis cell contains the same solution, but the electrodes are strips of Ni foil. In each case, a current of 0.10 A flows through the cell for 10 hours.
(a) Write a balanced equation for the chemical reaction that occurs at the anode in each cell.
(b) Calculate the mass, in grams, of the product formed at the anode in each cell. (The product may be a gas, a solid, or an ionic species in solution.)

90. A potential difference of 2.0 V is impressed across a pair of inert electrodes (e.g., platinum) that are immersed in a 0.050 M aqueous KBr solution. What are the products that form at the anode and the cathode?

*91. An aqueous solution is simultaneously 0.10 M in $SnCl_2$ and in $CoCl_2$.
(a) If the solution is electrolyzed, which metal will appear first?
(b) At what decomposition potential will that metal first appear?
(c) As the electrolysis proceeds, the concentration of the metal being reduced will drop and the voltage will change. How complete a separation of the metals using electrolysis is theoretically possible? In other words, at the point where the second metal begins to form, what fraction of the first metal is left in solution?

## CUMULATIVE PROBLEMS

92. A 1.0 M solution of NaOH is electrolyzed, generating $O_2(g)$ at the anode. A current of 0.15 A is passed through the cell for a period of 75 minutes. Calculate the volume of (wet) oxygen generated in this period if the temperature is held at 25°C and the total pressure is 0.985 atm. (*Hint:* Use the vapor pressure of water at this temperature from Table 5.1.)

93. Use standard entropies from Appendix D to predict whether the standard voltage of the $Cu|Cu^{2+} \parallel Ag^+|Ag$ cell (diagrammed in Figure 12.4) will increase or decrease if the temperature is raised above 25°C.

94. About 50,000 kJ of electrical energy is required to produce 1.0 kg of Al from its $Al(OH)_3$ ore. The major energy cost in

recycling aluminum cans is the melting of the aluminum. The enthalpy of fusion of Al(s) is 10.7 kJ mol$^{-1}$. Compare the energy cost for making new aluminum with that for recycling.

95. Use the following half-reactions and their reduction potentials to calculate the $K_{sp}$ of AgBr:

$$Ag^+(aq) + e^- \longrightarrow Ag(s) \qquad \mathcal{E}° = 0.7996 \text{ V}$$

$$AgBr(s) + e^- \longrightarrow Ag(s) + Br^-(aq) \qquad \mathcal{E}° = 0.0713 \text{ V}$$

   (b) Estimate the solubility of AgBr in 0.10 M NaBr(aq).

96. Amounts of iodine dissolved in aqueous solution, $I_2(aq)$, can be determined by titration with thiosulfate ion ($S_2O_3^{2-}$). The thiosulfate ion is oxidized to $S_4O_6^{2-}$ while the iodine is reduced to iodide ion. Starch is used as an indicator because it has a strong blue color in the presence of dissolved iodine.

   (a) Write a balanced equation for this reaction.
   (b) If 56.40 mL of 0.100 M $S_2O_3^{2-}$ solution is used to reach the endpoint of a titration of an unknown amount of iodine, calculate the number of moles of iodine originally present.
   (c) Combine the appropriate half-cell potentials from Appendix E with thermodynamic data from Appendix D for the equilibrium

$$I_2(s) \rightleftharpoons I_2(aq)$$

   to calculate the equilibrium constant at 25°C for the reaction in part (a).

# U N I T 4 RATES OF CHEMICAL AND PHYSICAL PROCESSES

Zinc reacts readily with aqueous hydrochloric acid to give hydrogen gas and aqueous zinc chloride. *(Charles D. Winters)*

Thermodynamics explains *why* chemical reactions occur by identifying the driving force toward chemical equilibrium. Thermodynamics is a very powerful tool for understanding the nature of chemical equilibrium, but it gives no information about the crucial question of how rapidly that equilibrium is achieved.

Chemical kinetics explains *how* reactions occur by studying their rates and mechanisms. Chemical kinetics explains how the speeds of different chemical reactions vary from explosive rapidity to glacial sluggishness and how slow reactions can be accelerated by introducing materials called catalysts. Chemical kinetics has enormous practical importance because it provides the basis for modifying conditions to carry out chemical reactions with reasonable speed, under proper control.

The central conceptual method in chemical kinetics is to relate the rate of a reaction to the amount of reactants present. Once this connection is established, the influence of external conditions—principally the temperature—can be explained.

Similar concepts are used to describe nuclear reactions through which one element is transformed into another. Historically, the conceptual development of nuclear physics and chemistry relied heavily upon analogies with chemical kinetics. Consequently these subjects are presented together in Unit 4.

## CHAPTER 13
Chemical Kinetics

## CHAPTER 14
Nuclear Chemistry

THE GOALS OF UNIT 4 ARE   **(1)** To relate the rate of a chemical reaction to the instantaneous concentration(s) of reactants by developing the rate law and the rate constant for the reaction; **(2)** To describe the influence of temperature on the reaction rate by identifying the activation energy for the reaction; **(3)** To explain the mechanism of a complex reaction by identifying the separate elementary reaction steps through which it proceeds; **(4)** To explain the role of catalysts in manipulating reaction rates; **(5)** To develop an elementary description of the rates of nuclear reactions, emphasizing the half-life of radioactive species; and **(6)** To survey the applications and consequences of nuclear reactions in medicine, biology, energy production, and the environment.

# CHAPTER
# 13

# Chemical Kinetics

*Illustration*
Rust and corrosion of iron on the seashore. *(Copyright 1990 Diane Hirsch/ Fundamental Photographs)*

Why do some chemical reactions proceed with lightning speed, when others require days, months, or even years to produce detectable amounts of products? How do catalysts increase the rates of chemical reactions? Why do small changes in temperature often have such large effects on the cooking rate of food? How does a study of the rate of a chemical reaction inform us about the way in which molecules combine to form products? All of these questions involve chemical kinetics.

Chemical kinetics is a complex subject, and our mastery of it is not nearly as complete as that of thermodynamics. There are many reactions whose equilibrium constants are known accurately, but whose rates and detailed reaction pathways remain poorly understood. This is particularly true of reactions in which many species participate in the overall process that leads from reactants to products. An example is the reaction

$$5 \ Fe^{2+}(aq) + MnO_4^-(aq) + 8 \ H_3O^+(aq) \longrightarrow 5 \ Fe^{3+}(aq) + Mn^{2+}(aq) + 12 \ H_2O(\ell)$$

The equilibrium constant for this reaction is straightforward to measure (from the voltage of a galvanic cell, for example), and from it the equilibrium concentrations can be calculated for arbitrary initial conditions. The exact way in which the reaction goes from reactants to products is considerably harder to determine. It certainly does *not* involve the simultaneous collision of five $Fe^{2+}$ ions and one $MnO_4^-$ ion with eight $H_3O^+$ ions, because such a collision would be exceedingly rare. Instead, it proceeds through a series of elementary steps involving two or at most three ions, such as

$$Fe^{2+} + MnO_4^- \longrightarrow Fe^{3+} + MnO_4^{2-}$$

$$MnO_4^{2-} + H_3O^+ \longrightarrow HMnO_4^- + H_2O$$

$$HMnO_4^- + Fe^{2+} \longrightarrow HMnO_4^{2-} + Fe^{3+}$$

Other postulated steps take the process to its final products. Some of these steps are slow, others fast; taken together they constitute the **reaction mechanism.** A major goal of chemical kinetics is to discern the way the rate depends on the concentrations of the reacting species and to deduce the mechanism of a reaction from experimental knowledge of that rate.

---

## 13.1
### RATES OF CHEMICAL REACTIONS

The speed of a reaction depends on many factors. Concentrations of reacting species certainly play a major role in speeding up or slowing down a particular reaction (Fig. 13.1). As we will see in Section 13.5, many reactions are extremely sensitive to temperature, a fact that makes careful control of temperature critical for quantitative measurements in chemical kinetics. Finally, the physical forms of the reactants are often crucial to the observed rate. An iron nail oxidizes only very slowly in dry air to iron oxide, but steel wool burns spectacularly in oxygen (Fig. 13.2). The quantitative study of heterogeneous reactions (those involving two or more phases, such as a solid and a gas) is difficult, so we begin with homogeneous reactions (those that take place entirely within the gas phase or solution). In Section 13.7 we turn briefly to some important aspects of heterogeneous reactions.

### Measuring Reaction Rates

A kinetics experiment measures the rate at which the concentration of a substance taking part in a chemical reaction changes with time. How can a changing concentration be monitored experimentally? If the reaction is slow enough, we can allow

**FIGURE 13.1** The rate of reaction of zinc with aqueous sulfuric acid depends on the concentration of the acid. The dilute solution reacts slowly (left), and the more concentrated solution reacts rapidly (right). *(Charles Steele)*

it to proceed for a measured time and then abruptly "quench" (effectively stop) it by rapidly cooling the reaction mixture to a low enough temperature. At that low temperature there is time to carry out a chemical analysis for some particular reactant or product. This procedure is not useful for rapid reactions, especially those involving gas mixtures, which are difficult to cool quickly. An alternative is to use the absorption of light as a probe. As will be seen in Chapter 16, molecules differ in the wavelengths of light they absorb. If a wavelength is absorbed by only one particular reactant or product, then measuring the amount of light absorbed at that wavelength by the reaction mixture determines the concentration of the absorbing species at a series of times. Often a flash of light can also be used to initiate a very fast reaction.

The average rate of a reaction is analogous to the average speed of a car. If the average position of a car is noted at two different times, then

$$\text{average speed} = \frac{\text{distance traveled}}{\text{elapsed time}} = \frac{\text{change in location}}{\text{change in time}}$$

In the same way, the **average reaction rate** is obtained by dividing the change in concentration of a reactant or product by the time interval over which that change occurs:

$$\text{average reaction rate} = \frac{\text{change in concentration}}{\text{change in time}}$$

If concentration is measured in mol $L^{-1}$ and time in seconds, then the rate of a reaction has units of mol $L^{-1}$ $s^{-1}$.

Consider a specific example. In the gas-phase reaction

$$NO_2(g) + CO(g) \longrightarrow NO(g) + CO_2(g)$$

$NO_2$ and CO are consumed as NO and $CO_2$ are produced. If a probe can measure the NO concentration, the average rate of reaction can be estimated from the ratio of the change in NO concentration $\Delta[NO]$ to the time interval $\Delta t$:

$$\text{average rate} = \frac{\Delta[NO]}{\Delta t} = \frac{[NO]_f - [NO]_i}{t_f - t_i}$$

**FIGURE 13.2** Steel wool burning in oxygen. *(Leon Lewandowski)*

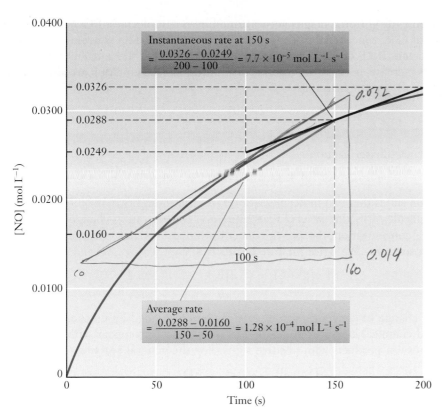

**FIGURE 13.3** A graph of the concentration of NO against time in the reaction $NO_2 + CO \rightarrow NO + CO_2$. The average rate during the time interval from 50 to 150 seconds is obtained by dividing the change in NO concentration by the duration of the interval (green box). Note that the average rate (green line) underestimates the true production rate over the time interval. The instantaneous rate 150 seconds after the start of the reaction is found by calculating the slope of the line tangent to the curve at that point (red box).

$$\frac{\Delta y}{\Delta x} = \frac{0.032 - 0.014}{160 - 10}$$

$$= 1.2 \times 10^{-4}$$

This estimate depends on the time interval $\Delta t$ that is selected, because the rate at which NO is produced changes with time. From the data in Figure 13.3, the average rate of reaction during the first 50 s is

$$\text{average rate} = \frac{\Delta[NO]}{\Delta t} = \frac{(0.0160 - 0)\ \text{mol L}^{-1}}{(50 - 0)\ \text{s}} = 3.2 \times 10^{-4}\ \text{mol L}^{-1}\,\text{s}^{-1}$$

During the second 50 s, the average rate is $1.6 \times 10^{-4}\ \text{mol L}^{-1}\,\text{s}^{-1}$, and during the third 50 s it is $9.6 \times 10^{-5}\ \text{mol L}^{-1}\,\text{s}^{-1}$. Clearly, this reaction slows down as it progresses, and its average rate indeed depends on the time interval chosen. Figure 13.3 shows a graphical method for determining average rates. The average rate is the slope of the straight line connecting the concentrations at the initial and final points of a time interval.

The **instantaneous rate** of a reaction is obtained by considering smaller and smaller time increments $\Delta t$ (with correspondingly smaller values of $\Delta[NO]$). As $\Delta t$ approaches 0, the rate becomes the slope of the line tangent to the curve at time $t$ (see Fig. 13.3). This slope is written as the derivative of [NO] with respect to time:

$$\text{instantaneous rate} = \text{limit}_{\Delta t \rightarrow 0} \frac{[NO]_{t+\Delta t} - [NO]_t}{\Delta t} = \frac{d[NO]}{dt}$$

Throughout the rest of this book, the instantaneous rate will be referred to simply as the *rate*. The instantaneous rate of a reaction at the moment that it begins (at $t = 0$) is the **initial rate** of that reaction.

The rate of this sample reaction could just as well have been measured by monitoring changes in the concentration of $CO_2$, $NO_2$, or $CO$ instead of NO. Because every molecule of NO produced is accompanied by one molecule of $CO_2$, the rate of increase of $CO_2$ concentration is the same as that of NO. The concentrations of the two reactants, $NO_2$ and CO, *decrease* at the same rate that the concentrations of the products increase, because the coefficients in the balanced equation are also both equal to 1. This is summarized as

$$\text{rate} = -\frac{d[NO_2]}{dt} = -\frac{d[CO]}{dt} = \frac{d[NO]}{dt} = \frac{d[CO_2]}{dt}$$

Another gas-phase reaction is

$$2\ NO_2(g) + F_2(g) \longrightarrow 2\ NO_2F(g)$$

This equation states that two molecules of $NO_2$ disappear and two of $NO_2F$ appear for each molecule of $F_2$ that reacts. Thus, the $NO_2$ concentration changes twice as fast as the $F_2$ concentration; the $NO_2F$ concentration also changes twice as fast and has the opposite sign. We write the rate in this case as

$$\text{rate} = -\frac{1}{2}\frac{d[NO_2]}{dt} = -\frac{d[F_2]}{dt} = \frac{1}{2}\frac{d[NO_2F]}{dt}$$

The rate of change of concentration of each species is divided by its coefficient in the balanced chemical equation. Rates of change of reactants appear with negative signs and those of products with positive signs. For the general reaction

$$a\text{A} + b\text{B} \longrightarrow c\text{C} + d\text{D}$$

the rate is

$$\text{rate} = -\frac{1}{a}\frac{d[A]}{dt} = -\frac{1}{b}\frac{d[B]}{dt} = \frac{1}{c}\frac{d[C]}{dt} = \frac{1}{d}\frac{d[D]}{dt} \qquad \textbf{[13.1]}$$

These relations hold true provided there are no transient intermediate species or, if there are intermediates, their concentrations are independent of time for most of the reaction period.

---

## 13.2

## RATE LAWS

In discussing chemical equilibrium we stressed that both forward and reverse reactions can occur; once products are formed, they can react back to give the original reactants. The net rate is the difference:

$$\text{net rate} = \text{forward rate} - \text{reverse rate}$$

Strictly speaking, measurements of concentration give the net rate rather than simply the forward rate. Near the beginning of a reaction that starts from pure reactants, however, the concentrations of reactants are far higher than those of products, and the reverse rate can be neglected. In addition, many reactions go to "completion" ($K \gg 1$) and so have a measurable rate only in the forward direction, or else the experiment can be arranged in such a way that the products are removed as they are formed. In this section, the focus is on forward rates exclusively.

## Order of a Reaction

The forward rate of a chemical reaction depends on the concentrations of the reactants. As an example, consider the decomposition of dinitrogen pentaoxide ($N_2O_5$). This compound is a white solid that is stable below 0°C but decomposes when vaporized:

$$N_2O_5(g) \longrightarrow 2\,NO_2(g) + \tfrac{1}{2}\,O_2(g)$$

The rate of the reaction depends on the concentration of $N_2O_5(g)$. In fact, as Figure 13.4 shows, a graph of rate versus concentration is a straight line that can be extrapolated to pass through the origin, and so the rate can be written

$$\text{rate} = k[N_2O_5]$$

This relation between the rate of a reaction and concentration is called a **rate expression** or **rate law,** and the proportionality constant $k$ is called the **rate constant** for the reaction. Like an equilibrium constant, a rate constant is independent of concentration but depends on temperature, as we will see with greater detail in Section 13.5.

For many (but not all) reactions with a single reactant, the rate is proportional to the concentration of that reactant raised to a power. That is, the rate expression for

$$a\text{A} \longrightarrow \text{products}$$

frequently has the form

$$\text{rate} = k[\text{A}]^n$$

It is important to note that the power $n$ in the rate expression has no direct relation to the coefficient $a$ in the balanced chemical equation. For the decomposition of ethane at high temperatures and low pressures,

$$C_2H_6(g) \longrightarrow 2\,CH_3(g)$$

the rate expression has the form

$$\text{rate} = k[C_2H_6]^2$$

so that $n = 2$ even though the coefficient in the chemical equation is 1.

The power to which the concentration is raised is called the **order** of the reaction with respect to that reactant. Thus, the decomposition of $N_2O_5$ is **first order,** whereas that of $C_2H_6$ is **second order.** Some processes are **zeroth order** over a range of concentrations. Because $[\text{A}]^0 = 1$, such reactions have rates that are independent of concentration:

$$\text{rate} = k \qquad \text{(zeroth-order kinetics)}$$

The order of a reaction does not have to be an integer; fractional powers are sometimes found. At 450 K, the decomposition of acetaldehyde ($CH_3CHO$) is described by the rate expression

$$\text{rate} = k[CH_3CHO]^{3/2}$$

The lesson from these examples is that reaction order is an experimental property that cannot be predicted from the form of the chemical equation.

The following example illustrates how the order of a reaction can be deduced from experimental data.

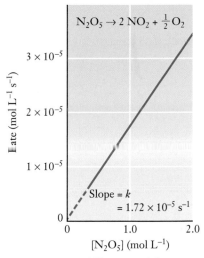

**FIGURE 13.4** The rate of decomposition of $N_2O_5(g)$ at 25°C is proportional to its concentration. The slope of this line is equal to the rate constant $k$ for the reaction.

## EXAMPLE 13.1

At elevated temperatures, HI reacts according to the chemical equation

$$2\ HI(g) \longrightarrow H_2(g) + I_2(g)$$

At 443°C, the rate of the reaction increases with concentration of HI as follows:

| [HI] (mol L$^{-1}$) | 0.0050 | 0.010 | 0.020 |
| Rate (mol L$^{-1}$ s$^{-1}$) | $7.5 \times 10^{-4}$ | $3.0 \times 10^{-3}$ | $1.2 \times 10^{-2}$ |

(a) Determine the order of the reaction and write the rate expression.
(b) Calculate the rate constant, and give its units.
(c) Calculate the reaction rate for a 0.0020 M concentration of HI.

### Solution

(a) The rate expressions at two different concentrations $[HI]_1$ and $[HI]_2$ are

$$rate_1 = k([HI]_1)^n$$

$$rate_2 = k([HI]_2)^n$$

After dividing the second equation by the first, the rate constant $k$ drops out, leaving the reaction order $n$ as the only unknown quantity.

$$\frac{rate_2}{rate_1} = \left(\frac{[HI]_2}{[HI]_1}\right)^n$$

We can now substitute any two sets of data into this equation and solve for $n$. Taking the first two sets with $[HI]_1 = 0.0050$ M and $[HI]_2 = 0.010$ M gives

$$\frac{3.0 \times 10^{-3}}{7.5 \times 10^{-4}} = \left(\frac{0.010}{0.0050}\right)^n$$

which simplifies to

$$4 = (2)^n$$

By inspection, $n = 2$, and so the reaction is second order in HI. In a case where the solution of the equation is less obvious, we can take the logarithms of both sides, giving (in this case)

$$\log_{10} 4 = n \log_{10} 2$$

$$n = \frac{\log_{10} 4}{\log_{10} 2} = \frac{0.602}{0.301} = 2$$

The rate expression has the form

$$rate = k[HI]^2$$

(b) The rate constant $k$ is calculated by inserting any of the sets of data into the rate expression. Taking the first set, for example, gives

$$7.5 \times 10^{-4}\ mol\ L^{-1}\ s^{-1} = k(0.0050\ mol\ L^{-1})^2$$

Solving for $k$ gives

$$k = 30\ L\ mol^{-1}\ s^{-1}$$

(c) Finally, the rate is calculated for [HI] = 0.0020 M:

$$rate = k[HI]^2 = (30\ L\ mol^{-1}\ s^{-1})(0.0020\ mol\ L^{-1})^2 = 1.2 \times 10^{-4}\ mol\ L^{-1}\ s^{-1}$$

So far, each reaction rate has depended only on a single concentration. In reality, many rates depend on the concentrations of two or more different chemical species, and the rate expression is written in a form such as

$$\text{rate} = -\frac{1}{a}\frac{d[A]}{dt} = k[A]^m[B]^n$$

Again the exponents $m$ and $n$ do not derive from the coefficients in the balanced equation for the reaction; they must be determined experimentally and are usually integers or half-integers.

The exponents $m$, $n$, ... give the order of the reaction, just as in the simpler case where only one concentration appeared in the rate expression. For example, the preceding reaction is said to be $m$th order in A, meaning that a change in the concentration of A by a certain factor leads to a change in the rate by that factor raised to the $m$th power. The reaction is $n$th order in B, and the **overall reaction order** is $m + n$. For the reaction

$$\text{H}_2\text{PO}_2^-(aq) + \text{OH}^-(aq) \longrightarrow \text{HPO}_3^{2-}(aq) + \text{H}_2(g)$$

the experimentally determined rate expression is

$$\text{rate} = k[\text{H}_2\text{PO}_2^-][\text{OH}^-]^2$$

so that the reaction is said to be first order in $\text{H}_2\text{PO}_2^-(aq)$ and second order in $\text{OH}^-(aq)$, with an overall reaction order of 3. The units of $k$ depend on the reaction order. If all concentrations are expressed in mol $\text{L}^{-1}$ and if $p = m + n + \cdots$ is the overall reaction order, then $k$ has units of $\text{mol}^{-(p-1)}\ \text{L}^{p-1}\ \text{s}^{-1}$.

## EXAMPLE 13.2

Use the preceding rate expression to determine the effect of the following changes on the rate of decomposition of $\text{H}_2\text{PO}_2^-(aq)$:
**(a)** Tripling the concentration of $\text{H}_2\text{PO}_2^-(aq)$ at constant pH
**(b)** Changing the pH from 13 to 14 at a constant concentration of $\text{H}_2\text{PO}_2^-(aq)$

### Solution
**(a)** Because the reaction is first order in $\text{H}_2\text{PO}_2^-(aq)$, tripling this concentration will triple the reaction rate.
**(b)** A change in pH from 13 to 14 corresponds to an increase in the $\text{OH}^-(aq)$ concentration by a factor of 10. Because the reaction is second order in $\text{OH}^-(aq)$ (i.e., this term is squared in the rate expression), this will increase the reaction rate by a factor of $10^2$, or 100.

**Related Problems: 5, 6**

Rate expressions that depend on more than one concentration are more difficult to obtain experimentally than those that depend on only one. One way to proceed is to find the instantaneous initial rates of reaction for several values of one of the concentrations, holding the other initial concentrations fixed from one run to the next. Then the experiment can be repeated, changing one of the other concentrations. The following example illustrates this procedure.

### EXAMPLE 13.3

The reaction of $NO(g)$ with $O_2(g)$ gives $NO_2(g)$:

$$2\ NO(g) + O_2(g) \longrightarrow 2\ NO_2(g)$$

From the dependence of the initial rate $(-\frac{1}{2}d[NO]/dt)$ on the initial concentrations of NO and $O_2$, determine the rate expression and the value of the rate constant.

| [NO] (mol L$^{-1}$) | [O$_2$] (mol L$^{-1}$) | Initial Rate (mol L$^{-1}$ s$^{-1}$) |
|---|---|---|
| $1.0 \times 10^{-4}$ | $1.0 \times 10^{-4}$ | $2.8 \times 10^{-6}$ |
| $1.0 \times 10^{-4}$ | $3.0 \times 10^{-4}$ | $8.4 \times 10^{-6}$ |
| $2.0 \times 10^{-4}$ | $3.0 \times 10^{-4}$ | $3.4 \times 10^{-5}$ |

### Solution

When $[O_2]$ is multiplied by 3 (with [NO] constant), the rate is also multiplied by 3 (from $2.8 \times 10^{-6}$ to $8.4 \times 10^{-6}$), and so the reaction is first order in $O_2$. When [NO] is multiplied by 2 (with $[O_2]$ constant), the rate is multiplied by

$$\frac{3.4 \times 10^{-5}}{8.4 \times 10^{-6}} \approx 4 = 2^2$$

so the reaction is second order in NO. Thus, the form of the rate expression is

$$\text{rate} = k[O_2][NO]^2$$

To evaluate $k$, any set of data can be inserted into the equation. From the first set,

$$2.8 \times 10^{-6}\ \text{mol L}^{-1}\ \text{s}^{-1} = k(1.0 \times 10^{-4}\ \text{mol L}^{-1})(1.0 \times 10^{-4}\ \text{mol L}^{-1})^2$$

$$k = 2.8 \times 10^6\ \text{L}^2\ \text{mol}^{-2}\ \text{s}^{-1}$$

**Related Problems: 7, 8**

## Integrated Rate Laws

Measuring an initial rate involves determining small changes in concentration $\Delta[A]$ that occur during a short time interval $\Delta t$. Sometimes it can be difficult to obtain sufficiently precise experimental data for these small changes. An alternative is to use an **integrated rate law,** which expresses the concentration of a species directly as a function of the time. For any given simple rate expression, a corresponding integrated rate law can be obtained.

### First-Order Reactions

As an example, consider again the reaction

$$N_2O_5(g) \longrightarrow 2\ NO_2(g) + \tfrac{1}{2}\ O_2(g)$$

whose rate law has been determined experimentally to be

$$\text{rate} = -\frac{d\,[N_2O_5]}{dt} = k\,[N_2O_5]$$

This is a first-order reaction. If we let $[N_2O_5] = c$, a function of time, we have

$$\frac{dc}{dt} = -kc$$

We seek a function whose slope at every time is proportional to the value of the function itself. This function can be found through the use of calculus. Separating the variables (with concentration $c$ on the left and time $t$ on the right) gives

$$\frac{1}{c}\, dc = -k\, dt$$

Integrating from an initial concentration $c_0$ at time $t = 0$ to a concentration $c$ at time $t$ (see Appendix C, Section C.5) gives

$$\int_{c_0}^{c} \frac{1}{c}\, dc = -k \int_{0}^{t} dt$$

$$\ln c - \ln c_0 = -kt$$

$$\ln (c/c_0) = -kt$$

$$c = c_0\, e^{-kt} \qquad \text{[13.2]}$$

The concentration will fall off exponentially with time. For a first-order reaction, a plot of $\ln c$ against $t$ should be a straight line with slope $-k$ (Fig. 13.5).

A useful concept in discussions of first-order reactions is the **half-life** $t_{1/2}$—the time it takes for the original concentration $c_0$ to be reduced to one-half its value, $c_0/2$. Setting $c = c_0/2$ gives

$$-kt_{1/2} = \ln\left(\frac{c}{c_0}\right) = \ln\left(\frac{c_0/2}{c_0}\right) = -\ln 2$$

$$t_{1/2} = \frac{\ln 2}{k} = \frac{0.6931}{k} \qquad \text{[13.3]}$$

If $k$ has units of $s^{-1}$, $(\ln 2)/k$ is the half-life in seconds. During each half-life, the concentration of A falls to half its value again (Fig. 13.6).

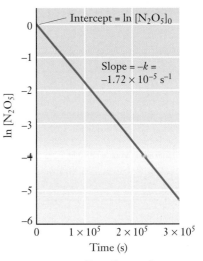

**FIGURE 13.5** In a first-order reaction such as the decomposition of $N_2O_5$, a graph of the natural logarithm of the concentration against time is a straight line, the negative of whose slope gives the rate constant for the reaction.

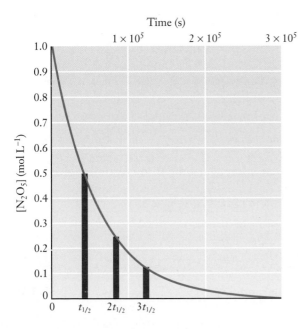

**FIGURE 13.6** The same data as in Figure 13.5 are graphed in a concentration-versus-time picture. The half-life $t_{1/2}$ is the time it takes for the concentration to be reduced to one-half its initial value. In two half-lives, the concentration falls to one quarter of its initial value.

## EXAMPLE 13.4

(a) What is the rate constant $k$ for the first-order decomposition of $N_2O_5(g)$ at 25°C if the half-life of $N_2O_5(g)$ at that temperature is $4.03 \times 10^4$ s?
(b) What percentage of the $N_2O_5$ molecules will *not* have reacted after one day?

### Solution

(a)
$$t_{1/2} = \frac{\ln 2}{k} = 4.03 \times 10^4 \text{ s}$$

Solving for the rate constant $k$ gives

$$k = \frac{\ln 2}{t_{1/2}} = \frac{0.6931}{4.03 \times 10^4 \text{ s}} = 1.72 \times 10^{-5} \text{ s}^{-1}$$

(b) From the integrated rate law for a first-order reaction,

$$\frac{c}{c_0} = e^{-kt}$$

Putting in the value for $k$ and setting $t$ to 1 day $= 8.64 \times 10^4$ s gives

$$\frac{c}{c_0} = \exp\left[-(1.72 \times 10^{-5} \text{ s}^{-1})(8.64 \times 10^4 \text{ s})\right]$$

$$= e^{-1.49} = 0.226$$

Therefore, 22.6% of the molecules will not have reacted after one day.

**Related Problems: 11, 12**

### Second-Order Reactions

Integrated rate laws can be obtained for reactions of other orders. The observed rate of the reaction

$$2 \text{ NO}_2 \longrightarrow 2 \text{ NO} + O_2$$

is second order in $[NO_2]$:

$$\text{rate} = -\frac{1}{2} \frac{d[NO_2]}{dt} = k[NO_2]^2$$

Writing $[NO_2] = c$ gives

$$\frac{dc}{dt} = -2kc^2$$

$$\frac{1}{c^2} dc = -2k \, dt$$

Integrating this from the initial concentration $c_0$ at time 0 to $c$ at time $t$ gives

$$\int_{c_0}^{c} \frac{1}{c^2} dc = -2k \int_0^t dt$$

$$-\frac{1}{c} + \frac{1}{c_0} = -2kt$$

$$\frac{1}{c} = \frac{1}{c_0} + 2kt \qquad\qquad [13.4]$$

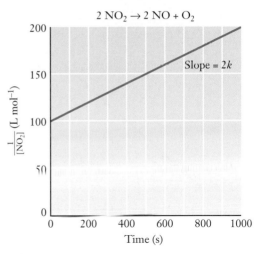

$2 NO_2 \rightarrow 2 NO + O_2$

Slope = $2k$

$\frac{1}{[NO_2]}$ (L mol$^{-1}$)

Time (s)

**FIGURE 13.7** For a second-order reaction such as $2 NO_2 \rightarrow 2 NO + O_2$, a graph of the reciprocal of the concentration against time is a straight line with slope $2k$.

For such a second-order reaction, a plot of $1/c$ against $t$ is linear (Fig. 13.7). The factor 2 multiplying $kt$ in this expression is a consequence of the stoichiometric coefficient 2 for $NO_2$ in the balanced equation for the specific example reaction. For other second-order reactions with different stoichiometric coefficients for the reactant (see for example the thermal decomposition of ethane described on page 451), the integrated rate law must be modified accordingly.

The concept of half-life has little use for second-order reactions. Setting $[NO_2]$ equal to $[NO_2]_0/2$ in the preceding equation and solving for $t$ gives

$$\frac{2}{[NO_2]} = 2kt_{1/2} + \frac{1}{[NO_2]_0}$$

$$t_{1/2} = \frac{1}{2k\,[NO_2]_0}$$

For second-order reactions, the half-life is not a constant; it depends on the initial concentration.

## EXAMPLE 13.5

The dimerization of tetrafluoroethylene ($C_2F_4$) to octafluorocyclobutane ($C_4F_8$) is second order in the reactant $C_2F_4$, and at 450 K its rate constant is $k = 0.0448$ L mol$^{-1}$ s$^{-1}$. If the initial concentration of $C_2F_4$ is 0.100 mol L$^{-1}$, what will its concentration be after 205 s?

### Solution

For this second-order reaction,

$$\frac{1}{c} - \frac{1}{c_0} = 2kt$$

Solving for the concentration $c$ after a time $t = 205$ s gives

$$\frac{1}{c} = (2)(0.0448 \text{ L mol}^{-1} \text{ s}^{-1})(205 \text{ s}) + \frac{1}{0.100 \text{ mol L}^{-1}} = 28.4 \text{ L mol}^{-1}$$

$$c = 3.53 \times 10^{-2} \text{ mol L}^{-1}$$

**Related Problems: 15, 16**

**FIGURE 13.8** For the reaction in Example 13.5, (a) plotting the logarithm of the concentration of $C_2F_4$ against time tests for first-order kinetics, and (b) plotting the reciprocal of the concentration of $C_2F_4$ against time tests for second-order kinetics. It is clear that the assumption of first-order kinetics does not fit the data as well; no straight line will pass through the data.

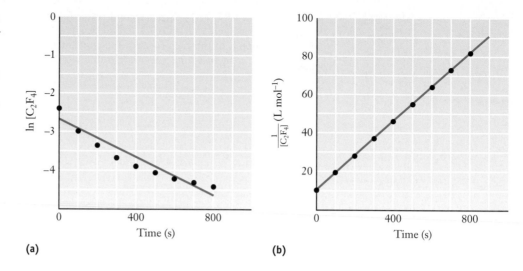

(a)                                          (b)

In a real experimental study, of course, the reaction order may be unknown. In such a case, it is necessary to try several plots to find the one that best fits the data (Fig. 13.8).

---

## 13.3

# REACTION MECHANISMS

Many reactions do not occur in a single step, but proceed through a series of steps. Each step is called an **elementary reaction** and is directly caused by the collisions of atoms, ions, or molecules. The rate expression for an overall reaction cannot be derived from the stoichiometry of the balanced equation, as we have emphasized, and must be determined experimentally. The rate of an elementary reaction, on the other hand, *is* directly proportional to the product of the concentrations of the reacting species, each raised to a power equal to its coefficient in the balanced elementary equation.

## Elementary Reactions

A **unimolecular** elementary reaction involves only a single reactant molecule. An example is the dissociation of energized $N_2O_5$ molecules in the gas phase:

$$N_2O_5^* \longrightarrow NO_2 + NO_3$$

The asterisk indicates that the $N_2O_5$ molecules have far more than thermal energy. This step is unimolecular and has the rate expression

$$\text{rate} = k[N_2O_5^*]$$

An important class of unimolecular reactions is the decay of radioactive nuclei, considered in Chapter 14.

The most common type of elementary reaction involves the collision of two atoms, ions, or molecules and is called **bimolecular.** An example is the reaction

$$NO(g) + O_3(g) \longrightarrow NO_2(g) + O_2(g)$$

The frequency at which a given NO molecule collides with ozone molecules is proportional to the concentration of ozone: if there are twice as many ozone molecules per unit volume, each NO molecule will undergo twice as many collisions as it moves through space, and ozone will react twice as rapidly. The rate of collisions of *all* the NO molecules in the container is proportional to the concentration of NO as well, so the rate law of a bimolecular reaction like this one has the form

$$\text{rate} = k[\text{NO}][\text{O}_3]$$

A **termolecular** reaction step involves the simultaneous collision of three molecules, which is a much less likely event. An example is the recombination of iodine atoms in the gas phase to form iodine molecules. So much energy is released in forming the I—I bond that the molecule would simply fly apart as soon as it was formed if the event were a binary collision. A third atom or molecule is necessary to take away some of the excess energy. If iodine recombination takes place in the presence of a sufficiently high concentration of an inert gas such as argon, termolecular reactions

$$\text{I} + \text{I} + \text{Ar} \longrightarrow \text{I}_2 + \text{Ar}$$

occur in which the argon atom leaves with more kinetic energy than it had initially. The rate law for this termolecular reaction is

$$\text{rate} = k[\text{I}]^2[\text{Ar}]$$

Elementary reactions involving collisions of four or more molecules are not observed, and even termolecular collisions are rare if other pathways are possible.

Elementary reactions in liquid solvents involve encounters of solute species with one another. If the solution is ideal, the rates of these processes are proportional to the product of the concentrations of the solute species involved. Solvent molecules are always present and may affect the reaction, even though they do not appear in the rate expression because the solvent concentration cannot be varied appreciably. A reaction such as the recombination of iodine atoms occurs readily in a liquid. It appears to be second order with rate law

$$\text{rate} = k[\text{I}]^2$$

only because the third body involved is a solvent molecule. In the same way, a reaction between a solvent molecule and a solute molecule appears to be unimolecular, and only the concentration of solute molecules enters the rate expression for that step.

## Reaction Mechanisms

A **reaction mechanism** is a detailed series of elementary reactions, with their rates, that are combined to yield the overall reaction. It is often possible to write several reaction mechanisms, each of which is consistent with a given overall reaction. One of the goals of chemical kinetics is to use the observed rate of a reaction to choose among various conceivable reaction mechanisms.

The gas-phase reaction of nitrogen dioxide with carbon monoxide provides a good example of a reaction mechanism. The generally accepted mechanism at low temperatures has two steps, both bimolecular:

$$\text{NO}_2 + \text{NO}_2 \longrightarrow \text{NO}_3 + \text{NO} \qquad \text{(slow)}$$
$$\text{NO}_3 + \text{CO} \longrightarrow \text{NO}_2 + \text{CO}_2 \qquad \text{(fast)}$$

For any reaction mechanism, combining the steps must give the overall reaction. When each elementary step occurs the same number of times in the course of the reaction, the chemical equations can simply be added. (If one step occurs twice as often as the others, it must be multiplied by 2 before the elementary reactions are added.) In this case, we add the two chemical equations to give

$$2\ NO_2 + NO_3 + CO \longrightarrow NO_3 + NO + NO_2 + CO_2$$

Canceling out the $NO_3$ and one molecule of $NO_2$ from each side leads to

$$NO_2 + CO \longrightarrow NO + CO_2$$

A **reaction intermediate** (here, $NO_3$) is a chemical species that is formed and consumed in the reaction but does not appear in the overall balanced chemical equation. One of the major challenges in chemical kinetics is to identify intermediates, which are frequently so short-lived that they are difficult to detect directly.

## EXAMPLE 13.6

Consider the following reaction mechanism:

$$Cl_2 \longrightarrow 2\ Cl$$

$$Cl + CHCl_3 \longrightarrow HCl + CCl_3$$

$$CCl_3 + Cl \longrightarrow CCl_4$$

**(a)** What is the molecularity of each elementary step?
**(b)** Write the overall equation for the reaction.
**(c)** Identify the reaction intermediate(s).

### Solution
**(a)** The first step is unimolecular, and the other two are bimolecular.
**(b)** Adding the three steps gives

$$Cl_2 + 2\ Cl + CHCl_3 + CCl_3 \longrightarrow 2\ Cl + HCl + CCl_3 + CCl_4$$

The two species that appear in equal amounts on both sides cancel out to leave

$$Cl_2 + CHCl_3 \longrightarrow HCl + CCl_4$$

**(c)** The two reaction intermediates are $Cl$ and $CCl_3$.

**Related Problems: 21, 22**

## Kinetics and Chemical Equilibrium

There is a direct connection between the rates for the elementary steps in a chemical reaction mechanism and the overall equilibrium constant $K$. To explore it, consider the reaction

$$2\ NO(g) + 2\ H_2(g) \rightleftharpoons N_2(g) + 2\ H_2O(g)$$

This reaction is believed to occur in a three-step process involving $N_2O_2$ and $N_2O$ as intermediates:

$$NO + NO \xrightleftharpoons[k_{-1}]{k_1} N_2O_2$$

$$N_2O_2 + H_2 \xrightleftharpoons[k_{-2}]{k_2} N_2O + H_2O$$

$$N_2O + H_2 \xrightleftharpoons[k_{-3}]{k_3} N_2 + H_2O$$

These elementary reactions are shown as equilibria, and so the reverse reactions (from products to reactants) are included here as well; $k_1$, $k_2$, and $k_3$ are the rate constants for the forward elementary steps, and $k_{-1}$, $k_{-2}$, and $k_{-3}$ are the rate constants for the corresponding reverse reactions.

We now invoke the principle of **detailed balance**, which states that at equilibrium the rate of *each* elementary process is balanced by (equal to) the rate of its reverse process. For the preceding mechanism we conclude that

$$k_1[NO]_{eq}^2 = k_{-1}[N_2O_2]_{eq}$$

$$k_2[N_2O_2]_{eq}[H_2]_{eq} = k_{-2}[N_2O]_{eq}[H_2O]_{eq}$$

$$k_3[N_2O]_{eq}[H_2]_{eq} = k_{-3}[N_2]_{eq}[H_2O]_{eq}$$

The equilibrium constants[1] $K_1$, $K_2$, and $K_3$ for the elementary reactions are equal to the ratio of the forward and reverse reaction rate constants:

$$K_1 = \frac{[N_2O_2]_{eq}}{[NO]_{eq}^2} = \frac{k_1}{k_{-1}}$$

$$K_2 = \frac{[N_2O]_{eq}[H_2O]_{eq}}{[N_2O_2]_{eq}[H_2]} = \frac{k_2}{k_{-2}}$$

$$K_3 = \frac{[N_2]_{eq}[H_2O]_{eq}}{[N_2O]_{eq}[H_2]_{eq}} = \frac{k_3}{k_{-3}}$$

The steps of the mechanism are now added together to obtain the overall reaction. Recall from Section 9.3 that when reactions are added, their equilibrium constants are multiplied. Therefore, the overall equilibrium constant $K$ is

$$K = K_1 K_2 K_3 = \frac{k_1 k_2 k_3}{k_{-1} k_{-2} k_{-3}} = \frac{[N_2O_2]_{eq}[N_2O]_{eq}[H_2O]_{eq}[N_2]_{eq}[H_2O]_{eq}}{[NO]_{eq}^2[N_2O_2]_{eq}[H_2]_{eq}[N_2O]_{eq}[H_2]_{eq}}$$

$$= \frac{[H_2O]_{eq}^2[N_2]_{eq}}{[NO]_{eq}^2[H_2]_{eq}^2}$$

The concentrations of the intermediates $N_2O_2$ and $N_2O$ cancel out, giving the usual expression of the mass-action law.

This result can be generalized to any reaction mechanism. The product of the forward rate constants for the elementary reactions divided by the product of the

---

[1] Thermodynamic equilibrium constants are dimensionless because they are expressed in terms of activities rather than partial pressure or concentration. The convention in chemical kinetics is to use concentrations rather than activities, even for gaseous species. Therefore, the equilibrium constants $K_1$, $K_2$, and $K_3$ introduced here are the empirical equilibrium constants $K_c$ described briefly in Sections 9.1 and 9.3. These constants are not dimensionless and must be multiplied by the concentration of the reference state, $c_{ref} = RT/P_{ref}$, raised to the appropriate power to be made equal to the thermodynamic equilibrium constant. Nevertheless, to maintain consistency with the conventions of chemical kinetics, such constants as $K_1$, $K_2$, and $K_3$ will be referred to as equilibrium constants in this section and will be written without the subscript $c$.

reverse rate constants is always equal to the equilibrium constant of the overall reaction. If there are several possible mechanisms for a given reaction (which might involve intermediates other than $N_2O_2$ and $N_2O$), their forward and reverse rate constants will all be consistent in this way with the equilibrium constant of the overall reaction.

---

## 13.4

## REACTION MECHANISMS AND RATE

In many reaction mechanisms, one step is significantly slower than all the others; it is called the **rate-determining step.** Because an overall reaction can occur only as fast as its slowest step, that step is crucial in determining the rate of the reaction. This is analogous to the flow of automobile traffic on a highway on which there is a slowdown at some point. The rate at which cars can complete a trip down the full length of the highway (in cars per minute) is approximately equal to the rate at which they pass through the bottleneck.

If the rate-determining step is the first one, the analysis is particularly simple. An example is the reaction

$$2 NO_2 + F_2 \longrightarrow 2 NO_2F$$

for which the experimental rate law is

$$\text{rate} = k_{obs}[NO_2][F_2]$$

A possible mechanism for the reaction is

$$NO_2 + F_2 \xrightarrow{k_1} NO_2F + F \quad \text{(slow)}$$

$$NO_2 + F \xrightarrow{k_2} NO_2F \quad \text{(fast)}$$

The first step is slow and determines the rate, $k_1[NO_2][F_2]$, in agreement with the observed rate expression. The subsequent fast step does not affect the reaction rate, because fluorine atoms react with $NO_2$ almost as soon as they are produced.

Mechanisms in which the rate-determining step occurs after one or more fast steps are often signaled by a reaction order greater than 2, by a nonintegral reaction order, or by an inverse concentration dependence on one of the species taking part in the reaction. An example is the reaction

$$2 NO + O_2 \longrightarrow 2 NO_2$$

for which the experimental rate law is

$$\text{rate} = k_{obs}[NO]^2[O_2]$$

One possible mechanism would be a single-step termolecular reaction of two NO molecules with one $O_2$ molecule. This would be consistent with the form of the rate expression, but termolecular collisions are quite rare, and if there is an alternative pathway it is usually followed.

One such alternative is the two-step mechanism

$$NO + NO \underset{k_{-1}}{\overset{k_1}{\rightleftharpoons}} N_2O_2 \quad \text{(fast equilibrium)}$$

$$N_2O_2 + O_2 \xrightarrow{k_2} 2 NO_2 \quad \text{(slow)}$$

Because the slow step determines the overall rate, we can write

$$\text{rate} = k_2[\text{N}_2\text{O}_2][\text{O}_2]$$

The concentration of a reactive intermediate such as $\text{N}_2\text{O}_2$ cannot be varied at will, however. Because the $\text{N}_2\text{O}_2$ reacts only slowly with $\text{O}_2$, the reverse reaction (to 2 NO) is possible and must be taken into account. In fact, it is reasonable to assume that all of the elementary reactions that occur *before* the rate-determining step are in equilibrium, with the forward and reverse reactions occurring at the same rate. In this case, we have

$$\frac{[\text{N}_2\text{O}_2]}{[\text{NO}]^2} = \frac{k_1}{k_{-1}} = K_1$$

$$[\text{N}_2\text{O}_2] = K_1[\text{NO}]^2$$

$$\text{rate} = k_2 K_1[\text{NO}]^2[\text{O}_2]$$

This result is consistent with the observed reaction order, with $k_2 K_1 = k_{\text{obs}}$.

## EXAMPLE 13.7

In basic aqueous solution the reaction

$$\text{I}^- + \text{OCl}^- \longrightarrow \text{Cl}^- + \text{OI}^-$$

follows a rate law that is consistent with the following mechanism:

$$\text{OCl}^-(aq) + \text{H}_2\text{O}(\ell) \underset{k_{-1}}{\overset{k_1}{\rightleftharpoons}} \text{HOCl}(aq) + \text{OH}^-(aq) \qquad \text{(fast equilibrium)}$$

$$\text{I}^-(aq) + \text{HOCl}(aq) \xrightarrow{k_2} \text{HOI}(aq) + \text{Cl}^-(aq) \qquad \text{(slow)}$$

$$\text{OH}^-(aq) + \text{HOI}(aq) \xrightarrow{k_3} \text{H}_2\text{O}(\ell) + \text{OI}^-(aq) \qquad \text{(fast)}$$

What rate law is predicted by this mechanism?

### Solution
The rate is determined by the slowest elementary step, the second one:

$$\text{rate} = k_2[\text{I}^-][\text{HOCl}]$$

However, the HOCl is in equilibrium with $\text{OCl}^-$ and $\text{OH}^-$ due to the first step:

$$\frac{[\text{HOCl}][\text{OH}^-]}{[\text{OCl}^-]} = K_1 = \frac{k_1}{k_{-1}}$$

Solving this for [HOCl] and inserting it into the previous expression gives the prediction

$$\text{rate} = k_2 K_1 \frac{[\text{I}^-][\text{OCl}^-]}{[\text{OH}^-]}$$

which is, in fact, the experimentally observed rate law.

**Related Problems: 25, 26, 27, 28, 29, 30**

The rate law of the foregoing example showed an inverse dependence on the concentration of $\text{OH}^-$ ion. Such a form is often a clue that a rapid equilibrium occurs in the first steps of a reaction, preceding the rate-determining step. Fractional orders of reaction provide a similar clue, as in the reaction of $H_2$ with $Br_2$ to form HBr,

$$H_2 + Br_2 \longrightarrow 2\ HBr$$

for which the initial reaction rate (before very much HBr builds up) is

$$\text{rate} = k_{obs}[H_2][Br_2]^{1/2}$$

How can such a fractional power appear? One reaction mechanism that predicts this rate law is

$$Br_2 + M \underset{k_{-1}}{\overset{k_1}{\rightleftarrows}} Br + Br + M \qquad \text{(fast equilibrium)}$$

$$Br + H_2 \xrightarrow{k_2} HBr + H \qquad \text{(slow)}$$

$$H + Br_2 \xrightarrow{k_3} HBr + Br \qquad \text{(fast)}$$

Here M stands for a second molecule that does not react but that supplies the energy to break up the bromine molecules. For such a mechanism the reaction rate is determined by the slow step:

$$\text{rate} = k_2[Br][H_2]$$

However, [Br] is fixed by the establishment of equilibrium in the first reaction,

$$\frac{[Br]^2}{[Br_2]} = K_1 = \frac{k_1}{k_{-1}}$$

so that

$$[Br] = K_1^{1/2}[Br_2]^{1/2}$$

The rate expression predicted by this mechanism is thus

$$\text{rate} = k_2 K_1^{1/2}[H_2][Br_2]^{1/2}$$

This is in accord with the observed fractional power in the rate law. On the other hand, the simple bimolecular mechanism

$$H_2 + Br_2 \xrightarrow{k_1} 2\ HBr \qquad \text{(slow)}$$

predicts a rate law:

$$\text{rate} = k_1[H_2][Br_2]$$

This disagrees with the observed rate law, and so it can be ruled out as the major contributor to the measured rate.

It should be clear from our discussion that deducing a rate law from a proposed mechanism is relatively straightforward, but doing the reverse is much harder. In fact, several competing mechanisms often give rise to the same rate law, and only some independent type of measurement can help one to choose between them. A reaction mechanism can never be proved from an experimental rate law; it can only be disproved if it is inconsistent with the experimental behavior.

A classic example is the reaction

$$H_2 + I_2 \longrightarrow 2\ HI$$

for which the observed rate law is

$$\text{rate} = k_{obs}[H_2][I_2]$$

(Contrast this with the rate law already given for the analogous reaction of $H_2$ with $Br_2$.) This is one of the earliest and most extensively studied reactions in chemical kinetics, and until 1967 it was widely believed to occur as a one-step elementary reaction. At that time J. H. Sullivan investigated the effect of illuminating the reacting sample with light, which splits some of the $I_2$ molecules into iodine atoms. If the mechanism we proposed is correct, the effect of the light on the reaction should be small because it leads only to a small decrease in the $I_2$ concentration.

Instead, Sullivan observed a dramatic *increase* in the rate of reaction under illumination, which could be explained only by the participation of iodine *atoms* in the reaction mechanism. One such mechanism is

$$I_2 + M \underset{k_{-1}}{\overset{k_1}{\rightleftharpoons}} I + I + M \qquad \text{(fast equilibrium)}$$

$$H_2 + I + I \overset{k_2}{\longrightarrow} 2\,HI \qquad\qquad \text{(slow)}$$

for which the rate law is

$$\text{rate} = k_2[H_2][I]^2 = k_2 K_1[H_2][I_2]$$

This mechanism gives the same rate law that is observed experimentally, but it is now consistent with the effect of light on the reaction. The other reaction mechanism also appears to contribute significantly to the overall rate.

This example illustrates the hazards of trying to determine reaction mechanisms from rate laws: several mechanisms can fit any given empirical rate law, and it is always possible that a new piece of information that suggests a different mechanism will be found. The problem is that, under ordinary conditions, reaction intermediates cannot be isolated and studied like the reactants and products. This situation is changing, however, with the development of experimental techniques that allow the direct study of the transient intermediates that form in small concentration during the course of a chemical reaction.

## The Steady-State Approximation

In some reaction mechanisms there is no single step that is much slower than the others, and the methods discussed so far cannot be used to predict the rate law. In this case, the **steady-state approximation** is helpful, in which it is assumed that the concentrations of reactive intermediates remain nearly constant through most of the reaction.

To illustrate this approximation, examine the mechanism proposed by F. A. Lindemann for the dissociation of molecules in the gas phase. A molecule such as $N_2O_5$ undergoes collisions with neighboring molecules M, where M can stand for another $N_2O_5$ molecule or for an inert gas such as argon if the latter is present. Through such collisions the $N_2O_5$ molecule can become excited (or activated) to a state indicated by $N_2O_5^*$:

$$N_2O_5 + M \underset{k_{-1}}{\overset{k_1}{\rightleftharpoons}} N_2O_5^* + M$$

The reverse process, with rate constant $k_{-1}$, is also indicated because the activated molecule can be deactivated by collisions with other molecules. The second step is the unimolecular decomposition of $N_2O_5^*$:

$$N_2O_5^* \overset{k_2}{\longrightarrow} NO_3 + NO_2$$

Subsequent reaction steps to form $O_2$ and $NO_2$ from $NO_3$ occur rapidly and do not affect the measured rate:

$$NO_3 + NO_2 \xrightarrow{k_3} NO + NO_2 + O_2 \qquad \text{(fast)}$$

$$NO_3 + NO \xrightarrow{k_4} 2\,NO_2 \qquad\qquad \text{(fast)}$$

The $N_2O_5^*$ is a reactive intermediate; it is produced at a rate $k_1[N_2O_5][M]$ from collisions of $N_2O_5$ molecules with other molecules and is lost at a rate $k_{-1}[N_2O_5^*][M]$ due to deactivation and at a rate $k_2[N_2O_5^*]$ due to dissociation. The net rate of change of $[N_2O_5^*]$ is then

$$\frac{d[N_2O_5^*]}{dt} = k_1[N_2O_5][M] - k_{-1}[N_2O_5^*][M] - k_2[N_2O_5^*]$$

At the beginning of the reaction, $[N_2O_5^*]$ is 0, but this concentration builds up after a short time to a small value. The steady-state approximation consists of the assumption that after this short time the rates of production and loss of $N_2O_5^*$ become equal, and

$$\frac{d[N_2O_5^*]}{dt} = 0$$

The steady-state concentration of $[N_2O_5^*]$ persists practically unchanged throughout most of the course of the reaction.

Setting the net rate of change of the $N_2O_5^*$ concentration to 0 gives

$$\frac{d[N_2O_5^*]}{dt} = 0 = k_1[N_2O_5][M] - k_{-1}[N_2O_5^*][M] - k_2[N_2O_5^*]$$

Solving for $[N_2O_5^*]$ gives

$$[N_2O_5^*]\{k_2 + k_{-1}[M]\} = k_1[N_2O_5][M]$$

$$[N_2O_5^*] = \frac{k_1[N_2O_5][M]}{k_2 + k_{-1}[M]}$$

The rate of the overall reaction $N_2O_5 \longrightarrow 2\,NO_2 + \frac{1}{2}\,O_2$ is

$$\text{rate} = \frac{1}{2}\frac{d[NO_2]}{dt} = k_2[N_2O_5^*] = \frac{k_1 k_2[N_2O_5][M]}{k_2 + k_{-1}[M]}$$

This expression has two limiting cases:

1. *Low pressure*  When $[M]$ is small enough, $k_2 \gg k_{-1}[M]$ and we can use the approximation

$$\text{rate} = k_1[N_2O_5][M] \qquad \text{(second order)}$$

This same result would be found by assuming the first step to be rate-determining.
2. *High pressure*  When $[M]$ is large enough, $k_{-1}[M] \gg k_2$ and we can use the approximation

$$\text{rate} = \frac{k_1}{k_{-1}}\,k_2[N_2O_5] \qquad \text{(first order)}$$

This same result would be found by assuming the second step to be rate-determining.

The steady-state approximation is a more general approach than those considered earlier and can be used when no single step is rate-determining.

## Chain Reactions

A **chain reaction** is one that proceeds through a series of elementary steps, some of which are repeated many times. Such chain reactions consist of three stages: (1) **initiation,** in which two or more reactive intermediates are generated; (2) **propagation,** in which products are formed but reactive intermediates are continuously regenerated; and (3) **termination,** in which two intermediates combine to give a stable product.

An example of a chain reaction is the reaction of methane with fluorine to give $CH_3F$ and HF:

$$CH_4(g) + F_2(g) \longrightarrow CH_3F(g) + HF(g)$$

Although in principle this reaction could occur through a one-step bimolecular process, that route turns out to be too slow to contribute significantly under normal reaction conditions. Instead, the mechanism involves a chain reaction of the following type:

| | |
|---|---|
| $CH_4 + F_2 \longrightarrow CH_3 + HF + F$ | (initiation) |
| $CH_3 + F_2 \longrightarrow CH_3F + F$ | (propagation) |
| $CH_4 + F \longrightarrow CH_3 + HF$ | (propagation) |
| $CH_3 + F + M \longrightarrow CH_3F + M$ | (termination) |

In the initiation step, two reactive intermediates ($CH_3$ and F) are produced. During the propagation steps, these intermediates are not used up while reactants ($CH_4$ and $F_2$) are being converted to products ($CH_3F$ and HF). The propagation steps can be repeated again and again, until eventually two reactive intermediates come together in a termination step. As we will see in Chapter 25, chain reactions are important in building up long-chain molecules called polymers.

The chain reaction just considered proceeds at a constant rate, because each propagation step both uses up and produces a reactive intermediate. The concentrations of the reactive intermediates remain approximately constant and are determined by the rates of chain initiation and termination. However, another type of chain reaction is possible in which the number of reactive intermediates increases during one or more propagation steps. This is called a **branching chain reaction.** An example is the reaction of oxygen with hydrogen. The mechanism is complex and can be initiated in various ways, leading to the formation of several reactive intermediates such as O, H, and OH. Some propagation steps are of the type already seen for $CH_4$ and $F_2$, such as

$$OH + H_2 \longrightarrow H_2O + H$$

in which one reactive intermediate (OH) is used up and one (H) is produced. Other propagation steps are branching, however:

$$H + O_2 \longrightarrow OH + O$$
$$O + H_2 \longrightarrow OH + H$$

In these steps, each reactive intermediate that is used up causes the generation of two others. This leads to a rapid growth in the number of reactive species, speeding the rate further and possibly causing an explosion. Branching chain reactions are critical in the fission of uranium (Chapter 14).

---

13.5

---

## EFFECT OF TEMPERATURE ON REACTION RATES

The first four sections of this chapter discussed the experimental determination of rate laws and their relation to assumed mechanisms for chemical reactions. Little was said, however, about the actual magnitudes of rate constants (either for elementary reactions or for overall rates of multistep reactions), nor was the effect of temperature on reaction rates discussed. To consider these matters, it is necessary to establish the connection between collision rates and the rates of chemical reactions. Only gases, for which the kinetic theory of Chapter 4 is applicable, are considered here.

### Gas-Phase Reaction Rate Constants

In Section 4.6, the kinetic theory of gases was used to estimate the frequency of collisions in a gas between a given molecule and other molecules. In Example 4.11, this frequency was calculated to be $4.1 \times 10^9$ s$^{-1}$ under room conditions for a typical small molecule such as oxygen. If every collision led to reaction, the reaction would be practically complete in a period of about $10^{-9}$ s. Some reactions do proceed at almost this rate. An example is the bimolecular reaction between two $CH_3$ radicals to give ethane, $C_2H_6$,

$$2 \, CH_3 \longrightarrow C_2H_6$$

for which the observed rate constant is $1 \times 10^{10}$ L mol$^{-1}$ s$^{-1}$. If the initial pressure of $CH_3$ were near 1 atm, then at 25°C the concentration initially would be about 0.04 M. According to the second-order integrated rate law from Section 13.2, after a period of $10^{-9}$ s the concentration would have dropped to 0.02 M. Much more commonly, however, reactions proceed at rates that are far lower—by factors of $10^{12}$ or more. The naive idea that "to collide is to react" clearly must be modified if such rates are to be understood.

A clue can be found in the observed temperature dependence of reaction rate constants. The rates of many reactions increase extremely rapidly with increases in temperature; typically, a 10°C rise in temperature may double the rate. In 1889, Svante Arrhenius suggested that rate constants vary exponentially with inverse temperature,

$$k = Ae^{-E_a/RT} \qquad \text{[13.5]}$$

where $E_a$ is a constant with dimensions of energy and $A$ is a constant with the same dimensions as $k$. Taking the natural logarithm of this equation gives

$$\ln k = \ln A - \frac{E_a}{RT} \qquad \text{[13.6]}$$

so that a plot of $\ln k$ against $1/T$ should be a straight line with slope $-E_a/R$ and intercept $\ln A$. Many rate constants do show just this kind of temperature dependence (Fig. 13.9).

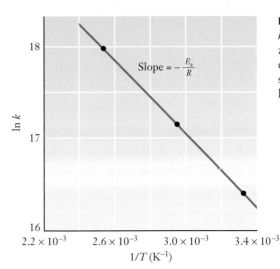

**FIGURE 13.9** An Arrhenius plot of $\ln k$ against $1/T$ for the reaction of benzene vapor with oxygen atoms. An extrapolation to $1/T = 0$ gives the constant $\ln A$ from the intercept of this line.

## EXAMPLE 13.8

The decomposition of hydroxylamine ($NH_2OH$) in the presence of oxygen follows the rate law

$$-\frac{d\,[NH_2OH]}{dt} = k_{obs}[NH_2OH][O_2]$$

where $k_{obs}$ is $0.237 \times 10^{-4}$ L mol$^{-1}$ s$^{-1}$ at 0°C and $2.64 \times 10^{-4}$ L mol$^{-1}$ s$^{-1}$ at 25°C. Calculate $E_a$ and the factor $A$ for this reaction.

### Solution

Let us write the Arrhenius equation at two different temperatures $T_1$ and $T_2$:

$$\ln k_1 = \ln A - \frac{E_a}{RT_1} \qquad \text{and} \qquad \ln k_2 = \ln A - \frac{E_a}{RT_2}$$

If the first equation is subtracted from the second, the term $\ln A$ cancels out, leaving

$$\ln k_2 - \ln k_1 = \ln \frac{k_2}{k_1} = -\frac{E_a}{R}\left(\frac{1}{T_2} - \frac{1}{T_1}\right)$$

which can be solved for $E_a$. In the present case, $T_1 = 273$ K and $T_2 = 298$ K, and so

$$\ln \frac{2.64 \times 10^{-4}}{0.237 \times 10^{-4}} = \frac{-E_a}{8.315 \text{ J K}^{-1} \text{ mol}^{-1}}\left(\frac{1}{298 \text{ K}} - \frac{1}{273 \text{ K}}\right)$$

$$2.410 = \frac{E_a}{8.315 \text{ J K}^{-1} \text{ mol}^{-1}}(3.07 \times 10^{-4} \text{ K}^{-1})$$

$$E_a = 6.52 \times 10^4 \text{ J mol}^{-1} = 65.2 \text{ kJ mol}^{-1}$$

Now that $E_a$ is known, the constant $A$ can be calculated by using data at either temperature. At 273 K,

$$\ln A = \ln k_1 + \frac{E_a}{RT}$$

$$= \ln (0.237 \times 10^{-4}) + \frac{6.52 \times 10^4 \text{ J mol}^{-1}}{8.315 \text{ J K}^{-1} \text{ mol}^{-1} (273 \text{ K})}$$

$$= -10.65 + 28.73 = 18.08$$

$$A = e^{18.08} = 7.1 \times 10^7 \text{ L mol}^{-1} \text{ s}^{-1}$$

A more accurate determination of $E_a$ and $A$ would use measurements at a series of temperatures rather than at only two, with a fit to a plot such as that in Figure 13.9.

**Related Problems: 35, 36**

---

Arrhenius believed that for molecules to react upon collision they must become "activated," and the parameter $E_a$ became known as the **activation energy.** His ideas were refined by scientists who followed, and in 1915 A. Marcelin pointed out that, although molecules make many collisions, not all collisions are reactive. Only those collisions for which the collision energy (i.e., the relative translational kinetic energy of the colliding molecules) exceeds some critical energy result in reaction. Thus, Marcelin gave a dynamic interpretation to the activation energy inferred from reaction rates.

The strong temperature dependence of rate constants, described by the Arrhenius law, can be related to the Maxwell–Boltzmann distribution of molecular energies (Fig. 13.10). If $E_a$ is the critical relative collision energy that a pair of molecules must have for a reaction to occur, only a small fraction of the molecules will possess this much energy (or more) when the temperature is sufficiently low. This fraction corresponds to the area under the Maxwell–Boltzmann distribution curve between $E_a$ and $\infty$. As the temperature is increased, the distribution function spreads out toward higher energies. The fraction of molecules that exceed the critical energy $E_a$ increases exponentially as $\exp(-E_a/RT)$, as Arrhenius's law and experiment require. The reaction rate is then proportional to $\exp(-E_a/RT)$, and both the strong temperature dependence and the order of magnitude of the experimental rate constants can be understood.

## The Reaction Coordinate and the Activated Complex

Why should there be a critical collision energy $E_a$ for the reaction between two molecules? To understand this, let us consider the physical analogy of marbles rolling on a hilly surface. As a marble rolls up a hill, its potential energy increases and its kinetic energy decreases (it gets higher and slows down). If it can reach the top of the hill, it will fall down the other side, with its kinetic energy increasing and its potential energy decreasing. Not every marble will make it over the hill, however. If its initial speed (and therefore kinetic energy) is too small, a marble will roll part of the way up and then fall back. Only those marbles with initial kinetic energy higher than a critical threshold will pass over the hill.

We can translate this physical model into a description of molecular collisions and reactions. As two reactant molecules, atoms, or ions approach each other along a **reaction path,** their potential energy increases as the bonds within them distort. At some maximum potential energy they are associated in an unstable entity called an **activated complex** or **transition state.** The activated complex is the cross-over

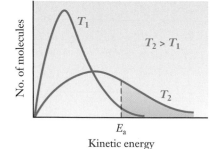

**FIGURE 13.10** The Maxwell–Boltzmann distribution of molecular kinetic energies, showing the effect of temperature on the fraction of molecules having large enough kinetic energies to react. This figure shows only translational energy; internal vibrational and rotational energy also promote reactions.

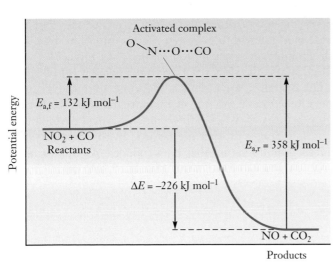

**FIGURE 13.11** The energy profile along the reaction coordinate for the reaction $NO_2 + CO \rightarrow NO + CO_2$. This direct reaction dominates the kinetics at high temperatures (above about 500 K).

stage where the smooth ascent in potential energy as the reactants come together becomes a smooth descent as the product molecules separate. As in the case of the marbles, not all pairs of reacting bodies react. Only those pairs with sufficient kinetic energy can stretch bonds and rearrange atoms sufficiently to reach the transition state that separates reactants from products. If the barrier is too high, almost all colliding pairs of reactant molecules separate from each other without reacting. The height of the barrier is close to the measured activation energy for the reaction.

Figure 13.11 is a graph of potential energy versus position along the reaction path for the reaction

$$NO_2(g) + CO(g) \longrightarrow NO(g) + CO_2(g)$$

Two activation energies are shown in Figure 13.11: $E_{a,f}$ is the activation energy for the forward reaction, and $E_{a,r}$ is that for the reverse reaction, in which NO takes an oxygen atom from $CO_2$ to form $NO_2$ and CO. The difference between the two is $\Delta E$, the change in internal energy of the chemical reaction:

$$\Delta E = E_{a,f} - E_{a,r}$$

Although $\Delta E$ is a thermodynamic quantity, obtainable from calorimetric measurements, $E_{a,f}$ and $E_{a,r}$ must be found from the temperature dependence of the rate constants for the forward and reverse reactions. In this reaction the forward and reverse activation energies are 132 and 358 kJ mol$^{-1}$, respectively, and $\Delta E$ from thermodynamics is $-226$ kJ mol$^{-1}$.

The activation energy for an elementary reaction is always positive (although in some cases it can be quite small) because there is always some energy barrier to surmount. Rates of elementary reactions therefore increase with increasing temperature. This is not necessarily true for rates of overall reactions consisting of more than one elementary reaction. These sometimes have "negative activation energies" so that the overall reaction rate slows at higher temperature. How can this be? Let us examine a specific example: the reaction of NO with oxygen,

$$2\ NO(g) + O_2(g) \longrightarrow 2\ NO_2(g)$$

for which the observed rate law is

$$\text{rate} = k_{obs}[NO]^2[O_2]$$

where $k_{obs}$ *decreases* with increasing temperature. In Section 13.4 we saw that this rate expression could be accounted for with a two-step mechanism. The first step is a rapid equilibrium (with equilibrium constant $K_1$) of two NO molecules with their dimer, $N_2O_2$. The second step is the slow reaction (rate constant $k_2$) of $N_2O_2$ with $O_2$ to form products. The overall rate constant is therefore the product of $k_2$ and $K_1$. Although $k_2$ is the rate constant for an elementary reaction (and thus increases with increasing temperature), $K_1$ is an *equilibrium* constant. Provided that the corresponding reaction is sufficiently exothermic (as it is in this case), $K_1$ will decrease so rapidly with increasing temperature that the product $k_2K_1$ will decrease as well. This accounts for the observation of negative activation energies in some overall chemical reactions.

## A  DEEPER  LOOK...

### 13.6   Reaction Dynamics

In this section, we take a closer look at the dynamics of gas-phase reactions and then compare them with the corresponding dynamics in solution. The starting point is a result from Section 4.6, that the rate of collisions of a single molecule with other molecules of the same type is

$$Z_1 = \sqrt{2}\,\pi\,d^2\,\bar{u}\,\frac{N}{V} = 4\,d^2\,\sqrt{\frac{\pi RT}{\mathcal{M}}}\,\frac{N}{V}$$

Here $d$ is the molecular diameter, $\bar{u}$ the average speed, $\mathcal{M}$ the molar mass, and $N/V$ the number density of molecules in the gas. If there are $N$ molecules in the volume, the *total* number of collisions per unit time is $\frac{1}{2}N \times Z_1$. (The factor $\frac{1}{2}$ comes from the fact that the collision of two A molecules counts as only one collision—not as one collision for the first A molecule with the second, plus one for the second with the first.) The rate of collisions *per unit volume* is then this result divided by $V$, or

rate of collisions per unit volume $= Z_{AA} = 2d^2\,\sqrt{\dfrac{\pi RT}{\mathcal{M}}}\left(\dfrac{N}{V}\right)^2$

The next step is to relate $Z_{AA}$ to the second-order rate constant for the reaction

$$A + A \longrightarrow \text{products}$$

If the activation energy for this reaction is $E_a$, then only a fraction $\exp(-E_a/RT)$ of these collisions will have sufficient energy to overcome the barrier leading to products. Each such *effective* collision of a pair of A molecules leads to a decrease in the number of A molecules in the reaction mixture by two, and so the rate of change of the number of A molecules per unit volume is

$$\frac{d(N/V)}{dt} = -2Z_{AA}e^{-E_a/RT}$$

$$= -2 \times 2\,d^2\,\sqrt{\frac{\pi RT}{\mathcal{M}}}\,e^{-E_a/RT}\left(\frac{N}{V}\right)^2$$

Rate constants involve the number of *moles* of A per unit volume, [A], not the number of molecules. The two are related by

$$N_0[A] = \frac{N}{V}$$

where $N_0$ is Avogadro's number. Substituting this equation and bringing a factor of $-\frac{1}{2}$ back to the left side (as in our usual definition of rate), we find

$$\text{rate} = -\frac{1}{2}\frac{d\,[A]}{dt} = 2\,d^2\,N_0\,\sqrt{\frac{\pi RT}{\mathcal{M}}}\,e^{-E_a/RT}\,[A]^2$$

The predicted rate constant can be identified as

$$k = 2\,d^2\,N_0\,\sqrt{\frac{\pi RT}{\mathcal{M}}}\,e^{-E_a/RT} \qquad \textbf{[13.7]}$$

How well does this simple theory agree with experiment? By fitting data on gas-phase elementary reaction rates to the Arrhenius form, the activation energy and the factor A can be obtained and the latter compared with the theory, once the molecular diameter is estimated. For the elementary reaction

$$2\,NOCl(g) \longrightarrow 2\,NO(g) + Cl_2(g)$$

the measured rate constant is 0.16 times the calculated rate constant. What this indicates is that not all collisions lead to reaction, even if the molecules have great enough relative kinetic energy. This result makes sense, because the relative orientations of the colliding molecules certainly should play a role in determining whether a particular collision results in a reaction. It seems obvious that, for a $Cl_2$ molecule to split off, the two NOCl molecules must approach each other in such a way that the chlorine atoms are close together (Fig. 13.12). The calculated collision frequency must therefore be multiplied by a **steric factor** $P$ (0.16 in

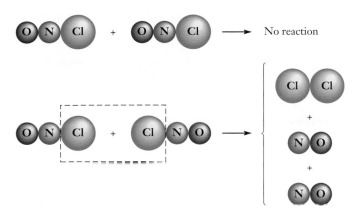

**FIGURE 13.12** The steric effect on the probability of a reaction. The two NOCl molecules must approach each other in such a way that the two chlorine atoms are close together, if the encounter is to produce $Cl_2(g)$ and $NO(g)$.

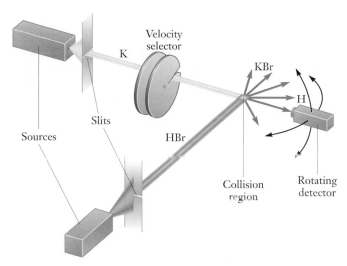

**FIGURE 13.13** Diagram of a crossed molecular beam experiment. The reaction being studied is $K + HBr \rightarrow KBr + H$.

this case) to account for the fact that only a fraction of the collisions occur with the proper orientation to lead to reaction.

The collision-theory calculation of the rate constant can be extended to bimolecular reactions of two species, A and B. Comparison of observed and predicted rate constants leads to the values of $P$ shown in Table 13.1. As the colliding molecules become larger and more complex, $P$ becomes smaller, because a smaller fraction of collisions is effective in causing reaction. The steric factor is an empirical correction that can be predicted only in especially simple cases.

## Molecular Beams

The kinetics experiments considered so far have involved changing reactant concentrations and temperatures, which are macroscopic properties of a reacting mixture. An alternative way to study kinetics is to employ the crossed molecular beam. In this device, two beams of molecules are crossed in a chamber under high vacuum, and the reaction products are determined (Fig. 13.13).

Such molecular beams allow a high degree of selection of reactant molecules. For example, a velocity selector (Fig. 4.14) can be used to keep only those molecules that have velocities within

a small range. This permits much finer tuning of reaction energies than can be achieved simply by changing the temperature. If electric and magnetic fields or laser light is used, molecules can be further selected according to how fast they rotate and how much they vibrate. The products can also be analyzed for velocity, and their angular distribution relative to the directions of the initial beams can be determined. This allows a much more detailed examination of the way in which molecules collide and react. For example, if the activated complex rotates many times before it finally breaks up, the angular distribution of products should be uniform, but if it breaks up before it has a chance to rotate, the angular distribution should be very nonuniform and should depend on the directions of the original beams. This type of measurement therefore gives information on the lifetime of an activated complex.

Molecular beams are limited to reactions that are carried out in the gas phase, for which well-defined beams of reactant molecules can be prepared. They cannot be used, for example, to study kinetics in liquid solvents. However, the detailed information that they give for certain reactions allows more precise testing of models and theories for the ways in which chemical reactions occur in the gas phase.

## Reaction Kinetics in Liquids

In a gas, atoms or molecules move in straight lines between occasional collisions (Fig. 4.18). In a liquid, the concept of collision has no meaning because molecules interact continuously with not one but many neighbors. Their trajectories can nonetheless be followed; these consist of rattling motions in a temporary "cage" formed by neighbors, superimposed upon a random and erratic diffusive displacement. In a time interval $t$, molecules undergo various net displacements $\Delta r$, but the average of the *square* of their displacements is found experimentally to be proportional to the time interval:

| **TABLE 13.1** | |
| --- | --- |

### *Steric Factors for Gas-Phase Reactions*

| Reaction | Steric Factor P |
| --- | --- |
| $2\ NOCl \rightarrow 2\ NO + Cl_2$ | 0.16 |
| $2\ NO_2 \rightarrow 2\ NO + O_2$ | $5.0 \times 10^{-2}$ |
| $2\ ClO \rightarrow Cl_2 + O_2$ | $2.5 \times 10^{-3}$ |
| $H_2 + C_2H_4 \rightarrow C_2H_6$ | $1.7 \times 10^{-6}$ |

Adapted from P. W. Atkins, *Physical Chemistry*. New York: W. H. Freeman, 1994, p. C30.

$$\overline{(\Delta r)^2} = 6Dt$$

This is the same law that was given for gases in Section 4.6, and it applies also to solids. The magnitude of the diffusion constant $D$ varies widely from one phase to another, however. Typical values of $D$ are on the order of $10^{-9}$ m$^2$ s$^{-1}$ for liquids (four orders of magnitude smaller than for a typical gas at atmospheric pressure), so that in a time interval of 1 s the mean-square displacement through diffusion is of order $10^{-8}$ m$^2$ and the root-mean-square displacement is of order $10^{-4}$ m (0.1 mm). For larger particles the diffusion coefficient can be much smaller and the resulting displacement smaller, as well.

The motion of molecules in a liquid has a significant effect on the kinetics of chemical reaction in solution. Molecules must diffuse together before they can react, and so their diffusion constants affect the rate of reaction. If the intrinsic reaction rate of two molecules that come into contact is fast enough (that is, if almost every encounter leads to reaction), then diffusion is the rate-limiting step. Such **diffusion-controlled reactions** have a maximum bimolecular rate constant on the order of $10^{10}$ L mol$^{-1}$ s$^{-1}$ in aqueous solution for the reaction of two neutral species. If the two species have opposite charges, the reaction rate can be even higher. One of the fastest known reactions in aqueous solution is the neutralization of hydronium ion ($H_3O^+$) by hydroxide ion ($OH^-$):

$$H_3O^+(aq) + OH^-(aq) \longrightarrow 2\ H_2O(\ell)$$

for which the diffusion-controlled rate constant is greater than $10^{11}$ L mol$^{-1}$ s$^{-1}$.

---

## 13.7

### KINETICS OF CATALYSIS

A **catalyst** is a substance that takes part in a chemical reaction and speeds it up, but itself undergoes no permanent chemical change. Catalysts therefore do not appear in the overall balanced chemical equation, but their presence very much affects the rate law, modifying and speeding existing pathways or, more commonly, providing completely new pathways by which a reaction can occur (Fig. 13.14). Catalysts exert significant effects on reaction rates even when they are present in very small amounts. In industrial chemistry, great effort is devoted to finding catalysts that will accelerate particular desired reactions without increasing the generation of undesired products.

Catalysis can be classified into two types: homogeneous and heterogeneous. In **homogeneous catalysis,** the catalyst is present in the same phase as the reactants, as when a gas-phase catalyst speeds up a gas-phase reaction, or a species dissolved in solution speeds up a reaction in solution. An example of homogeneous catalysis is the effect of chlorofluorocarbons and oxides of nitrogen on the depletion of ozone in the stratosphere, which will be examined in Section 16.9. A second example is the catalysis of the oxidation–reduction reaction

$$Tl^+(aq) + 2\ Ce^{4+}(aq) \longrightarrow Tl^{3+}(aq) + 2\ Ce^{3+}(aq)$$

by silver ions in solution. The direct reaction of Tl$^+$ with a single Ce$^{4+}$ ion to give Tl$^{2+}$ as an intermediate is slow. The reaction can be speeded by adding Ag$^+$ ion, which takes part in a reaction mechanism of the form

$$Ag^+ + Ce^{4+} \underset{k_{-1}}{\overset{k_1}{\rightleftharpoons}} Ag^{2+} + Ce^{3+} \qquad \text{(fast)}$$

$$Tl^+ + Ag^{2+} \xrightarrow{k_2} Tl^{2+} + Ag^+ \qquad \text{(slow)}$$

$$Tl^{2+} + Ce^{4+} \xrightarrow{k_3} Tl^{3+} + Ce^{3+} \qquad \text{(fast)}$$

**FIGURE 13.14**   The decomposition of hydrogen peroxide, $H_2O_2$, to water and oxygen is catalyzed by solid MnO$_2$. Here the water evolves as steam because of the heat given off in the reaction. (*Charles D. Winters*)

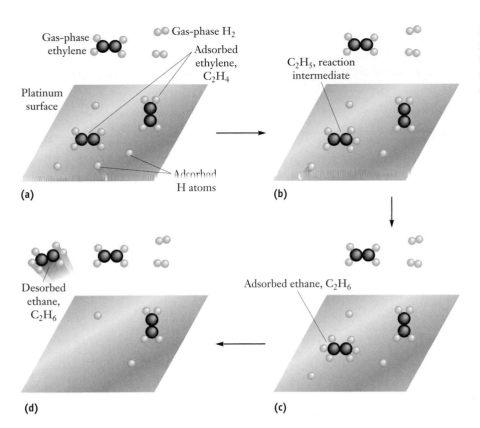

Gas-phase ethylene

Gas-phase $H_2$

Adsorbed ethylene, $C_2H_4$

Platinum surface

Adsorbed H atoms

(a)

$C_2H_5$, reaction intermediate

(b)

Adsorbed ethane, $C_2H_6$

(c)

Desorbed ethane, $C_2H_6$

(d)

**FIGURE 13.15** Platinum catalyzes the reaction $H_2 + C_2H_4$ by providing a surface that promotes the dissociation of $H_2$ to H atoms, which can then add to the $C_2H_4$ stepwise to give ethane, $C_2H_6$.

The $Ag^+$ ions are not permanently transformed by this reaction, because those used up in the first step are regenerated in the second; they play the role of catalyst in significantly speeding the rate of the overall reaction.

In **heterogeneous catalysis,** the catalyst is present as a distinct phase. The most important case is the catalytic action of certain solid surfaces on gas-phase and solution-phase reactions. A critical step in the production of sulfuric acid, for example, involves the use of a solid oxide of vanadium ($V_2O_5$) as catalyst. Many other solid catalysts are used in industrial processes. One of the best studied of such reactions is the addition of hydrogen to ethylene to form ethane:

$$C_2H_4(g) + H_2(g) \longrightarrow C_2H_6(g)$$

The process occurs extremely slowly in the gas phase, but is catalyzed by a platinum surface (Fig. 13.15).

A familiar type of catalyst is used in the exhaust streams of automobile engines to reduce the emission of pollutants such as unburned hydrocarbons, carbon monoxide, and nitrogen oxides (Fig. 13.16). A **catalytic converter** is designed to simultaneously *oxidize* hydrocarbons and CO

$$CO, C_xH_y, O_2 \xrightarrow{\text{Catalyst}} CO_2, H_2O$$

and *reduce* nitrogen oxides:

$$NO, NO_2 \xrightarrow{\text{Catalyst}} N_2, O_2$$

**FIGURE 13.16** A catalytic converter used to reduce automobile pollution, cut open to reveal the platinum, palladium, and rhodium catalysts. In this new design, a steel-alloy heating element raises the temperature to 400°C in seconds, activating the catalysts and reducing the pollution emitted in the first minutes after the car is started. *(Courtesy of Corning, Inc., Corning, NY)*

**FIGURE 13.17**  The most important way in which catalysts speed reactions is by reducing the activation energy. Both the uncatalyzed (blue) and catalyzed (red) reaction coordinates are shown.

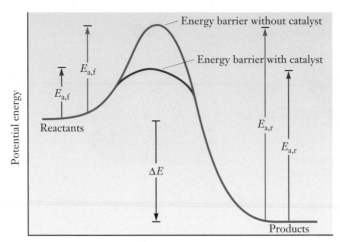

Clearly, the best catalyst for one process may not be the best for another, so a combination of two catalysts is employed. The noble metals, although expensive, are particularly useful. Typically, platinum and rhodium are deposited on a fine honeycomb mesh of alumina ($Al_2O_3$), giving a large surface area that increases the contact time of the exhaust gas with the catalysts. The platinum serves primarily as an oxidation catalyst and the rhodium as a reduction catalyst. Catalytic converters can be poisoned with certain metals that block their active sites and reduce their effectiveness. Lead is one of the most serious such poisons; as a result, unleaded fuel must be employed in automobiles with catalytic converters.

A catalyst speeds up the rate of a reaction by increasing the Arrhenius factor $A$ or, more often, by lowering the activation energy $E_a$ by providing a new activated complex of lower potential energy (Fig. 13.17). A corollary of this fact is that the same catalyst will speed both the forward *and* reverse reactions, because it lowers both the forward and reverse activation energies equally. It is important to remember that a catalyst has no effect on the thermodynamics of the overall reaction. Because the free energy is a function of state, $\Delta G$ is independent of the path followed and therefore the equilibrium constant is not changed by the presence of a catalyst. Reaction products that are not favored thermodynamically will still not form. The role of catalysts is to speed up the rate of production of products that *are* allowed by thermodynamics.

An **inhibitor** plays an opposite role to that of a catalyst. It slows the rate of a reaction, frequently by increasing the activation energy. Inhibitors are also important industrially because they can be used to reduce the rates of undesirable side reactions, allowing desired products to form in greater yield.

## Enzyme Catalysis

Many chemical reactions in living systems are carried out with enzymes acting as catalysts. An **enzyme** is a large protein molecule (typically of molar mass 20,000 g $mol^{-1}$ or more) that has a structure capable of carrying out a specific reaction or series of reactions. One or more reactant molecules (called **substrates**) bind to an enzyme at its **active sites.** These are regions on the surface of the enzyme whose structures and chemical properties are such that a given substrate will bind to them

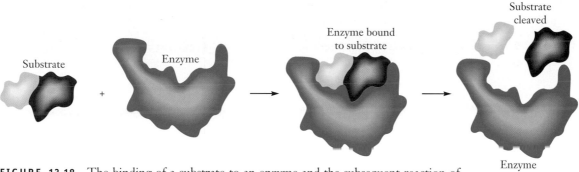

**FIGURE 13.18**  The binding of a substrate to an enzyme and the subsequent reaction of the substrate. As this schematic figure suggests, the size and shape of the active site play a role in determining which substrates bind. Equally important are the strengths of the intermolecular forces between nearby groups on the enzyme and substrate.

and its chemical transformation can be carried out (Fig. 13.18). Many enzymes are quite specific in their active sites. The enzyme urease, for example, catalyzes the hydrolysis of urea, $(NH_2)_2CO$,

$$H_3O^+(aq) + (NH_2)_2CO(aq) + H_2O(\ell) \xrightarrow{\text{Urease}} 2\,NH_4^+(aq) + HCO_3^-(aq)$$

but will not bind most other kinds of molecules, even those of similar structure. In some cases, however, a second species *does* bind to the enzyme and can act as an inhibitor, preventing the enzyme from carrying out its usual role as catalyst.

The kinetics of enzyme catalysis is indicated schematically by the reaction mechanism

$$E + S \underset{k_{-1}}{\overset{k_1}{\rightleftharpoons}} ES$$

$$ES \xrightarrow{k_2} E + P$$

Here E stands for the free enzyme, S for the substrate, ES for the complex they form when the substrate binds to the active site, and P for the product of the chemical transformation. The rate of formation of P can be derived for this mechanism by making a steady-state approximation for the concentration of the enzyme-substrate complex, [ES]. As shown in Section 13.4, this approximation involves setting the rate of change of [ES] to 0:

$$\frac{d[ES]}{dt} = 0 = k_1[E][S] - k_{-1}[ES] - k_2[ES]$$

This equation could be solved to find [ES] in terms of [E], but that is not quite what we want. Here [E] is the concentration of free enzyme (which does not have substrate bound to it), and [ES] is the concentration of bound enzyme. Their sum is $[E]_0$, the total amount of enzyme present:

$$[E]_0 = [E] + [ES]$$

It is $[E]_0$ that is accessible to experiment, so we replace [E] in the preceding steady-state equation with $[E]_0 - [ES]$. This gives

$$\frac{d[ES]}{dt} = 0 = k_1[E]_0[S] - k_1[ES][S] - k_{-1}[ES] - k_2[ES]$$

**FIGURE 13.19** The dependence of the rate of an enzyme reaction on substrate concentration. The dashed line gives the rate attained for very high [S].

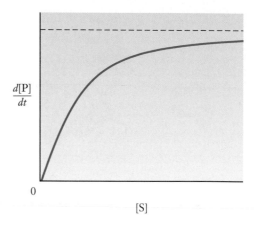

which can be solved for [ES] to give

$$[ES] = \frac{k_1[E]_0[S]}{k_1[S] + (k_{-1} + k_2)}$$

If $K_m$ is defined by

$$K_m = \frac{k_{-1} + k_2}{k_1}$$

then

$$[ES] = \frac{[E]_0[S]}{[S] + K_m}$$

The rate of formation of product is

$$\frac{d[P]}{dt} = k_2[ES] = \frac{k_2[E]_0[S]}{[S] + K_m} \qquad \text{[13.8]}$$

This is the **Michaelis–Menten equation,** which describes the kinetics of many enzyme-catalyzed reactions. Figure 13.19 shows a typical experimental curve for the dependence of the rate of such a reaction on substrate concentration [S]. For low substrate concentrations the rate is linear in [S], but at high concentrations it levels off. This behavior is easily understood on physical grounds. When [S] is small, the more substrate is added, the greater is the rate of formation of product. As the amount of substrate becomes large, all the active sites of the enzyme molecules become bound to substrate, and further addition of S does not affect the rate of the reaction. A saturation effect is seen, and the rate levels off.

A revealing form is obtained by taking the inverses of both sides of the Michaelis–Menten equation, to give

$$\frac{1}{d[P]/dt} = \frac{1}{k_2[E]_0} + \frac{K_m}{k_2[E]_0[S]}$$

A plot of the inverse of the rate of reaction against the inverse of the substrate concentration should be a straight line whose slope and intercept allow $k_2[E]_0$ and $K_m$ to be determined.

## CUMULATIVE EXERCISE

### Sulfite and Sulfate Kinetics

Sulfur dioxide dissolves in water droplets (fog, clouds, and rain) in the atmosphere and reacts according to the equation

$$SO_2(aq) + 2\,H_2O(\ell) \longrightarrow HSO_3^-(aq) + H_3O^+(aq)$$

The $HSO_3^-(aq)$ is then slowly oxidized by dissolved oxygen that is also in the droplets:

$$2\,HSO_3^-(aq) + O_2(aq) + 2\,H_2O(\ell) \longrightarrow 2\,SO_4^{2-}(aq) + 2\,H_3O^+(aq)$$

Although the second reaction has been studied for many years, only recently was the discovery made that it can proceed by the steps

$$2\,HSO_3^-(aq) + O_2(aq) \longrightarrow S_2O_7^{2-}(aq) + H_2O(\ell) \qquad \text{(fast)}$$

$$S_2O_7^{2-}(aq) + 3\,H_2O(\ell) \longrightarrow 2\,SO_4^{2-}(aq) + 2\,H_3O^+(aq) \qquad \text{(slow)}$$

The previously undetected intermediate, $S_2O_7^{2-}(aq)$, is well known in other reactions. It is the *disulfate* ion.

In an experiment at 25°C, a solution was mixed with the realistic initial concentrations of 0.270 M $HSO_3^-(aq)$ and 0.0135 M $O_2(aq)$. The initial pH was 3.90. The following table tells what happened in the solution, beginning at the moment of mixing.

| Time (s) | [HSO$_3^-$] (M) | [O$_2$] (M) | [S$_2$O$_7^{2-}$] (M) | [HSO$_4^-$] + [SO$_4^{2-}$] (M) |
|---|---|---|---|---|
| 0.000 | 0.270 | 0.0135 | 0.000 | 0.000 |
| 0.010 | 0.243 | 0.000 | $13.5 \times 10^{-3}$ | 0.000 |
| 10.0 | 0.243 | 0.000 | $11.8 \times 10^{-3}$ | $3.40 \times 10^{-3}$ |
| 45.0 | 0.243 | 0.000 | $7.42 \times 10^{-3}$ | $12.2 \times 10^{-3}$ |
| 90.0 | 0.243 | 0.000 | $4.08 \times 10^{-3}$ | $18.8 \times 10^{-3}$ |
| 150.0 | 0.243 | 0.000 | $1.84 \times 10^{-3}$ | $23.3 \times 10^{-3}$ |
| 450.0 | 0.243 | 0.000 | $0.034 \times 10^{-3}$ | $26.9 \times 10^{-3}$ |
| 600.0 | 0.243 | 0.000 | $0.005 \times 10^{-3}$ | $27.0 \times 10^{-3}$ |

(a) Determine the average rate of increase of the total of the concentrations of the sulfate plus hydrogen sulfate ions during the first 10 s of the experiment.
(b) Determine the average rate of disappearance of hydrogen sulfite ion during the first 0.010 s of the experiment.
(c) Explain why the hydrogen sulfite ion stops disappearing after 0.010 s.
(d) Plot the concentration of disulfate ion versus time on graph paper, and use the graph to estimate the instantaneous rate of disappearance of disulfate ion 90.0 s after the reaction starts.
(e) Determine the order with respect to the disulfate ion of the second step of the conversion, and the rate constant of that step.
(f) Determine the half-life of the second step of the conversion process.
(g) At 15°C, the rate constant of the second step of the conversion is only 62% of its value at 25°C. Compute the activation energy of the second step.
(h) The first step of the conversion occurs much faster when $1.0 \times 10^{-6}$ M $Fe^{2+}(aq)$ ion is added (but the rate of the second step is unaffected). What role does $Fe^{2+}$ play?
(i) Write a balanced equation for the overall reaction that gives sulfuric acid from $SO_2$ dissolved in water droplets in the air.

Crystals of sodium hydrogen sulfite, under polarized light. *(Copyright Stefan Eberhard/Fran Heyl Associates)*

**Answers**
(a) $3.40 \times 10^{-4}$ mol $L^{-1}$ $s^{-1}$
(b) $2.7$ mol $L^{-1}$ $s^{-1}$
(c) All of the oxygen is consumed, and so the first step of the process is over.
(d) $5.43 \times 10^{-5}$ mol $L^{-1}$ $s^{-1}$
(e) First order; $k = 0.0133$ $s^{-1}$
(f) 52 s
(g) 34 kJ $mol^{-1}$
(h) $Fe^{2+}$ acts as a catalyst.
(i) $2\ SO_2(aq) + O_2(aq) + 6\ H_2O(\ell) \rightarrow 2\ SO_4^{2-}(aq) + 4\ H_3O^+(aq)$

## CONCEPTS & SKILLS

*After studying this chapter and working the problems that follow, you should be able to*

1. Describe experimental methods of measuring average and instantaneous rates (Section 13.1, problems 1–2).
2. Deduce rate laws and reaction orders from experimental measurements of the dependence of reaction rates on concentrations (Section 13.2, problems 5–8).
3. Use the integrated rate laws for first- and second-order reactions to calculate the concentrations remaining after a certain elapsed time (Section 13.2, problems 9–18).
4. Describe the relationship between the equilibrium constant for a reaction and the corresponding forward and reverse rate constants (Section 13.3, problems 23–24).
5. Deduce the rate law from a mechanism characterized by a single rate-determining step (Section 13.4, problems 25–30).
6. Use the steady-state approximation to deduce rate laws in cases where no single step is rate-determining (Section 13.4, problems 31–34).
7. Calculate Arrhenius factors and activation energies from measurements of the temperature dependence of rate constants (Section 13.5, problems 35–40).
8. Discuss the connection between activation energy and the energy distribution of molecules, and relate the forward and reverse activation energies to each other through thermodynamics (Section 13.5, problems 41–42).
9. Outline the quantitative calculation of rate constants, using collision theory of gases (Section 13.6, problems 43–44).
10. Describe several types of catalysts and their effects on chemical reactions (Section 13.7).
11. Relate the rate of an enzyme-catalyzed reaction to the concentrations of substrate and enzyme in the reaction mixture (Section 13.7, problems 45–46).

## PROBLEMS

*Answers to problems whose numbers are boldface appear in Appendix G. Problems that are more challenging are indicated with asterisks.*

### Rates of Chemical Reactions

1. Use Figure 13.3 to estimate graphically the instantaneous rate of production of NO at $t = 200$ s.
2. Use Figure 13.3 to estimate graphically the instantaneous rate of production of NO at $t = 100$ s.

3. Give three related expressions for the rate of the reaction

$$N_2(g) + 3\ H_2(g) \longrightarrow 2\ NH_3(g)$$

assuming that the concentrations of any intermediates are constant and that the volume of the reaction vessel does not change.

4. Give four related expressions for the rate of the reaction

$$2\ H_2CO(g) + O_2(g) \longrightarrow 2\ CO(g) + 2\ H_2O(g)$$

assuming that the concentrations of any intermediates are constant and that the volume of the reaction vessel does not change.

## Rate Laws

**5.** Nitrogen oxide reacts with hydrogen at elevated temperatures according to the following chemical equation:

$$2\ NO(g) + 2\ H_2(g) \longrightarrow N_2(g) + 2\ H_2O(g)$$

It is observed that, when the concentration of $H_2$ is cut in half, the rate of the reaction is also cut in half. When the concentration of NO is multiplied by 10, the rate of the reaction increases by a factor of 100.
(a) Write the rate expression for this reaction, and give the units of the rate constant $k$.
(b) If [NO] were multiplied by 3 and [$H_2$] by 2, what change in the rate would be observed?

**6.** In the presence of vanadium oxide, $SO_2(g)$ reacts with an excess of oxygen to give $SO_3(g)$:

$$SO_2(g) + \tfrac{1}{2}O_2(g) \xrightarrow{\ V_2O_5\ } SO_3(g)$$

This reaction is an important step in the manufacture of sulfuric acid. It is observed that tripling the $SO_2$ concentration increases the rate by a factor of 3, but tripling the $SO_3$ concentration *decreases* the rate by a factor of $1.7 \approx \sqrt{3}$. The rate is insensitive to the $O_2$ concentration as long as an excess of oxygen is present.
(a) Write the rate expression for this reaction, and give the units of the rate constant $k$.
(b) If [$SO_2$] is multiplied by 2 and [$SO_3$] by 4 but all other conditions are unchanged, what change in the rate will be observed?

**7.** In a study of the reaction of pyridine ($C_5H_5N$) with methyl iodide ($CH_3I$) in a benzene solution, the following set of initial reaction rates was measured at 25°C for different initial concentrations of the two reactants:

| [$C_5H_5N$] (mol L$^{-1}$) | [$CH_3I$] (mol L$^{-1}$) | Rate (mol L$^{-1}$ s$^{-1}$) |
| --- | --- | --- |
| $1.00 \times 10^{-4}$ | $1.00 \times 10^{-4}$ | $7.5 \times 10^{-7}$ |
| $2.00 \times 10^{-4}$ | $2.00 \times 10^{-4}$ | $3.0 \times 10^{-6}$ |
| $2.00 \times 10^{-4}$ | $4.00 \times 10^{-4}$ | $6.0 \times 10^{-6}$ |

(a) Write the rate expression for this reaction.
(b) Calculate the rate constant $k$, and give its units.
(c) Predict the initial reaction rate for a solution in which [$C_5H_5N$] is $5.0 \times 10^{-5}$ M and [$CH_3I$] is $2.0 \times 10^{-5}$ M.

**8.** The rate for the oxidation of iron(II) by cerium(IV)

$$Ce^{4+}(aq) + Fe^{2+}(aq) \longrightarrow Ce^{3+}(aq) + Fe^{3+}(aq)$$

is measured at several different initial concentrations of the two reactants:

| [Ce$^{4+}$] (mol L$^{-1}$) | [Fe$^{2+}$] (mol L$^{-1}$) | Rate (mol L$^{-1}$ s$^{-1}$) |
| --- | --- | --- |
| $1.1 \times 10^{-5}$ | $1.8 \times 10^{-5}$ | $2.0 \times 10^{-7}$ |
| $1.1 \times 10^{-5}$ | $2.8 \times 10^{-5}$ | $3.1 \times 10^{-7}$ |
| $3.4 \times 10^{-5}$ | $2.8 \times 10^{-5}$ | $9.5 \times 10^{-7}$ |

(a) Write the rate expression for this reaction.
(b) Calculate the rate constant $k$, and give its units.
(c) Predict the initial reaction rate for a solution in which [Ce$^{4+}$] is $2.6 \times 10^{-5}$ M and [Fe$^{2+}$] is $1.3 \times 10^{-5}$ M.

**9.** The reaction $SO_2Cl_2(g) \rightarrow SO_2(g) + Cl_2(g)$ is first order, with a rate constant of $2.2 \times 10^{-5}$ s$^{-1}$ at 320°C. The partial pressure of $SO_2Cl_2(g)$ in a sealed vessel at 320°C is 1.0 atm. How long will it take for the partial pressure of $SO_2Cl_2(g)$ to fall to 0.50 atm?

**10.** The reaction $FClO_2(g) \rightarrow FClO(g) + O(g)$ is first order with a rate constant of $6.76 \times 10^{-4}$ s$^{-1}$ at 322°C.
(a) Calculate the half-life of the reaction at 322°C.
(b) If the initial partial pressure of $FClO_2$ in a container at 322°C is 0.040 atm, how long will it take to fall to 0.010 atm?

**11.** The decomposition of benzene diazonium chloride

$$C_6H_5N_2Cl \longrightarrow C_6H_5Cl + N_2$$

follows first-order kinetics with a rate constant of $4.3 \times 10^{-5}$ s$^{-1}$ at 20°C. If the initial partial pressure of $C_6H_5N_2Cl$ is 0.0088 atm, calculate its partial pressure after 10.0 hours.

**12.** At 600 K, the rate constant for the first-order decomposition of nitroethane

$$CH_3CH_2NO_2(g) \longrightarrow C_2H_4(g) + HNO_2(g)$$

is $1.9 \times 10^{-4}$ s$^{-1}$. A sample of $CH_3CH_2NO_2$ is heated to 600 K, at which point its initial partial pressure is measured to be 0.078 atm. Calculate its partial pressure after 3.0 hours.

**13.** Chloroethane decomposes at elevated temperatures according to the reaction

$$C_2H_5Cl(g) \longrightarrow C_2H_4(g) + HCl(g)$$

This reaction obeys first-order kinetics. After 340 s at 800 K, a measurement shows that the concentration of $C_2H_5Cl$ has decreased from 0.0098 mol L$^{-1}$ to 0.0016 mol L$^{-1}$. Calculate the rate constant $k$ at 800 K.

**14.** The isomerization reaction

$$CH_3NC \longrightarrow CH_3CN$$

obeys the first-order rate law

$$\text{rate} = -k[CH_3NC]$$

in the presence of an excess of argon. Measurements at 500 K reveal that in 520 s the concentration of $CH_3NC$ decreases to 71% of its original value. Calculate the rate constant $k$ of the reaction at 500 K.

**15.** At 25°C in CCl$_4$ solvent, the reaction

$$I + I \longrightarrow I_2$$

is second order in the concentration of the iodine atoms. The rate constant $k$ has been measured as $8.2 \times 10^9$ L mol$^{-1}$ s$^{-1}$.

Suppose the initial concentration of I atoms is $1.00 \times 10^{-4}$ M. Calculate their concentration after $2.0 \times 10^{-6}$ s.

16. $HO_2$ is a highly reactive chemical species that plays a role in atmospheric chemistry. The rate of the gas-phase reaction

$$HO_2(g) + HO_2(g) \longrightarrow H_2O_2(g) + O_2(g)$$

is second order in $[HO_2]$, with a rate constant at 25°C of $1.4 \times 10^9$ L mol$^{-1}$ s$^{-1}$. Suppose some $HO_2$ with an initial concentration of $2.0 \times 10^{-8}$ M could be confined at 25°C. Calculate the concentration that would remain after 1.0 s, assuming no other reactions take place.

17. The rate for the reaction

$$OH^-(aq) + NH_4^+(aq) \longrightarrow H_2O(\ell) + NH_3(aq)$$

is first order in both $OH^-$ and $NH_4^+$ concentrations, and the rate constant $k$ at 20°C is $3.4 \times 10^{10}$ L mol$^{-1}$ s$^{-1}$. Suppose 1.00 L of a 0.0010 M NaOH solution is rapidly mixed with the same volume of 0.0010 M $NH_4Cl$ solution. Calculate the time (in seconds) required for the $OH^-$ concentration to decrease to a value of $1.0 \times 10^{-5}$ M.

18. The rate for the reaction

$$OH^-(aq) + HCN(aq) \longrightarrow H_2O(\ell) + CN^-(aq)$$

is first order in both $OH^-$ and HCN concentrations, and the rate constant $k$ at 25°C is $3.7 \times 10^9$ L mol$^{-1}$ s$^{-1}$. Suppose 0.500 L of a 0.0020 M NaOH solution is rapidly mixed with the same volume of a 0.0020 M HCN solution. Calculate the time (in seconds) required for the $OH^-$ concentration to decrease to a value of $1.0 \times 10^{-4}$ M.

## Reaction Mechanisms

19. Identify each of the following elementary reactions as unimolecular, bimolecular, or termolecular, and write the rate expression.
    (a) $HCO + O_2 \rightarrow HO_2 + CO$
    (b) $CH_3 + O_2 + N_2 \rightarrow CH_3O_2 + N_2$
    (c) $HO_2NO_2 \rightarrow HO_2 + NO_2$

20. Identify each of the following elementary reactions as unimolecular, bimolecular, or termolecular, and write the rate expression.
    (a) $BrONO_2 \rightarrow BrO + NO_2$
    (b) $HO + NO_2 + Ar \rightarrow HNO_3 + Ar$
    (c) $O + H_2S \rightarrow HO + HS$

21. Consider the following reaction mechanism:

$$H_2O_2 \longrightarrow H_2O + O$$
$$O + CF_2Cl_2 \longrightarrow ClO + CF_2Cl$$
$$ClO + O_3 \longrightarrow Cl + 2\,O_2$$
$$Cl + CF_2Cl \longrightarrow CF_2Cl_2$$

(a) What is the molecularity of each elementary step?
(b) Write the overall equation for the reaction.
(c) Identify the reaction intermediate(s).

22. Consider the following reaction mechanism:

$$NO_2Cl \longrightarrow NO_2 + Cl$$
$$Cl + H_2O \longrightarrow HCl + OH$$
$$OH + NO_2 + N_2 \longrightarrow HNO_3 + N_2$$

(a) What is the molecularity of each elementary step?
(b) Write the overall equation for the reaction.
(c) Identify the reaction intermediate(s).

23. The rate constant of the elementary reaction

$$BrO(g) + NO(g) \longrightarrow Br(g) + NO_2(g)$$

is $1.3 \times 10^{10}$ L mol$^{-1}$ s$^{-1}$ at 25°C, and its equilibrium constant is $5.0 \times 10^{10}$ at this temperature. Calculate the rate constant at 25°C of the elementary reaction

$$Br(g) + NO_2(g) \longrightarrow BrO(g) + NO(g)$$

24. The compound $IrH_3(CO)(P(C_6H_5)_3)_2$ exists in two forms, the meridional ("mer") and facial ("fac"). At 25° in a nonaqueous solvent, the reaction mer → fac has a rate constant of $2.33$ s$^{-1}$, and the reaction fac → mer has a rate constant of $2.10$ s$^{-1}$. What is the equilibrium constant of the mer-to-fac reaction at 25°C?

## Reaction Mechanisms and Rate

25. Write the overall reaction and rate laws that correspond to the following reaction mechanisms. Be sure to eliminate intermediates from the answers.

(a) $A + B \underset{k_{-1}}{\overset{k_1}{\rightleftharpoons}} C + D$    (fast equilibrium)

   $C + E \overset{k_2}{\longrightarrow} F$    (slow)

(b)    $A \underset{k_{-1}}{\overset{k_1}{\rightleftharpoons}} B + C$    (fast equilibrium)

   $C + D \underset{k_{-2}}{\overset{k_2}{\rightleftharpoons}} E$    (fast equilibrium)

   $E \overset{k_3}{\longrightarrow} F$    (slow)

26. Write the overall reaction and the rate laws that correspond to the following reaction mechanisms. Be sure to eliminate intermediates from the answers.

(a) $2\,A + B \underset{k_{-1}}{\overset{k_1}{\rightleftharpoons}} D$    (fast equilibrium)

   $D + B \overset{k_2}{\longrightarrow} E + F$    (slow)

   $F \overset{k_3}{\longrightarrow} G$    (fast)

(b) $A + B \underset{k_{-1}}{\overset{k_1}{\rightleftharpoons}} C$    (fast equilibrium)

   $C + D \underset{k_{-2}}{\overset{k_2}{\rightleftharpoons}} F$    (fast equilibrium)

   $F \overset{k_3}{\longrightarrow} G$    (slow)

27. HCl reacts with propene ($CH_3CHCH_2$) in the gas phase according to the overall reaction

$$HCl + CH_3CHCH_2 \longrightarrow CH_3CHClCH_3$$

The experimental rate expression is

$$\text{rate} = k[\text{HCl}]^3[\text{CH}_3\text{CHCH}_2]$$

Which, if any, of the following mechanisms are consistent with the observed rate expression?

(a) $\text{HCl} + \text{HCl} \rightleftarrows \text{H} + \text{HCl}_2$ (fast equilibrium)
$\text{H} + \text{CH}_3\text{CHCH}_2 \rightarrow \text{CH}_3\text{CHCH}_3$ (slow)
$\text{HCl}_2 + \text{CH}_3\text{CHCH}_3 \rightarrow \text{CH}_3\text{CHClCH}_3 + \text{HCl}$ (fast)

(b) $\text{HCl} + \text{HCl} \rightleftarrows \text{H}_2\text{Cl}_2$ (fast equilibrium)
$\text{HCl} + \text{CH}_3\text{CHCH}_2 \rightleftarrows \text{CH}_3\text{CHClCH}_3^*$
(fast equilibrium)
$\text{CH}_3\text{CHClCH}_3^* + \text{H}_2\text{Cl}_2 \rightarrow \text{CH}_3\text{CHClCH}_3 + 2\ \text{HCl}$
(slow)

(c) $\text{HCl} + \text{CH}_3\text{CHCH}_2 \rightleftarrows \text{H} + \text{CH}_3\text{CHClCH}_2$
(fast equilibrium)
$\text{H} + \text{HCl} \rightleftarrows \text{H}_2\text{Cl}$ (fast equilibrium)
$\text{H}_2\text{Cl} + \text{CH}_3\text{CHClCH}_2 \rightarrow \text{HCl} + \text{CH}_3\text{CHClCH}_3$
(slow)

28. Chlorine reacts with hydrogen sulfide in aqueous solution

$$\text{Cl}_2(aq) + \text{H}_2\text{S}(aq) \longrightarrow \text{S}(s) + 2\ \text{H}^+(aq) + 2\ \text{Cl}^-(aq)$$

in a second-order reaction that follows the rate expression

$$\text{rate} = k[\text{Cl}_2][\text{H}_2\text{S}]$$

Which, if any, of the following mechanisms are consistent with the observed rate expression?

(a) $\text{Cl}_2 + \text{H}_2\text{S} \rightarrow \text{H}^+ + \text{Cl}^- + \text{Cl}^+ + \text{HS}^-$ (slow)
$\text{Cl}^+ + \text{HS}^- \rightarrow \text{H}^+ + \text{Cl}^- + \text{S}$ (fast)

(b) $\text{H}_2\text{S} \rightleftarrows \text{HS}^- + \text{H}^+$ (fast equilibrium)
$\text{HS}^- + \text{Cl}_2 \rightarrow 2\ \text{Cl}^- + \text{S} + \text{H}^+$ (slow)

(c) $\text{H}_2\text{S} \rightleftarrows \text{HS}^- + \text{H}^+$ (fast equilibrium)
$\text{H}^+ + \text{Cl}_2 \rightleftarrows \text{H}^+ + \text{Cl}^- + \text{Cl}^+$ (fast equilibrium)
$\text{Cl}^+ + \text{HS}^- \rightarrow \text{H}^+ + \text{Cl}^- + \text{S}$ (slow)

29. Nitryl chloride is a reactive gas with a normal boiling point of $-16°\text{C}$. Its decomposition to nitrogen dioxide and chlorine is described by the equation

$$2\ \text{NO}_2\text{Cl} \longrightarrow 2\ \text{NO}_2 + \text{Cl}_2$$

The rate expression for this reaction has the form

$$\text{rate} = k[\text{NO}_2\text{Cl}]$$

Which, if any, of the following mechanisms are consistent with the observed rate expression?

(a) $\text{NO}_2\text{Cl} \rightarrow \text{NO}_2 + \text{Cl}$ (slow)
$\text{Cl} + \text{NO}_2\text{Cl} \rightarrow \text{NO}_2 + \text{Cl}_2$ (fast)

(b) $2\ \text{NO}_2\text{Cl} \rightleftarrows \text{N}_2\text{O}_4 + \text{Cl}_2$ (fast equilibrium)
$\text{N}_2\text{O}_4 \rightarrow 2\ \text{NO}_2$ (slow)

(c) $2\ \text{NO}_2\text{Cl} \rightleftarrows \text{ClO}_2 + \text{N}_2\text{O} + \text{ClO}$ (fast equilibrium)
$\text{N}_2\text{O} + \text{ClO}_2 \rightleftarrows \text{NO}_2 + \text{NOCl}$ (fast equilibrium)
$\text{NOCl} + \text{ClO} \rightarrow \text{NO}_2 + \text{Cl}_2$ (slow)

30. Ozone in the upper atmosphere is decomposed by nitrogen oxide through the reaction

$$\text{O}_3 + \text{NO} \longrightarrow \text{O}_2 + \text{NO}_2$$

The experimental rate expression for this reaction is

$$\text{rate} = k[\text{O}_3][\text{NO}]$$

Which, if any, of the following mechanisms are consistent with the observed rate expression?

(a) $\text{O}_3 + \text{NO} \rightarrow \text{O} + \text{NO}_3$ (slow)
$\text{O} + \text{O}_3 \rightarrow 2\ \text{O}_2$ (fast)
$\text{NO}_3 + \text{NO} \rightarrow 2\ \text{NO}_2$ (fast)

(b) $\text{O}_3 + \text{NO} \rightarrow \text{O}_2 + \text{NO}_2$ (slow)

(c) $\text{NO} + \text{NO} \rightleftarrows \text{N}_2\text{O}_2$ (fast equilibrium)
$\text{N}_2\text{O}_2 + \text{O}_3 \rightarrow \text{NO}_2 + 2\ \text{O}_2$ (slow)

31. Consider the mechanism of problem 25(a). Suppose *no* assumptions are made about the relative rates of the steps. By making the steady-state approximation for the concentration of the intermediate (C), express the rate of production of the product (F) in terms of the concentrations of A, B, D, and E. In what limit does this reduce to the result of problem 25(a)?

32. Consider the mechanism of problem 25(b). Suppose *no* assumptions are made about the relative rates of the steps. By making a steady-state approximation for the concentrations of the intermediates (C and E), express the rate of production of the product (F) in terms of the concentrations of A, B, and D. In what limit does this reduce to the result of problem 25(b)?

33. The mechanism for the decomposition of $\text{NO}_2\text{Cl}$ is

$$\text{NO}_2\text{Cl} \underset{k_{-1}}{\overset{k_1}{\rightleftarrows}} \text{NO}_2 + \text{Cl}$$

$$\text{NO}_2\text{Cl} + \text{Cl} \overset{k_2}{\longrightarrow} \text{NO}_2 + \text{Cl}_2$$

By making a steady-state approximation for [Cl], express the rate of appearance of $\text{Cl}_2$ in terms of the concentrations of $\text{NO}_2\text{Cl}$ and $\text{NO}_2$.

34. A key step in the formation of sulfuric acid from dissolved $\text{SO}_2$ in acid precipitation is the oxidation of hydrogen sulfite ion by hydrogen peroxide:

$$\text{HSO}_3^-(aq) + \text{H}_2\text{O}_2(aq) \longrightarrow \text{HSO}_4^-(aq) + \text{H}_2\text{O}(\ell)$$

The mechanism involves peroxymonosulfurous acid, $\text{SO}_2\text{OOH}^-$:

$$\text{HSO}_3^-(aq) + \text{H}_2\text{O}_2(aq) \underset{k_{-1}}{\overset{k_1}{\rightleftarrows}} \text{SO}_2\text{OOH}^-(aq) + \text{H}_2\text{O}(\ell)$$

$$\text{SO}_2\text{OOH}^-(aq) + \text{H}_3\text{O}^+(aq) \overset{k_2}{\longrightarrow} \text{HSO}_4^-(aq) + \text{H}_3\text{O}^+(aq)$$

By making a steady-state approximation for the reactive intermediate concentration, $[\text{SO}_2\text{OOH}^-(aq)]$, express the rate of formation of $\text{HSO}_4^-(aq)$ in terms of the concentrations of $\text{HSO}_3^-(aq)$, $\text{H}_2\text{O}_2(aq)$, and $\text{H}_3\text{O}^+(aq)$.

## Effect of Temperature on Reaction Rates

35. The rate of the elementary reaction

$$\text{Ar} + \text{O}_2 \longrightarrow \text{Ar} + \text{O} + \text{O}$$

has been studied as a function of temperature between 5000 and 18,000 K. The following data were obtained for the rate constant $k$:

| Temperature (K) | $k$ (L mol$^{-1}$ s$^{-1}$) |
|---|---|
| 5,000 | $5.49 \times 10^6$ |
| 10,000 | $9.86 \times 10^8$ |
| 15,000 | $5.09 \times 10^9$ |
| 18,000 | $8.60 \times 10^9$ |

(a) Calculate the activation energy of this reaction.
(b) Calculate the factor $A$ in the Arrhenius equation for the temperature dependence of the rate constant.

36. The gas-phase reaction

$$H + D_2 \longrightarrow HD + D$$

is the exchange of isotopes of hydrogen of atomic mass 1 (H) and 2 (D, deuterium). The following data were obtained for the rate constant $k$ of this reaction:

| Temperature (K) | $k$ (L mol$^{-1}$ s$^{-1}$) |
|---|---|
| 299 | $1.56 \times 10^4$ |
| 327 | $3.77 \times 10^4$ |
| 346 | $7.6 \times 10^4$ |
| 440 | $1.07 \times 10^6$ |
| 549 | $8.7 \times 10^6$ |
| 745 | $8.7 \times 10^7$ |

(a) Calculate the activation energy of this reaction.
(b) Calculate the factor $A$ in the Arrhenius equation for the temperature dependence of the rate constant.

37. The rate constant of the elementary reaction

$$BH_4^-(aq) + NH_4^+(aq) \longrightarrow BH_3NH_3(aq) + H_2(g)$$

is $k = 1.94 \times 10^{-4}$ L mol$^{-1}$ s$^{-1}$ at 30.0°C, and the reaction has an activation energy of 161 kJ mol$^{-1}$.
(a) Compute the rate constant of the reaction at a temperature of 40.0°C.
(b) After equal concentrations of $BH_4^-(aq)$ and $NH_4^+(aq)$ are mixed at 30.0°C, $1.00 \times 10^4$ s is required for half of them to be consumed. How long will it take to consume half of the reactants if an identical experiment is performed at 40.0°C?

38. Dinitrogen tetraoxide ($N_2O_4$) decomposes spontaneously at room temperature in the gas phase:

$$N_2O_4(g) \longrightarrow 2\ NO_2(g)$$

The rate law governing the disappearance of $N_2O_4$ with time is

$$-\frac{d\,[N_2O_4]}{dt} = k\,[N_2O_4]$$

At 30°C, $k = 5.1 \times 10^6$ s$^{-1}$ and the activation energy for the reaction is 54.0 kJ mol$^{-1}$.
(a) Calculate the time (in seconds) required for the partial pressure of $N_2O_4(g)$ to decrease from 0.10 atm to 0.010 atm at 30°C.
(b) Repeat the calculation of part (a) at 300°C.

39. The activation energy for the isomerization reaction of $CH_3NC$ in problem 14 is 161 kJ mol$^{-1}$, and the reaction rate constant at 600 K is 0.41 s$^{-1}$.
(a) Calculate the Arrhenius factor $A$ for this reaction.
(b) Calculate the rate constant for this reaction at 1000 K.

40. Cyclopropane isomerizes to propylene according to a first-order reaction:

$$\text{cyclopropane} \longrightarrow \text{propylene}$$

The activation energy is $E_a = 272$ kJ mol$^{-1}$. At 500°C, the reaction rate constant is $6.1 \times 10^{-4}$ s$^{-1}$.
(a) Calculate the Arrhenius factor $A$ for this reaction.
(b) Calculate the rate constant for this reaction at 25°C.

41. The activation energy of the gas-phase reaction

$$OH(g) + HCl(g) \longrightarrow H_2O(g) + Cl(g)$$

is 3.5 kJ mol$^{-1}$, and the change in the internal energy in the reaction is $\Delta E = -66.8$ kJ mol$^{-1}$. Calculate the activation energy of the reaction

$$H_2O(g) + Cl(g) \longrightarrow OH(g) + HCl(g)$$

42. The compound HOCl is known, but the related compound HClO, with a different order for the atoms in the molecule, is not known. Calculations suggest that the activation energy for the conversion HOCl → HClO is 311 kJ mol$^{-1}$ and that for the conversion HClO → HOCl is 31 kJ mol$^{-1}$. Estimate $\Delta E$ for the reaction HOCl → HClO.

## A Deeper Look. . . Reaction Dynamics

43. Use collision theory to estimate the preexponential factor in the rate constant for the elementary reaction

$$NOCl + NOCl \longrightarrow 2\ NO + Cl_2$$

at 25°C. Take the average diameter of an NOCl molecule to be $3.0 \times 10^{-10}$ m and use the steric factor $P$ from Table 13.1.

44. Use collision theory to estimate the preexponential factor in the rate constant for the elementary reaction

$$NO_2 + NO_2 \longrightarrow 2\ NO + O_2$$

at 500 K. Take the average diameter of an $NO_2$ molecule to be $2.6 \times 10^{-10}$ m and use the steric factor $P$ from Table 13.1.

## Kinetics of Catalysis

45. Certain bacteria use the enzyme penicillinase to decompose penicillin and render it inactive. The Michaelis–Menten constants for this enzyme and substrate are $K_m = 5 \times 10^{-5}$ mol L$^{-1}$ and $k_2 = 2 \times 10^3$ s$^{-1}$.
(a) What is the maximum rate of decomposition of penicillin if the enzyme concentration is $6 \times 10^{-7}$ M?
(b) At what substrate concentration will the rate of decomposition be half that calculated in part (a)?

46. The conversion of dissolved carbon dioxide in blood to $HCO_3^-$ and $H_3O^+$ is catalyzed by the enzyme carbonic anhydrase. The Michaelis–Menten constants for this enzyme and substrate are $K_m = 8 \times 10^{-5}$ mol L$^{-1}$ and $k_2 = 6 \times 10^5$ s$^{-1}$.

(a) What is the maximum rate of reaction of carbon dioxide if the enzyme concentration is $5 \times 10^{-6}$ M?

(b) At what $CO_2$ concentration will the rate of decomposition be 30% of that calculated in part (a)?

## Additional Problems

47. Hemoglobin molecules in blood bind oxygen and carry it to cells, where it takes part in metabolism. The binding of oxygen

$$\text{hemoglobin}(aq) + O_2(aq) \longrightarrow \text{hemoglobin } O_2(aq)$$

is first order in hemoglobin and first order in dissolved oxygen, with a rate constant of $4 \times 10^7$ L mol$^{-1}$ s$^{-1}$. Calculate the initial rate at which oxygen will be bound to hemoglobin if the concentration of hemoglobin is $2 \times 10^{-9}$ M and that of oxygen is $5 \times 10^{-5}$ M.

* 48. Suppose 1.00 L of $9.95 \times 10^{-3}$ M $S_2O_3^{2-}$ is mixed with 1.00 L of $2.52 \times 10^{-3}$ M $H_2O_2$ at a pH of 7.0 and a temperature of 25°C. These species react by two competing pathways, represented by the balanced equations

$$S_2O_3^{2-} + 4 H_2O_2 \longrightarrow 2 SO_4^{2-} + H_2O + 2 H_3O^+$$

$$2 S_2O_3^{2-} + H_2O_2 + 2 H_3O^+ \longrightarrow S_4O_6^{2-} + 4 H_2O$$

At the instant of mixing, the thiosulfate ion ($S_2O_3^{2-}$) is observed to be disappearing at the rate of $7.9 \times 10^{-7}$ mol L$^{-1}$ s$^{-1}$. At the same moment, the $H_2O_2$ is disappearing at the rate of $8.8 \times 10^{-7}$ mol L$^{-1}$ s$^{-1}$.

(a) Compute the percentage of the $S_2O_3^{2-}$ that is, at that moment, reacting according to the first equation.

(b) It is observed that the hydronium ion concentration drops. Use the data and answer from part (a) to compute how many milliliters per minute of 0.100 M $H_3O^+$ must be added to keep the pH equal to 7.0.

49. $8.23 \times 10^{-3}$ mol of InCl(s) is placed in 1.00 L of 0.010 M HCl(aq) at 75°C. The InCl(s) dissolves quite quickly, and then the following reaction occurs:

$$3 In^+(aq) \longrightarrow 2 In(s) + In^{3+}(aq)$$

As this disproportionation proceeds, the solution is analyzed at intervals to determine the concentration of $In^+(aq)$ that remains.

| Time (s) | [In$^+$] (mol L$^{-1}$) |
|---|---|
| 0 | $8.23 \times 10^{-3}$ |
| 240 | $6.41 \times 10^{-3}$ |
| 480 | $5.00 \times 10^{-3}$ |
| 720 | $3.89 \times 10^{-3}$ |
| 1000 | $3.03 \times 10^{-3}$ |
| 1200 | $3.03 \times 10^{-3}$ |
| 10000 | $3.03 \times 10^{-3}$ |

(a) Plot ln [In$^+$] versus time, and determine the apparent rate constant for this first-order reaction.

(b) Determine the half-life of this reaction.

(c) Determine the equilibrium constant $K$ for the reaction under the experimental conditions.

* 50. A compound called di-$t$-butyl peroxide [abbreviation DTBP, formula $(CH_3)_3COOC(CH_3)_3$] decomposes to give acetone [$(CH_3)_2CO$] and ethane ($C_2H_6$):

$$(CH_3)_3COOC(CH_3)_3(g) \longrightarrow 2 (CH_3)_2CO(g) + C_2H_6(g)$$

The *total* pressure of the reaction mixture changes with time, as shown by the following data at 147.2°C:

| Time (min) | $P_{tot}$ (atm) | Time (min) | $P_{tot}$ (atm) |
|---|---|---|---|
| 0 | 0.2362 | 26 | 0.3322 |
| 2 | 0.2466 | 30 | 0.3449 |
| 6 | 0.2613 | 34 | 0.3570 |
| 10 | 0.2770 | 38 | 0.3687 |
| 14 | 0.2911 | 40 | 0.3749 |
| 18 | 0.3051 | 42 | 0.3801 |
| 20 | 0.3122 | 46 | 0.3909 |
| 22 | 0.3188 | | |

(a) Calculate the partial pressure of DTBP at each time from these data. Assume that at time 0, DTBP is the only gas present.

(b) Are the data better described by a first-order or a second-order rate expression with respect to DTBP concentration?

51. The reaction of OH$^-$ with HCN in aqueous solution at 25°C has a forward rate constant $k_f$ of $3.7 \times 10^9$ L mol$^{-1}$ s$^{-1}$. Using this information and the measured acid ionization constant of HCN (see Table 10.2), calculate the rate constant $k_r$ in the first-order rate law

$$\text{rate} = k_r[CN^-]$$

for the transfer of hydrogen ions to CN$^-$ from surrounding water molecules:

$$H_2O(\ell) + CN^-(aq) \longrightarrow OH^-(aq) + HCN(aq)$$

52. Carbon dioxide reacts with ammonia to give ammonium carbamate, a solid. The reverse reaction also occurs:

$$CO_2(g) + 2 NH_3(g) \rightleftharpoons NH_4OCONH_2(s)$$

The forward reaction is first order in $CO_2(g)$ and second order in $NH_3(g)$. Its rate constant is 0.238 atm$^{-2}$ s$^{-1}$ at 0.0°C (expressed in terms of partial pressures rather than concentrations). The reaction in the reverse direction is zero order, and its rate constant, at the same temperature, is $1.60 \times 10^{-7}$ atm s$^{-1}$. Experimental studies show that, at all stages in the progress of this reaction, the net rate is equal to the forward rate minus the reverse rate. Compute the equilibrium constant of this reaction at 0.0°C.

53. For the reactions

$$I + I + M \longrightarrow I_2 + M$$

$$Br + Br + M \longrightarrow Br_2 + M$$

the rate laws are

$$-\frac{d[I]}{dt} = k_I[I]^2[M]$$

$$-\frac{d[Br]}{dt} = k_{Br}[Br]^2[M]$$

The ratio $k_I/k_{Br}$ at 500°C is 3.0 when M is an Ar molecule. Initially, $[I]_0 = 2[Br]_0$, while [M] is the same for both reactions and is much greater than $[I]_0$. Calculate the ratio of the time required for [I] to decrease to half its initial value to the same time for [Br] at 500°C.

54. In some reactions there is a competition between kinetic control and thermodynamic control over product yields. Suppose compound A can undergo two elementary reactions to stable products:

$$A \underset{k_{-1}}{\overset{k_1}{\rightleftharpoons}} B \quad \text{or} \quad A \underset{k_{-2}}{\overset{k_2}{\rightleftharpoons}} C$$

For simplicity we assume first-order kinetics for both forward and reverse reactions. We take the numerical values $k_1 = 1 \times 10^8 \text{ s}^{-1}$, $k_{-1} = 1 \times 10^2 \text{ s}^{-1}$, $k_2 = 1 \times 10^9 \text{ s}^{-1}$, and $k_{-2} = 1 \times 10^4 \text{ s}^{-1}$.

(a) Calculate the equilibrium constant for the equilibrium

$$B \rightleftharpoons C$$

From this value, give the ratio of the concentration of B to that of C at equilibrium. This is an example of thermodynamic control.

(b) In the case of kinetic control, the products are isolated (or undergo additional reaction) before the back reactions can take place. Suppose the back reactions in the preceding example ($k_{-1}$ and $k_{-2}$) can be ignored. Calculate the concentration ratio of B to C reached in this case.

55. Compare and contrast the mechanisms for the two gas-phase reactions

$$H_2 + Br_2 \longrightarrow 2 HBr$$

$$H_2 + I_2 \longrightarrow 2 HI$$

56. In Section 13.4, the steady-state approximation was used to derive a rate expression for the decomposition of $N_2O_5(g)$:

$$\text{rate} = \frac{k_1 k_2[M][N_2O_5]}{k_2 + k_{-1}[M]} = k_{eff}[N_2O_5]$$

At 300 K, with an excess of nitrogen present, the following values of $k_{eff}$ as a function of total pressure were found:

| P (atm) | $k_{eff}$ (s$^{-1}$) | P (atm) | $k_{eff}$ (s$^{-1}$) |
|---------|--------|---------|--------|
| 9.21 | 0.265 | 0.625 | 0.116 |
| 5.13 | 0.247 | 0.579 | 0.108 |
| 3.16 | 0.248 | 0.526 | 0.104 |
| 3.03 | 0.223 | 0.439 | 0.092 |
|      |       | 0.395 | 0.086 |

Use the data to estimate the value of $k_{eff}$ at very high total pressure and the value of $k_1$ in L mol$^{-1}$ s$^{-1}$.

57. The decomposition of ozone by light can be described by the mechanism

$$O_3 + \text{light} \overset{k_1}{\longrightarrow} O_2 + O$$

$$O + O_2 + M \overset{k_2}{\longrightarrow} O_3 + M$$

$$O + O_3 \overset{k_3}{\longrightarrow} 2 O_2$$

with the overall reaction being

$$2 O_3 + \text{light} \longrightarrow 3 O_2$$

The rate constant $k_1$ depends on the light intensity and the type of light source used. By making a steady-state approximation for the concentration of oxygen atoms, express the rate of formation of $O_2$ in terms of the $O_2$, $O_3$, and M concentrations and the elementary rate constants. Show that only the ratio $k_3/k_2$, and not the individual values of $k_2$ and $k_3$, affects the rate.

*58. In Section 13.4 we considered the following mechanism for the reaction of $Br_2$ with $H_2$:

$$Br_2 + M \underset{k_{-1}}{\overset{k_1}{\rightleftharpoons}} Br + Br + M$$

$$Br + H_2 \overset{k_2}{\longrightarrow} HBr + H$$

$$H + Br_2 \overset{k_3}{\longrightarrow} HBr + Br$$

Although this is adequate for calculating the *initial* rate of reaction, before product HBr builds up, there is an additional process that can participate as the reaction continues:

$$HBr + H \overset{k_4}{\longrightarrow} H_2 + Br$$

(a) Write an expression for the rate of change of [H].
(b) Write an expression for the rate of change of [Br].
(c) As hydrogen and bromine atoms are both short-lived species, we can make the steady-state approximation and set the rates from parts (a) and (b) to 0. Express the steady-state concentrations [H] and [Br] in terms of concentrations of $H_2$, $Br_2$, HBr, and M. (*Hint:* Try adding the rate for part (a) to that for part (b).)
(d) Express the rate of production of HBr in terms of concentrations of $H_2$, $Br_2$, HBr, and M.

59. The gas-phase decomposition of acetaldehyde can be represented by the overall chemical equation

$$CH_3CHO \longrightarrow CH_4 + CO$$

It is thought to occur through the sequence of reactions

$$CH_3CHO \longrightarrow CH_3 + CHO$$

$$CH_3 + CH_3CHO \longrightarrow CH_4 + CH_2CHO$$

$$CH_2CHO \longrightarrow CO + CH_3$$

$$CH_3 + CH_3 \longrightarrow CH_3CH_3$$

Show that this reaction mechanism corresponds to a chain reaction, and identify the initiation, propagation, and termination steps.

60. Lanthanum(III) phosphate crystallizes as a hemihydrate, $LaPO_4 \cdot \frac{1}{2}H_2O$. When it is heated, it loses water to give anhydrous lanthanum(III) phosphate:

$$2(LaPO_4 \cdot \tfrac{1}{2}H_2O(s)) \longrightarrow 2\,LaPO_4(s) + H_2O(g)$$

This reaction is first order in the chemical amount of $LaPO_4 \cdot \frac{1}{2}H_2O$. The rate constant varies with temperature as follows:

| Temperature (°C) | k (s$^{-1}$) |
|---|---|
| 205 | $2.3 \times 10^{-4}$ |
| 219 | $3.69 \times 10^{-4}$ |
| 246 | $7.75 \times 10^{-4}$ |
| 260 | $12.3 \times 10^{-4}$ |

Compute the activation energy of this reaction.

61. The water in a pressure cooker boils at a temperature greater than 100°C because it is under pressure. At this higher temperature, the chemical reactions associated with the cooking of food take place at a greater rate.
    (a) Some food cooks fully in 5 min in a pressure cooker at 112°C and in 10 min in an open pot at 100°C. Calculate the average activation energy for the reactions associated with the cooking of this food.
    (b) How long will the same food take to cook in an open pot of boiling water in Denver, where the average atmospheric pressure is 0.818 atm and the boiling point of water is 94.4°C?

62. (a) A certain first-order reaction has an activation energy of 53 kJ mol$^{-1}$. It is run twice, first at 298 K and then at 308 K (10°C higher). All other conditions are identical. Show that, in the second run, the reaction occurs at double its rate in the first run.
    (b) The same reaction is run twice more at 398 K and 408 K. Show that the reaction goes 1.5 times as fast at 408 K as it does at 398 K.

*63. The gas-phase reaction between hydrogen and iodine

$$H_2(g) + I_2(g) \underset{k_r}{\overset{k_f}{\rightleftharpoons}} 2\,HI(g)$$

proceeds with a forward rate constant at 1000 K of $k_f = 240$ L mol$^{-1}$ s$^{-1}$ and an activation energy of 165 kJ mol$^{-1}$. By using this information and data from Appendix D, calculate the activation energy for the reverse reaction and the value of $k_r$ at 1000 K. Assume that $\Delta H$ and $\Delta S$ for the reaction are independent of temperature between 298 and 1000 K.

64. The following reaction mechanism has been proposed for a chemical reaction:

$$A_2 \underset{k_{-1}}{\overset{k_1}{\rightleftharpoons}} A + A \qquad \text{(fast equilibrium)}$$

$$A + B \underset{k_{-2}}{\overset{k_2}{\rightleftharpoons}} AB \qquad \text{(fast equilibrium)}$$

$$AB + CD \overset{k_3}{\longrightarrow} AC + BD \qquad \text{(slow)}$$

    (a) Write a balanced equation for the overall reaction.
    (b) Write the rate expression that corresponds to the preceding mechanism. Express the rate in terms of concentrations of reactants only ($A_2$, B, CD).
    (c) Suppose that the first two steps in the preceding mechanism are endothermic and the third one is exothermic. Will an increase in temperature increase the reaction rate constant, decrease it, or cause no change? Explain.

65. How would you describe the role of the $CF_2Cl_2$ in the reaction mechanism of problem 21?

66. In Section 13.7 we wrote a mechanism in which silver ions catalyze the reaction of $Tl^+$ with $Ce^{4+}$. Determine the rate law for this mechanism by making a steady-state approximation for the concentration of the reactive intermediate $Ag^{2+}$.

67. The rates of enzyme catalysis can be lowered by the presence of inhibitor molecules I, which bind to the active site of the enzyme. This adds the following additional step to the reaction mechanism considered in Section 13.7:

$$E + I \underset{k_{-3}}{\overset{k_3}{\rightleftharpoons}} EI \qquad \text{(fast equilibrium)}$$

Determine the effect of the presence of inhibitor at total concentration $[I]_0 = [I] + [EI]$ on the rate expression for formation of products derived at the end of this chapter.

68. The enzyme lysozyme kills certain bacteria by attacking a sugar called $N$-acetylglucosamine (NAG) in their cell walls. At an enzyme concentration of $2 \times 10^{-6}$ M, the maximum rate for substrate (NAG) reaction, found at high substrate concentration, is $1 \times 10^{-6}$ mol L$^{-1}$ s$^{-1}$. The rate is reduced by a factor of 2 when the substrate concentration is reduced to $6 \times 10^{-6}$ M. Determine the Michaelis–Menten constants $K_m$ and $k_2$ for lysozyme.

## CUMULATIVE PROBLEMS

69. The rate of the gas-phase reaction

$$H_2 + I_2 \longrightarrow 2\,HI$$

is given by

$$\text{rate} = -\frac{d\,[I_2]}{dt} = k\,[H_2][I_2]$$

with $k = 0.0242$ L mol$^{-1}$ s$^{-1}$ at 400°C. If the initial concentration of $H_2$ is 0.081 mol L$^{-1}$ and that of $I_2$ is 0.036 mol L$^{-1}$, calculate the initial rate at which heat is absorbed or emitted during the reaction. Assume that the enthalpy change at 400°C is the same as that at 25°C.

70. The rate of the reaction

$$2 \, ClO_2(aq) + 2 \, OH^-(aq) \longrightarrow$$
$$ClO_3^-(aq) + ClO_2^-(aq) + H_2O(\ell)$$

is given by

$$\text{rate} = k[ClO_2]^2[OH^-]$$

with $k = 230 \, L^2 \, mol^{-2} \, s^{-1}$ at 25°C. A solution is prepared that has initial concentrations $[ClO_2] = 0.020$ M, $[HCN] = 0.095$ M, and $[CN^-] = 0.17$ M. Calculate the initial rate of the reaction.

71. A gas mixture was prepared at 500 K with total pressure 3.26 atm and a mole fraction of 0.00057 of NO and 0.00026 of $O_3$. The elementary reaction

$$NO + O_3 \longrightarrow NO_2 + O_2$$

has a second-order rate constant of $7.6 \times 10^7 \, L \, mol^{-1} \, s^{-1}$ at this temperature. Calculate the initial rate of the reaction under these conditions.

72. The activation energy for the reaction

$$2 \, NO_2 \longrightarrow 2 \, N + O_2$$

is $E_a = 111 \, kJ \, mol^{-1}$. Calculate the root-mean-square velocity of an $NO_2$ molecule at 400 K and compare it to the velocity of an $NO_2$ molecule with kinetic energy $E_a/N_0$.

# Nuclear Chemistry

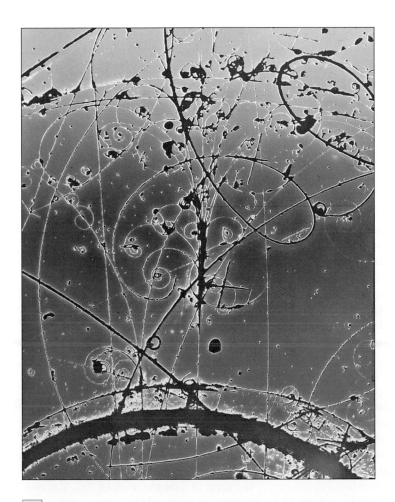

I n the preceding chapters, we have discussed chemical reactions essentially as rearrangements of atoms from one set of "bonding partners" to another set. We turn now to the subject of chemical reactions in which the elemental identity of the atomic nucleus changes in the course of the reaction. This is a fascinating subject that arose early in the twentieth century from experimental studies of the physical structure of the atom. (See Section 1.4).

*Illustration*
The tracks of subatomic particles in a bubble chamber. *(Science Photo Library/ Photo Researchers, Inc.)*

Two new features distinguish this field from the material discussed in earlier chapters.

First, the separate laws of conservation of matter and conservation of energy must be generalized. Nuclear reactions occur at energies sufficiently high that energy and matter are interconverted through Einstein's relation $E = mc^2$. The *sum* of energy and matter is conserved during nuclear reactions. This relation is the origin of the enormous amounts of energy that can be released during nuclear reactions.

Second, our point of view must shift from macroscopic or bulk properties of matter (masses of reactants and products in stoichiometry, pressures and concentrations in reacting mixtures at or away from equilibrium, and energy changes in physical and chemical processes) to a microscopic description of how the building blocks introduced in Section 1.4 (proton, neutron, and electron) fit together to give matter its properties and determine the magnitude of energy changes during reactions. In this chapter we consider the properties and reactions of the nucleus. In subsequent chapters the microscopic theme will continue as we study the structure of atoms, molecules, and the regular arrays of molecules that form solids.

---

## 14.1

## MASS–ENERGY RELATIONSHIPS IN NUCLEI

Recall from Section 1.4 that almost all the mass of an atom is concentrated in a very small volume in the nucleus. The small size of the nucleus (which occupies less than one *trillionth* of the space in the atom) and the strong forces between the protons and neutrons that make it up largely isolate its behavior from the outside world of electrons and other nuclei. This greatly simplifies our analysis of nuclear chemistry, allowing us to examine single nuclei without concern for the atoms, ions, or molecules in which they may be found.

As defined in Section 1.4, a nuclide is characterized by the number of protons, $Z$, and the number of neutrons, $N$, it contains. The atomic number $Z$ determines the charge $+Ze$ on the nucleus and therefore decides the identity of the element; the sum $Z + N = A$ is the mass number of the nuclide and is the integer closest to the relative atomic mass of the nuclide. Nuclides are designated by the symbol $^A_Z X$, where X is the chemical symbol for the element.

The absolute masses of nuclides and of elementary particles such as the proton, neutron, and electron are far too small to express in kilograms or grams without the constant use of 10 raised to some large negative exponent. A unit of mass that is better sized for measuring on the submicroscopic scale is needed. An **atomic mass unit** (abbreviated u) is defined as exactly $\frac{1}{12}$ the mass of a single atom of $^{12}C$. Because 1 mol of $^{12}C$ atoms weighs exactly 12 g, one atom weighs $12/N_0$ grams, and

$$1 \text{ u} = \frac{1}{12} \left( \frac{12 \text{ g mol}^{-1}}{6.022137 \times 10^{23} \text{ mol}^{-1}} \right)$$

$$= 1.660540 \times 10^{-24} \text{ g} = 1.660540 \times 10^{-27} \text{ kg}$$

The conversion factor between atomic mass units and grams is numerically equal to the inverse of Avogadro's number $N_0$, and the mass of a single atom in atomic

## TABLE 14.1

### Masses of Selected Elementary Particles and Atoms

| Elementary Particle | Symbol | Mass (u) | Mass (kg) |
|---|---|---|---|
| Electron, beta particle | $_{-1}^{0}e^-$ | 0.00054857990 | $9.109390 \times 10^{-31}$ |
| Positron | $_{1}^{0}e^+$ | 0.00054857990 | $9.109390 \times 10^{-31}$ |
| Proton | $_{1}^{1}p^+$ | 1.00727647 | $1.672623 \times 10^{-27}$ |
| Neutron | $_{0}^{1}n$ | 1.00866490 | $1.674929 \times 10^{-27}$ |

| Atom | Mass (u) | Atom | Mass (u) |
|---|---|---|---|
| $_{1}^{1}H$ | 1.00782505 | $_{11}^{23}Na$ | 22.989770 |
| $_{1}^{2}H$ | 2.0141079 | $_{12}^{24}Mg$ | 23.985045 |
| $_{1}^{3}H$ | 3.016049 | $_{14}^{30}Si$ | 29.973761 |
| $_{2}^{3}He$ | 3.01602930 | $_{15}^{30}P$ | 29.97832 |
| $_{2}^{4}He$ | 4.0026033 | $_{16}^{32}S$ | 31.972072 |
| $_{3}^{7}Li$ | 7.016005 | $_{17}^{35}Cl$ | 34.9688527 |
| $_{4}^{8}Be$ | 8.0053052 | $_{20}^{40}Ca$ | 39.962589 |
| $_{4}^{9}Be$ | 9.0121825 | $_{22}^{49}Ti$ | 48.947871 |
| $_{4}^{10}Be$ | 10.013535 | $_{35}^{81}Br$ | 80.91629 |
| $_{5}^{8}B$ | 8.024612 | $_{37}^{87}Rb$ | 86.909184 |
| $_{5}^{10}B$ | 10.012938 | $_{38}^{87}Sr$ | 86.908890 |
| $_{5}^{11}B$ | 11.009305 | $_{53}^{127}I$ | 126.904477 |
| $_{6}^{11}C$ | 11.011433 | $_{88}^{226}Ra$ | 226.025406 |
| $_{6}^{12}C$ | 12 exactly | $_{88}^{228}Ra$ | 228.03123 |
| $_{6}^{13}C$ | 13.003354 | $_{89}^{228}Ac$ | 228.03117 |
| $_{6}^{14}C$ | 14.003242 | $_{90}^{232}Th$ | 232.038054 |
| $_{7}^{14}N$ | 14.00307400 | $_{90}^{234}Th$ | 234.0436 |
| $_{8}^{16}O$ | 15.9949146 | $_{91}^{231}Pa$ | 231.035881 |
| $_{8}^{17}O$ | 16.999131 | $_{92}^{231}U$ | 231.0363 |
| $_{8}^{18}O$ | 17.9991594 | $_{92}^{234}U$ | 234.040947 |
| $_{9}^{19}F$ | 18.9984033 | $_{92}^{235}U$ | 235.043925 |
| $_{11}^{21}Na$ | 20.99764 | $_{92}^{238}U$ | 238.050786 |

mass units is numerically equal to the mass of 1 mol of atoms in grams. Thus, one atom of $^1H$ has a mass of 1.007825 u because 1 mol of $^1H$ has a mass of 1.007825 g. The *dalton* is a mass unit that is equivalent to the atomic mass unit and is used frequently in biochemistry.

Table 14.1 lists the masses of elementary particles and selected neutral atoms, obtained from mass spectrometry (Fig. 1.12). Note that each atomic mass includes the contribution from the surrounding $Z$ electrons as well as that from the nucleus.

## The Einstein Mass–Energy Relationship

Some nuclides are stable and can exist indefinitely, but others are unstable (radioactive) and decay spontaneously to form other nuclides in processes to be considered in Section 14.2. Figure 14.1 shows the distribution of the number of neutrons $N(= A - Z)$ against atomic number $Z$ (protons in the nucleus) for the known nuclides of the elements. For the light elements, $N/Z \approx 1$ for stable nuclides, and so nearly equal numbers of protons and neutrons are found in the nucleus. For the

**FIGURE 14.1** A plot of $N$ versus $Z$ for nuclides (represented by dots). Arrows show the directions of decay of unstable nuclides via alpha emission, beta emission, positron emission, or electron capture.

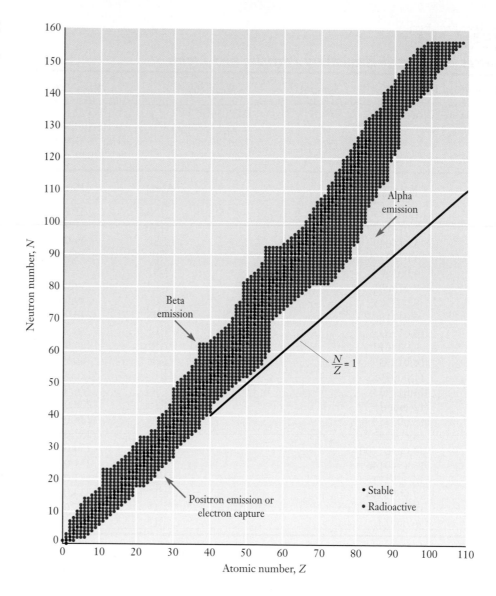

heavier elements, $N/Z > 1$ and there are progressively more neutrons than protons in the nucleus. One of the goals of nuclear chemistry is to understand the differing stabilities of the isotopes. Nuclear mass provides the key to this understanding.

Note from Table 14.1 that the neutron has a slightly greater mass than the proton. The neutron is stable inside a nucleus, but in free space it is unstable, decaying into a proton and an electron with a half-life of about 12 minutes:

$$\,^1_0n \longrightarrow \,^1_1p^+ + \,^0_{-1}e^-$$

Here $\,^1_0n$ and $\,^1_1p^+$ represent the neutron and the proton, respectively. We have written this as a **balanced nuclear equation.** To check such an equation for balance, first verify that the total mass numbers (the superscripts) are equal on the two sides. Then verify that the nuclear charges are balanced by checking the sum of the subscripts.

Mass is not conserved in the reaction just written. By subtracting the mass of the reactant from the masses of the products (using Table 14.1), the change in mass is calculated to be

$$\Delta m = -1.395 \times 10^{-30} \text{ kg} = -8.3985 10^{-4} \text{ u}$$

What has happened to the "lost" mass? It has been converted to energy—specifically, kinetic energy carried off by the electron and proton. According to Einstein's special theory of relativity, a change in mass always accompanies a change in energy:

$$E = mc^2$$

or, better,

$$\Delta E = c^2 \, \Delta m \qquad\qquad \text{[14.1]}$$

where $c$ is the speed of light in a vacuum and $\Delta m$ is the change in mass (mass of products minus mass of reactants). To express $\Delta E$ in joules, $c$ must have units of meters per second, and $\Delta m$, units of kilograms. In the decay of one neutron, for example,

$$\Delta E = c^2 \Delta m = (2.9979 \times 10^8 \text{ m s}^{-1})^2(-1.395 \times 10^{-30} \text{ kg}) = -1.254 \times 10^{-13} \text{ J}$$

In ordinary chemical reactions, the change in mass is negligibly small. In the combustion of 1 mol of carbon to $CO_2$, the energy change is about 400 kJ, corresponding to a change in mass of only $4 \times 10^{-9}$ g.

Changes in energy in nuclear reactions are nearly always expressed in more convenient energy units than the joule—namely, the **electron volt** (eV) and the **million electron volt** (MeV). The electron volt is defined as the energy given to an electron when it is accelerated through a potential difference of exactly 1 V:

$$\Delta E \text{ (joules)} = Q \text{ (coulombs)} \times \Delta\mathscr{E} \text{ (volts)}$$

$$1 \text{ eV} = 1.602177 \times 10^{-19} \text{ C} \times 1 \text{ V} = 1.602177 \times 10^{-19} \text{ J}$$

The million electron volt is 1 million times larger:

$$1 \text{ MeV} = 1.602177 \times 10^{-13} \text{ J}$$

A change in mass of $1 \text{ u} = 1.660540 \times 10^{-27}$ kg corresponds to

$$\Delta E = c^2 \Delta m = (2.9979246 \times 10^8 \text{ m s}^{-1})^2(1.660540 \times 10^{-27} \text{ kg})$$

$$= 1.492419 \times 10^{-10} \text{ J}$$

Converting to MeV,

$$\frac{1.492419 \times 10^{-10} \text{ J}}{1.602177 \times 10^{-13} \text{ J MeV}^{-1}} = 931.494 \text{ MeV}$$

We say that the **energy equivalent** of 1 u is 931.494 MeV. In the decay of a neutron,

$$\Delta E = (-8.3985 \times 10^{-4} \text{ u})(931.494 \text{ MeV u}^{-1}) = 0.782 \text{ MeV}$$

## Binding Energies of Nuclei

The **binding energy** $E_B$ of a nucleus is defined as the negative of the energy change $\Delta E$ that would occur if that nucleus were formed from its component protons and neutrons. For the $^4_2\text{He}$ nucleus, for example,

$$2 \, ^1_1p^+ + 2 \, ^1_0n \longrightarrow \, ^4_2\text{He}^{2+} \qquad\qquad \Delta E = ?$$

Binding energies of nuclei are calculated with Einstein's relationship and accurate data from a mass spectrometer. Most measurements, however, determine the mass of the atom with most or all of its electrons, not the mass of the bare nucleus (recall that it is the $^{12}C$ *atom*, not the $^{12}C$ nucleus, that is defined to have atomic mass 12). Therefore, we take the nuclear binding energy $E_B$ of the helium nucleus to be the negative of the energy change $\Delta E$ for the formation of an *atom* of helium from hydrogen *atoms* and neutrons.

$$2\,{}_1^1H + 2\,{}_0^1n \longrightarrow {}_2^4He$$

The correction for the differences in binding energies of electrons (which of course are present both in the hydrogen atoms and in the atom being formed) is a very small one that need not concern us here.

## EXAMPLE 14.1

Calculate the binding energy of $^4He$ from the data in Table 14.1, and express it both in joules and in million electron volts (MeV).

**Solution**
The mass change is

$$\Delta m = m[{}_2^4He] - 2m[{}_1^1H] - 2m[{}_0^1n]$$

$$= 4.0026033 - 2(1.00782505) - 2(1.0086649) = -0.0303766 \text{ u}$$

Einstein's relation then gives

$$\Delta E = (-0.0303766 \text{ u})(1.660540 \times 10^{-27} \text{ kg u}^{-1})(2.9979246 \times 10^8 \text{ m s}^{-1})^2$$

$$= -4.53346 \times 10^{-12} \text{ J}$$

$$E_B = 4.53346 \times 10^{-12} \text{ J}$$

If 1 *mole* of $^4_2He$ atoms were formed in this way, the energy change would be greater by a factor of Avogadro's number $N_0$, giving $\Delta E = -2.73 \times 10^{12}$ J mol$^{-1}$. This is an enormous quantity, seven orders of magnitude greater than typical energy changes in chemical reactions.

The energy change, in MeV, accompanying the formation of a $^4_2He$ atom is

$$\Delta E = (-0.0303766 \text{ u})(931.494 \text{ MeV u}^{-1}) = -28.2956 \text{ MeV}$$

**Related Problems: 3, 4**

There are four nucleons in the $^4_2He$ nucleus, and so the binding energy *per nucleon* is

$$\frac{E_B}{4} = \frac{28.2956 \text{ MeV}}{4} = 7.07390 \text{ MeV}$$

The binding energy per nucleon is a direct measure of the stability of the nucleus. It varies with atomic number among the stable elements, increasing to a maximum of about 8.8 MeV for iron and nickel, as Figure 14.2 illustrates.

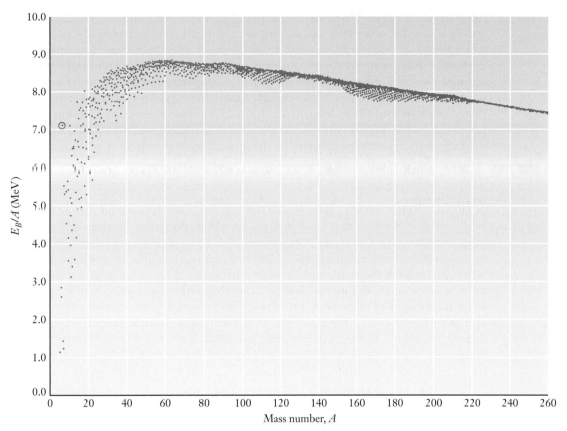

**FIGURE 14.2**  The variation of the binding energy per nucleon with mass number. A total of 2212 nuclides are shown, of which 274 are stable. The greatest nuclear stability is seen in the vicinity of $^{56}$Fe. Note that $^4$He (red circle) is unusually stable for its mass number.

---

## 14.2

## NUCLEAR DECAY PROCESSES

Why are some nuclei **radioactive,** decaying spontaneously, when others are stable? Thermodynamics gives the criterion $\Delta G < 0$ for a process to be spontaneous at constant $T$ and $P$. The energy change $\Delta E$ in a nuclear decay process is so great that the free energy change is essentially equal to it, and the thermodynamic criterion for spontaneous nuclear reaction consequently simplifies to

$$\Delta E < 0 \qquad \text{or} \qquad \Delta m < 0$$

The equivalence of these criteria follows from Einstein's mass–energy relationship. Spontaneous transformations of one nucleus into others can occur only if the combined mass of products is less than the mass of the original nuclide.

Before we address the details of nuclear decay, it is necessary to mention **antiparticles.** Each subatomic particle is thought to have an antiparticle of the same mass but opposite charge. Thus, the antiparticle of an electron is a **positron** ($e^+$), and the antiparticle of a proton is a negatively charged particle called an **antiproton.**

Antiparticles have only transient existences because matter and antimatter annihilate one another when they come together, emitting radiation carrying an equivalent amount of energy. In the case of an electron and a positron, two oppositely directed **gamma rays** (γ-rays; high-energy photons), called the **annihilation radiation,** are emitted. Another important particle–antiparticle pair is the neutrino (symbol $\nu$) and antineutrino ($\tilde{\nu}$), both of which appear to be massless, uncharged particles. (They differ only in parity, a quantum-mechanical symmetry upon reflection.) Neutrinos interact so weakly with matter that elaborate detectors are required to record the extremely rare events induced by them in selected nuclides.

## Beta Decay

If an unstable nuclide contains fewer protons than do stable isotopes of the same mass number, it is called "proton deficient." Such a nucleus can decay by transforming one of its neutrons into a proton and emitting a high-energy electron $_{-1}^{0}e^{-}$, also called a **beta particle,** and an antineutrino ($\tilde{\nu}$). The superscript 0 indicates a mass number of 0 because an electron contains neither protons nor neutrons and has a much smaller mass than a nucleus. The subscript $-1$ indicates the negative charge on the particle. It is not a true "atomic number," but writing the symbol in this way is helpful in balancing nuclear equations. The nuclide that results has the same mass number $A$, but its atomic number $Z$ is increased by 1 because a neutron has been transformed into a proton. Examples of **beta decay** of unstable nuclei are

$$^{14}_{6}C \longrightarrow {}^{14}_{7}N + {}^{0}_{-1}e^{-} + \tilde{\nu}_{11} \qquad {}^{24}_{11}Na \longrightarrow {}^{24}_{12}Mg + {}^{0}_{-1}e^{-} + \tilde{\nu}$$

The criterion $\Delta m < 0$ for spontaneous beta decay requires that

$$\underset{\text{(Parent nucleus)}}{m[^{A}Z]} > \underset{\text{(Daughter nucleus)}}{m[^{A}(Z+1)]} + m[_{-1}^{0}e^{-}]$$

because the antineutrino ($\tilde{\nu}$) has zero mass. This inequality refers to the masses of the parent and daughter *nuclei* and the emitted electron. It must be reformulated in terms of neutral *atom* masses because the mass spectrometer yields masses of atoms rather than nuclei. To do this, add the mass of $Z$ electrons to both sides, giving a total of $Z + 1$ electrons on the right side (because the emitted $_{-1}^{0}e^{-}$ is also counted). The inequality becomes

$$\underset{\text{(Parent atom)}}{m[^{A}Z]} > \underset{\text{(Daughter atom)}}{m[^{A}(Z+1)]}$$

A direct comparison of atomic masses (Table 14.1) lets us decide whether beta emission can take place.

The condition that energy be released follows from the Einstein mass–energy relation:

$$\Delta E = c^{2}\{m[^{A}(Z+1)] - m[^{A}Z]\} < 0$$

The liberated energy is carried off as kinetic energy by the beta particle (electron) and the antineutrino, because the nucleus produced is heavy enough that its recoil energy is small and can be neglected. The kinetic energy of the electron can fall anywhere in a continuous range from 0 up to $-\Delta E$ (Fig. 14.3), with the antineutrino carrying off the balance of the energy.

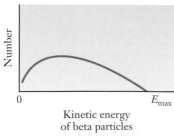

**FIGURE 14.3** Emitted beta particles have a distribution of kinetic energies up to a cutoff value of $E_{\text{max}}$.

## EXAMPLE 14.2

Calculate the maximum kinetic energy of the electron in the decay

$$^{14}_{6}C \longrightarrow {}^{14}_{7}N + {}^{0}_{-1}e^- + \tilde{\nu}$$

### Solution

Table 14.1 gives the masses of the relevant atoms. From them we calculate

$$\Delta m = m[^{14}_{7}N] - m[^{14}_{6}C] = 14.003074 - 14.003242 = -0.000168 \text{ u}$$

$$\Delta E = (-1.68 \times 10^{-4} \text{ u})(931.5 \text{ MeV u}^{-1}) = -0.156 \text{ MeV}$$

The maximum kinetic energy of the electron is 0.156 MeV.

**Related Problem: 8**

## Positron Emission

When a nucleus has too many protons for stability relative to the number of neutrons it contains, it may decay by emitting a positron. In this event, a proton is converted to a neutron, and a high-energy positron (symbolized ${}^{0}_{1}e^+$) and a neutrino ($\nu$) are emitted. As in beta decay, the mass number of the nuclide remains the same, but now the atomic number $Z$ *decreases* by 1. Examples of **positron emission** are

$$^{11}_{6}C \longrightarrow {}^{11}_{5}B + {}^{0}_{1}e^+ + \nu \qquad {}^{19}_{10}Ne \longrightarrow {}^{19}_{9}F + {}^{0}_{1}e^+ + \nu$$

The positron has the same mass as an electron but has a positive charge. The mass criterion $\Delta m < 0$ is different from that for beta decay. For the bare nuclei the criterion is

$$\underset{\text{(Parent nucleus)}}{m[^{A}Z]} > \underset{\text{(Daughter nucleus)}}{m[^{A}(Z-1)]} + m[^{0}_{1}e^+]$$

Adding the mass of $Z$ electrons to both sides gives the mass of the neutral atom $^{A}Z$ on the left, but only $Z - 1$ electrons are needed on the right side to form a neutral atom $^{A}(Z - 1)$. This leaves additional electron and positron masses on the right. These masses are the same, so the criterion for spontaneous positron emission is

$$\underset{\text{(Parent atom)}}{m[^{A}Z]} > \underset{\text{(Daughter atom)}}{m[^{A}(Z-1)]} + 2m[^{0}_{1}e^+]$$

The energy change in positron emission is

$$\Delta E = c^2\{m[^{A}(Z-1)] + 2m[^{0}_{1}e^+] - m[^{A}Z]\} < 0$$

The kinetic energy $(-\Delta E)$ is distributed between the positron and the neutrino.

## EXAMPLE 14.3

Calculate the maximum kinetic energy of the positron emitted in the decay

$$^{11}_{6}C \longrightarrow {}^{11}_{5}B + {}^{0}_{1}e^+ + \nu$$

### Solution

The change in mass is

$$\Delta m = m[^{11}_5B] + 2m[^0_1e^+] - m[^{11}_6C] = 11.009305 + 2(0.0005486) - 11.011433$$

$$= -0.001031 \text{ u}$$

The energy change is then

$$\Delta E = (-0.001031 \text{ u})(931.5 \text{ MeV u}^{-1}) = -0.960 \text{ MeV}$$

The maximum kinetic energy of the positron is $+0.960$ MeV.

**Related Problem: 7**

## Electron Capture

If $m[^AZ]$ exceeds $m[^A(Z-1)]$ (both referring to neutral atoms) but by less than $1.0972 \times 10^{-3}$ u (equivalent to $2m[^0_1e^+]$), positron emission cannot occur. An alternative process that *is* allowed is **electron capture,** in which an electron outside the nucleus of the parent atom is captured by the nucleus and a proton is converted to a neutron. As in positron emission, the mass number is unchanged and the atomic number decreases by 1. In this case, however, the only particle emitted is a neutrino. An example is

$$^{231}_{92}U + {}^0_{-1}e^- \longrightarrow {}^{231}_{91}Pa + \nu$$

The mass criterion for electron capture is simply that

$$m[^AZ] \quad > \quad m[^A(Z-1)]$$
$$\text{(Parent atom)} \quad \text{(Daughter atom)}$$

and so the energy change is

$$\Delta E = c^2\{m[^A(Z-1)] - m[^AZ]\} < 0$$

Electron capture is quite common for heavier neutron-deficient nuclei, for which the mass change with atomic number is too small for positron emission to be possible. In the preceding example,

$$\Delta m = m[^{231}_{91}Pa] - m[^{231}_{92}U] = 231.0359 - 231.0363 = -0.0004 \text{ u}$$

Because $\Delta m < 0$, this process can and does occur by electron capture, but because 0.0004 u is less than $2m[^0_1e^+]$, the analogous process could not occur through positron emission.

## Alpha Decay

The three forms of decay discussed so far all lead to changes in the atomic number $Z$ but not in the mass number $A$. Another process, **alpha decay,** involves the emission of an alpha particle ($^4_2He^{2+}$ ion) and a decrease in mass number $A$ by 4 (the atomic number $Z$ and the neutron number $N$ each decrease by 2). An example is

$$^{238}_{92}U \longrightarrow {}^{234}_{90}Th + {}^4_2He$$

Such alpha decay occurs chiefly for elements in the unstable region beyond bismuth ($Z = 83$) in the periodic table.

The criterion for alpha decay is once again $\Delta m < 0$. In the preceding example,

$$\Delta m = m[^{234}_{90}\text{Th}] + m[^{4}_{2}\text{He}] - m[^{238}_{92}\text{U}]$$
$$\text{(Atom)} \qquad \text{(Atom)} \qquad \text{(Atom)}$$

$$= 234.0436 + 4.0026 - 238.0508 \text{ u} = -0.0046 \text{ u} < 0$$

$$\Delta E = (-0.0046 \text{ u})(931.5 \text{ MeV u}^{-1}) = -4.3 \text{ MeV}$$

Most of the energy is carried away by the lighter helium atom, with a small fraction appearing as recoil energy of the heavy thorium atom.

## Other Modes of Decay

The arrows in Figure 14.1 show the directions of the types of decay discussed so far. Nuclei whose values of $N/Z$ are too great move toward the line of stability by beta emission, while those whose values are too small undergo positron emission or electron capture. Nuclei that are simply too massive can move to lower values of both $N$ and $Z$ by alpha emission. The range of stable nuclides is limited by the interplay of attractive nuclear forces and repulsive Coulomb forces between protons in the nucleus. Scientists believe that it is unlikely that additional stable elements will be found, although some hope that long-lived radioactive nuclides exist at still higher $Z$.

When nuclei are very proton-deficient or very neutron-deficient, an excess particle may "boil off," that is, be ejected directly from the nucleus. These decay modes are called **neutron emission** and **proton emission,** respectively, and move nuclides down or to the left in Figure 14.1. Finally, certain unstable nuclei undergo spontaneous **fission,** in which they split into two nuclei of roughly equal size. Nuclear fission will be discussed in more detail in Section 14.5.

## Detecting and Measuring Radioactivity

Many methods have been developed to detect, identify, and quantitatively measure the products of nuclear reactions. Some are quite simple, but others require complex electronic instrumentation. Perhaps the simplest radiation detector is the **photographic emulsion,** first employed by Henri Becquerel, the discoverer of radioactivity. In 1896 Becquerel reported his observation that potassium uranyl sulfate ($K_2UO_2(SO_4)_2 \cdot 2H_2O$) could expose a photographic plate even in the dark. Such detectors are still used today in the film badges that are worn to monitor exposure to penetrating radiation. The degree of darkening of the film is proportional to the quantity of radiation received.

Rutherford and his students used a screen coated with zinc sulfide to detect the arrival of alpha particles by the pinpoint scintillations of light they produce. That simple device has been developed into the modern **scintillation counter.** Instead of a ZnS screen, the modern scintillation counter employs a crystal of sodium iodide, in which a small fraction of the $Na^+$ ions have been replaced by thallium ($Tl^+$) ions. The crystal emits a pulse of light when it absorbs a beta particle or a gamma ray, and a photomultiplier tube detects and counts the light pulses.

The **Geiger counter** (Figure 14.4) consists of a cylindrical tube, usually of glass, coated internally with metal to provide a negative electrode and with a wire down the center for a positive electrode. The tube is filled to a total pressure of about 0.1 atm with a mixture of 90% argon and 10% ethyl alcohol vapor, and a potential

**FIGURE 14.4** In a Geiger tube, radiation ionizes gas in the tube, freeing electrons that are accelerated to the anode wire in a cascade. Their arrival creates an electrical pulse, which is detected by a ratemeter. The ratemeter displays the accumulated pulses as the number of ionization events per minute.

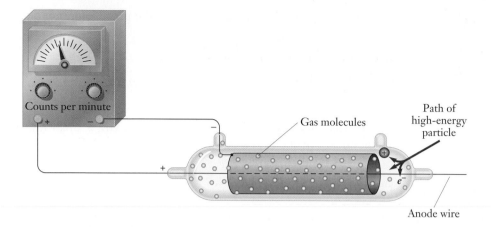

difference of about 1000 V is applied across the electrodes. When a high-energy electron (beta particle) enters the tube, it produces positive ions and electrons. The light electrons are quickly accelerated toward the positively charged wire. As they advance, they encounter and ionize other neutral atoms. An avalanche of electrons builds up, and a large electron current flows into the central wire. This causes a drop in the potential difference, which is recorded, and the multiplicative electron discharge is quenched by the alcohol molecules. In this way single beta particles produce electrical pulses that can be amplified and counted. Portable Geiger counters are widely used in uranium prospecting and to measure radiation in workplaces.

---

## 14.3

# KINETICS OF RADIOACTIVE DECAY

The decay of any given unstable nucleus is a random event and is independent of the number of surrounding nuclei that have decayed. When the number of nuclei is large, we can be confident that during any given period of time a definite fraction of the original number of nuclei will have undergone a transformation into another nuclear species. In other words, the rate of decay of a collection of nuclei is proportional to the number of nuclei present, showing that nuclear decay follows a first-order rate equation of the type discussed in Chapter 13. All the results developed there apply to the present situation; for example, the integrated rate law has the form

$$N = N_i \, e^{-kt}$$

where $N_i$ is the number of nuclei originally present at $t = 0$. The decay constant $k$ is related to a half-life $t_{1/2}$ through

$$t_{1/2} = \frac{\ln 2}{k} = \frac{0.6931}{k}$$

just as in the first-order gas-phase chemical kinetics of Section 13.2. The half-life is the time required for the nuclei in a sample to decay to one-half their initial number, and it can range from less than $10^{-21}$ s to more than $10^{24}$ years for unstable nuclei. Table 14.2 lists the half-lives and decay modes of some unstable nuclides.

## TABLE 14.2

### *Decay Characteristics of Some Radioactive Nuclei*

| Nuclide | $t_{1/2}$ | Decay Mode[†] | Daughter |
|---|---|---|---|
| $^3_1\text{H}$ (tritium) | 12.26 years | $e$ | $^3_2\text{He}$ |
| $^8_4\text{Be}$ | $\sim 1 \times 10^{-16}$ s | $\alpha$ | $^4_2\text{He}$ |
| $^{14}_6\text{C}$ | 5730 years | $e^-$ | $^{14}_7\text{N}$ |
| $^{22}_9\text{Na}$ | 2.601 years | $e^+$ | $^{22}_{10}\text{Ne}$ |
| $^{24}_8\text{Na}$ | 15.02 hours | $e^-$ | $^{24}_{15}\text{Mg}$ |
| $^{32}_{15}\text{P}$ | 14.28 days | $e^-$ | $^{32}_{16}\text{S}$ |
| $^{35}_{16}\text{S}$ | 87.2 days | $e^-$ | $^{35}_{17}\text{Cl}$ |
| $^{36}_{17}\text{Cl}$ | $3.01 \times 10^5$ years | $e^-$ | $^{36}_{18}\text{Ar}$ |
| $^{40}_{19}\text{K}$ | $1.28 \times 10^9$ years | $\begin{cases} e^- \ (89.3\%) \\ \text{E.C. } (10.7\%) \end{cases}$ | $^{40}_{20}\text{Ca}$ $^{40}_{18}\text{Ar}$ |
| $^{59}_{26}\text{Fe}$ | 44.6 days | $e^-$ | $^{59}_{27}\text{Co}$ |
| $^{60}_{27}\text{Co}$ | 5.27 years | $e^-$ | $^{60}_{28}\text{Ni}$ |
| $^{90}_{38}\text{Sr}$ | 29 years | $e^-$ | $^{90}_{39}\text{Y}$ |
| $^{109}_{48}\text{Cd}$ | 453 days | E.C. | $^{109}_{47}\text{Ag}$ |
| $^{125}_{53}\text{I}$ | 59.7 days | E.C. | $^{125}_{52}\text{Te}$ |
| $^{131}_{53}\text{I}$ | 8.041 days | $e^-$ | $^{131}_{54}\text{Xe}$ |
| $^{127}_{54}\text{Xe}$ | 36.41 days | E.C. | $^{127}_{53}\text{I}$ |
| $^{137}_{57}\text{La}$ | $\sim 6 \times 10^4$ years | E.C. | $^{137}_{56}\text{Ba}$ |
| $^{222}_{86}\text{Rn}$ | 3.824 days | $\alpha$ | $^{218}_{84}\text{Po}$ |
| $^{226}_{88}\text{Ra}$ | 1600 years | $\alpha$ | $^{222}_{86}\text{Rn}$ |
| $^{232}_{90}\text{Th}$ | $1.40 \times 10^{10}$ years | $\alpha$ | $^{228}_{88}\text{Ra}$ |
| $^{235}_{92}\text{U}$ | $7.04 \times 10^8$ years | $\alpha$ | $^{231}_{90}\text{Th}$ |
| $^{238}_{92}\text{U}$ | $4.468 \times 10^9$ years | $\alpha$ | $^{234}_{90}\text{Th}$ |
| $^{239}_{93}\text{Np}$ | 2.350 days | $e^-$ | $^{239}_{94}\text{Pu}$ |
| $^{239}_{94}\text{Pu}$ | $2.411 \times 10^4$ years | $\alpha$ | $^{235}_{92}\text{U}$ |

[†] E.C. stands for electron capture; $e^+$, for positron emission; $e^-$, for beta emission; $\alpha$, for alpha emission.

There is one important practical difference between chemical kinetics and nuclear kinetics. In chemical kinetics, the concentration of a reactant or product is monitored over time, and the rate of a reaction is then found from the rate of change of that concentration. In nuclear kinetics, the rate of occurrence of decay events, $-dN/dt$, is measured directly with a Geiger counter or other radiation detector. This decay rate—the average disintegration rate in numbers of nuclei per unit time—is called the **activity**.

$$\text{activity} = A = -\frac{dN}{dt} = kN \qquad \text{[14.2]}$$

Because the activity is proportional to the number of nuclei $N$, it also decays exponentially with time:

$$A = A_\text{i}\, e^{-kt} \qquad \text{[14.3]}$$

A plot of ln $A$ against time $t$ is linear with slope $-k = -(\ln 2)/t_{1/2}$, as Figure 14.5 shows. The activity $A$ is reduced to half its initial value in a time $t_{1/2}$. Once $A$ and

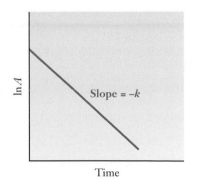

**FIGURE 14.5** A graph of the logarithm of the activity of a radioactive nuclide against time is a straight line with slope $-k$. The decay rate can also be measured as the ratio of the activity to the number of atoms of the radioisotope, if the latter is known.

$k$ are known, the number of nuclei $N$ at that time can be calculated from

$$N = \frac{A}{k} = \frac{At_{1/2}}{\ln 2} = \frac{At_{1/2}}{0.6931}$$

The S.I. unit of activity is the becquerel (Bq), defined as 1 radioactive disintegration per second. An older and much larger unit of activity is the curie (abbreviated Ci), which is defined as $3.7 \times 10^{10}$ disintegrations per second. The activity of 1 g of radium is 1 Ci.

---

## EXAMPLE 14.4

Tritium ($^3$H) decays by beta emission to $^3$He with a half-life of 12.26 years. A sample of a tritiated compound has an initial activity of 0.833 Bq. Calculate the number $N_i$ of tritium nuclei in the sample initially, the decay constant $k$, and the activity after 2.50 years.

### Solution

Convert the half-life to seconds:

$$t_{1/2} = (12.26 \text{ yr})(60 \times 60 \times 24 \times 365 \text{ s yr}^{-1}) = 3.866 \times 10^8 \text{ s}$$

The number of nuclei originally present was

$$N_i = \frac{A_i t_{1/2}}{\ln 2} = \frac{(0.883 \text{ s}^{-1})(3.866 \times 10^8 \text{ s})}{0.6931} = 4.65 \times 10^8 \text{ }^3\text{H nuclei}$$

The decay constant $k$ is calculated directly from the half-life:

$$k = \frac{\ln 2}{t_{1/2}} = \frac{0.6931}{3.866 \times 10^8 \text{ s}} = 1.793 \times 10^{-9} \text{ s}^{-1}$$

To find the activity after 2.50 years, convert this time to seconds ($7.884 \times 10^7$ s) and use

$$A = A_i \, e^{-kt} = (0.833 \text{ Bq}) \exp\,[-(1.793 \times 10^{-9} \text{ s}^{-1})(7.884 \times 10^7 \text{ s})] = 0.723 \text{ Bq}$$

**Related Problems: 21, 22**

---

## Radioactive Dating

The decay of radioactive nuclides with known half-lives enables geochemists to measure the ages of rocks from their isotopic compositions. Suppose that a uranium-bearing mineral was deposited some 2 billion ($2.00 \times 10^9$) years ago and has remained geologically unaltered to the present time. The $^{238}$U in the mineral decays with a half-life of $4.51 \times 10^9$ years to form a series of short-lived intermediates, ending in the stable lead isotope $^{206}$Pb (Fig. 14.6). The fraction of uranium remaining after $2.00 \times 10^9$ years should be

$$\frac{N}{N_i} = e^{-kt} = e^{-0.6931 \, t/t_{1/2}} = \exp\left(\frac{-0.691 \times 2.00 \times 10^9 \text{ yr}}{4.51 \times 10^9 \text{ yr}}\right) = 0.735$$

and the number of $^{206}$Pb atoms should be approximately

$$(1 - 0.735)N_i = 0.265 \, N_i(^{238}\text{U})$$

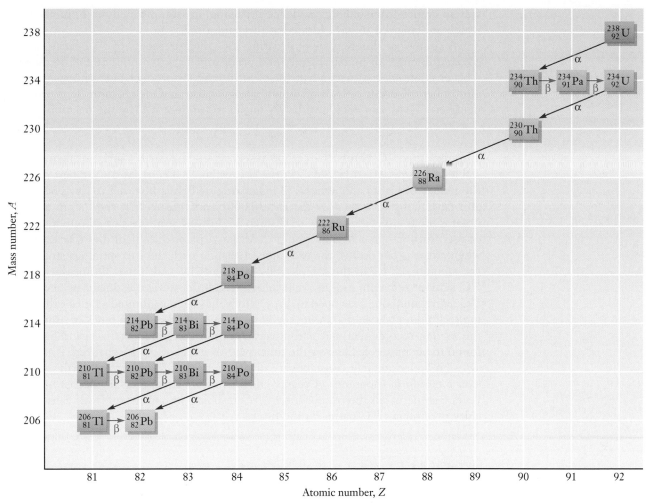

**FIGURE 14.6**   The radioactive nuclide $^{238}$U decays via a series of alpha and beta emissions to the stable nuclide $^{206}$Pb.

The ratio of abundances

$$\frac{N(^{206}\text{Pb})}{N(^{238}\text{U})} = \frac{0.265}{0.735} = 0.361$$

therefore is determined by the time elapsed since the deposit was originally formed—in this case, $2.00 \times 10^9$ years. Of course, to calculate the age of the mineral, we work backward from the measured $N(^{206}\text{Pb})/N(^{238}\text{U})$ ratio.

To use the method, it is necessary to be certain that the stable nuclide generated ($^{206}$Pb in this case) arises only from the parent species ($^{238}$U here) and that neither lead nor uranium has left or entered the rock over the course of geologic time. If possible, it is desirable to measure the ratios of several different isotopically paired species in the same rock sample. For example, $^{87}$Rb decays to $^{87}$Sr with a half-life of $4.9 \times 10^{10}$ years, and one mode of $^{40}$K decay (with a half-life of $1.28 \times 10^9$ years) is to $^{40}$Ar. Each pair should ideally yield the same age. Detailed analysis of a large number of samples suggests that the oldest surface rocks on earth are about 3.8 billion years old. An estimate of 4.5 billion years for the age of the earth and solar system comes from indirect evidence involving an isotopic analysis of meteorites, which are believed to have formed at the same time.

A somewhat different type of dating uses measurements of $^{14}C$ decay, which covers the range of human history and prehistory back to about 30,000 years ago. This unstable species (with a half-life of 5730 years) is produced continuously in the atmosphere. Cosmic rays of very high energy cause nuclear reactions that produce neutrons. These can collide with $^{14}_7N$ nuclei to produce $^{14}_6C$ by the reaction

$$^{14}_7N + ^1_0n \longrightarrow ^{14}_6C + ^1_1H$$

The resulting $^{14}C$ enters the carbon reservoir on the earth's surface, mixing with stable $^{12}C$ as dissolved $H^{14}CO_3^-$ in the oceans, as $^{14}CO_2$ in the atmosphere, and in the tissues of plants and animals. This mixing, which is believed to have occurred at a fairly constant rate over the last 50,000 years, means that the $^{14}C$ in a living organism has a specific activity of close to 15.3 disintegrations per minute per gram of total carbon—that is, 0.255 Bq g$^{-1}$. When a plant or animal dies (for example, when a tree is cut down), the exchange of carbon with the surroundings stops, and the amount of $^{14}C$ in the sample falls exponentially with time. By measuring the $^{14}C$ activity remaining in an archaeological sample, we can estimate its age. This $^{14}C$ dating method, developed by W. F. Libby, has been calibrated against other dating techniques (such as counting the annual rings of bristlecone pines or examining the written records that may accompany a carbon-containing artifact) and has been found to be quite reliable over the time span for which it can be checked. This indicates an approximately constant rate of production of $^{14}C$ near the earth's surface over a period of thousands of years. The burning of fossil fuels in the last hundred years has increased the proportion of $^{12}C$ in living organisms and will cause difficulty in applying $^{14}C$ dating in the future.

## EXAMPLE 14.5

A wooden implement has a specific activity of $^{14}C$ of 0.195 Bq g$^{-1}$. Estimate the age of the implement.

### Solution
The decay constant for $^{14}C$ is

$$k = \frac{0.6931}{5730 \text{ yr}} = 1.21 \times 10^{-4} \text{ yr}^{-1}$$

The initial specific activity was 0.255 Bq g$^{-1}$, and the measured activity now (after $t$ years) is 0.195 Bq g$^{-1}$, and so

$$A = A_i e^{-kt}$$

$$0.195 \text{ Bq g}^{-1} = 0.255 \text{ Bq g}^{-1} e^{-(1.21 \times 10^{-4})t}$$

$$\ln\left(\frac{0.195}{0.255}\right) = -(1.21 \times 10^{-4} \text{ yr}^{-1})t$$

$$t = 2200 \text{ yr}$$

The implement comes from a tree cut down approximately 2200 years ago.

**Related Problems: 25, 26**

## 14.4

# RADIATION IN BIOLOGY AND MEDICINE

Radiation has both harmful and beneficial effects for living organisms. All forms of radiation cause damage in direct proportion to the amount of energy they deposit in cells and tissues. The damage takes the form of chemical changes in cellular molecules, which alter their functions and lead either to uncontrolled multiplication and growth of cells or to their death. Alpha particles lose their kinetic energy over very short distances in matter (typically 10 cm in air or 0.05 cm in water or tissues), producing intense ionization in their wakes until they accept electrons and are neutralized to harmless helium atoms. Radium, for example, is an alpha emitter that substitutes for calcium in bone tissue and destroys its capacity to produce both red and white blood cells. Beta particles, gamma rays, and x-rays have greater penetrating power than alpha particles and so present a radiation hazard even when their source is well outside an organism.

The amount of damage produced in tissue by any of these kinds of radiation is proportional to the number of particles or photons and to their energy. A given activity of tritium causes less damage than the same activity of $^{14}$C, because the beta particles from tritium have a maximum kinetic energy of 0.0179 MeV, whereas those from $^{14}$C have an energy of 0.156 MeV. What is important is the amount of ionization produced or the quantity of energy deposited by radiation. For this purpose, several units are used. The *rad* (*r*adiation *a*bsorbed *d*ose) is defined as the amount of radiation that deposits $10^{-2}$ J of energy per kilogram of tissue. The damage produced in human tissue depends upon still other factors, such as the nature of the tissue, the kind of radiation, the total radiation dose, and the dose rate. To take all these into account, the *rem* (*r*oentgen *e*quivalent in *m*an) has been defined to measure the effective dosages of radiation received by humans. A physical dose of 1 rad of beta or gamma radiation translates into a human dose of 1 rem. Alpha radiation is more toxic; a physical dose of 1 rad of alpha radiation equals about 10 rems.

Exposure to radiation is unavoidable. The average person in the United States receives about 100 millirems (mrem) annually from natural sources that include cosmic radiation and radioactive nuclides such as $^{40}$K and $^{222}$Rn. Another 50 to 100 mrem (variable) comes from human activities (including dental and medical x-rays and airplane flights, which increase exposure to cosmic rays higher in the atmosphere). The safe level of exposure to radiation is a controversial issue among biologists; one group maintains that the effects of radiation are cumulative, another that a threshold dose is necessary for pathological change. The problem is made even more complex by the necessity to distinguish between tissue damage in an exposed individual and genetic damage, which may not become apparent for several generations. It is much easier to define the radiation level that, with high probability, will cause death in an exposed person. The $LD_{50}$ level in human beings (the level carrying a 50% probability that death will result within 30 days after a single exposure) is 500 rad.

## EXAMPLE 14.6

The beta decay of $^{40}$K that is a natural part of the body makes all human beings slightly radioactive. An adult weighing 70.0 kg contains about 170 g of potassium.

The relative natural abundance of $^{40}$K is 0.0118%, its half-life is $1.28 \times 10^9$ years, and its beta particles have an average kinetic energy of 0.55 MeV.

**(a)** Calculate the total activity of $^{40}$K in this person.

**(b)** Determine (in rad per year) the annual radiation absorbed dose arising from this internal $^{40}$K.

### Solution

**(a)** First calculate the decay constant of $^{40}$K in s$^{-1}$:

$$k = \frac{0.693}{t_{1/2}} = \frac{0.693}{(1.28 \times 10^9 \text{ yr})(365 \times 24 \times 60 \times 60 \text{ s yr}^{-1})}$$

$$= 1.72 \times 10^{-17} \text{ s}^{-1}$$

$$\text{number of } ^{40}\text{K atoms} = \frac{170 \text{ g}}{40.0 \text{ g mol}^{-1}} (1.18 \times 10^{-4})(6.02 \times 10^{23} \text{ mol}^{-1})$$

$$= 3.02 \times 10^{20}$$

$$A = -\frac{dN}{dt} = kN = (1.72 \times 10^{-17} \text{ s}^{-1})(3.02 \times 10^{20}) = 5.19 \times 10^3 \text{ s}^{-1}$$

**(b)** Each disintegration of $^{40}$K emits an average of 0.55 MeV of energy, and we assume that all of this energy is deposited within the body. From part **(a)**, $5.19 \times 10^3$ disintegrations occur per second, and we know how many seconds are in a year. The total energy deposited per year is then

$$5.19 \times 10^3 \text{ s}^{-1} \times (60 \times 60 \times 24 \times 365 \text{ s yr}^{-1}) \times 0.55 \text{ MeV} = 9.0 \times 10^{10} \text{ MeV yr}^{-1}$$

Next, because a rad is a centijoule (cJ) per kilogram of tissue, we express this answer in centijoules per year:

$$(9.0 \times 10^{10} \text{ MeV yr}^{-1})(1.602 \times 10^{-13} \text{ J MeV}^{-1})(10^2 \text{ cJ J}^{-1}) = 1.44 \text{ cJ yr}^{-1}$$

Each kilogram of body tissue receives 1/70.0 of this amount of energy per year, because the person weighs 70.0 kg. The dose is thus $21 \times 10^{-3}$ cJ kg$^{-1}$ yr$^{-1}$, which is equivalent to 21 mrad yr$^{-1}$. This is about a fifth of the annual background dosage received by a person.

**Related Problems: 35, 36**

**FIGURE 14.7** An advanced generator for $^{99m}$Tc, an excited state of $^{99}$Tc that decays to the nuclear ground state by emission of gamma rays. This isotope is used in nuclear medicine to study the heart. The $^{99m}$Tc is taken up by heart tissue; a gamma-ray detector then provides an image of the heart. (*DuPont Pharma*)

Although radiation can do great harm, it confers great benefits in medical applications (Fig. 14.7). The diagnostic importance of x-ray imaging hardly needs mention. Both x-rays and gamma rays are used selectively in cancer therapy to destroy malignant cells. The beta-emitting $^{131}$I nuclide finds use in the treatment of cancer of the thyroid because iodine is taken up preferentially by the thyroid gland. Heart pacemakers use the decay of tiny amounts of radioactive $^{238}$Pu, converted to electrical energy.

Positron emission tomography (PET) is an important diagnostic technique using radiation (Fig. 14.8). It employs radioisotopes such as $^{11}$C (half-life 20.3 min) or $^{15}$O (half-life 124 s) that emit positrons when they decay. These are incorporated (quickly, because of their short half-lives) into substances such as glucose, which are injected into the patient. By following the pattern of positron emission from the body, researchers can study blood flow and glucose metabolism in healthy and diseased individuals. Computer-reconstructed pictures of positron emissions from the brain

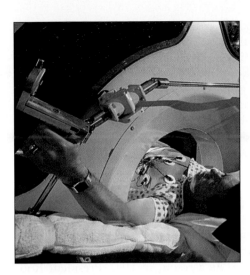

**FIGURE 14.8** A patient in a positron emission tomography (PET) scanner squeezes a handgrip to stimulate flow of blood to the heart. The electrodes on his chest record an electrocardiogram while an array of detectors around the scanner registers nuclear decays. *(GTE)*

are particularly useful, because the locations of glucose metabolism appear to differ between normal persons and patients suffering from ailments such as manic depression and schizophrenia.

Less direct benefits come from other applications. An example is the study of the mechanism of photosynthesis, in which carbon dioxide and water are combined to form glucose in the green leaves of plants.

$$6\ CO_2(g) + 6\ H_2O(\ell) \longrightarrow C_6H_{12}O_6(s) + 6\ O_2(g)$$

Exposure of plants to $CO_2$ containing a higher than normal proportion of $^{14}C$ as a tracer enables scientists to follow the mechanism of the reaction. At intervals, the plants are analyzed to find what compounds contain $^{14}C$ in their molecules and thereby to identify the intermediates in photosynthesis. Radioactive tracers are also widely used in medical diagnosis. The radioimmunoassay technique, invented by Nobel laureate Rosalind Yalow, determines the levels of drugs and hormones in body fluids.

---

## 14.5

## NUCLEAR FISSION

The nuclear reactions considered so far have been spontaneous, first-order decays of unstable nuclides. By the 1920s, physicists and chemists were using particle accelerators to bombard samples with high-energy particles in order to *induce* nuclear reactions. One of the first results of this program was the identification in 1932 by James Chadwick (a student of Rutherford) of the neutron as a product of the reaction of alpha particles with light nuclides such as $^9Be$:

$$^4_2He + {}^9_4Be \longrightarrow {}^1_0n + {}^{12}_6C$$

Shortly after Chadwick's discovery, a group of physicists in Rome, led by Enrico Fermi, began to study the interaction of neutrons with the nuclei of various elements. The experiments produced a number of radioactive species, and it was evident that the absorption of a neutron increased the $N:Z$ ratio in target nuclei above the stability line (Fig. 14.1). One of the targets used was uranium, the heaviest naturally occurring element. Several radioactive products resulted, none of which had chemical properties characteristic of the elements between $Z = 86$ (radon) and $Z = 92$ (uranium). It appeared to the Italian scientists in 1934 that several new transuranic elements ($Z > 92$) had been synthesized, and an active period of investigation followed.

In 1938 in Berlin, Otto Hahn and Fritz Strassmann sought to characterize the supposed transuranic elements. To their bewilderment, they found instead that barium ($Z = 56$) appeared among the products of the bombardment of uranium by neutrons. Hahn informed his former colleague Lise Meitner, and she conjectured that the products of the bombardment of uranium by neutrons were not transuranic elements but fragments of uranium atoms resulting from a process she termed **fission.** The implication of this phenomenon—the possible release of enormous amounts of energy—was immediately evident. When the outbreak of World War II appeared imminent in the summer of 1939, Albert Einstein wrote to President Franklin Roosevelt to inform him of the possible military uses of fission and of his concern that Germany might develop a nuclear explosive. As a result, in 1942 President Roosevelt authorized the Manhattan District Project, an intense, coordinated effort by a large number of physicists, chemists, and engineers to make a fission bomb of unprecedented destructive power.

The operation of the first atomic bomb hinged on the fission of uranium in a chain reaction induced by absorption of neutrons. The two most abundant isotopes of uranium are $^{235}U$ and $^{238}U$, whose natural relative abundances are 0.720% and 99.275%, respectively. Both undergo fission upon absorbing a neutron, the latter only with "fast" neutrons and the former with both "fast" and "slow" neutrons. In the early days of neutron research, it was not known that the absorption of neutrons by nuclei depends strongly upon neutron velocity. By accident, Fermi and his col-

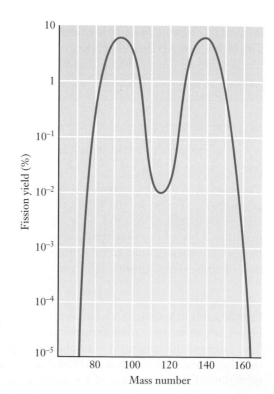

**FIGURE 14.9**   The distribution of nuclides produced in the fission of $^{235}U$ has two peaks. Nuclei having mass numbers in the vicinity of $A = 95$ and $A = 139$ are formed with the highest yield; those with $A \approx 117$ are produced with lower probability.

leagues discovered that experiments conducted on a wooden table led to a much higher yield of radioactive products than those performed on a marble-topped table. Fermi then repeated the irradiation experiments with a block of paraffin wax interposed between the radium–beryllium neutron source and the target sample, with the startling result that the induced level of radioactivity was greatly enhanced. Within hours Fermi had found the explanation: the high-energy neutrons emitted from the radium–beryllium source were reduced to thermal energy by collision with the low-mass nuclei of the paraffin molecules. Because of their lower energies, their probability of reaction with $^{235}U$ was greater and a higher yield was achieved. Hydrogen nuclei and the nuclei of other light elements such as $^{12}C$ (in graphite) are very effective in reducing the energies of high-velocity neutrons and are called **moderators.**

The fission of $^{235}U$ follows many different patterns, and some 34 elements have been identified among the fission products. In any single fission event two particular nuclides are produced together with two or three secondary neutrons; collectively, they carry away about 200 MeV of kinetic energy. Usually the daughter nuclei have different $Z$ and $A$ numbers, and so the fission process is asymmetric. Three of the many pathways are

$$
{}_{0}^{1}n + {}_{92}^{235}U \longrightarrow \begin{cases} {}_{30}^{72}Zn + {}_{62}^{162}Sm + 2\,{}_{0}^{1}n \\ {}_{38}^{80}Sr + {}_{54}^{153}Xe + 3\,{}_{0}^{1}n \\ {}_{36}^{94}Kr + {}_{56}^{139}Ba + 3\,{}_{0}^{1}n \end{cases}
$$

Figure 14.9 shows the distribution of the nuclides produced. The emission of more than one neutron per neutron absorbed in the fission process means that this is a branching chain reaction in which the number of neutrons grows exponentially with time (Fig. 14.10). Permitted to proceed unchecked, this reaction would quickly lead

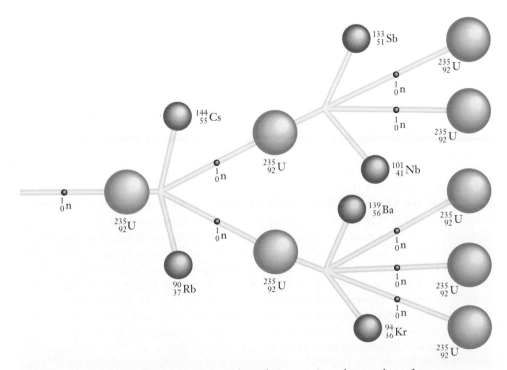

**FIGURE 14.10**  In a self-propagating nuclear chain reaction, the number of neutrons grows exponentially during fission. Uranium atoms are shown in green, and products are shown in blue.

to the release of enormous quantities of energy. Neutrons can be lost by various processes, however, so it is not obvious that a self-propagating reaction will occur. On December 2, 1942, Fermi and his associates demonstrated that a self-sustaining neutron chain reaction occurred in a uranium "pile" with a graphite moderator. They limited its power output to $\frac{1}{2}$ J s$^{-1}$ by inserting cadmium control rods to absorb neutrons, thereby balancing the rates of neutron generation and loss.

Fermi's work made two developments possible: (1) the exploitation of nuclear fission for the controlled generation of energy in nuclear reactors and (2) the production of $^{239}$Pu, a slow- and fast-neutron fissionable isotope of plutonium, as an alternative to $^{235}$U for the construction of atomic bombs. For the sudden release of energy that is required in an explosive, it was necessary to obtain $^{235}$U or $^{239}$Pu in a state free of neutron-absorbing impurities. Both alternatives were pursued simultaneously. The first required enriching $^{235}$U from its relative abundance of 0.72% in natural uranium. This was accomplished through gaseous diffusion (Section 4.6). For the second alternative, $^{239}$Pu was recovered from the partially spent uranium fuel of large nuclear reactors by means of redox reactions, precipitation, and solvent extraction.

Although Enrico Fermi and his associates were the first scientists to demonstrate a self-sustaining nuclear chain reaction, a natural uranium fission reactor "went critical" about 1.8 billion years ago in a place now called Oklo, in the Gabon Republic of equatorial Africa. In 1972 French scientists discovered that the $^{235}$U content of ore from a site in the open-pit mine at Oklo was only 0.7171%; the normal content of ore from other areas of the mine was 0.7207%. Although this deviation was not large, it was significant, and an investigation revealed that other elements were also present in the ore, in the exact proportions expected after nuclear fission. This discovery established that a self-sustaining nuclear reaction had occurred at Oklo. The geological age of the ore body was found to be about $1.8 \times 10^9$ years, and the original $^{235}$U concentration is calculated to have been about 3%. From the size of the active ore mass and the depletion of $^{235}$U, it is estimated that the reactor generated about 15,000 megawatt-years ($5 \times 10^{17}$ J) of energy over a period of about 100,000 years.

## Nuclear Power Reactors

Most **nuclear power reactors** in the United States (Fig. 14.11) use rods of $U_3O_8$ as fuel. The uranium is primarily $^{238}$U, but the amount of $^{235}$U is enriched above natural abundance to a level of about 3%. The moderator used to slow the neutrons (to increase the efficiency of the fission) is ordinary water in most cases, so these reactors are called "light-water" reactors. The controlled release of energy by nuclear fission in power reactors demands a delicate balance between neutron generation and neutron loss. As mentioned earlier, this is accomplished by means of steel control rods containing $^{112}$Cd or $^{10}$B, isotopes that have a very large neutron-capture probability. These rods are automatically inserted into or withdrawn from the fissioning system in response to a change in the neutron flux. As the nuclear reaction proceeds, the moderator (water) is heated and transfers its heat to a steam generator. The steam then goes to turbines that generate electricity (Fig. 14.12).

The power reactors discussed so far rely on the fission of $^{235}$U, an isotope in quite limited supply. An alternative is to convert the much more abundant $^{238}$U to fissionable plutonium ($^{239}$Pu) by neutron bombardment:

$$^{238}_{92}\text{U} + ^{1}_{0}n \longrightarrow ^{239}_{93}\text{Np} + _{-1}^{0}e^{-} \longrightarrow ^{239}_{94}\text{Pu} + 2\,_{-1}^{0}e^{-}$$

**FIGURE 14.11** A nuclear power plant. The large structure on the left is a cooling tower; the containment building is the smaller building on the right with the domed top. *(Jonathan Turk)*

Fissionable $^{233}$U can also be made from thorium:

$$^{232}_{90}\text{Th} + ^{1}_{0}n \longrightarrow ^{233}_{91}\text{Pa} + ^{0}_{-1}e^{-} \longrightarrow ^{233}_{92}\text{U} + 2 \, ^{0}_{-1}e^{-}$$

In a **breeder reactor,** in addition to heat being generated by fission, neutrons are absorbed in a blanket of uranium or thorium. This causes the preceding reactions to occur and generates additional fuel for the reactor to use. An advanced technology

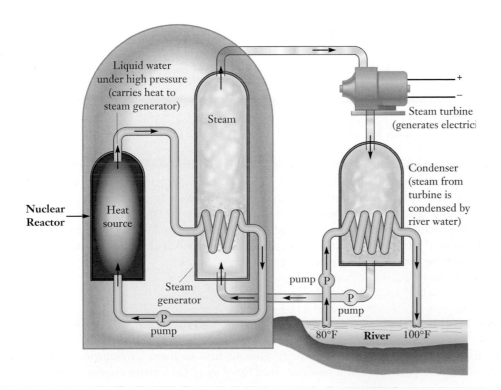

**FIGURE 14.12** A schematic diagram of a pressurized-water nuclear power reactor.

for breeder reactors uses liquid sodium instead of water as the coolant, allowing the use of faster neutrons than in water-cooled reactors. The faster neutrons cause more complete consumption of radioactive fuels, increasing efficiency and greatly reducing radioactive waste.

The risks associated with the operation of nuclear reactors are small but not negligible, as the failure of the Three Mile Island reactor in 1979 and the disaster at Chernobyl in 1987 demonstrated. If a reactor has to be shut down quickly, there is danger of a meltdown, in which the heat from the continuing fission processes melts the uranium fuel. Coolant must be circulated until heat from the decay of short-lived isotopes has been dissipated. The Three Mile Island accident resulted in a partial meltdown because some water coolant pumps were inoperative and others were shut down too soon, causing damage to the core and a slight release of radioactivity into the environment. The Chernobyl disaster was caused by a failure of the water-cooling system and a meltdown. The rapid and uncontrolled nuclear reaction that took place set the graphite moderator on fire and caused the reactor building to rupture, spreading radioactive nuclides with an activity estimated at $2 \times 10^{20}$ Bq into the atmosphere. A major part of the problem was that the reactor at Chernobyl, unlike those in the United States, was not in a massive containment building.

The safe disposal of the radioactive wastes from nuclear reactors is an important and controversial matter. A variety of proposals have been made, including the burial of radioactive waste in deep mines on either a recoverable or a permanent basis, burial at sea, and launching the waste into outer space. The first alternative is the only one that appears credible. The essential requirement is that the disposal site(s) be stable with respect to possible earthquakes or invasion by underground water. Spent nuclear fuel can be encased in blocks of borosilicate glass, packed in metal containers, and buried in stable rock formations. For a nuclide such as $^{239}$Pu, whose half-life is 24,000 years, a storage site that is stable over 240,000 years is needed before the activity drops to 0.1% of its original value. Some shorter lived isotopes are more hazardous over short periods of time, but their threat diminishes more quickly.

---

## 14.6

## Nuclear Fusion and Nucleosynthesis

**Nuclear fusion** is the union of two light nuclides to form a heavier nuclide with the release of energy. Fusion processes are often called **thermonuclear reactions,** because they require that the colliding particles possess very high kinetic energies, corresponding to temperatures of millions of degrees, before they are initiated. They are the processes that occur in the sun and other stars. In 1939 Hans Bethe (and, independently, Carl von Weizsäcker) proposed that in normal stars (main sequence) the following reactions take place:

$$^1_1\text{H} + ^1_1\text{H} \longrightarrow ^2_1\text{H} + ^0_1 e^+ + \nu$$

$$^2_1\text{H} + ^1_1\text{H} \longrightarrow ^3_2\text{He} + \gamma$$

$$^3_2\text{He} + ^3_2\text{He} \longrightarrow ^4_2\text{He} + 2\,^1_1\text{H}$$

In the first reaction, two high-velocity protons fuse to form a deuteron, with the emission of a positron and a neutrino that carry away (as kinetic energy) the additional 0.415 MeV of energy released. In the second reaction, a high-energy deuteron combines with a high-velocity proton to form a helium nucleus of mass 3 and a gamma ray. The third reaction completes the cycle with the formation of a normal helium nucleus ($_2^4$He) and the regeneration of two protons. Each of these reactions is exothermic, but up to 1.25 MeV is required to overcome the repulsive barrier between the positively charged nuclei. The overall result of the cycle is to convert hydrogen nuclei to helium nuclei, and the process is called **hydrogen burning.**

As such a star ages and accumulates helium, it begins to contract under the influence of its immense gravity. As it contracts, its helium core heats up, and when it reaches a temperature of about $10^8$ K, a stage of **helium burning** begins. The first reaction that occurs is

$$2 \; _2^4\text{He} \; \rightleftarrows \; _4^8\text{Be}$$

This reaction is written as an equilibrium because the $^8$Be quickly reverts to helium nuclei with a half-life of only $2 \times 10^{-16}$ s. Even with this short half-life, the $^8$Be nuclei are believed to occasionally react with alpha particles to form stable $^{12}$C:

$$_4^8\text{Be} + _2^4\text{He} \longrightarrow _6^{12}\text{C}$$

The overall effect of the helium-burning phase of a star's life is to convert three helium nuclei to a carbon nucleus, just as helium was formed from four hydrogen nuclei in the hydrogen-burning phase. The density of the core of a star that is burning helium is on the order of $10^5$ g cm$^{-3}$.

This process of **nucleosynthesis** continues beyond the formation of $^{12}$C to produce $^{13}$N, $^{13}$C, $^{14}$N, $^{15}$O, $^{15}$N, and $^{16}$O. The stars in this stage are classified as red giants. Similar cycles occur until the temperature of a star core is about $4 \times 10^9$ K, its density is about $3 \times 10^6$ g cm$^{-3}$, and the nuclei are $^{56}$Fe, $^{59}$Co, and $^{60}$Ni. These are the nuclei that have the maximum binding energy per nucleon (see Fig. 14.2). It is thought that the synthesis of still heavier nuclei occurs in the immense explosions of supernovae.

Heavy elements can also be produced in particle accelerators, which accelerate ions to high speeds, causing collisions that generate the new elements. Technetium, for example, is not found in nature but was first produced in 1937 when high-energy deuterons were directed at a molybdenum source:

$$_{42}^{96}\text{Mo} + _1^2\text{H} \longrightarrow _{43}^{97}\text{Tc} + _0^1 n$$

The first **transuranic element** was produced in 1940. Neptunium ($Z = 93$) results from the capture of a neutron by $^{238}$U, followed by beta decay. Subsequent work by Glenn Seaborg and others led to the production of plutonium ($Z = 94$) and heavier elements. In recent years, nuclides with $Z$ as high as 111 have been made, but in tiny quantities. These nuclides have very short half-lives.

Major efforts are now under way to achieve controlled nuclear fusion as a source of energy. One approach is based on the reaction between deuterium and tritium atoms,

$$_1^2\text{H} + _1^3\text{H} \longrightarrow _2^4\text{He} + _0^1 n$$

with a predicted energy release of 17.6 MeV. The neutrons produced in this reaction can be used to make the needed tritium by means of the reaction

$$_3^6\text{Li} + _0^1 n \longrightarrow _1^3\text{H} + _2^4\text{He}$$

in a lithium "blanket" that surrounds the region of the reaction. Temperatures on the order of $10^8$ K are required to initiate these reactions, by which point atoms are

**FIGURE 14.13** (a) The target chamber of Nova, a 16-foot-diameter aluminum sphere inside of which ten powerful laser beams converge. (b) The target itself is a tiny capsule (1 mm in diameter) of deuterium mixed with tritium. A laser pulse one billionth of a second in duration heats the target hotter than the sun's core, raising its pressure to more than 100 million atm, to initiate nuclear fusion. *(a, Phototake; b, Courtesy of the University of California, Lawrence Livermore National Laboratory, and the U.S. Department of Energy)*

**(a)**

**(b)**

stripped of all their electrons and the system is a **plasma** of nuclei, electrons, and neutrons. There are two major problems: heating the plasma to this high temperature and confining it. In a laser fusion reactor, pellets of deuterium–tritium fuel are raised to the thermonuclear ignition temperature by bursts of light fired from lasers (Fig. 14.13). Contact between the plasma and the wall of the confining vessel would immediately result in prohibitive energy losses and extinguish the fusion reaction. A plasma can be contained in an intense toroidal (doughnut-shaped) magnetic field. Research reactors using this confinement method have progressed and are reaching the point where they produce more energy than they consume. Formidable problems still must be solved, including the development of materials for containment vessels that can withstand the corrosive effects of the intense x-ray and neutron radiation that is present.

The rewards of a workable nuclear fusion process would be great. Fusion produces neither the long-lived radioactive nuclides that accompany nuclear fission (although tritium requires care in handling) nor the environmental pollutants released by the burning of fossil fuels. Although deuterium is present in only 1/6000 of the abundance of ordinary hydrogen, its separation from the latter by the electrolysis of water is readily accomplished, and the oceans contain a virtually unlimited quantity of deuterium.

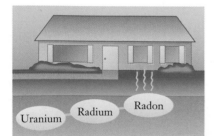

Radon most commonly enters houses through the foundation or basement walls.

## CUMULATIVE EXERCISE

### *Radon*

Radioactive $^{222}Rn$ and $^{220}Rn$ form constantly from the decay of uranium and thorium in rocks and soil and, being gaseous, seep out of the ground. The radon isotopes decay fairly quickly, but their products, which are also radioactive, are then in the air and attach themselves to dust particles. Thus, airborne radioactivity can accumulate to worrisome levels in poorly ventilated basements in ground that is rich in uranium and thorium.

(a) Describe the composition of an atom of $^{222}Rn$ and compare it with that of an atom of $^{220}Rn$.

(b) Although $^{222}Rn$ is a decay product of $^{238}U$, $^{220}Rn$ comes from $^{232}Th$. How many alpha particles are emitted in the formation of these radon isotopes from their uranium or thorium starting points? (*Hint:* Alpha decay changes the mass number $A$, but other decay processes do not.)

(c) Can alpha decay alone explain the formation of these radon isotopes from $^{238}U$ and $^{232}Th$? If not, state what other types of decay must occur.

(d) Can $^{222}_{86}Rn$ and $^{220}_{86}Rn$ decay by alpha particle emission? Write balanced nuclear equations for these two decay processes, and calculate the changes in mass that would result. The masses of $^{222}Rn$ and $^{220}Rn$ atoms are 222.0175 and 220.01140 u, respectively; those of $^{218}Po$ and $^{216}Po$ are 218.0089 and 216.00192 u, respectively.

(e) Calculate the energy change in the alpha decay of one $^{220}Rn$ nucleus, in million electron volts and in joules.

(f) The half-life of $^{222}Rn$ is 3.82 days. Calculate the initial activity of $2.00 \times 10^{-8}$ g of $^{222}Rn$, in disintegrations per second.

(g) What will be the activity of the $^{222}Rn$ from part (f) after 14 days?

(h) The half-life of $^{220}Rn$ is 54 s. Are the health risks of exposure to a given amount of radon for a given short length of time greater or smaller for $^{220}Rn$ than for $^{222}Rn$?

**Answers**

(a) An atom of $^{222}Rn$ has 86 electrons outside the nucleus. Inside the nucleus are 86 protons and $222 - 86 = 136$ neutrons. An atom of $^{220}Rn$ has the same number of electrons and protons, but only 134 neutrons in its nucleus.

(b) Four alpha particles are produced to make $^{222}Rn$ from $^{238}U$; three are produced to make $^{220}Rn$ from $^{232}Th$.

(c) If $^{238}_{92}U$ were to lose four alpha particles, $^{222}_{84}Po$ would result instead of $^{222}_{84}Rn$. Two $_{-1}^{0}e^{-}$ beta particles must be ejected from the nucleus along the way to raise the atomic number to $Z = 86$. The same is true of the production of $^{220}Rn$ from $^{232}Th$.

(d) $^{222}_{86}Rn \rightarrow {}^{218}_{84}Po + {}^{4}_{2}He$; $\Delta m = -0.0060$ u $< 0$; allowed. $^{220}_{86}Rn \rightarrow {}^{216}_{84}Po + {}^{4}_{2}He$; $\Delta m = -0.0069$ u $< 0$; allowed

(e) $\Delta E = -6.4$ MeV $= -1.03 \times 10^{-12}$ J

(f) $A = 1.14 \times 10^{8}$ s$^{-1}$

(g) $A = 9.0 \times 10^{6}$ s$^{-1}$

(h) Greater

## CONCEPTS & SKILLS

*After you have studied this chapter and worked the problems that follow, you should be able to*

1. Calculate the binding energy of a nucleus (Section 14.1, problems 3–6).
2. Write balanced nuclear equations for beta decay, positron emission, electron capture, and alpha decay processes and calculate the maximum kinetic energies of particles emitted (Section 14.2, problems 7–18).
3. Describe several methods that are used to detect the products of radioactive decay (Section 14.2).
4. Solve problems involving the half-life or decay constant of a radioactive sample and its activity (Section 14.3, problems 19–24).
5. Apply the kinetics of nuclear decay to the dating of rocks or artifacts (Section 14.3, problems 25–30).
6. Discuss the interactions of radiation with various kinds of matter and the measurement of radiation dosage (Section 14.4, problems 33–36).
7. Describe the processes of nuclear fission and fusion, and calculate the amounts of energy released when they occur (Sections 14.5 and 14.6, problems 37–47).
8. Explain the benefits and risks associated with the use of nuclear reactions for power generation (Sections 14.5 and 14.6).

## PROBLEMS

*Answers to problems whose numbers are boldface appear in Appendix G. Problems that are more challenging are indicated with asterisks.*

### Mass–Energy Relationships in Nuclei

1. Complete and balance the following equations for nuclear reactions that are thought to take place in stars:
   (a) $2 \, {}^{12}_{6}\text{C} \rightarrow ? + {}^{1}_{0}n$
   (b) $? + {}^{1}_{1}\text{H} \rightarrow {}^{12}_{6}\text{C} + {}^{4}_{2}\text{He}$
   (c) $2 \, {}^{3}_{2}\text{He} \rightarrow ? + 2 \, {}^{1}_{1}\text{H}$

2. Complete and balance the following equations for nuclear reactions that are used in particle accelerators to make elements beyond uranium:
   (a) ${}^{4}_{2}\text{He} + {}^{253}_{99}\text{Es} \rightarrow ? + 2 \, {}^{1}_{0}n$
   (b) ${}^{249}_{98}\text{Cf} + ? \rightarrow {}^{257}_{103}\text{Lr} + 2 \, {}^{1}_{0}n$
   (c) ${}^{238}_{92}\text{U} + {}^{12}_{6}\text{C} \rightarrow {}^{244}_{98}\text{Cf} + ?$

3. Calculate the total binding energy, in both kJ per mole and MeV per atom, and the binding energy per nucleon of the following nuclides, using the data from Table 14.1.
   (a) ${}^{40}_{20}\text{Ca}$   (b) ${}^{87}_{37}\text{Rb}$   (c) ${}^{238}_{92}\text{U}$

4. Calculate the total binding energy, in both kilojoules per mole and MeV per atom, and the binding energy per nucleon of the following nuclides, using the data from Table 14.1.
   (a) ${}^{10}_{4}\text{Be}$   (b) ${}^{35}_{17}\text{Cl}$   (c) ${}^{49}_{22}\text{Ti}$

5. Use the data from Table 14.1 to predict which is more stable: four protons, four neutrons, and four electrons organized as two ${}^{4}\text{He}$ atoms or as one ${}^{8}\text{Be}$ atom. What is the mass difference?

6. Use the data from Table 14.1 to predict which is more stable: 16 protons, 16 neutrons, and 16 electrons organized as two ${}^{16}\text{O}$ atoms or as one ${}^{32}\text{S}$ atom. What is the mass difference?

### Nuclear Decay Processes

7. The nuclide ${}^{8}_{5}\text{B}$ decays by positron emission to ${}^{8}_{4}\text{Be}$. What is the energy released (in MeV)?

8. The nuclide ${}^{10}_{4}\text{Be}$ undergoes spontaneous radioactive decay to ${}^{10}_{5}\text{B}$ with emission of a beta particle. Calculate the maximum kinetic energy of the emitted beta particle.

9. Write balanced equations that represent the following nuclear reactions.
   (a) Beta emission by ${}^{39}_{17}\text{Cl}$   (c) Alpha emission by ${}^{224}_{88}\text{Ra}$
   (b) Positron emission by ${}^{22}_{11}\text{Na}$   (d) Electron capture by ${}^{82}_{38}\text{Sr}$

10. Write balanced equations that represent the following nuclear reactions.
    (a) Alpha emission by ${}^{155}_{70}\text{Yb}$   (c) Electron capture by ${}^{65}_{30}\text{Zn}$
    (b) Positron emission by ${}^{26}_{14}\text{Si}$   (d) Beta emission by ${}^{100}_{41}\text{Nb}$

11. The stable isotopes of neon are ${}^{20}\text{Ne}$, ${}^{21}\text{Ne}$, and ${}^{22}\text{Ne}$. Predict the nuclides formed when ${}^{19}\text{Ne}$ and ${}^{23}\text{Ne}$ decay.

12. The two stable isotopes of carbon are ${}^{12}\text{C}$ and ${}^{13}\text{C}$. Predict the nuclides formed when ${}^{11}\text{C}$ and ${}^{14}\text{C}$ decay. Is alpha emission by ${}^{14}\text{C}$ possible?

13. The free neutron is an unstable particle that decays into a proton. What other particle is formed in neutron decay, and what is the maximum kinetic energy (in MeV) that it can possess?

14. The radionuclide ${}^{210}_{84}\text{Po}$ decays by alpha emission to a daughter nuclide. The atomic mass of ${}^{210}_{84}\text{Po}$ is 209.9829 u, and that of its daughter is 205.9745 u.
    (a) Identify the daughter, and write the nuclear equation for the radioactive decay process.
    (b) Calculate the total energy released per disintegration (in MeV).
    (c) Calculate the kinetic energy of the emitted alpha particle.

15. The natural abundance of ${}^{30}\text{Si}$ is 3.1%. Upon irradiation with neutrons, this isotope is converted to ${}^{31}\text{Si}$, which decays to the stable isotope ${}^{31}\text{P}$. This provides a way of introducing trace amounts of phosphorus into silicon in a much more uniform fashion than is possible by ordinary mixing of silicon and phosphorus and gives semiconductor devices the capability of handling much higher levels of power. Write balanced nuclear equations for the two steps in the preparation of ${}^{31}\text{P}$ from ${}^{30}\text{Si}$.

16. The most convenient way to prepare the element polonium is to expose bismuth (which is 100% ${}^{209}\text{Bi}$) to neutrons. Write balanced nuclear equations for the two steps in the preparation of polonium.

17. One convenient source of neutrons is the reaction of an alpha particle from an emitter such as polonium (${}^{210}\text{Po}$) with an atom of beryllium (${}^{9}\text{Be}$). Write nuclear equations for the reactions that occur.

18. Three atoms of element 111 were produced in 1994 by bombarding ${}^{209}\text{Bi}$ with ${}^{64}\text{Ni}$.
    (a) Write a balanced equation for this nuclear reaction. What other species is produced?
    (b) Write a balanced equation for the alpha decay process of this nuclide of element 111.

### Kinetics of Radioactive Decay

19. How many radioactive disintegrations occur per minute in a 0.0010-g sample of ${}^{209}\text{Po}$ that has been freshly separated from its decay products? The half-life of ${}^{209}\text{Po}$ is 103 years.

20. How many alpha particles are emitted per minute by a 0.0010-g sample of ${}^{238}\text{U}$ that has been freshly separated from its decay products? Assume that each decay emits one alpha particle. The half-life of ${}^{238}\text{U}$ is $4.47 \times 10^{9}$ years.

21. The nuclide ${}^{19}\text{O}$, prepared by neutron irradiation of ${}^{19}\text{F}$, has a half-life of 29 s.
    (a) How many ${}^{19}\text{O}$ atoms are in a freshly prepared sample if its decay rate is $2.5 \times 10^{4} \, \text{s}^{-1}$?
    (b) After 2.00 min, how many ${}^{19}\text{O}$ atoms remain?

22. The nuclide ${}^{35}\text{S}$ decays by beta emission with a half-life of 87.1 days.
    (a) How many grams of ${}^{35}\text{S}$ are in a sample that has a decay rate from that nuclide of $3.70 \times 10^{2} \, \text{s}^{-1}$?
    (b) After 365 days, how many grams of ${}^{35}\text{S}$ remain?

23. Astatine is the rarest naturally occurring element, with ${}^{219}\text{At}$ appearing as the product of a very minor side branch in the decay of ${}^{235}\text{U}$ (itself not a very abundant isotope). It is estimated that the mass of all the naturally occurring ${}^{219}\text{At}$ in the

upper kilometer of the earth's surface has a steady value of only 44 mg. Calculate the total activity (in disintegrations per second) caused by all the naturally occurring astatine in this part of the earth. The half-life of $^{219}$At is 54 s, and its atomic mass is 219.01 u.

24. Technetium has not been found in nature. It can be obtained readily as a product of uranium fission in nuclear power plants, however, and is now produced in quantities of many kilograms per year. One medical use relies on the tendency of $^{99m}$Tc (an excited nuclear state of $^{99}$Tc) to concentrate in abnormal heart tissue. Calculate the total activity (in disintegrations per second) caused by the decay of 1.0 μg of $^{99m}$Tc, which has a half-life of 6.0 hours.

25. The specific activity of $^{14}$C in the biosphere is 0.255 Bq g$^{-1}$. What is the age of a piece of papyrus from an Egyptian tomb if its beta counting rate is 0.153 Bq g$^{-1}$? The half-life of $^{14}$C is 5730 years.

26. The specific activity of an article found in the Lascaux Caves in France is 0.0375 Bq g$^{-1}$. Calculate the age of the article.

27. Over geological time, an atom of $^{238}$U decays to a stable $^{206}$Pb atom in a series of eight alpha emissions, each of which leads to the formation of one helium atom. A geochemist analyzes a rock and finds that it contains $9.0 \times 10^{-5}$ cm$^3$ of helium (at 0°C and atmospheric pressure) per gram and $2.0 \times 10^{-7}$ g of $^{238}$U per gram. Estimate the age of the mineral, given that $t_{1/2}$ of $^{238}$U is $4.47 \times 10^9$ years.

28. The isotope $^{232}$Th decays to $^{208}$Pb by the emission of six alpha particles, with a half-life of $1.39 \times 10^{10}$ years. Analysis of 1.00 kg of ocean sediment shows it to contain 7.4 mg of $^{232}$Th and $4.9 \times 10^{-3}$ cm$^3$ of gaseous helium at 0°C and atmospheric pressure. Estimate the age of the sediment, assuming no loss or gain of thorium or helium from the sediment since its formation and assuming that the helium arose entirely from the decay of thorium.

29. The half-lives of $^{235}$U and $^{238}$U are $7.04 \times 10^8$ years and $4.47 \times 10^9$ years, respectively, and the present abundance ratio is $^{238}$U/$^{235}$U = 137.7. It is thought that their abundance ratio was 1 at some time *before* our earth and solar system were formed about $4.5 \times 10^9$ years ago. Estimate how long ago the supernova occurred that supposedly produced all the uranium isotopes in equal abundance, including the two longest lived isotopes, $^{238}$U and $^{235}$U.

30. Using the result of problem 29 and the accepted age of the earth, $4.5 \times 10^9$ yr, calculate the $^{238}$U/$^{235}$U ratio at the time the earth was formed.

## Radiation in Biology and Medicine

31. Write balanced equations for the decays of $^{11}$C and $^{15}$O, both of which are used in positron emission tomography to scan the uptake of glucose in the body.

32. Write balanced equations for the decays of $^{13}$N and $^{18}$F, two other radioisotopes that are used in positron emission tomography. What is the ultimate fate of the positrons?

33. The positrons emitted by $^{11}$C have a maximum kinetic energy of 0.99 MeV, and those emitted by $^{15}$O have a maximum kinetic energy of 1.72 MeV. Calculate the ratio of the number of millirems of radiation exposure caused by ingesting a given fixed chemical amount (equal numbers of atoms) of each of these radioisotopes.

34. Compare the relative health risks of contact with a given amount of $^{226}$Ra, which has a half-life of 1622 years and emits 4.78-MeV alpha particles, with contact with the same chemical amount of $^{14}$C, which has a half-life of 5730 years and emits beta particles with energies of up to 0.155 MeV.

35. The nuclide $^{131}$I undergoes beta decay with a half-life of 8.041 days. Large quantities of this nuclide were released into the environment in the Chernobyl accident. A victim of radiation poisoning has absorbed $5.0 \times 10^{-6}$ g (5.0 μg) of $^{131}$I.
    (a) Compute the activity, in becquerels, of the $^{131}$I in this person, taking the atomic mass of the nuclide to equal 131 g mol$^{-1}$.
    (b) Compute the radiation absorbed dose, in millirads, caused by this nuclide during the first *second* after its ingestion. Assume that beta particles emitted by $^{131}$I have an average kinetic energy of 0.40 MeV, that all of this energy is deposited within the victim's body, and that the victim weighs 60 kg.
    (c) Is this dose likely to be lethal? Remember that the $^{131}$I diminishes its activity as it decays.

36. The nuclide $^{239}$Pu undergoes alpha decay with a half-life of $2.411 \times 10^4$ years. An atomic energy worker breathes in $5.0 \times 10^{-6}$ g (5.0 μg) of $^{239}$Pu, which lodges permanently in a lung.
    (a) Compute the activity, in becquerels, of the $^{239}$Pu ingested, taking the atomic mass of the nuclide to be 239 g mol$^{-1}$.
    (b) Determine the radiation absorbed dose, in millirads, during the first *year* after its ingestion. Assume that alpha particles emitted by $^{239}$Pu have an average kinetic energy of 5.24 MeV, that all of this energy is deposited within the worker's body, and that the worker weighs 60 kg.
    (c) Is this dose likely to be lethal?

## Nuclear Fission

37. Strontium-90 is one of the most hazardous products of atomic weapons testing because of its long half-life ($t_{1/2} = 28.1$ years) and its tendency to accumulate in the bones.
    (a) Write nuclear equations for the decay of $^{90}$Sr via the successive emission of two beta particles.
    (b) The atomic mass of $^{90}$Sr is 89.9073 u and that of $^{90}$Zr is 89.9043 u. Calculate the energy released per $^{90}$Sr atom, in MeV, in decaying to $^{90}$Zr.
    (c) What will be the initial activity of 1.00 g of $^{90}$Sr released into the environment, in disintegrations per second?
    (d) What activity will the material from part (c) show after 100 years?

38. Plutonium-239 is the fissionable isotope produced in breeder reactors; it is also produced in ordinary nuclear plants and in weapons tests. It is an extremely poisonous substance with a half-life of 24,100 years.
    (a) Write an equation for the decay of $^{239}$Pu via alpha emission.
    (b) The atomic mass of $^{239}$Pu is 239.05216 u and that of $^{235}$U is 235.04393 u. Calculate the energy released per $^{239}$Pu atom, in MeV, in decaying via alpha emission.

(c) What will be the initial activity, in disintegrations per second, of 1.00 g of $^{239}$Pu buried in a disposal site for radioactive wastes?

(d) What activity will the material from part (c) show after 100,000 years?

**39.** The three naturally occurring isotopes of uranium are $^{234}$U (half-life $2.5 \times 10^5$ years), $^{235}$U (half-life $7.0 \times 10^8$ years), and $^{238}$U (half-life $4.5 \times 10^9$ years). As time passes, will the average atomic mass of the uranium in a sample taken from nature increase, decrease, or remain constant?

**40.** Natural lithium consists of 7.42% $^6$Li and 92.58% $^7$Li. Much of the tritium ($^3_1$H) used in experiments with fusion reactions is made by the capture of neutrons by $^6$Li atoms.

(a) Write a balanced nuclear equation for the process. What is the other particle produced?

(b) After $^6$Li is removed from natural lithium, the remainder is sold for other uses. Is the molar mass of the leftover lithium greater or smaller than that of natural lithium?

**41.** Calculate the amount of energy released, in kilojoules per *gram* of uranium, in the fission reaction

$$^{235}_{92}U + ^0_1n \longrightarrow ^{94}_{36}Kr + ^{139}_{56}Ba + 3 \, ^1_0n$$

Use the atomic masses in Table 14.1. The atomic mass of $^{94}$Kr is 93.919 u and that of $^{139}$Ba is 138.909 u.

## Nuclear Fusion and Nucleosynthesis

**42.** Calculate the amount of energy released, in kilojoules per *gram* of deuterium ($^2$H), for the fusion reaction

$$^2_1H + ^2_1H \longrightarrow ^4_2He$$

Use the atomic masses in Table 14.1. Compare your answer with that from the preceding problem.

## Additional Problems

**43.** When an electron and a positron meet, they are replaced by two gamma rays, called the "annihilation radiation." Calculate the energies of these radiations, assuming that the kinetic energies of the particles are 0.

**44.** The nuclide $^{231}_{92}$U converts spontaneously to $^{231}_{91}$Pa.

(a) Write two balanced nuclear equations for this conversion, one if it proceeds by electron capture and the other if it proceeds by positron emission.

(b) Using the nuclidic masses in Table 14.1, calculate the change in mass for each process. Explain why electron capture can occur spontaneously in this case, but positron emission cannot.

**45.** The radioactive nuclide $^{64}_{29}$Cu decays by beta emission to $^{64}_{30}$Zn or by positron emission to $^{64}_{28}$Ni. The maximum kinetic energy of the beta particles is 0.58 MeV and that of the positrons is 0.65 MeV. The mass of the neutral $^{64}_{29}$Cu atom is 63.92976 u.

(a) Calculate the mass, in atomic mass units, of the neutral $^{64}_{30}$Zn atom.

(b) Calculate the mass, in atomic mass units, of the neutral $^{64}_{28}$Ni atom.

**46.** A puzzling observation that led to the discovery of isotopes was the fact that lead obtained from uranium-containing ores had an atomic mass lower by two full atomic mass units than lead obtained from thorium-containing ores. Explain this result, using the fact that decay of radioactive uranium and thorium to stable lead occurs via alpha and beta emission.

**47.** By 1913 the elements radium, actinium, thorium, and uranium had all been discovered, but element 91, between thorium and uranium in the periodic table, was not yet known. The approach used by Meitner and Hahn was to look for the parent that decays to form actinium. Alpha and beta emission are the most important decay pathways among the heavy radioactive elements. What elements would decay to actinium by each of these two pathways? If radium salts show no sign of actinium, what does this suggest about the parent of actinium? What is the origin of the name of element 91, discovered by Meitner and Hahn in 1918?

**48.** Working in Rutherford's laboratory in 1932, Cockcroft and Walton bombarded a lithium target with 700-keV protons and found that the following reaction occurred:

$$^7_3Li + ^1_1H \longrightarrow 2 \, ^4_2He$$

Each of the alpha particles was found to have a kinetic energy of 8.5 MeV. This research provided the first experimental test of Einstein's $\Delta E = c^2 \Delta m$ relationship. Discuss. Using the atomic masses from Table 14.1, calculate the value of $c$ needed to account for this result.

**49.** (a) Calculate the binding energy per nucleon in $^{30}_{15}$P.

(b) The radioactive decay of the $^{30}_{15}$P occurs through positron emission. Calculate the maximum kinetic energy carried off by the positron.

(c) The half-life for this decay is 150 s. Calculate the rate constant $k$ and the fraction remaining after 450 s.

**50.** Selenium-82 undergoes *double* beta decay:

$$^{82}_{34}Se \longrightarrow ^{82}_{36}Kr + 2 \, ^{\;\;0}_{-1}e^- + 2 \, \tilde{\nu}$$

This low-probability process occurs with a half-life of $3.5 \times 10^{27}$ s, one of the longest half-lives ever measured. Estimate the activity in an 82.0-g (1.00 mol) sample of this isotope. How many $^{82}$Se nuclei decay in a day?

**51.** Gallium citrate, which contains the radioactive nuclide $^{67}$Ga, is used in medicine as a tumor-seeking agent. Gallium-67 decays with a half-life of 77.9 hours. How much time is required for it to decay to 5.0% of its initial activity?

**52.** The nuclide $^{241}$Am is used in smoke detectors. As it decays (with a half-life of 458 years), the emitted alpha particles ionize the air. When combustion products enter the detector, the number of ions changes and with it the conductivity of the air, setting off an alarm. If the activity of $^{241}$Am in the detector is $3 \times 10^4$ Bq, calculate the mass of $^{241}$Am present.

**53.** The half-life of $^{14}$C is $t_{1/2} = 5730$ years, and 1.00 g of modern wood charcoal has an activity of 0.255 Bq.

(a) Calculate the number of $^{14}$C atoms per gram of carbon in modern wood charcoal.

(b) Calculate the fraction of carbon atoms in the biosphere that are $^{14}$C.

**54.** Carbon-14 is produced in the upper atmosphere by the reaction

$$^{14}_7N + ^1_0n \longrightarrow ^{14}_6C + ^1_1H$$

where the neutrons come from nuclear processes induced by cosmic rays. It is estimated that the steady-state $^{14}C$ activity in the biosphere is $1.1 \times 10^{19}$ Bq.

(a) Estimate the total mass of carbon in the biosphere, using the data in problem 53.

(b) The earth's crust has an average carbon content of 250 parts per million by mass, and the total crustal mass is $2.9 \times 10^{25}$ g. Estimate the fraction of the carbon in the earth's crust that is part of the biosphere. Speculate on the whereabouts of the rest of the carbon in the earth's crust.

55. Analysis of a rock sample shows that it contains 0.17 mg of $^{40}Ar$ for every 1.00 mg of $^{40}K$. Assuming that all the argon resulted from decay of the potassium and that neither element has left or entered the rock since its formation, estimate the age of the rock. (*Hint:* Use data from Table 14.2.) Note that not all the $^{40}K$ decays to $^{40}Ar$.

*56. Cobalt-60 and iodine-131 are used in treatments for some types of cancer. Cobalt-60 decays with a half-life of 5.27 years, emitting beta particles with a maximum energy of 0.32 MeV. Iodine-131 decays with a half-life of 8.04 days, emitting beta particles with a maximum energy of 0.60 MeV.

(a) Suppose a fixed small number of moles of each of these isotopes were to be ingested and remain in the body indefinitely. What is the *ratio* of the number of millirems of total lifetime radiation exposure that would be caused by the two radioisotopes?

(b) Now suppose that the contact with each of these isotopes is for a fixed short period, such as 1 hour. What is the ratio of millirems of radiation exposure for the two in this case?

57. Boron is used in control rods in nuclear power reactors because it is a good neutron absorber. When the isotope $^{10}B$ captures a neutron, an alpha particle (helium nucleus) is emitted. What other atom is formed? Write a balanced equation.

*58. The average energy released in the fission of a $^{235}U$ nucleus is about 200 MeV. Suppose the conversion of this energy to electrical energy is 40% efficient. What mass of $^{235}U$ is consumed in the fission process in a year's operation of a 1000-megawatt nuclear power station? Recall that 1 W is 1 J s$^{-1}$.

59. The energy released by a bomb is sometimes expressed in tons of TNT. When one ton of TNT (trinitrotoluene) explodes, $4 \times 10^9$ J of energy is released. The fission of one mole of uranium releases approximately $2 \times 10^{13}$ J of energy. Calculate the energy released by the fission of 1.2 kg of uranium in a small atomic bomb. Express your answer in tons of TNT.

60. The solar system abundances of the elements Li, Be, and B are four to seven orders of magnitude lower than those of the elements that immediately follow them: C, N, and O. Explain.

*61. The sun's distance from earth is approximately $1.50 \times 10^8$ km and the earth's radius is 6371 km. The earth receives radiant energy from hydrogen burning in the sun at a rate of 0.135 J s$^{-1}$ cm$^{-2}$. Using the data of Table 14.1, calculate the mass of hydrogen converted per second in the sun.

## CUMULATIVE PROBLEMS

62. Examine the ratio of atomic mass to atomic number for the elements with *even* atomic number through calcium. This ratio is approximately the ratio of the average mass number to the atomic number.

(a) Which two elements stand out as different in this set of ten?

(b) What would be the "expected" atomic mass of argon, based on the correlation considered here?

(c) Show how the anomaly in the ordering of natural atomic masses of argon and potassium can be accounted for by the formation of "extra" $^{40}Ar$ via decay of $^{40}K$ atoms.

63. Hydrazine, $N_2H_4(\ell)$, reacts with oxygen in a rocket engine to form nitrogen and water vapor:

$$N_2H_4(\ell) + O_2(g) \longrightarrow N_2(g) + 2\ H_2O(g)$$

(a) Calculate $\Delta H°$ for this highly exothermic reaction at 25°C, using data from Appendix D.

(b) Calculate $\Delta E°$ of this reaction at 25°C.

(c) Calculate the total change in mass (in grams) during the reaction of 1.00 mol of hydrazine.

64. The long-lived isotope of radium, $^{226}Ra$, decays by alpha particle emission to its daughter radon, $^{222}Rn$, with a half-life of 1622 years. The energy of the alpha particle is 4.79 MeV. Suppose 1.00 g of $^{226}Ra$, freed of all its radioactive progeny, were placed in a calorimeter that contained 10.0 g of water, initially at 25°C. Neglecting the heat capacity of the calorimeter and heat loss to the surroundings, calculate the tempera-

ture the water would reach after 1.00 hour. Take the specific heat of water to be 4.18 J K$^{-1}$ g$^{-1}$.

65. The radioactive nuclide $^{232}_{90}Th$ has a half-life of $1.39 \times 10^{10}$ years. It decays by a series of consecutive steps, the first two of which involve $^{228}_{88}Ra$ (half-life 6.7 years) and $^{228}_{89}Ac$ (half-life 6.13 hours).

(a) Write balanced equations for the first two steps in the decay of $^{232}Th$, indicating all decay products. Calculate the total kinetic energy carried off by the decay products.

(b) After a short initial time, the rate of formation of $^{228}Ra$ becomes equal to its rate of decay. Express the number of $^{228}Ra$ nuclei in terms of the number of $^{232}Th$ nuclei, using the steady-state approximation from Section 13.4.

66. Zirconium is used in the fuel rods of most nuclear power plants. The following half-cell reduction potential applies to aqueous acidic solution:

$$ZrO_2(s) + 4\ H_3O^+(aq) + 4e^- \longrightarrow Zr(s) + 6\ H_2O(\ell)$$
$$\mathscr{E}° = -1.43\ V$$

(a) Predict whether zirconium can reduce water to hydrogen. Write a balanced equation for the overall reaction.

(b) Calculate $\Delta\mathscr{E}°$ and $K$ for the reaction in part (a).

(c) Can your answer to part (b) explain the release of hydrogen in the Three Mile Island accident and the much greater release of hydrogen (which subsequently exploded) at Chernobyl?

# UNIT 5 QUANTAL DESCRIPTION OF ATOMIC AND MOLECULAR STRUCTURE

A "quantum corral" is formed by using a scanning tunneling microscope (STM) tip to arrange individual Fe atoms in a circular pattern on a clean Cu surface in vacuum. Electrons moving at the surface of the Cu are confined inside the corral. STM images show that inside the corral the density of surface electrons takes circular patterns that are explained only by quantum mechanics. *(Courtesy of Dr. Don Eigler, IBM Almaden Research Center, San Jose, CA)*

Rutherford's planetary model of atomic structure described in Section 1.4 ($Z$ electrons moving around a dense nucleus of charge $+Ze$) elegantly summarized experimental facts circa 1912 but was profoundly at odds with the theoretical understanding of the day. Maxwell's electromagnetic theory predicted that a planetary atom would collapse in a fraction of a second because the electrons would be accelerated toward the nucleus, lose energy continuously by radiation, and spiral into the nucleus. The stability of atoms and the existence of the chemical bond were explained only when a new theory called quantum mechanics was developed to replace Newtonian mechanics in describing the microscopic world of molecules, atoms, and fundamental particles. Quantum mechanics is based on two key observations: (1) Microscopic systems do not gain or lose energy in arbitrary amounts, but only in discrete amounts called quanta; and (2) microscopic particles do not follow definite trajectories, but are described by statistical equations that predict the probability of detecting a system at a particular location. These observations are foreign to ordinary human experience, making quantal systems appear strange and nonintuitive.

## CHAPTER 15
Quantum Mechanics and Atomic Structure

## CHAPTER 16
Quantum Mechanics and Molecular Structure

## CHAPTER 17
Bonding, Structure, and Reactions of Organic Molecules

## CHAPTER 18
Bonding in Transition Metals and Coordination Complexes

## CHAPTER 19
Structure and Bonding in Solids

THE GOALS OF UNIT 5 ARE **(1)** To convey the basic concepts and methods of quantum mechanics that describe the discrete energies and statistical behavior of microscopic systems and **(2)** To develop intuition for this behavior of systems and for the magnitudes of quantities described in quantum mechanics. Quantum mechanics will then be used **(3)** To describe the allowed energies and the probabilistic structures of atoms; **(4)** To explain the structure of the periodic table and periodic trends in the properties of atoms; **(5)** To describe covalent bond formation and structure of diatomic and small polyatomic molecules; **(6)** To describe bonding in more complex structures that include transition metal ions; and **(7)** To survey bonding types and their correlation with the properties of solids.

# Quantum Mechanics and Atomic Structure

**Illustration**
Calcium atoms heated in a flame emit light with a characteristic wavelength that is determined by the energy difference between two of their quantum states. *(Larry Cameron)*

S cience can advance in different ways. Usually the slow and steady accumulation of experimental results support and refine existing models, which leads to a more satisfactory description of natural phenomena. But occasionally new experiments give results that directly contradict previously accepted theories. In this case, a period of uncertainty ensues; it is resolved only through the eventual emergence of a new and more complete theory that explains both the previously understood results and the new experiments. This process is called a *scientific revolution*. In the first 25 years of the 20th century, a revolution in physics led to the development of quantum theory, which also profoundly affected the science of chemistry.

One of the fundamental assumptions of early science was that nature is continuous; that is, nature does not make "jumps." On a macroscopic scale, this seems true enough. We can measure out an amount of graphite (carbon) of mass 9 kg, or 8.23 kg, or 6.4257 kg, and it appears that the mass can have any value provided our balance is sufficiently accurate. On an atomic scale, however, this apparently continuous behavior breaks down. An analogy may be useful here. A sand beach from a distance appears smooth and continuous, but a close look reveals that it is made up of individual grains of sand. This same "graininess" is found in matter observed on the atomic scale. The mass of carbon ($^{12}C$) comes in "packets," each of which weighs $1.99265 \times 10^{-26}$ kg. Although in principle two, three, or any integral number of such packets can be "weighed out," we cannot obtain $1\frac{1}{2}$ packets. Carbon is not a continuous material but comes in chunks, each containing the minimum measurable mass of carbon: that of an atom. Similarly, electric charge comes in packets of size $e$, as shown in Section 1.4, and we cannot obtain nonintegral numbers of packets of charge in experiments conducted at chemical energies.

The central idea of quantum theory is that energy also is not continuous but comes in discrete packets. While discreteness of matter and charge on the microscopic scale seem entirely reasonable and familiar, discreteness of energy is an unsettling and unfamiliar notion. The soccer ball represented in Figure 1.15 can roll with arbitrary amounts of kinetic and potential energy; nothing in ordinary human experience suggests that the energy of a system should change by "jumps."

To understand the far-reaching nature of the quantum revolution, consider the state of physics at the end of the 19th century. The 200 years that followed the seminal work of Isaac Newton were the classical period in the study of mechanics, the branch of physics that predicts the motions of particles and the collections of particles that make up working mechanisms. By the end of that period, about 1900, physicists had achieved a deep understanding that successfully dealt with problems ranging from the motions of the planets in their orbits to the design of a bicycle. These attainments now make up the subject called *classical mechanics*, and they are eminently useful. In classical mechanics, a fundamental role is played by energy, which has two parts: kinetic energy, which arises from the motion of particles; and potential energy, which arises from the interactions of particles with one another or with an external field (such as gravity) and depends on the *positions* of the particles. Using classical mechanics, one can predict the future positions of a group of particles from their present positions and momenta if the forces among them are known. A major triumph of classical mechanics was the kinetic theory of gases outlined in Section 4.5.

At the end of the 19th century, it was naturally thought that the motion of elementary particles—such as the recently discovered electron—could be described by classical mechanics and that, once the correct laws of force were discovered, the properties of atoms and molecules could be predicted to any desired degree of accuracy by solving Newton's equations of motion. It was believed that all the fundamental laws of physics had been discovered. At the dedication of the Ryerson Physics Laboratory at the University of Chicago in 1894, A. A. Michelson said that "our future discoveries must be looked for in the sixth decimal place." Little did he imagine the far-reaching changes that would shake physics and chemistry during the next 30 years.

Central to those changes was not only the recognition that energy is quantized but also the discovery that all particles have wave-like properties. The effects of wave-like properties are most pronounced for small, low-mass particles such as electrons in atoms. The incorporation of wave and particle aspects of matter into a single comprehensive theory is the achievement of *quantum mechanics*, the great 20th-century extension of classical mechanics.

In this chapter we discuss the origins of the quantum theory, the implications of the theory for the structure of atoms, and its connection to the periodic table. In subsequent chapters we will consider the consequences of the quantum nature of matter for chemical bonds, coordination complexes, and the solid state.

---

## 15.1

# PRELIMINARIES: WAVE MOTION AND LIGHT

Many kinds of waves are studied in physics and chemistry. A familiar example is water waves stirred up by the winds over the oceans, set off by a stone dropped into a quiet pool, or created by a laboratory water-wave machine. Sound waves, periodic compressions of the air that move from a source to a detector such as the human ear, are another example. Light waves, as we will see, consist of oscillating electric and magnetic fields moving through space. Even some chemical reactions occur in such a way that waves of color pass through the sample as the reaction proceeds. Common to all these wave phenomena is the oscillatory variation of some property with time at a given fixed location in space (Table 15.1).

A snapshot of a water wave (Fig. 15.1) records the peaks and troughs present at some instant in time. The **amplitude** of the wave is the maximum displacement of the water surface over the undisturbed level of the water. The distance between two successive peaks (or troughs) is called the **wavelength** $\lambda$ (Greek lambda) of the wave, provided that this distance is reproducible from peak to peak. The **frequency** of a water wave can be measured by counting the number of peaks or troughs observed at a fixed point in space per second. The frequency, $\nu$ (Greek nu), has units of waves

**TABLE 15.1**

### *Kinds of Waves*

| Wave | Oscillating Quantity |
|------|---------------------|
| Water | Height of water surface |
| Sound | Density of air |
| Light | Electric and magnetic fields |
| Chemical | Concentrations of chemical species |

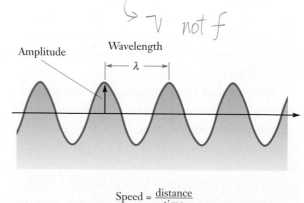

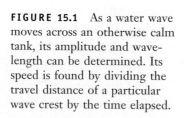

**FIGURE 15.1** As a water wave moves across an otherwise calm tank, its amplitude and wavelength can be determined. Its speed is found by dividing the travel distance of a particular wave crest by the time elapsed.

(or cycles) per second or simply $s^{-1}$. For example, if 12 water-wave peaks are observed at a certain point in 30 s, the frequency is

$$\text{frequency} = \nu = \frac{12}{30 \text{ s}} = 0.40 \text{ s}^{-1}$$

The wavelength and frequency of a wave are related through its speed—the rate at which a particular wave crest moves through the medium. As Figure 15.1 shows, in a time interval $\Delta t = \nu^{-1}$ the wave moves through one wavelength, and so the speed (the distance traveled divided by the time elapsed) is

$$\text{speed} = \frac{\text{distance traveled}}{\text{time elapsed}} = \frac{\lambda}{\nu^{-1}} = \lambda \nu$$

The speed of a wave is the product of its wavelength and its frequency.

## Electromagnetic Radiation

J. C. Maxwell had demonstrated by 1865 that light is **electromagnetic radiation.** A beam of light consists of oscillating electric and magnetic fields perpendicular to the direction in which the light is propagating (Fig. 15.2). The speed $c$ of light passing through a vacuum is equal to the product $\lambda \nu$:

$$c = \lambda \nu = 2.9979 \times 10^8 \text{ m s}^{-1} \qquad \textbf{[15.1]}$$

The speed $c$ is a universal constant,[1] the same for all types of light throughout the electromagnetic spectrum (Fig. 15.3). Regions of the spectrum are characterized by differing values of wavelength and frequency. The region visible to the eye, which is a very small fraction of the spectrum, comprises bands of colored light with particular ranges of wavelength and frequency. Green light has a range of frequencies near $5.7 \times 10^{14} \text{ s}^{-1}$ and wavelengths near $5.3 \times 10^{-7}$ m (530 nm). Red light has a lower frequency and a longer wavelength than green, and violet light has a higher frequency and shorter wavelength than green. A laser such as that shown in Figure 15.4 emits nearly monochromatic light (light with a single frequency and wavelength). White light contains the full range of visible wavelengths. White light can be resolved into its component wavelengths by passage through a prism.

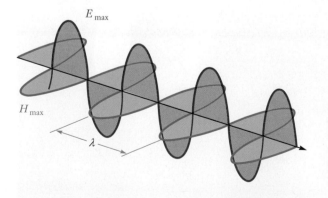

**FIGURE 15.2**   Light consists of waves of oscillating electric ($E$) and magnetic ($H$) fields that are perpendicular to each other and to the direction of propagation of the light.

[1] This constancy of $c$ has led to a new definition of the meter. One meter is now defined as the distance traveled by light in a vacuum during 1/299,792,458 second, and the speed of light is fixed at 299,792,458 m s$^{-1}$. The meter is no longer an independent unit but is defined in terms of the second.

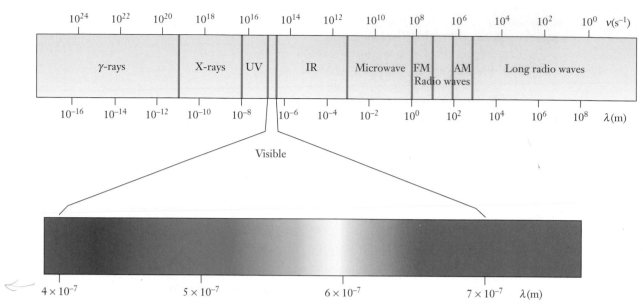

**FIGURE 15.3** The electromagnetic spectrum. Note the small fraction that is visible to the human eye.

Light outside the visible region is nonetheless familiar (Fig. 15.3). The warmth radiated from a stone pulled from a fire consists largely of infrared (IR) radiation, which has a wavelength longer than that of visible light. Microwave ovens employ radiation of greater than infrared wavelength, and still longer waves are employed in radio communication. The position of a radio station on the dial is given by its frequency in hertz (Hz), where 1 Hz is one cycle per second. Thus, FM stations typically broadcast at frequencies of tens to hundreds of megahertz ($1 \text{ MHz} = 10^6 \text{ s}^{-1}$); AM stations broadcast at lower frequencies, from hundreds to thousands of kilohertz ($1 \text{ kHz} = 10^3 \text{ s}^{-1}$). Light with wavelengths shorter than visible light includes ultraviolet (UV) light, x-rays, and gamma rays. These kinds of radiation, with their high frequencies and short wavelengths, penetrate more deeply than visible or infrared radiation.

**FIGURE 15.4** A laser emits a sharply focused beam of light with a very narrow range of wavelengths. The direction of motion of a laser beam can be manipulated by inserting mirrors in the beam's path. *(Copyright Richard Megna/Fundamental Photographs)*

**EXAMPLE 15.1**

Almost all commercially available microwave ovens employ radiation with a frequency of $2.45 \times 10^9$ s$^{-1}$. Calculate the wavelength of this radiation.

**Solution**

The wavelength is related to the frequency through

$$\lambda = \frac{c}{\nu} = \frac{3.00 \times 10^8 \text{ m s}^{-1}}{2.45 \times 10^9 \text{ s}^{-1}} = 0.122 \text{ m}$$

and so the wavelength is 12.2 cm.

**Related Problems: 3, 4**

---

## 15.2

### EXPERIMENTAL BASIS OF ENERGY QUANTIZATION: BLACKBODY RADIATION AND THE PHOTOELECTRIC EFFECT

The first rumblings of the quantum revolution occurred around the end of the 19th century when experimental results began to appear that could not be explained by classical physics. Most of these involved either the absorption of light by matter or the emission of light by matter. The quantum revolution was launched when Planck and Einstein introduced the radical concept of energy quantization to explain two of these results.

### Blackbody Radiation

Incandescence is the process by which radiation is emitted from the surface of a solid body. As a body—for example, the heating element on an electric kitchen range or the filament in an incandescent light bulb—is heated, it emits radiation and becomes first red, then orange, then yellow, then white as its temperature increases. The distribution of frequencies of radiated light changes with the temperature of the body. It is exactly this effect that gives stars of different temperatures their different colors.

At the end of the 19th century there was great interest in explaining the temperature dependence of the frequency distribution. Experiments had shown that the distribution depended only on temperature and not on the composition of the incandescent body. Methods of thermodynamics (see Chapters 7 and 8) could be applied to this problem but only if the emitted radiation was in thermodynamic equilibrium with the heated surface. This condition is met by a system made from perfectly absorbing material with a hollowed-out internal cavity whose walls are maintained at temperature $T$. Radiation in the cavity is constantly emitted and absorbed by the heated walls. The radiation inside this idealized model of incandescent systems is called **blackbody radiation.** A small hole allows a small amount of radiation to escape the system for analysis without disturbing the equilibrium in the cavity. Figure 15.5 shows typical experimental curves for the dependence of light intensity on wavelength at two different temperatures.

**FIGURE 15.5** The dependence of the intensity of blackbody radiation on wavelength for two temperatures: 5000 K (red) and 7000 K (blue). The sun has a blackbody temperature near 5780 K, and its light-intensity curve lies between the two shown. The classical theory (dashed curves) disagrees with observation at shorter wavelengths.

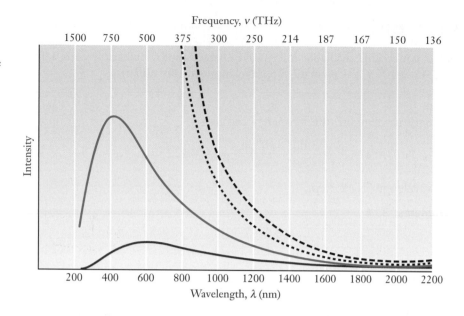

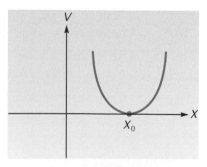

A collection of charged particles oscillating due to forces from "springs" connecting them together is a model for surface motions on a blackbody.

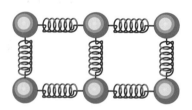

The potential energy curve for an oscillator is a consequence of the "restoring force" that always drives the oscillator toward its equilibrium position.

To explain this result, it was assumed that radiation of a given frequency was produced by a charged particle at the surface of the cavity undergoing simple harmonic motion at that same frequency around its equilibrium position (see Appendix B). In this motion the charged particles were modeled as balls of mass $m$ held in place by springs. As the particles move under influence of thermal energy in the cavity wall, the spring exerts a "restoring force" $f$ which returns the particle to its equilibrium position $x_0$. The restoring force is given by $f = -K(x-x_0)$, where the constant $K$ measures the "stiffness" of the spring. The particle oscillates about $x_0$ in a periodic motion of frequency $\nu = \dfrac{1}{2\pi}\left(\dfrac{K}{m}\right)^{1/2}$ and the associated potential energy is $V(x) = \dfrac{1}{2}K\left(x-x_0\right)^2$. A very broad range of oscillator frequencies must be included in the model since a very broad range of frequencies is detected in the radiation. Statistical thermodynamics (see Section 8.2) and classical mechanics were used to calculate the average energy in the oscillators at each frequency as a function of temperature. The resulting frequency distribution was

$$I = \left(\frac{8\pi R}{N_0}\right)\left(\frac{T}{\lambda^4}\right)$$

This calculated result, shown at 5000K and 7000K in Figure 15.5, agrees well with experiment at longer wavelength. It disagrees badly at short wavelength, where it overestimates the contributions from the high-frequency oscillators. This was referred to as the "ultraviolet catastrophe" because it predicts an infinite intensity at very short wavelengths.

The blackbody radiation paradox was resolved by Max Planck in 1900. Planck reasoned that the only way to eliminate the contributions from the very high frequency oscillators to the blackbody radiation was to guarantee that the thermal energy of the walls did not set them into motion, despite the fact that classical mechanics allows an oscillator to have any value of energy. Planck made the daring hypothesis that it was *not* possible to put an arbitrarily small amount of energy into an oscillator of frequency $\nu$. Instead, he postulated that the oscillator must gain and lose energy in "packets," or *quanta*, of magnitude $h\nu$, and that the total energy $\epsilon$ of an oscillator could take only discrete values that are integral multiples of $h\nu$:

$$\epsilon_{\text{osc}} = nh\nu \qquad n = 1, 2, 3, 4, \ldots \qquad \textbf{[15.2]}$$

In Planck's postulate, $h$ was a constant with physical units *energy* $\times$ *frequency*$^{-1}$ = *energy* $\times$ *time* but whose value was yet to be determined.

The dramatic contrast between Planck's postulate and the classical picture is illustrated by **energy level diagrams** in which horizontal lines represent the possible allowed energy values of a system. The distance of each line above the zero of energy represents the value of the total energy in that level.

Planck used statistical thermodynamics to calculate the average energy in these quantized oscillators at each frequency as a function of temperature. The resulting frequency distribution agreed with experimental results quite accurately and determined the value of the constant $h$, which has since been measured to very high precision. It is referred to as **Planck's constant** and has the value

$$h = 6.62608 \times 10^{-34} \text{ J s}$$

Planck initially applied his hypothesis to the charged particle oscillator model of blackbody radiation described previously. Because the measured frequency distribution depended only on the temperature and not on the properties of the wall, a more sophisticated classical description based on the oscillating electromagnetic waves within the cavity was developed by Lord Rayleigh and Sir James Jeans. When Planck's hypothesis was extended to quantization of the electromagnetic waves within the cavity, the calculated frequency distribution for the radiation was the same as Planck's original result.

The detailed calculation of the frequency distribution as a consequence of Planck's quantum hypothesis is beyond the scope of this book. Nonetheless, the reason Planck's quantum hypothesis led to the correct frequency distribution at both high and low frequencies can be explained qualitatively as follows. According to statistical thermodynamics, the probability that an oscillator of frequency $\nu$ contains its minimum allowed energy $h\nu$ and therefore contributes to the blackbody radiation is proportional to exp $[-h\nu/k_{\text{B}}T]$, where $k_{\text{B}}$ is Boltzmann's constant. (Compare this expression with the Maxwell–Boltzmann speed distribution for gas molecules (Equation 4.12) where the ratio of kinetic energy to $k_{\text{B}}T$ determines the statistical weight for each speed.) When $h\nu/k_{\text{B}}T$ is large, this probability is small. Oscillators with frequencies sufficiently high that $h\nu/k_{\text{B}}T \gg 1$ have negligible probability for having energy other than 0 and for contributing to the blackbody radiation at temperature $T$. Oscillators with frequencies sufficiently low that $h\nu/k_{\text{B}}T \ll 1$ have energy levels so closely spaced that they are in effect continuous and behave like classical systems. The magnitude of the energy quantum required to excite a quantized oscillator is determined by the spacing between its energy levels, $h\nu$; whether the thermal energy of the walls can provide these quanta is determined by the value of $k_{\text{B}}T$. This ratio $h\nu/k_{\text{B}}T$ sets a general pattern that recurs each time we consider whether a quantum system will absorb thermal energy from its surroundings.

## The Photoelectric Effect

A second conflict between experimental results and classical theory arose from the observation of the **photoelectric effect.** A beam of light shining onto a metal surface (called the *photocathode*) can eject electrons (called *photoelectrons*) and cause an

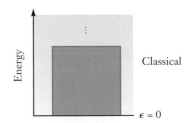

An oscillator obeying classical mechanics has continuous values of energy and can gain or lose energy in arbitrary amounts.

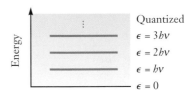

An oscillator described by Planck's postulate has discrete energy levels. It can gain or lose energy only in amounts that correspond to the difference between two energy levels.

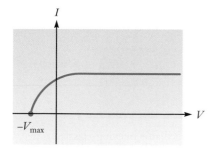

The current in a photocell depends on the potential between cathode and collector.

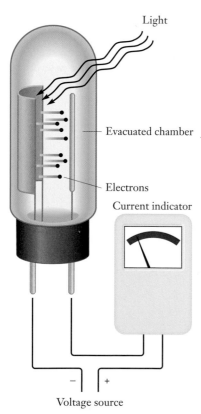

**FIGURE 15.6** In a photoelectric cell (photocell), light strikes a metal surface in an evacuated space and ejects electrons. The electrons are attracted to a positively charged collector, and a current flows through the cell.

electric current (called a *photocurrent*) to flow (Fig. 15.6). The photoelectrons are emitted with nonzero kinetic energy, demonstrated as follows. If the frequency and intensity of the beam are held constant, the magnitude of the photocurrent depends on the electrical potential of the collector relative to the photocathode. At sufficiently positive potential all the photoelectrons are attracted toward the collector. As the potential of the collector becomes more negative, photoelectrons with small kinetic energy values are repelled, and the photocurrent decreases. Only those photoelectrons with sufficient kinetic energy still arrive at the collector. As the collector is made still more negative, the photocurrent drops sharply to zero at $-V_{max}$, demonstrating a sharp maximum in the kinetic energy of the photoelectrons: $E_{max} = eV_{max}$. As an aside on physical units, recall that an electron accelerated through a potential difference of 1 V experiences a change in kinetic energy given by $\Delta E = eV = (1.602177 \times 10^{-19}\,C)(1\,V) = 1.602177 \times 10^{-19}\,J$. It is therefore convenient to define a unit of energy called the **electron volt (eV)** such that 1 eV = $1.602177 \times 10^{-19}$ J. Here the electrons are decelerated by the bias potential and lose their initial kinetic energy. The potential required to stop all electrons from arriving at the collector is thus a convenient measure of their maximum initial kinetic energy, expressed in units of eV. Assume that those photoelectrons with $E_{max}$ are emitted from atoms at the surface of the metal, while those with lower kinetic energy are emitted deeper in the metal and lose some kinetic energy through collisions with other metal atoms before escaping from the surface. Then, the value of $E_{max}$ should be directly related to the energy acquired by the photoelectron during the ejection process. Many measurements showed that while the total photocurrent collected at positive bias depended on light intensity, $E_{max}$ did not. $E_{max}$ was shown to depend on frequency, and for light frequencies less than a certain threshold frequency $\nu_0$, no electrons were ejected.

The photoelectric effect could not be explained by classical physics. According to classical theory, the energy associated with electromagnetic radiation depends only on the intensity and not on the frequency. Why, then, could a very weak beam of blue light eject electrons from sodium when an intense red beam had no effect (Fig. 15.7a)?

In 1905 Einstein used Planck's quantum hypothesis to explain the photoelectric effect. First, he suggested that a light wave of frequency $\nu$ consists of quanta of energy (later called **photons** by G. N. Lewis), each of which carries energy $\epsilon_{photon} = h\nu$. Second, he assumed that in the photoelectric effect an electron in the metal absorbs a photon of light and gains thereby the energy required to escape from the metal. A photoelecton emitted beneath the surface will lose energy of amount $L$ in collisions with other atoms and of amount $\Phi$ in escaping through the surface, after which it travels through the vacuum with kinetic energy $E$. Assuming conservation of energy leads to the relation $h\nu = L + \Phi + E$ for the process. Electrons with maximum kinetic energy are emitted at the surface, so that $L = 0$. The maximum kinetic energy of photoelectrons emitted by light of frequency $\nu$ is therefore given by

$$E_{max} = \frac{1}{2} m_e v^2 = h\nu - \Phi \qquad [15.3]$$

where $\Phi = h\nu_0$ is a constant characteristic of the metal.

Einstein's theory predicts that the maximum kinetic energy is a linear function of the frequency, which provides a means for testing the validity of the theory. Experiments conducted at several frequencies demonstrated that the relation between $E_{max}$ and frequency is indeed linear (Fig. 15.7b). The slope of the experimental data

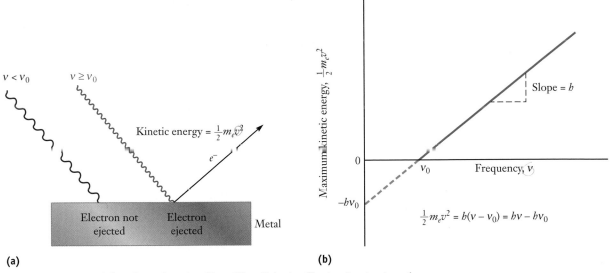

**(a)**                                                                    **(b)**

**FIGURE 15.7** (a) The photoelectric effect. Blue light is effective in ejecting electrons from the surface of this metal, but red light is not. (b) The maximum kinetic energy of the ejected electrons varies linearly with the frequency of light used.

determined the numerical value of $h$ to be identical to the value found by Planck in comparing his theory of blackbody radiation to experimental data. Einstein's interpretation also provided a means to obtain the value of the quantity $\Phi$ from the experimental data as the "energy intercept" of the linear graph. $\Phi$, called the **work function** of the metal, represents the binding energy or energy barrier that electrons must overcome to escape from the metal surface. $\Phi$ governs the extraction of electrons from metal surfaces by heat and by electric fields as well as by the photoelectric effect and is an essential parameter in the design of numerous electronic devices.

## EXAMPLE 15.2

Light with a wavelength of 400 nm strikes the surface of cesium in a photocell, and the maximum kinetic energy of the electrons ejected is $1.54 \times 10^{-19}$ J. Calculate the work function of cesium and the longest wavelength of light that is capable of ejecting electrons from that metal.

### Solution

The frequency of the light is

$$\nu = \frac{c}{\lambda} = \frac{3.00 \times 10^8 \text{ m s}^{-1}}{4.00 \times 10^{-7} \text{ m}} = 7.50 \times 10^{14} \text{ s}^{-1}$$

The binding energy $h\nu_0$ can be calculated from Einstein's formula:

$$(\tfrac{1}{2}m_e v^2)_{\text{max}} = h\nu - h\nu_0$$

$$1.54 \times 10^{-19} \text{ J} = (6.626 \times 10^{-34} \text{ J s})(7.50 \times 10^{14} \text{ s}^{-1}) - h\nu_0$$
$$= 4.97 \times 10^{-19} \text{ J} - h\nu_0$$

$$\Phi = h\nu_0 - (4.97 - 1.54) \times 10^{-19} \text{ J} = 3.43 \times 10^{-19} \text{ J}$$

The minimum frequency $\nu_0$ for the light to eject electrons is then

$$\nu_0 = \frac{3.43 \times 10^{-19} \text{ J}}{6.626 \times 10^{-34} \text{ J s}} = 5.18 \times 10^{14} \text{ s}^{-1}$$

From this, the maximum wavelength $\lambda_0$ is

$$\lambda_0 = \frac{c}{\nu_0} = \frac{3.00 \times 10^8 \text{ m s}^{-1}}{5.18 \times 10^{14} \text{ s}^{-1}} = 5.79 \times 10^{-7} \text{ m} = 579 \text{ nm}$$

**Related Problems: 13, 14**

The fact that two independent measurements of two totally different phenomena gave the same value of $h$ inspired great confidence in the validity of the quantum hypotheses of Planck and Einstein, despite their unsettling implications. Einstein's bold assertion that light consisted of a stream of bundles of energy that appeared to transfer their energy through collisions very like those of material particles was completely at odds with the classical wave representation of light, already amply confirmed by experimental studies. How could light be both a wave and a particle?

By 1930, these paradoxes had been resolved by quantum mechanics, which superseded Newtonian mechanics. The classical wave description of light is adequate under some circumstances, but the emission of light from matter and the absorption of light by matter are described by the particle-like photon picture. A hallmark of quantal, as opposed to classical, thinking is to ask not "What *is* light?" but instead "How does light behave under particular experimental conditions?" Thus, **wave–particle duality** is not a contradiction but rather part of the fundamental nature of light and of matter as well.

---

## 15.3

# EXPERIMENTAL DEMONSTRATIONS OF ENERGY QUANTIZATION IN ATOMS

The incompatibility of Rutherford's model of the atom with principles of classical physics, already mentioned at the beginning of Unit 5, was the most fundamental of the conceptual challenges facing physicists in the early 1900s. In atoms, as with blackbody radiation and the photoelectric effect, recognition that energy is quantized provided steps toward understanding the conceptual conflicts. This section describes three sets of experiments that demonstrate quantization of energy in free atoms in the gas phase. These experiments are interpreted through energy level diagrams that represent the discrete energy states of the atom. It remains for quantum theory to explain why energy is quantized and to predict the specific allowed values for each state. The presentation emphasizes physical insight and does not adhere to strict historical order.

## Atomic Spectra and Transitions between Energy States

Light that consists of a distribution of wavelengths (Fig. 15.8) can be resolved into its components by passing through a prism, because each wavelength is deflected through a different angle. One instrument used to separate light into its component

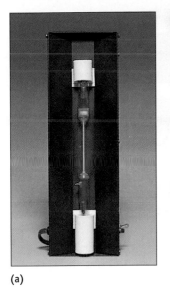
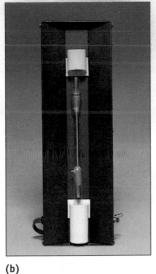
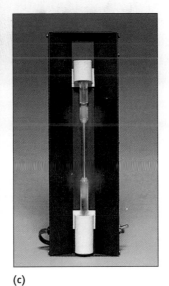

(a) (b) (c)

**FIGURE 15.8** When a gas is excited in an electrical discharge, it glows as it emits light. The colors of the light emitted by three gases—(a) neon, (b) argon, and (c) mercury—are shown. Each emission consists of several wavelengths of light, and the perceived color depends on which wavelength predominates. *(Leon Lewandowski)*

wavelengths is called a **spectrograph** (Fig. 15.9). The spectrograph is enclosed in a box-like container whose walls exclude stray light. The light to be analyzed enters through a narrow slit in the walls, passes to the prism where it is dispersed into its components, and then falls upon a photographic plate or other detector. The detector records the position and intensity of an image of the slit formed by each component wavelength. The recorded array of images is called the *spectrum* of the incoming light. If the light contains all frequencies, the spectrum is a continuous band as in Figure 15.3. Early experiments showed that light emitted from gaseous atoms excited in flames or in electrical discharges gave discrete spectra comprising arrays of parallel lines corresponding to slit images at specific wavelengths (Fig.

**FIGURE 15.9** (a) The emission spectrum of atoms or molecules is measured by passing the light emitted from an excited sample through a prism to separate it according to wavelength, then recording the image on photographic film or with another detector. (b) In absorption spectroscopy, white light from a source passes through the unexcited sample, which absorbs certain discrete wavelengths of light. Dark lines appear on a bright background.

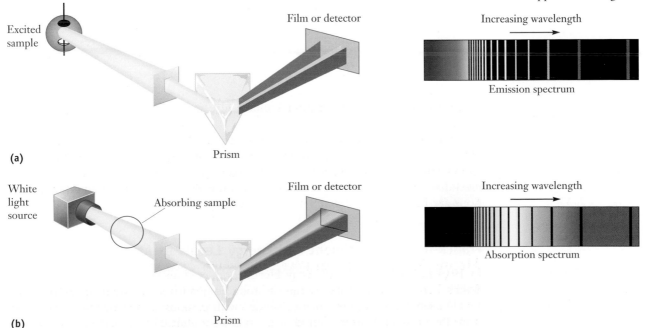

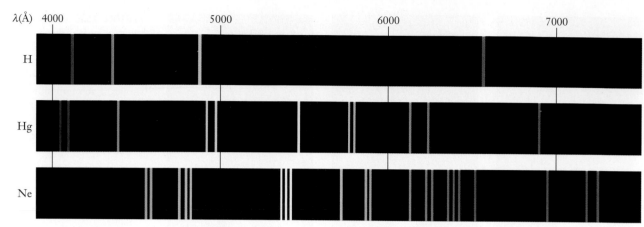

**FIGURE 15.10** Atoms of hydrogen, mercury, and neon emit light at discrete wavelengths. The pattern seen is characteristic of the element under study. 1 Å = $10^{-10}$ m.

15.9a). If white light is passed through gaseous atoms and then into the spectrograph, the absorption spectrum consists of dark slit images superposed on the continuous spectrum of white light. These experiments show that atoms emit and absorb light at a discrete set of frequencies characteristic of that element (Fig. 15.10). For example, in 1885 J. J. Balmer discovered that hydrogen atoms emit a series of lines in the visible region, with frequencies given by the simple formula

$$\nu = \left[\frac{1}{4} - \frac{1}{n^2}\right] \times 3.29 \times 10^{15}\ \text{s}^{-1} \qquad n = 3, 4, 5, \ldots$$

The H atom lines shown in Figure 15.10 fit this equation with $n = 3, 4, 5,$ and 6 in order red to blue.

The existence of discrete line spectra and various empirical equations that relate the frequencies of the lines challenged physicists for more than three decades. The first explanation was provided in 1913 by Niels Bohr. Bohr proposed a model of the H atom that allowed energy only in discrete amounts corresponding to specific energy states. Bohr proposed that emission of light would correspond to a **transition** of the atoms between two such states. The frequency of the emitted light would be connected to the energy of the initial and final states by the expression

$$\nu = \frac{E_i - E_f}{h}$$

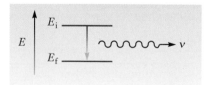

An atom makes a transition from state $E_i$ to $E_f$ and emits a photon of frequency $\nu = [E_i - E_f]/h$.

where $h$ is Planck's constant. Bohr's model also predicts the values of the discrete energy levels in H (Section 15.4).

Each atom can be represented by an energy level diagram in which the distance between two levels gives the frequency of a specific line in the experimental spectrum of the atom. Except for the simplest case of hydrogen, constructing the energy level diagram from the experimental spectrum is difficult because numerous transitions are involved.

## Franck–Hertz Experiment and Energy Levels of Atoms

In 1914 James Franck and Gustav Hertz conducted an experiment to test directly Bohr's hypothesis that the energy of atoms is quantized. In their apparatus (Fig. 15.11) electrons of known energy collided with gaseous atoms, and the energy lost from the electrons was measured. Electrons were emitted from the heated cathode

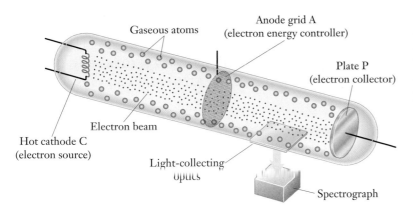

**FIGURE 15.11**  The apparatus of Franck and Hertz that demonstrates the quantization of energy in atoms. Gaseous atoms collide with electrons and gain energy by collisions only when the energy of the electron exceeds a certain threshold. Then, the atom emits a photon whose frequency is determined by the energy transferred during the collision. Gaseous atoms are distributed uniformly throughout the tube; for clarity, in this figure they are shown only near the walls.

C and accelerated toward the anode A. Holes in the anode allowed electrons to pass toward the collector plate P with known kinetic energy controlled by the accelerating voltage between C and A. The apparatus was filled to a low pressure with the gas to be studied. The current arriving at P was studied as a function of kinetic energy of the electrons by varying the accelerating voltage.

The experiment was begun at a very low voltage, and the current increased steadily as the accelerating voltage was increased. At a certain voltage $V_{thr}$, the current dropped sharply, nearly to zero. This fact implied that most of the electrons had lost their kinetic energy in collisions with the gas atoms and were unable to reach the collector. As the voltage was increased above $V_{thr}$, the current rose again, indicating that after losing energy to the gas atoms, the electrons were again accelerated sufficiently to reach the collector.

The abrupt fall in the plot of current versus voltage at $V_{thr}$ indicated that electrons required a *kinetic energy threshold* $eV_{thr}$ in order to transfer energy to the gas atoms. Therefore, the energy of the atoms must be quantized in discrete states. The **first excited state** must lie above the **ground state** (the state with lowest energy) by the amount $eV_{thr}$. Continuing the experiment with higher values of accelerating voltage reveals additional energy thresholds corresponding to excited states with higher energies.

To confirm their interpretation, Franck and Hertz used a spectrograph to analyze light emitted by the excited atoms. When the accelerating voltage was below $V_{thr}$, no light was observed. When the accelerating voltage was slightly above $V_{thr}$, a single emission line was observed with frequency very nearly equal to

$$\nu = \frac{\Delta E}{h} = \frac{eV_{thr}}{h}$$

At higher values of accelerating voltage an additional spectral emission line appeared as each additional excitation energy threshold was reached.

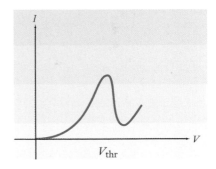

The current in the Franck–Hertz experiment shows a sharp change at a particular value of the accelerating voltage, corresponding to the threshold for energy transfer from the electron to a gaseous atom.

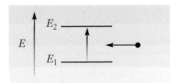

The Franck–Hertz experiment measures directly the separation between energy levels of the atom by measuring the energy lost by an electron colliding with the atom.

## EXAMPLE 15.3

The first two excitation voltage thresholds in the Franck–Hertz study of mercury vapor were found at 4.9 V and 6.7 V. Calculate the wavelength of light that should be emitted by Hg atoms after excitation past each of these thresholds.

## Solution

The emitted wavelength at a particular value of $V_{thr}$ is given by the equation

$$\lambda = \frac{hc}{\Delta E} = \frac{hc}{eV_{thr}} = \frac{(6.6261 \times 10^{-34}\,\text{J s})(2.9979 \times 10^8\,\text{m s}^{-1})}{(V_{thr}[\text{V}])(1.6022 \times 10^{-19}\,\text{C})}$$

$$= \frac{12.398 \times 10^{-7}\,\text{m}}{V_{thr}[\text{V}]}$$

The value of wavelength is calculated by substituting in a particular value of $V_{thr}$ expressed in units of volts (V). At $V_{thr} = 4.9$ V, the calculated wavelength is $\lambda = 2500$ Å. The wavelength actually observed above this threshold was 2537.0 Å. At $V_{thr} = 6.7$ V, the calculated wavelength is $\lambda = 1800$ Å. The wavelength actually observed above this threshold was 1849.6 Å. Values of energy separations measured by the Franck–Hertz method and by optical emission spectroscopy agree quite closely. The optical measurements are more precise.

**Related Problems: 21, 22**

---

The Franck–Hertz experiment demonstrated that atoms can acquire energy only in discrete, quantized amounts from collisions with electrons. This energy is then returned in discrete, quantized amounts by emission of light. This experiment provided dramatic confirmation of Bohr's hypothesis that the energy of atoms is quantized in discrete states. It also gave a direct method for measuring the energy differences between these states and constructing the energy level diagram starting at the ground state.

## Photoelectron Spectroscopy and Binding Energies

Atomic spectra and Franck–Hertz experiments measure the difference between energy states. In order to assign a specific value of energy to each state, it is necessary to define zero on the energy scale and to identify a measurement that probes the state of zero energy. A reasonable choice for zero energy of the atom is that state in which an electron has been removed from the atom, but has no kinetic energy. This is equivalent to a singly charged positive ion and an electron separated by infinite distance, so the Coulomb potential energy between them is zero (see Appendix B). Relative to this (arbitrary but reasonable) choice for zero, the bound states of the atom have negative values of energy. Unbound states with positive values of energy correspond to a free electron moving with nonzero kinetic energy in the vicinity of the ion.

The binding energy of a discrete state is most easily measured directly by an important experimental technique called **photoelectron spectroscopy,** which determines the amount of energy required to remove electrons from the atom. Photoelectron spectroscopy is simply the photoelectric effect of Section 15.2 applied not to metals but to free atoms. If radiation of sufficiently high frequency $\nu$ (in the ultraviolet or x-ray region of the spectrum), strikes an atom, an electron will be ejected with kinetic energy $\frac{1}{2}m_e v^2$. The kinetic energy of the ejected electrons is determined by an energy analyzer, which measures the extent to which the electrons are deflected during passage between charged plates in vacuum. As the voltage between the plates is changed, electrons with different values of kinetic energy will be deflected to the detector, and the spectrum of kinetic energy values can be

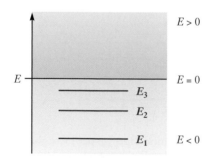

The zero of energy for an atom corresponds to complete separation of the electron. As the electron moves closer to the nucleus, the attractive potential causes the energy of the atom to decrease; bound state energies are thus negative. Positive energies correspond to free electrons passing by the nucleus.

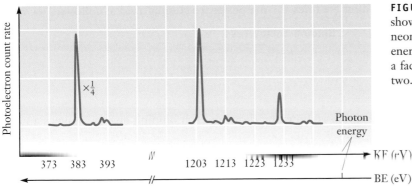

**FIGURE 15.12**   The photoelectron spectrum of neon shows three peaks, demonstrating that the electrons of neon are organized in three bonding states of distinct energy values. The peak at 383.4 eV has been reduced by a factor of four for display on the same scale as the other two.

recorded. Measuring the kinetic energy by deflection is analogous to the energy measurement in the photoelectric effect experiments, and the results are conveniently expressed in units of eV. Then the spectrum of binding energy, BE, is calculated by the principle of conservation of energy,

$$BE = h\nu_{photon} - \tfrac{1}{2}m_e v^2_{electron}$$

as shown in Section 15.2.

Figure 15.12 shows the measured photoelectron spectrum for neon excited by x-rays with wavelength $9.890 \times 10^{-11}$ m. Three peaks appear with kinetic energy of 383.4 eV, 1205.2 eV, and 1232.0 eV. The binding energy scale has been sketched beneath the kinetic energy scale. The peak at lowest binding energy (highest kinetic energy) is produced by the most weakly bound electrons (see preceding equation). This is the minimum amount of energy required to detach an electron from an atom and is the ionization energy $IE_1$ introduced in Section 3.1. Clearly there can be no signal in the photoelectron spectrum at binding energies less than this value. Peaks at higher binding energies correspond to electrons removed from more strongly bound states. This spectrum demonstrates that the 10 electrons of Ne are arranged in bonding states that produce three distinct discrete energy levels. Detailed interpretation of these bonding states requires the quantum theory of atomic structure (Section 15.8). For present purposes, this spectrum demonstrates that the energy of atoms is quantized at discrete values that can be measured relative to the selected zero of energy. Binding energies measured in this way are used to construct the energy level diagram for an atom.

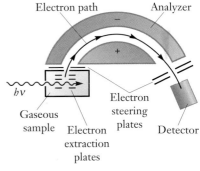

The energy of photoelectrons is determined by the extent of their deflection when passing between charged plates in vacuum.

*lowest binding energy ($BE_1$) is equal to the ionization energy.*

## EXAMPLE 15.4

Construct the energy level diagram for Ne from the data in Figure 15.12.

### Solution

The binding energy of each level is calculated as $BE = \epsilon_{photon} - (KE)_{electron}$. Since the measured kinetic energy values for the photoelectrons are reported in units of eV, the most convenient approach is to calculate the photon energy in eV and then subtract off the kinetic energy values. The energy of the photon is given by

$$\epsilon_{photon} = \frac{hc}{\lambda} = \frac{(6.6261 \times 10^{-34}\,\text{J s})(2.9979 \times 10^{8}\,\text{m s}^{-1})}{(0.9890 \times 10^{-10}\,\text{m})(1.6022 \times 10^{-19}\,\text{J eV}^{-1})} = 1253.6\,\text{eV}$$

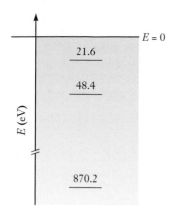

The energy level diagram of neon as determined by photoelectron spectroscopy.

The calculated binding energies for the peaks are summarized as follows:

| Kinetic Energy (eV) | Binding Energy (eV) |
| --- | --- |
| 383.4 | 870.2 |
| 1205.2 | 48.4 |
| 1232.0 | 21.6 |

The energy level diagram is drawn by showing each binding energy value as a negative number.

**Related Problems: 23, 24**

---

## 15.4

# THE BOHR MODEL: PREDICTING DISCRETE ENERGY LEVELS

### Bohr's Model of the Atom

In 1913 Niels Bohr developed a model for the atom that accounted for the striking regularities seen in the spectrum of the hydrogen atom and one-electron ions such as $He^+$, $Li^{2+}$, and $Be^{3+}$. Bohr supplemented Rutherford's *planetary model* of the atom with the assumption that an electron of mass $m_e$ moves in a circular orbit of radius $r$ about a fixed nucleus. According to Coulomb's law, the potential energy of interaction of two charges, $Q_1$ and $Q_2$, separated by a distance $r$ is

$$\text{potential energy} = \frac{Q_1 Q_2}{4\pi\epsilon_0 r}$$

where $\epsilon_0$, called the *permittivity of the vacuum*, is a proportionality constant with numerical value $8.854 \times 10^{-12}$ $C^2$ $J^{-1}$ $m^{-1}$. For a nucleus (of charge $+Ze$) and an electron (of charge $-e$) this potential energy is

$$\text{potential energy} = -\frac{Ze^2}{4\pi\epsilon_0 r}$$

where the minus sign indicates a net attractive interaction, giving an algebraically lower energy at shorter distances. The total energy (kinetic and potential) of the electron in the atom is

$$E = \tfrac{1}{2}m_e v^2 - \frac{Ze^2}{4\pi\epsilon_0 r} \qquad \text{[15.4]}$$

The Coulomb force of attraction on the electron, $F_{\text{coul}}$, is the derivative of the potential energy with respect to the separation $r$ and equals $Ze^2/4\pi\epsilon_0 r^2$. From Newton's law relating force and acceleration, $F_{\text{coul}} = m_e a$, and for uniform circular motion, the acceleration $a$ of the electron is $v^2/r$. Therefore,

$$F_{\text{coul}} = \frac{Ze^2}{4\pi\epsilon_0 r^2} = m_e a = m_e \frac{v^2}{r} \qquad \text{[15.5]}$$

As mentioned earlier, classical physics predicts that an accelerated electron will emit light of continuously increasing frequency and will spiral inward to the nucleus.

Bohr avoided this conflict by simply postulating that only certain discrete orbits (characterized by radius $r_n$ and energy $E_n$) are allowed and that light is emitted or absorbed only when the electron "jumps" from one stable orbit to another.

The *linear* momentum of an electron is the product of its mass and its velocity, $m_e v$. The *angular* momentum is a different quantity that describes rotational motion about an axis. An introduction to angular momentum is provided in Appendix B. For the circular paths of the Bohr model, the angular momentum of the electron is the product of its mass, its velocity, and the radius of the orbit ($m_e vr$). Bohr assumed that the angular momentum is quantized and must be an integral multiple of $h/2\pi$, where $h$ is Planck's constant:

$$\text{angular mometum} = m_e vr = n\frac{h}{2\pi} \qquad n = 1, 2, 3, \ldots \qquad \textbf{[15.6]}$$

Once this quantization is assumed, discrete orbits and energies follow as consequences. Equations 15.5 and 15.6 contain two unknowns, $v$ and $r$. Solving Equation 15.6 for $v$ ($= nh/2\pi m_e r$), inserting it into Equation 15.5, and solving for $r$ give the allowed values

$$r_n = \frac{\epsilon_0 n^2 h^2}{\pi Z e^2 m_e} = \frac{n^2}{Z} a_0 \qquad \textbf{[15.7]}$$

where $a_0$, the **Bohr radius**, has the numerical value $5.29 \times 10^{-11}$ m $= 0.529$ Å.

*$a_0$ is constant.*

The velocity $v_n$ corresponding to the orbit with radius $r_n$ is

$$v_n = \frac{nh}{2\pi m_e r_n} = \frac{Ze^2}{2\epsilon_0 nh}$$

The results for $r_n$ and $v_n$ can be substituted into Equation 15.4 to give the allowed values of energy:

$$E_n = \frac{-Z^2 e^4 m_e}{8\epsilon_0^2 n^2 h^2} = -(2.18 \times 10^{-18} \text{ J})\frac{Z^2}{n^2} \qquad n = 1, 2, 3, \ldots \qquad \textbf{[15.8a]}$$

To avoid the awkwardness of carrying these numerous fundamental constants through all calculations, it is convenient to express the allowed energy values in units of **rydbergs,** where 1 rydberg $= 2.18 \times 10^{-18}$ J. The energy level expression becomes

$$E_n = -\frac{Z^2}{n^2} \qquad \text{(rydberg)} \qquad n = 1, 2, 3, \ldots \qquad \textbf{[15.8b]}$$

As $n \to \infty$, the electron orbit is farther and farther from the nucleus ($r_n \to \infty$). The energy $E_n$ is 0 for the electron and nucleus at rest and at an infinite distance apart, and it is negative for smaller distances. Bohr's model thus predicts a discrete energy level diagram for the one-electron atom (Figs. 15.13 and 15.14). The ground state for the system of nucleus plus electron has $n = 1$; excited states have higher values of $n$ (see Fig. 15.14).

The **ionization energy** of an atom is the minimum energy needed to remove an electron from it (see Section 3.1). Removing an electron from a hydrogen atom in its ground state involves a transition from the $n = 1$ state to the $n = \infty$ state, in which $E_n = 0$. The associated energy change is

$$\Delta E = E_{\text{final}} - E_{\text{initial}} = 2.18 \times 10^{-18} \text{ J}$$

*energy to remove hydrogen's only electron is equal to Rydberg's Constant.*

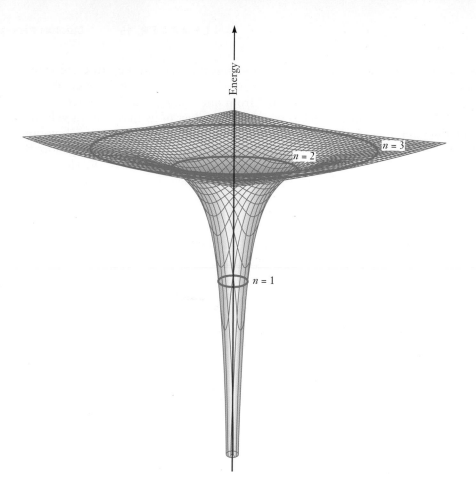

**FIGURE 15.13** The interaction between the electron and nucleus has its lowest energy when the electron is closest to the nucleus. The electron moving away from the nucleus can be seen as moving up the sides of a steep potential energy well. In the Bohr theory, it can "catch" and stick on the sides only at certain allowed values of $r$, the radius, and $E$, the energy. The first three of these are shown by rings.

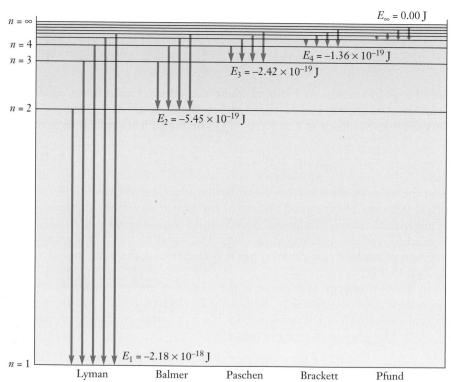

**FIGURE 15.14** In the energy levels of the hydrogen atom, the separated electron and proton are arbitrarily assigned zero energy, and all other energies are more negative than that. Atoms emit light as they fall from higher to lower energy levels (blue arrows). Each series of related transitions is named after its discoverer.

Multiplying this by Avogadro's number gives the ionization energy (*IE*) per mole of atoms:

$$IE = (6.022 \times 10^{23} \text{ mol}^{-1})(2.18 \times 10^{-18} \text{ J}) = 1.31 \times 10^6 \text{ J mol}^{-1} = 1310 \text{ kJ mol}^{-1}$$

This number agrees with the experimentally observed ionization energy of hydrogen atoms.

## EXAMPLE 15.5

Consider the $n = 2$ state of the $Li^{2+}$ ion. Using the Bohr model, calculate the radius of the electron orbit, the electron velocity, and the energy of the ion relative to the nucleus and electron separated by an infinite distance.

### Solution

Because $Z = 3$ for an $Li^{2+}$ ion (the nuclear charge is $+3e$) and $n = 2$, the radius is

$$r_2 = \frac{n^2}{Z} a_0 = \frac{4}{3} a_0 = \frac{4}{3} (0.529 \text{ Å}) = 0.705 \text{ Å}$$

The velocity is

$$v_2 = \frac{nh}{2\pi m_e r_2} = \frac{2(6.626 \times 10^{-34} \text{ J s})}{2\pi(9.11 \times 10^{-31} \text{ kg})(0.705 \times 10^{-10} \text{ m})} = 3.28 \times 10^6 \text{ m s}^{-1}$$

The energy is

$$E_2 = -\frac{(3)^2}{(2)^2} (2.18 \times 10^{-18} \text{ J}) = -4.90 \times 10^{-18} \text{ J}$$

**Related Problems: 25, 26**

## Atomic Spectra: Interpretation by Bohr's Model

When a one-electron atom or ion undergoes a transition from a state characterized by quantum number $n_i$ to a state lower in energy with quantum number $n_f$ ($n_i > n_f$), a photon is emitted to carry off the energy $h\nu$ lost by the atom. By conservation of energy, $E_i = E_f + h\nu$, so that

$$h\nu = \frac{Z^2 e^4 m_e}{8\epsilon_0^2 h^2} \left[ \frac{1}{n_f^2} - \frac{1}{n_i^2} \right] \qquad \text{(emission)}$$

*Bohr emission*

As $n_i$ and $n_f$ take on a succession of integral values, lines are seen in the emission spectrum (Fig. 15.14) with frequencies

$$\nu = \frac{Z^2 e^4 m_e}{8\epsilon_0^2 h^3} \left( \frac{1}{n_f^2} - \frac{1}{n_i^2} \right) = (3.29 \times 10^{15} \text{ s}^{-1}) Z^2 \left( \frac{1}{n_f^2} - \frac{1}{n_i^2} \right) \qquad \text{[15.9]}$$

$$n_i > n_f = 1, 2, 3, \ldots \qquad \text{(emission)}$$

On the other hand, an atom can *absorb* energy $h\nu$ from a photon as it undergoes a transition to a *higher* energy state ($n_f > n_i$). In this case, conservation of energy requires $E_i + h\nu = E_f$, and so the absorption spectrum shows a series of lines at frequencies

$$\nu = (3.29 \times 10^{15} \text{ s}^{-1})\, Z^2 \left(\frac{1}{n_i^2} - \frac{1}{n_f^2}\right)$$

$$n_f > n_i = 1, 2, 3, \ldots \quad \text{(absorption)}$$

[15.10]

For hydrogen, which has $Z = 1$, the predicted emission spectrum with $n_f = 2$ corresponds to the series of lines in the visible region described by Balmer and mentioned at the beginning of this section. A series of lines at higher frequencies (in the ultraviolet region) is predicted for $n_f = 1$ (the Lyman series), and other series are predicted at lower frequencies (in the infrared region) for $n_f = 3, 4, \ldots$. In fact, the predicted and observed spectra of hydrogen and one-electron ions are in excellent agreement—a major triumph of the Bohr theory.[2]

Despite these successes, Bohr's theory has a number of shortcomings. For one, it cannot predict the energy levels and spectra of atoms and ions with more than one electron. Also, fundamentally, Bohr's theory was an uncomfortable hybrid of classical and nonclassical concepts. The postulate of quantized angular momentum—which led to the circular orbits—had no fundamental basis and was grafted onto classical physics to force the predictions of classical physics to agree with experimental results. In 1926 Bohr's theory was replaced by modern quantum mechanics in which the quantization of energy and angular momentum is a natural consequence of the basic postulates and requires no extra assumptions. The circular orbits of the Bohr theory do not appear in quantum mechanics. Bohr's theory provided the conceptual bridge from classical theoretical physics to the new quantum mechanics. Its historical and intellectual importance cannot be exaggerated.

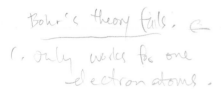

---

## 15.5

# WAVES, PARTICLES, AND THE SCHRÖDINGER EQUATION

Bohr's theory gave a prescription for obtaining the discrete energy levels of a one-level atom or ion, but did not explain the origin of energy quantization. A fundamental explanation was developed in 1926 by Schrödinger through an analogy with the theory of vibrations, where a form of quantization was already understood. A key step was the recognition by de Broglie that wave–particle duality, introduced by Einstein to describe photons, was equally a property of material particles like electrons.

### De Broglie Waves

Up to now we have considered only one type of wave, a **traveling wave.** Electromagnetic radiation (light, x-rays, and gamma rays) is described by such a traveling wave, moving through space at speed $c$. Another type of wave is a **standing wave,** of which a simple example is a guitar string with fixed ends. A plucked string vibrates, but only certain oscillations of the string are possible. Because the ends are fixed, the only oscillations that can persist are those in which a whole number of half-

---

[2] Strictly speaking, the electron mass $m_e$ in the expressions for the energy levels of one-electron atoms should be replaced by the *reduced mass* $\mu$, equal to $m_e m_N/(m_e + m_N)$ where $m_N$ is the nuclear mass; $\mu$ differs from $m_e$ by less than 0.1%.

wavelengths fits into the length of string $L$ (Fig. 15.15). The condition on the allowed wavelengths is

$$n\frac{\lambda}{2} = L \qquad n = 1, 2, 3, \ldots$$

It is impossible to create a wave with any other value of $\lambda$ if the ends of the string are fixed. The oscillation with $n = 1$ is called the **fundamental** or first harmonic, and higher values of $n$ correspond to higher harmonics. At certain points within the standing wave the amplitude of oscillation is 0; these points are called **nodes.** (The fixed ends are not counted as nodes.) The higher the number of the harmonic $n$, the more numerous the nodes, the shorter the wavelength, the higher the frequency and the higher the energy of the standing wave.

De Broglie realized that such standing waves are examples of quantization: only certain discrete states of vibration, characterized by the "quantum number" $n$, are allowed for the vibrating string. He suggested that the quantization of a one-electron atom might have the same origin and that the electron might be associated with a standing wave, in this case a *circular* standing wave about the nucleus of the atom (Fig. 15.16). For the wave's amplitude to be well defined (single valued and smooth), a whole number of wavelengths must fit into the circumference of the circle $2\pi r$. The condition on the allowed wavelengths for standing circular waves is

$$n\lambda = 2\pi r \qquad n = 1, 2, 3, \ldots \qquad \textbf{[15.11]}$$

Bohr's assumption about quantization of the angular momentum of the electron was

$$m_e vr = n\frac{h}{2\pi}$$

which can be rewritten as

$$2\pi r = n\left[\frac{h}{m_e v}\right] \qquad \textbf{[15.12]}$$

Comparison of de Broglie's equation (15.11) with Bohr's (15.12) shows that the wavelength of the standing wave is related to the linear momentum $p$ of the electron by the simple formula

$$\lambda = \frac{h}{m_e v} = \frac{h}{p} \qquad \textbf{[15.13]}$$

De Broglie showed with the theory of relativity that exactly the same relationship holds between the wavelength and momentum of a *photon*. De Broglie therefore proposed as a generalization that any particle moving with linear momentum $p$ has wave-like properties and a wavelength $\lambda = h/p$ associated with it.

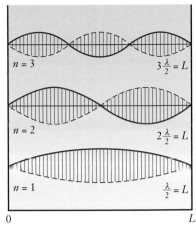

**FIGURE 15.15** A string of length $L$ with fixed ends can vibrate in only a restricted set of ways. The positions of largest amplitude for the first three harmonics are shown here. In a standing wave such as this, the whole string is in motion except at its ends and at the nodes.

$$n\lambda = n\left[\frac{h}{m_e v}\right]$$

$$\lambda = \frac{h}{m_e v} = \frac{h}{p}$$

since

$$m_e v = \text{momentum}$$
$$p$$

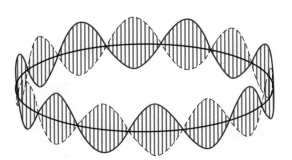

**FIGURE 15.16** A circular standing wave on a closed loop. The state shown has $n = 7$, with seven full wavelengths around the circle.

**EXAMPLE 15.6**

Calculate the de Broglie wavelengths of **(a)** an electron moving with velocity $1.0 \times 10^6$ m s$^{-1}$ and **(b)** a baseball of mass 0.145 kg, thrown with a velocity of 30 m s$^{-1}$.

**Solution**

**(a)**
$$\lambda = \frac{h}{p} = \frac{h}{m_e v} = \frac{6.626 \times 10^{-34} \text{ J s}}{(9.11 \times 10^{-31} \text{ kg})(1.0 \times 10^6 \text{ m s}^{-1})}$$
$$= 7.3 \times 10^{-10} \text{ m} = 7.3 \text{ Å}$$

**(b)**
$$\lambda = \frac{h}{mv} = \frac{6.626 \times 10^{-34} \text{ J s}}{(0.145 \text{ kg})(30 \text{ m s}^{-1})}$$
$$= 1.5 \times 10^{-34} \text{ m} = 1.5 \times 10^{-24} \text{ Å}$$

The latter wavelength is far too small to be observed. For this reason we do not see the wave-like properties of baseballs or other macroscopic objects, but on a microscopic level, electrons moving in atoms show wave-like properties that are essential for atomic structure.

**Related Problems: 31, 32**

Under what circumstances does the wave-like nature of particles become apparent? When waves from two sources pass through the same region of space, they *interfere* with each other. Consider water waves as an example. When two crests meet, *constructive* interference occurs and a higher crest (greater amplitude) appears; where a crest of one wave meets a trough of the other, *destructive* interference (smaller amplitude) occurs (Fig. 15.17). When x-rays are sent into a crystal, they are scattered by the atoms of the crystal (picture a water wave encountering a post embedded in the bottom of a lake). The x-rays scattered from each atom in the crystal interfere with one another to give a characteristic x-ray **diffraction pattern.** From de Broglie's generalization, such diffraction should also be observed when a beam of *electrons* with appropriate momentum (and therefore wavelength) is sent into a crystal. In 1927 C. Davisson and L. H. Germer showed that electrons are, in fact, diffracted by a crystal and that de Broglie's relationship correctly predicts their wavelength. Low Energy Electron Diffraction (LEED) is now widely used to study the atomic structure of solid surfaces (Fig. 15.18). These experiments provide a striking confirmation of de Broglie's hypothesis about the wave-like nature of matter. "Waves" and "particles" are idealized limits, and protons, electrons, and photons all possess both wave and particle aspects.

## The Heisenberg Uncertainty Principle

In classical physics, particles have well-defined positions and momenta and follow precise trajectories. On an atomic scale, however, positions and momenta cannot be determined with precision because of the **uncertainty principle** stated by Werner Heisenberg in 1927.

The uncertainty principle is directly connected with the process of measurement. How are the position and momentum of a macroscopic object such as a base-

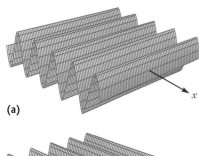

**(a)**

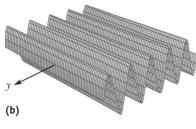

**(b)**

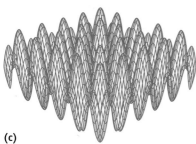

**(c)**

**FIGURE 15.17** When two waves (a and b) moving perpendicular to one another meet and overlap, their amplitudes add. Where two crests meet, constructive interference occurs and the resultant amplitude is greater than that of either wave alone (shown by the peaks in part c). Where a crest meets a trough, destructive interference gives an amplitude of zero.

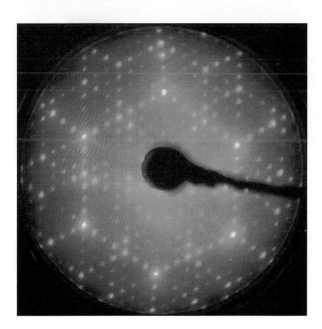

**FIGURE 15.18** LEED pattern of the same Si surface imaged by scanning tunneling microscopy (STM) in the figure opposite page 1 of this book. *(Courtesy of Dr. Gerard Parkinsen and Mr. William Gerace, OMICRON Vakuumphysik GMBH, Tanusstein, Germany)*

ball in motion determined? The simplest way is to take a series of snapshots of it at different times, with each picture a record of the light (photons) scattered by the baseball. To observe an object, we must scatter something (such as a photon) from it (Fig. 15.19). Scattering photons from a baseball does not change the baseball's trajectory appreciably, but scattering photons from an electron can strongly affect its trajectory. To locate the *position* of the object accurately, we must use light with a wavelength comparable to or shorter than the size of the object. By de Broglie's relationship, however, the shorter the wavelength, the greater the momentum of the photon. We cannot then make an accurate measurement of the object's *momentum*, because it will be knocked away from its original trajectory by the photon. Thus, we cannot measure to arbitrary accuracy both the position and momentum of an object.

A rough estimate takes the minimum uncertainty of the position measurement $\Delta x$ to be of order $\lambda$ and the uncertainty of the momentum measurement $\Delta p$ to be of order $h/\lambda$ (because that is the momentum carried by the photon). Their product is then $(\Delta x)(\Delta p) \geq h$. A fuller analysis gives the result

$$(\Delta x)(\Delta p) \geq h/4\pi \qquad [15.14]$$

The uncertainty principle places a fundamental limit on the accuracy to which the position and momentum of a particle can be known simultaneously. It reveals a fundamental defect in the Bohr model, which ascribes a definite position and momentum to the electron in its circular orbit around the nucleus.

## EXAMPLE 15.7

Suppose photons of green light (wavelength $5.3 \times 10^{-7}$ m) are used to locate the position of the baseball from Example 15.6 to an accuracy of one wavelength. Calculate the minimum uncertainty in the *speed* of the baseball.

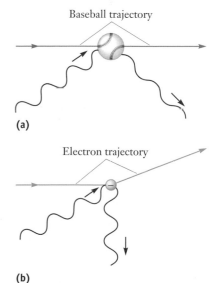

**FIGURE 15.19** A photon, which has a negligible effect on the trajectory of a baseball (a), significantly perturbs the trajectory of the far less massive electron (b).

**Solution**

The Heisenberg relation

$$(\Delta x)(\Delta p) \geq h/4\pi$$

gives

$$\Delta p \geq \frac{h}{4\pi\Delta x} = \frac{6.626 \times 10^{-34}\,\text{J s}}{4\pi(5.3 \times 10^{-7}\,\text{m})} = 9.9 \times 10^{-29}\,\text{kg m s}^{-1}$$

Because the momentum is just the mass (a constant) times the speed, the uncertainty in the speed is

$$\Delta v = \frac{\Delta p}{m} \geq \frac{9.9 \times 10^{-29}\,\text{kg m s}^{-1}}{0.145\,\text{kg}} = 6.8 \times 10^{-28}\,\text{m s}^{-1}$$

This is such a tiny fraction of the speed of the baseball ($30\,\text{m s}^{-1}$) that the uncertainty principle plays a negligible role in the measurement of the baseball's motion. Such is the case for all macroscopic objects.

**Related Problems: 33, 34**

## The Schrödinger Equation

De Broglie's work attributes wave-like properties to electrons in atoms, and the uncertainty principle shows that detailed trajectories of electrons cannot be defined. Consequently we must deal in terms of the *probability* of electrons having certain positions and momenta. These ideas are combined in the fundamental equation of quantum mechanics, the **Schrödinger equation,** discovered by the Austrian physicist Erwin Schrödinger in 1925.

Schrödinger, a recognized authority on the theory of vibrations and the associated "quantization" of standing waves, reasoned that an electron (or any other particle) with wave-like properties, should be described by a **wave function** that has a value at each position in space. (The value of a wave function at each position describes, for example, the height of a water wave or the amplitude of a classical electromagnetic wave.) This wave function is symbolized by the Greek letter $\psi$ (psi), and $\psi(x, y, z)$ is the "height" of the wave at the point in space defined by the set of Cartesian coordinates $(x, y, z)$. Schrödinger wrote down the equation satisfied by $\psi$ for a given set of interactions between particles. Time-independent solutions of this equation—called **stationary states**—exist only for certain discrete values of the energy, and so energy quantization is a natural consequence of the Schrödinger equation.

A wave function, like an electromagnetic wave, can take positive values in some regions of space (it is said to have **positive phase** in these regions) and negative values in other regions (where it has **negative phase**). The points at which the wave function passes through 0 and changes sign are called nodes, just as in our guitar-string model for standing waves.

Solutions of the Schrödinger equation are possible for more than one value of the energy. The solution that corresponds to the lowest energy is called the ground state (just as in the Bohr model), and higher energy solutions are called excited states. What is the physical meaning of the wave function $\psi$? We have no way of measuring $\psi$ directly, just as in classical wave optics we have no direct way of measuring the amplitudes of the electric and magnetic fields that constitute the light wave (Fig. 15.2). What *can* be measured in the latter case is the intensity of the light wave, which, according to the classical theory of electromagnetism, is proportional to the square of the amplitude of the electric field:

$$\text{intensity} \propto (E_{\text{max}})^2$$

In turn, this intensity is proportional to the density of photons at that point in space or, in other words, to the *probability density* of finding a photon there.

By analogy, we propose that the square of the wave function $\psi^2$ for a particle is a probability density for that particle. In other words, $\psi^2(x, y, z)\Delta x \, \Delta y \, \Delta z$ is the probability that the particle will be found in a small volume $\Delta x \, \Delta y \, \Delta z$ about the point $(x, y, z)$. This probabilistic interpretation of the wave function, proposed by the German physicist Max Born, is now generally accepted because it provides a consistent picture of particle motion on a microscopic scale. We are obliged to acknowledge that our information about the location of a particle is statistical.

Through Schrödinger's equation, quantum mechanics accurately describes the behavior of microscopic systems, which Newtonian mechanics could not do. Quantum mechanics is a more general theory and contains Newtonian mechanics as a limiting case, valid for macroscopic systems. To apply quantum mechanics to a particular system, one substitutes the potential energy of interaction for that system into Schrödinger's equation and solves the equation to obtain two principal results: the allowed energy values and the wave function. This procedure is illustrated in detail for a simplified model in the next section. In the remainder of the book quantum mechanics is used to analyze numerous systems. However, this prescription is not carried out in detail, and the two principal results are merely stated for each system under discussion. The predicted energy levels are used to describe gain or loss of energy by the system, and the square of the wave function is used to describe the statistics of locating the particles in the system at particular positions. Section 15.6 can be omitted without loss of continuity by readers who prefer to proceed directly to the results.

## A DEEPER LOOK...

## 15.6   THE PARTICLE IN A BOX

The Schrödinger equation cannot be derived (just as Newton's laws cannot). Nonetheless, it is worthwhile to explore some properties of its mathematical form. We begin by considering a particle moving freely in one dimension with classical momentum $p$. Such a particle is associated with a wave of wavelength $\lambda = h/p$. Two "wave functions" with this wavelength are

$$\psi(x) = A \sin \frac{2\pi x}{\lambda} \qquad \text{or} \qquad B \cos \frac{2\pi x}{\lambda}$$

where $A$ and $B$ are constants. What equation do these trigonometric functions satisfy? Let us take the sine function as an example. From differential calculus, the derivative (or slope) of $\psi(x)$ is

$$\frac{d\psi(x)}{dx} = A \frac{2\pi}{\lambda} \cos \frac{2\pi x}{\lambda}$$

The slope of *this* is the second derivative of $\psi$, written $\dfrac{d^2\psi(x)}{dx^2}$ and equal to

$$\frac{d^2\psi(x)}{dx^2} = -A\left(\frac{2\pi}{\lambda}\right)^2 \sin \frac{2\pi x}{\lambda}$$

This is just a constant, $-(2\pi/\lambda)^2$, multiplied by the original wave function $\psi(x)$:

$$\frac{d^2\psi(x)}{dx^2} = -\left(\frac{2\pi}{\lambda}\right)^2 \psi(x)$$

This is an equation (called a *differential equation*) that is satisfied by the function $\psi(x) = A \sin (2\pi x/\lambda)$. It is easy to verify that this equation is also satisfied by the function $\psi(x) = B \cos (2\pi x/\lambda)$.

Let us now replace the wavelength $\lambda$ with the momentum $p$ from the de Broglie relation:

$$\frac{d^2\psi(x)}{dx^2} = -\left(\frac{2\pi}{h} p\right)^2 \psi(x)$$

We can rearrange this into a suggestive form by multiplying both sides by $-h^2/8\pi^2 m$, giving

$$-\frac{h^2}{8\pi^2 m}\frac{d^2\psi(x)}{dx^2} = \frac{p^2}{2m}\,\psi(x) = E_K\psi(x)$$

where $E_K = p^2/2m$ is the kinetic energy of the particle. This shows that there must be a relation between the second derivative of the wave function and the kinetic energy $E_K$. When one is large, the other is large as well.

If external forces are present, a potential energy term $V(x)$ (due to walls enclosing the particle or the presence of fixed charges, for example) must be included. Writing the total energy as $E = E_K + V(x)$ and substituting in the previous equation give

$$-\frac{h^2}{8\pi^2 m}\frac{d^2\psi(x)}{dx^2} + V(x)\psi(x) = E\psi(x) \qquad \textbf{[15.15]}$$

This is the Schrödinger equation for a particle moving in one dimension. The calculation just given is not a derivation of this equation; rather, it indicates some useful intuitive connections to the de Broglie momentum–wavelength correspondence.

If $\psi^2(x)$ is to be a probability density, $\psi$ must have certain mathematical properties. It must be continuous, because the probability of finding a particle at every point in space must be defined and unique. Second, if all of space is subdivided into small intervals (in one dimension) or volumes (in three dimensions) and the probability of the particle's being in each volume is calculated, then the *sum* of those probabilities must be 1 because the particle is *somewhere*. Mathematically, this means

$$\int \psi^2(x)\,dx = 1 \qquad \textbf{[15.16]}$$

When these conditions are imposed on $\psi(x)$, for reasonable choices of the potential $V(x)$ there are solutions $\psi_n(x)$ of the Schrödinger equation only for certain values of the energy. Quantization of the energy arises as a natural consequence of the Schrödinger equation, rather than being introduced in an ad hoc fashion as in the Bohr model.

The simplest problem for which the Schrödinger equation can be solved is the so-called particle in a box. This consists of a particle confined to a certain region of space (the "box"), which in one dimension becomes the interval between 0 and $L$, where $L$ is the length of the box (Fig. 15.20a). If the particle is truly to be confined, the potential energy $V(x)$ must rise abruptly to an infinite value at the two end walls, to prevent even fast-moving particles from escaping. On the other hand, inside the box the particle's motion is free and $V(x) = 0$. The solution of the Schrödinger equation for this case illustrates the general methods used for other potential-energy functions.

Wherever the potential energy $V$ is infinite, the probability of finding the particle must be 0. Hence, $\psi(x)$ and $\psi^2(x)$ must be 0 in these regions:

$$\psi(x) = 0 \qquad \text{for} \qquad x \le 0 \qquad \text{or} \qquad x \ge L$$

Inside the box, where $V = 0$, the Schrödinger equation has the form

$$-\frac{h^2}{8\pi^2 m}\frac{d^2\psi(x)}{dx^2} = E\psi(x)$$

or, equivalently,

$$\frac{d^2\psi(x)}{dx^2} = -\frac{8\pi^2 mE}{h^2}\psi(x)$$

As already shown, the sine and cosine are two possible solutions to this equation, because their second derivative is a (negative) constant times the function itself. The cosine function can be eliminated in this case because, for the wave function to be continuous, $\psi(x)$ must be 0 at $x = 0$. This is true of the sine function but not of the cosine. Therefore,

$$\psi(x) = A \sin kx$$

If the wave function is also to be continuous at $x = L$, then

$$\psi(L) = 0$$

or

$$\psi(L) = A \sin kL = 0$$

**FIGURE 15.20** (a) The potential energy for a particle in a box of length $L$, with the first three energy levels marked. (b) Wave functions, showing the ground state $\psi_1$ and the first two excited states. The more numerous the nodes, the higher the energy of the state. (c) The squares of the wave functions from part b, equal to the probability density for finding the particle at a particular point in the box.

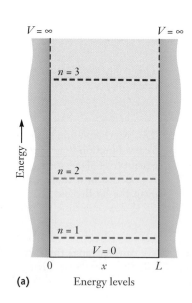

(a)   Energy levels

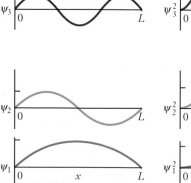
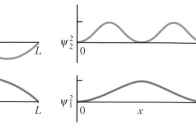

(b)   Wave functions

(c)   Probabilities

This can be true only if

$$kL = n\pi \qquad n = 1, 2, 3, \ldots$$

because $\sin(n\pi) = 0$, so that

$$\psi(x) = A \sin\left(\frac{n\pi}{L}x\right) \qquad n = 1, 2, 3, \ldots$$

For the total probability of finding the particle *somewhere* in the box to be unity,

$$\int_0^L \psi^2(x) \, dx = 1$$

or

$$A^2 \int_0^L \sin^2\left(\frac{n\pi}{L}x\right) dx = 1$$

The integral has the value $L/2$, and so

$$A^2\left(\frac{L}{2}\right) = 1$$

$$A = \sqrt{\frac{2}{L}}$$

The wave function is then

$$\psi_n(x) = \sqrt{\frac{2}{L}} \sin\left(\frac{n\pi}{L}x\right) \qquad \textbf{[15.17]}$$

To find the energy $E_n$ for a particle with wave function $\psi_n$, calculate the second derivative:

$$\frac{d^2\psi_n(x)}{dx^2} = \frac{d^2}{dx^2}\left[\sqrt{\frac{2}{L}} \sin\left(\frac{n\pi}{L}x\right)\right]$$

$$= -\left(\frac{n\pi}{L}\right)^2\left[\sqrt{\frac{2}{L}} \sin\left(\frac{n\pi}{L}x\right)\right]$$

$$= -\left(\frac{n\pi}{L}\right)^2 \psi_n(x)$$

This must be equal to $-\dfrac{8\pi^2 m E_n}{h^2}\,\psi_n(x)$. Setting the coefficients equal gives

$$\frac{8\pi^2 m E_n}{h^2} = \frac{n^2\pi^2}{L^2}$$

$$E_n = \frac{h^2 n^2}{8mL^2} \qquad \textbf{[15.18]}$$

This solution of the Schrödinger equation demonstrates that the particle in the box is quantized. An energy $E_n$ and a wave function $\psi_n(x)$ correspond to each quantum number $n$ (a positive integer). These are the only allowed stationary states of the particle in a box. The spectrum of the particle in a box consists of a series of frequencies given by

$$h\nu = |E_n - E_n'|$$

where $n$ and $n'$ are positive integers and the $E_n$ are given by Equation 15.18. The wave functions $\psi_n(x)$ plotted in Figure 15.20b for the quantum states $n$ are exactly the standing waves of Figure

15.15. The guitar string and the particle in the box are physically analogous. Both systems are quantized by imposing the condition that the amplitude of the wave function $\psi$ be 0 at each end of the string or at each wall of the box.

From the variation of the square of the wave function of the particle with position (Fig. 15.20c) following Born's interpretation, we infer that in its ground state the particle has maximum probability of being in the middle of the box and lower probability of being near either wall. In its first excited state ($n = 2$), the probability is a maximum when it is near $L/4$ and $3L/4$ and zero near 0, $L/2$, and $L$. Wherever there is a node in the wave function, the probability is small that the particle will be near there. The wave function $\psi_n$ has $n - 1$ nodes, and the more numerous the nodes, the higher the energy of the particle.

Can a particle in a box have zero energy? If $n$ could have the value 0, the equation $E_n = h^2 n^2/8mL^2$ would give $E_0 = 0$. But this is not possible. Setting $n$ equal to 0 in $\psi_n(x) = A \sin(n\pi x/L)$ makes $\psi_0(x)$ zero everywhere. In this case, $\psi_0^2(x)$ would be zero everywhere in the box, and therefore there could be no particle in the box. The same conclusion comes from the uncertainty principle. If the energy of the lowest state could be 0, the momentum of the particle would also be 0. Moreover, the uncertainty in the particle momentum, $\Delta p_x$, would be 0 and $\Delta x$ would be infinite, contradicting our condition that the particle be confined to a box of length $L$. Even at the absolute zero of temperature, where classical kinetic theory would suggest that all motion ceases, a finite quantity of energy remains for a bound system. It is required by the uncertainty principle and is called the **zero-point energy**.

The Schrödinger equation is readily generalized to a particle in three dimensions. The wave function $\psi$ and potential $V$ now depend on three coordinates, $(x, y, z)$, and derivatives with respect to each coordinate appear. The allowed energies for a particle in a three-dimensional rectangular box are

$$E_{n_x n_y n_z} = \frac{h^2}{8m}\left[\frac{n_x^2}{L_x^2} + \frac{n_y^2}{L_y^2} + \frac{n_x^2}{L_z^2}\right] \quad \begin{cases} n_x = 1, 2, 3, \ldots \\ n_y = 1, 2, 3, \ldots \\ n_z = 1, 2, 3, \ldots \end{cases}$$

where $L_x$, $L_y$, and $L_z$ are the side lengths of the box. Here the state is designated by a set of three quantum numbers, $(n_x, n_y, n_z)$. Each independently ranges over the positive integers.

### EXAMPLE 15.8

Consider the following two systems: **(a)** an electron in a one-dimensional box of length 1.0 Å and **(b)** a helium atom in a cube 30 cm on an edge. Calculate the energy difference between ground state and first excited state, expressing your answer in kJ mol$^{-1}$.

### Solution

**(a)** For a one-dimensional box,

$$E_{\text{ground state}} = \frac{h^2}{8mL^2}(1^2)$$

$$E_{\text{first excited}} = \frac{h^2}{8mL^2}(2^2)$$

Then, for one electron in the box,

$$\Delta E = \frac{3h^2}{8mL^2}$$

$$= \frac{3(6.626 \times 10^{-24}\ \text{J s})^2}{8(9.11 \times 10^{-31}\ \text{kg})(1.0 \times 10^{-10}\ \text{m})^2}$$

$$= 1.8 \times 10^{-17}\ \text{J}$$

Multiplying this by $10^{-3}\ \text{kJ J}^{-1}$ and by $N_0 = 6.022 \times 10^{23}\ \text{mol}^{-1}$ gives

$$\Delta E = 11{,}000\ \text{kJ mol}^{-1}$$

**(b)** For a three-dimensional cube, $L_x = L_y = L_z = L$, and

$$E_{\text{ground state}} = \frac{h^2}{8mL^2}\ (1^2 + 1^2 + 1^2)$$

$$E_{\text{first excited}} = \frac{h^2}{8mL^2}\ (2^2 + 1^2 + 1^2)$$

In this case, the three states (2, 1, 1), (1, 2, 1), and (1, 1, 2) have the same energy.

$$\Delta E = \frac{3h^2}{8mL^2}$$

$$= \frac{3(6.626 \times 10^{-34}\ \text{J s})^2}{8(4.0\ \text{u} \times 1.66 \times 10^{-27}\ \text{kg u}^{-1})(0.30\ \text{m})^2}$$

$$= 2.8 \times 10^{-40}\ \text{J}$$

$$= 1.7 \times 10^{-19}\ \text{kJ mol}^{-1}$$

The energy levels are so close together in the latter case (due to the much greater dimensions of the box) that they appear continuous, and quantum effects play no role. The properties are almost those of a classical particle.

**Related Problems: 35, 36**

---

**15.7**

## THE HYDROGEN ATOM

The hydrogen atom is the simplest example of a one-electron atom or ion; other examples are He$^+$, Li$^{2+}$, and other ions in which all but one electron have been stripped off. They differ in the charge $Ze$ on the nucleus and therefore in the attractive force felt by the electron.

The potential energy for the one-electron atom, discussed in Section 15.4, depends only on the distance of the electron from the nucleus and is independent of angular orientation (Fig. 15.13). Solution of the Schrödinger equation is most easily carried out in coordinates that reflect the natural symmetry of the potential energy function. For an isolated one-electron atom or ion, spherical coordinates are more appropriate than the more familiar Cartesian coordinates. Spherical coordinates are defined in Figure 15.21: $r$ is the distance of the electron at P from the nucleus at O, and the angles $\theta$ and $\phi$ are closely related to those used to locate points on the surface of the globe; $\theta$ is related to the latitude and $\phi$ to the longitude.

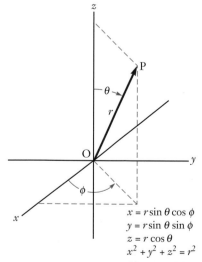

$$x = r\sin\theta\cos\phi$$
$$y = r\sin\theta\sin\phi$$
$$z = r\cos\theta$$
$$x^2 + y^2 + z^2 = r^2$$

**FIGURE 15.21** The relationship of spherical coordinates $(r, \theta, \phi)$ to Cartesian coordinates $(x, y, z)$. Here $\theta$ is the angle with respect to the Cartesian $z$ axis, and $\phi$ is the azimuthal angle (the angle between the $x$ axis and the projection onto the $xy$ plane of the arrow from the origin to P).

### Energy Levels

Solutions of the Schrödinger equation for the one-electron atom exist only for particular values of the energy

$$E = E_n = -\frac{Z^2 e^4 m_e}{8\epsilon_0^2 n^2 h^2} \qquad n = 1, 2, 3, \dots \qquad \textbf{[15.19a]}$$

In energy units of rydbergs, this becomes [see Equation (15.8b)]

$$E_n = -\frac{Z^2}{n^2} \qquad (\text{rydberg}) \qquad n = 1, 2, 3, \dots \qquad \textbf{[15.19b]}$$

The integer $n$, called the **principal quantum number**, indexes the individual energy levels. These are identical to the energy levels predicted by Bohr's theory. Here, however, quantization comes about naturally from the solution of the Schrödinger equation rather than through an arbitrary assumption about the angular momentum.

The energy of a one-electron atom depends only on the principal quantum number $n$, because the potential energy depends only on the radial distance. Schrödinger's equation also quantizes $L^2$, the square magnitude of the angular momentum as well as $L_z$, the projection of the angular momentum along the z-axis. (A review of elementary aspects of angular momentum in Appendix B provides useful background for the present discussion.) This requires two additional quantum numbers. The **angular momentum quantum number** $\ell$ may take on any integral value from 0 to $n - 1$, and the **angular momentum projection** quantum number $m$ may take on any integral value from $-\ell$ to $\ell$. $m$ is referred to as the **magnetic quantum number** because its value governs the behavior of the atom in an external magnetic field. The allowed values of angular momentum L and its z projection satisfy

$$L^2 = \ell(\ell + 1) \frac{h^2}{4\pi^2} \qquad \ell = 0, 1, ..., n - 1$$

$$L_z = m \frac{h}{2\pi} \qquad m = -\ell, -\ell + 1, ..., 0, ..., \ell - 1, \ell$$

For $n = 1$ (the ground state), the only allowed quantum numbers are ($\ell = 0, m = 0$). For $n = 2$, there are $n^2 = 4$ allowed sets of quantum numbers, which are

$$(\ell = 0, m = 0), \quad (\ell = 1, m = 1), \quad (\ell = 1, m = 0), \quad (\ell = 1, m = -1)$$

The restrictions on $\ell$ and $m$ give rise to $n^2$ sets of quantum numbers for every value of $n$. Each set $(n, \ell, m)$ identifies a specific **quantum state** of the atom in which the electron has energy equal to $E_n$, angular momentum equal to $\sqrt{\ell(\ell + 1)}h/2\pi$, and z-projection of angular momentum equal to $mh/2\pi$. When $n > 1$, a total of $n^2$ specific quantum states correspond to the single energy level $E_n$; consequently, this set of states is said to be **degenerate.**

It is conventional to label specific states by replacing the angular momentum quantum number with a letter; we signify $\ell = 0$ with $s$, $\ell = 1$ with $p$, $\ell = 2$ with $d$, $\ell = 3$ with $f$, $\ell = 4$ with $g$, and on through the alphabet. A state with $n = 1$ and $\ell = 0$ is thus called a 1s state, one with $n = 3$ and $\ell = 1$ is a 3p state, one with $n = 4$ and $\ell = 3$ is a 4f state, and so forth. The letters $s$, $p$, $d$, and $f$ come from early (pre-quantum mechanics) spectroscopy, in which certain spectral lines were referred to as sharp, principal, diffuse, and fundamental. These terms are not used in modern spectroscopy, but historical convention is maintained in the assignment of letters to the quantum number $\ell$. Figure 15.22 summarizes the allowed combinations of quantum numbers.

The energy levels, including the degeneracy due to $m$ and with states labeled by the spectroscopic notation, are conventionally displayed on a diagram as shown in Figure 15.22.

## Wave Functions

For each quantum state ($n \ell m$), solution of Schrödinger's equation provides a wave function written in the form

$$\psi_{n\ell m}(r, \theta, \phi) = R_{n\ell}(r) Y_{\ell m}(\theta, \phi) \qquad \text{[15.20]}$$

in which the total wave function is the product of a radial part $R_{n\ell}(r)$ and an angular part $Y_{\ell m}(\theta, \phi)$. This product form is a consequence of the spherically symmetric

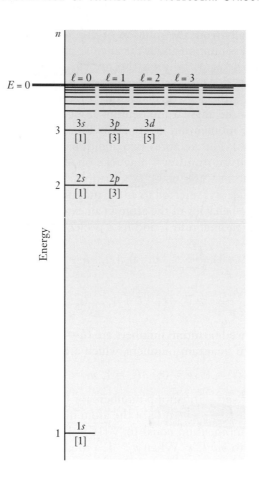

**FIGURE 15.22** The energy level diagram of the H atom predicted by quantum mechanics is arranged to show the degeneracy produced by $\ell$.

potential energy function and enables separate examination of the angular and radial contributions to the wavefunction. The functions $Y_{\ell m}$ are called **spherical harmonics.** They appear in many physical problems with spherical symmetry and were already well known before the advent of Schrödinger's equation. The angular motions of the electron described by $\theta$ and $\phi$ influence the shape of the wave function through the angular factor $Y_{\ell m}$, even though they do not influence the energy.

The wave function itself is not measured directly. It is to be viewed as an intermediate step toward calculating the physically significant quantity $\psi^2$, which gives the probability density for locating the electron at a particular point in the atom. More precisely,

$$(\psi_{n\ell m})^2 d\tau = [R_{n\ell}(r)]^2[Y_{\ell m}(\theta, \phi)]^2 d\tau \qquad \text{[15.21]}$$

gives the probability of locating the electron within a small three-dimensional volume $d\tau$ located at the position $(r, \theta, \phi)$ when it is known that the atom is in the state $(n, \ell, m)$. Specific examples will be presented in succeeding paragraphs.

A wave function $\psi_{n\ell m}(r, \theta, \phi)$ for a one-electron atom in the state $(n, \ell, m)$ is called an **orbital.** This term recalls the circular orbits of the Bohr atom, but there is no real resemblance. An orbital is *not* the trajectory traced by an individual electron. When the one-electron atom is in state $(n, \ell, m)$ it is conventional to say the electron is "in a $(n, \ell, m)$ orbital." This statement is merely another way of stating that the probability density of finding the electron at a particular point is given by $\psi_{n\ell m}^2$. The orbitals are labeled by the spectroscopic notation previously introduced.

## EXAMPLE 15.9

Give the names of all the orbitals with $n = 4$, and state how many $m$ values correspond to each type of orbital.

### Solution

The quantum number $\ell$ may range from 0 to $n - 1$, so its allowed values in this case are 0, 1, 2, and 3. The labels for the groups of orbitals are then

$$\ell = 0 \quad 4s \qquad \qquad \ell = 2 \quad 4d$$

$$\ell = 1 \quad 4p \qquad \qquad \ell = 3 \quad 4f$$

The quantum number $m$ ranges from $-\ell$ to $+\ell$, and so the number of $m$ values is $2\ell + 1$. This gives one $4s$ orbital, three $4p$ orbitals, five $4d$ orbitals, and seven $4f$ orbitals for a total of $16 = 4^2 = n^2$ orbitals with $n = 4$. They all have the same energy, but they differ in shape.

**Related Problems: 39, 40**

## Sizes and Shapes of Orbitals

The shapes and sizes of the hydrogen atom orbitals are important in chemistry because they provide the foundations for the quantal description of chemical bonding and the resulting shapes of molecules. Shapes and sizes are revealed by graphical analysis of the wave functions, of which the first few are given in Table 15.2. Note that the radial functions are written in terms of the dimensionless variable $\sigma$, which is the ratio of $Zr$ to $a_o$. For $Z = 1$, this is the radius of the first Bohr orbit of the H atom.

**s *orbitals*** Let us begin with **s orbitals,** corresponding to $\psi_{n\ell m}$ with $\ell = 0$ (and therefore $m = 0$, as well). For all $s$ orbitals, the angular part $Y$ is a constant (Table 15.2). Because $\psi$ depends on neither $\theta$ nor $\phi$, all $s$ orbitals are spherically symmetric about the nucleus. This means that the amplitude of an $s$ orbital (and therefore also the probability of finding the electron near some point in space) depends only on its distance $r$ from the nucleus and not on its direction in space. There are several ways to visualize the radial variation of the $ns$ orbitals with $n = 1, 2, 3, \ldots$. One way (Fig. 15.23a) is to make a "three-dimensional" picture in which the shading is heaviest where $\psi^2$ is largest and lighter where $\psi^2$ is smaller. A second way (Fig. 15.23b) is to plot the wave functions themselves: $\psi_{n00}(r) \propto R_{n0}(r)$. A third way (Fig. 15.23c) is to plot the **radial probability distribution** $r^2 R_{n0}^2(r)$. This is the probability density of finding the electron at any point in space at a distance $r$ from the nucleus for all angles $\theta$ and $\phi$. More precisely, the product $r^2 R_{n0}^2(r)dr$ gives the probability of finding the electron anywhere within a thin spherical shell of thickness $dr$, located at distance $r$ from the nucleus. The factor $r^2$ in front accounts for the increasing volume of spherical shells at greater distances from the nucleus. The radial probability distribution is small near the nucleus, where the shell volumes (proportional to $r^2$) are small, and has its maximum at the distance where the electron is most likely to be found.

What is meant by the *size* of an orbital? Strictly speaking, the wave function of an electron in an atom stretches out to infinity, and so an atom has no boundary

**TABLE 15.2**

### Angular and Radial Parts of Wave Functions for One-Electron Atoms

| Angular Part $Y(\theta, \phi)$ | Radial Part $R_{n\ell}(r)$ |
|---|---|
| | $R_{1s} = 2\left(\dfrac{Z}{a_0}\right)^{3/2} \exp(-\sigma)$ |
| $\ell = 0 \left\{ Y_s = \left(\dfrac{1}{4\pi}\right)^{1/2} \right.$ | $R_{2s} = \dfrac{1}{2\sqrt{2}}\left(\dfrac{Z}{a_0}\right)^{3/2} (2 - \sigma)\exp(-\sigma/2)$ |
| | $R_{3s} = \dfrac{2}{81\sqrt{3}}\left(\dfrac{Z}{a_0}\right)^{3/2}(27 - 18\sigma + 2\sigma^2)\exp(-\sigma/3)$ |
| $\ell = 1 \begin{cases} Y_{px} = \left(\dfrac{3}{4\pi}\right)^{1/2}\sin\theta\cos\phi \\ Y_{py} = \left(\dfrac{3}{4\pi}\right)^{1/2}\sin\theta\sin\phi \\ Y_{pz} = \left(\dfrac{3}{4\pi}\right)^{1/2}\cos\theta \end{cases}$ | $R_{2p} = \dfrac{1}{2\sqrt{6}}\left(\dfrac{Z}{a_0}\right)^{3/2}\sigma\exp(-\sigma/2)$ <br><br> $R_{3p} = \dfrac{4}{81\sqrt{6}}\left(\dfrac{Z}{a_0}\right)^{3/2}(6\sigma - \sigma^2)\exp(-\sigma/3)$ |
| $\ell = 2 \begin{cases} Y_{d_{z^2}} = \left(\dfrac{5}{16\pi}\right)^{1/2}(3\cos^2\theta - 1) \\ Y_{d_{xz}} = \left(\dfrac{15}{4\pi}\right)^{1/2}\sin\theta\cos\theta\cos\phi \\ Y_{d_{yz}} = \left(\dfrac{15}{4\pi}\right)^{1/2}\sin\theta\cos\theta\sin\phi \\ Y_{d_{xy}} = \left(\dfrac{15}{16\pi}\right)^{1/2}\sin^2\theta\sin 2\phi \\ Y_{d_{x^2-y^2}} = \left(\dfrac{15}{16\pi}\right)^{1/2}\sin^2\theta\cos 2\phi \end{cases}$ | $R_{3d} = \dfrac{4}{81\sqrt{30}}\left(\dfrac{Z}{a_0}\right)^{3/2}\sigma^2\exp(-\sigma/3)$ |

$$\sigma = \frac{Zr}{a_0} \qquad a_0 = \frac{\epsilon_0 h^2}{\pi e^2 m_e} = 0.529 \times 10^{-10}\ \text{m}$$

Skell

unless we set some arbitrary limit upon the extension of its wave function. We might define the size of an atom as that of the contour at which $\psi^2$ has fallen off to some particular numerical value, or as the extent of a "balloon skin" inside which the electron has a definite probability (for example, 90% or 99%) of being found. We have adopted the latter practice in pictures of orbitals for electrons. Figure 15.23a shows spheres that enclose 90% of the electron probability for the 1s, 2s, and 3s orbitals for hydrogen. This figure shows that the size of an orbital increases with increasing quantum number $n$. A 3s orbital is larger than a 2s orbital, which in turn is larger than a 1s orbital. This is the quantum mechanical analog of the increase in radius of the Bohr orbits with $n$.

Another measure of the size of an orbital is the most probable distance of the electron from the nucleus in that orbital. Figure 15.23c shows that the electron is most likely to be found farther from the nucleus in $ns$ orbitals for larger $n$. Nonetheless, there is a small but finite probability for finding the electron near the nucleus in both 2s and 3s orbitals.

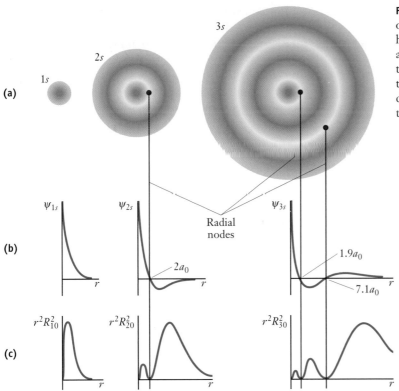

**FIGURE 15.23** Three representations of hydrogen $s$ orbitals. (a) An electron-density representation of a hydrogen atom in its $1s$, $2s$, and $3s$ states. The spheres are cut off at a radius that encloses 90% of the total electron density. (b) The wave functions plotted against distance from the nucleus, $r$. (c) The radial probability distribution, equal to the square of the radial wave function multiplied by $r^2$.

Finally, note that an $ns$ orbital has $n - 1$ **radial nodes** (a radial node is a sphere about the nucleus on which $\psi$ and $\psi^2$ are 0). The more numerous the nodes, the higher the energy of the corresponding state.

**p orbitals**   It is more difficult to draw pictures of orbitals with angular quantum numbers $\ell$ different from 0, because these are not spherically symmetric. In addition, when two or more wave functions have the same energy, linear combinations of them also satisfy the Schrödinger equation. This is the case for the three wave functions with $\ell = 1$ (the **p orbitals**), for which the allowed $m$ values are $-1$, 0, and $+1$. The usual convention is to take linear combinations that give three orbitals with the same shapes but different orientations, as shown (for $n = 2$) in Figure 15.24. The angular wave function for the $p_z$ orbital is proportional to $\cos \theta$ (or, in Cartesian coordinates, to $z$), and so such an orbital has maximum amplitude along the $z$ axis (where $\theta = 0$ or $\pi$) and a node in the $xy$ plane (where $\theta = \pi/2$, so that $\cos \theta = 0$). Therefore, it points along the $z$ axis, with positive phase on the side of the $xy$ plane where the $z$-axis is positive, and negative phase on the side where the $z$-axis is negative. An electron in a $p_z$ orbital has the highest probability of being found along the $z$ axis and has zero probability of being found in the $xy$ plane. This plane is a nodal plane or, more generally, an **angular node** across which the wave function changes sign. The $p_x$ orbital has the same shape but points along the $x$ axis, and the $p_y$ orbital points along the $y$ axis.

The radial parts of the $np$ wave functions (represented by $R_{n1}$) are illustrated in Figure 15.25. The $p$ orbitals, like the $s$ orbitals, may also have radial nodes, in which the electron density vanishes at certain distances from the nucleus irrespective of direction. From Figure 15.25, the $R_{2p}$ wave function possesses no nodes, the $R_{3p}$

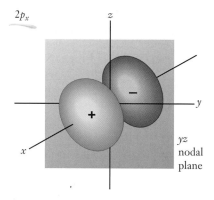

$2p_x$

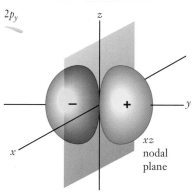

$2p_y$

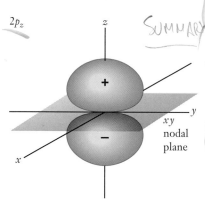

$2p_z$

**FIGURE 15.24**  The distribution of electron density in the three $2p$ orbitals, with phases and nodal planes indicated.

function has one node; the $R_{4p}$ function has two nodes. In general, the $R_{n1}$ radial wave functions have $n - 2$ nodes. Because the angular part of the $np$ wave function always has a nodal plane, the total wave function has $n - 1$ nodes ($n - 2$ radial and 1 angular), which is the same number as an $s$ orbital with the same principal quantum number. The $R_{21}(r)$ function (that is, $R_{2p}$) in Table 15.2 contains the factor $\sigma$, which is proportional to $r$ ($\sigma = Zr/a_0$) and causes it to vanish at the nucleus. This is true of all the radial wave functions except the $ns$ functions, and it means that the probability density is zero for the electron to be at the nucleus for all wave functions with $\ell > 0$ ($p, d, f, \ldots$).

**d *orbitals***  The spatial distribution of the five $d$ orbitals is somewhat more complicated (Fig. 15.26). The conventionally chosen linear combinations of these orbitals give four orbitals with the same shape but different orientations with respect to the Cartesian axes: $d_{xy}$, $d_{yz}$, $d_{xz}$, and $d_{x^2-y^2}$. A $d_{xy}$ orbital, for example, has four lobes, two with positive phase and two with negative phase; the maximum amplitude is at 45° to the $x$ and $y$ axes. The $d_{x^2-y^2}$ orbital has maximum amplitude along the $x$ and $y$ axes. The fifth orbital, $d_{z^2}$, has a different shape, with maximum amplitude along the $z$ axis and a little "doughnut" in the $xy$ plane. Each $d$ orbital has two angular nodes (for example, the $d_{xy}$ orbital has the $xz$ and $yz$ planes as its nodal surfaces). The radial functions $R_{n2}(r)$ have $n - 3$ radial nodes, giving once again $n - 1$ total nodes.

The wave functions for $f$ orbitals and orbitals of higher $\ell$ can be calculated, but they play a smaller role in chemistry than the $s$, $p$, and $d$ orbitals. We therefore stop at this point and summarize the important features of orbital shapes and sizes.

1. For a given value of $\ell$, an increase in $n$ leads to an increase in the average distance of the electron from the nucleus and therefore in the size of the orbital (Figs. 15.23 and 15.25).
2. An orbital with quantum numbers $n$ and $\ell$ has $\ell$ angular nodes and $n - \ell - 1$ radial nodes, giving a total of $n - 1$ nodes. For a one-electron atom or ion, the energy depends only on the number of nodes—that is, on $n$ but not on $\ell$ or $m$.
3. As $r$ approaches 0, $\psi(r, \theta, \phi)$ vanishes for all orbitals except $s$ orbitals; thus, only an electron in an $s$ orbital can "penetrate to the nucleus," that is, have a finite probability of being found right at the nucleus.

These general statements will prove important in the next section in determining the electronic structure of many-electron atoms even though they are deduced from the one-electron case.

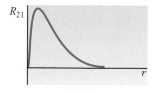

$R_{21}$

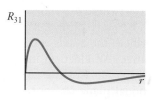

$R_{31}$

$r^2 R_{21}^2$

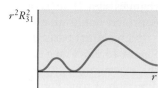

$r^2 R_{31}^2$

**FIGURE 15.25**  Radial wave functions $R_{n1}$ for $np$ orbitals and the corresponding radial probability distributions $r^2 R_{n1}^2$.

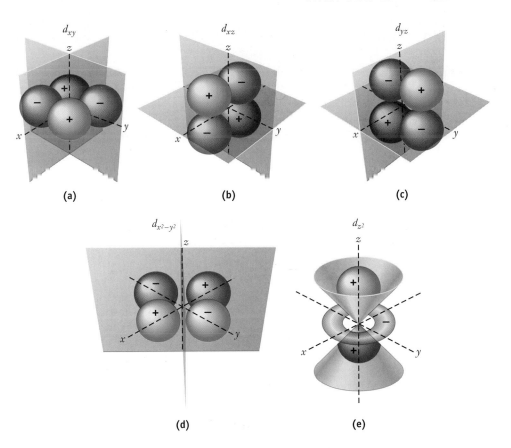

**FIGURE 15.26**   The shapes of the five 3$d$ orbitals, with phases and nodal surfaces indicated.

These features are especially well displayed by a quantitative plot showing $s$, $p$, and $d$ orbitals all on the same scale (Fig. 15.27). The best quantitative measure of the size of an orbital is $\bar{r}_{n\ell}$, the average value of the distance of the electron from the nucleus in that orbital. Quantum mechanics calculates $\bar{r}_{n\ell}$ as

$$\bar{r}_{n\ell} = \frac{n^2 a_0}{Z}\left\{1 + \frac{1}{2}\left[1 - \frac{\ell(\ell+1)}{n^2}\right]\right\} \qquad \textbf{[15.22]}$$

The leading term of this expression is the radius of the $n$th Bohr orbit (Equation 15.7). In Figure 15.27, the small arrow on each curve locates the value of $\bar{r}_{n\ell}$ in that orbital.

## EXAMPLE 15.10

Compare the 3$p$ and 4$d$ orbitals of a hydrogen atom with respect to **(a)** number of radial and angular nodes and **(b)** energy of the corresponding atom.

### Solution
**(a)** The 3$p$ orbital has a total of $n - 1 = 3 - 1 = 2$ nodes. Of these, one is angular ($\ell = 1$) and one is radial. The 4$d$ orbital has $4 - 1 = 3$ nodes. Of these, two are angular ($\ell = 2$) and one is radial.
**(b)** The energy of a one-electron atom depends only on $n$. The energy of an atom with an electron in a 4$d$ orbital is higher than that of an atom with an electron in a 3$p$ orbital, because $4 > 3$.

**Related Problems: 41, 42**

**FIGURE 15.27** Radial probability densities for hydrogen orbitals with $n = 1, 2, 3$. The small arrow on each curve locates the value of $\bar{r}_{n\ell}$ in that orbital.

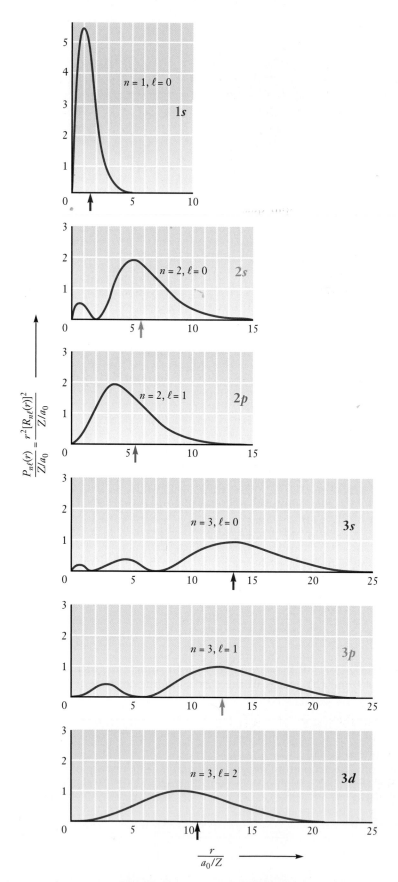

## Electron Spin

Our discussion of one-electron atoms has so far omitted one important point. If a beam of hydrogen atoms in their ground state (with $n = 1$, $\ell = 0$, $m = 0$) is sent through a spatially varying magnetic field, it splits into two beams, each containing half of the atoms (Fig. 15.28). The result of this experiment is explained by introducing a *fourth* quantum number, $m_s$, which can take on two values, conventionally chosen to be $+\frac{1}{2}$ and $-\frac{1}{2}$. Because the behavior in the magnetic field is what would be predicted by classical theory if the electrons were balls of charge spinning about their own axes, the fourth quantum number is referred to as the **spin quantum number.** When $m_s = +\frac{1}{2}$ the electron spin is said to be "up," and when $m_s = -\frac{1}{2}$ the spin is "down." We should not take the classical image of a spinning ball of charge too literally, however. The spin quantum number comes from relativistic effects that are not included in the Schrödinger equation. For most practical purposes in chemistry, it is sufficient simply to solve the ordinary Schrödinger equation and then associate with each electron a spin quantum number $m_s = \pm\frac{1}{2}$, which does not affect the spatial probability density of the electron. The spin does affect the number of different quantum states with energy $E_n$, however, doubling it from $n^2$ to $2n^2$. This fact will assume considerable importance when we consider many-electron atoms in the next section.

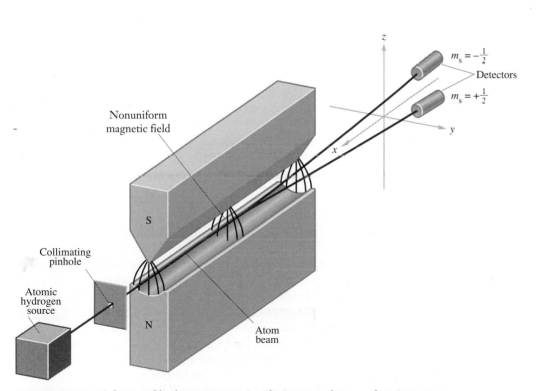

**FIGURE 15.28**   A beam of hydrogen atoms is split into two beams when it traverses a nonuniform magnetic field. Atoms with spin quantum number $m_s = +\frac{1}{2}$ follow one trajectory, and those with $m_s = -\frac{1}{2}$ follow another.

---

**15.8**

---

# MANY-ELECTRON ATOMS AND THE PERIODIC TABLE

As we move from one-electron to many-electron atoms, both the Schrödinger equation and its solutions become increasingly complicated. The simplest many-electron atom, helium, has two electrons and a nuclear charge of $+2e$. The positions of the two electrons in a helium atom can be described classically by two sets of Cartesian coordinates, $(x_1, y_1, z_1)$ and $(x_2, y_2, z_2)$. The wave function $\psi$ depends on all six of these variables: $\psi = \psi(x_1, y_1, z_1, x_2, y_2, z_2)$. Its square, $\psi^2(x_1, y_1, z_1, x_2, y_2, z_2)$, is the probability density of finding the first electron at point $(x_1, y_1, z_1)$ and, simultaneously, the second at $(x_2, y_2, z_2)$. The Schrödinger equation is now more complicated, and an explicit solution for helium is not possible. Nevertheless, modern computers have enabled us to solve this equation numerically to very high accuracy, and the predicted properties of helium are in excellent agreement with experiment.

Although these numerical calculations demonstrate conclusively the usefulness of the Schrödinger equation for predicting atomic properties, they suffer from two defects: first, they are somewhat difficult to interpret physically, and second, they become increasingly difficult to solve, even numerically, as the number of electrons increases. As a result, approximate approaches to the many-electron Schrödinger equation have been developed.

The **self-consistent field (SCF) orbital approximation method** developed by Douglas Hartree is especially well suited for applications in chemistry. Hartree's method generates a set of approximate one-electron orbitals and associated energy levels reminiscent of those for the H atom. The electronic structure of an atom with atomic number $Z$ is then "built up" by placing $Z$ electrons into these orbitals in accordance with certain rules to be described later. This approximate description rationalizes periodic trends in atomic properties and serves as a starting point for qualitative descriptions of chemical bond formation. The present section introduces Hartree's method and uses it to describe the electron arrangements in many-electron atoms and to survey periodic trends in atomic properties. The discussion proceeds through several steps.

## Hartree Orbitals: Shell Model of the Atom

For any atom, Hartree's method begins with the exact Schrödinger equation in which each electron is attracted to the nucleus and repelled by all the other electrons in accordance with the Coulomb potential. Two simplifying assumptions are made immediately: (1) each electron moves in an *effective field* due to all the other electrons, to be obtained by averaging over all the positions of the other electrons, and (2) the effective field is spherically symmetric; that is, it has no angular dependence. Under these two assumptions, each electron is then described by a one-electron orbital similar to those of the H atom. These two assumptions in effect convert the exact Schrödinger equation for the atom into a set of simultaneous equations for the unknown effective field and the unknown one-electron orbitals. These equations must be solved by iteration until a self-consistent solution is obtained. (In spirit, this approach is identical to the solution of complicated algebraic equations by the method of iteration described in Appendix C.) Like any other method for solving Schrödinger's equation, Hartree's method produces two principal results: energy levels and orbitals.

These Hartree orbitals resemble the atomic orbitals of hydrogen in many ways. Their angular dependence is identical, and so it is straightforward to associate quantum numbers $\ell$ and $m$ with each atomic orbital. The radial dependence of the orbitals in many-electron atoms differs from that of one-electron orbitals because the effective field differs from the Coulomb potential, but a principal quantum number $n$ can still be defined. The lowest energy orbital is a $1s$ orbital and has no radial nodes, the next lowest $s$ orbital is a $2s$ orbital and has one radial node, and so forth. Each electron in an atom has associated with it a set of four quantum numbers ($n$, $\ell$, $m$, $m_s$) that describes its spatial distribution and spin state. The allowed quantum numbers follow the same pattern as those for the hydrogen atom.

The spatial properties of Hartree orbitals are best conveyed through a specific example. It will be seen in the discussion that follows that the ground state of the argon atom has electrons in the $1s$, $2s$, $2p$, $3s$, and $3p$ Hartree orbitals. Figure 15.29 shows the radial probability distributions for these five occupied states calculated by Hartree's method. The distribution shown for $2p$ is the sum of the distributions for the $2p_x$, $2p_y$, and $2p_z$ orbitals; similar remarks hold for the $3p$ distribution. Comparing Figure 15.29 with Figure 15.27 reveals that each Hartree orbital for argon is "smaller" than the corresponding orbital for hydrogen in the sense that the region of maximum probability is closer to the nucleus. This difference occurs because the argon nucleus ($Z = 18$) has a much stronger attraction for electrons than does the hydrogen nucleus ($Z = 1$). A semiquantitative relation between orbital size and $Z$ will be presented in the next subsection.

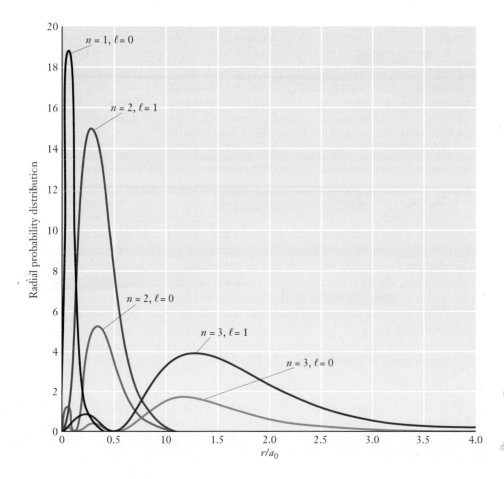

**FIGURE 15.29** The shell structure of an argon atom calculated by Hartree's method. The fact that the radial probability for all orbitals with the same value of $n$ have maxima very near one another suggests that the electrons can be viewed in "shells" made up of these orbitals.

The fact that Hartree orbitals with the same value of $n$ are large in the same narrow regions of space, despite their different values of $\ell$, has interesting consequences. If the radial distribution functions in Figure 15.29 are all added together, the result is the total radial probability distribution $P(r)$, which gives the probability of finding an electron in a thin spherical shell of radius $r$, regardless of the orbital to which the electron belongs. A plot of $P(r)$ on the same scale as Figure 15.29 would show three peaks at $r$ values approximately 0.1, 0.3, and 1.2 in units of $a_0$. The total electron density of the argon atom is thus concentrated in three concentric shells, where a **shell** is defined as all electrons with the same value of $n$. Each shell has a radius determined by the principal quantum number $n$. The shell model summarizes the coarse features of the electron density of an atom by averaging over all those local details not described by the principal quantum number $n$. Within each shell, a more detailed picture is provided by the **subshells,** which are defined as the set of orbitals with the same values of both $n$ and $\ell$.

The subshells, illustrated in Figure 15.29, determine the structure of the periodic table and the formation of chemical bonds. In preparation for discussing these connections it is necessary to describe the energy values for Hartree orbitals.

## Shielding Effects: Energy Sequence of Hartree Orbitals

The energy level diagrams calculated for many-electron atoms by Hartree's method resemble the diagram for the hydrogen atom (see Fig. 15.22) but differ in two important respects. First, the degeneracy of the $p$, $d$, and $f$ orbitals is removed. Because the effective field in Hartree's method is different from the Coulomb field in the hydrogen atom, the energy levels of Hartree orbitals depend on both $n$ and $\ell$. Second, the energy values, especially at smaller values of $n$, are distinctly shifted from the values of corresponding hydrogen orbitals because of the stronger attractive force exerted by nuclei with $Z > 1$. These two effects can be explained qualitatively as follows.

In a very useful simplified approximation to the Hartree description, the effective field experienced by each electron in a particular shell is given by

$$V_n^{eff}(r) = -\frac{Z_{eff}(n)e^2}{r} \qquad \textbf{[15.23]}$$

where $Z_{eff}(n)$ is the **effective nuclear charge** in that shell. To understand the origin and magnitude of $Z_{eff}(n)$, consider a particular electron $e_1$ in an atom. Inner electrons near the nucleus *shield* $e_1$ from the full charge $Z$ of the nucleus by effectively canceling some of the positive nuclear charge. $Z_{eff}(n)$ is the net reduced nuclear charge felt by a particular electron, due to the presence of the other electrons. For a neutral atom, $Z_{eff}(n)$ can range from a maximum value of $Z$ near the nucleus (no screening) to a minimum value of 1 far from the nucleus (complete screening by the other $Z-1$ electrons). Detailed Hartree calculations for argon show that $Z_{eff}(1) \sim 16$, $Z_{eff}(2) \sim 8$, and $Z_{eff}(3) \sim 2.5$. The effect of shielding on the energy and radius of a Hartree orbital is easily estimated in this simplified picture by using the hydrogen atom equations with $Z$ replaced by $Z_{eff}(n)$. Thus

$$E_n \approx -\frac{[Z_{eff}(n)]^2}{n^2} \qquad \text{(rydbergs)} \qquad \textbf{[15.24]}$$

and

$$\bar{r}_{n\ell} \approx \frac{n^2 a_0}{Z_{eff}(n)}\left\{1 + \frac{1}{2}\left[1 - \frac{\ell(\ell+1)}{n^2}\right]\right\} \qquad \textbf{[15.25]}$$

Thus electrons in inner shells (small $n$) are tightly bound to the nucleus and their average position is quite near the nucleus because they are only slightly shielded from the full nuclear charge $Z$. Electrons in outer shells are only weakly attracted to the nucleus and their average position is far from the nucleus because they are almost fully shielded from the nuclear charge $Z$.

## EXAMPLE 15.11

Estimate the energy and average of $r$ in the $1s$ orbital of argon. Compare the results with the corresponding values for hydrogen.

### Solution
Using Equation 15.24 and the value $Z_{eff}(1) \sim 16$ leads to $E_{1s} \sim -256$ rydbergs for Ar. This is more strongly bound than H($1s$) by a factor 256. (Compare Equation 15.19b for the H atom.)

Using Equation 15.25 and the value $Z_{eff}(1) \sim 16$ leads to $\bar{r}_{1s} = \dfrac{3a_0}{2 \cdot 16}$ for Ar. This is smaller than $r_{1s}$ for H by a factor 16.

---

The dependence of the energy on $\ell$ can be demonstrated qualitatively by considering the effective charge within a *subshell*. Figures 15.23 to 15.27 show that only the $s$ orbitals penetrate to the nucleus; both $p$ and $d$ orbitals have nodes at the nucleus. Consequently, the shielding will be smallest and the effective charge greatest in $s$ orbitals: $Z_{eff}(ns) > Z_{eff}(np) > Z_{eff}(nd)$. It follows from Equation 15.24 that

$$E_{ns} < E_{np} < E_{nd}$$

The approximate energy level diagram for Hartree orbitals showing dependence on both $n$ and $\ell$ is presented in Figure 15.30.

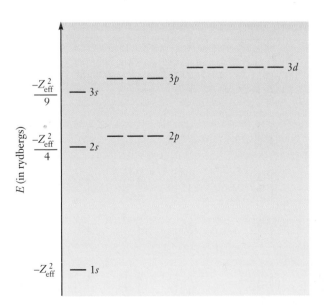

**FIGURE 15.30** Approximate energy level diagram for Hartree orbitals, estimated by incorporating values of $Z_{eff}$. Energy values are in units of rydbergs. The result of electron–electron repulsion is to remove the degeneracy of the H-atom levels with $\ell > 0$.

## The Aufbau ("Building Up") Principle

The ground-state electronic structures of atoms are built up by arranging the Hartree atomic orbitals in order of increasing energy and filling them one electron at a time, starting with the lowest energy orbital and heeding two additional restrictions at each step. (1) The **Pauli exclusion principle** states that no two electrons in an atom can have the same set of four quantum numbers ($n$, $\ell$, $m$, $m_s$). Another way of stating this principle is to say that each Hartree atomic orbital (characterized by a set of three quantum numbers, $n$, $\ell$, and $m$) holds at most two electrons, one with spin up and the other with spin down. (2) **Hund's rule** states that when electrons are added to Hartree orbitals of equal energy, a single electron enters each orbital before a second one enters any orbital. In addition, the spins remain parallel if possible.

## Building Up from Helium to Argon

Let us see how the aufbau principle works for the atoms from helium through neon. The lowest energy orbital is always the $1s$ orbital, and so helium has two electrons (with opposite spins) in that orbital. The ground-state **electron configuration** of

*[handwritten margin notes: 2 rules  1) Pauli Exclusion  2) Hund's Rule.]*

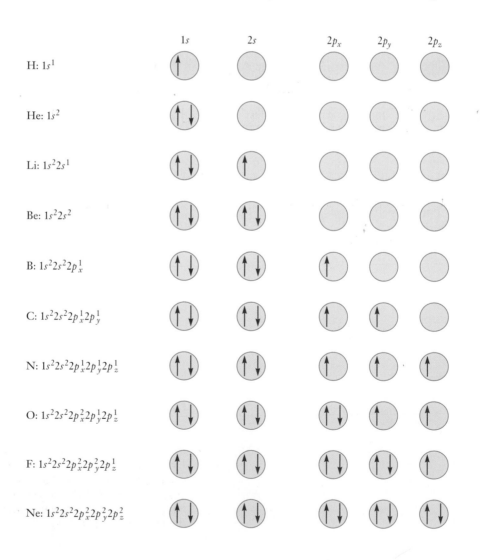

**FIGURE 15.31** The ground-state electron configurations of first- and second-period atoms. The arrows represent electrons with spin quantum numbers $m_s = +\frac{1}{2}$ or $m_s = -\frac{1}{2}$; each circle represents an orbital.

the helium atom is symbolized as $1s^2$ and is conveniently illustrated by a "pigeon-hole" diagram such as Figure 15.31. The $1s$ orbital in the helium atom is somewhat larger than the $1s$ orbital in the helium *ion* ($He^+$). In the ion the electron feels the full nuclear charge $+2e$, but in the atom each electron partially screens or shields the other electron from the nuclear charge. The orbital in the helium atom can be described by the approximate equations given previously with an "effective" nuclear charge $Z_{eff}$ of 1.7, which lies between $+1$ (the value for complete shielding by the other electron) and $+2$ (no shielding).

An atom of lithium has three electrons. The third electron does not join the first two in the $1s$ orbital. It occupies a different orbital because, by the Pauli principle, at most two electrons may occupy any orbital. The third electron goes into the $2s$ orbital, which is the next lowest in energy. The ground state of the lithium atom is therefore $1s^2 2s^1$, and $1s^2 2p^1$ is an excited state.

The next two elements present no difficulties. Beryllium has the ground-state configuration $1s^2 2s^2$. The $2s$ orbital is now filled, and boron, with five electrons, has the ground-state configuration $1s^2 2s^2 2p^1$. Because the three $2p$ orbitals of boron (for convenience, taken to be the $2p_x$, $2p_y$, and $2p_z$ orbitals) have the same energy, there is an equal chance for the electron to be in each one.

With carbon, the sixth element, a new question arises. Will the sixth electron go into the same orbital as the fifth (for example, both into a $2p_x$ orbital with opposite spins), or will it go into the $2p_y$ orbital, which has equal energy? The answer is found in the observation that two electrons occupying the same atomic orbital experience stronger electron–electron repulsion than they would if they occupied orbitals in different regions of space. Thus, putting the last two electrons of carbon into two different $p$ orbitals, such as the $2p_x$ and $2p_y$, in accordance with Hund's rules, leads to lower energy than putting them into the same $p$ orbital. The electron configuration of carbon is then $1s^2 2s^2 2p_x^1 2p_y^1$ or, more simply, $1s^2 2s^2 2p^2$. This configuration is shown in Figure 15.32.

The presence of unpaired electrons in atoms of carbon affects their behavior in a magnetic field. A substance is **paramagnetic** if it is attracted into a magnetic field. Any substance that has one or more unpaired electrons in the atoms, molecules, or ions that compose it is paramagnetic; ground-state carbon atoms are an example. A substance in which all the electrons are paired is weakly **diamagnetic:** it is pushed

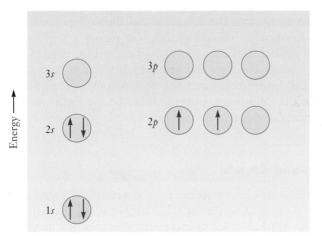

**FIGURE 15.32** For a many-electron atom such as carbon, orbitals with different $\ell$ values (such as the $2s$ and $2p$) have different energies (see Fig. 15.30). When two or more orbitals have the *same* energy (such as the three $2p$ orbitals here), electrons occupy different orbitals with parallel spins.

*out* of a magnetic field, although the force it experiences is much smaller in magnitude than the force that pulls a typical paramagnetic substance into a magnetic field. Of the atoms examined so far, H, Li, B, and C are paramagnetic, but He and Be have all of their electron spins paired and so are diamagnetic.

The electron configurations from N through Ne follow from the stepwise filling of the 2*p* orbitals. The six elements from boron to neon are **p-block** elements because they appear during filling of *p* orbitals in the building-up process. The four elements that precede them (hydrogen through beryllium) are **s-block** elements.

The build-up of the third period, from sodium to argon, is an echo of what happened in the second; first the one 3*s* orbital is filled and then the three 3*p* orbitals. As the number of electrons in an atom reaches 15 or 20, it is frequently the practice to explicitly include only those electrons added in the building up beyond the last preceding noble-gas element. The configuration of that noble gas is then represented by its chemical symbol enclosed in brackets. The ground-state configuration of silicon, for example, is written $[Ne]3s^2 3p^2$ under this system.

---

### EXAMPLE 15.12

Write the ground-state electron configurations for magnesium and sulfur. Are the gaseous atoms of these elements paramagnetic or diamagnetic?

### Solution

The noble-gas element preceding both elements is neon. Magnesium has two electrons beyond the neon core, which must be placed in the 3*s* orbital, the next higher in energy, to give the ground-state configuration $[Ne]3s^2$. A magnesium atom is diamagnetic because all its electrons are paired in orbitals.

Sulfur has six electrons beyond the neon core; the first two of these are in the 3*s* orbital, and the next four are in the 3*p* orbital. Sulfur's ground-state configuration is $[Ne]3s^2 3p^4$. When four electrons are put into three *p* orbitals, two electrons must occupy one of the orbitals, and the other two occupy different orbitals to reduce electron–electron repulsion. By Hund's rules, the electrons' spins are parallel, and the sulfur atom is paramagnetic.

### Related Problems: 45, 46, 47, 48

---

The electron configuration for an atom is a concise shorthand-notation that represents a great deal of information about the structure and the energy levels of the atom. Each configuration corresponds to wavefunction comprising a product of occupied Hartree orbitals. Each orbital has energy levels, given by Equation 15.24 and Figure 15.30, and average radius, given by Equation 15.25. The orbitals are grouped into subshells with radial distribution functions as shown in Figure 15.29. This information will be used in the following chapter to describe the formation of chemical bonds by atoms in particular electron configurations.

## Transition-Metal Elements and Beyond

After the three 3*p* orbitals are filled up with six electrons, the natural next step is a continuation of the building-up process using the 3*d* subshell. However, the aufbau principle requires that orbitals be filled in order of increasing energies. For neutral atoms at the beginning of the fourth period, it is an experimental fact that the $4s$ orbital ($n = 4$, $\ell = 0$) actually has a lower energy than the $3d$ orbital ($n = 3$, $\ell = 2$), although the two energies are very close. The $4s$ orbital is therefore filled first. The resulting ground-state configurations are $[Ar]4s^1$ for potassium and $[Ar]4s^2$ for calcium. The energy of the $3d$ orbitals does decline relative to the $4s$ orbital with increasing nuclear charge, however. In scandium, the very next element, the $3d$ orbital has a lower energy than the $4s$ orbital, and so an electron is removed more easily from the $4s$ orbital. The $Sc^+$ ion therefore has the ground-state configuration $[Ar]3d^14s^1$ rather than $[Ar]4s^2$, and $Sc^{2+}$ has the configuration $[Ar]3d^1$. The drop in energy of the $3d$ orbitals relative to the $4s$ continues throughout the transition-metal series from scandium to zinc, with the two $4s$ electrons removed more easily than the $3d$ electrons. The energy drop has important chemical consequences: the elements copper and zinc rarely have oxidation states higher than $+2$, and in the elements beyond zinc the $3d$ electrons are held so tightly that they play no chemical role. The ten elements from scandium to zinc are ***d*-block** elements because they appear with the filling of a $d$ orbital in the building-up process.

Experimental evidence shows that two of the transition-metal elements in the fourth period have ground-state electron configurations that are contrary to expectation based on the configurations that precede them. In its ground state, chromium has the configuration $[Ar]3d^54s^1$ rather than $[Ar]3d^44s^2$, and copper has the configuration $[Ar]3d^{10}4s^1$ rather than $[Ar]3d^94s^2$. Similar anomalies occur in the transition-metal atoms of the fifth period, and some other anomalies occur there as well, such as the ground-state configuration $[Kr]4d^75s^1$ that is observed for ruthenium in place of the expected $[Kr]4d^65s^2$.

In the sixth period, the filling of the $4f$ orbitals (and the generation of the ***f*-block** elements) begins as the rare-earth (lanthanide) elements from lanthanum to ytterbium are reached. The energies of the $4f$, $5d$, and $6s$ orbitals are comparable over much of the sixth period, and so their order of filling is erratic. (See Appendix F for the ground-state configurations of all the elements.) The sixth period ends with the radioactive noble gas radon. All of the elements in the seventh period, including the actinides, the second set of $f$-block elements, are radioactive.

The periodic table of Figure 15.33 groups elements in periods according to the subshell that is being filled as the atomic number increases. Exceptions to the "standard" order of filling are shown.

## Shells and the Periodic Table

A shell has been defined as a set of orbitals that have the same principal quantum number, reflecting the fact that the average positions of the electron in each of these shells are close to each other, but far from those of orbitals with different $n$ values (Fig. 15.29). **Valence electrons** are those that occupy the outermost shell, beyond the immediately preceding noble-gas configuration. Electrons in inner shells are closer to the nucleus, on the average, and lower in energy (that is, more strongly bound) than valence electrons; they are called **core electrons**. A **closed-shell atom** is one whose outermost shell is completely filled with electrons so that any excitation (promotion of an electron to a higher energy orbital) requires a large amount of energy; the noble-gas atoms have closed shells, and the next higher unoccupied orbitals are separated from the occupied orbitals by a significant energy gap.

Two elements in the same group (column) of the periodic table have related valence electron configurations. An atom's bonding properties and therefore its chemical behavior flow from the configuration of its valence electrons. Sodium

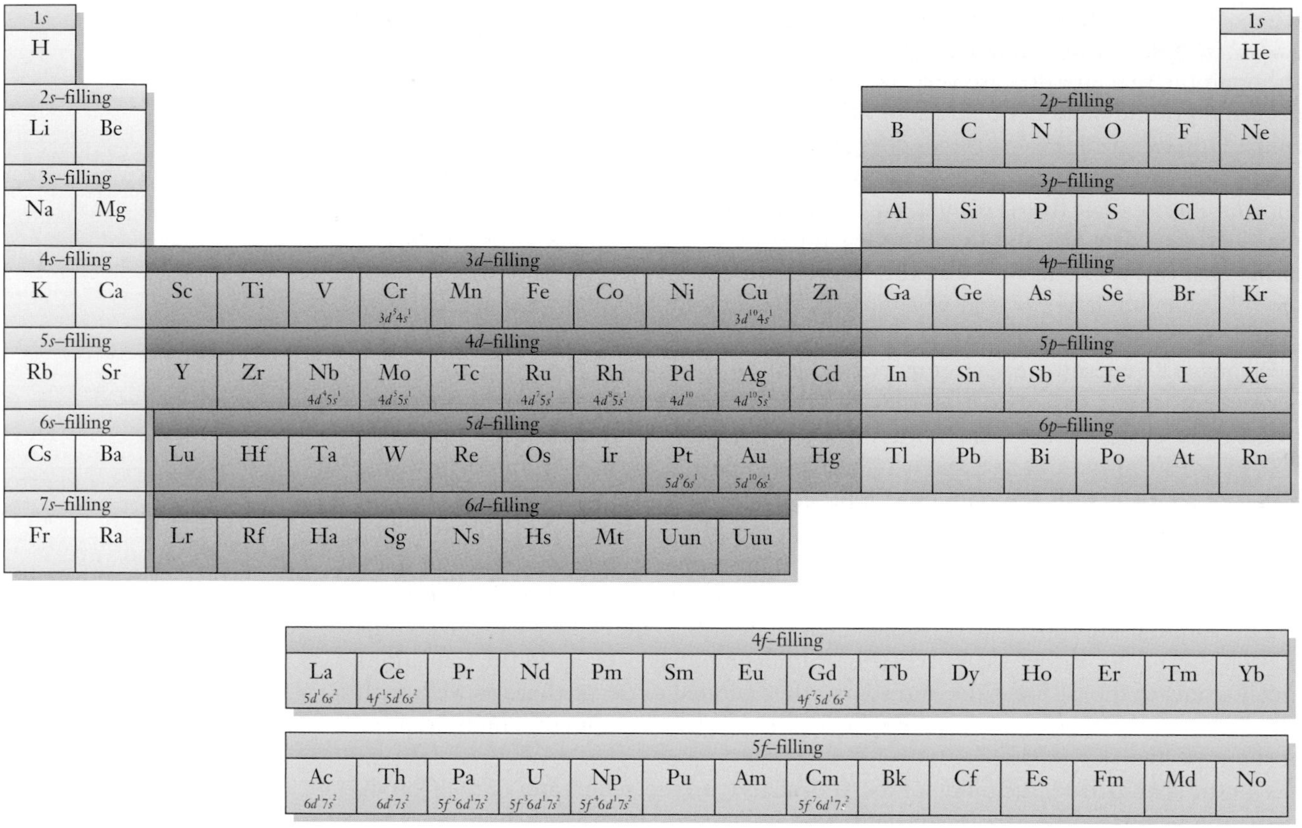

| 1s |  |  |  |  |  |  |  |  |  |  |  |  |  |  |  |  | 1s |
|---|---|---|---|---|---|---|---|---|---|---|---|---|---|---|---|---|---|
| H |  |  |  |  |  |  |  |  |  |  |  |  |  |  |  |  | He |

| 2s–filling | | | | | | | | | | | 2p–filling | | | | | |
|---|---|---|---|---|---|---|---|---|---|---|---|---|---|---|---|---|
| Li | Be |  |  |  |  |  |  |  |  |  | B | C | N | O | F | Ne |

| 3s–filling | | | | | | | | | | | 3p–filling | | | | | |
|---|---|---|---|---|---|---|---|---|---|---|---|---|---|---|---|---|
| Na | Mg |  |  |  |  |  |  |  |  |  | Al | Si | P | S | Cl | Ar |

| 4s–filling | | 3d–filling | | | | | | | | | | 4p–filling | | | | | |
|---|---|---|---|---|---|---|---|---|---|---|---|---|---|---|---|---|---|---|
| K | Ca | Sc | Ti | V | Cr $3d^5 4s^1$ | Mn | Fe | Co | Ni | Cu $3d^{10} 4s^1$ | Zn | Ga | Ge | As | Se | Br | Kr |

| 5s–filling | | 4d–filling | | | | | | | | | | 5p–filling | | | | | |
|---|---|---|---|---|---|---|---|---|---|---|---|---|---|---|---|---|---|---|
| Rb | Sr | Y | Zr | Nb $4d^4 5s^1$ | Mo $4d^5 5s^1$ | Tc | Ru $4d^7 5s^1$ | Rh $4d^8 5s^1$ | Pd $4d^{10}$ | Ag $4d^{10} 5s^1$ | Cd | In | Sn | Sb | Te | I | Xe |

| 6s–filling | | 5d–filling | | | | | | | | | | 6p–filling | | | | | |
|---|---|---|---|---|---|---|---|---|---|---|---|---|---|---|---|---|---|---|
| Cs | Ba | Lu | Hf | Ta | W | Re | Os | Ir | Pt $5d^9 6s^1$ | Au $5d^{10} 6s^1$ | Hg | Tl | Pb | Bi | Po | At | Rn |

| 7s–filling | | 6d–filling | | | | | | | |
|---|---|---|---|---|---|---|---|---|---|
| Fr | Ra | Lr | Rf | Ha | Sg | Ns | Hs | Mt | Uun | Uuu |

| 4f–filling | | | | | | | | | | | | | |
|---|---|---|---|---|---|---|---|---|---|---|---|---|---|
| La $5d^1 6s^2$ | Ce $4f^1 5d^1 6s^2$ | Pr | Nd | Pm | Sm | Eu | Gd $4f^7 5d^1 6s^2$ | Tb | Dy | Ho | Er | Tm | Yb |

| 5f–filling | | | | | | | | | | | | | |
|---|---|---|---|---|---|---|---|---|---|---|---|---|---|
| Ac $6d^1 7s^2$ | Th $6d^2 7s^2$ | Pa $5f^2 6d^1 7s^2$ | U $5f^3 6d^1 7s^2$ | Np $5f^4 6d^1 7s^2$ | Pu | Am | Cm $5f^7 6d^1 7s^2$ | Bk | Cf | Es | Fm | Md | No |

**FIGURE 15.33** The filling of shells and the structure of the periodic table. Only the anomalous electron configurations are shown.

(configuration [Ne]$3s^1$) and potassium (configuration [Ar]$4s^1$), for example, each have a single valence electron in an *s* orbital outside a closed shell; consequently, the two elements resemble each other closely in chemical properties. A major triumph of quantum mechanics is its ability to account for the periodic trends discovered by chemists many years earlier. Mendeleyev and others developed the periodic table of the elements purely on the basis of parallel chemical and physical properties (Section 1.5). Now we see that the structure of the periodic table can be derived from quantum mechanics. The ubiquitous octets in the Lewis electron-dot diagrams of second- and third-period atoms and ions arise from the eight available sites for electrons in the one *s* orbital and three *p* orbitals of the valence shell. The special properties of the transition-metal elements can be ascribed to the partial filling of their *d* orbitals, a feature to which we will return in Chapter 18.

## Sizes of Atoms and Ions

The sizes of atoms and ions influence how they interact in chemical compounds. Although atomic radius is not a precisely defined concept, these sizes can be estimated in several ways. If the electron density is known from theory or experiment, a contour surface of fixed electron density can be drawn, as demonstrated in Section 15.7 for one-electron atoms. Alternatively, if the atoms or ions in a crystal are assumed to be in contact with one another, a size can be defined from the measured distances between their centers (an approach to be explored in greater detail in Chapter 19.). These and other measures of size are reasonably consistent with each other

*See page 555*

and allow us to tabulate sets of atomic and ionic radii, many of which are listed in Appendix F.

Certain systematic trends appear in these radii. For a series of elements or ions in the same group (column) of the periodic table, the radius usually increases with increasing atomic number. This occurs mainly because the Pauli exclusion principle effectively excludes added electrons from the region occupied by the core electrons and so forces an increase in size as more distant electron shells are occupied. On the other hand, Coulomb (electrostatic) forces cause the radii of atoms to *decrease* with increasing atomic number across a period. As the nuclear charge increases steadily, electrons join the same valence shell and are ineffective in shielding each other from its attraction. This "incomplete shielding" of the added proton by the added electron as we go from atomic number Z to Z + 1 leads to an increase in $Z_{eff}$ *effective nuclear charge.* across a period.

Superimposed upon these broad trends are some subtler effects that have significant consequences in chemistry. One is shown dramatically in Figure 15.34. The radii of several sets of ions and atoms increase with atomic number in a given group, as already explained, but the *rate* of this increase changes considerably when the ions and atoms containing the same number of electrons as Ar are reached ($S^{2-}$, $Cl^-$, Ar, $K^+$, $Ca^{2+}$, $Sc^{3+}$, $Ti^{4+}$). The change in size from $Li^+$ to $Na^+$ to $K^+$ is quite great, for example, but the subsequent changes, to $Rb^+$ and $Cs^+$, are significantly smaller due to the filling of the *d* orbitals, which begins after $K^+$ is reached. Because atomic and ionic size decrease from left to right across a series of transition-metal elements (owing to increased effective nuclear charge), the radius of a main-group element when it is ultimately reached is smaller than it would have been had the transition series not intervened. A similar phenomenon occurs during the filling of

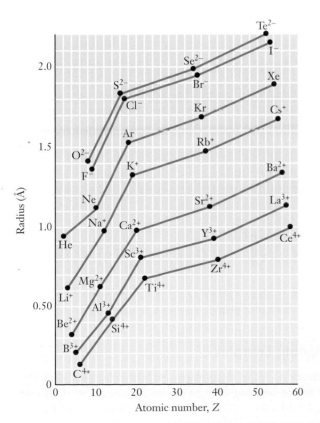

**FIGURE 15.34** Ionic and atomic radii plotted versus atomic number. Each line connects a set of atoms or ions having the same charge; all species have noble-gas configurations.

**FIGURE 15.35** The molar volumes (in cm³ mol⁻¹) of some elements in their solid states. Note the high peaks corresponding to the alkali metals.

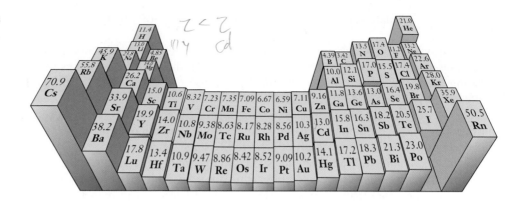

the 4*f* orbitals in the lanthanide series and is called the *lanthanide contraction*. Its effect on the sizes of transition-metal atoms is discussed in Section 18.1.

A different measure of atomic size is the volume occupied by a mole of atoms of the element in the solid phase. Figure 15.35 shows the pronounced periodicity of the molar volume, with maxima occurring for the alkali metals. Two factors affect the experimentally measured molar volume: the "size" of the atoms and the geometry of the bonding that connects them. The large molar volumes of the alkali metals stem from both the large size of the atoms and the fact that they are organized in a rather open, loosely packed structure in the solid.

---

**EXAMPLE 15.13**

Predict which atom or ion in each of the following pairs should be larger: **(a)** Kr or Rb, **(b)** Y or Cd, **(c)** F⁻ or Br⁻.

**Solution**
**(a)** Rb should be larger because it has an extra electron in a 5*s* orbital beyond the Kr closed shell.
**(b)** Y should be larger because the effective nuclear charge increases through the transition series from Y to Cd.
**(c)** Br⁻ should be larger because the extra outer electrons are excluded from the core.

**Related Problems: 55, 56, 57, 58**

---

**15.9**

## EXPERIMENTAL MEASURES OF ORBITAL ENERGIES

The order of the energies of atomic orbitals determines the building up of electron configurations. How are these energies measured?

### Direct Measurement of Binding Energies

Photoelectron spectroscopy (Section 15.3) determines orbital energies directly. Neon, for example, has a peak at the binding energy of the 2*p* electrons, then one for the *2s* electrons, and finally, at the highest binding energy, one for the *1s* electrons (Fig. 15.12). Figure 15.36 shows an experimental spectrum in which the binding energies of the valence electrons in Na and Cl are directly compared. Figure 15.37 shows measured orbital energies for neutral atoms. The sequence of orbital energies displayed in Figure 15.37 is used to build up electron configurations. The energies are reported in units of rydbergs.

## Periodic Trends in Ionization Energies

The ionization energy of an atom is defined as the minimum energy necessary to detach an electron from the neutral gaseous atom (see Section 3.1.). It can be obtained directly from the photoelectron spectrum of an atomic gas. Appendix F lists measured ionization energies of the elements, and Figure 15.38 shows the periodic trends in first and second ionization energies with atomic number.

Let us examine the periodic trends in the first ionization energy to learn about the stabilities of the various electron configurations. There is a large drop in $IE_1$ from helium to lithium for two reasons: first, a *2s* electron is much farther from the nucleus than a *1s* electron; second, the *1s* electrons screen the nucleus in lithium so effectively that the *2s* electron "sees" a net positive charge close to +1 rather than the larger charge seen by the electrons in helium. Beryllium shows an increase in $IE_1$ over Li because the effective nuclear charge has increased, but the electron being removed is still from a *2s* orbital.

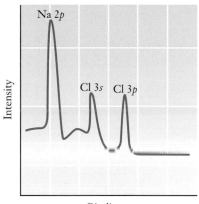

**FIGURE 15.36** Sodium chloride consists of $Na^+$ ions (with the $[He]2s^2 2p^6$ configuration) and $Cl^-$ ions (with the $[Ne]3s^2 3p^6$ configuration). The portion of the photoelectron spectrum of NaCl that covers the binding energies of the outer electrons is shown here. Binding energy increases from right to left. The core electrons on the two ions, which are more tightly held than the outer electrons, would lie farther to the left.

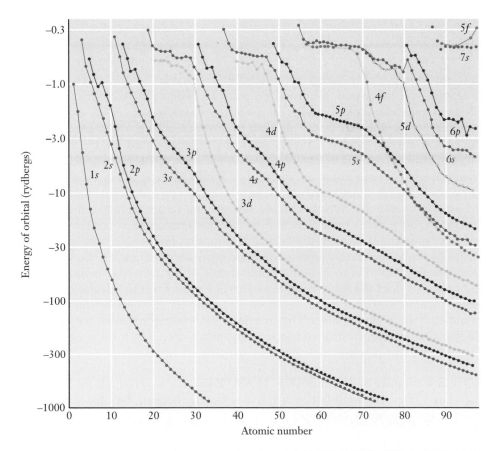

**FIGURE 15.37** The orbital energies of the first 97 elements, based on photoelectron spectroscopy experiments. Orbitals of the same principal quantum number *n*, such as the *2s* and *2p* orbitals, have similar energies and are well separated from orbitals of different *n*. Note, however, the *3d* orbitals. In elements 21 through 30 (scandium through zinc), they lie well above the *3s* and *3p* orbitals and only slightly below the *4s* orbitals. Their energy then drops rapidly as *Z* increases further. The *4d*, *5d*, and *4f* orbitals behave similarly. Note the logarithmic energy scale. One rydberg is $2.8 \times 10^{-18}$ J.

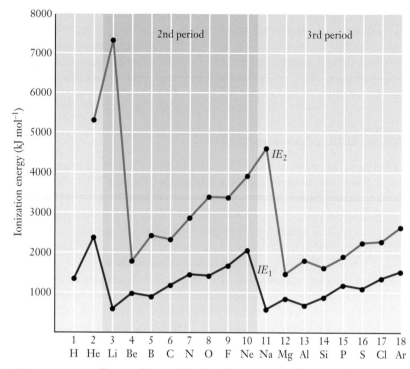

**FIGURE 15.38**  First and second ionization energies of atoms of the first three periods.

The $IE_1$ of boron is somewhat less than that of beryllium because the fifth electron is in a higher energy (and therefore less stable) $2p$ orbital. In carbon and nitrogen, the additional electrons go into $2p$ orbitals as the effective nuclear charge increases to hold the outer electrons more tightly; hence, $IE_1$ increases. The nuclear charge is higher in oxygen than in nitrogen, which would give it a higher ionization energy; however, this is more than compensated for by the fact that oxygen must accommodate two electrons in the same $2p$ orbital, leading to greater electron–electron repulsion and diminished binding. Oxygen has a lower $IE_1$ than nitrogen. Fluorine and neon have successively higher first ionization energies because of increasing effective nuclear charge. The general trends of increasing ionization energy across a given period, as well as the dips that occur at certain points, can thus be understood through the orbital description of many-electron atoms.

The ionization energy tends to decrease down a group in the periodic table (for example, from lithium to sodium to potassium). As the principal quantum number increases, so does the distance of the outer electrons from the nucleus. There are some exceptions to this trend, however, especially for the heavier elements in the middle of the periodic table. The first ionization energy of gold, for example, is higher than that of silver or of copper. This fact is crucial in making gold a "noble metal," one that is resistant to attack by oxygen.

Similar trends are observed in the second ionization energy but are shifted higher in atomic number by one unit (Fig. 15.38). Thus, $IE_2$ is very high for Li (because $Li^+$ has a filled $1s^2$ shell) but relatively low for Be (because $Be^+$ has a single electron in the outermost $2s$ orbital).

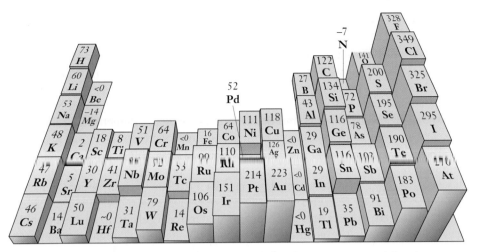

**FIGURE 15.39** The electron affinities (in kJ mol$^{-1}$) of gaseous atoms of the elements. Some of the atoms have negative electron affinities; this includes the noble gases, which are not visible at the extreme right.

## Electron Affinity

The **electron affinity** *EA* of an atom is the energy *released* when an electron is added to it (see Section 3.1). Appendix F lists the electron affinities of the elements.

The periodic trends in electron affinity (Fig. 15.39) parallel those in ionization energy for the most part, except that they are shifted one unit *lower* in atomic number. The reason is clear. Attaching an electron to F gives F$^-$, with the configuration $1s^2 2s^2 2p^6$, the same as that of Ne. Fluorine has a large affinity for electrons because the resulting closed-shell configuration is very stable. In contrast, the noble gases have negative electron affinities for the same reason that the alkali metals have small ionization energies: the last electron resides in a new shell far from the nucleus and is almost totally screened from the nuclear charge.

---

## EXAMPLE 15.14

Consider the elements selenium (Se) and bromine (Br). Which has the higher first ionization energy and which the higher electron affinity?

### Solution

These two atoms are adjacent to each other in the periodic table. Bromine has one more electron in the 4$p$ subshell, and this electron should be more tightly bound than the 4$p$ electrons in Se because of the incomplete shielding and the extra unit of positive charge on the nucleus. Thus, $IE_1$ should be greater for Br.

Bromine has a greater electron affinity than selenium. Gaining the extra electron changes the Br atom into a Br$^-$ ion, which has a particularly stable closed-shell electron configuration (the same as that of the noble gas atom Kr). No such configuration is created when an Se atom gains an additional electron.

**Related Problems: 59, 60, 61, 62**

Clouds of gas surround very hot stars in these galactic clusters. The red color arises from hydrogen radiation. *(Copyright 1984 Anglo–Australian Observatory)*

## CUMULATIVE EXERCISE

### Interstellar Space

The vast stretches of space between the stars are by no means empty. They contain both gases and dust particles at very low concentrations. Interstellar space extends so far that these low-density species significantly affect the electromagnetic radiation arriving from distant stars and other sources, which is detected by telescopes. The gas in interstellar space consists primarily of hydrogen atoms (either neutral or ionized) at a concentration on the order of one atom per cubic centimeter. The dust (thought to be mostly solid water, methane, or ammonia) is even less concentrated, with typically only a few dust particles (each $10^{-4}$ to $10^{-5}$ cm in radius) per cubic kilometer.

(a) The hydrogen in interstellar space near a star is largely ionized by the high-energy photons from the star. Such regions are called H II regions. Suppose a ground-state hydrogen atom absorbs a photon with a wavelength of 65 nm. Calculate the kinetic energy of the electron ejected. (Note: This is the gas-phase analog of the photoelectric effect for solids.)

(b) What is the de Broglie wavelength of the electron from part (a)?

(c) Free electrons in H II regions can be recaptured by hydrogen nuclei. In such an event, the atom emits a series of photons of increasing energy as the electrons cascade down through the quantum states of the hydrogen atom. The particle densities are so low that extremely high quantum states can be detected in interstellar space. In particular, the transition from the state $n = 110$ to $n = 109$ for the hydrogen atom has been detected. What is the Bohr radius of an electron in the state $n = 110$ for hydrogen?

(d) Calculate the wavelength of light emitted as an electron undergoes a transition from level $n = 110$ to $n = 109$. In what region of the electromagnetic spectrum does this lie?

(e) H II regions also contain ionized atoms heavier than hydrogen. Calculate the longest wavelength of light that will ionize a ground-state helium atom. Use data from Appendix F.

(f) The regions farther from stars are called H I regions. Here almost all of the hydrogen atoms are neutral rather than ionized and are in the ground state. Will such hydrogen atoms absorb light in the Balmer series emitted by atoms in H II regions?

(g) We stated in Section 15.7 that the energy of the hydrogen atom depends only on the quantum number $n$. In fact, this is not quite true. The electron spin ($m_s$ quantum number) couples very weakly with the spin of the nucleus, making the ground state split into two states of almost equal energy. The radiation emitted in a transition from the upper to the lower of these levels has a wavelength of 21.2 cm and is of great importance in astronomy because it allows the H I regions to be studied. What is the energy difference between these two levels, both for a single atom and for a mole of atoms?

(h) The gas and dust particles between a star and the earth absorb the star's light more strongly in the blue region of the spectrum than in the red. As a result, stars appear slightly redder than they actually are. Will an estimate of the temperature of a star based on its apparent color give too high or too low a number?

**Answers**
(a) $8.8 \times 10^{-19}$ J
(b) 0.52 nm = 5.2 Å

(c) $6.40 \times 10^{-7}$ m = 6400 Å

(d) $5.98 \times 10^{-2}$ m = 5.98 cm, in the microwave region

(e) 50.4 nm

(f) No, because the lowest energy absorption for a ground-state hydrogen atom is in the ultraviolet region of the spectrum.

(g) $9.37 \times 10^{-25}$ J; 0.564 J mol$^{-1}$

(h) Too low, because red corresponds to emitted light of lower energy. Experience with blackbody radiation curves would assign a lower temperature to a star emitting lower energy light.

## CONCEPTS & SKILLS

*After studying this chapter and working the problems that follow, you should be able to*

1. Relate the frequency, wavelength, and speed of light and other waves (Section 15.1, problems 1–8).

2. Describe blackbody radiation and the photoelectric effect, and discuss how these paradoxes of classical physics were resolved by quantum mechanics (Section 15.2, problems 9–10).

3. Using the law of conservation of energy, relate the work function of a metal to the wavelength of light used to eject electrons in the photoelectric effect and the kinetic energy of those electrons (Section 15.2, problems 11–14).

4. Use experimental emission and absorption spectra to determine spacings between energy levels in atoms (Section 15.3, problems 15–20).

5. Use the Franck–Hertz method to determine spacings between adjacent energy levels in atoms (Section 15.3, problems 21–22).

6. Use photoelectron spectroscopy to determine the binding energies for atoms (Section 15.3, problems 23–24).

7. Use the Bohr model to calculate the energy levels of one-electron atoms and to find the frequencies and wavelengths of light emitted in transitions between energy levels (Section 15.4, problems 25–28).

8. Discuss the de Broglie relation and use it to calculate the wavelengths associated with particles in motion (Section 15.5, problems 31–32).

9. State the Heisenberg uncertainty principle and use it to establish bounds within which the position and momentum of a particle can be known (Section 15.5, problems 33–34).

10. Determine the energy levels for particles in rigid rectangular boxes (Section 15.6, problems 35–36).

11. Give the quantum numbers that characterize one-electron atoms, and discuss the shapes, sizes, and nodal properties of the corresponding orbitals (Section 15.7, problems 37–42).

12. Use the aufbau principle to predict electron configurations of atoms and ions and to account for the structure of the periodic table (Section 15.8, problems 45–54).

13. Discuss the factors that lead to systematic variation of sizes of atoms and ion through the periodic table (Section 15.8, problems 55–58).

14. Describe the trends in ionization energy and electron affinity across the periodic table and relate them to the electronic structure of atoms (Section 15.9, problems 59–62).

*Answers to problems whose numbers are boldface appear in Appendix G. Problems that are more challenging are indicated with asterisks.*

## Preliminaries: Wave Motion and Light

1. Some water waves reach the beach at a rate of one every 3.2 s, and the distance between their crests is 2.1 m. Calculate the speed of these waves.

2. The spacing between bands of color in a chemical wave from an oscillating reaction is measured to be 1.2 cm, and a new wave appears every 42 s. Calculate the speed of propagation of the chemical waves.

3. An FM radio station broadcasts at a frequency of $9.86 \times 10^7$ s$^{-1}$ (98.6 MHz). Calculate the wavelength of the radio waves.

4. The gamma rays emitted by $^{60}$Co are used in radiation treatment of cancer. They have a frequency of $2.83 \times 10^{20}$ s$^{-1}$. Calculate their wavelength, expressing your answer in meters and in angstroms.

5. Radio waves of wavelength $6.00 \times 10^2$ m can be used to communicate with spacecraft over large distances.
   (a) Calculate the frequency of these radio waves.
   (b) Suppose a radio message is sent home by the astronauts in a spaceship approaching the planet Mars at a distance of $8.0 \times 10^{10}$ m from earth. How long (in minutes) will it take for the message to travel from the spaceship to earth?

6. An argon ion laser emits light of wavelength 488 nm.
   (a) Calculate the frequency of the light.
   (b) Suppose a pulse of light from this laser is sent from earth, is reflected from a mirror on the moon, and returns to its starting point. Calculate the time elapsed for the round trip, taking the distance from earth to moon to be $3.8 \times 10^5$ km.

7. The speed of sound in dry air at 20°C is 343.5 m s$^{-1}$, and the frequency of the sound from the middle C note on a piano is 261.6 s$^{-1}$ (according to the American standard pitch scale). Calculate the wavelength of this sound and the time it will take to travel 30.0 m across a concert hall.

8. Ultrasonic waves have frequencies too high to be detected by the human ear but can be produced and detected by vibrating crystals. Calculate the wavelength of an ultrasonic wave of frequency $5.0 \times 10^4$ s$^{-1}$ that is propagating through a sample of water at a speed of $1.5 \times 10^3$ m s$^{-1}$. Explain why ultrasound can be used to probe the size and position of the fetus inside its mother's abdomen. Could audible sound with a frequency of 8000 s$^{-1}$ be used for this purpose?

## Experimental Basis of Energy Quantization: Blackbody Radiation and the Photoelectric Effect

9. Both blue and green light eject electrons from the surface of potassium. In which case do the ejected electrons have the higher average kinetic energy?

10. When an intense beam of green light is directed onto a copper surface, no electrons are ejected. What will happen if the green light is replaced with red light?

11. Cesium is frequently employed in photocells because its work function ($3.43 \times 10^{-19}$ J) is the lowest of all the elements. Such photocells are efficient because the broadest range of wavelengths of light can eject electrons. What colors of light will eject electrons from cesium? What colors of light will eject electrons from selenium, which has a work function of $9.5 \times 10^{-19}$ J?

12. Alarm systems employ the photoelectric effect. A beam of light strikes a piece of metal in the photocell, ejecting electrons continuously and causing a small electric current to flow. When someone steps into the light beam, the current is interrupted and the alarm is triggered. What is the maximum wavelength of light that can be used in such an alarm system if the photocell metal is sodium, with a work function of $4.41 \times 10^{-19}$ J?

13. Light having a wavelength of $2.50 \times 10^{-7}$ m falls upon the surface of a piece of chromium in an evacuated glass tube. If the work function of the chromium is $7.21 \times 10^{-19}$ J, determine (a) the maximum kinetic energy of the emitted photoelectrons and (b) the speed of photoelectrons having this maximum kinetic energy.

14. Calculate the maximum wavelength of electromagnetic radiation if it is to cause detachment of electrons from the surface of metallic tungsten, which has a work function of $7.29 \times 10^{-19}$ J. If the maximum speed of the emitted photoelectrons is to be $2.00 \times 10^6$ m s$^{-1}$, what should the wavelength of the radiation be?

## Experimental Demonstrations of Energy Quantization in Atoms

15. Excited lithium atoms emit light strongly at a wavelength of 671 nm. This emission predominates when Li atoms are excited in a flame. Predict the color of the flame.

16. Excited mercury atoms emit light strongly at a wavelength of 454 nm. This emission predominates when Hg atoms are excited in a flame. Predict the color of the flame.

17. Barium atoms in a flame emit light as they undergo transitions from one energy level to another that is $3.6 \times 10^{-19}$ J lower in energy. Calculate the wavelength of light emitted and, by referring to Figure 15.3, predict the color visible in the flame.

18. Potassium atoms in a flame emit light as they undergo transitions from one energy level to another that is $4.9 \times 10^{-19}$ J lower in energy. Calculate the wavelength of light emitted and, by referring to Figure 15.3, predict the color visible in the flame.

19. The sodium D-line is actually a pair of closely spaced spectroscopic lines seen in the emission spectrum of sodium atoms. The wavelengths are centered at 589.3 nm. The intensity of this emission makes it the major source of light (and causes the yellow color) in the sodium arc light.

(a) Calculate the energy change per sodium atom emitting a photon at the D-line wavelength.

(b) Calculate the energy change per mole of sodium atoms emitting photons at the D-line wavelength.

(c) If a sodium arc light is to produce 1.000 kilowatt ($1000 \, J \, s^{-1}$) of radiant energy at this wavelength, how many moles of sodium atoms must emit photons per second?

20. The power output of a laser is measured by its wattage, the number of joules of energy it radiates per second (1 W = $1 \, J \, s^{-1}$). A 10-W laser produces a beam of green light with a wavelength of 520 nm ($5.2 \times 10^{-7}$ m).

(a) Calculate the energy carried by each photon.

(b) Calculate the number of photons emitted by the laser per second.

21. In a Franck–Hertz experiment on Na atoms, the first excitation threshold occurs at 2.103 eV. Calculate the wavelength of emitted light expected just above this threshold. (Note: Sodium vapor lamps used in street lighting emit spectral lines with wavelengths 5891.8 Å and 5889.9 Å.)

22. In a Franck–Hertz experiment on H atoms, the first two excitation thresholds occur at 10.1 eV and 11.9 eV. Three optical emission lines are associated with these levels. Sketch an energy level diagram for H atoms based on this information. Identify the three transitions associated with these emission lines. Calculate the wavelength of each emitted line.

23. Photoelectron spectra of Hg atoms acquired with radiation from a He lamp at 584.4 Å show a peak in which the photoelectrons have kinetic energy of 11.7 eV. Calculate the binding energy of electrons in that level.

24. Quantum mechanics predicts that the binding energy of the ground state of the H atom is $-13.6$ eV. Insight into the magnitude of this quantity is gained by considering several methods by which it can be measured. (a) Calculate the maximum wavelength of light that will ionize H atoms in their ground state. (b) Assume the atom is ionized by collision with an electron that transfers all its kinetic energy to the atom in the ionization process. Calculate the speed of the electron before the collision. Express your answer in m $s^{-1}$ and miles hour$^{-1}$. (c) Calculate the temperature required to ionize a H atom in its ground state by thermal excitation. (*Hint:* Recall the criterion for thermal excitation of an oscillator in Planck's theory of blackbody radiation.)

## The Bohr Model: Predicting Discrete Energy Levels

25. Use the Bohr model to calculate the radius and the energy of the $B^{4+}$ ion in the $n = 3$ state. How much energy would be required to remove the electrons from one mole of $B^{4+}$ ions in this state? What frequency and wavelength of light would be emitted in a transition from the $n = 3$ to the $n = 2$ state of this ion? Express all results in SI units.

26. $He^+$ ions are observed in stellar atmospheres. Use the Bohr model to calculate the radius and the energy of the $He^+$ ion in the $n = 5$ state. How much energy would be required to remove the electrons from one mole of $He^+$ in this state? What frequency and wavelength of light would be emitted in a transition from the $n = 5$ to the $n = 3$ state of this ion? Express all results in SI units.

27. The radiation emitted in the transition from $n = 3$ to $n = 2$ in a neutral hydrogen atom has a wavelength of 656.1 nm. What would be the wavelength of radiation emitted from a doubly ionized lithium atom ($Li^{2+}$) if a transition occurred from $n = 3$ to $n = 2$? In what region of the spectrum does this radiation lie?

28. The $Be^{3+}$ ion has a single electron. Calculate the frequencies and wavelengths of light in the ion's emission spectrum for the first three lines of each of the series that are analogous to the Lyman and the Balmer series of neutral hydrogen. In what region of the spectrum does this radiation lie?

## Waves, Particles, and the Schrödinger Equation

29. A guitar string with fixed ends has a length of 50 cm.
(a) Calculate the wavelengths of its fundamental mode of vibration (i.e., its first harmonic) and its third harmonic.
(b) How many nodes does the third harmonic have?

30. Suppose we picture an electron in a chemical bond as being a wave with fixed ends. Take the length of the bond to be 1.0 Å.
(a) Calculate the wavelength of the electron wave in its ground state and in its first excited state.
(b) How many nodes does the first excited state have?

31. Calculate the de Broglie wavelength of the following:
(a) An electron moving at a speed of $1.00 \times 10^3$ m $s^{-1}$
(b) A proton moving at a speed of $1.00 \times 10^3$ m $s^{-1}$
(c) A baseball with a mass of 145 g, moving at a speed of 75 km hr$^{-1}$

32. Calculate the de Broglie wavelength of the following:
(a) Electrons that have been accelerated to a kinetic energy of $1.20 \times 10^7$ J mol$^{-1}$
(b) A helium atom moving at a speed of 353 m $s^{-1}$ (the root-mean-square speed of helium atoms at 20 K)
(c) A krypton atom moving at a speed of 299 m $s^{-1}$ (the root-mean-square speed of krypton atoms at 300 K)

33. (a) The position of an electron is known to be within 10 Å ($1.0 \times 10^{-9}$ m). What is the minimum uncertainty in its velocity?
(b) Repeat the calculation of part (a) for a helium atom.

34. No object can travel faster than the speed of light, so it would seem evident that the uncertainty in the speed of any object is at most $3 \times 10^8$ m $s^{-1}$.
(a) What is the minimum uncertainty in the position of an electron, given that we know nothing about its speed except that it is less than the speed of light?
(b) Repeat the calculation of part (a) for the position of a helium atom.

## A Deeper Look . . . The Particle in a Box

35. In Chapter 3 we introduced the concept of a double bond between carbon atoms, represented by C=C, with a length near 1.34 Å. The motion of an electron in such a bond can be treated very crudely as motion in a one-dimensional box. Calculate the energy of an electron in each of its three lowest

allowed states if it is confined to move in a one-dimensional box of length 1.34 Å. Calculate the wavelength of light necessary to excite the electron from its ground state to the first excited state.

36. When metallic sodium is dissolved in liquid sodium chloride, electrons are released into the liquid. These dissolved electrons absorb light with a wavelength near 800 nm. Suppose we treat the positive ions surrounding an electron very crudely as defining a three-dimensional cubic box of edge $L$, and we assume that the absorbed light excites the electron from its ground state to the first excited state. Calculate the edge length $L$ in this simple model.

## The Hydrogen Atom

37. Which of the following combinations of quantum numbers are allowed for an electron in a one-electron atom and which are not?
    (a) $n = 2$, $\ell = 2$, $m = 1$, $m_s = \frac{1}{2}$
    (b) $n = 3$, $\ell = 1$, $m = 0$, $m_s = -\frac{1}{2}$
    (c) $n = 5$, $\ell = 1$, $m = 2$, $m_s = \frac{1}{2}$
    (d) $n = 4$, $\ell = -1$, $m = 0$, $m_s = \frac{1}{2}$

38. Which of the following combinations of quantum numbers are allowed for an electron in a one-electron atom and which are not?
    (a) $n = 3$, $\ell = 2$, $m = 1$, $m_s = 0$
    (b) $n = 2$, $\ell = 0$, $m = 0$, $m_s = -\frac{1}{2}$
    (c) $n = 7$, $\ell = 2$, $m = -2$, $m_s = \frac{1}{2}$
    (d) $n = 3$, $\ell = -3$, $m = -2$, $m_s = -\frac{1}{2}$

39. Label the orbitals described by each of the following sets of quantum numbers.
    (a) $n = 4$, $\ell = 1$
    (b) $n = 2$, $\ell = 0$
    (c) $n = 6$, $\ell = 3$

40. Label the orbitals described by each of the following sets of quantum numbers.
    (a) $n = 3$, $\ell = 2$
    (b) $n = 7$, $\ell = 4$
    (c) $n = 5$, $\ell = 1$

41. How many radial nodes and how many angular nodes does each of the orbitals in problem 39 have?

42. How many radial nodes and how many angular nodes does each of the orbitals in problem 40 have?

43. Use the mathematical expression for the $2p_z$ wave function of a one-electron atom (see Table 15.2) to show that the probability of finding an electron in that orbital anywhere in the $xy$ plane is 0. What are the nodal planes for a $d_{xz}$ orbital and for a $d_{x^2-y^2}$ orbital?

44. (a) Use the radial wave function for the $3p$ orbital of a hydrogen atom (see Table 15.2) to calculate the value of $r$ for which a node exists.
    (b) Find the values of $r$ for which nodes exist for the $3s$ wave function of the hydrogen atom.

## Many-Electron Atoms and the Periodic Table

45. Give the ground-state electron configurations of the following elements.
    (a) C        (b) Se        (c) Fe

46. Give the ground-state electron configurations of the following elements.
    (a) P        (b) Tc        (c) Ho

47. Write ground-state electron configurations for the ions $Be^+$, $C^-$, $Ne^{2+}$, $Mg^+$, $P^{2+}$, $Cl^-$, $As^+$, and $I^-$. Which do you expect will be paramagnetic due to the presence of unpaired electrons?

48. Write ground-state electron configurations for the ions $Li^-$, $B^+$, $F^-$, $Al^{3+}$, $S^-$, $Ar^+$, $Br^+$, and $Te^-$. Which do you expect to be paramagnetic due to the presence of unpaired electrons?

49. Identify the atom or ion corresponding to each of the following descriptions:
    (a) An atom with ground-state electron configuration $[Kr]4d^{10}5s^25p^1$
    (b) An ion with charge $-2$ and ground-state electron configuration $[Ne]3s^23p^6$
    (c) An ion with charge $+4$ and ground-state electron configuration $[Ar]3d^3$

50. Identify the atom or ion corresponding to each of the following descriptions:
    (a) An atom with ground-state electron configuration $[Xe]4f^{14}5d^66s^2$
    (b) An ion with charge $-1$ and ground-state electron configuration $[He]2s^22p^6$
    (c) An ion with charge $+5$ and ground-state electron configuration $[Kr]4d^6$

51. Predict the atomic number of the (as yet undiscovered) element in the seventh period that is a halogen.

52. (a) Predict the atomic number of the (as yet undiscovered) alkali-metal element in the eighth period.
    (b) Suppose the eighth-period alkali-metal atom turned out to have atomic number 137. What explanation would you give for such a high atomic number (recall that the atomic number of francium is only 87)?

53. Suppose that the spin quantum number did not exist, so that only one electron could occupy each orbital of a many-electron atom. Give the atomic numbers of the first three noble-gas atoms in this case.

54. Suppose that the spin quantum number had three allowed values ($m_s = 0$, $+\frac{1}{2}$, $-\frac{1}{2}$). Give the atomic numbers of the first three noble-gas atoms in this case.

55. For each of the following pairs of atoms or ions, state which you expect to have the larger radius.
    (a) Na or K          (d) K or Ca
    (b) Cs or $Cs^+$     (e) $Cl^-$ or Ar
    (c) $Rb^+$ or Kr

56. For each of the following pairs of atoms or ions, state which you expect to have the larger radius.
    (a) Sm or $Sm^{3+}$  (d) Ge or As
    (b) Mg or Ca         (e) $Sr^+$ or Rb
    (c) $I^-$ or Xe

57. Predict the larger ion in each of the following pairs. Give reasons for your answers.
    (a) $O^-$, $S^{2-}$      (c) $Mn^{2+}$, $Mn^{4+}$
    (b) $Co^{2+}$, $Ti^{2+}$  (d) $Ca^{2+}$, $Sr^{2+}$

58. Predict the larger ion in each of the following pairs. Give reasons for your answers.

(a) $S^{2-}$, $Cl^-$     (c) $Ce^{3+}$, $Dy^{3+}$
(b) $Tl^+$, $Tl^{3+}$     (d) $S^-$, $I^-$

## Experimental Measures of Orbital Energies

59. For each of the following pairs of atoms, state which you expect to have the higher first ionization energy.
    (a) Rb or Sr     (c) Xe or Cs
    (b) Po or Rn     (d) Ba or Sr

60. For each of the following pairs of atoms, state which you expect to have the higher first ionization energy.
    (a) Bi or Xe     (c) Rb or Y
    (b) Se or Te     (d) K or Ne

61. For each of the following pairs of atoms, state which you expect to have the greater electron affinity.
    (a) Xe or Cs     (c) Ca or K
    (b) Pm or F     (d) Po or At

62. For each of the following pairs of atoms, state which you expect to have the higher electron affinity.
    (a) Rb or Sr     (c) Ba or Te
    (b) I or Rn     (d) Bi or Cl

63. The cesium atom has the lowest ionization energy of all the neutral atoms in the periodic table, 375.7 kJ mol$^{-1}$. What is the longest wavelength of light that could ionize a cesium atom? In which region of the electromagnetic spectrum does this light fall?

64. Until recently, it was thought that the $Ca^-$ ion was unstable, so that the calcium atom had a negative electron affinity. Some new experiments have now measured an electron affinity of +2.0 kJ mol$^{-1}$ for Ca. What is the longest wavelength of light that could remove an electron from a $Ca^-$ ion? In which region of the electromagnetic spectrum does this light fall?

## Additional Problems

65. A piano tuner uses a tuning fork that emits sound with a frequency of 440 s$^{-1}$. Calculate the wavelength of the sound from this tuning fork and the time the sound takes to travel 10.0 m across a large room. Take the speed of sound in air to be 343 m s$^{-1}$.

66. The distant galaxy called Cygnus A is one of the strongest sources of radio waves reaching the earth. The distance of this galaxy from the earth is $3 \times 10^{24}$ m. How long (in years) does it take a radio wave of wavelength 10 m to reach the earth? What is the frequency of this radio wave?

67. Hot objects can emit blackbody radiation that appears red, orange, white, or bluish white but never green. Explain.

68. Compare the energy (in joules) carried by an x-ray photon (wavelength $\lambda = 0.20$ nm) with that carried by an AM radio wave photon ($\lambda = 200$ m). Calculate the energy of 1.00 mol of each type of photon. What effect do you expect each type of radiation to have in terms of inducing chemical reactions in the substances through which it passes?

69. When ultraviolet light of wavelength 131 nm strikes a polished nickel surface, the maximum kinetic energy of ejected electrons is measured to be $7.04 \times 10^{-19}$ J. Calculate the work function of nickel.

70. Express the velocity of the electron in the Bohr model in terms of fundamental constants ($m_e$, $e$, $h$, $\epsilon_0$), the nuclear charge $Z$, and the quantum number $n$. Evaluate the velocity of an electron in the ground states of an $He^+$ ion and a $U^{91+}$ ion. Compare these velocities with the speed of light $c$. As the velocity of an object approaches the speed of light, relativistic effects become important. In which kinds of atoms do you expect relativistic effects to be greatest?

71. Photons are emitted in the Lyman series as hydrogen atoms undergo transitions from various excited states to the ground state. If ground-state $He^+$ ions are present in the same gas (near stars, for example), can they absorb these photons? Explain.

*72. Name a transition in the $C^{5+}$ ion that will lead to the absorption of green light.

73. The energies of macroscopic objects as well as those of microscopic ones are quantized, but the effects of the quantization are not seen because the difference in energy between adjacent states is so small. Apply Bohr's quantization of angular momentum to the revolution of the earth (mass $6.0 \times 10^{24}$ kg), which moves with a velocity of $3.0 \times 10^4$ m s$^{-1}$ in a circular orbit (radius $1.5 \times 10^{11}$ m) about the sun. The sun can be treated as fixed. Calculate the value of the quantum number $n$ for the present state of the earth–sun system. What would be the effect of an increase in $n$ by 1?

74. Sound waves, like light waves, can interfere with each other, giving maximum and minimum levels of sound. Suppose a listener standing directly between two loudspeakers hears the same tone being emitted from both. This listener observes that, when one of the speakers is moved 0.16 m farther away, the perceived intensity of the tone decreases from a maximum to a minimum.
    (a) Calculate the wavelength of the sound.
    (b) Calculate its frequency, using 343 m s$^{-1}$ as the speed of sound.

75. (a) If the kinetic energy of an electron is known to lie between $1.59 \times 10^{-19}$ J and $1.61 \times 10^{-19}$ J, what is the smallest distance within which it can be known to lie?
    (b) Repeat the calculation of part (a) for a helium atom instead of an electron.

76. It is interesting to speculate on the properties of a universe with different values for the fundamental constants.
    (a) In a universe in which Planck's constant had the value $h = 1$ J s, what would be the de Broglie wavelength of a 145-g baseball moving at a speed of 20 m s$^{-1}$?
    (b) Suppose the velocity of the ball from part (a) is known to lie between 19 and 21 m s$^{-1}$. What is the smallest distance within which it can be known to lie?
    (c) Suppose that in this universe the mass of the electron is 1 g and the charge on the electron is 1 C. Calculate the Bohr radius of the hydrogen atom in this universe.

77. The normalized wave function for a particle in a one-dimensional box, in which the potential energy is zero, is $\psi(x) = \sqrt{2/L} \sin(n\pi x/L)$, where $L$ is the length of the box. What is the probability that the particle will lie between $x = 0$ and $x = L/4$ if the particle is in its $n = 2$ state?

78. A particle of mass $m$ is placed in a three-dimensional rectangular box with edge lengths $2L$, $L$, and $L$. Inside the box the potential energy is zero and outside it is infinite, and so the wave function goes smoothly to zero at the sides of the box. Calculate the energies, and give the quantum numbers of the ground state and the first five excited states (or sets of states of equal energy) for the particle in the box.

79. How does the $3d_{xy}$ orbital of an electron in an $O^{7+}$ ion resemble, and how does it differ from, the $3d_{xy}$ orbital of an electron in a hydrogen atom?

\* 80. The wave function of an electron in the lowest (i.e., ground) state of the hydrogen atom is

$$\psi(r) = \left(\frac{1}{\pi a_0^3}\right)^{1/2} \exp\left(-\frac{r}{a_0}\right)$$

$$a_0 = 0.529 \times 10^{-10} \text{ m}$$

(a) What is the probability of finding the electron inside a sphere of volume 1.0 $pm^3$, centered at the nucleus (1 pm $= 10^{-12}$ m)?

(b) What is the probability of finding the electron in a volume of 1.0 $pm^3$ at a distance of 52.9 pm from the nucleus, in a fixed but arbitrary direction?

(c) What is the probability of finding the electron in a spherical shell of 1.0 pm thickness, at a distance of 52.9 pm from the nucleus?

81. An atom of sodium has the electron configuration [Ne]$6s^1$. Explain how this is possible.

\* 82. (a) The nitrogen atom has one electron in each of the $2p_x$, $2p_y$, and $2p_z$ orbitals. By using the form of the angular wave functions, show that the total electron density, $\psi^2(2p_x) + \psi^2(2p_y) + \psi^2(2p_z)$, is spherically symmetric (that is, it is independent of the angles $\theta$ and $\phi$). The neon atom, which has *two* electrons in each $2p$ orbital, is also spherically symmetric.

(b) The same result as in part (a) applies to $d$ orbitals, so that a filled or half-filled subshell of $d$ orbitals is spherically symmetric. Identify the spherically symmetric atoms or ions among the following: $F^-$, Na, Si, $S^{2-}$, $Ar^+$, Ni, Cu, Mo, Rh, Sb, W, Au.

83. Chromium(IV) oxide is used in making magnetic recording tapes because it is paramagnetic. It can be described as a solid made up of $Cr^{4+}$ and $O^{2-}$ ions. Give the electron configuration of $Cr^{4+}$ in $CrO_2$, and determine the number of unpaired electrons on each chromium ion.

84. Use the data from Appendix F to graph the variation of atomic radius with atomic number for the rare-earth elements from lanthanum to lutetium.

(a) What is the general trend in these radii? How do you account for it?

(b) Which two elements in the series present exceptions to the trend?

85. Arrange the following six atoms or ions in order of size, from the smallest to the largest: K, $F^+$, Rb, $Co^{25+}$, Br, F, $Rb^-$.

86. Which is higher, the third ionization energy of lithium or the energy required to eject a $1s$ electron from a lithium atom in a photoelectron spectroscopy experiment? Explain.

87. The outermost electron in an alkali-metal atom is sometimes described as resembling an electron in the corresponding state of a one-electron atom. Compare the first ionization energy of Li with the binding energy of a $2s$ electron in a one-electron atom that has nuclear charge $Z_{eff}$, and determine the value of $Z_{eff}$ that is necessary for the two energies to agree. Repeat the calculation for the $3s$ electron of Na and the $4s$ electron of K.

\* 88. In two-photon ionization spectroscopy, the combined energies carried by two different photons are used to remove an electron from an atom or molecule. In such an experiment a potassium atom in the gas phase is to be ionized by two different light beams, one of which has wavelength 650 nm. What is the maximum wavelength for the second beam that will cause two-photon ionization?

---

## CUMULATIVE PROBLEMS

\* 89. Because of its particle nature, light can exert pressure on a surface. Suppose a 1.0-W laser is focused on a circular area 0.10 mm in radius. The wavelength of the laser light is 550 nm. Assume that the photons travel perpendicular to the surface and are perfectly reflected, so that their momenta change sign after reflection from the wall. Using results from the kinetic theory of gases (see Section 4.5), calculate the pressure exerted by the light on the surface. The momentum $p$ of a photon is related to its energy through $E = pc$.

90. As stated in Example 15.1, microwave ovens use radiation with a frequency of $2.45 \times 10^9$ $s^{-1}$. Calculate the number of such photons that must be absorbed to raise the temperature of 100.0 g of water from 15°C to its boiling point, 100°C. Take the specific heat capacity of water to be 4.2 J $K^{-1}$ $g^{-1}$.

91. The standard energy change (at 25°C and atmospheric pressure) for the process

$$K(g) + Cl(g) \longrightarrow KCl(s)$$

has been measured to be $-653$ kJ. Use this fact, together with data on ionization energies and electron affinities from Appendix F, to calculate the lattice energy per mole of the KCl lattice, defined as the energy change required to separate the solid into its component ions:

$$KCl(s) \longrightarrow K^+(g) + Cl^-(g)$$

\* 92. Chapter 14 stated that the nuclei of atoms contain protons and neutrons rather than protons and electrons.

(a) Show by a calculation using the Heisenberg uncertainty principle, $\Delta p \, \Delta x \geq h/4\pi$, that an electron cannot be confined within a nucleus. Assume the radius of a nucleus to be $\sim 1 \times 10^{-15}$ m.

(b) Repeat the calculation for a proton or neutron inside the nucleus.

# Quantum Mechanics and Molecular Structure

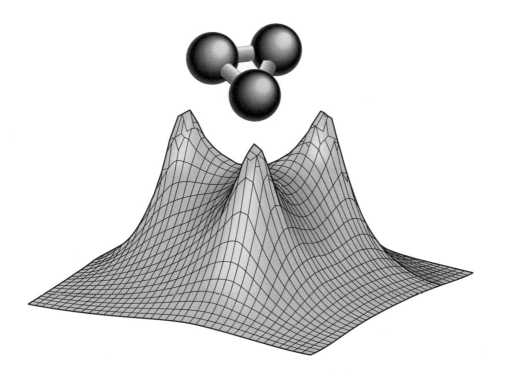

**M**odern quantum mechanics embraces the disquieting notion that matter and energy have dual nature—part particle and part wave—and from this notion it correctly predicts the sizes, energy levels, and spectra of atoms and ions. This is a substantial achievement. But if quantum mechanics could do no more, that subject would hold little interest for chemists, because chemistry is concerned with the atomic interactions that form molecules and extended solid structures with useful properties. In fact, quantum mechanics goes far beyond atoms and gives a firm conceptual foundation to the simple ideas of bonding used in the Lewis electron-dot model.

In Chapter 3 we posed the question why some combinations of atoms form molecules of great stability but others join together only for fleeting moments or not at all. The Lewis electron dot model rationalized the formation of stable compounds based on the octet rule and the relative positions of the participating atoms in the periodic table, but it provided no fundamental explanation of bond formation in terms of forces and energy. The origin of molecular shapes, that is, the three-dimensional arrangement of atoms in space, was explored by supplementing the

*Illustration*
The electron density in a delocalized three-center chemical bond for $H_3^+$ calculated by quantum mechanics. (*Courtesy of Dr. Richard P. Muller and Professor William A. Goddard III, California Institute of Technology.*)

Lewis model with VSEPR theory. Again, the answer based on the Lewis model rationalized experimental results but did not provide a fundamental explanation. A third question, why oxygen is paramagnetic and reactive when nitrogen is diamagnetic and relatively inert, is unanswered by the simple Lewis dot model.

In this chapter we revisit those questions and answer new ones as well by extending the quantal concepts and methods introduced in Chapter 15 to molecular structure. *Molecular* orbitals emerge as a natural extension of atomic orbitals and help to explain the stability of molecules and the nature of their bonds. We develop the electron configurations of molecules and present an aufbau principle analogous to that for atoms. The rearrangement of electrons in polyatomic molecules leads to the VSEPR model for molecular geometry. Our overall conceptual goal is to understand how changes in the wave functions of electrons in atoms lead to the formation of chemical bonds.

The second half of the chapter is devoted to experimental studies of molecular structure, which test the predictions of the quantal description. Developed as an extension of atomic spectroscopy, molecular spectroscopy exploits the interaction of light with molecules to measure molecular bond energies, bond lengths, and bond angles.

The chapter concludes with two illustrations that require both theoretical and experimental studies of molecular structure. Conjugated systems include intensely colored dye molecules, molecules of biological significance, and the recently discovered class of carbon materials called fullerenes. Atmospheric photochemistry deals with important environmental problems caused by interaction of sunlight with substances released into the atmosphere by living species and industrial processes.

---

## 16.1

## MOLECULAR ORBITALS IN DIATOMIC MOLECULES

Section 3.4 introduced a simple classical model of the covalent bond by pointing out that electron pairs between nuclei exert electrostatic forces that pull the nuclei together. Chapter 15 showed that electrons must be described by quantum mechanics. In particular, the Heisenberg uncertainty principle requires that an electron cannot simply be "at" some point in space but must be described as occupying an orbital with a characteristic wave function whose square is the probability density for finding the electron at a point. In a molecule, such an orbital is spread out, or *delocalized*, over more than one atom and is called a **molecular orbital.** This section discusses the molecular orbitals of diatomic molecules, and the following section considers polyatomic molecules.

### Molecular Orbitals and Covalent Bonding

The simplest bonded species is the one-electron molecular ion $H_2^+$. The wave function of the electron is the solution of the Schrödinger equation for an electron bound in the electrostatic field of *two* nuclei (protons, here) separated by a distance $R$ (see

Fig. 3.3). A full calculation can be carried out numerically, but for many purposes a simpler description is sufficient and provides useful insight. When the electron is very close to nucleus A, it experiences a potential that is not very different from that in an isolated hydrogen atom. The ground-state wave function for the electron near A should therefore resemble a $1s$ atomic wave function $\psi_{1s}^A$. Near B, the wave function should resemble $\psi_{1s}^B$. A simple way to construct a molecular wave function with these properties is to take it to be a linear combination of both atomic wave functions, obtained by adding or subtracting the two with constant coefficients:

$$\psi_{mol} = C_A \psi_{1s}^A + C_B \psi_{1s}^B$$

This wave function is an approximate molecular orbital for the electron in $H_2^+$. The constants $C_A$ and $C_B$ give the relative weights of the two atomic orbitals. If $C_A$ were greater in magnitude than $C_B$, the $\psi_{1s}^A$ orbital would be more heavily weighted and the electron would be more likely to be found near nucleus A and vice versa. Because the two nuclei in the $H_2^+$ ion are identical, however, the electron should have the same probability density near one as near the other. Therefore, the magnitudes of $C_A$ and $C_B$ must be equal and either $C_A = C_B$ or $C_A = -C_B$. For these two choices $(\psi_{mol})^2$ is symmetric in the two nuclei.

In this way two molecular orbitals are constructed, designated $\sigma_{1s}$ and $\sigma_{1s}^*$, with the Greek letter $\sigma$ (sigma) indicating that the electron density is distributed symmetrically about the bond axis:

$$\sigma_{1s} = C_1[\psi_{1s}^A + \psi_{1s}^B]$$

$$\sigma_{1s}^* = C_2[\psi_{1s}^A - \psi_{1s}^B]$$

where $C_1$ and $C_2$ are proportionality constants chosen to ensure that the total probability of finding the electron *somewhere* is unity. The distributions of electron probability are obtained by squaring the wave functions:

$$[\sigma_{1s}]^2 = C_1^2[(\psi_{1s}^A)^2 + (\psi_{1s}^B)^2 + 2\psi_{1s}^A\psi_{1s}^B]$$

$$[\sigma_{1s}^*]^2 = C_2^2[(\psi_{1s}^A)^2 + (\psi_{1s}^B)^2 - 2\psi_{1s}^A\psi_{1s}^B]$$

These can be compared with the electron probability distribution for a noninteracting (n.i.) system (obtained by averaging the probabilities for $H_A + H_B^+$ and $H_A^+ + H_B$), which is

$$\psi^2(\text{n.i.}) = C_3^2[(\psi_{1s}^A)^2 + (\psi_{1s}^B)^2]$$

The values of $(\psi_{mol})^2$ along the internuclear axis are plotted for these various forms in Figure 16.1 (right side). An electron in a $\sigma$ orbital has an enhanced probability of being found where the two atomic orbitals overlap (between the nuclei), and so $\sigma$ is a **bonding orbital.** In contrast, an electron in a $\sigma^*$ orbital has a *reduced* probability of being found between the nuclei; thus, $\sigma^*$ is an **antibonding orbital.** Note that the $\sigma^*$ orbital has a node between the two nuclei, and so the probability density of finding the electron at that point vanishes. Figure 16.2 shows the potential energy of the $H_2^+$ ion in its $\sigma_{1s}$ and $\sigma_{1s}^*$ states. The force between the nuclei in the antibonding state is everywhere repulsive, but in the bonding state the nuclei are attracted to each other and form a bound state at the distance corresponding to the lowest potential energy. The antibonding orbital has a higher potential energy because it has a node, and so in the ground state of $H_2^+$ the electron occupies the $\sigma_{1s}$ molecular orbital.

A molecular orbital, like an atomic orbital, can hold two electrons, one with spin up and the other with spin down. The ground-state $H_2$ molecule therefore

**FIGURE 16.1** Antibonding and bonding molecular orbitals of $H_2^+$ along the internuclear axis. For comparison, the red lines show the independent atomic orbitals and the electron probability distribution $\psi^2$(n.i.) for a noninteracting system. Using this reference system, the bonding orbital shows *increased* probability density between the nuclei, but the antibonding orbital shows *decreased* probability density in this region.

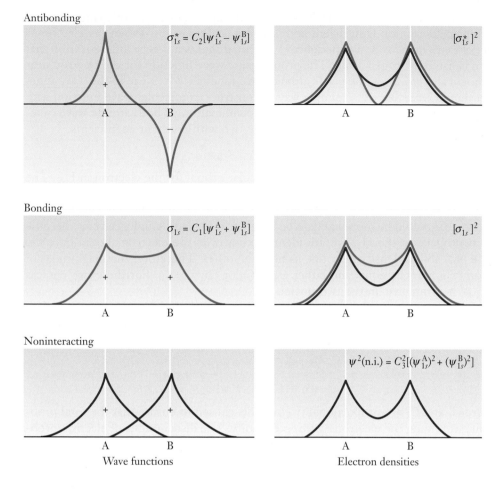

Antibonding

$$\sigma_{1s}^* = C_2[\psi_{1s}^A - \psi_{1s}^B]$$

$$[\sigma_{1s}^*]^2$$

Bonding

$$\sigma_{1s} = C_1[\psi_{1s}^A + \psi_{1s}^B]$$

$$[\sigma_{1s}]^2$$

Noninteracting

$$\psi^2(\text{n.i.}) = C_3^2[(\psi_{1s}^A)^2 + (\psi_{1s}^B)^2]$$

Wave functions                     Electron densities

accommodates two electrons with opposite spins in a $\sigma_{1s}$ bonding molecular orbital. In bonding molecular orbitals, covalent bonding arises from the sharing of electrons (most often electron pairs with opposite spins). The average electron density is greatest between the nuclei and tends to pull them together. Electron sharing alone is not sufficient for chemical bond formation. Electrons that are shared in an *antibonding* molecular orbital tend to force the nuclei apart, reducing the bond strength.

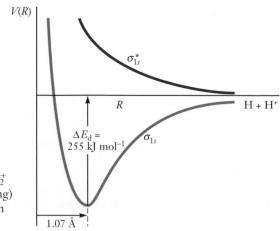

**FIGURE 16.2** Potential energy of $H_2^+$ in a $\sigma_{1s}$ (bonding) and $\sigma_{1s}^*$ (antibonding) molecular orbital, shown as a function of internuclear separation $R$.

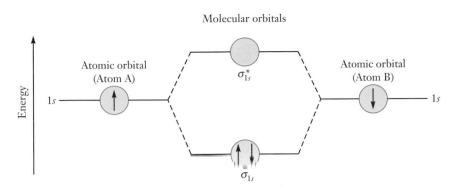

**FIGURE 16.3**  Correlation diagram for first-period diatomic molecules. The red arrows indicate the electron filling for the $H_2$ molecule. All of the atomic electrons are pooled and used to fill the molecular orbitals using the aufbau principle. In the molecules, electrons no longer have an atomic identity.

The energy difference between the two molecular orbitals is illustrated by a **correlation diagram** (Fig. 16.3), in which two $1s$ atomic orbitals are combined to give a lower energy $\sigma_{1s}$ molecular orbital and a higher energy $\sigma_{1s}^*$ molecular orbital. We can then apply an aufbau principle, analogous to that for atomic orbitals, in which electrons available from the two atoms are "fed" into the molecular orbitals, starting with the one of lowest energy. At most two electrons can occupy each molecular orbital. In the Lewis theory of chemical bonding, a shared pair of electrons corresponds to a single bond, two shared pairs to a double bond and so forth. In the molecular orbital picture, the ground-state $H_2$ molecule, which has a pair of electrons in a bonding molecular orbital, has a single bond (i.e., a **bond order** of 1). Electrons in antibonding orbitals reduce the bond strength and therefore also the bond order. In summary, the bond order can be defined as follows:

$$\text{bond order} = \tfrac{1}{2}(\text{number of electrons in bonding molecular orbitals}$$
$$- \text{ number of electrons in antibonding molecular orbitals})$$

## EXAMPLE 16.1

Give the ground-state electron configuration and the bond order of the $He_2^+$ molecular ion.

### Solution

The $He_2^+$ ion has three electrons, which are placed in molecular orbitals to give the ground-state configuration $(\sigma_{1s})^2(\sigma_{1s}^*)^1$, indicating that the ion has a doubly occupied $\sigma_{1s}$ orbital (bonding) and a singly occupied $\sigma_{1s}^*$ orbital (antibonding). The bond order is

$$\text{bond order} = \tfrac{1}{2}(2 \text{ electrons in } \sigma_{1s} - 1 \text{ electron in } \sigma_{1s}^*) = \tfrac{1}{2}$$

This should be a weaker bond than that in $H_2$.

## Homonuclear Diatomic Molecules

Table 16.1 lists the molecular orbital configurations of **homonuclear** diatomic molecules and molecular ions made from first-period elements. These configurations are simply the occupied molecular orbitals in order of increasing energy, together with the number of electrons in each orbital. Higher bond order corresponds to

**TABLE 16.1**

**Configurations and Bond Orders for First-Row Diatomic Molecules**

| Species | Electron Configuration | Bond Order | Bond Energy (kJ mol$^{-1}$) | Bond Length (Å) |
|---------|------------------------|------------|------------------------------|-----------------|
| $H_2^+$ | $(\sigma_{1s})^1$ | $\frac{1}{2}$ | 255 | 1.06 |
| $H_2$ | $(\sigma_{1s})^2$ | 1 | 431 | 0.74 |
| $He_2^+$ | $(\sigma_{1s})^2(\sigma_{1s}^*)^1$ | $\frac{1}{2}$ | 251 | 1.08 |
| $He_2$ | $(\sigma_{1s})^2(\sigma_{1s}^*)^2$ | 0 | ~0 | Large |

higher bond energies and shorter bond lengths. The species $He_2$ has bond order 0 and does not form a true chemical bond.

A general prescription for obtaining a molecular orbital description from atomic orbitals can now be stated:

**1.** Form linear combinations of atomic orbitals to give molecular orbitals. The total number of molecular orbitals formed in this way must equal the number of atomic orbitals used.
**2.** Place the molecular orbitals in order from lowest to highest energy.
**3.** Put in electrons (at most two per molecular orbital), starting from the orbital of lowest energy. Apply Hund's rules where appropriate.

This procedure is readily applied to second-period homonuclear diatomic molecules. The $2s$ atomic orbitals of the two atoms are combined in the same fashion as $1s$ orbitals, giving a $\sigma_{2s}$ bonding orbital and a $\sigma_{2s}^*$ antibonding orbital. The $2p$ orbitals form different molecular orbitals depending on whether they are parallel or perpendicular to the internuclear (bond) axis. The $z$ axis is taken to lie along the bond. Then the $2p_z$ orbitals of the two atoms combine to give either a bonding $\sigma_{2p_z}$ or an antibonding $\sigma_{2p_z}^*$ molecular orbital (Fig. 16.4). The bonding orbital has increased electron density between the nuclei, whereas the antibonding orbital has a node.

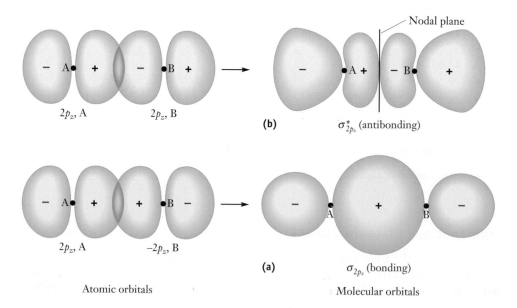

**FIGURE 16.4** Formation of (a) $\sigma_{2p_z}$ bonding and (b) $\sigma_{2p_z}^*$ antibonding molecular orbitals from $2p_z$ orbitals on atoms A and B. The signs refer to phases of the wave function.

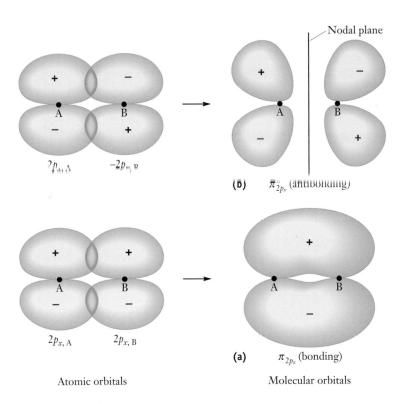

**FIGURE 16.5**   Formation of (a) $\sigma_{2p_x}$ bonding and (b) $\sigma^*_{2p_x}$ antibonding molecular orbitals from $2p_x$ orbitals on atoms A and B.

Nodal plane

**(b)**   $\pi^*_{2p_x}$ (antibonding)

$2p_{x, A}$   $-2p_{x, B}$

$2p_{x, A}$   $2p_{x, B}$

**(a)**   $\pi_{2p_x}$ (bonding)

Atomic orbitals          Molecular orbitals

These orbitals are given the designation "$\sigma$" because, like the $\sigma_{1s}$ and $\sigma^*_{1s}$ orbitals, their electron density is symmetrically distributed about the internuclear axis.

The two $2p_x$ orbitals, which are perpendicular to the bond axis, also combine to form a bonding molecular orbital and an antibonding molecular orbital (Fig. 16.5). These orbitals have a nodal plane containing the internuclear axis (in this case, the $yz$ plane) and are designated by the letter $\pi$ rather than $\sigma$. The $\pi_{2p_x}$ orbital is bonding, and the $\pi^*_{2p_x}$ is antibonding. In the same way, $\pi_{2p_y}$ and $\pi^*_{2p_y}$ orbitals can be formed from the $2p_y$ atomic orbitals. Their lobes project above and below the $xz$ nodal plane, which is the plane of the page in Figure 16.5.

It is important to remember that the + and − signs in Figures 16.4 and 16.5 refer not to positive and negative *charges* but to positive and negative *phases* of the wave function. The relative phases of the two atomic orbitals are critical in determining whether the resulting molecular orbital is bonding or antibonding. Bonding orbitals form from the overlap of wave functions with the same phase; antibonding orbitals form from the overlap of wave functions with opposite phase.

The next step is to determine the energy ordering of the molecular orbitals. In general, that would require a calculation beyond the scope of this book, but we can apply some simple qualitative rules:

1. The average energy of a bonding–antibonding pair of molecular orbitals lies approximately at the energy of the original atomic orbitals.
2. The energy difference between a bonding–antibonding pair becomes greater as the overlap of the atomic orbitals increases.

In second-period diatomic molecules, the $1s$ orbitals of the two atoms scarcely overlap; because the $\sigma_{1s}$ bonding and $\sigma^*_{1s}$ antibonding orbitals are both doubly

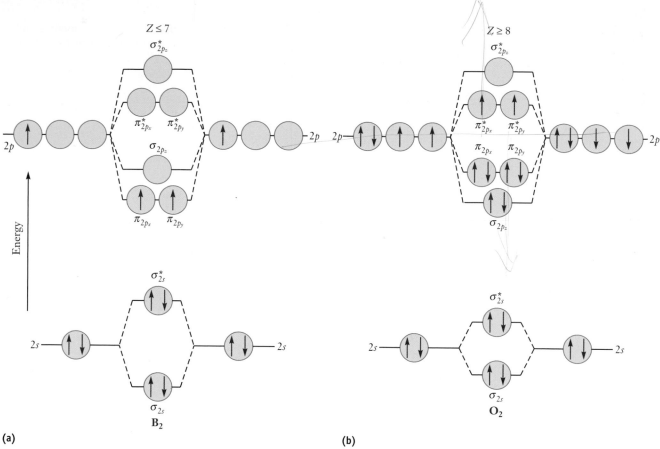

**FIGURE 16.6**  Correlation diagrams for second-period diatomic molecules. In each case, the atomic orbitals at the sides combine to give the molecular orbitals in the center. The arrows represent the valence electrons in $B_2$ and $O_2$, both paramagnetic molecules.

occupied, they have little net effect on bonding properties and need not be considered. For this reason, chemistry is dominated by valence electron behavior. Figure 16.6 comprises correlation diagrams for the molecular orbitals formed from the $2s$ and $2p$ orbitals. There are two different energy orderings for diatomic molecules formed from second-period elements. The first (Fig. 16.6a) applies to the molecules with atoms Li through N (i.e., the first part of the period) and their positive and negative ions. The second (Fig. 16.6b) applies to the later elements, O, F, and Ne, and their positive and negative ions. Usually the splitting between $\sigma_{2p}$ and $\sigma_{2p}^*$ orbitals is larger than that between $\pi_{2p}$ and $\pi_{2p}^*$ because $2p$ orbitals directed along the bond axis have greater overlap than those perpendicular to it, giving the result shown in Figure 16.6b. The different pattern of Figure 16.6a arises largely from electron–electron repulsions. The electron density from the filled $\sigma_{2s}$ and $\sigma_{2s}^*$ orbitals is concentrated along the bond axis. The $\pi_{2p}$ orbitals are filled before the $\sigma_{2p}$ orbitals because the electron density in a $\pi$ orbital is concentrated away from the bond axis, leading to reduced electron–electron repulsion. Toward the end of the second period (beginning with oxygen) the effective nuclear charge is great enough that the $\sigma_{2s}$ and $\sigma_{2s}^*$ electrons stay close to the nuclei and interact less strongly with electrons in a $\sigma_{2p_z}$ orbital.

The valence electrons are now placed into the molecular orbitals of Figure 16.6 to determine the electron configurations, and the bond orders can be calculated as before.

## EXAMPLE 16.2

Determine the ground-state electron configuration and bond order of the $F_2$ molecule.

### Solution

Each atom of fluorine has 7 valence electrons, and so 14 electrons are placed in the molecular orbitals to represent bonding in the $F_2$ molecule. The correlation diagram of Figure 16.6b gives the electron configuration

$$(\sigma_{2s})^2(\sigma_{2s}^*)^2(\sigma_{2p_z})^2(\pi_{2p})^4(\pi_{2p}^*)^4$$

Because there are eight valence electrons in bonding orbitals and six in antibonding orbitals, the bond order is

$$\text{bond order} = \tfrac{1}{2}(8 - 6) = 1$$

and the $F_2$ molecule has a single bond.

**Related Problems: 1, 2, 3, 4**

---

Table 16.2 summarizes the properties of second-period homonuclear diatomic molecules. Note the close relationship among bond order, bond length, and bond energy and the fact that the bond orders calculated from the molecular orbitals agree completely with the results of the Lewis electron-dot model.

An important prediction comes from the two correlation diagrams in Figure 16.6. In the ground states of both $B_2$ and $O_2$, the last two electrons should follow Hund's rules and have parallel spins. Thus, the molecular orbital approach predicts paramagnetism for $B_2$ and $O_2$ because they contain unpaired electrons. This paramagnetism is exactly what is found experimentally (Fig. 16.7). In contrast, in the Lewis electron-dot diagram for $O_2$,

$$:\overset{..}{O}::\underset{..}{O}:$$

## TABLE 16.2

### Molecular Orbitals of Homonuclear Diatomic Molecules

| Species | Number of Valence Electrons | Valence Electron Configuration | Bond Order | Bond Length (Å) | Bond Energy (kJ mol$^{-1}$) |
|---|---|---|---|---|---|
| $H_2$ | 2 | $(\sigma_{1s})^2$ | 1 | 0.74 | 431 |
| $He_2$ | 4 | $(\sigma_{1s})^2(\sigma_{1s}^*)^2$ | 0 | Observed only at very low temperatures | |
| $Li_2$ | 2 | $(\sigma_{2s})^2$ | 1 | 2.67 | 105 |
| $Be_2$ | 4 | $(\sigma_{2s})^2(\sigma_{2s}^*)^2$ | 0 | 2.45 | 9 |
| $B_2$ | 6 | $(\sigma_{2s})^2(\sigma_{2s}^*)^2(\pi_{2p})^2$ | 1 | 1.59 | 289 |
| $C_2$ | 8 | $(\sigma_{2s})^2(\sigma_{2s}^*)^2(\pi_{2p})^4$ | 2 | 1.24 | 599 |
| $N_2$ | 10 | $(\sigma_{2s})^2(\sigma_{2s}^*)^2(\pi_{2p})^4(\sigma_{2p_z})^2$ | 3 | 1.10 | 942 |
| $O_2$ | 12 | $(\sigma_{2s})^2(\sigma_{2s}^*)^2(\sigma_{2p_z})^2(\pi_{2p})^4(\pi_{2p}^*)^2$ | 2 | 1.21 | 494 |
| $F_2$ | 14 | $(\sigma_{2s})^2(\sigma_{2s}^*)^2(\sigma_{2p_z})^2(\pi_{2p})^4(\pi_{2p}^*)^4$ | 1 | 1.41 | 154 |
| $Ne_2$ | 16 | $(\sigma_{2s})^2(\sigma_{2s}^*)^2(\sigma_{2p_z})^2(\pi_{2p})^4(\pi_{2p}^*)^4(\sigma_{2p_z}^*)^2$ | 0 | Observed only at very low temperatures | |

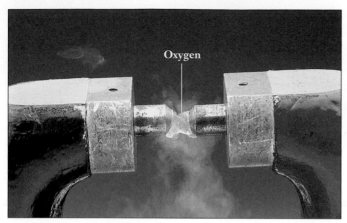

**(a)**

**(b)**

**FIGURE 16.7**  (a) Oxygen is paramagnetic; liquid oxygen poured between the pole faces of a magnet is attracted and held there.
(b) When the experiment is repeated with liquid nitrogen, which is diamagnetic, the liquid pours straight through. *(Larry Cameron)*

all the electrons appear to be paired. Moreover, the high chemical reactivity of molecular oxygen can be rationalized as resulting from the readiness of the two $\pi^*$ electrons, unpaired and in different regions of space, to find additional bonding partners in other molecules.

## Heteronuclear Diatomic Molecules

Diatomic molecules such as CO and NO, formed from atoms of two different elements, are called **heteronuclear.** To construct molecular orbitals from the $2s$ atomic orbital of carbon and the $2s$ atomic orbital of oxygen, for example, the atomic wave functions combine linearly. The combinations produce a bonding molecular orbital (without a node)

$$\sigma_{2s} = C_A \psi_{2s}^A + C_B \psi_{2s}^B$$

and an antibonding molecular orbital (with a node)

$$\sigma_{2s}^* = C_A' \psi_{2s}^A - C_B' \psi_{2s}^B$$

where A and B refer to the two different atoms in the molecule. In the homonuclear case we argued that $C_A = C_B$ and $C_A' = C_B'$ because the electron should spend the same amount of time near one nucleus as near the other. If the two nuclei are different, however, such reasoning does not apply. If atom B is more electronegative than atom A, then the lower energy $\sigma_{2s}$ orbital will have $C_B > C_A$ (so that the electron spends more time on the electronegative atom); the higher energy $\sigma_{2s}^*$ orbital will have $C_A' > C_B'$ and more closely resemble a $2s_A$ atomic orbital.

Molecular orbital correlation diagrams show the energy levels of the more electronegative atom displaced *downward*, because that atom attracts valence electrons more strongly than does the less electronegative atom. Figure 16.8 shows the result for many heteronuclear diatomic molecules of second-period elements (those in which the electronegativity difference is not too great). This diagram can now be filled with the valence electrons to obtain the molecular orbital configuration. As an example, consider NO, which has 11 valence electrons (5 from N, 6 from O). Its ground-state configuration is

$$(\sigma_{2s})^2 (\sigma_{2s}^*)^2 (\pi_{2p})^4 (\sigma_{2p_z})^2 (\pi_{2p}^*)^1$$

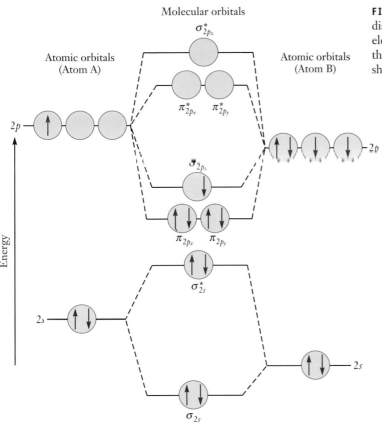

Molecular orbitals

**FIGURE 16.8** Correlation diagram for heteronuclear diatomic molecules, AB. The atomic orbitals for the more electronegative atom (B) are displaced downward because they have lower energies than those for A. The orbital filling shown is that for BO.

From the fact that eight electrons are in bonding orbitals and three are in antibonding orbitals, NO has a bond order of $\frac{1}{2}(8 - 3) = 2\frac{1}{2}$ and is paramagnetic. Its bond energy is predicted to be (and is) smaller than the bond energy of CO, which has one fewer electron and a bond order of 3.

Atomic orbitals mix significantly to form molecular orbitals only if they are fairly close in energy and have appropriate symmetries. In the HF molecule, for example, the $1s$ orbital of the F atom is far too low in energy to mix with the hydrogen $1s$ orbital, and the same is true of the $2s$ F orbital (Fig. 16.9a). The net overlap of the $1s$ orbital of hydrogen with the $2p_x$ or $2p_y$ F orbital is zero (Fig. 16.9b), because the positive and negative phases of the combined wave function sum to zero. This leaves only the $2p_z$ orbital, which combines with the hydrogen $1s$ orbital to give both $\sigma$ and $\sigma^*$ orbitals (Fig. 16.9c and d). Figure 16.10 is the correlation diagram for HF. The $2s$, $2p_x$, and $2p_y$ orbitals of fluorine do not mix with the $1s$ of hydrogen and so remain as atomic (nonbonding) states. Electrons in these orbitals do not contribute significantly to the chemical bonding. Because fluorine is more electronegative than hydrogen, its $2p$ orbitals lie below the $1s$ hydrogen orbital in energy. The $\sigma$ orbital then contains more fluorine $2p_z$ character, and the $\sigma^*$ orbital more closely resembles a hydrogen $1s$ atomic orbital. When the eight valence electrons are put in for HF, the result is the molecular orbital configuration

$$(2s_F)^2(\sigma)^2(2p_F)^4$$

The net bond order is 1 because electrons in nonbonding atomic orbitals do not affect bond order. The electrons in the $\sigma$ orbital are more likely to be found near

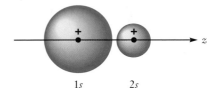

**(a)** Nonbonding
(negligible overlap)

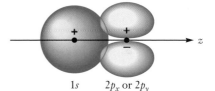

**(b)** Nonbonding
(phases cancel)

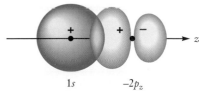

**(c)** Bonding, $\sigma$
(significant overlap)

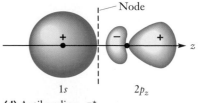

**(d)** Antibonding, $\sigma^*$

**FIGURE 16.9** Overlap of atomic orbitals in HF.

**FIGURE 16.10** Correlation diagram for HF. The $2s$, $2p_x$, and $2p_y$ atomic orbitals of fluorine do not mix with the $1s$ atomic orbital of hydrogen and so remain nonbonding.

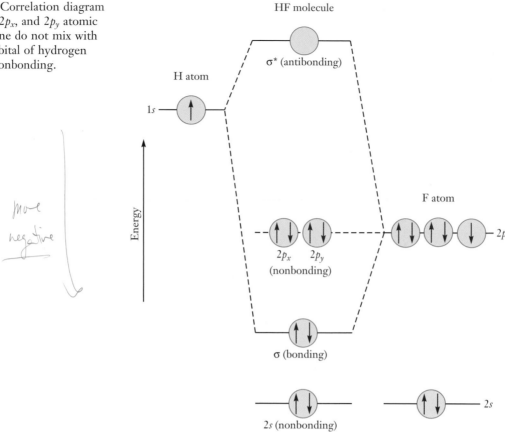

the fluorine atom than near the hydrogen, and so HF has the dipole moment $H^{\delta+}F^{\delta-}$.

If a more electropositive atom (such as Na or K) is substituted for H, the energy of its outermost $s$ orbital will be higher than that of the hydrogen atom, because its ionization energy is lower. In this case, the $\sigma$ orbital will resemble a fluorine $2p_z$ orbital even more (that is, the coefficient $C_F$ of the fluorine wave function will be close to 1, and $C_A$ for the alkali atom will be very small). In this limit the molecule can be described as having the valence electron configuration $(2s_F)^2(2p_F)^6$, which corresponds to the ionic species $Na^+F^-$ or $K^+F^-$. The magnitudes of the coefficients in the molecular orbital wave function are thus closely related to the ionic–covalent character of the bonding and to the dipole moment.

---

### 16.2

## MOLECULAR ORBITALS IN POLYATOMIC MOLECULES

There are two ways to describe bonding in polyatomic molecules. The first employs *delocalized* molecular orbitals, with electron wave functions that are spread out over the entire molecule. These molecular orbitals are formed as linear combinations of valence atomic orbitals, then placed in order of increasing energy, and finally filled with the available valence electrons. The electrons that lead to bonding are spread out over all the atoms in the molecule. One problem with this approach is of a fundamentally chemical nature. Bonds have fairly universal characteristics such as energy and length (see Section 3.3). If electrons are described by molecular orbitals that are spread out over the entire molecule, then why should the properties of a bond be so weakly dependent on the particular molecule involved? Out of such chemical considerations a second approach was developed that employs molecular orbitals that are localized, either as bonding orbitals between particular pairs of atoms or as lone-pair orbitals on individual atoms. This approach is related to the qualitative Lewis model, in which a similar localization of electrons is assumed.

In this section, localized orbitals are used for $\sigma$ bonds and, wherever possible, for $\pi$ bonds. In some molecules, however, delocalized molecular orbitals are useful to describe $\pi$ bonds correctly. This approach gives physical insight with relatively few mathematical complications. The following section describes more general approaches that are useful for quantitatively accurate calculations.

## Hybridization

The $BeH_2$ molecule is readily described by the localized orbital method. This molecule has four valence electrons (two from the Be and one from each H atom), all of which are involved in bonding in the Lewis model. From the VSEPR theory, the steric number is 2, and so the molecule is predicted to be linear, with one hydrogen atom on each side of the central beryllium atom. In the localized orbital method, atomic orbitals of the central atom are mixed to form new, singly occupied *atomic* orbitals, each of which can pair with a singly occupied orbital of another atom to form a $\sigma$ bond along the internuclear axis. In $BeH_2$, the $2s$ and $2p_z$ orbitals of Be are mixed to form two **$sp$ hybrid atomic orbitals** on the central Be atom:

$$sp^-_{Be} = 2s - 2p_z$$

$$sp^+_{Be} = 2s + 2p_z$$

Figure 16.11a shows the form of these orbitals. An electron in a hybrid $sp^-_{Be}$ orbital is localized in that it is much more likely to be found on the left side of the nucleus

**FIGURE 16.11** (a) The two $sp$ hybrid orbitals formed from the $2s$ and $2p_z$ orbitals on the Be atom. (b) The two $\sigma$ bonds that form from the overlap of the $sp$ hybrid orbitals with the hydrogen $1s$ orbitals, making two single bonds in the $BeH_2$ molecule.

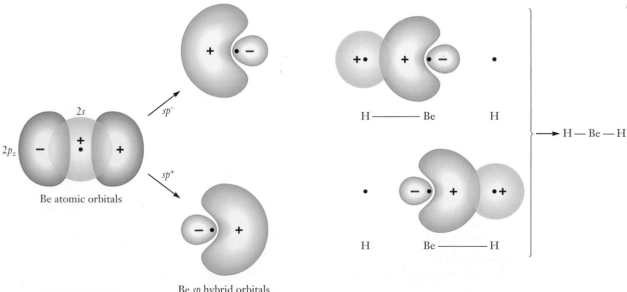

(a)    Be atomic orbitals    Be $sp$ hybrid orbitals    (b)

than on the right; the opposite is true of an electron in the $sp_{Be}^{+}$ orbital. Once the hybrid atomic orbitals of the central atom form, the two hydrogen atoms can be brought near, and each shares its electron with the corresponding hybrid orbital to form two single $\sigma$ bonds (Fig. 16.11b).

What happens in the hydrides of the other second-period elements? The boron atom has valence electron configuration $2s^2 2p$, and from the VSEPR theory $BH_3$ would have steric number 3, corresponding to a trigonal planar structure. Promotion of one of the $2s$ electrons creates the excited-state configuration $2s^1 2p_x^1 2p_y^1$, and these three atomic orbitals can be mixed to form the three equivalent **$sp^2$ hybrid atomic orbitals** shown in Figure 16.12. They lie in a plane with an angle of 120° between them. Experimentally, $BH_3$ molecules turn out to be unstable and react rapidly to form $B_2H_6$ or other higher compounds called "boranes." However, the closely related $BF_3$ molecule has the trigonal planar geometry characteristic of $sp^2$ hybridization.

In $CH_4$, VSEPR theory assigns a steric number of 4 to the central carbon atom. The $2s$ and three $2p$ orbitals of the central carbon atom can be combined to form four equivalent **$sp^3$ hybrid atomic orbitals,** which point toward the vertices of a tetrahedron (Fig. 16.13). Each can overlap a $1s$ orbital of one of the hydrogen atoms to give an overall tetrahedral structure for $CH_4$.

Lone-pair electrons can occupy hybrid orbitals. The nitrogen atom in $NH_3$ also has steric number 4, and its bonding can be described in terms of $sp^3$ hybridization. Of the eight valence electrons in $NH_3$, six are involved in $\sigma$ bonds between nitrogen and hydrogen, and the other two occupy the fourth $sp^3$ hybrid orbital as a lone pair. Oxygen in $H_2O$ likewise has steric number 4 and can be described with $sp^3$

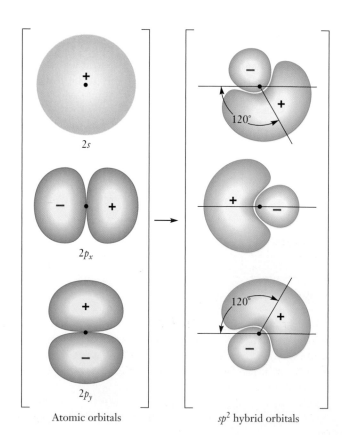

$2s$

$2p_x$

$2p_y$

Atomic orbitals

$sp^2$ hybrid orbitals

**FIGURE 16.12**  Shapes and relative dispositions of the three $sp^2$ hybrid orbitals, formed from the $2s$, $2p_x$, and $2p_y$ atomic orbitals on a single atom.

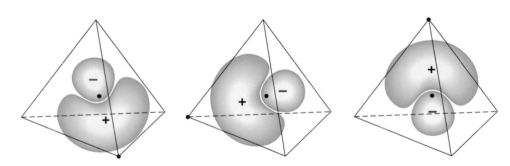

hybridization, with two lone pairs in $sp^3$ orbitals. This hybrid orbital description for $NH_3$ and $H_2O$ is consistent with bond angles close to the central angle of a tetrahedron (109.5°).

There is a close relationship between the VSEPR theory of Section 3.6 and the hybrid orbital approach, with steric numbers of 2, 3, and 4 corresponding to $sp$, $sp^2$, and $sp^3$ hybridization, respectively. Both theories are based on the energy savings resulting from the reduction of electron–electron repulsion. The physical importance of hybrid orbitals should not be overemphasized, though; they are really no more than a device to rationalize observed bond angles and equivalent bonds in some molecules.

## EXAMPLE 16.3

Predict the structure of hydrazine ($H_2NNH_2$) by writing down its Lewis diagram and using VSEPR theory. What is the hybridization of the two nitrogen atoms?

### Solution

The Lewis diagram is

$$
\begin{array}{c}
\phantom{H:}\overset{\displaystyle H}{\phantom{.}} \\
H : \overset{..}{\underset{..}{N}} : \overset{..}{\underset{..}{N}} : H \\
\phantom{H:}\underset{\displaystyle H}{\phantom{.}}
\end{array}
$$

Both nitrogen atoms have steric number 4 and are $sp^3$ hybridized, with H—N—H and H—N—N angles of approximately 109.5°. The extent of rotation about the N—N bond cannot be predicted from the VSEPR theory or the hybrid orbital model. Figure 16.14 shows the full three-dimensional structure of hydrazine.

**Related Problems: 15, 16, 17, 18**

## Triatomic Nonhydrides

Many triatomic molecules and molecular ions can be formed from atoms of elements in the second and third periods. Some are linear (examples are $CO_2$, $N_2O$, $OCS$, and $NO_2^+$), and others are bent ($NO_2$, $O_3$, $NF_2$, and $NO_2^-$). We use localized orbitals for the $\sigma$ bonds and delocalized orbitals for the $\pi$ bonds of these molecules.

For linear molecules it is logical to use $sp$ hybridization to describe the central atom. This leaves two $p$ orbitals ($p_x$ and $p_y$) perpendicular to the bond axis. Hybridization of the outer atoms is not necessary to treat such molecules; the outer atoms are

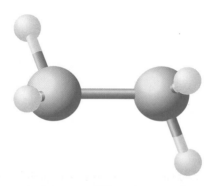

**FIGURE 16.14** The structure of hydrazine, $N_2H_4$.

**FIGURE 16.15** Pi bonding in linear triatomic molecules. From three *p* orbitals lying perpendicular to the bond axis can be constructed one bonding, one antibonding, and one nonbonding molecular orbital. A second group of molecular orbitals can be constructed from the *p* orbitals that lie perpendicular to the plane of the page.

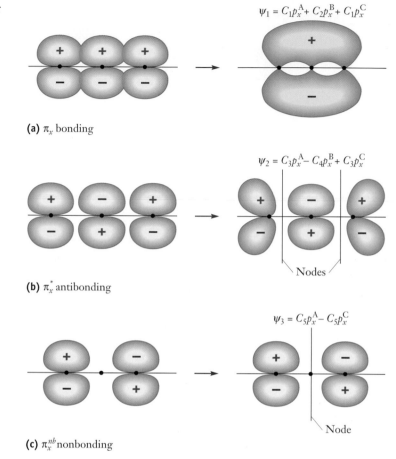

$$\psi_1 = C_1 p_x^A + C_2 p_x^B + C_1 p_x^C$$

**(a)** $\pi_x$ bonding

$$\psi_2 = C_3 p_x^A - C_4 p_x^B + C_3 p_x^C$$

Nodes

**(b)** $\pi_x^*$ antibonding

$$\psi_3 = C_5 p_x^A - C_5 p_x^C$$

Node

**(c)** $\pi_x^{nb}$ nonbonding

most simply described as having fully occupied *s* orbitals not involved in bonding, a $p_z$ orbital that takes part in $\sigma$ bonding, and $p_x$ and $p_y$ orbitals that form delocalized $\pi$ bonds with the other atoms. The $\sigma$ bonds (which extend along the bond axes) can be obtained by combining the $p_z$ orbital of each of the outer atoms with the *sp* hybrid orbital of the central atom that points toward it to form a localized bonding orbital. These two localized $\sigma$ bonds are filled by two pairs of electrons.

Linear combinations of the $p_x$ and $p_y$ orbitals of all three atoms make up the delocalized molecular orbitals involved in $\pi$ bonding. How do three $p_x$ orbitals combine to form three molecular orbitals? Two of the combinations are obvious. One (Fig. 16.15a) is fully bonding, with no nodes, and another (Fig. 16.15b) is fully antibonding, with two nodes, one between each pair of atoms. The third linear combination is less obvious. Mathematical analysis shows that it has the form of Figure 16.15c, with a single node and with the coefficient of the wave function on the central atom being zero. This third orbital is called nonbonding ($\pi_x^{nb}$) because there is neither a node nor a region of increased electron density between the central atom and its neighbors. There is, however, weak antibonding between the outer atoms. The three $p_y$ orbitals can be combined to give the same types of molecular orbitals as the $p_x$. Figure 16.16 is the resulting correlation diagram for the $\pi$ levels only, with the nonbonding orbitals lying between the bonding and the antibonding orbitals in energy. As before, the energy increases with the number of nodes.

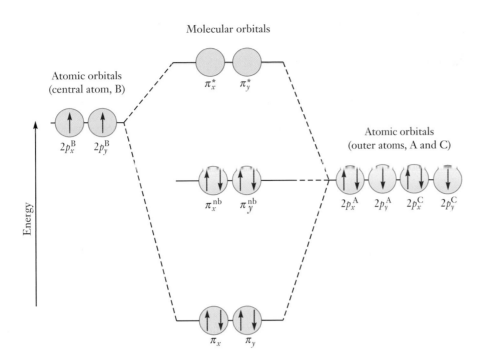

Molecular orbitals

Atomic orbitals
(central atom, B)

$2p_x^B$   $2p_y^B$

$\pi_x^*$   $\pi_y^*$

Atomic orbitals
(outer atoms, A and C)

$\pi_x^{nb}$   $\pi_y^{nb}$

$2p_x^A$   $2p_y^A$   $2p_x^C$   $2p_y^C$

$\pi_x$   $\pi_y$

Energy

**FIGURE 16.16** Correlation diagram for $\pi$ electrons in linear triatomic molecules. The orbital filling shown is that for $CO_2$.

Consider the $CO_2$ molecule. It has 16 valence electrons (4 from the carbon and 6 from each oxygen). Two electrons on each oxygen are nonbonding electrons in $2s$ atomic orbitals. Of the remaining 12 electrons, 4 take part in localized $\sigma$ bonds, 2 between each oxygen and the carbon. This leaves 8 electrons for the $\pi$ system. Placing them in the correlation diagram of Figure 16.16 gives two electron pairs in $\pi$ orbitals and two pairs in $\pi^{nb}$ orbitals. The total bond order for the molecule is 4, because there are two pairs of electrons in $\sigma$ bonding orbitals (from the $sp$ hybrids) and two pairs in $\pi$ bonding orbitals (the nonbonding $\pi^{nb}$ and atomic $2s_{oxygen}$ orbitals neither increase nor decrease the bond order). Each bond is then of order 2, in agreement with the Lewis diagram result for $CO_2$. The meaning is slightly different here, however, in that some of the bonding electrons (those in the $\pi$ orbital system) are delocalized over the full molecule rather than shared by only two of the atoms.

Nonlinear triatomic molecules can be described through $sp^2$ hybridization of the central atom. If the molecule lies in the $xy$ plane, then the $s$, $p_x$, and $p_y$ orbitals of the central atom can be combined to form three $sp^2$ hybrid orbitals with an angle close to 120° between each pair. One of these orbitals holds a lone pair of electrons, and the other two take part in $\sigma$ bonds with the outer atoms. The fourth orbital, a $p_z$ orbital, takes part in delocalized $\pi$ bonding (Fig. 16.17a). On the outer atoms, the $p$ orbital pointing toward the central atom takes part in a localized $\sigma$ bond and the $p_z$ orbital takes part in $\pi$ bonding; the third $p$ orbital and the $s$ orbital are atomic orbitals that do not participate in bonding. The three $p_z$ atomic orbitals can be combined into bonding, nonbonding, and antibonding $\pi$ orbitals much as in the linear molecule case (Fig. 16.17b). Here there is only one of each type of orbital ($\pi$, $\pi^{nb}$, $\pi^*$) rather than two as for linear molecules.

Consider a specific example, $NO_2^-$, with 18 electrons. Two electrons are placed in each oxygen $2s$ orbital and two more in each nonbonding oxygen $2p$ orbital, so that a total of eight electrons are localized on oxygen atoms. Two electrons also are

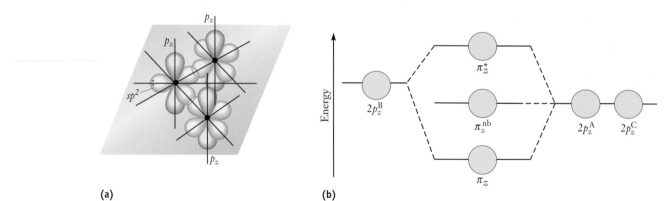

(a)

(b)

**FIGURE 16.17** (a) Molecular orbitals for bent triatomic molecules. The central atom has three $sp^2$ hybrid orbitals lying in the plane of the molecule (green). From the three $p_z$ orbitals perpendicular to this plane (blue), three $\pi$ orbitals can be constructed. (b) Correlation diagram for the $\pi$ orbitals.

placed as a lone pair in the third $sp^2$ orbital of the nitrogen atom. Of the remaining eight electrons, four are involved in the two $\sigma$ bonds between nitrogen and the oxygen atoms. The last four are placed into the $\pi$ electron system: two into the bonding $\pi$ orbital and two into the nonbonding $\pi^{nb}$ orbital. Because a total of six electrons are in bonding orbitals and none are in antibonding, the net bond order for the molecule is 3, or $1\frac{1}{2}$ per bond. In the Lewis model, two resonance forms are needed to represent $NO_2^-$. The awkwardness of the resonance model is avoided by treating the electrons in the bonding $\pi$ molecular orbital as delocalized over the three atoms in the molecule.

Experiment shows that triatomic nonhydrides with 16 or fewer valence electrons generally are linear, whereas those with 17 to 20 valence electrons generally are bent. This is consistent with the VSEPR theory, because the former have steric number 2 (giving linear molecules), whereas the latter have steric number 3 or 4 (giving bent molecules with one or two lone-pair orbitals on the central atom). For molecules involving larger atoms later in the periodic table, both the molecular orbital and hybridization methods become more complicated and less useful for qualitative interpretation. The success of the simple VSEPR theory in predicting molecular geometries is all the more striking for these heavier atoms.

# A  DEEPER  LOOK...

## 16.3   Quantum Calculations for Molecules

The wide availability of high-speed computers has greatly expanded the range of calculations of molecular properties using the principles of quantum mechanics. The simple molecular orbital approaches described up to now are not only useful for gaining qualitative insight; they also provide a starting point for highly accurate predictions of the properties of small molecules.

### The Variational Principle

The total potential energy of a molecule includes three terms:

$$V = V_{el-el} + V_{el-nuc} + V_{nuc-nuc}$$

These are the repulsion of each electron by each other electron, the attraction of each electron by each nucleus, and the mutual

repulsions of the nuclei. To proceed, let us take the nuclei to have fixed positions (which can be varied later), making the last of these terms an easily calculated constant. As discussed in Section 15.8, the wave function for an $N$-electron system depends on the positions of all $N$ electrons and can be written as $\psi(x_1, y_1, z_1, \ldots, x_N, y_N, z_N)$. The Schrödinger equation can be written in the symbolic form

$$\mathcal{H}\psi = E\psi$$

where $E$ is the energy of the molecule and $\mathcal{H}$ is an **operator** that acts on the wave function. This equation is a generalization of Equation 15.15 to many electrons in three spatial dimensions. The operator $\mathcal{H}$ involves second derivatives with respect to electron positions and multiplication by the total potential energy.

As stated in Chapter 15, exact solutions of the Schrödinger equation are not available for many-electron systems. Suppose, however, that an approximate solution $\psi_{app}$ is known. Then the corresponding approximate energy $E_{app}$ can be calculated according to the following prescription from quantum mechanics. Make the operator $\mathcal{H}$ act on $\psi_{app}$, multiply the result by $\psi_{app}$, and integrate over all the electron positions:

$$E_{app} = \int (\mathcal{H}\psi_{app})\psi_{app} \, dx_1 \ldots dz_N$$

Mathematical analysis shows that this approximate energy is always greater than or equal to the exact energy:

$$E_{app} \geq E$$

with the equals sign applying only if the exact wave function is used in the integral.

This result is known as the **variational principle** in quantum mechanics and is of fundamental importance in calculating molecular properties. It enables us to systematically improve an approximate wave function by varying one or more of its parameters. The "best" wave function is the one that gives the lowest total energy, and that approximate energy lies above the exact energy of the system.

## The Self-Consistent Field Approximation

If each electron in a molecule could be treated separately, calculation would be easy. However, the electron–electron repulsion term $V_{el-el}$ in the total potential energy couples the electrons in a complicated fashion. In the **self-consistent field (SCF) approximation** (a generalization of Hartree's method for atoms described in Section 15.8), the total wave function for the molecule is taken to be a product of single-electron wave functions. Each of these,

in turn, is written as a linear combination of atomic orbitals; the coefficients are determined by minimizing the energy, using the variational principle. In the SCF theory, each electron is taken to interact with the *average* field of the other $N - 1$ electrons, calculated from their single-electron wave functions. The process is iterative, with a series of one-electron variational calculations that use the average field exerted by the remaining electrons. Each newly calculated wave function is reinserted and used to calculate an improved average field for the next electron. The process is continued until the wave functions stop changing; they are then said to be "self-consistent."

Large-scale SCF calculations (employing linear combinations of atomic orbitals) have been carried out for small molecules, with the bond lengths and angles further varied to find the stable (lowest energy) structures. Reasonably accurate predictions of molecular properties usually result. One example of a quantitative SCF calculation is the one for ammonia, $NH_3$. The calculated H—N—H bond angle for this molecule is 107.2° (experiment: 106.7°), and the equilibrium N—H bond lengths are 1.00 Å (experiment: 1.012 Å).

*Orbital* shapes and energies can also be compared with those from simple bonding theories. Figure 16.18 is the orbital energy correlation diagram for ammonia. The lowest energy ($a_1$) and $e$ orbitals are bonding molecular orbitals between the nitrogen atom and the three hydrogen atoms. They contain altogether six electrons, corresponding to three single bonds. The higher energy $a_1$ orbital, also doubly occupied, is largely a nonbonding lone-pair orbital on the nitrogen. Notice that this description of orbital energies is quite different from the result of the hybrid orbital picture. In that case, one would expect $sp^3$ hybridization to give three identical bonding orbitals and one lone pair. The discrepancy illustrates one limitation of the hybrid orbital picture. The

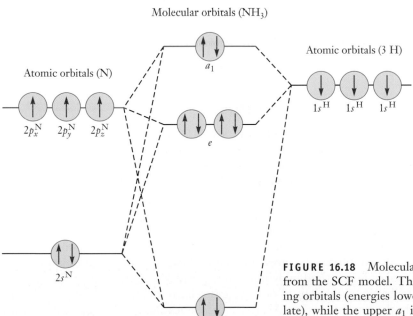

Molecular orbitals ($NH_3$)

Atomic orbitals (N)

Atomic orbitals (3 H)

$a_1$

$2p_x^N$    $2p_y^N$    $2p_z^N$

$1s^H$    $1s^H$    $1s^H$

$e$

$2s^N$

$a_1$

**FIGURE 16.18** Molecular orbital correlation diagram for $NH_3$, calculated from the SCF model. The lower energy $a_1$ orbital and the $e$ orbitals are bonding orbitals (energies lower than the atomic orbitals with which they correlate), while the upper $a_1$ is largely a nonbonding lone pair. Three higher energy antibonding orbitals (unoccupied in $NH_3$) are not shown.

total bonding electron distribution in $NH_3$ is evenly distributed among the three N—H bonds, but it arises from three molecular orbitals of which only two have equal energies.

Calculations on molecules such as ammonia are not limited to the equilibrium geometry. Suppose the H atoms are moved upward until the N atom lies in a plane with them. The H atoms can then "pop through" to the other side, giving an inverted geometry in a process analogous to the inversion of an umbrella in a strong wind. The inverted molecule has the same symmetry, structure, and energy as the original molecule, but the intermediate planar state has a higher energy. With the SCF method, this energy is calculated to lie 21 kJ mol$^{-1}$ above the stable pyramidal state (the experimental value is 24 kJ mol$^{-1}$). SCF calculations give potential energy surfaces for chemical reactions (of the type used to study reaction kinetics in Section 13.6), but they become less accurate in cases in which bonds are made or broken.

## Electron Correlation

The SCF method is a useful approximation, but it ignores one important fact: electrons actually respond not only to the average charge density of other electrons but to the instantaneous positions of those electrons. In other words, a **correlation** exists between different electron positions, and it is not correct to take the wave function to be a simple product of one-electron wave functions. One of the effects of correlation is to produce a long-range attractive force (see Section 4.7) between atoms and molecules, even when they are nonpolar. In a pair of argon atoms, for example, there is an instantaneous correlation between the positions of electrons on the two atoms, giving a fluctuating pair of dipole moments (Fig. 16.19) that interact with each other, lowering the energy of the pair and causing the two atoms to attract each other.

Electrons in molecules are also correlated. In the simple SCF picture that fact is ignored, and it is possible for both electrons in an electron-pair bond to be instantaneously located on one of

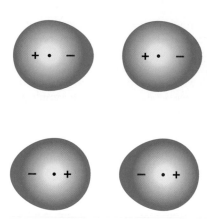

**FIGURE 16.19** The fluctuating charge distribution in an atom can give rise to an instantaneous dipole moment. When two atoms are close to each other, their fluctuating dipole moments are correlated, leading to attraction that lowers the potential energy of the pair.

the two bonded atoms. There are several ways to incorporate electron correlation into quantal calculations. One is to work with wave functions that depend explicitly on the distance between a pair of electrons—a difficult method for all but two-electron atoms and molecules. A second way is to choose wave functions that never allow a pair of bonding electrons to occupy the same atom, which is the basis of the **valence bond method.** In its simplest form, this approach tends to exaggerate the importance of electron correlation, but it serves as a starting point for more accurate approaches. A third way to include electron correlation is through **configuration interaction.** In this approach, the SCF wave function is supplemented with many other wave functions that represent excited states in the one-electron picture. Variational calculations then lead to highly accurate predictions of molecular properties.

---

### 16.4

## General Aspects of Molecular Spectroscopy

In Chapter 15 we discussed atomic spectroscopy and showed how the frequencies of the light absorbed or emitted by an atom are related to the energy differences between the atom's quantum states. In particular, the spectrum of the hydrogen atom can be interpreted in terms of solutions of the Schrödinger equation for that atom. Since molecules include several nuclei as well as electrons, their internal motions are more complicated than those of atoms, and their energy level diagrams and spectra consequently exhibit new features not seen in those for atoms. For example, rotation of the molecule as a rigid body leads to molecular transitions in the microwave region, and vibrational motions of one nucleus relative to another lead to molecular transitions in the infrared region. Analysis of these features gives useful information

## TABLE 16.3

### Spectroscopic Experiments

| Spectral Region | Frequency ($s^{-1}$) | Energy Levels Involved | Information Obtained |
|---|---|---|---|
| Radio waves | $10^7–10^9$ | Nuclear spin states | Electronic structure near the nucleus |
| Microwave, far infrared | $10^9–10^{12}$ | Rotational | Bond lengths and bond angles |
| Near infrared | $10^{12}–10^{14}$ | Vibrational | Stiffness of bonds |
| Visible, ultraviolet | $10^{14}–10^{17}$ | Valence electrons | Electron configuration |
| X-ray | $10^{17}–10^{19}$ | Core electrons | Core electron energies |

about molecular structure, bond lengths, and bond energies (Table 16.3). In fact, molecular spectroscopy is the main source of experimental information about chemical bonding and the structure and shape of molecules.

Molecular absorption spectra are recorded by *spectrophotometers* (Fig. 16.20), which differ from the spectrographs illustrated with atomic spectra in Figure 15.9. Light from the source is directed to a prism or grating to select a specific wavelength $\lambda$, which then is passed through the sample confined in a cell. The intensity of light transmitted through the sample cell, $I_S$, is measured. To remove spurious effects due to absorption or scattering of light at the cell walls, the incoming beam is actually split into two parts, one of which is passed through a reference cell identical to that confining the sample. The spectrophotometer can be calibrated to record a graph of either the transmission $T = I_S/I_R$ or the absorbance $A = \log [I_R/I_S]$ versus $\lambda$ as wavelength is scanned over the range of interest. The magnitude of the signal is related to properties of the sample as follows. The transmittance $T$ decreases (and the absorbance $A$ increases) as cell width $\ell$ increases:

$$-\log \frac{I_S}{I_R} = a\ell = A$$

The parameter $a$ is called the absorption, or extinction, coefficient. If the sample is a solution, then $T$ decreases and $A$ increases as the concentration $c$ (in mol $L^{-1}$) of the light-absorbing solute increases, as described by the Beer-Lambert law:

$$-\log \frac{I_S}{I_R} = c\epsilon\ell = A$$

**FIGURE 16.20** Schematic of double-beam spectrophotometer. Incoming light passes through a reference cell identical to the sample cell.

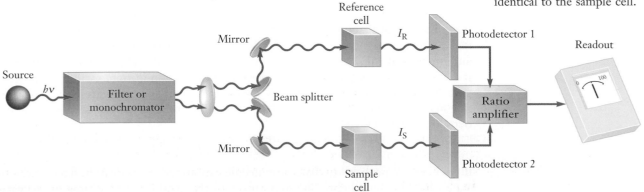

The parameter $\epsilon$, called the molar extinction coefficient, is a property of the light-absorbing solute in the solution; it measures the extent to which that species can absorb light at a particular wavelength.

Peaks in the graph at particular values of $\lambda$ correspond to transitions between molecular energy levels $E_i$ and $E_f$ that satisfy the relation $\Delta E = E_f - E_i = h\nu = hc/\lambda$. Positions of the peaks correlate with some feature of molecular structure associated with the molecular energy levels $E_i$ and $E_f$. In principle, these can be identified by solving Schrödinger's equation for the molecule. In practice, they are identified by comparison with extensive tables of spectral data already compiled; the peaks serve as "fingerprints" for identifying structural features. Representative spectra are shown in Figures 16.26, 16.32, and 16.37 and will be discussed later.

The area under each peak reflects the concentration of molecules present, as well as the *strength* of the absorption governed by $\epsilon$. This fact is illustrated by the solutions shown in Figure 9.1a and c, in which merely 2% concentration of the blue tetrachloro Co(II) species changes the color of the solution from pink to violet due to its very large molecular extinction coefficient.

The magnitude of $\epsilon$ is determined by the magnitude of the interaction between the absorbing molecule and the light wave and the difference in populations of the initial and final states involved in the transition. Detailed discussion of the interaction of light with molecules requires quantal concepts beyond the scope of this book. The essential physical effect is that the changing dipole moment of the molecule couples with the oscillating electric field of the light wave (see Fig. 15.2) and functions as a "molecular antenna" to absorb energy from the electric field. In order for energy to be absorbed at a particular wavelength, the number of molecules in the initial state with energy $E_i$ must be greater than the number in the final state with energy $E_f$.

The thermal population of the various molecular quantum states is determined by the **Boltzmann distribution,** which is similar to energy distribution functions seen earlier. Recall from Section 4.5 that molecules have a distribution of speeds, which is described by the Maxwell–Boltzmann distribution for thermal equilibrium at absolute temperature $T$. This corresponds to a distribution $P(E)$ of kinetic energies that falls off exponentially:

$$P(E) \propto \exp\left(-E/k_B T\right)$$

where $k_B$ is the Boltzmann constant, equal to $1.38 \times 10^{-23}$ J K$^{-1}$. (If energies per *mole* are used, $k_B$ is replaced by the gas constant $R$.) This distribution gives rise to the Arrhenius dependence of gas-phase reaction rates on temperature (see Section 13.5). The Boltzmann distribution

$$P_i \propto \exp\left(-E_i/k_B T\right)$$

describes the probability that at thermal equilibrium a gas-phase molecule will be in a particular quantum state $i$ with energy $E_i$, except that now $i$ is a discrete label, and not all energies are allowed the molecule by quantum mechanics. Equivalently, if a sample contains a great many molecules, the fraction in state $i$ will be $P_i$ at temperature $T$. This exponential falloff means that very few molecules have energies that are large compared to $k_B T$ at thermal equilibrium. The *ratio* of occupation numbers for two states, $i$ and $f$,

$$P_i/P_f = \exp[-(E_i - E_f)/k_B T]$$

influences the probability that the molecule can undergo a transition from state $i$ to state $f$ by absorbing light. The importance of this factor will be estimated for various specific examples of molecular transitions in the following paragraphs.

---
**16.5**

---

# NUCLEAR MAGNETIC RESONANCE SPECTROSCOPY

In **nuclear magnetic resonance (NMR) spectroscopy,** low-energy radio waves (photons carrying energies between 0.00002 and 0.00020 kJ mol$^{-1}$) "tickle" the nuclei in a molecule. Some nuclei have spins (just as electrons have spin), and the energies of different spin states are split apart by the magnetic field employed in a magnetic resonance spectrometer. Hydrogen ($^1$H) nuclei, for example, have "spin-up" and "spin-down" states in which the proton behaves as if it were a tiny bar magnet. In a magnetic field, the protons with spin up align parallel to the field (lower energy configuration), and those with spin down align antiparallel to the field (higher energy configuration). The analogy with bar magnets makes it clear that these two states will differ in energy by an amount determined by the strength of the magnetic field $H$. The magnitude of the energy splitting is $\Delta E = g_N \beta_N H$ where $g_N$ is the "nuclear g-factor" for protons and $\beta_N$ is a constant called the nuclear magneton (Fig. 16.21). These nuclei can change spin state by absorbing or emitting photons of magnitude $h\nu = g_N \beta_N H$, which fall in the FM radio frequency (rf) region of the electromagnetic spectrum.

The NMR spectrum can be recorded in various ways. The earliest commercial NMR spectrometers operated in continuous-wave mode, in which the sample is irradiated at constant frequency $\nu$ while the magnetic field is swept through a range of values. The rf power absorbed by the sample is recorded at each value of $H$. When the value of $H$ satisfies the resonance condition, a peak appears in the spectrum. Newer instruments rely on **Fourier transform (FT NMR) spectroscopy,** in which a sample held in a fixed magnetic field is irradiated with a short, intense burst of rf power spanning a frequency range broad enough to stimulate many transitions. The specific frequencies of all the many transitions are measured simultaneously and stored in computer memory. A complete spectrum can be acquired within a few seconds. The measurement can be repeated many times and the results averaged in computer memory to reduce the effects of random background noise. FT NMR spectrometers detect numerous nuclei besides $^1$H; $^{13}$C holds special interest in organic chemistry and biochemistry.

Nuclear spins are sensitive to the chemical environment of a nucleus because electrons moving near the nucleus establish an internal magnetic field that modifies the local effective field felt by each proton to a value different from that of the externally applied field. The resulting **chemical shift** causes protons within different structural units of the molecule to show nmr peaks at different values of magnetic field. All protons in chemically equivalent environments will contribute to a single absorption peak in the spectrum. The relative area under each absorption peak is proportional to the number of protons within each equivalent group. In order to standardize procedures, chemical shift values are recorded relative to the selected reference compound trimethylsilane (TMS) by adding a very small amount of TMS to the sample. The chemical shift for each proton in the spectrum is then defined in units of parts per million (ppm) as

$$\delta = \frac{H_s - H_r}{H_r} \times 10^6 \qquad \leftarrow \text{FOURIER.}$$

where $H_s$ is the value of the magnetic field at the peak in the sample spectrum and $H_r$ is the value at the TMS peak. Extensive tables of chemical shift values relative to TMS aid the interpretation of spectra. For example, these considerations predict that the nmr spectrum of ethanol, $CH_3$—$CH_2$—$OH$ should have three peaks of

**FIGURE 16.21** (a) Energy level diagram for proton nmr measurements. The magnitude of the energy gap between the spin states depends on the strength of the magnetic field. (b) Energy is absorbed when the magnetic field and the rf radiation satisfy the resonance condition.

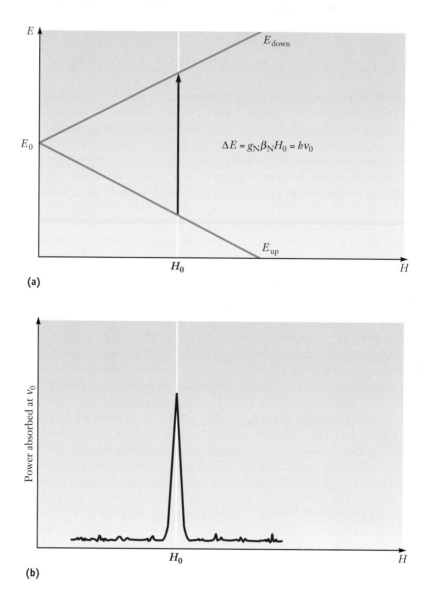

(a)

(b)

relative area 3:2:1. This is exactly the result obtained when the spectrum is acquired with low resolution (Fig. 16.22a).

Nuclear spins are also sensitive to other nearby nuclear spins. Since each spin behaves as if it were a small bar magnet, each influences the local value of magnetic field felt by others nearby. The result is **spin-spin splitting,** which breaks the broad peak of each chemically equivalent group of protons into a multiplet of narrower, sharper peaks. So long as the chemical shift between chemically nonequivalent groups of protons is large, the multiplet patterns can be interpreted by qualitative rules that will not be described in this book. For example, when applied to ethanol, these rules predict splitting of the $CH_3$— protons into a triplet and splitting of the $CH_2$— protons into a quartet, as seen in high-resolution experimental spectra (Fig. 16.22b). Through chemical shifts and spin-spin splitting, nuclear magnetic resonance spectroscopy offers a way to identify the bonding groups in a molecule and interpret molecular structure.

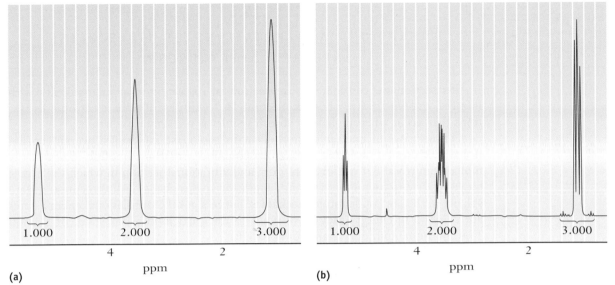

**FIGURE 16.22** Proton nmr spectrum for ethanol. (a) The low-resolution spectrum shows a single broad peak for each chemically equivalent group of protons. (b) In high resolution, spin–spin splitting separates the peak for each chemically equivalent group of protons into a multiplet. *(Courtesy of Dr. Jane Strouse, UCLA)*

An adaptation of NMR spectroscopy used in medical diagnosis is called **magnetic resonance imaging (MRI).** It relies on emission by the protons in the water contained in the body's organs. The patient is placed in the opening of a large magnet (Fig. 16.23), and a radio transmitter raises the proton spins in the relevant part of the body to their high-energy state. The radio-frequency photons subsequently emitted are then detected by a radio receiver coil. The amplitude of the signal indicates the density of water present. Thus, MRI can identify tumors by the excess water in their cells. The time delay until emission occurs is related to the type of tissue. Figure 16.24 shows an MRI brain scan.

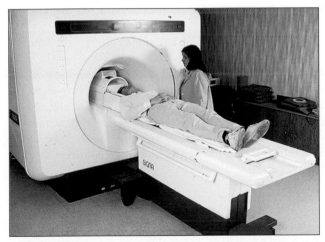

**FIGURE 16.23** A magnetic resonance imaging (MRI) machine. The patient is placed on the platform and rolled forward into the magnet opening. *(Peter Arnold)*

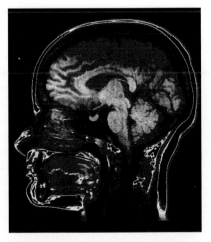

**FIGURE 16.24** Computer-enhanced MRI scan of a normal human brain with the pituitary gland highlighted. *(Scott Camazine/Photo Researchers, Inc.)*

---
## 16.6
---

# VIBRATIONS AND ROTATIONS OF MOLECULES: INFRARED SPECTROSCOPY

In this section we consider the interaction of *nuclear* motion in molecules with infrared and microwave photons. Section 16.7 will describe the electronic excitations that occur when molecules absorb visible or ultraviolet radiation.

Three types of nuclear motion occur in gas-phase molecules: overall translational motion of a molecule through its container, rotational motion in which the molecule turns about one or more axes, and vibrational motion in which the nuclei move relative to each other as bond lengths or angles change. All three motions are subject to the laws of quantum mechanics, but in a gas the translational energy states are so close in energy (they correspond to the particle-in-a-box states of Section 15.6) that quantum effects are not apparent.

## Rotations of Molecules

Rotational energies depend on the molecular **moments of inertia.** (See Appendix B for background on rotational motion.) For a diatomic molecule, only a single moment of inertia $I$ enters, defined by

$$I = \mu R_e^2$$

where $\mu$ is the **reduced mass** of the molecule:

$$\mu = \frac{m_1 m_2}{m_1 + m_2}$$

with $m_1$ and $m_2$ being the masses of the two atoms and $R_e$ the bond length. Polyatomic molecules have up to three different moments of inertia, corresponding to rotations about three axes. Quantum mechanics demonstrates that rotational motion is quantized, and only certain discrete rotational energy levels are permitted. In a linear molecule, for example, the rotational energy can take on only the values

$$E_{rot,J} = \frac{h^2}{8\pi^2 I} J(J + 1) \qquad J = 0, 1, 2, \ldots$$

where $J$ is the rotational quantum number. Typical rotational energy differences range from 0.001 to 1 kJ mol$^{-1}$.

Radiation in the far (long-wavelength) infrared and microwave regions of the electromagnetic spectrum excites rotational states of molecules. From gas-phase absorption spectra in these regions, the moments of inertia can be calculated and (because the atomic masses are known) bond lengths and angles can be determined. For a heteronuclear diatomic molecule, absorption of light is possible only between states that differ by 1 in rotational quantum number ($\Delta J = 1$). The allowed absorption frequencies are

$$\nu = \frac{\Delta E}{h} = \frac{h}{8\pi^2 I} [J_f(J_f + 1) - J_i(J_i + 1)]$$

$$= \frac{h}{8\pi^2 I} [(J_i + 1)(J_i + 2) - J_i(J_i + 1)]$$

$$= \frac{h}{4\pi^2 I} (J_i + 1) \qquad J_i = 0, 1, 2, \ldots$$

because the quantum number of the final rotational state $J_f$ must be 1 greater than that of the initial rotational state $J_i$. The rotational absorption spectrum of such a molecule shows a series of equally spaced lines with a frequency separation of $h/4\pi^2 I$. The spectra for nonlinear polyatomic molecules are more complex, but their interpretation has enabled chemists to determine to high accuracy the molecular geometries for many small polyatomic molecules.

## Vibrations of Molecules

Consider first the vibrational motion of a diatomic molecule. If the bond is stretched, the two atoms experience a restoring force that tends to bring them back to their original separation $R_e$, just as a restoring force acts on the ends of a spring that has been stretched. For a small change in bond length $R - R_e$, the force is proportional to that change:

$$F = -k(R - R_e)$$

where the **force constant** $k$ determines the stiffness of the bond. In SI units, $k$ is measured in joules per square meter, equivalent to kilograms per square second. According to classical mechanics, when a stretched bond is released, the atoms will oscillate back and forth about their average separation, much as two balls connected by a spring do. The oscillation frequency $\nu$ is given by

$$\nu = \frac{1}{2\pi}\sqrt{\frac{k}{\mu}}$$

where $\mu$ is again the reduced mass of the harmonic oscillator. In classical mechanics, any vibrational energy is permitted for the oscillator, but in quantum mechanics, the Schrödinger equation gives a discrete set of vibrational energy levels:

$$E_{\text{vib},v} = h\nu(v + \tfrac{1}{2}) \qquad v = 0, 1, 2, \ldots$$

where $h$ is Planck's constant and $v$ is the vibrational quantum number. Note that even in the ground state ($v = 0$) the vibrational energy is not 0. The residual energy $\frac{1}{2}h\nu$ is the **zero-point energy;** it arises from the quantum nature of vibrational motion. Typical energy differences between the ground ($v = 0$) and first excited ($v = 1$) vibrational states range from 2 to 40 kJ mol$^{-1}$; they are greater than rotational energy differences but less than electronic energy differences (Table 16.3). These energy differences correspond to the absorption of radiation in the infrared region of the electromagnetic spectrum, with wavelengths longer than those of visible light.

In a polyatomic molecule, several types of vibrational motion are possible (Fig. 16.25). Each has a different frequency, and each gives rise to a series of allowed quantum vibrational states. Infrared absorption spectra provide useful information about vibrational frequencies and force constants in molecules. For larger molecules, vibrational spectra are useful for identifying compounds, because each functional group (such as a C=O double bond or a C—OH bond) has its own characteristic

**FIGURE 16.25** The three types of vibrational motion that are possible for a bent triatomic molecule. Arrows show the displacement of each atom during each type of vibration.

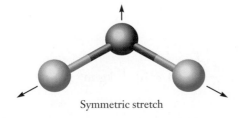

Symmetric stretch

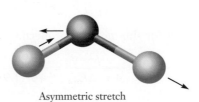

Asymmetric stretch

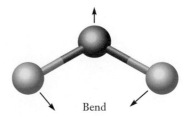

Bend

**FIGURE 16.26** The infrared absorption spectrum of vitamin C (ascorbic acid, $C_6H_8O_6$). Strong absorptions at 1 (6100 nm), 2 (7600 nm), 3 (8800 nm), and 4 (9600 nm) are characteristic of particular features of the molecule. For example, band 1 is primarily due to a C=C bond stretch, and band 4, to a C—OH stretch.

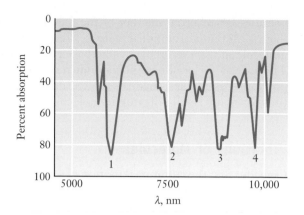

frequency in the spectrum, largely independent of the mass, geometry, and other features of the molecule of which it is a part. The spectrum thus provides information about the functional groups in a given molecule (Fig. 16.26). Strong absorption is observed only for transitions between states that differ by 1 in vibrational quantum number ($\Delta v = 1$).

## EXAMPLE 16.4

The spectrum of gaseous NaH (isotope: $^{23}Na^1H$) was determined via absorption spectroscopy. The photon wavelength required to excite the molecules rotationally from the state $J = 0$ to $J = 1$ was measured to be $1.02 \times 10^{-3}$ m, and the wavelength needed to excite them vibrationally from the state $v = 0$ to $v = 1$ was $8.53 \times 10^{-6}$ m. Calculate **(a)** the bond length and **(b)** the vibrational force constant of the NaH molecule. Use isotope atomic masses from Table 14.1.

### Solution

**(a)** The reduced mass is

$$\mu = \frac{m_{Na}m_H}{m_{Na} + m_H} = \frac{(22.9898 \text{ u})(1.0078 \text{ u})}{22.9898 + 1.0078 \text{ u}}$$

$$= 0.9655 \text{ u} = 1.603 \times 10^{-27} \text{ kg}$$

The energy change is

$$\Delta E = h\nu = \frac{hc}{\lambda}$$

which in this case is

$$\Delta E_{rot} = \frac{(6.626 \times 10^{-34} \text{ J s})(2.998 \times 10^8 \text{ m s}^{-1})}{1.02 \times 10^{-3} \text{ m}} = 1.95 \times 10^{-22} \text{ J}$$

$$= \frac{h^2}{8\pi^2 I}[(1)(2) - (0)(1)] = \frac{h^2}{4\pi^2 I}$$

Solving for the moment of inertia $I$ gives

$$I = \frac{h^2}{4\pi^2 \Delta E_{rot}} = \frac{(6.626 \times 10^{-34} \text{ J s})^2}{4\pi^2(1.95 \times 10^{-22} \text{ J})} = 5.70 \times 10^{-47} \text{ kg m}^2$$

The moment of inertia $I$ is related to the bond length $R_e$ by $I = \mu R_e^2$, and so

$$R_e^2 = \frac{I}{\mu} = \frac{5.70 \times 10^{-47} \text{ kg m}^2}{1.603 \times 10^{-27} \text{ kg}} = 3.56 \times 10^{-20} \text{ m}^2$$

$$R_e = 1.89 \times 10^{-10} \text{ m} = 1.89 \text{ Å}$$

**(b)** The vibrational frequency is

$$\nu = \frac{c}{\lambda} = \frac{2.998 \times 10^8 \text{ m s}^{-1}}{8.53 \times 10^{-6}} = 3.515 \times 10^{13} \text{ s}^{-1} = \frac{1}{2\pi}\sqrt{\frac{k}{\mu}}$$

Solving for the constant $k$ gives

$$k = \mu(2\pi\nu)^2 = (1.603 \times 10^{-27} \text{ kg})(2\pi)^2(3.515 \times 10^{13} \text{ s}^{-1})^2$$

$$= 78.2 \text{ kg s}^{-2} = 78.2 \text{ J m}^{-2}$$

**Related Problems: 29, 30, 31, 32, 33, 34**

## Molecular Energy Levels and Their Thermal Occupation

The electrons, being much less massive, move much more rapidly than the nuclei in a molecule. This observation permits a simplified approximate discussion of the quantum mechanics of molecular energy states, due originally to Max Born and J. Robert Oppenheimer. In a particular electronic state, the energy of the rapidly moving electrons in effect provides a potential energy function that governs the motions (vibrations and rotations) of the more sluggish nuclei. Therefore, each quantized electronic energy state will have an associated set of quantized vibrational and rotational energy states. The separation between electronic states, to be considered in the next section, is much greater than that for rotational and vibrational states. Figure 16.27 schematically illustrates the Born–Oppenheimer picture for the energy levels of a typical diatomic molecule. Two electronic states with numerous rotational and vibrational states are shown. A polyatomic molecule would have an even more complicated set of energy levels, each characterized by several vibrational quantum numbers and by up to three rotational quantum numbers.

To an excellent level of approximation, the total energy of a molecule can be viewed as the sum of the energies associated with each of these separate motions:

$$E_{tot} = E_{trans} + E_{rot} + E_{vib} + E_{el}$$

When the total energy of a molecule changes during absorption or emission of radiation, the transition usually involves changes in more than one kind of energy. For example, a line in the visible region of the emission spectrum may connect two energy levels that differ in rotational, vibrational, and electronic energy (see the red arrow in Figure 16.27). The intensity of the line will depend on the strength of the transition (Section 16.4) and on the population ratio of the initial and final states. The population ratio for two states, $i$ and $f$, is

$$P_i/P_f = \exp[-(E_i - E_f)/k_B T]$$

The labels $i$ and $f$ stand for the complete set of rotational and vibrational quantum numbers, in addition to designating the electronic states under consideration.

At room temperature, $k_B T$ has the value $4 \times 10^{-21}$ J (equivalent to 2.5 kJ mol$^{-1}$). This is large compared to the typical spacing of rotational levels, indicating that at

**FIGURE 16.27** An energy-level diagram for a diatomic molecule, showing the electronic, vibrational, and rotational levels schematically. The arrow indicates one of the many possible transitions; in this case the rotational, vibrational, and electronic states of the molecule all change.

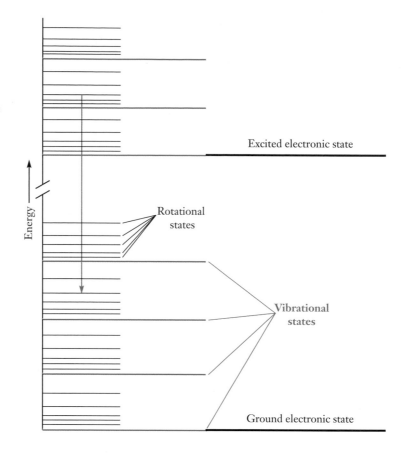

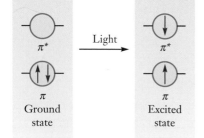

**FIGURE 16.28** Occupancy of the $\pi$ molecular orbitals in the ground state and first excited state of ethylene, $C_2H_4$. Ethylene in the excited state is produced by irradiation of ground-state ethylene with ultraviolet light in the appropriate frequency range.

room temperature many rotational states are occupied by a collection of molecules. This value is also a little smaller than typical vibrational level spacings, and so most small molecules at room temperature are in their ground vibrational states, but a measurable fraction occupy excited states. Finally, $k_B T$ is small compared to typical electronic energy spacings, and so only the ground electronic state is occupied at room temperature. When molecules are heated, the occupation of more highly excited levels increases.

---

16.7

# EXCITED ELECTRONIC STATES

Excited states of atoms result from the promotion of one or more electrons into orbitals of higher energy. Excited electronic states in molecules arise similarly and can be described in the framework of molecular orbital theory. As an example, consider the ethylene molecule ($C_2H_4$). The bonding of the two carbon atoms in ethylene can be described through $sp^2$ hybridization, with the two remaining carbon $2p_z$ orbitals combined into a $\pi$ (bonding) molecular orbital and a $\pi^*$ (antibonding) molecular orbital (Fig. 16.28). The lowest energy state of ethylene has two electrons in the $\pi$ orbital, but an excited state can be attained by removing one electron from the $\pi$ orbital and placing it in the $\pi^*$ orbital. An ethylene molecule in such an excited state has quite different properties from one in the ground state. The electron in the $\pi^*$ antibonding orbital cancels the effect of the electron in the $\pi$ bonding orbital, and the net bond order is reduced from 2 to 1 (the $\sigma$ bond). Consequently, the molecule in the excited state has a lower bond energy and a longer C—C bond length. The observed wavelength for this $\pi \rightarrow \pi^*$ transition in ethylene is 162 nm, which lies in the ultraviolet region of the spectrum. Because ethylene does not absorb strongly in the visible region, visible light passes almost unaffected through samples of the substance, which is therefore colorless.

## The Fate of Excited Electronic States

What happens to molecules after they absorb radiation? Some reemit a photon almost immediately and return to the ground electronic state, although they usually end up in rotational and vibrational states different from the original ones. This process is called **fluorescence** and typically occurs within about $10^{-9}$ s. A second type of light emission occurs much more slowly, taking seconds (or in some cases hours), and is called **phosphorescence.** To understand phosphorescence, it is necessary to consider the spins of the electrons in an excited state. The absorption of light (shown in Fig. 16.28) does not change the electron spin and so gives an excited state (called a *singlet*) with one spin up and one down. Another possible excited electronic state (called a *triplet*) has both spins up (Fig. 16.29) and, by Hund's rule, has a lower energy than the singlet. Although the triplet state cannot be reached directly by absorption of light, there is a small probability that an excited molecule will cross over to the triplet state as it evolves in time or undergoes collisions. Once there, it remains for a long time because emission from a triplet state to a ground singlet state is very slow.

Figure 16.30 schematically illustrates these two pathways for loss of energy by radiation from an electronically excited molecule, along with one nonradiative pathway. The molecule may cross over to the ground electronic state and gradually lose

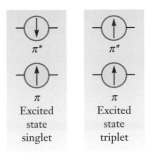

**FIGURE 16.29** The first excited singlet and triplet states have the same orbital occupancy but different spin states. The triplet is generally lower in energy.

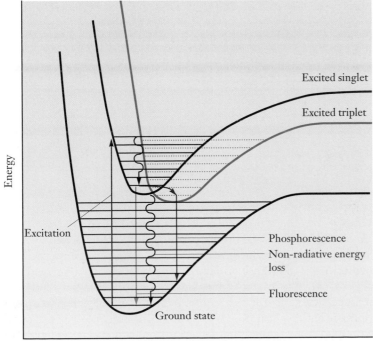

**FIGURE 16.30** A schematic picture of the loss of energy by a molecule in an excited electronic state. After the molecule absorbs a photon (purple arrow), it cascades down through a series of levels to the bottom of the excited singlet state (short wavy black arrow). Then three outcomes are possible. The molecule can return to the ground state by fluorescence (green arrow), it can cross over to the triplet state and return from there to the ground state by phosphorescence (red arrow), or it can cross over to the ground state and cascade down without radiating (long wavy black arrow).

its energy as heat to the surroundings as it cascades down to lower vibrational and rotational levels.

It is possible that the energy of the excited molecule is high enough that bonds break, producing a chemical reaction. Molecules in excited states have enhanced reactivity and can often fragment or rearrange to give new molecules. **Photochemistry,** the study of the chemical reactions that follow the excitation of molecules to higher electronic states through absorption of photons, is an active area of research.

Some compounds are so photochemically sensitive that they must be stored in the dark, because they react rapidly to form products when exposed to light. An example is anhydrous hydrogen peroxide, which reacts explosively in light to give water and oxygen:

$$H_2O_2 \longrightarrow 2\ OH \longrightarrow H_2O + \tfrac{1}{2}\ O_2$$

In a process called **photodissociation,** when the energy of a visible photon is added to the molecule, the rather weak O—O bond in $H_2O_2$ breaks apart.

## New Developments in Spectroscopy

Modern molecular spectroscopy aims not only to determine energy levels (and, from them, information about molecular structure and bonding) but to give us an understanding of the dynamics of excited molecules—a detailed series of events that may include nuclear motion, light emission, and electron transfer. Such an understanding requires **time-resolved spectroscopy,** in which short light pulses are used to excite molecules and the emitted photons are identified not only by wavelength but also by the time delay until each appears. Light pulses with durations of picoseconds ($10^{-12}$ s) and even femtoseconds ($10^{-15}$ s) can be made with modern lasers, and the very rapid evolution of excited molecules is monitored by measuring the absorption of a second light pulse, which arrives at a variable time after the initial pulse, or by counting photons emitted with time (Fig. 16.31).

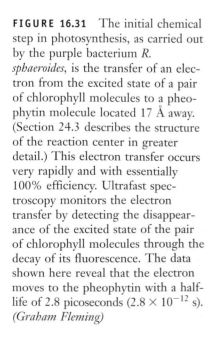

**FIGURE 16.31** The initial chemical step in photosynthesis, as carried out by the purple bacterium *R. sphaeroides*, is the transfer of an electron from the excited state of a pair of chlorophyll molecules to a pheophytin molecule located 17 Å away. (Section 24.3 describes the structure of the reaction center in greater detail.) This electron transfer occurs very rapidly and with essentially 100% efficiency. Ultrafast spectroscopy monitors the electron transfer by detecting the disappearance of the excited state of the pair of chlorophyll molecules through the decay of its fluorescence. The data shown here reveal that the electron moves to the pheophytin with a half-life of 2.8 picoseconds ($2.8 \times 10^{-12}$ s). *(Graham Fleming)*

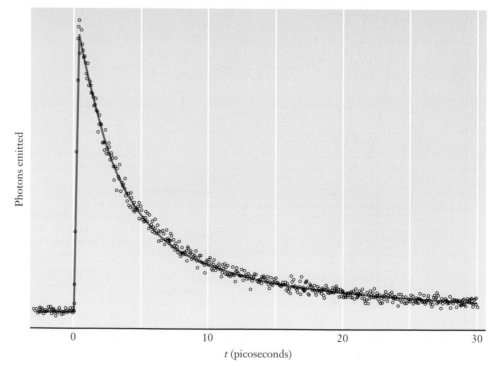

A promising approach to studying the spectra of complex molecules involves cooling them in supersonic jets. A typical, moderate-size molecule in solution has a rather broad and featureless spectrum in the visible and ultraviolet regions (Fig. 16.32, red curve) because there are many vibrational and rotational levels, and interactions with solvent molecules at different distances shift their energies, giving a blurred overall picture. (The same blurring of rotational levels within a particular vibration state produced the broad bands in Fig. 16.26.) If a molecule can be vaporized into the gas state, the effects of solvent are removed and additional structure appears (Fig. 16.32, blue curve); however, molecular collisions and the numerous thermally occupied energy levels still give bands that are too broad to be interpreted quantitatively. To simplify the spectrum, it would be desirable to cool the molecule to low temperatures so that very few levels were occupied. In general, however, the gas would then condense to a liquid or solid and defeat the purpose of isolating the molecules. In a supersonic jet, molecules, along with a carrier gas such as helium, are forced through a nozzle at supersonic speeds. The gas expands adiabatically, which cools the molecules (see Section 7.6). Collisions redistribute molecules into lower energy levels, until downstream in the jet the effective temperatures are quite low and much more structure appears in the spectra (Fig. 16.32, inset).

Surprising as it may seem, another way to simplify spectra is to examine one molecule at a time. Chemists have studied the spectra of molecules that are present in low concentrations in a low-temperature solid. Working with thin samples and focused laser beams, they can probe a region of the sample with a relatively small

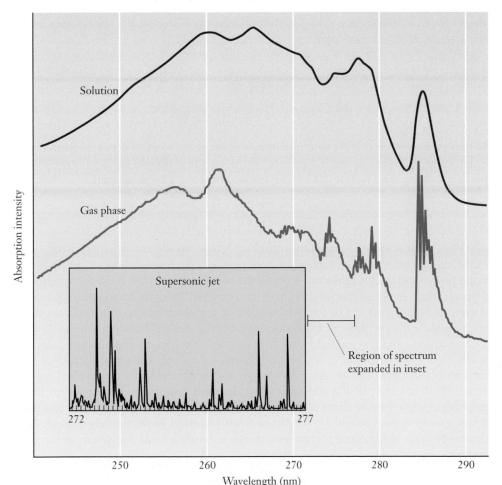

**FIGURE 16.32** A portion of the absorption spectrum of indole, $C_8H_7N$, in the ultraviolet region. In heptane solution (red curve) the spectrum has broad and featureless bands, while in the gas phase (blue curve) additional structure appears. Cooling indole molecules in a supersonic jet reveals a large number of sharp absorption lines (see inset, which covers a very small range of wavelengths). *(D. Levy, H. Strickland)*

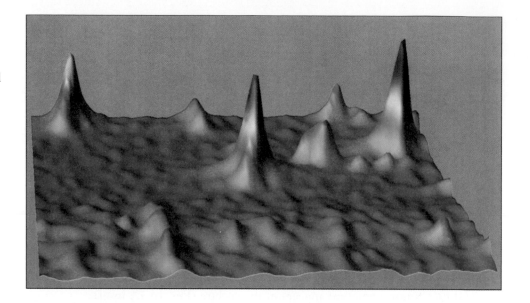

number of those molecules. Using extraordinarily sensitive detectors, they measure signals from one molecule at a time (Fig. 16.33). Chemists can even see how a signal changes with time; as the molecules near the probed molecule move, the frequency of the probed molecule's transition shifts as well. Thus, chemists not only can "see" single atoms with scanning tunneling microscopes (see Fig. 1.14), they can also probe the spectra of single molecules.

---

## 16.8

# ILLUSTRATION: CONJUGATED SYSTEMS

When two or more double or triple bonds occur close to each other in a molecule, a delocalized molecular orbital picture of the bonding should be used. One example is 1,3-butadiene ($CH_2CHCHCH_2$), which has the Lewis diagram

$$
\begin{array}{cccc}
H & H & H & H \\
\ddot{} & \ddot{} & \ddot{} & \ddot{} \\
H : \ddot{C} : : \ddot{C} : \ddot{C} : : \ddot{C} : H
\end{array}
$$

All four carbon atoms have steric number 3, and so all are $sp^2$ hybridized. The remaining $p_z$ orbitals have maximum overlap when the four carbon atoms lie in the same plane, and so this molecule is predicted to be planar. From these four $p_z$ atomic orbitals, four molecular orbitals can be constructed by combining their phases as shown in Figure 16.34. The four electrons that remain after the $\sigma$ orbitals are filled are placed in the two lowest $\pi$ orbitals. The first of these is bonding among all four carbon atoms, and the second is bonding between the outer carbon atom pairs but antibonding between the central pair. Therefore, 1,3-butadiene has stronger and shorter bonds between the outer carbon pairs than between the two central carbon atoms. It is an example of a **conjugated $\pi$ electron system,** in which two or more double or triple bonds alternate with single bonds. Such conjugated systems have lower energies than would be predicted from localized bond models and are best described with delocalized molecular orbitals extending over the entire $\pi$ electron system.

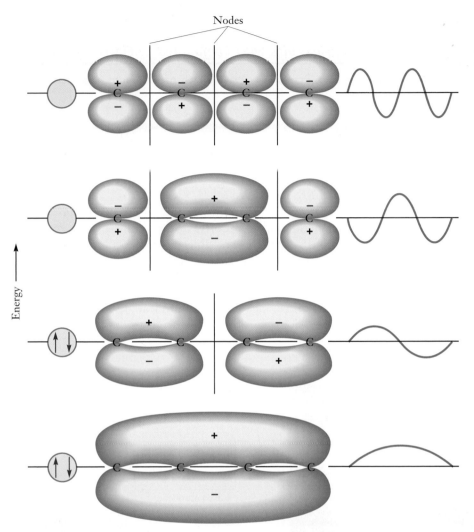

**FIGURE 16.34** The four $\pi$ molecular orbitals formed from four $2p_z$ atomic orbitals in 1,3 butadiene, viewed from the side. Planar nodes are indicated by red lines. Note the similarity in nodal properties to the first four harmonics of a vibrating string (right). Only the two lowest energy orbitals are occupied in the molecule in the ground state (left).

Now consider a molecule in which several double bonds alternate with single bonds in a conjugated $\pi$ bonding system such as that discussed for butadiene. Observe that the lowest-energy $\pi$ orbital in Figure 16.34 extends over the full conjugated $\pi$ system and resembles a standing wave with a wavelength proportional to the length of the molecule. As more alternating single and double bonds are added, the characteristic wavelengths become longer and the frequencies and energies become lower (Table 16.4). For a sufficiently long chain, the first absorption is shifted into the visible region of the spectrum, and the substance takes on color. The perceived color is related to the absorption spectrum of the material, but indirectly. What we see is the light transmitted through the material, not the light absorbed. In other words, the color perceived is **complementary** to the color most strongly absorbed by the molecule (Fig. 16.35a and b). Examples are the dyes indigo and

**TABLE 16.4**

**Absorption of Light by Molecules with Conjugated π Electron Systems**

| Molecule | Number of C=C Bonds | Wavelength of Maximum Absorption (nm) |
|---|---|---|
| $C_2H_4$ | 1 | 162 |
| $C_4H_6$ | 2 | 217 |
| $C_6H_8$ | 3 | 251 |
| $C_8H_{10}$ | 4 | 304 |

beta-carotene (Fig. 16.36). Figure 16.37 plots the strength of the absorption of these two dyes at different wavelengths. Indigo absorbs in the yellow-orange region of the spectrum, and so to an observer it appears violet-blue (the colors complementary to yellow and orange). Beta-carotene absorbs at shorter wavelengths in the blue and violet regions and appears orange or yellow. Indigo is used to dye blue jeans; beta-carotene is responsible for the orange color of carrots, the colors of some processed foods, and the yellow and orange in certain bird feathers. Most bird colors, however, arise from diffraction, not absorption, of light.

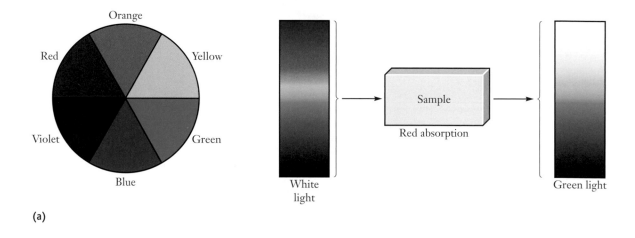

(a)

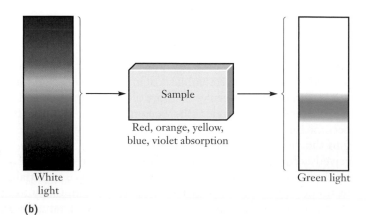

(b)

**FIGURE 16.35**  Two ways in which the color green can be produced. (a) If the sample strongly absorbs red light, but transmits yellow and blue light, the sample appears green. (b) If all visible light *except* green is absorbed by the sample, the sample also appears green. The color wheel in (a) shows complementary colors opposite each other.

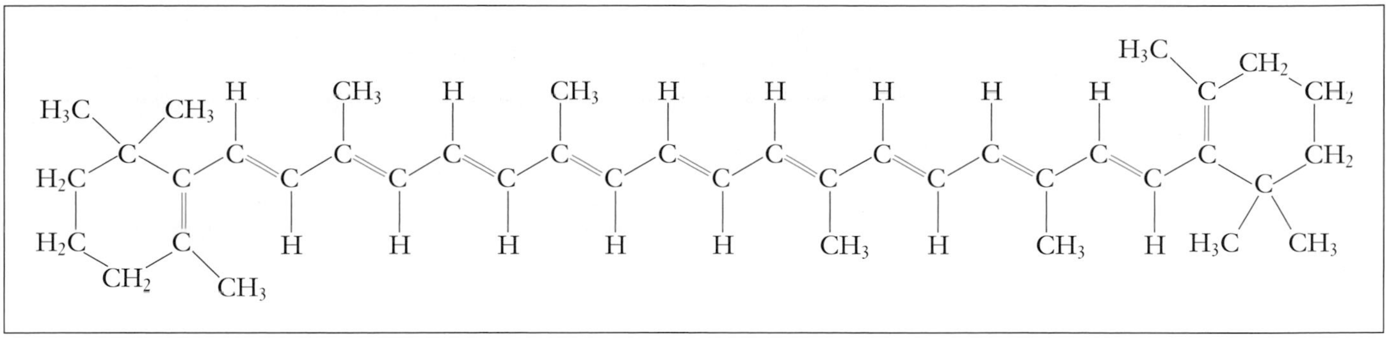

**FIGURE 16.36** The molecular structure of the dye beta-carotene. Eleven double bonds alternate with single bonds in this conjugated structure.

## EXAMPLE 16.5

Suppose you set out to design a new green dye. Over what range of wavelengths would you want your trial compound to absorb light?

### Solution

A good green dye must transmit green light and absorb other colors of light. This can be achieved by a molecule that absorbs in the violet-blue *and* orange-red regions of the spectrum. Thus, you would want a dye with strong absorptions in both these regions.

The naturally occurring substance chlorophyll, which is responsible for the green colors of grass and leaves, absorbs light over just these wavelength ranges, converting solar energy to chemical energy for the growth of the plant. It is used commercially as a green dye.

**Related Problems: 39, 40**

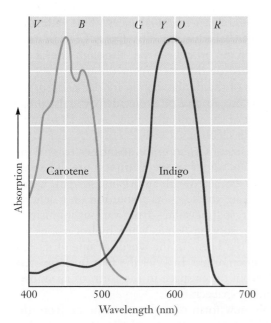

**FIGURE 16.37** The absorption spectra of the dyes beta-carotene and indigo differ in the visible region and hence have different colors. The letters stand for the colors of the light at particular wavelengths (violet, blue, green, yellow, orange, and red).

**FIGURE 16.38** (a) The structure of $C_{60}$, buckminsterfullerene. Note the pattern of hexagons and pentagons. (b) The design on the surface of a soccer ball has the same pattern as the structure of $C_{60}$. (*b, George Semple*)

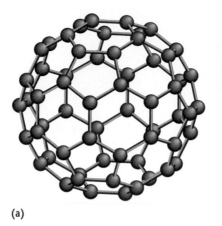

(a)

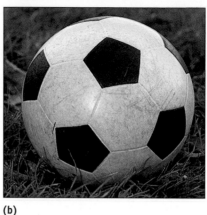

(b)

## Fullerenes

Among the most interesting conjugated $\pi$ electron systems is a molecule discovered in 1985: buckminsterfullerene, $C_{60}$. Previously, only two forms of carbon (diamond and graphite) were known. In 1985, Harold Kroto, Robert Curl, and Richard Smalley were studying certain long-chain carbon molecules that had been discovered spectroscopically by radioastronomers in the vicinity of red giant stars. They sought to duplicate the conditions near those stars by vaporizing a graphite target with a laser beam. Analysis of the products by mass spectrometry revealed not only the hoped-for molecules but a large proportion of molecules of molar mass 720 g mol$^{-1}$, which corresponds to the molecular formula $C_{60}$. Although the amounts of $C_{60}$ present were far too small to isolate for direct determination of molecular structure, Kroto, Curl, and Smalley correctly suggested the cage structure shown in Figure 16.38a and named the molecule *buckminsterfullerene* after the architect Buckminster Fuller, the inventor of the geodesic dome—which the molecular structure of $C_{60}$ resembles.

Molecules of $C_{60}$ have a highly symmetric structure: 60 carbon atoms arranged in a closed net with 20 hexagonal faces and 12 pentagonal faces. The pattern is exactly the design on the surface of a soccer ball (Fig. 16.38b). Every C atom has a steric number of 3; all 60 atoms are accordingly $sp^2$ hybridized, although one of the three bond angles at each carbon atom must be distorted from the usual 120° $sp^2$ bond angle down to 108°. The $\pi$ electrons of the double bonds are delocalized: the 60 $p$ orbitals (one from each carbon atom) mix to give 60 molecular orbitals spread over both sides of the entire closed surface of the molecule. The lowest 30 of these molecular orbitals are occupied by the 60 $\pi$ electrons.

In 1990 scientists succeeded in synthesizing $C_{60}$ in gram quantities by striking an electric arc between two carbon rods held under an inert atmosphere. The carbon vapors condensed to a soot, which was extracted with an organic solvent. Using chromatography, the scientists could separate the $C_{60}$—as a solution of a delicate magenta hue (Fig. 16.39) and finally as a crystalline solid—from various impurities. Since 1990, $C_{60}$ has been found to form in sooting flames when hydrocarbons are burned. Thus, the newest form of carbon has been (in the words of one of its discoverers) "under our noses since time immemorial." In 1994 the first buckminsterfullerene molecules were brought back from outer space in the form of an impact crater from a tiny meteorite on an orbiting spacecraft.

Buckminsterfullerene is not the only new form of carbon to emerge from the chaos of carbon vapor condensing at high temperature. Synthesis of $C_{60}$ simulta-

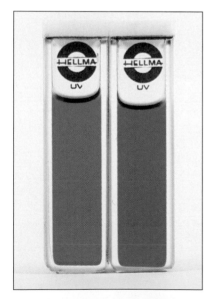

**FIGURE 16.39** (Left) A solution of $C_{60}$ in an organic solvent has a delicate magenta color. (Right) The related fullerene $C_{70}$ is distinctly orange in the same solvent. (*Hellma Cello, Inc.*)

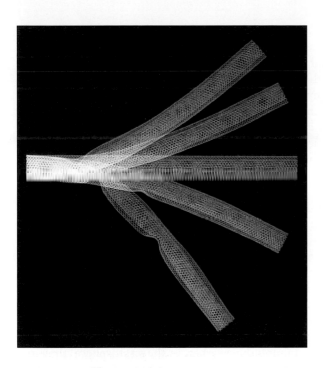

**FIGURE 16.40** A computer simulation of the oscillations experienced by a multiwalled carbon nanotube 300 Å in length after bending to the point of buckling. *(Courtesy of Daniel H. Robertson of IUPUI and Carter T. White of the Naval Research Laboratory)*

neously produces a whole family of closed-cage carbon molecules called **fullerenes.** All the fullerenes have even numbers of atoms, with formulas ranging up to $C_{400}$ and higher. These materials offer exciting prospects for technical applications. For example, because $C_{60}$ readily accepts and donates electrons to or from its $\pi$ molecular orbitals, it has possible applications in batteries. It forms compounds (such as $Rb_3C_{60}$) that are superconducting (have zero resistance to the passage of an electric current) up to 30 K. Fullerenes also can encapsulate foreign atoms that are present during synthesis. If a graphite disk is soaked with a solution of $LaCl_3$, dried, and used as a laser target, the substance $La@C_{60}$ forms, where the symbol @ means that the La atom is trapped within the 60-atom carbon cage.

Proper conditions during condensation of the carbon vapor favor formation of *nanotubes*, which consist of seamless, cylindrical shells of thousands of $sp^2$ hybridized carbon atoms arranged in hexagons. The ends of the tubes are capped by introduction of pentagons into the hexagonal network. These structure all have a delocalized $\pi$-electron system that covers the inner and outer surfaces of the cage or cylinder. Nanotubes also offer exciting prospects for materials science and technological applications. For example, the mechanical properties of nanotubes suggest applications as high-strength fibers (Fig. 16.40).

---

### 16.9

## ILLUSTRATION: GLOBAL WARMING AND THE GREENHOUSE EFFECT

The photochemical reactions in green plants and photosynthetic bacteria are not the only reactions induced by sunlight. The ability of light to cause chemical reactions is also apparent in the earth's atmosphere. In fact, photosynthesis and atmospheric chemistry are intimately connected. Reactions in the atmosphere determine the intensities and wavelengths of the light that reach the earth's surface to be harvested by living species. At the same time, oxygen, the gas produced by green-plant photosynthesis, has transformed the atmosphere; before photosynthesis began, there was almost no free oxygen at the earth's surface.

Although the chemical composition summarized in Table 4.1 describes the average makeup of the portion of the atmosphere closest to the earth's surface, it does not do justice to the variation in chemical properties with height, to the dramatic role of local fluctuations in trace gases, and to the dynamics underlying observed average concentrations. The atmosphere is a complex chemical system that is far from equilibrium. Its properties are determined by an intricate combination of thermodynamic and kinetic factors. It is a multilayered structure, bathed in radiation from the sun and interacting at the bottom with the oceans and land masses. At least four layers are identifiable, each with a characteristic variation of temperature (Fig. 16.41). In the outer two layers (the **thermosphere** and the **mesosphere**) atmospheric density is low, and the intense radiation from the sun causes extensive ionization of the particles that are present. The third layer, the **stratosphere,** is the region from 12 to 50 km (approximately) above the earth's surface. The **troposphere** is the lowest region, extending 12 km out from the earth's surface. In the troposphere, warmer air lies beneath cooler air. This is a dynamically unstable situation because warm air is less dense and tends to rise, and so convection takes place, mixing the gases in the troposphere and determining the weather. In the stratosphere, the temperature increases with height and there is little vertical mixing from convection. Mixing across the borders between the layers is also slow, and so most chemical processes in each layer can be described separately.

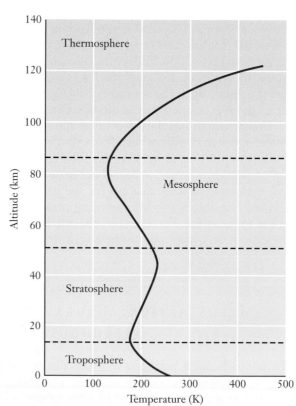

**FIGURE 16.41** The variation of the temperature of the atmosphere with altitude, showing the layered structure of the atmosphere.

Atmospheric chemistry dates back to the 18th century. Cavendish, Priestley, Lavoisier, and Ramsay were the first scientists to study the composition of the atmosphere. In recent years, atmospheric chemistry has developed in two different but related directions. First, the sensitivity of chemical analysis has greatly improved, and analyses for substances at concentrations below the part-per-billion (ppb) level are now carried out routinely. Airplanes and satellites enable scientists to map the global distributions of trace substances. Second, advances in gas-phase chemical kinetics have led to a better quantitative understanding of the ways in which substances in the atmosphere react with one another and with light. Much of the impetus for these studies of atmospheric chemistry comes from concern about the effect of air pollution on life.

## Stratospheric Chemistry

As Figure 15.3 shows, the sun emits light over a broad range of wavelengths, with the highest intensity at about 500 nm, in the visible region of the spectrum. The intensities at wavelengths down to 100 nm in the ultraviolet region are quite substantial, and because the energy $h\nu$ carried by a photon is inversely proportional to the wavelength $\lambda$ ($h\nu = hc/\lambda$), the ultraviolet photons carry much more energy than do photons of visible light. If these photons could penetrate to the earth's surface in substantial numbers, they could do a great deal of damage to living organisms. The outer portions of the atmosphere (especially the thermosphere) play a crucial role in preventing this penetration through the photodissociation of oxygen molecules:

$$O_2 + h\nu \longrightarrow 2\ O$$

where "$h\nu$" symbolizes a photon. This reaction reduces the number of high-energy photons reaching the lower parts of the atmosphere, especially those with wavelengths less than 200 nm.

## EXAMPLE 16.6

The bond dissociation energy of $O_2$ is 496 kJ mol$^{-1}$. Calculate the maximum wavelength of light that can photodissociate an oxygen molecule.

### Solution

Because 496 kJ dissociates 1.00 mol of $O_2$ molecules, the energy to dissociate one molecule is found by dividing by Avogadro's number:

$$\frac{496 \times 10^3\ \text{J mol}^{-1}}{6.022 \times 10^{23}\ \text{mol}^{-1}} = 8.24 \times 10^{-19}\ \text{J}$$

A photon carrying this energy has a wavelength $\lambda$, given by

$$8.24 \times 10^{-19}\ \text{J} = h\nu = \frac{hc}{\lambda}$$

so that

$$\lambda = \frac{hc}{8.24 \times 10^{-19}\ \text{J}} = \frac{(6.626 \times 10^{-34}\ \text{J s})(2.998 \times 10^8\ \text{m s}^{-1})}{8.24 \times 10^{-19}\ \text{J}}$$

$$= 2.41 \times 10^7\ \text{m} = 241\ \text{nm}$$

Any photons with wavelengths *shorter* than this are energetic enough to dissociate oxygen molecules. Those with wavelengths shorter than 200 nm are the most efficient in causing photodissociation.

**Related Problems: 45, 46**

---

Rather few photons with wavelengths less than 200 nm can penetrate to the stratosphere, but those that do establish a small concentration of oxygen atoms in that layer. These atoms can collide with the much more prevalent oxygen molecules to form highly excited molecules of ozone, symbolized by $O_3^*$:

$$O + O_2 \rightleftharpoons O_3^*$$

The excited-state $O_3^*$ can dissociate in a unimolecular reaction back to O and $O_2$, as indicated by the reverse arrow in this equilibrium. Alternatively, if another atom or molecule collides with it soon enough, some of its excess energy can be transferred to that atom or molecule, a process represented by

$$O_3^* + M \longrightarrow O_3 + M$$

where M stands for the atom or molecule with which the $O_3^*$ collides (the most likely are oxygen and nitrogen molecules, because they are the most abundant species in the atmosphere). The net effect of these two reactions is to produce a small concentration of ozone in the stratosphere.

Under laboratory conditions, ozone is an unstable compound. Its conversion to oxygen is thermodynamically spontaneous

$$O_3(g) \longrightarrow \tfrac{3}{2} O_2(g) \qquad\qquad \Delta G° = -163 \text{ kJ}$$

but takes place quite slowly in the absence of light. In the stratosphere, ozone photodissociates readily to $O_2$ and O in a reaction that requires about 106 kJ per mole of ozone, much less than the dissociation of $O_2$:

$$O_3 + h\nu \longrightarrow O_2 + O$$

This process occurs most efficiently for wavelengths between 200 and 350 nm. The energy of light in this range of wavelengths is too small to be absorbed by molecular oxygen but quite large enough to damage organisms at the earth's surface. The ozone layer shields the earth's surface from 200- to 350-nm ultraviolet radiation coming from the sun. The balance between formation and photodissociation leads to a steady-state concentration of more than $10^{15}$ molecules of ozone per liter in the stratosphere.

It is important to know whether molecules being released in the lower atmosphere can reach the stratosphere and affect its amount of ozone. Certain types of air pollution give rise to *radicals* that catalyze ozone depletion. A radical is a chemical species that contains an odd (unpaired) electron, and it is usually formed by the breaking of a covalent bond to form a pair of neutral species. One pressing concern involves chlorofluorocarbons (CFCs)—compounds of chlorine, fluorine, and carbon that are used as refrigerants and as propellants in some aerosol sprays. CFCs are nonreactive at sea level but can photodissociate in the stratosphere:

$$CCl_2F_2 + h\nu \longrightarrow CClF_2 + Cl$$

The atomic chlorine thus released soon reacts with O atoms or $O_3$ to give ClO:

$$Cl + O \longrightarrow ClO$$
$$Cl + O_3 \longrightarrow ClO + O_2$$

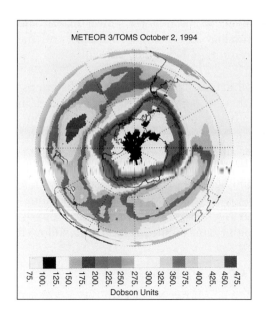

**FIGURE 16.42** This color image shows total ozone amounts in the Southern Hemisphere for October 2, 1994. The Dobson unit measures the stratospheric concentration of ozone. Preliminary minimum ozone values for this day were below 100 Dobson units, while the ozone hole area was at its maximum extent in a region near the South Pole. The total ozone values were measured by the Total Ozone Mapping Spectrometer (TOMS) instrument aboard the Russian Meteor-3 satellite, launched in August 1991. The color scale at the bottom of the figure shows the total ozone values. The Antarctic ozone hole is located in the center of the figure and is indicated by all colors contained within the blue boundary. Prior to 1975, values of total ozone between 250 and 300 Dobson units (blue to green) were normal, indicating a decrease of about 170 Dobson units (NASA.)

The ClO radical is the immediate culprit in the destruction of stratospheric ozone. Local increases in ClO concentration are directly correlated with decreases in $O_3$ concentration. ClO catalyzes the destruction of ozone, probably by the mechanism

$$2\ ClO + M \longrightarrow ClOOCl + M$$

$$ClOOCl + h\nu \longrightarrow ClOO + Cl$$

$$ClOO + M \longrightarrow Cl + O_2 + M$$

$$\underline{2 \times (Cl + O_3 \longrightarrow ClO + O_2)}$$

$$\text{net reaction:}\quad 2\ O_3 \longrightarrow 3\ O_2$$

where M stands for $N_2$ and $O_2$ molecules. This catalytic cycle is ordinarily disrupted rather quickly as Cl reacts with other stratospheric species to form less reactive "reservoir molecules" such as HCl and $ClONO_2$. The lack of mixing in the stratosphere keeps the reservoir molecules around for long periods, however.

It is believed that atomic Cl escapes its reservoirs through heterogeneous reactions on the polar stratospheric clouds that form during the intense Antarctic winter. The escaped Cl gives rise to the annual "ozone holes" above the Antarctic. During these episodes, more than 70% of the total column ozone is depleted before the values rise again (Fig. 16.42). Researchers have demonstrated that the extent of depletion above the Antarctic increases each year (Fig. 16.43), and measurements have shown smaller but still serious depletions over other parts of the globe. International agreement (the Montreal Protocol) has led to major reductions in the production of chlorofluorocarbons, with a complete phase-out planned by 1996. Even if this goal is met, stratospheric ozone concentrations will continue to decrease for some time to come as already-released chlorofluorocarbons mix upward into the stratosphere.

## Tropospheric Chemistry

The troposphere is the part of the atmosphere in contact with the earth's surface. It is therefore most directly and immediately influenced by human activities, especially

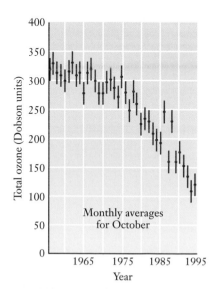

**FIGURE 16.43** Worsening ozone depletion over Halley Bay, Antarctica. These measurements were all taken in the Antarctic spring (October) when depletion is at its worst.

by the gases or small particles put into the air by automobiles, power plants, and factories. Some pollutants have long lifetimes and are spread fairly evenly over the earth's surface; others attain large concentrations only around particular cities or industrial areas.

The oxides of nitrogen are major air pollutants. Their persistence in the atmosphere demonstrates the importance of kinetics, as opposed to thermodynamics, in the chemistry of the atmosphere. All of the oxides of nitrogen are thermodynamically unstable with respect to the elements at 25°C, as shown by their positive free energies of formation. They form whenever air is heated to high enough temperatures, either in an industrial process or in the engine of a car. They accumulate to concentrations much higher than equilibrium because they decompose slowly. The interconversion among NO, $NO_2$, and $N_2O_4$ is rapid and strongly temperature-dependent, and so they are generally grouped together as "$NO_x$" in pollution reports. Photochemical smog is formed by the action of light on nitrogen dioxide, followed by reaction to produce ozone:

$$NO_2 + h\nu \longrightarrow NO + O$$

$$O + O_2 + M \longrightarrow O_3 + M \qquad (M = N_2 \text{ or } O_2)$$

It may seem paradoxical that concern about ozone in the stratosphere involves its potential *depletion*, whereas concern about the troposphere involves the *production* of ozone. Although ozone is beneficial in preventing radiation from penetrating to the earth's surface, it is quite harmful in direct contact with organisms because of its strong oxidizing power; levels of 10 to 15 ppm are sufficient to kill small mammals, and a concentration as low as 3 ppm is enough to trigger an "ozone alert." In addition, ozone reacts with incompletely oxidized organic compounds from gasoline and with nitrogen oxides in the air to produce harmful irritants such as methyl nitrate ($CH_3NO_3$).

The oxides of sulfur create global pollution problems because they have longer lifetimes in the atmosphere than the oxides of nitrogen. Some of the $SO_2$ and $SO_3$ in the air originates from biological processes and from volcanoes, but much comes from the oxidation of sulfur in petroleum and in coal that is burned for fuel. If the sulfur is not removed from the fuel or the exhaust gas, $SO_2$ enters the atmosphere as a stable but reactive pollutant. Further oxidation by radicals leads to sulfur trioxide:

$$SO_2 + OH \longrightarrow SO_2OH$$

$$SO_2OH + O_2 \longrightarrow SO_3 + OOH$$

$$OOH \longrightarrow O + OH$$

The atomic oxygen thus produced can react with molecular oxygen to form ozone. **Acid rain** results from the reaction of $NO_2$ and $SO_3$ with hydroxyl radical and water vapor in the air to form nitric acid ($HNO_3$) and sulfuric acid ($H_2SO_4$). The acids dissolve in water and return to the earth in the rain.

## The Greenhouse Effect

As we have seen, most of the short-wavelength photons from the sun are absorbed by the outer atmosphere and do not reach the earth's surface. The radiation that does reach the earth is in balance with reradiation into space and helps to maintain a livable temperature. Certain gases in the troposphere play a crucial role in this balance, because they absorb infrared radiation emitted by the warm surface of the

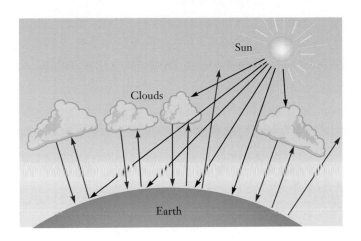

**FIGURE 16.44** Visible light from the sun strikes the earth's surface and heats it. Much energy is radiated back into space in the longer wavelength infrared region of the spectrum. Molecules in the atmosphere and in clouds can block some of this loss, keeping the lower atmosphere warmer than it would be without these molecules.

earth rather than letting it pass out to space (Fig. 16.44). In winter, the temperature does not fall as low on cloudy nights as it does on clear nights because the water vapor in the clouds provides a thermal blanket, absorbing outgoing radiation with wavelengths near 20,000 nm and reradiating it to the earth's surface.

Two other gases that absorb infrared radiation to significant extents are carbon dioxide and methane. Both are uniformly distributed in fairly low concentrations through the troposphere, but their concentrations have increased steadily over the 200 to 300 years since the beginning of the industrial revolution (Fig. 16.45). The tremendous increase in the burning of fossil fuels for energy is a major contributor to this trend (Fig 16.46). It is estimated that by the year 2050 the concentration of $CO_2$ in the atmosphere will be double the premodern value. Chlorofluorocarbons also absorb infrared radiation, especially over some key wavelength ranges left open by $CO_2$ and water vapor.

Such changes are viewed with alarm because of a phenomenon called the **greenhouse effect**—a global increase in average surface temperature that occurs if heat given off by the earth's surface is prevented from escaping by higher concentrations of gases that absorb and reradiate infrared radiation. An increase in average surface temperature of 2 to 5°C over the next century would have major repercussions, including the melting of some polar ice, a consequent increase in the level of the oceans, and the conversion of arable land to desert. To prevent these undesirable

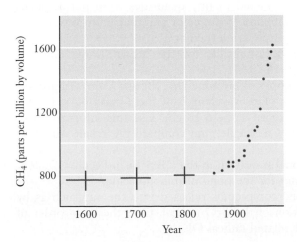

**FIGURE 16.45** The concentration of methane in the lower atmosphere has doubled since 1600. Note the dramatic increase in the last 100 years.

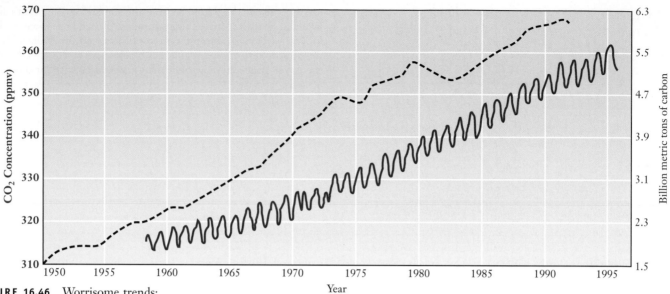

**FIGURE 16.46** Worrisome trends: annual worldwide $CO_2$ emissions from fossil fuel combustion, 1950–1992 (black); average monthly concentration of atmospheric $CO_2$ at Mauna Loa Hawaii, 1958–1996 (red). The latter varies seasonally.

A chemical plant uses bromine to manufacture brominated flame retardants for plastics and fibers. (*Ethyl Corporation*)

changes, it will be necessary to develop new energy sources not based on the burning of fossil fuels by early in the 21st century.

## CUMULATIVE EXERCISE

### Bromine

Elemental bromine is a brownish red liquid that was first isolated in 1826. It is currently produced by oxidation of bromide ion in natural brines with elemental chlorine.

(a) What is the ground-state configuration of the valence electrons of bromine molecules ($Br_2$)? Is bromine paramagnetic or diamagnetic?

(b) What is the electron configuration of the $Br_2^+$ molecular ion? Is its bond stronger or weaker than that in $Br_2$? What is its bond order?

(c) Bromine compounds have been known and used for centuries. The deep purple color that symbolized imperial power in ancient Rome originated with the compound dibromoindigo, which was extracted in tiny quantities from purple snails (about 8000 snails per gram of compound). What color and maximum wavelength of *absorbed* light would give a deep purple (violet) color?

(d) What excited electronic state is responsible for the brownish red color of bromine? Refer to Figures 16.6 and 16.28.

(e) The two naturally occurring isotopes of bromine are $^{79}Br$ and $^{81}Br$, with masses of 78.918 and 80.916 u, respectively. The wavelength of the $J = 0$ to $J = 1$ rotational transition in $^{79}Br^{81}Br$ is measured to be 6.18 cm. Use this information to calculate the bond length in the $Br_2$ molecule, and compare the result with that listed in Table 3.2.

(f) The wavelength of the vibrational transition in the $^{79}Br^{81}Br$ molecule is $3.09 \times 10^{-5}$ m. Calculate the force constant for the bond in this molecule.

(g) The action of light on bromine compounds released into the air (such as by leaded gasoline) causes the formation of the BrO radical. Give the bond order of this species by comparing it to the related radical OF.

(h) There is concern that synthetic bromine-containing compounds, in addition to chlorofluorocarbons, are helping to destroy ozone in the stratosphere. The BrO (see part (e) can take part with ClO in the following catalytic cycle:

$$Cl + O_3 \longrightarrow ClO + O_2$$

$$Br + O_3 \longrightarrow BrO + O_2$$

$$ClO + BrO \longrightarrow Cl + Br + O_2$$

Write the overall equation for this cycle.

## Answers

(a) $(\sigma_{4s})^2(\sigma_{4s}^*)^2(\sigma_{4p_z})^2(\pi_{4p})^4(\pi_{4p}^*)^4$; diamagnetic
(b) $(\sigma_{1s})^2(\sigma_{1s}^*)^2(\sigma_{1p_z})^2(\pi_{4p})^4(\pi_{4p}^*)^3$; stronger; bond order is $\frac{3}{2}$ versus 1
(c) Yellow light, near 530 nm
(d) The lowest energy excited state, which arises from the excitation of an electron from the filled $\pi_{4p}^*$ orbital to the unfilled $\sigma_{4p_z}^*$ orbital
(e) 2.28 Å (from Table 3.2: 2.286 Å)
(f) 247 J m$^{-2}$
(g) $\frac{3}{2}$ order
(h) Overall: $2\ O_3 \rightarrow 3\ O_2$

## CONCEPTS & SKILLS

*After studying this chapter and working the problems that follow, you should be able to:*

1. Show how molecular orbitals can be constructed from the atomic orbitals of two atoms that form a chemical bond, and explain how the electron density between the atoms is related to the molecular orbital (Section 16.1).

2. Construct correlation diagrams for diatomic molecules formed from second- and third-period main-group elements. From these diagrams, give the electron configurations, work out the bond orders, and comment on their magnetic properties (Section 16.1, problems 1–14).

3. Describe the hybrid-atomic-orbital basis for representing the structures of molecules (Section 16.2, problems 15–22).

4. Discuss the delocalization of $\pi$ electrons in polyatomic molecules (Section 16.2, problems 23–26).

5. Discuss the SCF approach to calculating molecular properties, and comment on the role played by electron correlation (Section 16.3).

6. Outline the principles of magnetic resonance spectroscopy (Section 16.5).

7. Relate the moments of inertia, bond lengths, and vibrational force constants of diatomic molecules to their rotational and vibrational spectra (Section 16.6, problems 27–34).

8. Discuss the Boltzmann distribution of population among possible molecular quantum states (Section 16.6, problems 35, 36).

9. Describe the preparation of excited electronic states by absorption of radiation and the subsequent flow of energy (Section 16.7, problems 37–42).

10. Discuss several new techniques in molecular spectroscopy and the purposes they serve (Section 16.7).

11. Describe the bonding in conjugated molecules and determine their geometries (Section 16.8, problems 43, 44).

12. Indicate processes in the atmosphere that are beneficial and those that are potentially damaging to the ecosystem (Section 16.9, problems 45–48).

## Molecular Orbitals in Diatomic Molecules

**1.** If an electron is removed from a fluorine molecule, an $F_2^+$ molecular ion forms.
(a) Give the molecular electron configurations for $F_2$ and $F_2^+$.
(b) Give the bond order of each species.
(c) Predict which species should be paramagnetic.
(d) Predict which species has the greater bond dissociation energy.

**2.** When one electron is added to an oxygen molecule, a super-oxide ion ($O_2^-$) is formed. The addition of two electrons gives a peroxide ion ($O_2^{2-}$). Removal of an electron from $O_2$ leads to $O_2^+$.
(a) Construct the correlation diagram for $O_2^-$.
(b) Give the molecular electron configuration for each of the following species: $O_2^+$, $O_2$, $O_2^-$, $O_2^{2-}$.
(c) Give the bond order of each species.
(d) Predict which species are paramagnetic.
(e) Predict the order of increasing bond dissociation energy among the species.

**3.** Predict the valence electron configuration and the total bond order for the molecule $S_2$, which forms in the gas phase when sulfur is heated to a high temperature. Will $S_2$ be paramagnetic or diamagnetic?

**4.** Predict the valence electron configuration and the total bond order for the molecule $I_2$. Will $I_2$ be paramagnetic or diamagnetic?

**5.** For each of the following valence electron configurations of a homonuclear diatomic molecule or molecular ion, identify the element X, Q, or Z, and determine the total bond order.
(a) $X_2$: $(\sigma_{2s})^2(\sigma_{2s}^*)^2(\sigma_{2p_z})^2(\pi_{2p})^4(\pi_{2p}^*)^4$
(b) $Q_2^+$: $(\sigma_{2s})^2(\sigma_{2s}^*)^2(\pi_{2p})^4(\sigma_{2p_z})^1$
(c) $Z_2^-$: $(\sigma_{2s})^2(\sigma_{2s}^*)^2(\sigma_{2p_z})^2(\pi_{2p})^4(\pi_{2p}^*)^3$

**6.** For each of the following valence electron configurations of a homonuclear diatomic molecule or molecular ion, identify the element X, Q, or Z, and determine the total bond order.
(a) $X_2$: $(\sigma_{2s})^2(\sigma_{2s}^*)^2(\sigma_{2p_z})^2(\pi_{2p})^4(\pi_{2p}^*)^4$     _12 - not F._
(b) $Q_2^-$: $(\sigma_{2s})^2(\sigma_{2s}^*)^2(\pi_{2p})^3$
(c) $Z_2^{2+}$: $(\sigma_{2s})^2(\sigma_{2s}^*)^2(\sigma_{2p_z})^2(\pi_{2p})^4(\pi_{2p}^*)^2$

**7.** For each of the electron configurations in problem 5, determine whether the molecule or molecular ion is paramagnetic or diamagnetic.

**8.** For each of the electron configurations in problem 6, determine whether the molecule or molecular ion is paramagnetic or diamagnetic.

**9.** Following the pattern of Figure 16.8, work out the correlation diagram for the CN molecule, showing the relative energy levels of the atoms and the bonding and antibonding orbitals of the molecule. Indicate the occupation of the molecular orbitals with arrows. State the order of the bond, and comment upon the magnetic properties of CN.

**10.** Following the pattern of Figure 16.8, work out the correlation diagram for the BeN molecule, showing the relative energy levels of the atoms and the bonding and antibonding orbitals of the molecule. Indicate the occupation of the molecular orbitals with arrows. State the order of the bond, and comment upon the magnetic properties of BeN.

**11.** The bond length of the transient diatomic molecule CF is 1.291 Å; that of the molecular ion $CF^+$ is 1.173 Å. Explain why the CF bond shortens with the loss of an electron. Refer to the proper molecular orbital correlation diagram.

**12.** The compound nitrogen oxide (NO) forms when the nitrogen and oxygen in air are heated. Predict whether the nitrosyl ion ($NO^+$) will have a shorter or a longer bond than the NO molecule. Will $NO^+$ be paramagnetic like NO, or diamagnetic?

**13.** What would be the electron configuration for an $HeH^-$ molecular ion? What bond order would you predict? How stable should such a species be?

**14.** The molecular ion $HeH^+$ has an equilibrium bond length of 0.774 Å. Draw an electron correlation diagram for this ion, indicating the occupied molecular orbitals. Is $HeH^+$ paramagnetic? When the $HeH^+$ ion dissociates, is a lower energy state reached by forming $He + H^+$ or $He^+ + H$?

## Molecular Orbitals in Polyatomic Molecules

**15.** Formulate a localized bond picture for the amide ion ($NH_2^-$). What hybridization do you expect the central nitrogen atom to have, and what geometry do you predict for the molecular ion?

**16.** Formulate a localized bond picture for the hydronium ion ($H_3O^+$). What hybridization do you expect the central oxygen atom to have, and what geometry do you predict for the molecular ion?

**17.** Write a Lewis electron-dot diagram for each of the following molecules and ions. Formulate the hybridization for the central atom in each case, and give the molecular geometry.
(a) $CCl_4$     (d) $CH_3^-$
(b) $CO_2$      (e) $BeH_2$
(c) $OF_2$

**18.** Write a Lewis electron-dot diagram for each of the following molecules and ions. Formulate the hybridization for the central atom in each case, and give the molecular geometry.
(a) $BF_3$      (d) $CS_2$
(b) $BH_4^-$    (e) $CH_3^+$
(c) $PH_3$

**19.** Describe the hybrid orbitals on the chlorine atom in the $ClO_3^+$ and $ClO_2^+$ molecular ions. Sketch the expected geometries of these ions.

**20.** Describe the hybrid orbitals on the chlorine atom in the $ClO_4^-$ and $ClO_3^-$ molecular ions. Sketch the expected geometries of these ions.

**21.** The sodium salt of the unfamiliar orthonitrate ion ($NO_4^{3-}$) has been prepared. What hybridization is expected on the N atom at the center of this ion? Predict the geometry of the $NO_4^{3-}$ ion.

22. Describe the hybrid orbitals used by the carbon atom in N≡C—Cl. Predict the geometry of the molecule.

23. The azide ion ($N_3^-$) is a weakly bound molecular ion. Formulate its molecular orbital structure in terms of localized $\sigma$ bonds and delocalized $\pi$ bonds. Do you expect $N_3$ and $N_3^+$ to be bound as well? Which of the three species do you expect to be paramagnetic?

24. Formulate the molecular orbital structure of $NO_2^+$ in terms of localized $\sigma$ bonds and delocalized $\pi$ bonds. Is it linear or nonlinear? Do you expect it to be paramagnetic? Repeat the analysis for $NO_2$ and for $NO_2^-$.

25. Discuss the nature of the bonding in the nitrite ion ($NO_2^-$). Draw the possible Lewis resonance diagrams for this ion. Use the VSEPR theory to determine the steric number, the hybridization of the central N atom, and the geometry of the ion. Show how the use of resonance structures can be avoided by introducing a delocalized $\pi$ molecular orbital. What bond order does the molecular orbital model predict for the N—O bonds in the nitrite ion?

26. Discuss the nature of the bonding in the nitrate ion ($NO_3^-$). Draw the possible Lewis resonance diagrams for this ion. Use the VSEPR theory to determine the steric number, the hybridization of the central N atom, and the geometry of the ion. Show how the use of resonance structures can be avoided by introducing a delocalized $\pi$ molecular orbital. What bond order is predicted by the molecular orbital model for the N—O bonds in the nitrate ion?

## General Aspects of Molecular Spectroscopy

27. Use data from Tables 14.1 and 3.2 to predict the energy spacing between the ground state and the first excited rotational state of the $^{14}N^{16}O$ molecule.

28. Use data from Tables 14.1 and 3.2 to predict the energy spacing between the ground state and the first excited rotational state of the $^1H^{19}F$ molecule.

29. The first three absorption lines in the pure rotational spectrum of gaseous $^{12}C^{16}O$ are found to have the frequencies $1.15 \times 10^{11}$, $2.30 \times 10^{11}$, and $3.46 \times 10^{11}$ s$^{-1}$. Calculate:
   (a) The moment of inertia $I$ of CO (in kg m$^2$)
   (b) The energies of the $J = 1$, $J = 2$, and $J = 3$ rotational levels of CO, measured from the $J = 0$ state (in joules)
   (c) The C—O bond length (in angstroms)

30. Four consecutive absorption lines in the pure rotational spectrum of gaseous $^1H^{35}Cl$ are found to have the frequencies $2.50 \times 10^{12}$, $3.12 \times 10^{12}$, $3.74 \times 10^{12}$, and $4.37 \times 10^{12}$ s$^{-1}$. Calculate:
   (a) The moment of inertia $I$ of HCl (in kg m$^2$)
   (b) The energies of the $J = 1$, $J = 2$, and $J = 3$ rotational levels of HCl, measured from the $J = 0$ state (in joules)
   (c) The H—Cl bond length (in angstroms)
   (d) The initial and final $J$ states for the observed absorption lines

31. The $Li_2$ molecule ($^7Li$ isotope) shows a very weak infrared line in its vibrational spectrum at a wavelength of $2.85 \times 10^{-5}$ m. Calculate the force constant for the $Li_2$ molecule.

32. The $Na_2$ molecule ($^{23}Na$ isotope) shows a very weak infrared line in its vibrational spectrum at a wavelength of $6.28 \times 10^{-5}$ m. Calculate the force constant for the $Na_2$ molecule, and compare your result with that of problem 31. Give a reason for any difference.

33. The "signature" infrared absorption that indicates the presence of a C—H stretching motion in a molecule occurs at wavelengths near $3.4 \times 10^{-6}$ m. Use this information to estimate the force constant of the C—H stretch. Take the reduced mass in this motion to be approximately equal to the mass of the hydrogen atom (a good approximation when the H atom is attached to a heavy group).

34. Repeat the calculation of the preceding problem for the N—H stretch, where absorption occurs near $2.9 \times 10^{-6}$ m. Which bond is stiffer: N—H or C—H?

35. Estimate the ratio of the number of molecules in the first excited vibrational state of the molecule $N_2$ to the number in the ground state, at a temperature of 450 K. The vibrational frequency of $N_2$ is $7.07 \times 10^{13}$ s$^{-1}$.

36. The vibrational frequency of the ICl molecule is $1.15 \times 10^{13}$ s$^{-1}$. For every million ($1.00 \times 10^6$) molecules in the ground vibrational state, how many will be in the first excited vibrational state at a temperature of 300 K?

## Excited Electronic States

37. Suppose that an ethylene molecule gains an additional electron to give the $C_2H_4^-$ ion. Will the bond order of the carbon–carbon bond increase or decrease? Explain.

38. Suppose that an ethylene molecule is ionized by a photon to give the $C_2H_4^+$ ion. Will the bond order of the carbon–carbon bond increase or decrease? Explain.

39. The color of the dye "indanthrene brilliant orange" is evident from its name. In what wavelength range would you expect to find the maximum in the absorption spectrum of this molecule? Refer to the color spectrum in Figure 15.3.

40. In what wavelength range would you expect to find the maximum in the absorption spectrum of the dye "crystal violet"?

## Illustration: Conjugated Systems

41. Use data from Table 3.2 to give an upper bound on the wavelengths of light that are capable of dissociating a molecule of ClF.

42. Use data from Table 3.2 to give an upper bound on the wavelengths of light that are capable of dissociating a molecule of ICl.

43. To satisfy the octet rule, fullerene must have double bonds. How many? Give a simple rule for one way of placing them in the structure shown in Figure 16.38a.

44. It has been suggested that a compound of formula $C_{12}B_{24}N_{24}$ might exist and have a structure like that of $C_{60}$ (buckminsterfullerene).
   (a) Explain the logic of this suggestion by comparing the number of valence electrons in $C_{60}$ and $C_{12}B_{24}N_{24}$.
   (b) Propose the most symmetric pattern of C, B, and N atoms in $C_{12}B_{24}N_{24}$ to occupy the 60 atom sites in the

buckminsterfullerene structure. Where could the double bonds be placed in such a structure?

## Illustration: Global Warming and the Greenhouse Effect

45. The bond dissociation energy of a typical C—F bond in a chlorofluorocarbon is approximately 440 kJ mol$^{-1}$. Calculate the maximum wavelength of light that can photodissociate a molecule of $CCl_2F_2$, breaking such a C—F bond.

46. The bond dissociation energy of a typical C—Cl bond in a chlorofluorocarbon is approximately 330 kJ mol$^{-1}$. Calculate the maximum wavelength of light that can photodissociate a molecule of $CCl_2F_2$, breaking such a C—Cl bond.

47. Draw a Lewis diagram(s) for the ozone molecule ($O_3$). Determine the steric number and hybridization of the central oxygen atom, and identify the molecular geometry. Describe the nature of the $\pi$ bonds and give the bond order of the O—O bonds in ozone.

48. The compounds carbon dioxide ($CO_2$) and sulfur dioxide ($SO_2$) are formed by the burning of coal. Their apparently similar formulas mask underlying differences in molecular structure. Determine the shapes of these two types of molecules, identify the hybridization at the central atom of each, and compare the natures of their $\pi$ bonds.

## Additional Problems

49. (a) Sketch the occupied molecular orbitals of the valence shell for the $N_2$ molecule. Label the orbitals as $\sigma$ or $\pi$ orbitals, and specify which are bonding and which are antibonding.
    (b) If one electron is removed from the highest occupied orbital of $N_2$, will the equilibrium N—N distance become longer or shorter? Explain briefly.

50. Calcium carbide ($CaC_2$) is an intermediate in the manufacture of acetylene ($C_2H_2$). It is the calcium salt of the carbide (also called acetylide) ion ($C_2^{2-}$). What is the electron configuration of this molecular ion? What is its bond order?

51. Show how the fact that the $B_2$ molecule is paramagnetic indicates that the energy ordering of the orbitals in this molecule is given by Figure 16.6a rather than 16.6b.

52. The $Be_2$ molecule has been detected experimentally. It has a bond length of 2.45 Å and a bond dissociation enthalpy of 9.46 kJ mol$^{-1}$. Write the ground-state electron configuration of $Be_2$ and predict its bond order, using the theory developed in the text. Compare the experimental bonding data on $Be_2$ with those recorded for $B_2$, $C_2$, $N_2$, and $O_2$ in Table 16.2. Is the prediction that stems from the simple theory seriously incorrect?

* 53. (a) The ionization energy of molecular hydrogen ($H_2$) is *greater* than that of atomic hydrogen (H), but that of molecular oxygen ($O_2$) is *lower* than that of atomic oxygen (O). Explain. (*Hint:* Think about the stability of the molecular ion that forms, in relation to bonding and antibonding electrons.)
    (b) What prediction would you make for the relative ionization energies of atomic and molecular fluorine (F and $F_2$)?

54. The molecular ion HeH$^+$ has an equilibrium bond length of 0.774 Å. Draw an electron correlation diagram for this molecule, indicating the occupied molecular orbitals. If the lowest energy molecular orbital has the form $C_1\psi_{1s}^H + C_2\psi_{1s}^{He}$, do you expect $C_2$ to be larger or smaller than $C_1$?

* 55. The molecular orbital of the ground state of a heteronuclear diatomic molecule AB is

$$\psi_{mol} = C_A\psi_A + C_B\psi_B$$

If a bonding electron spends 90% of its time in an orbital $\psi_A$ on atom A and 10% of its time in $\psi_B$ on atom B, what are the values of $C_A$ and $C_B$? (Neglect the overlap of the two orbitals.)

56. The stable molecular ion $H_3^+$ is triangular, with H—H distances of 0.87 Å. Sketch the molecule and indicate the region of greatest electron density of the lowest energy molecular orbital.

* 57. According to recent spectroscopic results, nitramide

$$\begin{matrix} H & & & O \\ & \diagdown & & \diagup \\ & & N{-}N \\ & \diagup & & \diagdown \\ H & & & O \end{matrix}$$

is a nonplanar molecule. It was previously thought to be planar.
    (a) Predict the bond order of the N—N bond in the nonplanar structure.
    (b) If the molecule really were planar after all, what would be the bond order of the N—N bond?

58. *Trans*-tetrazene ($N_4H_4$) consists of a chain of four nitrogen atoms with each of the two end atoms bonded to two hydrogen atoms. Use the concepts of steric number and hybridization to predict the overall geometry of the molecule. Give the expected structure of *cis*-tetrazene.

59. What are the moments of inertia of $^1H^{19}F$ and $^1H^{81}Br$, expressed in kg m$^2$? Compute the spacings $\nu = \Delta E/h$ of the rotational states, in s$^{-1}$, between $J = 0$ and 1 and between $J = 1$ and 2. Explain, in one sentence, why the large change in mass from 19 to 81 causes only a small change in rotational energy differences.

60. The average bond length of a molecule can change slightly with vibrational state. In $^{23}Na^{35}Cl$, the frequency of light absorbed in a change from the $J = 1$ to the $J = 2$ rotational state in the ground vibrational state ($v = 0$) was measured to be $\nu = 2.60511 \times 10^{10}$ s$^{-1}$, and that for a change from $J = 1$ to $J = 2$ in the first excited vibrational state ($v = 1$) was $\nu = 2.58576 \times 10^{10}$ s$^{-1}$. Calculate the average bond lengths of NaCl in these two vibrational states, taking the relative atomic mass of $^{23}Na$ to be 22.9898 and that of $^{35}Cl$ to be 34.9689.

61. The vibrational frequencies of $^{23}Na^1H$, $^{23}Na^{35}Cl$, and $^{23}Na^{127}I$ are $3.51 \times 10^{13}$ s$^{-1}$, $1.10 \times 10^{13}$ s$^{-1}$, and $0.773 \times 10^{13}$ s$^{-1}$, respectively. Their bond lengths are 1.89, 2.36, and 2.71 Å. What are their reduced masses? What are their force constants? If NaH and NaD have the same force constant, what is the vibrational frequency of NaD? D is $^2H$.

62. Recall that nuclear spin states in nuclear magnetic resonance are typically separated by energies of $2 \times 10^{-5}$ to $2 \times 10^{-4}$ kJ mol$^{-1}$. What are the ratios of occupation probability between a pair of such levels at thermal equilibrium and a temperature of 25°C?

63. The vibrational temperature of a molecule prepared in a supersonic jet can be estimated from the observed populations of its vibrational levels, assuming a Boltzmann distribution. The vibrational frequency of HgBr is $5.58 \times 10^{12}$ s$^{-1}$, and the ratio of the number of molecules in the $v = 1$ state to the number in the $v = 0$ state is 0.127. Estimate the vibrational temperature under these conditions.

64. An electron in the $\pi$ orbital of ethylene (C$_2$H$_4$) is excited by a photon to the $\pi^*$ orbital. Do you expect the equilibrium bond length in the excited ethylene molecule to be greater or less than that in ground-state ethylene? Will the vibrational frequency in the excited state be higher or lower than in the ground state? Explain your reasoning.

65. One isomer of retinal is converted to a second isomer by the absorption of a photon:

This process is a key step in the chemistry of vision. Although free retinal (in the form shown to the left of the arrow) has an absorption maximum at 376 nm, in the ultraviolet region of the spectrum this absorption shifts into the visible range when the retinal is bound in a protein, as it is in the eye.

(a) How many of the C=C double bonds are *cis* and how many are *trans* in each of the preceding structures? (When assigning labels, consider the relative positions of the two largest groups attached at each double bond.) Describe the motion that takes place upon absorption of a photon.

(b) If the ring and the —CHO group in retinal were replaced by CH$_3$ groups, would the absorption maximum in this molecule shift to longer or shorter wavelengths?

* 66. The ground-state electron configuration of the H$_2^+$ molecular ion is $(\sigma_{1s})^1$.

(a) A molecule of H$_2^+$ absorbs a photon and is excited to the $\sigma_{1s}^*$ molecular orbital. Predict what happens to the molecule.

(b) Another molecule of H$_2^+$ absorbs even more energy in an interaction with a photon and is excited to the $\sigma_{2s}$ molecular orbital. Predict what happens to this molecule.

* 67. (a) Draw a Lewis diagram for formaldehyde (H$_2$CO), and decide the hybridization of the central carbon atom.

(b) Formulate the molecular orbitals for the molecule.

(c) A strong absorption is observed in the ultraviolet region of the spectrum and is attributed to a $\pi \rightarrow \pi^*$ transition. Another, weaker transition is observed at lower frequencies. What electronic excitation causes the weaker transition?

68. Write balanced chemical equations that describe the formation of nitric acid and sulfuric acid in rain, starting with the sulfur in coal and the oxygen, nitrogen, and water vapor in the atmosphere.

69. Compare and contrast the roles of ozone (O$_3$) and nitrogen dioxide (NO$_2$) in the stratosphere and in the troposphere.

70. Describe the greenhouse effect and its mechanism of operation. Give three examples of energy sources that contribute to increased CO$_2$ in the atmosphere and three that do not.

## CUMULATIVE PROBLEMS

71. At thermal equilibrium, is the rate as a molecule is excited from $v = 0$ to the $v = 1$ level greater than or less than the rate for the reverse process? What is the ratio of the rate constants? (*Hint:* Think of the analogy with the chemical equilibrium between two species.)

72. The standard enthalpy of combustion of C$_{60}$(s) has been found to be $-25,891$ kJ mol$^{-1}$. Compute the standard enthalpy of formation of C$_{60}$(s).

73. The hydroxyl radical has been referred to as the "chief cleanup agent in the troposphere." Its concentration is approximately zero at night and becomes as high as $1 \times 10^7$ molecules per cm$^3$ in highly polluted air.

(a) Calculate the maximum mole fraction and partial pressure of OH in polluted air at 25°C and atmospheric pressure.

(b) Write an equation for the reaction of HO with NO$_2$ in the atmosphere. How does the oxidation state of nitrogen change in this reaction? What is the ultimate fate of the product of this reaction?

74. In unpolluted air at 300 K, the hydroxyl radical OH reacts with CO with a bimolecular rate constant of $1.6 \times 10^{11}$ L mol$^{-1}$ s$^{-1}$ and with CH$_4$ with a rate constant of $3.8 \times 10^9$ L mol$^{-1}$ s$^{-1}$. Take the partial pressure of CO in air to be constant at $1.0 \times 10^{-7}$ atm and that of CH$_4$ to be $1.7 \times 10^{-6}$ atm, and assume that these are the primary mechanisms by which OH is consumed in the atmosphere. Calculate the half-life of OH under these conditions.

# Bonding, Structure, and Reactions of Organic Molecules

*Illustration*

A petroleum refining tower. *(Courtesy of Ashland Oil, Inc.)*

arbon is unique among the elements in the large number of compounds it forms and in the variety of their structures. In combination with hydrogen, it forms molecules with single, double, and triple bonds; chains; rings; branched structures; and cages (Fig. 17.1). The thousands of stable hydrocarbons are in sharp contrast with the mere two stable compounds between oxygen and hydrogen (water and hydrogen peroxide). Even the rather versatile elements nitrogen and oxygen form only six nitrogen oxides.

The unique properties of carbon relate to its position in the periodic table. As a second-period element, carbon has relatively small atoms, so it can easily form the double and triple bonds that are rare in the compounds of related elements, such as silicon. As a Group IV element, carbon can form four bonds, more than the other second-period elements; this characteristic gives it wide scope for structural elaboration. Finally, as an element of intermediate electronegativity, carbon forms covalent compounds both with relatively electronegative elements, such as oxygen, nitrogen, and the halogens, and with relatively electropositive elements, such as hydrogen and the heavy metals mercury and lead.

The study of the compounds of carbon is the discipline traditionally called **organic chemistry,** although the chemistry of carbon is intimately bound up with that of inorganic elements and with biochemistry. This chapter begins with a look at the variety of bonding and structures evidenced by organic molecules, made possible by the properties of the element carbon described in the previous paragraph. The description is based on the general principles of covalent bonding in polyatomic molecules presented in Section 16.2. The relation between molecular structure and properties of organic substances is illustrated by examining the composition, refining, and chemical processing of petroleum, the primary starting material for the production of hydrocarbons and their derivatives. The chapter concludes with an introduction to the types of compounds that result when elements such as chlorine, oxygen, and nitrogen combine with carbon and hydrogen, and with a survey of their importance in synthetic reactions.

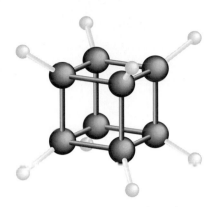

**FIGURE 17.1**   One simple and unusual hydrocarbon is cubane ($C_8H_8$), in which the eight carbon atoms are arranged at the corners of a cube.

---

## 17.1

# BONDING AND STRUCTURE IN ORGANIC MOLECULES

## Bonding in Organic Molecules

The structures of many organic molecules (compounds of carbon) can be described through the simple molecular orbital picture presented in Section 16.2. A helpful first step in analyzing molecular bonding is to write the Lewis electron-dot diagram. As an example, consider the molecule 2-butene ($CH_3CHCHCH_3$). Its Lewis diagram is

$$
\begin{array}{ccccc}
H & H & & H & H \\
\ddot{} & \ddot{} & & \ddot{} & \ddot{} \\
H:\overset{\displaystyle H}{\underset{\displaystyle H}{\ddot{C}}}:\overset{\displaystyle H}{\ddot{C}}: & :\overset{\displaystyle H}{\underset{\displaystyle H}{\ddot{C}}}:\ddot{C}:H
\end{array}
$$

From the VSEPR theory, the steric number of the two outer carbon atoms is 4 (and so they have $sp^3$ hybridization), and that of the two central carbon atoms is 3 (giving $sp^2$ hybridization). The bonding about the outer carbon atoms will be tetrahedral, and that about the central ones will be trigonal planar. Electron pairs are placed into each localized ($\sigma$) molecular orbital, giving a single bond between each pair of bonding atoms. In the case of 2-butene, these placements use 22 out of the 24 available valence electrons, forming a total of 11 single bonds.

Next, the remaining $p$ orbitals that were not involved in hybridization are combined to form $\pi$ molecular orbitals. The $p_z$ orbitals from the two central carbon

**FIGURE 17.2** As the 2-butene molecule is twisted about the C=C bond, the overlap of the two *p* orbitals decreases, giving a higher energy.

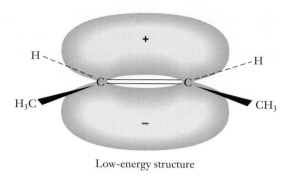

Low-energy structure

High-energy structure

atoms can be mixed to form a $\pi$ (bonding) molecular orbital and a $\pi^*$ (antibonding) molecular orbital. The remaining two valence electrons are placed into the $\pi$ orbital, giving a double bond between the central carbon atoms. If the $p_z$ orbital of one of these atoms is rotated about the central C—C axis, its overlap with the $p_z$ orbital of the other carbon atom changes (Fig. 17.2). The overlap is greatest, and the energy lowest, when the two $p_z$ orbitals are parallel to each other. In the most stable molecular geometry, the hydrogen atoms on the central carbon atoms lie in the same plane as the C—C—C—C carbon skeleton. This prediction is verified by experiment.

## Structural Isomerism

In 2-butene there are two possible placements for the outer $CH_3$ groups relative to the double bond (Fig. 17.3). In one, called *cis*-2-butene, the two methyl ($CH_3$) groups lie on the same side of the double bond; in the other, *trans*-2-butene, they lie on opposite sides. Converting one form to the other requires breaking the central $\pi$ bond (by rotating the two $p_z$ orbitals 180° with respect to each other, as in Fig. 17.2), then re-forming it in the other configuration. Because breaking a $\pi$ bond costs a significant amount of energy, both *cis* and *trans* forms are stable at room temperature, and interconversion between the two is very slow. The change in molecular structure from *trans* to *cis* can be accomplished by photochemistry without complete breaking of bonds (see Section 16.7). Molecules such as those of *trans*-2-butene can absorb ultraviolet light by the excitation of an electron from a $\pi$ to a $\pi^*$ molecular orbital, just as ethylene does (see Fig. 16.28). In the excited electronic state of *trans*-2-butene, the carbon–carbon double bond is effectively reduced to a single bond, and the one $CH_3$ group can rotate relative to the other to form *cis*-2-butene. Molecules that have the same chemical formula but different structures are called **geometrical isomers.** Absorption of ultraviolet light can cause interconversion of one geometrical isomer into another.

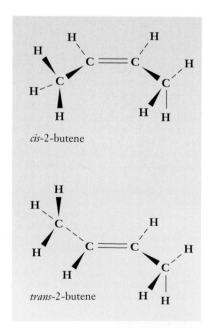

*cis*-2-butene

*trans*-2-butene

**FIGURE 17.3** The two *cis-trans* isomers of 2-butene.

**EXAMPLE 17.1**

The Lewis diagram for propyne ($CH_3CCH$) is

$$
\begin{array}{c}
\text{H} \\
\overset{\displaystyle ..}{\underset{\displaystyle ..}{\text{H}:\text{C}}}\text{—C}\equiv\text{C}:\text{H} \\
\text{H}
\end{array}
$$

Discuss its bonding and predict its geometry.

### Solution

The leftmost carbon atom in the structure is $sp^3$ hybridized, and the other two carbon atoms are $sp$ hybridized. The atoms in the molecule are on a single straight line with the exception of the three hydrogen atoms on the leftmost carbon atom, which point outward toward three of the vertices of a tetrahedron. There is a $\sigma$ bond between each pair of bonded atoms. The $p_x$ and $p_y$ orbitals on carbon atoms 2 and 3 combine to form two $\pi$ orbitals and two $\pi^*$ orbitals; only the former are occupied in the ground-state electron configuration.

**Related Problems: 1, 2**

A second type of isomerism characteristic of organic molecules is **optical isomerism, or chirality.** A carbon atom that forms single bonds to four different atoms or groups of atoms can exist in two forms that are mirror images of each other but that cannot be interconverted without breaking and re-forming bonds (Fig. 17.4). If a mixture of the two forms is resolved into its optical isomers, the two forms rotate the plane of polarized light in different directions, and so such molecules are said to be "optically active." Although paired optical isomers have identical physical properties, their chemical properties can differ when they interact with other optically active molecules. As we will see in Section 25.3, proteins and other biomolecules are optically active. One goal of pharmaceutical research is to prepare particular optical isomers of carbon compounds for medicinal use. In many cases one optical isomer is beneficial and the other is useless or even harmful.

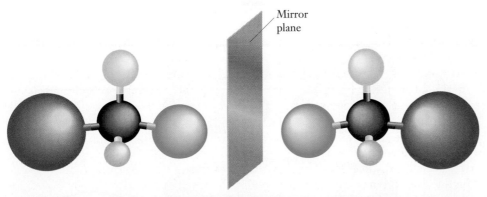

**FIGURE 17.4**  A molecule such as $CHBrClF$, which has four different atoms or groups of atoms bonded to a single carbon atom, exists in two mirror-image forms that cannot be superimposed by rotation. Such pairs of molecules are optical isomers; the carbon atom is called a chiral center.

## Conjugated Molecules

When two or more double or triple bonds occur close to each other in a molecule, a delocalized molecular orbital picture of the bonding should be used. One example is 1,3-butadiene ($CH_2CHCHCH_2$), which has the Lewis diagram

$$
\begin{array}{cccc}
\text{H} & \text{H} & \text{H} & \text{H} \\
\ddots & \ddots & \ddots & \ddots \\
\end{array}
$$

$$
\text{H}\!:\!\ddot{\text{C}}\!:\!:\!\ddot{\text{C}}\!:\!\ddot{\text{C}}\!:\!:\!\ddot{\text{C}}\!:\!\text{H}
$$

All four carbon atoms have steric number 3, and so all are $sp^2$ hybridized. The remaining $p_z$ orbitals have maximum overlap when the four carbon atoms lie in the same plane, and so this molecule is predicted to be planar. From these four $p_z$ atomic orbitals, four molecular orbitals can be constructed by combining their phases as shown in Figure 17.5. The four electrons that remain after the $\sigma$ orbitals are filled are placed in the two lowest $\pi$ orbitals. The first of these is bonding among all four carbon atoms, and the second is bonding between the outer carbon atom pairs but antibonding between the central pair. Therefore, 1,3-butadiene has stronger and shorter bonds between the outer carbon pairs than between the two central carbon atoms. It is an example of a **conjugated $\pi$ electron system,** in which two or more

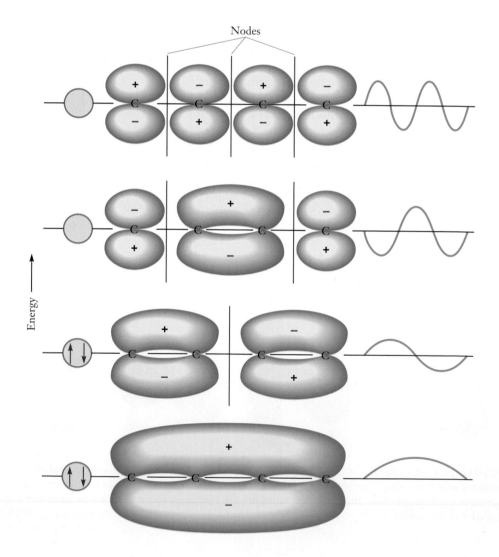

**FIGURE 17.5** The four $\pi$ molecular orbitals formed from four $2p_z$ atomic orbitals in 1,3-butadiene, viewed from the side. Planar nodes are indicated by red lines. Note the similarity in nodal properties to the first four harmonics of a vibrating string (right). Only the two lowest energy orbitals are occupied in the molecule in the ground state (left).

double or triple bonds alternate with single bonds. Such conjugated systems have lower energies than would be predicted from localized bond models and are best described with delocalized molecular orbitals extending over the entire $\pi$ electron system.

Benzene is a ring molecule with the formula $C_6H_6$. In the language of Chapter 3, benzene is represented as a resonance hybrid of two Lewis diagrams:

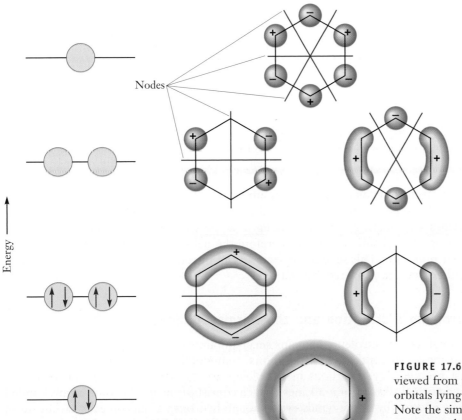

The modern view is that the molecule does not jump between two structures but rather has a single, time-independent electron distribution in which the $\pi$ bonding is described through delocalized molecular orbitals. Each carbon atom has $sp^2$ hybridization, and the remaining six $p_z$ orbitals combine to give six molecular orbitals delocalized over the entire molecule. Figure 17.6 shows the $\pi$ orbitals and their energy level diagram. The lowest energy $\pi$ orbital has no nodes, the next two have two nodes, the next two have four nodes, and the highest energy orbital has six nodes.

Energy

Nodes

**FIGURE 17.6**  The six $\pi$ molecular orbitals for benzene, viewed from the top, formed from the six $2p_z$ atomic orbitals lying perpendicular to the plane of the molecule. Note the similarity in nodal properties to the standing waves on a loop shown in Figure 15.16. Only the three lowest energy orbitals are occupied in molecules of benzene in the ground state.

The $C_6H_6$ molecule has 30 valence electrons, of which 24 occupy $sp^2$ hybrid orbitals and form $\sigma$ bonds. When the six remaining valence electrons are placed in the three lowest energy $\pi$ orbitals, the resulting electron distribution is the same in all six carbon–carbon bonds. As a result, benzene has six carbon–carbon bonds of equal length with properties intermediate between those of single and double bonds.

---

## 17.2

## PETROLEUM REFINING AND THE HYDROCARBONS

When the first oil well was drilled in 1859 near Titusville, Pennsylvania, the future effects of the exploitation of petroleum on everyday life could not have been anticipated. Today, the petroleum and petrochemical industries span the world and touch the most minute details of our daily lives. In the early years of this century, the development of the automobile, fueled by low-cost gasoline derived from petroleum, dramatically changed many people's lifestyles. The subsequent use of gasoline and oil to power trains and planes, tractors and harvesters, pumps and coolers, transformed travel, agriculture, and industry. Natural gas and heating oil warm the vast majority of homes in the United States. Finally, the spectacular growth of the petrochemical industry since 1945 has led to the introduction of innumerable new products ranging from pharmaceuticals to plastics and synthetic fibers. More than half of the chemical compounds produced in greatest volumes stem directly or indirectly from petroleum.

As we look forward to the next century, the prospects for the continued enjoyment of cheap petroleum and petrochemicals are clouded. Many wells have been drained, and the remaining petroleum is relatively difficult and costly to extract. Petroleum is not easy to make. It originated from the deposition and decay of organic matter (of animal or vegetable origin) in oxygen-poor marine sediments. Subsequently, petroleum migrated to the porous sandstone rocks from which it is extracted today. In the last 100 years, we have consumed a significant fraction of the petroleum accumulated in the earth over many millions of years. The imperative for the future is to save the remaining reserves for uses for which few substitutes are available (such as the manufacture of specialty chemicals) while finding other sources of heat and energy.

Isolating individual hydrocarbon substances from the complex mixtures obtained as petroleum is therefore an industrial process of central importance. Moreover, it provides a fascinating story illustrating how the structure of molecules determines the properties of substances and therefore the behavior of those substances in particular processes. This section presents a brief introduction to this story, emphasizing the structure-property correlations.

### Petroleum Distillation and the Normal Alkanes

Although crude petroleum contains small amounts of oxygen, nitrogen, and sulfur, its major constituents are **hydrocarbons,** compounds of carbon and hydrogen. The most prevalent hydrocarbons in petroleum are the **straight-chain alkanes** (also called normal alkanes, or $n$-alkanes), which consist of chains of carbon atoms bonded to one another by single bonds, with enough hydrogen atoms on each carbon atom to bring it to the maximum bonding capacity of four. These alkanes have the generic formula $C_nH_{2n+2}$; Table 17.1 lists the names and formulas of the first few. The ends

## TABLE 17.1

### Straight-Chain Alkanes

| Name | Formula |
| --- | --- |
| Methane | $CH_4$ |
| Ethane | $C_2H_6$ |
| Propane | $C_3H_8$ |
| Butane | $C_4H_{10}$ |
| Pentane | $C_5H_{12}$ |
| Hexane | $C_6H_{14}$ |
| Heptane | $C_7H_{16}$ |
| Octane | $C_8H_{18}$ |
| Nonane | $C_9H_{20}$ |
| Decane | $C_{10}H_{22}$ |
| Undecane | $C_{11}H_{24}$ |
| Dodecane | $C_{12}H_{26}$ |
| Tridecane | $C_{13}H_{28}$ |
| Tetradecane | $C_{14}H_{30}$ |
| Pentadecane | $C_{15}H_{32}$ |
| ⋮ | ⋮ |
| Triacontane | $C_{30}H_{62}$ |

of the molecules are methyl (—CH$_3$) groups, with methylene (—CH$_2$—) groups between. We could write pentane (C$_5$H$_{12}$) as CH$_3$CH$_2$CH$_2$CH$_2$CH$_3$ to indicate the structure more explicitly or, in abbreviated fashion, as CH$_3$(CH$_2$)$_3$CH$_3$.

Each carbon atom in a straight-chain alkane forms four single bonds, which point to the corners of a tetrahedron (exhibiting *sp*$^3$ hybridization). Although bond *lengths* do not change much through vibration, rotation about a C—C single bond occurs quite easily (Fig. 17.7), and so a given hydrocarbon molecule in a gas or liquid constantly changes its conformation. The term "straight chain" refers only to the bonding pattern in which each carbon atom is bonded to the next one in a sequence; it does not mean that the carbon atoms are positioned along a straight line. An alkane molecule with 10 to 20 carbon atoms looks quite different from its curled-up form when its bonds are extended to give a "stretched" molecule (Fig. 17.8). These two extreme conformations and many others interconvert rapidly at room temperature.

Figure 17.9 shows the melting and boiling points of the straight-chain alkanes; both increase with the number of carbon atoms and thus with molecular mass. This is a consequence of the increasing strength of dispersion forces between heavier molecules, as discussed in Section 4.7. Methane, ethane, propane, and butane are all gases at room temperature, but the hydrocarbons that follow them in the alkane series are liquids. Alkanes beyond about C$_{17}$H$_{36}$ are waxy solids at 20°C, with melting points that increase with the number of carbon atoms present. Paraffin wax, a low-melting solid, is a mixture of alkanes that have 20 to 30 carbon atoms per molecule. Petrolatum (petroleum jelly, or Vaseline) is a different mixture that is semi-solid at room temperature.

A mixture of hydrocarbons such as petroleum does not boil at a single, sharply defined temperature. Instead, as such a mixture is heated, the compounds with lower boiling points (the most volatile) boil off first and, as the temperature rises, more and more of the material vaporizes. The existence of a boiling-point range permits components of a mixture to be separated by distillation, as Section 6.5 discussed. The earliest petroleum distillation was a simple batch process: the crude oil was heated in a still, the volatile fractions were removed at the top and condensed to

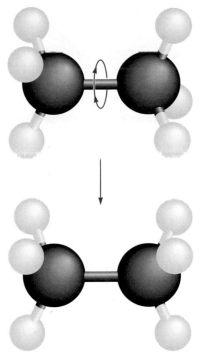

**FIGURE 17.7** The two —CH$_3$ groups in ethane rotate easily about the bond that joins them.

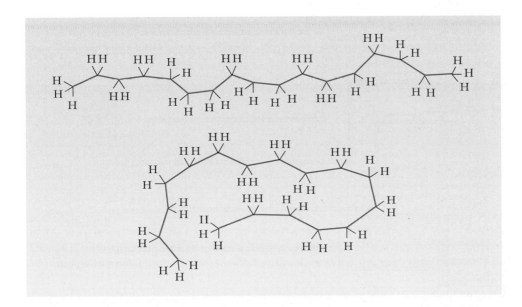

**FIGURE 17.8** Two of the very many possible conformations of the alkane C$_{17}$H$_{36}$. The carbon atoms are not shown explicitly, but they lie at every intersection of four lines.

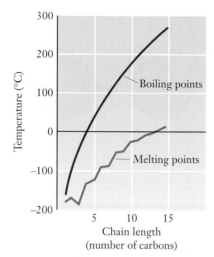

**FIGURE 17.9** The melting and boiling points of the straight-chain alkanes increase with chain length $n$. Note the alternation in the melting points: alkanes with $n$ odd tend to have lower melting points because they are more difficult to pack into a crystal lattice.

**FIGURE 17.10** In the distillation of petroleum, the lighter, more volatile hydrocarbon fractions are removed from higher up the column and the heavier fractions from lower down.

gasoline, and the still was cleaned for another batch. Modern petroleum refineries use much more sophisticated and efficient distillation methods, in which crude oil is added continuously and fractions of different volatility are tapped off at various points up and down the distillation column (Fig. 17.10). To save on energy costs, heat exchangers are employed to capture the heat from condensation of the liquid products.

Distillation allows hydrocarbons to be separated by boiling point and thus by molecular mass. The gases that emerge from the top of the column resemble the natural gas that collects in rock cavities above petroleum deposits. The gas mixtures can be separated further by redissolving the ethane, propane, and butane in a liquid solvent such as hexane. The methane-rich mixture of gases that remains is used for chemical synthesis or is shipped by pipeline to heat homes. Redistilling the hexane and its dissolved gases allows their separation and their use as chemical starting materials. Propane and butane are also bottled under pressure as liquefied petroleum gas (LPG), which is used for fuel in rural areas. After the gases, the next fraction to emerge from the petroleum distillation column is **naphtha,** which is used primarily in the manufacture of gasoline. Subsequent fractions of successively higher molecular mass are employed for jet and diesel fuel, heating oil, and machine lubricating oil. The heavy, involatile sludge that remains at the bottom of the distillation unit is pitch or asphalt, which is used for roofing and paving.

## Cyclic and Branched-Chain Alkanes

The straight-chain alkanes are not the only hydrocarbons in petroleum. Two other important classes of compounds, the cyclic and branched-chain alkanes, are also represented. A **cycloalkane** consists of at least one chain of carbon atoms attached at

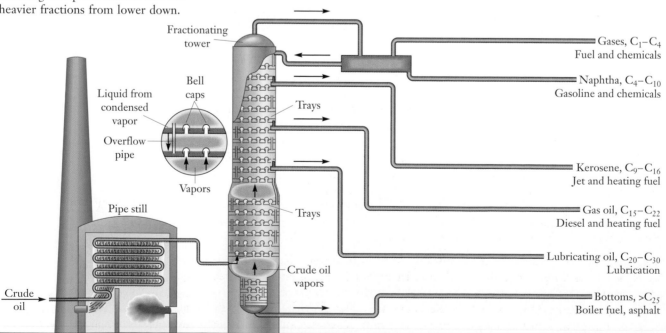

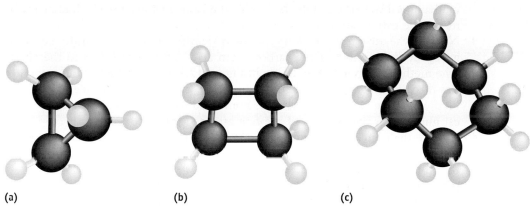

**FIGURE 17.11**    Three cyclic hydrocarbons. (a) Cyclopropane, $C_3H_6$. (b) Cyclobutane, $C_4H_8$. (c) Cyclohexane, $C_6H_{12}$.

the ends to form a closed loop. In the formation of this additional C—C bond, two hydrogen atoms must be eliminated, and so the general formula for cycloalkanes with one ring is $C_nH_{2n}$ (Fig. 17.11). The smallest cycloalkanes, cyclopropane and cyclobutane, are very strained compounds because the C—C—C bond angle is 60° (in $C_3H_6$) or 90° (in $C_4H_8$), far less than the normal tetrahedral angle of 109.5°. As a result, these compounds are more reactive than the heavier cycloalkanes or their straight-chain analogs, propane and butane.

**Branched-chain alkanes** are hydrocarbons that contain only C—C and C—H single bonds but in which the carbon atoms are no longer arranged in a straight chain. One or more carbon atoms in each molecule are bonded to three or four other carbon atoms rather than to only one or two as in the normal alkanes or cycloalkanes. The simplest branched-chain molecule (Fig. 17.12b) is that of 2-methylpropane, sometimes referred to as isobutane. This molecule has the same molecular formula as butane ($C_4H_{10}$) but a different bonding structure in which the central carbon atom is bonded to three —$CH_3$ groups and only one hydrogen atom. The compounds butane and 2-methylpropane are **isomers.** Their molecules have the same formula but different three-dimensional structures that can be interconverted only by breaking and re-forming chemical bonds.

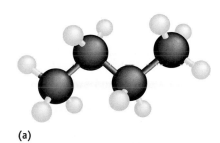

**(a)**

The number of possible isomers increases rapidly with increasing numbers of carbon atoms in the hydrocarbon molecule. Thus, butane and 2-methylpropane are the only two isomers of chemical formula $C_4H_{10}$, but there are three isomers of $C_5H_{12}$, five of $C_6H_{14}$, nine of $C_7H_{16}$, and millions of $C_{30}H_{62}$. A systematic procedure for naming these isomers has been codified by the International Union of Pure and Applied Chemistry (IUPAC). The following set of rules is a part of that procedure.

1. Find the longest continuous chain of carbon atoms in the molecule. The molecule is named as a derivative of this alkane. In Figure 17.12b a chain of three carbon atoms can be found, and so the molecule is a derivative of propane.
2. The hydrocarbon groups attached to the chain are called alkyl groups. Their names are obtained by dropping the ending *-ane* from the corresponding alkane and replacing it with *-yl* (Table 17.2). The methyl group, —$CH_3$, is derived from

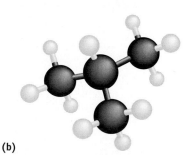

**(b)**

**FIGURE 17.12**    Two isomeric hydrocarbons with the molecular formula $C_4H_{10}$. (a) Butane. (b) 2-methylpropane.

## TABLE 17.2

### Alkyl Side Groups

| Name | Formula |
|------|---------|
| Methyl | $-CH_3$ |
| Ethyl | $-CH_2CH_3$ |
| Propyl | $-CH_2CH_2CH_3$ |
| Isopropyl | $-CH(CH_3)_2$ |
| Butyl | $-CH_2CH_2CH_2CH_3$ |

methane ($CH_4$), for example. Note also the isopropyl group, which attaches by its middle carbon atom.

3. Number the carbon atoms along the chain identified in rule 1. Identify each alkyl group by the number of the carbon atom at which it is attached to the chain. The methyl group in the molecule in Figure 17.12b is attached to the second of the three carbon atoms in the propane chain, and so the molecule is called 2-methylpropane. The carbon chain is numbered from the end that gives the lowest number for the position of the first attached group.

4. If more than one alkyl group of the same type is attached to the chain, use the prefixes *di-* (two), *tri-* (three), *tetra-* (four), *penta-* (five), and so forth to specify the total number of such attached groups in the molecule. Thus, 2,2,3-trimethylbutane has two methyl groups attached to the second carbon atom and one to the third carbon atom of the four-atom butane chain. It is an isomer of heptane ($C_7H_{16}$).

5. If several types of alkyl groups appear, name them in alphabetical order. Ethyl is listed before methyl, which appears before propyl.

## EXAMPLE 17.2

Name the following branched-chain alkane:

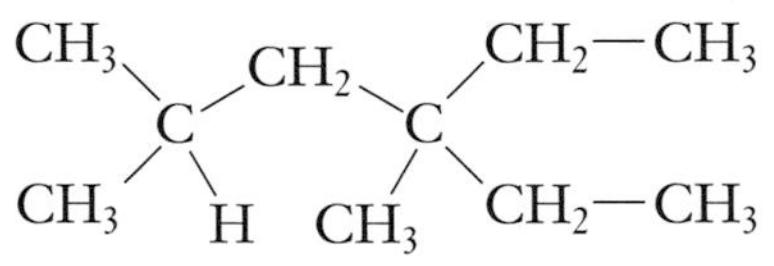

### Solution

The longest continuous chain of carbon atoms is six, and so this is a derivative of hexane. Number the carbon atoms starting from the left.

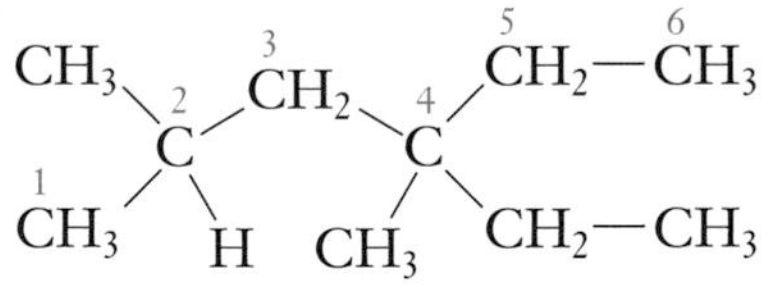

Methyl groups are attached to carbon atoms 2 and 4, and an ethyl group is attached to atom 4. The name is thus 4-ethyl-2,4-dimethylhexane. Note that if we had started numbering from the right, the higher number 3 would have appeared for the position of the first methyl group, and so the numbering from the left is preferred.

**Related Problems: 13, 14**

The fraction of branched-chain alkanes in a gasoline affects how it burns in an engine. Gasoline composed entirely of straight-chain alkanes burns very unevenly, causing "knocking" that can damage the engine. Blends that are richer in branched-chain and cyclic alkanes burn with much less knocking. Smoothness of combustion is rated quantitatively via the **octane number** of the gasoline, which was defined in 1927 by selecting two compounds that lie at extremes in the knocking they cause. Pure 2,2,4-trimethylpentane (commonly known as isooctane) burns very smoothly and was assigned an octane number of 100. Of the compounds examined at the time, pure heptane caused the most knocking and was assigned octane number 0. Mixtures of heptane and isooctane cause intermediate amounts of knocking. Standard mixtures of these two compounds define a scale for evaluating the knocking caused by real gasolines, which are complex mixtures of branched-chain and straight-chain hydrocarbons. If a gasoline sample produces the same amount of knocking in a test engine as a mixture of 90% (by volume) 2,2,4-trimethylpentane and 10% heptane, it is assigned the octane number 90.

Certain additives increase the octane rating of gasoline. The least expensive of these is tetraethyllead, $Pb(C_2H_5)_4$, a compound that has very weak bonds between the central lead atom and the ethyl carbon atoms. It readily releases ethyl radicals ($C_2H_5$) into the gasoline during combustion; these reactive species speed and smooth the combustion process, reducing knocking and giving better fuel performance. The lead is released into the atmosphere and causes serious long-term health hazards, however. As a result, the use of lead in gasoline is being phased out, and other low-cost additives are being sought to increase octane number. Chemical processing to make branched-chain compounds from straight-chain compounds represents another solution to the problem, although only a partial one.

## Alkenes and Alkynes

The hydrocarbons discussed so far are referred to as **saturated,** because all the carbon–carbon bonds are single bonds. Hydrocarbons that have double and triple carbon–carbon bonds are referred to as **unsaturated** (Fig. 17.13). Ethylene ($C_2H_4$) has a double bond between its carbon atoms and is called an **alkene.** The simplest **alkyne** is acetylene ($C_2H_2$), which has a triple bond between its carbon atoms. In naming these compounds, the *-ane* ending of the corresponding alkane is replaced by *-ene* when a double bond is present and by *-yne* when a triple bond is present. Ethene is thus the systematic name for ethylene, and ethyne for acetylene, although we will continue to use their more common names. For any compound with a carbon backbone of four or more carbon atoms, it is necessary to specify the location of the double or triple bond. This is done by numbering the carbon–carbon bonds and putting the number of the lower numbered carbon involved in the multiple bond before the name of the alkene or alkyne. Thus, the two different isomeric alkynes with the formula $C_4H_6$ are

$$HC\equiv C-CH_2-CH_3 \qquad \text{1-butyne}$$

$$CH_3-C\equiv C-CH_3 \qquad \text{2-butyne}$$

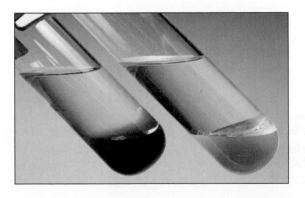

**FIGURE 17.13** One way to distinguish alkanes from alkenes is by their reactions with aqueous $KMnO_4$. This strong oxidizing agent undergoes no reaction with hexane and retains its purple color (left). When $KMnO_4$ is placed in contact with 1-hexene, however, a redox reaction takes place in which the brown solid $MnO_2$ forms (right) and —OH groups are added to both sides of the double bond in the 1-hexene, giving a compound with the formula $CH_3(CH_2)_3CH(OH)CH_2OH$. *(Charles Steele)*

There is yet another subtlety in alkenes. We recall from Section 17.1 that rotation does not take place as readily about a carbon–carbon double bond as it does about a single bond. Many alkenes therefore exist in contrasting isomeric forms, depending on whether bonding groups are on the same side or on opposite sides of the double bond. There is only a single isomer of 1-butene but two of 2-butene:

*cis*-2-butene          *trans*-2-butene

These compounds differ in melting and boiling points, density, and other physical and chemical properties. Compounds with two double bonds are called dienes, those with three are trienes, and so forth. The compound 1,3-pentadiene, for example, is a derivative of pentane with two double bonds:

$$CH_2{=}CH{-}CH{=}CH{-}CH_3$$

In such **polyenes,** each double bond may lead to *cis* and *trans* conformations, depending on its neighboring groups, and so there may be several isomers with the same bonding patterns but different molecular geometries and physical properties.

Alkenes are not present to a significant extent in crude petroleum. They are key compounds for the synthesis of organic chemicals and polymers, however, and so their production from alkanes is of great importance. One way to achieve this production is by the **cracking** of the petroleum by heat or with catalysts. In **catalytic cracking,** the heavier fractions from the distillation column (compounds of $C_{12}$ or higher) are passed over a silica–alumina catalyst at temperatures of 450 to 550°C. Reactions such as

$$CH_3(CH_2)_{12}CH_3 \longrightarrow CH_3(CH_2)_4{-}CH{=}CH_2 + C_7H_{16}$$

occur to break the long chain into fragments. This type of reaction accomplishes two purposes. First, the shorter chain hydrocarbons have lower boiling points and can be added to gasoline. Second, the alkenes that result have higher octane numbers than the corresponding alkanes and perform better in the engine. Moreover, these alkenes can react with alkanes to give the more highly branched alkanes that are desirable in gasoline. **Thermal cracking** uses higher temperatures of 850 to 900°C and no catalyst. It produces shorter chain alkenes, such as ethylene and propylene, through reactions such as

$$CH_3(CH_2)_{10}CH_3 \longrightarrow CH_3(CH_2)_8CH_3 + CH_2{=}CH_2$$

The short-chain alkenes are too volatile to be good components of gasoline but are among the most important starting materials for chemical synthesis.

## Aromatic Hydrocarbons

One last group of hydrocarbons in crude petroleum is the **aromatic hydrocarbons,** of which benzene is the simplest example. The benzene molecule consists of a six-carbon ring in which delocalized $\pi$ electrons significantly increase the molecular stability (see Sections 16.8 and 17.1). Benzene is sometimes represented by its chemical formula $C_6H_6$ and sometimes (to show structure) by a hexagon with a circle inside it:

The six points of the hexagon represent the six carbon atoms, with the hydrogen atoms omitted for simplicity. The circle represents the delocalized $\pi$ electrons, which are spread out evenly over the ring. The molecules of other aromatic compounds contain benzene rings with various side groups or two (or more) benzene rings linked by alkyl chains or fused side by side, as in naphthalene ($C_{10}H_8$):

Besides benzene, the most prevalent aromatic compounds in petroleum are toluene, in which one hydrogen atom on the benzene ring is replaced by a methyl group, and the xylenes, in which two such replacements are made:

| Toluene | o-Xylene | m-Xylene | p-Xylene |
|---------|----------|----------|----------|

This set of compounds is referred to as BTX (for *b*enzene-*t*oluene-*x*ylene). The BTX in petroleum is very important to polymer synthesis, as we will see in Section 25.1. These components also significantly increase octane number and are used to make high-performance fuels with octane numbers above 100, as are required in modern aviation.

A major advance in petroleum refining has been the development of **reforming reactions,** which produce BTX aromatics from straight-chain alkanes containing the same numbers of carbon atoms. A fairly narrow distillation fraction containing only $C_6$ to $C_8$ alkanes is taken as the starting material. The reactions employ high temperatures and transition-metal catalysts such as platinum or rhenium on alumina supports, and their detailed mechanisms are not fully understood. Apparently a normal alkane such as hexane is cyclized and dehydrogenated to give benzene as the primary product (Fig. 17.14). Heptane yields mostly toluene, and octane yields a mixture of xylenes. Toluene is of increasing importance as a solvent because tests on laboratory animals show it to be far less carcinogenic than benzene. Benzene is more important for chemical synthesis, however, and so a large fraction of the toluene produced is converted to benzene by **hydrodealkylation:**

$$\text{Toluene} + H_2 \longrightarrow \text{Benzene} + CH_4$$

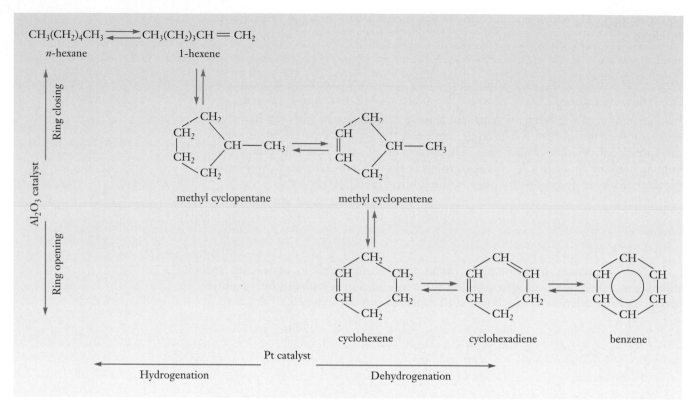

**FIGURE 17.14**  The reforming reaction that produces benzene from hexane uses a catalyst of platinum on alumina. The platinum facilitates the removal of hydrogen, and the alumina ($Al_2O_3$) facilitates the opening and closing of rings. One typical multistep sequence is shown; other intermediates can form, as well.

This reaction is carried out at high temperatures (550–650°C) and pressures of 40 to 80 atm.

---

### 17.3

# FUNCTIONAL GROUPS AND ORGANIC SYNTHESIS

The second section of this chapter described the diverse compounds formed from the two elements carbon and hydrogen and the winning of these compounds from petroleum. We now consider the effect of the attachment on or insertion into a hydrocarbon of additional chemical elements, such as oxygen, nitrogen, or a halogen. In doing so, we shift our attention from structures of entire molecules to the properties of **functional groups,** which consist of the noncarbon atoms plus the portions of the molecule immediately adjacent to them. Functional groups tend to be the reactive sites in organic molecules, and their chemical properties depend only rather weakly on the natures of the hydrocarbons to which they are attached. This fact permits us to regard an organic molecule as a hydrocarbon frame, which mainly governs size and shape, to which are attached functional groups that mainly determine the molecule's chemistry. Table 17.3 shows some of the most important functional groups.

## TABLE 17.3

### Common Functional Groups

| Functional Group[†] | Type of Compound | Examples |
|---|---|---|
| R—F, —Cl, —Br, —I | Alkyl or aryl halide | $CH_3CH_2Br$ (bromoethane) |
| R—OH | Alcohol | $CH_3CH_2OH$ (ethanol) |
| | Phenol | (phenol) |
| R—O—R′ | Ether | $CH_3—O—CH_3$ (dimethyl ether) |
| R—C(=O)H | Aldehyde | $CH_3CH_2CH_2—C(=O)—H$ (butyraldehyde, or butanal) |
| R,R′C=O | Ketone | $CH_3—C(=O)—CH_3$ (propanone, or acetone) |
| R—C(=O)OH | Carboxylic acid | $CH_3COOH$ (acetic acid, or ethanoic acid) |
| R—C(=O)O—R′ | Ester | $CH_3—C(=O)O—CH_3$ (methyl acetate) |
| R—NH₂ | Amine | $CH_3NH_2$ (methylamine) |
| R—C(=O)(R′)N—R″ | Amide | $CH_3—C(=O)NH_2$ (acetamide) |

[†] The symbols R, R′, R″ stand for hydrocarbon radicals. In some cases they may represent H.

A small number of hydrocarbon building blocks from petroleum and natural gas (methane, ethylene, propylene, benzene, and xylene) are the starting points for the synthesis of most of the high-volume organic chemicals produced today. How are these chemicals synthesized?

## Halides

One of the simplest functional groups consists of a single halogen atom, which we take to be chlorine for illustrative purposes. **Alkyl halides** form when mixtures of alkanes and halogens (except iodine) are heated or exposed to light.

$$CH_4 + Cl_2 \xrightarrow{\text{250–400°C or light}} CH_3Cl + HCl$$

Methane    Chlorine    Chloromethane    Hydrogen chloride

The mechanism is a chain reaction, as described in Section 13.4. Ultraviolet light initiates the reaction by dissociating a small number of chlorine molecules into highly reactive atoms. These take part in linked reactions of the form

$$Cl\cdot + CH_4 \longrightarrow HCl + \cdot CH_3 \qquad \text{(propagation)}$$

$$\cdot CH_3 + Cl_2 \longrightarrow CH_3Cl + Cl\cdot \qquad \text{(propagation)}$$

Chloromethane (also called methyl chloride) is used in synthesis to add methyl groups to organic molecules. If sufficient chlorine is present, more highly chlorinated methanes form. This phenomenon provides a method for the industrial synthesis of dichloromethane ($CH_2Cl_2$, also called methylene chloride), trichloromethane ($CHCl_3$, chloroform), and tetrachloromethane ($CCl_4$, carbon tetrachloride). All three are used as solvents and have vapors with anesthetic or narcotic effects.

A more important industrial route to alkyl halides than the free-radical reactions just described is the addition of chlorine to C=C double bonds. Billions of kilograms of 1,2-dichloroethane (commonly called ethylene dichloride) are manufactured each year, making this compound the largest volume derived organic chemical. It is made by adding chlorine to ethylene over an iron(III) oxide catalyst at moderate temperatures (40–50°C), either in the vapor phase or in a solution of 1,2-dibromoethane:

$$CH_2{=}CH_2 + Cl_2 \longrightarrow ClCH_2CH_2Cl$$

Almost all the 1,2-dichloroethane produced is used to make chloroethylene (vinyl chloride, $CH_2{=}CHCl$). This is accomplished by heating the 1,2-dichloroethane to 500°C over a charcoal catalyst to abstract HCl:

$$ClCH_2CH_2Cl \longrightarrow CH_2{=}CHCl + HCl$$

The HCl can be recovered and converted to $Cl_2$ (Section 22.1) for further production of 1,2-dichloroethane from ethylene. Vinyl chloride has a much lower boiling point than 1,2-dichloroethane (−13°C compared with 84°C), and so the two are easily separated by fractional distillation. Vinyl chloride is used in the production of polyvinyl chloride plastic (see Section 25.2).

## Alcohols and Phenols

**Alcohols** have the —OH functional group attached to an alkyl chain. The simplest alcohol is methanol ($CH_3OH$), which is made from synthesis gas, as will be described in Section 20.2. The next higher alcohol, ethanol ($CH_3CH_2OH$), can be produced from the fermentation of sugars. Although fermentation is the major source of ethanol for alcoholic beverages and for "gasohol" (automobile fuel made up of 90% gasoline and 10% ethanol), it is not significant for industrial production, which uses the direct hydration of ethylene:

$$CH_2{=}CH_2 + H_2O \longrightarrow CH_3CH_2OH$$

Temperatures of 300 to 400°C and pressures of 60 to 70 atm are employed, with a phosphoric acid catalyst. Both methanol and ethanol are widely used as solvents and as intermediates for further chemical synthesis.

Two three-carbon alcohols exist, depending on whether the —OH group is attached to a terminal carbon atom or the central carbon atom. They are 1-propanol and 2-propanol:

$$CH_3CH_2CH_2OH \qquad CH_3CHCH_3$$
$$\qquad\qquad\qquad\qquad |$$
$$\qquad\qquad\qquad\qquad OH$$

1-Propanol        2-Propanol

The two are frequently referred to as propyl alcohol and isopropyl alcohol, respectively. The systematic names of alcohols are obtained by replacing the *-ane* ending of the corresponding alkane with *-anol* and using a numerical prefix, where necessary, to identify the carbon atom to which the —OH group is attached.[1] Isopropyl alcohol is made from propylene by means of an interesting hydration reaction that is catalyzed by sulfuric acid. The first step is addition of $H^+$ to the double bond,

$$CH_3{-}CH{=}CH_2 + H^+ \longrightarrow CH_3{-}\overset{\oplus}{C}H{-}CH_3$$

producing a transient charged species in which the positive charge is centered on the central carbon atom. Attack by negative $HSO_4^-$ ions occurs at this positive site and leads to a neutral intermediate that can be isolated:

$$CH_3{-}CH{-}CH_3$$
$$\qquad\quad |$$
$$\qquad\quad OSO_3H$$

Further reaction with water then causes the replacement of the —OSO₃H group with an —OH group and the regeneration of the sulfuric acid:

$$H_2O + CH_3{-}CH{-}CH_3 \longrightarrow CH_3{-}CH{-}CH_3 + H_2SO_4$$
$$\qquad\qquad\quad |\qquad\qquad\qquad\qquad\quad |$$
$$\qquad\qquad\quad OSO_3H\qquad\qquad\qquad OH$$

This mechanism explains why only 2-propanol, and no 1-propanol, forms.

The compound 1-propanol is a **primary alcohol:** the carbon atom to which the —OH group is bonded has exactly one other carbon atom attached to it. The isomeric compound 2-propanol is a **secondary alcohol** because the carbon atom to which the —OH group is attached has two carbon atoms (in the two methyl groups) attached to it. The simplest **tertiary alcohol** (in which the carbon atom attached to the —OH group is also bonded to three other carbon atoms) is 2-methyl-2-propanol:

$$OH$$
$$\quad |$$
$$CH_3{-}C{-}CH_3$$
$$\quad |$$
$$\quad CH_3$$

Primary, secondary, and tertiary alcohols differ in chemical properties.

**Phenols** are compounds in which an —OH group is attached directly to an aromatic ring. The simplest example is phenol itself ($C_6H_5OH$). Phenols differ from alcohols in both physical and chemical properties. One of the most important differences is acidity. Phenol (also called carbolic acid) has an acid ionization constant

---

[1] Contrast the names of alcohols with the corresponding names of alkyl halides. If the —OH group were replaced by a chlorine atom, the names of these compounds would be 1-chloropropane and 2-chloropropane.

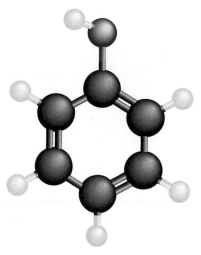

The structure of phenol, $C_6H_5OH$.

of $1 \times 10^{-10}$, much greater than the $K_a$ values of typical alcohols, which range from $10^{-16}$ to $10^{-18}$. The reason for this difference is the greater stability of the conjugate base (the phenoxide ion, $C_6H_5O^-$) due to the spreading out of the negative charge over the aromatic ring. Phenol, although not a strong acid, does react readily with sodium hydroxide to form the salt sodium phenoxide:

$$C_6H_5OH + NaOH \longrightarrow C_6H_5O^-Na^+ + H_2O$$

The corresponding reaction between NaOH and alcohols does not occur to a significant extent, although sodium ethoxide can be prepared by reaction of sodium metal with anhydrous ethanol:

$$Na + C_2H_5OH \longrightarrow C_2H_5ONa + \tfrac{1}{2} H_2$$

The manufacture of phenols uses quite different types of reactions from those employed to make alcohols. One method, introduced in 1924 and still employed to a small extent today, involves the chlorination of the benzene ring followed by reaction with sodium hydroxide:

$$C_6H_6 + Cl_2 \longrightarrow HCl + C_6H_5Cl$$

$$C_6H_5Cl + 2\, NaOH \longrightarrow C_6H_5O^-Na^+ + NaCl + H_2O$$

This approach illustrates a characteristic difference between the reactions of aromatics and alkenes. When chlorine reacts with an alkene, it *adds* across the double bond (as just shown in the production of 1,2-dichloroethane). When an aromatic ring is involved, substitution of chlorine for hydrogen occurs instead, and the aromatic $\pi$-bonding structure is preserved.

The method used to make almost all phenol today is a different one, however. It involves, first, the acid-catalyzed reaction of benzene with propylene to give cumene, or isopropyl benzene:

$$C_6H_6 + \underset{CH_3 \quad\; H}{\overset{\overset{\textstyle CH_2}{\underset{\textstyle \|}{C}}}{\diagdown\;\diagup}} \xrightarrow{\;H^+\;} \underset{\underset{\textstyle CH_3}{|}}{\overset{\overset{\textstyle CH_3}{|}}{C_6H_5{-}C{-}H}}$$

Cumene

As in the production of 2-propanol (see p. 649), the first step is the addition of $H^+$ to propylene to give $CH_3{-}CH^+{-}CH_3$. This ion then attaches to the benzene ring through its central carbon atom to give the cumene and regenerate the $H^+$ ion. Subsequent reaction of cumene with oxygen (Fig. 17.15) gives phenol and acetone, an important compound to be discussed later in this section. The ultimate products of the manufacture of phenol are mostly polymers and aspirin (see Fig. 20.2).

## Ethers

**Ethers** are characterized by the —O— functional group, in which an oxygen atom provides a link between two separate alkyl or aromatic groups. One important ether is diethyl ether, often called simply ether, in which two ethyl groups are linked to the same oxygen atom:

$$C_2H_5{-}O{-}C_2H_5$$

This is a useful solvent for organic reactions and was formerly used as an anesthetic. It can be produced by a **condensation reaction** (a reaction in which a small mol-

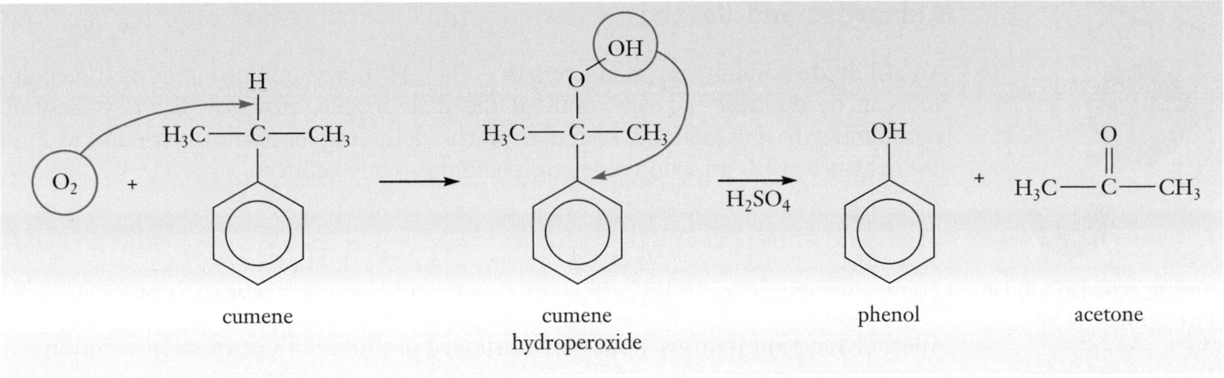

The production of phenol and acetone from cumene (reaction scheme):

cumene → cumene hydroperoxide → phenol + acetone

cumene

cumene hydroperoxide

phenol

acetone

**FIGURE 17.15** The production of phenol and acetone from cumene is a two-step process involving insertion of $O_2$ to make a peroxide, followed by acid-catalyzed migration of the —OH group to form the products.

ecule such as water is split out) between two molecules of ethanol in the presence of concentrated sulfuric acid as a dehydrating agent:

$$CH_3CH_2OH + HOCH_2CH_3 \xrightarrow{H_2SO_4} CH_3CH_2-O-CH_2CH_3 + H_2O$$

Another ether of growing importance is methyl *t*-butyl ether (MTBE):

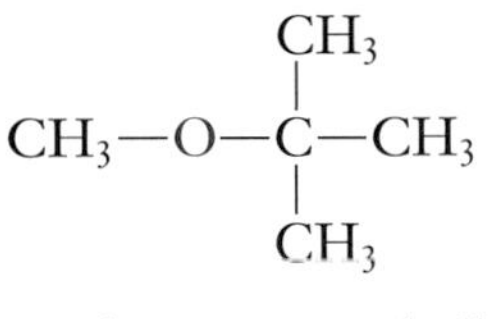

$$CH_3-O-\underset{\underset{CH_3}{|}}{\overset{\overset{CH_3}{|}}{C}}-CH_3$$

This compound is increasingly replacing tetraethyllead as an additive to gasolines to increase their octane ratings. The bond between the oxygen and the *t*-butyl group is weak, and it breaks to form radicals that assist the smooth combustion of gasoline.

In a cyclic ether, oxygen forms part of a ring with carbon atoms, as in the common solvent tetrahydrofuran (Fig. 17.16). The smallest such ring has two carbon atoms bonded to each other and to the oxygen atom; it occurs in ethylene oxide,

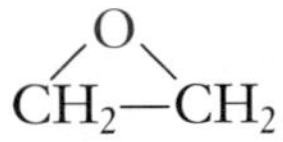

$$\overset{O}{\overset{/\backslash}{CH_2-CH_2}}$$

which is made by direct oxidation of ethylene over a silver catalyst:

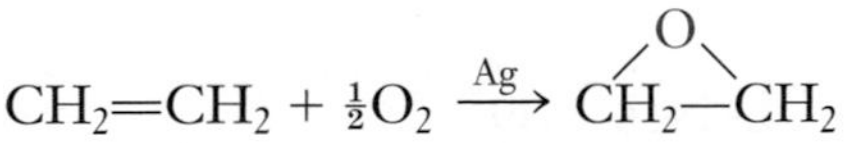

$$CH_2{=}CH_2 + \tfrac{1}{2}O_2 \xrightarrow{Ag} \overset{O}{\overset{/\backslash}{CH_2-CH_2}}$$

Such ethers with three-membered rings are called **epoxides.** The major use of ethylene oxide is in the preparation of ethylene glycol:

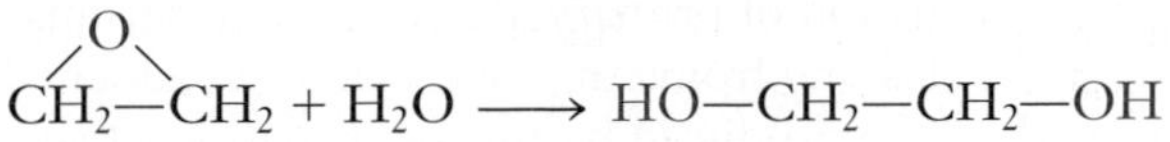

$$\overset{O}{\overset{/\backslash}{CH_2-CH_2}} + H_2O \longrightarrow HO-CH_2-CH_2-OH$$

This reaction is carried out either at 195°C under pressure or at lower temperatures (50–70°C) with sulfuric acid as a catalyst. Ethylene glycol is a dialcohol, or **diol,** in which two —OH groups are attached to adjacent carbon atoms. Its primary use is as antifreeze: when it is added to water, it substantially lowers the freezing point. It is commonly combined with water in automobile radiators for this reason.

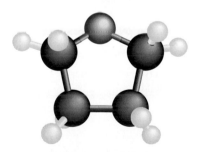

**FIGURE 17.16** The structure of tetrahydrofuran, $C_4H_8O$.

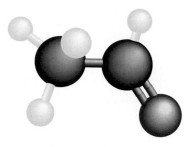

**FIGURE 17.17**  The structure of acetaldehyde, $CH_3CHO$.

## Aldehydes and Ketones

An **aldehyde** contains the characteristic $-\overset{\displaystyle O}{\overset{\displaystyle \|}{C}}-H$ functional group in its molecules and can be described as the result of the dehydrogenation of a primary alcohol. Formaldehyde, for example, results from the dehydrogenation of methanol at high temperatures with an iron oxide–molybdenum oxide catalyst:

$$CH_3OH \longrightarrow \overset{H}{\underset{H}{>}}C=O + H_2$$

Another reaction that gives the same primary product is the oxidation reaction

$$CH_3OH + \tfrac{1}{2}O_2 \longrightarrow \overset{H}{\underset{H}{>}}C=O + H_2O$$

Formaldehyde is readily soluble in water, and a 40% aqueous solution called formalin is used to preserve biological specimens. It is a component of wood smoke and helps to preserve smoked meat and fish, probably by reacting with nitrogen-containing groups in the proteins of attacking bacteria. Its major use is in making polymer adhesives and insulating foam.

The next aldehyde in the series is acetaldehyde, the dehydrogenation product of ethanol (Fig. 17.17). Industrially produced acetaldehyde is made not from ethanol but by the oxidation of ethylene, using a $PdCl_2$ catalyst.

**Ketones** have the $>C=O$ functional group in which a carbon atom forms a double bond to an oxygen atom and single bonds to two separate alkyl or aromatic groups. Such compounds can be described as the products of dehydrogenation or oxidation of secondary alcohols, just as aldehydes come from primary alcohols. The simplest ketone, acetone, is in fact made by the dehydrogenation of 2-propanol over a copper oxide or zinc oxide catalyst at 500°C:

$$CH_3-\underset{\underset{OH}{|}}{CH}-CH_3 \longrightarrow CH_3-\underset{\underset{O}{\|}}{C}-CH_3 + H_2$$

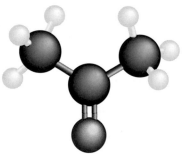

The structure of acetone, $CH_3COCH_3$.

Acetone is also produced (in greater volume) as the coproduct with phenol of the oxidation of cumene, as discussed earlier. It is a widely used solvent and is the starting material for the synthesis of a number of polymers.

## Carboxylic Acids and Esters

**Carboxylic acids** contain the $-\overset{\displaystyle O}{\overset{\displaystyle \|}{C}}-OH$ functional group (also written as $-COOH$). They are the products of the oxidation of aldehydes, just as aldehydes are the products of the oxidation of primary alcohols. (The turning of wine to vinegar is a two-step oxidation leading from ethanol through acetaldehyde to acetic acid.) Industrially, acetic acid can be produced by the air oxidation of acetaldehyde over a manganese acetate catalyst at 55 to 80°C:

$$CH_3C\overset{\displaystyle \nearrow O}{\underset{\displaystyle \searrow H}{}} + \tfrac{1}{2}O_2 \xrightarrow{\text{Mn}(CH_3COO)_2} CH_3C\overset{\displaystyle \nearrow O}{\underset{\displaystyle \searrow OH}{}}$$

The reaction now preferred for acetic acid production, for reasons of economy, is the combination of methanol with carbon monoxide (both derived from natural gas) over a catalyst containing rhodium and iodine. The overall reaction is

$$CH_3OH(g) + CO(g) \xrightarrow{Rh,I_2} CH_3COOH(g)$$

and can be described as a **carbonylation,** or the insertion of CO into the methanol C—O bond. Figure 18.27 shows the mechanism of the reaction.

Acetic acid is a member of a series of carboxylic acids with formulas H—$(CH_2)_n$—COOH. Before acetic acid (with $n = 1$) comes the very simplest of these carboxylic acids, formic acid (HCOOH), in which $n = 0$. This compound was first isolated from extracts of the crushed bodies of ants, and its name stems from the Latin word *formica*, meaning "ant." Formic acid is the strongest acid of the series, and acid strength decreases with increasing length of the hydrocarbon chain. The longer chain carboxylic acids are called fatty acids. Sodium stearate, the sodium salt of stearic acid, $CH_3(CH_2)_{16}COOH$, is a typical component of soap. It cuts grime by simultaneously interacting with grease particles at its hydrocarbon tail and with water at its carboxylate ion end group. This makes the grease soluble in water.

Carboxylic acids react with alcohols or phenols to give **esters,** forming water as the coproduct. An example is the condensation of acetic acid with methanol to give methyl acetate:

$$CH_3\overset{\displaystyle O}{\overset{\|}{C}}\boxed{—OH + H}OCH_3 \longrightarrow CH_3\overset{\displaystyle O}{\overset{\|}{C}}—OCH_3 + H_2O$$

Esters are named by stating the name of the alkyl group of the alcohol (the methyl group in this case), followed by the name of the carboxylic acid with the ending *-ate* (acetate). One of the most important esters in commercial production is vinyl acetate, with the structure

$$CH_3—\overset{\displaystyle O}{\overset{\|}{C}}\underset{O—CH—CH_2}{}$$

It is prepared by the reaction of acetic acid not with an alcohol but with ethylene and oxygen over a catalyst such as $CuCl_2$ and $PdCl_2$:

$$CH_3\overset{\displaystyle O}{\overset{\|}{C}}—OH + CH_2{=}CH_2 + \tfrac{1}{2}O_2 \xrightarrow{CuCl_2} CH_3\overset{\displaystyle O}{\overset{\|}{C}}—O—CH{=}CH_2 + H_2O$$

Esters are colorless, volatile liquids that frequently have pleasant odors. Many occur naturally in flowers and fruits. Isoamyl acetate (Fig. 17.18a) is generated in apples as they ripen and contributes to the flavor and odor of the fruit. Benzyl acetate, the ester formed from acetic acid and benzyl alcohol (Fig. 17.18b), is a major component of oil of jasmine and is used in the preparation of perfumes.

## Amines and Amides

The **amines** are derivatives of ammonia with the general formula $R_3N$, where R can represent a hydrocarbon group or hydrogen. If only one hydrogen atom of

**FIGURE 17.18** The structures of (a) isoamyl acetate, $CH_3COO(CH_2)_2CH(CH_3)_2$, and (b) benzyl acetate, $CH_3COOCH_2C_6H_5$.

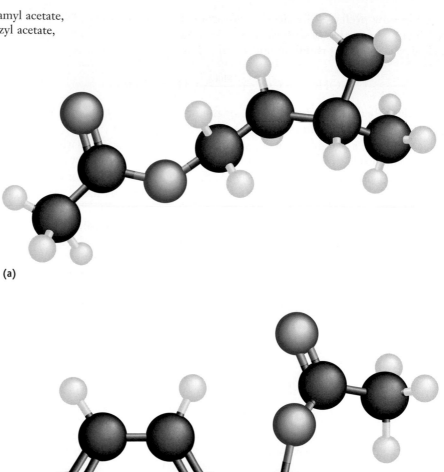

(a)

(b)

ammonia is replaced by a hydrocarbon group, the result is a **primary amine.** Examples are ethylamine and aniline:

Ethylamine   Aniline   Dimethylamine   Trimethylamine

If two hydrocarbon groups replace hydrogen atoms in the ammonia molecule, the compound is a **secondary amine** (such as dimethylamine), and three replacements make a **tertiary amine** (trimethylamine). Amines are bases because the lone electron pair on the nitrogen can accept a hydrogen ion in the same way that the lone pair on the nitrogen in ammonia does.

A primary or secondary amine (or ammonia itself) can react with a carboxylic acid to form an **amide.** This is another condensation reaction and is analogous to the formation of an ester from reaction of an alcohol with a carboxylic acid. An example of amide formation is

$$CH_3C\!\!-\!\!\overline{OH + H}\!\!-\!\!N(CH_3)_2 \longrightarrow CH_3C\!\!-\!\!N(CH_3)_2 + H_2O$$

If ammonia is the reactant, an $-NH_2$ group replaces the $-OH$ group in the carboxylic acid as the amide is formed:

$$CH_3C\!\!-\!\!OH + NH_3 \longrightarrow CH_3C\!\!-\!\!NH_2 + H_2O$$
Acetamide

Amide linkages are present in the backbone of every protein molecule and therefore have great importance in biochemistry (see Section 25.3).

## CONCEPTS & SKILLS

*After studying this chapter and working the problems that follow, you should be able to*

1. Describe the bonding in organic molecules and determine their geometries (Section 17.1, problems 1–4).
2. Identify important hydrocarbons in crude petroleum and describe how they are separated (Section 17.2, problems 5–8 ).
3. Write names and structural formulas for hydrocarbons (Section 17.2, problems 9–14).
4. Identify important functional groups and outline chemical processes by which important chemical compounds are synthesized (Section 17.3, problems 17–22).

## PROBLEMS

*Answers to problems whose numbers are boldface appear in Appendix G. Problems that are more challenging are indicated with asterisks.*

### Bonding and Structure in Organic Molecules

1. Acetic acid can be made by the oxidation of acetaldehyde ($CH_3CHO$). Molecules of acetaldehyde have a $-CH_3$ group, an oxygen atom, and a hydrogen atom attached to a carbon atom. Draw the Lewis diagram for this molecule, give the hybridization of each carbon atom, and describe the $\pi$ orbitals and the number of electrons that occupy each one. Draw the three-dimensional structure of the molecule, showing all angles.
2. Acrylic fibers are polymers made from a starting material called acrylonitrile, $H_2C(CH)CN$. In acrylonitrile, a $-C\equiv N$ group replaces an H atom on ethylene. Draw the Lewis diagram for this molecule, give the hybridization of each carbon atom, and describe the $\pi$ orbitals and the number of electrons that occupy each one. Draw the three-dimensional structure of the molecule, showing all angles.
3. Compare the bonding in formic acid (HCOOH) with that in its conjugate base formate ion ($HCOO^-$). Each molecule has a central carbon atom bonded to the two oxygen atoms and to a hydrogen atom. Draw Lewis diagrams, determine the steric numbers and hybridization of the central carbon atom, and give the molecular geometries. How do the $\pi$ orbitals differ in formic acid and the formate molecular ion? The bond lengths of the C—O bonds in HCOOH are 1.23 Å (for the bond to the lone oxygen) and 1.36 Å (for the bond to the oxygen with a hydrogen atom attached). In what range of lengths do you predict the C—O bond length in the formate ion to lie?

4. In Section 17.1, we showed that the compound 2-butene exists in two isomeric forms, which can be interconverted only by breaking a bond (in that case, the central double bond). How many possible isomers correspond to each of the following chemical formulas? Remember that a simple rotation of an entire molecule does not give a different isomer. Each molecule contains a central C=C double bond.
   (a) $C_2H_2Br_2$
   (b) $C_2H_2BrCl$
   (c) $C_2HBrClF$

## Petroleum Refining and the Hydrocarbons

5. (a) Use average bond enthalpies from Table 7.3 to estimate the standard enthalpy of formation of cyclopropane ($C_3H_6$).
   (b) The measured enthalpy change for the combustion of cyclopropane

   $$C_3H_6(g) + \tfrac{9}{2} O_2(g) \longrightarrow 3\ CO_2(g) + 3\ H_2O(g)$$

   is $\Delta H° = -1959$ kJ. Use this, together with data from Appendix D, to calculate the standard enthalpy of formation of cyclopropane.
   (c) By comparing your answers from (a) and (b), estimate the "strain enthalpy" associated with forming a ring of three carbon atoms (with C—C—C bond angles of only 60°) in cyclopropane.

6. (a) Use average bond enthalpies from Table 7.3 to estimate the standard enthalpy of formation of cyclobutane ($C_4H_8$).
   (b) The measured enthalpy change for the combustion of cyclobutane

   $$C_4H_8(g) + 6\ O_2(g) \longrightarrow 4\ CO_2(g) + 4\ H_2O(g)$$

   is $\Delta H° = -2568$ kJ. Use this, together with data from Appendix D, to calculate the standard enthalpy of formation of cyclobutane.
   (c) By comparing your answers from (a) and (b), estimate the "strain enthalpy" associated with forming a ring of four carbon atoms (with C—C—C bond angles of only 90°) in cyclobutane.

7. (a) Write a chemical equation involving structural formulas for the catalytic cracking of decane into an alkane and an alkene that contain equal numbers of carbon atoms. Assume that both products have straight chains of carbon atoms.
   (b) Draw and name one other isomer of the alkene.

8. (a) Write an equation involving structural formulas for the catalytic cracking of 2,2,3,4,5,5-hexamethylhexane. Assume that the cracking occurs between carbon atoms 3 and 4.
   (b) Draw and name one other isomer of the alkene.

9. Write structural formulas for the following:
   (a) 2,3-dimethylpentane
   (b) 3-ethyl-2-pentene
   (c) Methylcyclopropane

   (d) 2,2-dimethylbutane
   (e) 3-propyl-2-hexene
   (f) 3-methyl-1-hexene
   (g) 4-ethyl-2-methylheptane
   (h) 4-ethyl-2-heptyne

10. Write structural formulas for the following:
    (a) 2,3-dimethyl-1-cyclobutene
    (b) 2-methyl-2-butene
    (c) 2-methyl-1,3-butadiene
    (d) 2,3-dimethyl-3-ethylhexane
    (e) 4,5-diethyloctane
    (f) Cyclooctene
    (g) Propadiene
    (h) 2-pentyne

11. Write structural formulas for *trans*-3-heptene and *cis*-3-heptene.

12. Write structural formulas for *cis*-4-octene and *trans*-4-octene.

13. Name the following hydrocarbons.

    (a)
    $$\underset{\phantom{x}}{CH_2}=C=\overset{\displaystyle H}{\underset{\displaystyle |}{C}}-CH_2-CH_2-CH_3$$

    (b)
    $$H_2C=\overset{\displaystyle H}{\underset{\displaystyle |}{C}}-\overset{\displaystyle H}{\underset{\displaystyle |}{C}}=C\overset{\displaystyle }{\underset{\displaystyle }{}}-\overset{\displaystyle }{\underset{\displaystyle H}{C}}=CH_2$$
    with H below the third and fourth carbons

    (c)
    $$H_2C=\overset{\displaystyle CH_3}{\underset{\displaystyle |}{C}}-CH_2-CH_2-CH_2-CH_3$$

    (d) $CH_3-CH_2-C\equiv C-CH_2-CH_3$

14. Name the following hydrocarbons.
    (a)
    $$CH_2=\underset{\displaystyle CH_3}{C}-\underset{\displaystyle CH_3}{C}=CH_2$$

    (b)
    $$CH_3-\underset{\displaystyle H}{C}=\underset{\displaystyle H}{C}-\underset{\displaystyle H}{C}=\underset{\displaystyle H}{C}-CH_3$$

    (c)
    $$CH_3-\overset{\displaystyle CH_3}{\underset{\displaystyle CH_3}{C}}-CH_2-CH_3$$

    (d)
    $$CH_3-\overset{\displaystyle }{\underset{\displaystyle CH_2}{C}}-CH_3$$

15. State the hybridization of each of the carbon atoms in the hydrocarbon structures in problem 13.

16. State the hybridization of each of the carbon atoms in the hydrocarbon structures in problem 14.

## Functional Groups and Organic Synthesis

17. Write balanced equations for the following reactions. Use structural formulas to represent the organic compounds.

(a) The production of butyl acetate from butanol and acetic acid

(b) The conversion of ammonium acetate to acetamide and water

(c) The dehydrogenation of 1-propanol

(d) The complete combustion (to $CO_2$ and $H_2O$) of heptane

18. Write balanced equations for the following reactions. Use structural formulas to represent the organic compounds.
    (a) The complete combustion (to $CO_2$ and $H_2O$) of cyclopropanol
    (b) The reaction of isopropyl acetate with water to give acetic acid and isopropanol
    (c) The dehydration of ethanol to give ethylene
    (d) The reaction of 1-iodobutane with water to give 1-butanol

19. Outline, using chemical equations, the synthesis of the following from easily available petrochemicals and inorganic starting materials.
    (a) Vinyl bromide ($CH_2$=CHBr)
    (b) 2-butanol
    (c) Acetone ($CH_3COCH_3$)

20. Outline, using chemical equations, the synthesis of the following from easily available petrochemicals and inorganic starting materials.
    (a) Vinyl acetate ($CH_3COOCH$=$CH_2$)
    (b) Formamide ($HCONH_2$)
    (c) 1,2-difluoroethane

21. Write a general equation (using R to represent a general alkyl group) for the formation of an ester by the condensation of a tertiary alcohol with a carboxylic acid.

22. Explain why it is impossible to form an amide by the condensation of a tertiary amine with a carboxylic acid.

23. In a recent year, the United States produced $6.26 \times 10^9$ kg of ethylene dichloride (1,2-dichloroethane) and $15.87 \times 10^9$ kg of ethylene. Assuming that all significant quantities of ethylene dichloride were produced from ethylene, what fraction of the ethylene production went into making ethylene dichloride? What mass of chlorine was required for this conversion?

24. In a recent year, the United States produced $6.26 \times 10^9$ kg of ethylene dichloride (1,2-dichloroethane) and $3.73 \times 10^9$ kg of vinyl chloride. Assuming that all significant quantities of vinyl chloride were produced from ethylene dichloride, what fraction of the ethylene dichloride production went into making vinyl chloride? What mass of hydrogen chloride was generated as a byproduct?

## Additional Problems

* 25. The pyridine molecule ($C_5H_5N$) is obtained by replacing one C—H group in benzene with a nitrogen atom. Because nitrogen is more electronegative than the C—H group, orbitals with electron density on nitrogen are lower in energy. How do you expect the $\pi$ molecular orbitals and energy levels of pyridine to differ from those of benzene?

* 26. For each of the following molecules, construct the $\pi$ molecular orbitals from the $2p_z$ atomic orbitals perpendicular to the plane of the carbon atoms.

(a) Cyclobutadiene

$$\begin{array}{ccc} HC{=}CH & & HC{-}CH \\ | \quad | & \longleftrightarrow & \| \quad \| \\ HC{=}CH & & HC{-}CH \end{array}$$

(b) Allyl radical

$$H{-}\overset{.}{C}{-}\overset{.}{C}{-}\overset{.}{C}{-}H$$
$$\begin{array}{ccc} | & | & | \\ H & H & H \end{array}$$

Indicate which, if any, of these orbitals have identical energies on the basis of symmetry considerations. Show the number of electrons occupying each $\pi$ molecular orbital in the ground state, and indicate whether either or both of the molecules are paramagnetic. (*Hint:* Refer to Figures 17.5 and 17.6.)

27. *Trans*-cyclodecene boils at 193°C, but *cis*-cyclodecene boils at 195.6°C. Write structural formulas for these two compounds.

28. Consider the following proposed structures for benzene, each of which is consistent with the molecular formula $C_6H_6$.

(i)

(ii)

(iii)

(iv) $CH_3{-}C{\equiv}C{-}C{\equiv}C{-}CH_3$

(v) $CH_2{=}CH{-}C{\equiv}C{-}CH{=}CH_2$

(a) When benzene reacts with chlorine to give $C_6H_5Cl$, only one isomer of that compound forms. Which of the five proposed structures for benzene are consistent with this observation?

(b) When $C_6H_5Cl$ reacts further with chlorine to give $C_6H_4Cl_2$, exactly three isomers of the latter compound form. Which of the five proposed structures for benzene are consistent with this observation?

29. How does *dehydration* differ from *dehydrogenation*? Give an example of each.

30. When an ester forms from an alcohol and a carboxylic acid, an oxygen atom links the two parts of each ester molecule. This atom could have come originally from the alcohol, from the carboxylic acid, or randomly from either. Propose an experiment using isotopes to determine which is the case.

31. Hydrogen can be added to a certain unsaturated hydrocarbon in the presence of a platinum catalyst to form hexane.

When the same hydrocarbon is oxidized with $KMnO_4$, it yields acetic acid and butanoic acid. Identify the hydrocarbon and write balanced chemical equations for the reactions.

32. (a) It is reported that ethylene is released when pure ethanol is passed over alumina ($Al_2O_3$) that is heated to 400°C, but diethyl ether is obtained at a temperature of 230°C. Write balanced equations for both of these dehydration reactions.

    (b) If the temperature is raised well above 400°C, an aldehyde forms. Write a chemical equation for this reaction.

---

## CUMULATIVE PROBLEMS

33. The standard enthalpy of formation of heptane is $-187.82$ kJ $mol^{-1}$, and the standard enthalpy of formation of isooctane (2,2,4-trimethylpentane) is $-224.13$ kJ $mol^{-1}$.

    (a) Compute the standard enthalpy of combustion of 1.00 mol of each of the two if they are burned to $CO_2(g)$ and $H_2O(g)$ in an internal-combustion engine.

    (b) Which compound delivers more heat per gallon? (Isooctane has a density of 5.77 lb per U.S. gallon, and heptane has a density of 5.71 lb per U.S. gallon.)

34. Ethylene ($C_2H_4$) is an organic compound of great economic significance that is often transported by pipeline. It has a critical temperature of 282.65 K and a critical pressure of 50.096 atm.

    (a) Express the critical temperature and pressure of ethylene in degrees Fahrenheit and in pounds per square inch.

    (b) A pipeline contains ethylene at a pressure of 54 atm and a temperature of 15°C. Discuss the physical state of the ethylene within the pipe.

    (c) The pipeline is 100 km long and has an interior diameter of 250 mm. Estimate the mass of ethylene in the pipeline, assuming it is an ideal gas.

    (d) The density of the ethylene within the pipe from part (c) is 0.20 g $cm^{-3}$. Compute the actual mass of ethylene in the pipe.

    (e) If the temperature of the entire pipeline and its contents falls to 0°C on a cold day, suggest what the operators could do to make sure that the ethylene in the pipe does not change its physical state.

35. (a) Use thermodynamic data from Appendix D to calculate the enthalpy change when 1.00 mol of gaseous benzene is formed from carbon and hydrogen *atoms*, all in the gas phase.

    (b) Compare this result with your calculated enthalpy change for forming the structure represented by one of the Lewis resonance diagrams for benzene. Use the bond enthalpies of C=C, C—C, and C—H in Table 7.3.

    (c) The additional lowering of the enthalpy found experimentally (the more negative value of $\Delta H$) is due to *resonance stabilization*, the delocalization of the electrons as they are free to move over the whole carbon ring. How great is the enthalpy change due to resonance stabilization in benzene?

36. Use bond enthalpies from Table 7.3 to estimate the barrier to rotation about the central C=C bond in 2-butene (see Fig. 17.2). (*Hint:* Think about what bonds are broken in the process.) Do you expect that conversion of *cis*-2-butene to *trans*-2-butene will be faster or slower than the inversion of ammonia described in Section 16.3?

37. The structure of the molecule cyclohexene is

Does the absorption of ultraviolet light by cyclohexene occur at longer or at shorter wavelengths than the absorption by benzene? Explain.

38. The naphthalene molecule has a structure that corresponds to two benzene molecules fused together:

The $\pi$ electrons in this molecule are delocalized over the entire molecule. The wavelength of maximum absorption in benzene is 255 nm. Will the corresponding wavelength in naphthalene be shorter or longer than 255 nm?

# Bonding in Transition Metals and Coordination Complexes

The partially filled *d* electron shells of the transition-metal elements manifest a range of physical properties and chemical reactions that contrast with those of the main-group elements. The presence of unpaired electrons in the transition-metal elements and their compounds, the availability of low-lying unoccupied orbitals, and the facility with which transition-metal oxidation states change all give rise to a fascinating and rich chemistry. Of particular interest in this chapter are the magnetic properties and colors of transition-metal compounds, which are variable and striking, and the wide variety of geometrical arrangements in which ions and small molecules position themselves around metal atoms. We describe how some of

***Illustration***
The purple color of this amethyst results from the absorption of light by $[FeO_4]^{4-}$ color center defects in quartz. *(Charles D. Winters)*

these properties vary through the transition-metal series and present bonding models that seek to explain the observed properties. We also describe some organometallic compounds, which contain bonds between metal and carbon atoms. Organometallic substances play an important role in the catalysis of reactions in organic chemistry and biochemistry.

---

## 18.1
## CHEMISTRY OF THE TRANSITION METALS

Let us begin by surveying some key physical and chemical properties of the transition-metal elements, with particular emphasis on the fourth-period elements (those from scandium through zinc in which the $3d$ shell is progressively filled) and by exploring connections to the quantum theory of atomic structure discussed in Section 15.8.

### Physical Properties

Table 18.1 lists some key physical properties of the fourth-period transition elements, mostly taken from Appendix F. As the nuclear charge increases across the period, the first and second ionization energies tend to increase, although this process is by no means smooth. The energies of the $4s$ and $3d$ orbitals are so close that details of the neutral atom and ion electron configurations are not easy to predict from a qualitative model of atomic structure.

The increasing nuclear charge also causes electrons to be held closer and the atomic radius to decrease across each period, as Table 18.1 and Figure 18.1 show. Near the end of each period, however, the increased electron–electron repulsion in a nearly filled $d$ shell outweighs the effect of nuclear charge and forces the atom to expand. Atoms of the fifth-period elements (from yttrium to cadmium) are larger than those of the fourth-period elements, but almost no further increase occurs from

## TABLE 18.1
### Properties of the Fourth-Period Transition Elements

| Element | Sc | Ti | V | Cr | Mn | Fe | Co | Ni | Cu | Zn |
|---|---|---|---|---|---|---|---|---|---|---|
| $IE_1$ (kJ mol$^{-1}$) | 631 | 658 | 650 | 653 | 717 | 759 | 758 | 737 | 745 | 906 |
| $IE_2$ (kJ mol$^{-1}$) | 1235 | 1310 | 1414 | 1592 | 1509 | 1562 | 1648 | 1753 | 1958 | 1733 |
| Radius (Å) | 1.61 | 1.45 | 1.31 | 1.25 | 1.37 | 1.24 | 1.25 | 1.25 | 1.28 | 1.34 |
| $\Delta H_f^\circ(M(g))$ (kJ mol$^{-1}$) | 378 | 470 | 514 | 397 | 281 | 416 | 425 | 430 | 338 | 131 |
| Boiling point (°C) | 2831 | 3287 | 3380 | 2672 | 1962 | 2750 | 2870 | 2732 | 2567 | 907 |
| Melting point (°C) | 1541 | 1660 | 1890 | 1857 | 1244 | 1535 | 1495 | 1453 | 1083 | 420 |
| $M^{2+}$ configuration | $d^1$ | $d^2$ | $d^3$ | $d^4$ | $d^5$ | $d^6$ | $d^7$ | $d^8$ | $d^9$ | $d^{10}$ |
| $\Delta H_{hyd}(M^{2+})^\dagger$ (kJ mol$^{-1}$) | | | | −2799 | −2740 | −2839 | −2902 | −2985 | −2989 | −2937 |
| $\mathscr{E}°(M^{2+}\vert M)$ (V) | | −1.63 | −1.2 | −0.56 | −1.03 | −0.41 | −0.28 | −0.23 | +0.34 | −0.76 |

$^\dagger$ Defined as $\Delta H_f^\circ(M^{2+}(aq)) - \Delta H_f^\circ(M^{2+}(g))$. Because aqueous species are defined relative to $\Delta H_f^\circ(H^+(aq)) = 0$, this is not an absolute enthalpy of hydration.

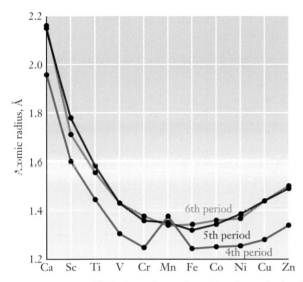

**FIGURE 18.1**   Variation of atomic radius through the fourth-, fifth-, and sixth-period transition elements. Symbols shown are for the fourth-period elements.

the fifth to the sixth period (Fig. 18.1) because of the lanthanide contraction: the filling of the $4f$ shell prior to the sixth-period transition elements makes the outer $d$ electrons feel a greater effective charge and causes the atom to be smaller than it would be otherwise. (A parallel effect, shown in Figure 15.34 for main-group atoms and ions, arose from the filling of the $3d$ shell.) The atomic and ionic radii of hafnium, in the sixth period, are essentially the same as those of zirconium, in the fifth period. Because these elements are similar in both valence configuration *and* size, they have quite similar properties and are difficult to separate from each other.

The binding strength of atoms in an elemental metal has a maximum in the middle of each transition series and falls off on both sides. This pattern shows up in the variation of the enthalpy of formation of the gaseous atoms $\Delta H_f^\circ(M(g))$ and in the boiling points of the fourth-period elements (Table 18.1). It is also evident in the melting points of the transition metals (Fig. 18.2). All three properties correlate roughly with the number of unpaired electrons in the involved elements, which reaches a maximum in the middle of the corresponding series. Another approach is to think in terms of the formation of delocalized orbitals in the solid from outer $d$ orbitals on the atoms. Lower energy orbitals in the solid are primarily bonding (and are progressively filled through the first half of each transition series), and higher energy antibonding orbitals become filled after that, giving a lower bond strength. Tungsten, near the middle of the sixth period, has a very high melting point (3410°C), which makes it useful in light-bulb filaments; mercury, at the end of the same period, has a melting point well below room temperature (−39°C).

Table 18.1 and Figure 18.3 illustrate the variation of the enthalpy change of hydration for transition-metal ions $M^{2+}$. This variation is connected to the heat released in the process

$$M^{2+}(g) \longrightarrow M^{2+}(aq)$$

or (recognizing the existence of hydrated complex ions of the transition-metal ions)

$$M^{2+}(g) + 6\,H_2O(\ell) \longrightarrow [M(H_2O)_6]^{2+}(aq)$$

**FIGURE 18.2** Variation of melting point through the three periods of transition elements.

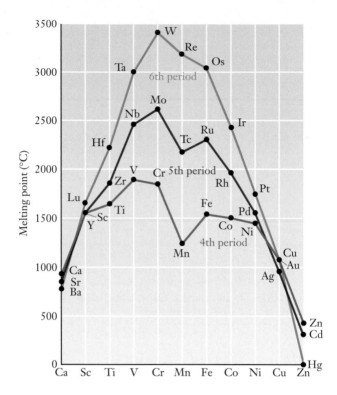

The enthalpy of hydration becomes more negative later in the fourth period, but superimposed on this general trend is a secondary trend that gives a maximum near $Mn^{2+}$, a $d^5$ species. Similar anomalies near Mn appear in Figures 18.1 and 18.2 and are related to the fact that Mn contains a half-filled $d$ shell. Section 18.4 will discuss the origin of the trend shown in Figure 18.3.

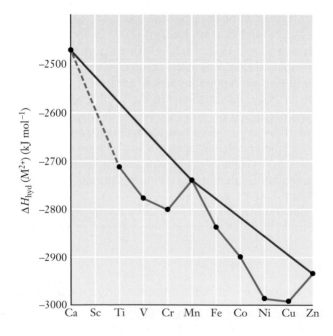

**FIGURE 18.3** Enthalpies of hydration of $M^{2+}$ ions, defined as $\Delta H_f^\circ(M^{2+}(aq)) - \Delta H_f^\circ(M^{2+}(g))$. The crystal field stabilization energy (discussed in Section 18.4) stabilizes certain ions, lowering $\Delta H_{hyd}$ from a line representing a linear change with increasing atomic number (red line) to the experimental (blue) line.

## Redox Chemistry of the Transition Elements

Figure 18.4 shows the characteristic oxidation states of compounds of the transition metals and should be compared with Figure 3.15 for main-group elements. The early members of each period have maximum oxidation numbers that correspond to the participation of all the outer $s$ and $d$ electrons in ionic or covalent bonds. Thus, elements in the scandium group have only the +3 oxidation state, and manganese has a maximum oxidation state of +7, as do the other elements in its group, forming compounds such as $HReO_4$ and the dark red liquid $Mn_2O_7$. An interesting trend in Figure 18.4 is the increasing tendency toward higher oxidation states among the heavier transition elements, which is the opposite of the trend among main-group elements. Thus, the chemistry of iron is dominated by the +2 and +3 oxidation states, as in the common oxides FeO and $Fe_2O_3$, but the +8 state, which is nonexistent for iron, is important for the later members of the iron group, ruthenium and osmium. The oxide $OsO_4$, for example, is a volatile yellow solid that melts at 41°C and boils at 131°C. Its selective reaction with C=C double bonds makes it useful in organic synthesis and as a biological stain. The chemistry of nickel is almost entirely that of the +2 oxidation state, but the later elements in the nickel group, palladium and platinum, have chemistries increasingly dominated by the +4 state. For example, $NiF_2$ is the only stable fluoride of nickel, but both $PdF_2$ and $PdF_4$ exist; $PtF_2$ is not found, but both $PtF_4$ and $PtF_6$ have been prepared.

Compounds in which the transition-metal element has a high oxidation number tend to be relatively covalent, whereas those with lower oxidation numbers are more ionic. Oxides are an example: $Mn_2O_7$ is a covalent compound that is a liquid at room temperature (crystallizing only at 6°C), but $Mn_3O_4$ is an ionic compound, containing both Mn(II) and Mn(III), that melts at 1564°C. Covalent oxides tend to be acid anhydrides, whereas ionic oxides are basic, just as happens in the main-group elements (compare Fig. 10.2). Table 18.2 shows that the oxides of two transition elements, chromium and manganese, demonstrate these tendencies.

The transition elements show such a range of oxidation states because their partially filled $d$ orbitals can accept or donate electrons in chemical reactions. This same property makes many of their compounds effective catalysts, both homogeneous and heterogeneous. Because an element such as iron can exist as either $Fe^{2+}$ or $Fe^{3+}$ in

**FIGURE 18.4** Some of the oxidation states found in compounds of the transition-metal elements.

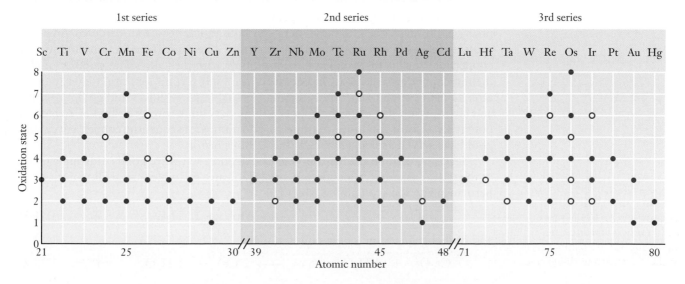

**TABLE 18.2**

## Acid–Base Character of Oxides of Chromium and Manganese

| Oxidation State | Chromium | | | Manganese | | |
|---|---|---|---|---|---|---|
| | Oxide | Hydroxide | Acidic/Basic | Oxide | Hydroxide | Acidic/Basic |
| +2 | CrO<br>Black | $Cr(OH)_2$ | Basic | MnO<br>Green | $Mn(OH)_2$ | Basic |
| +3 | $Cr_2O_3$<br>Green | $Cr(OH)_3$ | Amphoteric | $Mn_2O_3$<br>Black | $Mn(OH)_3$ | Weakly basic |
| +4 | | | | $MnO_2$<br>Black | $H_2MnO_3$ | Amphoteric |
| +6 | $CrO_3$<br>⇅<br>Dark red | $H_2CrO_4$<br><br>$H_2Cr_2O_7$ | Weakly acidic<br><br>Acidic | $MnO_3$<br>⇅<br>Reddish | $H_2MnO_4$ | Acidic |
| +7 | | | | $Mn_2O_7$<br><br>Dark red | $HMnO_4$ | Strongly acidic |

solution, it can catalyze electron-transfer reactions by shuttling back and forth between oxidized and reduced forms without ever leaving solution, unlike an ion such as $K^+$ that has only one oxidation state in solution. Iron oxides are also used as heterogeneous catalysts for the production of ammonia from the elements and solid $V_2O_5$ catalyzes the production of $SO_3$ from $SO_2$. (Chapter 21 will discuss both of these processes.) Many enzymes (biological catalysts) contain transition-metal ions that undergo redox reactions at their active sites.

Table 18.1 lists the standard reduction potentials to the metallic states of $M^{2+}$ transition elements. All but copper have negative reduction potentials, indicating that the metals should be oxidized by 1 M acidic solutions; several metals (Ti, V, and Mn) are strong enough reducing agents that they should even react with water. In practice, however, each of these metals in bulk rapidly gains an oxide coating that protects (passivates) it against further reaction. The positive reduction potential of copper ions means that copper is oxidized only with difficulty. The stability of elemental copper arises at least in part from the high second ionization energy of its atoms relative to the other transition metals (Table 18.1). Silver and gold, the fifth- and sixth-period metals in the copper group, have ions with even higher reduction potentials and are still more resistant to oxidation. Consequently, copper, silver, and gold have served as coinage metals throughout history.

Several transition elements in higher oxidation states are good oxidizing agents, as their large positive reduction potentials indicate. Examples are acidic solutions of permanganate ion (containing Mn(VII))

$$MnO_4^- + 8\ H_3O^+ + 5\ e^- \longrightarrow Mn^{2+} + 12\ H_2O \qquad \mathscr{E}° = +1.51\ V$$

and of dichromate ion (containing Cr(VI)):

$$Cr_2O_7^{2-} + 14\ H_3O^+ + 6\ e^- \longrightarrow 2\ Cr^{3+} + 21\ H_2O \qquad \mathscr{E}° = +1.33\ V$$

Dichromate ion is isoelectronic with manganese(VII) oxide, the explosive red liquid. The chromium-containing species that is isoelectronic with $MnO_4^-$ is the chromate ion, $CrO_4^{2-}$. Chromate and dichromate ion take part in an equilibrium that depends on pH:

$$2\,CrO_4^{2-} + 2\,H_3O^+ \rightleftharpoons Cr_2O_7^{2-} + 3\,H_2O \qquad K = 4.2 \times 10^{14}$$

Low pH (acidic conditions) favors formation of the orange dichromate ion, and high pH (basic conditions) favors the yellow chromate ion.

---

## 18.2

## THE FORMATION OF COORDINATION COMPLEXES

One of the key features of the transition metals is their ability to form complexes with small molecules and ions. For example, solid copper(II) sulfate is made by reacting copper and hot concentrated sulfuric acid ("oil of vitriol"). Its common name, "blue vitriol," recalls this origin and reports the color that is its most obvious property. There is more to this compound than copper and sulfate, however; it contains water as well. The water in blue vitriol is important, because when it is driven away by strong heat, the blue color vanishes, leaving greenish white anhydrous copper(II) sulfate (Fig. 18.5). The blue of blue vitriol comes from a **coordination complex** in which $H_2O$ molecules bond directly to $Cu^{2+}$ ions to form composite ions with the formula $[Cu(H_2O)_4]^{2+}$. As a Lewis acid, the $Cu^{2+}$ ion *coordinates* four water molecules into a group by accepting electron density from a lone pair on each. By acting as electron-pair donors and sharing electron density with the $Cu^{2+}$ ion, the four water molecules, which in this interaction are called **ligands,** come into the **coordination sphere** of the ion. Blue vitriol has the chemical formula $[Cu(H_2O)_4]SO_4 \cdot H_2O$; the fifth water molecule is not coordinated directly to copper.

The positive ions of every metal in the periodic table accept electron density to some degree and can therefore coordinate surrounding electron donors, even if only weakly. The solvation of the $K^+$ ion by $H_2O$ molecules in aqueous solution (see Fig. 6.3) is an example of weak coordination. The ability to make fairly strong, *directional* bonds by accepting electron pairs from neighboring molecules or ions is characteristic of the transition-metal elements. Coordination occupies a middle place

**FIGURE 18.5** Hydrated copper(II) sulfate, $CuSO_4 \cdot 5H_2O$, is blue (left), but the anhydrous compound, $CuSO_4$, is greenish white (right). A structural study of the solid compound reveals that four of the water molecules are closely associated with the copper and the fifth is not. Thus, a better representation of the hydrated compound is $[Cu(H_2O)_4]SO_4 \cdot H_2O$. *(Leon Lewandowski)*

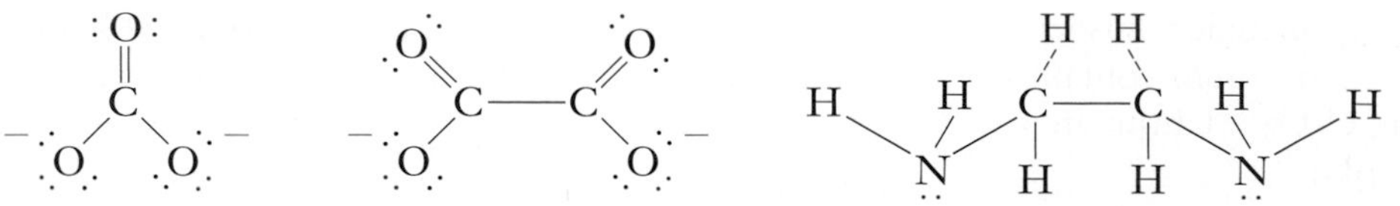

**FIGURE 18.6**   These three ligands are bidentate; each is capable of donating two pairs of electrons.

Carbonate ion, $CO_3^{2-}$

Oxalate ion, $C_2O_4^{2-}$

Ethylenediamine, $NH_2CH_2CH_2NH_2$

energetically between the weak intermolecular attractions in solids (see Chapter 19) and the stronger covalent and ionic bonding (see Chapters 3 and 16). Thus, heating blue vitriol disrupts the $Cu—H_2O$ bonds at temperatures well below those required to break the covalent bonds in the $SO_4^{2-}$ group. The enthalpy change in breaking a $+2$ transition-metal ion away from a coordinated water molecule falls in the range of 170 to 210 kJ mol$^{-1}$. This is far less than the enthalpy changes for the strongest chemical bonds, but it is by no means small. Metal ions with $+3$ charges have still stronger coordinate bonds with water.

The total number of metal-to-ligand bonds in a complex (usually two to six) is the **coordination number** of the metal. Some common ligands are the halide ions ($F^-$, $Cl^-$, $Br^-$, $I^-$), ammonia ($NH_3$), carbon monoxide (CO), and water (Table 18.3). Each of these is capable of forming only a single bond to a central metal atom. They are called *monodentate* (from Latin *mono*, meaning "one," plus *dens*, meaning "tooth," indicating that they bind at only one point). Other ligands can form two or more such bonds and are referred to as *bidentate*, *tridentate*, and so forth. Ethylenediamine ($NH_2CH_2CH_2NH_2$), in which two $NH_2$ groups are held together by a carbon backbone, is a particularly important bidentate ligand. Both nitrogen atoms in ethylenediamine have lone electron pairs to share. If all the nitrogen donors of three ethylenediamine molecules bind to a single $Co^{3+}$ ion, then that ion has a coordination number of 6, and the formula of the resulting complex is $[Co(en)_3]^{3+}$ (where "en" is the abbreviation for ethylenediamine). Complexes in which a ligand coordinates via two or more donors to the same central atom are called **chelates** (from Greek *chele*, meaning "claw," because the ligand grabs onto the central atom like a pincers). Figure 18.6 shows the structures of some important chelating ligands.

Brackets are used in chemical formulas of coordination complexes to group together the symbols of the central atom and its coordinated ligands. In the formula $[Pt(NH_3)_6]Cl_4$, the portion in brackets represents a positively charged coordination complex in which Pt coordinates six $NH_3$ ligands. The brackets emphasize that a complex is a distinct chemical entity with its own properties. Within the brackets, the symbol of the central atom comes first. The electric charge on a coordination complex is the sum of the oxidation number of the metal ion and the charges of the ligands that surround it. Thus, the complex of copper(II) ($Cu^{2+}$) with four $Br^-$ ions is an anion with a $-2$ charge, $[CuBr_4]^{2-}$.

## TABLE 18.3

### Common Ligands and Their Names

| Ligand | Name |
|---|---|
| †$:NO_2^-$ | Nitro |
| $:OCO_2^{2-}$ | Carbonato |
| $:ONO^-$ | Nitrito |
| $:CN^-$ | Cyano |
| $:SCN^-$ | Thiocyanato |
| $:NCS^-$ | Isothiocyanato |
| $:OH^-$ | Hydroxo |
| $:OH_2$ | Aqua |
| $:NH_3$ | Ammine |
| $:CO$ | Carbonyl |
| $:NO^+$ | Nitrosyl |

† The ligating atom is indicated by a pair of dots (:) to show a lone pair of electrons. In the $CO_3^{2-}$ ligand, either one or two of the oxygen atoms can donate a lone pair to a metal.

## EXAMPLE 18.1

Determine the oxidation state of the coordinated metal atom in each of the following compounds:

**(a)** $K[Co(NH_3)_2(CN)_4]$, **(b)** $[Os(CO)_5]$, **(c)** $Na[Co(H_2O)_3(OH)_3]$.

## Solution

**(a)** The oxidation state of K is known to be $+1$, and so the complex in brackets is an anion with a $-1$ charge, $[Co(NH_3)_2(CN)_4]^-$. The charge on the two $NH_3$ ligands is 0, and the charge on each of the four $CN^-$ ligands is $-1$. The oxidation state of the Co must then be $+3$, because $4 \times -1$ (for the $CN^-$) $+ 2 \times 0$ (for the $NH_3$) $+ 3$ (for Co) equals the required $-1$.

**(b)** The ligand CO has zero charge, and the complex has zero charge as well. Therefore, the oxidation state of the osmium is 0.

**(c)** There are three neutral ligands (the water molecules) and three ligands with $-1$ charges (the hydroxide ions). The $Na^+$ ion contributes only $+1$, and so the oxidation state of the cobalt must be $+2$.

**Related Problems: 9, 10**

---

Coordination modifies the chemical and physical properties of both the central atom and the ligands. Consider the chemistry of aqueous cyanide ($CN^-$) and iron(II) ($Fe^{2+}$) ions. The first reacts immediately with acid to generate gaseous HCN, a deadly poison, and the second, when mixed with aqueous base, instantly precipitates a gelatinous hydroxide. Reaction between the two ions gives the complex ion $[Fe(CN)_6]^{4-}(aq)$, which undergoes neither these two reactions nor any others considered diagnostic of simple $CN^-$ and $Fe^{2+}$. Because coordination changes a ligand's chemical behavior, a given ligand may be present in multiple forms in the same compound. The two $Cl^-$ ions in $[Pt(NH_3)_3Cl]Cl$ differ chemically, because one is coordinated and one is not. Treatment of an aqueous solution of this substance with $Ag^+$ ion immediately precipitates the uncoordinated $Cl^-$ as $AgCl(s)$, but *not* the coordinated $Cl^-$.

Ionic coordination complexes of opposite charges can combine with each other—just as any positive ion can combine with a negative ion—to form a salt. For example, the cation $[Pt(NH_3)_4]^{2+}$ and the anion $[PtCl_4]^{2-}$ form a doubly complex ionic compound of formula $[Pt(NH_3)_4][PtCl_4]$. This compound and the four compounds

$$[Pt(NH_3)_2Cl_2] \qquad [Pt(NH_3)_3Cl][Pt(NH_3)Cl_3]$$

$$[Pt(NH_3)_3Cl]_2[PtCl_4] \qquad [Pt(NH_3)_4][Pt(NH_3)Cl_3]_2$$

all contain Pt, $NH_3$, and Cl in the ratio of 1 to 2 to 2; that is, they have the same percentage composition. Two pairs even have the same molar mass. Yet the five compounds differ in structure and in physical and chemical properties. The concept of coordination organizes an immense number of chemical compositions as combinations of ligands linked in varied ratios with central metal atoms or ions. An otherwise bewildering collection of information on chemical reactivity is rationalized in terms of one ligand substituting for another.

## Naming Coordination Compounds

Up to now, we have used only the chemical symbols for coordination compounds, but for many purposes names are also useful. Some substances have names that were given to them before their structures were known. Thus, $K_3[Fe(CN)_6]$ was called potassium ferricyanide, and $K_4[Fe(CN)_6]$ was potassium ferrocyanide (these are complexes of $Fe^{3+}$ (ferric) and $Fe^{2+}$ (ferrous) ions, respectively). The older names still find some use but are being replaced by systematic names based on the following set of rules:

**1.** The name of a coordination complex is written as a single word built from the names of the ligands, a prefix for each to indicate how many are present, and a name for the central metal.

**2.** If a compound contains a coordination complex, it is named as a simple ionic compound is named: the positive ion is named first, followed (after a space) by the name of the negative ion, regardless of which is the complex ion.

**3.** The name of an anionic ligand is obtained by replacing the usual ending with the suffix *-o*. The names of neutral ligands are unchanged. Exceptions to the latter rule are *aqua* (for water), *ammine* (for $NH_3$), and *carbonyl* (for CO). See Table 18.3.

**4.** Greek prefixes (*di-*, *tri-*, *tetra-*, *penta-*, *hexa-*) are used to indicate the number of ligands of a given type attached to the central ion, if there is more than one. The prefix *mono-* (for one) is not used. If the name of the ligand itself contains a term such as *mono-* or *di-* (as in ethylene*di*amine), then the name of the ligand is placed in parentheses and the prefixes *bis-*, *tris-*, and *tetrakis-* are used instead of *di-*, *tri-*, and *tetra-*.

**5.** The ligands are listed in alphabetical order, without regard for the prefixes that tell how often each type of ligand occurs in the coordination sphere.

**6.** A Roman numeral enclosed in parentheses immediately following the name of the metal gives the oxidation state of the central metal atom. If the complex ion has a net negative charge, the ending *-ate* is added to the stem of the name of the metal.

The following list of complexes illustrates systematic naming.

| Complex | Systematic Name |
|---|---|
| $K_3[Fe(CN)_6]$ | Potassium hexacyanoferrate(III) |
| $K_4[Fe(CN)_6]$ | Potassium hexacyanoferrate(II) |
| $Fe(CO)_5$ | Pentacarbonyliron(0) |
| $[Co(NH_3)_5CO_3]Cl$ | Pentaamminecarbonatocobalt(III) chloride |
| $K_3[Co(NO_2)_6]$ | Potassium hexanitrocobaltate(III) |
| $[Cr(H_2O)_4Cl_2]Cl$ | Tetraaquadichlorochromium(III) chloride |
| $[Pt(NH_2CH_2CH_2NH_2)_3]Br_4$ | Tris(ethylenediamine)platinum(IV) bromide |
| $K_2[CuCl_4]$ | Potassium tetrachlorocuprate(II) |

## EXAMPLE 18.2

Interpret the names and write the formulas of these coordination compounds: **(a)** sodium tricarbonatocobaltate(III), **(b)** diamminediaquadichloroplatinum(IV) bromide, **(c)** sodium tetranitratoborate(III).

### Solution

**(a)** In the anion, three carbonate ligands (with $-2$ charges) are coordinated to a cobalt atom in the $+3$ oxidation state. Because the complex ion thus has an overall charge of $-3$, three sodium cations are required, and the correct formula is $Na_3[Co(CO_3)_3]$.

**(b)** The ligands coordinated to a Pt(IV) include two ammonia molecules, two water molecules, and two chloride ions. Ammonia and water are electrically neutral, but the two chloride ions contribute a total charge of $2 \times (-1) = -2$ that adds with the

+4 of the platinum and gives the complex ion a +2 charge. Two bromide anions are required to balance this, and so the formula is $[Pt(NH_3)_2(H_2O)_2Cl_2]Br_2$.

**(c)** The complex anion has four −1 nitrate ligands coordinated to a central boron(III). This gives a net charge of −1 on the complex ion and requires one sodium ion in the formula $Na[B(NO_3)_4]$.

**Related Problems: 11, 12**

---

## Ligand Substitution Reactions

If the yellow crystalline solid nickel(II) sulfate is exposed to moist air at room temperature, it takes up six water molecules per formula unit. These water molecules coordinate the nickel ions to form a bright green complex:

$$NiSO_4(s) + 6\ H_2O(g) \longrightarrow [Ni(H_2O)_6]SO_4(s)$$
Yellow          Colorless                  Green

Heating the green hexaaquanickel(II) sulfate to a temperature well above the boiling point of water drives off the water and regenerates the yellow $NiSO_4$ in a reversal of the reaction just shown. A similar coordination reaction generates a product with completely different color when yellow $NiSO_4(s)$ is exposed to gaseous ammonia, $NH_3(g)$. This time, the product is a blue-violet complex:

$$NiSO_4(s) + 6\ NH_3(g) \longrightarrow [Ni(NH_3)_6]SO_4(s)$$
Yellow          Colorless                  Blue-violet

Heating the blue-violet product drives off ammonia, and the color of the solid then changes back to yellow. Given these facts, it is not hard to explain the observation that a green $[Ni(H_2O)_6]^{2+}(aq)$ solution turns blue-violet when treated with $NH_3(aq)$ (Fig. 18.7). The $NH_3$ must be displacing the $H_2O$ from the coordination sphere.

$$[Ni(H_2O)_6]^{2+} + 6\ NH_3(aq) \longrightarrow [Ni(NH_3)_6]^{2+} + 6\ H_2O$$
Green          Colorless                  Blue-violet          Colorless

Complexes that undergo rapid substitution of one ligand for another are **labile.** Those in which substitution proceeds slowly or not at all are **inert.** In an inert complex, a high activation energy for ligand substitution prevents rapid reaction even

**FIGURE 18.7**  When ammonia is added to the green solution of nickel(II) sulfate on the left (which contains $[Ni(H_2O)_6]^{2+}$ ions), ligand substitution occurs to give the blue-violet solution on the right (which contains $[Ni(NH_3)_6]^{2+}$ ions). *(Leon Lewandowski)*

though there may be a thermodynamic tendency to proceed. In the substitution reaction

$$[Co(NH_3)_6]^{3+}(aq) + 6\,H_3O^+(aq) \longrightarrow [Co(H_2O)_6]^{3+} + 6\,NH_4^+(aq)$$

the products are favored thermodynamically by an enormous amount (the equilibrium constant is about $10^{64}$). Yet the inert $[Co(NH_3)_6]^{3+}$ complex ion lasts for weeks in acidic solution because no low-energy path for the reaction exists. The $[Co(NH_3)_6]^{3+}$ ion is thermodynamically unstable relative to $[Co(H_2O)_6]^{3+}$ yet kinetically stable (that is, inert). The closely related cobalt(II) complex, $[Co(NH_3)_6]^{2+}$, undergoes a similar substitution reaction in a few seconds:

$$[Co(NH_3)_6]^{2+}(aq) + 6\,H_3O^+(aq) \longrightarrow [Co(H_2O)_6]^{2+} + 6\,NH_4^+(aq)$$

The hexaaminecobalt(II) complex is thermodynamically unstable and also labile.

Substitution of one ligand for another can proceed in stages in the coordination sphere of a metal. By controlling the reaction conditions, substitution can usually be stopped at intermediate stages. For example, all of the possible four-coordinate compositions of Pt(II) with the two ligands $NH_3$ and $Cl^-$

$$[Pt(NH_3)_4]^{2+} \quad [Pt(NH_3)_3Cl]^+ \quad [Pt(NH_3)_2Cl_2] \quad [Pt(NH_3)Cl_3]^- \quad [PtCl_4]^{2-}$$

can be prepared. Such mixed-ligand complexes add greatly to the variety and richness of coordination chemistry.

---

## 18.3

### STRUCTURES OF COORDINATION COMPLEXES

The Alsatian-Swiss chemist Alfred Werner pioneered the field of coordination chemistry in the late 19th century. At that time, a number of compounds of cobalt(III) chloride with ammonia were known. They had the following chemical formulas and colors:

| | |
|---|---|
| Compound 1: $CoCl_3 \cdot 6NH_3$ | Orange-yellow |
| Compound 2: $CoCl_3 \cdot 5NH_3$ | Purple |
| Compound 3: $CoCl_3 \cdot 4NH_3$ | Green |
| Compound 4: $CoCl_3 \cdot 3NH_3$ | Green |

The fact that treatment of these compounds with aqueous hydrochloric acid did not remove the ammonia suggested that the ammonia was somehow closely bound with the cobalt ions. Treatment with aqueous silver nitrate at 0°C, on the other hand, gave interesting results. With compound 1, all of the chloride present precipitated as solid AgCl. With compound 2, only two thirds of the chloride precipitated, and with compound 3, only one third. Compound 4 did not react at all with the silver nitrate. Werner accounted for these facts by positing the existence of coordination complexes with six ligands (either chloride ions or ammonia molecules) attached to each $Co^{3+}$ ion. Specifically, he wrote the formulas for compounds 1 through 4 as

Compound 1: $[Co(NH_3)_6]^{3+}\,(Cl^-)_3$
Compound 2: $[Co(NH_3)_5Cl]^{2+}\,(Cl^-)_2$
Compound 3: $[Co(NH_3)_4Cl_2]^+\,(Cl^-)$
Compound 4: $[Co(NH_3)_3Cl_3]$

Only those chloride ions that were *not* ligands attached directly to cobalt were precipitated upon addition of cold aqueous silver nitrate.

Werner realized that with his proposed complexes came predictions about the electrical conductivity of aqueous solutions of the salts of these complex ions. Compound 1, for example, should have a molar conductivity close to that of $Al(NO_3)_3$, which also forms one 3+ ion and three 1− ions per formula unit dissolved in water. His experiments confirmed this resemblance and showed that compound 2 resembled $Mg(NO_3)_2$ and compound 3 resembled $NaNO_3$ in conductivity. Compound 4 behaved as a nonelectrolyte, with a very low electrical conductivity, as expected from the absence of ions in its formula.

Werner and other chemists studied a variety of other coordination complexes, using both physical and chemical techniques. Their research has shown that the most common coordination number by far is 6, as in the cobalt complexes just discussed. Coordination numbers ranging from 2 to 12 have been observed, however, including coordination numbers 2 (as in $[Ag(NH_3)_2]^+$), 4 (as in $[PtCl_4]^{2-}$), and 5 (as in $[Ni(CN)_5]^{3-}$).

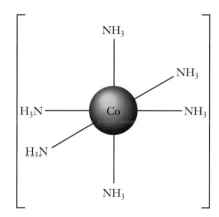

**FIGURE 18.8** The octahedral structure of the $Co(NH_3)_6^{3+}$ ion. All six corners of the octahedron are equivalent.

## Three-Dimensional Structures

If the central cobalt ion in $[Co(NH_3)_6]^{3+}$ is bonded to six ammonia ligands, what is the geometrical structure of the complex? This question naturally occurred to Werner, who suggested that the arrangement should be the simplest and most symmetric possible, with ligands at the six vertices of a regular octahedron (Fig. 18.8). Modern methods of x-ray diffraction (Section 19.1) enable us to make very precise determinations of atomic positions in crystals and completely confirm Werner's hypothesis of octahedral coordination for this complex. Such techniques were not available a century ago, however, and so Werner turned to a study of the properties of substituted complexes to test his hypothesis.

If one ammonia ligand is replaced with a chloride ion, the resulting complex has the formula $[Co(NH_3)_5Cl]^{2+}$, with one vertex of the octahedron occupied by $Cl^-$ and the other five by $NH_3$. Only one structure of this type is possible, because all six vertices of a regular octahedron are equivalent and the various singly substituted complexes $[MA_5B]$ (where $A = NH_3$, $B = Cl^-$, $M = Co^{3+}$, for example) can be superimposed upon one another. Now suppose a second $NH_3$ ligand is replaced with $Cl^-$. The second $Cl^-$ can either lie in one of the four positions closest to the first $Cl^-$ (in the plane perpendicular to the Co–$Cl^-$ line; see Fig. 18.9a) or lie in the sixth position, on the opposite side of the central metal atom (Fig. 18.9b). The

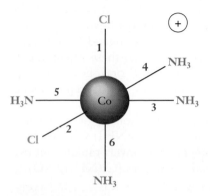

**(a)** cis-$[Co(NH_3)_4Cl_2]^+$

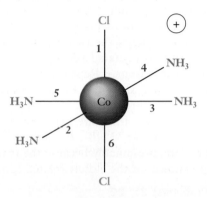

**(b)** trans-$[Co(NH_3)_4Cl_2]^+$

**FIGURE 18.9** The cis-$[Co(NH_3)_4Cl_2]^+$ and trans-$[Co(NH_3)_4Cl_2]^+$ ions. The cis complex is purple in solution, but the trans is green.

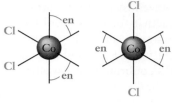

**FIGURE 18.10** The complex ion $[CoCl_2(en)_2]^+$ is an octahedral complex that has *cis* and *trans* isomers, according to the relative positions of the two $Cl^-$ ligands. Salts of the *cis* isomers are purple, and salts of the *trans* isomers are green. *(Charles Steele)*

former arrangement, in which the two $Cl^-$ ligands are closer to each other, is *cis*-$[Co(NH_3)_4Cl_2]^+$, and the latter, with the two $Cl^-$ ligands farther apart, is *trans*-$[Co(NH_3)_4Cl_2]^+$. The octahedral structure model predicts that exactly two different ions with the chemical formula $[Co(NH_3)_4Cl_2]^+$ exist. Such species are called **geometrical isomers.** When Werner began his work, only the green *trans* form was known, but by 1907 he had prepared the *cis* isomer and shown that it differed from the *trans* isomer in color (it was violet rather than green) and other physical properties. The isolation of two, and only two, geometrical isomers of this ion was good (although not conclusive) evidence that the octahedral structure was correct. Similar isomerism is displayed by the complex ion $[CoCl_2(en)_2]^+$, which also has a purple *cis* form and a green *trans* form (Fig. 18.10).

**EXAMPLE 18.3**

How many geometrical isomers exist for the octahedral coordination compound $[Co(NH_3)_3Cl_3]$?

**Solution**

We begin with the two isomers of $[Co(NH_3)_4Cl_2]^+$ and see how many different structures can be made by replacing one more ammonia molecule with $Cl^-$.

Starting with the *trans* form (Fig. 18.9b), it is clear that replacement of any of the four $NH_3$ ligands at site 2, 3, 4, or 5 gives an equivalent structure that can be superimposed by rotation. Figure 18.11a shows this isomer. What isomers can be made from the *cis* form of Figure 18.9a? If either the ammonia ligand at site 4 or that at site 6 (i.e., one of the two that are *trans* to existing $Cl^-$ ligands) is replaced, the result is simply a rotated version of Figure 18.11a, and so these replacements do *not* give another isomer. Replacement of the ligand at site 3 or 5, however, gives a different structure, shown in Figure 18.11b. (Note that the 1, 2, 3 replacements give the same structure as the 1, 2, 5 replacements.)

We conclude that there are two, and only two, possible isomers of the octahedral complex structure $MA_3B_3$. In fact, only one form of $[Co(NH_3)_3Cl_3]$ has been prepared to date, presumably because the two isomers interconvert rapidly, but two isomers are known for the closely related coordination complex $[Cr(NH_3)_3(NO_2)_3]$.

**Related Problems: 21, 22**

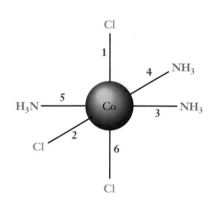

**(a)**

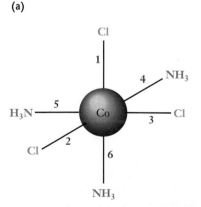

**FIGURE 18.11** Two structural isomers of the coordination compound $Co(NH_3)_3Cl_3$.

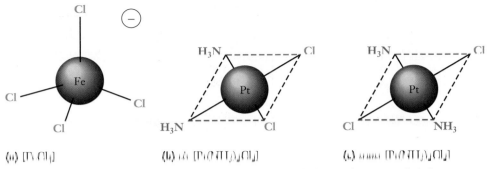

(a) [FeCl₄] (b) cis-[Pt(NH₃)₂Cl₂] (c) trans-[Pt(NH₃)₂Cl₂]

**FIGURE 18.12**  Four-coordinate complexes. (a) Tetrahedral, $[FeCl_4]^-$. (b and c) Square planar, illustrating (b) the *cis* and (c) the *trans* forms of $[Pt(NH_3)_2Cl_2]$.

## Square-Planar, Tetrahedral, and Linear Structures

Complexes with coordination numbers of 4 typically are either tetrahedral or square planar. The tetrahedral geometry (Fig. 18.12a) predominates for four-coordinate complexes of the early transition metals (those toward the left side of the *d* block of elements in the periodic table). Geometrical isomerism is not possible for tetrahedral complexes of the general form $MA_2B_2$, because all such structures are superimposable.

The square-planar geometry (Fig. 18.12b and c) is common for four-coordinate complexes of $Au^{3+}$, $Ir^+$, and $Rh^+$ and, most especially, ions with the $d^8$ valence electron configurations: $Ni^{2+}$, $Pd^{2+}$, and $Pt^{2+}$. The $Ni^{2+}$ ion forms a few tetrahedral structures, but four-coordinate $Pd^{2+}$ and $Pt^{2+}$ are nearly exclusively square planar. Square-planar complexes of the type $MA_2B_2$ can have isomers, as Figure 18.12b and c illustrate for *cis*- and *trans*-$[Pt(NH_3)_2Cl_2]$. The *cis* form of this compound is a potent and widely used anticancer drug called cisplatin, but the *trans* form has no therapeutic properties.

Finally, linear complexes with coordination numbers of 2 exist, especially for ions with $d^{10}$ configurations such as $Cu^+$, $Ag^+$, $Au^+$, and $Hg^{2+}$. The central silver atom in a complex such as $[Ag(NH_3)_2]^+$ in aqueous solution strongly attracts several water molecules as well, however, and so its actual coordination number under these circumstances may be greater than 2.

## Chiral Structures

The complex ion $[Pt(en)_3]^{4+}$ displays a type of isomerism that differs from the geometrical isomerism discussed to this point. The two structures shown in Figure 18.13 are mirror-image pairs that cannot be superimposed via rotation. Section 17.1 discussed such optical isomers in connection with tetrahedrally bonded carbon atoms (recall Fig. 17.4); like such carbon atoms, the central atom in an optical isomer is called a chiral center.

**EXAMPLE 18.4**

Suppose that the complex ion $[Co(NH_3)_2(H_2O)_2Cl_2]^+$ is synthesized with the two ammine ligands *cis* to each other, the two aqua ligands *cis* to each other, and the two chloro ligands *cis* to each other (Fig. 18.14a). Is this complex optically active?

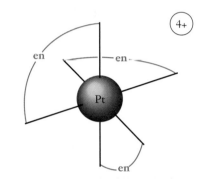

Mirror plane

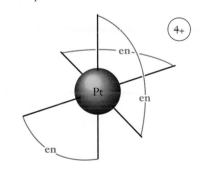

en = $H_2\ddot{N}$–$CH_2CH_2$–$\ddot{N}H_2$

**FIGURE 18.13**  Enantiomers of the $[Pt(en)_3]^{4+}$ ion. Reflection through the mirror plane transforms one enantiomer into the other. The two cannot be superimposed by simple rotation.

**FIGURE 18.14**  The structure of (a) the all-*cis* $[Co(NH_3)_2(H_2O)_2Cl_2]^+$ complex ion, together with (b) its mirror image.

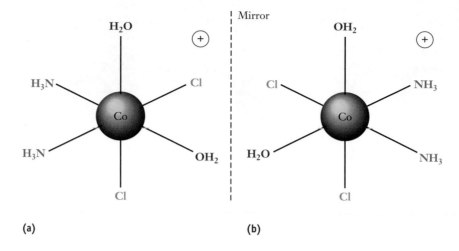

(a)                                    (b)

### Solution

We indicate a mirror with a dashed line and create the mirror image by making each point in it lie the same distance from the dashed line as does its generating point in the original structure (Fig. 18.14). Comparison of the original and the mirror image (Fig. 18.14b) shows that *cis, cis*-$[Co(NH_3)_2(H_2O)_2Cl_2]^+$ is chiral, because the two structures cannot be superimposed even after they are turned. As this example proves, nonchelates can be chiral.

**Related Problems: 23, 24**

---

The hexadentate ligand EDTA (ethylenediaminetetraacetate ion) forms chiral complexes. Figure 18.15 shows the structure of this chelating ligand, as coordinated to a $Co^{3+}$ ion. The central metal ion is literally "enveloped" by the ligand as it coordinates simultaneously at the six corners of the coordination octahedron. Such a chelating agent has a very strong affinity for certain metal ions and can **sequester** them effectively in solution. EDTA solubilizes the scummy precipitates that $Ca^{2+}$ ion forms with anionic constituents of soap by forming a stable complex with $Ca^{2+}$.

**FIGURE 18.15**  The chelation complex of EDTA with cobalt(III). Each EDTA ion has six donor sites at which it can bind (by donating lone pair electrons) to the central metal ion.

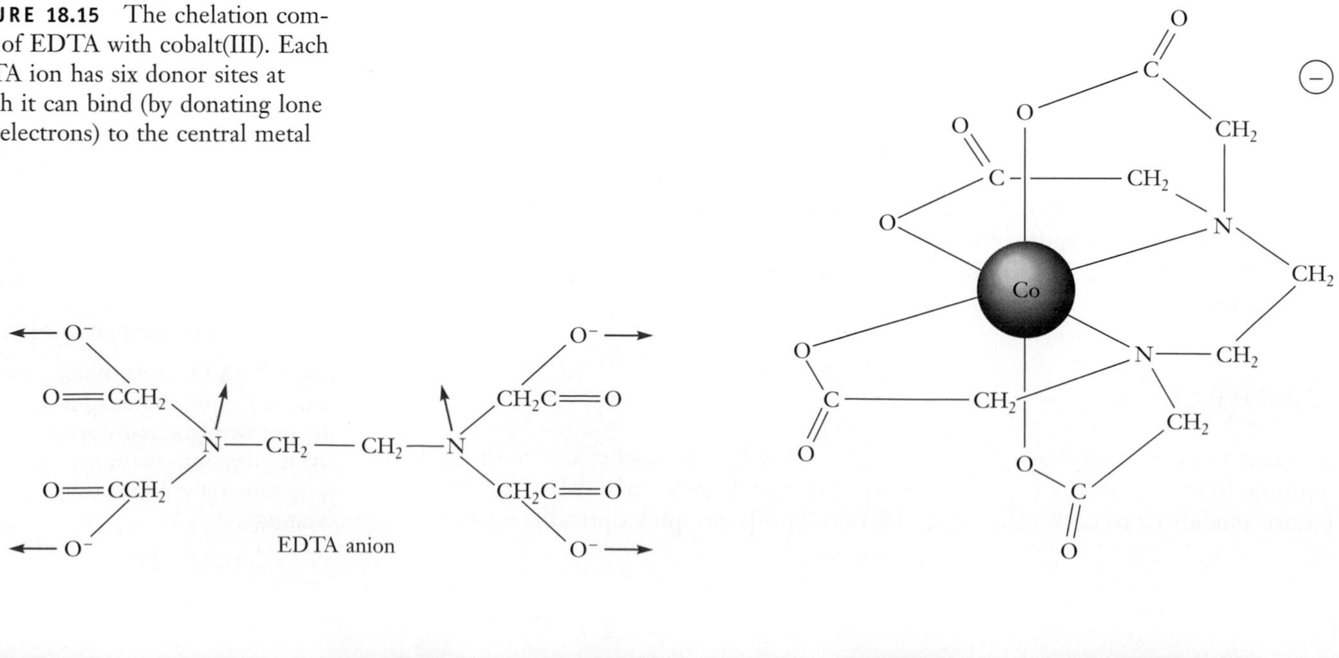

Thus, it breaks up the main contributor to bathtub rings and is a "miracle ingredient" in some bathtub cleaners. EDTA is used to recover trace contaminants from water (some metal ions, especially heavy ones, are toxic). It has been used as an antidote for lead poisoning because of its great affinity for $Pb^{2+}$ ions. Complexes of EDTA with iron in plant foods permit a slow release of iron to the plant. EDTA also sequesters copper and nickel ions in edible fats and oils. Because these metal ions catalyze the oxidation reactions that turn oils rancid, EDTA preserves freshness.

---

## 18.4

## CRYSTAL FIELD THEORY AND MAGNETIC PROPERTIES

What is the nature of the bonding in coordination complexes of the transition metals that leads to their special properties? Why does Pt(IV) form only octahedral complexes while Pt(II) forms square-planar ones, and under what circumstances does Ni(II) form octahedral, square-planar, and tetrahedral complexes? Can trends in the length and strength of metal to ligand bonds be understood? To answer these questions, we need a theoretical description of bonding in coordination complexes.

### Crystal Field Theory

A simple and useful model for the energies of coordination complexes is **crystal field theory,** which is based on an ionic description of the metal–ligand bonds. It treats the complex as a central metal ion perturbed by the approach of negatively charged ligands. In an octahedral complex, negative charges from the lone pairs of six ligands are brought up along the $\pm x$, $\pm y$, and $\pm z$ coordinate axes toward a metal atom or ion at the origin. In an atom or ion of a transition metal in free space, the energy of an electron is the same in each of the five $d$ orbitals, but when the external charges are brought up (Fig. 18.16), the energies of the $d$ orbitals change by dif-

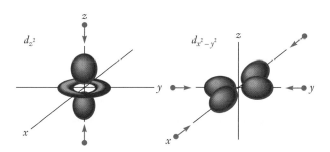

Large repulsions

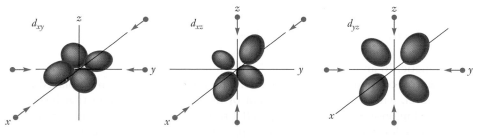

Smaller repulsions

**FIGURE 18.16**  The basis for octahedral crystal field splitting of $3d$ orbital energies by ligands. As the external charges approach the five $3d$ orbitals, the largest repulsions arise in the $d_{z^2}$ and $d_{x^2-y^2}$ orbitals, which point directly at two or four of the approaching charges.

**EXAMPLE 18.6**

Predict which of the following octahedral complexes should have the shortest wavelength of absorption in the visible spectrum: $[FeF_6]^{3-}$, $[Fe(CN)_6]^{3-}$, $[Fe(H_2O)_6]^{3+}$.

### Solution

The $[Fe(CN)_6]^{3-}$ has the strongest field ligands of the three complexes, and so its energy levels are split by the greatest amount. The frequency of light absorbed should be greatest, and the absorption wavelength should be the shortest, for this ion.

The observed color of $[Fe(CN)_6]^{3-}$ solutions is red, indicating absorption of blue and violet light. Solutions of $[Fe(H_2O)_3]^{3+}$ are a very pale violet, due to weak absorption of red light, and $[FeF_6]^{3-}$ solutions are colorless, indicating that the absorption lies beyond the long wavelength limit of the visible spectrum.

**Related Problems: 37, 38, 39, 40**

## Ligand Field Theory

The ordering of the spectrochemical series cannot be explained within the crystal field model and indeed points up an unsatisfactory feature of that model. The bonding in coordination compounds *cannot* be fully ionic, as assumed in the crystal field description, because neutral ligands such as $NH_3$ and CO give *larger* splittings than those induced by a small negative ion such as $F^-$. The crystal field theory has been modified to include covalent as well as ionic aspects of coordination. The resulting **ligand field theory** gives a powerful description of transition-metal complexes.

Ligand field theory begins by constructing molecular orbitals from the valence *d*, *s*, and *p* orbitals of the central metal atom and from the six ligand orbitals that point along the metal–ligand bonds in an octahedral complex. The ligand orbital would be the $sp^3$ orbital containing a lone pair in $NH_3$, or the $p_z$ orbital from a $Cl^-$ ion lying on the z axis. Because the phases of the wave function alternate in its different lobes, there is no net overlap between the $d_{xy}$, $d_{yz}$, or $d_{xz}$ orbital of the metal atom and a ligand orbital directed along the bond axis (this is analogous to the absence of net overlap between a *p* orbital perpendicular to the bond axis and an *s* orbital on the second atom, shown in Figure 16.9b). As a result, these three *d* orbitals are nonbonded orbitals on the metal atom, and their energy shifts neither up nor down. The other two metal *d* orbitals ($d_{x^2-y^2}$ and $d_{z^2}$), together with the metal *s* and *p* valence orbitals (a total of six atomic orbitals), *can* be combined with the six ligand orbitals to form a total of twelve molecular orbitals, six of which are σ bonding orbitals and six σ* antibonding orbitals. Figure 18.22 is the resulting orbital correlation diagram.

The molecular orbitals, beginning with those of lowest energy, are then filled with electrons. Twelve electrons are available from the bonding electron pairs on the six ligands, and these fill the six σ bonding orbitals. The metal *d* electrons then are placed in the nonbonding $d_{xy}$, $d_{xz}$, and $d_{yz}$ orbitals ($t_{2g}$) and the pair of antibonding σ* $d_{x^2-y^2}$ and σ* $d_{z^2}$ orbitals that lie above them ($e_g$). These five orbitals have exactly the same level structure as in the crystal field theory (see Fig. 18.17), so that all of the results from that theory can be carried over to this one. The magnitude of the splitting $\Delta_o$ arises here from the increase in energy of the antibonding σ* molecular orbitals relative to the nonbonding metal-atom *d* orbitals. The

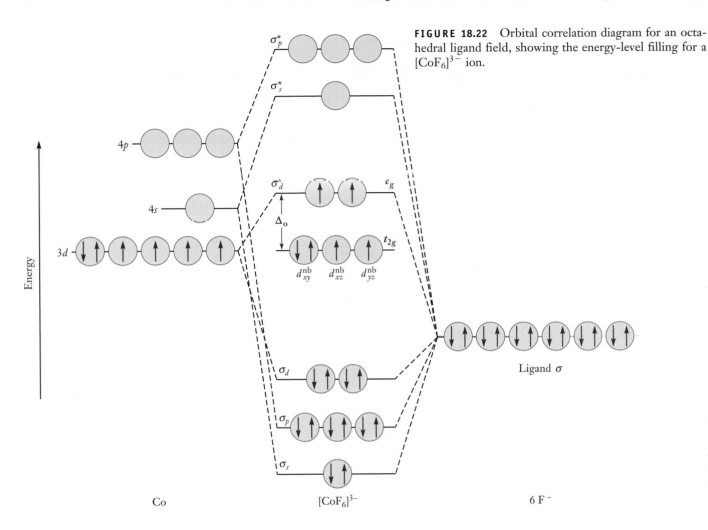

**FIGURE 18.22** Orbital correlation diagram for an octahedral ligand field, showing the energy-level filling for a $[CoF_6]^{3-}$ ion.

ligand field picture allows for partial covalent and partial ionic character in the bonding of the complex. The lower the energy of the original ligand orbitals relative to the metal $d$ orbitals, the more closely the $\sigma$ bonding orbitals resemble the pure ligand orbitals. In this limit the bonding becomes purely ionic. The situation closely parallels the discussion in Section 16.1 of the effect of electronegativity on the ionic–covalent character of bonding in heteronuclear diatomic molecules (see Fig. 16.8).

Ligand field theory helps to explain the order of bonding strengths for different ligands. Consider a nonbonded metal-atom $d$ orbital and the way in which it interacts with the electrons in ligand $p$ orbitals *perpendicular* to the metal–ligand axis. An electron in a $d_{xy}$ orbital, for example, is repelled by an electron in a $p_y$ orbital of a $Cl^-$ ion lying along the $x$ axis (Fig. 18.23). This increases the energy of the $t_{2g}$ orbitals and decreases the splitting $\Delta_o$. This electrostatic crystal field effect dominates for spherical ionic ligands such as $I^-$, $Br^-$, $Cl^-$, $F^-$, and $OH^-$, which are therefore called **weak-field ligands** (small $\Delta_o$). The larger, less electronegative ions in the group, such as $I^-$ and $Br^-$, give the greatest increase in energy of the $t_{2g}$ orbitals because they hold their electrons least tightly. As a result, $\Delta_o$ is smallest for these ions and is somewhat larger for $F^-$.

At the opposite extreme are ligands such as CO and $CN^-$. They increase $\Delta_o$ by *lowering* the energy of the $d_{xy}$, $d_{yz}$, and $d_{xz}$ orbitals through a mechanism called $\pi$

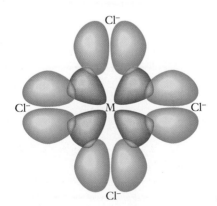

**FIGURE 18.23** Repulsion of an electron in a $d_{xy}$ orbital by $2p_x$ and $2p_y$ electrons on neighboring chlorine ligands. All the orbitals are occupied.

**FIGURE 18.24** The effect of $\pi$ bonding on metal carbonyl complexes. (a) The slight overlap of an occupied $\pi$ orbital of CO with a $d_{xy}$, $d_{xz}$, or $d_{yz}$ metal orbital raises $t_{2g}$ slightly. (b) The large overlap of an unoccupied $\pi^*$ orbital of CO with a $d_{xy}$, $d_{xz}$, or $d_{yz}$ metal orbital. Backbonding lowers $t_{2g}$ significantly.

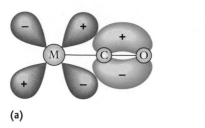

(a)

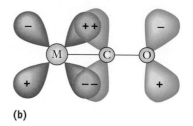

(b)

**back-bonding.** To understand this mechanism, recall that CO and $CN^-$ are diatomic species that have occupied $\pi$ bonding orbitals and unoccupied $\pi^*$ antibonding orbitals. Figure 18.24 shows the way in which these orbitals overlap with the $t_{2g}$ orbitals of the metal. The $\pi$ bonding orbital in CO is concentrated in the region between the C and O nuclei (see Fig. 18.24a) rather than on the C atom, and so the repulsive overlap with a $d_{xy}$ orbital is less than that shown in Figure 18.23 for a chloride ion. On the other hand, the unoccupied CO $\pi^*$ orbital extends *toward* the metal atom and overlaps strongly with the $d_{xy}$ orbital (Fig. 18.24b), allowing the electron in that orbital to be partially delocalized over the CO molecule and lowering its energy ($\pi$ back-bonding). Recall that strong overlap with a doubly occupied orbital raises the energy through electron repulsion and Pauli exclusion, but strong overlap with an empty orbital lowers the energy because the electron is free to move over a larger volume. The first effect is small and the second is large for ligands with $\pi$ bonds, increasing $\Delta_o$ and making these **strong-field ligands.**

Ligands such as $H_2O$ and $NH_3$ give neither strong repulsive overlap with $p$ orbitals, as in $Cl^-$, nor strong back-bonding into $\pi^*$ orbitals, as in CO, and so are **intermediate-field ligands.** Ligand field theory, with $\pi$ orbital contributions included, thus explains the observed strengths of bonding in coordination complexes. Although our description of these effects has been qualitative, quantum-mechanical calculations that support it can be carried out.

---

### 18.6

## ORGANOMETALLIC COMPOUNDS AND CATALYSIS

An **organometallic compound** is one that contains bonds between metal and carbon atoms. Transition metals with carbonyl (CO) ligands are an important example. Nickel forms a tetrahedral carbonyl compound, $Ni(CO)_4$, which is the basis of the Mond process for purifying the element. Crude metallic nickel reacts with carbon monoxide to form nickel tetracarbonyl:

$$Ni(s) + 4\,CO(g) \xrightarrow{60°C} Ni(CO)_4(g)$$

This volatile compound can be separated easily from impurities originally present in the crude metal. Strong heating then decomposes the $Ni(CO)_4$,

$$Ni(CO)_4(g) \xrightarrow{200°C} Ni(s) + 4\,CO(g)$$

and pure metallic nickel is recovered. The structures of metal carbonyl compounds are by no means restricted to the few types considered in Section 18.3. For example, iron forms a pentacarbonyl, $Fe(CO)_5$, with a trigonal-bipyramidal structure like that shown for $PF_5$ in Figure 3.11. In organometallic compounds, special stability occurs when the central metal atom is surrounded by 18 valence electrons, which give it a closed shell (10 $d$, 6 $p$, and 2 $s$ electrons). The iron atom in $Fe(CO)_5$ is surrounded by 8 of its own valence electrons plus 10 from the five CO ligand molecules, giving a total of 18 electrons. Other novel structures (Fig. 18.25) include metal–metal bonds and three-center bridge bonds formed by carbonyl groups. In molecules of $Mn_2(CO)_{10}$, each Mn atom is surrounded by 7 of its own valence electrons, 10 electrons from five CO molecules, and 1 from the electron-pair bond formed with the second Mn atom, for a total of 18. Some compounds do not fit the "rule of 18," however; molecules of the blue-green crystalline compound $V(CO)_6$ have a central vanadium atom that shares only 17 valence electrons. The odd number of electrons makes this compound paramagnetic.

Other organometallic compounds are "sandwich compounds," in which a transition-metal atom is situated between two planar hydrocarbon rings with delocalized $\pi$ electrons. One of the most stable sandwich compounds is ferrocene, $Fe(C_5H_5)_2$, in which an $Fe^{2+}$ ion is coordinated to two cyclopentadienyl anions $(C_5H_5^-)$, giving the structure shown in Figure 18.26. The $C_5H_5^-$ anion itself is a resonance hybrid of five Lewis diagrams, two of which are

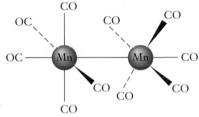

**(a)** $Mn_2(CO)_{10}$

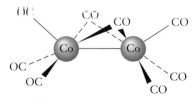

**(b)** $Co_2(CO)_8$

**FIGURE 18.25** The structures of two binuclear carbonyls. In part b, note the bridge bonding by CO molecules.

Molecules of ferrocene have 18 electrons about the iron ion: 6 $d$ electrons from $Fe^{2+}$ and 6 $\pi$ electrons from each $C_5H_5^-$ ion. The compound is remarkably stable, forming orange crystals that melt at 173°C to an orange liquid that boils at 249°C. The compound is thermally stable to about 500°C. A related, more weakly bound 18-electron compound has molecules with a chromium atom sandwiched between two benzene rings, $Cr(C_6H_6)_2$.

One reason to take particular interest in metal–carbon bonds is the catalytic role transition metals often play in making organic compounds. An example is the Monsanto process, which is now the predominant industrial route to acetic acid. The overall reaction is

$$CH_3OH + CO \longrightarrow CH_3COOH$$

with both starting materials (methanol and carbon monoxide) produced economically through the partial oxidation of methane from natural gas. An organometallic complex of rhodium(II) and iodine catalyzes the synthesis. In the presence of this compound, the production of acetic acid proceeds smoothly under mild conditions (a pressure of 1 atm and a temperature of 175°C). The mechanism of the reaction is believed to be that shown in Figure 18.27. Methanol reacts with hydrogen iodide (formed from the iodine) to give methyl iodide and water (see central box of figure):

$$CH_3OH + HI \longrightarrow CH_3I + H_2O$$

The methyl iodide adds to a four-coordinated rhodium complex to give a six-coordinated complex with a Rh—CH$_3$ bond (step $a$). In the crucial step of the mechanism, one of the carbonyl groups already attached to the Rh atom swings up and inserts itself into the Rh—CH$_3$ bond (step $b$). Another CO molecule then coordinates to the Rh atom to replace the CO lost to this insertion (step $c$). Upon addition of water, the —COCH$_3$ group detaches from the rhodium to give acetic acid (step $d$).

**FIGURE 18.26** The structure of ferrocene, $Fe(C_5H_5)_2$, in the gaseous state. Each flat pentagon represents a $C_5H_5$ ring.

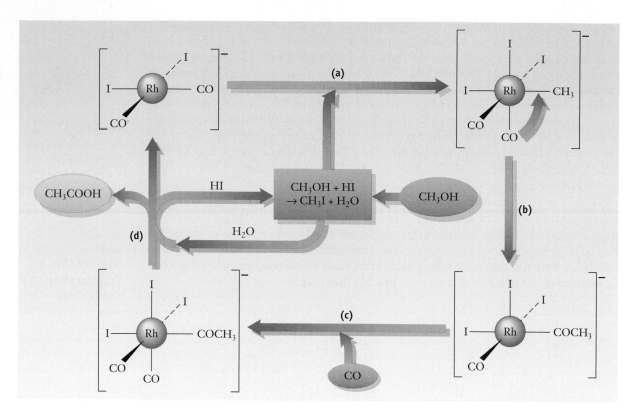

**FIGURE 18.27**    The Monsanto process for making acetic acid (shown in green) from methanol and carbon monoxide (red). This cyclic process proceeds clockwise around the diagram. The rhodium coordination complex (upper left) returns to its initial state after each cycle. The crucial step is *b*, in which a CO ligand swings up (colored arrow) and inserts itself into the Rh—CH₃ bond.

The iodine atom leaves at the same time, regenerating hydrogen iodide, and the rhodium complex is returned to its original state, ready to start another catalytic cycle.

## Coordination Complexes in Biology

Coordination complexes, particularly chelates, play fundamental roles in the biochemistry of both plants and animals. Trace amounts of at least nine transition elements are essential to life—vanadium, chromium, manganese, iron, cobalt, nickel, copper, zinc, and molybdenum.

Several of the most important complexes are based on *porphine*, a compound of carbon, nitrogen, and hydrogen whose structure is shown in Figure 24.8a. Removal of two central hydrogen ions leaves four nitrogen atoms that can bind to a metal ion $M^{2+}$ and form a tetradentate chelate structure. Section 24.3 will describe the key role of such structures in photosynthesis, the process through which green plants, algae, and certain bacteria capture light and transform it into energy for use in chemical reactions. Such a structure, modified by the addition of several side groups, gives a complex with $Fe^{2+}$ ions called *heme* (see Fig. 24.8c). *Hemoglobin*, the compound that transports oxygen in the blood, contains four such heme groups. The fifth coordination site of each iron(II) ion binds *globin* (a high-molar-mass protein) via nitrogen atoms on that protein, and the sixth site is occupied by water or molecular oxygen. Oxygen is a strong-field ligand that makes the $d^6$ $Fe^{2+}$ diamagnetic and

The structure diagram shows vitamin B$_{12}$ with labels: NH$_2$, O, H, H$_3$C, H$_3$C, NH$_2$, H$_2$N, O, H, H$_2$N, CN, N, H, CH$_3$, H$_2$N, O, H$_3$C, H$_3$C, N, CH$_3$, Co$^+$, N, N, H, H, CH$_3$, H$_2$N, O, H, CH$_3$, O, NH$_2$, N, CH$_3$, HN, O, CH$_3$, H, HO, N, H$_3$C, O, P, O, O, O, O, H, CH$_2$OH

**FIGURE 18.28** The structure of vitamin B$_{12}$. Carbon atoms are understood to lie at the intersections of bonds where no other atoms are indicated.

gives the short-wavelength absorption that is responsible for the red color of blood in the arteries. After the oxygen is delivered to the cells, it is replaced by the weaker field ligand water, and a paramagnetic complex forms that absorbs at longer wavelengths, producing the bluish tint of blood in the veins. In cases of carbon monoxide poisoning, CO molecules occupy the sixth coordination site and block the binding and transport of oxygen; the equilibrium constant for CO binding is 200 times that for O$_2$ binding.

Vitamin B$_{12}$ is useful in the treatment of pernicious anemia and other diseases. Its structure (Fig. 18.28) has certain similarities to that of heme. Again, a metal ion is coordinated by a planar tetradentate ligand, with two other donors completing the coordination octahedron by occupying *trans* positions. In vitamin B$_{12}$, the metal ion is cobalt and the planar ring is *corrin*, which is somewhat similar to porphine. In the human body, enzymes derived from vitamin B$_{12}$ accelerate a range of important reactions, including those involved in producing red blood cells.

# CUMULATIVE EXERCISE

## Platinum

The precious metal platinum was first used by South American Indians, who found impure, native samples of it in the gold mines of what is now Ecuador and used the samples to make small items of jewelry. Platinum's high melting point (1772°C) makes it harder to work than gold (m.p. 1064°C) and silver (m.p. 962°C), but this same property and a high resistance to chemical attack make platinum suitable as a material for high-temperature crucibles. Although platinum is a noble metal, in the +4 and +2 oxidation states it forms a wide variety of compounds, many of which

Crystals of potassium hexachloroplatinate(IV) (K$_2$PtCl$_6$). (ASTRA)

are coordination complexes. Its coordinating abilities make it an important catalyst for organic and inorganic reactions.

(a) The anticancer drug cisplatin, *cis*-[Pt(NH$_3$)$_2$Cl$_2$] (Fig. 18.12b), can be prepared from K$_2$PtCl$_6$ via reduction with N$_2$H$_4$ (hydrazine), giving K$_2$PtCl$_4$, followed by replacement of two chloride ion ligands with ammonia. Give systematic names to the three platinum complexes referred to in this statement.

(b) The coordination compound diamminetetracyanoplatinum(IV) has been prepared, but salts of the hexacyanoplatinate(IV) ion have not. Write the chemical formulas of these two species.

(c) Platinum forms organometallic complexes quite readily. In one of the simplest of these, Pt(II) is coordinated to two chloride ions and two molecules of ethylene (C$_2$H$_4$) to give an unstable yellow crystalline solid. Can this complex have more than one isomer? If so, describe the possible isomers.

(d) Platinum(IV) is readily complexed by ethylenediamine. Draw the structures of both enantiomers of the complex ion *cis*-[Pt(Cl)$_2$(en)$_2$]$^{2+}$. In this compound the Cl$^-$ ligands are *cis* to one another.

(e) In platinum(IV) complexes, the octahedral crystal field splitting $\Delta_o$ is relatively large. Is K$_2$PtCl$_6$ diamagnetic or paramagnetic? What is its $d$ electron configuration?

(f) Is cisplatin diamagnetic or paramagnetic?

(g) The salt K$_2$[PtCl$_4$] is red, but [Pt(NH$_3$)$_4$]Cl$_2$·H$_2$O is colorless. In what regions of the spectrum do the dominant absorptions for these compounds lie?

(h) When the two salts from part (g) are dissolved in water and the solutions mixed, a green precipitate called Magnus's green salt forms. Propose a chemical formula for this salt, and assign the corresponding systematic name.

**Answers**

(a) *cis*-diamminedichloroplatinum(II), potassium hexachloroplatinate(IV), and potassium tetrachloroplatinate(II)

(b) [Pt(NH$_3$)$_2$(CN)$_4$] and [Pt(CN)$_6$]$^{2-}$

(c) Pt(II) forms square-planar complexes. There are two possible forms, arising from *cis* and *trans* placement of the ethylene molecules. These are analogous to the two isomers shown in Figure 18.12b and c.

(e) The six $d$ electrons in Pt(IV) are all in the lower $t_{2g}$ level in a low-spin, large-$\Delta_o$ complex. All are paired, and so the compound is diamagnetic.

(f) Diamagnetic, with the bottom four levels in the square-planar configuration (all but $d_{x^2-y^2}$) occupied

(g) Red transmission corresponds to absorption of green light by K$_2$[PtCl$_4$]. A colorless solution has absorptions at either higher or lower frequency than visible. Because Cl$^-$ is a weaker field ligand than NH$_3$, the absorption frequency should be higher for the Pt(NH$_3$)$_4^{2+}$ complex, putting it in the ultraviolet region of the spectrum.

(h) [Pt(NH$_3$)$_4$][PtCl$_4$], tetraammineplatinum(II) tetrachloroplatinate(II)

# Concepts & Skills

*After studying this chapter and working the problems that follow, you should be able to*

1. Discuss the systematic variation of physical and chemical properties of the transition elements through the periodic table (Section 18.1, problems 1–6).

2. Define the terms "coordination compound," "coordination number," "ligand," and "chelation" (Section 18.2).

**3.** Name coordination compounds, given their molecular formulas (Section 18.2, problems 9–14).

**4.** Draw geometrical isomers of octahedral, tetrahedral, and square-planar complexes (Section 18.3, problems 21–24).

**5.** Use crystal field theory to interpret the magnetic properties of coordination compounds in terms of the electron configurations of their central ions (Section 18.4, problems 25–28).

**6.** Relate the colors of coordination compounds to their crystal field splitting energies and crystal field stabilization energies (Section 18.5, problems 33–40).

**7.** Use ligand field theory to order the energy levels in coordination compounds and to account for the spectrochemical series (Section 18.5).

**8.** Describe several organometallic compounds, including compounds that have roles in biology (Section 18.6, problems 41–44).

---

## PROBLEMS

*Answers to problems whose numbers are boldface appear in Appendix G. Problems that are more challenging are indicated with asterisks.*

### Chemistry of the Transition Metals

**1.** Of the compounds $PtF_4$ and $PtF_6$, predict (a) which is more soluble in water and (b) which is more volatile.

**2.** The melting point of $TiCl_4$ ($-24°C$) lies below those of $TiF_4$ ($284°C$) and $TiBr_4$ ($38°C$). Explain why by considering the covalent–ionic nature of these compounds and the intermolecular forces in each case.

**3.** The decavanadate ion is a complex species with chemical formula $V_{10}O_{28}^{6-}$. It reacts with an excess of acid to form the dioxovanadium ion, $VO_2^+$, and water. Write a balanced chemical equation for this reaction. In what oxidation state is vanadium before and after this reaction? What vanadium oxide has the same oxidation state?

**4.** What is the chemical formula of vanadium(III) oxide? Do you predict this compound to be more basic or more acidic than the vanadium oxide of problem 3? Write a balanced chemical equation for the reaction of vanadium(III) oxide with a strong acid.

**5.** Titanium(III) oxide is prepared by reaction of titanium(IV) oxide with hydrogen at high temperature. Write a balanced chemical equation for this reaction. Which oxide do you expect to have more strongly basic properties?

**6.** Treatment of cobalt(II) oxide with oxygen at high temperatures gives $Co_3O_4$. Write a balanced chemical equation for this reaction. What is the oxidation state of cobalt in $Co_3O_4$?

### The Formation of Coordination Complexes

**7.** Will methylamine ($CH_3NH_2$) be a monodentate or a bidentate ligand? With which of its atoms will it bind to a metal ion?

**8.** Show how the glycinate ion ($H_2N—CH_2—COO^-$) can act as a bidentate ligand. (Draw a Lewis diagram if necessary.) Which atoms in the glycinate ion will bind to a metal ion?

**9.** Determine the oxidation state of the metal in each of the following coordination complexes: $[V(NH_3)_4Cl_2]$, $[Mo_2Cl_8]^{4-}$, $[Co(H_2O)_2(NH_3)Cl_3]^-$, $[Ni(CO)_4]$.

**10.** Determine the oxidation state of the metal in each of the following coordination complexes: $Mn_2(CO)_{10}$, $[Re_3Br_{12}]^{3-}$, $[Fe(H_2O)_4(OH)_2]^+$, $[Co(NH_3)_4Cl_2]^+$.

**11.** Give the chemical formula that corresponds to each of the following compounds.
(a) Sodium tetrahydroxozincate(II)
(b) Dichlorobis(ethylenediamine)cobalt(III) nitrate
(c) Triaquabromoplatinum(II) chloride
(d) Tetraamminedinitroplatinum(IV) bromide

**12.** Give the chemical formula of each of the following compounds.
(a) Silver hexacyanoferrate(II)
(b) Potassium tetraisothiocyanatocobaltate(II)
(c) Sodium hexafluorovanadate(III)
(d) Potassium trioxalatochromate(III)

**13.** Assign a systematic name to each of the following chemical compounds.
(a) $NH_4[Cr(NH_3)_2(NCS)_4]$
(b) $[Tc(CO)_5]I$
(c) $K[Mn(CN)_5]$
(d) $[Co(NH_3)_4(H_2O)Cl]Br_2$

**14.** Give the systematic name for each of the following chemical compounds.
(a) $[Ni(H_2O)_4(OH)_2]$
(b) $[HgClI]$
(c) $K_4[Os(CN)_6]$
(d) $[FeBrCl(en)_2]Cl$

**15.** Describe the behavior of a coordination complex that is both thermodynamically stable and labile.

**16.** Describe the behavior of a coordination complex that is both thermodynamically unstable and inert.

**17.** Calculate the standard free energy change for the dissociation of tetraamminecopper(II) ion to $Cu^{2+}(aq)$ and $NH_3(aq)$ in basic aqueous solution, using data from Appendix D. Is this complex thermodynamically stable with respect to this reaction under standard conditions? Why does it dissociate under acidic conditions? Write a balanced equation for the dissociation reaction at pH = 0, and calculate the standard free energy change for that reaction.

18. The $\Delta G_f^\circ$ of the tetracyanonickelate(II) anion is +472.1 kJ mol$^{-1}$ in aqueous solution. Use this information, together with data from Appendix D, to calculate the equilibrium constant at 25°C for the dissociation

$$[Ni(CN)_4]^{2-}(aq) \longrightarrow Ni^{2+}(aq) + 4\ CN^-(aq)$$

Isotopically labeled Na$^{13}$CN is added to a solution of K$_2$Ni(CN)$_4$. Subsequent examination of the Ni(CN)$_4^{2-}$ present reveals extensive incorporation of $^{13}$CN$^-$ into the complex ion. Comment on these observations, and discuss the difference between lability and thermodynamic stability as applied to complexes of nickel(II) with the cyanide ion.

## Structures of Coordination Complexes

19. Suppose 0.010 mol of each of the following compounds is dissolved (separately) in 1.0 L of water: KNO$_3$, [Co(NH$_3$)$_6$]Cl$_3$, Na$_2$[PtCl$_6$], [Cu(NH$_3$)$_2$Cl$_2$]. Rank the resulting four solutions in order of conductivity, from lowest to highest.

20. Suppose 0.010 mol of each of the following compounds is dissolved (separately) in 1.0 L of water: BaCl$_2$, K$_4$[Fe(CN)$_6$], [Cr(NH$_3$)$_4$Cl$_2$]Cl, [Fe(NH$_3$)$_3$Cl$_3$]. Rank the resulting four solutions in order of conductivity, from lowest to highest.

21. Draw the structures of all possible isomers for the following complexes. Indicate which isomers are enantiomer pairs.
    (a) Diamminebromochloroplatinum(II) (square-planar)
    (b) Diaquachlorotricyanocobaltate(III) ion (octahedral)
    (c) Trioxalatovanadate(III) ion (octahedral)

22. Draw the structures of all possible isomers for the following complexes. Indicate which isomers are enantiomer pairs.
    (a) Bromochloro(ethylenediamine)platinum(II) (square-planar)
    (b) Tetraamminedichloroiron(III) ion (octahedral)
    (c) Amminechlorobis(ethylenediamine)iron(III) ion (octahedral)

23. Iron(III) forms octahedral complexes. Sketch the structures of all the distinct isomers of [Fe(en)$_2$Cl$_2$]$^+$, indicating which pairs of structures are mirror images of each other.

24. Platinum(IV) forms octahedral complexes. Sketch the structures of all the distinct isomers of [Pt(NH$_3$)$_2$Cl$_2$F$_2$], indicating which pairs of structures are mirror images of each other.

## Crystal Field Theory and Magnetic Properties

25. For each of the following ions, draw diagrams like those in Figure 18.17 to show orbital occupancies in both weak and strong octahedral fields. Indicate the total number of unpaired electrons in each case.
    (a) Mn$^{2+}$    (c) Cr$^{3+}$    (e) Fe$^{2+}$
    (b) Zn$^{2+}$    (d) Mn$^{3+}$

26. Repeat the work of the preceding problem for the following ions.
    (a) Cr$^{2+}$    (c) Ni$^{2+}$    (e) Co$^{2+}$
    (b) V$^{3+}$     (d) Pt$^{4+}$

27. Experiments can measure not only whether a compound is paramagnetic but also the number of unpaired electrons. It is found that the octahedral complex ion [Fe(CN)$_6$]$^{3-}$ has fewer unpaired electrons than the octahedral complex ion [Fe(H$_2$O)$_6$]$^{3+}$. How many unpaired electrons are present in each species? Explain. In each case, express the crystal field stabilization energy in terms of $\Delta_o$.

28. The octahedral complex ion [MnCl$_6$]$^{3-}$ has more unpaired spins than the octahedral complex ion [Mn(CN)$_6$]$^{3-}$. How many unpaired electrons are present in each species? Explain. In each case, express the crystal field stabilization energy in terms of $\Delta_o$.

29. Explain why octahedral coordination complexes with three and eight $d$ electrons on the central metal atom are particularly stable. Under what circumstances would you expect complexes with five or six $d$ electrons on the central metal atom to be particularly stable?

30. The following reduction potentials have been measured in acidic aqueous solution:

$$Mn^{3+} + e^- \longrightarrow Mn^{2+} \qquad \mathscr{E}^\circ = 1.51\ V$$
$$Fe^{3+} + e^- \longrightarrow Fe^{2+} \qquad \mathscr{E}^\circ = 0.77\ V$$
$$Co^{3+} + e^- \longrightarrow Co^{2+} \qquad \mathscr{E}^\circ = 1.84\ V$$

Explain why the reduction potential for Fe$^{3+}$ lies below those for the +3 oxidation states of the elements on both sides of it in the periodic table.

## The Spectrochemical Series and Ligand Field Theory

31. An aqueous solution of zinc nitrate contains the [Zn(H$_2$O)$_6$]$^{2+}$ ion and is colorless. What conclusions can be drawn about the absorption spectrum of the [Zn(H$_2$O)$_6$]$^{2+}$ complex ion?

32. An aqueous solution of sodium hexaiodoplatinate(IV) is black. What conclusions can be drawn about the absorption spectrum of the [PtI$_6$]$^{2-}$ complex ion?

33. Estimate the wavelength of maximum absorption for the octahedral ion hexacyanoferrate(III) from the fact that light transmitted by a solution of it is red. Estimate the crystal field splitting energy $\Delta_o$ (in kJ mol$^{-1}$).

34. Estimate the wavelength of maximum absorption for the octahedral ion hexaaquanickel(II) from the fact that its solutions are colored green by transmitted light. Estimate the crystal field splitting energy $\Delta_o$ (in kJ mol$^{-1}$).

35. Estimate the crystal field stabilization energy for the complex in problem 33. (*Note:* This is a high-field (low-spin) complex.)

36. Estimate the crystal field stabilization energy for the complex in problem 34.

37. The chromium(III) ion in aqueous solution has a blue-violet color.
    (a) What is the complementary color to blue-violet?
    (b) Estimate the wavelength of maximum absorption for a Cr(NO$_3$)$_3$ solution.
    (c) Will the wavelength of maximum absorption increase or decrease if cyano ligands are substituted for the coordinated water? Explain.

38. An aqueous solution containing the hexaamminecobalt(III) ion has a yellow color.
    (a) What is the complementary color to yellow?
    (b) Estimate this solution's wavelength of maximum absorption in the visible spectrum.

39. (a) An aqueous solution of $Fe(NO_3)_3$ has only a pale color, but an aqueous solution of $K_3[Fe(CN)_6]$ is bright red. Do you expect a solution of $K_3[FeF_6]$ to be brightly colored or pale? Explain your reasoning.
    (b) Would you predict a solution of $K_2[HgI_4]$ to be colored or colorless? Explain.

40. (a) An aqueous solution of $Mn(NO_3)_2$ has a very pale pink color, but an aqueous solution of $K_4[Mn(CN)_6]$ is deep blue. Explain why the two differ so much in the intensities of their colors.
    (b) Predict which of the following compounds would be colorless in aqueous solution: $K_2[Co(NCS)_4]$, $Zn(NO_3)_2$, $[Cu(NH_3)_4]Cl_2$, $CdSO_4$, $AgClO_3$, $Cr(NO_3)_2$.

## Organometallic Compounds and Catalysis

41. Give the number of valence electrons surrounding the central transition-metal ion or atom in each of the following: $[Cr(CO)_4]$, $[Os(CO)_5]$, $[H_2Fe(CO)_4]$, $K_3[Fe(CN)_5CO]$. The hydrogen atoms are covalently bonded to the central iron atom in the third case.

42. Give the number of valence electrons surrounding the central transition-metal ion or atom in each of the following: $[HMn(CO)_5]$, $[V(CO)_6]$, $Na[Co(CO)_4]$, $[HTc(CO)_5]$. The hydrogen atoms are covalently bonded to the central metal atom in each case.

43. Vitamin $B_{12s}$ is 4.43% cobalt by mass. Determine the molar mass of vitamin $B_{12s}$ if each molecule of it contains one atom of cobalt.

44. Hemoglobin, the compound that transports oxygen in the bloodstream, has a molar mass of $6.8 \times 10^4 \ g \ mol^{-1}$. It is also 0.33% iron by mass. Determine how many iron atoms are in a single hemoglobin molecule.

## Additional Problems

45. Of the ten fourth-period transition metal elements in Table 18.1, which one has particularly low melting and boiling points? How can you explain this in terms of the electronic configuration of this element?

*46. Although copper lies between nickel and zinc in the periodic table, the reduction potential of $Cu^{2+}$ is above that for both $Ni^{2+}$ and $Zn^{2+}$ (see Table 18.1). Use other data from Table 18.1 to account for this observation. (*Hint:* Think of a multistep process to convert a metal atom in the solid to a metal ion in solution. For each step, compare the relevant energy or enthalpy changes for Cu with those for Ni or Zn.)

47. A reference book lists five different values for the electronegativity of molybdenum, a different value for each oxidation state from +2 through +6. Predict which electronegativity is highest and which is lowest.

48. If *trans*-$[Cr(en)_2(NCS)_2]SCN$ is heated, it forms gaseous ethylenediamine and solid $[Cr(en)_2(NCS)_2][Cr(en)(NCS)_4]$. Write a balanced chemical equation for this reaction. What are the oxidation states of the chromium ions in the reactant and in the two complex ions in the product?

49. A coordination complex has the molecular formula $[Ru_2(NH_3)_6Br_3](ClO_4)_2$. Determine the oxidation state of ruthenium in this complex.

50. Heating 2.0 mol of a coordination compound gives 1.0 mol $NH_3$, 2.0 mol $H_2O$, 1.0 mol HCl, and 1.0 mol $(NH_4)_3[Ir_2Cl_9]$. Write the formula of the original (six-coordinate) coordination compound and name it.

51. Explain why ligands are usually negative or neutral in charge and only rarely positive.

52. Match each compound in the group on the left with the compound on the right most likely to have the same electrical conductivity per mole in aqueous solution.
    (a) $[Fe(H_2O)_5Cl]CO_3$     HCN
    (b) $[Mn(H_2O)_6]Cl_3$     $Fe_2(SO_4)_3$
    (c) $[Zn(H_2O)_3(OH)]Cl$     NaCl
    (d) $[Fe(NH_3)_6]_2(CO_3)_3$     $MgSO_4$
    (e) $[Cr(NH_3)_3Br_3]$     $Na_3PO_4$
    (f) $K_3[Fe(CN)_6]$     $GaCl_3$

53. Three different compounds are known to have the empirical formula $CrCl_3 \cdot 6H_2O$. When exposed to a dehydrating agent, compound 1 (which is dark green) loses 2 mol of water per mole of compound, compound 2 (light green) loses 1 mol of water, and compound 3 (violet) loses no water. What are the probable structures of these compounds? If an excess of silver nitrate solution is added to 100.0 g of each of these compounds, what mass of silver chloride will precipitate in each case?

54. The octahedral structure is not the only possible six-coordinate structure. Other possibilities include a planar hexagonal structure and a triangular prism structure. In the latter, the ligands are arranged in two parallel triangles, one lying above the metal atom and the other below the metal atom with its corners directly in line with the corners of the first triangle. Show that the existence of two and only two isomers of $[Co(NH_3)_4Cl_2]^+$ is evidence against both of these possible structures.

55. Cobalt(II) forms more tetrahedral complexes than any other ion except zinc(II). Draw the structure(s) of the tetrahedral complex $[CoCl_2(en)]$. Could this complex exhibit geometrical or optical isomerism? If one of the $Cl^-$ ligands is replaced by $Br^-$, what kinds of isomerism, if any, are possible in the resulting compound?

56. Is the coordination compound $[Co(NH_3)_6]Cl_2$ diamagnetic or paramagnetic?

57. A coordination compound has the empirical formula $PtBr(en)(SCN)_2$ and is diamagnetic.
    (a) Examine the *d* electron configurations on the metal atoms, and explain why the formulation $[Pt(en)_2(SCN)_2]$ $[PtBr_2(SCN)_2]$ is preferred for this substance.
    (b) Name this compound.

58. We used crystal field theory to order the energy-level splittings induced in the five *d* orbitals. The same procedure could

be applied to $p$ orbitals. Predict the level splittings (if any) induced in the three $p$ orbitals by octahedral and square-planar crystal fields.

59. The three complex ions $[Mn(CN)_6]^{5-}$, $[Mn(CN)_6]^{4-}$, and $[Mn(CN)_6]^{3-}$ have all been synthesized, and all are low-spin octahedral complexes. For each complex, determine the oxidation number of Mn, the configuration of the $d$ electrons (how many $t_{2g}$ and how many $e_g$), and the number of unpaired electrons present.

60. Is the $Mn^{2+}(aq)$ ion a strong reducing agent? Is it a strong oxidizing agent? (Use Appendix E.) What can you conclude about the stability of $Mn^{2+}(aq)$?

61. The following ionic radii (in angstroms) are estimated for the $+2$ ions of selected elements of the first transition-metal series, based on the structures of their oxides: $Ca^{2+}(0.99)$, $Ti^{2+}(0.71)$, $V^{2+}(0.64)$, $Mn^{2+}(0.80)$, $Fe^{2+}(0.75)$, $Co^{2+}(0.72)$, $Ni^{2+}(0.69)$, $Cu^{2+}(0.71)$, $Zn^{2+}(0.74)$. Draw a graph of ionic radius versus atomic number in this series, and account for its shape. The oxides take the rock salt structure. Are these solids better described as high-spin or low-spin transition-metal complexes?

62. The coordination geometries of $[Mn(NCS)_4]^{2-}$ and $[Mn(NCS)_6]^{4-}$ are tetrahedral and octahedral, respectively. Explain why the two have the same room-temperature molar magnetic susceptibility.

63. The complex ion $CoCl_4^{2-}$ has a tetrahedral structure. How many $d$ electrons are on the Co? What is its electronic configuration? Why is the tetrahedral structure stable in this case?

64. In the coordination compound $(NH_4)_2[Fe(H_2O)F_5]$, the Fe is octahedrally coordinated.
    (a) Based on the fact that $F^-$ is a weak-field ligand, predict whether this compound is diamagnetic or paramagnetic. If it is paramagnetic, tell how many unpaired electrons it has.
    (b) By comparison with other complexes discussed in the chapter, discuss the likely color of this compound.
    (c) Determine the $d$ electron configuration of the iron in this compound.
    (d) Name this compound.

65. The compound $Cs_2[CuF_6]$ is bright orange and paramagnetic. Determine the oxidation number of copper in this compound, the most likely geometry of the coordination around the copper, and the possible configurations of the $d$ electrons of the copper.

66. In what directions do you expect the bond length and vibrational frequency of a free CO molecule to change when it becomes a CO ligand in a $Ni(CO)_4$ molecule? Explain your reasoning.

67. Give the number of valence electrons surrounding the central transition-metal ion in each of the following known organometallic compounds or complex ions: $[Co(C_5H_5)_2]^+$, $[Fe(C_5H_5)(CO)_2Cl]$, $[Mo(C_5H_5)_2Cl_2]$, $[Mn(C_5H_5)(C_6H_6)]$.

68. Molecular nitrogen ($N_2$) can act as a ligand in certain coordination complexes. Predict the structure of $[V(N_2)_6]$, which is isolated by condensing V with $N_2$ at 25 K. Is this compound diamagnetic or paramagnetic? What is the formula of the carbonyl compound of vanadium that has the same number of electrons?

69. How is the rule of 18, discussed in Section 18.6, related to the ligand field theory of Section 18.5? What levels are occupied in a complex such as hexacarbonylchromium(0)? Are any electrons placed into antibonding orbitals that are derived from the chromium $d$ orbitals?

* 70. The compound $WH_2(C_5H_5)_2$ acts as a base, but $TaH_3(C_5H_5)_2$ does not. Explain.

71. Discuss the role of transition-metal complexes in biology. Consider such aspects as their absorption of light, the existence of many different structures, and the possibility of multiple oxidation states.

## CUMULATIVE PROBLEMS

72. Predict the volume of hydrogen generated at 1.00 atm and 25°C by the reaction of 4.53 g of scandium with excess aqueous hydrochloric acid.

* 73. Mendeleev's early periodic table placed manganese and chlorine in the same group. Discuss the chemical evidence for these placements, focusing on the oxides of the two elements and their acid–base and redox properties. Is there a connection between the electronic structures of their atoms? In what ways are the elements different?

74. A 0.10 M solution of $Na_2Cr_2O_7$ is prepared, and the pH is adjusted to 7.00. Calculate the equilibrium concentration of $CrO_4^{2-}$ that results.

* 75. Methylamine ($NH_2CH_3$) is a monodentate ligand, whereas ethylenediamine ($H_2NCH_2CH_2NH_2$) is bidentate. For the two reactions

$$Cd^{2+}(aq) + 4 NH_2CH_3(aq) \longrightarrow Cd(NH_2CH_3)_4^{2+}(aq)$$

$$Cd^{2+} + 2 H_2NCH_2CH_2NH_2(aq) \longrightarrow Cd(en)_2^{2+}(aq)$$

the energy and enthalpy changes are quite similar, because the bonds formed in the complexes are much the same. Nevertheless, the formation constants for the two complexes differ by more than four orders of magnitude. Explain these facts by predicting which reaction will have the less negative entropy change. Which complex is more stable? This effect of chelation on the entropy change is sometimes referred to as the *chelate effect*.

76. An orange-yellow osmium carbonyl compound is heated to release CO and leave elemental osmium behind. Treatment of 6.79 g of the compound releases 1.18 L of CO(g) at 25°C and 2.00 atm pressure. What is the empirical formula of this compound? Propose a possible structure for it by comparing with the metal carbonyls shown in Figure 18.25.

# Structure and Bonding in Solids

A lthough the symmetry and beauty of crystals have always excited curiosity and wonder, the science of **crystallography** did not begin until the latter part of the 18th century. In those closing days of the Age of Enlightenment, when Lavoisier led the modern approach to chemistry, another brilliant French thinker established the fundamental laws of crystallography. Réné Just Haüy was struck by the fact that when he accidentally dropped a crystal of calcite (a form of calcium carbonate), it fractured into smaller crystals that had the same interfacial angles

*Illustration*
The structure of diamond consists of a network of interlocking tetrahedral carbon atoms. *(Charles D. Winters)*

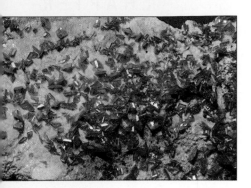

**FIGURE 19.1** Microcrystals of realgar, $As_4S_4$. *(Brian Parker/Tom Stack & Associates)*

between their plane surfaces as did the original crystal. The constancy of the interfacial angles observed when crystals are cleaved is known as Haüy's law. Haüy drew the conclusion that the outward symmetry of crystals (Fig. 19.1) implies an internal regularity and the existence of a smallest crystal unit. His inferences were correct; what distinguishes the crystalline state from the gaseous and liquid states is the nearly perfect positional order of the atoms, ions, or molecules in crystals.

This chapter begins with a look at the microscopic structure of a perfect crystal. We then examine the sources of cohesion in matter, identifying the forces that hold together different kinds of solids. More disordered condensed phases of matter—defective crystals and amorphous solids—are then considered. The chapter closes with a look at liquid crystals, states of matter that are intermediate between solids and liquids in their physical properties.

---

## 19.1

### CRYSTAL SYMMETRY AND THE UNIT CELL

Crystals can be classified according to the natures of the bonds between the atoms, ions, or molecules that compose them, and such classification will later prove useful (Section 19.3). A more fundamental classification of crystals employs the numbers and kinds of their **symmetry elements.** When the result of rotation, reflection, or inversion of an object can be exactly superimposed on (matched point for point to) the original object, the structure is said to contain the corresponding symmetry element—an axis of rotation, a plane of reflection (mirror plane), or a central point (inversion center), as shown in Figure 19.2. These symmetry operations can be applied to geometrical shapes, to physical objects, and to molecular structures.

Consider a cube as an example. Suppose the center of the cube is placed at the origin of its coordinate system and the symmetry operations that transform it into identity with itself are counted (Fig. 19.3). The $x$, $y$, and $z$ coordinate axes are 4-fold axes of rotational symmetry, denoted by $C_4$, because a cube that is rotated

**FIGURE 19.2** Three types of symmetry operation: rotation about an $n$-fold axis, reflection in a plane, and inversion through a point.

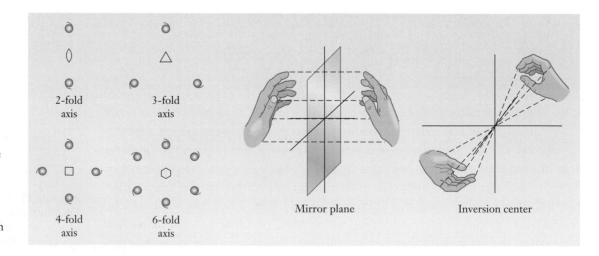

2-fold axis    3-fold axis    4-fold axis    6-fold axis    Mirror plane    Inversion center

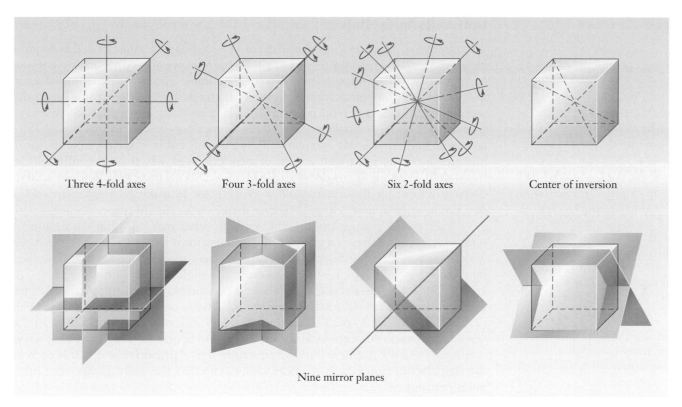

Three 4-fold axes        Four 3-fold axes        Six 2-fold axes        Center of inversion

Nine mirror planes

**FIGURE 19.3**   Symmetry operations acting on a cube.

through a multiple of 90° (= 360°/4) about any one of these axes is indistinguishable from the original cube. Similarly, a cube has four 3-fold axes of rotational symmetry, designated $C_3$, that are the body diagonals of the cube, connecting opposite vertices. In addition, a cube has six 2-fold rotational axes of symmetry, defined by the six axes that pass through the centers of edges and through the coordinate origin. Next, the cube has nine mirror planes of symmetry (designated by the symbol *m*), which reflect any point in one half of the cube into an equivalent point in the other half. Finally, a cube has a center of inversion (reflection through a point, designated *i*).

## EXAMPLE 19.1

Identify the symmetry elements of the ammonia molecule ($NH_3$).

### Solution

If the ammonia molecule is drawn as a pyramid with the nitrogen atom at the top (Fig. 19.4), then the only axis of rotational symmetry is a 3-fold axis passing downward through the N atom. Three mirror planes intersect at this 3-fold axis.

**Related Problems: 3, 4**

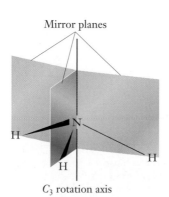

Mirror planes

$C_3$ rotation axis

**FIGURE 19.4**   An ammonia molecule has a 3-fold axis of rotation and three mirror planes of symmetry.

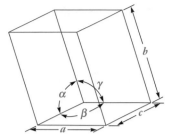

**FIGURE 19.5**  Unit cells always have three pairs of mutually parallel faces. Only six pieces of information are required to construct a scale model of a unit cell: the three cell edges ($a$, $b$, and $c$) and the three angles between the edges ($\alpha$, $\beta$, and $\gamma$). By convention, $\gamma$ is the angle between edges $a$ and $b$, $\alpha$ the angle between $b$ and $c$, and $\beta$ the angle between $a$ and $c$.

## Unit Cells in Crystals

The symmetry operations just described can be applied to crystals as well as to individual molecules or shapes. Identical sites within a crystal recur regularly because of long-range order in the organization of the atoms. The three-dimensional array made up of all the points within a crystal that have the same environment in the same orientation is a **crystal lattice.** Such a lattice is an abstraction "lifted away" from a real crystal, embodying the scheme of repetition at work in that crystal. The lattice of highest possible symmetry is that of the **cubic system.** This lattice is obtained by filling space with a series of identical cubes, which are the **unit cells** in the system. A single unit cell contains all structural information about its crystal, because in principle the crystal could be constructed by making a great many copies of a single original unit cell and stacking them. Unit cells fill space.

Other crystal systems can be achieved with other unit cells, but constraints of symmetry permit only seven types of three-dimensional lattices. Each has a unit cell with the shape of a parallelepiped (Fig. 19.5), whose size and shape are fully described by three edge lengths ($a$, $b$, and $c$) and the three angles between those edges ($\alpha$, $\beta$, and $\gamma$). These lengths and angles are the **cell constants.** The symmetry that defines each of the seven crystal systems imposes conditions on the shape of the unit cell (Table 19.1 and Fig. 19.6). The unit cell chosen is the smallest unit that has all the symmetry elements of the crystal lattice; there is no benefit in using large cells when small ones will do. A unit cell of the minimum size is **primitive** and shares each of the eight **lattice points** at its corners with seven other unit cells, giving one lattice point per unit cell.

Other cells with the same volume (an infinite number, in fact) could be constructed, and each could generate the macroscopic crystal by repeated elementary translations, but only those shown in Figure 19.6 possess the symmetry elements of their crystal systems. Figure 19.7 illustrates a few of the infinite number of cells that can be constructed for a two-dimensional rectangular lattice. Only the rectangular cell B in the figure has three 2-fold rotation axes and two mirror planes. Although the other cells all have the same area, each of them has only one 2-fold axis and no mirror planes; they are therefore not acceptable unit cells.

**TABLE 19.1**

### *The Seven Crystal Systems*

| Crystal System | Minimum Essential Symmetry | Conditions on Unit-Cell Edges and Angles |
|---|---|---|
| Hexagonal | One 6-fold rotation | $a = b$; $\alpha = \beta = 90°$, $\gamma = 120°$ |
| Cubic | Four independent 3-fold rotations† | $a = b = c$; $\alpha = \beta = \gamma = 90°$ |
| Tetragonal | One 4-fold rotation | $a = b$; $\alpha = \beta = \gamma = 90°$ |
| Trigonal | One 3-fold rotation | $a = b = c$; $\alpha = \beta = \gamma \neq 90°$ |
| Orthorhombic | Three mutually perpendicular 2-fold rotations | $\alpha = \beta = \gamma = 90°$ |
| Monoclinic | One 2-fold rotation | $\alpha = \gamma = 90°$ |
| Triclinic | No symmetry required | None |

† Each of these axes makes 70.53° angles with the other three.

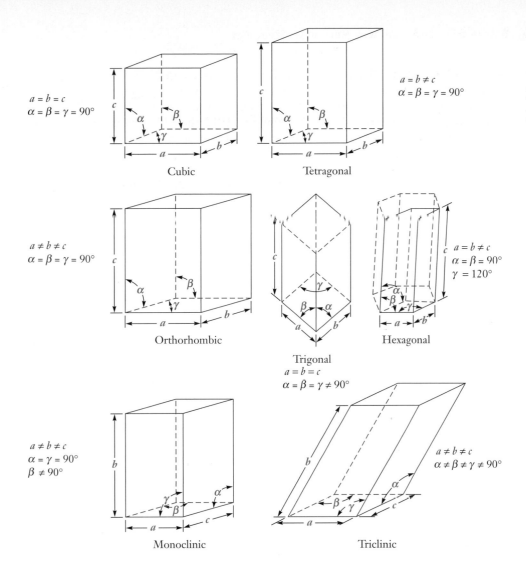

FIGURE 19.6 Shapes of the unit cells in the seven crystal systems. The more symmetric crystal systems have more symmetric cells.

Cubic

$a = b = c$
$\alpha = \beta = \gamma = 90°$

Tetragonal

$a = b \neq c$
$\alpha = \beta = \gamma = 90°$

Orthorhombic

$a \neq b \neq c$
$\alpha = \beta = \gamma = 90°$

Trigonal
$a = b = c$
$\alpha = \beta = \gamma \neq 90°$

Hexagonal

$a = b \neq c$
$\alpha = \beta = 90°$
$\gamma = 120°$

Monoclinic

$a \neq b \neq c$
$\alpha = \gamma = 90°$
$\beta \neq 90°$

Triclinic

$a \neq b \neq c$
$\alpha \neq \beta \neq \gamma \neq 90°$

Sometimes the smallest, or primitive, unit cell does not have the full symmetry of the crystal lattice. If so, a larger *nonprimitive* unit cell that does have the characteristic symmetry is deliberately chosen (Fig. 19.8). Only three types of nonprimitive cells are commonly used in the description of crystals: **body-centered, face-centered,** and **side-centered.** They are shown in Figure 19.9.

## Scattering of X-Rays by Crystals

In the 19th century, crystallographers could classify crystals into the seven crystal systems only on the basis of their external symmetries. They could not measure the dimensions of unit cells or the positions of atoms within them. Wilhelm Roentgen's discovery of x-rays in 1895 provided a tool of enormous power for determining the

FIGURE 19.7 Cells with and without mirror planes in a two-dimensional lattice. Each cell has the same area, but only cell B possesses mirror planes in addition to the 2-fold rotation axes required by the lattice.

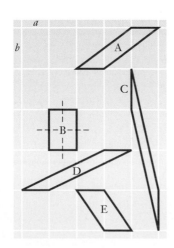

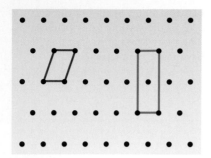

**FIGURE 19.8** In this two-dimensional lattice, every lattice point is at the intersection of a horizontal mirror line and a vertical mirror line. It is possible to draw a primitive unit cell (red), but the larger, centered unit cell (blue) is preferred because it also has two mirror lines, the full symmetry of the lattice. A similar argument applies to the choice of unit cells on three-dimensional lattices.

structures of crystals. Max von Laue made the original suggestion that crystals might serve as three-dimensional gratings for the diffraction of electromagnetic radiation with a wavelength comparable to the distance between planes of atoms. Friedrich and Knipping demonstrated experimentally in 1912 that this was indeed the case, and von Laue was awarded the Nobel prize in physics in 1914 for his theory of the diffraction of x-rays by crystals. At about the same time, W. H. Bragg and W. L. Bragg (father and son) at Cambridge University also demonstrated the diffraction of x-rays by crystals and shared the Nobel prize in physics the following year. (W. L. Bragg was 22 and still a student at Cambridge when he discovered the diffraction law.) The formulation of the diffraction law proposed by the Braggs is equivalent to von Laue's suggestion and somewhat simpler to visualize. We will follow an approach similar to theirs.

When electromagnetic radiation passes through matter, it interacts with the electrons in atoms, and some of it is scattered in all directions. Suppose that x-radiation strikes two neighboring scattering centers. The expanding spheres of scattered waves soon encounter each other and interfere. In some directions, the waves are in phase and reinforce each other, or interfere *constructively* (Fig. 19.10a); in others they are out of phase and cancel each other out, or interfere *destructively* (Fig. 19.10b). Constructive interference occurs when the paths traversed by two waves differ in length by a whole number of wavelengths. The amplitudes of waves that interfere constructively add to one another, and the intensity of the scattered radiation in that direction is proportional to the square of the total amplitude.

Figure 19.11 illustrates the constructive interference of x-rays scattered by the electrons in atoms in equally spaced planes separated by the distance *d*. A parallel bundle of coherent x-rays of a single known wavelength is allowed to fall on the

no central atom —

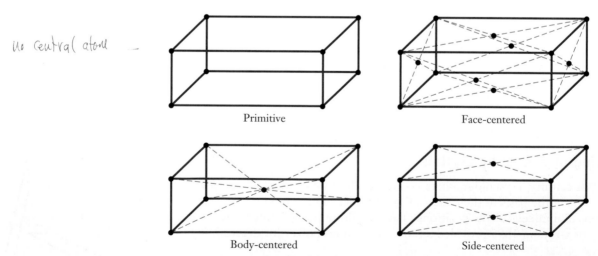

Primitive

Face-centered

Body-centered

Side-centered

**FIGURE 19.9** Centered lattices, like all lattices, have lattice points at the eight corners of the unit cell. A body-centered lattice has an additional lattice point at the center of the cell, a face-centered lattice has additional points at the centers of the six faces, and a side-centered lattice has points at the centers of two parallel sides of the unit cell. (*Note:* The dots in these diagrams are not atoms but lattice points.)

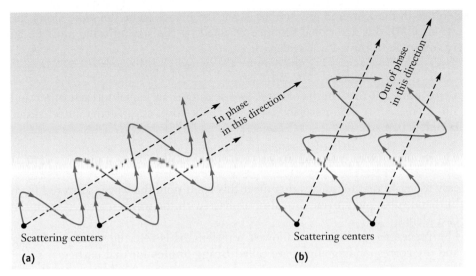

Scattering centers

(a)

Scattering centers

(b)

**FIGURE 19.10**   A beam of x-rays (not shown) is striking two scattering centers, which emit scattered radiation. The difference in the lengths of the paths followed by the scattered waves determines whether they interfere (a) constructively or (b) destructively. This path difference depends on both the distance between the centers and the direction in which the scattered waves are moving.

surface of a crystal, making an angle $\theta$ with a set of parallel planes of atoms in the crystal. The scattering angle $2\theta$ is then varied by rotating the crystal about an axis perpendicular to the plane of the figure. Line AD in Figure 19.11 represents a wave front of waves that are in phase as they approach the crystal. The wave that is scattered at B follows the path ABC, and the one that is scattered at F follows the path DFH. The second wave travels a greater distance than the first, and the difference in path length is the sum of the two segments EF and FG. To achieve constructive

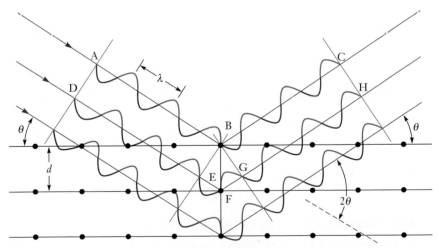

**FIGURE 19.11**   Constructive interference of x-rays scattered by atoms in lattice planes. Three beams of x-rays, scattered by atoms in three successive layers of a simple cubic crystal, are shown. Note that the phases of the waves are the same along the line CH, indicating constructive interference at this scattering angle $2\theta$.

interference in the scattered waves (that is, for the phases to be the same along the wave front CH), this additional distance traveled by the second wave must be an integral multiple of the x-ray wavelength $\lambda$:

$$EF + FG = n\lambda \qquad n = 1, 2, 3, \ldots$$

From trigonometry, the lengths of these two segments are equal to each other and to $d \sin \theta$, where $d$ is the interplanar spacing. Therefore, constructive interference occurs only when

$$n\lambda = 2d \sin \theta \qquad n = 1, 2, 3, \ldots \qquad \text{[19.1]}$$

It is easy to verify that for angles that meet this condition, the waves scattered from the third and subsequent planes are also in phase with the waves scattered from the first two planes.

The preceding condition on allowed wavelengths is referred to as the **Bragg law,** and the corresponding angles are called **Bragg angles** for that particular set of parallel planes of atoms. It appears as though the beam of x-rays has been reflected symmetrically from those crystal planes, and we often speak colloquially of the "Bragg reflection" of x-rays. The x-rays have not been reflected, however, but have undergone constructive interference, more commonly called diffraction. The case $n = 1$ is called first-order Bragg diffraction, $n = 2$ is second-order, and so forth.

We now possess a tool of immense value for determining the interplanar spacings of crystals. If a crystal is turned through different directions, other parallel sets of planes with different separations are brought into the Bragg condition. The symmetry of the resulting diffraction pattern identifies the crystal system, and the Bragg angles determine the cell constants. Moreover, the *intensities* of the diffracted beams permit the locations of the atoms in the unit cell to be determined.

Analogous scattering techniques use beams of neutrons. Here the scattering interaction is between the magnetic moments of neutrons and nuclei, but the principles are the same as for x-ray diffraction. Of course, it is the wave character of neutrons (in particular, their de Broglie wavelength; see Chapter 15) that is relevant for neutron diffraction. Recall that the de Broglie wavelength and neutron momentum are related by $\lambda = h/p$.

---

### EXAMPLE 19.2

A diffraction pattern of aluminum is obtained by using x-rays with wavelength $\lambda = 0.709$ Å. The second-order Bragg diffraction from the parallel faces of the cubic unit cells is observed at the angle $2\theta = 20.2°$. Calculate the lattice parameter $a$.

**Solution**
From the Bragg condition for $n = 2$,

$$2\lambda = 2d \sin \theta$$

the spacing between planes, which is the lattice parameter, is

$$d = \frac{\lambda}{\sin \theta} = \frac{0.709 \text{ Å}}{\sin (10.1°)} = 4.04 \text{ Å} = a$$

**Related Problems: 5, 6**

## 19.2

## CRYSTAL STRUCTURE

Certain elements crystallize in particularly simple solid structures, in which an atom is situated at each point of the lattice. Polonium is the only element known to crystallize in the **simple cubic lattice,** with its atoms at the intersections of three sets of equally spaced planes that meet at right angles. Each unit cell contains one Po atom, separated from each of its six nearest neighbors by 3.35 Å.

The alkali metals crystallize in the **body-centered cubic (b.c.c.) structure** at atmospheric pressure (Fig. 19.12). A unit cell of this structure contains two lattice points, one at the center of the cube and the other at any one of the eight corners. A single alkali-metal atom is associated with each lattice point. An alternative way to visualize this is to realize that each of the eight atoms that lie at the corners of a b.c.c. unit cell is shared by the eight unit cells that meet at those corners. The contribution of the atoms to one unit cell is therefore $8 \times \frac{1}{8} = 1$ atom, to which is added the atom that lies wholly within that cell at its center. *∞ B.c.c = 2 atoms*

The metals aluminum, nickel, copper, and silver, among others, crystallize in the **face-centered cubic (f.c.c.) structure** shown in Figure 19.13. This unit cell contains four lattice points, with a single atom associated with each point. No atom lies wholly within the unit cell; there are atoms at the centers of its six faces, each of which is shared with another cell (contributing $6 \times \frac{1}{2} = 3$ atoms), and an atom at each corner of the cell (contributing $8 \times \frac{1}{8} = 1$ atom), for a total of four atoms per unit cell.   *F.CC. = 4 atoms*

The volume of a unit cell is given by the formula

$$V_c = abc \sqrt{1 - \cos^2\alpha - \cos^2\beta - \cos^2\gamma + 2 \cos\alpha \cos\beta \cos\gamma} \qquad \text{[19.2]}$$

When the angles are all 90° (so that their cosines are 0), this formula reduces to the simple result $V = abc$ for the volume of a rectangular box. If the mass of the unit cell contents is known, the theoretical cell density can be computed. This density must come close to the measured density of the crystal, a quantity that can be determined by entirely independent experiments. For an *element* whose crystal contains $n_c$ atoms per unit cell, the calculated cell density is

$$\text{density} = \rho = \frac{\text{mass}}{\text{volume}} = \frac{n_c \dfrac{\mathcal{M}}{N_0}}{V_c} = \frac{n_c \mathcal{M}}{N_0 V_c}$$

where we have used the fact that the molar mass $\mathcal{M}$ of the element divided by Avogadro's number $N_0$ is the mass of a single atom. This equation can also be used to calculate Avogadro's number from the measured density and cell constants, as the following example illustrates.

### EXAMPLE 19.3

Sodium has a density of $\rho = 0.9700$ g cm$^{-3}$ at 20°C, and its lattice parameter is $a = 4.2856$ Å. What is the value of Avogadro's number, given that the molar mass of sodium is 22.9898 g mol$^{-1}$?

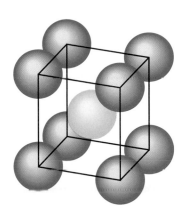

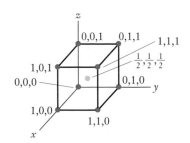

**FIGURE 19.12** The b.c.c. structure. An atom is located at the center of each cubic cell (yellow) as well as at each corner of the cube (blue). The atoms are reduced slightly in size to make their positions clear.

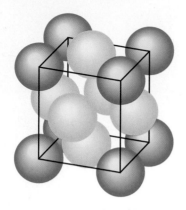

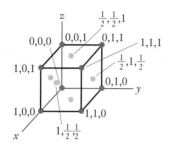

**FIGURE 19.13** The f.c.c. structure. Atoms are located at the centers of the faces (yellow) as well as at the corners of the cube (blue). The atoms are reduced slightly in size to make their positions clear.

## Solution

Sodium has two atoms per unit cell because its structure is b.c.c., and the volume of the unit cell is $a^3$. Solve the foregoing equation for Avogadro's number:

$$N_0 = \frac{n_c \mathcal{M}}{\rho a^3} = \frac{(2)(22.9898 \text{ g mol}^{-1})}{(0.9700 \text{ g cm}^{-3})(4.2856 \times 10^{-8} \text{ cm})^3} = 6.022 \times 10^{23} \text{ mol}^{-1}$$

**Related Problems: 15, 16**

It is useful to define the locations of atoms in the unit cell with a set of three numbers measured in units of the lattice parameter(s). For this purpose, one corner of the unit cell is taken to be at the origin of the coordinate axes appropriate to the crystal system, and an atom at that lattice point has the coordinates (0, 0, 0). Equivalent atoms at the seven remaining corners of the cell then have the coordinates (1, 0, 0), (0, 1, 0), (0, 0, 1), (1, 0, 1), (1, 1, 0), (0, 1, 1), and (1, 1, 1). These atom positions are generated from an atom at (0, 0, 0) by successive translations through unit distances along the three axes. An atom in a body-centered site has coordinates $(\frac{1}{2}, \frac{1}{2}, \frac{1}{2})$, and the prescription for locating it is to proceed from the coordinate origin at (0, 0, 0) a distance $a/2$ along $a$, then a distance $b/2$ along $b$, and finally $c/2$ along $c$. In the same way, an atom in a face-centered site has coordinates such as $(\frac{1}{2}, \frac{1}{2}, 0)$, $(\frac{1}{2}, 0, \frac{1}{2})$, or $(0, \frac{1}{2}, \frac{1}{2})$.

So far, we have considered only cubic metals in which one atom corresponds to each lattice point. More complicated structures also occur, even for the elements, in which atoms occupy positions that are not lattice points. Diamond, for example, has an f.c.c. structure with eight atoms (not four) per unit cell. Boron has a tetragonal structure with a very complex unit cell that contains 50 atoms. In molecular crystals, the number of atoms per unit cell can be still greater; for example, in a protein with a molecular mass of $10^5$, there may be tens of thousands of atoms per unit cell. The protein shown in Figure 24.9 is an example of a complex structure revealed by modern x-ray crystallography.

## Atomic Packing in Crystals

We have already discussed two measures of the "size" of an atom or molecule. In Section 4.7, the van der Waals parameter $b$ was related to the volume excluded per mole of molecules, so that $b/N_0$ is one measure of molecular size. In Section 15.8, an approximate radius of an atom was obtained by using the distance at which the electron density had fallen off to a particular value, or the distance at which a certain fraction of the total electron density lay within a sphere of that radius. A third, related, measure of atomic size is based on the interatomic separations in a crystal. The radius of a noble-gas or metallic atom, for example, can be approximated as half the distance between the center of an atom and the center of its nearest neighbor in the crystal. Crystal structures are pictured as resulting from the packing of spheres in which nearest neighbors are in contact.

The nearest neighbor separation in a simple cubic crystal is equal to the lattice parameter $a$, and so the atomic radius in that case is $a/2$. For the b.c.c. lattice, the central atom in the unit cell "touches" each of the eight atoms at the corners of the cube, but those at the corners do not touch one another, as Figure 19.12 shows. The nearest neighbor distance is calculated from the Cartesian coordinates of the atom at the origin (0, 0, 0) and that at the cell center ($a/2$, $a/2$, $a/2$). By the Pythagorean

## TABLE 19.2

### Structural Properties of Cubic Lattices

| | Simple Cubic | Body-Centered Cubic | Face-Centered Cubic |
|---|---|---|---|
| Lattice points per cell | 1 | 2 | 4 |
| Number of nearest neighbors | 6 | 8 | 12 |
| Nearest-neighbor distance | $a$ | $a\sqrt{3}/2 = 0.866a$ | $a\sqrt{2}/2 = 0.707a$ |
| Atomic radius | $a/2$ | $a\sqrt{3}/4 = 0.433a$ | $a\sqrt{2}/4 = 0.354a$ |
| Packing fraction | $\dfrac{\pi}{6} = 0.524$ | $\dfrac{\sqrt{3}\,\pi}{8} = 0.680$ | $\dfrac{\sqrt{2}\,\pi}{6} = 0.740$ |

theorem, the distance between these points is $\sqrt{(a/2)^2 + (a/2)^2 + (a/2)^2} = a\sqrt{3}/2$, so that the atomic radius is $a\sqrt{3}/4$. Figure 19.13 shows that in an f.c.c. crystal the atom at the center of a face [such as at $(0, a/2, a/2)$] touches each of the neighboring corner atoms [such as at $(0, 0, 0)$], so that the nearest-neighbor distance is $\sqrt{(a/2)^2 + (a/2)^2} = a\sqrt{2}/2$ and the atomic radius is $a\sqrt{2}/4$. Table 19.2 summarizes the results for cubic lattices.

What is the densest crystal packing that can be achieved? To answer this question, construct a crystal by first putting down a plane of atoms with the highest possible density, shown in Figure 19.14a. Each sphere is in contact with six other spheres in the plane. Then put down a second close-packed plane on top of the first one (Fig. 19.14b) in such a way that each sphere in the second plane is in contact with three spheres in the plane below it; that is, each sphere in the second plane forms a tetrahedron with three spheres beneath. This can be accomplished by placing the second plane of spheres above the sites marked $b$ in Figure 19.14a. When the third plane is laid down, there are two possibilities. The spheres can be placed in the depressions marked $c$ or in those marked $a$, which lie directly above the spheres in the first layer.

Clearly, there are two choices for placing each plane, and an infinite number of crystal structures can be generated that have the same atomic packing density. The

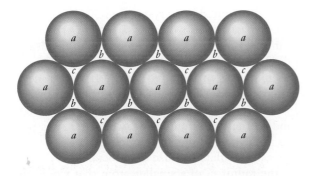

(a)

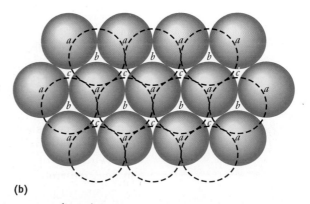

(b)

**FIGURE 19.14** Close-packing of spheres. (a) One layer, with atoms centered on sites labeled $a$. (b) Two layers, with the atoms of the second layer centered on sites labeled $b$. The third layer can be placed on the sites labeled $c$ (giving cubic close-packing) or over those marked $a$ (giving hexagonal close-packing).

two simplest such structures correspond to the periodic layer sequences *abcabcab* ... and *abababa*. ... The first of these is the face-centered cubic structure already discussed, and the second is a close-packed structure in the hexagonal crystal system termed **hexagonal close-packed (h.c.p.).** In each of these simple structures, atoms occupy 74.0% of the unit cell volume, as the following example shows. (Atoms that crystallize in the b.c.c. structure occupy only 68.0% of the crystal volume, and the packing fraction for a simple cubic array is only 52.4%.)

## EXAMPLE 19.4

Calculate **(a)** the atomic radius of an aluminum atom and **(b)** the fraction of the volume of aluminum that is occupied by its atoms.

### Solution
**(a)** Aluminum crystallizes in the f.c.c. crystal system, and its unit cell therefore contains four atoms. Because a face-centered atom touches each of the atoms at the corners of its face, the atomic radius $r_1$ (see the triangle in Fig. 19.15) can be expressed by

$$4r_1 = a\sqrt{2}$$

Using the value $a = 4.04$ Å derived from x-ray diffraction (Example 19.2) and solving for $r_1$ give

$$r_1 = 1.43 \text{ Å}$$

**(b)** The ratio of the volume occupied by four atoms to the volume of a unit cell is

$$f = \frac{4\left[\frac{4}{3}\pi r_1^3\right]}{a^3} = \frac{4\frac{4}{3}\pi\left[\frac{a\sqrt{2}}{4}\right]^3}{a^3} = 0.740$$

**Related Problem: 23**

## Interstitial Sites

The ways in which the empty volume is distributed in a crystal are both interesting and important. For the close-packed f.c.c. structure, two types of **interstitial sites,** upon which the free volume in the unit cell is centered, are identifiable. An **octahedral site** is surrounded at equal distances by six nearest neighbor atoms. Figure 19.15 shows that such sites lie at the midpoints of the edges of the f.c.c. unit cell. A cell has 12 edges, each of which is shared by four unit cells, so that the edges contribute three octahedral interstitial sites per cell. In addition, the site at the center of the unit cell is also octahedral, and so the total number of octahedral sites per f.c.c. unit cell is four, the same as the number of atoms in the unit cell.

With a bit of simple geometry we can calculate the size of an octahedral site in an f.c.c. structure or, more precisely, the radius $r_2$ of a smaller atom that would fit in the site without overlapping its neighboring atoms. Figure 19.15 represents a cell

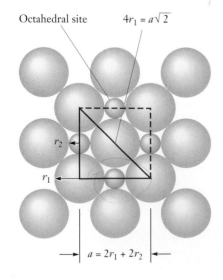

Octahedral site     $4r_1 = a\sqrt{2}$

$r_2$

$r_1$

$a = 2r_1 + 2r_2$

**FIGURE 19.15**   Octahedral sites in an f.c.c. lattice. The geometric procedure for relating the site radius $r_2$ to the atom radius $r_1$ is shown.

face in which the length of the diagonal is $4r_1$ and the length of the cell edge is $2r_1 + 2r_2$, where $r_1$ is the radius of the host atoms and $r_2$ is the radius of the octahedral site. From the figure,

$$r_1 = a \frac{\sqrt{2}}{4}$$

$$a = 2r_2 + 2r_1 = 2r_2 + 2a \frac{\sqrt{2}}{4}$$

$$r_2 = \frac{a}{2} - a \frac{\sqrt{2}}{4} = 0.146a$$

and the ratio of the octahedral-site radius to the host-atom radius is

$$\frac{r_2}{r_1} = \frac{0.146a}{a\sqrt{2}/4} = 0.414$$

The second type of interstitial site, known as a **tetrahedral site,** lies at the center of the space defined by four touching spheres. In an f.c.c. cell, a tetrahedral site occurs in the volume between a corner atom and the three face-centered atoms nearest to it. Geometrical reasoning like that used for the octahedral site gives the following ratio of the radius of a tetrahedral site to that of a host atom:

$$\frac{r_2}{r_1} = 0.225$$

The f.c.c. unit cell contains eight tetrahedral sites, twice the number of atoms in the cell.

Interstitial sites are important when a crystal contains atoms of several kinds with considerably different radii. We will return to this shortly when we consider the structures of ionic crystals.

---

## 19.3

### COHESION IN SOLIDS

At the beginning of this chapter, we stated that the nature of the bonding forces between atoms might provide a useful way to classify solids. Such a classification does indeed lead to an understanding of the remarkable differences in the chemical and physical properties of different materials. For this purpose, we now consider crystals with bonding interactions classified as primarily molecular, ionic, metallic, or covalent.

### Molecular Crystals

Molecular crystals include the noble gases; oxygen; nitrogen; the halogens; compounds such as carbon dioxide; metal halides of low ionicity such as $Al_2Cl_6$, $FeCl_3$, and $BiCl_3$; and the vast majority of organic compounds. All these molecules are held in their lattice sites by the intermolecular forces discussed in Sections 4.7 and 5.2. The trade-off between attractive and repulsive forces among even small molecules in a molecular crystal is quite complex because so many atoms are involved. A useful simplification is to picture a molecule as a set of fused spheres centered at each nucleus. The radius of each sphere is the van der Waals radius of the element

**FIGURE 19.16** The van der Waals radii of the carbon and nitrogen atoms superimposed on an outline of the molecular structure of cyanuric triazide, $C_3N_{12}$, to show the volume of space from which each molecule excludes the others. Van der Waals forces in the molecular crystal hold the molecules in contact in a pattern that minimizes empty space. The thin black lines emphasize the 3-fold symmetry of the pattern.

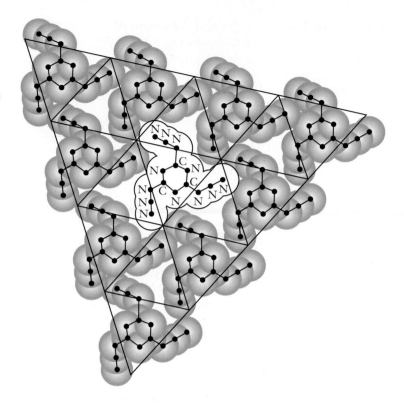

involved. In molecular crystals, such shapes pack together so that no molecules overlap but empty space is minimized. Figure 19.16 depicts such a "space-filling model" of cyanuric triazide ($C_3N_{12}$), showing how nature solves the problem of efficiently packing many copies of the rather complicated molecular shape of $C_3N_{12}$ in a single layer. In the three-dimensional molecular crystal of $C_3N_{12}$, many such layers stack up with a slight offset that minimizes unfilled space between layers.

Van der Waals forces are much weaker than the forces that operate in ionic, metallic, and covalent crystals. Consequently, molecular crystals typically have low melting points and are soft and easily deformed. Although at atmospheric pressure the noble-gas elements crystallize in the highly symmetric face-centered cubic lattice shown in Figure 19.13, molecules (especially those with complex geometries) more often form crystals of low symmetry in the monoclinic or triclinic systems. Few display the striking collective physical properties of electrical conductivity and magnetism that are characteristic of metals. The interesting properties of these crystals are primarily the properties of their molecular units.

Molecular crystals are of great scientific value. If proteins and other macromolecules are obtained in the crystalline state, their structures can be determined by x-ray diffraction. Knowing the three-dimensional structures of biological molecules is the starting point for understanding their functions.

## Ionic Crystals

Compounds formed by atoms with significantly different electronegativities are largely ionic, and to a first approximation the ions can be treated as hard, charged spheres that occupy positions on the crystal lattice (see the ionic radii in Appendix

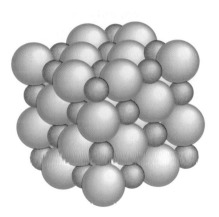

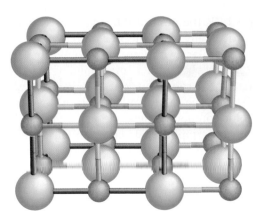

**FIGURE 19.17** The sodium chloride, or rock-salt, structure. On the left, the sizes of the Na$^+$ and Cl$^-$ ions are shown to scale. On the right, the crystal lattice is face-centered cubic, as can be seen from the fact that translating each atom by half a unit cell (shown by red lines) in the direction of the face diagonal superimposes identical atoms. There are eight ions (4 Na$^+$ and 4 Cl$^-$) per unit cell.

F). All the elements of Groups I and II of the periodic table react with Group VI and VII elements to form ionic compounds, the great majority of which crystallize in the cubic system. Specifically, the alkali-metal halides (except for the cesium halides), the ammonium halides, and the oxides and sulfides of the alkaline-earth metals all crystallize in the **rock-salt,** or **sodium chloride, structure** shown in Figure 19.17. It may be viewed as an f.c.c. lattice of anions whose octahedral sites are all occupied by cations or, equivalently, as an f.c.c. lattice of cations whose octahedral sites are all occupied by anions. Either way, each ion is surrounded by six equidistant ions of the opposite charge. The rock-salt structure is a stable crystal structure when the cation–anion radius ratio lies between 0.414 and 0.732, if cations and anions are assumed to behave as incompressible charged spheres.

When the hard-sphere cation–anion radius ratio exceeds 0.732, as it does for the cesium halides, a different crystal structure called the **cesium chloride structure,** is more stable. It may be looked on as two interpenetrating simple cubic lattices, one of anions and the other of cations, as shown in Figure 19.18. When the cation–anion radius is less than 0.414, the **zinc blende,** or **sphalerite, structure** (named after the structure of ZnS) results. This crystal consists of a f.c.c. lattice of S$^{2-}$ ions, with Zn$^{2+}$ ions occupying half of the available tetrahedral sites in alternation, as Figure 19.19a illustrates. **Fluorite** (CaF$_2$) has yet another structure; the unit cell is based on an f.c.c. lattice of Ca$^{2+}$ ions. The F$^-$ ions occupy all eight of the tetrahedral sites, and so the unit cell contains four Ca$^{2+}$ and eight F$^-$ ions (Fig. 19.19b). The radius ratios (0.414 and 0.732) at which crossovers from one type of crystal to another occur are not accidental numbers. Recall from Section 19.2 that

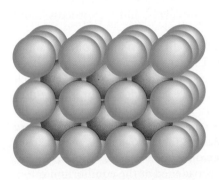

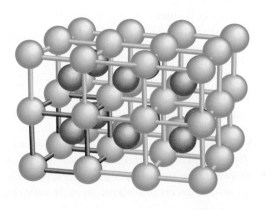

**FIGURE 19.18** The cesium chloride structure. This crystal lattice is simple cubic, with one Cs$^+$ and one Cl$^-$ ion per unit cell.

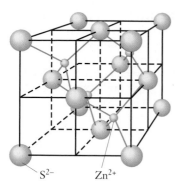

Sphalerite (ZnS)

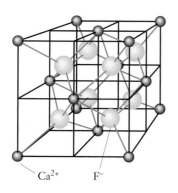

Fluorite (CaF$_2$)

**FIGURE 19.19** Two ionic lattices in the f.c.c. system. A single (nonprimitive) cubic unit cell of each is shown.

0.414 is the ratio of the octahedral-site radius to the host-atom radius for an f.c.c. lattice; only when this size ratio is exceeded does the ion inserted into that site come into contact with ions of the opposite sign in the rock-salt structure. The number 0.732 comes from a corresponding calculation of the radius ratio of the interstitial site at the center of a simple cubic unit cell (see problem 25). It is important to realize that the radius-ratio criterion for the stability limits of the structures of binary ionic compounds depends on the ions' being incompressible and the wave functions' not overlapping. The criterion fails when these approximations are not met.

The strength and range of the electrostatic attractions make ionic crystals hard, high-melting, brittle solids that are electrical insulators. Melting an ionic crystal, however, disrupts the lattice and sets the ions free to move, and so ionic liquids are good electrical conductors.

## Metallic Crystals

The characteristic property of metals is their good ability to conduct electricity and heat. Both phenomena are due to the ease with which valence electrons move; electrical conduction is a result of the flow of electrons from regions of high potential energy to those of low potential energy, and heat conduction is a result of the flow of electrons from high-temperature regions (where their kinetic energies are high) to low-temperature regions (where their kinetic energies are low). Why are electrons so mobile in a metal but so tightly bound to atoms in an insulating solid, such as diamond or sodium chloride?

The answer lies in the type of bonding found in metals, which is quite different from that in other crystals. The valence electrons in a metal are delocalized in huge molecular orbitals that extend over the entire crystal. To understand the origin of these molecular orbitals, suppose just two sodium atoms are brought together, each in its electronic ground state with the configuration $1s^2 2s^2 2p^6 3s^1$. As the atoms approach each other, the wave functions of their $3s$ electrons combine to form two molecular orbitals—one in which their phases are symmetric ($\sigma_{3s}$) and another in which they are antisymmetric ($\sigma_{3s}^*$). Solving the Schrödinger equation yields two energy states, one above and the other below the energy of the atomic $3s$ levels. (This is analogous to the formation of a hydrogen molecule from two hydrogen atoms, considered in Chapter 16.) If both valence electrons are put into the level of lower energy with spins opposed, the result is an Na$_2$ molecule. If a third sodium atom is added, the $3s$ atomic levels of the atoms split into three sublevels (Fig. 19.20). Two electrons occupy the lowest level with their spins opposed, and the third electron occupies the middle level. The three energy levels and the three electrons belong collectively to the three sodium atoms. A fourth sodium atom could be added so that there would be four closely spaced energy sublevels, and this process could be carried on without limit.

The foregoing is not a mere "thought experiment." Sodium vapor contains about 17% Na$_2$ molecules at its normal boiling point. Larger sodium clusters have been produced in molecular beam experiments, and mass spectrometry shows that such clusters can contain any desired number of atoms. For each added Na atom, another energy sublevel is added. Because the sublevels are so very closely spaced in a solid (with, say, $10^{23}$ atoms), the manifold of sublevels can be regarded as a **band**. Figure 19.20 depicts the generation of bands of sublevels that broaden as the spacing between the nuclear centers decreases. The electrons that belong to the $1s$, $2s$, and $2p$ atomic levels of sodium are only very slightly broadened at the equilibrium inter-

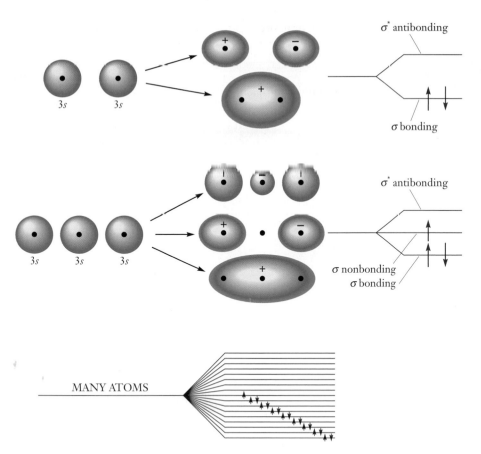

**FIGURE 19.20**  As sodium atoms are brought together, the molecular orbitals formed from their 3s atomic orbitals spread out into a band of levels, half occupied by electrons.

nuclear separation of the crystal, but even they would broaden into bands if the spacing were diminished further.

Now suppose a small electric potential difference is applied across a sodium crystal. The spin-paired electrons that lie deep in the band cannot be accelerated by a weak electric field, because there are occupied levels just above them. They have no place to go (recall the Pauli principle, which states that an energy level contains at most two electrons). At the top of the "sea" of occupied levels, however, there is an uppermost electron-occupied or half-occupied level called the **Fermi level.** Electrons that lie near that level have the highest kinetic energy of all the valence electrons in the crystal and can be accelerated by the electric field to occupy the levels above. They are free to migrate in response to the electric field so that they conduct an electric current. These same electrons at the Fermi level are responsible for the high thermal conductivities of metals. They are also the electrons freed by the photoelectric effect, when a photon gives them sufficient kinetic energy to escape from the metal (Section 15.2).

A natural question arises: Why are the alkaline earth elements metals, given the argument just presented? A metal such as magnesium contains two 3s electrons, and one might expect the band derived from the broadened 3s level to be completely filled. The answer to this question is that the energies of the 3s orbitals and 3p orbitals for magnesium are not greatly different. When the internuclear separation becomes small enough, the 3p band overlaps the 3s band, and many unoccupied sublevels are then available above the highest filled level.

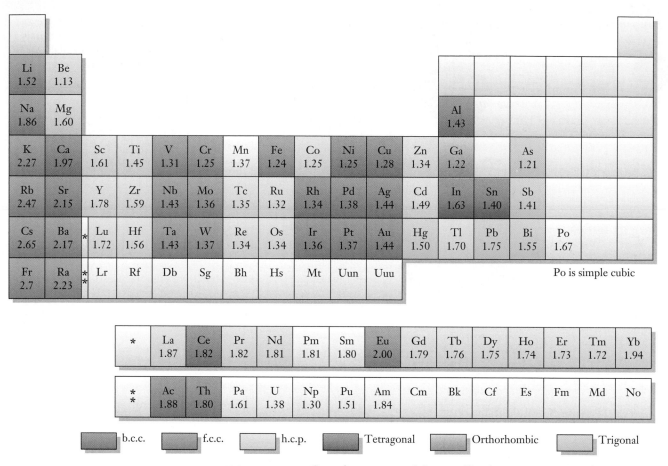

Po is simple cubic

b.c.c. | f.c.c. | h.c.p. | Tetragonal | Orthorhombic | Trigonal

**FIGURE 19.21** Crystal structures of the metallic elements at 25°C and 1 atm pressure. Atomic radii (in angstroms) are calculated from the hard sphere contact model, as in Example 19.4.

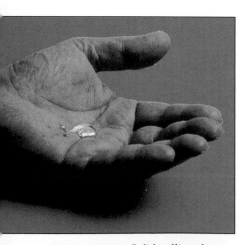

**FIGURE 19.22** Solid gallium has a low melting point, low enough to melt from the heat of the body. *(Leon Lewandowski)*

Most metals have crystal structures of high symmetry and crystallize in b.c.c., f.c.c., or h.c.p. lattices (Fig. 19.21). Relatively few metals (Ga, In, Sn, Sb, Bi, and Hg) have more complex crystal structures. Many metals undergo phase transitions to other structures when the temperature or pressure is changed. Liquids as well as crystals can be metals; in fact, the conductivity usually drops by only a small amount when a metal melts. The electron sea provides very strong binding in most metals, as shown by their high boiling points. Metals also have a very large range of melting points. Gallium melts at 29.78°C (Fig. 19.22), and mercury stays liquid at temperatures that freeze water. On the other hand, many transition metals require temperatures in excess of 1000°C to melt, and tungsten, the highest melting elemental metal, melts at 3410°C (Section 18.1).

Chapter 20 will consider the production and uses of metals in more detail.

## Covalent Crystals

We turn finally to a class of crystalline solids whose atoms are linked by covalent bonds rather than by the electrostatic attractions of ions or the valence-electron "glue" in a metal. The archetype of the covalent crystal is diamond, which belongs to the cubic system. The ground-state electron configuration of a carbon atom is $1s^2 2s^2 2p^2$, and, as shown in Sections 16.2 and 17.1, its bonding can be described by four hybrid $sp^3$ orbitals that are directed to the four corners of a regular tetrahedron. Each of the equivalent hybrid orbitals contains one electron that can spin-pair with the electron in one of the $sp^3$ orbitals of another carbon atom. Each carbon atom can thus link covalently to four others to yield the space-filling network shown in Figure 19.23. Covalent crystals are also called "network crystals," for obvious reasons. In a sense, every atom in a covalent crystal is part of one giant molecule that is the crystal itself. These crystals have very high melting points because of the strong attractions between covalently bound atoms. They are hard and brittle. Chapters 23 and 24 will discuss the mechanical and electrical properties of covalent crystals.

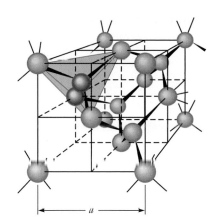

**FIGURE 19.23**  The structure of diamond. Each carbon atom has four nearest neighbors surrounding it at the corners of a tetrahedron.

## Structures of the Elements

The elements provide examples of three of the four classes of crystalline solids described in this section. Only ionic solids are excluded, because a single element cannot have the two types of atoms of different electronegativities needed to form an ionic material. We have already discussed some of the structures formed by metallic elements, which are sufficiently electropositive that their atoms readily give up electrons to form the electron sea of metallic bonding. The nonmetallic elements are more complex in their structures, reflecting a competition between inter- and intramolecular bonding and producing molecular or covalent solids with varied properties.

Each halogen atom has seven valence electrons and can react with one other halogen atom to form a diatomic molecule. Once this single bond forms, there is no further bonding capacity; the halogen diatomic molecules interact with one another only through relatively weak van der Waals forces and form molecular solids with low melting and boiling points.

The Group VI elements oxygen, sulfur, and selenium display dissimilar structures in the solid state. Each oxygen atom (with six valence electrons) can form one double or two single bonds. Except in ozone, its high-free-energy form, oxygen uses up all its bonding capacity with an intramolecular double bond, forming a molecular liquid and a molecular solid that are only weakly bound. In contrast, diatomic sulfur molecules (S=S) are relatively rare, being encountered only in high-temperature vapors. The favored forms of sulfur involve the bonding of every atom to two other sulfur atoms. This leads to either rings or chains, and both are observed. The stable form of sulfur at room temperature consists of $S_8$ molecules, with eight sulfur atoms arranged in a puckered ring (Fig. 19.24). The weak interactions between $S_8$

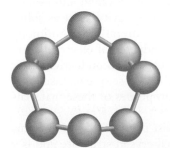

**FIGURE 19.24**  The structure of the sulfur molecule. The orthorhombic unit cell of rhombic sulfur, the most stable form of elemental sulfur at room temperature, is large and contains 16 of these $S_8$ molecules for a total of 128 atoms of sulfur.

**FIGURE 19.25**  Structures of elemental phosphorus.

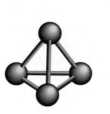

**(a)** White phosphorus

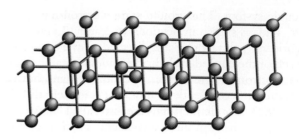

**(b)** Black phosphorus

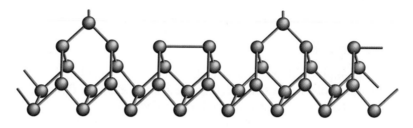

**(c)** Red phosphorus

molecules make elemental sulfur a rather soft molecular solid. Above 160°C, the rings in molten sulfur break open and relink to form long, tangled chains, producing a highly viscous liquid. An unstable ring form of selenium ($Se_8$) is known, but the thermodynamically stable form of this element is a gray crystal of metallic appearance that consists of very long spiral chains with weak interchain interaction. Crystalline tellurium has a similar structure. The Group VI elements thus show a trend (moving down the periodic table) away from the formation of multiple bonds and toward the chains and rings characteristic of atoms that each form two single bonds.

A similar trend is evident in Group V. Only nitrogen forms diatomic molecules with triple bonds, in which all the bonding capacity is used between pairs of atoms. Elemental phosphorus exists in three forms, in all of which each phosphorus atom forms three single bonds rather than one triple bond. White phosphorus (Fig. 19.25a) consists of tetrahedral $P_4$ molecules, which interact with each other through weak van der Waals forces. Black phosphorus and red phosphorus (Fig. 19.25b and c) are higher melting network solids in which the three bonds formed by each atom connect it directly or indirectly with all the other atoms in the sample. Unstable solid forms of arsenic and antimony that consist of $As_4$ or $Sb_4$ tetrahedra like those in white phosphorus can be prepared by rapid cooling of the vapor. The stable forms of these elements have structures related to that of black phosphorus.

The elements considered so far lie on the border between covalent and molecular solids. Other elements, those of intermediate electronegativity, exist as solids on the border between metallic and covalent and are called **metalloids.** Antimony has a metallic luster, for example, but is a rather poor conductor of electricity and heat. Silicon and germanium are **semiconductors,** with electrical conductivities far lower than those of metals but still significantly higher than those of true insulators such as diamond. Section 24.1 will examine the special properties of these materials more closely.

Some elements of intermediate electronegativity exist in two crystalline forms with very different properties. White tin has a tetragonal crystal structure and is a metallic conductor. Below 13°C it crumbles slowly to form a powder of gray tin (with the diamond structure) that is a poor conductor. Its formation at low temperature is known as the "tin disease" and can be prevented by the addition of small amounts of bismuth or antimony. The thermodynamically stable form of carbon at room conditions is not the insulator diamond, but graphite. Graphite consists of sheets of fused hexagonal rings with only rather weak interactions between layers (Fig. 19.26). Each carbon atom shows $sp^2$ hybridization, with its remaining $p$ orbital (perpendicular to the graphite layers) taking part in extended $\pi$-bonding interactions over the whole plane. Graphite can be pictured as a series of interlocked benzene rings, with $\pi$-electron delocalization contributing significantly to its stability. The delocalized electrons give graphite a conductivity in the planes of fused hexagons approaching that of the metallic elements. The conductivity and relative chemical inertness of graphite make it useful for electrodes in electrochemistry.

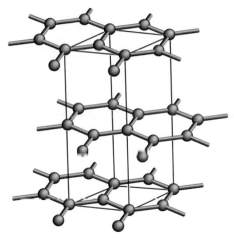

**FIGURE 19.26**  The structure of graphite.

# A DEEPER LOOK...

## 19.4  LATTICE ENERGIES OF CRYSTALS

The **lattice energy** of a crystal is the energy required to separate the crystal into its component atoms, molecules, or ions at 0 K. In this section we examine the calculation and measurement of lattice energies for molecular and ionic crystals.

### Lattice Energy of a Molecular Crystal

The lattice energy of a molecular crystal can be estimated by using the simple Lennard-Jones potential of Section 4.7:

$$V_{LJ}(R) = 4\epsilon\left[\left(\frac{\sigma}{R}\right)^{12} - \left(\frac{\sigma}{R}\right)^{6}\right]$$

Table 4.4 lists the values of $\epsilon$ and $\sigma$ for various atoms and molecules. To obtain the total potential energy for 1 mol, sum over all pairs of atoms or molecules:

$$V_{tot} = \frac{1}{2}\sum_{i=1}^{N_0}\sum_{j=1}^{N_0}V_{LJ}(R_{ij})$$

where $R_{ij}$ is the distance between atom $i$ and atom $j$. The factor $\frac{1}{2}$ arises because each interaction between a pair of atoms should be counted only once, not twice. For a crystal of macroscopic size, this can be rewritten as

$$V_{tot} = \frac{N_0}{2}\sum_{j=1}^{N_0}V_{LJ}(R_{ij})$$

where $i$ is taken to be some atom in the middle of the crystal. If the nearest neighbor distance is $R_0$, then we define a ratio of distances $p_{ij} = R_{ij}/R_0$ and rewrite $V_{tot}$ for the Lennard-Jones potential as

$$V_{tot} = \frac{N_0}{2}(4\epsilon)\left[\sum_{j}\left(\frac{\sigma}{p_{ij}R_0}\right)^{12} - \sum_{j}\left(\frac{\sigma}{p_{ij}R_0}\right)^{6}\right]$$

$$= 2\epsilon N_0\left[\left(\frac{\sigma}{R_0}\right)^{12}\sum_{j}(p_{ij})^{-12} - \left(\frac{\sigma}{R_0}\right)^{6}\sum_{j}(p_{ij})^{-6}\right]$$

The two summations are dimensionless properties of the lattice structure, and accurate values can be obtained by summing over the first few sets of nearest neighbors (Table 19.3). The resulting total energy for the f.c.c. lattice is

$$V_{tot} = 2\epsilon N_0\left[12.132\left(\frac{\sigma}{R_0}\right)^{12} - 14.454\left(\frac{\sigma}{R_0}\right)^{6}\right]$$

The equilibrium atomic spacing at $T = 0$ K should be close to the one that gives a minimum in $V_{tot}$, which can be calculated by differentiating the preceding expression with respect to $R_0$ and setting the derivative to 0. The result is

$$R_0 \approx 1.09\sigma$$

and the value of $V_{tot}$ at this value of $R_0$ is

$$V_{tot} = -8.61\epsilon N_0$$

The corresponding potential energy when the atoms or molecules are completely separated from one another is zero. The lattice energy is the difference between these quantities and is a positive number:

$$\text{lattice energy} = -V_{tot} = 8.61\epsilon N_0$$

## TABLE 19.3

### Lattice Sums for Molecular Crystals (f.c.c. Structure)

|  | Number, $n$ | $p_{ij}$ | $n(p_{ij})^{-12}$ | $n(p_{ij})^{-6}$ |
|---|---|---|---|---|
| Nearest neighbors | 12 | 1 | 12 | 12 |
| Second nearest neighbors | 6 | $\sqrt{2}$ | 0.0938 | 0.750 |
| Third nearest neighbors | 24 | $\sqrt{3}$ | 0.0329 | 0.889 |
| Fourth nearest neighbors | 12 | 2 | 0.0029 | 0.188 |
| Fifth nearest neighbors | 24 | $\sqrt{5}$ | 0.0015 | 0.192 |
|  | $\vdots$ | $\vdots$ | $\vdots$ | $\vdots$ |
| Total |  |  | 12.132 | 14.454 |

This overestimates the true lattice energy because of the quantum-mechanical effect of zero-point energy (Section 16.6). When a quantum correction is applied, the binding energy is reduced by 28%, 10%, 6%, and 4% for Ne, Ar, Kr, and Xe, respectively. Table 19.4 shows the resulting predictions for crystal lattice energies and nearest neighbor distances. The agreement with experiment is quite reasonable, considering the approximations inherent in the use of a Lennard–Jones potential derived entirely from gas-phase data. For helium the amplitude of zero-point motion is so great that if a crystal did form, it would immediately melt. Consequently, helium remains liquid down to absolute zero at atmospheric pressure.

## Lattice Energy of an Ionic Crystal

In Section 3.2 we calculated the potential energy of a gaseous diatomic ionic molecule relative to the separated ions by means of Coulomb's law:

$$V = \frac{Q_1 Q_2}{4\pi\epsilon_0 R_0}$$

where $R_0$ is the equilibrium internuclear separation. With a few modifications, the same considerations apply to a calculation of the lattice energies of ionic compounds in the crystalline state.

For simplicity, consider a hypothetical one-dimensional crys-

## TABLE 19.4

### Properties of Noble-Gas Crystals[†]

|  | $R_0$ (Å) | | Lattice Energy (kJ mol$^{-1}$) | |
|---|---|---|---|---|
|  | Predicted | Observed | Predicted | Observed |
| Ne | 3.00 | 3.13 | 1.83 | 1.88 |
| Ar | 3.71 | 3.76 | 7.72 | 7.74 |
| Kr | 3.92 | 4.01 | 11.5 | 11.2 |
| Xe | 4.47 | 4.35 | 15.2 | 16.0 |

[†] All data are extrapolated to 0 K and zero pressure.

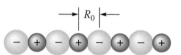

**FIGURE 19.27** Lattice energy for a one-dimensional ionic crystal.

tal (Fig. 19.27), in which ions of charge $+e$ and $-e$ alternate with an internuclear separation of $R_0$. One ion, selected to occupy an arbitrary origin, will interact attractively with all ions of opposite sign to make the following contribution to the crystal energy:

$$-\frac{e^2}{4\pi\epsilon_0 R_0}[2(1) + 2(\tfrac{1}{3}) + 2(\tfrac{1}{5}) + \cdots]$$

Here the factors of 2 come from the fact that there are *two* ions of opposite sign at a distance $R_0$ from a given ion, two at a distance $3R_0$, two at $5R_0$, and so forth. The negative sign occurs because the ions that occupy odd-numbered sites have a charge opposite that of the ion at the origin, and their interaction with the ion at the origin is attractive. The ion at the origin also interacts repulsively with all ions of the same sign to make the following contribution to the crystal energy:

$$+\frac{e^2}{4\pi\epsilon_0 R_0}[2(\tfrac{1}{2}) + 2(\tfrac{1}{4}) + 2(\tfrac{1}{6}) + \cdots]$$

The net interaction of $N_0$ such ions of each sign with one another is:

$$-\frac{N_0 e^2}{4\pi\epsilon_0 R_0}\left[2 - \frac{2}{2} + \frac{2}{3} - \frac{2}{4} + \frac{2}{5} - \frac{2}{6} + \cdots\right]$$

We must be very careful with factors of 2. To obtain the potential energy for the interaction of $N_0$ positive ions with $N_0$ negative ions, it is necessary to multiply the total potential energy of a given ion due to all others by $2N_0$ and then divide by 2 to avoid counting the interaction of a given pair of ions twice. This gives the preceding result.

If such a calculation is carried out for a real three-dimensional crystal, the result is a series (such as that just given in brackets) whose value sums to a dimensionless number that depends upon

## TABLE 19.5

### *Madelung Constants*

| Lattice | M |
|---|---|
| Rock Salt | 1.7476 |
| CsCl | 1.7627 |
| Zinc Blende | 1.6381 |
| Fluorite | 2.5194 |

the crystal structure. That number is called the **Madelung constant,** $M$, and its value is independent of the unit-cell dimensions. Table 19.5 lists the values of the Madelung constant for several crystal structures. The lattice energy is again the opposite of the total potential energy. Expressed in terms of the Madelung constant, it is

$$\text{lattice energy} = \frac{N_0 e^2}{4\pi\epsilon_0 R_0} M$$

## EXAMPLE 19.5

Calculate the electrostatic part of the lattice energy of sodium chloride, given that the internuclear separation between $Na^+$ and $Cl^-$ ions is 2.82 Å.

### Solution

Using the Madelung constant of 1.7476 for this structure gives

$$\frac{(6.02 \times 10^{23} \text{ mol}^{-1})(1.602 \times 10^{-19} \text{ C})^2 (1.7476)}{(4\pi)(8.854 \times 10^{-12} \text{ C}^2 \text{ J}^{-1} \text{ m}^{-1})(2.82 \times 10^{-10} \text{ m})}$$

$$= 8.61 \times 10^5 \text{ J mol}^{-1} = 861 \text{ kJ mol}^{-1}$$

**Related Problems: 31, 32**

Ionic lattice energies can be measured experimentally by means of a thermodynamic cycle developed by Max Born and Fritz Haber. The **Born–Haber cycle** is an application of Hess's law (the first law of thermodynamics). It is illustrated by a calculation of the lattice energy of sodium chloride, which is $\Delta E$ for the reaction

$$NaCl(s) \longrightarrow Na^+(g) + Cl^-(g) \qquad \Delta E = ?$$

This reaction can be represented as a series of steps, each with a measurable energy or enthalpy change. In the first step, the ionic solid is converted to the elements in their standard states:

$$NaCl(s) \longrightarrow Na(s) + \tfrac{1}{2} Cl_2(g)$$
$$\Delta E_1 \approx \Delta H = -\Delta H_f^\circ(NaCl(s)) = +411.2 \text{ kJ}$$

In the second step, the elements are transformed into gas-phase atoms:

$$Na(s) \longrightarrow Na(g) \qquad \Delta E \approx \Delta H = \Delta H_f^\circ(Na(g))$$
$$= +107.3 \text{ kJ}$$
$$\tfrac{1}{2}Cl_2(g) \longrightarrow Cl(g) \qquad \Delta E \approx \Delta H = \Delta H_f^\circ(Cl(g))$$
$$= +121.7 \text{ kJ}$$

$$Na(s) + \tfrac{1}{2} Cl_2(g) \longrightarrow Na(g) + Cl(g) \qquad \Delta E_2 = +229.0 \text{ kJ}$$

Finally, in the third step, electrons are transferred from the sodium atoms to the chlorine atoms to give ions:

$$Na(g) \longrightarrow Na^+(g) + e^- \qquad \Delta E = IE_1(Na) = 496 \text{ kJ}$$
$$Cl(g) + e^- \longrightarrow Cl^-(g) \qquad \Delta E = -EA(Cl) = -349 \text{ kJ}$$

$$Na(g) + Cl(g) \longrightarrow Na^+(g) + Cl^-(g) \qquad \Delta E_3 = +147 \text{ kJ}$$

Here $EA(Cl)$ is the electron affinity of Cl, and $IE_1(Na)$ is the first ionization energy of Na. The total energy change is

$$\Delta E = \Delta E_1 + \Delta E_2 + \Delta E_3 = 411 + 229 + 147 = +787 \text{ kJ}$$

The small differences between $\Delta E$ and $\Delta H$ were neglected in this calculation. If their difference is taken into account (using $\Delta H = \Delta E + RT\Delta n_g$, where $\Delta n_g$ is the change in the number of moles of gas molecules in each step of the reaction), then $\Delta E_2$ is decreased by $\tfrac{3}{2}RT$ and $\Delta E_1$ by $\tfrac{1}{2}RT$, giving a net decrease of $2RT$ and changing $\Delta E$ to $+782$ kJ mol$^{-1}$. Comparing this experimental lattice energy with the energy calculated in Example 19.5 shows that the latter is approximately 10% greater, presumably because short-range repulsive interactions and zero-point energy were not taken into account in the lattice energy calculation.

## 19.5

# DEFECTS AND AMORPHOUS SOLIDS

Although real crystals display beautiful symmetries to the eye, they are not perfect. As a practical matter, it is impossible to rid a crystal of all impurities or to ensure that it contains perfect periodic ordering. Amorphous solids are states of matter in which so many defects are present that crystalline order is destroyed.

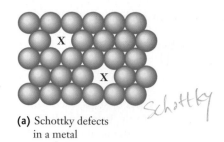

**(a)** Schottky defects
in a metal

*Schottky*

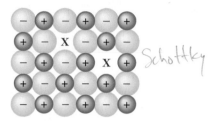

**(b)** Schottky defect in
an ionic crystal

*Schottky*

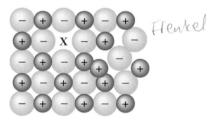

**(c)** Frenkel defect in
an ionic crystal

*Frenkel*

**FIGURE 19.28**  Point imperfections
in a lattice. The red Xs denote
vacancies.

*Causes colour*

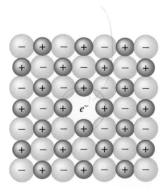

**FIGURE 19.29**  An F-center in a
crystal.

## Point Defects  *Was covered in Learn.*

Point defects in a pure crystalline substance include **vacancies,** in which atoms are missing from lattice sites, and **interstitials,** in which atoms are inserted in sites different from their normal sites. In real crystals, a small fraction of the normal atom sites remain unoccupied. Such vacancies are called **Schottky defects,** and their concentration depends on temperature:

$$N = N_s \exp(-\Delta G/RT)$$

where $N$ is the number of lattice vacancies per unit volume, $N_s$ is the number of atom sites per unit volume, and $\Delta G$ is the molar free energy of formation of vacancies. Figure 19.28a illustrates Schottky defects in the crystal structure of a metal or noble gas. Schottky defects also occur in ionic crystals but with the restriction that the imperfect crystals remain electrically neutral. Thus, in sodium chloride, for every missing $Na^+$ ion there must also be a missing $Cl^-$ ion (Fig. 19.28b).

In certain kinds of crystals, atoms or ions are displaced from their regular lattice sites to interstitial sites, and the crystal defect consists of the lattice vacancy plus the interstitial atom or ion. Figure 19.28c illustrates this type of lattice imperfection, known as a **Frenkel defect.** The silver halides (AgCl, AgBr, AgI) are examples of crystals in which Frenkel disorder is extreme. The crystal structures of these compounds are established primarily by the anion lattice, and the silver ions occupy highly disordered, almost random, sites. The rate of diffusion of silver ions in these solids is exceptionally high, as studies employing radioactive isotopes of silver have shown. Both Frenkel and Schottky defects in crystals are mobile, jumping from one lattice site to a neighboring site with frequencies that depend on the temperature and the strengths of the atomic forces. Diffusion in crystalline solids is due largely to the presence and mobility of point defects; it is a thermally activated process, just like the rates of chemical reactions considered in Chapter 13. The coefficient of self-diffusion has the form

$$D = D_0 \exp[-E_a/RT]$$

where $E_a$ is the activation energy. The rates of diffusive motion in crystalline solids vary enormously from one substance to another. In a crystal of a low-melting metal such as sodium, an average atom undergoes about $10^8$ diffusive jumps per second at 50°C, whereas in a metal such as tungsten that melts at 3410°C, an average atom jumps to another lattice site less than once per year at 1000°C!

If an alkali halide crystal such as NaCl is irradiated with x-rays, ultraviolet radiation, or high-energy electrons, some $Cl^-$ ions may lose an electron:

$$Cl^- + h\nu \longrightarrow Cl + e^-$$

The resulting Cl atom, being uncharged and much smaller than a $Cl^-$ ion, is no longer strongly bound in the crystal and can diffuse to the surface and escape. The electron can migrate through the crystal quite freely until it encounters an anion vacancy and is trapped in the Coulomb field of the surrounding cations (Fig. 19.29). This crystal defect is called an **F-center** (from the German word *Farbenzentrum,* meaning **color center**). It is the simplest of a family of electronic crystal defects. As the name suggests, it imparts a color to ionic crystals (Fig. 19.30).

## Nonstoichiometric Compounds

As Chapter 1 emphasized, the law of definite proportions was one of the principal pieces of evidence that led to the acceptance of Dalton's atomic theory. It is now recognized that a great many solid-state binary compounds do *not* have fixed and unvarying compositions but exist over a range of compositions in a single phase. Thus, FeO (wüstite) has the composition range $Fe_{0.85}O_{1.00}$ to $Fe_{0.95}O_{1.00}$ and is never found with its nominal 1:1 composition. The compounds NiO and $Cu_2S$ also deviate considerably from their nominal stoichiometries.

The explanation depends on the existence of more than one oxidation state for the metal. In wüstite, iron can exist in either the +2 or the +3 oxidation state. Suppose a solid were to begin at the hypothetical composition of $Fe_{1.00}O_{1.00}$, with iron entirely in the +2 oxidation state. For every two $Fe^{3+}$ ions introduced, three $Fe^{2+}$ ions must be removed to maintain overall charge neutrality. The total number of moles of iron is then less than that in the ideal FeO stoichiometry. The departure from the nominal stoichiometry can be far more extreme than that found in wüstite. The composition of "TiO" ranges from $Ti_{0.75}O$ to $Ti_{1.45}O$. Nickel oxide varies only from $Ni_{0.97}O$ to NiO in composition, but the variation is accompanied by a dramatic change in properties. When the compound is prepared in the 1:1 composition, it is pale green and is an electrical insulator. When it is prepared in an excess of oxygen, it is black and conducts electricity fairly well. In the black material, a small fraction of $Ni^{2+}$ ions are replaced by $Ni^{3+}$ ions, and compensating vacancies occur at some nickel atom sites in the crystal.

**FIGURE 19.30** Pure calcium fluoride, $CaF_2$, is white, but the natural sample of calcium fluoride (fluorite) shown here is a rich purple because F-centers are present. These are lattice sites at which the $F^-$ anion is replaced by an electron only. *(Leon Lewandowski)*

## EXAMPLE 19.6

The composition of a sample of wüstite is $Fe_{0.930}O_{1.00}$. What percentage of the iron is in the form of iron(III)?

**Solution**

For every 1.00 mol of oxygen atoms in this sample, there is 0.930 mol of iron atoms. Suppose $y$ mol of the iron is in the +3 oxidation state and $0.930 - y$ is in the +2 oxidation state. Then the total positive charge from the iron (in moles of electron charge) is

$$+3y + 2(0.930 - y)$$

This positive charge must exactly balance the 2 mol of negative charge carried by the mole of oxygen atoms (recall that each oxygen atom has oxidation number $-2$). We conclude that

$$3y + 2(0.930 - y) = +2$$

Solving this equation for $y$ gives

$$y = 0.140$$

The percentage of iron in the form of $Fe^{3+}$ is then the ratio of this to the total number of moles of iron, 0.930, multiplied by 100%:

$$\% \text{ iron in form of } Fe^{3+} = \frac{0.140}{0.930} \times 100\% = 15.1\%$$

**Related Problems: 37, 38**

## Alloys

The nonstoichiometric compounds just described are ionic materials with compositional disorder. A related type of disorder is exhibited by an **alloy,** a mixture of elements that displays metallic properties.

*SUBSTITUTION OF COMPARABLE SIZE ATOMS.*

*Many Application*

There are two types of alloys. In a **substitutional alloy,** some of the metal atoms in a crystal lattice are replaced by other atoms (usually of comparable size). Examples are brass, in which approximately one third of the atoms in a copper crystal are replaced by zinc atoms, and pewter, an alloy of tin that contains 7% copper, 6% bismuth, and 2% antimony. In an **interstitial alloy,** atoms of one or more additional elements enter the interstitial sites of the host metal lattice. An example is steel, in which carbon atoms occupy interstitial sites of an iron crystal, making the material stronger and harder than pure iron. Mild steel contains less than 0.2% C and is used for nails, whereas high-carbon steels can contain up to 1.5% C and are used in specialty applications such as tools and springs. *Alloy steels* are both substitutional and interstitial, as atoms from metals such as chromium and vanadium substitute for iron atoms, with carbon remaining in interstitial sites. Alloy steels have a variety of specialized purposes, ranging from cutlery to bicycle frames.

## Amorphous Solids and Glasses

The atoms, ions, or molecules in crystalline solids exhibit high degrees of spatial order. Now let us briefly consider solids that lack this characteristic. **Amorphous solids,** commonly called **glasses,** resemble crystalline solids in many respects; they may have chemical compositions, mechanical properties such as hardness and elasticity, and electrical and magnetic properties that are similar to those of crystals. Like crystals, glasses may have molecular, ionic, covalent, or metallic bonding. On an atomic scale, however, amorphous solids lack the regular periodic structure of crystals. They are states of matter in which so many defects are present that crystalline order is destroyed.

Some substances have a strong tendency to solidify as glasses. The best example is the material used in common window panes, with the approximate chemical formula $Na_2O \cdot CaO \cdot (SiO_2)_6$. This is a partly ionic, partly covalent material with $Na^+$ and $Ca^{2+}$ ions distributed through a covalently bonded Si—O network. Glass-forming ability is not restricted to a few special materials, however. If a substance can be liquefied, it can almost certainly be prepared in an amorphous state. Even metals, which are known primarily in the crystalline state, have been made into amorphous solids. The trick is to bypass crystallization by cooling molten material very fast. One technique involves shooting a jet of liquid metal at a rapidly rotating cold cylinder, which produces a continuous ribbon of amorphous metal at a rate up to 2 km per minute.

On the molecular level, a strong tendency to form glasses is associated with the presence of long or irregularly shaped molecules that can easily become tangled and disordered. Even slowly cooling a liquid assembly of such molecules may not afford enough time for them to organize into a crystalline lattice before solidification. Instead of a sharp liquid-to-crystal transition, such glass-formers transform continuously, over a range of temperature, into amorphous solids. They lend themselves to fabrication into articles of every conceivable shape, because the flow properties of the work piece can be managed by controlling its temperature. This plasticity is the reason that glass has played an indispensable role in science, industry, and the arts.

One of the most exciting new uses of a glass is the transmission of voice messages, television images, and data as light pulses. Tens of thousands of audio messages can be transmitted simultaneously through glass fibers no greater in diameter than a human hair by encoding the audio signal into electronic impulses that modulate light from a laser source. The light passes down the glass fiber as though it were a tube. Chemical control of the glass composition reduces light loss and permits messages to travel many kilometers without amplification (Fig. 19.31).

**FIGURE 19.31** Extremely thin glass fibers of specialized composition can conduct a beam of light for miles without substantial loss. (*Copyright 1988 Paul Silverman/Fundamental Photographs*)

---

## 19.6
# LIQUID CRYSTALS

**Liquid crystals** constitute an interesting state of matter with properties intermediate between those of true liquids and those of crystals. Unlike glasses, liquid-crystal states are thermodynamically stable. Many organic materials do not show a single solid-to-liquid transition but rather a cascade of transitions involving new intermediate phases. In recent years, liquid crystals have been used in a wide variety of practical applications, ranging from temperature sensors to displays on calculators and other electronic devices.

## The Structure of Liquid Crystals

Substances that form liquid crystals are usually characterized by molecules with elongated, rod-like shapes. An example is terephthal-bis-(4-*n*-butylaniline), called TBBA, whose molecular structure can be represented as

$$H_9C_4 - \bigcirc - N{=}C - \bigcirc - C{=}N - \bigcirc - C_4H_9$$

with hydrocarbon groups at the ends separated by a relatively rigid backbone of benzene rings and $N{=}C$ double bonds. Such rod-like molecules tend to line up even in the liquid phase, as Figure 19.32a shows. Ordering in this phase persists only over small distances, however, and on average a given molecule is equally likely to take any orientation.

The simplest type of liquid-crystal phase is the **nematic phase** (Fig. 19.32b). In a nematic liquid crystal, the molecules display a preferred orientation in a particular direction, but their centers are distributed at random, as they would be in an ordinary liquid. Although liquid-crystal phases are characterized by a net orientation of molecules over large distances, not all the molecules point in exactly the same direction. There are fluctuations in the orientation of each molecule, and only *on average* do the molecules have a greater probability of pointing in a particular direction.

Some liquid crystals form one or more **smectic phases.** These display a variety of microscopic structures that are indicated by the letters A, B, C, and so forth. Figure 19.32c shows one of them, the smectic A structure; the molecules continue to display net orientational ordering, but now, unlike in the nematic phase, the centers of the molecules also tend to lie in layers. Within each layer, however, these centers are distributed at random as in an ordinary liquid.

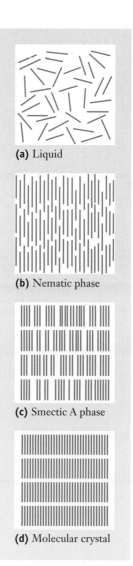

**(a)** Liquid

**(b)** Nematic phase

**(c)** Smectic A phase

**(d)** Molecular crystal

**FIGURE 19.32** Different states of structural order for rod-shaped molecules. The figure is only schematic; in a real sample, the lining up of the molecules would not be so nearly perfect.

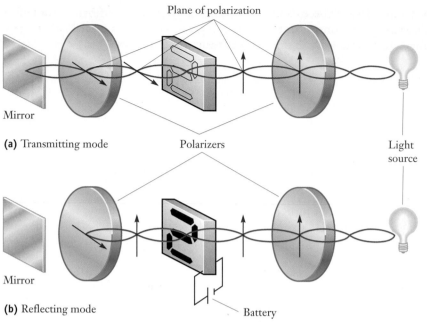

**FIGURE 19.33**   The mode of operation of a liquid-crystal display device. (a) The light has a polarization that permits it to pass through the second polarizing filter and strike the mirror, giving a bright display. (b) Imposition of a potential difference across some portion of the display causes the liquid-crystal molecules to rotate, creating a different polarization of light. Because the "rotated" light is blocked by the second filter, it does not reach the mirror, and that part of the display appears black.

At low enough temperatures, a liquid crystal freezes into a crystalline solid (Fig. 19.32d) in which the molecules' orientations are ordered and their centers lie on a regular three-dimensional lattice. The progression of structures in Figure 19.32 illustrates the meaning of the term "liquid crystal." Liquid crystals are solid-like in showing orientational ordering but liquid-like in the random distribution of the centers of their molecules.

In taking a particular orientation, a liquid crystal is very sensitive both to the positions and natures of the surfaces with which it is in contact and to small electric or magnetic fields. This sensitivity is the basis for the use of nematic liquid crystals in electronic display devices such as digital watches and calculators (Fig. 19.33) as well as in large-screen liquid-crystal displays.

## CUMULATIVE EXERCISE

### *Phosphorus*

Solid elemental phosphorus appears in a rich variety of forms, with crystals in all seven crystal systems reported under various conditions of temperature, pressure, and sample preparation.

(a) The thermodynamically stable form of phosphorus under room conditions is black phosphorus. Its unit cell is orthorhombic with edges of lengths 3.314, 4.376,

and 10.48 Å. Calculate the volume of one unit cell, and determine the number of phosphorus atoms per unit cell, if the density of this form of phosphorus is 2.69 g cm$^{-3}$.
(b) The form of phosphorus that is easiest to prepare from the liquid or gaseous state is white phosphorus, which consists of P$_4$ molecules in a cubic lattice. When x-rays of wavelength 2.29 Å are scattered from the parallel faces of its unit cells, the first-order Bragg diffraction is observed at an angle $2\theta$ of 7.10°. Calculate the length of the unit-cell edge for white phosphorus. At what angle will third-order Bragg diffraction be seen?
(c) Amorphous red phosphorus has been reported to convert to monoclinic, triclinic, tetragonal, and cubic red forms with different heat treatments. Identify the changes in the shape of the unit cell as a cubic lattice is converted first to tetragonal, then monoclinic, then triclinic.
(d) A monoclinic form of red phosphorus has been studied that has cell edge lengths 9.21, 9.15, and 22.60 Å, with an angle $\beta$ of 106.1°. Each unit cell contains 84 atoms of phosphorus. Estimate the density of this form of phosphorus.
(e) Phosphorus forms many compounds with other elements. Describe the nature of the bonding in the solids white elemental phosphorus (P$_4$), black elemental phosphorus, sodium phosphate (Na$_3$PO$_4$), and phosphorus trichloride (PCl$_3$).

Two forms of elemental phosphorus: white and red. *(Charles D. Winters)*

### Answers

(a) Volume is 152.0 Å$^3$; eight atoms per unit cell
(b) 18.5 Å; angle $2\theta = 21.4°$
(c) Cubic to tetragonal: one cell edge is stretched or shrunk. Tetragonal to monoclinic: a second cell edge is stretched or shrunk and the angles between two adjacent faces (and their opposite faces) are changed from 90°. Monoclinic to triclinic: the remaining two angles between faces are deformed from 90°.
(d) Density is 2.36 g cm$^{-3}$
(e) P$_4$(white) and PCl$_3$ are molecular solids; P(black) is covalent; Na$_3$PO$_4$ is ionic.

## CONCEPTS & SKILLS

*After studying this chapter and working the problems that follow, you should be able to*

1. Identify the symmetry elements of different crystal systems (Section 19.1, problems 1–4).
2. Explain how x-rays and neutrons are diffracted by crystals, and use information from such experiments to calculate lattice spacings (Section 19.1, problems 5–10).
3. Describe the packing of atoms in simple crystal lattices (Section 19.2, problems 11–24).
4. Compare the natures of the forces that hold atoms or molecules in their lattice sites in molecular, ionic, metallic, and covalent crystals (Section 19.3, problems 27–28).
5. Calculate lattice energies of molecular and ionic crystals (Section 19.4, problems 31–34).
6. Describe the kinds of equilibrium defects that are present in crystalline solids and the properties of amorphous solids (Section 19.5).
7. Determine the oxidation states present in nonstoichiometric solids (Section 19.5, problems 37–38).
8. Explain what a liquid crystal is, and state how nematic and smectic phases differ from ordinary liquids and crystalline solids (Section 19.6, problems 39–40).

---

## PROBLEMS

*Answers to problems whose numbers are boldface appear in Appendix G. Problems that are more challenging are indicated with asterisks.*

### Crystal Symmetry and the Unit Cell

**1.** Which of the following has 3-fold rotational symmetry? Explain.
  (a) An isosceles triangle
  (b) An equilateral triangle
  (c) A tetrahedron
  (d) A cube

2. Which of the following has 4-fold rotational symmetry? Explain.
  (a) A cereal box (exclusive of the writing on the sides)
  (b) A stop sign (not counting the writing)
  (c) A tetrahedron
  (d) A cube

**3.** Identify the symmetry elements of the $CCl_2F_2$ molecule (see Fig. 7.15).

4. Identify the symmetry elements of the $PF_5$ molecule (see Fig. 3.11a).

**5.** The second-order Bragg diffraction of x-rays with $\lambda = 1.660$ Å from a set of parallel planes in copper occurs at an angle $2\theta = 54.70°$. Calculate the distance between the scattering planes in the crystal.

6. The second-order Bragg diffraction of x-rays with $\lambda = 1.237$ Å from a set of parallel planes in aluminum occurs at an angle $2\theta = 35.58°$. Calculate the distance between the scattering planes in the crystal.

**7.** The distance between members of a set of equally spaced planes of atoms in crystalline lead is 4.950 Å. If x-rays with $\lambda = 1.936$ Å are diffracted by this set of parallel planes, calculate the angle $2\theta$ at which fourth-order Bragg diffraction will be observed.

8. The distance between members of a set of equally spaced planes of atoms in crystalline sodium is 4.28 Å. If x-rays with $\lambda = 1.539$ Å are diffracted by this set of parallel planes, calculate the angle $2\theta$ at which second-order Bragg diffraction will be observed.

**9.** The members of a series of equally spaced parallel planes of ions in crystalline LiCl are separated by 2.570 Å. Calculate all the angles $2\theta$ at which diffracted beams of various orders may be seen, if the x-ray wavelength used is 2.167 Å.

10. The members of a series of equally spaced parallel planes in crystalline vitamin $B_{12}$ are separated by 16.02 Å. Calculate all the angles $2\theta$ at which diffracted beams of various orders may be seen, if the x-ray wavelength used is 2.294 Å.

**11.** A crucial protein at the photosynthetic reaction center of the purple bacterium *Rhodopseudomonas viridis* (Section 24.3) has been separated from the organism, crystallized, and studied by x-ray diffraction. This substance crystallizes with a primitive unit cell in the tetragonal system. The cell dimensions are $a = b = 223.5$ Å and $c = 113.6$ Å.

(a) Determine the volume, in cubic angstroms, of this cell.
(b) One of the crystals in this experiment was box-shaped, with dimensions $1 \times 1 \times 3$ mm. Compute the number of unit cells in this crystal.

12. Compute the volume (in cubic angstroms) of the unit cell of potassium hexacyanoferrate(III) ($K_3Fe(CN)_6$), a substance that crystallizes in the monoclinic system with $a = 8.40$ Å, $b = 10.44$ Å, and $c = 7.04$ Å and with $\beta = 107.5°$.

**13.** The compound $Pb_4In_3B_{17}S_{18}$ crystallizes in the monoclinic system with a unit cell having $a = 21.021$ Å, $b = 4.014$ Å, $c = 18.898$ Å, and the only non-90° angle equal to 97.07°. There are two molecules in every unit cell. Compute the density of this substance.

14. Strontium chloride hexahydrate ($SrCl_2 \cdot 6H_2O$) crystallizes in the trigonal system in a unit cell with $a = 8.9649$ Å and $\alpha = 100.576°$. The unit cell contains three formula units. Compute the density of this substance.

**15.** At room temperature, the edge length of the cubic unit cell in elemental silicon is 5.431 Å, and the density of silicon at the same temperature is 2.328 g cm$^{-3}$. Each cubic unit cell contains eight silicon atoms. Using only these facts, perform the following operations.
  (a) Calculate the volume (in cubic centimeters) of one unit cell.
  (b) Calculate the mass (in grams) of silicon present in a unit cell.
  (c) Calculate the mass (in grams) of an atom of silicon.
  (d) The mass of an atom of silicon is 28.0855 u. Estimate Avogadro's number to four significant figures.

16. One form of crystalline iron has a body-centered cubic lattice with an iron atom at every lattice point. Its density at 25°C is 7.86 g cm$^{-3}$. The length of the edge of the cubic unit cell is 2.87 Å. Use these facts to estimate Avogadro's number.

**17.** Sodium sulfate ($Na_2SO_4$) crystallizes in the orthorhombic system in a unit cell with $a = 5.863$ Å, $b = 12.304$ Å, and $c = 9.821$ Å. The density of these crystals is 2.663 g cm$^{-3}$. Determine how many $Na_2SO_4$ formula units are present in the unit cell.

18. The density of turquoise, $CuAl_6(PO_4)_4(OH)_8(H_2O)_4$, is 2.927 g cm$^{-3}$. This gemstone crystallizes in the triclinic system with cell constants $a = 7.424$ Å, $b = 7.629$ Å, $c = 9.910$ Å, $\alpha = 68.61°$, $\beta = 69.71°$, and $\gamma = 65.08°$. Calculate the volume of the unit cell, and determine how many copper atoms are present in each unit cell of turquoise.

**19.** An oxide of rhenium has a structure with a Re atom at each corner of the cubic unit cell and an O atom at the center of each edge of the cell. What is the chemical formula of this compound?

20. The mineral perovskite has a calcium atom at each corner of the unit cell, a titanium atom at the center of the unit cell, and an oxygen atom at the center of each face. What is the chemical formula of this compound?

**21.** Iron has a body-centered cubic structure with a density of $7.86$ g cm$^{-3}$.
   (a) Calculate the nearest neighbor distance in crystalline iron.
   (b) What is the lattice parameter for the cubic unit cell of iron?
   (c) What is the atomic radius of iron?

22. The structure of aluminum is face-centered cubic and its density is $\rho = 2.70$ g cm$^{-3}$.
   (a) How many Al atoms belong to a unit cell?
   (b) Calculate $a$, the lattice parameter, and $d$, the nearest neighbor distance.

**23.** Sodium has the body-centered cubic structure, and its lattice parameter is $4.28$ Å.
   (a) How many Na atoms does a unit cell contain?
   (b) What fraction of the volume of the unit cell is occupied by Na atoms, if they are represented by spheres in contact with one another?

24. Nickel has a face-centered cubic structure with a density of $8.90$ g cm$^{-3}$.
   (a) Calculate the nearest neighbor distance in crystalline nickel.
   (b) What is the atomic radius of nickel?
   (c) What is the radius of the largest atom that could fit into the interstices of a nickel lattice, approximating the atoms as spheres?

**25.** Calculate the ratio of the maximum radius of an interstitial atom at the center of a simple cubic unit cell to the radius of the host atom.

26. Calculate the ratio of the maximum radius of an interstitial atom at the center of each face of a body-centered cubic unit cell to the radius of the host atom.

## Cohesion in Solids

**27.** Classify each of the following solids as molecular, ionic, metallic, or covalent.
   (a) BaCl$_2$    (b) SiC    (c) CO    (d) Co
28. Classify each of the following solids as molecular, ionic, metallic, or covalent.
   (a) Rb    (b) C$_5$H$_{12}$    (c) B    (d) Na$_2$HPO$_4$
**29.** By examining Figure 19.18, determine the number of nearest neighbors, second nearest neighbors, and third nearest neighbors of a Cs$^+$ ion in crystalline CsCl. The nearest neighbors of the Cs$^+$ ion are Cl$^-$ ions, and the second nearest neighbors are Cs$^+$ ions.
30. Repeat the determinations of the preceding problem for the NaCl crystal, referring to Figure 19.17.

## A Deeper Look ... Lattice Energies of Crystals

**31.** Calculate the energy needed to dissociate 1.00 mol of crystalline RbCl into its gaseous ions if the Madelung constant for its structure is 1.7476 and the radii of Rb$^+$ and Cl$^-$ are $1.48$ Å and $1.81$ Å, respectively. Assume that the repulsive

energy reduces the lattice energy by 10% from the pure Coulomb energy.
32. Repeat the calculation of problem 31 for CsCl, taking the Madelung constant from Table 19.5 and taking the radii of Cs$^+$ and Cl$^-$ to be $1.67$ Å and $1.81$ Å.
**33.** (a) Use the Born–Haber cycle, with data from Appendices D and F, to calculate the lattice energy of LiF.
   (b) Compare the result of part (a) with the Coulomb energy calculated by using an Li—F separation of $2.014$ Å in the LiF crystal, which has the rock-salt structure.
34. Repeat the calculations of problem 33 for crystalline KBr, which has the rock-salt structure with a K—Br separation of $3.298$ Å.

## Defects and Amorphous Solids

**35.** Will the presence of Frenkel defects change the measured density of a crystal?
36. What effect will the (unavoidable) presence of Schottky defects have on the determination of Avogadro's number via the method described in problems 15 and 16?
**37.** Iron(II) oxide is nonstoichiometric. A particular sample was found to contain 76.55% iron and 23.45% oxygen by mass.
   (a) Calculate the empirical formula of the compound (four significant figures).
   (b) What percentage of the iron in this sample is in the $+3$ oxidation state?
38. A sample of nickel oxide contains 78.23% Ni by mass.
   (a) What is the empirical formula of the nickel oxide to four significant figures?
   (b) What fraction of the nickel in this sample is in the $+3$ oxidation state?

## Liquid Crystals

**39.** Compare the natures and extents of order in the smectic liquid-crystal and isotropic liquid phases of a substance. Which has the higher entropy? Which has the higher enthalpy?
40. Nematic liquid crystals form when a liquid of long rod-like molecules is cooled. What *additional* types of intermolecular interactions would you expect to favor the formation of a *smectic* phase?

## Additional Problems

**41.** Some water waves with a wavelength of $3.0$ m are diffracted by an array of evenly spaced posts in the water. If the rows of posts are separated by $5.0$ m, calculate the angle $2\theta$ at which the first-order "Bragg diffraction" of these water waves will be seen.
42. A crystal scatters x-rays of wavelength $\lambda = 1.54$ Å at an angle $2\theta$ of $32.15°$. Calculate the wavelength of the x-rays in another experiment if this same diffracted beam from the same crystal is observed at an angle $2\theta$ of $34.46°$.
43. The number of beams diffracted by a single crystal depends on the wavelength $\lambda$ of the x-rays used and on the volume

associated with one lattice point in the crystal—that is, on the volume $V_p$ of a primitive unit cell. An approximate formula is

$$\text{number of diffracted beams} = \frac{4}{3}\pi\left(\frac{2}{\lambda}\right)^3 V_p$$

(a) Compute the volume of the conventional unit cell of crystalline sodium chloride. This cell is cubic and has an edge length of 5.6402 Å.

(b) The NaCl unit cell contains four lattice points. Compute the volume of a primitive unit cell for NaCl.

(c) Use the formula given in this problem to estimate the number of diffracted rays that will be observed if NaCl is irradiated with x-rays of wavelength 2.2896 Å.

(d) Use the formula to estimate the number of diffracted rays that will be observed if NaCl is irradiated with x-rays having the shorter wavelength 0.7093 Å.

44. If the wavelength $\lambda$ of the x-rays is too large relative to the spacing of planes in the crystal, no Bragg diffraction will be seen because $\sin\theta$ would be larger than 1 in the Bragg equation, even for $n = 1$. Calculate the longest wavelength of x-rays that can give Bragg diffraction from a set of planes separated by 4.20 Å.

45. The crystal structure of diamond is face-centered cubic, and the atom coordinates in the unit cell are $(0, 0, 0)$, $(\frac{1}{2}, \frac{1}{2}, 0)$, $(\frac{1}{2}, 0, \frac{1}{2})$, $(0, \frac{1}{2}, \frac{1}{2})$, $(\frac{1}{4}, \frac{1}{4}, \frac{1}{4})$, $(\frac{3}{4}, \frac{1}{4}, \frac{3}{4})$, $(\frac{3}{4}, \frac{3}{4}, \frac{1}{4})$, and $(\frac{1}{4}, \frac{3}{4}, \frac{3}{4})$. The lattice parameter is $a = 3.57$ Å. What is the C—C bond distance in diamond?

46. Polonium is the only element known to crystallize in the simple cubic lattice.

(a) What is the distance between nearest neighbor polonium atoms if the first-order diffraction of x-rays with $\lambda = 1.785$ Å from the parallel faces of its unit cells appears at an angle of $2\theta = 30.96°$ from these planes?

(b) What is the density of polonium in this crystal (in g cm$^{-3}$)?

47. At room temperature, monoclinic sulfur has the unit-cell dimensions $a = 11.04$ Å, $b = 10.98$ Å, $c = 10.92$ Å, and $\beta = 96.73°$. Each cell contains 48 atoms of sulfur.

(a) Explain why it is not necessary to give the values of the angles $\alpha$ and $\gamma$ in this cell.

(b) Compute the density of monoclinic sulfur in units of g cm$^{-3}$.

48. A compound contains three elements: sodium, oxygen, and chlorine. It crystallizes in a cubic lattice. The oxygen atoms are at the corners of the unit cells, the chlorine atoms are at the centers of the unit cells, and the sodium atoms are at the centers of the faces of the unit cells. What is the formula of the compound?

*49. Show that the radius of the largest sphere that can be placed in a tetrahedral interstitial site in a f.c.c. lattice is $0.225r_1$, where $r_1$ is the radius of the atoms making up the lattice. (*Hint:* Consider a cube with the centers of four spheres placed at alternate corners, and visualize the tetrahedral site at the center of the cube. What is the relationship between $r_1$ and the length of a diagonal of a face? The length of a body diagonal?)

50. What is the closest packing arrangement possible for a set of thin circular discs lying in a plane? What fraction of the area of the plane is occupied by the discs? Show how the same reasoning can be applied to the packing of infinitely long, straight cylindrical fibers.

51. Name two elements that form molecular crystals, two that form metallic crystals, and two that form covalent crystals. What generalizations can you make about the portions of the periodic table where each type is found?

52. The nearest-neighbor distance in crystalline LiCl (rock-salt structure) is 2.570 Å; the bond length in a gaseous LiCl molecule is significantly shorter, 2.027 Å. Explain.

53. (a) Using the data of Table 4.4, estimate the lattice energy and intermolecular separation of nitrogen in its solid state, assuming a f.c.c. structure for the solid lattice.

(b) The density of cubic nitrogen is 1.026 g cm$^{-3}$. Calculate the lattice parameter $a$ and the nearest neighbor distance. Compare your answer with that from part (a).

*54. Solid $CuI_2$ is unstable relative to CuI at room temperature, but $CuBr_2$, $CuCl_2$, and $CuF_2$ are all stable relative to the copper(I) halides. Explain by considering the steps in the Born–Haber cycle for these compounds.

55. A crystal of sodium chloride has a density of 2.165 g cm$^{-3}$ in the absence of defects. Suppose a crystal of NaCl is grown in which 0.15% of the sodium ions and 0.15% of the chloride ions are missing. What is the density in this case?

56. The activation energy for the diffusion of sodium atoms in the crystalline state is 42.22 kJ mol$^{-1}$, and $D_0 = 0.145$ cm$^2$ s$^{-1}$.

(a) Calculate the diffusion constant $D = D_0 \exp(-E_a/RT)$ of sodium in the solid at its melting point (97.8°C).

(b) What is the root-mean-square displacement of an average sodium atom from an arbitrary origin after the lapse of 1.0 hour at $t = 97.8°C$? (*Hint:* Use Equation 4.17 in Chapter 4.)

57. A compound of titanium and oxygen contains 28.31% oxygen by mass.

(a) If the compound's empirical formula is $Ti_xO$, calculate $x$ to four significant figures.

(b) The nonstoichiometric compounds $Ti_xO$ can be described as having a $Ti^{2+}$—$O^{2-}$ lattice in which certain $Ti^{2+}$ ions are missing or are replaced by $Ti^{3+}$ ions. Calculate the fraction of $Ti^{2+}$ sites in the nonstoichiometric compound that are vacant and the fraction that are occupied by $Ti^{3+}$ ions.

58. Classify the bonding in the following amorphous solids as molecular, ionic, metallic, or covalent.

(a) Amorphous silicon, used in photocells to collect light energy from the sun

(b) Polyvinyl chloride, a plastic of long-chain molecules composed of —$CH_2CHCl$— repeating units, used in pipes and siding

(c) Soda–lime–silica glass, used in windows

(d) Copper–zirconium glass, an alloy of the two elements with approximate formula $Cu_3Zr_2$, used for its high strength and good conductivity

## CUMULATIVE PROBLEMS

59. Sodium hydride (NaH) crystallizes in the rock-salt structure, with four formula units of NaH per cubic unit cell. A beam of monoenergetic neutrons, selected to have a velocity of $2.639 \times 10^3$ m s$^{-1}$, is scattered in second order through an angle of $2\theta = 36.26°$ by the parallel faces of the unit cells of a sodium hydride crystal.
    (a) Calculate the wavelength of the neutrons.
    (b) Calculate the edge length of the cubic unit cell.
    (c) Calculate the distance from the center of an Na$^+$ ion to the center of a neighboring H$^-$ ion.
    (d) If the radius of a Na$^+$ ion is 0.98 Å, what is the radius of an H$^-$ ion, assuming the two ions are in contact?

60. Chromium(III) oxide has a structure in which chromium ions occupy two thirds of the octahedral interstitial sites in a hexagonal close-packed lattice of oxygen ions. What is the $d$-electron configuration on the chromium ion?

61. A useful rule of thumb is that in crystalline compounds every nonhydrogen atom occupies 18 Å$^3$, and the volume occupied by hydrogen atoms can be neglected. Using this rule, estimate the density of ice (in g cm$^{-3}$). Explain why the answer is so different from the observed density of ice.

62. Estimate, for the F-centers in CaF$_2$, the wavelength of maximum absorption in the visible region of the spectrum that will give rise to the color shown in Figure 19.30.

# UNIT 6 CHEMICAL PROCESSES

The smelting of copper involves a series of chemical reactions to convert the ore into elemental copper. Here, molten copper streams from an oven. *(Charles D. Winters)*

The first five Units of this book introduced the basic principles of chemistry. Unit 1 presented fundamental concepts and methods, and Units 2 to 4 emphasized macroscopic aspects of chemistry. Unit 5 took a microscopic approach, showing how electrons and nuclei combine to form atoms, which join together through chemical bonds to make molecules and complex ions. The further interactions between molecules give rise to the observed properties of bulk matter: gases, liquids, and solids.

In Unit 6, we combine these pieces to see how reactions fit together in overall chemical processes that transform starting materials into products. Such chemical processes existed in the geochemical and biochemical spheres long before the appearance of humankind. The challenge for modern chemists is to learn from such naturally occurring processes in order to improve the efficiency and speed of synthesis of new chemical substances.

**CHAPTER 20**
Chemical Processes for the Recovery of Pure
Substances

**CHAPTER 21**
Chemical Processes Based on Sulfur, Phosphorus,
and Nitrogen

**CHAPTER 22**
Chemical Processes Based on the Halogens and the
Noble Gases

THE GOALS OF UNIT 6 ARE **(1)** To discuss the chemical processes involved in preparing selected metallic and nonmetallic elements from their naturally occurring sources and then **(2)** To consider the uses of these elements and their further reactions to form other compounds.

# Chemical Processes for the Recovery of Pure Substances

**Illustration**
A catalytic converter, used to allow wood stoves to burn more efficiently.
*(Charles D. Winters)*

I n Chapter 20 we introduce a **process-based approach** to chemistry by first examining the nature of the chemical industry, its major raw materials and products, and the types of chemical processes used for the recovery of pure substances. We next take a close look at the production and uses of the lightest chemical element, hydrogen, as an illustration of the approach followed in the remaining chapters of the book. Finally, we describe the technologically important approaches used to extract and purify several important metals from their naturally occurring ores.

---
**20.1**
---

# THE CHEMICAL INDUSTRY

A firm grasp of chemical principles is essential to the design of clean and efficient manufacturing processes for chemical products. These processes are continually developed and changed to improve product yield, reduce waste, and minimize costs. Chemical plants are usually expensive to construct, and a new process invariably passes through several **pilot-plant stages** after its development in the research laboratory. In these scaling-up operations, the process is optimized with respect to such variables as temperature, pressure, and time of reaction. Almost all laboratory-scale chemical processes are **batch processes,** in which the reactants are mixed in a container and allowed to react, and the products are removed in a sequence of operations. Many industrial-scale processes, in contrast, are **continuous processes,** designed so that reactants are fed in and products are removed continuously. It is rarely a simple matter to convert a process from batch to continuous operation, but the latter type is generally more economical.

We can represent processes in the chemical industry schematically (Fig. 20.1) as

$$\text{starting materials} \longrightarrow \text{desired products} + \text{byproducts}$$

Energy, often a great deal of it, plays a part in such transformations. Sometimes energy must be put in to carry out the desired reaction; in other cases it is produced, usually as heat, and must be recycled or safely discarded. Let us examine each of the terms in this equation in light of basic principles of chemistry.

In the ideal case, the starting materials for a chemical process are easily obtainable and easily transportable to the site where they will be used. Starting materials are often in thermodynamic states of rather low free energy, simply because compounds with high free energy tend to react spontaneously to give products of lower free energy. Over the long history of the earth, compounds of high free energy have mostly disappeared. Thus, aluminum is very abundant in the earth's crust but is never found in elemental form. Rather, it is tied up in minerals of much lower free energy such as kaolinite, a clay with the chemical formula $Al_2Si_2O_5(OH)_4$ and a large negative free energy of formation ($\Delta G_f^\circ = -3800 \text{ kJ mol}^{-1}$). Some compounds of high free energy remain in nature because kinetics is involved as well as thermodynamics; states of high free energy can persist over long periods if the rate of reaction is sufficiently low. The compounds of high free energy that do occur in nature are particularly valuable resources for chemistry. Hydrocarbons such as octane occur in petroleum deposits and have high free energies relative to their combustion products:

$$C_8H_{18}(g) + \tfrac{25}{2}\, O_2(g) \longrightarrow 9\, H_2O(\ell) + 8\, CO_2(g) \qquad \Delta G^\circ = -5272 \text{ kJ}$$

In many cases, a nonspontaneous reaction (one with $\Delta G > 0$) can be caused to proceed by linking it to another reaction for which $\Delta G < 0$—for instance, the combustion of a fossil fuel such as coal, oil, or natural gas. In many chemical processes, energy is supplied as electricity. The powerful oxidizing (bleaching) agent sodium chlorate, for example, is made by the overall reaction

$$NaCl(aq) + 3\, H_2O(\ell) \longrightarrow NaClO_3(aq) + 3\, H_2(g) \qquad \Delta G^\circ = +835 \text{ kJ}$$

This reaction becomes possible only with an external source of electrical work. The electricity itself is generated by burning fossil fuels or harnessing the energy of a

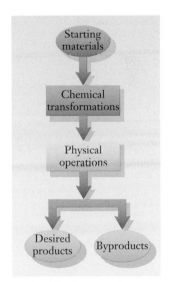

**FIGURE 20.1** Schematic diagram of a chemical process, employing the color conventions that were used in Figure 18.27 and that will be followed in subsequent chapters. Starting materials (red) are converted to products (green) via a series of chemical reactions (blue) and physical operations (orange). Filtering and evaporation are examples of physical operations.

nuclear reaction so that, again, coupling to a spontaneous reaction affects a non-spontaneous reaction. Many chemical transformations are practical only if inexpensive energy sources are available. It is very desirable to use the energy from the sun directly to carry out chemical reactions. Nature, of course, does just this in photosynthesis (see Section 24.3), the process in which light is harvested by chlorophyll and used to convert water and carbon dioxide (low-free-energy states) to glucose and other sugars (high-free-energy states). Much current research is aimed at understanding photosynthesis and copying its principles to carry out other chemical transformations.

In an ideal chemical process, the desired product would be obtained (1) in 100% yield at room temperature and pressure, (2) at a convenient and controllable rate, and (3) already pure or in a condition that allows easy separation and purification. In real processes, compromises are necessary. The yield of a chemical reaction is fundamentally limited by thermodynamics, and so conditions of pressure and temperature must be adjusted to optimize it, often by expensive equipment. The rate of a reaction depends on pressure and temperature but is also sensitive to the presence of catalysts. The control of reaction rates is so important in the chemical industry that researchers constantly search for catalysts that will speed the formation of desired products without leading to unwanted byproducts. Finally, the importance of the last point—separability of the desired product—cannot be overemphasized. A process that yields a product contaminated with large amounts of impurities (or with dangerous impurities) may be abandoned even if it gives good yields at a convenient rate.

This brings us to the last term in the schematic equation for industrial processes: the byproducts. In some cases the byproducts are relatively harmless, but the question of what to do with them remains. Each year, a small mountain of $CaSO_4$ is produced in the synthesis of phosphate fertilizers; although $CaSO_4$ is harmless in itself (it is used in gypsum wallboard), large volumes of it create awkward and expensive disposal problems. Other byproducts are toxic to plant and animal life, ranging from slightly irritating to deadly. In the past, many of these toxic wastes were buried in landfills. This practice carries the risk of contamination of water sources and direct exposure of nearby residents and is now thoroughly regulated. Important social, economic, and legal incentives to eliminate or reduce worthless and hazardous byproducts already exist. Parts of the chemical industry have responded by modifying their processes to minimize the generation of wastes. Where this has not been possible, other measures have been taken to diminish the effects of production on the environment. Recycling of materials such as glass and paper and incineration of flammable wastes are the methods of choice, with adequate precautions taken to ensure complete combustion and with the use of alkaline scrubbers to remove acidic compounds from the effluent gas stream. In this way, even very hazardous organic compounds are converted to carbon dioxide and water vapor. Wastes that are not combustible can generally be made nonhazardous through chemical treatment. The relatively small fraction of wastes that cannot be incinerated or treated chemically can be put in a secure landfill.

The ideal way to dispose of byproducts is to find a use for them. The cost of waste disposal has encouraged research into ways to recycle the wastes or employ them differently. One example comes from the uranium fuel processing industry, in which scrap uranium contaminated with dirt and grease is oxidized by nitric acid to uranyl nitrate, $UO_2(NO_3)_2$, which is soluble in water and can therefore be purified. The oxidation reaction gives off the nitrogen oxides NO and $NO_2$, which are pollutants. Vented as waste, they would contribute to the formation of acid rain. Instead, these gases are caused to react first with oxygen and then with water to give nitric acid:

$$NO(g) + \tfrac{1}{2} O_2(g) \longrightarrow NO_2(g)$$

$$3\ NO_2(g) + H_2O(\ell) \longrightarrow 2\ HNO_3(aq) + NO(g)$$

The use of these reactions offers two advantages: reduction of the amount of nitrogen oxide released to the atmosphere and recycling of the nitric acid product for use in the oxidation reaction. A cost savings also results.

The picture of chemical processing painted so far has been a simplified description of raw materials being converted directly to final products. In fact, such conversions are usually multistep processes, with the involvement of many chemicals that do not appear in the finished product. Figure 20.2 shows how aspirin is made from a set of raw materials that can be traced all the way back to air, water, salt, sulfur, petroleum, and natural gas. Numerous chemicals appear only as intermediates; there is no sulfuric acid in the aspirin that is eventually produced, but sulfuric acid plays a key role in several steps of the manufacturing process. The figure also shows the interdependency built into the chemical industry. An aspirin manufacturer does not purchase salt, sulfur, and the other primary materials but rather buys intermediate chemicals that may have applications in a wide variety of other fields. Instead of manufacturing sulfuric acid in small quantities, the aspirin company buys it from a firm that makes far larger quantities—for use by steel mills, for example.

Table 20.1 shows the amounts of the chemicals produced in the greatest volumes in the United States during 1995. Relatively few are used *directly* as consumer products. Household ammonia, for example, is a negligible fraction of the total produced. Most ammonia undergoes further chemical transformations for use as fertilizer in forms that include ammonium nitrate, urea, and ammonium phosphate. Some

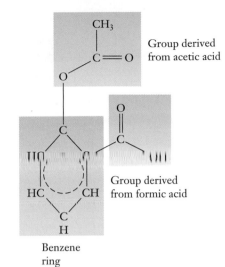

Group derived from acetic acid

Group derived from formic acid

Benzene ring

The structure of acetylsalicylic acid, the active component of aspirin. Note the substructures, which are those of three simpler compounds: benzene, formic acid, and acetic acid.

## TABLE 20.1

### Chemicals Produced in Largest Volume in the United States (1995)[†]

| Rank | Name | Formula | U.S. Production (billions of kg) | Principal End Use |
|---|---|---|---|---|
| 1 | Sulfuric acid | $H_2SO_4$ | 42.9 | Fertilizers, chemicals, processing |
| 2 | Nitrogen | $N_2$ | 30.6 | Fertilizers |
| 3 | Oxygen | $O_2$ | 24.1 | Steel, welding |
| 4 | Ethylene | $C_2H_4$ | 21.1 | Plastics, antifreeze |
| 5 | Lime | $CaO$ | 18.5 | Paper, chemicals, cement |
| 6 | Ammonia | $NH_3$ | 16.0 | Fertilizers |
| 7 | Phosphoric acid | $H_3PO_4$ | 11.8 | Fertilizers |
| 8 | Sodium hydroxide | $NaOH$ | 11.8 | Chemical processing, aluminum production, soap |
| 9 | Propylene | $C_3H_6$ | 11.6 | Gasoline, plastics |
| 10 | Chlorine | $Cl_2$ | 11.3 | Bleaches, plastics, water purification |
| 11 | Sodium carbonate | $Na_2CO_3$ | 10.0 | Glass |
| 12 | Methyl *tert*-butyl ether | $(CH_3)_3COCH_3$ | 7.9 | Gasoline additive |
| 13 | Ethylene dichloride | $C_2H_4Cl_2$ | 7.8 | Plastics, drycleaning |
| 14 | Nitric acid | $HNO_3$ | 7.8 | Fertilizers, explosives |
| 15 | Ammonium nitrate | $NH_4NO_3$ | 7.2 | Fertilizers, mining |
| 16 | Benzene | $C_6H_6$ | — | Chemicals, processing, solvent |

[†] Not included as "chemicals" are steel and concrete. Adapted from *Chemical and Engineering News*, April 10, 1996, p. 17.

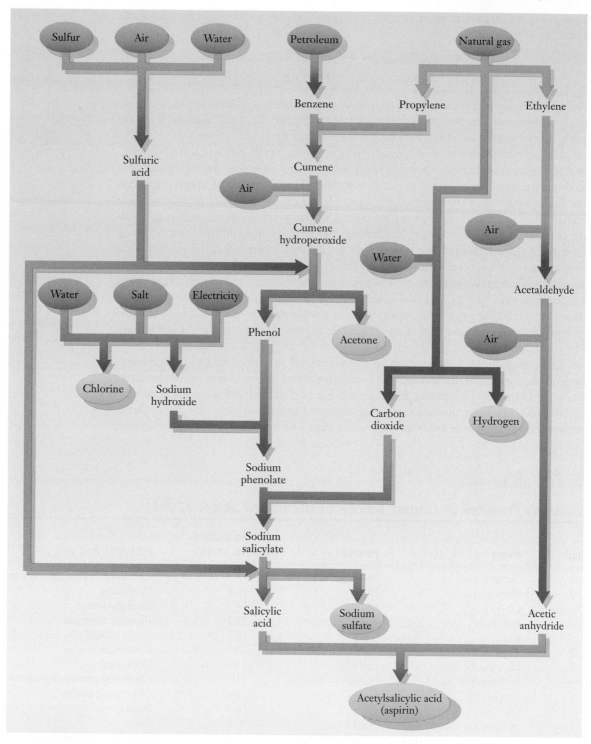

**FIGURE 20.2**  Process flow diagram for the synthesis of aspirin. The ultimate precursors of aspirin are air, water, salt, sulfur, petroleum, and natural gas.

chemicals that are not made in huge quantities are nevertheless important. Pharmaceuticals, for example, are vital to health care even though they are produced in relatively small amounts.

## Raw Materials for Chemistry

The raw materials of the heavy-chemicals industry are relatively few in number—chiefly air, water, limestone, salt, coal, natural gas and petroleum, sulfur, and phosphate rock—simply because these are abundantly available at low cost. Table 20.2 lists the sources for the important raw materials of the chemical industry. The atmosphere is the envelope of gases that surrounds the earth. The lithosphere includes the outer parts of the solid earth. The hydrosphere is the earth's water in its many forms. The biosphere comprises living organisms, their immediate surroundings, and their products.

The raw materials in chemistry that come from the atmosphere, hydrosphere, and lithosphere are the products of geochemical transformations over many millions of years, as will be discussed in greater depth in Section 23.2. The classifications of Table 20.2 are not absolute, because many materials move between hydrosphere and lithosphere as they dissolve and reprecipitate, or they are shaped by both biological and geochemical processes. Sodium chloride, for example, is extracted from salt brines (concentrated aqueous solutions of mixed salts) in addition to occurring in mineral form. Calcium carbonate, which is found most extensively in limestone deposits, also forms the shells of many marine animals and is obtained economically from them. It is the source of lime (CaO), the cheapest base available and a chemical of major importance.

The fossil fuels are important chemical raw materials produced by biological activity. The decay of plant matter through the action of bacterial reducing agents creates first peat, then lignite, and finally bituminous coal. Crude oil (petroleum) is thought to have formed from the decomposition of the remains of small marine

## TABLE 20.2

### Sources of Major Raw Materials for Chemistry

| Atmosphere | Hydrosphere |
|---|---|
| Oxygen ($O_2$) | Water ($H_2O$) |
| Nitrogen ($N_2$) | Sodium chloride (NaCl) |
| Noble gases | Sodium bromide (NaBr) |
| | Sodium iodide (NaI) |
| | Magnesium chloride ($MgCl_2$) |

| Lithosphere | Biosphere |
|---|---|
| Silica ($SiO_2$) | Coal (C) |
| Limestone ($CaCO_3$) | Petroleum |
| Sodium chloride (NaCl) | Natural gas ($CH_4$) |
| Potassium chloride (KCl) | Sulfur (S) |
| Sodium carbonate ($Na_2CO_3$) | Phosphate rock ($Ca_5(PO_4)_3F$) |
| Sodium sulfate ($Na_2SO_4$) | Organic natural products |
| Metal ores | |

The foxglove plant produces digitalis, an important drug in the treatment of heart disease. (*John Gerlach/ Tom Stack & Associates*)

organisms under reducing conditions. Hydrocarbons of low molar mass that formed in this way are gases rather than liquids. Pockets of natural gas (primarily methane, with smaller amounts of hydrogen, ethane, and propane) are found in association with deposits of petroleum. Although all of these fossil fuels have biological origins, geological transformations have brought them to the forms in which they are found today.

A chemical raw material with a somewhat more unexpected biological origin is sulfur. Although sulfur is widely distributed in compounds, its most convenient commercial source is the deposits of elemental sulfur often associated with salt domes under the surface of the earth. Sulfur is thought to have precipitated in these settings as calcium sulfate, in contact with natural gas and carbon dioxide. Reaction of these substances to give elemental sulfur and calcium carbonate was then catalyzed by enzymes secreted by certain bacteria.

$$4 \ CaSO_4(s) + 3 \ CH_4(g) + CO_2(g) \longrightarrow 4 \ CaCO_3(s) + 4 \ S(s) + 6 \ H_2O(\ell)$$

The origins of phosphate rock, which has the approximate chemical formula $Ca_5(PO_4)_3F$, are less clear. The rock is found in large deposits in certain parts of the world, notably Florida and North Africa. Phosphate ions are present in rather low concentration in seawater, and the solubility of calcium phosphate is comparable to that of calcium carbonate. Therefore, it is surprising that the phosphate is found in such concentration rather than scattered through the world's limestone deposits. The concentrated deposits of phosphate rock may have resulted from extensive dissolution and reprecipitation of phosphate–carbonate mixtures, perhaps in a small but critical pH range in which the phosphate precipitated and the carbonate remained in solution. It is more likely, however, that the phosphate rock resulted from accumulation by marine organisms. Phosphate rock is the source of phosphorus and phosphate-based fertilizers. As such, it is the raw material used in greatest volume in the chemical industry, as Chapter 21 will discuss further.

Other natural products obtained from plants and animals play a large role in chemistry even though they may not be employed in billion-kilogram quantities. So many natural products are in common use that it is impossible to list even a small fraction of them. Penicillin, for example, was first produced from the accidental contamination of a bacterial culture by mold. The terpenes are a class of compounds found in almost all plants; they are a major component of turpentine, which is obtained from the resin of pine trees. Many terpenes are used directly as artificial scents in the perfume industry; others serve as precursors in the manufacture of vitamin A (Fig. 20.3).

**FIGURE 20.3** (a) The compound β-ionone, $C_{13}H_{20}O$, is a terpene that is used extensively in perfumes. (b) Its more significant use is as a precursor in the synthesis of vitamin A. Note the close connection between the structure of vitamin A and that of retinal, a molecule crucial for vision (see problem 67 in Chapter 16).

(a) β-ionone

(b) Vitamin A

The distinction between chemical synthesis and biological synthesis has in fact blurred with time. On the one hand, living agents such as specialized bacteria are increasingly used to carry out the synthesis of complex products that are difficult to make by conventional chemical means. On the other hand, certain natural products are more easily obtained by synthesizing them from other starting materials than by extracting them from existing sources. Methanol, for example, used to be obtained by the destructive distillation of wood—hence the name "wood alcohol"—but it is now produced mainly from synthesis gas (discussed in the next section).

## 20.2

### HYDROGEN

Hydrogen is the most abundant atomic species in the universe; at the earth's surface it ranks third in abundance, behind only oxygen and silicon. This section describes how elemental hydrogen is produced and used.

### Production of Hydrogen

Hydrogen constitutes 11.2% of the mass of water and is therefore abundantly available nearly everywhere on the earth's surface. From the point of view of easy access to starting materials, the ideal method for producing elemental hydrogen would be to split water:

$$H_2O(\ell) \longrightarrow H_2(g) + \tfrac{1}{2} O_2(g)$$

This reaction can be carried out in an electrolytic cell, as Section 12.7 discusses (see also Fig. 1.5). Commercial cells use solutions of sodium hydroxide, rather than pure water, to permit the passage of large currents. Although electrolysis is useful when very pure hydrogen is needed, it is not a major source of hydrogen today because of the prohibitive cost of the electricity. Considerable research is being devoted to the possibility of using sunlight directly to split water in the presence of a catalyst, but trial processes of this type are far from being practical for the large amounts of hydrogen needed.

The reaction of a strong acid with iron produces gaseous hydrogen:

$$H_2SO_4(aq) + Fe(s) \longrightarrow H_2(g) + FeSO_4(aq)$$

The French scientist Jacques Charles used this reaction in 1783 to generate hydrogen for the first human flight in a hydrogen balloon (Fig. 20.4). The similar reaction of zinc with acid is a useful laboratory-scale reaction for the production of hydrogen. Neither of these reactions is useful for making hydrogen on a large scale, however.

Currently, the major source for hydrogen is the methane in natural gas, and Figure 20.5 shows the method of production. There are two crucial steps in this process. The first is the **reforming reaction:**

$$CH_4(g) + H_2O(g) \longrightarrow CO(g) + 3\,H_2(g) \qquad \Delta H° = +206 \text{ kJ}$$

This reaction is endothermic and proceeds with an increase in the chemical amount of gas from 2 mol to 4. Accordingly, it is carried out at high temperatures (750 to 1000°C) and low total pressures to increase yield. An excess of steam is added to drive the reaction toward completion, and a nickel catalyst is used. The gas mixture

**FIGURE 20.4**   The French chemist Jacques Charles took part in the first hydrogen balloon ascent from Paris on December 1, 1783. (*The Granger Collection*)

**FIGURE 20.5** The reforming and shift reactions produce hydrogen and carbon dioxide from methane and steam.

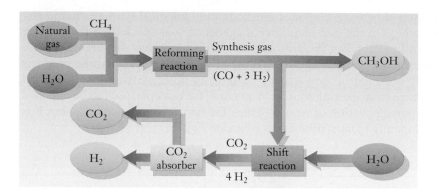

produced, called **synthesis gas,** is used directly to make methanol ($CH_3OH$) and other chemicals. If a feedstock other than methane is used, synthesis gas is obtained with different ratios of carbon monoxide to hydrogen. With propane, for example, the reforming reaction is

$$C_3H_8(g) + 3\ H_2O(g) \longrightarrow 3\ CO(g) + 7\ H_2(g)$$

The second step in the production of hydrogen is called the **shift reaction.** In it, additional steam reacts with the carbon monoxide from the first step:

$$CO(g) + H_2O(g) \longrightarrow CO_2(g) + H_2(g) \qquad\qquad \Delta H° = -41\ kJ$$

This reaction is exothermic and so occurs to a greater extent at low temperatures. As is frequently the case with exothermic processes, however, the temperature cannot be lowered too far because the reaction then becomes too slow. The best compromise comes at a temperature of about 350°C, with an iron oxide catalyst. Two things are accomplished with the shift reaction: (1) additional hydrogen beyond that from the reforming reaction is obtained, and (2) carbon dioxide is obtained as a byproduct. The shift reaction is the major commercial source of carbon dioxide, which has many uses. Carbon dioxide is frozen and sold for refrigeration (Dry Ice). In liquid form it is used to fill fire extinguishers and to make carbonated water for sparkling beverages. Gaseous $CO_2$ is used in chemical synthesis, such as in the production of salicylic acid for aspirin (see Fig. 20.2).

At present, natural gas is the primary raw material for making hydrogen and carbon dioxide. As reserves of natural gas are consumed, other sources must be developed. One older process that is attracting new attention is the reaction of coal with steam:

$$C(s) + H_2O(g) \longrightarrow CO(g) + H_2(g) \qquad\qquad \Delta H° = +131\ kJ$$

This mixture of product gases is called **water gas** and burns with a blue flame because of its carbon monoxide content. It has less hydrogen proportionately than the synthesis gas produced from methane or higher hydrocarbons. Water gas can react further, in the shift reaction, to give additional hydrogen and carbon dioxide. Once a mixture of CO and $H_2$ is prepared with the appropriate proportions the reforming reaction just described can be reversed to make methane for use as a fuel; the overall process is referred to as the gasification of coal.

## The Uses of Hydrogen

Hydrogen does not appear on the list of the top 16 chemicals in Table 20.1, or even on a list of the top 50 in a given year. One reason is that it is the lightest element, and so it ranks low on a mass basis. A second reason is that the list refers to the sales or shipments of chemicals and does not include amounts that are **captive**— that is, used directly after production to make other chemical end-products. This is very much the case with hydrogen, which is a far more important chemical than its commercial ranking indicates. Its major use is for reaction with nitrogen to make ammonia for fertilizers, a process Section 21.3 considers in greater detail. Chemical plants that produce ammonia use natural gas, steam, and nitrogen as starting materials, and hydrogen plays only the role of an intermediate. Another captive use for hydrogen is in making methanol. As already mentioned, this synthesis is carried out with the product from the reforming reaction, synthesis gas:

$$CO(g) + 2\ H_2(g) \longrightarrow CH_3OH(g) \qquad\qquad \Delta H° = -90\ kJ$$

Low temperatures (300°C) and high pressures (250–300 atm) are used. In most cases, the same plant makes ammonia, methanol, carbon dioxide, and hydrogen.

One final use of hydrogen is its addition to carbon–carbon double bonds (which are present in what are called **unsaturated hydrocarbons** because they are capable of taking up more hydrogen) to convert them to single bonds **(saturated compounds).** The simplest example is the reduction of ethylene to ethane:

$$H_2C{=}CH_2 + H_2 \longrightarrow H_3C{-}CH_3$$

Hydrogen is used extensively in petroleum refining to eliminate carbon–carbon double bonds. It is also employed in food processing to convert liquid vegetable oils to solids. For example, a large proportion of sunflower oil is an oily liquid composed of molecules with the structural formula

$$
\begin{array}{c}
\quad\quad\quad\quad\quad\quad\quad\quad\quad\quad\quad\quad\quad\quad\quad\overset{\textstyle O}{\overset{\|}{\phantom{C}}} \\
CH_3(CH_2)_4CH{=}CHCH_2CH{=}CH(CH_2)_7C{-}OCH_2 \\
\quad\quad\quad\quad\quad\quad\quad\quad\quad\quad\quad\quad\quad\quad\quad\quad\quad\quad\quad | \\
CH_3(CH_2)_4CH{=}CHCH_2CH{=}CH(CH_2)_7COOCH_2 \\
\quad\quad\quad\quad\quad\quad\quad\quad\quad\quad\quad\quad\quad\quad\quad\quad\quad\quad\quad | \\
CH_3(CH_2)_4CH{=}CHCH_2CH{=}CH(CH_2)_7C{-}OCH_2 \\
\quad\quad\quad\quad\quad\quad\quad\quad\quad\quad\quad\quad\quad\quad\quad\overset{\|}{\underset{\textstyle O}{\phantom{C}}}
\end{array}
$$

This is a **polyunsaturated oil** because there are six C=C double bonds per molecule. **Hydrogenation** with 6 mol of $H_2$ in the presence of a catalyst saturates this oil, and the product has a high enough melting point to make it a solid at room conditions. The use of solid fats (or solidified oils) offers advantages in food processing and preservation, so that "hydrogenated vegetable oil" is an ingredient in many foodstuffs. Heavy consumption of saturated fats has been linked to diseases of the heart and circulatory system, however, and the presence of *un*hydrogenated (polyunsaturated) oils in foods is now heavily advertised.

---

## 20.3

## EXTRACTIVE METALLURGY

The recovery of metals from their sources in the earth is the science of **extractive metallurgy,** a discipline that draws on chemistry, physics, and engineering for its methods. As a science it is a comparatively recent subject, but its beginnings, which were evidently in the Near East about 6000 years ago, marked the emergence of humanity from the Stone Age. The earliest known metals were undoubtedly gold, silver, and copper, because they could be found in their native (elemental) states

**FIGURE 20.6** A specimen of native copper. *(Charles D. Winters)*

(Fig. 20.6). Gold and silver were valued for their ornamental uses, but they are too soft to have been made into tools. Iron was also found in elemental form (although rarely) in meteorites.

Most metals are combined with other elements, such as oxygen and sulfur, in ores, and chemical processes are required to free them. As Table 20.3 shows, the free energies of formation of most metal oxides are negative, indicating that the reverse reactions, which yield the free metal and oxygen, have positive free energy changes. Scientists can carry out such a reverse reaction only by coupling it with a second, spontaneous chemical reaction. The greater the cost in free energy, the more difficult the production of the free metal. Thus, silver and gold (at the bottom of Table 20.3) exist in nature as elements, and mercury can be released from its oxide or sulfide ore (cinnabar) merely by moderate heating (see Fig. 1.6). Winning copper, zinc, and iron requires more stringent conditions; the ores of these metals are reduced in chemical reactions at high temperatures, collectively called **pyrometallurgy**. These reactions are carried out in huge furnaces, in which a fuel such as coke (coal from which volatile components have been expelled) serves both as the reducing agent and as the source of heat to maintain the required high temperatures. The process, called **smelting**, involves both chemical change and melting. The combustion of carbon

$$C(s) + O_2(g) \longrightarrow CO_2(g) \qquad \Delta G° = -394 \text{ kJ mol}^{-1}$$

provides the driving force for the overall reaction. Even smelting is not sufficient for the metals at the top of Table 20.3, which have the most negative free energies of formation for their oxides (and for their sulfides as well). Electrometallurgical methods, to be considered in Section 20.4, are needed.

---

## TABLE 20.3

### Metal Oxides Arranged According to Ease of Reduction

| Metal Oxide | Metal | $n^†$ | $\Delta G_f°/n^†$ (kJ mol$^{-1}$) | A Method of Production of the Metal |
|---|---|---|---|---|
| MgO | Mg | 2 | −285 | Electrolysis of MgCl$_2$ |
| Al$_2$O$_3$ | Al | 6 | −264 | Electrolysis |
| TiO$_2$ | Ti | 4 | −222 | Reaction with Mg |
| Na$_2$O | Na | 2 | −188 | Electrolysis of NaCl |
| Cr$_2$O$_3$ | Cr | 6 | −176 | Electrolysis, reduction by Al |
| ZnO | Zn | 2 | −159 | Smelting of ZnS |
| SnO$_2$ | Sn | 4 | −130 | Smelting |
| Fe$_2$O$_3$ | Fe | 6 | −124 | Smelting |
| NiO | Ni | 2 | −106 | Smelting of nickel sulfides |
| PbO | Pb | 2 | −94 | Smelting of PbS |
| CuO | Cu | 2 | −65 | Smelting of CuFeS$_2$ |
| HgO | Hg | 2 | −29 | Moderate heating of HgS |
| Ag$_2$O | Ag | 2 | −6 | Found in elemental form |
| Au$_2$O$_3$ | Au | 6 | >0 | Found in elemental form |

$^†$ The standard free energies of formation of the metal oxides in kJ mol$^{-1}$ are adjusted for fair comparison by dividing them by $n$, the total decrease in oxidation state required to reduce the metal atoms contained in the oxide to oxidation states of 0. Thus, the reduction of Cr$_2$O$_3$ involves a change in the oxidation state of two chromium atoms from +3 to 0, so that $n = 2 \times 3 = 6$.

When the oxide of a metal is reduced to elemental form, the molar free energies of formation in Table 20.3 are divided by the total change in the oxidation number. This allows a fair comparison of the relative ease of reduction of two compounds that have the metals in different oxidation states. Reducing 1 mol of $Fe_2O_3$ to elemental iron, for example, requires oxidizing three times as much carbon (to $CO_2$) as reducing 1 mol of ZnO, and so its free energy of formation is divided by 6 in the table; the free energy of formation of ZnO is divided by only 2. Rank in the table tells which metals can reduce which metal oxides. The standard free energy change for the reduction

$$Cr_2O_3(s) + 3\ Mg(s) \longrightarrow 2\ Cr(s) + 3\ MgO(s)$$

is

$$\Delta G° = 3\ \Delta G_f°(MgO) - \Delta G_f°(Cr_2O_3) = 6 \times (\tfrac{1}{2}\ \Delta G_f°(MgO) - \tfrac{1}{6}\ \Delta G_f°(Cr_2O_3))$$

Because MgO lies above $Cr_2O_3$ in the table, the difference enclosed in parentheses is negative, and this reaction is spontaneous. A similar computation for any pair of entries in the table confirms that a metal will spontaneously reduce the oxide of any element below it in the table. Thus, magnesium is used commercially to produce titanium from its ore, rutile ($TiO_2$). Once the reaction is initiated, aluminum reacts spectacularly with iron(III) oxide in the thermite reaction (see Fig. 7.11) to give iron and $Al_2O_3$. The metals at the top of the table are the most difficult to produce in terms of thermodynamics and, once made, are the most reactive. It is no accident that a correlation exists between the reactivity of a metal and its date of first isolation in elemental form.

The remainder of this section is devoted to the production and uses of two of the most important metals produced by chemical reduction, copper and iron. Section 20.4 discusses two other important metals, aluminum and magnesium, which require electrochemical processing. Four other metals have been introduced through the Cumulative Exercises earlier in the book: titanium (see Chapter 2), nickel (see Chapter 8), manganese (see Chapter 12), and platinum (see Chapter 18).

## Copper and Its Alloys

Copper occurs in a variety of minerals in the form of veins that are frequently exposed as outcrops at the earth's surface. Chalcopyrite ($CuFeS_2$), chalcocite ($Cu_2S$), and bornite ($Cu_5FeS_4$) are chief among the copper-bearing ores. On exposure to the weathering action of the atmosphere, they are oxidized to beautiful green or blue basic copper carbonates such as malachite, $Cu_2CO_3(OH)_2$, and azurite, $Cu_3(CO_3)_2(OH)_2$, both of which are semiprecious stones (Fig. 20.7). We speculate that metallurgy began when a potter introduced an ore such as malachite into the kiln, perhaps as a colored glaze, and found a globule of a red metal after firing the ore to high temperatures in the reducing atmosphere generated by the burning of carbon (charcoal):

$$Cu_2CO_3(OH)_2(s) + C(s) \longrightarrow 2\ Cu(s) + 2\ CO_2(g) + H_2O(g)$$

This would not have been pure copper, of course. Pure copper is not hard enough to make serviceable cutting tools, although to some extent it can be "work-hardened" by hammering. We cannot be sure whether the use of tin in bronze originated by design or from the accidental use of ores that contained minor concentrations of tin as an impurity. But its appearance, at first in trace amounts and increasing to about 10% as the centuries passed, surely marks the birth of a remarkably sophisticated

**FIGURE 20.7** Several gems from copper minerals. Clockwise from top: malachite with azurite, malachite, azurite with chalcopyrite, and turquoise. *(Leon Lewandowski)*

**FIGURE 20.8** A bronze ax-head from Luristan (Persia) that dates from about 1200 B.C. *(Leon Lewandowski)*

industry. Bronze is an alloy that contains about 10% tin and 90% copper (Fig. 20.8). It is hard without being brittle, can be cast in a mold, and melts at a temperature (950°C) substantially below the melting point of pure copper (1084°C).

Brass is a second important alloy of copper, containing up to 42% zinc along with copper. Brass was first discovered by the Romans, and to this day it is prized as a material for precision instruments because it is hard, easy to machine, and resistant to corrosion.

The modern high demand for copper stems largely from its use in wires for conducting electricity and has led to worldwide exploitation of copper ores, of which the most abundant is chalcopyrite ($CuFeS_2$). Open-pit mining operations on a mammoth scale (Fig. 20.9) remove and crush tons of rock that may contain less than 1% copper. **Froth flotation** is used to concentrate the copper ores (Fig. 20.10). In this process, crushed low-grade ore is mixed with water, a special oil, and a detergent. A stream of compressed air produces a foamy mixture in which oil-wetted particles of ore are buoyed up by clinging to air bubbles, while water-wetted earth and rock sink to the bottom. The copper-rich froth is skimmed off, the oil is separated for reuse, and (after drying) a solid mass enriched in $CuFeS_2$ (typically to 25–30% copper) remains.

The next step is the **roasting** of this enriched ore—reaction with air at high temperatures to convert the iron to its oxide while leaving the copper as a sulfide:

$$6\ CuFeS_2(s) + 13\ O_2(g) \longrightarrow 3\ Cu_2S(s) + 2\ Fe_3O_4(s) + 9\ SO_2(g)$$

In the next stage, the mixture of $Cu_2S$ and $Fe_3O_4$ is put in a furnace at 1100°C; limestone ($CaCO_3$) and silica ($SiO_2$) are added as a flux to remove the iron:

$$CaCO_3(s) + Fe_3O_4(s) + 2\ SiO_2(s) \longrightarrow CO_2(g) + CaO \cdot FeO \cdot Fe_2O_3 \cdot (SiO_2)_2(\ell)$$
$$\text{(Slag, composition variable)}$$

Because $Cu_2S(\ell)$ is not soluble in the calcium–iron slag and exceeds it in density, it settles to the bottom, and the molten slag layer can be poured off. The $Cu_2S(\ell)$ is run into another furnace through which additional air is blown to cause the redox reaction

$$Cu_2S(\ell) + O_2(g) \longrightarrow 2\ Cu(\ell) + SO_2(g)$$

The liquid copper is then cooled and cast into large slabs of "blister copper" for further purification. The final refinement routinely employs an electrochemical method (Section 20.4) to recover the small amounts of gold and silver that have accompanied the copper to this stage.

**FIGURE 20.9** An open-pit copper mine. The ore contains less than 1% copper, but the metal can nevertheless be extracted economically, depending on fluctuations in the world market price. *(Byron Augustin/ Tom Stack & Associates)*

Three major problems afflict the smelting of copper: the pollution that can be caused by the $SO_2$ byproduct, the disposal of the calcium–iron–silicate slag that can contain toxic heavy-metal impurities such as cadmium, and the energy costs associated with operating high-temperature furnaces. Emission control regulations have required new processes in which nearly all of the $SO_2$ is captured and converted to commercially important sulfuric acid (see Chapter 21). Careful attention to recycling water in newer smelters has greatly reduced the amount of heavy metals leached out from the slag. Finally, energy costs are reduced by new **hydrometallurgical** methods for separating the copper and iron in the ore. When an aqueous mixture of copper(II) chloride and iron(III) chloride is added to the ore, the copper-bearing minerals are reduced in the reactions

$$CuFeS_2(s) + 3\ CuCl_2(aq) \longrightarrow 4\ CuCl(s) + FeCl_2(aq) + 2\ S(s)$$
$$CuFeS_2(s) + 3\ FeCl_3(aq) \longrightarrow CuCl(s) + 4\ FeCl_2(aq) + 2\ S(s)$$

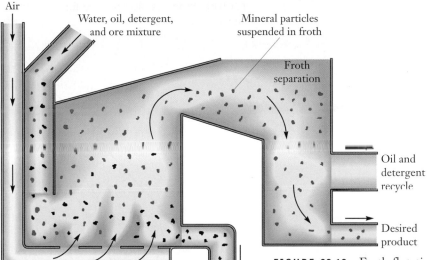

**FIGURE 20.10** Froth flotation for the enrichment of ore. Frothing agents are added to a water suspension of the pulverized ore. As compressed air is bubbled in, the desired mineral ($CuFeS_2$, in the case of copper production) is concentrated at the surface of a foam and borne upward while the unwanted rocky material settles to the bottom.

Excess aqueous sodium chloride is added to the solids that form to convert the $CuCl(s)$ to the soluble complex ion $[CuCl_2]^-(aq)$, which is separated from the sulfur and converted to copper and $CuCl_2$ in a disproportionation reaction:

$$2\ CuCl_2^-(aq) \longrightarrow Cu(s) + CuCl_2(aq) + 2\ Cl^-(aq)$$

The copper(II) chloride is recycled to reduce more ore.

## Iron and Steel

The chief sources of iron are the minerals hematite ($Fe_2O_3$), magnetite ($Fe_3O_4$), goethite ($FeO(OH)$), limonite ($FeO(OH)\cdot nH_2O$), and siderite ($FeCO_3$). The metallurgy of iron requires considerably higher temperatures than that of copper, and even though forced drafts were used to fan the fires, the molten iron produced in early times merely formed globules interspersed in a semiliquid slag. By repeatedly heating and hammering the red-hot mass, the more fluid slag could be squeezed out to yield a product known as wrought iron. Such iron was fairly soft and usually inferior to bronze, until artisans learned that prolonged heating of iron in contact with charcoal caused it to absorb carbon. When iron containing several percent of dissolved carbon was quenched in water, it produced an extremely hard metal, a form of steel. This technology, developed by the Hittites around 1200 B.C., was further refined in Greco-Roman times and then was little changed until the invention of the **blast furnace** in the late 14th and early 15th centuries.

Figure 20.11 illustrates the design of a blast furnace for smelting iron ore. It consists of a cylindrical steel shell some 10 m in diameter at its widest part and about 30 m high, lined with a refractory "fire brick." Designed for continuous operation, it is run for about five years, after which its lining must be torn out and replaced. A charge of iron ore, coke, and limestone ($CaCO_3$) or dolomite ($CaMg(CO_3)_2$) enters the top of the stack through a hopper at intervals and works its way downward as the reaction proceeds. Preheated air enters the stack through nozzles called **tuyeres** above the **hearth,** where the molten iron collects.

**FIGURE 20.11** A blast furnace for smelting iron ore.

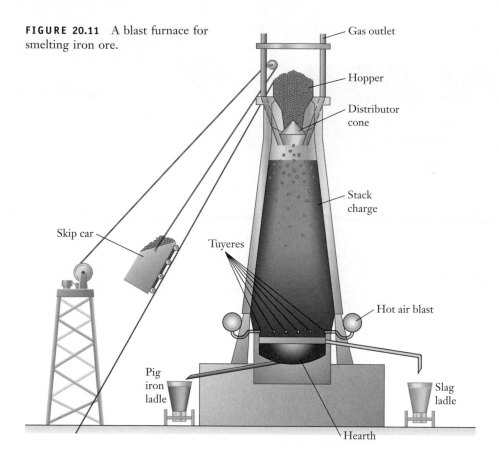

High-grade iron ore is not pure $Fe_2O_3$ (70% Fe) but contains up to about 27% silica, alumina, clay, and minor impurities that must be converted to a fluid slag for the process to be continuous. That is why limestone or dolomite is added with the charge; each converts silica and alumina (melting points 1700° and 2050°C, respectively) to lower melting calcium aluminum silicates (m.p. approximately 1550°C):

$$CaCO_3(s) \longrightarrow CaO(s) + CO_2(g)$$

$$CaO(s) + Al_2O_3(s) + 2\ SiO_2(s) \longrightarrow CaO \cdot Al_2O_3 \cdot (SiO_2)_2(\ell)$$
$$\text{(Composition variable)}$$

Hematite is reduced not by direct reaction with carbon itself but by carbon monoxide in the hot gases as they pass up through the charge under forced draft:

$$Fe_2O_3(s) + 3\ CO(g) \longrightarrow 2\ Fe(s) + 3\ CO_2(g)$$

The carbon monoxide is produced in highly exothermic reactions between hot coke and the preheated air that enters through the tuyeres:

$$2\ C(s) + O_2(g) \longrightarrow 2\ CO(g)$$

At this level in the blast furnace, the temperature is a maximum (about 1875°C). At this high temperature, the equilibrium between carbon dioxide and carbon monoxide

$$C(s) + CO_2(g) \rightleftharpoons 2\ CO(g)$$

lies strongly to the right. The equilibrium constant of this reaction at 1875°C is readily calculated from data in Appendix D (if we assume temperature-independent enthalpies and entropies) to be about $1 \times 10^5$ (see problem 39). The reduction of hematite to iron occurs as a heterogeneous gas–solid reaction fairly high in the stack, where the temperature is still well below the melting point of pure iron (1535°C). The result is a spongy form of iron interspersed with the other components of the charge. As the mass slowly descends through the stack, it encounters increasingly higher temperatures until, just above the tuyeres, both the iron and the slag components liquefy and trickle down into the hearth of the furnace. Molten calcium aluminum silicate (slag) is less dense than molten iron and floats on it. Periodically, the two liquids are separately drawn off into ladle cars, the iron to be further refined and the slag to be solidified into clinker, which is ground up for use in cement or crushed for use as road ballast. The iron produced in the blast furnace, called **pig iron,** contains a total of 5 to 10% carbon, silicon, phosphorus, and other impurities. A modern blast furnace produces about 1500 metric tons of pig iron and 750 metric tons of slag per day while consuming about 1200 metric tons of coke.

Steel is iron mixed with a small amount of carbon (up to about 1.7% by mass) and other alloying elements. It is made from pig iron by removing silicon, phosphorus, manganese, sulfur, and other impurities and lowering the carbon content to a target value that depends on the intended end use. Modern methods of conversion rely on the oxidation of the unwanted elements in the molten pig iron. The oxides escape as gases or react with a flux, which is added for this purpose, to form a slag phase that separates from the hot metal. The greatest quantities of steel are produced in **basic oxygen converters**—barrel-shaped furnaces open at one end and lined with refractory bricks. A converter stands 10 meters high, has a diameter of $5\frac{1}{2}$ meters, and can be tilted to allow the starting materials to be tipped in or product to be poured out. Such a converter is charged with some 150 to 250 tons of molten pig iron, which is transferred in insulated ladles from a blast furnace (Fig. 20.12). Cold scrap steel is also often added. The converter is then turned upright under a hood (to collect gases) and a high-velocity jet of pure oxygen is blown down onto the charge through nozzles in a water-cooled retractable lance positioned above the surface of the molten charge (Fig. 20.13). More oxygen is blown in through bottom-mounted tuyeres. Each tuyere is protected from burning by injection of a small amount of propane gas through an annular space at its end. The propane forms a sheath around the tuyeres. Its endothermic decomposition

$$C_3H_8 \longrightarrow 3\ C + 4\ H_2$$

cools and protects the tuyeres.

Within seconds after blowing starts, strong oxidation commences in the pig iron. Some iron is oxidized to FeO and is rapidly distributed throughout the charge:

$$2\ Fe + O_2 \longrightarrow 2\ FeO$$

Silicon burns directly to silicon dioxide, $SiO_2$, or reacts with FeO to give the same product:

$$Si + O_2 \longrightarrow SiO_2$$

$$2\ FeO + Si \longrightarrow 2\ Fe + SiO_2$$

The acidic $SiO_2$ then reacts with the basic flux (typically lime, CaO, which has been blown into the mixture along with the oxygen) and forms a product that separates from the hot metal as slag:

$$SiO_2 + CaO \xrightarrow[\text{In slag}]{} CaO \cdot SiO_2(\ell)$$

**FIGURE 20.12** Molten iron being charged into a basic oxygen furnace. The basic oxygen process for steel-making is an *oxygen* process because it employs a blast of oxygen, rather than air, to oxidize impurities. It is a *basic* process because the flux contains a base, lime, to capture the acidic oxides (such as $SO_2$ and $P_4O_{10}$) of those impurities. (*Courtesy of Bethlehem Steel Corporation*)

**FIGURE 20.13** The operation of a basic oxygen converter. Blasts of pure oxygen from above (the lance) and below (the tuyeres) refine pig iron into steel. After processing, the converter is tipped on its side and the molten steel is tapped off through the taphole. The impurity-containing slag floats atop the steel and is easily separated.

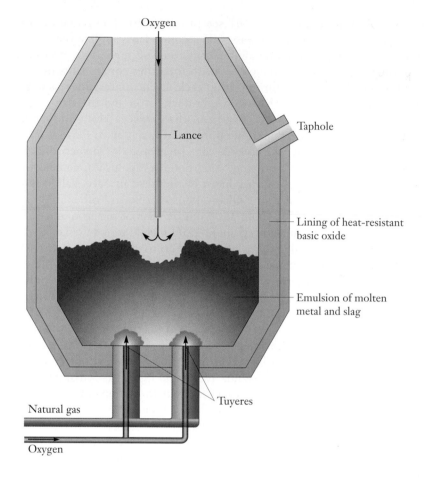

Oxidation of the phosphorus and its removal into the slag proceed similarly, the overall equation being

$$5\ FeO + 2\ P + 3\ CaO \xrightarrow[\text{In slag}]{} 5\ Fe + (CaO)_3 \cdot P_2O_5(\ell)$$

The slag also takes up MnO (as it forms by oxidation of manganese in the pig iron) and some FeO; both of these metal oxides are acidic. Carbon, the major impurity in the pig iron, is oxidized to gaseous carbon monoxide:

$$C + FeO \longrightarrow Fe + CO$$
$$C + \tfrac{1}{2}\ O_2 \longrightarrow CO$$

Under the conditions in the converter, essentially no carbon dioxide ($CO_2(g)$) is produced. The escape of $CO(g)$ from the melt causes a boiling action that mixes the slag and metal phases thoroughly and hastens the transfer of oxides into the slag. All of the oxidation reactions are exothermic. The rising temperature in the converter melts the scrap steel and keeps the slag molten. A probe plunged into the melt measures the temperature (a good indicator of carbon content) and oxygen content of the steel. The data are used under automatic computer control to adjust the blowing rate, the rate of addition of flux, and other operating conditions to reach the desired carbon content and temperature simultaneously. The process is complete in 25 to 30 minutes.

Oxygen, hydrogen, and nitrogen are soluble in molten steel and must be removed before the steel can be cast. Otherwise, the escape of the dissolved gases would cause large void spaces (blowholes) in the casting. Dissolved oxygen also can react with alloying elements to form particles of oxides that remain in the finished steel as internal and surface defects. The next stage of steel refining is therefore **vacuum degassing.** Large vacuum pumps reduce the pressure above the steel as argon gas, which is not soluble in molten steel, is blown through the melt from the bottom to speed the escape of dissolved gases. Toward the end of the degassing, aluminum is added. This further reduces the level of oxygen in the steel by the reactions

$$2 \ Al + \tfrac{3}{2} \ O_2 \longrightarrow Al_2O_3$$

$$2 \ Al + 3 \ FeO \longrightarrow Al_2O_3 + 3 \ Fe$$

The insoluble aluminum oxide forms an easily separated layer. Any desired alloying elements are also added during degassing.

The carbon content of fully refined steel is far lower than that of the original pig iron. For example, in steel intended for automobile body sheets, carbon is reduced to ultralow levels (0.002% C by mass). Ultralow-carbon steels have better press-forming characteristics than ordinary low-carbon or mild steel (0.1 to 0.4% C by mass). The basic oxygen process allows the production, at short intervals, of large batches of steel of controlled chemical make-up. It has major advantages over other ways of making steel, particularly when the steel is taken directly for the continuous casting of sheets. In recent years, the basic oxygen process has entirely supplanted the formerly dominant open-hearth process, in which pig iron was cooked for several hours in oxygen-enriched air with a basic flux.

---

## 20.4

## ELECTROMETALLURGY

The methods of pyrometallurgy and hydrometallurgy considered in Section 20.3 are unsuitable for production of metals such as aluminum and the alkali and alkaline-earth elements from their ore sources. These metals have particularly high free energies relative to their compounds in readily available ores. **Electrometallurgy,** or electrolytic production, provides the best ways of winning such elements from their ores. Electrochemical cells are also used to purify the metals produced by the techniques of pyrometallurgy.

### Aluminum

Aluminum is the third most abundant element in the earth's crust (after oxygen and silicon), accounting for 8.2% of the total mass. It occurs most commonly in association with silicon in the aluminosilicates of feldspars and micas and in clays, the products of weathering of these rocks. The most important ore for aluminum production is bauxite, a hydrated aluminum oxide that contains 50 to 60% $Al_2O_3$; 1 to 20% $Fe_2O_3$; 1 to 10% silica; minor concentrations of titanium, zirconium, vanadium, and other transition-metal oxides; and the balance (20 to 30%) water. Bauxite is purified via the **Bayer process,** which takes advantage of the fact that the amphoteric oxide alumina is soluble in strong bases but iron(III) oxide is not. Crude bauxite is dissolved in sodium hydroxide

$$Al_2O_3(s) + 2\ OH^-(aq) + 3\ H_2O(\ell) \longrightarrow 2\ Al(OH)_4^-(aq)$$

and separated from hydrated iron oxide and other insoluble impurities by filtration. Pure hydrated aluminum oxide precipitates when the solution is cooled to super-saturation and seeded with crystals of the product:

$$2\ Al(OH)_4^-(aq) \longrightarrow Al_2O_3 \cdot 3H_2O(s) + 2\ OH^-(aq)$$

The water of hydration is removed by calcining at high temperature (1200°C).

Compared with copper, iron, gold, and lead, which were known in antiquity, aluminum is a relative newcomer. Sir Humphry Davy obtained it as an alloy of iron and proved its metallic nature in 1809. It was first prepared in relatively pure form by H. C. Oersted in 1825, by reduction of aluminum chloride with an amalgam of potassium dissolved in mercury,

$$AlCl_3(s) + 3\ K(Hg)_x(\ell) \longrightarrow 3\ KCl(s) + Al(Hg)_{3x}(\ell)$$

after which the mercury was removed by distillation. Aluminum remained largely a laboratory curiosity until 1886, when Charles Hall in the United States (then a 21-year-old graduate of Oberlin College) and Paul Héroult (a Frenchman of the same age) independently invented an efficient process for its production. In the 1990s the worldwide production of aluminum by the **Hall–Héroult process** was approximately $1.5 \times 10^7$ metric tons.

The Hall–Héroult process involves the cathodic deposition of aluminum, from molten cryolite ($Na_3AlF_6$) containing dissolved $Al_2O_3$, in electrolysis cells (Fig. 20.14). Each cell consists of a rectangular steel box some 6 m long, 2 m wide, and 1 m high, which serves as the cathode, and massive graphite anodes that extend through the roof of the cell into the molten cryolite bath. Enormous currents (50,000 to 100,000 A) are passed through the cell, and as many as 100 such cells may be connected in series.

Molten cryolite, which is completely dissociated into $Na^+$ and $AlF_6^{3-}$ ions, is an excellent solvent for aluminum oxide, giving rise to an equilibrium distribution of ions such as $Al^{3+}$, $AlF^{2+}$, $AlF_2^+$, ..., $AlF_6^{3-}$, and $O^{2-}$ in the electrolyte. Cryolite melts at 1000°C, but its melting point is lowered by dissolved aluminum oxide, so that the operating temperature of the cell is about 950°C. Compared with the melt-

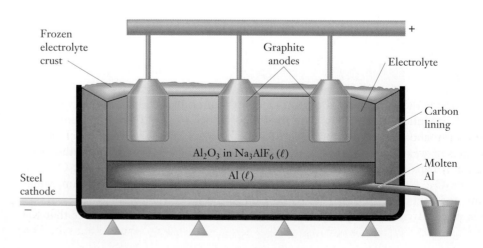

**FIGURE 20.14** An electrolytic cell used in the Hall–Héroult process for production of aluminum.

ing point of pure $Al_2O_3$ (2050°C), this is a low temperature, and it is the reason the Hall–Héroult process has succeeded. Molten aluminum is somewhat denser than the melt at 950°C and so collects at the bottom of the cell, from which it is tapped periodically. Oxygen is the primary anode product, but it reacts with the graphite electrode to produce carbon dioxide. The overall cell reaction is

$$2\ Al_2O_3 + 3\ C \longrightarrow 4\ Al + 3\ CO_2$$

Aluminum and its alloys have a tremendous variety of applications. Many of these make use of aluminum's low density (Table 20.4), an advantage over iron or steel when weight savings are desirable—such as in the transportation industry, which uses aluminum in vehicles from automobiles to satellites. Aluminum's high electrical conductivity and low density make it useful for electrical transmission lines. For structural and building applications, its resistance to corrosion is an important feature, as is the fact that it becomes stronger at subzero temperatures. (Steel and iron sometimes become brittle under these circumstances.) Household products that contain aluminum include foil, soft drink cans, and cooking utensils.

## Magnesium

Like aluminum, magnesium is a very abundant element on the surface of the earth, but it is not easy to prepare in elemental form. Although ores such as dolomite $(CaMg(CO_3)_2)$ and carnallite $(KCl \cdot MgCl_2 \cdot 6H_2O)$ exist, the major commercial source of magnesium and its compounds is seawater. Magnesium forms the second most abundant positive ion in the sea, and scientists separate $Mg^{2+}$ from the other cations in seawater ($Na^+$, $Ca^{2+}$, and $K^+$, in particular) by taking advantage of the fact that magnesium hydroxide is the least soluble hydroxide of the group. Economical recovery of magnesium requires a low-cost base to treat large volumes of seawater and efficient methods for separating the $Mg(OH)_2(s)$ that precipitates from the solution. One base that is used in this way is calcined dolomite, prepared by heating dolomite to high temperatures to drive off carbon dioxide:

$$CaMg(CO_3)_2(s) \longrightarrow CaO \cdot MgO(s) + 2\ CO_2(g)$$

The greater solubility of calcium hydroxide ($K_{sp} = 5.5 \times 10^{-6}$) relative to magnesium hydroxide ($K_{sp} = 1.2 \times 10^{-11}$) leads to the reaction

$$CaO \cdot MgO(s) + Mg^{2+}(aq) + 2\ H_2O(\ell) \longrightarrow 2\ Mg(OH)_2(s) + Ca^{2+}(aq)$$

The magnesium hydroxide that is produced in this process includes not only the magnesium from the seawater but also that from the dolomite.

An interesting alternative to dolomite as a base for magnesium production is a process used off the coast of Texas (Fig. 20.15). Oyster shells (composed largely of $CaCO_3$) are calcined to give lime (CaO), which is added to the seawater to yield magnesium hydroxide. The $Mg(OH)_2$ slurry (a suspension in water) is washed and filtered in huge nylon filters.

After purification, $Mg(OH)_2$ can be reacted with carbon dioxide to give magnesium carbonate, used for coating sodium chloride in table salt to prevent caking and for antacid remedies. Another alternative is to add hydrochloric acid to the magnesium hydroxide to neutralize it and yield hydrated magnesium chloride:

$$Mg(OH)_2(s) + 2\ HCl(aq) \longrightarrow MgCl_2(aq) + 2\ H_2O(\ell)$$

After the water is evaporated, the solid magnesium chloride is melted (m.p. 708°C) in a large steel electrolysis cell that holds as much as 10 tons of the molten salt. The

**TABLE 20.4**

### Densities of Selected Metals

| Metal | Density (g cm$^{-3}$) at Room Conditions |
|---|---|
| Li | 0.534 |
| Na | 0.971 |
| Mg | 1.738 |
| Al | 2.702 |
| Ti | 4.54 |
| Zn | 7.133 |
| Fe | 7.874 |
| Ni | 8.902 |
| Cu | 8.96 |
| Ag | 10.50 |
| Pb | 11.35 |
| U | 18.95 |
| Au | 19.32 |
| Pt | 21.45 |

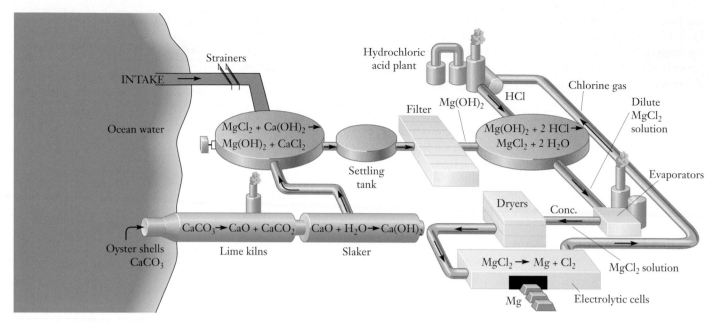

**FIGURE 20.15** The production of magnesium hydroxide starts with the addition of lime (CaO) to seawater. Reaction of the magnesium hydroxide with hydrochloric acid produces magnesium chloride, which, after drying, is electrolyzed to give magnesium.

steel in the cell acts as the cathode during electrolysis, with graphite anodes suspended from the top. The cell reaction is

$$MgCl_2(\ell) \longrightarrow Mg(\ell) + Cl_2(g)$$

The molten magnesium liberated at the cathode floats to the surface and is dipped out periodically, while the chlorine released at the anodes is collected and reacted with steam at high temperatures to produce hydrochloric acid. This is recycled for further reaction with magnesium hydroxide.

Until 1918, elemental magnesium was used mainly in fireworks and flashbulbs, which took advantage of its great reactivity with the oxygen in air and the bright light given off in that reaction (see Fig. 6.6). Since that time, many further uses for the metal and its alloys have been developed. Magnesium is even less dense than aluminum and is used in alloys with aluminum to lower its density and improve its resistance to corrosion under basic conditions. As discussed in Section 12.6 magnesium is used as a sacrificial anode to prevent the oxidation of another metal with which it is in contact. It is also used as a reducing agent to produce other metals such as titanium, uranium, and beryllium from their compounds.

## Electrorefining and Electroplating

Metals such as copper, silver, nickel, and tin that have been produced by pyrometallurgical methods are too impure for many purposes, and **electrorefining** is used to purify them further. Crude metallic copper, for example, is cast into slabs, which are used as anodes in electrolysis cells that contain a solution of $CuSO_4$ in aqueous $H_2SO_4$. Thin sheets of pure copper serve as cathodes, and the copper that dissolves at the anodes is deposited in purer form on the cathodes (Fig. 20.16). Impurities that are more easily oxidized than copper, such as nickel, dissolve along with the copper but remain in solution; elements such as silver and gold that are less easily oxidized do not dissolve but fall away from the anode as a metallic slime. Periodi-

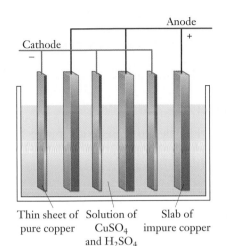

**FIGURE 20.16** In the electrolytic refining of copper, many slabs of impure copper, which serve as anodes, alternate with thin sheets of pure copper (the cathodes). Both are dipped into a dilute acidic solution of copper. As the copper is oxidized from the impure anodes, it enters the solution and migrates to the cathodes, where it plates out in purer form. *(ASARCO)*

cally the anode slime and the solution are removed and further processed for recovery of the elements they contain.

A related process is **electroplating,** in which electrolysis is used to plate out a thin layer of a metal on another material, frequently a second metal. In chrome plating, the piece of metal to be plated is placed in a hot bath of sulfuric acid and chromic acid ($H_2CrO_4$) and is made the cathode in an electrolytic cell. As current passes through the cell, chromium is reduced from the +6 oxidation state in chromic acid to elemental chromium and plates out on the cathode. A decorative chromium layer can be as thin as $2.5 \times 0^{-5}$ cm (corresponding to 2 g of Cr per square meter of surface). Thicker layers ranging up to $10^{-2}$ cm are found in hard chromium plate, prized for its resistance to wear and used in automobile trim. Steel can be plated with cadmium to improve its resistance to corrosion in marine environments. Gold and silver are used both for decorative plating and (because they are good conductors of electricity) on electronic devices.

## EXAMPLE 20.1

Suppose a layer of chromium $3.0 \times 10^{-3}$ cm thick is to be plated onto an automobile bumper with a surface area of $2.0 \times 10^3$ cm$^2$. If a current of 250 A is used, how long must current be passed through the cell to achieve the desired thickness? The density of chromium is 7.2 g cm$^{-3}$.

### Solution

The volume of the Cr is the product of the thickness of the layer and the surface area:

$$\text{volume} = (3.0 \times 10^{-3}\ \text{cm})(2.0 \times 10^3\ \text{cm}^2) = 6.0\ \text{cm}^3$$

The mass of chromium is the product of this volume and the density:

$$\text{mass Cr} = (6.0\ \text{cm}^3)(7.2\ \text{g cm}^{-3}) = 43.2\ \text{g}$$

From this, the chemical amount of Cr that must be reduced is

$$\frac{43.2\ \text{g}}{52.00\ \text{g mol}^{-1}} = 0.831\ \text{mol Cr}$$

Because Cr is being reduced from oxidation state $+6$ in $H_2CrO_4$ to 0 in the elemental form, six electrons are required for each atom of Cr deposited. The number of moles of electrons is then

$$0.831 \text{ mol Cr} \times \left( \frac{6 \text{ mol } e^-}{1 \text{ mol Cr}} \right) = 4.98 \text{ mol } e^-$$

$$\text{total charge} = (4.98 \text{ mol})(96{,}485 \text{ C mol}^{-1}) = 4.81 \times 10^5 \text{ C}$$

The required electrolysis time is the total charge divided by the current (in amperes):

$$\text{time} = \frac{4.81 \times 10^5 \text{ C}}{250 \text{ C s}^{-1}} = 1.9 \times 10^3 \text{ s} = 32 \text{ min}$$

**Related Problems: 29, 30**

## CONCEPTS & SKILLS

*After studying this chapter and working the problems that follow, you should be able to*

1. Describe how the chemical industry operates, and the variety of considerations that must be taken into account in the design of a new chemical process (Section 20.1, problems 1–2).
2. Outline the sources of important raw materials for chemistry (Section 20.1, problems 3–4).
3. Describe methods for the production of hydrogen and the uses of this element (Section 20.2, problems 5–12).
4. Outline the chemistry involved in the reduction of copper and iron ores in smelters and blast furnaces (Section 20.3, problems 19–22).
5. Describe the Hall–Héroult process for the production of aluminum (Section 20.4, problem 25).
6. Summarize the methods used to recover magnesium compounds from seawater and the compounds' subsequent conversion to elemental magnesium (Section 20.4, problem 26).
7. Use Faraday's laws to relate current to metal deposited in electrorefining and electroplating operations (Section 20.4, problems 29–30).

## PROBLEMS

*Answers to problems whose numbers are boldface appear in Appendix G. Problems that are more challenging are indicated with asterisks.*

### The Chemical Industry

1. Define the terms "batch process" and "continuous process." How do the designs of the two types of processes differ?
2. Define the terms "byproducts" and "waste products." Is there a distinction between the two?
3. Identify a useful source of each of the following chemical elements as specifically as you can, based on information in the chapter.

(a) Sulfur    (c) Phosphorus
(b) Carbon    (d) Calcium

4. In what ways are biochemical processes analogous to and different from processes in chemical industry? Draw a process diagram for the production of sulfur, beginning with calcium sulfate, carbon dioxide, and marine bacteria.

### Hydrogen

5. You need a small supply of hydrogen for a chemical reaction. Describe a good way to prepare it in the laboratory, using a balanced chemical equation.

6. When water is electrolyzed on the laboratory scale to produce hydrogen and oxygen, a small amount of sulfuric acid is added to the water.
   (a) Why is the acid added?
   (b) How could you prove that the hydrogen and oxygen come primarily from the water and not from the acid?

7. The production of synthesis gas from methane in natural gas and steam

$$CH_4(g) + H_2O(g) \longrightarrow CO(g) + 3\ H_2(g)$$

in an endothermic process, and so it is favored at moderately high temperatures. Estimate the temperature at which the equilibrium constant for the reaction is equal to 1 (that is, at which $\Delta G = 0$).

8. The shift reaction

$$CO(g) + H_2O(g) \longrightarrow CO_2(g) + H_2(g)$$

is an exothermic process, and so it is favored at sufficiently low temperatures. Estimate the temperature at which the equilibrium constant for the reaction is 1 (that is, at which $\Delta G = 0$).

9. One procedure for generating gaseous hydrogen involves the action of steam on iron to give iron oxide ($Fe_3O_4$).
   (a) Write a balanced chemical equation for this reaction.
   (b) Calculate the enthalpy change $\Delta H°$ for this reaction, and determine whether the equilibrium yield would be enhanced or reduced by an increase in temperature.

10. Iron oxide produced as described in the preceding problem is subsequently reduced to iron by the carbon monoxide present in synthesis gas.
    (a) Write a balanced chemical equation for this step.
    (b) Calculate the enthalpy change $\Delta H°$ for this second step. What will be the effect of an increase in temperature on the equilibrium yield of this step? An increase in pressure?

11. Calculate the volume of hydrogen at 0°C and 1.00 atm that is required to convert 500.0 g of linoleic acid ($C_{18}H_{32}O_2$) to stearic acid ($C_{18}H_{36}O_2$).

12. A chemist determines that 4.20 L of hydrogen at 298 K and a pressure of 1.00 atm is required to completely hydrogenate 48.5 g of the unsaturated compound oleic acid to stearic acid ($C_{18}H_{36}O_2$). How many units of unsaturation (where a unit of unsaturation is one double bond) are in a molecule of oleic acid?

## Extractive Metallurgy

13. The principal ore of manganese is pyrolusite ($MnO_2$), which can be reduced to elemental manganese with aluminum. Write a balanced equation for the reaction and calculate the standard enthalpy and free energy changes per mole of manganese produced from pyrolusite using data in Appendix D.

14. Uranium can be prepared in elemental form by reducing $U_3O_8$ with calcium. Write a balanced equation for the reaction, and calculate the standard enthalpy and free energy changes per mole of uranium produced from $U_3O_8$, using the following thermodynamic data:

| | $\Delta H_f°$ (kJ mol$^{-1}$) | $S°$ (J K$^{-1}$ mol$^{-1}$) |
|---|---|---|
| $U_3O_8(s)$ | $-3575$ | 282.6 |
| $U(s)$ | 0 | 50.2 |
| $CaO(s)$ | $-635$ | 39.8 |
| $Ca(s)$ | 0 | 41.4 |

15. As mentioned in the text, mercury(II) oxide (HgO) decomposes to mercury and oxygen when heated. Use data from Appendix D to estimate the temperature at which this change becomes spontaneous—that is, at which the free energy change ($\Delta G$) becomes negative.

16. Use data from Appendix D to predict the temperature above which CuO decomposes spontaneously to Cu(s) and $\frac{1}{2} O_2(g)$ at atmospheric pressure.

17. Explain why it is possible to reduce tungsten(VI) oxide ($WO_3$) to metal at elevated temperature with hydrogen. At what temperature does the equilibrium constant become unity? Use data from Appendix D.

18. $WO_3$ can be reduced to tungsten if it is heated with carbon in an electric furnace:

$$2\ WO_3(s) + 3\ C(s) \longrightarrow 2\ W(s) + 3\ CO_2(g)$$

   (a) Calculate the standard free energy change $\Delta G°$ for the reduction of $WO_3$ to W, and comment on the feasibility of the process.
   (b) What must be done to make the process thermodynamically feasible, assuming that $\Delta H$ and $\Delta S$ do not vary much with temperature?

19. Analysis of a copper-bearing seam of rock shows it to contain three copper-containing substances: 1.1% chalcopyrite ($CuFeS_2$), 0.42% covellite (CuS), and 0.51% bornite ($Cu_5FeS_4$) by mass. Calculate the total mass of copper present in 1.0 metric ton ($1.0 \times 10^3$ kg) of this copper-bearing ore.

20. Calculate (a) the mass of copper and (b) the volume of sulfur dioxide at 0°C and 1 atm that are produced in the smelting of 1.0 metric ton ($1.0 \times 10^3$ kg) of chalcopyrite ($CuFeS_2$).

21. (a) Write balanced equations for the reduction with carbon monoxide of hematite ($Fe_2O_3$), magnetite ($Fe_3O_4$), and siderite ($FeCO_3$) to iron.
    (b) Calculate the standard free energy change $\Delta G°$ per mole of iron produced from each of the three reactions in part (a). From your results, predict which iron compound should be easiest to reduce from the standpoint of thermodynamics.

22. Calculate $\Delta G°$ for each of the following reactions:

$$Fe_2O_3(s) + \tfrac{3}{2} C(s) \longrightarrow 2\ Fe(s) + \tfrac{3}{2} CO_2(g)$$
$$Fe_2O_3(s) + 3\ C(s) \longrightarrow 2\ Fe(s) + 3\ CO(g)$$
$$Fe_2O_3(s) + 3\ CO(g) \longrightarrow 2\ Fe(s) + 3\ CO_2(g)$$

From your results, decide which is the better reducing agent—C (being oxidized to CO or $CO_2$) or CO.

## Electrometallurgy

**23.** In the Downs process, molten sodium chloride is electrolyzed to produce sodium. A valuable byproduct is chlorine. Write equations representing the processes taking place at the anode and at the cathode in the Downs process.

**24.** The first element to be prepared by electrolysis was potassium. In 1807, Humphry Davy, then 29 years old, passed an electric current through molten potassium hydroxide (KOH), obtaining liquid potassium at one electrode and water and oxygen at the other. Write equations to represent the processes taking place at the anode and at the cathode.

**25.** A current of 55,000 A is passed through a series of 100 Hall–Héroult cells for a period of 24 hours. Calculate the maximum theoretical mass of aluminum that can be recovered.

**26.** A current of 75,000 A is passed through an electrolysis cell containing molten $MgCl_2$ for a period of 7.0 days. Calculate the maximum theoretical mass of magnesium that can be recovered.

**27.** An important use for magnesium is to make titanium. In the Kroll process, magnesium reduces titanium(IV) chloride to elemental titanium in a sealed vessel at 800°C. Write a balanced chemical equation for this reaction. What mass of magnesium is needed, in theory, to produce 100 kg of titanium from titanium(IV) chloride?

**28.** Calcium is used to reduce vanadium(V) oxide to elemental vanadium in a sealed steel vessel. Vanadium is used in vanadium steel alloys for jet engines, high-quality knives, and tools. Write a balanced chemical equation for this process. What mass of calcium is needed, in theory, to produce 20.0 kg of vanadium from vanadium(V) oxide?

**29.** Galvanized steel consists of steel with a thin coating of zinc to reduce corrosion. The zinc can be deposited electrolytically, by making the steel object the cathode and a block of zinc the anode in an electrochemical cell containing a dissolved zinc salt. Suppose a steel garbage can is to be galvanized and requires that a total mass of 7.32 g of zinc be coated to the required thickness. For how long should a current of 8.50 A be passed through the cell to achieve this?

**30.** In the electroplating of a silver spoon, the spoon acts as the cathode and a piece of pure silver as the anode. Both dip into a solution of silver cyanide (AgCN). Suppose that a current of 1.5 A is passed through such a cell for a period of 22 min and that the spoon has a surface area of 16 cm². Calculate the average thickness of the silver layer deposited on the spoon, taking the density of silver to be 10.5 g cm$^{-3}$.

## Additional Problems

**31.** Identify the state of lower free energy in each of the following pairs.
 (a) Aspirin versus: coal, hydrogen, air
 (b) Seawater versus: pure water, sodium chloride, magnesium carbonate
 (c) Vitamin A, oxygen versus: carbon dioxide, water

**32.** Calcium hydride ($CaH_2$) is used as a convenient source of gaseous hydrogen in the laboratory. Calculate the volume of hydrogen produced at 25°C and a pressure of 1.00 atm when 12.21 g of $CaH_2$ is added to a large excess of water.

**33.** Explain why reacting iron or zinc with sulfuric acid is not a practical means for the large-scale generation of hydrogen.

**34.** As the technical chief in a soft-drink factory, you buy a tank that contains 900 lb of liquid carbon dioxide and set up a machine that uses the carbon dioxide to carbonate the company's product. After 800 lb of $CO_2$ has been taken out of the tank, the pressure gauge on the cylinder still reads 64 atm as it did on the day it arrived. The boss concludes that the gauge must be broken and tells you to buy a costly replacement gauge. Use Figure 5.21 to explain to the boss why the gauge is almost certainly *not* broken. What is a reliable way to measure the amount of $CO_2$ that remains in the cylinder?

**35.** In a recent year, the total consumption of carbonated beverages in the United States corresponded to 63 *billion* 12-oz cans or bottles. If a 12-oz can has a volume of 355 mL, and the concentration of dissolved $CO_2$ is 0.15 M, calculate the total mass of $CO_2$ produced and used for making carbonated beverages during that year.

**36.** Compute the oxidation number of carbon in ethylene and in ethane. Is hydrogen oxidized or reduced when it is used to saturate the multiple bonds in a hydrocarbon?

**37.** Classify the following as redox, dissolution, precipitation, or acid–base reactions.
 (a) $4\,CaSO_4(s) + 3\,CH_4(g) + CO_2(g) \rightarrow$
 $\qquad\qquad 4\,CaCO_3(s) + 4\,S(s) + 6\,H_2O(\ell)$
 (b) $CO(g) + 2\,H_2(g) \rightarrow CH_3OH(g)$
 (c) $H_2C{=}CH_2 + H_2 \rightarrow H_3C{-}CH_3$
 (d) $H_2SO_4(aq) + Fe(s) \rightarrow H_2(g) + FeSO_4(aq)$

**38.** (a) On a single graph, plot $\Delta G$ as a function of temperature from 0°C to 1500°C for the reactions

$$2\,C(s, \text{gr}) + O_2(g) \longrightarrow 2\,CO(g)$$
$$2\,Fe(s) + O_2(g) \longrightarrow 2\,FeO(s)$$
$$2\,Ni(s) + O_2(g) \longrightarrow 2\,NiO(s)$$
$$\tfrac{4}{3}\,Al(s) + O_2(g) \longrightarrow \tfrac{2}{3}\,Al_2O_3(s)$$

Assume that $\Delta H$ and $\Delta S$ for these reactions are independent of temperature, and approximate FeO by the nonstoichiometric compound wüstite.
 (b) Use your graph to predict the temperature range over which coke can be used to reduce NiO and FeO to the elements.
 (c) Can $Al_2O_3$ be reduced to elemental aluminum by coke over this temperature range?

**39.** Compute the equilibrium constant of the reaction

$$CO_2(g) + C(s) \rightleftharpoons 2\,CO(g)$$

at room temperature (25°C) and at 1875°C.

**40.** The principal ore of chromium is the mineral chromite ($FeCr_2O_4$), which has a standard free energy of formation of

($\Delta G_f^\circ$) of $-1344$ kJ mol$^{-1}$. Because the main use of chromium is in chromium-containing steels, the ore is generally reduced directly to the alloy ferrochrome by the reaction

$$FeCr_2O_4(s) + 4\ C(s) \rightarrow Fe(s) + 2\ Cr(s) + 4\ CO(g)$$

Calculate the standard free energy change for the production of ferrochrome according to this equation, using Appendix D for other information you may need. (*Note:* The free energy change in mixing iron and chromium in the alloy is small and can be neglected.)

41. Compare and contrast the smelting of copper with that of iron. Include in your discussion the differences in starting materials, in temperatures and reducing agents employed, and in byproducts and their disposal.

42. The radius of the earth is 6370 km. Estimate the amount of aluminum (in kilograms) in the top 1 km of the earth's crust, assuming the average density of the crust to be approximately 2.8 g cm$^{-3}$.

43. Compare and contrast the production of aluminum with that of magnesium. Include in your discussion the differences in starting materials, in reactions preparatory to the electrolysis, and in reaction byproducts.

44. Use data from Appendix D to predict whether elemental magnesium should react with water on purely thermodynamic grounds. What is observed in practice? Explain.

45. If pure chromium is to be obtained from its chromite ore, $FeCr_2O_4(s)$, the process commonly used is to fuse it with sodium carbonate. The unbalanced equation for this reaction is

$$FeCr_2O_4(s) + Na_2CO_3(\ell) + O_2(g) \longrightarrow$$
$$Fe_2O_3(s) + Na_2CrO_4(\ell) + CO_2(g)$$

(a) Balance the preceding chemical equation.
(b) Calculate the standard enthalpy change of the reaction, assuming that the enthalpies of fusion of $Na_2CO_3$ and $Na_2CrO_4$ are approximately equal. The standard molar enthalpy of formation of $FeCr_2O_4(s)$ is $-1445$ kJ mol$^{-1}$, and that of $Na_2CrO_4(s)$ is $-1342$ kJ mol$^{-1}$. Use Appendix D for the additional data you require.
(c) The sodium chromate produced in the reaction is soluble in water and can be leached from the solidified melt. Suggest a procedure for converting it to elemental chromium.

46. A 55.5-kg slab of crude copper from a smelter has a copper content of 98.3%. Estimate the time required to purify it electrochemically if it is used as the anode in a cell that has acidic copper(II) sulfate as its electrolyte and a current of $2.00 \times 10^3$ A is passed through the cell.

47. Sheet iron can be galvanized by passing a direct current through a cell containing a solution of zinc sulfate between a graphite anode and the iron sheet. Zinc plates out on the iron. The process can be made continuous if the iron sheet is a coil that unwinds as it passes through the electrolysis cell and coils up again after it emerges from a rinse bath. Calculate the cost of the electricity required to deposit a 0.250-mm-thick layer of zinc on both sides of an iron sheet that is 1.00 m wide and 100 m long, if a current of 25 A at a voltage of 3.5 V is used and the energy efficiency of the process is 90%. The cost of electricity is 10 cents per kilowatt–hour (1 kWh = 3.6 MJ). Consult Appendix F for data on zinc.

# Chemical Processes Based on Sulfur, Phosphorus, and Nitrogen

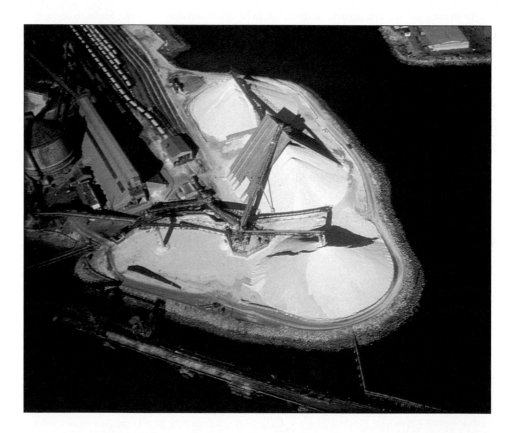

*Illustration*
Aerial view of sulfur piled in the harbor in Vancouver, British Columbia, Canada. *(Copyright 1990 Stuart McCall)*

Sulfur, phosphorus, and nitrogen are three elements of intermediate electronegativity that are of fundamental importance in industrial and biological processes. The "workhorse" acids of chemical industry—sulfuric, phosphoric, and nitric, together with hydrochloric acid—are produced in quantities of tens of billions of kilograms each year and are used directly or indirectly in virtually every chemical manufacturing process. These three elements also play a central role in the production of chemical fertilizers.

# SULFURIC ACID AND ITS USES

Sulfuric acid is the chemical produced in greatest amount in the world. Its uses range from metal treatment to the production of pharmaceuticals to fertilizer manufacture. How did this strongly acidic, extremely reactive chemical come to have such an important place in chemistry and in modern industry? In fact its role was established between 1750 and 1900. In 1750 sulfuric acid was produced and used on a very modest scale in metal assaying and treatment and by some doctors for cures that had no scientific basis. New uses for the acid stimulated the search for new ways to produce it; as the price went down, further uses were discovered and exploited. From 1750 to 1900, the price of sulfuric acid declined steadily, and the amount produced grew explosively. This mutual stimulation between new uses and new processes is the hallmark of the chemical industry even today, and the story of sulfuric acid is a good way to illustrate it.

Sulfuric acid was probably first generated by alchemists who heated crystalline green vitriol, or iron(II) sulfate heptahydrate, in a retort:

$$FeSO_4 \cdot 7H_2O(s) \longrightarrow H_2SO_4(\ell) + FeO(s) + 6\ H_2O(g)$$

By the 17th century it was produced on a limited commercial scale from iron pyrite ores. The $FeS_2$ in the ores was converted to $FeSO_4$, then oxidized to solid iron(III) sulfate, $Fe_2(SO_4)_3$. This solid was broken up and strongly heated in small clay vessels, a treatment that decomposed it into iron(III) oxide and sulfur trioxide:

$$Fe_2(SO_4)_3(s) \longrightarrow Fe_2O_3(s) + 3\ SO_3(g)$$

The gas produced was then absorbed in water to make sulfuric acid:

$$SO_3(g) + H_2O(\ell) \longrightarrow H_2SO_4(\ell)$$

Production of sulfuric acid was stimulated by the discovery in 1744 that the valued dyestuff indigo, used for many years to dye cotton, could also be used as a wool dye after the wool was treated with concentrated sulfuric acid.

## The Lead-Chamber Process

The lead-chamber process, which was developed in the second half of the 18th century, probably also had its origins in the laboratories of alchemists, who burned sulfur in earthenware vessels. The small amounts of $SO_3$ that were produced (along with the $SO_2$ that was the main product) were condensed and absorbed into water to make sulfuric acid. An accidental discovery revealed that the addition of sodium nitrate or potassium nitrate improves the yield of $SO_3$. These salts decompose to give nitrogen dioxide, which reacts with $SO_2$ to give $SO_3$:

$$SO_2(g) + NO_2(g) \longrightarrow SO_3(g) + NO(g)$$

In 1736 Joshua Ward took the next important step by replacing the earthenware vessels in which the sulfur was burned with large glass bottles arranged in series, to speed up the process.

The development of the room-size **lead chamber,** first employed by John Roebuck in 1746, dramatically expanded the manufacture of sulfuric acid. The output of one of the older earthenware vessels was several ounces, and Ward's glass bottles had outputs on the order of several pounds. By contrast, the lead chamber allowed sulfuric acid to be produced in amounts of hundreds of pounds to tons, reducing the price of the product through economy of scale and lower labor costs. In the lead-chamber process, a mixture of sulfur and potassium nitrate was placed in a ladle and ignited inside a large chamber lined with lead, the floor of which was covered with water. The gases condensed on the walls and were absorbed in the water. After the process was repeated several times, the dilute sulfuric acid was removed and boiled down to concentrate it further. Later developments included blowing in steam to speed up the reaction with water and disperse the gases and separating the burning chamber from the absorption chamber.

Joseph Gay-Lussac took a significant step forward in 1835 when he built a tower to recover the NO that had previously been vented and to convert it back to $NO_2$ by reaction with oxygen. More precisely, in a Gay-Lussac tower, NO was converted to nitrous acid ($HNO_2$) dissolved in aqueous sulfuric acid:

$$2 \, NO(g) + \tfrac{1}{2} \, O_2(g) + H_2O(\ell) \longrightarrow 2 \, HNO_2(aq)$$

The nitrous acid then reacted in a second tower—named after its developer, John Glover—to oxidize sulfur dioxide:

$$2 \, HNO_2(aq) + SO_2(g) \longrightarrow H_2SO_4(aq) + 2 \, NO(g)$$

The overall reaction from these steps is simply

$$SO_2(g) + \tfrac{1}{2} \, O_2(g) + H_2O(\ell) \longrightarrow H_2SO_4(aq)$$

Recycling the oxides of nitrogen greatly reduced the consumption of sodium nitrate or potassium nitrate, which was now needed only to make up for losses in the process. In addition, the Glover tower produced more concentrated sulfuric acid—75 to 85% $H_2SO_4$ by mass compared with 60 to 70% obtained by earlier methods. Figure 21.1 is a flow diagram for the full process, after the development of the Glover tower in 1859.

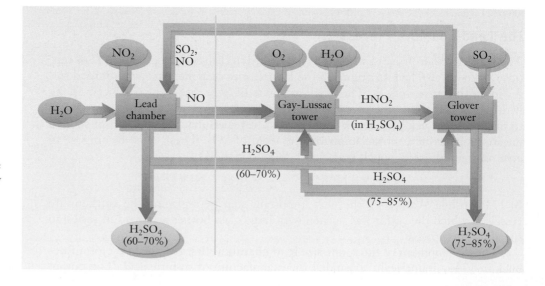

**FIGURE 21.1** The Glover–Gay-Lussac process for making sulfuric acid. The portion of the diagram to the left of the yellow line shows the original lead-chamber process, which produced a weaker acid and consumed and emitted more nitrogen oxides than the full process as it later developed.

## The Contact Process

As early as 1831, the Englishman Peregrine Phillips observed that platinum could catalyze the conversion of $SO_2$ to $SO_3$, the step at the heart of the production of sulfuric acid. Little attention was paid to this discovery until the 1870s, when the growth of the German dye industry stimulated a search for a method of producing sulfuric acid more concentrated than that from the Glover tower. Platinum catalysts were rediscovered and patented but initially had limited usefulness because they were poisoned and rendered ineffective by impurities in the sulfur dioxide feed. As a result, the early applications of this method used, somewhat paradoxically, sulfuric acid from the Glover tower as a raw material. The sulfuric acid was decomposed by heating,

$$H_2SO_4(aq) \longrightarrow SO_2(g) + H_2O(\ell) + \tfrac{1}{2} O_2(g)$$

and the relatively pure $SO_2$ was converted to $SO_3$ over the catalyst and then back to $H_2SO_4$. In this way, a highly concentrated acid was obtained but at great expense. Over the next 30 years, researchers recognized the role of arsenic and other impurities in poisoning the catalyst and took steps to remove the impurities from the $SO_2$ feed, which allowed direct production of concentrated sulfuric acid without the intermediate lead-chamber step. Different catalysts were studied as well, leading eventually to the selection of an oxide of vanadium ($V_2O_5$) that is the primary catalyst in use today.

The catalytic production of sulfuric acid from $SO_2$ is called the **contact process** and is outlined in Figure 21.2. The portion to the right of the yellow line represents a process used in some modern plants to reduce air pollution from residual $SO_2$, which will be discussed later. In the reaction of $SO_2$ with oxygen,

$$SO_2(g) + \tfrac{1}{2} O_2(g) \longrightarrow SO_3(g)$$

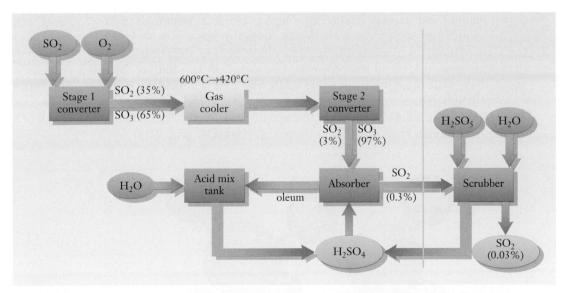

**FIGURE 21.2** Flow chart of the contact process for making sulfuric acid. The portion to the right of the yellow line represents a process used in some modern plants to reduce air pollution from residual $SO_2$.

the number of moles of gas decreases. This suggests running the reaction under high total pressures to increase the yield of the product. The small advantage gained by using high pressures in this reaction is more than offset by the greater cost of the necessary equipment, however, and so the reaction is carried out at atmospheric pressure. A thermodynamic analysis gives

$$\Delta H° = -98.9 \text{ kJ} \qquad \Delta S° = -94.0 \text{ J K}^{-1}$$

Because the reaction is exothermic, the lower the temperature, the higher the equilibrium constant and the greater the degree of conversion to products at equilibrium. The change in free energy is 0 (and the equilibrium constant 1) when

$$\Delta G = \Delta H - T \Delta S \approx \Delta H° - T \Delta S° = 0$$

or

$$T = \frac{\Delta H°}{\Delta S°} = \frac{-98,900 \text{ J}}{-94.0 \text{ J K}^{-1}} = 1050 \text{ K}$$

or approximately 780°C. The temperature must be maintained substantially below this point to achieve significant yield of product.

The problem is that at lower temperature the reaction becomes slower, although the catalyst does help to speed it up. A two- to four-stage process is typically used. The entering $SO_2(g)$ reaches the first catalyst at a temperature of 420°C. As the reaction begins, heat is evolved and the temperature of the reacting gas mixture rises. After a few seconds, the mixture has reached equilibrium at about 600°C with conversion of 60 to 70% of the $SO_2$. The gases are then cooled back to 420°C and allowed to react one or two more times over the catalyst, with lower temperatures and longer exposure periods used. The result is conversion of about 97% of the $SO_2$ to $SO_3$. For even greater conversion, the gases are then passed into a tower where $SO_3$ dissolves in sulfuric acid. This removes the reaction product, so that the reaction again shifts to the right when the unreacted $SO_2$ is passed over the catalyst for a last time. The $SO_3$ from this last stage is then absorbed, giving an overall yield of about 99.7% of the $SO_2$ introduced originally. Careful consideration of thermodynamics and kinetics has led to a highly efficient process. Almost all sulfuric acid manufactured today is made by the contact process.

If $SO_3$ were absorbed in water rather than in sulfuric acid, the product would be more dilute and less $SO_3$ would be absorbed. In addition, the direct reaction of $SO_3$ with water produces a fine acidic mist that is difficult to condense. Absorption of $SO_3$ into sulfuric acid gives fuming sulfuric acid, or **oleum,** that can be used directly or diluted with water to the desired strength. An equimolar amount of $SO_3$ dissolved in $H_2SO_4$ gives disulfuric acid ($H_2S_2O_7$) (Fig. 21.3).

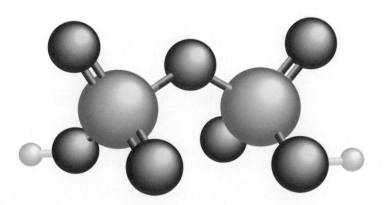

**FIGURE 21.3**  Disulfuric acid, $H_2S_2O_7$.

The amount of $SO_2$ released by the contact process to pollute the air is very small, but further removal of objectionable $SO_2$ from the tail gases can be achieved (at additional expense) in another step. Some $H_2SO_4$, which is available in abundance, of course, is electrolytically oxidized to peroxydisulfuric acid ($H_2S_2O_8$):

$$2\ H_2SO_4(aq) + 2\ H_2O(\ell) \longrightarrow H_2S_2O_8(aq) + 2\ H_3O^+(aq) + 2\ e^- \qquad \text{(anode)}$$

$$2\ H_3O^+(aq) + 2\ e^- \longrightarrow H_2(g) + 2\ H_2O(\ell) \qquad \text{(cathode)}$$

The $H_2S_2O_8$ quickly reacts:

$$H_2O(\ell) + H_2S_2O_8(aq) \longrightarrow H_2SO_4(aq) + H_2SO_5(aq)$$

The $H_2SO_5$ is called peroxymonosulfuric acid. The exiting gases are passed into a scrubber to mingle with a solution of this powerful oxidizing agent (see Fig. 21.2). In the scrubber, the reaction

$$SO_2(g) + H_2SO_5(aq) + H_2O(\ell) \longrightarrow 2\ H_2SO_4(aq)$$

converts $SO_2(g)$ to sulfuric acid. More than 90% of the already small amount of residual $SO_2(g)$ is removed in this way. The product, dilute sulfuric acid, is cycled back into the main process.

## Sources for Sulfur and Sulfur Dioxide

The primary raw material for making sulfuric acid is sulfur (Fig. 21.4) or sulfur dioxide. Sources for these chemicals have changed over the years, driven by considerations of price and the desirability of reducing atmospheric pollution. Centuries ago, the sulfur mines of Sicily were the major source of elemental sulfur. Increases in price, especially after a cartel was established to exploit the mines, led to the search for other sources. As the use of metals increased in the 19th century, sulfur was obtained in the form of $SO_2$ as a byproduct of the roasting of sulfide ores of zinc, iron, or copper, through reactions such as

$$ZnS(s) + \tfrac{3}{2}\ O_2(g) \longrightarrow ZnO(s) + SO_2(g)$$

Capture of sulfur dioxide and its conversion to sulfuric acid created a cheap starting material and reduced air pollution from $SO_2$.

In the late 1890s, a new development shifted interest from sulfide ores back to elemental sulfur as a starting material, at least in the United States. This was the discovery by Herman Frasch of a new method to extract sulfur from underground deposits. The salt domes of Louisiana, Texas, and Mexico contained substantial reserves of sulfur in porous limestones, but extraction of the sulfur had proved difficult. The **Frasch process** (Fig. 21.5) is an ingenious mining method in which sulfur is liquefied by steam injected into the deposit and pumped to the surface. The sulfur is usually shipped as a liquid in heated vessels to sulfuric acid plants, where it is used to make sulfuric acid.

Since the 1970s, the Frasch process has declined in importance as a source of sulfur because of a competing process that recovers sulfur impurities in natural gas and petroleum. The latter process was developed to reduce sulfur oxide pollution from the burning of oil and gas. Hydrogen sulfide and other sulfide impurities are acidic and so are preferentially extracted into solutions of basic potassium carbonate:

$$H_2S(g) + CO_3^{2-}(aq) \longrightarrow HS^-(aq) + HCO_3^-(aq)$$

**FIGURE 21.4** A sulfur crystal. (*Charles D. Winters*)

**FIGURE 21.5** The Frasch process uses three concentric pipes driven into the ground. Superheated water (at a temperature of 160°C) is pumped under pressure through the outermost pipe into the sulfur-bearing rock formation. This heats the rock above the melting point of sulfur, 119°C. The molten sulfur is heavier than water and collects in a pool. Heated, compressed air pumped through the innermost pipe works the sulfur in the pool into a froth that rises to the surface through the third pipe.

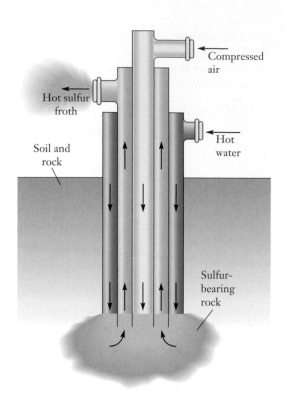

After removal from this solution, a portion of the $H_2S$ is then burned in air at 1000°C over an aluminum oxide catalyst to give sulfur dioxide:

$$H_2S(g) + \tfrac{3}{2}\,O_2(g) \longrightarrow SO_2(g) + H_2O(g)$$

This can be used directly to make sulfuric acid or converted back to elemental sulfur by reaction with additional $H_2S$ in the **Claus process,**

$$SO_2(g) + 2\,H_2S(g) \longrightarrow 3\,S(\ell) + 2\,H_2O(g)$$

which uses an $Fe_2O_3$ catalyst. The Claus process greatly reduces pollution from the burning of natural gas and petroleum. New processes to reduce emission of sulfur oxides from the burning of coal are needed now.

## Uses of Sulfuric Acid and Its Derivatives

Pure sulfuric acid is a colorless, viscous liquid that freezes at 10.4°C and boils at 279.6°C. It reacts avidly with water (Fig. 21.6) and with organic compounds. It may be mixed with water in any proportion, with provisions for the large amounts of heat that are liberated. Despite its corrosive power, sulfuric acid is easily handled and transported in steel drums. This fact, together with its acid strength and low cost, has given it a tremendous range of uses. In metals processing, sulfuric acid is used to leach copper, uranium, and vanadium from their ores and to "pickle," or descale, steel. In pickling, layers of oxides on the surfaces of the metal are dissolved away by reaction with the acid. Much sulfuric acid is used as a **dehydrating agent** (Fig. 21.7) in the synthesis of organic chemicals and in the processing of petrochemicals. Lesser but still important amounts are used in making hydrochloric and hydrofluoric acids and $TiO_2$ pigments for paint. As will be discussed, the largest sin-

**FIGURE 21.6** Sulfuric acid should always be added to water, rather than the reverse. The dissolution reaction is quite exothermic; water can boil and splatter acid out if water is added to acid because the heat is generated in the region where the less dense water touches the dense acid. Adding acid to water mixes the two much more effectively. Here, the indicator in the water, methyl orange, is changing color from yellow to red as acid is added. *(Leon Lewandowski)*

gle use of sulfuric acid is in the fertilizer industry. All of these uses are *indirect*, however, in the sense that the sulfur from the sulfuric acid does not become incorporated as an ingredient in products. Rather, it ends up either as sulfate wastes or as **spent acid,** acid that has been diluted and contaminated by its use in processing. The *direct* uses of sulfuric acid are surprisingly few.

An important chemical made from sulfuric acid is sodium sulfate ($Na_2SO_4$). It is produced in the **Hargreaves process,** in which the immediate reactant is $SO_2$ rather than $H_2SO_4$:

$$4 \, NaCl + 2 \, SO_2 + 2 \, H_2O + O_2 \longrightarrow 2 \, Na_2SO_4 + 4 \, HCl$$

The $SO_2$ for this reaction often comes from the decomposition of spent sulfuric acid. Sodium sulfate is also a byproduct of the manufacture of rayon (Section 25.2). In addition, there are major natural salt sources for sodium sulfate in California and Texas. Sodium sulfate is used as a phosphate substitute in detergents and in the manufacture of paper products.

## Wood Pulp and Paper Processing

One of the most important applications of sulfur chemistry is in processing wood to wood pulp for use in paper and cardboard. Wood has three major components: cellulose, lignin, and oils and resins. Cellulose is a fibrous polymeric compound of carbon, oxygen, and hydrogen that provides a supporting structure (see Fig. 25.11). Lignin, which is also a complex polymeric material, acts as a matrix to bind the fibers of cellulose together. To produce wood pulp, the cellulose must be separated from the lignin. The other materials in the wood are removed by solvent extraction and in many cases have important uses (for example, as turpentine, pine oil, rosin, soaps, and tannin).

There are four steps in making paper from trees:

1. Logging and milling operations to collect and debark trees and convert the logs to wood chips
2. Digestion, the chemical process by which lignin is separated from cellulose
3. Bleaching to lighten the color of the cellulose
4. Production of paper or cardboard from the cellulose fibers

The first and last steps are mechanical operations and will not concern us here. The next chapter will discuss bleaching. We discuss only the second step, digestion, here. The chemical reaction involved is a hydrolysis, whereby the addition of water under proper conditions breaks down the lignin polymer into alcohols and organic acids containing long chains of carbon atoms. The hydrolysis proceeds only in a slow digestive process in the presence of other chemicals. The two primary methods used are the **sulfate (Kraft) process** and the **sulfite process.**

The sulfate process (Fig. 21.8) is the more widely used of these two methods because it works for almost all kinds of wood and produces the strongest paper (*Kraft* is German for "strength"). The digestion liquor is an aqueous solution of NaOH and $Na_2S$, the active agents in the hydrolysis of lignin. After digestion for several hours at 175°C, the pulp (now primarily cellulose) is separated by filtration, and the liquor is treated to extract any desired organic compounds. The next and crucial step is the removal and recycling of the inorganic components from the spent liquor. First, sodium sulfate is added to make up for the inevitable losses in the process. Then the liquor is concentrated and its organic component is burned, generating

**FIGURE 21.7** Concentrated sulfuric acid is an excellent dehydrating agent. It can remove water from a carbohydrate (here, sucrose, $C_{12}H_{22}O_{11}$), leaving mostly carbon. *(Charles D. Winters)*

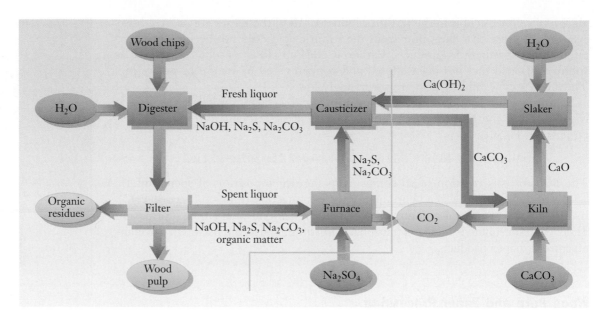

**FIGURE 21.8** The sulfate process for pulping wood. The overall reaction is wood chips + water → wood pulp + organic residues. The entire part of the flow diagram to the right of the yellow line serves only to make up losses in the process.

heat to run the paper mill. During the burning, the sodium sulfate is reduced by the organic matter in the liquor, a process indicated schematically as

$$Na_2SO_4(s) + 2\,[C] \longrightarrow Na_2S(s) + 2\,CO_2(g)$$

The sodium hydroxide in the liquor reacts with the carbon dioxide produced by the oxidation of the organic matter, giving sodium carbonate:

$$2\,NaOH(s) + CO_2(g) \longrightarrow Na_2CO_3(s) + H_2O(g)$$

To regenerate the sodium hydroxide for a new cycle of digestion, the solid residue from the burning is dissolved in water. Limestone ($CaCO_3$) is decomposed to lime (CaO) and carbon dioxide in a separate lime kiln, and the lime is hydrated with water to produce slaked lime, $Ca(OH)_2$. The addition of this base to the dissolved residue precipitates the carbonate ion that was produced in the preceding step and generates hydroxide ion:

$$CO_3^{2-}(aq) + Ca(OH)_2(s) \longrightarrow 2\,OH^-(aq) + CaCO_3(s)$$

The sodium hydroxide is thus restored to the digestion liquor. The calcium carbonate is itself separated and recycled to the kiln to produce additional lime. In this process the only material that is consumed chemically is water; sodium sulfate and limestone are added only as needed to make up for losses from what would ideally be a closed cycle. The major problem with the Kraft process is the occurrence of side reactions that lead to the formation of $H_2S$ and other sulfur compounds, which smell vile and cause serious air and water pollution if they are not captured.

The sulfate process involves a basic digestion medium, but the sulfite process uses a mildly acidic medium, with pH near 4.5. It is limited to certain soft woods such as spruce and hemlock, and it produces a lighter pulp that requires less bleaching and is suitable for making high-quality writing paper. In the digestion liquor, gaseous sulfur dioxide is added to a solution containing hydroxide ion (generally prepared from MgO) to produce hydrogen sulfite ions:

$$SO_2(g) + OH^-(aq) \longrightarrow HSO_3^-(aq)$$

The action of the hydrogen sulfite ion at 60°C over 6 to 12 hours hydrolyzes the cellulose–lignin complex and dissolves the lignin under conditions milder than those required for the sulfate process. After the process is complete, the pulp is recovered by filtration and the liquor is evaporated. The organic residues are again burned to recover heat; during this procedure the $Mg(HSO_3)_2$ decomposes according to the equation

$$Mg(HSO_3)_2(s) \longrightarrow MgO(s) + H_2O(g) + 2\ SO_2(g)$$

Both the $SO_2$ and MgO are recovered and recycled through the process.

---

## 21.2

## PHOSPHORUS CHEMISTRY

Until rather recently, no single use of sulfuric acid dominated the others. The situation has changed with the growing production of phosphate fertilizers (Fig. 21.9), which are processed with sulfuric acid and which now account for well over half of the annual consumption of sulfuric acid in the United States. Phosphorus occurs in nature primarily as the $PO_4^{3-}$ ion in phosphate rock, the main component of which is fluorapatite, $Ca_5(PO_4)_3F$ (Fig. 21.10). Originally, the rock was simply ground up and applied to fields to make them more fertile. A more active fertilizer is made by treating a slurry of ground-up fluorapatite with sulfuric acid to produce **super-phosphate,** a mixture of calcium dihydrogen phosphate and gypsum:

$$2\ Ca_5(PO_4)_3F(s) + 7\ H_2SO_4(aq) + 17\ H_2O(\ell) \longrightarrow$$
$$3\ Ca(H_2PO_4)_2 \cdot H_2O(s) + 7\ CaSO_4 \cdot 2H_2O(s) + 2\ HF(g)$$

Superphosphate is more effective because the essential phosphorus nutrient is in a more soluble form. This mixture also provides the important secondary nutrients calcium and sulfur, as well as phosphorus, to growing crops.

**FIGURE 21.9**  The world's largest linear disc reclaimer, reclaiming phosphate shale at an elemental phosphorus plant in Idaho. *(Manley Prim/FMC Corporation)*

**FIGURE 21.10** Apatite on quartzite. *(Copyright Mark A. Schneider/Visuals Unlimited)*

A variant of the superphosphate process, the **wet-acid process,** uses an excess of sulfuric acid to produce phosphoric acid. The chemical reaction in this case is

$$Ca_5(PO_4)_3F(s) + 5\ H_2SO_4(aq) + 10\ H_2O(\ell) \longrightarrow$$
$$3\ H_3PO_4(aq) + 5\ CaSO_4 \cdot 2H_2O(s) + HF(g)$$

The hydrogen fluoride that is liberated is carried to an absorption tower containing $SiF_4$. In this tower it reacts to give $H_2SiF_6$, which is used in aqueous solution to fluoridate drinking water. The solid gypsum ($CaSO_4 \cdot 2H_2O$) is filtered out, and the dilute solution of phosphoric acid ($H_3PO_4$) is concentrated by evaporation. The phosphoric acid made by this process is not pure because of residual gypsum and impurities from the phosphate rock. The major use for wet-process phosphoric acid is again fertilizer manufacture, where it is substituted for sulfuric acid to produce **triple superphosphate** fertilizer, which does not contain gypsum.

$$Ca_5(PO_4)_3F + 7\ H_3PO_4 + 5\ H_2O \longrightarrow 5\ Ca(H_2PO_4)_2 \cdot H_2O + HF(g)$$

This fertilizer contains a greater percentage of phosphorus, which reduces transportation costs.

The use of triple superphosphate fertilizer has fallen in recent years. The major phosphorus-containing fertilizer now is ammonium phosphate, which is obtained by the reaction of phosphoric acid with ammonia:

$$H_3PO_4(aq) + 3\ NH_3(aq) \longrightarrow (NH_4)_3PO_4(aq)$$

This supplies nitrogen, a necessary nutrient, as well as phosphorus. The next section will discuss the manufacture of ammonia and other nitrogen-based fertilizers.

An alternative to the wet-acid process, used to make purer phosphoric acid, is called the **furnace process.** Phosphate rock, $SiO_2$ (in the form of sand), and elemental carbon (in the form of coke) are fed continuously into an electric-arc furnace that operates at 2000°C. The overall reaction is approximately

$$12\ Ca_5(PO_4)_3F(\ell) + 43\ SiO_2(\ell) + 90\ C(s) \longrightarrow$$
$$90\ CO(g) + 60\ CaO \cdot \tfrac{2}{3}SiO_2(\ell) + 3\ SiF_4(g) + 9\ P_4(g)$$

The elemental white phosphorus is burned in air to tetraphosphorus decaoxide ($P_4O_{10}$) (Fig. 21.11):

$$P_4(g) + 5\ O_2(g) \longrightarrow P_4O_{10}(s)$$

This is a powerful drying agent that dehydrates $HNO_3$ to $N_2O_5$ and even $H_2SO_4$ to $SO_3$. The ultimate product of its reaction with water is phosphoric acid:

$$P_4O_{10}(s) + 6\ H_2O(\ell) \longrightarrow 4\ H_3PO_4(aq)$$

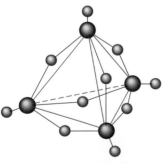

**FIGURE 21.11** The structure of $P_4O_{10}$ is closely related to the tetrahedral $P_4$ structure in white phosphorus.

## Derivatives of Phosphoric Acid

The major nonfertilizer use of wet-process phosphoric acid is in the manufacture of detergents and cleaning agents that are sodium salts of phosphoric acid and its derivatives. Phosphoric acid in excess reacts with sodium carbonate to give sodium dihydrogen phosphate, a weak acid:

$$2\ H_3PO_4(aq) + Na_2CO_3(aq) \longrightarrow 2\ NaH_2PO_4(aq) + CO_2(aq) + H_2O(\ell)$$

When the sodium carbonate is in excess, the product is sodium hydrogen phosphate, a weak base:

$$H_3PO_4(aq) + 2\ Na_2CO_3(aq) \longrightarrow 2\ NaHCO_3(aq) + Na_2HPO_4(aq)$$

Sodium carbonate is too weak a base to remove the last hydrogen ion from $HPO_4^{2-}$, so the stronger (and more expensive) base NaOH must be added:

$$NaOH(aq) + Na_2HPO_4(aq) \longrightarrow Na_3PO_4(aq) + H_2O(\ell)$$

Sodium phosphate ($Na_3PO_4$), also called trisodium phosphate, or TSP, gives strongly basic solutions that are excellent for cleaning.

Heating solid sodium hydrogen phosphate ($Na_2HPO_4$) expels one molecule of water for every two formula units:

The product, sodium diphosphate ($Na_4P_2O_7$), is the sodium salt of diphosphoric acid ($H_4P_2O_7$). This is the phosphorus analog of the disulfuric acid ($H_2S_2O_7$) mentioned in Section 21.1. In diphosphates, each phosphorus atom is tetrahedrally coordinated to four oxygen atoms (as in the original $PO_4^{3-}$ ion), but two tetrahedra now share a single corner. A third tetrahedron can be linked to the other two through the high-temperature reaction

$$2\ Na_2HPO_4(s) + NaH_2PO_4(s) \longrightarrow Na_5P_3O_{10}(s) + 2\ H_2O(g)$$

with one molecule of water expelled for each linkage that forms. The product sodium tripolyphosphate ($Na_5P_3O_{10}$), or STPP (Fig. 21.12), is used as a **builder** in detergents. It helps to sequester $Ca^{2+}$ and $Mg^{2+}$ ions, to maintain the pH in an appropriate range, to keep dirt in suspension, and to increase the efficiency of the detergent itself. The use of detergents that contain phosphates has diminished in recent years because phosphates released into the groundwater fertilize algae in lakes. Excessive algal growth can destroy a lake through **eutrophication.** When the algae die, their bodies fall to the bottom of the lake. As the lake becomes shallower, bottom-rooted plants can take hold; eventually the lake becomes a marsh and, finally, a meadow. This phenomenon has inspired legislation to limit the use of phosphates in detergents.

The purer phosphoric acid from the furnace process is used widely in the food industry. Its acidity gives a pleasing sour taste to soft drinks (called "phosphates" when they were first introduced). Calcium dihydrogen phosphate monohydrate, $Ca(H_2PO_4)_2 \cdot H_2O$, is mixed with $NaHCO_3$ in some baking powders. When water

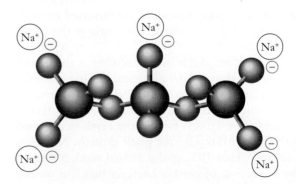

**FIGURE 21.12** The structure of sodium tripolyphosphate.

is added, these ionic compounds dissolve, and the acidic dihydrogen phosphate ions react with the basic hydrogen carbonate ions to generate carbon dioxide:

$$H_2PO_4^-(aq) + HCO_3^-(aq) \longrightarrow HPO_4^{2-}(aq) + CO_2(g) + H_2O(\ell)$$

The liberated gas makes baked goods rise. Another phosphoric acid derivative used in the food industry is sodium hydrogen phosphate ($Na_2HPO_4$), which is used as an emulsifier to distribute butterfat uniformly through pasteurized processed cheese.

## 21.3

## NITROGEN FIXATION

Nitrogen at the earth's surface exists almost entirely (99.9%) as gaseous diatomic molecules ($N_2$), which make up 78% by volume of the atmosphere. The N—N bond is a triple bond, and its enthalpy of dissociation is 945 kJ mol$^{-1}$, making it the largest of any diatomic molecule except carbon monoxide. The nitrogen molecule is remarkably unreactive in comparison with other triple-bonded molecules. This is apparent in a simple thermochemical comparison. In each of the two parallel reactions,

$$N{\equiv}N(g) + 3\ H_2(g) \longrightarrow 2\ NH_3(g) \qquad \Delta H° = -92.2\ kJ$$

$$HC{\equiv}CH(g) + 3\ H_2(g) \longrightarrow 2\ CH_4(g) \qquad \Delta H° = -376.4\ kJ$$

a triple bond and three H—H bonds are broken and six X—H single bonds are formed (X is N or C). The enthalpy changes in the formation of an N—H bond and a C—H bond are roughly equal, and so the large difference in the $\Delta H°$ values of the reactions must mean that the triple bond in $N_2$ is substantially stronger than the triple bond in $C_2H_2$. The great stability of the $N_2$ molecule diminishes its reactivity and means that an $N_2$ atmosphere can be used to prevent air oxidation in metallurgical, chemical, and food-processing applications. For most biological and industrial purposes, however, what is needed is not molecular nitrogen but compounds of nitrogen with other elements. The formation of such compounds is referred to as **nitrogen fixation.**

### Natural Sources of Fixed Nitrogen

Figure 21.13 illustrates how nitrogen moves through a worldwide cycle involving biological, geological, and atmospheric processes. Herbivores and omnivores use nitrogen from the plants and vegetable matter they have consumed in a variety of essential biological functions, including the synthesis of proteins. Eventually, the nitrogen is excreted and returned to the soil in forms such as urea, $(NH_2)_2CO$. A portion is then reused by other plants as they grow, but some is converted to molecular nitrogen by bacteria and returned to the atmosphere. To continue the cycle, a way is needed to fix the nitrogen again.

Nature has at least two ways of fixing atmospheric nitrogen. One is the action of lightning on air, in which the high temperature of the electrical discharge causes the reaction of nitrogen with oxygen to form nitrogen oxide (NO). The nitrogen oxide is subsequently oxidized to $NO_2$, which reacts with hydroxyl radicals (OH) in the atmosphere to form dilute nitric acid. The $HNO_3$ falls to the earth in rain and provides fixed nitrogen (in the form of the nitrate ion) for plant growth. Nitrogen fixed in this way is thought to amount to at least 10% of the annual total. A second way in which nitrogen is fixed is by the action of certain bacteria that live on the roots of clover and alfalfa plants.

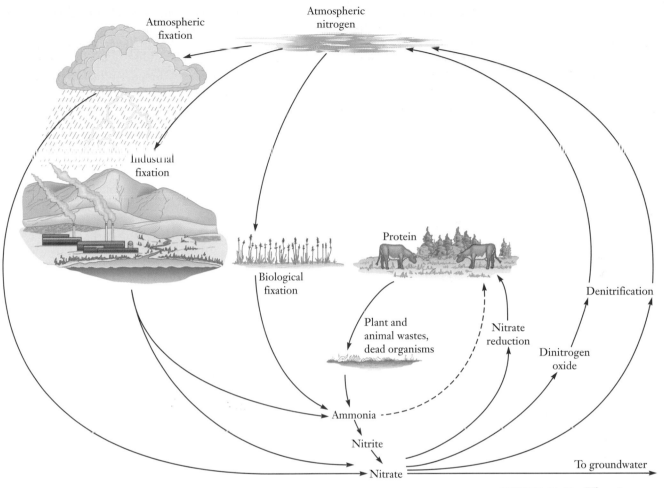

**FIGURE 21.13** The nitrogen cycle.

By the late 18th century, the population of Europe had grown sufficiently that methods of increasing crop yields were needed. The first step was to reuse animal wastes efficiently. Sailing ships went as far as the coastal regions of South America to bring back guano, the excrement dropped by seagulls, because it was rich in nitrogen and easily mined. Mineral deposits of sodium nitrate in Chile were exploited for fertilizer. By the end of the 19th century some fixed nitrogen was being recovered from the waste liquors of natural gas plants in the form of ammonia. It was generally converted to ammonium sulfate, $(NH_4)_2SO_4$, by reaction with sulfuric acid, and it was used in this form as a fertilizer.

## Early Industrial Processes

In 1898, in a presidential address to the British Association, Sir William Crookes said,

> *England and all civilized nations stand in deadly peril of not having enough to eat. As mouths multiply, food resources dwindle....It is the chemist who must come to the rescue of the threatened communities. It is through the laboratory that starvation may ultimately be turned to plenty.*[1]

[1] Quoted in L. F. Haber, *The Chemical Industry, 1900–1930*. Oxford: Clarendon Press, 1971, p. 84.

He concluded by adding that "there is a gleam of light amid this darkness and despondency…the fixation of atmospheric nitrogen," which would be "one of the great discoveries awaiting the ingenuity of chemists." Within ten years of his prediction, three methods had been developed for fixing atmospheric nitrogen.

The first was the **electric-arc process,** which was based on the observation by Cavendish in the 1780s that a spark causes the combination of nitrogen and oxygen in air. A century later, electrical technology was sufficiently advanced to create electric arcs in which the temperature was raised to 2000 to 3000°C. The NO that formed was cooled, oxidized to $NO_2$, and passed into an absorption tower where it reacted with water to make dilute nitric acid ($HNO_3$). This process closely resembled the fixing of nitrogen by lightning discharges. The energy costs for running the electric arc were high, and it was difficult to continuously obtain large amounts of fixed nitrogen. As a result, the electric-arc process was never a significant factor in nitrogen fixation.

More promising was the **cyanamide process,** also called the **Frank–Caro process** after its developers. This method uses calcium carbide ($CaC_2$) as a starting material, and the discovery of this compound has an interesting history itself. In 1892 T. L. Willson in North Carolina was trying to make elemental calcium from lime (CaO) and tar in an electric-arc furnace. He obtained a product that was clearly not calcium and threw it into a stream, where, to his surprise, it reacted with water to liberate large amounts of a combustible gas. The compound was calcium carbide, and its reaction with water gives acetylene ($C_2H_2$):

$$CaC_2(s) + 2\ H_2O(\ell) \longrightarrow C_2H_2(g) + Ca(OH)_2(aq)$$

Calcium carbide is now made from lime and coke in an electric-arc furnace at 2000 to 2200°C:

$$CaO(s) + 3\ C(s) \longrightarrow CaC_2(s) + CO(g) \qquad \Delta H° = +465\ kJ$$

In the cyanamide process, calcium carbide then reacts (at 1100°C in an electric furnace) with nitrogen, obtained from liquefying and distilling air, to give calcium cyanamide:

$$CaC_2(s) + N_2(g) \longrightarrow CaCN_2(s) + C(s) \qquad \Delta H° = -291\ kJ$$

Steam is added to convert this to ammonia:

$$CaCN_2(s) + 4\ H_2O(g) \longrightarrow Ca(OH)_2(s) + CO_2(g) + 2\ NH_3(g)$$

Between 1900 and 1930, calcium cyanamide was itself used as a fertilizer because it furnishes the soil with ammonia and calcium. The major current use of $CaCN_2$ is as a starting material for making plastics.

## The Haber–Bosch Process

Both the electric-arc and the cyanamide processes have been superseded by the **Haber–Bosch process,** which is the predominant source of industrial fixed nitrogen today. The German chemists Fritz Haber and Walther Nernst carried out studies from 1900 to 1910 on the effects of pressure and temperature on the equilibrium between ammonia and its component elements:

$$N_2(g) + 3\ H_2(g) \rightleftharpoons 2\ NH_3(g)$$

The practical ammonia synthesis Haber developed with chemical engineer Kurt Bosch required both deep chemical insight and equipment capable of operating under conditions of high temperature and pressure. At 298 K, the standard free energy and standard enthalpy of formation of 2 mol of ammonia from its elements are

$$\Delta G° = -33.0 \text{ kJ} \qquad \text{and} \qquad \Delta H° = -92.2 \text{ kJ}$$

and the value for the equilibrium constant is

$$\ln K_{298} = \frac{-\Delta G°}{RT} = \frac{33{,}000 \text{ J mol}^{-1}}{(8.315 \text{ J K}^{-1} \text{ mol}^{-1})(298 \text{ K})} = 13.32$$

$$K_{298} = e^{13.32} = 6 \times 10^5$$

which is large relative to 1; the formation of ammonia is favored on thermodynamic grounds. Unfortunately, the reaction is very slow at 298 K, and the temperature must be raised to speed it. Because the process is exothermic, increased temperature lowers the equilibrium constant and decreases product yield. Thus, competition occurs between thermodynamic factors, which favor a high yield at low temperature, and kinetic factors, which favor the use of a high temperature.

The compromise adopted in industrial production is a temperature of 700 to 900 K. At 800 K, for example, the equilibrium constant can be estimated from the van't Hoff equation (Section 9.5):

$$\ln \frac{K_{800}}{K_{298}} = -\frac{\Delta H°}{R}\left(\frac{1}{T_2} - \frac{1}{T_1}\right)$$

$$= \frac{92{,}200 \text{ J mol}^{-1}}{8.315 \text{ J K}^{-1} \text{ mol}^{-1}}\left(\frac{1}{800 \text{ K}} - \frac{1}{298 \text{ K}}\right) = -23.35$$

Solving this for the equilibrium constant at 800 K gives

$$K_{800} = (7 \times 10^{-11}) K_{298} = 4 \times 10^{-5}$$

This value is only approximate because $\Delta H$ changes with temperature between 298 and 800 K. The qualitative result, however, is that the product yield is greatly reduced at the higher temperature. To increase the yield, high total pressures are used, which favor the product in this process because 2 mol of gaseous product is formed from 4 mol of gaseous reactants (recall Fig. 9.9). At a total pressure of 200 atm, the reaction yields 15 to 30% of the amount of $NH_3$ that would be produced by complete consumption of the reactants.

A catalyst increases the rate of this reaction. Haber used osmium, a rare and expensive metal, in his original work. Later he developed a less expensive catalyst based on partially oxidized iron containing small amounts of aluminum, which has been used ever since. The ammonia produced is liquefied to separate it from the unreacted nitrogen and hydrogen, which are recycled.

Although the principles of the Haber–Bosch process were developed before World War I, its implementation in Germany was greatly speeded by the need for fixed nitrogen to make military explosives when the supplies of nitrates from Chile were cut off. It is one of the ironies of history that the commercialization of this process, which more than any other has expanded the food-producing capacity of the world, began as a military project that probably prolonged World War I by at least a year.

Despite the significant role the Haber–Bosch process has played since that time, its use to fix nitrogen has some problems. Fertilizer factories consume large amounts of energy and require expensive structural materials to operate at high pressure and high temperature. Dependence on hydrogen as a reactant in the Haber–Bosch process is another problem. Because most hydrogen is currently obtained from petroleum and natural gas (Chapter 20), both nonrenewable resources, an alternative hydrogen production process that uses coal or, ultimately, water will eventually have to be found. Scientists are also trying to design chemical catalysts that imitate the action of nitrogen-fixing bacteria as an economical way of converting atmospheric nitrogen to ammonia or to some other fixed form.

---

## 21.4
## CHEMICALS FROM AMMONIA

Once molecular nitrogen has been fixed in the form of ammonia, further reactions to produce many other compounds are possible. Reactions with sulfuric, phosphoric, and nitric acid give ammonium sulfate, ammonium phosphate, and ammonium nitrate, respectively. All three have been used as chemical fertilizers, with the latter two being particularly popular in recent years. Ammonium nitrate mixed with small amounts of fuel oil is also used as an industrial explosive. Ammonium nitrate decomposes nonexplosively above 200°C,

$$NH_4NO_3(s) \longrightarrow N_2O(g) + 2\ H_2O(g) \qquad \Delta G° = -169\ kJ$$

providing a commercial source of dinitrogen oxide (nitrous oxide), which is used as an anesthetic ("laughing gas") and an aerosol propellant. When the mass of ammonium nitrate exceeds a critical value (many tons) or when the ammonium nitrate contains chloride impurities, it may detonate:

$$NH_4NO_3(s) \longrightarrow N_2(g) + 2\ H_2O(g) + \tfrac{1}{2}\ O_2(g) \qquad \Delta G° = -273\ kJ$$

After a shipload of ammonium nitrate blew up in the harbor of Texas City, Texas, in 1947, devastating the town, the U.S. government set limits on the quantity of ammonium nitrate that may be stored in one place.

Some ammonia is also converted to urea, $(NH_2)_2CO$. Modern methods for producing urea use the reaction of $CO_2$ with ammonia in hot aqueous solution (180 to 200°C) under pressure to form the intermediate ammonium carbamate $(NH_2COONH_4)$, which is then decomposed to urea by heat:

$$2\ NH_3(aq) + CO_2(aq) \longrightarrow NH_2COONH_4(aq)$$

$$NH_2COONH_4(aq) \longrightarrow (NH_2)_2CO(s) + H_2O(g)$$

Urea is a widely used solid-nitrogen fertilizer and is an ingredient in other products ranging from glues to skin creams to disinfectants. Urea has historical significance as well, having been first synthesized by Friedrich Wöhler in 1828, from ammonia and cyanic acid (HCNO). Wöhler's work was important in demonstrating that an "organic" compound, formed in human and animal metabolism (and excreted in urine), could be synthesized from strictly inorganic starting materials.

### Nitric Acid

One of the most important products made from ammonia is nitric acid, which is generated by the overall reaction

$$NH_3 + 2\ O_2 \longrightarrow HNO_3 + H_2O$$

Urea, $(NH_2)_2CO$, is an important solid nitrogen fertilizer.

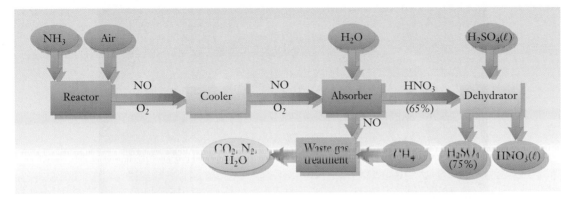

**FIGURE 21.14** Flow diagram for the synthesis of nitric acid from ammonia. Both the second step (oxidation of NO to NO₂) and the third step (absorption into water) described in the text take place in the absorber.

This reaction is deceptively simple as written, because direct combination of ammonia with oxygen will *not* produce nitric acid. The most commonly used industrial process, the **Ostwald process** (Fig. 21.14), involves three steps that, added together, give the preceding overall reaction. Let us examine these steps.

The first step is the partial oxidation of ammonia in air (Fig. 21.15):

$$4\,NH_3(g) + 5\,O_2(g) \longrightarrow 4\,NO(g) + 6\,H_2O(g) \qquad \Delta H° = -905.5 \text{ kJ}$$

This is a strongly exothermic reaction, but it occurs very slowly at room temperature. In addition, once some NO is produced, a competing reaction occurs,

$$4\,NH_3(g) + 6\,NO(g) \longrightarrow 5\,N_2(g) + 6\,H_2O(g)$$

which is undesirable because it returns nitrogen to its elemental state. To obtain practical yields of NO, the reaction is carried out at elevated temperatures (800–950°C) and a catalyst, consisting of a fine gauze of noble metals such as platinum and rhodium or gold and palladium, is used (Fig. 21.16). Such a catalyst promotes the desired reaction in preference to the competing one, and this first step occurs with extraordinary speed: only about $3 \times 10^{-4}$ s of contact time between gases and the gauze is needed to effect excellent conversion.

The second, and rate-determining, step is the reaction of nitrogen oxide with oxygen to produce nitrogen dioxide:

$$2\,NO(g) + O_2(g) \longrightarrow 2\,NO_2(g) \qquad \Delta H° = -114.1 \text{ kJ}$$

Example 13.3 and Section 13.5 discussed the kinetics of this reaction, which requires no catalyst, and showed that the rate law has the form

$$\text{rate} = k[NO]^2[O_2]$$

and the rate constant $k$ *decreases* with increasing temperature. Because the reaction occurs faster at low temperatures, this step is carried out at the lowest temperature conveniently reached by the available cooling water, 10 to 40°C.

**FIGURE 21.15** When a heated platinum wire is placed just above a solution of concentrated aqueous ammonia, it glows for as long as half an hour as it catalyzes the continued exothermic oxidation of NH₃ to NO by oxygen from the air. *(Leon Lewandowski)*

**FIGURE 21.16** This gold–palladium gauze catalyst is used to make nitric acid from ammonia. (*Johnson Matthey*)

The third step is the absorption of the $NO_2(g)$ into water:

$$3\ NO_2(g) + 3\ H_2O(\ell) \longrightarrow 2\ H_3O^+(aq) + 2\ NO_3^-(aq) + NO(g)$$
$$\Delta H° = -113.5\ kJ$$

The $NO(g)$ produced reacts again in the second step to give more $NO_2(g)$. The nitric acid has a concentration in the range of 50 to 65% by mass. Distilling off the water does not increase the concentration beyond 69% $HNO_3$, which is the "concentrated nitric acid" in normal laboratory use. Addition of a powerful dehydrating agent (concentrated sulfuric acid) and distillation separate more water to give a solution that contains 95 to 98% nitric acid (fuming nitric acid).

We have analyzed the production of nitric acid in some detail not only because nitric acid is an important chemical but also because the Ostwald process illustrates some of the problems involved in practical chemistry. The original overall reaction is quite simple and is strongly favored thermodynamically:

$$\frac{[H_3O^+][NO_3^-]}{P_{NH_3}(P_{O_2})^2} = K_{298} = 10^{52}$$

Nevertheless, kinetic effects and the presence of competing equilibria require the careful design of a multistep process to obtain practical yields.

## Hydrazine

Another important product made from ammonia is hydrazine ($N_2H_4$). In principle, many reactions are capable of producing hydrazine from ammonia, such as

$$2\ NH_3 + \tfrac{1}{2}\ O_2 \longrightarrow N_2H_4 + H_2O$$

This and other direct pathways fail as practical methods, however. Competing reactions such as

$$2\ NH_3 + \tfrac{3}{2}\ O_2 \longrightarrow N_2 + 3\ H_2O$$

are thermodynamically favored because the $N_2$ molecule is so stable. The method used to make hydrazine commercially is the reaction of ammonia with hypochlorite ion ($OCl^-$) in aqueous solution, known as the **Raschig synthesis:**

$$2\ NH_3(aq) + OCl^-(aq) \longrightarrow N_2H_4(aq) + H_2O(\ell) + Cl^-(aq)$$

Pure liquid hydrazine resembles hydrogen peroxide in many of its physical properties, boiling at $114°C$ and freezing at $2°C$ ($H_2O_2$ boils at $152°C$ and freezes at $-1.7°C$). Hydrazine has had some use as a rocket fuel because its reaction with oxygen is extremely exothermic:

$$N_2H_4(\ell) + O_2(g) \longrightarrow N_2(g) + 2\ H_2O(g) \qquad \Delta H° = -534\ kJ$$

A small mass of liquid hydrazine produces a tremendous thrust as the hot gaseous products rush out at high speed from the rocket engine.

Nitrogen in hydrazine has an oxidation state of $-2$, intermediate between those of molecular nitrogen (0) and nitrogen in ammonia ($-3$). As a result, hydrazine (like hydrogen peroxide) can act either as a reducing agent or as an oxidizing agent in aqueous solution. In basic solution (at pH 14), the standard reduction potential for the half-reaction

$$N_2(g) + 4\ H_2O(\ell) + 4\ e^- \longrightarrow N_2H_4(aq) + 4\ OH^-(aq)$$

is $\mathscr{E}° = -1.16$ V. This means that hydrazine tends to be easily oxidized to nitrogen under these conditions and so should act as a good reducing agent. In a strongly acidic medium, hydrazine is almost completely converted to its conjugate acid, the hydrazinium ion $N_2H_5^+$. The standard reduction potential $\mathscr{E}° = 1.275$ V for the half-reaction

$$N_2H_5^+(aq) + 3\ H_3O^+(aq) + 2\ e^- \longrightarrow 2\ NH_4^+(aq) + 3\ H_2O(\ell)$$

at pH 0 suggests that $N_2H_5^+$ should be easily reduced to $NH_4^+$ and therefore should serve as an excellent oxidizing agent under these conditions. This reaction is slow, however, and hydrazine is mainly known as a reducing agent.

Hydrazine is used to control boiler corrosion through the reaction

$$6\ Fe_2O_3(s) + N_2H_4(aq) \longrightarrow 4\ Fe_3O_4(s) + N_2(g) + 2\ H_2O(\ell)$$

The red $Fe_2O_3$ oxide (ordinary rust) is reduced to $Fe_3O_4$ that forms a protective black layer to hinder further rusting. Hydrazine is also used to remove certain metal ions from the wastewaters of chemical plants. Chromate ion ($CrO_4^{2-}$), for example, is reduced to $Cr^{3+}$ in the reaction

$$20\ H_3O^+(aq) + 4\ CrO_4^{2-}(aq) + 3\ N_2H_4(aq) \longrightarrow$$
$$4\ Cr^{3+}(aq) + 3\ N_2(g) + 36\ H_2O(\ell)$$

The $Cr^{3+}$ is then precipitated by adding base:

$$Cr^{3+}(aq) + 3\ OH^-(aq) \longrightarrow Cr(OH)_3(s)$$

Hydrazine also removes halogens ($X_2$, X = F, Cl, Br, I) from industrial wastewater by the reaction

$$N_2H_4(aq) + 2\ X_2(aq) \longrightarrow N_2(g) + 4\ HX(aq)$$

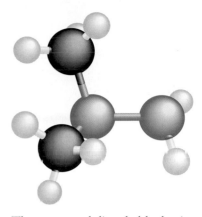

The compound dimethyl hydrazine, $(CH_3)_2NNH_2$, fueled the lunar excursion modules in their departures from the surface of the moon. The oxidant was the nitrogen oxide $N_2O_4$. Ignition was a certainty because the two burst into flame upon contact.

## 21.5

### OXOACIDS OF SULFUR, PHOSPHORUS, AND NITROGEN

The production of the oxoacids of sulfur, phosphorus, and nitrogen was discussed earlier in this chapter. Here we compare two facets of their aqueous chemistry: their ability to act as oxidizing (or reducing) agents and their acid strengths.

Figure 21.17 shows the reduction potentials for these oxoacids and their conjugate bases. The species listed are those that predominate at the pH given (0 or 14). In acidic solution, nitrous acid exists largely as $HNO_2$ molecules, for example, but in basic solution almost all these molecules have ionized to form $NO_2^-$ ions. Nitric acid, on the other hand, is a strong acid and is almost fully dissociated to nitrate and hydronium ions, even at pH 0. Because hydronium ion or hydroxide ion participates in the reduction half-reactions, the half-cell potentials depend strongly on pH, as is seen by comparing the left half of the figure with the right half.

The reduction potentials of nitrogen, sulfur, and phosphorus oxoacids on the left side of the figure show that the phosphorus oxoacids are the most difficult to reduce and therefore are the least effective oxidizing agents. The oxoacids of sulfur are intermediate in this regard, and nitric acid is the strongest oxidizing acid listed. In concentrated form, it will attack and dissolve many metals—such as copper (see Fig. 4.8)—that phosphoric acid will not attack. In applications needing a strong acid that is *not* a strong oxidizing agent, phosphoric acid is preferable to nitric or sulfuric acid (Fig. 21.18).

The right side of Figure 21.17 shows that the oxidizing strengths of all of these oxoacids are drastically curtailed when they are replaced by their basic forms at pH

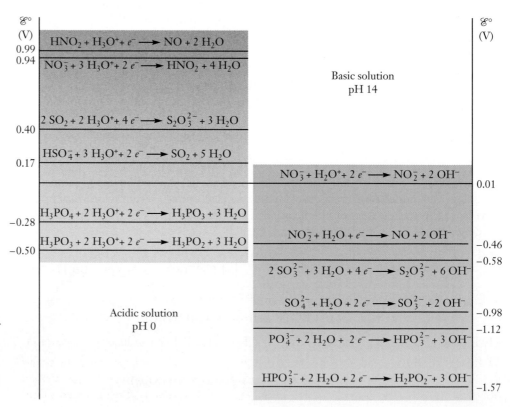

**FIGURE 21.17** The standard reduction potentials for oxides of nitrogen, sulfur, and phosphorus are higher in acidic solution at pH 0 (left) than in basic solution at pH 14 (right).

14. The highest reduction potential achieved (that for nitrate ion) is only $+0.01$ V at pH 14, and so basic conditions should be avoided if oxidation is desired. On the other hand, the lower oxidation states of the sulfur and phosphorus compounds can be excellent *reducing* agents at pH 14. Both sulfite ($SO_3^{2-}$) and thiosulfate ($S_2O_3^{2-}$) ions are fairly strong reducing agents in basic solution, as shown by their appearance on the right sides of half-equations with negative reduction potentials ($-0.98$ and $-0.58$ V, respectively). The bases of the lower oxoacids of phosphorus ($H_2PO_2^-$ and $HPO_3^{2-}$) are even stronger reducing agents.

From Figure 21.17 we can identify which species of intermediate oxidation state are unstable with respect to disproportionation under the conditions given. Nitrous acid, for example, disproportionates spontaneously in acidic solution according to

$$3\ HNO_2 \longrightarrow NO_3^- + 2\ NO + H_3O^+ \qquad \Delta\mathscr{E}° = 0.99 - 0.94 = +0.05\ V$$

In basic solution, however, the nitrite ion is stable against disproportionation to NO and $NO_3^-$ because the hypothetical reaction

$$3\ NO_2^- + H_2O \longrightarrow NO_3^- + 2\ NO + 2\ OH^-$$

has $\Delta\mathscr{E}° = -0.46 - 0.01 = -0.47\ V < 0$, and so the equilibrium lies strongly to the left. The $NO_2^-$ ion is still thermodynamically unstable with respect to $NO_3^-$ and $N_2O$ or $N_2$, but these reactions occur so slowly that $NO_2^-$ can be kept in basic solution indefinitely.

One might guess at first that having an element in a very high oxidation state would make a compound a strong oxidizing agent—as is the case with permanganate solutions, which contain Mn(VII) and certainly do have great oxidizing power. Figure 21.17 shows that this is not true in general, however. Nitrous acid (N oxidation state $= +3$) would be a stronger oxidizing agent than nitric acid ($+5$) at pH 0, were it stable against disproportionation. Aqueous sulfur dioxide (S oxidation state $= +4$) is a stronger oxidizing agent than hydrogen sulfate ion ($+6$). The important factor is not the oxidation number itself but rather the relative stabilities of the redox pair of chemical species. Nitrous acid is a strong oxidizing agent precisely because it is much less stable than its reduced form, nitrogen oxide. Thermodynamic stability changes with oxidation state in an irregular fashion.

Oxidation state also affects acid strength for the oxoacids of S, P, and N as well as those of other nonmetals. Higher oxidation states should correspond to more covalent X—O bonds and thus higher acid ionization constants (recall Fig. 10.6). An even better correlation than oxidation state, however, is a related quantity: the number of lone oxygen atoms attached to the central atom. The rule is that the strength of oxoacids with a given central element increases with the number of lone oxygen atoms attached to the central atom. If the formula of these acids is written as $XO_n(OH)_m$, the corresponding acid strengths fall into distinct classes according to the value of $n$, the number of lone oxygen atoms (see Table 21.1). Each increase of 1 in $n$ increases the acid ionization constant $K_a$ by a factor of about $10^5$. Another way to describe this effect is to focus on the stability of the conjugate base, $XO_{n+1}(OH)_{m-1}^-$, of the oxoacid. The greater the number of lone oxygen atoms attached to the central atom, the more easily the net negative charge can be spread out over the ion, and therefore the more stable the base. This leads to a larger $K_a$.

An unusual and interesting structural result can be obtained from Table 21.1. Figure 21.19a shows the simplest Lewis diagram for phosphorous acid ($H_3PO_3$) in which each atom achieves an octet configuration. Such a diagram could also be written $P(OH)_3$ and would be analogous to $As(OH)_3$, which has no lone oxygen atoms bonded to the central atom ($n = 0$). On the basis of this analogy, the expected value

**FIGURE 21.18** Concentrated nitric acid reacts with sugar after some time (left), whereas sulfuric acid (center) reacts immediately. By contrast, sugar and phosphoric acid (right) form a viscous solution, but no reaction is seen. (*Leon Lewandowski*)

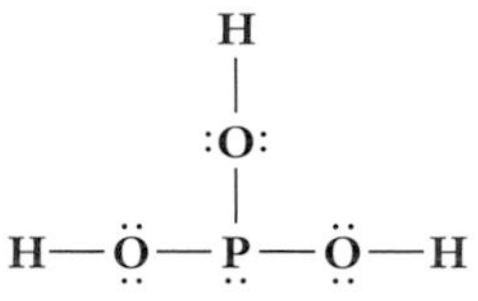

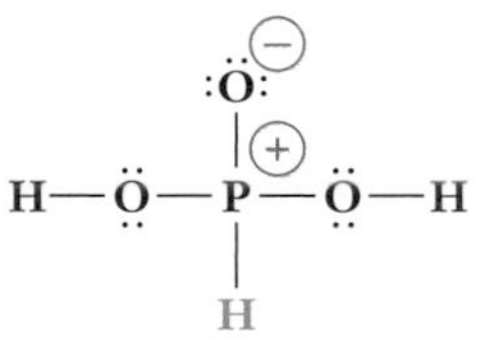

**FIGURE 21.19** (a) The simplest Lewis diagram that can be drawn for $H_3PO_3$ gives an incorrect structure. This acid would be triprotic, like $H_3PO_4$. (b) The observed structure of $H_3PO_3$ requires assigning formal charge to the P atom and the lone O atom. The hydrogen atom attached to the P is not released into acidic solution, and so the acid is diprotic.

## TABLE 21.1

### Acid Ionization Constants for Oxoacids of the Nonmetals

| $X(OH)_m$ Very Weak | $K_a$ | $XO(OH)_m$ Weak | $K_a$ | $XO_2(OH)_m$ Strong | $K_a$ | $XO_3(OH)_m$ Very Strong | $K_a$ |
|---|---|---|---|---|---|---|---|
| Cl(OH) | $3 \times 10^{-8}$ | $H_2PO(OH)$ | $8 \times 10^{-2}$ | $SeO_2(OH)_2$ | $10^3$ | $ClO_3(OH)$ | $2 \times 10^7$ |
| Te(OH)$_6$ | $2 \times 10^{-8}$ | IO(OH)$_5$ | $2 \times 10^{-2}$ | $ClO_2(OH)$ | $5 \times 10^2$ | | |
| Br(OH) | $2 \times 10^{-9}$ | SO(OH)$_2$ | $2 \times 10^{-2}$ | $SO_2(OH)_2$ | $1 \times 10^2$ | | |
| As(OH)$_3$ | $6 \times 10^{-10}$ | ClO(OH) | $1 \times 10^{-2}$ | $NO_2(OH)$ | $2 \times 10^1$ | | |
| B(OH)$_3$ | $6 \times 10^{-10}$ | HPO(OH)$_2$ | $1 \times 10^{-2}$ | $IO_2(OH)$ | $1.6 \times 10^{-1}$ | | |
| Ge(OH)$_4$ | $4 \times 10^{-10}$ | PO(OH)$_3$ | $8 \times 10^{-3}$ | | | | |
| Si(OH)$_4$ | $2 \times 10^{-10}$ | AsO(OH)$_3$ | $5 \times 10^{-3}$ | | | | |
| I(OH) | $4 \times 10^{-11}$ | SeO(OH)$_2$ | $3 \times 10^{-3}$ | | | | |
| | | TeO(OH)$_2$ | $3 \times 10^{-3}$ | | | | |
| | | NO(OH) | $5 \times 10^{-4}$ | | | | |

of $K_a$ for $P(OH)_3$ is on the order of $10^{-9}$ (a very weak acid). In fact, however, $H_3PO_3$ is only a moderately weak acid ($K_a = 1 \times 10^{-2}$) and fits better into the class of acids with one lone oxygen atom bonded to the central atom. X-ray diffraction measurements support this analysis based on chemical properties. The true structure of $H_3PO_3$ is best represented by Figure 21.19b and corresponds either to a Lewis diagram with more than eight electrons around the central phosphorus atom or to one with formal charges on the central phosphorus and lone oxygen atoms. The formula of this acid is written as $HPO(OH)_2$ in Table 21.1. Unlike phosphoric acid ($H_3PO_4$), which is a triprotic acid, $H_3PO_3$ is a diprotic acid. The third hydrogen atom, the one directly bonded to the phosphorus atom, does not ionize even in strongly basic aqueous solution.

## CONCEPTS & SKILLS

*After studying this chapter and working the problems that follow, you should be able to*

1. Compare the lead-chamber and contact processes for the manufacture of sulfuric acid (Section 21.1, problems 1–6).

2. Outline the role of sulfur in the manufacture of paper pulp from wood (Section 21.1, problem 7).

3. Describe methods for making phosphoric acid and its uses in fertilizer manufacture and consumer chemistry (Section 21.2, problems 9–12).

4. Give thermodynamic analyses of different ways of fixing nitrogen in the form of ammonia (Section 21.3, problems 13–18).

5. Describe the stages in the industrial synthesis of nitric acid and the uses of this chemical (Section 21.4, problems 19–20).

6. Discuss the structures and chemistries of nitrogen compounds with oxidation states intermediate between ammonia and elemental nitrogen (Section 21.4, problems 23–29).

7. Compare the redox and acid–base chemistries of the aqueous oxoacids of S, P, and N (Section 21.5, problems 33–34).

## PROBLEMS

*Answers to problems whose numbers are boldface appear in Appendix G. Problems that are more challenging are indicated with asterisks.*

### Sulfuric Acid and Its Uses

1. Propose several ways in which predictions from Le Châtelier's principle could be used to improve the yield of $SO_3(g)$ from $SO_2(g)$ and air.

2. Give a thermodynamic analysis of the Claus process; that is, determine the $\Delta H°$ and $\Delta S°$ of the main reaction and predict conditions that will favor a high yield of sulfur. Take $S(s,$ rhombic) as the physical form of sulfur.

3. Residual $SO_2$ can be removed from the effluent gases of sulfuric acid plants by oxidation with $H_2SO_5$, as explained in the text. Suppose that the electrolysis system that supplies the $H_2SO_5$ failed. Would it be thermodynamically feasible to substitute an aqueous solution of $H_2O_2$ as the oxidant for the $SO_2$? Consult Appendix D for necessary data.

4. Amounts of sulfur in a sample can be measured by burning the sample in oxygen and then collecting the oxides of sulfur in a dilute solution of hydrogen peroxide. What is the product of this reaction? Write a balanced equation. Suggest a good way to determine quantitatively the amount of product that is formed.

5. Write a balanced overall chemical equation for the production of sulfuric acid from zinc sulfide ore.

6. Write a balanced overall chemical equation for the production of sulfuric acid from hydrogen sulfide in natural gas.

7. Describe (via chemical equations) how sodium sulfide for use in the Kraft process can be prepared starting from sulfuric acid and other easily accessible materials.

8. Predict the product(s) of the reaction of disulfuric acid $(H_2S_2O_7)$ with water.

### Phosphorus Chemistry

9. Hot concentrated sulfuric acid reacts with phosphorus to give sulfur dioxide, phosphoric acid, and water. Write a balanced chemical equation for this reaction.

10. Describe (via chemical equations) how sodium tripolyphosphate for detergents can be prepared starting from phosphoric acid and other easily accessible materials.

11. Phosphoric acid $(H_3PO_4)$ can be kept pure only in the crystalline state. Above its melting point of 42.4°C, pairs of $H_3PO_4$ units gradually lose water. What is the other product of this dehydration reaction? What is its structure?

12. The P—O—P linkages in diphosphoric acid are extended by adding other phosphate groups (through reaction with $H_3PO_4$ and elimination of water). The resulting long-chain phosphate is called "metaphosphoric acid." Determine its empirical formula, assuming that the chain is very long.

### Nitrogen Fixation

13. A minor source of fixed nitrogen is emission from the inside of the earth in volcanic eruptions. One estimate of the total amount of fixed nitrogen per year from this source is $2 \times 10^8$ kg. How many atoms of fixed nitrogen are emitted per second, on the average, by volcanoes?

14. Observations of lightning suggest that the average lightning stroke produces $10^{27}$ molecules of $NO$. The rate of lightning strokes worldwide is about 100 per second. According to these data, how much nitrogen (in metric tons) is fixed annually by lightning strokes?

15. Write an overall equation to represent the production of $NH_3$ by the cyanamide process as described in the text. Assume that the starting materials are limestone $(CaCO_3)$, coke (C), nitrogen $(N_2)$, and water. Determine the $\Delta H°$, per mole of $NH_3$ produced, of the process represented by your equation.

16. Suppose that, in the preceding problem, the calcium hydroxide and carbon dioxide produced were recycled to give the calcium carbonate starting material. Write the overall equation in this case, and calculate the $\Delta H°$ per mole of $NH_3$ produced.

17. Some chemical plants manufacture ammonia via the Haber–Bosch process using pressures as high as 700 atm rather than the 200 atm cited in the text. What are the advantages and disadvantages of working at these higher pressures? For a given yield of product, should a higher or lower temperature be used if the pressure is raised? Will the reaction then be faster or slower?

18. Suppose a catalyst were developed that allowed the rapid production of ammonia from the elements at 600 K. Would this be useful in the Haber–Bosch process? Estimate the equilibrium constant for the formation of 2 mol of $NH_3$ at this temperature using the van't Hoff equation.

### Chemicals from Ammonia

19. For each of the three steps in the Ostwald process for making nitric acid from ammonia, state whether a high yield of product will be favored by low or high temperature.

20. For each of the three steps in the Ostwald process for making nitric acid from ammonia, state whether a high yield of product will be favored by low or high total pressure.

21. Write balanced equations to represent the reaction of concentrated nitric acid with $Cr(s)$ and with $Fe(s)$. Assume that the nitric acid is reduced to $NO_2(g)$ and that Cr and Fe are oxidized to the +3 oxidation state.

22. Write balanced equations to represent the reaction of concentrated nitric acid with $Ni(s)$ and with $Tl(s)$, assuming that the nitric acid is reduced to $NO(g)$ and that Ni and Tl are oxidized to the +2 and +3 oxidation states, respectively.

23. *trans*-Tetrazene has the formula $N_4H_4$. It consists of a chain of four nitrogen atoms with the two terminal nitrogen atoms

bonded to two hydrogen atoms each. Draw a Lewis diagram for this compound.

24. Diimine ($N_2H_2$) has not been isolated, but there is evidence that it exists in solution.
    (a) Draw a Lewis diagram for this compound.
    (b) Would the conversion of hydrazine to diimine in solution require an oxidizing agent or a reducing agent?

**25.** Two reduction half-reactions involving hydrazine ($N_2H_4$) are

$$N_2H_4 + 2\ H_2O + 2\ e^- \longrightarrow 2\ NH_3 + 2\ OH^-$$
$$\mathscr{E}° = 0.1\ V$$

$$N_2 + 4\ H_2O + 4\ e^- \longrightarrow N_2H_4 + 4\ OH^-$$
$$\mathscr{E}° = -1.16\ V$$

Determine whether hydrazine is thermodynamically stable against disproportionation to $N_2$ and $NH_3$ at pH 14.

26. In our discussion of conversions among $N_2$, $N_2H_4$, and $NH_3$, we did not mention a compound of nitrogen named hydroxylamine ($NH_2OH$).
    (a) What is the oxidation number of nitrogen in $NH_2OH$?
    (b) Two reduction half-reactions involving $NH_2OH$ are

$$2\ NH_2OH + 2\ e^- \longrightarrow N_2H_4 + 2\ OH^-$$
$$\mathscr{E}° = -0.73\ V$$

$$N_2 + 4\ H_2O + 2\ e^- \longrightarrow 2\ NH_2OH + 2\ OH^-$$
$$\mathscr{E}° = -1.59\ V$$

Determine whether hydroxylamine is thermodynamically stable against disproportionation to $N_2$ and $N_2H_4$ at pH 14.

**27.** The suggestion is made that the reaction between liquid hydrazine and liquid nitric acid be used to boost a specialized type of rocket because the products, $N_2(g)$ and $H_2O(\ell)$, are completely innocuous in the atmosphere. Write a balanced equation, and compute $\Delta H°$ for this reaction.

28. Write a balanced chemical equation for a reaction between hydrazine and dinitrogen tetraoxide that gives only nitrogen and water. Compute $\Delta H°$ for this reaction.

**29.** Hydrazine ($N_2H_4$) is less basic than ammonia, and $NF_3$ is not basic at all. Formulate a general principle about the effect of substituents attached to nitrogen on the nitrogen's ability to accept hydrogen ions (or, equivalently, to donate electrons). Do you expect hydroxylamine ($NH_2OH$) to be more basic or less basic than ammonia?

30. Suppose the hydrogen atoms in phosphine ($PH_3$) are replaced by chlorine atoms to give $PCl_3$. Do you expect $PCl_3$ to be a stronger or a weaker Lewis base than $PH_3$? Explain your reasoning.

**31.** Describe a practical reaction sequence by which $N_2O_4(s)$ could be obtained from the elements.

32. Describe a practical reaction sequence by which $N_2O(g)$ could be obtained from the elements.

## Oxoacids of Sulfur, Phosphorus, and Nitrogen

**33.** Figure 21.17 can be extended to include species in which the oxidation states of sulfur, phosphorus, and nitrogen are 0 or

negative. In basic solution, for example, the following addition reduction potentials are measured at pH 14:

$$P_4(s) + 12\ H_2O(\ell) + 12\ e^- \longrightarrow 4\ PH_3(g) + 12\ OH^-$$
$$\Delta\mathscr{E}° = -0.89\ V$$

$$4\ H_2PO_2^- + 4\ e^- \longrightarrow P_4(s) + 8\ OH^- \qquad \Delta\mathscr{E}° = -2.05\ V$$

(a) Will $P_4$ disproportionate spontaneously in basic solution?
(b) Which is the stronger reducing agent at pH 14—$P_4$ or $PH_3$?

34. Extending Figure 21.17, the following additional reduction potentials are measured at pH 14:

$$S(s) + H_2O + 2\ e^- \longrightarrow HS^- + OH^- \qquad \Delta\mathscr{E}° = -0.51\ V$$

$$S_2O_3^{2-} + 3\ H_2O(\ell) + 4\ e^- \longrightarrow 2\ S(s) + 6\ OH^-$$
$$\Delta\mathscr{E}° = -0.74\ V$$

(a) Will sulfur disproportionate spontaneously under basic conditions?
(b) Which is the stronger reducing agent—S or $HS^-$?

## Additional Problems

35. Use thermodynamics to compare oxidation of $SO_2$ in the absence of $NO_2$

$$SO_2(g) + \tfrac{1}{2}\ O_2(g) \longrightarrow SO_3(g)$$

with oxidation of $SO_2$ by $NO_2$:

$$SO_2(g) + NO_2(g) \longrightarrow SO_3(g) + NO(g)$$

Which has the greater thermodynamic driving force ($\Delta G°$) at 25°C? At low temperatures, does the $NO_2$ act by increasing the yield or by speeding up the reaction?

36. In the lead-chamber process, the sulfuric acid was contained in a lead container. Using the information in Table 11.2, explain the great resistance of lead to corrosion by sulfuric acid.

37. The standard reduction potential for the half-reaction

$$S_2O_8^{2-} + 2\ e^- \longrightarrow 2\ SO_4^{2-}$$

is 2.0 V. Use the Nernst equation to compute the minimum voltage required to operate an electrolysis unit that converts 1.0 M $SO_4^{2-}$ to 0.50 M $S_2O_8^{2-}$ and hydrogen at a pressure of 0.10 atm if the pH is 0.

38. The world production of pure $H_2SO_4$ in 1991 was $5.51 \times 10^{10}$ kg. Suppose that all locations converted sulfur to $H_2SO_4$ at 99.97% efficiency but lost the rest of the sulfur as $SO_2$ to the atmosphere. What, then, was the total mass of emitted sulfur dioxide (in kilograms)?

39. Write a chemical equation for (a) a precipitation reaction, (b) an acid–base reaction, and (c) an oxidation–reduction reaction in which aqueous sulfuric acid is a reactant.

40. The $Na_2S$ for the Kraft process can be made by heating $Na_2SO_4$ with coke (C). Write a balanced chemical equation for this reaction. Can this reaction be used to supply heat for the digestion of the wood, or does it consume heat?

41. Most sulfuric acid (other than that used in making fertilizer) does not form part of the final product. It lingers inconveniently as spent sulfuric acid that is both diluted and contaminated. Spent acid must not be dumped in the environment. You propose to regenerate $H_2SO_4$ by heating spent acid to high temperatures (about 1000 K) in the presence of natural gas ($CH_4$) to make $SO_2(g)$ by the reaction

$$4\ SO_3(g) + CH_4(g) \longrightarrow CO_2(g) + 4\ SO_2(g) + 2\ H_2O(g)$$

and then recycling the $SO_2(g)$ to make more $H_2SO_4$.
   (a) Estimate $\Delta G$ for your proposed reaction.
   (b) A critic asserts that the reaction will probably proceed as follows:

$$SO_3(g) + CH_4(g) \longrightarrow CO_2(g) + H_2O(g) + H_2S(g)$$

   Estimate $\Delta G$ at 1000 K for the critic's proposed reaction.
   (c) How can you tell which reaction will occur? Explain.
42. Suppose a fertilizer plant produces phosphoric acid through the wet-acid process and uses it to make triple superphosphate fertilizer $Ca(H_2PO_4)_2 \cdot H_2O$. Write a balanced equation for the overall reaction. How many moles of gypsum are generated per mole of $Ca(H_2PO_4)_2 \cdot H_2O$?
43. In fertilizer manufacture, it is conventional to measure relative amounts of phosphorus in different fertilizers by arbitrarily assuming that the phosphorus is present as $P_2O_5$, dividing the mass of $P_2O_5$ that would be present by the total mass, and expressing that mass as a percentage. For example, the formula of triple superphosphate is rewritten as

$$Ca(H_2PO_4)_2 \cdot H_2O - (P_2O_5)(CaO)(H_2O)_3$$

   The ratio of the mass of $P_2O_5$ to the total mass is then the ratio of the molar mass of $P_2O_5$ to the molar mass of the formula unit, or $141.94/252.07 = 0.563$, and so this fertilizer is 56.3% $P_2O_5$. Carry out the corresponding calculation for ordinary superphosphate fertilizer, which contains 7 mol of gypsum for every 3 mol of $Ca(H_2PO_4)_2 \cdot H_2O$. Account for the origin of the word "triple" in "triple superphosphate."
44. The disposal of byproduct gypsum ($CaSO_4 \cdot 2H_2O(s)$) is a significant problem for fertilizer manufacturers.
   (a) A typical fertilizer plant may produce 100 metric tons of phosphoric acid per day by the wet-acid process (1 metric ton = $10^3$ kg). What mass of gypsum is produced per *year*?
   (b) The density of crystalline gypsum is 2.32 g cm$^{-3}$. What volume of gypsum is generated per year? (In fact, the volume is several times *larger* because of the porosity of the gypsum produced.)
45. During the early 1970s, when the price of petroleum and natural gas rose rapidly, the price of nitrogen-based fertilizers increased rapidly as well. Explain why the two commodities are so closely linked.
46. A piece of hot platinum foil is inserted into a container filled with a mixture of ammonia and oxygen. Brown fumes appear near the surface of the foil. Write balanced equations for the reactions taking place.

47. Nitrogen oxides ($NO_x$) from power stations, steam generators, and other fossil-fuel-burning sources are serious air pollutants. One promising technology (called NOxOUT) for removing $NO_x$ from stack emissions injects an aqueous solution of urea, $(NH_2)_2CO$, into various points of the combustion system. Urea reacts with the nitrogen oxides to give $N_2$, $CO_2$, and $H_2O$.
   (a) Write a balanced chemical equation to represent the reaction of urea with $NO_2$.
   (b) Write a second balanced chemical equation to represent the reaction of urea with NO.
48. The concentration of $NO_3^-(aq)$ in a solution can be determined by adding sulfuric acid and mercury and then collecting the gaseous NO that the reaction generates:

$$2\ NO_3^-(aq) + 8\ H_3O^+(aq) + 3\ SO_4^{2-}(aq) + 6\ Hg(\ell) \longrightarrow$$
$$3\ Hg_2SO_4(s) + 12\ H_2O(\ell) + 2\ NO(g)$$

   A 357-mL sample of a solution is treated in this way and generates 11.46 mL of NO(g), measured at a pressure of 0.965 atm and a temperature of 18.8°C. Compute the concentration of nitrate ion in the solution.
49. The formation of hydrazine in the Raschig synthesis is thought to be a two-step process that involves $NH_2Cl$ as an intermediate. Write balanced equations for the two steps in the Raschig synthesis.
50. Molecules of hydrazine ($N_2H_4$) have a sizable dipole moment. Sketch a structure that this information definitely *eliminates* as the structure of the hydrazine molecule.
51. (a) Compute the $\Delta G°$ of the reaction

$$6\ Fe_2O_3(s) + N_2H_4(aq) \longrightarrow$$
$$4\ Fe_3O_4(s) + N_2(g) + 2\ H_2O(\ell)$$

   (b) As explained in the text, this reaction is used to prevent further rusting of the walls of boilers. A boilermaker asks you about a proposal to improve this procedure by making the solution of hydrazine strongly acidic (to "burn off the worst of the rust as the hydrazine simultaneously seals off the rest"). Evaluate this idea.
52. Oxoacids can be formed that involve several central atoms of the same chemical element. An example is $H_3P_3O_9$, which can be written $P_3O_6(OH)_3$. (Sodium salts of these polyphosphoric acids are used as "builders" in detergents to improve their cleaning power.) In such a case, we would expect acid strength to correlate approximately with the *ratio* of the number of lone oxygen atoms to the number of central atoms (this ratio is 6:3 for $H_3P_3O_9$, for example). Rank the following in order of increasing acid strength: $H_3PO_4$, $H_3P_3O_9$, $H_4P_2O_6$, $H_4P_2O_7$, $H_5P_3O_{10}$. Assume that no hydrogen atoms are directly bonded to phosphorus in these compounds.
53. The first acid ionization constant of the oxoacid $H_3PO_2$ is $8 \times 10^{-2}$. What molecular structure do you predict for $H_3PO_2$? Will this acid be monoprotic, diprotic, or triprotic in aqueous solution?

# Chemical Processes Based on Halogens and the Noble Gases

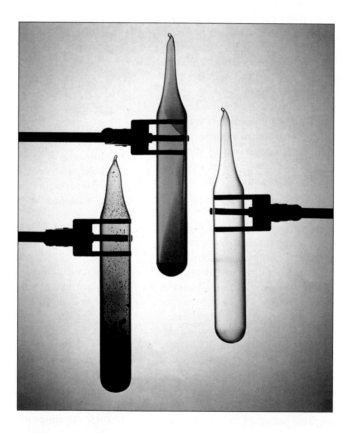

*Illustration*
Halogens. Left, Iodine ($I_2$) is a grayish black solid that sublimes to form violet vapor. Center, Bromine ($Br_2$) is a reddish brown liquid that forms vapor of the same color. Right, Chlorine ($Cl_2$) is a greenish yellow gas at room temperature. *(Copyright 1990 Richard Megna/ Fundamental Photographs)*

T he elements that make up the last two main groups of the periodic table are a study in contrasts. The halogens are among the most reactive of all the elements and are found terrestrially only in combination with other elements. Except for oxygen itself, they are the only elements that are classified as oxidizing agents. In contrast, the noble gases are so unreactive that they are found in nature only in elemental form.

This chapter begins with an historical survey of the industrial-scale preparation of chlorine and its linkage with the two important bases sodium carbonate and sodium hydroxide. After a comparative look at the chemistries of chlorine, bromine, and iodine, the special properties of fluorine are introduced. This element and some of its compounds are highly reactive and toxic, but other compounds are at the opposite extreme and are deliberately chosen for uses demanding inert, nontoxic materials. Under proper conditions, fluorine forms compounds with nearly all of the other elements, the only exceptions being helium, neon, and argon.

---

## 22.1

## CHEMICALS FROM SALT

By the second half of the 18th century, several important industries in Europe and North America faced a growing crisis in the supply of raw materials. The glass industry relied on soda ash (sodium carbonate, $Na_2CO_3$) or potash (potassium carbonate, $K_2CO_3$) as a flux to lower the viscosity of the molten glass. As their common names imply, the two carbonates were derived from the leachings of ashes: soda ash from the ashes of sea plants, and potash from the ashes of forest trees and shrubs. By 1750, demand for glass was exceeding the supplies from traditional sources, which were of variable quality in any case.

Soapmakers faced a supply problem of their own, because they needed a cheap base to react with vegetable oils and animal fats. One base, lime, was readily available. It was produced by high-temperature decomposition of the limestone that was abundantly available throughout the world:

$$CaCO_3(s) \xrightarrow{850°C} CaO(s) + CO_2(g)$$

In the 18th century, lime kilns were among the largest industrial structures in Europe. Quicklime (CaO) could be **slaked** by adding water to give a sparingly soluble hydroxide:

$$CaO(s) + H_2O(\ell) \longrightarrow Ca(OH)_2(s)$$

In this form, it served well to neutralize acids but had no use in the soap industry because the calcium salts that formed from its reaction with oils were insoluble in water. A soap must be water-soluble.

An expanding demand for fine clothing greatly increased the need to bleach cotton and linen in the mid-18th century. In early bleaching methods, the cloth or fibers were repeatedly exposed to lactic acid from sour milk, bases from wood ashes, and the bright light of the sun. Cotton took up to three months to bleach, and linen up to six, and the entire process was very dependent on the weather and the supply of milk. In the closing years of the 18th century, the lead-chamber process (Section 21.1) began to supply sulfuric acid that could replace the sour milk and greatly hasten that part of the process. The need for a cheap base and a bleaching agent more effective than sunlight created major bottlenecks.

This section describes how salt (sodium chloride, NaCl) provided the answer to these problems. Processes were developed to convert salt to sodium carbonate and sodium hydroxide, and the byproducts of that conversion included compounds of chlorine that could be made into effective bleaches. Several competing processes rose and fell, with utilization of waste materials to achieve economies playing a central role. These processes, together with the production of sulfuric acid (examined in Chapter 21), dominated the 19th-century chemical industry and continue to play an important role today.

## The Leblanc Process

In 1775 the French Academy of Sciences offered a prize for a satisfactory process to make sodium carbonate from sodium chloride. A French physician and amateur chemist, Nicolas Leblanc, solved the problem during the next 15 years, although he did not receive credit for the discovery within his lifetime. Figure 22.1 outlines the process named after him.

The first step in the **Leblanc process** is the production of sodium sulfate ($Na_2SO_4$). The process Leblanc employed was the standard one at the time and involved the addition of sulfuric acid to sodium chloride at temperatures of 800 to 900°C:

$$2\ NaCl(s) + H_2SO_4(aq) \longrightarrow Na_2SO_4(aq) + 2\ HCl(g)$$

The sodium sulfate was ground up, mixed with limestone and charcoal or powdered coal, and put in trays in a furnace. The reactions that took place can be summarized in two equations:

$$Na_2SO_4(s) + 2\ C(s) \longrightarrow Na_2S(s) + 2\ CO_2(g)$$
$$Na_2S(s) + CaCO_3(s) \longrightarrow Na_2CO_3(s) + CaS(s)$$

**FIGURE 22.1** In the original version of the Leblanc process, HCl, $CO_2$, and CaS were all waste products generated along with the desired $Na_2CO_3$. Later versions of the process (below the yellow line) converted HCl to useful $Cl_2$ and recovered sulfur from the CaS to make sulfuric acid.

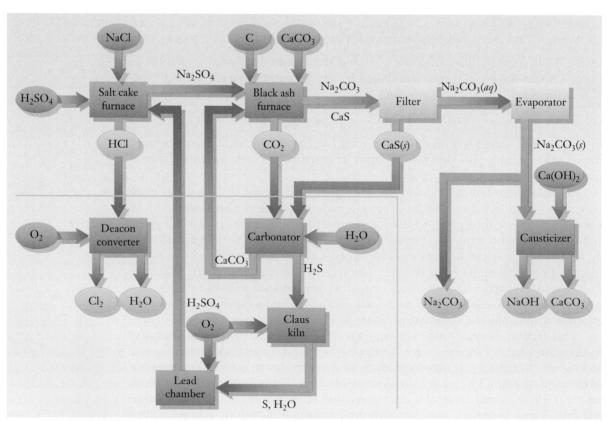

The result was a porous gray-black mixture called black ash, which consisted primarily of calcium sulfide and sodium carbonate, with some unreacted calcium carbonate and other impurities. Filtering systems were developed to leach out the soluble sodium carbonate, leaving behind the solid calcium sulfide as a byproduct. The liquid was then burned in a furnace to remove sulfur and carbon impurities as $SO_2(g)$ and $CO_2(g)$ and leave behind the desired sodium carbonate. If sodium hydroxide was needed, the sodium carbonate was boiled with slaked lime:

$$Na_2CO_3(aq) + Ca(OH)_2(aq) \longrightarrow CaCO_3(s) + 2\ NaOH(aq)$$

The precipitation of the insoluble calcium carbonate left a solution of sodium hydroxide that could be concentrated further or evaporated to dryness.

The use of the Leblanc process spread rapidly, especially in Great Britain, and increased the demand for sulfuric acid as a starting material. A look at the overall equation for the process

$$2\ NaCl + H_2SO_4 + 2\ C + CaCO_3 \longrightarrow Na_2CO_3 + 2\ CO_2 + CaS + 2\ HCl$$

shows that in addition to the primary product, sodium carbonate, and the carbon dioxide, two significant byproducts result—calcium sulfide and hydrogen chloride. In the 19th century, calcium sulfide was dumped at sea or left in piles on land, where it weathered to produce noxious $H_2S$ and $SO_2$ gases. Hydrogen chloride was initially vented as an acidic gas, which laid waste to vegetation for miles around the sodium carbonate plants. Chimneys hundreds of feet high were built, but they only dispersed the HCl farther. The next remedy passed the gaseous hydrogen chloride into towers in which water trickled down over a packing of coke. The hydrogen chloride dissolved to create a dilute solution of hydrochloric acid that was run off into streams and rivers, substituting water pollution for air pollution. In 1863 the Alkali Act prohibited such pollution in Great Britain. Although some condemned the law as excessive interference by the government in private business, it stimulated the search for ways of using the hydrochloric acid. Efficient processes were developed to convert it to bleaching powder, making a noxious waste into a useful product.

Before we consider hydrochloric acid conversion processes, let us return to a somewhat earlier period to examine the role of chlorine and its compounds as bleaching agents.

## Chlorine and Bleaches

The Swedish chemist Karl Wilhelm Scheele discovered chlorine in 1774 and prepared it in its elemental form through the reaction of hydrochloric acid with pyrolusite, a mineral containing $MnO_2$:

$$4\ HCl(aq) + MnO_2(s) \longrightarrow Cl_2(g) + MnCl_2(aq) + 2\ H_2O(\ell)$$

Scheele did not realize that the greenish yellow gas he had produced was an element, and it remained for Humphry Davy to identify it as such in 1811 and to name it (from the Greek word *chloros*, meaning "green"). In the meantime, Berthollet and de Saussure had already described chlorine's bleaching properties in 1786. The chlorine was unsatisfactory in several ways, however: it would disintegrate the cloth unless monitored carefully, it was difficult to handle, and at the time it could not be transported. The Scottish chemist Charles Tennant made a significant advance in

Addition of aqueous HBr from the funnel to solid $MnO_2$ produces gaseous bromine; a similar reaction generates chlorine from HCl.

1799 when he patented a material that he called **bleaching powder,** which was made by saturating slaked lime with chlorine:

$$Ca(OH)_2(s) + Cl_2(g) \longrightarrow CaCl(OCl)(s) + H_2O(\ell)$$

Together with sulfuric acid, bleaching powder greatly reduced the time needed to bleach cotton and linen, so that by the 1830s one week (rather than several months) was sufficient to bleach cotton goods.

Bleaching powder acts in aqueous solution by releasing hypochlorite ion ($OCl^-$),

$$CaCl(OCl)(s) \longrightarrow Ca^{2+}(aq) + Cl^-(aq) + OCl^-(aq)$$

which in turn can carry out the bleaching. One disadvantage of bleaching powder is that, on standing, it decomposes to calcium chloride and oxygen:

$$CaCl(OCl)(s) \longrightarrow CaCl_2(s) + \tfrac{1}{2} O_2(g)$$

This can be avoided by preparing calcium hypochlorite, $Ca(OCl)_2$, which also produces twice as much hypochlorite ion in solution, making it more effective on the basis of mass. Modern household bleach uses the sodium salt of hypochlorite ion ($NaOCl$) rather than the calcium salt. Sodium chlorite ($NaClO_2$) is an even stronger bleach that is used in industrial applications such as paper bleaching.

For almost a century after chlorine's discovery in 1774, the major method of making chlorine for bleach was the original reaction used by Scheele. This is an extremely wasteful method, because all of the manganese and much of the chlorine is lost as $MnCl_2$. By the mid-19th century, hydrochloric acid (the noxious byproduct of the Leblanc process) was in extensive use in bleach manufacture, and a less wasteful method of oxidizing it was needed.

Between 1868 and 1874, the British chemist and industrialist Henry Deacon developed a process to convert gaseous hydrogen chloride into chlorine over a copper chloride catalyst:

$$2\ HCl(g) + \tfrac{1}{2} O_2(g) \longrightarrow Cl_2(g) + H_2O(g)$$

The **Deacon process** was widely adopted, and hydrochloric acid from the Leblanc process became the major source of chlorine for bleaches. As time went on, in fact, the sodium carbonate from the Leblanc process became the less important product economically and was sold at a loss. The Leblanc process yielded profits only because bleaching powder was co-produced from hydrogen chloride, the former waste product. This turnabout illustrates the way in which the chemical industry adapts to changing circumstances.

## The Solvay Process

In the mid-19th century, a new process for making sodium carbonate was being developed in Belgium. Figure 22.2 is a flow diagram for the **Solvay process.** The first and most important reaction is

$$2\ NaCl(aq) + 2\ NH_3(aq) + 2\ CO_2(g) + 2\ H_2O(\ell) \longrightarrow$$
$$2\ NaHCO_3(s) + 2\ NH_4Cl(aq)$$

This reaction is carried out in a carbonating tower, a tall reactor in which brine (concentrated aqueous NaCl) is saturated with ammonia and carbon dioxide. The sodium hydrogen carbonate ($NaHCO_3$) that forms is relatively insoluble and is

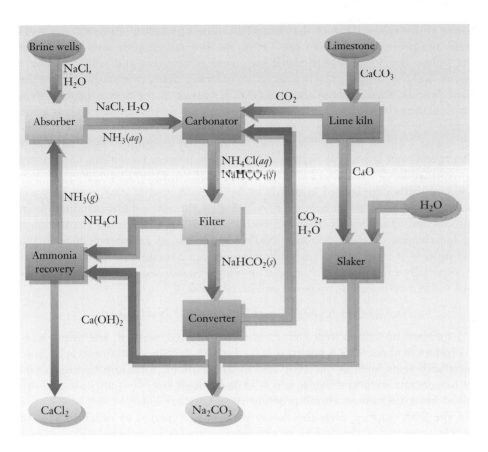

**FIGURE 22.2** The net reaction in the Solvay process is the production of sodium carbonate and calcium chloride from salt and calcium carbonate. Ammonia plays a crucial role in making the process work but is not consumed (except for inevitable small losses).

removed by filtration on huge rotating drum filters. It is then decomposed thermally in a rotary dryer above 270°C to give sodium carbonate as the product:

$$2\ NaHCO_3(s) \longrightarrow Na_2CO_3(s) + H_2O(g) + CO_2(g)$$

The carbon dioxide freed in this reaction is returned to the carbonating tower, supplemented by $CO_2$ generated in a lime kiln:

$$CaCO_3(s) \longrightarrow CaO(s) + CO_2(g)$$

The lime is slaked to $Ca(OH)_2(aq)$,

$$CaO(s) + H_2O(\ell) \longrightarrow Ca(OH)_2(aq)$$

which is then used in a final step to recover ammonia for reuse in the carbonating tower:

$$2\ NH_4Cl(aq) + Ca(OH)_2(aq) \longrightarrow 2\ NH_3(aq) + CaCl_2(aq) + 2\ H_2O(\ell)$$

The sum of these five steps gives the remarkably simple overall reaction for the Solvay process:

$$2\ NaCl(aq) + CaCO_3(s) \longrightarrow Na_2CO_3(s) + CaCl_2(aq)$$

Although this "ammonia–soda" process had been described and studied by many chemists beginning in 1800, it had languished because it was uneconomical. For many years, the stumbling block was the inefficient recovery of the ammonia. The high cost of this material meant that losses had to be kept below about 5% per cycle to make the process feasible. In the 1860s, the Belgian chemist and industrialist Ernest Solvay developed and patented improved carbonators and stills to recover ammonia, and the process grew rapidly in importance, taking over markets from the Leblanc process. Raw material and fuel costs are lower in the Solvay process than in the Leblanc process (lower temperatures are required, and sulfuric acid is not a reactant). The Solvay process also produces a much purer grade of sodium carbonate in higher yield. It has the final advantage of being a continuous rather than a batch process, and it was the first such continuous process to achieve widespread commercial success.

Sodium carbonate is used primarily in the manufacture of glass, especially the soda-lime glass discussed in Section 23.4. Another use for sodium carbonate is in the production of sodium hydrogen carbonate ($NaHCO_3$), which has the common name bicarbonate of soda. Although $NaHCO_3$ appears as an intermediate in the Solvay process, it is not taken out for direct use because it contains too many impurities, especially ammonia, which gives it an odor. Instead, a saturated aqueous solution of sodium carbonate is prepared and carbonated:

$$Na_2CO_3(aq) + CO_2(g) + H_2O(\ell) \longrightarrow 2\ NaHCO_3(s)$$

The suspension of sodium hydrogen carbonate is filtered, washed, and dried, leaving a product of about 99.9% purity. It is used in the kitchen as baking soda because it reacts with acids, such as the lactic acid in sour milk or potassium hydrogen tartrate, to generate carbon dioxide, which makes dough rise. It is also used in dry chemical fire extinguishers (recall problem 14 in Chapter 11).

In the 20th century, even the Solvay process has yielded to other sources of bases. One reason for the decline of the Solvay process is its byproduct, calcium chloride. This salt is used for dust control in summer and to melt ice on streets in winter (a saturated $CaCl_2$ solution freezes at $-50°C$, as compared with $-20°C$ for a saturated NaCl solution). Beyond these uses, it has little commercial value and poses a difficult disposal problem. Although the Solvay process remains important worldwide, it has been almost completely supplanted in the United States by the mining of a double salt of sodium carbonate and sodium hydrogen carbonate called **trona,** with the approximate composition $2Na_2CO_3 \cdot NaHCO_3 \cdot 2H_2O$. Heating this ore gives sodium carbonate.

## Electrolysis and the Chlor-Alkali Industry

In the 1890s, a new process for the production of bases and chlorine was developed and grew rapidly in importance: the electrolytic generation of sodium hydroxide and chlorine from salt brines. The dominant technique for electrolysis in the United States is the **diaphragm cell** (Fig. 22.3). If an electric current is passed through a sufficiently concentrated solution of sodium chloride, the anode reaction is

$$2\ Cl^-(aq) \longrightarrow Cl_2(g) + 2\ e^-$$

and the cathode reaction is

$$2\ H_2O(\ell) + 2\ e^- \longrightarrow H_2(g) + 2\ OH^-(aq)$$

The overall reaction is the sum of these half-reactions:

$$2\ H_2O(\ell) + 2\ Cl^-(aq) \longrightarrow 2\ OH^-(aq) + H_2(g) + Cl_2(g) \qquad \Delta G° = +422\ kJ$$

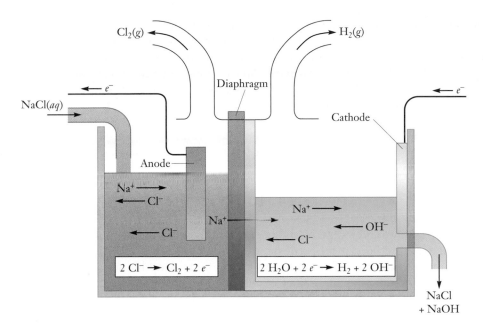

$Cl_2(g)$

$H_2(g)$

$e^-$

NaCl(*aq*)

Diaphragm

Cathode

$e^-$

Anode

Na$^+$
Cl$^-$
Na$^+$
Cl$^-$

Na$^+$
Cl$^-$
OH$^-$

$$2\ Cl^- \rightarrow Cl_2 + 2\ e^-$$

$$2\ H_2O + 2\ e^- \rightarrow H_2 + 2\ OH^-$$

NaCl
+ NaOH

**FIGURE 22.3**   In a diaphragm cell, the cathode is a perforated steel box. Note that the level of the solution is deliberately made higher in the anode compartment than in the cathode to minimize migration of OH$^-$ through the diaphragm to the anode, where it could react with the chlorine being generated.

The gaseous hydrogen and chlorine bubble out and are collected and dried, while the hydroxide ion forms part of a mixed solution of sodium hydroxide and sodium chloride. The less soluble sodium chloride precipitates upon evaporation and can be removed by filtration, leaving a solution containing 50% by mass of sodium hydroxide in water. This solution may be used directly or evaporated further to make solid sodium hydroxide.

Early diaphragm cells used graphite anodes, but these had to be replaced frequently and have been superseded by anodes made of titanium coated with platinum, ruthenium, or iridium. The cathodes are steel boxes with perforated sides. The diaphragm prevents the mixing of the H$_2$ and Cl$_2$ gases but allows ions to pass. Asbestos has been the most widely used material for these diaphragms, but new diaphragms are being developed based on fluorinated polymers. Another approach is to replace the diaphragm with an ion-exchange membrane, which permits only sodium ions to pass to the cathode (Fig. 22.4). By preventing the passage of chloride ions, this method generates pure aqueous sodium hydroxide at the cathode.

The electrolytic production of sodium hydroxide and chlorine was proposed in the mid-19th century, but the only source of electricity at the time was the battery. This made the process so expensive that it was little more than a laboratory curiosity. By the 1890s, however, the development of the dynamo (generator) changed the situation dramatically. The electrolysis industry grew up rapidly around sources of cheap electricity from water power—in Norway and near Niagara Falls in the United States. This industry's first effect was to doom the Leblanc process, which had survived so long only from the profits in the co-production of chlorine with sodium carbonate. With competing chlorine produced in greater purity electrolytically, the Leblanc process quickly disappeared. Over a longer period, the electrolytic chlor-alkali process has reduced the role of the Solvay process as well, by yielding a base (NaOH) that competes effectively with sodium carbonate in many applications.

The laws of chemistry require that each mole of chlorine produced by the electrolysis of brine be accompanied by two moles of sodium hydroxide. Because the

**FIGURE 22.4**   State-of-the-art membrane technology installed in a Washington State chlor-alkali plant that produces chlorine and sodium hydroxide. (*Occidental Petroleum*)

demand for these two products is rarely in ideal balance, the prices of both commodities fluctuate. A byproduct of the reaction is hydrogen. It can react directly with chlorine to give gaseous hydrogen chloride of high purity, although there are less expensive sources for this chemical. Alternatively, the chlor-alkali process can be coupled with a fertilizer plant, and the hydrogen can be reacted with nitrogen to make ammonia (see Section 21.3).

---

## 22.2

## THE CHEMISTRY OF CHLORINE, BROMINE, AND IODINE

Chlorine, bromine, and iodine resemble one another much more closely than they do fluorine, whose special properties are discussed in Section 22.3. Section 22.1 described the production of chlorine and its role in making bleaches. That role is now overshadowed by the importance of chlorine in the manufacture of chlorinated hydrocarbons for use as solvents, pesticides, and plastics such as polyvinyl chloride (Sec. 23.3). Growing concern about the environmental effects and toxic properties of chlorinated hydrocarbons (see Fig. 22.5) has slowed the growth of the industry. Another major use for chlorine is the purification of titanium dioxide. This substance occurs in impure form; it reacts above 900°C with chlorine and coke to give titanium tetrachloride:

$$TiO_2(s) + 2\ C(s) + 2\ Cl_2(g) \longrightarrow TiCl_4(\ell) + 2\ CO(g)$$

Because $TiCl_4$ is a volatile liquid (b.p. 136°C), it can easily be purified by distillation and then oxidized to pure titanium dioxide, regenerating chlorine:

$$TiCl_4(\ell) + O_2(g) \longrightarrow TiO_2(s) + 2\ Cl_2(g)$$

Titanium dioxide is the major white pigment used in paint. It is more opaque than other pigments, and so paints containing it have high covering power.

Bromine is obtained largely from naturally occurring brines, where it occurs as aqueous bromide ion ($Br^-$). Treatment of these brines with chlorine (Fig. 22.6) results in the reaction

$$2\ Br^-(aq) + Cl_2(aq) \longrightarrow 2\ Cl^-(aq) + Br_2(aq)$$

This equilibrium lies to the right because chlorine is a stronger oxidizing agent than bromine. The bromine liberated is swept out with steam, distilled, and purified. Until recently, the most important use for bromine was in the production of dibro-

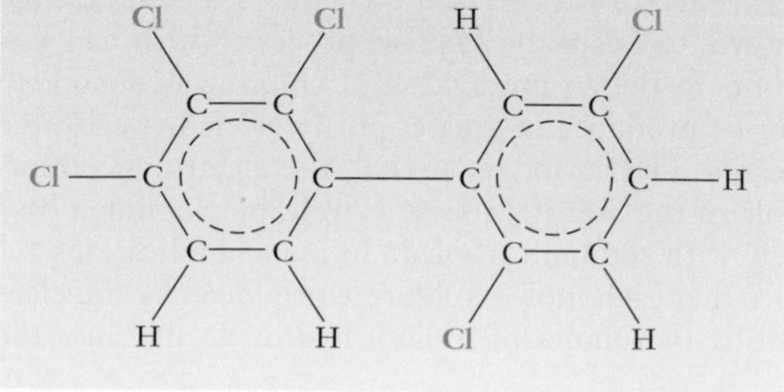

**FIGURE 22.5** One of the many polychlorinated biphenyls (PCBs), which were once used as insulating fluids in electrical transformers. These compounds are being phased out because of their toxic effect on marine life.

moethane ($C_2H_4Br_2$), which was used in gasoline to scavenge the lead deposited from the breakdown of tetraethyl lead; the use of this "antiknock" additive is decreasing, however, because of concerns about contamination of the environment with lead. A major current use for bromine is in the production of flame-retardant organic compounds. Aqueous solutions of $CaBr_2$ and $ZnBr_2$ are also employed as high-density fluids to recover oil from deep wells.

Iodine occurs in seawater to the extent of only $6 \times 10^{-7}\%$, but it is concentrated in certain species of kelp, from whose ash its recovery is commercially feasible. Iodine is present in the growth-regulating hormone thyroxin, produced by the thyroid gland. Much of the table salt sold has 0.01% NaI added to prevent goiter, the enlargement of the thyroid gland. Silver iodide is used in high-speed photographic film, whereas silver bromide and silver chloride are used in slower speed film and in photographic printing paper.

The hydrogen halides HCl, HBr, and HI are all gases at room conditions, and all three dissolve in water to form strong acids. Hydrogen chloride is usually prepared by the direct combination of hydrogen and chlorine over a platinum catalyst or as a byproduct of organic chemicals processing. Its aqueous solution, hydrochloric acid, is a major industrial acid, used extensively for the cleaning of metal surfaces. Hydrogen bromide and hydrogen iodide have fewer uses. Their aqueous solutions are prepared on a laboratory scale by the hydrolysis of phosphorus tribromide or phosphorus triiodide:

$$PBr_3(\ell) + 3\ H_2O(\ell) \longrightarrow 3\ HBr(aq) + H_3PO_3(aq)$$

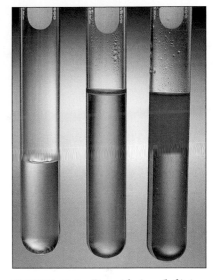

**FIGURE 22.6** Bromide ion (left) is colorless. Oxidation by chlorine yields bromine (center), which is colored in aqueous solution. The color is even more intense in an organic solvent (upper layer on right). (*Charles D. Winters*)

## Solution Chemistry of the Halogens

When chlorine, bromine, or iodine is added to water, the halogen disproportionates in part to form its halo- and hypohalous acids according to the equilibrium

$$X_2(aq) + 2\ H_2O(\ell) \rightleftharpoons HOX(aq) + X^-(aq) + H_3O^+(aq)$$

The equilibrium constant for this reaction is very small, but the reaction can be shifted to the right in basic solution, where HOX is neutralized and the overall reaction is

$$X_2(aq) + 2\ OH^-(aq) \rightleftharpoons OX^-(aq) + X^-(aq) + H_2O(\ell)$$

No complications arise with chlorine, and stable solutions of chloride and hypochlorite ions result at 25°C. In this way, a solution of sodium hypochlorite (NaOCl)—used as a laundry bleach—can be prepared. The hypochlorite ion is thermodynamically unstable with respect to disproportionation into chloride and chlorate ions,

$$3\ ClO^-(aq) \longrightarrow 2\ Cl^-(aq) + ClO_3^- \qquad \Delta G° = -90.7\ kJ$$

but at ordinary temperatures this reaction is very slow. When the temperature is increased to about 75°C, however, the rate of disproportionation is rapid.

The equilibrium constant for the disproportionation of bromine in water to bromide and hypobromite ions is even smaller than that for the disproportionation of chlorine; if the concentration of base is increased, the further disproportionation of hypobromite to bromide and bromate ions occurs rapidly even at 25°C. The same is true of iodine, but to an even greater extent. The solubility of iodine in water at 25°C is only $1.34 \times 10^{-3}$ M, but in the presence of iodide ion, much more dissolves as triiodide ion forms:

$$I^-(aq) + I_2(s) \rightleftharpoons I_3^-(aq) \qquad K\ (25°C) = 714$$

The $I_3^-$ ion has a linear structure that is symmetric in solution and in certain crystalline salts but unsymmetric (unequal I—I distances) in others.

Among the possible acids of formula $HXO_3$, only iodic acid exists in anhydrous form, a crystalline solid that is stable up to its melting point. The alkali chlorates, bromates, and iodates have the generic formula $MXO_3$, and all are well-characterized salts that can be crystallized from aqueous solution. Acidic solutions of the salts are powerful oxidizing agents that react quantitatively with their corresponding halide ions to yield the halogen:

$$IO_3^-(aq) + 5\ I^-(aq) + 6\ H_3O^+(aq) \longrightarrow 3\ I_2(aq) + 9\ H_2O(\ell)$$

When potassium chlorate is carefully heated in the absence of catalysts, it is converted to potassium perchlorate:

$$4\ KClO_3(s) \longrightarrow 3\ KClO_4(s) + KCl(s)$$

Perchloric acid can be formed by heating potassium perchlorate with concentrated sulfuric acid, but this is a somewhat hazardous reaction that is prone to explosions. A safer procedure is to oxidize an aqueous solution of an alkali chlorate at the anode of an electrolytic cell:

$$ClO_3^-(aq) + 3\ H_2O(\ell) \longrightarrow ClO_4^-(aq) + 2\ H_3O^+(aq) + 2\ e^-$$

Perchloric acid is a colorless liquid that boils at 82°C. In its anhydrous form it is a powerful oxidizing agent, but in aqueous solution at moderate concentration it does not display oxidizing ability. Aqueous perbromic acid, on the other hand, is a powerful oxidizing agent that is also formed electrolytically.

Aqueous periodic acid is a more useful oxidizing agent than either perbromic or perchloric acid, because its behavior is more controllable. Periodic acid exists in several forms, the simplest of which is metaperiodic acid ($HIO_4$). Like iodic acid, metaperiodic acid is a crystalline solid at room temperature. It is possible to expand the coordination sphere of iodine(VII) from four to six ligands by the addition of two water molecules, and the resulting orthoperiodic acid has the formula $H_5IO_6$. This substance is a crystalline solid that decomposes at 140°C.

## Halogen Oxides

Chlorine, bromine, and iodine form several oxides, most of which are rather unstable. Dichlorine monoxide is prepared by reaction of chlorine with mercury(II) oxide:

$$2\ Cl_2(g) + 2\ HgO(s) \longrightarrow Cl_2O(g) + HgCl_2 \cdot HgO(s)$$

In basic solution it forms hypochlorites, as it is the anhydride of hypochlorous acid:

$$Cl_2O(g) + 2\ OH^-(aq) \longrightarrow 2\ OCl^-(aq) + H_2O(\ell)$$

Chlorine dioxide is a very unstable gas that is prepared by reaction of chlorine with silver chlorate:

$$Cl_2(g) + 2\ AgClO_3(s) \longrightarrow 2\ ClO_2(g) + 2\ AgCl(s) + O_2(g)$$

It is a very powerful oxidizing agent that is used to bleach wheat flour. Like nitrogen dioxide, it consists of "odd-electron" molecules, but unlike that molecule, it shows little tendency to dimerize. In basic solution it disproportionates to form chlorite and chlorate ions and can be regarded as a mixed anhydride of chlorous and chloric acids:

$$2\ ClO_2(g) + 2\ OH^-(aq) \longrightarrow ClO_2^-(aq) + ClO_3^-(aq) + H_2O(\ell)$$

**FIGURE 22.7** Iodine reacts with starch to form a dark-colored complex. *(Leon Lewandowski)*

The most stable of the chlorine oxides is dichlorine heptaoxide ($Cl_2O_7$), a colorless liquid that boils without decomposing at 82°C. It is prepared by dehydration of perchloric acid with $P_4O_{10}$:

$$12\ HClO_4(\ell) + P_4O_{10}(s) \longrightarrow 6\ Cl_2O_7(\ell) + 4\ H_3PO_4(\ell)$$

The oxides of bromine, $Br_2O$ and $BrO_2$, are prepared by methods that parallel those for the corresponding oxides of chlorine. The most important of the iodine oxides is diiodine pentaoxide, a white crystalline solid that is prepared by thermally dehydrating iodic acid:

$$2\ HIO_3(s) \longrightarrow I_2O_5(s) + H_2O(g)$$

Diiodine pentaoxide is useful in analytical chemistry for the detection and quantitative determination of carbon monoxide, which it readily oxidizes:

$$5\ CO(g) + I_2O_5(s) \longrightarrow I_2(s) + 5\ CO_2(g)$$

The iodine released in this reaction is detected visually by means of the deep blue-black colored complex it forms with starch (Fig. 22.7).

---

## 22.3

# FLUORINE AND ITS COMPOUNDS

Fluorine ranks 13th in abundance among the elements in the rocks of the earth's crust, and its compounds were used as early as 1670 for the decorative etching of glass. Yet heroic efforts to isolate it as a free element failed for many years, until 1886. The difficulty of preparing elemental fluorine results from the fact that is among the strongest oxidizing agents. The other halogens are moderate to very good at accepting electrons in aqueous media, as shown by their standard reduction potentials, 0.535 V (for $I_2$), 1.065 V (for $Br_2$), and 1.358 V (for $Cl_2$). Fluorine is *avid* for electrons:

$$F_2(g) + 2\ e^- \longrightarrow 2\ F^-(aq) \qquad\qquad \mathscr{E}° = 2.87\ V$$

Because fluor*ine* is so good at gaining electrons, fluor*ides* strongly resist oxidation.

The major source of fluorine is the ore fluorspar, which contains fluorite ($CaF_2$) (Fig. 19.30). Figure 19.19b showed the cubic structure of $CaF_2$. Another source of fluorine is fluorapatite, $Ca_5(PO_4)_3F$, which is mined in great quantity to make phosphate fertilizer (Section 21.2). At one time, fluorine-containing byproducts (HF, $SiF_4$, and $H_2SiF_6$) from fertilizer plants were valueless and created environmental pollution comparable to the hydrogen chloride pollution caused by the Leblanc process (Section 22.1). Now the fluorine from fluorapatite has economic value in making fluorine compounds.

## Hydrogen Fluoride

Almost half of the fluorspar that is mined is used in steelmaking to increase the fluidity of the slag. Most of the remainder is reacted with concentrated sulfuric acid to make hydrogen fluoride:

$$CaF_2(s) + H_2SO_4(aq) \longrightarrow 2\ HF(g) + CaSO_4(s)$$

The resulting gaseous HF is condensed and redistilled to free it from sulfur oxides. Anhydrous hydrogen fluoride is a colorless liquid that freezes at $-83.37°C$ and boils at $19.54°C$, near room temperature. The boiling point far exceeds what the trend among the other hydrogen halides predicts and is strong evidence for association of the HF molecules (by hydrogen bonds) in the liquid. In this and other physical properties, HF is closer to water and ammonia than to HCl, HBr, or HI. As pure liquids, both water and hydrogen fluoride contain polar molecules but few ions; both are poor electrical conductors. Liquid hydrogen fluoride undergoes autoionization much as water does:

$$2\ HF(\ell) \rightleftharpoons H^+ + FHF^-$$

The reaction has an equilibrium constant (at $19.5°C$) of $2.0 \times 10^{-14}$, a value close to the room-temperature autoionization constant of water ($1.0 \times 10^{-14}$).

Hydrogen fluoride in aqueous solution is a weak acid ($K_a = 6.6 \times 10^{-4}$ at $25°C$), quite unlike HCl, HBr, and HI, which are strong acids. Its weakness as an acid is due both to the strength of the H—F bond and to the extensive hydrogen bonding in its aqueous solutions. Fluoride ions interact so strongly with hydronium ions, $H_3O^+$, that they remain bound in $F^-\cdots H_3O^+$ complexes rather than separating as chloride, bromide, and iodide ions do.

Anhydrous hydrogen fluoride is quite poisonous, causes serious burns, and corrodes silicate glass by the reaction

$$4\ HF(\ell) + SiO_2(s) \longrightarrow SiF_4(g) + 2\ H_2O(\ell)$$

This reaction underlies one long-standing use of aqueous hydrogen fluoride: etching decorative patterns in glass (Fig. 22.8). Hydrogen fluoride does not attack Teflon and certain other fluorine-containing plastics. Their availability in recent years has made it easier to handle hydrogen fluoride safely. Another major use of HF(g) is in the aluminum industry. It reacts with $NaAlO_2(aq)$, itself prepared from bauxite ore ($Al_2O_3(s)$) and $NaOH(aq)$, to make synthetic cryolite according to the equation

$$NaAlO_2(aq) + 2\ NaOH(aq) + 6\ HF(g) \longrightarrow Na_3AlF_6(aq) + 4\ H_2O(\ell)$$

The very important Hall–Héroult process to produce aluminum (Fig. 20.16) depends on the electrolysis of solutions of $Al_2O_3$ in molten cryolite.

**FIGURE 22.8** Glass can be etched in intricate patterns with aqueous hydrogen fluoride. The geometric patterns and background were formed by protecting portions of the glass and exposing the unprotected areas to HF(aq). *(Copyright Christie's Images)*

## The Preparation of Fluorine

Elemental fluorine is produced from hydrogen fluoride by a method similar to that first used by Moissan—the electrolysis of molten potassium hydrogen fluoride ($KF \cdot 2HF$), a solution of KF in liquid HF. In the solution, $F^-$ ion associates strongly with HF to form $FHF^-$ (the hydrogen difluoride ion), the species that actually loses the electrons:

$$FHF^- \longrightarrow F_2(g) + H^+ + 2\,e^- \qquad \text{(anode, oxidation)}$$

$$2\,H^+ + 2\,e^- \longrightarrow H_2(g) \qquad \text{(cathode, reduction)}$$

$$2\,HF \longrightarrow H^+ + FHF^- \longrightarrow F_2(g) + H_2(g) \qquad \text{(overall)}$$

This process resembles the electrolytic production of $Cl_2(g)$ (Section 22.1). As fluorine and hydrogen evolve, $HF(g)$ is fed in continuously to keep the electrolyte composition constant at $KF \cdot 2HF$. A commercial fluorine cell uses a steel cathode and carbon anode. The cells are carefully engineered to prevent contact between the product gases, which react at catastrophic speed.

Fluorine is a very pale yellow gas. It condenses to a canary-yellow liquid at $-188.14°C$ and solidifies at $-219.62°C$. It is an exceedingly toxic substance. Concentrations exceeding 25 ppm in air quickly damage the eyes, nose, lungs, and skin. Fluorine's pungent, irritating odor makes it first detectable in air at a concentration of about 3 ppm, well below the level of acute poisoning but above the recommended safe working level of 0.1 ppm. The ferocious reactivity of fluorine demands the utmost care in its use, and handling procedures that minimize the hazards have been developed. Certain metals (nickel, copper, steel, and the nickel–copper alloy Monel metal) resist attack by fluorine by forming a layer of a fluoride salt on their surfaces. Reactions can be carried out at room temperature in vessels made of these metals. The pure element is regularly packaged and shipped as a compressed gas in special cylinders, but it is also often generated at the point of use.

The role of elemental fluorine in the chemical industry has grown to major proportions since 1945. Before then, elemental fluorine was mostly a laboratory curiosity. The motive for finding ways to prepare, purify, and handle it was the need to synthesize a volatile uranium compound that could separate the $^{235}U$ and $^{238}U$ isotopes by gaseous diffusion, as described in Section 4.6. Uranium hexafluoride met that requirement, and gaseous diffusion is still used for preparing $^{235}U$-enriched reactor fuel. At present, more than half of the world's production of $F_2(g)$ is used to make $UF_6$ from uranium(IV) oxide by these reactions:

$$UO_2(g) + 4\,HF(g) \longrightarrow UF_4(s) + 2\,H_2O(g)$$

$$UF_4(s) + F_2(g) \longrightarrow UF_6(g)$$

Most of the remaining fluorine production goes to make sulfur hexafluoride by the reaction

$$S(g) + 3\,F_2(g) \longrightarrow SF_6(g)$$

This unreactive gas is a good electrical insulator. Its presence within heavy-duty electrical switches minimizes arcing or sparking when the switches are thrown. Electrical discharges that do occur in an $SF_6$ atmosphere decompose the gas to an extent (into sulfur and fluorine), but the dissociation products rapidly recombine. Some triple-pane windows are manufactured with $SF_6$ sealed in the "dead-air" spaces between the panes because it cuts heat transmission and muffles sound better than air.

## Fluorine and the Other Halogens

The differences between fluorine and the other halogens are extreme enough to place fluorine in a class by itself. It is much more reactive than chlorine, bromine, and iodine; indeed, it is by far the most reactive of all the elements. In forming a compound, a fluorine atom either shares electrons in a single covalent bond or gains one electron to form the fluoride ion ($F^-$). In either case, it always attains a noble-gas electron configuration ($1s^2 2s^2 2p^6$). The other halogens tend to behave similarly but tolerate numerous exceptions; fluorine does not. The rules for assigning oxidation numbers (Section 3.7) recognize the special character of fluorine by setting its oxidation number in compounds always equal to $-1$. Chlorine, bromine, and iodine may have positive oxidation numbers.

The chemical and physical properties of fluorine (and its compounds) deviate from the values that would be predicted by extrapolation of the trends among the other halogens. Figure 22.9 presents some data from Appendix F that show this deviation. The bond dissociation energy of $F_2$ is *smaller* than would be expected from the trend among the other halogens, but the first ionization energy of F is *greater*. Despite the position of F as the most electronegative of the elements, electron-affinity data show that $F(g)$ accepts another electron (to form the gaseous anion) *less* readily than $Cl(g)$ and only slightly more readily than $Br(g)$. The unexpectedly low electron affinity of a fluorine atom is related to the contracted size of the atom. As Figure 22.9b shows, $F(g)$ and $F^-(g)$ are smaller than an extrapolation of the trends among the other halogens would predict, causing crowding and greater mutual repulsion among the electrons in their $n = 2$ quantum levels.

Strong bonds, both covalent and ionic, characterize the interaction of fluorine with most other elements. Figure 22.10a shows that the H—F, C—F, and Na—F bond dissociation enthalpies exceed those of H, C, and Na combined with the other halogens. Fluorine's strong bonds with other elements and weak bond with itself (in the $F_2$ molecule the bond dissociation enthalpy is only $158 \ kJ \ mol^{-1}$) together explain its terrific reactivity. Both factors are traceable to the small size of the fluorine atom. The two atoms in the fluorine molecule are only 1.417 Å apart. This puts their lone-

**FIGURE 22.9** The measured physical properties of fluorine often do not fall on the trend lines defined by the behavior of the other halogens. (a) Here, the ionization energy of fluorine is unexpectedly high, and its electron affinity and bond dissociation enthalpy are unexpectedly low. (b) The sizes of the fluorine atom and the fluoride ion are both smaller than predicted by comparison with the other halogens.

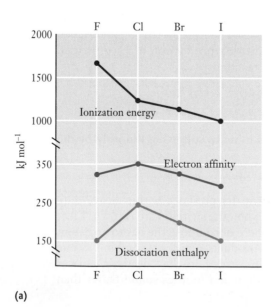

(a)

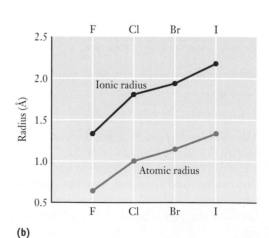

(b)

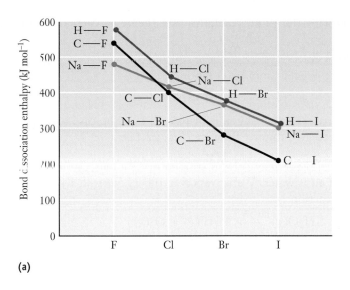

(a)

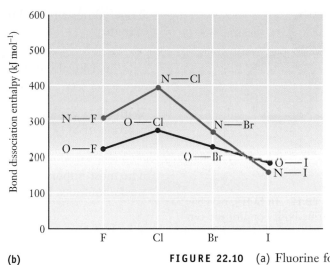

(b)

FIGURE 22.10  (a) Fluorine forms strong bonds with more electropositive elements such as C, H, and Na. (b) In contrast, fluorine forms unexpectedly weak bonds with the electronegative elements N and O in comparison to the bonds formed by the other halogens.

pair electrons closer to each other than those in the other halogens: repulsion among them lowers the bond dissociation energy. Small fluorine makes short bonds to nonfluorine atoms, as well. Lone-pair repulsion of the type just described becomes important only with elements that, like fluorine itself, are small and electronegative. Hence, O—F and N—F bonds are, like the F—F bond, weaker than would otherwise be expected (Fig. 22.10b).

Because fluorine atoms are small, many of them fit around a given central atom before crowding each other. Therefore, atoms often attain higher coordination numbers with fluorine than with other elements. In many fluorides, such as $K_2[NiF_6]$, $Cs[AuF_6]$, and $PtF_6$, the central element has an unexpectedly high oxidation number. The other halogens offer no comparable compounds.

## Reactions of Fluorine

In general, fluorine reacts with other elements and compounds by oxidizing them. It displaces the other, less reactive, halogens from compounds, just as chlorine displaces bromine and iodine, the elements located below it in the periodic table, and bromine displaces iodine. An example is the reaction

$$2\ NaCl(s) + F_2(g) \longrightarrow 2\ NaF(s) + Cl_2(g)$$

Fluorine sometimes breaks this pattern by going much further. Thus, if $F_2(g)$ is heated with potassium chloride, the reaction is

$$KCl(s) + 2\ F_2(g) \longrightarrow KClF_4(s)$$

Here the chlorine is not displaced from the compound but oxidized to the +3 state and surrounded by fluorine atoms in the "tetrafluorochlorate(III)" ion. Further heating of $KClF_4(s)$ with $F_2$ eventually does displace chlorine from potassium but in an unexpected form, $ClF_5(g)$:

$$KClF_4(s) + F_2(g) \longrightarrow KF(s) + ClF_5(g)$$

In a variation, fluorine oxidizes iodine in potassium iodide and also displaces it:

$$KI(s) + F_2(s) \longrightarrow KF(s) + IF(g)$$

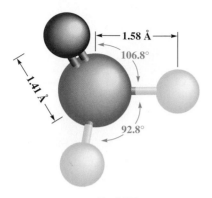

**FIGURE 22.11** In SOF$_2$, oxygen and fluorine both bond to the central sulfur atom. The VSEPR model correctly predicts this geometry.

Under different conditions with the same reactants, the theme is carried to an extreme:

$$\text{KI}(s) + 4\ \text{F}_2(g) \longrightarrow \text{KF}(s) + \text{IF}_7(g)$$

Fluorine also can displace oxygen from its compounds:

$$2\ \text{CaO}(s) + 2\ \text{F}_2(g) \longrightarrow 2\ \text{CaF}_2(s) + \text{O}_2(g)$$

With other oxides, only a portion of the oxygen is displaced, as is the case with the reaction of SO$_2$($g$) and F$_2$($g$) to give thionyl fluoride (Fig. 22.11):

$$2\ \text{SO}_2(g) + 2\ \text{F}_2(g) \longrightarrow 2\ \text{SOF}_2(g) + \text{O}_2(g)$$

The most important oxide of all is water. Water vapor at 100°C *burns* in F$_2$ with a weak, luminous flame. The reaction is

$$2\ \text{F}_2(g) + 2\ \text{H}_2\text{O}(g) \longrightarrow 4\ \text{HF}(g) + \text{O}_2(g)$$

As is often the case, different conditions favor different products. An excess of fluorine in contact with liquid water or ice gives hydrogen fluoride and highly unstable hydrogen oxyfluoride:

$$\text{F}_2(g) + \text{H}_2\text{O}(s) \longrightarrow \text{HF}(\ell) + \text{HOF}(\ell)$$

Oxygen difluoride also forms,

$$2\ \text{F}_2(g) + \text{H}_2\text{O}(s) \longrightarrow 2\ \text{HF}(\ell) + \text{OF}_2(g)$$

and the reaction that gives O$_2$ and HF still takes place to a limited extent. In all three of these reactions fluorine disrupts stable H$_2$O, taking oxygen from the $-2$ to the $0$ or $+2$ state. Few materials other than fluorine can oxidize water in this way.

## Metal Fluorides

The alkali and alkaline-earth metals form ionic salts with fluorine of the expected stoichiometry. Except for lithium fluoride, the alkali fluorides are all quite soluble in water. The differences between the alkaline-earth fluorides (BeF$_2$, MgF$_2$, CaF$_2$, SrF$_2$, and BaF$_2$) and the corresponding chlorides, bromides, and iodides are striking: BeF$_2$ is ionic and soluble in water, but BeCl$_2$ is covalent enough to sublime; CaF$_2$ is insoluble in water, but the other calcium halides are quite soluble in water. All these fluorides are prepared most easily by reaction of the corresponding hydroxide or carbonate with aqueous hydrofluoric acid. In their lower oxidation states, the transition metals and Group III and IV metals also form fluoride salts with a range of solubilities; ZnF$_2$, SnF$_2$, and PbF$_2$ are examples.

Low concentrations (about 1 mg L$^{-1}$) of fluoride ion in drinking water help to prevent tooth decay. Fluoride ion substitutes for hydroxide ion in tooth enamel, changing some of the hydroxyapatite, Ca$_5$(PO$_4$)$_3$OH, in the enamel to fluorapatite, Ca$_5$(PO$_4$)$_3$F. The replacement works because the F$^-$ and OH$^-$ ions have the same charge ($-1$) and similar radii (1.33 Å for F$^-$ versus approximately 1.2 Å for OH$^-$). Fluorapatite is less soluble than hydroxyapatite in the acidic oral environment fostered by the consumption of sweets. The deliberate fluoridation of civic water supplies to cut down tooth decay is a common public health measure in the United States. In addition, fluoride salts such as SnF$_2$ or NaF are often added to toothpaste.

Fluorine forms salts with most metals in their lower oxidation states but is capable of oxidizing many metals to exceptionally high oxidation states as well. The chemistry of cobalt centers on the $+2$ oxidation state, for example, except in complex ions, where the $+3$ state is not uncommon. Cobalt(III) fluoride, obtained by direct fluorination of the metal, is one of the very few simple ionic salts of that element in which the cation has an oxidation number of +3. Platinum is directly oxidized to platinum(IV) by fluorine and (at more extreme temperatures) to platinum(VI) fluoride. The latter is a dark red solid with a low melting point (56.7°C), suggesting its largely covalent nature. Other heavy metals that form volatile hexafluorides include osmium, iridium, molybdenum, tungsten, and uranium. Fluorides of lower oxidation state are also known for all of these metals.

Normally, one thinks of silver as having only an oxidation state of +1, and in the great majority of its compounds this is the case. Silver in the +2 oxidation state, however, can be formed by heating silver(I) fluoride in a stream of fluorine:

$$2\ AgF(s) + F_2(g) \longrightarrow 2\ AgF_2(s)$$

Even the +3 state of silver can be formed in a ternary compound when potassium fluoride is mixed with silver fluoride and heated in fluorine:

$$KF(s) + AgF(s) + F_2(g) \longrightarrow KAgF_4(s)$$

In these reactions the $4d$ subshell of the silver ion must be opened, at a cost of 2073 and 3360 kJ mol$^{-1}$, respectively, to form silver in the +2 and +3 oxidation states.

## Covalent Fluorides

Fluorine forms compounds with all the elements in Groups III through VII. The Group III elements (boron, aluminum, gallium, indium, and thallium) all form trifluorides, and thallium forms TlF as well. Boron trifluoride is one of the most powerful Lewis acids, readily forming the complex ion $[BF_4]^-$ in solutions that contain F$^-$. Boron trifluoride is a gas at room temperature, but the trifluorides of aluminum and the higher members of Group III are solids that melt near 1000°C. They are intermediate between ionic and covalent in their bonding, and in the crystalline state their structures consist of extended arrays of metal atoms in sixfold coordination with fluorine atoms.

The transitional character of bonding is even more evident in the fluorides of Group IV. At room temperature, carbon tetrafluoride and silicon tetrafluoride are gases whose molecules are covalent; germanium tetrafluoride is a molecular liquid, but the tetrafluorides of tin and lead are solids with considerable ionic character. As already mentioned, the difluorides of tin and lead are ionic salts. This fact illustrates a general rule: the halides of the heavier transition and posttransition metals are more covalent when the oxidation number of the metal is high and more ionic when it is low.

The fluorides of the Group V elements are just as varied in composition and behavior. Nitrogen forms fluorides with the compositions $N_2F_2$ (dinitrogen difluoride), $N_2F_4$ (dinitrogen tetrafluoride), and $NF_3$ (nitrogen trifluoride). All are gases at room temperature. Nitrogen trifluoride is rather inert and exhibits no electron-donor properties; like $CF_4$, it is not hydrolyzed by water. It fluorinates certain metals when heated:

$$2\ NF_3(g) + Cu(s) \longrightarrow N_2F_4(g) + CuF_2(s)$$

Dinitrogen tetrafluoride (Fig. 22.12) is much more reactive than nitrogen trifluoride and undergoes hydrolysis in water. The trifluoride and pentafluoride of phosphorus are also both gases at room temperature. Unlike $NF_3$, both react readily with water:

$$PF_3(g) + 3\ H_2O(\ell) \longrightarrow H_3PO_3(aq) + 3\ HF(aq)$$

$$PF_5(g) + 4\ H_2O(\ell) \longrightarrow H_3PO_4(aq) + 5\ HF(aq)$$

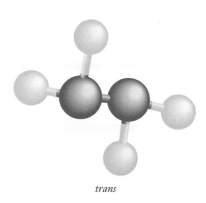

*trans*

*gauche*

**FIGURE 22.12**  Molecules of $N_2F_4$ exist in two isomeric forms.

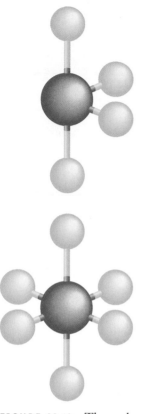

The PF₃ molecule has a trigonal pyramidal structure like NH₃, as the VSEPR model predicts. Unlike ammonia, however, it does not donate its lone electron pair readily, presumably because the highly electronegative fluorine atoms withdraw electron density from the central atom. Like BF₃, PF₅ is a very powerful Lewis acid and is useful as a catalyst for the polymerization of hydrocarbons. The remaining Group V elements (arsenic, antimony, and bismuth) resemble phosphorus in forming both trifluorides and pentafluorides that readily hydrolyze in aqueous solution. They are powerful fluorinating agents, capable of introducing fluorine into other molecules by substitution reactions.

The highly toxic gas oxygen difluoride ($OF_2$) can be made by passing fluorine through aqueous alkaline solutions:

$$2\ F_2(g) + 2\ OH^-(aq) \longrightarrow OF_2(g) + 2\ F^-(aq) + H_2O(\ell)$$

It is worthwhile to compare $OF_2$ with the analogous compound formed between chlorine and oxygen. Both $Cl_2O$ and $OF_2$ are gases under room conditions, and both have oxygen atoms bonded to two halogens. In both compounds the steric number of the oxygen atom is 4, with two bonds, two lone pairs on the central oxygen atom, and a bent shape. Despite these similarities, $Cl_2O$ and $OF_2$ differ profoundly in reactivity. Oxygen difluoride contains "positive oxygen" (note the +2 oxidation number for oxygen), because fluorine is highly electronegative. It reacts with water by oxidizing it to liberate $O_2$:

$$OF_2(g) + H_2O(\ell) \longrightarrow O_2(g) + 2\ HF(aq)$$

Dichlorine oxide contains ordinary "negative oxygen" and behaves quite differently with water. It reacts to give hypochlorous acid without oxidation or reduction:

$$Cl_2O(g) + H_2O(\ell) \longrightarrow 2\ HOCl(aq)$$

**FIGURE 22.13**  The molecular structures of SF₄ and SF₆.

The chemistries of the compounds diverge along the same lines in many other reactions.

The later Group VI elements also form fluorides. The two most important fluorides of sulfur are quite dissimilar in their properties. Both sulfur tetrafluoride and sulfur hexafluoride (Fig. 22.13) are gases at room temperature; the former is highly reactive, and the latter is extremely inert. Both selenium and tellurium form a tetrafluoride and a hexafluoride with fluorine. These compounds, like the fluorides of their Group V neighbors arsenic and antimony, are excellent fluorinating agents. Their strikingly enhanced reactivity compared with sulfur hexafluoride is not easy to explain, except by noting that the larger central atoms are not as well shielded by their fluorine atoms.

Fluorine reacts with the other halogens to form several **interhalogen** compounds; they are liquids at room temperature except for the chlorofluorides, which are gases. In these compounds, fluorine atoms, as always in the −1 oxidation state, surround a central chlorine, bromine, or iodine atom as XF, XF₃, XF₅, where X is Cl, Br, or I. The only XF₇ compound occurs with iodine. The VSEPR model provides accurate predictions of the structures of these compounds. In IF₇ iodine has a steric number of 7, and the structure is based on the pentagonal bipyramid (Fig. 22.14). As liquids, the interhalogens have low electrical conductivities, reflecting their largely covalent character (see, however, problem 27).

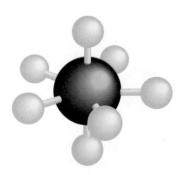

**FIGURE 22.14**  The molecular structure of IF₇ is based on the pentagonal bipyramid structure. The two axial fluorines are 1.786 Å away from the central iodine; the five equatorial fluorines are farther away at 1.858 Å, presumably so as to relieve crowding in the equatorial plane.

## Fluorinated Hydrocarbons and Chlorofluorocarbons

Fluorine can replace hydrogen in hydrocarbons, giving rise to an important class of compounds called fluorinated hydrocarbons. The C—F bond is short and strong, an average of 1.2 times stronger than the C—H bond and 1.4 to 2.2 times stronger than the bonds between carbon and the other halogens; it is the strongest single bond. Fluorination of hydrocarbons with no double or triple bonds therefore increases thermal and chemical stability. The reverse is true in the reactions of the other halogens with hydrocarbons. When all the hydrogen atoms in a hydrocarbon are replaced by fluorine atoms, compounds called **fluorocarbons (or perfluoro carbons**) result. The electron density on the fluorine atoms in fluorocarbons is tightly held and is not easily distorted by approaching atoms or ions. This reduces the reactivity of these compounds and weakens their intermolecular forces. The compounds have low melting and boiling points and low enthalpies of vaporization.

One provocative use of fluorine compounds is in the preparation of blood substitutes. Many fluorocarbons are excellent solvents for oxygen. A milliliter of liquid $C_7F_{16}$, for example, dissolves 1900 times more oxygen at room conditions than a milliliter of water. Small laboratory animals can survive total immersion in oxygenated fluorocarbon liquids, as their lungs successfully extract the oxygen they need from solution. Fluorocarbons are insoluble in water. For use in the bloodstream, they are prepared as fine-grained emulsions in water. Rats have survived for days without apparent harm after their blood was completely replaced with a concentrated emulsion of $C_8F_{17}Br$ in water. Over time, the composition of the rats' blood reverted to normal as the blood substitute was excreted and new red blood cells grew. Practical fluorocarbon blood substitutes would need no matching of blood types, could be stored for emergency use, and would not transmit disease.

If an excess of elemental fluorine is mixed with a hydrocarbon under ordinary conditions, the result is usually a fire or explosion to give HF and $CF_4$. Controlled cryogenic (very cold) conditions can avoid such outcomes, and the treatment of hydrocarbons with elemental fluorine can give reasonable yields of partially and completely fluorinated products. The best-known perfluorocarbon is the solid polytetrafluoroethylene (Teflon), formulated as $-(CF_2-CF_2)-_n$, where $n$ is a large number. Chemically, this compound is nearly completely inert, resisting attack from boiling sulfuric acid, molten potassium hydroxide, gaseous fluorine, and other aggressive chemicals. Physically, it has excellent heat stability (a working temperature up to 260°C), is a very good electrical insulator, and has a low coefficient of friction that makes it useful for bearing surfaces in machines as well as coating frying pans ("no-stick" Teflon). In Teflon the carbon atoms lie in a long chain that is encased by tightly bound fluorine atoms (Fig. 22.15). Even reactants with a strong innate ability to disrupt C—C bonds (such as fluorine itself) fail to attack Teflon at observable rates because there is no route of attack past the surrounding fluorine atoms and their tightly held electrons.

**Chlorofluorocarbons** (often called CFCs) contain one or more carbon atoms with both fluorine and chlorine atoms attached as side groups. Chlorofluorocarbons are chemically inert, nontoxic gases or low-boiling liquids (Fig. 22.16). They are relatively cheap to make and have boiling points conveniently near room temperature. These properties, together with their lack of reactivity, fit them to serve as

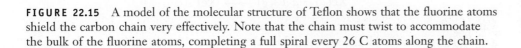

**FIGURE 22.15**   A model of the molecular structure of Teflon shows that the fluorine atoms shield the carbon chain very effectively. Note that the chain must twist to accommodate the bulk of the fluorine atoms, completing a full spiral every 26 C atoms along the chain.

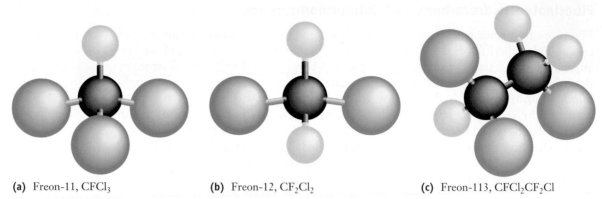

(a) Freon-11, $CFCl_3$     (b) Freon-12, $CF_2Cl_2$     (c) Freon-113, $CFCl_2CF_2Cl$

**FIGURE 22.16** Some chlorofluoro-carbons (Freons) formerly of commercial importance.

propellants in aerosol sprays and as working fluids in air-conditioning machinery. Soon after their discovery, chlorofluorocarbons replaced $SO_2$ and $NH_3$ in the latter application. The inertness of chlorofluorocarbons has drawbacks as well, because the molecules that escape into the air persist unchanged. In time they rise into the stratosphere, where ultraviolet radiation from the sun finally breaks them down. Chlorine atoms generated in this decomposition catalyze the destruction of ozone ($O_3$) in the upper atmosphere, as Section 16.9 discussed in detail. In response to this danger, the developed nations have ceased production of CFCs and have found replacements that will satisfy the applications described with lesser risk to the atmosphere.

---

## 22.4

# COMPOUNDS OF FLUORINE AND THE NOBLE GASES

For many years the noble gases were believed to be chemically inert. They failed to react with the strongest oxidizing agents available, and an early (1895) attempt by Moissan to make argon react with fluorine also failed. In 1933 Linus Pauling suggested on theoretical grounds that xenon and fluorine should form compounds. Soon thereafter, chemists subjected a mixture of the two gases to an electric discharge but saw no evidence of reaction. This result, other negative results, and the dangers of working with elemental fluorine discouraged further experiments. It seemed clear that if fluorine could not force a noble gas into combination, then nothing could. Interest in the subject was further stifled because generally accepted simple bonding theory appeared to ratify the complete chemical inertness of the noble gases. Then, in 1962, Neil Bartlett treated gaseous xenon with the powerful fluorinating agent $PtF_6$, and a yellow-orange solid compound of platinum, fluorine, and xenon resulted. In 1965 an irony emerged: the same fluorine–xenon mixtures that did not react in the 1933 electric discharge experiment *did* give $XeF_2(s)$ simply upon exposure to sunlight!

Bartlett's discovery spurred research on the reactions of the noble gases with fluorine, and soon afterward the synthesis of xenon tetrafluoride by direct combination of the elements at high pressure was reported:

$$Xe(g) + 2\ F_2(g) \longrightarrow XeF_4(s)$$

Varying the relative amounts of xenon and fluorine permitted synthesis of $XeF_2(s)$ and $XeF_6(s)$, as well. All three xenon fluorides are colorless crystalline solids at room temperature (Fig. 22.17). The VSEPR theory correctly predicts the linear structure of $XeF_2$ molecules and the square-planar structure of $XeF_4$ molecules. Because the steric number of the central Xe is 7 and not 6, $XeF_6$ molecules are not octahedral; however, the VSEPR theory does not correctly predict their complex structure.

The three xenon fluorides are moderate to powerful fluorinating agents but are thermodynamically stable with respect to decomposition into xenon and fluorine at room temperature. Their behavior toward water is quite diverse. Xenon difluoride hydrolyzes only very slowly in water, and solutions more concentrated than 0.1 M can be prepared at low temperatures. On the other hand, $XeF_4$ and $XeF_6$ react vigorously with water to form xenon trioxide ($XeO_3$).

$$3\ XeF_4(s) + 6\ H_2O(\ell) \longrightarrow XeO_3(aq) + 2\ Xe(g) + 12\ HF(aq) + \tfrac{3}{2}\ O_2(g)$$

$$XeF_6(s) + 3\ H_2O(\ell) \longrightarrow XeO_3(aq) + 6\ HF(aq)$$

Xenon trioxide dissolves in aqueous solution without ionizing, but in basic solution it acts as a Lewis acid, accepting $OH^-$:

$$XeO_3(aq) + OH^- \longrightarrow HXeO_4^-(aq)$$

The hydrogen xenate ion then disproportionates to form xenon and the perxenate ion, in which the oxidation number of xenon is +8:

$$2\ HXeO_4^-(aq) + 2\ OH^- \longrightarrow XeO_6^{4-}(aq) + Xe(g) + O_2(g) + 2\ H_2O(\ell)$$

Only one simple krypton halide has been prepared. Irradiating a mixture of krypton and fluorine at low temperature with electrons forms krypton difluoride, a thermodynamically unstable substance. It is a volatile, colorless, crystalline solid. Experiment confirms the linear structure predicted by the VSEPR theory. Krypton difluoride forms salt-like compounds with strong fluoride ion acceptors such as $AsF_5$ and $SbF_5$:

$$KrF_2 + AsF_5 \longrightarrow [KrF]^+[AsF_6]^-$$

Krypton difluoride is a stronger oxidizing agent than fluorine itself. It has a total bond enthalpy (for *both* Kr—F bonds) of only 96 kJ mol$^{-1}$; this is substantially less than the bond enthalpy of $F_2(g)$, 158 kJ mol$^{-1}$. Krypton difluoride oxidizes Xe(g) to $XeF_6(s)$ at room temperature and oxidizes $I_2$ directly to $IF_7$. A solution of $KrF_2$ in liquid HF even attacks gold:

$$7\ KrF_2(g) + 2\ Au(s) \longrightarrow 2\ KrF[AuF_6] + 5\ Kr$$

Heating the $KrF[AuF_6]$ to 60°C generates orange $AuF_5$, in which gold has a +5 oxidation number. By way of comparison, $F_2(g)$ itself requires a temperature of 250°C to act on powdered gold and then generates only $AuF_3$.

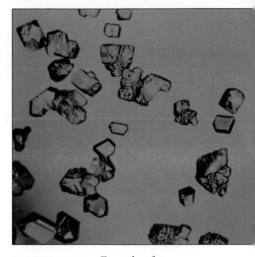

**FIGURE 22.17** Crystals of xenon tetrafluoride, formed in the creation of the first binary noble gas compound. *(Argonne National Laboratory)*

## CONCEPTS & SKILLS

*After studying this chapter and working the problems that follow, you should be able to*

1. Outline the historical development of the chlor-alkali industry, including the linkage between production of chlorine and the bases sodium carbonate and sodium hydroxide (Section 22.1, problems 1–6).

2. Describe the production of and uses for important compounds of chlorine, bromine, and iodine (Section 22.2, problems 7–12).

3. Compare and contrast the chemical properties of the oxoacids and oxides of chlorine, bromine, and iodine (Section 22.2, problems 13–16).

4. Discuss the preparation of elemental fluorine and the uses of this element and its compounds (Section 22.3).
5. Compare the properties of fluorine and its compounds with those of the other halogens (Section 22.3, problems 17–20).
6. Discuss the physical and chemical properties of fluorides of the elements through the periodic table (Section 22.3, problems 21–32).
7. Describe the chemical reactions undergone by the heavier noble-gas elements (Section 22.4, problems 33–38).

## PROBLEMS

*Answers to problems whose numbers are boldface appear in Appendix G. Problems that are more challenging are indicated with asterisks.*

### Chemicals from Salt

1. Calculate the equilibrium constant at 25°C for the production of sodium hypochlorite bleach from dichlorine oxide and sodium hydroxide:

$$Cl_2O(g) + 2\ OH^-(aq) \rightleftharpoons 2\ OCl^-(aq) + H_2O(\ell)$$

   Relevant data can be found in Appendix D.

2. Calculate the equilibrium constant at 25°C for the preparation of chlorine by the reaction of hydrochloric acid with pyrolusite, used by Scheele in 1774:

$$2\ Cl^-(aq) + 4\ H_3O^+(aq) + MnO_2(s) \rightleftharpoons$$
$$Cl_2(g) + Mn^{2+}(aq) + 6\ H_2O(\ell)$$

   Relevant data can be found in Appendix D.

3. Calculate the standard enthalpy change per mole of $Na_2CO_3$ produced by the Leblanc process, using data from Appendix D. From your calculation, what do you conclude about the overall heat requirements of that process?

4. Calculate the standard enthalpy change per mole of $Na_2CO_3$ produced by the Solvay process, using data from Appendix D. From your calculation, what do you conclude about the overall heat requirements of that process?

5. A recent magazine article about the chlor-alkali process in the United States reported that "production of chlorine in 1987 . . . total[ed] about 10.9 million tons. . . . Caustic soda production as usual [was] 4 to 5 percent higher than chlorine production." Explain the use of the phrase "as usual" by computing the theoretical ratio of the yields of these two chemicals in the process.

6. Calculate the minimum theoretical external cell voltage to drive a chlor-alkali cell in which all reactants and products are in their standard states.

### The Chemistry of Chlorine, Bromine, and Iodine

7. At 25°C, 33.6 g of bromine will dissolve in 1 L of water. Use this fact to estimate $\Delta G°$ for the reaction

$$Br_2(\ell) \longrightarrow Br_2(aq)$$

Compare your estimate with the result calculated from data in Appendix D.

8. At 25°C, the solubility of $I_2$ in water is only 0.33 g $L^{-1}$. Estimate $\Delta G°$ for the reaction

$$I_2(s) \longrightarrow I_2(aq)$$

Compare your estimate with the result calculated from data in Appendix D.

9. Calculate the equilibrium constant at 25°C for the reaction

$$2\ Br^-(aq) + Cl_2(g) \rightleftharpoons Br_2(g) + 2\ Cl^-(aq)$$

Use your result to comment on the feasibility of using cheap chlorine gas to extract bromine from seawater.

10. Hypobromite ion in aqueous solution disproportionates into bromide ion and bromate ion according to the equation

$$3\ BrO^-(aq) \rightleftharpoons 2\ Br^-(aq) + BrO_3^-(aq)$$

Calculate the equilibrium constant at 25°C for this reaction, given the following thermodynamic data:

|             | $\Delta H_f°$ (kJ mol$^{-1}$) | $S°$ (J K$^{-1}$ mol$^{-1}$) |
|-------------|--------------|-------------|
| $Br^-(aq)$  | −122         | 82          |
| $BrO^-(aq)$ | −94          | 42          |
| $BrO_3^-(aq)$ | −67        | 162         |

11. Gaseous hydrogen bromide can be produced by the action of a nonoxidizing acid such as phosphoric acid on a solution of sodium bromide.
    (a) Write a balanced chemical equation for this reaction.
    (b) What volume of hydrogen bromide at 0°C and 1.00 atm will be produced when an excess of phosphoric acid is added to 100 mL of a solution with an NaBr concentration of 0.050 M?

12. When chlorine is bubbled through a basic solution containing iodide ions, chloride ions and iodate ions are produced.
    (a) Write a balanced chemical equation for this reaction.
    (b) Suppose 0.216 L of chlorine at 0°C and 1.00 atm is needed to react completely with 50.0 mL of the iodide solution in part (a). Determine the initial concentration of iodide in that solution.

**13.** Given the reduction potentials,

$$ClO_2 + e^- \longrightarrow ClO_2^- \qquad \mathscr{E}° = 0.954 \text{ V}$$

$$ClO_3^- + H_2O + e^- \longrightarrow ClO_2 + 2\ OH^- \qquad \mathscr{E}° = -0.25 \text{ V}$$

decide whether $ClO_2$ is stable with respect to disproportionation in basic aqueous solution.

**14.** Given the following reduction potentials

$$ClO^- + H_2O + 2\ e^- \longrightarrow Cl^- + 2\ OH^- \qquad \mathscr{E}° = 0.90 \text{ V}$$

$$ClO_3^- + 2\ H_2O + 4\ e^- \longrightarrow ClO^- + 4\ OH^- \qquad \mathscr{E}° = 0.48 \text{ V}$$

decide whether hypochlorite ion ($ClO^-$) is stable with respect to disproportionation in basic aqueous solution under standard conditions.

**15.** The reduction potential diagram for iodine in aqueous solution at pH 14 and 25°C has the form

$$IO_3^- \xrightarrow{\ 0.14\text{ V}\ } IO^- \xrightarrow{\ 0.45\text{ V}\ } I_2 \xrightarrow{\ 0.535\text{ V}\ } I^-$$
$$\underset{0.49\text{ V}}{\underline{\hspace{6cm}}}$$

(a) Calculate the stable products that result when $I_2$ is dissolved in water at pH 14 and 25°C. Explain your answer.

(b) Calculate the standard half-cell potential at pH 14 for the half-reaction

$$IO_3^-(aq) + 3\ H_2O(\ell) + 6\ e^- \longrightarrow I^-(aq) + 6\ OH^-(aq)$$

**16.** The reduction potential diagram for bromine in aqueous solution at pH 14 and 25°C has the form

$$BrO_3^- \xrightarrow{\ 0.54\text{ V}\ } BrO^- \xrightarrow{\ 0.45\text{ V}\ } Br_2 \xrightarrow{\ 1.06\text{ V}\ } Br^-$$
$$\underset{0.76\text{ V}}{\underline{\hspace{6cm}}}$$

(a) Calculate the stable products that result when $Br_2$ is dissolved in water at pH 14 and 25°C. Explain your answer.

(b) Calculate the standard half-cell potential at pH 14 for the half-reaction

$$BrO_3^-(aq) + 3\ H_2O(\ell) + 6\ e^- \longrightarrow Br^-(aq) + 6\ OH^-(aq)$$

## Fluorine and Its Compounds

**17.** Write balanced equations for the reaction of elemental fluorine with the following:
(a) $SrO(s)$    (b) $O_2(g)$    (c) $UF_4(s)$

**18.** Write balanced equations for the reaction of elemental fluorine with the following:
(a) $NaCl(s)$    (b) $MgO(s)$    (c) $Na(s)$

**19.** The alkali metals form one or more acid fluorides, with formulas MF·HF, MF·2HF, and MF·3HF (where M stands for the alkali metal). In these compounds, additional HF molecules link with $F^-$ ions by means of hydrogen bonds, in a manner similar to the incorporation of water of hydration into crystalline hydrates. Nothing similar is observed with the other alkali-metal halides.

Suppose a compound is found to contain 2.48 g of fluorine for every gram of sodium. What is the empirical formula

of this compound? Assume that any hydrogen present is not detected in this analysis.

**20.** There are ways to combine carbon and fluorine beyond those mentioned in the text. When carbon in the form of graphite is treated with elemental fluorine, F atoms occupy interstitial spaces between the layers of C atoms, and compounds of empirical formula $C_4F$, $C_2F$, and CF result. The bonds between carbon and fluorine in all three of these materials are covalent. Graphite fluorides are used as electrodes in advanced primary batteries and as lubricants.

Suppose a graphite fluoride is analyzed to contain 0.53 g of carbon for every gram of fluorine. What is the empirical formula of this compound?

**21.** A gaseous binary compound of chlorine and fluorine has a density at 0°C and 1.00 atm of 4.13 g L$^{-1}$ and is 38.35% chlorine by mass. Determine its molecular formula.

**22.** A compound has the molecular formula $ClO_3F$. Estimate the volume occupied by 225 g of this compound at 0°C and 1.00 atm.

**23.** Predict the structures of the following fluorine-containing molecules.
(a) $OF_2$    (c) $BrF_3$    (e) $IF_7$
(b) $BF_3$    (d) $BrF_5$    (f) $SeF_6$

**24.** Predict the structures of the following fluorine-containing molecular ions.
(a) $SiF_6^{2-}$    (c) $IF_4^+$    (e) $BF_4^-$
(b) $ClF_2^+$    (d) $AsF_6^-$    (f) $BrF_6^+$

**25.** Thionyl and selenyl difluoride have the compositions $SOF_2$ and $SeOF_2$, with the oxygen and fluorine atoms directly linked to the central sulfur or selenium atom. They can be prepared by reaction of a strong fluorinating agent, such as $PF_5$, with sulfur dioxide or selenium dioxide:

$$SO_2(g) + PF_5(g) \longrightarrow SOF_2(g) + POF_3(g)$$

(a) Use the VSEPR theory to predict the structures of $SOF_2$ and $SeOF_2$. Draw the structures.

(b) Thionyl difluoride can coordinate with $BF_3$ through its lone pair. Does it act as a Lewis acid or as a Lewis base in this reaction?

**26.** When two oxygen atoms take the place of four fluorine atoms in $SF_6$, the compound has the composition $SO_2F_2$ and is called "sulfuryl difluoride." Like $SF_6$, $SO_2F_2$ is comparatively unreactive and hydrolyzes with difficulty.

(a) Use the VSEPR theory to predict the structure of $SO_2F_2$. Draw this structure.

(b) In the conversion of thionyl difluoride ($SOF_2$) to sulfuryl difluoride,

$$2\ SOF_2 + O_2 \longrightarrow 2\ SO_2F_2$$

does the thionyl difluoride act as a Lewis acid or as a Lewis base?

**27.** Although $BrF_3$ is a covalent compound, $BrF_3(\ell)$ is slightly conductive due to the autoionization of $BrF_3$ molecules. This is analogous to the small conductivity of water arising from autoionization:

$$2\ H_2O(\ell) \rightleftharpoons H_3O^+(aq) + OH^-(aq)$$

Write a balanced chemical equation for the autoionization of $BrF_3$. Can this reaction be described as a Lewis acid–base reaction?

28. The molecules in liquid $BrF_3$ are linked through "fluorine bonds" that are analogous to the hydrogen bonds in water. Would you expect the entropy of vaporization of $BrF_3(\ell)$ to be larger or smaller than the Trouton's rule value, $88 \ J \ K^{-1} \ mol^{-1}$?

29. Rank the following compounds in order of normal boiling point, from lowest to highest: $CaF_2$, $PtF_6$, $PtF_4$.

30. For each of the following pairs of fluorine compounds, predict which will have the higher melting point.
   (a) $NaF$ or $PF_3$
   (b) $AsF_3$ or $AsF_5$
   (c) $CF_4$ or $C_5F_{12}$

31. One of the hazards in making Teflon is that the starting material, tetrafluoroethylene, $C_2F_4(g)$, can explode, giving $C(s,$ graphite$)$ and $CF_4(g)$. The standard enthalpy of formation of $C_2F_4(g)$ is $-651 \ kJ \ mol^{-1}$. Use this fact, together with data from Appendix D, to estimate the amount of heat that would be released if a tank containing 1.00 kg of $C_2F_4$ were to explode.

32. Compute the $\Delta G°$ of the reaction

$$CF_4(g) + 2 \ H_2O(\ell) \longrightarrow CO_2(aq) + 4 \ HF(aq)$$

using data from Appendix D. Explain how $CF_4$ spontaneously decomposes according to the preceding equation, yet in practice is stable up to 500°C.

## Compounds of Fluorine and the Noble Gases

33. The standard enthalpy change $\Delta H°$ for the reaction

$$XeF_6(g) + 3 \ H_2(g) \longrightarrow Xe(g) + 6 \ HF(g)$$

is $-1282 \ kJ$. Using this fact and information from Appendix D, calculate the following:
   (a) The standard molar enthalpy of formation $\Delta H_f°$ of $XeF_6(g)$
   (b) The average enthalpy of an Xe—F bond in $XeF_6(g)$

34. The standard enthalpy change $\Delta H°$ for the reaction

$$XeF_4(g) + 2 \ H_2(g) \longrightarrow Xe(g) + 4 \ HF(g)$$

is $-887 \ kJ$. Using this fact and information from Appendix D, calculate the following:
   (a) The standard molar enthalpy of formation $\Delta H_f°$ of $XeF_4(g)$
   (b) The average enthalpy of an Xe—F bond in $XeF_4(g)$

35. Xenon oxotetrafluoride ($XeOF_4$) is a liquid at room temperature. It can be prepared by the controlled hydrolysis of xenon hexafluoride:

$$XeF_6(g) + H_2O(\ell) \longrightarrow XeOF_4(\ell) + 2 \ HF(g)$$

The standard enthalpies of formation of the noble-gas compounds in this reaction are

$$\Delta H_f°(XeF_6(g)) = -298 \ kJ \ mol^{-1}$$

$$\Delta H_f°(XeOF_4(\ell)) = +148 \ kJ \ mol^{-1}$$

Calculate $\Delta H°$ for the hydrolysis reaction. Use Appendix D for additional data.

36. The energy of explosion of $XeO_3(s)$ into its gaseous elements was measured in a constant-volume calorimeter.

$$XeO_3(s) \longrightarrow Xe(g) + \tfrac{3}{2}O_2(g)$$

   (a) A $2.763 \times 10^{-4}$ mol sample released 112 J of heat when it was exploded. Calculate the standard energy of formation $(\Delta E_f°)$ of $XeO_3(s)$.
   (b) Will the standard *enthalpy* of formation of $XeO_3(s)$ be larger or smaller than your answer to part (a)?

37. The reduction of perxenate ion ($XeO_6^{4-}$) to xenon is so favored thermodynamically in acidic solution that it will even oxidize $Mn^{2+}$ to $MnO_4^-$, one of the strongest oxidizing agents ordinarily encountered, with evolution of oxygen. Write balanced half-equations and an equation for the overall reaction in this process.

38. The following standard reduction potentials have been measured for the oxides of xenon in acidic aqueous solution:

$$H_4XeO_6(aq) + 2 \ H_3O^+(aq) + 2 \ e^- \longrightarrow$$
$$XeO_3(aq) + 5 \ H_2O(\ell) \qquad \mathscr{E}° = +2.36 \ V$$

$$XeO_3(aq) + 6 \ H_3O^+(aq) + 6 \ e^- \longrightarrow$$
$$Xe(g) + 9 \ H_2O(\ell) \qquad \mathscr{E}° = +2.12 \ V$$

   (a) Would you classify perxenic acid ($H_4XeO_6$) as an oxidizing agent or as a reducing agent?
   (b) In the preceding, $XeO_3$ acts as an oxidizing agent in one half-reaction and as a reducing agent in the other. At pH 0, is $XeO_3$ stronger as an oxidizing agent or as a reducing agent?
   (c) Is $XeO_3$ stable with respect to disproportionation to $Xe(g)$ and $H_4XeO_6(aq)$ in acidic aqueous solution?

## Additional Problems

39. Chlorine-containing bleaches act by oxidizing. By referring to Appendix E, predict whether bromine will be a stronger or a weaker bleach than chlorine.

40. In the Deacon process, an equilibrium is reached among chlorine, steam, hydrogen chloride, and oxygen at 450°C.
   (a) Use the $\Delta H_f°$ and $\Delta S_f°$ values from Appendix D to estimate an equilibrium constant at 450°C for the reaction

$$2 \ HCl(g) + \tfrac{1}{2} \ O_2(g) \rightleftharpoons Cl_2(g) + H_2O(g)$$

   (b) Suppose that bromine replaces chlorine in the preceding reaction. After referring to the appropriate bond energies in Table 3.2, state whether the equilibrium constant at 450°C will be larger or smaller. Explain.

41. Suppose a chlor-alkali plant has 250 cells, through each of which a current of 100,000 A passes continuously.
   (a) Calculate the mass of chlorine that this plant can produce per 24-hour day.
   (b) If the cell voltage is 3.5 V, calculate the total electrical energy consumed by the plant in one 24-hour day, and express it both in joules and in kilowatt-hours.

(c) If electricity costs $0.05 per kilowatt-hour, calculate the cost for the electricity needed to run the cells for one 24-hour day.

42. The solubility of iodine in water at 25°C is 0.033 g per 100 g. Calculate the solubility of iodine in 1.0 M KI solution at this temperature, taking into account the equilibrium

$$I^-(aq) + I_2(aq) \rightleftharpoons I_3^-(aq)$$

Relevant data can be found in Appendix D.

43. Astatine is a little-studied element whose longest lived isotope, $^{210}At$, has a half-life of only 8.3 hours.
    (a) Predict whether HAt will be a stronger or a weaker acid than HI.
    (b) The $At^-$ ion is produced when elemental astatine reacts in aqueous solution with zinc, but no reaction is seen with 1 M $Fe^{2+}(aq)$. What range of reduction potentials for conversion of $At_2$ to $At^-$ is consistent with these observations?
    (c) If $Cl_2(g)$ is bubbled through a solution containing $At^-(aq)$, what reaction will result?
    (d) Write a balanced equation for the reaction that should occur when $At_2(s)$ is dissolved in basic aqueous solution.

44. Solid iodine has a small electrical conductivity at room temperature, which increases with temperature. By considering the neighbors of iodine in the periodic table, explain how this may be true. Do you predict that the conductivity of solid bromine at −20°C will be greater than or less than that of solid iodine at the same temperature? Why?

45. The reaction

$$K_2[NiF_6] + TiF_4(s) \longrightarrow K_2[TiF_6](s) + NiF_2(s) + F_2(g)$$

generates elemental fluorine. Does $TiF_4(s)$ play the role of a Lewis acid or a Lewis base in this reaction? Explain.

46. A commercial fluorine cell generates 3.3 kg of fluorine per hour by electrolysis of KF·2HF. Compute the average current passing through this cell.

47. Suppose an aqueous solution of NaF is electrolyzed between platinum electrodes. Will $F_2(g)$ be produced? Why or why not?

48. (a) Predict the vapor density of HF(g) at 1.00 atm pressure and its normal boiling point, 20°C, assuming ideal gas behavior.
    (b) The observed vapor density under the conditions in part (a) is 3.11 g $L^{-1}$. What is the explanation for the large discrepancy with the calculated result of part (a)?

49. The reaction

$$4\,HF(\ell) + SiO_2(s) \longrightarrow SiF_4(g) + 2\,H_2O(\ell)$$

can be used to release gold that is distributed in certain quartz veins of hydrothermal origin. Let us say the quartz contains $1.0 \times 10^{-3}$% gold by mass and the gold has a market value of $350 per troy ounce. Will the process be economically feasible if commercial (50% by mass) aqueous hydrogen

fluoride (density 1.17 g $cm^{-3}$) costs $0.25 per liter? (1 troy ounce = 31.3 g.)

50. Ingestion of 5 to 10 g of NaF is lethal to a 70-kg man. Smaller doses can be treated with calcium therapy, which restores $Ca^{2+}$ levels in the body. Write a chemical equation for the effect of NaF on the body.

51. Dioxygen difluoride, $O_2F_2$, is a particularly potent fluorinating agent, made by irradiating a mixture of $O_2$ and $F_2$ at the temperature of liquid nitrogen. The O-to-O distance in $O_2F_2$ is nearly as short as the distance in $O_2$, and the O-to-F distances are quite long. Draw the Lewis structure for this molecule, determine the bond order of all bonds, and describe its geometry. What is the name of the analogous compound of oxygen and hydrogen?

52. The fluorinating agent dioxygen difluoride, $O_2F_2$ (see the preceding problem), is important because it converts the plutonium in almost any plutonium-containing material to $PuF_6$ under mild conditions. The volatility of $PuF_6$ then allows the separation of radioactive Pu from various impurities. Write balanced chemical equations for the reaction of $PuO_2$ with $O_2F_2$ and the reaction of Pu with $O_2F_2$.

53. When $SF_4(g)$ reacts with $CsF(s)$, the $SF_5^-$ ion is formed. Use the VSEPR theory to predict its geometry.

54. The compound $S_2F_{10}$ was not mentioned in the chapter.
    (a) Taking the known bond-forming tendencies of fluorine into account, suggest a probable Lewis structure for this compound.
    (b) Use the VSEPR model to predict the molecular geometry of $S_2F_{10}$.

55. The result of the accidental first preparation of Teflon was immediately subjected to elemental analysis for both chlorine and fluorine. The report on the new, mysterious white powder was that it contained no chlorine and 48.4% fluorine by mass. Compare these first values to the actual percentages of fluorine and chlorine in Teflon.

56. The ionization energy of $O_2$ is 1180 kJ $mol^{-1}$. In 1962 Neil Bartlett reported that it reacts with $PtF_6(g)$ to form the solid ionic compound $O_2^+PtF_6^-$. By referring to data in Appendix F, show why it occurred to Bartlett that xenon might also form a compound with $PtF_6$.

57. Calculate the equilibrium constant for the reaction

$$Xe(g) + 2\,F_2(g) \rightleftharpoons XeF_4(s)$$

at 25°C. The standard molar free energy of formation of $XeF_4(s)$ is −134 kJ $mol^{-1}$.

58. Xenon difluoride reacts with arsenic pentafluoride to give an ionic compound:

$$2\,XeF_2 + AsF_5 \longrightarrow [Xe_2F_3]^+[AsF_6]^-$$

Identify which species are Lewis acids in this reaction and which are Lewis bases.

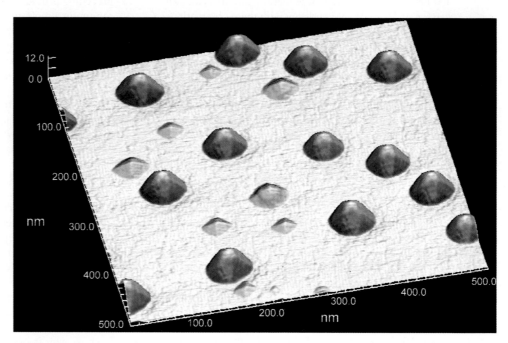

Ge islands on an Si surface undergo a "shape transition" from pyramids to domes at a certain stage of growth. *(Image courtesy of Dr. Gilberto Medeiros-Ribeiro and Dr. R. Stanley Williams, Hewlett-Packard Research Laboratories, Palo Alto, CA)*

Throughout history, the discovery of new materials from which to fashion the structures, machines, and devices of everyday life has set off great change in human affairs. Modern science and engineering—with chemistry in the central role—provide routes for modifying properties of materials to meet specific applications and for synthesizing and processing new materials designed from the beginning to have specific properties. One very exciting contemporary approach is to create "quantum dots" of materials, small samples less than about 25 nm in extent, whose properties are governed by quantum mechanics and depend more strongly on the size of the dots than on their composition.

THE GOAL OF UNIT 7 IS   to survey three essential classes of materials—ceramics, optical and electronic materials, and polymers—to illustrate the role of modern chemistry in determining properties of and identifying applications for both natural and manufactured materials.

# CHAPTER
# 23

# Ceramic Materials

**Illustration**
Azurite is a basic copper carbonate with chemical formula $Cu_3(CO_3)_2(OH)_2$. *(Charles D. Winters)*

I n this chapter we focus on **ceramics,** defined as synthetic materials whose essential components are inorganic, nonmetallic materials. Along with metals, whose properties were described in Chapter 19 and whose recovery from ores was discussed in Chapter 20, ceramics are one of the oldest classes of materials prepared by mankind. New discoveries in ceramics are occurring at a startling rate, and major new technological advances will certainly come from these discoveries. Ceramics have value both as structural materials, the role emphasized in this chapter, and for their wide range of electronic and optical properties (Chapter 24).

The chapter begins with a survey of the naturally occurring inorganic, non-metallic minerals from which ceramics are manufactured.

---
## 23.1

# NATURALLY OCCURRING MINERALS: THE LITHOSPHERE

The study of the solid earth involves techniques from chemistry, physics, and geology. Seismic data (shock-wave propagation velocities from earthquakes and man-made explosions) show that the earth consists of three distinct parts: the **core,** the **mantle,** and the **crust** (Table 23.1). The core of the earth is believed to consist primarily of iron, probably with smaller amounts of nickel and iron sulfide (FeS). This would be consistent with the composition of meteorites, which are thought to resemble the material from which the earth formed. The outer part of the core is probably liquid, and there may be a solid region at the center. Outside the core is the mantle, whose seismic properties are consistent with a composition largely of magnesium-iron silicates.

The crust of the earth has an average thickness of only 17 km, and so it contributes relatively little to the total mass. The crust shows greater variation in rock types than does the mantle. In contrast with the core and the mantle, the crust of the earth (also called the **lithosphere**) is in a state of continual transformation. Heat and matter flow from its hot interior; it is subjected to the chemical action of its atmosphere and hydrosphere; and it is modulated by its interaction with the biosphere (living matter on the surface of the earth). Contact of the lithosphere with these zones outside it has a profound effect on the distribution of the elements within it.

This section surveys the types of silicate compounds that make up much of the lithosphere and the chemical transformations they undergo at and beneath the surface of the earth.

## Silicates

Silicon and oxygen make up most of the earth's crust, with oxygen accounting for 47% and silicon for 28% of its mass. The silicon–oxygen bond is strong and partially ionic. It forms the basis for a class of minerals called **silicates,** which make up

---
## TABLE 23.1

### The Shell Structure of the Earth

| | Thickness (km) | Volume ($10^{20}$ m$^3$) | Mean Density (g cm$^{-3}$) | Mass ($10^{24}$ kg) | Mass (percent) |
|---|---|---|---|---|---|
| Atmosphere | — | — | — | 0.000005 | 0.00009 |
| Hydrosphere | 3.8 | 0.0137 | 1.03 | 0.00141 | 0.024 |
| Crust | 17 | 0.08 | 2.8 | 0.024 | 0.4 |
| Mantle | 2883 | 8.99 | 4.5 | 4.016 | 67.2 |
| Core | 3471[†] | 1.75 | 11.0 | 1.936 | 32.4 |
| Whole earth | 6371[†] | 10.83 | 5.52 | 5.976 | 100 |

[†] Radius.

**TABLE 23.2**

*Silicate Structures*

| Structure | Figure | Corners Shared at Each Si | Repeat Unit | Si:O Ratio | Example |
|---|---|---|---|---|---|
| Tetrahedra | 23.1a | 0 | $SiO_4^{4-}$ | 1:4 | Olivines |
| Pairs of tetrahedra | 23.1b | 1 | $Si_2O_7^{6-}$ | $1:3\frac{1}{2}$ | Thortveitite |
| Closed rings | 23.1c | 2 | $SiO_3^{2-}$ | 1:3 | Beryl |
| Infinite single chains | 23.1d | 2 | $SiO_3^{2-}$ | 1:3 | Pyroxenes |
| Infinite double chains | 23.1e | $2\frac{1}{2}$ | $Si_4O_{11}^{6-}$ | $1:2\frac{3}{4}$ | Amphiboles |
| Infinite sheets | 23.1f | 3 | $Si_2O_5^{2-}$ | $1:2\frac{1}{2}$ | Talc |
| Infinite network | 23.1g | 4 | $SiO_2$ | 1:2 | Quartz |

the bulk of the rocks, clays, sand, and soils in the earth's crust. From time immemorial, silicates have provided the ingredients for the building materials such as bricks, cement, concrete, and glass considered later in this chapter.

The structure-building properties of silicates (Table 23.2) originate in the tetrahedral orthosilicate anion ($SiO_4^{4-}$) in which the negative charge of the silicate ion is balanced by the compensating charge of one or more cations. The simplest silicates consist of individual $SiO_4^{4-}$ anions (Fig. 23.1a), with cations arranged around

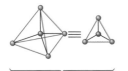

**(a)** Tetrahedral unit

**(b)** Disilicate

**(c)** Cyclosilicate

**(d)** Infinite single chain

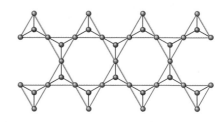

**(e)** Infinite double chain

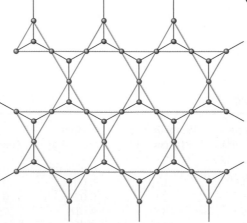

**(f)** Infinite sheet

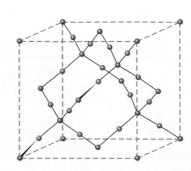

**(g)** Infinite network (cristobalite, $SiO_2$)

**FIGURE 23.1** The classes of silicate structures. Part a shows the symbol used for the $SiO_4^{4-}$ tetrahedron. Bear in mind that all of these structures are actually three-dimensional. Planar projections are used here for convenience of representation.

them on a regular crystalline lattice. Such silicates are properly called **orthosili-cates.** Examples are forsterite ($Mg_2SiO_4$) and fayalite ($Fe_2SiO_4$), which are the extreme members of a class of minerals called **olivines,** $[Mg,Fe]_2SiO_4$. There is a continuous range of proportions of magnesium and iron in the olivines.

Other silicate structures form when two or more $SiO_4^{4-}$ tetrahedra link and share oxygen vertices. The simplest such minerals are the **disilicates**—such as thortveitite, $Sc_2(Si_2O_7)$—in which two $SiO_4^{4-}$ tetrahedra are linked (Fig. 23.1b). Additional linkages of tetrahedra create the ring, chain, sheet, and network structures shown in Figure 23.1 and listed in Table 23.2. In each of these the fundamental tetrahedron is readily identified, but the Si:O ratio is no longer 4 because oxygen atoms are shared at the linkages.

## EXAMPLE 23.1

By referring to Table 23.2, predict the structural class in which the mineral Egyptian blue ($CaCuSi_4O_{10}$) belongs. Give the oxidation state of each of its atoms.

### Solution

Because the Si:O ratio is 4:10, or $1:2\frac{1}{2}$, this mineral should have an infinite sheet structure with the repeating unit $Si_2O_5^{2-}$. The oxidation states of Si and O are +4 and −2, as usual, and that of Ca is +2. In order that the total oxidation number per formula unit sum to 0, the oxidation state of Cu must be +2.

**Related Problems: 3, 4**

**FIGURE 23.2**  The mineral quartz is one form of silica, $SiO_2$. *(Leon Lewandowski)*

The physical properties of the silicates correlate closely with their structures. An example of an infinite layered structure (Fig. 23.1f) is talc, $Mg_3(Si_4O_{10})(OH)_2$. In talc, all of the bonding interactions among the atoms occur in a single layer. Layers of talc sheets are attracted to one another only by van der Waals interactions, which (being weak) permit one layer to slip easily across another. This accounts for the slippery feel of talc (called talcum powder). When all four vertices of each tetrahedron are linked to other tetrahedra (Fig. 23.1g), three-dimensional network structures such as quartz result (Fig. 23.2). Note that the quartz network carries no charge; consequently, there are no cations in its structure. Three-dimensional network silicates such as quartz are much stiffer and harder than the linear and layered silicates, and they resist deformation well.

**Asbestos** is a generic term for a group of naturally occurring hydrated silicates that can be processed mechanically into long fibers (Fig. 23.3). Some of these silicates, such as tremolite, $Ca_2Mg_5(Si_4O_{11})_2(OH)_2$, show the infinite double-chain structure of Figure 23.1e. Another kind of asbestos mineral is chrysotile, $Mg_3(Si_2O_5)(OH)_4$. As the formula indicates, this mineral has a sheet structure (Fig. 23.1f), but the sheets are rolled into long tubes. Asbestos minerals are fibrous because the bonds along the strand-like tubes are stronger than those that hold different tubes together. Asbestos is an excellent thermal insulator that does not burn, resists acids, and is very strong. For many years, it was used in cement for pipes and ducts and woven into fabric to make fire-resistant roofing paper and floor tiles. Its use has decreased significantly in recent years because inhalation of its small fibers can cause the lung disease asbestosis. The risk comes with breathing asbestos dust that is raised during mining and manufacturing processes or is released in buildings in which asbestos-containing materials are fraying, crumbling, or being removed.

**FIGURE 23.3**  The fibrous structure of asbestos is apparent in this sample. *(Leon Lewandowski)*

**FIGURE 23.4** Naturally occurring muscovite mica. The mechanical properties of crystals of mica are quite anisotropic. Thin sheets can be peeled off a crystal of mica by hand, but the sheets resist stresses in other directions more strongly. Transparent, thin sheets of mica, sometimes called isinglass, have been used for heat-resistant windows in stoves or in place of window glass. *(Copyright Doug Sokell/Visuals Unlimited)*

## Aluminosilicates

An important class of minerals called **aluminosilicates** results from the replacement of some of the silicon atoms in silicates with aluminum atoms. Aluminum is the third most abundant element in the earth's crust (8% by mass), where it occurs largely in the form of aluminosilicates. Aluminum in minerals can be a simple cation ($Al^{3+}$) or it can replace silicon in tetrahedral coordination. When it replaces silicon, it contributes only three electrons to the bonding framework in place of the four electrons of silicon atoms. The additional required electron is supplied by the ionization of a metal atom such as sodium or potassium; the resulting alkali-metal ions occupy nearby sites in the aluminosilicate structure.

The most abundant and important of the aluminosilicate minerals in the earth's surface are the **feldspars**, which result from the substitution of aluminum for silicon in three-dimensional silicate networks such as quartz. The aluminum ions must be accompanied by other cations such as sodium, potassium, or calcium to maintain overall charge neutrality. Albite is a feldspar with chemical formula $NaAlSi_3O_8$. In the high-temperature form of this mineral, the aluminum and silicon atoms are distributed at random (in 1:3 proportion) over the tetrahedral sites available to them. At lower temperatures, other crystal structures become thermodynamically stable, with partial ordering of the Al and Si sites.

If one of the four silicon atoms in the structural unit of talc, $Mg_3(Si_4O_{10})(OH)_2$, is replaced by an aluminum atom, and a potassium atom is furnished to supply the fourth electron needed for bonding in the tetrahedral silicate framework, the result is the composition $KMg_3(AlSi_3O_{10})(OH)_2$, which belongs to the family of **micas** (Fig. 23.4). Mica is harder than talc, and its layers slide less readily over one another, although the crystals still cleave easily into sheets. The cations occupy sites between the infinite sheets, and the van der Waals bonding that holds adjacent sheets together in talc is augmented by an ionic contribution. The further replacement of the three $Mg^{2+}$ ions in $KMg_3(AlSi_3O_{10})(OH)_2$ with two $Al^{3+}$ ions gives the mineral muscovite, $KAl_2(AlSi_3O_{10})(OH)_2$. Writing its formula in this way indicates that there are aluminum atoms in two kinds of sites in the structure: one Al atom per formula unit occupies a tetrahedral site, substituting for an Si atom, and the other two Al atoms are between the two adjacent layers. The formulas that mineralogists and crystallographers use convey more information than the usual empirical chemical formula of a compound.

## Clay Minerals

**Clays** are minerals produced by the weathering action of water and heat on primary minerals. Their compositions can vary widely as a result of the replacement of one element with another. Invariably they are microcrystalline or powdered in form and are usually hydrated. Often they are used as supports for catalysts, as fillers in paint, and as ion-exchange vehicles. The clays that readily absorb water and swell are used as lubricants and bore-hole sealers in the drilling of oil wells.

The derivation of clays from talcs and micas provides a direct way to understand the clays' structures. The infinite-sheet mica pyrophyllite, $Al_2(Si_4O_{10})(OH)_2$, serves as an example. If one out of six $Al^{3+}$ ions in the pyrophyllite structure is replaced by one $Mg^{2+}$ ion and one $Na^+$ ion (which together carry the same charge), a type of clay called montmorillonite, $MgNaAl_5(Si_4O_{10})_3(OH)_6$, results. This clay readily absorbs water, which infiltrates between the infinite sheets and hydrates the $Mg^{2+}$ and $Na^+$ ions there, causing the montmorillonite to swell (Fig. 23.5).

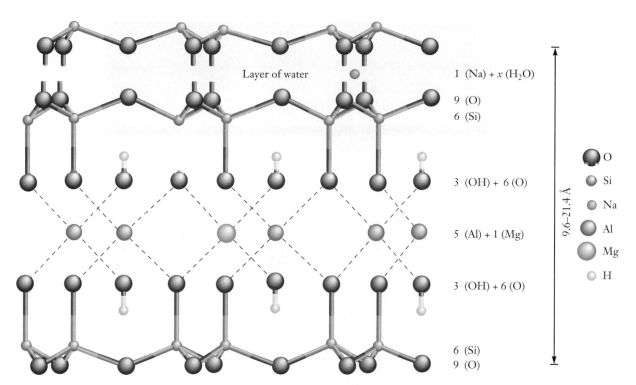

Layer of water

1 (Na) + $x$ (H$_2$O)

9 (O)
6 (Si)

3 (OH) + 6 (O)

5 (Al) + 1 (Mg)

3 (OH) + 6 (O)

6 (Si)
9 (O)

○ O
○ Si
○ Na
○ Al
○ Mg
○ H

9.6–21.4 Å

**FIGURE 23.5**  The structure of the clay mineral montmorillonite. Insertion of variable amounts of water causes the distance between layers to swell from 9.6 Å to more than 20 Å. When one Al$^{3+}$ ion is replaced by an Mg$^{2+}$ ion, an additional ion such as Na$^+$ is introduced into the water layers to maintain overall charge neutrality.

A different clay derives from the layered mineral talc, $Mg_3(Si_4O_{10})(OH)_2$. If iron(II) and aluminum replace magnesium and silicon in varying proportions, and if water molecules are allowed to take up positions between the layers, the swelling clay vermiculite results. When heated, vermiculite pops like popcorn, as the steam generated by the vaporization of water between the layers puffs the flakes up into a light, fluffy material with air inclusions. Because of its porous structure, vermiculite is used for thermal insulation or as an additive to loosen soils.

## Zeolites

**Zeolites** are a class of three-dimensional aluminosilicates. Like the feldspars, they carry a negative charge on the aluminosilicate framework that is compensated by neighboring alkali-metal or alkaline-earth cations. Zeolites differ from feldspars in having much more open structures consisting of polyhedral cavities connected by tunnels (Fig. 23.6). Many zeolites are found in nature, but they can also be synthesized under conditions controlled to favor cavities of very uniform size and shape. Most zeolites accommodate water molecules in their cavities, where they provide a mobile phase for the migration of the charge-compensating cations. This enables zeolites to serve as ion-exchange materials (in which one kind of positive ion can be readily exchanged for another) and is the key to their ability to soften water. Water "hardness" arises from soluble calcium and magnesium salts such as $Ca(HCO_3)_2$ and $Mg(HCO_3)_2$. Such salts are converted to insoluble carbonates (boiler scale) when

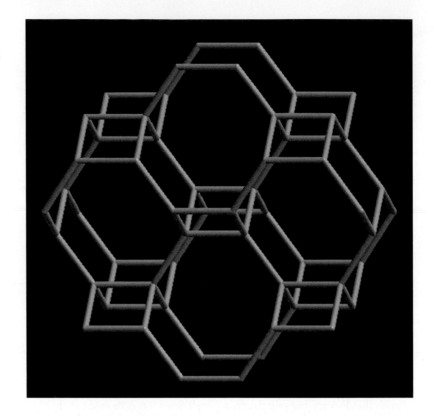

the water is heated and form objectionable precipitates (bathtub ring) with soaps. When hard water is passed through a column packed with a zeolite that has sodium ions in its structure, the calcium and magnesium ions exchange with the sodium ions and are removed from the water phase:

$$2\ NaZ(s) + Ca^{2+}(aq) \rightleftharpoons CaZ_2(s) + 2\ Na^+(aq)$$

When the ion-exchange capacity of the zeolite is exhausted, this reaction can be reversed by passing a concentrated solution of sodium chloride through the zeolite to regenerate it in the sodium form.

A second use of zeolites derives from the ease with which they adsorb small molecules. Their sponge-like affinity for water makes them useful as drying agents; they are put between the panes of double-pane glass windows to prevent moisture from condensing on the inner surfaces. The pore size of zeolites can be selected to allow molecules that are smaller than a certain size to pass through but hold back larger molecules. Such zeolites serve as "molecular sieves"; they have been used to capture nitrogen molecules in a gas stream while permitting oxygen molecules to pass through.

Perhaps the most exciting use of zeolites is as catalysts. Molecules of varying sizes and shapes have different rates of diffusion through a zeolite; this feature enables chemists to enhance the rates and yields of desired reactions and suppress unwanted reactions. The most extensive applications of zeolites at present are in the catalytic cracking of crude oil, a process that involves breaking down long-chain hydrocarbons and re-forming them into branched-chain molecules of lower molecular mass for use in high-octane unleaded gasoline. A relatively new process employs "shape-selective" zeolite catalysts to convert methanol ($CH_3OH$) to high-quality gasoline.

Plants have been built to make gasoline by this process, using methanol derived from coal or from natural gas.

---

## 23.2
### GEOCHEMISTRY

**Geochemistry** is the study of the chemistry of the earth. It attempts to explain the distribution of elements and compounds throughout the earth and the migrations and chemical transformations they undergo. This section presents a brief introduction to this fascinating subject, based on the types of rocks found in the crust: igneous, sedimentary, and metamorphic. These rock types differ in both origin and chemical composition, and their study requires the application of almost every principle discussed in the first 19 chapters of this book, from acid–base equilibrium to thermodynamics to the structures of solids.

### Igneous Rocks

**Igneous rocks** make up more than 95% of the earth's crust and contain a wide variety of minerals with the silicate structures described in Table 23.2. Such rocks form during the crystallization of a **magma,** a hot, molten silicate fluid. Although there is some variation in the compositions of magmas, the primary elements present are oxygen, silicon, and aluminum, with smaller amounts of iron, magnesium, calcium, sodium, and potassium. High-temperature melts of the resultant oxides have been studied in the laboratory, and the observed order of crystallization of minerals from these melts can be correlated with properties of igneous rocks.

The crystallization of a magma takes place in two different reaction series, as Figure 23.7 illustrates. In the continuous series, feldspars of compositions varying continuously from $CaAl_2Si_2O_8$ (anorthite) to $NaAlSi_3O_8$ (albite) crystallize successively.

**FIGURE 23.7** The reaction series of a magma. As the magma cools, minerals form along the two reaction series from top to bottom.

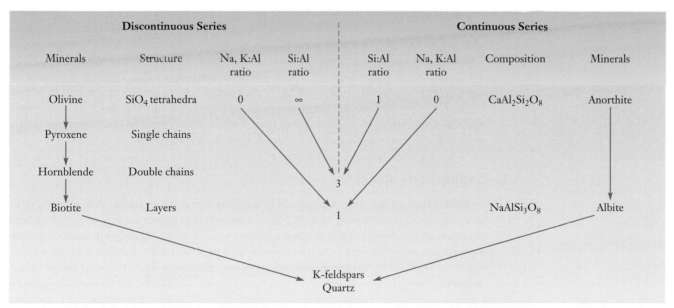

These two minerals form a nearly ideal solid solution, and the graph of temperature versus composition for the liquid–solid equilibrium is closely analogous to the graph for the vaporization of an ideal liquid mixture (see Fig. 6.15). The second reaction series is a discontinuous one that passes from silicates with low connectivities (olivines) through chain and layer structures to network crystals. This series is discontinuous because only a finite number of discrete structure types are formed, in contrast with the continuous composition change in the feldspar series.

As a magma cools, the first solids that crystallize are olivines from the discontinuous series and calcium-rich feldspars from the continuous series. At this stage, two possibilities exist. If these solids stay in contact with the remaining melt, they can react with it as the temperature falls, giving pyroxene and feldspars somewhat richer in sodium, the next minerals in the reaction series. On the other hand, if the solids are separated from the melt (a process called **fractionation**), the remaining liquid is deficient in silicon relative to aluminum and deficient in calcium relative to sodium; it goes on to crystallize into later members of the two series, such as biotite and albite. The greater the degree of fractionation, the farther the reaction series proceeds, leading eventually to a melt of silica and potassium feldspars, which crystallize at the lowest temperature. In a quantitative treatment of magma crystallization, both the effect of pressure on the equilibria and the catalytic effect of small amounts of water vapor must be taken into account. The latter effect is particularly dramatic, with a relatively small mass fraction of water depressing the freezing point of $SiO_2$ by 600°C and accelerating reaction rates by many orders of magnitude.

Besides the major elements in magmas discussed so far, other elements often enter the silicate structures in place of the more common ions. If a minor element has approximately the same charge and ionic radius as a major element, it is said to be "camouflaged" in the crystal containing the major element, and the ratio of abundances of the two elements is roughly constant throughout a whole rock series. Gallium ($Ga^{3+}$) is camouflaged in aluminum minerals, and cobalt ($Co^{2+}$) is camouflaged in magnesium compounds. A minor element ion with similar radius but higher charge than a major element ion, or of the same charge but a smaller radius, is said to be "captured" and generally appears in the early stages of crystallization of the corresponding major element, because it is held to the corresponding lattice site more tightly by electrostatic forces. Barium can replace potassium in potassium feldspars because of its similar ionic radius and higher charge. Ions whose sizes and charges do not permit them to substitute for major ions remain in the residual melt, a silica-rich liquid containing water and other volatile compounds such as $CO_2$, $H_2S$, HCl, and $N_2$. The pressure of the volatiles forces the residual liquid into cracks in the surrounding rock. In the cracks, it eventually crystallizes as **pegmatites,** which are rich in the less abundant elements, such as beryllium, molybdenum, tin, uranium, and the lanthanides. A knowledge of how ions fit into crystal lattices therefore helps in the search for elements that are economically important.

## Sedimentary Rocks

**Sedimentary rocks** are the products of weathering of other rocks. Water (especially moving water in streams) breaks rocks into smaller pieces and partially dissolves them. Carbon dioxide (which forms carbonic acid when dissolved) causes further dissolution and chemical reaction. Subsequent precipitation of solids leads to a variety of minerals with elemental abundances very different from those of the original material.

Resistance to weathering increases along the reaction series of Figure 23.7. Olivines and calcium feldspars erode and dissolve quite readily, but quartz (a network solid) is very resistant and is therefore left behind in the weathering process, often in the form of sandstone that contains as much as 99% $SiO_2$. Layered aluminosilicates such as biotite (a mica) absorb water readily to become clays. The compositions of clays depend strongly on the pH of the waters from which they form. Because aluminum ions are amphoteric, $Al_2O_3$ is more soluble in mildly acidic solutions (pH < 4) than is $SiO_2$, and so clays with high aluminum content tend to form at lower pH. Shales, which are rocks derived from layers of clay, are the most abundant sedimentary rocks. In some cases they contain extensive deposits of crude oil; as a result, they are of great interest to petroleum geologists.

Elements that do not remain in the silica residue and that are not incorporated into clays remain in aqueous solution. Certain ions are subject to oxidation or reduction, and the transformations they undergo can depend strongly on pH. These relationships are most clearly seen in a reduction potential–pH diagram, such as Figure 23.8 for the oxidation states of iron. The standard reduction potential for the half-reaction

$$Fe^{3+} + e^- \longrightarrow Fe^{2+}$$

is 0.770 V; it is shown as the corresponding intercept on the left side of the figure. Because $H_3O^+$ and $OH^-$ do not appear explicitly in this half-reaction, the half-cell potential does not change with a moderate increase of pH. As the pH increases

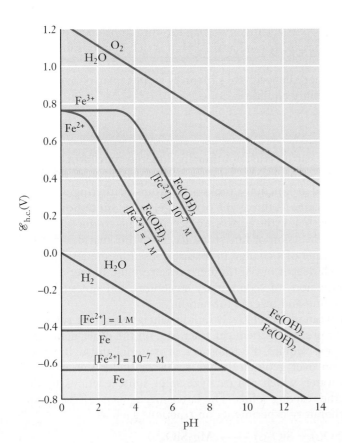

**FIGURE 23.8**  A potential-pH diagram for the ions and hydroxides of iron. Each line illustrates the dependence of half-cell voltage on pH for the redox couples indicated.

**FIGURE 23.9** Folded beds of hematite, $Fe_2O_3$. *(Copyright Glenn Oliver/Visuals Unlimited)*

further, however, hydroxides begin to form, and the half-cell reaction in basic solution is

$$Fe(OH)_3(s) + e^- \longrightarrow Fe(OH)_2(s) + OH^-(aq)$$

According to the Nernst equation, its reduction potential is

$$\mathscr{E} = \mathscr{E}°(Fe(OH)_3|Fe(OH)_2) - 0.0592 \log_{10} [OH^-]$$
$$= -0.56 \text{ V} + (0.0592 \text{ V})(14 - pH)$$

which changes linearly with pH at high pH. The region of intermediate pH is somewhat more complicated, because the complex ions $FeOH^{2+}$, $FeOH^+$, and $Fe(OH)_2^+$ form. A second curve in Figure 23.8 shows the pH dependence of the $Fe^{2+}|Fe$ half-cell equilibrium. In a strongly oxidizing environment, iron(III) is the stable oxidation state (top of diagram), but in a strongly reducing environment iron(0) would be stable (bottom of diagram). Between the two is the stability region of iron(II). Potential-pH diagrams help in analyzing what minerals will form under given conditions. Thus, hematite ($Fe_2O_3$) (Fig. 23.9) contains iron(III) and tends to form at high pH and under oxidizing conditions, whereas siderite ($FeCO_3$) contains iron(II) and forms at low pH and under reducing conditions.

Ions that do *not* react to form solids remain in solution until enough water evaporates that the volume decreases below the limit of their solubility. Ocean water is saturated with respect to calcium carbonate so that, when evaporation begins, $CaCO_3$ is the first solid to precipitate, usually in the form of limestone. Once the solution has been reduced to about one fifth of its original volume, calcium sulfate precipitates as gypsum ($CaSO_4·2H_2O$). After further evaporation of seawater (a reduction to about one-tenth the original volume), halite (NaCl) precipitates in great salt "domes," which are mined as the starting material for the chlor-alkali industry (Chapter 22). The brine that remains behind at this stage is rich in potassium, magnesium, and sulfate ions.

The formation of sedimentary deposits thus involves an interplay of oxidation–reduction, acid–base, and solubility equilibria.

## Metamorphic Rocks

The minerals that make up the earth's crust continually undergo transformations in structure and composition. As we have seen, igneous rocks form through the crystallization of fluid magmas, and sedimentary rocks arise from weathering and subsequent hydration and ion exchange. A third type of change, called **metamorphism,** is induced by heat and pressure below the earth's surface and leads to the formation of metamorphic rocks. The conversion of limestone to crystalline marble with the same composition is one example of a change in texture caused by metamorphism. Metamorphism involves much deeper burial (higher pressures) and higher temperatures than typical interconversions of sedimentary structures, but not such high temperatures as to cause melting and magma formation.

Purely structural changes can occur relatively rapidly when crystals are subjected to changes in temperature or pressure. Chemical reactions, on the other hand, lead to changes in composition and require the transport of material through a rock, either by solid-state diffusion over geological time scales or under the action of volatile substances such as water, which can catalyze solid–solid reactions by dissolving and reprecipitating the reacting solids. The latter mechanism can be important, as demonstrated by laboratory experiments on reactions such as

$$2 \text{ MgO}(s) + SiO_2(s) \longrightarrow Mg_2SiO_4(s)$$

This reaction occurs slowly under dry conditions, even at 1300°C, but much more rapidly in the presence of water vapor at only 600°C. The kinetics and mechanisms of such fluid-catalyzed reactions deserve further study because they can be a major contributor to metamorphism.

Because metamorphic rocks form below the earth's surface, pressure has an important effect, which can be analyzed in qualitative terms through Le Châtelier's principle: an increase in pressure at constant temperature displaces an equilibrium toward substances that have smaller volumes and greater densities. The silicates that are most characteristic of metamorphic rocks have chain and layer structures and are denser than those with the more open network structures, such as quartz and the feldspars. The denser minerals are stabilized at higher pressures.

A simple example of the effect of pressure on solid-solid equilibrium is a substance such as $CaCO_3$ that exists in two crystalline forms (calcite and aragonite) with different molar volumes. In a pressure–temperature phase diagram (see Section 5.6) the line of phase coexistence has a slope given by the **Clapeyron equation:**

$$\left(\frac{dp}{dT}\right)_{coex} = \frac{\Delta S}{\Delta V} = \frac{\Delta H}{T\,\Delta V} \qquad \text{[23.1]}$$

where $\Delta S$ is the entropy change for the transition between the two forms and $\Delta V$ is the change in volume (see problem 34). Because $\Delta G = 0$ along the coexistence curve, it must also be true that at each point $\Delta H = T\Delta S$, which gives the second equality in Equation 23.1. The denser (lower volume) form of the solid (which in the case of calcium carbonate is aragonite) lies on the high-pressure side of the coexistence curve.

Now consider a chemical reaction between solid compounds. At atmospheric pressure, the temperature at which $\Delta G = 0$ for such a reaction is $T = \Delta H°/\Delta S°$ (recall problem 37 in Chapter 8). At higher pressures, the temperature that divides the range of stable products from stable reactants shifts according to an equation closely analogous to Equation 23.1:

$$\left(\frac{dp}{dT}\right)_{\Delta G=0} = \frac{\Delta S}{\Delta V}.$$

where $\Delta S$ and $\Delta V$ are now the entropy change and volume change of reaction. If these depend weakly on pressure (as they do for most reactions involving only solids), the result is a straight line for the equilibrium curve between reactants and products ($\Delta G = 0$), as Figure 23.10a shows.

If gases are either reactants or products in a reaction, the effect of pressure on the equilibrium can be much greater. An example is the reaction

$$\underset{\text{Dolomite}}{CaMg(CO_3)_2(s)} + \underset{\text{Quartz}}{2\ SiO_2(s)} \longrightarrow \underset{\text{Diopside}}{CaMgSi_2O_6(s)} + \underset{\text{Carbon dioxide}}{2\ CO_2(g)}$$

Because the volume of the carbon dioxide is much greater than that of the solids taking part, an increase in pressure shifts the reaction to the left, decreasing the amounts of diopside and $CO_2$. The equilibrium curve for this reaction with $\Delta G = 0$ can be found by noting that $\Delta H$ is approximately independent of pressure, and $\Delta S$ depends on pressure primarily through the resulting compression of the $CO_2(g)$. If the carbon dioxide is at a pressure $P$, then

$$\Delta S \approx \Delta S° - (\Delta n_g)R \ln P$$

where $\Delta n_g$ is the change in the number of moles of gas in the chemical equation, and we have used the result from Section 8.5 that the entropy change on isothermal

**FIGURE 23.10** Equilibrium curves in metamorphism, showing the locus of states for which $\Delta G = 0$ for the transformations indicated.

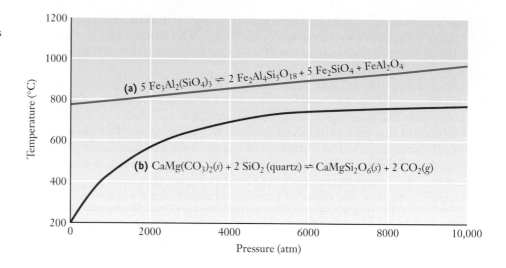

compression of $\Delta n_g$ mol of an ideal gas from 1 atm to $P$ atm is $-(\Delta n_g)R \ln P$. Setting $\Delta G$ to 0 gives

$$T = \frac{\Delta H}{\Delta S} = \frac{\Delta H°}{\Delta S° - (\Delta n_g)R \ln P} \qquad [23.2]$$

with the characteristic nonlinear dependence of $T$ on $P$ shown in Figure 23.10b.

---

## EXAMPLE 23.2

The standard enthalpies of formation of dolomite and diopside are $-2326.3$ and $-3206.2$ kJ mol$^{-1}$, respectively, and their standard entropies are 155.18 and 142.93 J K$^{-1}$ mol$^{-1}$. Use this information, together with data from Appendix D, to calculate the temperature at which $\Delta G = 0$ for the reaction of dolomite with quartz **(a)** at atmospheric pressure and **(b)** at $P = 1000$ atm.

### Solution
**(a)** $\Delta H° = 2\,\Delta H_f°(CO_2) + \Delta H_f°(CaMgSi_2O_6) - \Delta H_f°(CaMg(CO_3)_2) - 2\,\Delta H_f°(SiO_2)$
$\qquad = +155$ kJ

$\Delta S° = 2\,S°(CO_2) + S°(CaMgSi_2O_6) - S°(CaMg(CO_3)_2) - 2\,S°(SiO_2)$
$\qquad = +331$ J K$^{-1}$

$$T = \frac{\Delta H°}{\Delta S°} = \frac{155,000\text{ J}}{331\text{ J K}^{-1}} = 468\text{ K} = 195°C$$

At this temperature, $K = 1$ for the reaction.
**(b)** Using $\Delta n_g = 2$ mol in Equation 23.2 gives

$$T = \frac{155,000\text{ J}}{331\text{ J K}^{-1} - (2\text{ mol})(8.315\text{ J K}^{-1}\text{ mol}^{-1})\ln 1000} = 717\text{ K} = 444°C$$

This result is only approximate because we neglected pressure effects on the enthalpy and entropy of the solids and assumed ideal gas behavior even at very high pressure.

**Related Problems: 9, 10**

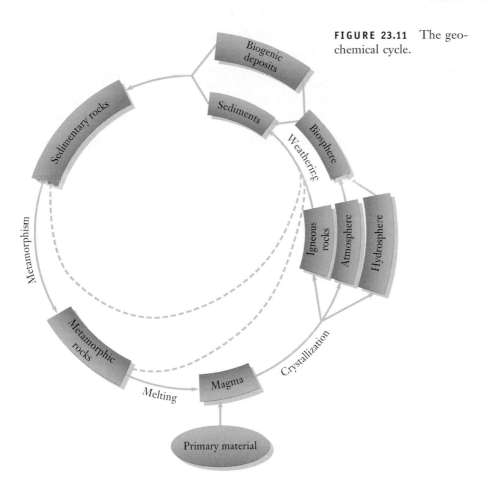

**FIGURE 23.11** The geo-chemical cycle.

As metamorphism proceeds, the elemental composition of the involved rocks becomes more uniform. High-grade metamorphism eventually merges (at high temperatures) with magma formation. In fact, a full geochemical cycle can be described in which a primary magma crystallizes to give igneous rocks, weathering yields sedimentary rocks, and finally metamorphism leads to remelting and magma formation once again (Fig. 23.11). The composition tends to become less uniform as the magma is converted to igneous and then sedimentary rocks and then, at later stages in the cycle, becomes more uniform through metamorphism.

---

## 23.3

# THE PROPERTIES OF CERAMICS

The term **"ceramics"** covers synthetic materials that have as their essential components inorganic, nonmetallic materials. This broad definition includes cement, concrete, and glass in addition to the more traditional fired clay products such as bricks, roof tiles, pottery, and porcelain. The use of ceramics predates recorded history; the emergence of civilization from a primitive state is chronicled in fragments of pottery. No one knows when small vessels were first shaped by human hands from moist clay and left to harden in the heat of the sun. Such containers held nuts, grains, and berries well, but lost their shape and slumped into formless mud when water was poured into them. Then someone discovered (perhaps by accident) that if clay was placed in the glowing embers of a fire, it became as hard as rock and withstood water well. Molded figures (found in what is now the Czech Republic) that were made 24,000 years ago are the earliest fired ceramic objects discovered so far, and fired clay vessels from the Near East date from 8000 B.C. With the action of fire on clay, the art and science of ceramics began.

Ceramics offer stiffness, hardness, resistance to wear, and resistance to corrosion (particularly by oxygen and water), even at high temperature. They are less dense than most metals, which makes them desirable metal substitutes when weight is a factor. Most are good electrical insulators at ordinary temperatures, a property that is exploited in electronics and power transmission. Ceramics retain their strength well at high temperatures. Several important structural metals soften or melt at temperatures a thousand degrees below the melting points of their chemical compounds in ceramics. Aluminum, for example, melts at 660°C, whereas aluminum oxide ($Al_2O_3$), an important compound in many ceramics, does not melt until a temperature of 2051°C is reached.

Against these advantages must be listed some serious disadvantages. Ceramics are generally brittle and low in tensile strength. They tend to have high thermal expansion but low thermal conductivity, making them subject to **thermal shock,** in which sudden local temperature change causes cracking or shattering. Metals and plastics dent or deform under stress, but ceramics cannot absorb stress in this way: they break instead. A major drawback of ceramics as structural materials is their tendency to fail unpredictably and catastrophically in use. Moreover, some ceramics lose mechanical strength as they age, an insidious and serious problem.

## Composition and Structure of Ceramics

Ceramics employ a wide variety of chemical compounds, and useful ceramic bodies are nearly always mixtures of several compounds. **Silicate ceramics,** which include the commonplace pots, dishes, and bricks, are made from aluminosilicate clay minerals. All contain the tetrahedral $SiO_4$ grouping discussed in Section 23.1. In **oxide ceramics,** silicon is a minor or nonexistent component. Instead, a number of metals combine with oxygen to give compounds such as alumina ($Al_2O_3$), magnesia (MgO), or yttria ($Y_2O_3$). **Nonoxide ceramics** contain compounds that are free of oxygen as principal components. Some important compounds in nonoxide ceramics are silicon nitride ($Si_3N_4$), silicon carbide (SiC), and boron carbide (approximate composition $B_4C$).

One important property of ceramics is their porosity. Porous ceramics have small openings into which fluids (typically air or water) can infiltrate. Fully dense ceramics have no channels of this sort. Two ceramic pieces can have the same chemical composition but quite different densities if the first is porous and the second is not.

A **ceramic phase** is any portion of the whole body that is physically homogeneous and bounded by a surface that separates it from other parts. Distinct phases are visible at a glance in coarse-grained ceramic pieces; in a fine-grained piece, phases can be seen with a microscope. When examined on a still finer scale, most ceramics, like metals, are microcrystalline, consisting of small crystalline grains cemented together (Fig. 23.12). The **microstructure** of such objects includes the sizes and shapes of the grains, the sizes and distribution of voids (openings between grains) and cracks, the identity and distribution of impurity grains, and the presence of stresses within the structure. Microstructural variations have enormous importance

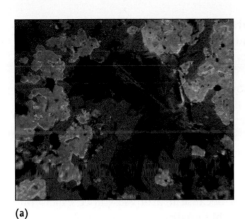

**(a)**                                    **(b)**

**FIGURE 23.12** Microstructures of aluminosilicate ceramics, viewed by the different colors of light emitted after bombardment by electrons. (a) Forsterite (red), spinel (green), and periclase (dark brown) grains. (b) Periclase (blue) and oldhamite (yellow) grains. *(Musa Karakus and Richard Hagni)*

in ceramics because slight changes at this level strongly influence the properties of individual ceramic pieces. This is less true for plastic and metallic objects.

The microstructure of a ceramic body depends markedly on the details of its fabrication. The techniques of forming and firing a ceramic piece are as important as its chemical composition in determining ultimate behavior because they confer a unique microstructure. This fact calls attention to the biggest problem with ceramics as structural materials: inconsistent quality. Ceramic engineers can produce parts that are stronger than steel, but not reliably so because of the difficulties of monitoring and controlling microstructure. Gas turbine engines fabricated of silicon nitride (Fig. 23.13), for example, run well at 1370°C, which is hot enough to soften or melt most metals. The higher operating temperature increases engine efficiency, and the ceramic turbines weigh less, which further boosts fuel economy. Despite these advantages, there is no commercial ceramic gas turbine. Acceptable ceramic turbines have to be built from selected, pretested components. The testing costs and rejection rates are so high that economical mass production has been impossible so far.

## Making Ceramics

The manufacture of most ceramics involves four steps: (1) the preparation of the raw material; (2) the forming of the desired shape, often achieved by mixing a powder with water or other binder and molding the resulting plastic mass; (3) the drying and firing of the piece, also called its **densification,** because pores (voids) in the dried ceramic fill in; and (4) the finishing of the piece by sawing, grooving, grinding, or polishing.

The raw materials for traditional ceramics are natural clays that come from the earth as powders or thick pastes and become plastic enough after adjustment of their water content to be formed freehand or on a potter's wheel. Special ceramics (both oxide and nonoxide) require chemically pure raw materials that are produced synthetically. Close control of the purity of the starting materials for these ceramics is essential to producing finished pieces with the desired properties. In addition to being formed by hand or in open molds, ceramic pieces are shaped by the squeezing (compacting) of the dry or semidry powders in a strong, closed mold of the desired shape, at either ordinary or elevated temperatures (hot pressing).

Firing a ceramic causes **sintering** to occur. In sintering, the fine particles of the ceramic start to merge together by diffusion at high temperatures. The density of

**FIGURE 23.13** A part freshly fabricated from silicon nitride for use in a gas turbine engine. *(GTE Labs)*

the material increases as the voids between grains are partially filled. Sintering occurs below the melting point of the material and shrinks the ceramic body. In addition to the merging of the grains of the ceramic, firing causes partial melting, chemical reactions among different phases, reactions with gases in the atmosphere of the firing chamber, and recrystallization of compounds with an accompanying growth in crystal size. All of these changes influence the microstructure of a piece and must be understood and controlled. Firing accelerates physical and chemical changes, of course, but thermodynamic equilibrium in a fired ceramic piece is rarely reached. Kinetic factors—including the rate of heating, the length of time at which each temperature is held, and the rate of cooling—influence microstructure. As a result, the use of microwave radiation (as in microwave ovens) rather than kilns to fire ceramics is under development in ceramic factories because it promises more exact control of the heating rate and thereby more reliable quality.

## 23.4

## SILICATE CERAMICS

The silicate ceramics include materials that vary widely in composition, structure, and uses. They range from simple earthenware bricks and pottery to cement, fine porcelain, and glass. Their structural strength is based on the same linking of silicate ion tetrahedra that gives structure to silicate minerals in nature.

### Pottery and Structural Clay Products

In Section 23.2 it was noted that aluminosilicate clays are the products of the weathering of primary minerals. When water is added to such clays in moderate amount, a thick paste results that is easily molded into different shapes. Clays expand as water invades the space between adjacent aluminosilicate sheets of the mineral, but they release most of this water to a dry atmosphere and shrink. A small fraction of the water or hydroxide ions remains rather tightly bound by ion–dipole forces to cations between the aluminosilicate sheets and is lost only when the clay is heated to a high temperature. The firing of aluminosilicate clays simultaneously causes irreversible chemical changes to take place. The clay kaolinite ($Al_2Si_2O_5(OH)_4$) undergoes the following reaction:

$$3 \ Al_2Si_2O_5(OH)_4 \longrightarrow Al_6Si_2O_{13}(s) + 4 \ SiO_2(s) + 6 \ H_2O(g)$$

     (kaolinite)          (mullite)     (silica)     (water)

The fired ceramic body is a mixture of two phases, mullite and silica. Mullite, a rare mineral in nature, takes the form of needle-like crystals that interpenetrate and confer strength on the ceramic. When the temperature is above 1470°C, the silica phase forms as minute grains of cristobalite, one of the several crystalline forms of $SiO_2$.

If chemically pure kaolinite is fired, the finished ceramic object is white. Such purified clay minerals are the raw material for fine china. As they occur in nature, clays contain impurities, such as transition-metal oxides, that affect the color of both the unfired clay and the fired ceramic object if they are not removed. The colors of the metal oxides arise from their absorption of light at visible wavelengths, as explained by crystal field theory (see Section 18.4). Common colors for ceramics are yellow or greenish yellow, brown, and red. Bricks are red when the clay used to make them has high iron content.

Before a clay is fired in a kiln it must first be freed of moisture by slowly heating to about 500°C. If a clay body dried at room temperature were to be placed directly into a hot kiln, it would literally explode from the sudden, uncontrolled expulsion of water. A fired ceramic shrinks somewhat as it cools, causing cracks to form. These imperfections limit the strength of the fired object and are undesirable. Their occurrence can be reduced by coating the surface of a partially fired clay object with a **glaze,** a thin layer that minimizes crack formation in the underlying ceramic by holding it in a state of tension as it cools. Glazes, as their name implies, are glasses that have no sharp melting temperature, but rather harden and develop resistance to shear stresses increasingly as the temperature of the high-fired clay object is gradually reduced. Glazes generally are aluminosilicates that have high aluminum content to raise their viscosity and thereby reduce the tendency to run off the surface during firing. They also provide the means of coloring the surfaces of fired clays and imparting decorative designs to them. Transition-metal oxides (particularly those of Ti, V, Cr, Mn, Fe, Co, Ni, and Cu) are responsible for the colors. The oxidation state of the transition metal in the glaze is critical in determining the color produced and is controlled by regulating the composition of the atmosphere in the kiln; atmospheres rich in oxygen give high oxidation states, and those poor or lacking in oxygen give low oxidation states.

## Glass

Glassmaking probably originated in the Near East about 3500 years ago. It is one of the oldest domestic arts but its beginnings, like those of metallurgy, are obscure. Both required high-temperature charcoal fueled ovens and vessels made of materials that did not easily melt, in order to initiate and contain the necessary chemical reactions. In the early period of glassmaking, desired shapes were fabricated by sculpting them from solid chunks of glass. At a later date, molten glass was poured in successive layers over a core of sand. A great advance was the invention of glassblowing, which probably occurred in the first century B.C. A long iron tube was dipped into molten glass and a rough ball of viscous material was caused to accumulate on its end by rotating the tube. Blowing into the iron tube forced the soft glass to take the form of a hollow ball (Fig. 23.14) that could be further shaped into

**FIGURE 23.14** Handcrafted glassware is trimmed after being blown into the desired shape. *(Hank Morgan/Photo Researchers, Inc.)*

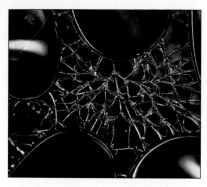

**FIGURE 23.15** The strains associated with internal stresses in rapidly cooled glass can be made visible by viewing the glass in polarized light. They appear as colored regions. *(David Guizak/Phototake)*

a vessel and severed from the blowing tube with a blade. Artisans in the Roman Empire developed glassblowing to a high degree, but with the decline of that civilization the skill of glassmaking in Europe deteriorated until the Venetians redeveloped the lost techniques a thousand years later.

Glasses are amorphous solids of widely varying composition whose physical properties were discussed in Section 19.5. In this chapter, the term "glass" is used in the restricted and familiar sense to refer to materials formed from silica, usually in combination with metal oxides. The absence of long-range order in glasses has the consequence that they are **isotropic**—that is, their physical properties are the same in all directions. This has advantages in technology, including the fact that glasses expand uniformly in all directions with an increase in temperature. The mechanical strength of glass is intrinsically very high, exceeding the tensile strength of steel, provided the surface is free of scratches and other imperfections. Flaws in the surface provide sites where fractures can start when the glass is stressed. When a glass object of intricate shape and nonuniform thickness is suddenly cooled, internal stresses are locked in that may be relieved catastrophically when the object is heated or struck (Fig. 23.15). Slowly heating a strained object to a temperature somewhat below its softening point and holding it there for a while before allowing it to cool slowly is called **annealing;** it gives short-range diffusion of atoms a chance to occur and to eliminate internal stresses.

The softening and annealing temperatures of a glass and other properties, such as density, depend upon its chemical composition (Table 23.3). Silica itself ($SiO_2$)

## TABLE 23.3

### *Composition and Properties of Various Glasses*

| | Silica Glass | Soda–Lime Glass | Borosilicate Glass | Aluminosilicate Glass | Leaded Glass |
|---|---|---|---|---|---|
| **Composition** | $SiO_2$, 99.9% | $SiO_2$, 73% | $SiO_2$, 81% | $SiO_2$, 63% | $SiO_2$, 56% |
| | $H_2O$, 0.1% | $Na_2O$, 17% | $B_2O_3$, 13% | $Al_2O_3$, 17% | $PbO$, 29% |
| | | $CaO$, 5% | $Na_2O$, 4% | $CaO$, 8% | $K_2O$, 9% |
| | | $MgO$, 4% | $Al_2O_3$, 2% | $MgO$, 7% | $Na_2O$, 4% |
| | | $Al_2O_3$, 1% | | $B_2O_3$, 5% | $Al_2O_3$, 2% |
| **Coefficient of Linear Thermal Expansion ($°C^{-1} \times 10^7$)[†]** | | | | | |
| | 5.5 | 93 | 33 | 42 | 89 |
| **Softening Point (°C)** | | | | | |
| | 1580 | 695 | 820 | 915 | 630 |
| **Annealing Point (°C)** | | | | | |
| | 1050 | 510 | 565 | 715 | 435 |
| **Density ($g\ cm^{-3}$)** | | | | | |
| | 2.20 | 2.47 | 2.23 | 2.52 | 3.05 |
| **Refractive Index[‡] at $\lambda = 589$ nm** | | | | | |
| | 1.459 | 1.512 | 1.474 | 1.530 | 1.560 |

[†] The coefficient of linear thermal expansion is defined as the fractional increase in length of a body when its temperature is increased by 1°C.

[‡] The refractive index is a vital property of glass for optical applications. It is defined by $n = \sin \theta_i / \sin \theta_r$ where $\theta_i$ is the angle of incidence of a ray of light on the surface of the glass and $\theta_r$ is the angle of refraction of the ray of light in the glass.

forms a glass if it is heated above its melting point and then cooled rapidly to avoid crystallization. The resulting vitreous (glassy) silica has limited use because the high temperatures required to shape it make it quite expensive. Sodium silicate glasses are formed in the high-temperature reaction of silica sand with anhydrous sodium carbonate (soda ash, $Na_2CO_3$):

$$Na_2CO_3(s) + nSiO_2(s) \longrightarrow Na_2O \cdot (SiO_2)_n(s) + CO_2(g)$$

The melting point of the nonvolatile product is about 900°C, and the glassy state results if cooling through that temperature is rapid. The product, called "water glass," is water-soluble and thus unsuitable for making vessels. Its aqueous solutions, however, are used in some detergents and as adhesives for sealing cardboard boxes.

An insoluble glass with useful structural properties results if lime (CaO) is added to the sodium carbonate–silica starting materials. **Soda-lime glass** is the resulting product, with the approximate composition $Na_2O \cdot CaO \cdot (SiO_2)_6$. Soda–lime glass is easy to melt and shape and is used in applications ranging from bottles to window glass. It accounts for over 90% of all the glass manufactured today. The structure of this ionic glass is shown schematically in Figure 23.16. It is a three-dimensional network of the type discussed in Section 23.1, but with random coordination of the silicate tetrahedra. The network is a giant "polyanion," with $Na^+$ and $Ca^{2+}$ ions distributed in the void spaces to compensate for the negative charge on the network.

Replacing lime and some of the silica in a glass by other oxides ($Al_2O_3$, $B_2O_3$, $K_2O$, or PbO) modifies its properties noticeably. For example, the thermal conductivity of ordinary (soda-lime) glass is quite low, and its coefficient of thermal expansion is high. This means that internal stresses are set up when its surface is subjected locally to extreme heat or cold; it may shatter. The coefficient of thermal expansion is appreciably lower in certain borosilicate glasses, in which many of the silicon sites are occupied by boron. Pyrex, the most familiar of these glasses, has a coefficient of linear expansion about one third that of ordinary soda-lime glass and is the material of choice for laboratory glassware and household ovenware. Vycor has an even smaller coefficient of expansion (approximating that of fused silica) and is made by chemical treatment of a borosilicate glass to leach out its sodium. This leaves a porous structure that is densified by raising the temperature and shrinking the glass to its final volume.

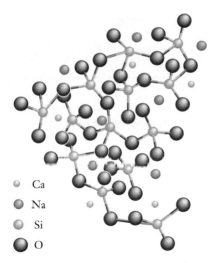

○ Ca

● Na

○ Si

⬤ O

**FIGURE 23.16** The structure of a soda–lime glass. Note the tetrahedral coordination of oxygen atoms around each silicon atom.

## Cements

Hydraulic cement was first developed by the Romans, who found that a mixture of lime (CaO) and dry volcanic ash reacts slowly with water, even at low temperatures, to form a durable solid. They used this knowledge to build the Pantheon in Rome, a circular building whose concrete dome, spanning 143 ft without internal support, still stands nearly 2000 years after its construction! The knowledge of cement making was lost for centuries after the fall of the Roman Empire. It was rediscovered in 1824 by an English bricklayer, Joseph Aspdin, who patented a process for calcining a mixture of limestone and clay. He called the product **portland cement** because, when mixed with water, it hardened to a material that resembled a kind of limestone found on the Isle of Portland. Portland cement is now manufactured in every major country on earth, and annual worldwide production is currently about 800,000,000 metric tons, exceeding the production of all other materials. Portland cement opened up a new age in the methods of constructing highways and buildings: rock could be crushed and then molded in cement, rather than shaped with cutting tools.

**TABLE 23.4**

## Composition of Portland Cement

| Oxide | Percentage by Mass |
|---|---|
| Lime (CaO) | 61–69 |
| Silica ($SiO_2$) | 18–24 |
| Alumina ($Al_2O_3$) | 4–8 |
| Iron(III) oxide ($Fe_2O_3$) | 1–8 |
| Minor oxides (MgO, Na$_2$O, K$_2$O, SO$_3$) | 2–4 |

Portland cement is a finely ground powdered mixture of compounds produced by the high-temperature reaction of lime, silica, alumina, and iron oxide. The lime (CaO) may come from limestone or chalk deposits, and the silica ($SiO_2$) and alumina ($Al_2O_3$) are often obtained in clays or slags. The blast furnaces of steel mills are a common source of slag, which is a byproduct of the smelting of iron ore (Section 20.4). The composition of slag varies, but it can be represented as a calcium aluminum silicate of approximate formula $CaO \cdot Al_2O_3 \cdot (SiO_2)_2$. Molten slag solidifies into "blast furnace clinkers" upon quenching in water. This material is crushed and ground to a fine powder, blended with lime in the correct proportion, and burned again in a horizontal rotary kiln at temperatures up to 1500°C to produce "cement clinker." A final stage of grinding and the addition of about 5% gypsum ($CaSO_4 \cdot 2H_2O$) to lengthen the setting time completes the process of manufacture. The composition of a typical portland cement is shown in Table 23.4. The table gives percentages of the separate oxides; these simple materials, which are the "elements" of cement making, combine in the cement in more complex compounds such as tricalcium silicate, $(CaO)_3 \cdot SiO_2$, and tricalcium aluminate, $(CaO)_3 \cdot Al_2O_3$.

Cement *sets* when the semiliquid slurry first formed by the addition of water to the powder becomes a solid of low strength. Subsequently, it gains strength in a slower *hardening* process. Setting and hardening involve a complex group of exothermic reactions in which several hydrated compounds form. Portland cement is a **hydraulic cement,** because it hardens not by loss of admixed water but by chemical reactions that incorporate water into the final body. The main reaction during setting is the hydration of the tricalcium aluminate, which can be approximated by the equation

$$(CaO)_3 \cdot Al_2O_3(s) + 3(CaSO_4 \cdot 2H_2O)(s) + 26\ H_2O(\ell) \longrightarrow (CaO)_3 \cdot Al_2O_3 \cdot (CaSO_4)_3 \cdot 32H_2O(s)$$

The product forms after 5 or 6 hours as a microscopic forest of long crystalline needles that lock together to solidify the cement. Later, the calcium silicates react with water to harden the cement. An example is

$$6(CaO)_3 \cdot SiO_2(s) + 18\ H_2O(\ell) \longrightarrow (CaO)_5 \cdot (SiO_2)_6 \cdot 5H_2O(s) + 13\ Ca(OH)_2(s)$$

The hydrated calcium silicates develop as strong tendrils that coat and enclose unreacted grains of cement, each other, and other particles that may be present, binding them in a robust network. Most of the strength of cement comes from these entangled networks, which in turn depend ultimately for strength on chains of O—Si—O—Si silicate bonds. Hardening is slower than setting; it may take as long as a year for the final strength of a cement to be attained.

Portland cement is rarely used alone. Generally, it is combined with sand, water, and lime to make **mortar,** which is applied with a trowel to bond bricks or stone together in an assembled structure. When portland cement is mixed with sand and aggregate (crushed stone or pebbles) in the proportions of $1 : 3.75 : 5$ by volume, the mixture is called **concrete.** Concrete is outstanding in its resistance to compressive forces, and therefore is the primary material in use for the foundations of buildings and the construction of dams, in which the compressive loads are enormous. The stiffness (resistance to bending) of concrete is high, but its fracture toughness (resistance to impact) is substantially lower, and its tensile strength is relatively poor. For this reason, concrete is usually reinforced with steel rods when it is used in structural elements such as beams that are subject to transverse or tensile stresses.

As excess water evaporates from cement during hardening, pores form that typically comprise 25 to 30% of the volume of the solid. This porosity weakens concrete, and recent research has shown that the fracture strength is related inversely to the size of the largest pores in the cement. A new material, called "macro defect free" (MDF) cement, has been developed in which the size of the pores is reduced from about a millimeter to a few micrometers by the addition of water-soluble polymers that make a dough-like "liquid" cement that is moldable with the use of far less water. Unset MDF cement is mechanically kneaded and extruded into the desired shape. The final result possesses substantially increased bending resistance and fracture toughness. MDF cement can even be molded into springs and shaped on a conventional lathe. When it is reinforced with organic fibers, its toughness is further increased. The development of MDF cement is a good illustration of the way in which chemistry and engineering collaborate to furnish new materials.

---

### 23.5

## NONSILICATE CERAMICS

Many useful ceramics exist that are *not* based on the Si—O bond and the $SiO_4$ tetrahedron. They have important uses in electronics, optics, and the chemical industry. Some of these materials are oxides, but others contain neither silicon nor oxygen.

### Oxide Ceramics

Oxide ceramics are materials that contain oxygen in combination with any of a number of metals. These materials are named by adding an *-ia* ending to the stem of the name of the metallic element. Thus, if the main chemical component of an oxide ceramic is $Be_2O_3$, it is a *beryllia* ceramic; if the main component is $Y_2O_3$, it is an *yttria* ceramic; and if it is MgO, it is a *magnesia* ceramic. As Table 23.5 shows, the melting points of these and other oxides are substantially higher than the melting points of the elements themselves. Such high temperatures are hard to achieve and maintain, and the molten oxides corrode most container materials. Oxide ceramic bodies are therefore not shaped by melting the appropriate oxide and pouring it into a mold. Instead, these ceramics are fabricated by sintering, like the silicate ceramics.

    **Alumina** ($Al_2O_3$) is the most important nonsilicate ceramic material. It melts at a temperature of 2051°C and retains strength even at temperatures of 1500 to 1700°C. Alumina has a large electrical resistivity and withstands both thermal shock and corrosion well. These properties make it a good material for spark plug insulators, and most spark plugs now use a ceramic that is 94% alumina.

---

**TABLE 23.5**

### *Melting Points of Some Metals and Their Oxides*

| Metal | Melting Point (°C) | Oxide | Melting Point (°C) |
|-------|--------------------|-------|--------------------|
| Be | 1287 | BeO | 2570 |
| Mg | 651 | MgO | 2800 |
| Al | 660 | $Al_2O_3$ | 2051 |
| Si | 1410 | $SiO_2$ | 1723 |
| Ca | 865 | CaO | 2572 |
| Y | 1852 | $Y_2O_3$ | 2690 |

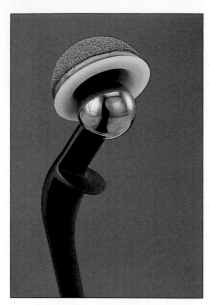

**FIGURE 23.17**   The socket in this artificial hip is made of high-density alumina. *(Copyright SIU/Photo Researchers, Inc.)*

High-density alumina is fabricated in such a way that open pores between the grains are nearly completely eliminated; the grains are small, having an average diameter as low as 1.5 $\mu$m. Unlike most others, this ceramic has good mechanical strength against impact, which has led to its use in armor plating. The ceramic absorbs the energy of an impacting projectile by breaking, so penetration does not occur. High-density alumina is also used in high-speed cutting tools for machining metals. The temperature resistance of the ceramic allows much faster cutting speeds, and a ceramic cutting edge has no tendency to weld to the metallic workpiece, as metallic tools do. These properties make alumina cutting tools superior to metallic tools, as long as they do not break too easily. High-density alumina is also used in artificial joints (Fig. 23.17).

If $Al_2O_3$ doped with a small percentage of MgO is fired in a vacuum or a hydrogen atmosphere (instead of air) at a temperature of 1800 to 1900°C, even very small pores, which scatter light and make the material white, are removed. The resulting ceramic is translucent. This material is used to contain the sodium in high-intensity sodium discharge lamps. With envelopes of high-density alumina, these lamps can be operated at temperatures of 1500°C to give a whiter and more intense light. Old-style sodium-vapor lamps with glass envelopes were limited to a temperature of 600°C because the sodium vapor reacted with the glass at higher temperatures. At low temperature the light from a sodium-vapor lamp has an undesirable yellow color.

**Magnesia** (MgO) is mainly used as a **refractory**—a ceramic material that withstands a temperature of more than 1500°C without melting (MgO melts at 2800°C). A major use of magnesia is as insulation in electrical heating devices, because it combines high thermal conductivity with excellent electrical resistance. Magnesia is prepared from magnesite ores, which consist of $MgCO_3$ and a variety of impurities. When purified magnesite is heated to 800–900°C, carbon dioxide is driven off to form MgO($s$) in fine grains:

$$MgCO_3(s) \longrightarrow MgO(s) + CO_2(g)$$

After cooling, fine-grained MgO reacts avidly with water to form magnesium hydroxide:

$$MgO(s) + H_2O(\ell) \longrightarrow Mg(OH)_2(s)$$

Heating fine-grained MgO($s$) to 1700°C causes the MgO grains to sinter, giving a "dead-burned magnesia" that consists of large crystals and does not react with water. The $\Delta G°$ of the reaction between MgO and $H_2O$ does not change when magnesia is dead burned. The altered microstructure (larger crystals versus small) makes the reaction with water exceedingly slow, however.

## Superconducting Ceramics

The oxide ceramics discussed so far all consisted of single chemical compounds, except for minor additives. A natural idea for new ceramics is to make materials containing two (or more) oxides in equal or nearly equal molar amounts. Thus, if $BaCO_3$ and $TiO_2$ are mixed and heated to high temperature, they react to give the ceramic barium titanate:

$$BaCO_3(s) + TiO_2(s) \longrightarrow BaTiO_3(s) + CO_2(g)$$

Barium titanate, which has many novel properties, is a **mixed oxide ceramic.** It has the same structure as the mineral **perovskite,** $CaTiO_3$ (see Fig. 23.18), except, of course, that Ba replaces Ca. Perovskites typically have two metal atoms for every three oxygen atoms, giving them the general formula $ABO_3$, where A stands for a metal atom at the center of the unit cube and B for an atom of a different metal at the cube corners.

Recently, interest in perovskite ceramic compositions has grown explosively following the discovery that some of them become **superconducting** at relatively high temperatures. A superconducting material offers no resistance whatever to the flow of an electric current. The phenomenon was discovered by the Dutch physicist Heike Kamerlingh-Onnes in 1911 when he cooled mercury below its superconducting transition temperature of 4 K. Such very low temperatures are hard to achieve and maintain, but if practical superconductors could be made to work at higher temperatures (or even at room temperature!), then power transmission, electronics, transportation, medicine, and many other aspects of human life would be transformed. More than 60 years of research with metallic systems culminated in 1973 in the discovery of a niobium–tin alloy with a world-record superconducting transition temperature of 23.3 K. Progress toward higher temperature superconductors then stalled until 1986, when K. Alex Müller and J. Georg Bednorz, who had had the inspired notion to look for higher transition temperatures among perovskite ceramics, found a Ba—La—Cu—O perovskite phase having a transition temperature of 35 K. This result motivated other scientists, who soon discovered another rare-earth-containing perovskite ceramic that became a superconductor at 90 K. This result was particularly exciting because 90 K exceeds the boiling point of liquid nitrogen (77 K), a relatively cheap refrigerant (Fig. 23.19). This 1-2-3 compound (so called because its formula $YBa_2Cu_3O_{(9-x)}$ has one yttrium, two barium, and three copper atoms per formula unit) is not an ideal perovskite because it has fewer than nine oxygen atoms in combination with its six metal atoms. The deficiency makes $x$ in the formula somewhat greater than 2, depending on the exact method of preparation. The structure of this nonstoichiometric solid is given in Figure 23.20. More recently, the maximum superconducting transition temperature has risen to 125 K in another class of ceramics that does not contain rare-earth elements.

A crucial concern in the application of superconducting ceramics is to devise ways to fabricate the new materials in desired shapes such as wires. This will be quite a challenge since these superconductors are ceramics and have the brittleness and fragility typical of ceramic materials.

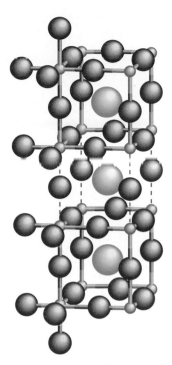

**FIGURE 23.18** The structure of perovskite ($CaTiO_3$). A stack of three unit cells is shown, with some additional O atoms from neighboring cells. Each Ti (blue) is surrounded by six O atoms; each Ca (green) has eight O atoms as nearest neighbors.

**FIGURE 23.19** The levitation of a small magnet above a disk of superconducting material. A superconducting substance cannot be penetrated by an external magnetic field. At room temperature, the magnet rests on the ceramic disk. When the disk is cooled with liquid nitrogen, it becomes superconducting and excludes the magnet's field, forcing the magnet into the air. *(Copyright D.O.E./Science Source/Photo Researchers, Inc.)*

**FIGURE 23.20** The structure of $YBa_2Cu_3O_{9-x}$ is a layered perovskite. The layers differ because one third of them contain yttrium (orange) and two thirds of them contain barium (green). The structure should be compared to that of perovskite in Figure 23.18. Every third Ba in the hypothetical "$BaCuO_3$" is replaced by a Y, and the oxygen atoms in the layer occupied by the Y are removed. Note the puckering in the Cu—O layers, as well.

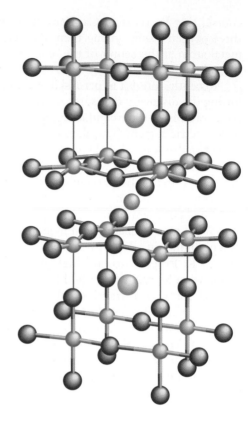

## Nonoxide Ceramics

In nonoxide ceramics, nitrogen or carbon takes the place of oxygen in combination with silicon or boron. Specific substances are boron nitride (BN), boron carbide ($B_4C$), the silicon borides ($SiB_4$ and $SiB_6$), silicon nitride ($Si_3N_4$), and silicon carbide (SiC). All of these compounds possess strong, short covalent bonds. They are hard and strong, but brittle. Table 23.6 lists the enthalpies of the chemical bonds in these compounds.

Much research has aimed at making gas-turbine and other engines from ceramics. Of the oxide ceramics, only alumina and zirconia ($ZrO_2$) are strong enough, but both resist thermal shock too poorly for this application. Attention has therefore turned to the nonoxide **silicon nitride** ($Si_3N_4$). In this network solid (Fig. 23.21), every silicon atom bonds to four nitrogen atoms that surround it at the corners of a tetrahedron; these tetrahedra link into a three-dimensional network by sharing corners. The Si—N bond is covalent and strong (the bond enthalpy is 439 kJ $mol^{-1}$). The similarity to the joining of $SiO_4$ units in silicate minerals (Section 23.1) is clear.

At first, silicon nitride seems chemically unpromising as a high-temperature structural material. It is unstable in contact with water because the reaction

$$Si_3N_4(s) + 6\ H_2O(\ell) \longrightarrow 3\ SiO_2(s) + 4\ NH_3(g)$$

has a negative $\Delta G°$. In fact, this reaction causes finely ground $Si_3N_4$ powder to give off an odor of ammonia in moist air at room temperature. Silicon nitride is also thermodynamically unstable in air, reacting spontaneously with oxygen:

$$Si_3N_4(s) + 3\ O_2(g) \longrightarrow 3\ SiO_3(s) + 2\ N_2(g)$$

### TABLE 23.6

#### Bond Enthalpies in Nonoxide Ceramics

| Bond | Bond Enthalpy (kJ $mol^{-1}$) |
|------|------|
| B—N | 389 |
| B—C | 448 |
| Si—N | 439 |
| Si—C | 435 |
| B—Si | 289 |
| C—C | 350 |

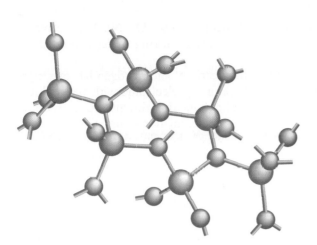

**FIGURE 23.21** The structure of silicon nitride ($Si_3N_4$). Each Si atom is bonded to four nitrogen atoms and each N atom is bonded to three Si atoms. The result is a strong network.

The reaction has a $\Delta G°$ of $-1927$ kJ mol$^{-1}$. In practice, neither reaction occurs at a perceptible rate when $Si_3N_4$ is in bulk form. Initial contact of oxygen or water with $Si_3N_4(s)$ forms a surface film of $SiO_2(s)$ that protects the bulk of the $Si_3N_4(s)$ from further attack. When *strongly* heated in air (to about 1900°C), $Si_3N_4$ does decompose, violently, but until that temperature is reached, it resists attack.

Fully dense silicon nitride parts are stronger than metallic alloys at high temperatures. Ball bearings made of dense silicon nitride work well without lubrication at temperatures up to 700°C, for example, lasting longer than steel ball bearings. Because the strength of silicon nitride increases with the density attained in the production process, the trick is to form a dense piece of silicon nitride in the desired shape. One method for making useful shapes of silicon nitride is "reaction bonding." Powdered silicon is compacted in molds, removed, and then fired under an atmosphere of nitrogen at 1250 to 1450°C. The reaction

$$3\ Si(s) + 2\ N_2(g) \longrightarrow Si_3N_4(s)$$

forms the ceramic. The parts neither swell nor shrink significantly during the chemical conversion from Si to $Si_3N_4$, making it possible to fabricate complex shapes reliably. Unfortunately, reaction-bonded $Si_3N_4$ is still somewhat porous and is not strong enough for many applications. In the "hot-pressed" forming process, $Si_3N_4$ powder is prepared in the form of exceedingly small particles by reaction of silicon tetrachloride with ammonia:

$$3\ SiCl_4(g) + 4\ NH_3(g) \longrightarrow Si_3N_4(s) + 12\ HCl(g)$$

The solid $Si_3N_4$ forms as a smoke. It is captured as a powder, mixed with a carefully controlled amount of MgO additive, placed in an enclosed mold, and sintered at 1850°C under a pressure of 230 atm. The resulting ceramic shrinks to nearly full density (no pores). Because the material does not flow well (to fill a complex mold completely), only simple shapes are possible. Hot-pressed silicon nitride is impressively tough and can be machined only with great difficulty, with diamond tools.

In the silicate minerals of Section 23.1, $AlO_4^{5-}$ units routinely substitute for $SiO_4^{4-}$ tetrahedra as long as positive ions of some type are present to balance the electric charge. This fact suggests that in silicon nitride some $Si^{4+}$ ions, which lie at the centers of tetrahedra of nitrogen atoms, could be replaced by $Al^{3+}$ if a compensating replacement of $O^{2-}$ for $N^{3-}$ were simultaneously made. Experiments show that ceramic alloying of this type works well, giving many new ceramics with great potential called **sialons** (named for the four elements Si—Al—O—N). These ceramics illustrate the way a structural theme in a naturally occurring material guided the search for new materials.

Boron has one fewer valence electron than carbon, and nitrogen has one more valence electron. **Boron nitride** (BN) is therefore isoelectronic with $C_2$, and it is not surprising that it has two structural modifications that resemble the structures of graphite and diamond. In hexagonal boron nitride, the boron and nitrogen atoms take alternate places in an extended "chicken-wire" sheet in which the B—N distance is 1.45 Å. The sheets stack in such a way that each boron atom has a nitrogen atom directly above it and directly below it and vice versa. The cubic form of boron nitride has the diamond structure, is comparable in hardness to diamond, and resists oxidation better. Boron nitride is often prepared by **chemical vapor deposition,** a method used in fabricating several other ceramics as well. In this method, a controlled chemical reaction of gases on a contoured, heated surface gives a solid product of the desired shape. If a cup made of BN is needed, a cup-shaped mold is heated to a temperature exceeding 1000°C, and a mixture of $BCl_3(g)$ and $NH_3(g)$ is passed over its surface. The reaction

$$BCl_3(g) + NH_3(g) \longrightarrow BN(s) + 3\ HCl(g)$$

deposits a cup-shaped layer of BN(s). Boron nitride cups and tubes are used to contain and evaporate molten metals.

**Silicon carbide** (SiC) is diamond in which half of the C atoms are replaced by Si atoms. Also known by its trade name of Carborundum, silicon carbide was originally developed as an abrasive, but is now used primarily as a refractory and as an additive in steel manufacture. It is formed and densified by methods similar to those used with silicon nitride. Silicon carbide is often produced in the form of small plates or whiskers in order to reinforce other ceramics. Fired silicon carbide whiskers are quite small (0.5 $\mu$m in diameter and 50 $\mu$m long), but very strong. They are mixed with a second ceramic material before that material is formed. Firing then gives a **composite ceramic.** Such composites are stronger and tougher than unreinforced bodies of the same primary material. Whiskers serve to reinforce the main material by stopping cracks; they either deflect advancing cracks or soak up their energy, and a widening crack must dislodge them to proceed (Fig. 23.22).

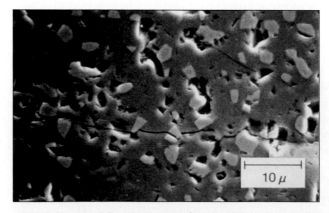

**FIGURE 23.22**  Microstructure of a composite ceramic. The light-colored areas are $TiB_2$, the gray areas are SiC, and the dark areas are voids. The $TiB_2$ acts to toughen the SiC matrix. Note that the crack shown passing through from left to right is forced to deflect around the $TiB_2$ particles. *(Carborundum Company)*

## CONCEPTS & SKILLS

*After studying this chapter and working the problems that follow, you should be able to*

1. Show how the fundamental silicate tetrahedral unit ($SiO_4^{4-}$) links to other silicate tetrahedra to form rings, chains, double chains, sheets, and space-filling crystalline networks (Section 23.1, problems 1–4).
2. Describe the chemical compositions and structures of aluminosilicates, clays, and zeolites (Section 23.1, problems 5–6).
3. Outline the crystallization sequences of silicate minerals from magma and their subsequent weathering to form sedimentary rocks (Section 23.2).
4. Apply the Clapeyron relationship to calculate the effects of pressure and temperature on the phase equilibria of minerals (Section 23.2, problems 9–12).
5. Describe the structure of ceramic materials and the ways in which they are formed (Section 23.3).
6. Outline the properties of pottery, glass, and cement and the chemical reactions that give them structural strength (Section 23.4, problems 13–20).
7. List several important oxide and mixed oxide ceramics and give some of their uses (Section 23.5, problems 21–22).
8. Discuss the special properties of nonoxide ceramics and the kinetic and thermodynamic factors that make them useful (Section 23.5, problems 23–26).

---

## PROBLEMS

*Answers to problems whose numbers are boldface appear in Appendix G. Problems that are more challenging are indicated with asterisks.*

### Naturally Occurring Minerals: The Lithosphere

1. Draw a Lewis electron-dot diagram for the disilicate ion ($Si_2O_7^{6-}$). What changes in this structure would be necessary to produce the structure of the pyrophosphate ion ($P_2O_7^{4-}$) and the pyrosulfate ion ($S_2O_7^{2-}$)? What is the analogous compound of chlorine?

2. Draw a Lewis electron-dot diagram for the cyclosilicate ion $Si_6O_{18}^{12-}$, which forms part of the structures of beryl and emerald.

3. By referring to Table 23.2, predict the structure of each of the following silicate minerals (network, sheets, double chains, and so forth). Give the oxidation state of each atom.
   (a) Andradite, $Ca_3Fe_2(SiO_4)_3$
   (b) Vlasovite, $Na_2ZrSi_4O_{10}$
   (c) Hardystonite, $Ca_2ZnSi_2O_7$
   (d) Chrysotile, $Mg_3Si_2O_5(OH)_4$

4. By referring to Table 23.2, predict the structure of each of the following silicate minerals (network, sheets, double chains, and so forth). Give the oxidation state of each atom.
   (a) Tremolite, $Ca_2Mg_5(Si_4O_{11})_2(OH)_2$
   (b) Gillespite, $BaFeSi_4O_{10}$
   (c) Uvarovite, $Ca_3Cr_2(SiO_4)_3$
   (d) Barysilate, $MnPb_8(Si_2O_7)_3$

5. By referring to Table 23.2, predict the structure of the following aluminosilicate minerals (network, sheets, double chains, and so forth). In each case, the aluminum atoms grouped with the silicon and oxygen in the formula substitute for Si in tetrahedral sites. Give the oxidation state of each atom.
   (a) Keatite, $Li(AlSi_2O_6)$
   (b) Muscovite, $KAl_2(AlSi_3O_{10})(OH)_2$
   (c) Cordierite, $Al_3Mg_2(AlSi_5O_{18})$

6. By referring to Table 23.2, predict the structure of each of the following aluminosilicate minerals (network, sheets, double chains, and so forth). In each case, the aluminum atoms grouped with the silicon and oxygen in the formula substitute for Si in tetrahedral sites. Give the oxidation state of each atom.
   (a) Amesite, $Mg_2Al(AlSiO_5)(OH)_4$
   (b) Phlogopite, $KMg_3(AlSi_3O_{10})(OH)_2$
   (c) Thomsonite, $NaCa_2(Al_5Si_5O_{20})\cdot6H_2O$

### Geochemistry

7. Sketch the dependence of cell potential on pH for the half-reactions corresponding to the reduction of $Co^{2+}$ at 1 M concentration (or $Co(OH)_2$) to $Co(s)$. Use data from Appendix E.

8. Sketch the dependence of cell potential on pH for the half-reactions corresponding to the reduction of $Pb^{2+}$ at 1 M concentration (or PbO) to $Pb(s)$. Use data from Appendix E.

9. Calcite ($CaCO_3$) reacts with quartz to form wollastonite ($CaSiO_3$) and carbon dioxide.
   (a) Write a balanced equation for this reaction.
   (b) Use the data in Appendix D to calculate $\Delta H°$ and $\Delta S°$ for this reaction.
   (c) Calculate the temperature at which $\Delta G = 0$ ($K = 1$) for this reaction at atmospheric pressure.

(d) Calculate the temperature at which $\Delta G = 0$ for this reaction at $P = 500$ atm, assuming ideal gas behavior for the carbon dioxide.

10. Tremolite, $Ca_2Mg_5Si_8O_{22}(OH)_2$, reacts with calcite ($CaCO_3$) and quartz to give diopside ($CaMgSi_2O_6$), water vapor, and carbon dioxide.
   (a) Write a balanced chemical equation for this reaction.
   (b) The standard enthalpies of formation of tremolite and diopside are $-12360$ and $-3206.2$ kJ mol$^{-1}$, respectively, and their standard entropies are 548.9 and 142.93 J K$^{-1}$ mol$^{-1}$. Use this information, together with data from Appendix D, to calculate $\Delta H°$ and $\Delta S°$ for this reaction.
   (c) Calculate the temperature at which $\Delta G = 0$ ($K = 1$) for this reaction at atmospheric pressure.
   (d) Calculate the temperature at which $\Delta G = 0$ for this reaction at $P = 600$ atm, assuming ideal gas behavior for the water vapor and carbon dioxide.

11. Calcium carbonate exists in two stable forms, calcite (molar volume 36.94 cm$^3$ mol$^{-1}$) and aragonite (molar volume 34.16 cm$^3$ mol$^{-1}$).
   (a) Use data from Appendix D to calculate $\Delta H°$, $\Delta S°$, and $\Delta G°$ for the conversion of calcite to aragonite at 25°C and 1 atm pressure.
   (b) Which form is thermodynamically stable under these conditions?
   (c) Which form will be stable at very low temperature?
   (d) Which form will be stable at high pressures?
   (e) Use the Clapeyron equation to calculate the slope of the coexistence curve for this reaction at $P = 1$ atm.

12. Silica exists in several stable forms, including quartz (molar volume 22.69 cm$^3$ mol$^{-1}$) and cristobalite (molar volume 25.74 cm$^3$ mol$^{-1}$).
   (a) Use data from Appendix D to calculate $\Delta H°$, $\Delta S°$, and $\Delta G°$ for the conversion of quartz to cristobalite at 25°C and 1 atm pressure.
   (b) Which form is thermodynamically stable under these conditions?
   (c) Which form will be stable at very high temperatures?
   (d) Which form will be stable at high pressures?
   (e) Use the Clapeyron equation to calculate the slope of the coexistence curve for this reaction at $P = 1$ atm.

## Silicate Ceramics

13. A ceramic that has been much used by artisans and craftsmen for the carving of small figurines is based upon the mineral steatite (commonly known as soapstone). Steatite is a hydrated magnesium silicate that has the composition $Mg_3Si_4O_{10}(OH)_2$. It is remarkably soft—a fingernail can scratch it. When heated in a furnace to about 1000°C, chemical reaction transforms it into a hard, two-phase composite of magnesium silicate ($MgSiO_3$) and quartz in much the same way that clay minerals are converted into mullite ($Al_6Si_2O_{13}$) and cristobalite ($SiO_2$) on firing. Write a balanced chemical equation for this reaction.

14. A clay mineral that is frequently used together with or in place of kaolinite is pyrophyllite ($Al_2Si_4O_{10}(OH)_2$). Write a balanced chemical equation for the production of mullite and cristobalite on the firing of pyrophyllite.

15. Calculate the volume of carbon dioxide produced at STP when a sheet of ordinary glass of mass 2.50 kg is made from its starting materials—sodium carbonate, calcium carbonate, and silica. Take the composition of the glass to be $Na_2O \cdot CaO \cdot (SiO_2)_6$.

16. Calculate the volume of steam produced when a 4.0-kg brick made from pure kaolinite is completely dehydrated at 600°C and a pressure of 1.00 atm.

17. A sample of soda-lime glass for tableware is analyzed and found to contain the following percentages by mass of oxides: $SiO_2$, 72.4%; $Na_2O$, 18.1%; $CaO$, 8.1%; $Al_2O_3$, 1.0%; $MgO$, 0.2%; $BaO$, 0.2%. (The elements are not actually present as binary oxides, but this is the way compositions are usually given.) Calculate the chemical amounts of silicon, sodium, calcium, aluminum, magnesium, and barium atoms per mole of oxygen atoms in this sample.

18. A sample of portland cement is analyzed and found to contain the following percentages by mass of oxides: $CaO$, 64.3%; $SiO_2$, 21.2%; $Al_2O_3$, 5.9%; $Fe_2O_3$, 2.9%; $MgO$, 2.5%; $SO_3$, 1.8%; $Na_2O$, 1.4%. Calculate the chemical amounts of calcium, silicon, aluminum, iron, magnesium, sulfur, and sodium atoms per mole of oxygen atoms in this sample.

19. The most important contributor to the strength of hardened portland cement is tricalcium silicate, $(CaO)_3 \cdot SiO_2$, for which the measured standard enthalpy of formation is $-2929.2$ kJ mol$^{-1}$. Calculate the standard enthalpy change for the production of 1.00 mol of tricalcium silicate from quartz and lime.

20. One of the simplest of the heat-generating reactions that take place when water is added to cement is the production of calcium hydroxide (slaked lime) from lime. Write a balanced chemical equation for this reaction, and use data from Appendix D to calculate the amount of heat generated by the reaction of 1.00 kg of lime with water at room conditions.

## Nonsilicate Ceramics

21. Calculate the average oxidation number of the copper in $YBa_2Cu_3O_{9-x}$ if $x = 2$. Assume that the rare-earth element yttrium is in its usual $+3$ oxidation state.

22. The mixed oxide ceramic $Tl_2Ca_2Ba_2Cu_3O_{10+x}$ has zero electrical resistance at 125 K. Calculate the average oxidation number of the copper in this compound if $x = 0.50$ and thallium is in the $+3$ oxidation state.

23. Silicon carbide ($SiC$) is made by the high-temperature reaction of silica sand (quartz) with coke; the byproduct is carbon monoxide.
   (a) Write a balanced chemical equation for this reaction.
   (b) Calculate the standard enthalpy change per mole of $SiC$ produced.
   (c) Predict (qualitatively) the following physical properties of silicon carbide: conductivity, melting point, and hardness.

24. Boron nitride (BN) is made by the reaction of boron trichloride with ammonia.
    (a) Write a balanced chemical equation for this reaction.
    (b) Calculate the standard enthalpy change per mole of BN produced, given that the standard molar enthalpy of formation of BN(s) is $-254.4$ kJ mol$^{-1}$.
    (c) Predict (qualitatively) the following physical properties of boron nitride: conductivity, melting point, and hardness.

25. The standard free energy of formation of cubic silicon carbide (SiC) is $-62.8$ kJ mol$^{-1}$. Determine the standard free energy change when 1.00 mol of SiC reacts with oxygen to form SiO$_2$ (s, quartz) and CO$_2(g)$. Is silicon carbide thermodynamically stable in the air at room conditions?

26. The standard free energy of formation of boron carbide (B$_4$C) is $-71$ kJ mol$^{-1}$. Determine the standard free energy change when 1.00 mol of B$_4$C reacts with oxygen to form B$_2$O$_3(s)$ and CO$_2(g)$. Is boron carbide thermodynamically stable in the air at room conditions?

## Additional Problems

27. Predict the structure of each of the following silicate minerals (network, sheets, double chains, and so forth). Give the oxidation state of each atom.
    (a) Apophyllite, KCa$_4$(Si$_8$O$_{20}$)F$\cdot$8H$_2$O
    (b) Rhodonite, CaMn$_4$(Si$_5$O$_{15}$)
    (c) Margarite, CaAl$_2$(Al$_2$Si$_2$O$_{10}$)(OH)$_2$

28. By referring to Table 23.2, predict the kind of structure formed by manganpyrosmalite, a silicate mineral with chemical formula Mn$_{12}$FeMg$_3$(Si$_{12}$O$_{30}$)(OH)$_{10}$Cl$_{10}$. Give the oxidation state of each atom in this formula unit.

29. A reference book lists the chemical formula of one form of vermiculite as

    [(Mg$_{2.36}$Fe$_{0.48}$Al$_{0.16}$)(Si$_{2.72}$Al$_{1.28}$)O$_{10}$(OH)$_2$][Mg$_{0.32}$(H$_2$O)$_{4.32}$]

    Determine the oxidation state of the iron in this mineral.

30. The most common feldspars are those containing potassium, sodium, and calcium cations. They are called, respectively, orthoclase (KAlSi$_3$O$_8$), albite (NaAlSi$_3$O$_8$), and anorthite (CaAl$_2$Si$_2$O$_8$). The solid solubility of orthoclase in albite is limited, and its solubility in anorthite is almost negligible. Albite and anorthite, however, are completely miscible at high temperatures and show complete solid solution. Offer an explanation for these observations, based on the tabulated radii of the K$^+$, Na$^+$, and Ca$^{2+}$ ions from Appendix F.

31. The clay mineral kaolinite (Al$_2$Si$_2$O$_5$(OH)$_4$) is formed by the weathering action of water containing dissolved carbon dioxide on the feldspar mineral anorthite (CaAl$_2$Si$_2$O$_8$). Write a balanced chemical equation for the reaction that occurs. The CO$_2$ forms H$_2$CO$_3$ as it dissolves. As the pH is lowered, will the weathering occur to a greater or a lesser extent?

32. Certain kinds of zeolite have the general formula M$_2$O$\cdot$Al$_2$O$_3\cdot y$SiO$_2\cdot w$H$_2$O where M is an alkali metal such as

Na or K, $y$ is 2 or more, and $w$ is any integer. Compute the mass percentage of Al in a zeolite that has M = K, $y$ = 4, and $w$ = 6.

33. (a) Use data from Tables 10.2 and 11.3 to calculate the solubility of CaCO$_3$ in water at pH 7.
    (b) Will the solubility increase or decrease if the pH is lowered and the water becomes more acidic?
    (c) Calculate the maximum amount of limestone (primarily calcium carbonate) that could dissolve per year in a river at pH 7 with an average flow rate of $1.0 \times 10^6$ m$^3$ per hour.

* 34. Show how the Clapeyron equation follows from the second law of thermodynamics, expressed as

$$dG = V\,dP - S\,dT$$

35. Talc, Mg$_3$Si$_4$O$_{10}$(OH)$_2$, reacts with forsterite (Mg$_2$SiO$_4$) to form enstatite (MgSiO$_3$) and water vapor.
    (a) Write a balanced chemical equation for this reaction.
    (b) If the water pressure is equal to the total pressure, will formation of products be favored or disfavored with increasing total pressure?
    (c) The entropy change for this reaction is positive. Will the slope of the coexistence curve (pressure plotted against temperature) be positive or negative?

36. In what ways does soda–lime glass resemble and in what ways does it differ from a pot made from the firing of kaolinite? Include the following aspects in your discussion: composition, structure, physical properties, and method of preparation.

37. Iron oxides are red when the average oxidation state of iron is high and black when it is low. To impart each of these colors to a pot made from clay that contains iron oxides, would you employ an air-rich or a smoky atmosphere in the kiln? Explain.

38. Refractories can be classified as acidic or basic, depending on the properties of the oxides in question. A basic refractory must not be used in contact with acid, and an acidic refractory must not be used in contact with a base. Classify magnesia and silica as acidic or basic refractories.

39. Dolomite bricks are used in the linings of furnaces in the cement and steel industries. Pure dolomite contains 45.7% MgCO$_3$ and 54.3% CaCO$_3$ by mass. Determine the empirical formula of dolomite.

40. A book states correctly that beryllia (BeO) ceramics have some use but possess only poor resistance to strong acids and bases. Write likely chemical equations for the reaction of BeO with a strong acid and with a strong base.

41. Silicon nitride resists all acids except hydrofluoric, with which it reacts to give silicon tetrafluoride and ammonia. Write a balanced chemical equation for this reaction.

42. Compare oxide ceramics like alumina and magnesia, which have significant ionic character, with covalently bonded nonoxide ceramics like SiC and B$_4$C (problems 25 and 26) with respect to thermodynamic stability at ordinary conditions.

# CHAPTER
# 24

# Optical and Electronic Materials

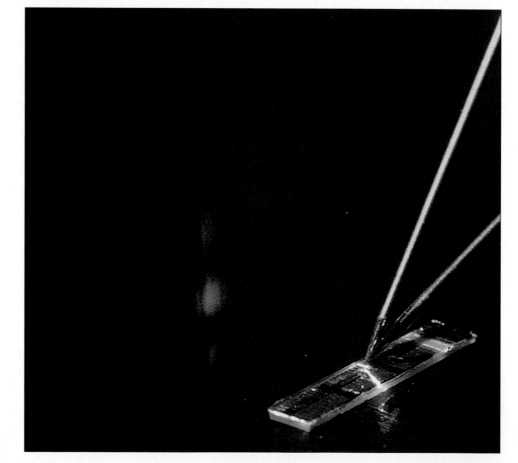

*Illustration*

The blue light emitted from a GaN-InN diode laser depends on the wide bandgaps of Group III nitride semiconductors. Active materials were deposited directly on a sapphire substrate. Electric power is provided via two metallic probes. *(Courtesy of Professors S. Den Baars and J. Speck, College of Engineering, University of California—Santa Barbara)*

The silicate-bearing minerals of Chapter 23 have done far more than contribute the structural materials discussed there; they also provide the source for the element (silicon) that lies at the heart of the electronics and computer revolution of the second half of the twentieth century. For these purposes, it is the electronic properties of this element (in combination with elements from neighboring groups) that lead to its critical applications. This chapter begins with a discussion of the role of synthetic ceramics as electronic materials and then turns to their interaction with light; the absorption of light gives rise to the bright colors of inorganic pigments and to the conversion of solar energy into electrical energy in solar cells. The chapter closes with a look at a natural form of light harvesting through photosynthesis in living plants.

---

## 24.1

## SEMICONDUCTORS

As a Group IV element, silicon bears a resemblance to carbon in its chemistry. Both it and carbon exhibit valences of 4 and the capacity to form extended molecular structures. As elements, both crystallize in the cubic system and possess the same structure. The bonding in solid silicon can be described by $sp^3$ hybrid orbitals directed to the four corners of a regular tetrahedron. Each atom is covalently linked by a single bond to four others in a space-filling "giant molecule" network just like that of diamond (see Fig. 19.23).

### The Band Picture of Bonding in Silicon

An alternative to the localized orbital explanation of the bonding in silicon is to consider the electrons as occupying "molecular" orbitals that are spread out over the entire silicon crystal. An analogous procedure was employed in Figure 19.20 to construct a band of half-occupied levels in metallic sodium. In the case of silicon, the $4N_0$ valence atomic orbitals from 1 mol ($N_0$ atoms) of silicon split into *two* bands in the silicon crystal, each containing $2N_0$ very closely spaced levels (Fig. 24.1). The lower band is called the **valence band** and the upper one the **conduction band.** Between the top of the valence band and the bottom of the conduction band is an energy region that is forbidden to electrons. The magnitude of the separation in energy between the valence band and the lowest level of the conduction band is called the **band gap** $E_g$, which for pure silicon is $1.94 \times 10^{-19}$ J. This is the amount of energy that an electron must gain to be excited from the top of the valence band to the bottom of the conduction band. For 1 mol of electrons to be excited in this way, the energy is larger by a factor of Avogadro's number $N_0$, giving 117 kJ mol$^{-1}$. (Another unit used for band gaps is the *electron volt*, defined in Section 15.2 as $1.60218 \times 10^{-19}$ J. In this unit, the band gap in Si is 1.21 eV.)

Each silicon atom in the crystal contributes four valence electrons to the bands of orbitals in Figure 24.1, for a total of $4N_0$ per mole. This is a sufficient number to place two electrons in each level of the valence band (with opposing spins) and leave the conduction band empty. There are no low-lying energy levels for those electrons at the top of the valence band to enter if given a small increment in their energy. The band gap means that an electron at the top of the filled valence band must acquire an energy of at least $1.94 \times 10^{-19}$ J, equivalent to 117 kJ mol$^{-1}$, to jump to the lowest empty level of the conduction band. This is a large amount of

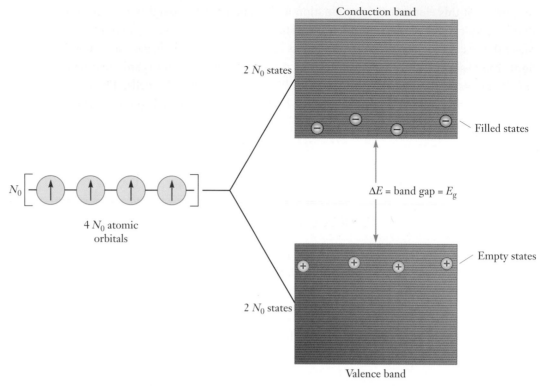

**FIGURE 24.1** The valence orbitals of the silicon atom combine in crystalline silicon to give two bands of very closely spaced levels. The valence band is almost completely filled, and the conduction band is almost empty.

energy. If it were to be supplied by a thermal source, the temperature of the source would have to be on the order of

$$T = \frac{\Delta E}{R} = \frac{117{,}000 \text{ J mol}^{-1}}{8.315 \text{ J K}^{-1} \text{ mol}^{-1}} = 14{,}000 \text{ K}$$

At room temperature only a few electrons per mole in the extreme tail of the Boltzmann distribution have enough energy to jump the gap, so the conduction band in pure silicon is very sparsely populated with electrons. The result is that silicon is not a good conductor of electricity; good electrical conductivity requires a net motion of many electrons under the impetus of a small electric potential difference. Silicon has an electrical conductivity that is 11 orders of magnitude smaller than that of a typical metal (copper) at room temperature. It is called a **semiconductor,** because its electrical conductivity, although smaller than that of a metal, is far greater than that of an **insulator** like diamond, which has a larger band gap. The conductivity of a semiconductor is increased by increasing the temperature, because more electrons are then excited into levels in the conduction band. Another way to increase the conductivity of a semiconductor is to irradiate it with a beam of electromagnetic radiation having a frequency high enough to excite electrons from the valence band to the conduction band. This process resembles the photoelectric effect described in Chapter 15, with the difference that now the electrons are not removed from the material but only moved into the conduction band, ready to conduct a current if a potential difference is imposed.

| EXAMPLE 24.1 |

Calculate the longest wavelength of light that can excite electrons from the valence to the conduction band in silicon. In what region of the spectrum does this wavelength fall?

### Solution

The energy carried by a photon is $h\nu = hc/\lambda$, where $h$ is Planck's constant, $\nu$ the photon frequency, $c$ the speed of light, and $\lambda$ the photon wavelength. For a photon to just excite an electron across the band gap, this energy must be equal to that band gap energy, $E_g = 1.94 \times 10^{-19}$ J.

$$\frac{hc}{\lambda} = E_g$$

Solving for the wavelength gives

$$\lambda = \frac{hc}{E_g} = \frac{(6.626 \times 10^{-34} \text{ J s}) \times \left(2.998 \times 10^8 \frac{\text{m}}{\text{s}}\right)}{1.94 \times 10^{-19} \text{ J}}$$

$$= 1.02 \times 10^{-6} \text{ m} = 1020 \text{ nm}$$

This wavelength falls in the infrared region of the spectrum. Photons with shorter wavelengths (for example, visible light) carry more than enough energy to excite electrons to the conduction band in silicon.

**Related Problems: 1, 2**

## Doped Semiconductors

Silicon in very high purity is said to display its **intrinsic properties.** When certain other elements are added to pure silicon in a process called **doping,** it acquires interesting electronic properties. For example, if atoms of a Group V element such as arsenic or antimony are diffused into silicon, they substitute for silicon atoms in the network. Such atoms have five valence electrons, so each one introduces one more electron into the silicon crystal than is needed for bonding. The extra electrons occupy energy levels just below the lowest level of the conduction band. Very little energy is required to promote electrons from such a **donor impurity** level into the conduction band, and the electrical conductivity of the silicon crystal is increased without the necessity of raising the temperature. Silicon doped with atoms of a Group V element is called an **n-type semiconductor** to indicate that the charge carrier is negative.

On the other hand, if a Group III element such as gallium is used as a dopant, there is one fewer electron in the valence band per dopant atom because Group III elements have only three valence electrons, not four. This situation corresponds to creating one **hole** in the valence band, with an effective charge of +1, for each Group III atom added. If a voltage difference is impressed across a crystal that is doped in this way, it causes the positively charged holes to move toward the negative source of potential. Equivalently, a valence band electron next to the (positive) hole will move in the opposite direction (that is, toward the positive source of potential). Whether we think of holes or electrons as the mobile charge carrier in the

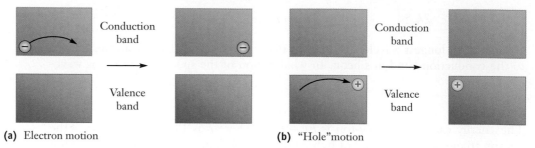

**(a)** Electron motion          **(b)** "Hole" motion

**FIGURE 24.2** (a) In an *n*-type semiconductor, a small number of electrons occupy levels in the conduction band. When an electric field is imposed, each electron moves (black arrows) into one of the numerous nearby vacant energy levels in the conduction band. (b) In a *p*-type semiconductor, a small number of levels in the valence band are unoccupied. Conduction occurs as electrons in the numerous occupied levels of the valence band jump into the sparsely distributed unoccupied levels. The arrow shows the motion of an electron to occupy a previously empty level. The process can also be described as the motion of positively charged holes in the opposite direction.

valence band, the result is the same; we are simply using different words to describe the same physical phenomenon (Fig. 24.2). Silicon that has been doped with a Group III element is called a **p-type semiconductor** to indicate that the carrier has an effective positive charge.

A different class of semiconductors is based not on silicon, but on equimolar compounds of Group III with Group V elements. Gallium arsenide, for example, is isoelectronic to the Group IV semiconductor germanium. When GaAs is doped with the Group VI element tellurium, an *n*-type semiconductor is produced; doping with zinc, which has one *fewer* valence electron than gallium, gives a *p*-type semiconductor. Other III–V combinations have different band gaps and are useful in particular applications. Indium antimonide (InSb), for example, has a small enough band gap that absorption of infrared radiation causes electrons to be excited from the valence to the conduction band, and an electric current then flows when a small potential difference is applied. This compound is therefore used as a detector of infrared radiation. Still other compounds formed between the zinc group (zinc, cadmium, and mercury) and Group VI elements such as sulfur also have the same average number of valence electrons per atom as silicon and make useful semiconductors.

### p–n Junctions and Device Performance

Of what value is it to have a semiconductor that can conduct an electric current by the flow of electrons if it is *n*-type or by the flow of holes if it is *p*-type? Many electronic functions can be fulfilled by semiconductors that possess these properties, but the simplest is **rectification**: the conversion of alternating current into direct current. Suppose thin crystals of *n*-type and *p*-type silicon are placed in contact with one another and connected to a battery. In Figure 24.3a the positive pole of the battery is connected to the *n*-type silicon, and the *p*-type silicon is connected to the negative pole. Only a very small transient current can flow through the circuit in this case, because when the electrons in the conduction band of the *n*-type silicon have flowed out to the positive pole of the battery, there are none to take their place and current ceases to flow. However, if the negative pole of the battery is connected

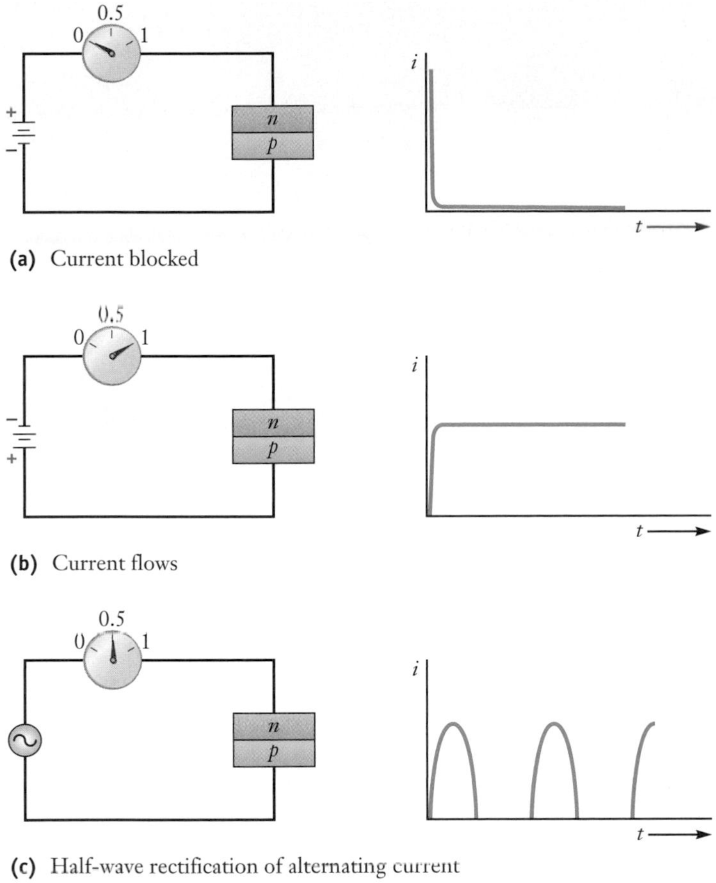

**(a)** Current blocked

$i$

$t \longrightarrow$

**(b)** Current flows

$i$

$t \longrightarrow$

**(c)** Half-wave rectification of alternating current

$i$

$t \longrightarrow$

**FIGURE 24.3** Current rectification by a *p–n* transistor.

to the *n*-type silicon and the positive pole to the *p*-type silicon (Fig. 24.3b), a steady current flows because electrons and holes move in opposite directions and recombine at the *n–p* junction. In effect, electrons flow toward the *n–p* junction in the *n*-type material, holes flow toward the *n–p* junction in the *p*-type material, and the junction is a "sink" where electrons fill the holes in the valence band and neutralize one another. If, instead of a galvanic cell, an alternating current source were connected to the *n–p* rectifier, current would flow in one direction only, creating pulsed direct current (Fig. 24.3c).

Semiconductors perform a wide range of electronic functions that formerly required the use of vacuum tubes. Vacuum tubes occupy much more space, generate large amounts of heat, and require considerably more energy to operate than **transistors,** their semiconductor counterparts. More importantly, semiconductors can be built into integrated circuits (Fig. 24.4) and made to store information and to process it at great speeds.

Solar cells based on silicon or gallium arsenide provide a way to convert the radiant energy of the sun directly into electrical work by a technology that is virtually nonpolluting (Fig. 24.5). The high capital costs of solar cells make them uncompetitive with conventional fossil fuel sources of energy at this time, but as reserves of fossil fuels dwindle, solar energy will become an important option.

**FIGURE 24.4** A microchip, or integrated circuit, contains many thousands of electrical components squeezed onto a thin sliver of silicon less than 0.4 inch (1 cm) square. The microchip is set in a plastic base and is connected to two rows of pins that conduct electricity. *(Copyright 1992 Martin Bough/Fundamental Photographs)*

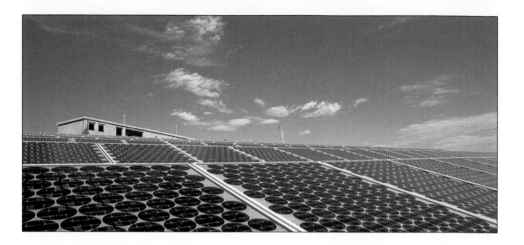

---

## 24.2

## PIGMENTS AND PHOSPHORS: OPTICAL DISPLAYS

The band gap of an insulator or semiconductor has a significant effect on its color. Pure diamond has a large band gap, so even blue light does not have enough energy to excite electrons from the valence band to the conduction band. As a result, light passes through it without being absorbed, and diamond is colorless. Cadmium sulfide (Fig. 24.6) has a band gap of $4.2 \times 10^{-19}$ J, which corresponds to a wavelength of 470 nm, in the visible region of the spectrum. CdS therefore absorbs violet and blue light but strongly transmits yellow, giving it a deep yellow color. Cadmium sulfide is the pigment called cadmium yellow. Cinnabar (HgS, Fig. 24.7) has a smaller band gap of $3.2 \times 10^{-19}$ J and absorbs all light except red. It has a deep red color and is the pigment vermilion. Semiconductors with band gaps of less than $2.8 \times 10^{-19}$ J absorb all wavelengths of visible light and appear black. These include silicon (Example 24.1), germanium, and gallium arsenide.

Doping silicon brings donor levels close enough to the conduction band or acceptor levels close enough to the valence band that thermal excitation can cause

FIGURE 24.6 Mixed crystals of two semiconductors with different band gaps, CdS (yellow) and CdSe (black), show a range of colors, illustrating a decrease in the band gap energy as the composition of the mixture becomes richer in Se. (*Kurt Nassau/ Phototake*)

electrons to move into a conducting state. The corresponding doping of *insulators* or wide band-gap semiconductors can bring donor or acceptor states into positions where *visible light* can be absorbed or emitted. This changes the colors and optical properties of the materials. Nitrogen doped in diamond gives a donor impurity level in the band gap. Transitions to this level can absorb some blue light, giving the diamond an (undesirable) yellowish color. On the other hand, boron doped into diamond gives an acceptor level that absorbs red light most strongly and gives the highly prized and rare "blue diamond."

**Phosphors** are wide band-gap materials with dopants selected to create new levels such that particular colors of light are emitted. Electrons in these materials are excited by light of other wavelengths, or by electrons hitting their surfaces, and light is then emitted as they return to lower energy states. A fluorescent lamp, for example, is a mercury-vapor lamp in which the inside of the tube has been coated with phosphors. These absorb the violet and ultraviolet light emitted by mercury vapor and emit at lower energies and longer wavelengths, giving a nearly white light that is more desirable than the bluish light that comes from a mercury-vapor lamp without the phosphors.

Phosphors are also used in television screens. The picture is formed by scanning a beam of electrons (from an electron gun) over the screen. The electrons strike the phosphors coating the screen, exciting their electrons and causing them to emit light. In a black-and-white television tube, the phosphors are a mixture of silver doped into $ZnS$, which gives blue light, and silver doped into $Zn_xCd_{1-x}S$, which gives yellow light. The combination of the two provides a reasonable approximation of white. A color television uses three different electron guns, with three corresponding types of phosphor on the screen. Silver doped in $ZnS$ gives blue; manganese doped in $Zn_2SiO_4$ is used for green; and europium doped in $YVO_4$ gives red light. Masks are used to ensure that each electron beam encounters only the phosphors corresponding to the desired color.

**FIGURE 24.7**  Crystalline cinnabar, $HgS$. (*Charles D. Winters*)

---

## 24.3

## PHOTOSYNTHESIS

Even the most efficient manufactured solar collectors described in Section 24.1 fall far short of Nature in their ability to convert solar energy from the sun into other useful forms of energy. Living species harvest light and store its energy by carrying out chemical reactions with positive free energy changes. For this purpose they use compounds dominated by carbon rather than by silicon, in which metal ions play a critical role. This process of **photosynthesis** has transformed our planet and permitted the evolution of human life. Before we consider green plants (which provide most of the photosynthetic energy storage today), let us examine the reaction mechanisms in existent photosynthetic bacteria that are thought to have evolved earlier, and are considerably simpler, than plants. These bacteria have recently yielded some important secrets in experimental investigations.

The purple photosynthetic bacteria include the species *Rhodobacter sphaeroides* and *Rhodopseudomonas viridis*. Although these two differ in the details of their chemical makeup, the broad features of their mechanisms appear to be closely related. Their purple color reflects their absorption of light predominantly in the red region of the spectrum, with wavelengths up to 925 nm. *R. sphaeroides* and *R. viridis* use a variety of pigment molecules, including derivatives of chlorophyll (Fig. 24.8b), as

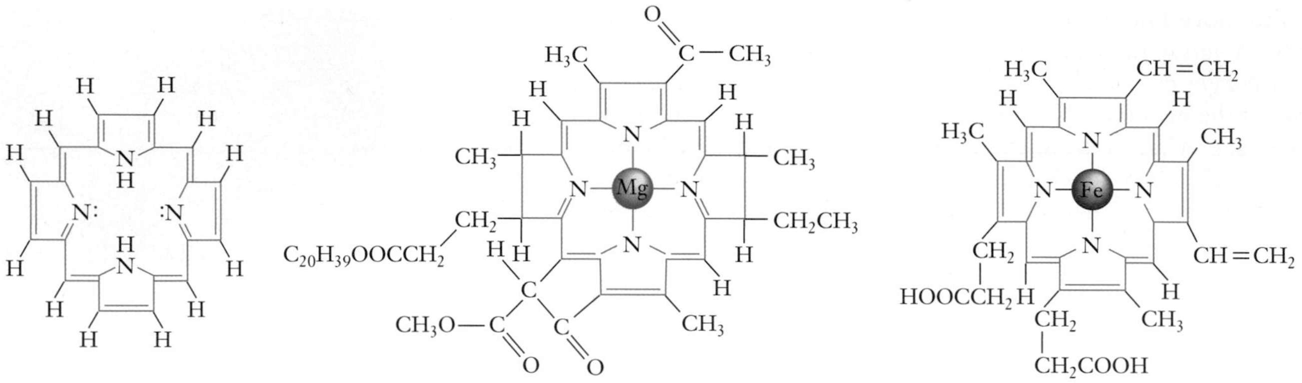

**(a)** Porphine        **(b)** Bacteriochlorophyll        **(c)** Heme

**FIGURE 24.8** Structures of (a) porphine, (b) bacteriochlorophyll, and (c) heme. Carbon atoms are understood to lie at intersections of bonds where no other atom is indicated. In bacteriopheophytin, the central $Mg^{2+}$ ion shown in part b is replaced by two $H^+$ ions, producing a ring like that in part a. Cytochromes are proteins containing heme groups at their centers.

"antennas" to absorb photons that reach their surface. The antenna molecules store the light energy through promotion of electrons to excited singlet states (see Fig. 16.28), then transfer this energy rapidly to other molecules nearby. In such a transfer, the initially excited molecule returns to its ground state while a neighboring molecule reaches an excited singlet state. Within about $10^{-10}$ s (100 picoseconds), the energy migrates to the **reaction center,** where it is trapped and is ready to be used to carry out chemical reactions.

The initial trapping site in the reaction center consists of a **special pair** of bacteriochlorophyll molecules. The existence of this special pair was postulated in the 1960s by scientists who used magnetic resonance and absorption spectroscopy to probe the dynamics of the reaction center. Dramatic confirmation came in 1984 when three German chemists (Hartmut Michel, Johann Deisenhofer, and Robert Huber) succeeded in preparing crystals of the membrane-bound reaction-center protein in *R. viridis* and determining its chemical structure by x-ray diffraction (Section 19.1). Not only did they locate the bacteriochlorophyll molecules that form the special pair, they also identified the positions of the molecules involved in the subsequent steps in photosynthesis.

Figure 24.9 shows the reaction center of the related bacterium *R. sphaeroides*; its structure is much more evident when the surrounding protein is stripped away. This complex consists of four bacteriochlorophyll molecules (the special pair and two others), two bacteriopheophytin molecules (which are bacteriochlorophyll molecules in which the central Mg ion is replaced by two hydrogen ions), two ubiquinone molecules (Fig. 24.10), and an iron(II) ion. Interestingly, these molecules are arranged in an almost perfectly symmetric fashion, with two branches (called A and B). However, the surrounding protein breaks the symmetry of the branches and causes the energy flow to pass almost entirely through the A branch (on the right in Fig. 24.9). Any energy passing through the B branch is too little to detect.

Time-resolved spectroscopy (see Fig. 16.31) has shown that within about 2.8 picoseconds ($2.8 \times 10^{-12}$ s) an electron is transferred from the special pair to the bacteriopheophytin molecule in the A branch, creating an anion on that site and leaving a cation on the special pair. Within 200 picoseconds, the electron reaches the ubiquinone molecule in the A branch and then is transferred across to the other ubiquinone molecule, in the B branch. The special pair picks up an electron from the iron atom in the heme group (Fig. 24.8c) on a cytochrome protein outside the

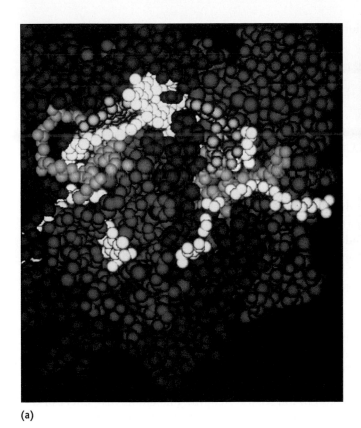

(a)

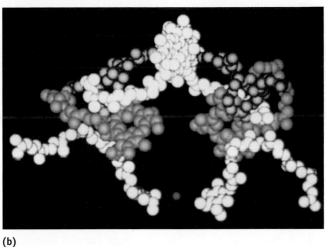

(b)

**FIGURE 24.9** Computer-generated diagrams of the reaction center of the purple bacterium *R. sphaeroides*, based on x-ray diffraction data. (a) The reaction center has been sliced through the middle to show the donor-acceptor molecules surrounded by protein (dark green, red, and blue portions). (b) The protein has been removed entirely. The four bacteriochlorophyll molecules are shown in white (for the special pair) and purple (for the other two); bacteriopheophytin molecules are light green; ubiquinone molecules are yellow. An iron(II) ion is shown between the ubiquinone molecules. The A branch is on the right, and the B branch is on the left. *(Copyright J. Norris)*

membrane and is reduced back to its original state. It then absorbs a second photon and transfers a second electron to the same B-branch ubiquinone molecule. This doubly reduced ubiquinone is less tightly bound to the reaction-center protein and moves away, being replaced by an unreduced ubiquinone molecule. The reduced ubiquinone gives up its electrons and hydrogen ions to another protein, which releases $H^+$ ions outside the cell membrane. The resulting hydrogen ion (proton) gradient across the cell wall stores energy (it is equivalent to the concentration cell considered in Chapter 12) and can carry out chemical reactions needed by the cell, ultimately transferring hydrogen ions back into the cell to close the cycle.

**FIGURE 24.10** The structures of (a) ubiquinone (UQ) and (b) its reduced form, $UQH_2$. Note the tail that contains a $C_5H_8$ group repeated ten times.

The chemical reactions at the reaction center can be summarized by the following six steps:

**1.** $(BChl)_2 + UQ \xrightarrow{\text{light}} (BChl)_2^+ + UQ^-$

**2.** $(BChl)_2^+ + Cyt \longrightarrow (BChl)_2 + Cyt^+$

**3.** $(BChl)_2 + UQ^- \xrightarrow{\text{light}} (BChl)_2^+ + UQ^{2-}$

**4.** $(BChl)_2^+ + Cyt \longrightarrow (BChl)_2 + Cyt^+$

**5.** $UQ^{2-} + 2\,H_{in}^+ \longrightarrow UQH_2$

**6.** $UQH_2 + 2\,H_{in}^+ + 2\,Cyt^+ \longrightarrow UQ + 4\,H_{out}^+ + 2\,Cyt$

Overall: $4\,H_{in}^+ \longrightarrow 4\,H_{out}^+$

Here $(BChl)_2$ stands for the special pair of bacteriochlorophyll molecules, UQ for ubiquinone, and Cyt for the cytochrome protein. Steps 1 and 3 involve excitation of bacteriochlorophyll and transfer of a pair of electrons to a ubiquinone molecule; steps 2 and 4 restore the special pair to its initial state; steps 5 and 6 transfer hydrogen ions outside the membrane wall and restore the cytochrome to its reduced form. The net reaction is the light-driven movement of hydrogen ions from inside to outside the cell.

Purple bacteria have only a single type of reaction center, but green plants have two types, referred to as photosystem I (PS I) and photosystem II (PS II). These absorb light at somewhat different wavelengths, allowing plants to use light energy from the sun more efficiently than purple bacteria do. Moreover, the two photosystems are linked to one another, so that the excitation of PS II is followed by the excitation of PS I with a photon of longer wavelength. The cumulative effect of the two photons absorbed in series is the storage of a larger amount of energy than is available to purple bacteria. Whereas the bacteria use $H_2S$ and other reduced compounds as sources of hydrogen, green plants are able to split abundant water molecules and in the process give off oxygen, which is, of course, vital to the existence and survival of animal species on the planet.

Chemists are studying the structure and kinetics of the photosynthetic reaction center both to understand the fundamentals of this important natural process and to design new materials that mimic the ability of Nature to harvest light energy at such high efficiency. Artificial photosynthesis may lead to carbon-based materials that will replace the silicon collectors in solar cells in the 21st century. This will help to reduce human dependence on stored fossil fuels as energy sources in the future.

## CONCEPTS & SKILLS

*After studying this chapter and working the problems that follow, you should be able to*

**1.** Describe the mechanism of action of intrinsic, *n*-type, and *p*-type semiconductors (Section 24.1, problems 1–6).

**2.** Describe how silicon-based solar energy collectors operate (Section 24.1).

**3.** Relate the band gap of a semiconductor or phosphor to the frequencies of electromagnetic radiation absorbed or emitted when electrons make transitions between the valence and conduction bands (Section 24.2, problems 7–10).

**4.** Explain the mechanism by which photosynthetic bacteria and green plants convert light energy to chemical energy (Section 24.3, problems 11–12).

**PROBLEMS**

*Answers to problems whose numbers are boldface appear in Appendix G.*

## Semiconductors

**1.** Electrons in a semiconductor can be excited from the valence band to the conduction band through the absorption of photons with energies exceeding the band gap. At room temperature, indium phosphide (InP) is a semiconductor that absorbs light only at wavelengths less than 920 nm. Calculate the band gap in InP.

**2.** Both GaAs and CdS are semiconductors that are being studied for possible use in solar cells to generate electric current from sunlight. Their band gaps are $2.29 \times 10^{-19}$ J and $3.88 \times 10^{-19}$ J, respectively, at room temperature. Calculate the longest wavelength of light that is capable of exciting electrons across the band gap in each of these substances. In which region of the electromagnetic spectrum do these wavelengths fall? Use this result to explain why CdS-based sensors are employed in some cameras to estimate the proper exposure conditions.

**3.** The number of electrons excited to the conduction band per cubic centimeter in a semiconductor can be estimated from the equation

$$n_e = (4.8 \times 10^{15} \text{ cm}^{-3} \text{ K}^{-3/2}) \, T^{3/2} \, e^{-E_g/(2\,RT)}$$

where $T$ is the temperature in kelvins and $E_g$ the band gap in joules *per mole*. The band gap of diamond at 300 K is $8.7 \times 10^{-19}$ J. How many electrons are thermally excited to the conduction band at this temperature in a 1.00-cm$^3$ diamond crystal?

**4.** The band gap of pure crystalline germanium is $1.1 \times 10^{-19}$ J at 300 K. How many electrons are excited from the valence band to the conduction band in a 1.00-cm$^3$ crystal of germanium at 300 K? Use the equation given in the preceding problem.

**5.** Describe the nature of electrical conduction in (a) Si doped with P and (b) InSb doped with Zn.

**6.** Describe the nature of electrical conduction in (a) Ge doped with In and (b) CdS doped with As.

**7.** In a light-emitting diode (LED), which is used in displays on electronic equipment, watches, and clocks, a voltage is imposed across an *n–p* semiconductor junction. The electrons on the *n* side combine with the holes on the *p* side and emit light at the frequency of the band gap. This process can also be described as the emission of light as electrons fall from levels in the conduction band to empty levels in the valence band. It is the reverse of the production of electric current by illumination of a semiconductor.

Many LEDs are made from semiconductors having the general composition GaAs$_{1-x}$P$_x$. When $x$ is varied between 0 and 1, the band gap changes, and with it the color of light emitted by the diode. When $x = 0.4$, the band gap is $2.9 \times 10^{-19}$ J. Determine the wavelength and color of the light emitted by this LED.

**8.** When the LED described in the previous problem has the composition GaAs$_{0.14}$P$_{0.86}$ (that is, $x = 0.86$), the band gap has increased to $3.4 \times 10^{-19}$ J. Determine the wavelength and color of the light emitted by this LED.

## Pigments and Phosphors: Optical Displays

**9.** The pigment zinc white (ZnO) turns bright yellow when heated, but the white color returns when the sample is cooled. Does the band gap increase or decrease when the sample is heated?

**10.** Mercury(II) sulfide (HgS) exists in two different crystalline forms. In cinnabar, the band gap is $3.2 \times 10^{-19}$ J; in metacinnabar, it is $2.6 \times 10^{-19}$ J. In some old paintings with improperly formulated paints, the pigment vermilion (cinnabar) has transformed to metacinnabar upon exposure to light. Describe the color change that results.

## Photosynthesis

**11.** One way in which photosynthetic bacteria store chemical energy is through the conversion of a compound called adenosine diphosphate (ADP), together with hydrogen phosphate ion, to adenosine triphosphate (ATP):

$$\text{ADP}^{3-} + \text{HPO}_4^{2-} + \text{H}_3\text{O}^+ \longrightarrow \text{ATP}^{4-} + 2\,\text{H}_2\text{O}$$
$$\Delta G = +34.5 \text{ kJ (pH 7)}$$

Suppose some chlorophyll molecules absorb 1.00 mol of photons of blue light with wavelength 430 nm. If *all* this energy could be used to convert ADP to ATP at room conditions and pH 7, how many molecules of ATP would be produced per photon absorbed? (The actual number is smaller because the conversion is not 100% efficient.)

**12.** Repeat the calculation of the preceding problem for red light with wavelength 700 nm.

## Additional Problems

**13.** Compare the hybridization of silicon atoms in Si(*s*) with that of carbon atoms in graphite (see Fig. 19.26). If silicon were to adopt the graphite structure, would its electrical conductivity be high or low?

**14.** Describe how the band gap varies from a metal to a semiconductor to an insulator.

**15.** Suppose some people are sitting in a row at a movie theater, with a single empty seat on the left end of the row. Every five minutes, a person moves into a seat on his or her left if it is empty. In what direction and with what speed does the empty seat "move" along the row? Comment on the connection with hole motion in *p*-type semiconductors.

16. A sample of silicon doped with antimony is an *n*-type semiconductor. Suppose a small amount of gallium is added to such a semiconductor. Describe how the conduction properties of the solid will vary with the amount of gallium added.

17. Draw a schematic diagram of the steps in bacterial photosynthesis, numbering them in sequence and showing the approximate spatial relations of the involved molecules.

18. Do you expect the energy of the special pair of bacteriochlorophyll molecules to be higher or lower than the energy of an isolated bacteriochlorophyll? (*Hint:* Think about the analogy between the mixing of atomic orbitals to make molecular orbitals and the mixing of molecular orbitals on two nearby molecules.)

# Polymeric Materials

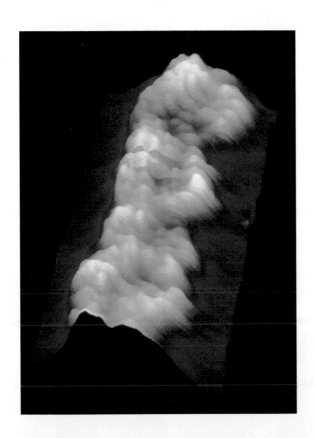

The organic compounds discussed in Chapter 17 were relatively small molecules, ranging from those with four or five atoms (such as methane or formaldehyde) to long-chain hydrocarbons, with up to 30 carbon atoms and relative molecular masses of several hundred. In addition to these smaller molecules, carbon atoms string together in stable chains of essentially unlimited length. Such chains provide the backbones of truly huge molecules that may contain hundreds of thousands or even millions of atoms. Such compounds, called **polymers,** are formed by the linking of numerous separate small **monomer units** in strands and webs.

*Illustration*
A false-color scanning tunneling micrograph (STM) of a DNA double-helix molecule adsorbed on a graphite substrate. *(Driscoll, Youngquist, and Baldeschwieler, California Institute of Technology/Science Photo Library/Photo Researchers, Inc.)*

Although many polymers are based on the ability of carbon to form stable long-chain molecules with various functional groups attached, carbon is not unique in this ability. (Recall from Chapter 23 the chains, sheets, and networks found in natural silicates, in which the elements silicon and oxygen join together to form extended structures.) This chapter, however, focuses on organic polymers, whose chemical and physical properties depend on the bonding and functional group chemistry discussed in Section 17.3. We examine both synthetic polymers (built largely from the hydrocarbon raw materials discussed in Section 17.2) and naturally occurring biopolymers such as starch, proteins, and nucleic acids (built from products of biological synthesis).

---

## 25.1

# POLYMERIZATION REACTIONS FOR SYNTHETIC POLYMERS

To construct a polymer, very many monomers must add to a growing polymer molecule, and the reaction must not falter after the first few molecules have reacted. This is achieved by having the polymer molecule retain highly reactive functional groups at all times during its synthesis. The two major types of polymer growth are addition polymerization and condensation polymerization.

In **addition polymerization,** monomers react to form a polymer chain without net loss of atoms. The most common type of addition polymerization involves the free-radical chain reaction of molecules that have C=C double bonds. As in the chain reactions considered in Section 13.4, the overall process consists of three steps: initiation, propagation (repeated many times to build up a long chain), and termination. As an example, consider the polymerization of vinyl chloride (chloroethene, $CH_2$=CHCl) to polyvinyl chloride (Fig. 25.1). This process can be initiated by a small concentration of molecules that have bonds weak enough to be broken by the

(a)

(b)

**FIGURE 25.1** (a) This chemical plant in Texas produces several billion kilograms of polyvinyl chloride each year. The spherical tanks store gaseous raw material. (b) A pipefitting of polyvinyl chloride. *(a, Occidental Petroleum; b, Leon Lewandowski)*

action of light or heat, giving radicals. An example of such an **initiator** is a peroxide, which can be represented as R—O—O—R′, where R and R′ represent alkyl groups. The weak O—O bonds break

$$R—\overset{..}{\underset{..}{O}}—\overset{..}{\underset{..}{O}}—R' \longrightarrow R—\overset{..}{\underset{..}{O}}\cdot + \cdot\overset{..}{\underset{..}{O}}—R' \quad \text{(initiation)}$$

to give radicals, whose oxygen valence shells are incomplete. The radicals remedy this by reacting avidly with vinyl chloride, accepting electrons from the C=C double bonds to reestablish a closed-shell electron configuration on the oxygen atoms:

$$R—\overset{..}{\underset{..}{O}}\cdot + CH_2{=}CHCl \longrightarrow R—\overset{..}{\underset{..}{O}}—CH_2—\overset{\overset{\textstyle H}{|}}{\underset{\underset{\textstyle Cl}{|}}{C}}\cdot \quad \text{(propagation)}$$

One of the two $\pi$ electrons in the vinyl chloride double bond has been used to form a single bond with the R—O· radical. The other remains on the second carbon atom, leaving it as a seven-valence-electron atom that will react with another vinyl chloride molecule:

$$R—O—CH_2—\overset{\overset{\textstyle H}{|}}{\underset{\underset{\textstyle Cl}{|}}{C}}\cdot + CH_2{=}\overset{\overset{\textstyle H}{|}}{\underset{\underset{\textstyle Cl}{|}}{C}} \longrightarrow R—O—CH_2—\overset{\overset{\textstyle H}{|}}{\underset{\underset{\textstyle Cl}{|}}{C}}—CH_2—\overset{\overset{\textstyle H}{|}}{\underset{\underset{\textstyle Cl}{|}}{C}}\cdot \quad \text{(propagation)}$$

At each stage, the end group of the lengthening chain is one electron short of a valence octet and remains quite reactive. The reaction can continue, building up long-chain molecules of high molecular mass. The vinyl chloride monomers always attach to the growing chain with their $CH_2$ group, because the odd electron is more stable on a CHCl end group. This gives the polymer a regular alternation of —$CH_2$— and —CHCl— groups. Its chemical formula is $+CH_2CHCl\frac{}{}_n$.

Termination occurs when the radical end groups on two different chains encounter each other and the two chains couple to give a longer chain:

$$R—O+CH_2—CHCl\frac{}{}_m CH_2—\overset{\overset{\textstyle H}{|}}{\underset{\underset{\textstyle Cl}{|}}{C}}\cdot + \cdot\overset{\overset{\textstyle H}{|}}{\underset{\underset{\textstyle Cl}{|}}{C}}—CH_2+CHCl—CH_2\frac{}{}_n O—R' \longrightarrow$$

$$R—O+CH_2—CHCl\frac{}{}_m CH_2—\overset{\overset{\textstyle H}{|}}{\underset{\underset{\textstyle Cl}{|}}{C}}—\overset{\overset{\textstyle H}{|}}{\underset{\underset{\textstyle Cl}{|}}{C}}—CH_2+CHCl—CH_2\frac{}{}_n O—R'$$

Alternatively, a hydrogen atom may transfer from one end group to the other:

$$R—O+CH_2—CHCl\frac{}{}_m\overset{\overset{\textstyle \textcircled{H}\ H}{|\ \ |}}{\underset{\underset{\textstyle H\ \ Cl}{|\ \ |}}{C—C}}\cdot + \cdot\overset{\overset{\textstyle H}{|}}{\underset{\underset{\textstyle Cl}{|}}{C}}—CH_2+CHCl—CH_2\frac{}{}_n O—R' \longrightarrow$$

$$R—O+CH_2—CHCl\frac{}{}_m CH{=}CHCl + CH_2Cl—CH_2+CHCl—CH_2\frac{}{}_n O—R'$$
$$\text{(termination)}$$

The latter termination step leaves a double bond on one chain end and a —$CH_2Cl$ group on the other. When the polymer molecules are long, the exact natures of the end groups have little effect on the physical and chemical properties of the material.

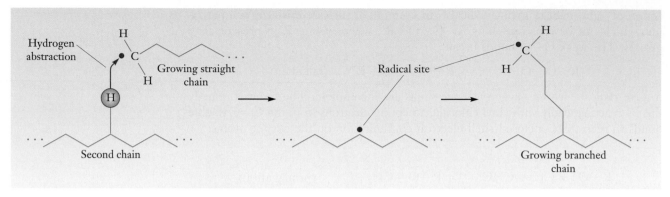

**FIGURE 25.2** A growing branched chain is created during free-radical polymerization when a radical at one end of a growing straight chain extracts a hydrogen atom from the middle of a second chain.

A different type of hydrogen transfer step often has a much greater effect on the properties of the resulting polymer. Suppose that a hydrogen atom transfers not from the monomer unit on the *end* of a second chain but from a monomer unit in the middle of that chain (Fig. 25.2). Then the first chain stops growing, but the radical site moves to the middle of the second chain, and growth resumes from that point, forming a *branched* polymeric chain with very different properties.

Addition polymerization can be initiated by ions as well as by free radicals. An example is the polymerization of acrylonitrile:

$$n \ CH_2{=}CH \longrightarrow \left[ CH_2CH \right]_n$$
$$\quad\quad\quad\;\; |\quad\quad\quad\quad\quad |$$
$$\quad\quad\quad C{\equiv}N \quad\quad\quad\quad C{\equiv}N$$

A suitable initiator for this process is butyl lithium, $(CH_3CH_2CH_2CH_2)^-Li^+$. The butyl anion (abbreviated $Bu^-$) reacts with the end carbon atom in a molecule of acrylonitrile to give a new anion:

$$Bu^{\ominus}Li^{\oplus} + CH_2{=}CH \longrightarrow Bu{-}CH_2{-}CH^{\ominus}Li^{\oplus} \quad \text{(initiation)}$$
$$\quad\quad\quad\quad\quad\quad | \quad\quad\quad\quad\quad\quad\quad\quad\quad\quad |$$
$$\quad\quad\quad\quad\quad\quad C{\equiv}N \quad\quad\quad\quad\quad\quad\quad\quad C{\equiv}N$$

The new anion then reacts with an additional molecule of acrylonitrile:

$$Bu{-}CH_2{-}CH^{\ominus}Li^{\oplus} + CH_2{=}CH \longrightarrow$$
$$\quad\quad\quad\quad | \quad\quad\quad\quad\quad\quad\quad |$$
$$\quad\quad\quad\quad C{\equiv}N \quad\quad\quad\quad\quad C{\equiv}N$$

$$Bu{-}CH_2{-}CH{-}CH_2{-}CH^{\ominus}Li^{\oplus} \quad \text{(propagation)}$$
$$\quad\quad\quad\quad\quad\quad | \quad\quad\quad\quad |$$
$$\quad\quad\quad\quad\quad C{\equiv}N \quad\quad\; C{\equiv}N$$

The process continues, building up a long-chain polymer.

Ionic polymerization differs from free-radical polymerization because the negatively charged end groups repel one another, ruling out termination by the coupling of two chains. The ionic group at the end of the growing polymer is stable at each stage. Once the supply of monomer has been used up, the polymer can exist indefinitely with its ionic end group, in contrast with the free-radical case, in which some reaction must take place to terminate the process. Ion-initiated polymers are called "living" polymers because, when additional monomer is added (even months later), they resume growth and increase in molecular mass. Termination can be achieved by adding water to replace the $Li^+$ with a hydrogen ion:

$$-(CH_2-CH)_n CH_2-CH^{\ominus}Li^{\oplus} + H_2O \longrightarrow$$
$$\quad\quad\quad |\quad\quad\quad\quad |$$
$$\quad\quad\quad C\equiv N\quad\quad C\equiv N$$

$$-(CH_2-CH)_n CH_2-CH_2 \; + \; Li^{\oplus} \; + \; OH^{\ominus} \quad\quad \text{(termination)}$$
$$\quad\quad\quad |\quad\quad\quad\quad |$$
$$\quad\quad\quad C\equiv N\quad\quad C\equiv N$$

A second important mechanism of polymerization is **condensation polymerization,** in which a small molecule (frequently water) is split off as each monomer unit is attached to the growing polymer.[1] An example is the polymerization of 6-aminohexanoic acid. The first two molecules react upon heating according to

$$\underset{HO}{\overset{O}{\|}}{C}-(CH_2)_5-N\overset{H}{\underset{\boxed{H\quad HO}}{}} \;+\; \overset{O}{\|}{C}-(CH_2)_5-NH_2 \longrightarrow$$

$$\underset{HO}{\overset{O}{\|}}{C}-(CH_2)_5-N\overset{\overset{O}{\|}}{\underset{H}{C}}(CH_2)_5-NH_2 + H_2O$$

An amide linkage and water form from the reaction of an amine with a carboxylic acid. The new molecule still has an amine group on one end and a carboxylic acid group on the other, and so it can react with two more molecules of 6-aminohexanoic acid. The process repeats to build up a long-chain molecule. For each monomer unit added, one molecule of water is split off. The final polymer in this case is called nylon 6 and is used in fiber-belted radial tires and in carpets.

Both addition and condensation polymerization can be carried out with mixtures of two or more types of monomers present in the reaction mixture. The result is a **random copolymer** that incorporates both types of monomers in an irregular sequence along the chain. For example, a 1:6 molar ratio of styrene to butadiene monomers is used to make styrene–butadiene rubber (SBR) for automobile tires and a 2:1 ratio gives a copolymer that is an ingredient in latex paints.

## Cross-Linking: Nonlinear Synthetic Polymers

If every monomer forming a polymer has only two reactive sites, then only chains and rings can be made. The 6-aminohexanoic acid used in making nylon 6, for example, has one amine group and one carboxylic acid group per molecule. When both functional groups react, one link is forged in the polymer chain, but that link cannot react further. If some or all of the monomers in a polymer have three or more reactive sites, however, then cross-linking to form sheets or networks is possible.

One important example of cross-linking involves phenol-formaldehyde copolymers (Fig. 25.3). When these two compounds are mixed (with the phenol in excess

---

[1] Condensation reactions have appeared several times outside the context of polymer synthesis. For example, two molecules of $H_2SO_4$ condense to form disulfuric acid ($H_2S_2O_7$) (see Fig. 21.3), and a carboxylic acid condenses with an alcohol to form an ester (see Section 17.3).

**FIGURE 25.3** When a mixture of phenol ($C_6H_5OH$) and formaldehyde ($CH_2O$) dissolved in acetic acid is treated with concentrated hydrochloric acid, a phenol-formaldehyde polymer grows. *(Charles D. Winters)*

in the presence of an acid catalyst), straight-chain polymers form. The first step is the addition of formaldehyde to phenol to give methylolphenol:

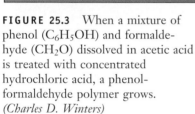

Molecules of methylolphenol then undergo condensation reactions (releasing water) to form a linear polymer called novalac:

If, on the other hand, the reaction is carried out with an excess of formaldehyde, di- and trimethylolphenols form:

Each of these monomers has more than two reactive sites and can react with up to three others to form a cross-linked polymer that is much stronger and more impact-resistant than the linear polymer. The very first synthetic plastic, Bakelite, was made in 1907 from cross-linked phenol and formaldehyde. Modern phenol-formaldehyde polymers are used as adhesives for plywood; more than a billion kilograms are produced per year in the United States.

Cross-linking is often desirable because it leads to a stronger material. Sometimes cross-linking agents are added deliberately to form additional bonds between polymer chains. Polybutadiene, for example, contains double bonds that can be linked upon addition of appropriate oxidizing agents. One especially important kind of cross-linking occurs through sulfur chains in rubber, as we will see.

---

## 25.2
## APPLICATIONS FOR SYNTHETIC POLYMERS

The three largest uses for polymers are in fibers, plastics, and elastomers (rubbers). We can distinguish between these three types of materials on the basis of their physical properties, especially their resistance to stretching. A typical fiber strongly resists stretching and can be elongated by less than 10% before breaking. Plastics are intermediate in their resistance to stretching and elongate 20 to 100% before breaking. Finally, elastomers stretch readily, with elongations of 100 to 1000% (that is, some types of rubber can be stretched by a factor of 10 without breaking). The remainder of this section examines the major kinds of polymers and their uses.

### Fibers

Many important fibers, including cotton and wool, are naturally occurring polymers. The first commercially successful synthetic polymers were made not by polymerization reactions but through the chemical regeneration of the natural polymer cellulose, a condensation polymer of the sugar glucose that is made by plants:

**FIGURE 25.4** Filter paper (cellulose) will dissolve in a concentrated ammonia solution containing $[Cu(NH_3)_4]^{2+}$ ions. When the solution is extruded into aqueous sulfuric acid, a dark blue thread of rayon (regenerated cellulose) precipitates. (*Leon Lewandowski*)

In the viscose rayon process, still employed today, cellulose is digested in a concentrated solution of NaOH to convert the —OH groups to $—O^- Na^+$ ionic groups. Reaction with $CS_2$ leads to the formation of about one "xanthate" group for every two glucose monomer units:

$$—\overset{|}{\underset{|}{C}}—O^{\ominus}Na^{\oplus} + CS_2 \longrightarrow —\overset{|}{\underset{|}{C}}—O—\overset{\overset{S}{\|}}{C}—S^{\ominus}Na^{\oplus}$$

Xanthate

Such substitutions reduce the hydrogen-bond forces holding polymer chains together. In the ripening step, some of these xanthate groups are removed with regeneration of $CS_2$, and others migrate to the —$CH_2OH$ groups from the ring —OH groups. Afterward, sulfuric acid is added to neutralize the NaOH and to remove the remaining xanthate groups. At the same time, the viscose rayon is spun out to form fibers (Fig. 25.4) while new hydrogen bonds form.

Rayon is a "semisynthetic" fiber because it is prepared from a natural polymeric starting material. The first truly synthetic polymeric fiber was nylon, developed in the 1930s by Wallace Carothers at Du Pont Company. He knew of the condensation of an amine with a carboxylic acid to form an amide linkage (see Section 17.3) and noted that, if each molecule had *two* amine or carboxylic acid functional groups, long-chain polymers could form. The specific starting materials upon which Carothers settled, after numerous attempts, were adipic acid and hexamethylenediamine:

$$\underset{\text{Adipic acid}}{\text{HO}-\overset{\overset{\text{O}}{\|}}{\text{C}}-(\text{CH}_2)_4-\overset{\overset{\text{O}}{\|}}{\text{C}}-\text{OH}} \qquad \underset{\text{Hexamethylenediamine}}{\text{H}_2\text{N}-(\text{CH}_2)_6-\text{NH}_2}$$

The two react with loss of water, according to the equation

$$\text{HO}-\overset{\overset{\text{O}}{\|}}{\text{C}}-(\text{CH}_2)_4-\overset{\overset{\text{O}}{\|}}{\text{C}}-\boxed{\text{OH}+\text{H}}-\underset{\overset{|}{\text{H}}}{\text{N}}-(\text{CH}_2)_6-\text{NH}_2 \longrightarrow$$

$$\text{HO}-\overset{\overset{\text{O}}{\|}}{\text{C}}-(\text{CH}_2)_4-\overset{\overset{\text{O}}{\|}}{\text{C}}-\underset{\overset{|}{\text{H}}}{\text{N}}-(\text{CH}_2)_6-\text{NH}_2 + \text{H}_2\text{O}$$

The resulting molecule has a carboxylic acid group on one end (which can react with another molecule of hexamethylenediamine) and an amine group on the other end (which can react with another molecule of adipic acid). The process can continue indefinitely, leading to a polymer with the formula

$$\left[ -\overset{\overset{\text{O}}{\|}}{\text{C}}-(\text{CH}_2)_4-\overset{\overset{\text{O}}{\|}}{\text{C}}-\underset{\overset{|}{\text{H}}}{\text{N}}-(\text{CH}_2)_6-\underset{\overset{|}{\text{H}}}{\text{N}}- \right]_n$$

called nylon 66 (Fig. 25.5). The nylon is extruded as a thread or spun as a fiber from the melt. The combination of well-aligned polymer molecules and N—H · · · O hydrogen bonds between chains makes nylon one of the strongest materials known. The designation "66" indicates that this nylon has six carbon atoms on the starting carboxylic acid and six on the diamine. Other nylons can be made with different numbers of carbon atoms.

Just as a carboxylic acid reacts with an amine to give an amide, so it reacts with an alcohol to give an ester. This suggests the possible reaction of a dicarboxylic acid and a glycol (dialcohol) to form a polymer. The polymer produced most extensively in this way is polyethylene terephthalate, which is built up from terephthalic acid (a benzene ring with —COOH groups on both ends) and ethylene glycol. The first two molecules react according to

**FIGURE 25.5** Hexamethylenediamine is dissolved in water (lower layer) and adipyl chloride, a derivative of adipic acid, is dissolved in hexane (upper layer). At the interface between the layers, nylon forms and is drawn out onto the stirring bar. (*Charles D. Winters*)

**TABLE 25.1**

*Fibers*

| Name | Structural Units | Properties | Sample Uses |
|------|-----------------|-----------|-------------|
| Rayon | Regenerated cellulose | Absorbent, soft, easy to dye, poor wash and wear | Dresses, suits, coats, curtains, blankets |
| Acetate | Acetylated cellulose | Fast drying, supple, shrink-resistant | Dresses, shirts, draperies, upholstery |
| Nylon | Polyamide | Strong, lustrous, easy to wash, smooth, resilient | Carpeting, upholstery, tents, sails, hosiery, stretch fabrics, rope |
| Dacron | Polyester | Strong, easy to dye, shrink-resistant | Permanent-press fabrics, rope, sails, thread |
| Acrylic (Orlon) | $-(CH_2-CH)_n-$ <br> $\quad\quad\;\; \vert$ <br> $\quad\quad C\equiv N$ | Warm, lightweight, resilient, quick-drying | Carpeting, sweaters, baby clothes, socks |

Adapted from P. J. Chenier, *Survey of Industrial Chemistry.* New York: John Wiley & Sons, 1986, Table 18.4.

Further reaction then builds up the polymer, which is called polyester and sold under trade names such as Dacron. The planar benzene rings in this polymer make it stiffer than nylon, which has no aromatic groups in its backbone, and help make polyester fabrics crush-resistant. The same polymer formed in a thin sheet rather than a fiber becomes Mylar, a very strong film used for audio and video tapes.

Table 25.1 summarizes the structures, properties, and uses of some important fibers.

## Plastics

**Plastics** are loosely defined as polymeric materials that can be molded or extruded into desired shapes and that harden upon cooling or solvent evaporation. Rather than being spun into threads in which their molecules are aligned, as in fibers, plastics are cast into three-dimensional forms or spread into films for packaging applications. Although celluloid articles were fabricated by plastic processing by the late 1800s, the first important synthetic plastic was Bakelite, the phenol-formaldehyde resin whose cross-linking was discussed earlier in this section. Table 25.2 lists some of the most important plastics and their properties.

Ethylene ($CH_2=CH_2$) is the simplest monomer that will polymerize. Through free-radical-initiated addition polymerization at high pressures (1000–3000 atm) and temperatures (300–500°C), it forms polyethylene:

$$n\; CH_2=CH_2 \longrightarrow -(CH_2-CH_2)_n-$$

The polyethylene formed in this way is not the perfect linear chain implied by this simple equation. Free radicals frequently abstract hydrogen from the middles of chains in this synthesis, and so the polyethylene is heavily branched with hydrocarbon side chains of varying length. It is **low-density polyethylene** (LDPE) because the difficulty of packing the irregular side chains gives it a lower density ($<0.94\ \text{g cm}^{-3}$) than that of perfectly linear polyethylene. This irregularity also makes it relatively

**TABLE 25.2**

*Plastics*

| Name | Structural Units | Properties | Sample Uses |
|---|---|---|---|
| Polyethylene | $+CH_2—CH_2\frac{}{}_n$ | High density: hard, strong, stiff | Molded containers, lids, toys, pipe |
| | | Low density: soft, flexible, clear | Packaging, trash bags, squeeze bottles |
| Polypropylene | $+CH_2—CH\frac{}{}_n$ <br> \|  <br> $CH_3$ | Stiffer, harder than high-density polyethylene, higher melting point | Containers, lids, carpeting, luggage, rope |
| Polyvinyl chloride | $+CH_2—CH\frac{}{}_n$ <br> \|  <br> $Cl$ | Nonflammable, resistant to chemicals | Water pipes, roofing, credit cards, records |
| Polystyrene | $+CH_2—CH\frac{}{}_n$ <br> ⬡ | Brittle, flammable, not resistant to chemicals, easy to process and dye | Furniture, toys, refrigerator linings, insulation |
| Phenolics | Phenol-formaldehyde copolymer | Resistant to heat, water, chemicals | Plywood adhesive, Fiberglas binder, circuit boards |

Adapted from P. J. Chenier, *Survey of Industrial Chemistry.* New York: John Wiley & Sons, 1986, pp. 252–264.

soft, and so its primary uses are in coatings, plastic packaging, trash bags, and squeeze bottles in which softness is an advantage, not a drawback.

A major breakthrough occurred in 1954, when the German chemist Karl Ziegler showed that ethylene could also be polymerized with a catalyst consisting of $TiCl_4$ and an organoaluminum compound (for example, $Al(C_2H_5)_3$). The addition of ethylene takes place at each stage within the coordination sphere of the titanium atom, so that monomers can add only at the end of the growing chain. The result is linear polyethylene, also called **high-density polyethylene** (HDPE) because of its density ($0.96 \text{ g cm}^{-3}$). Because its linear chains are regular, HDPE contains large crystalline regions, which make it much harder than LDPE and thus suitable for molding into plastic bowls, lids, and toys.

A third kind of polyethylene introduced in the late 1970s is called **linear low-density polyethylene** (LLDPE). It is made by the same metal-catalyzed reactions as HDPE, but it is a deliberate copolymer with other 1-alkenes such as 1-butene. It has some side groups (which reduce the crystallinity and density), but they are of a controlled short length as opposed to the irregular, long side branches in LDPE. LLDPE is stronger and more rigid than LDPE; it is also less expensive because lower pressures and temperatures are used in its manufacture.

If one of the hydrogen atoms of the ethylene monomer unit is replaced with a different type of atom or functional group, the plastics that form upon polymerization have different properties. Substitution of a methyl group (that is, the use of propylene as monomer) leads to polypropylene:

$$\left[ CH_2—CH \atop \quad\quad | \atop \quad\quad CH_3 \right]_n$$

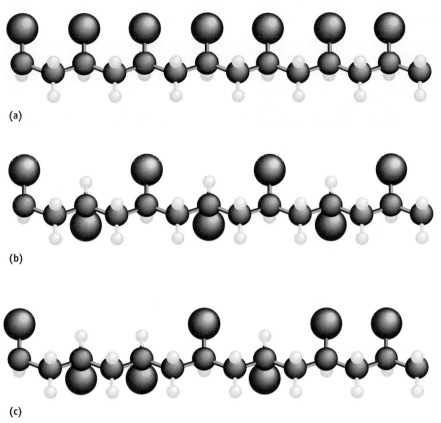

(a)

(b)

(c)

**FIGURE 25.6**  The structures of (a) isotactic, (b) syndiotactic, and (c) atactic polypropylene. In these structures the purple spheres represent —CH$_3$ (methyl) side groups on the long chain.

This reaction cannot be carried out successfully by free-radical polymerization, however. It was first achieved in 1953–1954 by Ziegler and the Italian chemist Giulio Natta, who used the Ziegler catalyst later employed in making HDPE. In polypropylene, the methyl groups attached to the carbon backbone can be arranged in different conformations (Fig. 25.6). In the **isotactic form,** all the methyl groups are arranged on the same side, whereas in the **syndiotactic form** they alternate in a regular fashion. The **atactic form** shows a random positioning of methyl groups. Natta showed that the Ziegler catalyst led to isotactic polypropylene, and he developed another catalyst, using VCl$_4$, that gave the syndiotactic form. Polypropylene plastic is stiffer and harder than HDPE and has a higher melting point, and so it is particularly useful in applications requiring high temperatures (such as the sterilization of medical instruments).

In polystyrene, a benzene ring replaces one hydrogen atom of each ethylene monomer unit. Because such a ring is bulky, atactic polystyrene does not crystallize to any significant extent. The most familiar application of this polymer is in the polystyrene foam used in disposable containers for food and drinks and as insulation. A volatile liquid or a compound that dissociates to gaseous products on heating is added to the molten polystyrene. It forms bubbles that remain as the polymer is cooled and molded. The gas-filled pockets in the final product make it a good thermal insulator.

## Rubber

An **elastomer** is a polymer that can be deformed to a great extent and still recover its original form when the deforming stress is removed. The term "rubber" was introduced by Joseph Priestley, who observed that such materials can be used to rub out pencil marks. Natural rubber is a polymer of isoprene (2-methylbutadiene). The isoprene molecule contains two double bonds of which polymerization removes only one, and so natural rubber is unsaturated, containing one double bond per isoprene unit. In polymeric isoprene, the geometry at each double bond can be either *cis* or *trans* (Fig. 25.7). Natural rubber is all-*cis* polyisoprene. The all-*trans* form also occurs in nature in the sap of certain trees and is called gutta-percha. This material is used to cover golf balls because it is particularly tough. Isoprene can be polymerized by free-radical addition polymerization, but the resulting polymer contains a mixture of *cis* and *trans* double bonds and is useless as an elastomer.

Even pure natural rubber is of limited usefulness because it melts, is soft, and does not fully spring back to its original form after being stretched. In 1839 the American inventor Charles Goodyear discovered that if sulfur is added to rubber and the mixture is heated, the rubber hardens, becomes more resilient, and does not melt. This process is referred to as **vulcanization** and involves the formation of sulfur bridges between the methyl side groups on different chains. Small amounts of sulfur (<5%) yield an elastic material in which sulfur links between chains remain after stretching and enable the rubber to regain its original form when the external force is removed. Large amounts of sulfur give the very hard, nonelastic material ebonite.

Research on synthetic substitutes for natural rubber began in the United States and Europe before World War II. Attention focused on copolymers of butadiene with styrene (now called SBR rubber) and with acrylonitrile (NBR rubber). The Japanese occupation of the rubber-producing countries of Southeast Asia sharply curtailed the supply of natural rubber to the Allied nations, and rapid steps were taken to increase production of synthetic rubber. The initial production goal was 40,000 tons per year of SBR. By 1945, U.S. production had reached an incredible total of more than 600,000 tons per year. During those few years, many advances were made in production techniques, quantitative analysis, and basic understanding of rubber elasticity. Styrene–butadiene rubber production continued after the war and in 1950 SBR exceeded natural rubber in overall production volume for the first time. More recently, several factors have favored natural rubber: the increasing cost of the hydrocarbon feed stock for synthetic rubber, gains in productivity of natural rubber, and the growing preference for belted radial tires, which use more natural rubber.

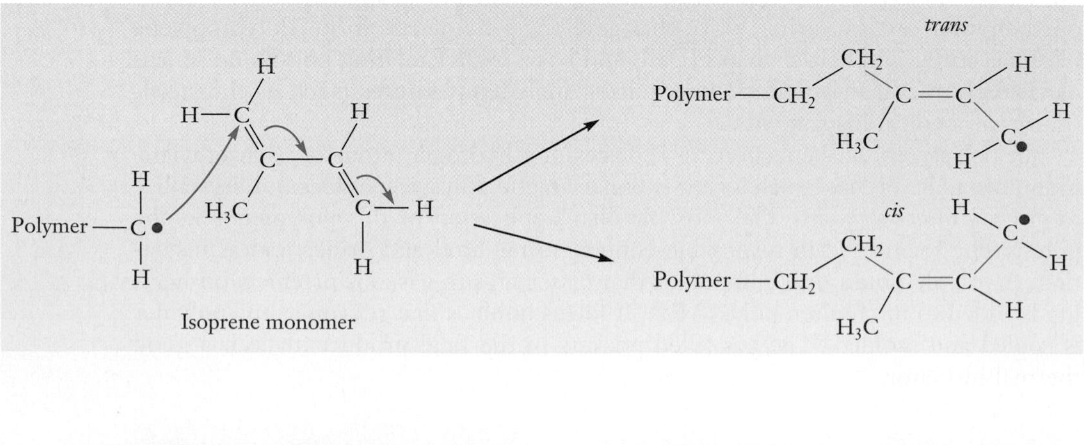

**FIGURE 25.7** In the polymerization of isoprene, a *cis* or *trans* configuration can form at each double bond in the polymer. The blue arrows show the redistribution of the electrons upon bond formation.

The development of the Ziegler–Natta catalysts has affected rubber production as well. First, it facilitated the synthesis of all-*cis* polyisoprene and the demonstration that its properties were nearly identical to those of natural rubber. (A small amount of "synthetic natural rubber" is produced today.) Second, a new kind of synthetic rubber was developed: all-*cis* polybutadiene. It now ranks second in production after styrene–butadiene rubber.

---

## 25.3

# NATURAL POLYMERS

All the products of human ingenuity in the design of polymers pale beside the products of nature. Plants and animals employ a tremendous variety of long-chain molecules with different functions: some for structural strength, others to act as catalysts, and still others to provide instructions for the synthesis of vital components of the cell. In this section we discuss these three important classes of natural polymers: polysaccharides, proteins, and nucleic acids.

## Carbohydrates and Polysaccharides

**Carbohydrates** form a class of compounds of carbon with hydrogen and oxygen. The name comes from the chemical formulas of these compounds, which can be written $C_n(H_2O)_m$, suggesting a "hydrate" of carbon. Simple **sugars,** or **monosaccharides,** are carbohydrates with the chemical formula $C_nH_{2n}O_n$. Sugars with three, four, five, and six carbon atoms are called trioses, tetroses, pentoses, and hexoses, respectively.

Glucose is a hexose sugar that exists in several forms in solution (Fig. 25.8). There is a rapid equilibrium between a straight-chain form (a six-carbon molecule with five —OH groups and one aldehyde —CHO group) and a cyclic form, in which the ring is composed of five carbon atoms and one oxygen, with four —OH side groups and one —CH$_2$OH side group. In the straight-chain form, four of the carbon atoms (those numbered 2 through 5) are chiral centers, with four different groups bonded to them. As discussed in Section 17.1 (see Fig. 17.4), each such carbon atom can exist in two configurations, giving rise to $2^4 = 16$ distinct hexose sugars.

**FIGURE 25.8** D-glucose exists in two ring forms in solution (a and c), which interconvert via an open-chain form (b). The two rings differ in the placement of the —OH and —H groups on carbon atom 1.

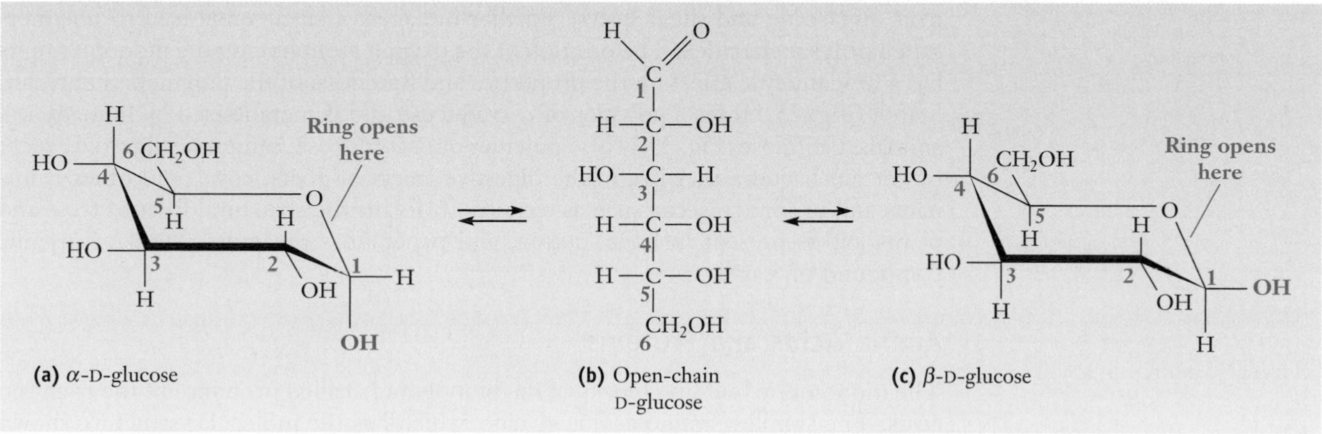

(a) $\alpha$-D-glucose

(b) Open-chain D-glucose

(c) $\beta$-D-glucose

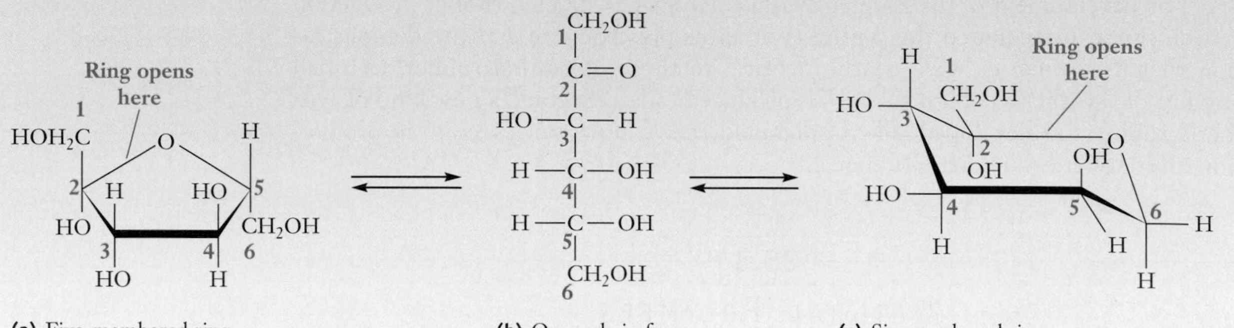

**(a)** Five-membered ring          **(b)** Open-chain form          **(c)** Six-membered ring

**FIGURE 25.9**   In aqueous solutions of the sugar D-fructose, an equilibrium exists among a five-atom ring, an open chain, and a six-atom ring. In addition to the $\beta$ isomers shown here, both ring forms have $\alpha$ isomers, in which the —CH$_2$OH and —OH on carbon 2 are exchanged.

The glucose formed in plant photosynthesis always has the chirality shown in Figure 25.8b. Of the 15 other straight-chain hexose sugars, the only ones found in nature are D-galactose (in the milk sugar lactose) and D-mannose (a plant sugar).

Figure 25.8 shows that glucose actually has two different ring forms, depending on whether the —OH group created from the aldehyde by the closing of the ring lies above or below the plane of the ring. Another way to see this is to note that closing the ring creates a fifth chiral carbon atom. The two ring forms of D-glucose are called $\alpha$-D-glucose (Fig. 25.8a) and $\beta$-D-glucose (Fig. 25.8c). In aqueous solution, these two forms interconvert rapidly via the open-chain glucose form and cannot be separated. They can be isolated separately in crystalline form, however. D-fructose, a common sugar found in fruit and honey, has the same molecular formula as D-glucose but is a member of a class of hexose sugars that are ketones rather than aldehydes. In their straight-chain forms, these sugars have the C=O double bond at carbon atom 2 rather than carbon atom 1 (Fig. 25.9).

Many plant cells do not stop the synthesis process with simple sugars such as glucose, but continue by linking sugars together to form more complex carbohydrates. **Disaccharides** are composed of two simple sugars linked together by a condensation reaction with the elimination of water. Examples shown in Figure 25.10 are the milk sugar lactose and the plant sugar sucrose (ordinary table sugar, extracted from sugarcane and sugar beets). Further linkages of sugar units lead to polymers called **polysaccharides.** The position of the oxygen atom linking the monomer units has a fundamental effect on the properties and functions of the polymers that result. Starch (Fig. 25.11a) is a polymer of $\alpha$-D-glucose and is metabolized by humans and animals. Cellulose (Fig. 25.11b), a polymer of $\beta$-D-glucose, cannot be digested except by certain bacteria that live in the digestive tracts of goats, cows, and other ruminants and in some insects, such as termites. It forms the structural fiber of trees and plants and is present in linen, cotton, and paper. It is the most abundant organic compound on earth.

## Amino Acids and Proteins

The monomeric building blocks of the biopolymers called proteins are the $\alpha$-amino acids. The simplest amino acid is glycine, which has the molecular structure shown

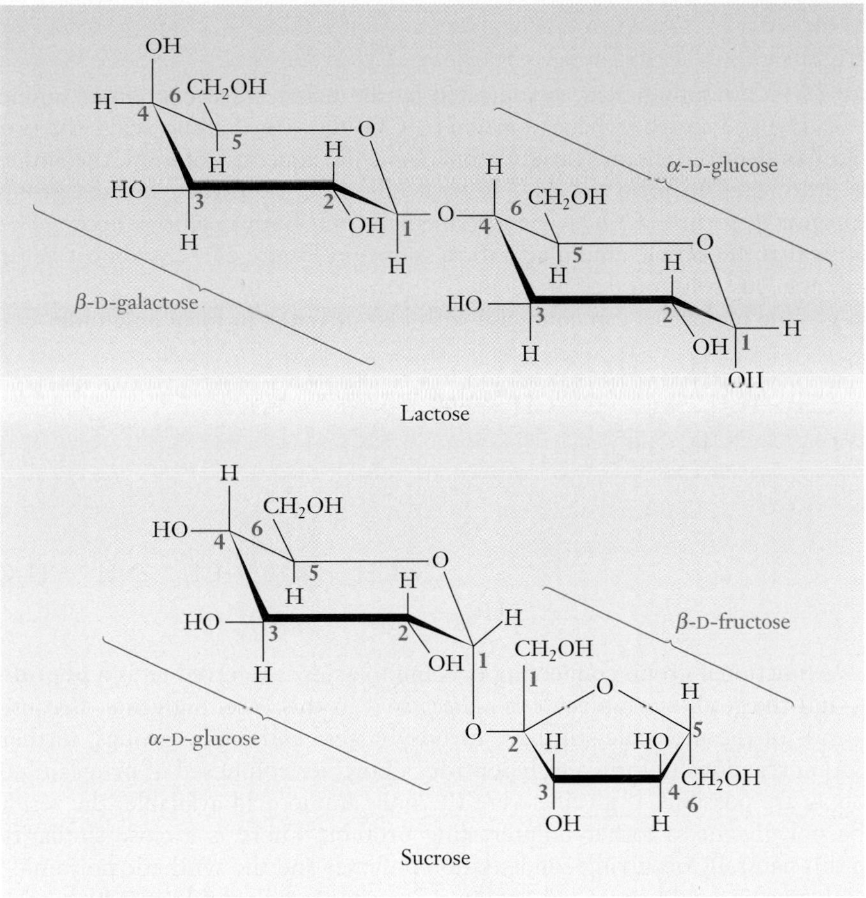

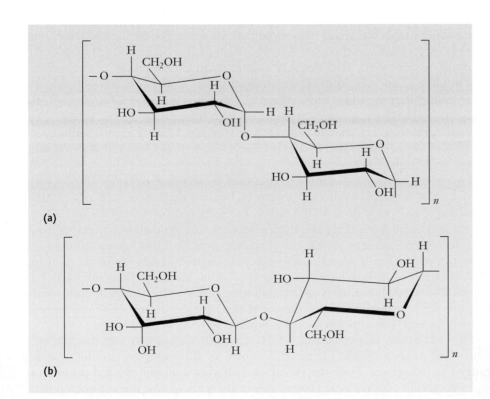

**FIGURE 25.10** Two disaccharides. Their derivations from monosaccharide building blocks are shown.

$\beta$-D-galactose

$\alpha$-D-glucose

Lactose

$\alpha$-D-glucose

$\beta$-D-fructose

Sucrose

**(a)**

**(b)**

**FIGURE 25.11** Both starch (a) and cellulose (b) are polymers of glucose. In starch, all the cyclic glucose units are $\alpha$-D-glucose. In cellulose, all the monomer units are $\beta$-D-glucose.

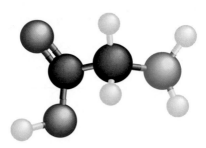

**FIGURE 25.12** The structure of glycine. On the left side is the carboxylic acid group (—COOH), and on the right is the amine group (—NH₂).

in Figure 25.12. An amino acid, as indicated by the name, must contain an amine group (—NH₂) and a carboxylic acid group (—COOH). In α-amino acids, the two groups are bonded to the same carbon atom. In acidic aqueous solution, the amine group is protonated to form —NH₃⁺; in basic solution, the carboxylic acid group loses a proton to form —COO⁻. At intermediate pH, both reactions occur. The net result is that the simple amino acid form shown in Figure 25.12 is almost never present in aqueous solution (see problem 27).

Two glycine molecules can condense with loss of water to form an amide:

$$\underset{HO}{\overset{O}{\|}}C-CH_2-N\overset{H}{\underset{\boxed{H \quad HO}}{}} + \underset{}{\overset{O}{\|}}C-CH_2-NH_2 \longrightarrow$$

$$\underset{HO}{\overset{O}{\|}}C-CH_2-N\underset{H}{|}-\overset{O}{\overset{\|}{C}}-CH_2-NH_2 + H_2O$$

The amide functional group connecting two amino acids is referred to as a **peptide linkage,** and the resulting molecule is a *dipeptide*—in this case, diglycine. Because the two ends of the molecule still have carboxylic acid and amine groups, further condensation reactions to form a **polypeptide,** a polymer composed of many amino acid groups, are possible. If glycine were the only amino acid available, the result would be polyglycine, a rather uninteresting protein. There is a close similarity between this naturally occurring condensation polymer and the synthetic polyamide nylon. Polyglycine could be called "nylon 2," a simple polyamide in which each repeating unit contains two carbon atoms.

Nature does not stop with glycine as a monomer unit. Instead, any of 20 different α-amino acids are found in most natural polypeptides. In each of these, one of the hydrogen atoms on the central carbon atom of glycine is replaced by another side group. Alanine is the next simplest α-amino acid after glycine; it has a —CH₃ group in place of an —H atom. This substitution has a profound consequence. In alanine, four different groups are attached to a central carbon: —COOH, —NH₂, —CH₃, and —H. There are two ways in which four different groups can be arranged in a tetrahedral structure about a central atom (see Fig. 17.4). The two optical isomers of alanine are designated by the prefixes L- and D- for *levo* and *dextro* (Latin for "left" and "right," respectively).

If a mixture of L- and D-alanine were caused to polymerize, nearly all the polymer molecules would have different structures, because their sequences of D-alanine and L-alanine monomer units would differ. To create polymers with definite structures for particular roles, there is only one recourse: to build all polypeptides from one of the optical isomers so that the properties will be reproducible from molecule to molecule. Nearly all naturally occurring α-amino acids are of the L form, and most earthly organisms have no use for D-α-amino acids in making polypeptides. Terrestrial life could presumably have begun equally well using mainly D-amino acids (all biomolecules would be mirror images of their present forms). The mechanism by which the established preference was initially selected is not known.

The —H group of glycine and —CH₃ group of alanine give just the first two amino acid building blocks. Table 25.3 shows all 20 important α-amino acids, arranged by side group. Note the variety in their chemical and physical properties. Some side groups contain basic groups; others are acidic. Some are compact; others are bulky. Some can take part in hydrogen bonds; others can complex readily with metal ions to form coordination complexes.

This variety in properties of the $\alpha$-amino acids leads to even more variety in the polymers derived from them, called **proteins.** The term "protein" is usually applied to polymers with more than about 50 amino acid groups; large proteins may contain many thousand such groups. Given the fact that any one of 20 $\alpha$-amino acids may appear at each point in the chain, the number of possible sequences of amino acids in even small proteins is staggering. Moreover, the amino acid sequence describes only one aspect of the molecular structure of a protein. It contains no information about the three-dimensional conformation adopted by the protein. The carbonyl group and the amine group in each amino acid along the protein chain are

---

## TABLE 25.3

### $\alpha$-Amino Acid Side Groups

| | Symbol | Structure of Side Group |
|---|---|---|
| **Hydrogen "Side Group"** | | |
| Glycine | Gly | —H |
| **Alkyl Side Groups** | | |
| Alanine | Ala | —CH$_3$ |
| Valine | Val | 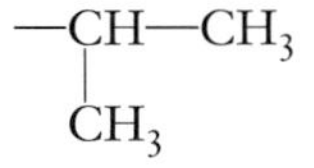 —CH—CH$_3$ <br>     CH$_3$ |
| Leucine | Leu | 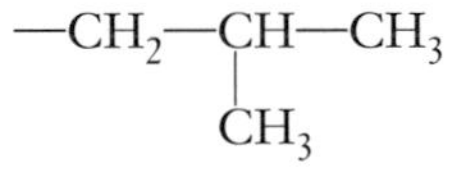 —CH$_2$—CH—CH$_3$ <br>        CH$_3$ |
| Isoleucine | Ile | 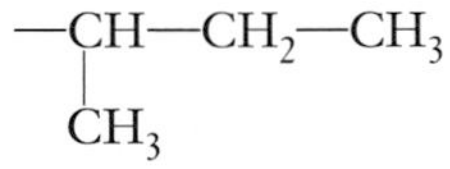 —CH—CH$_2$—CH$_3$ <br>   CH$_3$ |
| Proline | Pro <br> (structure of entire amino acid) | 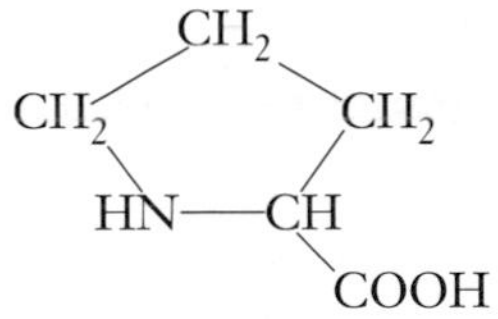 |
| **Aromatic Side Groups** | | |
| Phenylalanine | Phe | 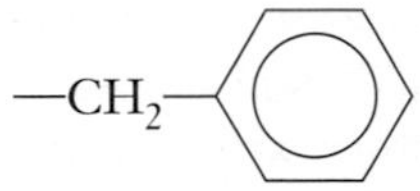 |
| Tyrosine | Tyr |  |
| Tryptophan | Trp | 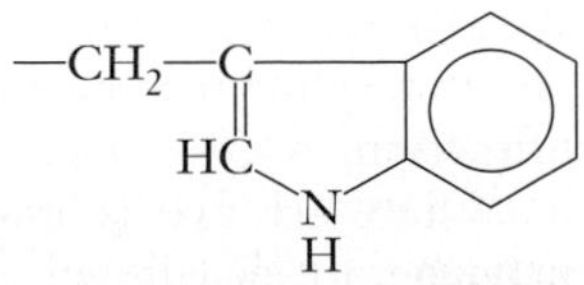 |

*(Continued)*

## TABLE 25.3

### α-Amino Acid Side Groups (continued)

|  | Symbol | Structure of Side Group |
|---|---|---|
| **Alcohol-Containing Side Groups** | | |
| Serine | Ser | $-CH_2OH$ |
| Threonine | Thr | $-\overset{\displaystyle OH}{\underset{\displaystyle CH_3}{CH}}$ |
| **Basic Side Groups** | | |
| Lysine | Lys | $-CH_2CH_2CH_2CH_2NH_2$ |
| Arginine | Arg | $-CH_2CH_2CH_2NH-C\overset{\nearrow NH}{\searrow NH_2}$ |
| Histidine | His | $-CH_2-C=CH$ with ring $HN$, $N$, $\underset{H}{C}$ |
| **Acidic Side Groups** | | |
| Aspartic acid | Asp | $-CH_2COOH$ |
| Glutamic acid | Glu | $-CH_2CH_2COOH$ |
| **Amide-Containing Side Groups** | | |
| Asparagine | Asn | $-CH_2\overset{\displaystyle O}{\overset{\|}{C}}-NH_2$ |
| Glutamine | Gln | $-CH_2CH_2\overset{\displaystyle O}{\overset{\|}{C}}-NH_2$ |
| **Sulfur-Containing Side Groups** | | |
| Cysteine | Cys | $-CH_2-SH$ |
| Methionine | Met | $-CH_2CH_2-S-CH_3$ |

potential sites for hydrogen bonds, which may also involve functional groups on the amino acid side chains. Also, the cysteine side groups ($-CH_2-SH$) can react with one another, with loss of hydrogen, to form $-CH_2-S-S-CH_2-$ disulfide bridges between different cysteine groups in a single chain or between neighboring chains (the same kind of cross-linking by sulfur occurs in the vulcanization of rubber). As a result of these strong intrachain interactions, the molecules of a given protein have a rather well-defined conformation even in solution, as compared with the much more varied range of conformations available to a simple alkane chain (see Fig. 17.8). The three-dimensional structures of many proteins have been determined by x-ray diffraction.

There are two primary categories of proteins: fibrous and globular. **Fibrous proteins** are usually structural materials and consist of polymer chains linked in sheets or twisted in long fibers. Silk is a fibrous protein in which the monomer units

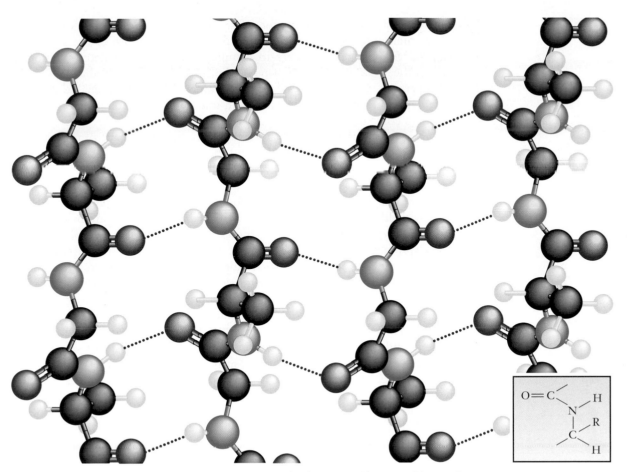

**FIGURE 25.13**    The structure of silk. The amino acid side groups (shown as R) must be small to fit into a sheet-like structure. Glycine (R=H) and alanine (R=CH$_3$) predominate. The repeating unit is shown as an inset.

are primarily glycine and alanine, with smaller amounts of serine and tyrosine. The protein chains are cross-linked by hydrogen bonds to form sheet-like structures (Fig. 25.13) that are arranged so that the nonhydrogen side groups all lie on one side of the sheet; the sheets then stack in layers. The relatively weak forces between sheets give silk its characteristic smooth feel. The amino acids in wool and hair have side chains that are larger, bulkier, and less regularly distributed than those in silk, and so sheet structures do not form. Instead, the protein molecules twist into a right-handed coil called an **$\alpha$-helix** (Fig. 25.14). In this structure, each carbonyl group is hydrogen-bonded to the amine group of the fourth amino acid farther along the chain; the bulky side groups jut out from the helix and do not interfere with one another.

The second type of protein is the **globular protein.** Globular proteins include the carriers of oxygen in the blood (hemoglobin) and in cells (myoglobin). They have irregular folded structures (Fig. 25.15) and typically consist of 100 to 1000 amino acid groups in one or more chains. Globular proteins frequently have parts of their structures in $\alpha$-helices and sheets, with other portions in more disordered forms. Hydrocarbon side groups tend to cluster in regions that exclude water, whereas charged and polar side groups tend to remain in close contact with water.

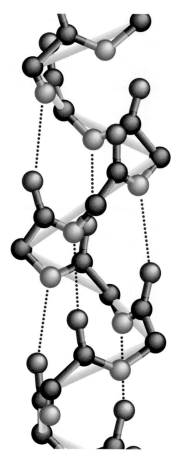

The sequences of amino acid units for many such proteins have been worked out by cleaving them into smaller pieces and analyzing the structure of the fragments. It took Frederick Sanger ten years to complete the first such determination of sequence for the 51 amino acids in bovine insulin, an accomplishment that earned him the Nobel prize in chemistry in 1958. Now, automated procedures enable scientists to rapidly determine amino acid sequences in much longer protein molecules.

**Enzymes** constitute a very important class of globular proteins. They catalyze particular reactions in the cell, such as the synthesis and breakdown of proteins, the transport of substances across cell walls, and the recognition and resistance of foreign bodies. Enzymes act by lowering the activation barrier for a reaction, and they must be selective so as to act only on a restricted group of substrates.

Let us examine the enzyme carboxypeptidase A, whose structure has been determined by x-ray diffraction. It removes amino acids one at a time from the carboxylic acid end of a polypeptide. Figure 25.16 shows the structure of the active site (with a peptide chain in place, ready to be cleaved). A special feature of this enzyme is the role played by the zinc ion, which is coordinated to two histidine residues in the enzyme and to a carboxylate group on a nearby glutamic acid residue. The zinc ion helps to remove electrons from the carbonyl group of the peptide linkage, making it more positive and thereby more susceptible to attack by water or by the carboxylate group of a second glutamic acid residue. The side chain on the outer amino acid of the peptide being cleaved is positioned in a hydrophobic cavity, which favors large aromatic or branched side chains (such as that in tyrosine) over smaller hydrophilic side chains (such as that in aspartic acid). Carboxypeptidase A is thus selective in the rates with which it cleaves peptide chains.

The molecular "engineering" that lies behind nature's design of carboxypeptidase A and other enzymes is truly remarkable. The amino acid residues that form the active site and determine its catalytic properties are *not* adjacent to one another

**FIGURE 25.14** The structure of an α-helix for a stretch of linked glycine monomer units. The superimposed yellow line highlights the helical structure, which is maintained by hydrogen bonds (red dotted lines). The hydrogen atoms themselves are omitted for clarity.

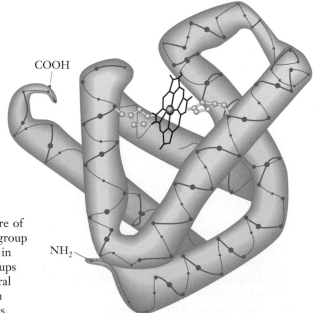

**FIGURE 25.15** A computer-generated model of the structure of myoglobin. The central heme group (red) is shown in greater detail in Figure 24.8c; two histidine groups (green) extend toward the central iron atom. Much of the protein (blue tube) is coiled in α-helices.

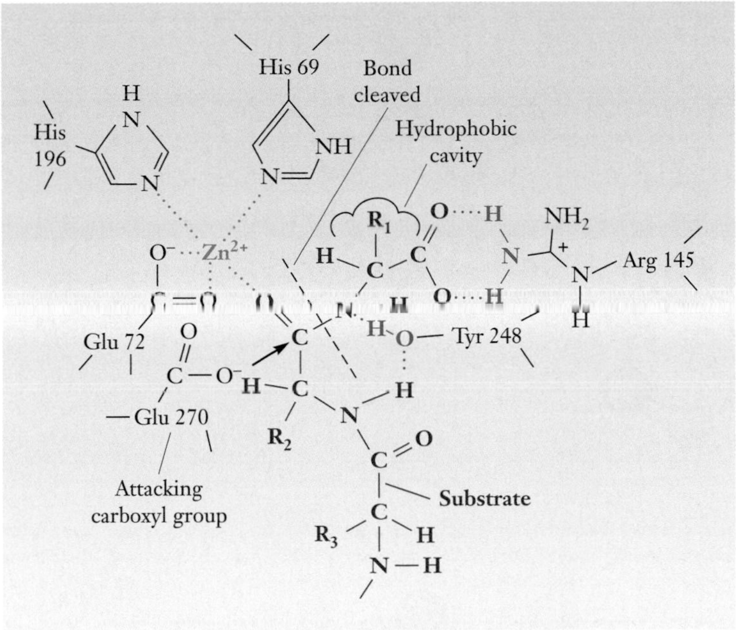

**FIGURE 25.16** The active site of carboxypeptidase A. Shown in red is a substrate polypeptide that is being cleaved by the enzyme. Green is used to show the role of $Zn^{2+}$ as a complexing ion, and blue is used to show the hydrogen bonds that maintain the geometry.

in the protein chain. As indicated by the numbers after the residues in Figure 25.16, the two glutamic acid residues are the 72nd and 270th amino acids along the chain. The enzyme adopts a conformation in which the key residues, distant from one another in terms of chain position, are nonetheless quite close in three-dimensional space, allowing the enzyme to carry out its specialized function.

## Nucleotides and Nucleic Acids

We have seen that proteins are copolymers made up typically of 20 types of monomer units. Simply mixing the amino acids and letting them dehydrate to form polymer chains at random would never lead to the particular structures needed by living cells. How does the cell preserve information about the amino acid sequences that make up its proteins, and how does it transmit this information to daughter cells through the reproductive process? These questions lie in the field of molecular genetics, an area in which chemistry plays the central role.

The primary genetic material is deoxyribonucleic acid (DNA). This biopolymer is made up of four types of monomer units called **nucleotides.** Each nucleotide is composed of three parts:

1. One molecule of a pyrimidine or purine base. The four bases are thymine, cytosine, adenine, and guanine (Fig. 25.17a).
2. One molecule of the sugar D-deoxyribose ($C_5H_{10}O_4$). D-ribose is a pentose sugar with a five-membered ring.
3. One molecule of phosphoric acid ($H_3PO_4$).

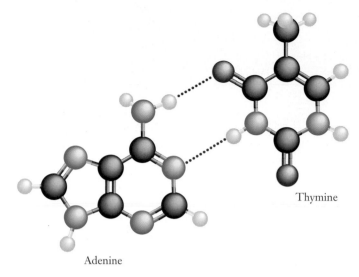

Thymine

Adenine

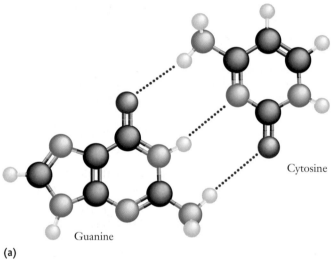

Cytosine

Guanine

(a)

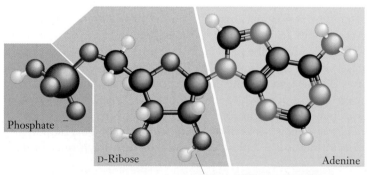

Phosphate

D-Ribose

Adenine

This —OH group is replaced
by —H in the deoxy form
found in DNA.

(b)

**FIGURE 25.17**    (a) The structures of the purine and pyrimidine bases. Hydrogen bonding between pairs of bases is indicated by red dots. (b) The structure of the nucleotide adenosine monophosphate (AMP).

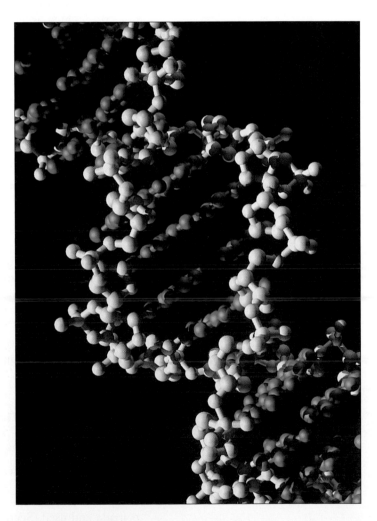

**FIGURE 25.18** The double-helix structure of DNA. *(Right, Copyright Kenneth Edward/Biografx/Photo Researchers, Inc.)*

The cyclic sugar molecule links the base to the phosphate group, undergoing two condensation reactions, with loss of water, to form the nucleotide (Fig. 25.17b). The first key to discovering the structure of DNA was the observation that, although the proportions of the four bases in DNA from different organisms were quite variable, the chemical amount of cytosine (C) was always approximately equal to that of guanine (G), and the chemical amount of adenine (A) was always approximately equal to that of thymine (T). This suggested some type of base pairing in DNA that could lead to association of C with G and A with T. The second crucial observation was an x-ray diffraction study by Rosalind Franklin and Maurice Wilkins that suggested the presence of helical structures of more than one chain in DNA.

James Watson and Francis Crick put together these two pieces of information in their famous 1953 proposal of a double-helix structure for DNA. They concluded that DNA consists of two interacting helical strands of nucleic acid polymer (Fig. 25.18), with each cytosine on one strand linked through hydrogen bonds to a guanine on the other and each adenine to a thymine. This accounted for the observed molar ratios of the bases, and it also provided a model for the replication of the molecule, which is crucial for passing on information during the reproductive process. One DNA strand serves as a template upon which a second DNA strand is synthesized. A DNA molecule reproduces by starting to unwind at one end. As it does so, new nucleotides are guided into position opposite the proper bases on each of the two strands. If the nucleotide does not fit the template, it cannot link to the polymeric strand under construction. The result of the polymer synthesis is two double-helix molecules, each containing one strand from the original and one new strand that is identical to the original in every respect.

Information is encoded in DNA in the sequence of the base pairs. Subsequent research has broken this genetic code and established the connection between the base sequence in a segment of DNA and the amino acid sequence of the protein synthesized according to the directions in that segment. The code in a nucleic acid is read as consecutive, nonoverlapping triplets of bases, with each triplet standing for a particular amino acid. Thus, a nucleic acid strand consisting of pure cytosine gives a polypeptide of pure proline, meaning that the triplet CCC codes for proline. The nucleic acid strand AGAGAGAG... is read as the alternating triplets AGA and GAG and gives a polypeptide consisting of alternating arginine (coded by AGA) and glutamic acid (coded by GAG) monomer units. There are 64 ($4^3$) possible triplets, and so typically more than one code exists for a particular amino acid. Some triplets serve as signals to terminate a polypeptide chain. Remarkably, the genetic code appears to be universal, independent of the species of plant or animal, a finding that suggests a common origin for all terrestrial life.

## CONCEPTS & SKILLS

*After studying this chapter and working the problems that follow, you should be able to*

1. Contrast the methods of addition and condensation polymerization (Section 25.1, problems 1–6).
2. Give several examples of fibers, plastics, and rubbers and describe how they are made and used (Section 25.2, problems 7–10).
3. Describe the formation of polysaccharides from sugars, proteins from amino acids, and DNA from nucleotides and the roles of these biopolymers in living cells (Section 25.3, problems 11–18).

*Answers to problems whose numbers are boldface appear in Appendix G. Problems that are more challenging are indicated with asterisks.*

## Polymerization Reactions for Synthetic Polymers

1. Write a balanced chemical equation to represent the addition polymerization of 1,1-dichloroethylene. The product of this reaction is Saran, used as a plastic wrap.

2. Write a balanced chemical equation to represent the addition polymerization of tetrafluoroethylene. The product of this reaction is Teflon.

3. A polymer produced by addition polymerization consists of $-(CH_2-O)-$ groups joined in a long chain. What was the starting monomer?

4. The polymer polymethyl methacrylate is used to make Plexiglas. It has the formula

$$
\left[ -CH_2-\underset{\underset{O \quad\quad OCH_3}{\underset{\|}{C}}}{\overset{CH_3}{\underset{|}{C}}}- \right]_n
$$

Draw the structural formula of the starting monomer.

5. The monomer glycine ($NH_2-CH_2-COOH$) can undergo condensation polymerization to form polyglycine, in which the structural units are joined by amide linkages.
   (a) What molecule is split off in the formation of polyglycine?
   (b) Draw the structure of the repeat unit in polyglycine.

6. The polymer $-(NH-CH(CH_3)-\overset{\overset{O}{\|}}{C})_n$ forms upon condensation polymerization with loss of water. Draw the structure of the starting monomer.

7. Determine the mass of adipic acid and the mass of hexamethylenediamine needed to make $1.00 \times 10^3$ kg of nylon 66 fiber.

8. Determine the mass of terephthalic acid and the mass of ethylene glycol needed to make 10.0 kg of polyester fiber.

9. In a recent year, 4.37 billion kilograms of low-density polyethylene was produced in the United States. What volume of gaseous ethylene at 0°C and 1.00 atm would give this amount?

10. In a recent year, 2.84 billion kilograms of polystyrene was produced in the United States. Polystyrene is the addition polymer formed from the styrene monomer, $C_6H_5CH=CH_2$. How many styrene monomer units were incorporated in that 2.84 billion kilograms of polymer?

## Natural Polymers

11. By referring to Figure 25.10a, draw the structure of the ring form of $\beta$-D-galactose. How many asymmetric carbon atoms (chiral centers) are there in the molecule?

12. By referring to Figure 25.10b, draw the structure of the ring form of D-ribose. How many asymmetric carbon atoms (chiral centers) are there in the molecule?

13. How many tripeptides can be synthesized using just three different species of $\alpha$-amino acids?

14. How many different polypeptides, each containing ten amino acids, can be made from the amino acids listed in Table 25.3? How many different polypeptides, each containing 100 amino acids, can be made?

15. Draw the structure of the pentapeptide alanine–leucine–phenylalanine–glycine–isoleucine. Assume that the free $-NH_2$ group is at the alanine end of the peptide chain. Would this compound be more likely to dissolve in water or in octane? Explain.

16. Draw the structure of the pentapeptide aspartic acid–serine–lysine–glutamic acid–tyrosine. Assume that the free $-NH_2$ group is at the aspartic acid end of the peptide chain. Would this compound be more likely to dissolve in water or in octane? Explain.

17. Suppose a long-chain polypeptide is constructed entirely from phenylalanine monomer units. What is its empirical formula? How many amino acids does it contain if its molar mass is 17,500 g mol$^{-1}$?

18. A typical bacterial DNA has a molar mass of $4 \times 10^9$ g mol$^{-1}$. Approximately how many nucleotides does it contain?

## Additional Problems

19. In the addition polymerization of acrylonitrile, a very small amount of butyl lithium causes a reaction that can consume hundreds of pounds of the monomer; however, the butyl lithium is called an initiator, not a catalyst. Explain why.

20. Based on the fact that the free-radical polymerization of ethylene is spontaneous and the fact that polymer molecules are less disorganized than the starting monomers, decide whether the polymerization reaction is exothermic or endothermic. Explain.

21. Approximately 950 million lb of ethylene dichloride was exported from the United States in a recent year, according to a trade journal. The article states that "between 500 million and 550 million pounds of PVC could have been made from that ethylene dichloride." Compute the range of percentage yields of PVC from ethylene dichloride that is implied by these figures.

22. The complete hydrogenation of natural rubber (the addition of $H_2$ to all double bonds) gives a product that is indistinguishable from the product of the complete hydrogenation of gutta-percha. Explain how this strengthens the conclusion that these two substances are isomers of each other.

23. A reducing solution breaks S—S bonds in proteins, whereas an oxidizing solution allows them to re-form. Discuss how such solutions might be used to carry out the permanent waving of hair.

24. L-sucrose tastes sweet, but it is not metabolized. It has been suggested as a potential nonnutritive sweetener. Draw the molecular structure of L-sucrose, using Figure 25.10b as a starting point.

25. Polypeptides are synthesized from a 50:50 mixture of L-alanine and D-alanine. How many different isomeric molecules containing 22 monomer units are possible?

26. An osmotic pressure measurement taken on a solution containing hemoglobin shows that the molar mass of that protein is approximately 65,000 g mol$^{-1}$. A chemical analysis shows it to contain 0.344% of iron by mass. How many iron atoms does each hemoglobin molecule contain?

\* 27. At very low pH, alanine is a diprotic acid that can be represented as $H_3N^+$—$CH(CH_3)$—$COOH$. The $pK_a$ of the carboxyl group is 2.3, and the $pK_a$ of the —$NH_3^+$ group is 9.7.
   (a) At pH 7, what fraction of the amino acid molecules dissolved in an aqueous solution will have the form $H_3N^+$—$CH(CH_3)$—$COO^-$?
   (b) What fraction of the molecules at this pH will have the form $H_2N$—$CH(CH_3)$—$COOH$?

28. The sequence of bases in one strand of DNA reads ACTTGACCG. Write the sequence of bases in the complementary strand.

# Appendices

A crystal of elemental bismuth. *(Charles D. Winters)*

A.1

# Scientific Notation and Experimental Error

---

## A.1

### SCIENTIFIC NOTATION

Very large and very small numbers are common in chemistry. Repeatedly writing such numbers in the ordinary way (for example, the important number 602,213,700,000,000,000,000,000) would be tedious and would engender errors. **Scientific notation** offers a better way. A number in scientific notation is expressed as a number from 1 to 10 multiplied by 10 raised to some power. Any number can be represented in this way, as the following examples show.

$$643.8 = 6.438 \times 10^2$$

$$-19,000,000 = -1.9 \times 10^7$$

$$0.0236 = 2.36 \times 10^{-2}$$

$$602,213,700,000,000,000,000,000 = 6.022137 \times 10^{23}$$

A simple rule of thumb is that the power to which 10 is raised is $n$ if the decimal point is moved $n$ places to the left and is $-n$ if the decimal is moved $n$ places to the right.

When two or more numbers written in scientific notation are to be added or subtracted, they should first be expressed as multiples of the *same* power of 10:

$$
\begin{array}{rcr}
6.431 \times 10^4 & \longrightarrow & 6.431 \times 10^4 \\
+\ 2.1\ \ \times 10^2 & \longrightarrow & +\ 0.021 \times 10^4 \\
+\ 3.67\ \ \times 10^3 & \longrightarrow & +\ 0.367 \times 10^4 \\
\hline
? & & 6.819 \times 10^4
\end{array}
$$

When two numbers in scientific notation are multiplied, the coefficients are multiplied and then the powers of 10 are multiplied (by adding the exponents):

$$1.38 \times 10^{-16} \times 8.80 \times 10^3 = (1.38 \times 8.80) \times 10^{(-16+3)}$$

$$= 12.1 \times 10^{-13} = 1.21 \times 10^{-12}$$

We divide one number by a second by dividing the coefficients and then multiplying by 10 raised to the first exponent minus the second (exponents are subtracted):

$$\frac{6.63 \times 10^{-27}}{2.34 \times 10^{-16}} = \frac{6.63}{2.34} \times \frac{10^{-27}}{10^{-16}}$$

$$= 2.83 \times 10^{[-27-(-16)]} = 2.83 \times 10^{-11}$$

Any calculator or computer equipped to perform scientific and engineering calculations can accept and display numbers in scientific notation. It cannot determine whether the input has an error, however, or whether the answer makes sense. That is your responsibility! Develop the habit of mentally estimating the order of magnitude of the answer as a rough check on your calculator's result.

---

## A.2

## EXPERIMENTAL ERROR

Chemistry is an experimental science in which every quantitative measurement is subject to some degree of error. We can seek to reduce error by carrying out additional measurements or by changing our experimental apparatus, but we can never eliminate error altogether. It is important, therefore, to be able to assess the results of an experiment quantitatively to establish the limits of the experiment's validity. Errors are of two types: random (lack of precision) and systematic (lack of accuracy).

### Precision and Random Errors

**Precision** refers to the degree of agreement in a collection of experimental results and is estimated by repeating the measurement under conditions as nearly identical as possible. If the conditions are truly identical, then differences among the trials are due to random error. As a specific example, consider some actual results of an early, important experiment by American physicist Robert Millikan in 1909, to measure the charge $e$ on the electron. The experiment (discussed in greater detail in Chapter 1) involved a study of the motion of charged oil drops suspended in air in an electric field. Millikan made hundreds of measurements on many different oil drops, but we shall consider only a set of results for $e$ found for one particular drop (Table A.1). The values he found ranged from 4.894 to $4.941 \times 10^{-10}$ esu. What do we choose to report as the best estimate for $e$? The proper procedure is to first examine the data to see whether any of the results are especially far from the rest (a value above $5 \times 10^{-10}$ esu would fall into this category). Such values are likely to result from some mistake in carrying out or reporting that particular measurement and therefore are excluded from further consideration (although there have been cases in science where just such exceptional results have led to significant breakthroughs). In Millikan's data, no such points should be excluded. To obtain our best estimate for $e$, we calculate the **mean,** or **average value,** by adding up the values found and dividing by the number of measurements. We can write the average value of any property after a series of $N$ measurements $x_1, x_2, \ldots, x_N$ as

$$\bar{x} = \frac{1}{N}(x_1 + x_2 + \cdots + x_N) = \frac{1}{N}\sum_{i=1}^{N} x_i$$

## TABLE A.1

| Measurement Number $e$ ($10^{-10}$ esu) | | | | | | | | | | | | |
|---|---|---|---|---|---|---|---|---|---|---|---|---|
| 1 | 2 | 3 | 4 | 5 | 6 | 7 | 8 | 9 | 10 | 11 | 12 | 13 |
| 4.915 | 4.920 | 4.937 | 4.923 | 4.931 | 4.936 | 4.941 | 4.902 | 4.927 | 4.900 | 4.904 | 4.897 | 4.894 |

From R. A. Millikan, *Phys. Rev.* 32:349, 1911. [1 esu = $3.3356 \times 10^{-10}$ C]

where a capital Greek sigma ($\Sigma$) is introduced to indicate a summation of $x_i$ over values of $i$ from 1 to $N$. In the present case, this gives an average for $e$ of $4.917 \times 10^{-10}$ esu.

This average by itself does not convey any estimate of uncertainty. If all of the measurements had given results between 4.91 and $4.92 \times 10^{-10}$ esu, the uncertainty would be less than if the results had ranged from $4 \times 10^{-10}$ to $6 \times 10^{-10}$ esu. Furthermore, an average of 100 measurements should have less uncertainty than an average of 5. How are these ideas made quantitative? A statistical measure of the spread of data, called the **standard deviation** $\sigma$, is useful in this regard. It is given by the formula

$$\sigma = \sqrt{\frac{(x_1 - \bar{x})^2 + (x_2 - \bar{x})^2 + \cdots + (x_N - \bar{x})^2}{N - 1}}$$

$$= \sqrt{\frac{1}{N-1} \sum_{i=1}^{N} (x_i - \bar{x})^2}$$

The standard deviation is found by adding up the squares of the deviations of the individual data points from the average value $\bar{x}$, dividing by $N - 1$, and taking the square root. Table A.2 shows how $\sigma$ is used quantitatively. A **confidence limit** is defined as

$$\text{confidence limit} = \pm \frac{t\sigma}{\sqrt{N}}$$

The table gives the factor $t$ for various numbers of measurements, $N$, and for various levels of confidence.

For Millikan's data, $N = 13$ and $\sigma = 0.017 \times 10^{-10}$. For 95% confidence with 13 measurements, the table shows $t = 2.18$ and the confidence limit is

$$\text{confidence limit} = \pm \frac{(2.18)(0.017 \times 10^{-10})}{\sqrt{13}} = \pm 0.010 \times 10^{-10} \text{ esu}$$

Thus, a 95% probability exists that the *true* average (obtained by repeating the experiment under the same conditions an infinite number of times) will lie within

### TABLE A.2

| N (Number of Observations) | Factor $t$ for Confidence Interval of | | | |
|---|---|---|---|---|
| | 80% | 90% | 95% | 99% |
| 2 | 3.08 | 6.31 | 12.7 | 63.7 |
| 3 | 1.89 | 2.92 | 4.30 | 9.92 |
| 4 | 1.64 | 2.35 | 3.18 | 5.84 |
| 5 | 1.53 | 2.13 | 2.78 | 4.60 |
| 6 | 1.48 | 2.02 | 2.57 | 4.03 |
| 7 | 1.44 | 1.94 | 2.45 | 3.71 |
| 8 | 1.42 | 1.90 | 2.36 | 3.50 |
| 9 | 1.40 | 1.86 | 2.31 | 3.36 |
| 10 | 1.38 | 1.83 | 2.26 | 3.25 |
| 11 | 1.37 | 1.81 | 2.23 | 3.17 |
| 12 | 1.36 | 1.80 | 2.20 | 3.11 |
| 13 | 1.36 | 1.78 | 2.18 | 3.06 |
| 14 | 1.35 | 1.77 | 2.16 | 3.01 |
| 15 | 1.34 | 1.76 | 2.14 | 2.98 |
| $\infty$ | 1.29 | 1.64 | 1.96 | 2.58 |

$\pm 0.010 \times 10^{-10}$ esu of the average $4.917 \times 10^{-10}$ esu. Within this 95% confidence level, our best estimate for $e$ is written as

$$(4.917 \pm 0.010) \times 10^{-10} \text{ esu}$$

For other confidence levels and other numbers of measurements, the factor $t$ and therefore the confidence limit change.

## Accuracy and Systematic Errors

The charge $e$ on the electron has been measured by several different techniques since Millikan's day. The current best estimate for $e$ is

$$e = (4.8032068 \pm 0.0000015) \times 10^{-10} \text{ esu}$$
$$= (1.60217733 \pm 0.00000049) \times 10^{-19} \text{ C}$$

This value lies outside the range of uncertainty we estimated from Millikan's original data. In fact, it lies well below the smallest of the 13 measurements of $e$. Why?

To understand this discrepancy, we need to remember that there is a second source of error in any experiment: *systematic* error that causes a shift in the measured values from the true value and reduces the **accuracy** of the result. By making more measurements, we can reduce the uncertainty due to *random* errors and improve the *precision* of our result, but if systematic errors are present, the average value will continue to deviate from the true value. Such systematic errors may result from a miscalibration of the experimental apparatus or from a fundamental inadequacy in the technique for measuring a property. In the case of Millikan's experiment, the then-accepted value for the viscosity of air (used in calculating the charge $e$) was subsequently found to be wrong. This caused his results to be systematically too high.

Error thus arises from two sources. Lack of precision (random errors) can be estimated by a statistical analysis of a series of measurements. Lack of accuracy (systematic errors) is much more problematical. If a systematic error is known to be present, we should do our best to correct for it before reporting the result. (For example, if our apparatus has not been calibrated correctly, it should be recalibrated.) The problem is that systematic errors of which we have no knowledge may be present. In this case, the experiment should be repeated with different apparatus to eliminate the systematic error caused by a particular piece of equipment; better still, a different and independent way to measure the property might be devised. Only after enough independent experimental data are available can we be convinced of the accuracy of a result—that is, how closely it approximates the true result.

---

## A.3

## SIGNIFICANT FIGURES

The number of **significant figures** is the number of digits used to express a measured or calculated quantity, excluding zeros that may precede the first nonzero digit. Suppose the mass of a sample of sodium chloride is measured to be 8.241 g and the uncertainty is estimated to be $\pm 0.001$ g. The mass is said to be given to four significant figures because we are confident of the first three digits (8, 2, 4) and the uncertainty appears in the fourth (1), which nevertheless is still significant. Writing additional digits beyond the 1 would not be justified, however, unless the accuracy of the weighing could be improved. When we record a volume as 22.4 L, we imply that the uncertainty in the measurement is in the last digit written ($V = 22.4 \pm 0.3$ L, for example). A volume written as 22.43 L, on the other hand, implies that the uncertainty is far less and appears only in the fourth significant figure.

In the same way, writing 20.000 m is quite different from writing 20.0 m. The second measurement (with three significant figures) could easily be made with a common meterstick. The first (with five significant figures) would require a more precise method. We should avoid reporting results such as "700 m," however, because the two trailing zeros may or may not be significant. The uncertainty in the measurement could be of order $\pm 1$ m or $\pm 10$ m or perhaps $\pm 100$ m; it is impossible to tell which without further information. To avoid this ambiguity, we can write such measurements using the scientific notation described in Section A.1. The measurement "700 m" translates into any of the following:

$$7.00 \times 10^2 \text{ m} \qquad \text{Three significant figures}$$

$$7.0 \times 10^2 \text{ m} \qquad \text{Two significant figures}$$

$$7 \times 10^2 \text{ m} \qquad \text{One significant figure}$$

Frequently it is necessary to combine several different experimental measurements to obtain a final result. Some operations involve addition or subtraction, and others entail multiplication or division. These operations affect the number of significant figures that should be retained in the calculated result. Suppose, for example, that a weighed sample of 8.241 g of sodium chloride is dissolved in 160.1 g of water. What will be the mass of the solution that results? It is tempting to simply write $160.1 + 8.241 = 168.341$ g, but this is *not* correct. In saying that the mass of water is 160.1 g, we imply that there is some uncertainty about the number of tenths of a gram measured. This uncertainty must also apply to the sum of the masses, and so the last two digits in the sum are not significant and should be **rounded off**, leaving 168.3 as the final result.

> Following addition or subtraction, round off the result to the leftmost decimal place that contained an uncertain digit in the original numbers.

Rounding off is a straightforward operation. It consists of first discarding the digits that are not significant and then adjusting the last remaining digit. If the first discarded digit is less than 5, the remaining digits are left as they are (for example, 168.341 is rounded down to 168.3 because the first discarded digit, 4, is less than 5). If the first discarded digit is greater than 5, or if it is equal to 5 and is followed by one or more nonzero digits, then the last digit is increased by 1 (for example, 168.364 and 168.3503 both become 168.4 when rounded off to four digits). Finally, if the first digit discarded is 5 and all subsequent digits are zeros, the last digit remaining is rounded to the nearest even digit (for example, both 168.35 and 168.45 would be rounded to 168.4). This last rule is chosen so that, on the average, as many numbers are rounded up as down. Other conventions are sometimes used.

In multiplication or division it is not the number of decimal places that matters (as in addition or subtraction) but the number of significant figures in the least precisely known quantity. Suppose, for example, the measured volume of a sample is 4.34 cm$^3$ and its mass is 8.241 g. The density, found by dividing the mass by the volume on a calculator, for example, is

$$\frac{8.241 \text{ g}}{4.34 \text{ cm}^3} = 1.89884\ldots \text{g cm}^{-3}$$

How many significant figures should we report? Because the volume is the less precisely known quantity (three significant figures as opposed to four for the mass), it controls the precision that may properly be reported in the answer. Only three significant figures are justified, and so the result is rounded to 1.90 g cm$^{-3}$.

> The number of significant figures in the result of a multiplication or division is the smallest of the numbers of significant figures used as input.

It is best to carry out the arithmetical operations and *then* round the final answer to the correct number of significant figures, rather than round off the input data first. The difference is usually small, but this recommendation is nevertheless worth following. For example, the correct way to add the three distances 15 m, 6.6 m, and 12.6 m is

| | | |
|---|---|---|
| 15   m | | 15   m $\longrightarrow$ 15 m |
| +  6.6 m | | 6.6 m $\longrightarrow$  7 m |
| + 12.6 m | rather than | 12.6 m $\longrightarrow$ 13 m |
| 34.2 m $\longrightarrow$ 34 m | | 35 m |

For the same reason, we frequently carry extra digits through the intermediate steps of a worked example and round off only for the final answer. If a calculation is done entirely on a scientific calculator or a computer, several extra digits are usually carried along automatically. Before the final answer is reported, however, it is important to round off to the proper number of significant figures.

Sometimes pure constant appear in expressions. In this case, the accuracy of the result is determined by the accuracy of the other factors. The uncertainty in the volume of a sphere, $\frac{4}{3}\pi r^3$, depends only on the uncertainty in the radius $r$; 4 and 3 are pure constants (4.000 . . . and 3.000 . . . , respectively), and $\pi$ can be given to as many significant figures (3.14159265 . . .) as are warranted by the radius.

---

## PROBLEMS

*Answers to problems whose numbers are boldface appear in Appendix G.*

### Scientific Notation

**1.** Express the following in scientific notation.
  (a) 0.0000582
  (b) 1402
  (c) 7.93
  (d) −6593.00
  (e) 0.002530
  (f) 1.47

2. Express the following in scientific notation.
  (a) 4579
  (b) −0.05020
  (c) 2134.560
  (d) 3.825
  (e) 0.0000450
  (f) 9.814

3. Convert the following from scientific notation to decimal form.
  (a) $5.37 \times 10^{-4}$
  (b) $9.390 \times 10^{6}$
  (c) $-2.47 \times 10^{-3}$
  (d) $6.020 \times 10^{-3}$
  (e) $2 \times 10^{4}$

4. Convert the following from scientific notation to decimal form.
  (a) $3.333 \times 10^{-3}$
  (b) $-1.20 \times 10^{7}$
  (c) $2.79 \times 10^{-5}$
  (d) $3 \times 10^{1}$
  (e) $6.700 \times 10^{-2}$

**5.** A certain chemical plant produces $7.46 \times 10^8$ kg of polyethylene in one year. Express this amount in decimal form.

**6.** A microorganism contains 0.0000046 g of vanadium. Express this amount in scientific notation.

## Experimental Error

**7.** A group of students took turns using a laboratory balance to weigh the water contained in a beaker. The results they reported were 111.42 g, 111.67 g, 111.21 g, 135.64 g, 111.02 g, 111.29 g, and 111.42 g.
   (a) Should any of the data be excluded before the average is calculated?
   (b) From the remaining measurements, calculate the average value of the mass of the water in the beaker.
   (c) Calculate the standard deviation $\sigma$ and, from it, the 95% confidence limit.

**8.** By measuring the sides of a small box, a group of students made the following estimates for its volume: $544 \text{ cm}^3$, $590 \text{ cm}^3$, $523 \text{ cm}^3$, $560 \text{ cm}^3$, $519 \text{ cm}^3$, $570 \text{ cm}^3$, and $578 \text{ cm}^3$.
   (a) Should any of the data be excluded before the average is calculated?
   (b) Calculate the average value of the volume of the box from the remaining measurements.
   (c) Calculate the standard deviation $\sigma$ and, from it, the 90% confidence limit.

**9.** Of the measurements in problems 7 and 8, which is more precise?

**10.** A more accurate determination of the mass in problem 7 (using a better balance) gives the value 104.67 g, and an accurate determination of the volume in problem 8 gives the value $553 \text{ cm}^3$. Which of the two measurements in problems 7 and 8 is more accurate, in the sense of having the smaller systematic error relative to the actual value?

## Significant Figures

**11.** State the number of significant figures in each of the following measurements.
   (a) 13.604 L
   (b) $-0.00345°C$
   (c) 340 lb
   (d) $3.40 \times 10^2$ miles
   (e) $6.248 \times 10^{-27}$ J

**12.** State the number of significant figures in each of the following measurements.
   (a) $-0.0025$ in
   (b) 7000 g
   (c) 143.7902 s

   (d) $2.670 \times 10^7$ Pa
   (e) $2.05 \times 10^{-19}$ J

**13.** Round off each of the measurements in problem 11 to two significant figures.

**14.** Round off each of the measurements in problem 12 to two significant figures.

**15.** Round off the measured number 2,997,215.548 to nine significant digits.

**16.** Round off the measured number in problem 15 to eight, seven, six, five, four, three, two, and one significant digits.

**17.** Express the results of the following additions and subtractions to the proper number of significant figures. All of the numbers are measured quantities.
   (a) $67.314 + 8.63 - 243.198 =$
   (b) $4.31 + 64 + 7.19 =$
   (c) $3.1256 \times 10^{15} - 4.631 \times 10^{13} =$
   (d) $2.41 \times 10^{-26} - 7.83 \times 10^{-25} =$

**18.** Express the results of the following additions and subtractions to the proper number of significant figures. All of the numbers are measured quantities.
   (a) $245.876 + 4.65 + 0.3678 =$
   (b) $798.36 - 1005.7 + 129.652 =$
   (c) $7.98 \times 10^{17} + 6.472 \times 10^{19} =$
   (d) $(4.32 \times 10^{-15}) + (6.257 \times 10^{-14}) - (2.136 \times 10^{-13}) =$

**19.** Express the results of the following multiplications and divisions to the proper number of significant figures. All of the numbers are measured quantities.
   (a) $\dfrac{-72.415}{8.62} =$
   (b) $52.814 \times 0.00279 =$
   (c) $(7.023 \times 10^{14}) \times (4.62 \times 10^{-27}) =$
   (d) $\dfrac{4.3 \times 10^{-12}}{9.632 \times 10^{-26}} =$

**20.** Express the results of the following multiplications and divisions to the proper number of significant figures. All of the numbers are measured quantities.
   (a) $129.587 \times 32.33 =$
   (b) $\dfrac{4.7791}{3.21 \times 5.793} =$
   (c) $\dfrac{10566.9}{3.584 \times 10^{29}} =$
   (d) $(5.247 \times 10^{13}) \times (1.3 \times 10^{-17}) =$

**21.** Compute the area of a triangle (according to the formula $A = \frac{1}{2}ba$) if its base and altitude are measured to equal 42.07 cm and 16.0 cm, respectively. Justify the number of significant figures in the answer.

**22.** An inch is defined as exactly 2.54 cm. The length of a table is measured as 505.16 cm. Compute the length of the table in inches. Justify the number of significant figures in the answer.

# SI Units, Unit Conversions, and Physics for General Chemistry

This appendix reviews essential concepts of physics, as well as systems of units of measure, essential for work in general chemistry.

---
## B.1
## SI UNITS AND UNIT CONVERSIONS
---

Scientific work requires measurement of quantities or properties observed in the laboratory. Results are expressed not as pure numbers but rather as **dimensions** that reflect the nature of the property under study. For example, mass, time, length, area, volume, energy, and temperature are fundamentally distinct quantities, each of which is characterized by its unique dimension. The magnitude of each dimensioned quantity can be expressed in various **units** (for example, feet or centimeters for length). Several systems of units are available for use, and facility at conversion among them is essential for scientific work. Over the course of history, different countries evolved different sets of units to express length, mass, and many other physical dimensions. Gradually, these diverse sets of units are being replaced by international standards that facilitate comparison of measurements made in different localities and that help avoid complications and confusion. The unified system of units recommended by international agreement is called SI, which stands for "Système International d'Unités," or the International System of Units. In this section, we outline the use of SI units and discuss interconversions with other systems of units.

The SI uses seven **base units,** which are listed in Table B.1. All other units can be written as combinations of the base units. In writing the units for a measurement, we abbreviate them (see Table B.1), and we use exponential notation to denote the power to which a unit is raised; a minus sign appears in the exponent when the

---
### TABLE B.1

**Base Units in the International System of Units**

| Quantity | Unit | Symbol |
|---|---|---|
| Length | meter | m |
| Mass | kilogram | kg |
| Time | second | s |
| Temperature | kelvin | K |
| Chemical amount (of substance) | mole | mol |
| Electric current | ampere | A |
| Luminous intensity | candela | cd |

## TABLE B.2

### Derived Units in SI

| Quantity | Unit | Symbol | Definition |
|---|---|---|---|
| Energy | joule | J | $kg\ m^2\ s^{-2}$ |
| Force | newton | N | $kg\ m\ s^{-2} = J\ m^{-1}$ |
| Power | watt | W | $kg\ m^2\ s^{-3} = J\ s^{-1}$ |
| Pressure | pascal | Pa | $kg\ m^{-1}\ s^{-2} = N\ m^{-2}$ |
| Electric charge | coulomb | C | $A\ s$ |
| Electric potential difference | volt | V | $kg\ m^2\ s^{-3}\ A^{-1} = J\ C^{-1}$ |

unit is in the denominator. For example, velocity is a quantity with dimensions of length divided by time, and so in SI it is expressed in meters per second, or $m\ s^{-1}$. Some derived units that are used frequently have special names. Energy, for example, is the product of mass and the square of the velocity. Therefore, it is measured in units of kilogram square meters per square seconds ($kg\ m^2\ s^{-2}$), and $1\ kg\ m^2\ s^{-2}$ is called a *joule*. Other derived units, such as the pascal for measuring pressure, appear in Table B.2. Although these names provide a useful shorthand, it is important to remember their meanings in terms of the base units.

Because scientists work on scales ranging from the microscopic to the astronomical, there is a tremendous range in the magnitudes of measured quantities. Consequently, a set of **prefixes** has been incorporated into the International System of Units to simplify the description of small and large quantities (Table B.3). The prefixes specify various powers of 10 times the base and derived units. Some of them are quite familiar in everyday use: the kilometer, for example, is $10^3$ m. Others may sound less familiar—for instance, the femtosecond ($1\ fs = 10^{-15}$ s) or the gigapascal ($1\ GPa = 10^9$ Pa).

In addition to base and derived SI units, several units that are not officially sanctioned are used in this book. The first is the liter (abbreviated L), a very convenient size for volume measurements in chemistry; a liter is $10^{-3}\ m^3$, or 1 cubic decimeter ($dm^3$):

$$1\ L = 1\ dm^3 = 10^{-3}\ m^3 = 10^3\ cm^3$$

## TABLE B.3

### Prefixes in SI

| Fraction | Prefix | Symbol | | Factor | Prefix | Symbol |
|---|---|---|---|---|---|---|
| $10^{-1}$ | deci- | d | | 10 | deca- | da |
| $10^{-2}$ | centi- | c | | $10^2$ | hecto- | h |
| $10^{-3}$ | milli- | m | | $10^3$ | kilo- | k |
| $10^{-6}$ | micro- | $\mu$ | | $10^6$ | mega- | M |
| $10^{-9}$ | nano- | n | | $10^9$ | giga- | G |
| $10^{-12}$ | pico- | p | | $10^{12}$ | tera- | T |
| $10^{-15}$ | femto- | f | | | | |
| $10^{-18}$ | atto- | a | | | | |

Second, we use the *angstrom* (abbreviated Å) as a unit of length for atoms and molecules:

$$1 \text{ Å} = 10^{-10} \text{ m} = 100 \text{ pm} = 0.1 \text{ nm}$$

This unit is used simply because most atomic sizes and chemical bond lengths fall in the range of one to several angstroms, and the use of either picometers or nanometers is slightly awkward. Next, we use the *atmosphere* (abbreviated atm) as a unit of pressure. It is not a simple power of 10 times the SI unit of the pascal, but rather is defined as follows:

$$1 \text{ atm} = 101,325 \text{ Pa} = 0.101325 \text{ MPa}$$

This unit is used because most chemistry procedures are carried out at pressures near atmospheric pressure, for which the pascal is too small a unit to be convenient. In addition, expressions for equilibrium constants (see Chapter 9) are simplified when pressures are expressed in atmospheres.

The non-SI temperature units require special mention. The two most important temperature scales in the United States are the Fahrenheit scale and the Celsius scale, which employ the Fahrenheit degree (°F) and the Celsius degree (°C), respectively. The *size* of the Celsius degree is the same as that of the SI temperature unit, the kelvin, but the two scales are shifted relative to each other by 273.15°C:

$$T_K = \frac{1 \text{ K}}{1°C}(t°C) + 273.15 \text{ K}$$

The size of the degree Fahrenheit is 5/9 the size of the Celsius degree. The two scales are related by

$$t°C = \frac{5°C}{9°F}(t°F - 32°F) \qquad \text{or} \qquad t°F = \frac{9°F}{5°C}(t°C) + 32°F$$

The advantage of a unified system of units is that if all the quantities in a calculation are expressed in SI units, the final result must come out in SI units. Nevertheless, it is important to become familiar with the ways in which units are interconverted because units other than SI base units often appear in calculations. The **unit conversion method** provides a systematic approach to this problem.

As a simple example, suppose the mass of a sample is measured to be 64.3 g. If this is to be used in a formula involving SI units, it should be converted to kilograms (the SI base unit of mass). To do this, we use the fact that 1 kg = 1000 g and write

$$\frac{64.3 \text{ g}}{1000 \text{ g kg}^{-1}} = 0.0643 \text{ kg}$$

Note that this is, in effect, division by 1; because 1000 g = 1 kg, 1000 g kg$^{-1}$ = 1, and we cancel units between numerator and denominator to obtain the final result. This unit conversion could also be written as

$$64.3 \text{ g}\left(\frac{1 \text{ kg}}{1000 \text{ g}}\right) = 0.0643 \text{ kg}$$

In this book we use the more compact first version of the unit conversion. Instead of *dividing* by 1000 g kg$^{-1}$, we can equally well *multiply* by 1 = 10$^{-3}$ kg g$^{-1}$:

$$(64.3 \text{ g})(10^{-3} \text{ kg g}^{-1}) = 0.0643 \text{ kg}$$

Other unit conversions may involve more than just powers of 10, but they are equally easy to carry out. For example, to express 16.4 inches in meters, we use the fact that 1 inch = 0.0254 m (or 1 = 0.0254 m inch$^{-1}$), so that

$$(16.4 \text{ inches})(0.0254 \text{ m inch}^{-1}) = 0.417 \text{ m}$$

More complicated combinations are possible. For example, to convert from liter-atmospheres to joules (the SI unit of energy), two separate unit conversions are used:

$$(1 \text{ L atm})(10^{-3} \text{ m}^3 \text{ L}^{-1})(101{,}325 \text{ Pa atm}^{-1}) = 101.325 \text{ kg m}^2 \text{ s}^{-2}$$
$$= 101.325 \text{ Pa m}^3 = 101.325 \text{ J}$$

When doing chemical calculations, it is very important to write out units explicitly and cancel units in intermediate steps in order to obtain the correct units for the final result. This practice is a way of checking to make sure that units have not been incorrectly mixed without unit conversions or that an incorrect formula has not been used. If a result that is supposed to be a temperature comes out with units of m$^3$ s$^{-2}$, then a mistake has been made!

---

## B.2

# BACKGROUND IN PHYSICS

Although physics and chemistry are distinct sciences with distinct objectives and methods of investigation, concepts from one can aid investigations in the other. One important case in which physical reasoning aids chemical understanding is the recognition that applying forces causes chemical systems to move (or be changed in size and shape) and, consequently, to change their energy. The resulting changes in energy can have chemical consequences. Two specific examples come readily to mind. First, the total energy content of a system can be increased as the system is compressed by externally applied pressure. Second, when significant forces exist between molecules, the mutual energy of a pair of molecules can be changed as the molecules are forced closer together. Such intermolecular forces have profound influence on the organization of matter into solid, liquid, and gaseous states, as well as on the effectiveness of molecular collisions in causing chemical reactions. Specific illustrations appear at many places in this book. The purpose of this section is to review general aspects of motion, forces, and energy as background for these specific applications. For simplicity, the discussion will be restricted to a **point mass,** which is an object characterized only by its total mass $m$. We do not inquire into the internal structure of the object. In various contexts, this model is applied to planets, projectiles, molecules, atoms, nuclei, or electrons in predicting their response to applied forces.

### Motion of an Object

Our goal here is to provide a precise description of the familiar observation that the location of an object can change and that change can occur at various rates. Several definitions are required. One defines the position of the object precisely by stating its **displacement** $x$ from a selected reference point. For example, an automobile could be located 1.5 miles north of the intersection of Wilshire and Westwood boulevards in Los Angeles. The electron in a hydrogen atom could be located $5 \times 10^{-11}$ m from the nucleus. Displacement has dimensions of length (L). The rate of change of location is given by the **average velocity** $v$, defined over the time interval $t_1$ to $t_2$ as $v = [x_2 - x_1]/[t_2 - t_1]$. Both the displacement and the average velocity are defined relative to the selected reference point and therefore possess direction as well as magnitude. The **average speed** $s$ gives the magnitude of the average velocity but not its direction. For example, an automobile can have average velocity of 35 mph southbound from Wilshire and Westwood, while its average speed is 35 mph. Both velocity and speed have dimension of length per time ($L\ t^{-1}$). The **momentum** $p$ of a body is defined as $p = mv$. The momentum indicates the ability of a moving body to exert impact on another body upon collision. For example, a slow-moving automobile "packs a bigger wallop" than a fast-moving bicycle. Momentum has dimensions of $M\ L\ t^{-1}$. The rate at which the velocity changes is given by the **acceleration** $a$, defined as $a = [v_2 - v_1]/[t_2 - t_1]$. Acceleration has direction as well as magnitude and has dimensions of $L\ t^{-2}$.

## Forces

Force is defined as the agent that causes changes in the motion of an object. This restatement of our daily experience that pushing or pulling an object causes it to move provides the basis of Sir Isaac Newton's first law of motion:

> Every object persists in a state of rest or of uniform motion in a straight line unless compelled to change that state by forces impressed upon it.

The properties of force are determined experimentally in the laboratory by measuring the consequences of applying the force. Suppose we have arbitrarily selected a standard test object of known mass $m_1$. We can apply a force $f_1$ and, by making distance and time measurements described earlier, determine the acceleration $a_{1,1}$ imparted to the object by the force. Experience shows that a different force $f_2$ imparts different acceleration $a_{1,2}$ to the test object. Forces can be ranked by the magnitude of the acceleration that they impart to the standard test object. Now suppose we choose a second test object of mass $m_2$. Applying the original force $f_1$ to this object produces acceleration $a_{2,1}$, which is different from the acceleration $a_{1,1}$ that it gave to the first test object. The results of such experiments are summarized in Newton's second law of motion:

> The acceleration imparted to an object by an applied force is proportional to the magnitude of the force, parallel to the direction of the force, and inversely proportional to the mass of the object.

In mathematical form this statement becomes

$$f = ma \qquad \text{or} \qquad a = f/m$$

This equation demonstrates that force must have dimensions of $M\ L\ t^{-2}$. In SI units, force is expressed in newtons (N).

One familiar example is the force caused by gravity at the surface of the Earth. Measurements show that this force produces a downward acceleration (toward the center of the Earth) of constant magnitude $9.80665\ \text{m s}^{-2}$, conventionally denoted by $g$, the gravitational constant. The gravitational force exerted on a body of mass $m$ is

$$f = mg$$

Another familiar example is the restoring force exerted on an object by a spring. In chemistry this serves as a useful model of the binding forces that keep atoms together in a chemical bond or near their home positions in a solid lattice. Imagine an object of mass $m$ located on a smooth table top. The object is connected to one end of a coiled metallic spring; the other end of the spring is anchored to a post in the table top. At rest, the object is located at position $x_0$. Now suppose that the object is pulled away from the post in a straight line to a new position $x$ beyond $x_0$ so that the spring is stretched. Measurements show that at the stretched position $x$ the magnitude of the force exerted on the object by the spring is directly proportional to the displacement from the rest position:

$$|f| = K(x - x_0)$$

In the preceding equation, $K$ is a constant whose magnitude must be determined empirically in each case studied. Clearly this force is directed back toward the rest position, and when the object is released the recoiling spring accelerates the object back toward the rest position. Since the displacement is directed *away* from the rest position and the restoring force is directed *toward* the rest position, a minus sign is inserted in the equation, giving the restoring force at each position $x$:

$$f(x) = -K(x - x_0)$$

When studying the dependence of force on position, one must pay careful attention to how displacement is defined in each particular problem. The force and the acceleration at any point are parallel to each other, but the displacement (defined relative to some convenient origin of coordinates in each particular case) may point in a different direction. In applications concerned with the height of an object above the Earth, the gravitational force is usually written as

$$f(y) = -mg$$

to emphasize that the vertical displacement $y$ is positive and pointing away from the surface of the Earth, while the gravitational force is clearly directed toward the Earth.

Once the force has been determined, the motion of the object under that force can be predicted from Newton's second law. If a constant force (that is, a force with constant acceleration) is applied to an object initially at rest for a period of time starting at $t_1$ and ending at $t_2$, the velocity at $t_2$ is given by

$$v(t_2) = a\,(t_2 - t_1)$$

and the position at $t_2$ is given by

$$x(t_2) = x(t_1) + \tfrac{1}{2}a(t_2 - t_1)^2$$

If the force depends on position, predicting the motion is slightly more complicated and requires the use of calculus. We merely quote the results for the restoring force. The object governed by the restoring force oscillates about the rest position $x_0$ with a period $\tau$ given by

$$\frac{1}{\tau} = \frac{1}{2\pi}\left(\frac{K}{m}\right)^{1/2}$$

The frequency $\nu$ of the oscillation (the number of cycles per unit time) is given by $\nu = \tau^{-1}$. If the motion is described in terms of the angular frequency $\omega = 2\pi\nu$, which has dimensions radians s$^{-1}$, then the position of the oscillating object is given by

$$x\,(t) - x_0 = A\cos{(\omega t + \delta)}$$

where the amplitude $A$ and the phase factor $\delta$ are determined by the initial position and velocity of the object.

## Potential Energy, Force, and Stability

Experience shows that applying a force to a moving object changes the kinetic energy of that object. If the applied force is in the same direction as the velocity of the object, then the kinetic energy is increased; if the force is opposed to the velocity, then the kinetic energy is decreased. Experience also shows that when a force causes an object to move to a new position, the potential energy of the object is changed. The concepts *energy*, *kinetic energy*, and *potential energy* are discussed in Section 1.7 of this book, which should be read at this point.

One very important application of the potential energy function in this book is to provide a procedure for qualitatively predicting the motion of an object without detailed solution of Newton's second law. The method determines the direction of the force applied to an object from knowledge of the potential energy of the object; it relies on the concept of potential energy curves introduced in Figure 1.15 of this book.

The value of the potential energy at a general point $x$ is always stated relative to its value at a specially selected reference point $x_0$. The value $V(x_0)$ is a constant and is usually assigned the value zero, and the point $x_0$ is frequently selected as the origin of coordinates for specifying the location of the object.

In the presence of a particular force $f$, the potential energy of the object at $x$ is defined as the work required to move the object from $x$ to the reference position $x_0$. Since the work done by a constant force in moving an object is defined as (force) $\times$ (displacement), the definition of potential energy in mathematical terms is

$$V(x) - V(x_0) = f \times (x_0 - x)$$

This is a more precise version of the equation on page 32 of the main text. Note carefully that the displacement referred to in the definition of potential energy starts at $x$ and ends at $x_0$. For example, it gives the potential energy due to gravity of an object of mass $m$ at height $h$ above the surface of the Earth as

$$V(h) - V(x_0) = -mg(x_0 - h) = mgh - mgx_0 = mgh$$

where the value $V(x_0)$ has been set to zero. For a variable force, the definition of potential energy becomes slightly more complicated and requires explicit use of calculus. The potential energy for the linear restoring force is

$$V(x) = \tfrac{1}{2}K(x)^2$$

where the value $V(x_0)$ has been set to 0 and $x_0$ has been selected as the origin of coordinates.

It is instructive to plot these two potential energy functions as graphs (see Appendix C). The gravitational potential energy function gives a straight line through the origin with slope $mg$, while the potential energy function for the linear restoring force yields a parabola. In each case, the potential energy of the object increases as the distance from the reference point $x_0$ increases. As explained on page 33, the net force on the object lies in the direction of decreasing slope of the potential energy function. For both the gravitational and restoring forces, the object is subject to negative forces that attract it back to the center of force. In general, we identify **attractive forces** as those that correspond to positive slope of the potential energy curve; **repulsive forces** operate in regions where the slope of the potential energy curve is negative.

The minimum of the potential energy curve shown in Figure 1.15 is called a point of **stable equilibrium** because the net force (that is, the slope of the potential energy curve) at that point is zero. As the ball tries to climb the wall on either side of the minimum, the restoring force always drives it back toward this position of stable equilibrium. This qualitative interpretation predicts that the ball will oscillate about the equilibrium position, as predicated by the exact solution to Newton's second law quoted earlier.

## Electrical Forces

The concepts summarized so far in this section also are used to describe the mutual interactions and the motions of electrically charged particles. The only additional features are to identify the proper force to represent electrical interactions and to obtain the corresponding potential energy function. Positive and negative charges and the electrical forces between them were first quantified by Charles Coulomb late in the 18th century. In his honor the unit of charge in SI units is the **coulomb** (C). Electrical charge is fundamentally quantized in units of the charge carried by a single electron $e$, which is equal to $1.60206 \times 10^{-19}$ C. The coulomb is thus an inconveniently large unit for chemical reasoning. Nonetheless, for consistency and for quantitative accuracy, physical equations involving charge employ SI units.

Suppose a charge of magnitude $q_1$ is held at the origin of coordinates and another charge of magnitude $q_2$ is brought near it, at the distance $r$ from the origin. The magnitude of the acceleration imparted to the charge $q_2$ by the charge $q_1$ fixed at the origin can be determined as described earlier for uncharged objects. From such measurements Coulomb determined that the magnitude of the force was directly proportional to the magnitudes of the two charges and inversely proportional to the distance between them:

$$f \propto \frac{q_1 q_2}{r^2}$$

The radial displacement variable $r$ has its origin at the same location as charge $q_1$, and its value increases in the outward direction. If the charges were of the same sign, then the acceleration pushed them apart in the same direction as $r$. Consequently, the force between charges is **positive** and **repulsive.** If the charges were of opposite sign, then the acceleration pulled them together in the direction opposite to the displacement variable $r$. In this case the force between charges is **negative** and **attractive.** It is instructive to sketch the results graphically in these two cases. The quantitative form of Coulomb's law in which force is expressed in newtons (N) and charge in coulombs (C) is

$$f = \frac{q_1 q_2}{4 \pi \epsilon_0 r^2}$$

where $\epsilon_0$, called the permittivity of the vacuum, is a constant with value $8.854 \times 10^{-12}$ $C^2 J^{-1} m^{-1}$. In this and related equations, the symbol $q$ for each charge represents the magnitude and carries implicitly the sign of the charge.

The **Coulomb potential energy** corresponding to this force is

$$V(r) = \frac{q_1 q_2}{4 \pi \epsilon_0 r}$$

Note that the force varies as $r^{-2}$, while the potential energy varies as $r^{-1}$. Note also that if the charges have the same sign, then the potential energy is positive and repulsive; if the charges have opposite sign, then the potential energy is negative and attractive. From knowledge of the potential energy curve between two charged particles, one can predict the direction of their relative motion.

The Coulomb potential energy function is of great importance in chemistry for examining the structure of atoms and molecules. In 1912, Ernest Rutherford proposed that an atom of atomic number $Z$ comprises a dense, central nucleus of positive charge with magnitude $Ze$ surrounded by a total of $Z$ individual electrons moving around the nucleus. Thus, each individual electron has a potential energy of interaction with the nucleus as given by

$$V(r) = \frac{Ze^2}{4\pi\epsilon_0 r}$$

which is clearly negative and attractive. Each electron has a potential energy of interaction with every other electron in the atom as given by

$$V(r) = \frac{e^2}{4\pi\epsilon_0 r}$$

which is clearly positive and repulsive. Rutherford's model of the atom, firmly based on experimental results, was completely at odds with the physical theories of the day. Attempts to reconcile these results with theory led to the development of the new theory called quantum mechanics.

## Circular Motion and Angular Momentum

An object executing uniform circular motion about a point (for example, a ball on the end of a rope being swung in circular motion) is represented in terms of position, velocity, speed, momentum, and force, just as described earlier for objects in linear motion. It is most convenient to describe this motion in polar coordinates $r$ and $\theta$, which represent, respectively, the distance of the object from the center and its angular displacement from the $x$-axis in ordinary cartesian coordinates. Since $r$ is constant for circular motion, the motion variables depend directly on $\theta$. The angular velocity during the time interval from $t_1$ to $t_2$ is given by

$$\omega = \frac{\theta_2 - \theta_1}{t_2 - t_1}$$

and the angular acceleration is given by

$$\alpha = \frac{\omega_2 - \omega_1}{t_2 - t_1}$$

In linear motion, we are concerned with the momentum $p = mv$ of an object as it heads toward a particular point; the linear momentum measures the impact that the object can transfer in a collision as it arrives at the point. To extend this concept to circular motion, we define the **angular momentum** of an object as it revolves around a point as $L = mvr$. This is in effect the *moment* of the linear momentum over the distance $r$, and it is the measure of the torque acting on the object as it executes angular motion. The angular momentum of an electron around a nucleus is a crucial feature of atomic structure, which is discussed in Chapter 15.

## PROBLEMS

*Answers to problems whose numbers are boldface appear in Appendix G.*

### SI Units and Unit Conversions

1. Rewrite the following in scientific notation, using only the base units of Table B.1, without prefixes.
   (a) 65.2 nanograms
   (b) 88 picoseconds
   (c) 5.4 terawatts
   (d) 17 kilovolts

2. Rewrite the following in scientific notation, using only the base units of Table B.1, without prefixes.
   (a) 66 $\mu$K
   (b) 15.9 MJ
   (c) 0.13 mg
   (d) 62 GPa

3. Express the following temperatures in degrees Celsius.
   (a) 9001°F
   (b) 98.6°F (the normal body temperature of human beings)
   (c) 20°F above the boiling point of water at 1 atm pressure
   (d) −40°F

4. Express the following temperatures in degrees Fahrenheit.
   (a) 5000°C
   (b) 40.0°C
   (c) 212°C
   (d) −40°C

5. Express the temperatures given in problem 3 in kelvins.

6. Express the temperatures given in problem 4 in kelvins.

7. Express the following in SI units, either base or derived.
   (a) 55.0 miles per hour (1 mile = 1609.344 m)
   (b) 1.15 g cm$^{-3}$
   (c) $1.6 \times 10^{-19}$ C Å
   (d) 0.15 mol L$^{-1}$
   (e) $5.7 \times 10^3$ L atm day$^{-1}$

8. Express the following in SI units, either base or derived.
   (a) 67.3 atm
   (b) $1.0 \times 10^4$ V cm$^{-1}$
   (c) 7.4 Å year$^{-1}$
   (d) 22.4 L mol$^{-1}$
   (e) 14.7 lb inch$^{-2}$ (1 inch = 2.54 cm; 1 lb = 453.59 g)

9. The kilowatt-hour (kWh) is a common unit in measurements of the consumption of electricity. What is the conversion factor between the kilowatt-hour and the joule? Express 15.3 kWh in joules.

10. A car's rate of fuel consumption is often measured in miles per gallon (mpg). Determine the conversion factor between miles per gallon and the SI unit of meters per cubic decimeter (1 gallon = 3.785 dm$^3$, and 1 mile = 1609.344 m). Express 30.0 mpg in SI units.

11. A certain V-8 engine has a displacement of 404 in$^3$. Express this volume in cubic centimeters (cm$^3$) and in liters.

12. Light travels in a vacuum at a speed of $3.00 \times 10^8$ m s$^{-1}$.
    (a) Convert this speed to miles per second.
    (b) Express this speed in furlongs per fortnight, a little-used unit of speed. (A furlong, a distance used in horse racing, is 660 ft; a fortnight is exactly 2 weeks.)

# Mathematics for General Chemistry

Mathematics is an essential tool in chemistry. This appendix reviews some of the most important mathematical techniques for general chemistry.

## C.1
### USING GRAPHS

In many situations in science, we are interested in how one quantity (measured or predicted) depends on another quantity. The position of a moving car depends on the time, for example, or the pressure of a gas depends on the volume of the gas at a given temperature. A very useful way to show such a relation is through a **graph,** in which one quantity is plotted against another.

The usual convention in drawing graphs is to use the horizontal axis for the variable over which we have control and use the vertical axis for the measured or calculated quantity. After a series of measurements, the points on the graph frequently lie along a recognizable curve, and that curve can be drawn through the points. Because any experimental measurement involves some degree of uncertainty, there will be some scatter in the points, so there is no purpose in drawing a curve that passes precisely through every measured point. If there is a relation between the quantities measured, however, the points will display a systematic trend, and a curve can be drawn that represents that trend.

The curves plotted on graphs can have many shapes. The simplest and most important shape is a straight line. A straight line is a graph of a relation such as

$$y = 4x + 7$$

or, more generally,

$$y = mx + b$$

where the variable $y$ is plotted along the vertical axis and the variable $x$ along the horizontal axis (Fig. C.1). The quantity $m$ is the **slope** of the line that is plotted, and $b$ is the **intercept**—the point at which the line crosses the $y$ axis. This can be demonstrated by setting $x$ equal to 0 and noting that $y$ is then equal to $b$. The slope is a measure of the steepness of the line; the greater the value of $m$, the steeper the line. If the line goes up and to the right, the slope is positive; if it goes down and to the right, the slope is negative.

The slope of a line can be determined from the coordinates of two points on the line. Suppose, for example, that when $x = 3, y = 5$, and when $x = 4, y = 7$. These two points can be written in shorthand notation as (3, 5) and (4, 7). The slope of the line is then defined as the "rise over the run"—the change in the $y$ coordinate divided by the change in the $x$ coordinate:

$$\text{slope} = m = \frac{\Delta y}{\Delta x} = \frac{7 - 5}{4 - 3} = \frac{2}{1} = 2$$

**FIGURE C.1**  A straight-line, or linear, relationship between two experimental quantities is a very desirable result because it is easy to graph and easy to represent mathematically. The equation of this straight line ($y = 2x - 1$) fits the general form $y = mx + b$. The line's $y$ intercept is $-1$ and its slope is 2.

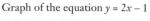

Graph of the equation $y = 2x - 1$

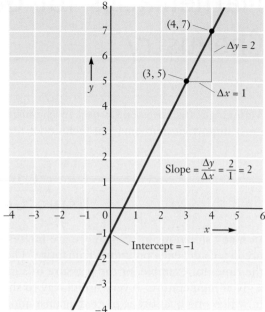

The symbol $\Delta$ (capital Greek delta) indicates the change in a quantity—the final value minus the initial value. In chemistry, if the two quantities begin graphed have dimensions, the slope has dimensions as well. If a graph of distance traveled (in meters) against time (in seconds) is a straight line, its slope has dimensions of meters per second (m s$^{-1}$).

---

## C.2

## SOLUTION OF ALGEBRAIC EQUATIONS

Frequently in chemistry it is necessary to solve an algebraic equation for an unknown quantity, such as a concentration or a partial pressure in an equilibrium-constant expression. Let us represent the unknown quantity with the symbol $x$. If the equation is linear, the method of solution is straightforward:

$$5x + 9 = 0$$

$$5x = -9$$

$$x = -\frac{9}{5}$$

or, more generally, if $ax + b = 0$, then $x = -b/a$.

Nonlinear equations are of many kinds. One of the most common in chemistry is the quadratic equation, which can always be rearranged into the form

$$ax^2 + bx + c = 0$$

where each of the constants ($a$, $b$, and $c$) may be positive, negative, or 0. The two solutions to a quadratic equation are given by the quadratic formula:

$$x = \frac{-b \pm \sqrt{b^2 - 4ac}}{2a}$$

As an example, suppose that the equation

$$x = 3 + \frac{7}{x}$$

arises in a chemistry problem. Multiplying by $x$ and rearranging the terms gives

$$x^2 - 3x - 7 = 0$$

Inserting $a = 1$, $b = -3$, and $c = -7$ into the quadratic formula gives

$$x = \frac{-(-3) \pm \sqrt{(-3)^2 - 4(1)(-7)}}{2} = \frac{3 \pm \sqrt{9 + 28}}{2} = \frac{3 \pm \sqrt{37}}{2}$$

The two roots of the equation are

$$x = 4.5414 \qquad \text{and} \qquad x = -1.5414$$

In a chemistry problem, the choice of the proper root can frequently be made on physical grounds. If $x$ corresponds to a concentration, for example, the negative root is unphysical and can be discarded.

The solution of cubic or higher-order algebraic equations (or more complicated equations involving sines, cosines, logarithms, or exponentials) becomes more difficult, and approximate or numerical methods must be used. As an illustration, consider the equation

$$x^2 \left( \frac{2.00 + x}{3.00 - x} \right) = 1.00 \times 10^{-6}$$

If $x$ is assumed to be small relative to both 2.00 and 3.00, we obtain the simpler approximate equation

$$x^2 \left( \frac{2.00}{3.00} \right) \approx 1.00 \times 10^{-6}$$

which leads to the roots $x = \pm 0.00122$. We immediately confirm that the solutions obtained in this way *are* small compared with 2.00 (and 3.00) and that our approximation was a good one. When a quantity ($x$ in this case) is added to or subtracted from a larger quantity in a complicated equation, it is usually worthwhile, in solving the equation, to simply neglect the occurrence of the small quantity. Note that this tactic works only for addition and subtraction, never for multiplication or division.

Suppose now that the equation is changed to

$$x^2 \left( \frac{2.00 + x}{3.00 - x} \right) = 1.00 \times 10^{-2}$$

In this case, the equation that comes from neglecting $x$ compared with 2.00 and 3.00 is

$$x^2 \left( \frac{2.00}{3.00} \right) \approx 1.00 \times 10^{-2}$$

which is solved by $x \approx \pm 0.122$. The number 0.122 is smaller than 2.00 and 3.00, but not so small that it can be ignored. In this case, more precise results can be obtained by **iteration.** Let us simply add and subtract the approximate positive root $x \approx +0.122$ as specified inside the parentheses and solve again for $x$:

$$x^2 \left( \frac{2.00 + 0.122}{3.00 - 0.122} \right) = 1.00 \times 10^{-2}$$

$$x = 0.116$$

This new value can again be inserted into the original equation and the process repeated:

$$x^2 \left( \frac{2.00 + 0.116}{3.00 - 0.116} \right) = 1.00 \times 10^{-2}$$

$$x = 0.117$$

Once the successive values of $x$ agree to within the desired accuracy, the iteration can be stopped. Another root of the equation is obtained in a similar fashion if the iterative procedure starts with $x = -0.122$.

In some cases, iteration fails. Suppose, for example, we have the equation

$$x^2 \left( \frac{2.00 + x}{3.00 - x} \right) = 10.0$$

There is no particular reason to believe that $x$ should be small compared with 2.00 or 3.00, but if we nevertheless assume that it is, we find $x = \pm\sqrt{15} = \pm 3.873$. Putting $x = +3.873$ back in for an iterative cycle leads to the equation

$$x^2 = -1.49$$

which has no real solutions. Starting with $x = -3.873$ succeeds no better; it gives

$$x^2 = -36.7$$

One way to overcome these difficulties is to solve the original equation graphically. We plot the left side of the equation against $x$ and see at which values it becomes equal to 10 (the right side). We might initially calculate

| $x$ | $x^2 \left( \dfrac{2.00 + x}{3.00 - x} \right)$ |
|---|---|
| 0 | 0 |
| 1 | 1.5 |
| 2 | 16 |

We observe that for $x = 1$ the left side is less than 10, whereas for $x = 2$ it is greater than 10. Somewhere in between there must be an $x$ for which the left side is *equal to* 10. We can pinpoint it by selecting values of $x$ between 1 and 2; if the left side is less than 10, $x$ should be increased, and if it is greater than 10, $x$ should be decreased.

| $x$ | $x^2 \left( \dfrac{2.00 + x}{3.00 - x} \right)$ |
|---|---|
| 1.5 | 5.25 |
| Increase to 1.8 | 10.26 |
| Decrease to 1.75 | 9.19 |
| Increase to 1.79 | 10.04 |

Thus, 1.79 is quite close to a solution of the equation. Improved values are easily obtained by further adjustments of this type.

---

## C.3

### POWERS AND LOGARITHMS

Raising a number to a power and the inverse operation, taking a logarithm, are important in many chemical problems. Although the ready availability of electronic calculators makes the mechanical execution of these operations quite routine, it remains important to understand what is involved in such "special functions."

The mathematical expression $10^4$ implies multiplying 10 by itself 4 times to give 10,000. Ten, or any other number, when raised to the power 0 always gives 1:

$$10^0 = 1$$

Negative powers of 10 give numbers less than 1 and are equivalent to raising 10 to the corresponding *positive* power and then taking the reciprocal:

$$10^{-3} = 1/10^3 = 0.001$$

We can extend the idea of raising to a power to include powers that are not whole numbers. For example, raising to the power $\frac{1}{2}$ (or 0.5) is the same as taking the square root:

$$10^{0.5} = \sqrt{10} = 3.1623\ldots$$

Scientific calculators have a $10^x$ (or INV LOG) key that can be used for calculating powers of 10 in cases where the power is not a whole number.

Numbers other than 10 can be raised to powers, as well; these numbers are referred to as **bases.** Many calculators have a $y^x$ key that lets any positive number $y$ be raised to any power $x$. One of the most important bases in scientific problems is the transcendental number called $e$ (2.7182818...). The $e^x$ (or INV LN) key on a calculator is used to raise $e$ to any power $x$. The quantity $e^x$, also denoted as exp $(x)$, is called the **exponential** of $x$. A key property of powers is that a base raised to the sum of two powers is equivalent to the product of the base raised separately to these powers. Thus, we can write

$$10^{21+6} = 10^{21} \times 10^6 = 10^{27}$$

The same type of relationship holds for any base, including $e$.

Logarithms also occur frequently in chemistry problems. The logarithm of a number is the exponent to which some base has to be raised to obtain the number. The base is almost always either 10 or the transcendental number $e$. Thus,

$$B^a = n \qquad \text{and} \qquad \log_B n = a$$

where $a$ is the logarithm, $B$ is the base, and $n$ is the number.

**Common logarithms** are base-10 logarithms; that is, they are powers to which 10 has to be raised in order to give the number. For example, $10^3 = 1000$, and so $\log_{10} 1000 = 3$. We shall frequently omit the 10 when showing common logarithms and write this equation as log 1000 = 3. Only the logarithms of 1, 10, 100, 1000, and so on are whole numbers; the logarithms of other numbers are decimal fractions. The decimal point in a logarithm divides it into two parts. To the left of the decimal point is the *characteristic;* to the right is the *mantissa.* Thus, the logarithm in the equation

$$\log (7.310 \times 10^3) = 3.8639$$

has a characteristic of 3 and a mantissa of 0.8639. As may be verified with a calculator, the base-10 logarithm of the much larger (but closely related) number $7.310 \times 10^{23}$ is 23.8639. As this case illustrates, the characteristic is determined solely by the location of the decimal point in the number and not by the number's precision, so it is *not* included when counting significant figures. The mantissa should be written with as many significant figures as the original number.

A logarithm is truly an exponent and as such follows the same rules of multiplication and division as other exponents. In multiplication and division we have

$$\log (n \times m) = \log n + \log m$$

$$\log \left(\frac{n}{m}\right) = \log n - \log m$$

Furthermore,

$$\log n^m = m \log n$$

and so the logarithm of $3^5 = 243$ is

$$\log 3^5 = 5 \log 3 = 5 \times 0.47712 = 2.3856$$

There is no such thing as the logarithm of a negative number, because there is no power to which 10 (or any other base) can be raised to give a negative number.

A frequently used base for logarithms is the number $e$ ($e = 2.7182818...$). The logarithm to the base $e$ is called the **natural logarithm** and is indicated by $\log_e$ or ln. Base-$e$ logarithms are related to base-10 logarithms by the formula

$$\ln n = 2.3025851 \log n$$

As already stated, calculations of logarithms and powers are inverse operations. Thus, if we want to find the number for which 3.8639 is the common logarithm, we simply calculate

$$10^{3.8639} = 7.310 \times 10^3$$

If we need the number for which the natural logarithm is 2.108, we calculate

$$e^{2.108} = 8.23$$

As before, the number of significant digits in the answer should correspond to the number of digits in the *mantissa* of the logarithm.

---

## C.4

## SLOPES OF CURVES AND DERIVATIVES

Very frequently in science, one measured quantity depends on a second one. If the property $y$ depends on a second property $x$, we can write $y = f(x)$, where $f$ is a function that expresses the dependence of $y$ on $x$. Often we are interested in the effect of a small change $\Delta x$ on the dependent variable $y$. If $x$ changes to $x + \Delta x$, then $y$ will change to $y + \Delta y$. How is $\Delta y$ related to $\Delta x$? Suppose we have the simple linear relation between $y$ and $x$

$$y = mx + b$$

where $m$ and $b$ are constants. If we substitute $y + \Delta y$ and $x + \Delta x$, we find

$$y + \Delta y = m(x + \Delta x) + b$$

Subtracting the first equation from the second leaves

$$\Delta y = m \, \Delta x$$

or

$$\frac{\Delta y}{\Delta x} = m$$

The change in $y$ is proportional to the change in $x$, with a proportionality constant equal to the slope of the line in the graph of $y$ against $x$.

Suppose now we have a slightly more complicated relationship such as

$$y = ax^2$$

where $a$ is a constant. If we substitute $y = y + \Delta y$ and $x = x + \Delta x$ here, we find

$$y + \Delta y = a(x + \Delta x)^2$$
$$= ax^2 + 2ax \, \Delta x + (\Delta x)^2$$

Subtracting as before leaves

$$\Delta y = 2ax \, \Delta x + (\Delta x)^2$$

Here we have a more complicated, nonlinear relationship between $\Delta y$ and $\Delta x$. If $\Delta x$ is small enough, however, the term $(\Delta x)^2$ will be small relative to the term proportional to $\Delta x$, and we may write

$$\Delta y \approx 2ax \, \Delta x \qquad (\Delta x \text{ small})$$

$$\frac{\Delta y}{\Delta x} \approx 2ax \qquad (\Delta x \text{ small})$$

How do we indicate this graphically? The graph of $y$ against $x$ is no longer a straight line, and so we need to generalize the concept of slope. We define a **tangent line** at the point $x_0$ as the line that touches the graph of $f(x)$ at $x = x_0$ without crossing it.[1] If $\Delta y$ and $\Delta x$ are small (Fig. C.2), the slope of the tangent line is

$$\frac{\Delta y}{\Delta x} = 2ax_0 \qquad \text{at} \qquad x = x_0$$

for the example just discussed. Clearly the slope is not constant but changes with $x_0$. If we define the slope at each point on the curve by the slope of the corresponding tangent line, we obtain a *new* function that gives the slope of the curve $f(x)$ at each point $x$. We call this new function the **derivative** of $f(x)$ and represent it with $df/dx$. We have already calculated the derivatives of two functions:

$$f(x) = mx + b \implies \frac{df}{dx} = m$$

$$f(x) = ax^2 \implies \frac{df}{dx} = 2ax$$

Table C.1 shows derivatives of several other functions that are important in basic chemistry.

[1] This is an intuitive, rather than a rigorously mathematical, definition of a tangent line.

**TABLE C.1**

### Derivatives of Simple Functions

| Function $f(x)$ | Derivative $\dfrac{df}{dx}$ |
|---|---|
| $mx + b$ | $m$ |
| $ax^2$ | $2ax$ |
| $\dfrac{a}{x} = ax^{-1}$ | $-\dfrac{a}{x^2} = -ax^{-2}$ |
| $ax^n$ | $nax^{n-1}$ |
| $e^{ax}$ | $ae^{ax}$ |
| $\ln ax$ | $\dfrac{1}{x}$ |
| $\sin ax$ | $a \cos ax$ |
| $\cos ax$ | $-a \sin ax$ |

The constants $a$, $b$, $m$, and $n$ may be positive or negative.

**FIGURE C.2** The slope of a curve at a point.

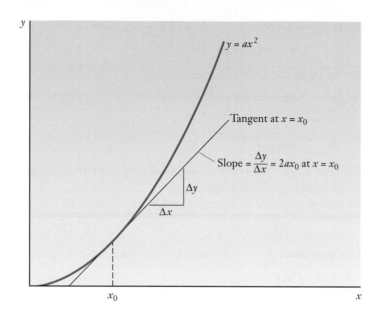

The derivative gives the slope of the tangent line to $f(x)$ at each point $x$ and can be used to approximate the response to a small perturbation $\Delta x$ in the independent variable $x$:

$$\Delta y \approx \frac{df}{dx} \Delta x$$

As an example, consider 1.00 mol of an ideal gas at 0°C that obeys the law

$$P\ (\text{atm}) = \frac{22.414\ \text{L atm}}{V\ (\text{L})}$$

Suppose $V$ is 20.00 L. If the volume $V$ is changed by a small amount $\Delta V$ at fixed $T$ and number of moles, what will be the corresponding change $\Delta P$ in the pressure? We write

$$P = f(V) = \frac{22.414}{V} = \frac{a}{V}$$

$$\frac{dP}{dV} = \frac{df}{dV} = -\frac{a}{V^2} = -\frac{22.414}{V^2} \qquad \text{(from Table C.1)}$$

$$\Delta P \approx \frac{dP}{dV} \Delta V = -\frac{22.414}{V^2} (\Delta V)$$

If $V = 20.00$ L, then

$$\Delta P \approx -\frac{22.414\ \text{L atm}}{(200.00\ \text{L})^2} \Delta V = -(0.0560\ \text{atm L}^{-1})\ \Delta V$$

---

## C.5

# AREAS UNDER CURVES AND INTEGRALS

Another mathematical operation that arises frequently in science is the calculation of the area under a curve. Some areas are those of simple geometric shapes and are easy to calculate. If the function $f(x)$ is a constant,

$$f(x) = a$$

then the area under a graph of $f(x)$ between the two points $x_1$ and $x_2$ is that of a rectangle (Fig. C.3a) and is easily calculated as

$$\text{area} = \text{height} \times \text{base} = a(x_2 - x_1) = [ax_2] - [ax_1]$$

If $f(x)$ is a straight line that is neither horizontal nor vertical,

$$f(x) = mx + b$$

then the area is that of a trapezoid (Fig. C.3b) and is equal to

$$\text{area} = \text{average height} \times \text{base}$$
$$= \tfrac{1}{2}(mx_1 + b + mx_2 + b)(x_2 - x_1)$$
$$= [\tfrac{1}{2}mx_2^2 + bx_2] - [\tfrac{1}{2}mx_1^2 + bx_1]$$

Suppose now that $f(x)$ is a more complicated function, such as that shown in Figure C.3c. We can estimate the area under this graph by approximating it with a series of small rectangles of width $\Delta x$ and varying heights. If the height of the $i$th rectangle is $y_i = f(x_i)$, then we have, approximately,

$$\text{area} \approx f(x_1)\, \Delta x + f(x_2)\, \Delta x + \cdots$$
$$= \sum_i f(x_i)\, \Delta x$$

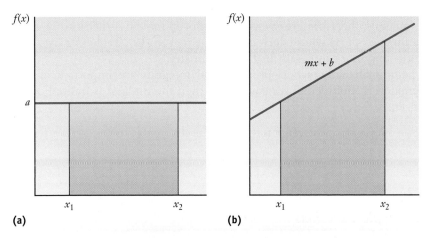

(a)    (b)

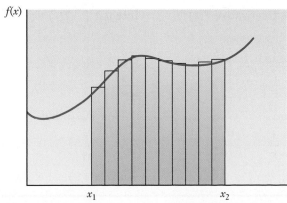

(c)

**FIGURE C.3**  The integral as the area under a curve.

As the widths of the rectangles $\Delta x$ become small, this becomes a better approximation. We *define* the area under a curve $f(x)$ between two points as the limiting value of this sum as $\Delta x$ approaches 0, called the **definite integral:**

$$\text{area} = \int_{x_1}^{x_2} f(x)\, dx$$

We have already worked out two examples of such integrals:

$$f(x) = a \implies \int_{x_1}^{x_2} a\, dx = [ax_2] - [ax_1]$$

$$f(x) = ax + b \implies \int_{x_1}^{x_2} (ax + b)\, dx = [\tfrac{1}{2}ax_2^2 + bx_2] - [\tfrac{1}{2}ax_1^2 + bx_1]$$

These integrals have an interesting form, which can be generalized to other cases as well: some function $F$ evaluated at the upper limit ($x_2$) minus that function $F$ evaluated at the lower limit ($x_1$).

$$\int_{x_1}^{x_2} f(x)\, dx = F(x_2) - F(x_1)$$

$F$ is called the **antiderivative** of $f$ because, for reasons we shall not go into here, it is obtained by the inverse of a derivative operation. In other words, if $F$ is the antiderivative of $f$, then $f$ is the derivative of $F$:

$$\frac{dF}{dx} = f(x)$$

To calculate integrals, therefore, we invert the results of Table C.1, because we need to find those functions $F(x)$ whose derivatives are equal to $f(x)$. Table C.2 lists several of the most important integrals for basic chemistry.

Several additional mathematical properties of integrals are important at this point. If a function is multiplied by a constant $c$, then the integral is multiplied by the same constant:

$$\int_{x_1}^{x_2} cf(x)\, dx = c \int_{x_1}^{x_2} f(x)\, dx$$

### TABLE C.2

#### Integrals of Simple Functions

| Function $f(x)$ | Antiderivative $F(x)$ | Integral $\int_{x_1}^{x_2} f(x)\, dx$ |
|---|---|---|
| $a$ | $ax$ | $ax_2 - ax_1$ |
| $mx + b$ | $\tfrac{1}{2}mx^2 + bx$ | $[\tfrac{1}{2}mx_2^2 + bx_2] - [\tfrac{1}{2}mx_1^2 - bx_1]$ |
| $x^n \ (n \neq -1)$ | $\dfrac{1}{n+1}x^{n+1}$ | $\dfrac{1}{n+1}(x_2^{n+1} - x_1^{n+1})$ |
| $\dfrac{1}{x} = x^{-1} \ (x > 0)$ | $\ln x$ | $(\ln x_2 - \ln x_1) = \ln \dfrac{x_2}{x_1}$ |
| $\dfrac{1}{x^2} = x^{-2}$ | $-\dfrac{1}{x}$ | $-\left(\dfrac{1}{x_2} - \dfrac{1}{x_1}\right)$ |
| $e^{ax}$ | $\dfrac{e^{ax}}{a}$ | $\dfrac{1}{a}(e^{ax_2} - e^{ax_1})$ |

The constants $a$, $b$, $m$, and $n$ may be positive or negative.

The reason is self-evident: if a function is multiplied everywhere by a constant factor, then the area under its graph must be increased by the same factor. Second, the integral of the sum of two functions is the sum of the separate integrals:

$$\int_{x_1}^{x_2} [f(x) + g(x)] \, dx = \int_{x_1}^{x_2} f(x) \, dx + \int_{x_1}^{x_2} g(x) \, dx$$

Finally, if the upper and lower limits of integration are reversed, the sign of the integral changes as well. This is easily seen from the antiderivative form:

$$\int_{x_2}^{x_1} f(x) \, dx = F(x_1) - F(x_2) = -[F(x_2) - F(x_1)] = -\int_{x_1}^{x_2} f(x) \, dx$$

## PROBLEMS

*Answers to problems whose numbers are boldface appear in Appendix G.*

### Using Graphs

**1.** A particular plot of distance traveled against time elapsed is found to be a straight line. After an elapsed time of 1.5 hours, the distance traveled was 75 miles, and after an elapsed time of 3.0 hours, the distance traveled was 150 miles. Calculate the slope of the plot of distance against time, and give its units.

2. The pressure of a gas in a rigid container is measured at several different temperatures, and it is found that a plot of pressure against temperature is a straight line. At 20.0°C the pressure is 4.30 atm, and at 100.0°C the pressure is 5.47 atm. Calculate the slope of the plot of pressure against temperature and give its units.

**3.** Rewrite each of the following linear equations in the form $y = mx + b$, and give the slope and intercept of the corresponding plot. Then draw the graph.
(a) $y = 4x - 7$
(b) $7x - 2y = 5$
(c) $3y + 6x - 4 = 0$

4. Rewrite each of the following linear equations in the form $y = mx + b$, and give the slope and intercept of the corresponding plot. Then draw the graph.
(a) $y = -2x - 8$
(b) $-3x + 4y = 7$
(c) $7y - 16x + 53 = 0$

**5.** Graph the relation

$$y = 2x^3 - 3x^2 + 6x - 5$$

from $y = -2$ to $y = +2$. Is the plotted curve a straight line?

6. Graph the relation

$$y = \frac{8 - 10x - 3x^2}{2 - 3x}$$

from $x = -3$ to $x = +3$. Is the plotted curve a straight line?

### Solution of Algebraic Equations

**7.** Solve the following linear equations for $x$.
(a) $7x + 5 = 0$
(b) $-4x + 3 = 0$
(c) $-3x = -2$

8. Solve the following linear equations for $x$.
(a) $6 - 8x = 0$
(b) $-2x - 5 = 0$
(c) $4x = -8$

**9.** Solve the following quadratic equations for $x$.
(a) $4x^2 + 7x - 5 = 0$
(b) $2x^2 = -3 - 6x$
(c) $2x + \dfrac{3}{x} = 6$

10. Solve the following quadratic equations for $x$.
(a) $6x^2 + 15x + 2 = 0$
(b) $4x = 5x^2 - 3$
(c) $\dfrac{1}{2 - x} + 3x = 4$

**11.** Solve each of the following equations for $x$ using the approximation of small $x$, iteration, or graphical solution, as appropriate.
(a) $x(2.00 + x)^2 = 2.6 \times 10^{-6}$
(b) $x(3.00 - 7x)(2.00 + 2x) = 0.230$
(c) $2x^3 + 3x^2 + 12x = -16$

12. Solve each of the following equations for $x$ using the approximation of small $x$, iteration, or graphical solution, as appropriate.
(a) $x(2.00 + x)(3.00 - x)(5.00 + 2x) = 1.58 \times 10^{-15}$
(b) $x\dfrac{(3.00 + x)(1.00 - x)}{2.00 - x} = 0.122$
(c) $12x^3 - 4x^2 + 35x = 10$

### Powers and Logarithms

**13.** Calculate each of the following expressions, giving your answers the proper numbers of significant figures.
(a) $\log (3.56 \times 10^4)$

(b) $e^{-15.69}$

(c) $10^{8.41}$

(d) $\ln(6.893 \times 10^{-22})$

14. Calculate each of the following expressions, giving your answers the proper numbers of significant figures.

(a) $10^{-16.528}$

(b) $\ln(4.30 \times 10^{13})$

(c) $e^{14.21}$

(d) $\log(4.983 \times 10^{-11})$

15. What number has a common logarithm of 0.4793?

16. What number has a natural logarithm of −15.824?

17. Determine the common logarithm of $3.00 \times 10^{121}$. It is quite likely that your calculator will *not* give a correct answer. Explain why.

18. Compute the value of $10^{-107.8}$. It is quite likely that your calculator will *not* give the correct answer. Explain why.

19. The common logarithm of 5.64 is 0.751. Without using a calculator, determine the common logarithm of $5.64 \times 10^7$ and of $5.64 \times 10^{-3}$.

20. The common logarithm of 2.68 is 0.428. Without using a calculator, determine the common logarithm of $2.68 \times 10^{192}$ and of $2.68 \times 10^{-289}$.

21. Use the graphical method and a calculator to solve the equation

$$\log \ln x = -x$$

for $x$. Give $x$ to four significant figures.

22. Use a calculator to find a number that is equal to the reciprocal of its own natural logarithm. Report the answer to four significant figures.

## Slopes of Curves and Derivatives

23. Calculate the derivatives of the following functions.

(a) $y = 4x^2 + 4$

(b) $y = \sin 3x + 4 \cos 2x$

(c) $y = 3x + 2$

(d) $y = \ln 7x$

24. Calculate the derivatives of the following functions.

(a) $y = 6x^{19}$

(b) $y = 7x^2 + 6x + 2$

(c) $y = e^{-6x}$

(d) $y = \cos 2x + \dfrac{7}{x}$

## Areas Under Curves and Integrals

25. Calculate the following integrals.

(a) $\displaystyle\int_2^4 (3x + 1)\, dx$

(b) $\displaystyle\int_0^5 x^6\, dx$

(c) $\displaystyle\int_1^4 e^{-2x}\, dx$

26. Calculate the following integrals.

(a) $\displaystyle\int_{-1}^3 4\, dx$

(b) $\displaystyle\int_2^{100} \dfrac{1}{x}\, dx$

(c) $\displaystyle\int_2^4 \dfrac{5}{x^2}\, dx$

# Standard Chemical Thermodynamic Properties

This table lists standard enthalpies of formation $\Delta H_f^\circ$, standard third-law entropies $S^\circ$, standard free energies of formation $\Delta G_f^\circ$, and molar heat capacities at constant pressure, $C_P$, for a variety of substances, all at 25°C (298.15 K) and 1 atm. The table proceeds from the left side to the right side of the periodic table. Binary compounds are listed under the element that occurs to the left in the periodic table, except that binary oxides and hydrides are listed with the other element. Thus, KCl is listed with potassium and its compounds, but $ClO_2$ is listed with chlorine and its compounds.

Note that the *solution-phase* entropies are not absolute entropies but are measured relative to the arbitrary standard $S^\circ(H^+(aq)) = 0$. Consequently, some of them (as well as some of the heat capacities) are negative.

Most of the thermodynamic data in these tables were taken from the *NBS Tables of Chemical Thermodynamic Properties* (1982) and changed, where necessary, from a standard pressure of 0.1 MPa to 1 atm. The data for organic compounds $C_nH_m(n > 2)$ were taken from the *Handbook of Chemistry and Physics* (1981).

| Substance | $\Delta H_f^\circ$ (25°C) kJ mol$^{-1}$ | $S^\circ$ (25°C) J K$^{-1}$ mol$^{-1}$ | $\Delta G_f^\circ$ (25°C) kJ mol$^{-1}$ | $C_P$ (25°C) J K$^{-1}$ mol$^{-1}$ |
|---|---|---|---|---|
| H(g) | 217.96 | 114.60 | 203.26 | 20.78 |
| H$_2$(g) | 0 | 130.57 | 0 | 28.82 |
| H$^+$(aq) | 0 | 0 | 0 | 0 |
| H$_3$O$^+$(aq) | −285.83 | 69.91 | −237.18 | 75.29 |
| Li(s) | 0 | 29.12 | 0 | 24.77 |
| Li(g) | 159.37 | 138.66 | 126.69 | 20.79 |
| Li$^+$(aq) | −278.49 | 13.4 | −293.31 | 68.6 |
| LiH(s) | −90.54 | 20.01 | −68.37 | 27.87 |
| Li$_2$O(s) | −597.94 | 37.57 | −561.20 | 54.10 |
| LiF(s) | −615.97 | 35.65 | −587.73 | 41.59 |
| LiCl(s) | −408.61 | 59.33 | −384.39 | 47.99 |
| LiBr(s) | −351.21 | 74.27 | −342.00 | — |
| LiI(s) | −270.41 | 86.78 | −270.29 | 51.04 |
| Na(s) | 0 | 51.21 | 0 | 28.24 |
| Na(g) | 107.32 | 153.60 | 76.79 | 20.79 |
| Na$^+$(aq) | −240.12 | 59.0 | −261.90 | 46.4 |
| Na$_2$O(s) | −414.22 | 75.06 | −375.48 | 69.12 |
| NaOH(s) | −425.61 | 64.46 | −379.53 | 59.54 |
| NaF(s) | −573.65 | 51.46 | −543.51 | 48.86 |
| NaCl(s) | −411.15 | 72.13 | −384.15 | 50.50 |
| NaBr(s) | −361.06 | 86.82 | −348.98 | 51.38 |
| NaI(s) | −287.78 | 98.53 | −286.06 | 52.09 |
| NaNO$_3$(s) | −467.85 | 116.52 | −367.07 | 92.88 |
| Na$_2$S(s) | −364.8 | 83.7 | −349.8 | — |
| Na$_2$SO$_4$(s) | −1387.08 | 149.58 | −1270.23 | 128.20 |

*continued*

| Substance | $\Delta H_f^\circ$ (25°C) kJ mol$^{-1}$ | $S^\circ$ (25°C) J K$^{-1}$ mol$^{-1}$ | $\Delta G_f^\circ$ (25°C) kJ mol$^{-1}$ | $C_P$ (25°C) J K$^{-1}$ mol$^{-1}$ |
|---|---|---|---|---|
| NaHSO$_4$(s) | −1125.5 | 113.0 | −992.9 | — |
| Na$_2$CO$_3$(s) | −1130.68 | 134.98 | −1044.49 | 112.30 |
| NaHCO$_3$(s) | −950.81 | 101.7 | −851.1 | 87.61 |
| K(s) | 0 | 64.18 | 0 | 29.58 |
| K(g) | 89.24 | 160.23 | 60.62 | 20.79 |
| K$^+$(aq) | −252.38 | 102.5 | −283.27 | 21.8 |
| KO$_2$(s) | −284.93 | 116.7 | −239.4 | 77.53 |
| K$_2$O$_2$(s) | −494.1 | 102.1 | −425.1 | — |
| KOH(s) | −424.76 | 78.9 | −379.11 | 64.9 |
| KF(s) | −567.27 | 66.57 | −537.77 | 49.04 |
| KCl(s) | −436.75 | 82.59 | −409.16 | 51.30 |
| KClO$_3$(s) | −397.73 | 143.1 | −296.25 | 100.25 |
| KBr(s) | −393.80 | 95.90 | −380.66 | 52.30 |
| KI(s) | −327.90 | 106.32 | −324.89 | 52.93 |
| KMnO$_4$(s) | −837.2 | 171.71 | −737.7 | 117.57 |
| K$_2$CrO$_4$(s) | −1403.7 | 200.12 | −1295.8 | 145.98 |
| K$_2$Cr$_2$O$_7$(s) | −2061.5 | 291.2 | −1881.9 | 219.24 |
| Rb(s) | 0 | 76.78 | 0 | 31.06 |
| Rb(g) | 80.88 | 169.98 | 53.09 | 20.79 |
| Rb$^+$(aq) | −251.17 | 121.50 | −283.98 | — |
| RbCl(s) | −435.35 | 95.90 | −407.82 | 52.38 |
| RbBr(s) | −394.59 | 109.96 | −381.79 | 52.84 |
| RbI(s) | −333.80 | 118.41 | −328.86 | 53.18 |
| Cs(s) | 0 | 85.23 | 0 | 32.17 |
| Cs(g) | 76.06 | 175.49 | 49.15 | 70.79 |
| Cs$^+$(aq) | −258.28 | 133.05 | −292.02 | −10.5 |
| CsF(s) | −553.5 | 92.80 | −525.5 | 51.09 |
| CsCl(s) | −443.04 | 101.17 | −414.55 | 52.47 |
| CsBr(s) | −405.81 | 113.05 | −391.41 | 52.93 |
| CsI(s) | −346.60 | 123.05 | −340.58 | 52.80 |

| | Substance | $\Delta H_f^\circ$ (25°C) kJ mol$^{-1}$ | $S^\circ$ (25°C) J K$^{-1}$ mol$^{-1}$ | $\Delta G_f^\circ$ (25°C) kJ mol$^{-1}$ | $C_P$ (25°C) J K$^{-1}$ mol$^{-1}$ |
|---|---|---|---|---|---|
| II | Be(s) | 0 | 9.50 | 0 | 16.44 |
| | Be(g) | 324.3 | 136.16 | 286.6 | 20.79 |
| | BeO(s) | −609.6 | 14.14 | −580.3 | 25.52 |
| | Mg(s) | 0 | 32.68 | 0 | 24.89 |
| | Mg(g) | 147.70 | 148.54 | 113.13 | 20.79 |
| | Mg$^{2+}$(aq) | −466.85 | −138.1 | −454.8 | — |
| | MgO(s) | −601.70 | 26.94 | −569.45 | 37.15 |
| | MgCl$_2$(s) | −641.32 | 89.62 | −591.82 | 71.38 |
| | MgSO$_4$(s) | −1284.9 | 91.6 | −1170.7 | 96.48 |
| | Ca(s) | 0 | 41.42 | 0 | 25.31 |
| | Ca(g) | 178.2 | 154.77 | 144.33 | 20.79 |
| | Ca$^{2+}$(aq) | −542.83 | −53.1 | −553.58 | — |
| | CaH$_2$(s) | −186.2 | 42 | −147.2 | — |
| | CaO(s) | −635.09 | 39.75 | −604.05 | 42.80 |
| | CaS(s) | −482.4 | 56.5 | −477.4 | 47.40 |
| | Ca(OH)$_2$(s) | −986.09 | 83.39 | −898.56 | 87.49 |
| | CaF$_2$(s) | −1219.6 | 68.87 | −1167.3 | 67.03 |
| | CaCl$_2$(s) | −795.8 | 104.6 | −748.1 | 72.59 |
| | CaBr$_2$(s) | −682.8 | 130 | −663.6 | — |
| | CaI$_2$(s) | −533.5 | 142 | −528.9 | — |
| | Ca(NO$_3$)$_2$(s) | −938.39 | 193.3 | −743.20 | 149.37 |
| | CaC$_2$(s) | −59.8 | 69.96 | −64.9 | 62.72 |

*continued*

| Substance | $\Delta H_f^\circ$ (25°C) kJ mol$^{-1}$ | $S^\circ$ (25°C) J K$^{-1}$ mol$^{-1}$ | $\Delta G_f^\circ$ (25°C) kJ mol$^{-1}$ | $C_P$ (25°C) J K$^{-1}$ mol$^{-1}$ |
|---|---|---|---|---|
| $CaCO_3(s$, calcite) | −1206.92 | 92.9 | −1128.84 | 81.88 |
| $CaCO_3(s$, aragonite) | −1207.13 | 88.7 | −1127.80 | 81.25 |
| $CaSO_4(s)$ | −1434.11 | 106.9 | −1321.86 | 99.66 |
| $CaSiO_3(s)$ | −1634.94 | 81.92 | −1549.66 | 85.27 |
| $CaMg(CO_3)_2(s$, dolomite) | −2326.3 | 155.18 | −2163.4 | 157.53 |
| $Sr(s)$ | 0 | 52.3 | 0 | 26.4 |
| $Sr(g)$ | 164.4 | 164.51 | 130.9 | 20.79 |
| $Sr^{2+}(aq)$ | −545.80 | 32.6 | −559.48 | |
| $SrCl_2(s)$ | −828.9 | 114.85 | −781.1 | 75.60 |
| $SrCO_3(s)$ | −1220.0 | 97.1 | −1140.1 | 81.42 |
| $Ba(s)$ | 0 | 62.8 | 0 | 28.07 |
| $Ba(g)$ | 180 | 170.24 | 146 | 20.79 |
| $Ba^{2+}(aq)$ | −537.64 | 9.6 | −560.77 | — |
| $BaCl_2(s)$ | −858.6 | 123.68 | −810.4 | 75.14 |
| $BaCO_3(s)$ | −1216.3 | 112.1 | −1137.6 | 85.35 |
| $BaSO_4(s)$ | −1473.2 | 132.2 | −1362.3 | 101.75 |
| $Sc(s)$ | 0 | 34.64 | 0 | 25.52 |
| $Sc(g)$ | 377.8 | 174.68 | 336.06 | 22.09 |
| $Sc^{3+}(aq)$ | −614.2 | −255 | −586.6 | — |
| $Ti(s)$ | 0 | 30.63 | 0 | 25.02 |
| $Ti(g)$ | 469.9 | 180.19 | 425.1 | 24.43 |
| $TiO_2(s$, rutile) | −944.7 | 50.33 | −889.5 | 55.02 |
| $TiCl_4(\ell)$ | −804.2 | 252.3 | −737.2 | 145.18 |
| $TiCl_4(g)$ | −763.2 | 354.8 | −726.8 | 95.4 |
| $Cr(s)$ | 0 | 23.77 | 0 | 23.35 |
| $Cr(g)$ | 396.6 | 174.39 | 351.8 | 20.79 |
| $Cr_2O_3(s)$ | −1139.7 | 81.2 | −1058.1 | 118.74 |
| $CrO_4^{2-}(aq)$ | −881.15 | 50.21 | −727.75 | — |
| $Cr_2O_7^{2-}(aq)$ | −1490.3 | 261.9 | −1301.1 | — |
| $W(s)$ | 0 | 32.64 | 0 | 24.27 |
| $W(g)$ | 849.4 | 173.84 | 807.1 | 21.31 |
| $WO_2(s)$ | −589.69 | 50.54 | −533.92 | 56.11 |
| $WO_3(s)$ | −842.87 | 75.90 | −764.08 | 73.76 |
| $Mn(s)$ | 0 | 32.01 | 0 | 26.32 |
| $Mn(g)$ | 280.7 | 238.5 | 173.59 | 20.79 |
| $Mn^{2+}(aq)$ | −220.75 | −73.6 | −228.1 | 50 |
| $MnO(s)$ | −385.22 | 59.71 | −362.92 | 45.44 |
| $MnO_2(s)$ | −520.03 | 53.05 | −465.17 | 54.14 |
| $MnO_4^-(s)$ | −541.4 | 191.2 | −447.2 | −82.0 |
| $Fe(s)$ | 0 | 27.28 | 0 | 25.10 |
| $Fe(g)$ | 416.3 | 180.38 | 370.7 | 25.68 |
| $Fe^{2+}(aq)$ | −89.1 | −137.7 | −78.9 | — |
| $Fe^{3+}(aq)$ | −48.5 | −315.9 | −4.7 | — |
| $Fe_{0.947}O(s$, wüstite) | −266.27 | 57.49 | −245.12 | 48.12 |
| $Fe_2O_3(s$, hematite) | −824.2 | 87.40 | −742.2 | 103.85 |
| $Fe_3O_4(s$, magnetite) | −1118.4 | 146.4 | −1015.5 | 143.43 |
| $Fe(OH)_3(s)$ | −823.0 | 106.7 | −696.5 | — |

*continued*

| Substance | $\Delta H_f^\circ$ (25°C) kJ mol$^{-1}$ | $S^\circ$ (25°C) J K$^{-1}$ mol$^{-1}$ | $\Delta G_f^\circ$ (25°C) kJ mol$^{-1}$ | $C_P$ (25°C) J K$^{-1}$ mol$^{-1}$ |
|---|---|---|---|---|
| FeS(s) | −100.0 | 60.29 | −100.4 | 50.54 |
| FeCO$_3$(s) | −740.57 | 93.1 | −666.72 | 82.13 |
| Fe(CN)$_6^{3-}$(aq) | 561.9 | 270.3 | 729.4 | — |
| Fe(CN)$_6^{4-}$(aq) | 455.6 | 95.0 | 695.1 | — |
| Co(s) | 0 | 30.04 | 0 | 24.81 |
| Co(g) | 424.7 | 179.41 | 380.3 | 23.02 |
| Co$^{2+}$(aq) | −58.2 | −113 | −54.4 | — |
| Co$^{3+}$(aq) | 92 | −305 | 134 | — |
| CoO(s) | −237.94 | 52.97 | −214.22 | 55.23 |
| CoCl$_2$(s) | −312.5 | 109.16 | −269.8 | 78.49 |
| Ni(s) | 0 | 29.87 | 0 | 26.07 |
| Ni(g) | 429.7 | 182.08 | 384.5 | 25.36 |
| Ni$^{2+}$(aq) | −54.0 | −128.9 | −45.6 | — |
| NiO(s) | −239.7 | 37.99 | −211.7 | 44.31 |
| Pt(s) | 0 | 41.63 | 0 | 25.86 |
| Pt(g) | 565.3 | 192.30 | 520.5 | 25.53 |
| PtCl$_6^{2-}$(aq) | −668.2 | 219.7 | −482.7 | — |
| Cu(s) | 0 | 33.15 | 0 | 24.44 |
| Cu(g) | 338.32 | 166.27 | 298.61 | 20.79 |
| Cu$^+$(aq) | 71.67 | 40.6 | 49.98 | — |
| Cu$^{2+}$(aq) | 64.77 | −99.6 | 65.49 | — |
| CuO(s) | −157.3 | 42.63 | −129.7 | 42.30 |
| Cu$_2$O(s) | −168.6 | 93.14 | −146.0 | 63.64 |
| CuCl(s) | −137.2 | 86.2 | −119.88 | 48.5 |
| CuCl$_2$(s) | −220.1 | 108.07 | −175.7 | 71.88 |
| CuSO$_4$(s) | −771.36 | 109 | −661.9 | 100.0 |
| Cu(NH$_3$)$_4^{2+}$(aq) | −348.5 | 273.6 | −111.07 | — |
| Ag(s) | 0 | 42.55 | 0 | 25.35 |
| Ag(g) | 284.55 | 172.89 | 245.68 | 20.79 |
| Ag$^+$(aq) | 105.58 | 72.68 | 77.11 | 21.8 |
| AgCl(s) | −127.07 | 96.2 | −109.81 | 50.79 |
| AgNO$_3$(s) | −124.39 | 140.92 | −33.48 | 93.05 |
| Ag(NH$_3$)$_2^+$(aq) | −111.29 | 245.2 | −17.12 | — |
| Au(s) | 0 | 47.40 | 0 | 25.42 |
| Au(g) | 366.1 | 180.39 | 326.3 | 20.79 |
| Zn(s) | 0 | 41.63 | 0 | 25.40 |
| Zn(g) | 130.73 | 160.87 | 95.18 | 20.79 |
| Zn$^{2+}$(aq) | −153.89 | −112.1 | −147.06 | 46 |
| ZnO(s) | −348.28 | 43.64 | −318.32 | 40.25 |
| ZnS(s, sphalerite) | −205.98 | 57.7 | −201.29 | 46.0 |
| ZnCl$_2$(s) | −415.05 | 111.46 | −369.43 | 71.34 |
| ZnSO$_4$(s) | −982.8 | 110.5 | −871.5 | 99.2 |
| Zn(NH$_3$)$_4^{2+}$(aq) | −533.5 | 301 | −301.9 | — |
| Hg($\ell$) | 0 | 76.02 | 0 | 27.98 |
| Hg(g) | 61.32 | 174.85 | 31.85 | 20.79 |
| HgO(s) | −90.83 | 70.29 | −58.56 | 44.06 |
| HgCl$_2$(s) | −224.3 | 146.0 | −178.6 | — |
| Hg$_2$Cl$_2$(s) | −265.22 | 192.5 | −210.78 | — |
| III    B(s) | 0 | 5.86 | 0 | 11.09 |
| B(g) | 562.7 | 153.34 | 518.8 | 20.80 |

*continued*

| Substance | $\Delta H_f^\circ$ (25°C) kJ mol$^{-1}$ | $S^\circ$ (25°C) J K$^{-1}$ mol$^{-1}$ | $\Delta G_f^\circ$ (25°C) kJ mol$^{-1}$ | $C_P$ (25°C) J K$^{-1}$ mol$^{-1}$ |
|---|---|---|---|---|
| $B_2H_6(g)$ | 35.6 | 232.00 | 86.6 | 56.90 |
| $B_5H_9(g)$ | 73.2 | 275.81 | 174.9 | 96.78 |
| $B_2O_3(s)$ | −1272.77 | 53.97 | −1193.70 | 62.93 |
| $H_3BO_3(s)$ | −1094.33 | 88.83 | −969.02 | 81.38 |
| $BF_3(g)$ | −1137.00 | 254.01 | −1120.35 | 50.46 |
| $BF_4^-(aq)$ | −1574.9 | 180 | −1486.9 | — |
| $BCl_3(g)$ | −403.76 | 289.99 | −388.74 | 62.72 |
| $BBr_3(g)$ | −205.64 | 324.13 | −232.47 | 67.78 |
| $Al(s)$ | 0 | 28.33 | 0 | 24.35 |
| $Al(g)$ | 326.4 | 164.43 | 285.7 | 21.38 |
| $Al^{3+}(aq)$ | 531 | 321.7 | −485 | — |
| $Al_2O_3(s)$ | −1675.7 | 50.92 | −1582.3 | 79.04 |
| $AlCl_3(s)$ | −704.2 | 110.67 | −628.8 | 91.84 |
| $Ga(s)$ | 0 | 40.88 | 0 | 25.86 |
| $Ga(g)$ | 277.0 | 168.95 | 238.9 | 25.36 |
| $Tl(s)$ | 0 | 64.18 | 0 | 26.32 |
| $Tl(g)$ | 182.21 | 180.85 | 147.44 | 20.79 |

**IV**

| Substance | $\Delta H_f^\circ$ (25°C) kJ mol$^{-1}$ | $S^\circ$ (25°C) J K$^{-1}$ mol$^{-1}$ | $\Delta G_f^\circ$ (25°C) kJ mol$^{-1}$ | $C_P$ (25°C) J K$^{-1}$ mol$^{-1}$ |
|---|---|---|---|---|
| $C(s, graphite)$ | 0 | 5.74 | 0 | 8.53 |
| $C(s, diamond)$ | 1.895 | 2.377 | 2.900 | 6.11 |
| $C(g)$ | 716.682 | 157.99 | 671.29 | 20.84 |
| $CH_4(g)$ | −74.81 | 186.15 | −50.75 | 35.31 |
| $C_2H_2(g)$ | 226.73 | 200.83 | 209.20 | 43.93 |
| $C_2H_4(g)$ | 52.26 | 219.45 | 68.12 | 43.56 |
| $C_2H_6(g)$ | −84.68 | 229.49 | −32.89 | 52.63 |
| $C_3H_8(g)$ | −103.85 | 269.91 | −23.49 | 73.0 |
| $n\text{-}C_4H_{10}(g)$ | −124.73 | 310.03 | −15.71 | 97.5 |
| $C_4H_{10}(g, isobutane)$ | −131.60 | 294.64 | −17.97 | 96.8 |
| $n\text{-}C_5H_{12}(g)$ | −146.44 | 348.40 | −8.20 | 120 |
| $C_6H_6(g)$ | 82.93 | 269.2 | 129.66 | 81.6 |
| $C_6H_6(\ell)$ | 49.03 | 172.8 | 124.50 | 136 |
| $CO(g)$ | −110.52 | 197.56 | −137.15 | 29.14 |
| $CO_2(g)$ | −393.51 | 213.63 | −394.36 | 37.11 |
| $CO_2(aq)$ | −413.80 | 117.6 | −385.98 | — |
| $CS_2(\ell)$ | 89.70 | 151.34 | 65.27 | 75.7 |
| $CS_2(g)$ | 117.36 | 237.73 | 67.15 | 45.40 |
| $H_2CO_3(aq)$ | −699.65 | 187.4 | −623.08 | — |
| $HCO_3^-(aq)$ | −691.99 | 91.2 | −586.77 | — |
| $CO_3^{2-}(aq)$ | −677.14 | −56.9 | −527.81 | — |
| $HCOOH(\ell)$ | −424.72 | 128.95 | −361.42 | 99.04 |
| $HCOOH(aq)$ | −425.43 | 163 | −372.3 | — |
| $COOH^-(aq)$ | −425.55 | 92 | −351.0 | −87.9 |
| $CH_2O(g)$ | −108.57 | 218.66 | −102.55 | 35.40 |
| $CH_3OH(\ell)$ | −238.66 | 126.8 | −166.35 | 81.6 |
| $CH_3OH(g)$ | −200.66 | 239.70 | −162.01 | 43.89 |
| $CH_3OH(aq)$ | −245.93 | 133.1 | −175.31 | — |
| $H_2C_2O_4(s)$ | −827.2 | — | — | 117 |
| $HC_2O_4^-(aq)$ | −818.4 | 149.4 | −698.34 | — |
| $C_2O_4^{2-}(aq)$ | −825.1 | 45.6 | −673.9 | — |
| $CH_3COOH(\ell)$ | −484.5 | 159.8 | −390.0 | 124.3 |
| $CH_3COOH(g)$ | −432.25 | 282.4 | −374.1 | 66.5 |
| $CH_3COOH(aq)$ | −485.76 | 178.7 | −396.46 | — |
| $CH_3COO^-(aq)$ | −486.01 | 86.6 | −369.31 | −6.3 |

*continued*

| Substance | $\Delta H_f^\circ$ (25°C) kJ mol$^{-1}$ | $S°$ (25°C) J K$^{-1}$ mol$^{-1}$ | $\Delta G_f^\circ$ (25°C) kJ mol$^{-1}$ | $C_P$ (25°C) J K$^{-1}$ mol$^{-1}$ |
|---|---|---|---|---|
| $CH_3CHO(\ell)$ | −192.30 | 160.2 | −128.12 | — |
| $C_2H_5OH(\ell)$ | −277.69 | 160.7 | −174.89 | 111.46 |
| $C_2H_5OH(g)$ | −235.10 | 282.59 | −168.57 | 65.44 |
| $C_2H_5OH(aq)$ | −288.3 | 148.5 | −181.64 | — |
| $CH_3OCH_3(g)$ | −184.05 | 266.27 | −112.67 | 64.39 |
| $CF_4(g)$ | −925 | 261.50 | −879 | 61.09 |
| $CCl_4(\ell)$ | −135.44 | 216.40 | −65.28 | 131.75 |
| $CCl_4(g)$ | −102.9 | 309.74 | −60.62 | 83.30 |
| $CHCl_3(g)$ | −103.14 | 295.60 | −70.37 | 65.69 |
| $COCl_2(g)$ | −218.8 | 283.53 | −204.6 | 57.66 |
| $CH_2Cl_2(g)$ | −92.47 | 270.12 | −65.90 | 50.96 |
| $CH_3Cl(g)$ | −80.83 | 234.47 | −57.40 | 40.75 |
| $CBr_4(s)$ | 18.8 | 212.5 | 47.7 | 144.3 |
| $CH_3I(\ell)$ | −15.5 | 163.2 | 13.4 | 126 |
| $HCN(g)$ | 135.1 | 201.67 | 124.7 | 35.86 |
| $HCN(aq)$ | 107.1 | 124.7 | 119.7 | — |
| $CN^-(aq)$ | 150.6 | 94.1 | 172.4 | — |
| $CH_3NH_2(g)$ | −22.97 | 243.30 | 32.09 | 53.1 |
| $CO(NH_2)_2(s)$ | −333.51 | 104.49 | −197.44 | 93.14 |
| $Si(s)$ | 0 | 18.83 | 0 | 20.00 |
| $Si(g)$ | 455.6 | 167.86 | 411.3 | 22.25 |
| $SiC(s)$ | −65.3 | 16.61 | −62.8 | 26.86 |
| $SiO_2(s, \text{quartz})$ | −910.94 | 41.84 | −856.67 | 44.43 |
| $SiO_2(s, \text{cristobalite})$ | −909.48 | 42.68 | −855.43 | 44.18 |
| $Ge(s)$ | 0 | 31.09 | 0 | 23.35 |
| $Ge(g)$ | 376.6 | 335.9 | 167.79 | 30.73 |
| $Sn(s, \text{white})$ | 0 | 51.55 | 0 | 26.99 |
| $Sn(s, \text{gray})$ | −2.09 | 44.14 | 0.13 | 25.77 |
| $Sn(g)$ | 302.1 | 168.38 | 267.3 | 21.26 |
| $SnO(s)$ | −285.8 | 56.5 | −256.9 | 44.31 |
| $SnO_2(s)$ | −580.7 | 52.3 | −519.6 | 52.59 |
| $Sn(OH)_2(s)$ | −561.1 | 155 | −491.7 | — |
| $Pb(s)$ | 0 | 64.81 | 0 | 26.44 |
| $Pb(g)$ | 195.0 | 161.9 | 175.26 | 20.79 |
| $Pb^{2+}(aq)$ | −1.7 | 10.5 | −24.43 | — |
| $PbO(s, \text{yellow})$ | −217.32 | 68.70 | −187.91 | 45.77 |
| $PbO(s, \text{red})$ | −218.99 | 66.5 | −188.95 | 45.81 |
| $PbO_2(s)$ | −277.4 | 68.6 | −217.36 | 64.64 |
| $PbS(s)$ | −100.4 | 91.2 | −98.7 | 49.50 |
| $PbI_2(s)$ | −175.48 | 174.85 | −173.64 | 77.36 |
| $PbSO_4(s)$ | −919.94 | 148.57 | −813.21 | 103.21 |
| $N_2(g)$ | 0 | 191.50 | 0 | 29.12 |
| $N(g)$ | 472.70 | 153.19 | 455.58 | 20.79 |
| $NH_3(g)$ | −46.11 | 192.34 | −16.48 | 35.06 |
| $NH_3(aq)$ | −80.29 | 111.3 | −26.50 | — |
| $NH_4^+(aq)$ | −132.51 | 113.4 | −79.31 | 79.9 |
| $N_2H_4(\ell)$ | 50.63 | 121.21 | 149.24 | 98.87 |
| $N_2H_4(aq)$ | 34.31 | 138 | 128.1 | — |
| $NO(g)$ | 90.25 | 210.65 | 86.55 | 29.84 |
| $NO_2(g)$ | 33.18 | 239.95 | 51.29 | 37.20 |
| $NO_2^-(aq)$ | −104.6 | 123.0 | −32.2 | −97.5 |
| $NO_3^-(aq)$ | −205.0 | 146.4 | −108.74 | −86.6 |

*continued*

| Substance | $\Delta H_f^\circ$ (25°C) kJ mol$^{-1}$ | $S^\circ$ (25°C) J K$^{-1}$ mol$^{-1}$ | $\Delta G_f^\circ$ (25°C) kJ mol$^{-1}$ | $C_P$ (25°C) J K$^{-1}$ mol$^{-1}$ |
|---|---|---|---|---|
| $N_2O(g)$ | 82.05 | 219.74 | 104.18 | 38.45 |
| $N_2O_4(g)$ | 9.16 | 304.18 | 97.82 | 77.28 |
| $N_2O_5(s)$ | −43.1 | 178.2 | 113.8 | 143.1 |
| $HNO_2(g)$ | −79.5 | 254.0 | −46.0 | 45.6 |
| $HNO_3(\ell)$ | −174.10 | 155.49 | −80.76 | 109.87 |
| $NH_4NO_3(s)$ | −365.56 | 151.08 | −184.02 | 139.3 |
| $NH_4Cl(s)$ | −314.43 | 94.6 | −202.97 | 84.1 |
| $(NH_4)_2SO_4(s)$ | −1180.85 | 220.1 | −901.90 | 187.49 |
| P(s, white) | 0 | 41.09 | 0 | 23.84 |
| P(s, red) | −17.6 | 22.80 | −12.1 | 21.21 |
| $P(g)$ | 314.64 | 163.08 | 278.28 | 20.79 |
| $P_2(g)$ | 144.3 | 218.02 | 103.7 | 32.05 |
| $P_4(g)$ | 58.91 | 279.87 | 24.47 | 67.15 |
| $PH_3(g)$ | 5.4 | 210.12 | 13.4 | 37.11 |
| $H_3PO_4(s)$ | −1279.0 | 110.50 | −1119.2 | 106.06 |
| $H_3PO_4(aq)$ | −1288.34 | 158.2 | −1142.54 | — |
| $H_2PO_4^-(aq)$ | −1296.29 | 90.4 | −1130.28 | — |
| $HPO_4^{2-}(aq)$ | −1292.14 | −33.5 | −1089.15 | — |
| $PO_4^{3-}(aq)$ | −1277.4 | −222 | −1018.7 | — |
| $PCl_3(g)$ | −287.0 | 311.67 | −267.8 | 71.84 |
| $PCl_5(g)$ | −374.9 | 364.47 | −305.0 | 112.80 |
| As(s, gray) | 0 | 35.1 | 0 | 24.64 |
| $As(g)$ | 302.5 | 174.10 | 261.0 | 20.79 |
| $As_2(g)$ | 222.2 | 239.3 | 171.9 | 35.00 |
| $As_4(g)$ | 143.9 | 314 | 92.4 | — |
| $AsH_3(g)$ | 66.44 | 222.67 | 68.91 | 38.07 |
| $As_4O_6(s)$ | −1313.94 | 214.2 | −1152.53 | 191.29 |
| Sb(s) | 0 | 45.69 | 0 | 25.33 |
| $Sb(g)$ | 262.3 | 180.16 | 222.1 | 20.79 |
| Bi(s) | 0 | 56.74 | 0 | 25.52 |
| $Bi(g)$ | 207.1 | 186.90 | 168.2 | 20.79 |

| | Substance | $\Delta H_f^\circ$ (25°C) kJ mol$^{-1}$ | $S^\circ$ (25°C) J K$^{-1}$ mol$^{-1}$ | $\Delta G_f^\circ$ (25°C) kJ mol$^{-1}$ | $C_P$ (25°C) J K$^{-1}$ mol$^{-1}$ |
|---|---|---|---|---|---|
| VI | $O_2(g)$ | 0 | 205.03 | 0 | 29.36 |
| | $O(g)$ | 249.17 | 160.95 | 231.76 | 21.91 |
| | $O_3(g)$ | 142.7 | 238.82 | 163.2 | 39.20 |
| | $OH^-(aq)$ | −229.99 | −10.75 | −157.24 | −148.5 |
| | $H_2O(\ell)$ | −285.83 | 69.91 | −237.18 | 75.29 |
| | $H_2O(g)$ | −241.82 | 188.72 | −228.59 | 35.58 |
| | $H_2O_2(\ell)$ | −187.78 | 109.6 | −120.42 | 89.1 |
| | $H_2O_2(aq)$ | −191.17 | 143.9 | −134.03 | — |
| | S(s, rhombic) | 0 | 31.80 | 0 | 22.64 |
| | S(s, monoclinic) | 0.30 | 32.6 | 0.096 | — |
| | $S(g)$ | 278.80 | 167.71 | 238.28 | 23.67 |
| | $S_8(g)$ | 102.30 | 430.87 | 49.66 | 156.44 |
| | $H_2S(g)$ | −20.63 | 205.68 | −33.56 | 34.23 |
| | $H_2S(aq)$ | −39.7 | 121 | −27.83 | — |
| | $HS^-(aq)$ | −17.6 | 62.8 | 12.08 | — |
| | $SO(g)$ | 6.26 | 221.84 | −19.87 | 30.17 |
| | $SO_2(g)$ | −296.83 | 248.11 | −300.19 | 39.87 |
| | $SO_3(g)$ | −395.72 | 256.65 | −371.08 | 50.67 |
| | $H_2SO_3(aq)$ | −608.81 | 232.2 | −537.81 | — |
| | $HSO_3^-(aq)$ | −626.22 | 139.7 | −527.73 | — |
| | $SO_3^{2-}(aq)$ | −635.5 | −29 | −486.5 | — |

*continued*

| Substance | $\Delta H_f^\circ$ (25°C) kJ mol$^{-1}$ | $S°$ (25°C) J K$^{-1}$ mol$^{-1}$ | $\Delta G_f^\circ$ (25°C) kJ mol$^{-1}$ | $C_P$ (25°C) J K$^{-1}$ mol$^{-1}$ |
|---|---|---|---|---|
| $H_2SO_4(\ell)$ | −813.99 | 156.90 | −690.10 | 138.91 |
| $HSO_4^-(aq)$ | −887.34 | 131.8 | −755.91 | −84 |
| $SO_4^{2-}(aq)$ | −909.27 | 20.1 | −744.53 | −293 |
| $SF_6(g)$ | −1209 | 291.71 | −1105.4 | 97.28 |
| Se(s, black) | 0 | 42.44 | 0 | 25.36 |
| Se(g) | 227.07 | 176.61 | 187.06 | 20.82 |
| VII  $F_2(g)$ | 0 | 202.67 | 0 | 31.30 |
| F(g) | 78.99 | 158.64 | 61.94 | 22.74 |
| $F^-(aq)$ | −332.63 | −13.8 | −278.79 | −106.7 |
| HF(g) | −271.1 | 173.67 | −273.2 | 29.13 |
| HF(aq) | −320.08 | 88.7 | −296.82 | — |
| $XeF_4(s)$ | −261.5 | — | — | — |
| $Cl_2(g)$ | 0 | 222.96 | 0 | 33.91 |
| Cl(g) | 121.68 | 165.09 | 105.71 | 21.84 |
| $Cl^-(aq)$ | −167.16 | 56.5 | −131.23 | −136.4 |
| HCl(g) | −92.31 | 186.80 | −95.30 | 29.12 |
| $ClO^-(aq)$ | −107.1 | 42 | −36.8 | — |
| $ClO_2(g)$ | 102.5 | 256.73 | 120.5 | 41.97 |
| $ClO_2^-(aq)$ | −66.5 | 101.3 | 17.2 | — |
| $ClO_3^-(aq)$ | −103.97 | 162.3 | −7.95 | — |
| $ClO_4^-(aq)$ | −129.33 | 182.0 | −8.52 | — |
| $Cl_2O(g)$ | 80.3 | 266.10 | 97.9 | 45.40 |
| HClO(aq) | −120.9 | 142 | −79.9 | — |
| $ClF_3(g)$ | −163.2 | 281.50 | −123.0 | 63.85 |
| $Br_2(\ell)$ | 0 | 152.23 | 0 | 75.69 |
| $Br_2(g)$ | 30.91 | 245.35 | 3.14 | 36.02 |
| $Br_2(aq)$ | −2.59 | 130.5 | 3.93 | — |
| Br(g) | 111.88 | 174.91 | 82.41 | 20.79 |
| $Br^-(aq)$ | −121.55 | 82.4 | −103.96 | −141.8 |
| HBr(g) | −36.40 | 198.59 | −53.43 | 29.14 |
| $BrO_3^-(aq)$ | −67.07 | 161.71 | 18.60 | — |
| $I_2(s)$ | 0 | 116.14 | 0 | 54.44 |
| $I_2(g)$ | 62.44 | 260.58 | 19.36 | 36.90 |
| $I_2(aq)$ | 22.6 | 137.2 | 16.40 | — |
| I(g) | 106.84 | 180.68 | 70.28 | 20.79 |
| $I^-(aq)$ | −55.19 | 111.3 | −51.57 | −142.3 |
| $I_3^-(aq)$ | −51.5 | 239.3 | −51.4 | — |
| HI(g) | 26.48 | 206.48 | 1.72 | 29.16 |
| ICl(g) | 17.78 | 247.44 | −5.44 | 35.56 |
| IBr(g) | 40.84 | 258.66 | 3.71 | 36.44 |
| VIII  He(g) | 0 | 126.04 | 0 | 20.79 |
| Ne(g) | 0 | 146.22 | 0 | 20.79 |
| Ar(g) | 0 | 154.73 | 0 | 20.79 |
| Kr(g) | 0 | 163.97 | 0 | 20.79 |
| Xe(g) | 0 | 169.57 | 0 | 20.79 |

# Standard Reduction Potentials at 25°C

| Half-Reaction | $\mathscr{E}°$ (volts) |
|---|---|
| $F_2(g) + 2\,e^- \longrightarrow 2\,F^-$ | 2.07 |
| $H_2O_2 + 2\,H_3O^+ + 2\,e^- \longrightarrow 4\,H_2O$ | 1.776 |
| $PbO_2(s) + SO_4^{2-} + 4\,H_3O^+ + 2\,e^- \longrightarrow PbSO_4(s) + 6\,H_2O$ | 1.685 |
| $Au^+ + e^- \longrightarrow Au(s)$ | 1.68 |
| $MnO_4^- + 4\,H_3O^+ + 3\,e^- \longrightarrow MnO_2(s) + 6\,H_2O$ | 1.679 |
| $HClO_2 + 2\,H_3O^+ + 2\,e^- \longrightarrow HClO + 3\,H_2O$ | 1.64 |
| $HClO + H_3O^+ + e^- \longrightarrow \frac{1}{2}\,Cl_2(g) + 2\,H_2O$ | 1.63 |
| $Ce^{4+} + e^- \longrightarrow Ce^{3+}$ (1 M $HNO_3$ solution) | 1.61 |
| $2\,NO(g) + 2\,H_3O^+ + 2\,e^- \longrightarrow N_2O(g) + 3\,H_2O$ | 1.59 |
| $BrO_3^- + 6\,H_3O^+ + 5\,e^- \longrightarrow \frac{1}{2}\,Br_2(\ell) + 9\,H_2O$ | 1.52 |
| $Mn^{3+} + e^- \longrightarrow Mn^{2+}$ | 1.51 |
| $MnO_4^- + 8\,H_3O^+ + 5\,e^- \longrightarrow Mn^{2+} + 12\,H_2O$ | 1.491 |
| $ClO_3^- + 6\,H_3O^+ + 5\,e^- \longrightarrow \frac{1}{2}\,Cl_2(g) + 9\,H_2O$ | 1.47 |
| $PbO_2(s) + 4\,H_3O^+ + 2\,e^- \longrightarrow Pb^{2+} + 6\,H_2O$ | 1.46 |
| $Au^{3+} + 3\,e^- \longrightarrow Au(s)$ | 1.42 |
| $Cl_2(g) + 2\,e^- \longrightarrow 2\,Cl^-$ | 1.3583 |
| $Cr_2O_7^{2-} + 14\,H_3O^+ + 6\,e^- \longrightarrow 2\,Cr^{3+} + 21\,H_2O$ | 1.33 |
| $O_3(g) + H_2O + 2\,e^- \longrightarrow O_2 + 2\,OH^-$ | 1.24 |
| $O_2(g) + 4\,H_3O^+ + 4\,e^- \longrightarrow 6\,H_2O$ | 1.229 |
| $MnO_2(s) + 4\,H_3O^+ + 2\,e^- \longrightarrow Mn^{2+} + 6\,H_2O$ | 1.208 |
| $ClO_4^- + 2\,H_3O^+ + 2\,e^- \longrightarrow ClO_3 + 3\,H_2O$ | 1.19 |
| $Br_2(\ell) + 2\,e^- \longrightarrow 2\,Br^-$ | 1.065 |
| $NO_3^- + 4\,H_3O^+ + 3\,e^- \longrightarrow NO(g) + 6\,H_2O$ | 0.96 |
| $2\,Hg^{2+} + 2\,e^- \longrightarrow Hg_2^{2+}$ | 0.905 |
| $Ag^+ + e^- \longrightarrow Ag(s)$ | 0.7996 |
| $Hg_2^{2+} + 2\,e \longrightarrow 2\,Hg(\ell)$ | 0.7961 |
| $Fe^{3+} + e^- \longrightarrow Fe^{2+}$ | 0.770 |
| $O_2(g) + 2\,H_3O^+ + 2\,e^- \longrightarrow H_2O_2 + 2\,H_2O$ | 0.682 |
| $BrO_3^- + 3\,H_2O + 6\,e^- \longrightarrow Br^- + 6\,OH^-$ | 0.61 |
| $MnO_4^- + 2\,H_2O + 3\,e^- \longrightarrow MnO_2(s) + 4\,OH^-$ | 0.588 |
| $I_2(s) + 2\,e^- \longrightarrow 2\,I^-$ | 0.535 |
| $Cu^+ + e^- \longrightarrow Cu(s)$ | 0.522 |
| $O_2(g) + 2\,H_2O + 4\,e^- \longrightarrow 4\,OH^-$ | 0.401 |
| $Cu^{2+} + 2\,e^- \longrightarrow Cu(s)$ | 0.3402 |
| $PbO_2(s) + H_2O + 2\,e^- \longrightarrow PbO(s) + 2\,OH^-$ | 0.28 |
| $Hg_2Cl_2(s) + 2\,e^- \longrightarrow 2\,Hg(\ell) + 2\,Cl^-$ | 0.2682 |
| $AgCl(s) + e^- \longrightarrow Ag(s) + Cl^-$ | 0.2223 |
| $SO_4^{2-} + 4\,H_3O^+ + 2\,e^- \longrightarrow H_2SO_3 + 5\,H_2O$ | 0.20 |
| $Cu^{2+} + e^- \longrightarrow Cu^+$ | 0.158 |
| $S_4O_6^{2-} + 2\,e^- \longrightarrow 2\,S_2O_3^{2-}$ | 0.0895 |
| $NO_3^- + H_2O + 2\,e^- \longrightarrow NO_2^- + 2\,OH^-$ | 0.01 |
| $2\,H_3O^+ + 2\,e^- \longrightarrow H_2(g) + 2\,H_2O(\ell)$ | 0.000 exactly |
| $Pb^{2+} + 2\,e^- \longrightarrow Pb(s)$ | −0.1263 |
| $Sn^{2+} + 2\,e^- \longrightarrow Sn(s)$ | −0.1364 |
| $Ni^{2+} + 2\,e^- \longrightarrow Ni(s)$ | −0.23 |

*continued*

| | |
|---|---|
| $Co^{2+} + 2\,e^- \longrightarrow Co(s)$ | $-0.28$ |
| $PbSO_4(s) + 2\,e^- \longrightarrow Pb(s) + SO_4^{2-}$ | $-0.356$ |
| $Mn(OH)_3(s) + e^- \longrightarrow Mn(OH)_2(s) + OH^-$ | $-0.40$ |
| $Cd^{2+} + 2\,e^- \longrightarrow Cd(s)$ | $-0.4026$ |
| $Fe^{2+} + 2\,e^- \longrightarrow Fe(s)$ | $-0.409$ |
| $Cr^{3+} + e^- \longrightarrow Cr^{2+}$ | $-0.424$ |
| $Fe(OH)_3(s) + e^- \longrightarrow Fe(OH)_2(s) + OH^-$ | $-0.56$ |
| $PbO(s) + H_2O + 2\,e^- \longrightarrow Pb(s) + 2\,OH^-$ | $-0.576$ |
| $2\,SO_3^{2-} + 3\,H_2O + 4\,e^- \longrightarrow S_2O_3^{2-} + 6\,OH^-$ | $-0.58$ |
| $Ni(OH)_2(s) + 2\,e^- \longrightarrow Ni(s) + 2\,OH^-$ | $-0.66$ |
| $Co(OH)_2(s) + 2\,e^- \longrightarrow Co(s) + 2\,OH^-$ | $-0.73$ |
| $Cr^{3+} + 3\,e^- \longrightarrow Cr(s)$ | $-0.74$ |
| $Zn^{2+} + 2\,e^- \longrightarrow Zn(s)$ | $-0.7628$ |
| $2\,H_2O + 2\,e^- \longrightarrow H_2(g) + 2\,OH^-$ | $-0.8277$ |
| $Cr^{2+} + 2\,e^- \longrightarrow Cr(s)$ | $-0.905$ |
| $SO_4^{2-} + H_2O + 2\,e^- \longrightarrow SO_3^{2-} + 2\,OH^-$ | $-0.92$ |
| $Mn^{2+} + 2\,e^- \longrightarrow Mn(s)$ | $-1.029$ |
| $Mn(OH)_2(s) + 2\,e^- \longrightarrow Mn(s) + 2\,OH^-$ | $-1.47$ |
| $Al^{3+} + 3\,e^- \longrightarrow Al(s)$ | $-1.706$ |
| $Sc^{3+} + 3\,e^- \longrightarrow Sc(s)$ | $-2.08$ |
| $Ce^{3+} + 3\,e^- \longrightarrow Ce(s)$ | $-2.335$ |
| $La^{3+} + 3\,e^- \longrightarrow La(s)$ | $-2.37$ |
| $Mg^{2+} + 2\,e^- \longrightarrow Mg(s)$ | $-2.375$ |
| $Mg(OH)_2(s) + 2\,e^- \longrightarrow Mg(s) + 2\,OH^-$ | $-2.69$ |
| $Na^+ + e^- \longrightarrow Na(s)$ | $-2.7109$ |
| $Ca^{2+} + 2\,e^- \longrightarrow Ca(s)$ | $-2.76$ |
| $Ba^{2+} + 2\,e^- \longrightarrow Ba(s)$ | $-2.90$ |
| $K^+ + e^- \longrightarrow K(s)$ | $-2.925$ |
| $Li^+ + e^- \longrightarrow Li(s)$ | $-3.045$ |

All voltages are standard reduction potentials (relative to the standard hydrogen electrode) at 25°C and 1 atm pressure. All species are in aqueous solution unless otherwise indicated.

# Physical Properties of the Elements

## Hydrogen and the Alkali Metals (Group I Elements)

|  | Hydrogen | Lithium | Sodium | Potassium | Rubidium | Cesium | Francium |
|---|---|---|---|---|---|---|---|
| Atomic number | 1 | 3 | 11 | 19 | 37 | 55 | 87 |
| Atomic mass | 1.00794 | 6.941 | 22.98977 | 39.0983 | 85.4678 | 132.9054 | (223) |
| Melting point (°C) | $-259.14$ | 180.54 | 97.81 | 63.65 | 38.89 | 28.40 | 25 |
| Boiling point (°C) | $-252.87$ | 1347 | 903.8 | 774 | 688 | 678.4 | 677 |
| Density at 25°C (g cm$^{-3}$) | 0.070 ($-253$°C) | 0.534 | 0.971 | 0.862 | 1.532 | 1.878 | |
| Color | Colorless | Silver | Silver | Silver | Silver | Silver | |
| Ground-state electron configuration | $1s^1$ | $[\text{He}]2s^1$ | $[\text{Ne}]3s^1$ | $[\text{Ar}]4s^1$ | $[\text{Kr}]5s^1$ | $[\text{Xe}]6s^1$ | $[\text{Rn}]7s^1$ |
| Ionization energy[†] | 1312.0 | 520.2 | 495.8 | 418.8 | 403.0 | 375.7 | $\approx 400$ |
| Electron affinity[†] | 72.770 | 59.63 | 52.867 | 48.384 | 46.884 | 45.505 | est. 44 |
| Electronegativity | 2.20 | 0.98 | 0.93 | 0.82 | 0.82 | 0.79 | 0.70 |
| Ionic radius (Å) | 1.46(H$^-$) | 0.68 | 0.98 | 1.33 | 1.48 | 1.67 | $\approx 1.8$ |
| Atomic radius (Å) | 0.37 | 1.52 | 1.86 | 2.27 | 2.47 | 2.65 | $\approx 2.7$ |
| Enthalpy of fusion[†] | 0.1172 | 3.000 | 2.602 | 2.335 | 2.351 | 2.09 | |
| Enthalpy of vaporization[†] | 0.4522 | 147.1 | 97.42 | 89.6 | 76.9 | 67.8 | |
| Bond enthalpy of $M_2$[†] | 436 | 102.8 | 72.6 | 54.8 | 51.0 | 44.8 | |
| Standard reduction potential (volts) | 0 | $-3.045$ | $-2.7109$ | $-2.924$ | $-2.925$ | $-2.923$ | $\approx 2.9$ |
|  | H$^+$/H$_2$ | Li$^+$/Li | Na$^+$/Na | K$^+$/K | Rb$^+$/Rb | Cs$^+$/Cs | Fr$^+$/Fr |

[†] In kilojoules per mole.

## The Alkaline-Earth Metals (Group II Elements)

|  | Beryllium | Magnesium | Calcium | Strontium | Barium | Radium |
|---|---|---|---|---|---|---|
| Atomic number | 4 | 12 | 20 | 38 | 56 | 88 |
| Atomic mass | 9.01218 | 24.3050 | 40.078 | 87.62 | 137.327 | (226) |
| Melting point (°C) | 1283 | 648.8 | 839 | 769 | 725 | 700 |
| Boiling point (°C) | 2484 | 1105 | 1484 | 1384 | 1640 |  |
| Density at 25°C (g cm$^{-3}$) | 1.848 | 1.738 | 1.55 | 2.54 | 3.51 | 5 |
| Color | Gray | Silver | Silver | Silver | Silver-yellow | Silver |
| Ground-state electron configuration | [He]$2s^2$ | [Ne]$3s^2$ | [Ar]$4s^2$ | [Kr]$5s^2$ | [Xe]$6s^2$ | [Rn]$7s^2$ |
| Ionization energy[†] | 899.4 | 737.7 | 589.8 | 549.5 | 502.9 | 509.3 |
| Electron affinity[†] | < 0 | < 0 | 2.0 | 4.6 | 13.95 | > 0 |
| Electronegativity | 1.57 | 1.31 | 1.00 | 0.95 | 0.89 | 0.90 |
| Ionic radius (Å) | 0.31 | 0.66 | 0.99 | 1.13 | 1.35 | 1.43 |
| Atomic radius (Å) | 1.13 | 1.60 | 1.97 | 2.15 | 2.17 | 2.23 |
| Enthalpy of fusion[†] | 11.6 | 8.95 | 8.95 | 9.62 | 7.66 | 7.15 |
| Enthalpy of vaporization[†] | 297.6 | 127.6 | 154.7 | 154.4 | 150.9 | 136.7 |
| Bond enthalpy of $M_2$[†] | 9.46 |  |  |  |  |  |
| Standard reduction potential (volts) | −1.70 | −2.375 | −2.76 | −2.89 | −2.90 | −2.916 |
|  | $Be^{2+}$/Be | $Mg^{2+}$/Mg | $Ca^{2+}$/Ca | $Sr^{2+}$/Sr | $Ba^{2+}$/Ba | $Ra^{2+}$/Ra |

## Group III Elements

|  | Boron | Aluminum | Gallium | Indium | Thallium |
|---|---|---|---|---|---|
| Atomic number | 5 | 13 | 31 | 49 | 81 |
| Atomic mass | 10.811 | 26.98154 | 69.723 | 114.82 | 204.3833 |
| Melting point (°C) | 2300 | 660.37 | 29.78 | 156.61 | 303.5 |
| Boiling point (°C) | 3658 | 2467 | 2403 | 2080 | 1457 |
| Density at 25°C (g cm$^{-3}$) | 2.34 | 2.702 | 5.904 | 7.30 | 11.85 |
| Color | Yellow | Silver | Silver | Silver | Blue-white |
| Ground-state electron configuration | [He]$2s^2 2p^1$ | [Ne]$3s^2 3p^1$ | [Ar]$3d^{10}4s^2 4p^1$ | [Kr]$4d^{10}5s^2 5p^1$ | [Xe]$4f^{14}5d^{10}6s^2 6p^1$ |
| Ionization energy[†] | 800.6 | 577.6 | 578.8 | 558.3 | 589.3 |
| Electron affinity[†] | 26.7 | 42.6 | 29 | 29 | ≈20 |
| Electronegativity | 2.04 | 1.61 | 1.81 | 1.78 | 1.83 |
| Ionic radius (Å) | 0.23 (+3) | 0.51 (+3) | 0.62 (+3) | 0.81 (+3) | 0.95 (+3) |
| Atomic radius (Å) | 0.88 | 1.43 | 1.22 | 1.63 | 1.70 |
| Enthalpy of fusion[†] | 22.6 | 10.75 | 5.59 | 3.26 | 4.08 |
| Enthalpy of vaporization[†] | 508 | 291 | 272 | 243 | 182 |
| Bond enthalpy of $M_2$[†] | 295 | 167 | 116 | 106 | ≈63 |
| Standard reduction potential (volts) | −0.890 | −1.706 | −0.560 | −0.338 | 0.719 |
|  | B(OH)$_3$/B | $Al^{3+}$/Al | $Ga^{3+}$/Ga | $In^{3+}$/In | $Tl^{3+}$/Tl |

[†] In kilojoules per mole.

## Group IV Elements

| | Carbon | Silicon | Germanium | Tin | Lead |
|---|---|---|---|---|---|
| Atomic number | 6 | 14 | 32 | 50 | 82 |
| Atomic mass | 12.011 | 28.0855 | 72.61 | 118.710 | 207.2 |
| Melting point (°C) | 3550 | 1410 | 937.4 | 231.9681 | 327.502 |
| Boiling point (°C) | 4827 | 2355 | 2830 | 2270 | 1740 |
| Density at 25°C (g cm$^{-3}$) | 2.25 (gr) 3.51 (dia) | 2.33 | 5.323 | 5.75 (gray) 7.31 (white) | 11.35 |
| Color | Black (gr) Colorless (dia) | Gray | Gray-white | Silver | Blue-white |
| Ground-state electron configuration | [He]$2s^22p^2$ | [Ne]$3s^23p^2$ | [Ar]$3d^{10}4s^24p^2$ | [Kr]$4d^{10}5s^25p^2$ | [Xe]$4f^{14}5d^{10}6s^26p^2$ |
| Ionization energy[†] | 1086.4 | 786.4 | 762.2 | 708.6 | 715.5 |
| Electron affinity[†] | 121.85 | 133.6 | ≈120 | ≈120 | 35.1 |
| Electronegativity | 2.55 | 1.90 | 2.01 | 1.88 | 2.10 |
| Ionic radius (Å) | 0.15 (+4) 2.60 (−4) | 0.42 (+4) 2.71 (−4) | 0.53 (+4) 0.73 (+2) 2.72 (−4) | 0.71 (+4) 0.93 (+2) | 0.84 (+4) 1.20 (+2) |
| Atomic radius (Å) | 0.77 | 1.17 | 1.22 | 1.40 | 1.75 |
| Enthalpy of fusion[†] | 105.0 | 50.2 | 34.7 | 6.99 | 4.774 |
| Enthalpy of vaporization[†] | 718.9 | 359 | 328 | 302 | 195.6 |
| Bond enthalpy of M$_2$[†] | 178 | 317 | 280 | 192 | 61 |
| Standard reduction potential (volts) | | | −0.13 $H_2GeO_3,H^+$/Ge | −0.1364 $Sn^{2+}$/Sn | −0.1263 $Pb^{2+}$/Pb |

## Group V Elements

| | Nitrogen | Phosphorus | Arsenic | Antimony | Bismuth |
|---|---|---|---|---|---|
| Atomic number | 7 | 15 | 33 | 51 | 83 |
| Atomic mass | 14.00674 | 30.97376 | 74.92159 | 121.75 | 208.9804 |
| Melting point (°C) | −209.86 | 44.1 | 817 (28 atm.) | 630.74 | 271.3 |
| Boiling point (°C) | −195.8 | 280 | 613 (subl.) | 1750 | 1560 |
| Density at 25°C (g cm$^{-3}$) | 0.808 (−196°C) | 1.82 (white) 2.20 (red) 2.69 (black) | 5.727 | 6.691 | 9.747 |
| Color | Colorless | | Gray | Blue-white | White |
| Ground-state electron configuration | [He]$2s^22p^3$ | [Ne]$3s^23p^3$ | [Ar]$3d^{10}4s^24p^3$ | [Kr]$4d^{10}5s^25p^3$ | [Xe]$4f^{14}5d^{10}6s^26p^3$ |
| Ionization energy[†] | 1402.3 | 1011.7 | 947 | 833.7 | 703.3 |
| Electron affinity[†] | −7 | 72.03 | ≈80 | 103 | 91.3 |
| Electronegativity | 3.04 | 2.19 | 2.18 | 2.05 | 2.02 |
| Ionic radius (Å) | 1.71 (−3) | 0.44 (+3) 2.12 (−3) | 0.46 (+5) 0.58 (+3) 2.22 (−3) | 0.62 (+5) 0.76 (+3) 2.45 (−3 | 0.96 (+3) |
| Atomic radius (Å) | 0.70 | 1.10 | 1.21 | 1.41 | 1.55 |
| Enthalpy of fusion[†] | 0.720 | 6.587 | 27.72 | 20.91 | 10.88 |
| Enthalpy of vaporization[†] | 5.608 | 59.03 | 334 | 262.5 | 184.6 |
| Bond enthalpy of M$_2$[†] | 945 | 485 | 383 | 289 | 194 |
| Standard reduction potential (volts) | 0.96 $NO_3^-,H^+$/NO | −0.276 $H_3PO_4$/$H_3PO_3$ | 0.234 $As_2O_3,H^+$/As | 0.1445 $Sb_2O_3,H^+$/Sb | −0.46 $Bi_2O_3,OH^-$/Bi |

[†] In kilojoules per mole.

## The Chalcogens (Group VI Elements)

|  | Oxygen | Sulfur | Selenium | Tellurium | Polonium |
|---|---|---|---|---|---|
| Atomic number | 8 | 16 | 34 | 52 | 84 |
| Atomic mass | 15.9994 | 32.066 | 78.96 | 127.60 | (209) |
| Melting point (°C) | −218.4 | 119.0 (mon.) | 217 | 449.5 | 254 |
|  |  | 112.8 (rhom.) |  |  |  |
| Boiling point (°C) | −182.962 | 444.674 | 684.9 | 989.8 | 962 |
| Density at 25°C (g cm$^{-3}$) | 1.14 | 1.957 (mon.) | 4.79 | 6.24 | 9.32 |
|  | (−183°C) | 2.07 (rhom.) |  |  |  |
| Color | Pale blue ($\ell$) | Yellow | Gray | Silver | Silver-gray |
| Ground-state electron configuration | [He]$2s^22p^4$ | [Ne]$3s^23p^4$ | [Ar]$3d^{10}4s^24p^4$ | [Kr]$4d^{10}5s^25p^4$ | [Xe]$4f^{14}5d^{10}6s^26p^4$ |
| Ionization energy† | 1313.9 | 999.6 | 940.9 | 869.3 | 812 |
| Electron affinity† | 140.97676 | 200.4116 | 194.967 | 190.15 | ≈180 |
| Electronegativity | 3.44 | 2.58 | 2.55 | 2.10 | 2.00 |
| Ionic radius (Å) | 1.40 (−2) | 0.29 (+6) | 0.42 (+6) | 0.56 (+6) | 0.67 (+6) |
|  |  | 1.84 (−2) | 1.98 (−2) | 2.21 (−2) | 2.30 (−2) |
| Atomic radius (Å) | 0.66 | 1.04 | 1.17 | 1.43 | 1.67 |
| Enthalpy of fusion† | 0.4187 | 1.411 | 5.443 | 17.50 | 10 |
| Enthalpy of vaporization† | 6.819 | 238 | 207 | 195 | 90 |
| Bond enthalpy of M$_2$† | 498 | 429 | 308 | 225 |  |
| Standard reduction potential (volts) | 1.229 | −0.508 | −0.78 | −0.92 | ≈−1.4 |
|  | $O_2,H^+/H_2O$ | $S/S^{2-}$ | $Se/Se^{2-}$ | $Te/Te^{2-}$ | $Po/Po^{2-}$ |

## The Halogens (Group VII Elements)

|  | Fluorine | Chlorine | Bromine | Iodine | Astatine |
|---|---|---|---|---|---|
| Atomic number | 9 | 17 | 35 | 53 | 85 |
| Atomic mass | 18.998403 | 35.4527 | 79.904 | 126.90447 | (210) |
| Melting point (°C) | −219.62 | −100.98 | −7.25 | 113.5 | 302 |
| Boiling point (°C) | −188.14 | −34.6 | 58.78 | 184.35 | 337 |
| Density at 25°C (g cm$^{-3}$) | 1.108 | 1.367 | 3.119 | 4.93 |  |
|  | (−189°C) | (−34.6°C) |  |  |  |
| Color | Yellow | Yellow-green | Deep red | Violet-black |  |
| Ground-state electron configuration | [He]$2s^22p^5$ | [Ne]$3s^23p^5$ | [Ar]$3d^{10}4s^24p^5$ | [Kr]$4d^{10}5s^25p^5$ | [Xe]$4f^{14}5d^{10}6s^26p^5$ |
| Ionization energy† | 1681.0 | 1251.1 | 1139.9 | 1008.4 | ≈930 |
| Electron affinity† | 328.0 | 349.0 | 324.7 | 295.2 | ≈270 |
| Electronegativity | 3.98 | 3.16 | 2.96 | 2.66 | 2.20 |
| Ionic radius (Å) | 1.33 | 1.81 | 1.96 | 2.20 | ≈2.27 |
| Atomic radius (Å) | 0.64 | 0.99 | 1.14 | 1.33 | 1.40 |
| Enthalpy of fusion† | 0.511 | 6.410 | 10.55 | 15.78 | 23.9 |
| Enthalpy of vaporization† | 6.531 | 20.347 | 29.56 | 41.950 |  |
| Bond enthalpy of M$_2$† | 158 | 243 | 193 | 151 | 110 |
| Standard reduction potential (volts) | 2.87 | 1.358 | 1.065 | 0.535 | ≈0.2 |
|  | $F_2/F^-$ | $Cl_2/Cl^-$ | $Br_2/Br^-$ | $I_2/I^-$ | $At_2/At^-$ |

† In kilojoules per mole.

## The Noble Gases (Group VIII Elements)

|  | Helium | Neon | Argon | Krypton | Xenon | Radon |
|---|---|---|---|---|---|---|
| Atomic number | 2 | 10 | 18 | 36 | 54 | 86 |
| Atomic mass | 4.00260 | 20.1797 | 39.948 | 83.80 | 131.29 | (222) |
| Melting point (°C) | −272.2 | −248.67 | −189.2 | −156.6 | −111.9 | −71 |
|  | (26 atm) |  |  |  |  |  |
| Boiling point (°C) | −268.934 | −246.048 | −185.7 | −152.30 | −107.1 | −61.8 |
| Density at 25°C (g cm$^{-3}$) | 0.147 | 1.207 | 1.40 | 2.155 | 3.52 | 4.4 |
|  | (−270.8°C) | (−246.1°C) | (−186°C) | (−152.9°C) | (−109°C) | (−52°C) |
| Color | Colorless | Colorless | Colorless | Colorless | Colorless | Colorless |
| Ground-state electron configuration | $1s^2$ | $[He]2s^22p^6$ | $[Ne]3s^23p^6$ | $[Ar]3d^{10}4s^24p^6$ | $[Kr]4d^{10}5s^25p^6$ | $[Xe]4f^{14}5d^{10}6s^26p^6$ |
| Ionization energy[†] | 2372.3 | 2080.6 | 1520.5 | 1350.7 | 1170.4 | 1037.0 |
| Electron affinity[†] | <0 | <0 | <0 | <0 | <0 | <0 |
| Atomic radius (Å) | 0.32 | 0.69 | 0.97 | 1.10 | 1.30 | 1.45 |
| Enthalpy of fusion[†] | 0.02093 | 0.3345 | 1.176 | 1.637 | 2.299 | 2.9 |
| Enthalpy of vaporization[†] | 0.1005 | 1.741 | 6.288 | 9.187 | 12.643 | 18.4 |

## The Transition Elements

|  | Scandium | Yttrium | Lutetium | Titanium | Zirconium | Hafnium |
|---|---|---|---|---|---|---|
| Atomic number | 21 | 39 | 71 | 22 | 40 | 72 |
| Atomic mass | 44.95591 | 88.90585 | 174.967 | 47.88 | 91.224 | 178.49 |
| Melting point (°C) | 1541 | 1522 | 1656 | 1660 | 1852 | 2227 |
| Boiling point (°C) | 2831 | 3338 | 3315 | 3287 | 4504 | 4602 |
| Density at 25°C (g cm$^{-3}$) | 2.989 | 4.469 | 9.840 | 4.54 | 6.506 | 13.31 |
| Color | Silver | Silver | Silver | Silver | Gray-white | Silver |
| Ground-state electron configuration | $[Ar]3d^14s^2$ | $[Kr]4d^15s^2$ | $[Xe]4f^{14}5d^16s^2$ | $[Ar]3d^24s^2$ | $[Kr]4d^25s^2$ | $[Xe]4f^{14}5d^26s^2$ |
| Ionization energy[†] | 631 | 616 | 523.5 | 658 | 660 | 654 |
| Electron affinity[†] | 18.1 | 29.6 | ≈50 | 7.6 | 41.1 | ≈0 |
| Electronegativity | 1.36 | 1.22 | 1.27 | 1.54 | 1.33 | 1.30 |
| Ionic radius (Å) | 0.81 | 0.93 | 0.848 (+3) | 0.68 | 0.80 | 0.78 |
| Atomic radius (Å) | 1.61 | 1.78 | 1.72 | 1.45 | 1.59 | 1.56 |
| Enthalpy of fusion[†] | 11.4 | 11.4 | 19.2 | 18.62 | 20.9 | 25.5 |
| Enthalpy of vaporization[†] | 328 | 425 | 247 | 426 | 590 | 571 |
| Standard reduction potential (volts) | −2.08 | −2.37 | −2.30 | −0.86 | −1.43 | −1.57 |
|  | $Sc^{3+}/Sc$ | $Y^{3+}/Y$ | $Lu^{3+}/Lu$ | $TiO_2,H^+/Ti$ | $ZrO_2,H^+/Zr$ | $HfO_2,H^+/Hf$ |

[†] In kilojoules per mole.

*continued*

## The Transition Elements (cont.)

|  | Vanadium | Niobium | Tantalum | Chromium | Molybdenum | Tungsten |
|---|---|---|---|---|---|---|
| Atomic number | 23 | 41 | 73 | 24 | 42 | 74 |
| Atomic mass | 50.9415 | 92.90638 | 180.9479 | 51.9961 | 95.94 | 183.85 |
| Melting point (°C) | 1890 | 2468 | 2996 | 1857 | 2617 | 3410 |
| Boiling point (°C) | 3380 | 4742 | 5425 | 2672 | 4612 | 5660 |
| Density at 25°C (g cm$^{-3}$) | 6.11 | 8.57 | 16.654 | 7.18 | 10.22 | 19.3 |
| Color | Silver-white | Gray-white | Steel gray | Silver | Silver | Steel gray |
| Ground-state electron configuration | [Ar]$3d^34s^2$ | [Kr]$4d^45s^1$ | [Xe]$4f^{14}5d^36s^2$ | [Ar]$3d^54s^1$ | [Kr]$4d^55s^1$ | [Xe]$4f^{14}5d^46s^2$ |
| Ionization energy[†] | 650 | 664 | 761 | 652.8 | 684.9 | 770 |
| Electron affinity[†] | 50.7 | 86.2 | 31.1 | 64.3 | 72.0 | 78.6 |
| Electronegativity | 1.63 | 1.60 | 1.50 | 1.66 | 2.16 | 2.36 |
| Ionic radius (Å) | 0.59 (+5) | 0.69 (+5) | 0.68 (+5) |  | 0.62 (+6) | 0.62 (+6) |
|  | 0.63 (+4) | 0.74 (+4) |  | 0.63 (+3) | 0.70 (+4) | 0.70 (+4) |
|  | 0.74 (+3) |  |  | 0.89 (+2) |  |  |
|  | 0.88 (+2) |  |  |  |  |  |
| Atomic radius (Å) | 1.31 | 1.43 | 1.43 | 1.25 | 1.36 | 1.37 |
| Enthalpy of fusion[†] | 21.1 | 26.4 |  | 20.9 | 27.8 | 35.4 |
| Enthalpy of vaporization[†] | 512 | 722 | 781 | 394.7 | 589.2 | 819.3 |
| Standard reduction potential (volts) | −1.2 | −0.62 | −0.71 | −0.74 | 0.0 | −0.09 |
|  | $V^{2+}$/V | $Nb_2O_5,H^+$/Nb | $Ta_2O_5,H^+$/Ta | $Cr^{3+}$/Cr | $H_2MoO_4,H^+$/Mo | $WO_3,H^+$/W |

|  | Manganese | Technetium | Rhenium | Iron | Ruthenium | Osmium |
|---|---|---|---|---|---|---|
| Atomic number | 25 | 43 | 75 | 26 | 44 | 76 |
| Atomic mass | 54.93805 | (98) | 186.207 | 55.847 | 101.07 | 190.2 |
| Melting point (°C) | 1244 | 2172 | 3180 | 1535 | 2310 | 3045 |
| Boiling point (°C) | 1962 | 4877 | 5627 | 2750 | 3900 | 5027 |
| Density at 25°C (g cm$^{-3}$) | 7.21 | 11.50 | 21.02 | 7.874 | 12.41 | 22.57 |
| Color | Gray-white | Silver-gray | Silver | Gray | White | Blue-white |
| Ground-state electron configuration | [Ar]$3d^54s^2$ | [Kr]$4d^55s^2$ | [Xe]$4f^{14}5d^56s^2$ | [Ar]$3d^64s^2$ | [Kr]$4d^75s^1$ | [Xe]$4f^{14}5d^66s^2$ |
| Ionization energy[†] | 717.4 | 702 | 760 | 759.3 | 711 | 840 |
| Electron affinity[†] | < 0 | ≈53 | ≈14 | 15.7 | ≈100 | ≈106 |
| Electronegativity | 1.55 | 1.90 | 1.90 | 1.90 | 2.2 | 2.20 |
| Ionic radius (Å) | 0.80 (+2) |  | 0.56 (+7) | 0.60 (+3) | 0.67 (+4) | 0.69 (+6) |
|  |  |  | 0.27 (+4) | 0.72 (+2) |  | 0.88 (+4) |
| Atomic radius (Å) | 1.37 | 1.35 | 1.34 | 1.24 | 1.32 | 1.34 |
| Enthalpy of fusion[†] | 14.6 | 23.8 | 33.1 | 15.19 | 26.0 | 31.8 |
| Enthalpy of vaporization[†] | 279 | 585 | 778 | 414 | 649 | 678 |
| Standard reduction potential (volts) | −0.183 | 0.738 | 0.3 | −0.036 | 0.49 | 0.85 |
|  | $Mn^{3+}$/Mn | $TcO_4^-,H^+$/$TcO_2$ | $Re^{3+}$/Re | $Fe^{3+}$/Fe | $Ru^{4+}$/$Ru^{3+}$ | $OsO_4,H^+$/Os |

[†] In kilojoules per mole.

*continued*

## The Transition Elements (cont.)

| | Cobalt | Rhodium | Iridium | Nickel | Palladium | Platinum |
|---|---|---|---|---|---|---|
| Atomic number | 27 | 45 | 77 | 28 | 46 | 78 |
| Atomic mass | 58.93320 | 102.90550 | 192.22 | 58.69 | 106.42 | 195.08 |
| Melting point (°C) | 1459 | 1966 | 2410 | 1453 | 1552 | 1772 |
| Boiling point (°C) | 2870 | 3727 | 4130 | 2732 | 3140 | 3827 |
| Density at 25°C (g cm$^{-3}$) | 8.9 | 12.41 | 22.42 | 8.902 | 12.02 | 21.45 |
| Color | Steel gray | Silver | Silver | Silver | Steel white | Silver |
| Ground-state electron configuration | [Ar]3$d^7$4$s^2$ | [Kr]4$d^8$5$s^1$ | [Xe]4$f^{14}$5$d^7$6$s^2$ | [Ar]3$d^8$4$s^2$ | [Kr]4$d^{10}$ | [Xe]4$f^{14}$5$d^9$6$s^1$ |
| Ionization energy[†] | 758 | 720 | 880 | 736.7 | 805 | 868 |
| Electron affinity[†] | 63.8 | 110 | 151 | 111.5 | 51.8 | 205.1 |
| Electronegativity | 1.88 | 2.28 | 2.20 | 1.91 | 2.20 | 2.28 |
| Ionic radius (Å) | 0.63 (+3) | 0.68 (+3) | 0.68 (+4) | 0.69 (+2) | 0.65 (+4) | 0.65 (+4) |
| | 0.72 (+2) | | | | 0.80 (+2) | 0.80 (+2) |
| Atomic radius (Å) | 1.25 | 1.34 | 1.36 | 1.25 | 1.38 | 1.37 |
| Enthalpy of fusion[†] | 16.2 | 21.5 | 26.4 | 17.6 | 17.6 | 19.7 |
| Enthalpy of vaporization[†] | 373 | 557 | 669 | 428 | 353 | 564 |
| Standard reduction potential (volts) | −0.28 | 1.43 | 0.1 | −0.23 | 0.83 | 1.2 |
| | Co$^{2+}$/Co | Rh$^{4+}$/Rh$^{3+}$ | Ir$_2$O$_3$/Ir,OH$^-$ | Ni$^{2+}$/Ni | Pd$^{2+}$/Pd | Pt$^{2+}$/Pt |

| | Copper | Silver | Gold | Zinc | Cadmium | Mercury |
|---|---|---|---|---|---|---|
| Atomic number | 29 | 47 | 79 | 30 | 48 | 80 |
| Atomic mass | 63.546 | 107.8682 | 196.96654 | 65.39 | 112.411 | 200.59 |
| Melting point (°C) | 1083.4 | 961.93 | 1064.43 | 419.58 | 320.9 | −38.87 |
| Boiling point (°C) | 2567 | 2212 | 2807 | 907 | 765 | 356.58 |
| Density at 25°C (g cm$^{-3}$) | 8.96 | 10.50 | 19.32 | 7.133 | 8.65 | 13.546 |
| Color | Red | Silver | Yellow | Blue-white | Blue-white | Silver |
| Ground-state electron configuration | [Ar]3$d^{10}$4$s^1$ | [Kr]4$d^{10}$5$s^1$ | [Xe]4$f^{14}$5$d^{10}$6$s^1$ | [Ar]3$d^{10}$4$s^2$ | [Kr]4$d^{10}$5$s^2$ | [Xe]4$f^{14}$5$d^{10}$6$s^2$ |
| Ionization energy[†] | 745.4 | 731.0 | 890.1 | 906.4 | 867.7 | 1007.0 |
| Electron affinity[†] | 118.5 | 125.6 | 222.749 | < 0 | < 0 | < 0 |
| Electronegativity | 1.90 | 1.93 | 2.54 | 1.65 | 1.69 | 2.00 |
| Ionic radius (Å) | 0.72 (+2) | 0.89 (+2) | 0.85 (+2) | 0.74 (+2) | 0.97 (+2) | 1.10 (+2) |
| | 0.96 (+1) | 1.26 (+1) | 1.37 (+1) | | 1.14 (+1) | 1.27 (+1) |
| Atomic radius (Å) | 1.28 | 1.44 | 1.44 | 1.34 | 1.49 | 1.50 |
| Enthalpy of fusion[†] | 13.3 | 11.95 | 12.36 | 7.39 | 6.11 | 2.300 |
| Enthalpy of vaporization[†] | 304 | 285 | 365 | 131 | 112 | 59.1 |
| Standard reduction potential (volts) | 0.340 | 0.800 | 1.42 | −0.763 | −0.403 | 0.796 |
| | Cu$^{2+}$/Cu | Ag$^+$/Ag | Au$^{3+}$/Au | Zn$^{2+}$/Zn | Cd$^{2+}$/Cd | Hg$_2^{2+}$/Hg |

[†] In kilojoules per mole.

## The Lanthanide Elements

| | Lanthanum | Cerium | Praseodymium | Neodymium | Promethium | Samarium | Europium |
|---|---|---|---|---|---|---|---|
| Atomic number | 57 | 58 | 59 | 60 | 61 | 62 | 63 |
| Atomic mass | 138.9055 | 140.115 | 140.90765 | 144.24 | (145) | 150.36 | 151.965 |
| Melting point (°C) | 921 | 798 | 931 | 1010 | ≈1080 | 1072 | 822 |
| Boiling point (°C) | 3457 | 3257 | 3212 | 3127 | ≈2400 | 1778 | 1597 |
| Density at 25°C (g cm$^{-3}$) | 6.145 | 6.657 | 6.773 | 6.80 | 7.22 | 7.520 | 5.243 |
| Color | Silver | Gray | Silver | Silver | | Silver | Silver |
| Ground-state electron configuration | [Xe]$5d^16s^2$ | [Xe]$4f^15d^16s^2$ | [Xe]$4f^36s^2$ | [Xe]$4f^46s^2$ | [Xe]$4f^56s^2$ | [Xe]$4f^66s^2$ | [Xe]$4f^76s^2$ |
| Ionization energy[†] | 538.1 | 528 | 523 | 530 | 536 | 543 | 547 |
| Electron affinity[†] | 50 | | | est. 50 | | | |
| Electronegativity | 1.10 | 1.12 | 1.13 | 1.14 | | 1.17 | |
| Ionic radius (Å) | 1.15 | 0.92 (+4) 1.034 (+3) | 0.90 (+4) 1.013 (+3) | 0.995 (+3) | 0.979 (+3) | 0.964 (+3) | 0.950 (+3) 1.09 (+2) |
| Atomic radius (Å) | 1.87 | 1.82 | 1.82 | 1.81 | 1.81 | 1.80 | 2.00 |
| Enthalpy of fusion[†] | 5.40 | 5.18 | 6.18 | 7.13 | 12.6 | 8.91 | (10.5) |
| Enthalpy of vaporization[†] | 419 | 389 | 329 | 324 | | 207 | 172 |
| Standard reduction potential (volts) | −2.37 La$^{3+}$/La | −2.335 Ce$^{3+}$/Ce | −2.35 Pr$^{3+}$/Pr | −2.32 Nd$^{3+}$/Nd | −2.29 Pm$^{3+}$/Pm | −2.30 Sm$^{3+}$/Sm | −1.99 Eu$^{3+}$/Eu |

| | Gadolinium | Terbium | Dysprosium | Holmium | Erbium | Thulium | Ytterbium |
|---|---|---|---|---|---|---|---|
| Atomic number | 64 | 65 | 66 | 67 | 68 | 69 | 70 |
| Atomic mass | 157.25 | 158.92534 | 162.50 | 164.93032 | 167.26 | 168.93421 | 173.04 |
| Melting point (°C) | 1311 | 1360 | 1409 | 1470 | 1522 | 1545 | 824 |
| Boiling point (°C) | 3233 | 3041 | 2335 | 2720 | 2510 | 1727 | 1193 |
| Density at 25°C (g cm$^{-3}$) | 7.900 | 8.229 | 8.550 | 8.795 | 9.066 | 9.321 | 6.965 |
| Color | Silver | Silver-gray | Silver | Silver | Silver | Silver | Silver |
| Ground-state electron configuration | [Xe]$4f^75d^16s^2$ | [Xe]$4f^96s^2$ | [Xe]$4f^{10}6s^2$ | [Xe]$4f^{11}6s^2$ | [Xe]$4f^{12}6s^2$ | [Xe]$4f^{13}6s^2$ | [Xe]$4f^{14}6s^2$ |
| Ionization energy[†] | 592 | 564 | 572 | 581 | 589 | 596.7 | 603.4 |
| Electron affinity[†] | | | | est. 50 | | | |
| Electronegativity | 1.20 | | 1.22 | 1.23 | 1.24 | 1.25 | |
| Ionic radius (Å) | 0.938 (+3) | 0.84 (+4) 0.923 (+3) | 0.908 (+3) | 0.894 (+3) | 0.881 (+3) | 0.869 (+3) | 0.858 (+3) 0.93 (+2) |
| Atomic radius (Å) | 1.79 | 1.76 | 1.75 | 1.74 | 1.73 | 1.72 | 1.94 |
| Enthalpy of fusion[†] | 15.5 | 16.3 | 17.2 | 17.2 | 17.2 | 18.2 | 9.2 |
| Enthalpy of vaporization[†] | 301 | 293 | 165 | 285 | 280 | 240 | 165 |
| Standard reduction potential (volts) | −2.28 Gd$^{3+}$/Gd | −2.31 Tb$^{3+}$/Tb | −2.29 Dy$^{3+}$/Dy | −2.33 Ho$^{3+}$/Ho | −2.32 Er$^{3+}$/Er | −2.32 Tm$^{3+}$/Tm | −2.22 Yb$^{3+}$/Yb |

[†] In kilojoules per mole.

## The Actinide Elements

| | Actinium | Thorium | Protactinium | Uranium | Neptunium | Plutonium | Americium |
|---|---|---|---|---|---|---|---|
| Atomic number | 89 | 90 | 91 | 92 | 93 | 94 | 95 |
| Atomic mass | (227) | 232.0381 | 231.03588 | 238.0289 | (237) | (244) | (243) |
| Melting point (°C) | 1050 | 1750 | 1600 | 1132.3 | 640 | 624 | 994 |
| Boiling point (°C) | 3200 | 4790 | | 3818 | 2732 | 3232 | 2607 |
| Density at 25°C (g cm$^{-3}$) | 10.07 | 11.72 | 15.37 | 18.95 | 20.25 | 19.84 | 13.67 |
| Color | Silver | Silver | Silver | Silver | Silver | Silver | Silver |
| Ground-state electron configuration | [Rn]$6d^17s^2$ | [Rn]$6d^27s^2$ | [Rn]$5f^26d^17s^2$ | [Rn]$5f^36d^17s^2$ | [Rn]$5f^46d^17s^2$ | [Rn]$5f^67s^2$ | [Rn]$5f^77s^2$ |
| Ionization energy[†] | 499 | 587 | 568 | 587 | 597 | 585 | 578 |
| Electronegativity | 1.1 | 1.3 | 1.5 | 1.38 | 1.36 | 1.28 | 1.3 |
| Ionic radius (Å) | 1.11 (+3) | 0.99 (+4) | 0.89 (+5) | 0.80 (+6) | 0.71 (+7) | 0.90 (+4) | 0.89 (+4) |
| | | | 0.96 (+4) | 0.93 (+4) | 0.92 (+4) | 1.00 (+3) | 0.99 (+3) |
| | | | 1.05 (+3) | 1.03 (+3) | 1.01 (+3) | | |
| Atomic radius (Å) | 1.88 | 1.80 | 1.61 | 1.38 | 1.30 | 1.51 | 1.84 |
| Enthalpy of fusion[†] | 14.2 | 18.8 | 16.7 | 12.9 | 9.46 | 3.93 | 14.4 |
| Enthalpy of vaporization[†] | 293 | 575 | 481 | 536 | 337 | 348 | 238 |
| Standard reduction potential (volts) | −2.6 Ac$^{3+}$/Ac | −1.90 Th$^{4+}$/Th | −1.0 PaO$_2^+$,H$^+$/Pa | −1.8 U$^{3+}$/U | −1.9 Np$^{3+}$/U | −2.03 Pu$^{3+}$/Pu | −2.32 Am$^{3+}$/Am |

| | Curium | Berkelium | Californium | Einsteinium | Fermium | Mendelevium | Nobelium |
|---|---|---|---|---|---|---|---|
| Atomic number | 96 | 97 | 98 | 99 | 100 | 101 | 102 |
| Atomic mass | (247) | (247) | (251) | (252) | (257) | (258) | (259) |
| Melting point (°C) | 1340 | | | | | | |
| Boiling point (°C) | | | | | | | |
| Density at 25°C (g cm$^{-3}$) | 13.51 | 14 | | | | | |
| Color | Silver | Silver | Silver | Silver | | | |
| Ground-state electron configuration | [Rn]$5f^76d^17s^2$ | [Rn]$5f^97s^2$ | [Rn]$5f^{10}7s^2$ | [Rn]$5f^{11}7s^2$ | [Rn]$5f^{12}7s^2$ | [Rn]$5f^{13}7s^2$ | [Rn]$5f^{14}7s^2$ |
| Ionization energy[†] | 581 | 601 | 608 | 619 | 627 | 635 | 642 |
| Electronegativity | 1.3 | 1.3 | 1.3 | 1.3 | 1.3 | 1.3 | 1.3 |
| Ionic radius (Å) | 0.88 (+4) | 0.87 (+4) | 0.86 (+4) | 0.85 (+4) | 0.84 (+4) | 0.84 (+4) | 0.83 (+4) |
| | 1.01 (+3) | 1.00 (+3) | 0.99 (+3) | 0.98 (+3) | 0.97 (+3) | 0.96 (+3) | 0.95 (+3) |
| | 1.19 (+2) | 1.18 (+2) | 1.17 (+2) | 1.16 (+2) | 1.15 (+2) | 1.14 (+2) | 1.13 (+2) |
| Standard reduction potential (volts) | −2.06 Cm$^{3+}$/Cm | −1.05 Bk$^{3+}$/Bk | −1.93 Cf$^{3+}$/Cf | −2.0 Es$^{3+}$/Es | −1.96 Fm$^{3+}$/Fm | −1.7 Md$^{3+}$/Md | −1.2 No$^{3+}$/No |

[†] In kilojoules per mole.

## The Transactinide Elements[‡]

| | Lawrencium | Rutherfordium | Dubnium | Seaborgium | Bohrium | Hassium | Meitnerium |
|---|---|---|---|---|---|---|---|
| Atomic number | 103 | 104 | 105 | 106 | 107 | 108 | 109 |
| Atomic mass | (262) | (261) | (262) | (263) | (262) | (265) | (266) |
| Melting point (°C) | 1600 | | | | | | |
| Ground-state electron configuration | [Rn]$5f^{14}7s^27p^1$ | [Rn]$5f^{14}6d^27s^2$ | [Rn]$5f^{14}6d^37s^2$ | [Rn]$5f^{14}6d^47s^2$ | [Rn]$5f^{14}6d^57s^2$ | [Rn]$5f^{14}6d^67s^2$ | [Rn]$5f^{14}6d^77s^2$ |
| Ionization energy | | 490 | 640 | 730 | 660 | 750 | 840 |

[‡] All missing data are unknown.

# Answers to Odd-Numbered Problems

## CHAPTER 1

1. Mercury is an element; water and sodium chloride are compounds. The other materials are mixtures: seawater and air are homogeneous; table salt and wood are heterogeneous. Mayonnaise appears homogeneous to the naked eye, but under magnification reveals itself as water droplets suspended in oil.

3. Substances

5. 16.9 g

7. (a) 2.005, and 1.504 g    (b) 2.005/1.504 = 1.333 = 4/3. SiN (or a multiple).

9. 2, 3, 4, 5

11. (a) HO (or any multiple such as $H_2O_2$)    (b) All would give $H_2$ and $O_2$ in 1:1 ratio.

13. 2.0 L $N_2O$, 3.0 L $O_2$

15. (a) 145/94 = 1.54    (b) 94 electrons

17. 95 protons, 146 neutrons, 95 electrons

19. Melting point 1250°C (obs. 1541°C), boiling point 2386°C (obs. 2831°C), density 3.02 g cm$^{-3}$ (obs. 2.99)

21. $SbH_3$, HBr, $SnH_4$, $H_2Se$    23. 28.086    25. 11.01

27. $2.107300 \times 10^{-22}$ g

29. (a) 283.89    (b) 115.36    (c) 164.09    (d) 158.03    (e) 132.13

31. The total count of $2.52 \times 10^9$ atoms of gold has a mass of only $8.3 \times 10^{-13}$ g, far too small to detect with a balance.

33. 1.041 mol    35. 2540 cm$^3$ = 2.54 L    37. $7.03 \times 10^{23}$ atoms

39. $2 \times 10^2$ J

## CHAPTER 2

1. Pt: 47.06%, F: 36.67%, Cl: 8.553%, O: 7.720%    3. $N_4H_6$, $H_2O$, LiH, $C_{12}H_{26}$

5. 0.225%    7. $Zn_3P_2O_8$    9. $Fe_3Si_7$    11. BaN, $Ba_3N_2$

13. (a) 0.923 g C, 0.077 g H    (b) No    (c) 92.3% C, 7.7% H    (d) CH

15. $C_4F_8$

17. (a) 62.1    (b) 6    (c) 56    (d) Si (atomic mass 28.1), N (atomic mass 14.0)
    (e) $Si_2H_6$

19. (a) 3 $H_2$ + $N_2$ → 2 $NH_3$
    (b) 2 K + $O_2$ → $K_2O_2$
    (c) $PbO_2$ + Pb + 2 $H_2SO_4$ → 2 $PbSO_4$ + 2 $H_2O$
    (d) 2 $BF_3$ + 3 $H_2O$ → $B_2O_3$ + 6 HF
    (e) 2 $KClO_3$ → 2 KCl + 3 $O_2$
    (f) $CH_3COOH$ + 2 $O_2$ → 2 $CO_2$ + 2 $H_2O$
    (g) 2 $K_2O_2$ + 2 $H_2O$ → 4 KOH + $O_2$
    (h) 3 $PCl_5$ + 5 $AsF_3$ → 3 $PF_5$ + 5 $AsCl_3$

21. (a) 12.06 g    (b) 1.258 g    (c) 4.692 g

**23.** 7.83 g $K_2Zn_3[Fe(CN)_6]_2$ **25.** 0.134 g $SiO_2$ **27.** $1.18 \times 10^3$ g
**29.** 418 g KCl; 199 g $Cl_2$ **31.** (a) 58.8 (b) Probably Ni
**33.** 42.49% NaCl, 57.51% KCl **35.** 14.7 g $NH_4Cl$, 5.3 g $NH_3$
**37.** 303.0 g Fe; 83.93%

## CHAPTER 3

**1.** (a) $:\ddot{\underset{..}{Rn}}:$ $86e^-$ (8 valence, 78 core) (b) $Sr\cdot^+$ $37e^-$ (1 valence, 36 core)
(c) [illegible] (d) [illegible]

**3.** (a) CsCl, cesium chloride $Cs\cdot + :\ddot{\underset{..}{Cl}}\cdot \rightarrow (Cs^+)(:\ddot{\underset{..}{Cl}}:^-)$
(b) $CaAt_2$, calcium astatide $\dot{Ca}\cdot + 2:\ddot{A}\cdot \rightarrow (Ca^{2+})(:\ddot{At}:^-)_2$
(c) $Al_2S_3$, aluminum sulfide $2\cdot\dot{Al}\cdot + 3\ \cdot\ddot{\underset{..}{S}}\cdot \rightarrow (Al^{3+})_2(:\ddot{\underset{..}{S}}:^{2-})_3$
(d) $K_2Te$, potassium telluride $2\ K\cdot + \cdot\ddot{\underset{..}{Te}}\cdot \rightarrow (K^+)_2(:\ddot{\underset{..}{Te}}:^{2-})$

**5.** (a) Aluminum oxide (b) Rubidium selenide (c) Ammonium sulfide
(d) Calcium nitrate (e) Cesium sulfate (f) Potassium hydrogen carbonate

**7.** (a) AgCN (b) $Ca(OCl)_2$ (c) $K_2CrO_4$ (d) $Ga_2O_3$ (e) $KO_2$
(f) $Ba(HCO_3)_2$

**9.** $Na_3PO_4$, sodium phosphate **11.** 450 kJ $mol^{-1}$

**13.** The As—H bond length will lie between 1.42 and 1.71 Å (observed: 1.52 Å). $SbH_3$ will have the weakest bond.

**15.**

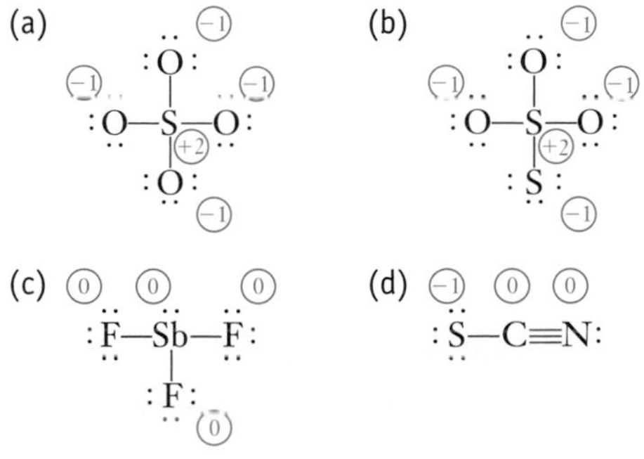

**17.** The first is favored. H—N=O (formal charges 0, 0, 0) H—O=N (formal charges 0, +1, −1)

**19.** (a) Group IV, $CO_2$ (b) Group VII, $Cl_2O_7$ (c) Group V, $NO_2^-$
(d) Group VI, $HSO_4^-$

**21.**

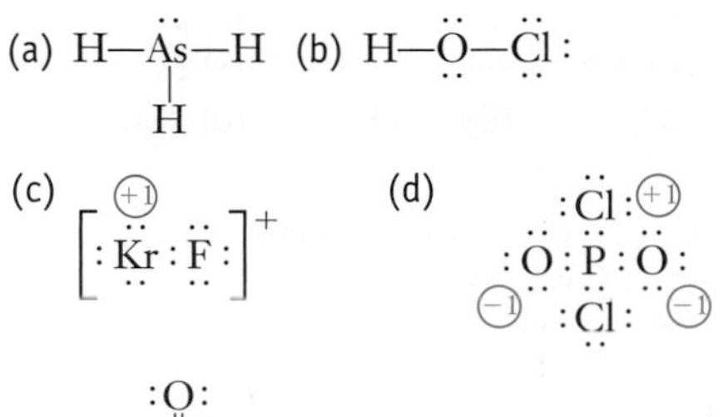

**23.**

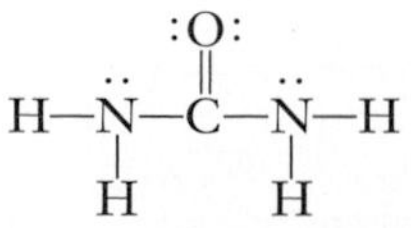

Bond lengths: N—H $1.01 \times 10^{-10}$ m, N—C $1.47 \times 10^{-10}$ m, C=O $1.20 \times 10^{-10}$ m

**25.** $:S-S-S:$ / $:S:$ ... $:S:$ / $:S-S-S:$ (ring of eight S atoms)

**27.** (a) $H-N-B-F$ structure with $H:F:$, $\ominus$ on B, $\oplus$ on N, $H$ and $:F:$ groups

(b) $H-C-C$ with $O$ groups; two resonance structures with $\ominus$ charges

(c) $H-O-C$ with $O$ and $O$ groups; two resonance structures with $\ominus$ charges

**29.** $:O-N=O \leftrightarrow O=N-O:$ (with $\ominus$ charges) Between $1.18 \times 10^{-10}$ m and $1.43 \times 10^{-10}$ m.

**31.** $H-C-N=C=O \leftrightarrow H-C-N\equiv C-O: \leftrightarrow H-C-N-C\equiv O:$ (three resonance structures with formal charges $\oplus$, $\ominus$)

**33.** (a) $PF_5$ structure with F atoms around P

(b) $F-S-F$ structure with F and O

(c) $XeO_2F_2$ structure

**35.** (a) $CI_4$    (b) $OF_2$    (c) $SiH_4$

**37.** ClO, 16%; KI, 74%; TlCl, 38%; InCl, 33%

**39.** HF, 40%; HCl, 19%; HBr, 14%; HI, 8%; CsF, 87%

**41.** (a) SN = 4, tetrahedral    (b) SN = 3, trigonal planar    (c) SN = 6, octahedral
(d) SN = 4, pyramidal    (e) SN = 5, distorted T-shape

**43.** (a) SN = 6, square planar    (b) SN = 4, bent, angle < 109.5°
(c) SN = 4, pyramidal, angle < 109.5°    (d) SN = 2, linear

**45.** (a) $SO_3$    (b) $NF_3$    (c) $NO_2^-$    (d) $CO_3^{2-}$

**47.** Only (d) and (e) are polar.

**49.** No, because VSEPR theory predicts a steric number of 3 and a bent molecule in both cases.

**51.** (a) Linear    (b) The N end

**53.** $SrBr_2$ ($+2-1$) $Zn(OH)_4^{2-}$ ($+2-2+1$) $SiH_4$ ($-4+1$) $CaSiO_3$ ($+2+4-2$) $Cr_2O_7^{2-}$ ($+6-2$) $Ca_5(PO_4)_3F$ ($+2$ $+5-2$ $-1$) $KO_2$ ($+1-\frac{1}{2}$) $CsH$ ($+1-1$)

**55.** (a) $SiO_2$    (b) $(NH_4)_2CO_3$    (c) $PbO_2$    (d) $P_2O_5$    (e) $CaI_2$
(f) $Fe(NO_3)_3$

**57.** (a) Copper(I) sulfide and copper(II) sulfide    (b) Sodium sulfate
(c) Tetraarsenic hexaoxide    (d) Zirconium(IV) chloride
(e) Dichlorine heptaoxide    (f) Gallium(I) oxide

# CHAPTER 4

**1.** $NH_4HS(s) \rightarrow NH_3(g) + H_2S(g)$

**3.** $NH_4Br(s) + NaOH(aq) \rightarrow NH_3(g) + H_2O(\ell) + NaBr(aq)$

**5.** 10.3 m    **7.** $1.40 \times 10^4$ ft    **9.** 1697.5 atm, $1.7200 \times 10^3$ bar

**11.** 0.857 atm       **13.** 8.00 L       **15.** 14.3 gill       **17.** 134 L       **19.** 35.2 psi  → get solution.

**21.** (a) 19.8 atm       (b) 23.0 atm

**23.** The mass of a gas in a given volume changes proportionately to the absolute tempera-ture. This statement will only be true at $-19.8°C$.

**25.** (a) $2 Na(s) + 2 HCl(g) \rightarrow 2 NaCl(s) + H_2(g)$       (b) 4.23 L

**27.** $3.0 \times 10^6$ L HCl       **29.** 24.2 L

**31.** (a) 932 L $H_2S$       (b) 1.33 kg, 466 L $SO_2$

**33.** $X_{SO_3} = 0.135$; $P_{SO_3} = 0.128$ atm       **35.** $X_{N_2} = 0.027$; $P_{N_2} = 1.6 \times 10^{-4}$ atm

**37.** (a) $X_{CO} = 0.444$       (b) $X_{CO} = 0.33$       **39.** (a) $5.8 \times 10^{17}$       (b) 520 L

**41.** (a) $1.93 \times 10^3$ m s$^{-1}$ = 1.93 km s$^{-1}$       (b) 226 m s$^{-1}$

**43.** 6100 m s$^{-1}$ (6000 K), 790 m s$^{-1}$ (100 K)

**45.** Greater. As $T$ increases, the Maxwell–Boltzmann distribution shifts to higher speeds.

**47.** $6.1 \times 10^{-9}$ m       **49.** 92.6 g mol$^{-1}$       **51.** 1830 stages

**53.** $7.4 \times 10^{-7}$ atm; 20 m$^2$ s$^{-1}$       **55.** 162 atm = $2.38 \times 10^3$ psi

**57.** (a) 27.8 atm       (b) 24.6 atm; attractive forces dominate.

## Chapter 5

**1.** Gas       **3.** (a) Condensed       (b) 10.3 cm$^3$ mol$^{-1}$       **5.** Condensed

**7.** The liquid water has vaporized to steam.

**9.** Harder, because forces binding the particles in NaCl are stronger and resist deforma-tion better.

**11.** In all three phases, the diffusion constant should decrease as its density is increased. At higher densities, molecules are closer to each other. In gases, they will collide more often and travel shorter distances between collisions. In liquids and solids, there will be less space for molecules to move around each other.

**13.** Both involve the interaction of an ion with a dipole. In the first case the dipole is per-manent (pre-existing) whereas in the second, it is induced by the approach of the ion. Induced dipole forces are weaker than ion–dipole forces. Examples: $Na^+$ with HCl (ion–dipole); $Na^+$ with $Cl_2$ (induced dipole).

**15.** (a) *Ion-ion*, dispersion       (b) *Dipole-dipole*, dispersion       (c) *Dispersion*
     (d) *Disperson*

**17.** Bromide ion       **19.** (a) $2.0 \times 10^{-10}$ m       (b) $2.5 \times 10^{-10}$ m

**21.** Heavier gases have stronger attractive forces, favoring the liquid and solid states.

**23.** Ne < NO < $NH_3$ < RbCl; nonpolar < polar < hydrogen bonded < ionic

**25.** An eight-membered ring of alternating H's and O's is reasonable. The ring is probably not planar.

**27.** The two have comparable molar masses, but the hydrogen bonding in hydrazine should give it a higher boiling point.

**29.** $6.7 \times 10^{25}$

**31.** 6.16 L; several times smaller than the volume of 1 mole at STP (22.4 L mol$^{-1}$).

**33.** $7.02 \times 10^{13}$ atoms cm$^{-3}$       **35.** 0.9345 g L$^{-1}$       **37.** 2.92 g $CaCO_3$

**39.** 0.69 atm; 31% lies below

**41.** Iridium. Its higher melting and boiling points indicate that the intermolecular forces in iridium are stronger than those in sodium.

**43.** No phase change will occur.

**47.** (a) Liquid       (b) Gas       (c) Solid       (d) Gas

**49.** (a) Above. If gas and solid coexist at $-84.0°C$, their coexistence must extend upward in temperature to the triple point.

(b) The solid will sublime at some temperature below $-84.0°C$.

**51.** The meniscus between gas and liquid phases will disappear at the critical temperature, 126.19 K.

## CHAPTER 6

**1.** (a) $5.53 \times 10^{-3}$ M     (b) $5.5 \times 10^{-3}$ molal     (c) 3.79 L

**3.** Molarity = 12.39 M in HCl; mole fraction of HCl = 0.2324; molality = 16.81 molal in HCl

**5.** 8.9665 molal     **7.** $0.00643$ g $H_2O$     **9.** (a) 1.33 g mL$^{-1}$     (b) 164 mL

**11.** 1.06 M in NaOH

**13.** (a) $Ag^+(aq) + Cl^-(aq) \rightarrow AgCl(s)$

(b) $K_2CO_3(s) + 2 H^+(aq) \rightarrow 2 K^+(aq) + CO_2(g) + H_2O(\ell)$

(c) $2 Cs(s) + 2 H_2O(\ell) \rightarrow 2 Cs^+(aq) + 2 OH^-(aq) + H_2(g)$

(d) $2 MnO_4^-(aq) + 16 H^+(aq) + 10 Cl^-(aq) \rightarrow 5 Cl_2(g) + 2 Mn^{2+}(aq) + 8 H_2O(\ell)$

**15.** 16.8 mL of 7.91 M $HNO_3$     **17.** $6.74 \times 10^3$ L $CO_2(g)$

**19.** (a) $Ca(OH)_2(aq) + 2 HF(aq) \rightarrow CaF_2(aq) + 2 H_2O(\ell)$

Hydrofluoric acid, calcium hydroxide, calcium fluoride

(b) $2 RbOH(aq) + H_2SO_4(aq) \rightarrow Rb_2SO_4(aq) + 2 H_2O(\ell)$

Sulfuric acid, rubidium hydroxide, rubidium sulfate

(c) $Zn(OH)_2(s) + 2 HNO_3(aq) \rightarrow Zn(NO_3)_2(aq) + 2 H_2O(\ell)$

Nitric acid, zinc hydroxide, zinc nitrate

(d) $KOH(aq) + CH_3COOH(aq) \rightarrow KCH_3COO(aq) + H_2O(\ell)$

Acetic acid, potassium hydroxide, potassium acetate

**21.** Sodium sulfide

**23.** (a) $PF_3 + 3 H_2O \rightarrow H_3PO_3 + 3 HF$

(b) $[H_3PO_3] = 0.0882$ M; $[HF] = 0.265$ M

**25.** $0.04841$ M $HNO_3$

**27.** (a) $2 \overset{+3}{P}F_2I(\ell) + 2 \overset{0}{Hg}(\ell) \rightarrow \overset{+2}{P_2}F_4(g) + \overset{+1}{Hg_2}I_2(s)$

(b) $2 K\overset{+5}{Cl}\overset{-2}{O_3}(s) \rightarrow 2 K\overset{-1}{Cl}(s) + 3 \overset{0}{O_2}(g)$

(c) $4 \overset{-3}{N}H_3(g) + 5 \overset{0}{O_2}(g) \rightarrow 4 \overset{+2}{N}\overset{-2}{O}(g) + 6 H_2\overset{-2}{O}(g)$

(d) $2 \overset{0}{As}(s) + 6 Na\overset{+1}{O}H(\ell) \rightarrow 2 Na_3\overset{+3}{As}O_3(s) + 3 \overset{0}{H_2}(g)$

**29.** $2 \overset{0}{Au}(s) + 6 H_2\overset{+6}{Se}O_4(aq) \rightarrow \overset{+3}{Au_2}(\overset{+6}{Se}O_4)_3(aq) + 3 H_2\overset{+4}{Se}O_3(aq) + 3 H_2O(\ell)$

Gold is oxidized and selenium (half of it) is reduced.

**31.** $7.175 \times 10^{-3}$ M $Fe^{2+}(aq)$     **33.** 0.2985 atm     **35.** 3.34 K kg mol$^{-1}$

**37.** 340 g mol$^{-1}$     **39.** $1.7 \times 10^2$ g mol$^{-1}$

**41.** $-2.8°C$. As the solution becomes more concentrated, its freezing point will decrease further.

**43.** 2.70 particles (complete dissociation of $Na_2SO_4$ gives 3 particles)

**45.** $1.88 \times 10^4$ g mol$^{-1}$     **47.** $7.46 \times 10^3$ g mol$^{-1}$

**49.** (a) 0.17 mol $CO_2$     (b) Because the partial pressure of $CO_2$ in the atmosphere is much less than 1 atm, the excess $CO_2$ will bubble out from the solution and escape when the cap is removed.

**51.** $4.13 \times 10^2$ atm     **53.** 0.774     **55.** (a) 0.491     (b) 0.250 atm     (c) 0.575

## CHAPTER 7

**1.** $-2.16 \times 10^4$ L atm $= -2.19 \times 10^6$ J      **3.** 86.6 m

**5.** 24.8, 28.3, 29.6, 31.0, 32.2 J K$^{-1}$ mol$^{-1}$. Extrapolating gives about 33.5 J K$^{-1}$ mol$^{-1}$ for Fr.

**7.** 26.1, 25.4, 25.0, 24.3, 25.4, 27.6 J K$^{-1}$ mol$^{-1}$

**9.** (a) $w$ is zero, $q$ is positive, and $\Delta E$ is positive.
   (b) $w$ is zero, $q$ is negative, and $\Delta E$ is negative.
   (c) $(w_1 + w_2)$ is zero. $(\Delta E_1 + \Delta E_2) = (q_1 + q_2)$. The latter two sums could be any of three possibilities: both positive, both negative, or both zero.

**11.** 0.468 J K$^{-1}$ g$^{-1}$

**13.** $q_1 = Mc_{s_1} \Delta T_1 = -q_2 = -Mc_{s_2} = \Delta T_2$   $\dfrac{C_{s_1}}{C_{s_2}} = \dfrac{\Delta T_2}{\Delta T_1}$

**15.** 314 J g$^{-1}$ (modern value is 333 J g$^{-1}$)

**17.** $w = -323$ J; $\Delta E = -394$ J; $q = -71$ J

**19.** (a) 38.3 L      (b) $w = -1.94 \times 10^3$ J; $q = 0$; $\Delta E = -1.94 \times 10^3$ J      (c) 272 K

**21.** $\Delta E = +11.2$ kJ; $q = 0$; $w = +11.2$ kJ

**23.** (a) $-6.68$ kJ      (b) $+7.49$ kJ      (c) $+0.594$ kJ

**25.** 41.3 kJ      **27.** $+513$ J      **29.** 10.4°C      **31.** $-623.5$ kJ

**33.** Diamond. This is recommended as a source of heat only as a last resort, however!

**35.** $-555.93$ kJ      **37.** (a) $-878.26$ kJ      (b) $-1.35 \times 10^7$ kJ of heat absorbed

**39.** (a) $-81.4$ kJ      (b) 55.1°C      **41.** $\Delta H_f^\circ = -152.3$ kJ mol$^{-1}$

**43.** (a) $C_{10}H_8(s) + 12\,O_2(g) \rightarrow 10\,CO_2(g) + 4\,H_2O(\ell)$      (b) $-5157$ kJ
   (c) $-5162$ kJ      (d) $+84$ kJ mol$^{-1}$

**45.** $-264$ kJ mol$^{-1}$      **47.** $-1.58 \times 10^3$ kJ mol$^{-1}$

**49.** Each side of the equation has three B—Br and three B—Cl bonds.

**51.** $w = -6.87$ kJ; $q = +6.87$ kJ; $\Delta E = \Delta H = 0$

**53.** $T_f = 144$ K; $w = \Delta E = -3.89$ kJ; $\Delta H = -6.48$ kJ

## CHAPTER 8

**1.** (a) The system is all the matter participating in the reaction $NH_4NO_3(s) \rightarrow NH_4^+(aq) +$ $NO_3^-(aq)$. This includes solid ammonium nitrate, the water in which it dissolves, and the aquated ions that are the products of the dissolution process. The surroundings include the flask or beaker in which the system is held, the air above the system, and other neighboring materials. The dissolution of ammonium nitrate is spontaneous after any physical separation (such as a glass wall or a space of air) between the water and the ammonium nitrate has been removed.
   (b) The system is all the matter participating in the reaction $H_2(g) + O_2(g) \rightarrow$ products. The surroundings are the walls of the bomb and other portions of its environment that might deliver heat or work or else absorb heat or work. The reaction of hydrogen with oxygen is spontaneous. Once hydrogen and oxygen are mixed in a closed bomb, no constraint exists to prevent their reaction. It is found experimentally that this system gives products quite slowly at room temperature (no immediate explosion). It explodes instantly at higher temperatures.
   (c) The system is the rubber band. The surroundings consist of the weight (visualized as attached to the lower end of the rubber band), a hanger at the top of the rubber band, and the air in contact with the rubber band. The change is spontaneous once a constraint such as a stand or support underneath the weight is removed.
   (d) The system is the gas contained in the chamber. The surroundings are the walls of the chamber and the moveable piston head. The process is spontaneous if the force

exerted by the weight on the piston exceeds the force exerted by the collisions of the molecules of the gas on the bottom of the piston. (The forces due to the mass of the piston itself and friction between the piston and the walls within which it moves are neglected.)

(e) The system is the drinking glass in the process glass $\rightarrow$ fragments. The surroundings are the floor, the air, and the other materials in the room. The change is spontaneous. It occurs when the constraint, which is whatever portion of the surroundings holds the glass above the floor, is removed.

**3.** (a) $6 \times 6 = 36$     (b) 1 in 36     **5.** The tendency for entropy to increase

**7.** $10^{-3.62 \times 10^{23}}$ (i.e., one part in $10^{3.62 \times 10^{23}}$)

**9.** (a) $\Delta S > 0$     (b) $\Delta S > 0$     (c) $\Delta S < 0$

**11.** (a) 0.333     (b) $q = -1000$ J     (c) $w = -500$ J

**13.** 9.61 J K$^{-1}$ mol$^{-1}$     **15.** 29 kJ mol$^{-1}$

**17.** $\Delta E = \Delta H = 0$; $w = -1.22 \times 10^4$ J; $q = +1.22 \times 10^4$ J; $\Delta S = +30.5$ J K$^{-1}$

**19.** $\Delta S_{sys} = +30.2$ J K$^{-1}$; $\Delta S_{surr} = -30.2$ J K$^{-1}$; $\Delta S_{univ} = 0$

**21.** $\Delta S_{Fe} = -8.24$ J K$^{-1}$; $\Delta S_{H2O} = +9.49$ J K$^{-1}$; $\Delta S_{tot} = +1.25$ J K$^{-1}$

**23.** (a) $-116.58$ J K$^{-1}$     (b) Lower (more negative)

**25.** $\Delta S° = -162.54, -181.12, -186.14, -184.72, -191.08$ J K$^{-1}$. The entropy changes in these reactions become increasingly negative with increasing atomic mass, except that the rubidium reaction is out of line.

**27.** $\Delta S_{surr}$ must be positive, and greater in magnitude than 44.7 J K$^{-1}$.

**29.** $\Delta S_{sys} > 0$ because the gas produced (oxygen) has many possible microstates.

**31.** (a) $+740$ J     (b) 2.65 kJ     (c) No     (d) 196 K

**33.** $q = +38.7$ kJ; $w = -2.92$ kJ; $\Delta E = +35.8$ kJ; $\Delta S_{sys} = +110$ J K$^{-1}$; $\Delta G = 0$

**35.** Overall reaction: 2 Fe$_2$O$_3$($s$) + 3 C($s$) $\rightarrow$ 4 Fe($s$) + 3 CO$_2$($g$)
$$\Delta G = +840 + 3(-400) = -360 < 0$$

**37.** (a) $0 < T < 3000$ K     (b) $0 < K < 1050$ K     (c) Spontaneous at all temperatures

**39.** WO$_3$($s$) + 3 H$_2$($g$) $\rightarrow$ W($s$) + 3 H$_2$O($g$)
$\Delta H° = 117.41$ kJ; $\Delta S° = 131.19$ J K$^{-1}$. $\Delta G < 0$ for $T > \Delta H°/\Delta S° = 895$ K.

# Chapter 9

**1.** (a) $\dfrac{P^2_{H_2O}}{P^2_{H_2}P_{O_2}} = K$     (b) $\dfrac{P_{XeF_6}}{P_{Xe}P^3_{F_2}} = K$     (c) $\dfrac{P^{12}_{CO_2}P^6_{H_2O}}{P^2_{C_6H_6}P^{15}_{O_2}} = K$

**3.** P$_4$($g$) + 6 Cl$_2$($g$) + 2 O$_2$($g$) $\rightleftharpoons$ 4 POCl$_3$($g$)     $\dfrac{P^4_{POCl_3}}{P_{P_4}P^6_{Cl_2}P^2_{O_2}} = K$

**5.** (a) $\dfrac{P_{CO_2}P_{H_2}}{P_{CO}P_{H2O}} = K$     (b) 0.056 atm     **7.** $\Delta G° = -550.23$ kJ; $K = 2.5 \times 10^{96}$

**9.** $2.6 \times 10^{12}$     $\dfrac{P_{SO_3}}{(P_{SO_2})(P_{O_2})^{1/2}} = K$     **11.** $1.04 \times 10^{-4}$     **13.** 14.6

**15.** (a) $P_{SO_2} = P_{Cl_2} = 0.58$ atm; $P_{SO_2Cl_2} = 0.14$ atm     (b) 2.4     **17.** $K_1 = (K_2)^3$

**19.** $K_2/K_1$     **21.** (a) 0.180 atm     (b) 0.756

**23.** $P_{PCl_5} = 0.078$ atm; $P_{PCl_3} = P_{Cl_2} = 0.409$ atm

**25.** $P_{Br_2} = 0.0116$ atm; $P_{I_2} = 0.0016$ atm; $P_{IBr} = 0.0768$ atm

**27.** $P_{N_2} = 0.52$ atm; $P_{O_2} = 0.70$ atm; $P_{NO} = 3.9 \times 10^{-16}$ atm

**29.** $P_{N_2} = 0.0148$ atm; $P_{H_2} = 0.0445$ atm; $P_{NH_3} = 0.941$ atm

**31.** $5.6 \times 10^{-5}$ mol L$^{-1}$     **33.** (a) $9.83 \times 10^{-4}$     (b) Net consumption

**35.** $K > 5.1$

**37.** (a) 0.80, left    (b) $P_{P_4} = 5.12$ atm, $P_{P_2} = 1.77$ atm    (c) Dissociation

**39.** (a) Shifts left    (b) Shifts right    (c) Shifts right    (d) The volume must have been increased to keep the total pressure constant; shifts left    (e) No effect

**41.** (a) Exothermic    (b) Decrease

**43.** Run the first step at low temperature and high pressure, and the second step at high temperature and low pressure.

**45.** Low temperature and high pressure    **47.** $-58$ kJ

**49.** (a) $-56.9$ kJ    (b) $\Delta H° = -55.6$ kJ, $\Delta S° = +4.2$ J K$^{-1}$    **51.** $4.3 \times 10^{-3}$

**53.** (a) 23.8 kJ mol$^{-1}$    (b) $T_b = 240$ K

# Chapter 10

**1.** (a) Cl$^-$ cannot act as a Brønsted–Lowry acid.    (b) SO$_4^{2-}$    (c) NH$_3$
(d) NH$_2^-$    (e) OH$^-$

**3.** HCO$_3^-(aq)$ serves as the base.

**5.** (a) CaO(s) + H$_2$O($\ell$) $\rightarrow$ Ca(OH)$_2$(s)
(b) The CaO acts as a Lewis base, donating a pair of electrons (located on the oxide ion) to one of the hydrogen ions (the Lewis acid) of the water molecule.

**7.** (a) Fluoride acceptor    (b) Acids: BF$_3$, TiF$_4$; bases: ClF$_3$O$_2$, KF

**9.** (a) Base, Mg(OH)$_2$    (b) Acid, HOCl    (c) Acid, H$_2$SO$_4$    (d) Base, CsOH

**11.** SnO(s) + 2 HCl(aq) $\rightarrow$ Sn$^{2+}$(aq) + 2 Cl$^-$(aq) + H$_2$O($\ell$);
SnO(s) + NaOH(aq) + H$_2$O($\ell$) $\rightarrow$ Sn(OH)$_3^-$(aq) + Na$^+$(aq)

**13.** 3.70

**15.** $3 \times 10^{-7}$ M $< [H_3O^+] < 3 \times 10^{-6}$ M; $3 \times 10^{-9}$ M $< [OH^-] < 3 \times 10^{-8}$ M

**17.** $[H_3O^+] = 1.0 \times 10^{-8}$ M; $[OH^-] = 1.7 \times 10^{-6}$ M

**19.** The first reaction is more likely to be correct. In the second case, the reactant H$_3$O$^+$ would be present at very low concentration ($10^{-7}$ M) and would give neither a fast nor a vigorous reaction.

**21.** (a) C$_{10}$H$_{15}$ON(aq) + H$_2$O($\ell$) $\rightleftharpoons$ C$_{10}$H$_{15}$ONH$^+$(aq) + OH$^-$(aq)    (b) $7.1 \times 10^{-11}$
(c) stronger

**23.** $K = 24$; HClO$_2$ is the stronger acid and NO$_2^-$ the stronger base.

**25.** (a) methyl orange    (b) 3.8 to 4.4    **27.** 2.35

**29.** (a) 2.45    (b) 0.72 mol    **31.** 1.16    **33.** $1.2 \times 10^{-6}$

**35.** 10.4    **37.** 0.083 M

**39.** The reaction gives a base of moderate strength, the acetate ion, in solution, so the pH $> 7$.

**41.** HCl $<$ NH$_4$Br $<$ KI $<$ NaCH$_3$COO $<$ NaOH    **43.** 8.08

**45.** (a) 4.36    (b) 4.63    **47.** *m*-chlorobenzoic acid    **49.** 639 mL

**51.** 13.88, 11.24, 7.00, 2.77    **53.** 2.86, 4.72, 8.71, 11.00

**55.** 11.89, 11.52, 10.79, 8.20, 6.05, 3.90, 1.95    **57.** 0.97 g

**59.** 0.0872 g, pH = 6.23 if no approximations are made, bromothymol blue

**61.** $4 \times 10^{-7}$

**63.** $[H_3AsO_4] = 8.0 \times 10^{-2}$ M; $[H_2AsO_4^-] = [H_3O^+] = 2.0 \times 10^{-2}$ M;
$[HAsO_4^{2-}] = 9.3 \times 10^{-8}$ M; $[AsO_4^{3-}] = 1.4 \times 10^{-17}$ M

**65.** $[PO_4^{3-}] = 0.020$ M; $[HPO_4^{2-}] = [OH^-] = 0.030$ M; $[H_2PO_4^-] = 1.61 \times 10^{-7}$ M;
$[H_3PO_4] = 7.1 \times 10^{-18}$ M

**67.** $[H_2CO_3] = 8.5 \times 10^{-6}$ M; $[HCO_3^-] = 1.5 \times 10^{-6}$ M; $[CO_3^{2-}] = 2.8 \times 10^{-11}$ M

**69.** 6.86    **71.** 1.51, 1.61, 2.07, 4.01, 6.07, 8.77, 9.29, 11.51

## CHAPTER 11

**1.** (a) $\dfrac{(P_{H_2S})^8}{(P_{H_2})^8} = K$  (b) $\dfrac{(P_{COCl_2})(P_{H_2})}{(P_{Cl_2})} = K$  (c) $P_{CO_2} = K$  (d) $\dfrac{1}{(P_{C_2H_2})^3} = K$

**3.** (a) $\dfrac{[Zn^{2+}]}{[Ag^+]^2} = K$  (b) $\dfrac{[VO_3(OH)^{2-}][OH^-]}{[VO_4^{3-}]} = K$  (c) $\dfrac{[HCO_3^-]^6}{[As(OH)_6^{3-}]^2 P_{CO_2}^6} = K$

**5.** (a) The graph is a straight line passing through the origin.
   (b) The experimental $K$'s range from $3.42 \times 10^{-2}$ to $4.24 \times 10^{-2}$ with a mean of $3.84 \times 10^{-2}$.

**7.** $2.0 \times 10^{-2}$    **9.** (a) 0.31 atm   (b) 1.65 atm, 0.15 atm

**11.** (a) $Q = 2.05$; reaction shifts to right.   (b) $Q = 3.27$; reaction shifts to left.

**13.** (a) $8.46 \times 10^{-5}$   (b) 0.00336 atm   **15.** 76

**17.** (a) $K_1 = 1.64 \times 10^{-2}$; $K_2 = 5.4$   (b) 330   **19.** 4.65 L

**21.** About 48°C

**23.** $Fe_2(SO_4)_3(s) \rightleftharpoons 2\,Fe^{3+}(aq) + 3\,SO_4^{2-}(aq)$   $[Fe^{3+}]^2[SO_4^{2-}]^3 = K_{sp}$

**25.** 0.0665 g per 100 mL water   **27.** $14.3$ g $L^{-1}$

**29.** $[I^-] = 6.2 \times 10^{-10}$ M, $[Hg_2^{2+}] = 3.1 \times 10^{-10}$ M   **31.** $1.9 \times 10^{-12}$

**33.** $1.6 \times 10^{-8}$   **35.** Yes. The initial reaction quotient is $6.2 \times 10^{-10} > K_{sp}$.

**37.** No   **39.** $[Pb^{2+}] = 2.3 \times 10^{-10}$ M; $[IO_3^-] = 0.033$ M

**41.** $[Ag^+] = 1.8 \times 10^{-2}$ M; $[CrO_4^{2-}] = 6.2 \times 10^{-9}$ M

**43.** $2.4 \times 10^{-8}$ mol $L^{-1}$   **45.** (a) $3.4 \times 10^{-6}$ M   (b) $1.6 \times 10^{-14}$ M

**47.** $9.1 \times 10^{-3}$ M   **49.** In pure water: $1.2 \times 10^{-4}$ M; at pH 7: 0.15 M

**51.** (a) Unchanged   (b) Increase   (c) Increase

**53.** (a) $8.6 \times 10^{-4}$ M, $Pb^{2+}$ in solid   (b) $3.1 \times 10^{-7}$

**55.** $[I^-] = 5.3 \times 10^{-4}$ M   **57.** $2 \times 10^{-13}$ M

**59.** pH = 2.4; $[Pb^{2+}] = 6 \times 10^{-11}$ M

**61.** $[Cu(NH_3)_4^{2+}] = 0.10$ M; $[Cu^{2+}] = 7 \times 10^{-14}$ M

**63.** $[K^+] = 0.0051$ M and $[Na^+] = 0.0076$ M

**65.** More will dissolve in 1 M NaCl ($3 \times 10^{-5}$ versus $1.3 \times 10^{-5}$ mol $L^{-1}$). Less will dissolve in 0.100 M NaCl ($3 \times 10^{-6}$ versus $1.3 \times 10^{-5}$ mol $L^{-1}$).

**67.** $Cu^{2+}(aq) + 2\,H_2O(\ell) \rightleftharpoons CuOH^+(aq) + H_3O^+(aq)$ or
$Cu(H_2O)_4^{2+}(aq) + H_2O(\ell) \rightleftharpoons Cu(H_2O)_3OH^+(aq) + H_3O^+(aq)$

**69.** 5.3   **71.** $K_a = 9.6 \times 10^{-10}$

**73.** In the first case, solid $Pb(OH)_2$ will precipitate; $[Pb^{2+}] = 4.2 \times 10^{-13}$ M, $[Pb(OH)_3^-] = 0.17$ M. In the second case, solid $Pb(OH)_2$ will not precipitate; $[Pb^{2+}] = 1 \times 10^{-13}$ M, $[Pb(OH)_3^-] = 0.050$ M.

## CHAPTER 12

**1.** (a) $2\,VO_2^+(aq) + SO_2(g) \rightarrow 2\,VO^{2+}(aq) + SO_4^{2-}(aq)$
   (b) $Br_2(\ell) + SO_2(g) + 6\,H_2O(\ell) \rightarrow 2\,Br^-(aq) + SO_4^{2-}(aq) + 4\,H_3O^+(aq)$
   (c) $Cr_2O_7^{2-}(aq) + 3\,Np^{4+}(aq) + 2\,H_3O^+(aq) \rightarrow 2\,Cr^{3+}(aq) + 3\,NpO_2^{2+}(aq) + 3\,H_2O(\ell)$
   (d) $5\,HCOOH(aq) + 2\,MnO_4^-(aq) + 6\,H_3O^+(aq) \rightarrow$
$$5\,CO_2(g) + 2\,Mn^{2+}(aq) + 14\,H_2O(\ell)$$
   (e) $3\,Hg_2HPO_4(s) + 2\,Au(s) + 8\,Cl^-(aq) + 3\,H_3O^+(aq) \rightarrow$
$$6\,Hg(\ell) + 3\,H_2PO_4^-(aq) + 2\,AuCl_4^-(aq) + 3\,H_2O(\ell)$$

**3.** (a) $2\,Cr(OH)_3(s) + 3\,Br_2(aq) + 10\,OH^-(aq) \rightarrow$
$$2\,CrO_4^{2-}(aq) + 6\,Br^-(aq) + 8\,H_2O(\ell)$$

(b) $ZrO(OH)_2(s) + 2 SO_3^{2-}(aq) \rightarrow Zr(s) + 2 SO_4^{2-}(aq) + H_2O(\ell)$

(c) $7 HPbO_2^-(aq) + 2 Re(s) \rightarrow 7 Pb(s) + 2 ReO_4^-(aq) + H_2O(\ell) + 5 OH^-(aq)$

(d) $4 HXeO_4^-(aq) + 8 OH^-(aq) \rightarrow 3 XeO_6^{4-}(aq) + Xe(g) + 6 H_2O(\ell)$

(e) $N_2H_4(aq) + 2 CO_3^{2-}(aq) \rightarrow N_2(g) + 2 CO(g) + 4 OH^-(aq)$

**5.** (a) $Fe^{2+}(aq) \rightarrow Fe^{3+}(aq) + e^-$    oxidation
$H_2O_2(aq) + 2 H_3O^+(aq) + 2 e^- \rightarrow 4 H_2O(\ell)$    reduction

(b) $5 H_2O(\ell) + SO_2(aq) \rightarrow HSO_4^-(aq) + 3 H_3O^+(aq) + 2 e^-$    oxidation
$MnO_4^-(aq) + 8 H_3O^+(aq) + 5 e^- \rightarrow Mn^{2+}(aq) + 12 H_2O(\ell)$    reduction

(c) $ClO_2^-(aq) \rightarrow ClO_2(g) + e^-$    oxidation
$ClO_2^-(aq) + 4 H_3O^+(aq) + 4 e^- \rightarrow Cl^-(aq) + 6 H_2O(\ell)$    reduction

**7.** $3 HNO_2(aq) \rightarrow NO_3^-(aq) + 2 NO(g) + H_3O^+(aq)$    **11.** 0.180 mol

**13.** (a) $Zn(s) + Cl_2(g) \rightarrow Zn^{2+}(aq) + 2 Cl^-(aq)$

(b) $1.20 \times 10^3$ C; $1.24 \times 10^{-2}$ mol $e^-$

(c) Decreases by 0.407 g    (d) 0.152 L $Cl_2$ consumed

**15.** +2

**17.** (a) Anode: $Cl^- \rightarrow \frac{1}{2}Cl_2(g) + e^-$; cathode: $K^+ + e^- \rightarrow K(\ell)$;
total: $Cl^- + K^+ \rightarrow K(\ell) + \frac{1}{2}Cl_2(g)$

(b) Mass K = 14.6 g; mass $Cl_2$ = 13.2 g

**19.** $\Delta G° = -921$ J; $w_{max} = +921$ J

**21.** (a) Anode: $Co \rightarrow Co^{2+} + 2 e^-$; cathode: $Br_2 + 2 e^- \rightarrow 2 Br^-$;
total: $Co + Br_2 \rightarrow Co^{2+} + 2 Br^-$

(b) 1.34 V

**23.** (a) Anode: $Zn \rightarrow Zn^{2+} + 2 e^-$; cathode: $In^{3+} + 3 e^- \rightarrow In$    (b) $-0.338$ V

**25.** Reducing agent    **27.** $Br_2$ less effective than $Cl_2$

**29.** (a) $BrO_3^-$    (b) Cr    (c) Co

**31.** (a) $\mathscr{E}° = -0.183$ V    (b) It will not disproportionate.

**33.** (a) No    (b) $Br^-$    **35.** 0.37 V    **37.** $-0.31$ V    **39.** 5.17

**41.** (a) 0.31 V    (b) $1 \times 10^{-8}$ M    **43.** $K = 3 \times 10^{31}$, orange

**45.** $3 \times 10^6$    **47.** pH = 2.53; $K_a = 0.0029$

**49.** (a) 1.065 V    (b) $1.3 \times 10^{-11}$ M    (c) $7.6 \times 10^{-13}$

**51.** 2.041 V; 12.25 V    **53.** (a) $9.3 \times 10^6$ C    (b) $1.1 \times 10^8$ J

**55.** No, because $H_2SO_4(aq)$ is not the only substance whose amount changes during discharge. The accumulated $PbSO_4$ must also be removed and replaced by Pb and $PbO_2$.

**57.** 7900 J $g^{-1}$

**59.** $\Delta\mathscr{E}° = -0.419$ V, not spontaneous under standard conditions (pH = 14). If $[OH^-]$ is small enough, the equilibrium will shift to the right and the reaction will become spontaneous.

**61.** According to its reduction potential, yes. In practice, however, the sodium would react instantly and explosively with the water and would therefore be useless for this purpose.

**63.** (a) Ni    (b) 21.9 g (provided the electrolyte volume is very large)    (c) $H_2$

## CHAPTER 13

**1.** $5.3 \times 10^{-5}$ mol $L^{-1}$ $s^{-1}$

**3.** rate $= -\dfrac{\Delta[N_2]}{\Delta t} = -\dfrac{1}{3}\dfrac{\Delta[H_2]}{\Delta t} = \dfrac{1}{2}\dfrac{\Delta[NH_3]}{\Delta t}$

**5.** (a) Rate $= k[NO]^2[H_2]$; $k$ has the units $L^2$ $mol^{-2}$ $s^{-1}$
(b) An increase by a factor of 18

**7.** Rate $= k[C_5H_5N][CH_3I]$    (b) $k = 75$ L $mol^{-1}$ $s^{-1}$    (c) $7.5 \times 10^{-8}$ mol $L^{-1}$ $s^{-1}$

**9.** $3.2 \times 10^4$ s     **11.** 0.0019 atm     **13.** $5.3 \times 10^{-3}$ s$^{-1}$

**15.** $2.34 \times 10^{-5}$ M     **17.** $2.9 \times 10^{-6}$ s

**19.** (a) Bimolecular, rate = k[HCO][$O_2$]
(b) Termolecular, rate = k[$CH_3$][$O_2$][$N_2$]
(c) Unimolecular, rate = k[$HO_2NO_2$]

**21.** (a) The first step is unimolecular; the others are bimolecular.
(b) $H_2O_2 + O_3 \rightarrow H_2O + 2\ O_2$     (c) O, ClO, $CF_2Cl$, Cl

**23.** 0.26 L mol$^{-1}$ s$^{-1}$

**25.** (a) $A + B + E \rightarrow D + F$     rate $= \dfrac{k_1 k_2}{k_{-1}} \dfrac{[A][B][E]}{[D]}$

(b) $A + D \rightarrow B + F$     rate $= \dfrac{k_1 k_2 k_3}{k_{-1} k_{-2}} \dfrac{[A][D]}{[B]}$

**27.** Only mechanism (b)     **29.** Only mechanism (a)

**31.** rate $= \dfrac{k_2 k_1 [A][B][E]}{k_2[E] + k_{-1}[D]}$     When $k_2[E] \ll k_{-1}[D]$

**33.** $\dfrac{d\,[Cl_2]}{dt} = \dfrac{k_1 k_2 [NO_2Cl]^2}{k_{-1}[NO_2] + k_2[NO_2Cl]}$

**35.** (a) $4.25 \times 10^5$ J     (b) $1.54 \times 10^{11}$ L mol$^{-1}$ s$^{-1}$

**37.** (a) $1.49 \times 10^{-3}$ L mol$^{-1}$ s$^{-1}$     (b) $1.30 \times 10^3$ s

**39.** (a) $4.3 \times 10^{13}$ s$^{-1}$     (b) $1.7 \times 10^5$ s$^{-1}$     **41.** 70.3 kJ mol$^{-1}$

**43.** $6.0 \times 10^9$ L mol$^{-1}$ s$^{-1}$     **45.** (a) $1.2 \times 10^{-3}$ mol L$^{-1}$ s$^{-1}$     (b) $5 \times 10^{-5}$ M

## CHAPTER 14

**1.** (a) $2\ ^{12}_{6}C \rightarrow\ ^{23}_{12}Mg +\ ^{1}_{0}n$     (b) $^{15}_{7}N +\ ^{1}_{1}H \rightarrow\ ^{12}_{6}C +\ ^{4}_{2}He$
(c) $2\ ^{3}_{2}He \rightarrow\ ^{4}_{2}He + 2\ ^{1}_{1}H$

**3.** (a) $3.300 \times 10^{10}$ kJ mol$^{-1}$ = 342.1 MeV total; 8.551 MeV per nucleon
(b) $7.312 \times 10^{10}$ kJ mol$^{-1}$ = 757.9 MeV total; 8.711 MeV per nucleon
(c) $1.738 \times 10^{11}$ kJ mol$^{-1}$ = 1801.7 MeV total; 7.570 MeV per nucleon

**5.** The two $^4$He atoms are more stable, with a mass lower than that of one $^8$Be atom by $9.86 \times 10^{-5}$ u.

**7.** 16.96 MeV

**9.** (a) $^{39}_{17}Cl \rightarrow\ ^{39}_{18}Ar +\ ^{0}_{-1}e^- + \tilde{\nu}$     (b) $^{22}_{11}Na \rightarrow\ ^{22}_{10}Ne +\ ^{0}_{1}e^+ + \nu$
(c) $^{224}_{88}Ra \rightarrow\ ^{220}_{86}Rn +\ ^{4}_{2}He$     (d) $^{82}_{38}Sr +\ ^{0}_{-1}e^- \rightarrow\ ^{82}_{37}Rb + \nu$

**11.** $^{19}$Ne decays to $^{19}$F by positron emission; $^{23}$Ne decays to $^{23}$Na by beta decay.

**13.** An electron, 0.78 MeV     **15.** $^{30}_{14}Si +\ ^{1}_{0}n \rightarrow\ ^{31}_{14}Si \rightarrow\ ^{31}_{15}P +\ ^{0}_{-1}e^- + \tilde{\nu}$

**17.** $^{210}_{84}Po \rightarrow\ ^{206}_{82}Pb +\ ^{4}_{2}He$     $^{4}_{2}He +\ ^{9}_{4}Be \rightarrow\ ^{12}_{6}C +\ ^{1}_{0}n$     (b) **19.** $3.7 \times 10^{10}$ min$^{-1}$

**21.** (a) $1.0 \times 10^6$     (b) $5.9 \times 10^4$     **23.** $1.6 \times 10^{18}$ disintegrations s$^{-1}$

**25.** 4200 years     **27.** $3.0 \times 10^9$ years     **29.** $6.0 \times 10^9$ years

**31.** $^{11}_{6}C \rightarrow\ ^{11}_{5}B +\ ^{0}_{1}e^+ + \nu;\ ^{15}_{8}O \rightarrow\ ^{15}_{7}N +\ ^{0}_{1}e^+ + \nu$

**33.** Exposure from $^{15}$O is greater by a factor of 1.72/0.99 = 1.74.

**35.** (a) $2.3 \times 10^{10}$ Bq     (b) 2.4 mrad     (c) Yes. A dose of 500 rad has a 50% chance of being lethal; this dose is greater than 850 rad in the first 8 days.

**37.** (a) $^{90}_{38}Sr \rightarrow\ ^{90}_{40}Zr + 2\ ^{0}_{-1}e^- + 2\,\tilde{\nu}$     (b) 2.8 MeV
(c) $5.23 \times 10^{12}$ disintegrations s$^{-1}$.     (d) $4.44 \times 10^{11}$ disintegrations s$^{-1}$.

**39.** Increase (assuming no light isotopes of U are products of the decay)

**41.** $7.59 \times 10^7$ kJ g$^{-1}$

## CHAPTER 15

**1.** $0.66$ m s$^{-1}$    **3.** $3.04$ m    **5.** (a) $5.00 \times 10^5$ s$^{-1}$    (b) $4.4$ min

**7.** $\lambda = 1.313$ m, time $= 0.0873$ s    **9.** Blue

**11.** Part of the yellow light, together with green and blue light, will eject electrons from Cs. No visible light will eject electrons from Se; ultraviolet light is required.

**13.** (a) $7.4 \times 10^{-20}$ J    (b) $4.0 \times 10^5$ m s$^{-1}$    **15.** Red

**17.** 550 nm, green

**19.** (a) $3.371 \times 10^{-19}$ J    (b) $203.0$ kJ mol$^{-1}$    (c) $4.926 \times 10^{-3}$ mol s$^{-1}$

**21.** $5.895 \times 10^{-7}$ m

**23.** $9.52$ eV or $1.52 \times 10^{-18}$ J

**25.** $r_3 = 0.0952$ nm, $E_3 = -6.06 \times 10^{-18}$ J, energy per mole $= 3.65 \times 10^3$ kJ mol$^{-1}$, $\nu = 1.14 \times 10^{16}$ s$^{-1}$, $\lambda = 2.63 \times 10^{-8}$ m $= 26.3$ nm

**27.** 72.90 nm, ultraviolet    **29.** (a) 100 cm, 33 cm    (b) 2 nodes

**31.** (a) $7.27 \times 10^{-7}$ m    (b) $3.96 \times 10^{-10}$ m    (c) $2.2 \times 10^{-34}$ m

**33.** (a) $5.8 \times 10^4$ m s$^{-1}$    (b) $7.9$ m s$^{-1}$

**35.** $E_1 = 3.36 \times 10^{-18}$ J; $E_2 = 1.34 \times 10^{-17}$ J; $E_3 = 3.02 \times 10^{-17}$ J; $\lambda = 1.97 \times 10^{-8}$ m

**37.** Only (b) is allowed.

**39.** (a) $4p$    (b) $2s$    (c) $6f$

**41.** (a) 2 radial, 1 angular    (b) 1 radial, 0 angular    (c) 2 radial, 3 angular

**43.** $R^2_{p_z} \propto \cos^2 \theta = 0$ for $\theta = \pi/2$; $d_{xz}$ nodal planes are the $yz$ and $xy$ planes; $d_{x^2-y^2}$ nodal planes are two planes containing the $z$ axis at $45°$ from the $x$ and $y$ axes.

**45.** (a) $1s^2 2s^2 2p^2$    (b) $[\text{Ar}]3d^{10}4s^2 4p^4$    (c) $[\text{Ar}]3d^6 4s^2$

**47.** Be$^+$: $1s^2 2s^1$; C$^-$: $1s^2 2s^2 2p^3$; Ne$^{2+}$: $1s^2 2s^2 2p^4$; Mg$^+$: $[\text{Ne}]3s^1$; P$^{2+}$: $[\text{Ne}]3s^2 3p^1$; Cl$^-$: $[\text{Ne}]3s^2 3p^6$; As$^+$: $[\text{Ar}]3d^{10}4s^2 4p^2$; I$^-$: $[\text{Kr}]4d^{10}5s^2 5p^6$. All except Cl$^-$ and I$^-$ are paramagnetic.

**49.** (a) In    (b) S$^{2-}$    (c) Mn$^{4+}$    **51.** 117

**53.** First "noble gases" at $Z = 1, 5, 9$

**55.** (a) K    (b) Cs    (c) Kr    (d) K    (e) Cl$^-$

**57.** (a) S$^{2-}$    (b) Ti$^{2+}$    (c) Mn$^{2+}$    (d) Sr$^{2+}$

**59.** (a) Sr    (b) Rn    (c) Xe    (d) Sr

**61.** (a) Cs    (b) F    (c) K    (d) At    **63.** 318.4 nm, near ultraviolet

## CHAPTER 16

**1.** (a) F$_2$: $(\sigma_{2s})^2(\sigma^*_{2s})^2(\sigma_{2p})^2(\pi_{2p})^4(\pi^*_{2p})^4$    F$_2^+$: $(\sigma_{2s})^2(\sigma^*_{2s})^2(\sigma_{2p})^2(\pi_{2p})^4(\pi^*_{2p})^3$
   (b) F$_2$: 1, F$_2^+$: $\frac{3}{2}$    (c) F$_2^+$ should be paramagnetic.    (d) F$_2 <$ F$_2^+$

**3.** $(\sigma_{3s})^2(\sigma^*_{3s})^2(\sigma_{3pz})^2(\pi_{3p})^4(\pi^*_{3p})^2$; bond order $= 2$; paramagnetic

**5.** (a) F, 1    (b) N, $2\frac{1}{2}$    (c) O, $1\frac{1}{2}$

**7.** (a) is diamagnetic; (b) and (c) are paramagnetic.    **9.** Bond order $2\frac{1}{2}$; paramagnetic

**11.** The outermost electron in CF is in a $\pi^*_{2p}$ molecular orbital. Removing it gives a stronger bond.

**13.** $(\sigma_{1s})^2(\sigma^*_{1s})^2$, bond order 0, should be unstable.

**15.** $sp^3$ hybridization of central N$^-$, bent molecular ion

**17.** (a) $sp^3$ on C, tetrahedral    (b) $sp$ on C, linear    (c) $sp^3$ on O, bent
   (d) $sp^3$ on C, pyramidal    (e) $sp$ on Be, linear

**19.** $sp^2$ hybrid orbitals. ClO$_3^+$ is trigonal planar, ClO$_2^+$ is bent.

**21.** $sp^3$ hybrid orbitals, tetrahedral

23. Sixteen electrons, so molecule is linear; $sp$ hybridization of central N gives two $\sigma$ bonds with $2p_z$ orbitals on outer N atoms (four electrons).

    Lone pairs on both $2s$ orbitals on outer N atoms (four atoms). $\pi$ system as in Figure 16.16: $(\pi)^4(\pi^{nb})^4$ with eight electrons. Total bond order = 4; bond order 2 per N—N bond. $N_3$ and $N_3^+$ should be bound. $N_3$ and $N_3^+$ are paramagnetic.

25. $\{\ddot{\text{O}}=\text{N}—\ddot{\ddot{\text{O}}}: \leftrightarrow :\ddot{\ddot{\text{O}}}—\text{N}=\ddot{\text{O}}\}$

    $SN = 3$, $sp^2$ hybridization, bent molecule. The $2p_z$ orbitals perpendicular to the plane of the molecule can be combined into a $\pi$ molecular orbital containing one pair of electrons. This orbital adds bond order $\frac{1}{2}$ to each N—O bond, for a total bond order of $\frac{3}{2}$ per bond.

27. $6.736 \times 10^{-23}$ J = 40.56 J mol$^{-1}$

29. (a) $1.45 \times 10^{-46}$ kg m$^2$    (b) $E_1 = 7.65 \times 10^{-23}$ J; $E_2 = 2.30 \times 10^{-22}$ J; $E_3 = 4.59 \times 10^{-22}$ J    (c) 1.13 Å

31. 25.4 kg s$^{-2}$    33. $5.1 \times 10^2$ kg s$^{-2}$    35. $5.3 \times 10^{-4}$

37. Decrease, because the electron will enter a $\pi^*$ antibonding orbital

39. In the blue range, around 450 nm    41. 474 nm

43. 30 double bonds; on the bonds shared by two hexagonal faces    45. $2.72 \times 10^{-7}$ m

47. $\{\ddot{\text{O}}=\ddot{\text{O}}—\ddot{\ddot{\text{O}}}: \leftrightarrow :\ddot{\ddot{\text{O}}}—\ddot{\text{O}}=\ddot{\text{O}}\}$

    $SN = 3$, $sp^2$ hybridization, bent molecule.

    Two electrons in a $\pi$ orbital formed from the three $2p_z$ orbitals perpendicular to the molecular plane, total bond order of $\frac{3}{2}$ for each O—O bond.

## CHAPTER 17

1. —$CH_3$ carbon $sp^3$, other carbon $sp^2$

   A $\pi$ orbital with two electrons bonds the second C atom with the O atom.

   The three groups around the second carbon form an approximately trigonal planar structure, with bond angles near 120°. The geometry about the first carbon atom is approximately tetrahedral, with angles near 109.5°.

3. The Lewis diagrams for HCOOH and HCOO$^-$ are

   One resonance form is given for HCOOH, but two are given for the formate anion HCOO$^-$. In formic acid, one oxygen atom is double bonded to the carbon atom, and the other is singly bonded. In the anion, there is some double-bond character in both C—O bonds. The carbon atom in HCOOH is $sp^2$ hybridized ($SN$ 3) and the OH oxygen atom is $sp^3$ hybridized ($SN$ 4). The immediate surroundings of the carbon atom have trigonal planar geometry, and the C—O—H group is bent. In the HCOO$^-$ ion, the carbon atom and both oxygen atoms are $sp^2$ hybridized ($SN$ 3), possessing a three-center four-electron $\pi$ system. In HCOOH, $\pi$ overlap occurs between orbitals on the carbon atom and only one oxygen atom. Both the C— to —O bond lengths in the formate ion should lie somethwere between the value for the single bond (1.36 Å) and the value for the double bond (1.23 Å).

5. (a) $-64$ kJ mol$^{-1}$    (b) $+53$ kJ mol$^{-1}$    (c) $+117$ kJ mol$^{-1}$

7. (a) $C_{10}H_{22} \rightarrow C_5H_{10} + C_5H_{12}$    (b) Two: 1-pentene and 2-pentene

**9. (a)**

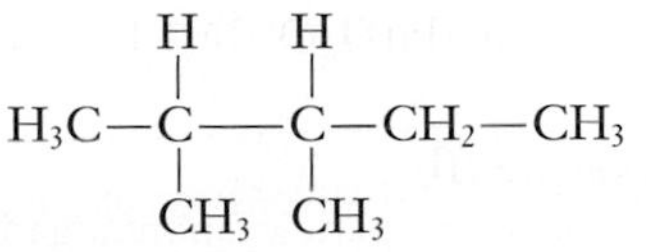

$$H_3C-\underset{\underset{CH_3}{|}}{\overset{\overset{H}{|}}{C}}-\underset{\underset{CH_3}{|}}{\overset{\overset{H}{|}}{C}}-CH_2-CH_3$$

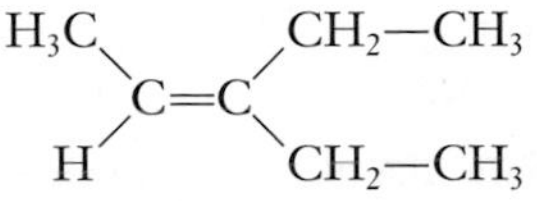

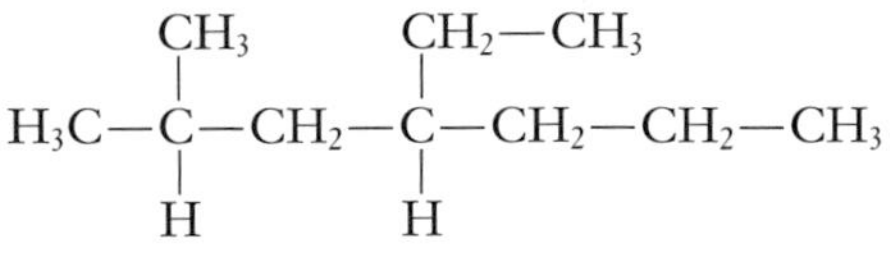

**(b)**

$$\underset{H}{\overset{H_3C}{\diagdown}}C=C\underset{CH_2-CH_3}{\overset{CH_2-CH_3}{\diagup}}$$

**(c)**

(cyclopropane ring with CH_3)

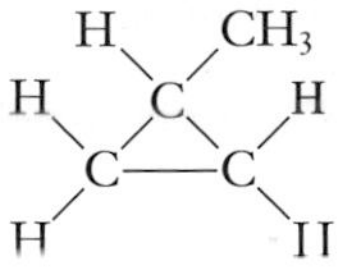

**(d)**

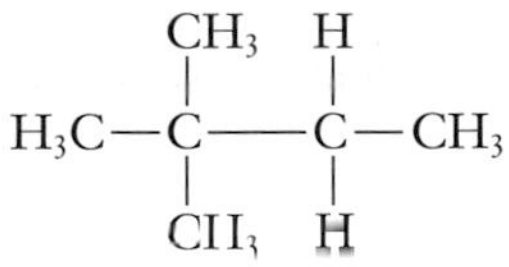

$$H_3C-\underset{\underset{CH_3}{|}}{\overset{\overset{CH_3}{|}}{C}}-\underset{\underset{H}{|}}{\overset{\overset{H}{|}}{C}}-CH_3$$

**(e)**

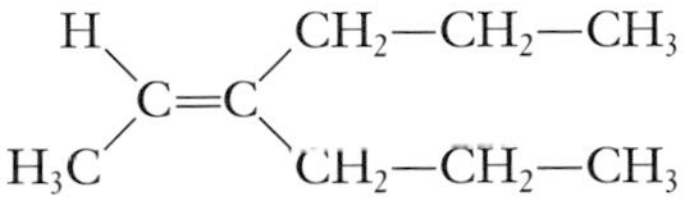

$$\underset{H_3C}{\overset{H}{\diagdown}}C=C\underset{CH_2-CH_2-CH_3}{\overset{CH_2-CH_2-CH_3}{\diagup}}$$

**(f)**

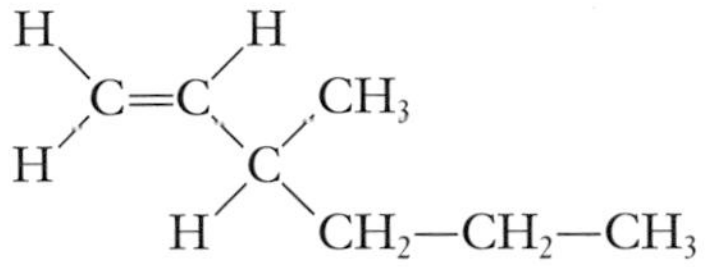

$$\underset{H}{\overset{H}{\diagdown}}C=C\underset{CH(CH_3)-CH_2-CH_2-CH_3}{\overset{H}{\diagup}}$$

**(g)**

$$H_3C-\underset{\underset{H}{|}}{\overset{\overset{CH_3}{|}}{C}}-CH_2-\underset{\underset{H}{|}}{\overset{\overset{CH_2-CH_3}{|}}{C}}-CH_2-CH_2-CH_3$$

**(h)**

$$H_3C-C\equiv C-\underset{\underset{H}{|}}{\overset{\overset{CH_2-CH_3}{|}}{C}}-CH_2-CH_2-CH_3$$

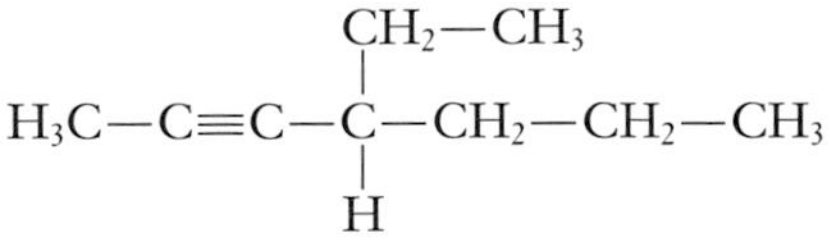

**11.**

$$\underset{H_3C-CH_2}{\overset{H}{\diagdown}}C=C\underset{H}{\overset{CH_2-CH_2-CH_3}{\diagup}}$$

$$\underset{H_3C-CH_2}{\overset{H}{\diagdown}}C=C\underset{H}{\overset{CH_2-CH_2-CH_3}{\diagup}}$$

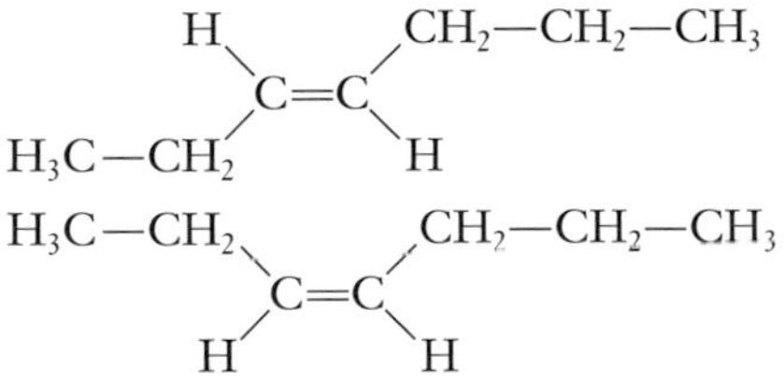

**13. (a)** 1,2-hexadiene    **(b)** 1,3,5-hexatriene    **(c)** 2-methyl-1-hexene
**(d)** 3-hexyne

**15. (a)** $sp^2$, $sp$, $sp^2$, $sp^3$, $sp^3$, $sp^3$    **(b)** All $sp^2$    **(c)** $sp^2$, $sp^2$; all others $sp^3$
**(d)** $sp^3$, $sp^3$, $sp$, $sp$, $sp^3$, $sp^3$

**17. (a)** $CH_3CH_2CH_2CH_2OH + CH_3COOH \rightarrow CH_3COOCH_2CH_2CH_2CH_3 + H_2O$
**(b)** $NH_4CH_3COO \rightarrow CH_3CONH_2 + H_2O$
**(c)** $CH_3CH_2CH_2OH \rightarrow CH_3CH_2CHO$ (propionaldehyde) $+ H_2$
**(d)** $CH_3(CH_2)_5CH_3 + 11\,O_2 \rightarrow 7\,CO_2 + 8\,H_2O$

**19. (a)** $CH_2{=}CH_2 + Br_2 \rightarrow CH_2BrCH_2Br$; $CH_2BrCH_2Br \rightarrow CH_2{=}CHBr + HBr$
**(b)** $CH_3CH_2CH{=}CH_2 + H_2O \rightarrow CH_3CH_2CH(OH)CH_3$ (using $H_2SO_4$)
**(c)** $CH_3CH{=}CH_2 + H_2O \rightarrow CH_3CH(OH)CH_3$ (using $H_2SO_4$);
$CH_3CH(OH)CH_3 \rightarrow CH_3COCH_3 + H_2$ (copper or zinc oxide catalyst)

**21.** $R-\overset{\overset{O}{\|}}{C}-OH + (R')_3C-OH \rightarrow R-\overset{\overset{O}{\|}}{C}-C(R')_3 + H_2O$

**23.** 11.2%; $4.49 \times 10^9$ kg

# CHAPTER 18

**1. (a)** $PtF_4$ **(b)** $PtF_6$    **3.** $V_{10}O_{28}^{6-} + 16\,H_3O^+ \rightarrow 10\,VO_2^+ + 24\,H_2O + 5$ state, $V_2O_5$

**5.** $2\,TiO_2 + H_2 \rightarrow Ti_2O_3 + H_2O$  Titanium(III) oxide

**7.** Monodentate, at the N-atom lone pair

**9.** +2, +2, +2, 0

**11.** (a) $Na_2[Zn(OH)_4]$  (b) $[CoCl_2(en)_2]NO_3$  (c) $[Pt(H_2O)_3Br]Cl$
(d) $[Pt(NH_3)_4(NO_2)_2]Br_2$

**13.** (a) Ammonium diamminetetraisothiocyanatochromate(III)
(b) Pentacarbonyltechnetium(I) iodide  (c) Potassium pentacyanomanganate(IV)
(d) Tetraammineaquachlorocobalt(III) bromide

**15.** Such a complex will persist in solution but will exchange ligands with other complexes, as can be verified by isotopic labeling experiments.

**17.** $\Delta G° = +70.56$ kJ, stable against dissociation. In acid,

$$[Cu(NH_3)_4]^{2+}(aq) + 4\,H_3O^+(aq) \rightarrow Cu^{2+}(aq) + 4\,NH_4^+(aq) + 4\,H_2O(\ell)$$

$\Delta G° = -140.68$ kJ, spontaneous

**19.** $[Cu(NH_3)_2Cl_2] < KNO_3 < Na_2[PtCl_6] < [Co(NH_3)_6]Cl_3$

**25.** (a) Strong: 1, weak: 5  (b) Strong or weak: 0  (c) Strong or weak: 3
(d) Strong: 2, weak: 4  (e) Strong: 0, weak: 4

**27.** $[Fe(CN)_6]^{3-}$: 1 unpaired, CFSE $= -2\,\Delta_o$; $[Fe(H_2O)_6]^{3+}$: 5 unpaired, CFSE $= 0$

**29.** $d^3$ gives half-filled and $d^8$ filled and half-filled shells for metal ions in an octahedral field. $d^5$ will be half filled and stable for high spin (small $\Delta_o$), and $d^6$ will be a filled subshell and stable for low spin (large $\Delta_o$).

**31.** This ion does not absorb any significant amount of light in the visible range.

**33.** $\lambda \approx 480$ nm; $\Delta_o \approx 250$ kJ mol$^{-1}$  **35.** $-500$ kJ mol$^{-1}$

**37.** (a) Orange-yellow  (b) Approximately 600 nm (actual: 575 nm)  (c) Decrease, because $CN^-$ is a strong-field ligand and will increase $\Delta_o$.

**39.** (a) Pale, because $F^-$ is an even weaker field ligand than $H_2O$, so it should be high-spin $d^5$. This is a half-filled shell and the complex will absorb only weakly.
(b) Colorless, because Hg(II) is a $d^{10}$ filled-subshell species

**41.** 14, 18, 18, 18  **43.** 1330 g mol$^{-1}$

# CHAPTER 19

**1.** (b), (c), and (d)  **3.** Two mirror planes, one 2-fold rotation axis

**5.** 3.613 Å  **7.** 102.9°  **9.** 49.87° and 115.0°

**11.** (a) $5.675 \times 10^6$ Å$^3$  (b) $5 \times 10^{14}$  **13.** 4.059 g cm$^{-3}$

**15.** (a) $1.602 \times 10^{-22}$ cm$^3$  (b) $3.729 \times 10^{-22}$ g  (c) $4.662 \times 10^{-23}$ g
(d) $6.025 \times 10^{23}$ (in error by 0.05%)

**17.** 8  **19.** $ReO_3$  **21.** (a) 2.48 Å  (b) 2.87 Å  (c) 1.24 Å

**23.** (a) 2  (b) 0.680  **25.** 0.732

**27.** (a) Ionic  (b) Covalent  (c) Molecular  (d) Metallic

**29.** 8, 6, 12  **31.** 664 kJ mol$^{-1}$  **33.** (a) 1041 kJ mol$^{-1}$  (b) 1205 kJ mol$^{-1}$

**35.** Not by a significant amount, because each vacancy is accompanied by an interstitial. In large numbers, such defects could cause a small bulging of the crystal and a decrease in its density.

**37.** (a) $Fe_{0.9352}O$  (b) 13.86% of the iron

**39.** The isotropic phase has higher entropy and higher enthalpy.

# CHAPTER 20

**1.** A batch process is carried out by mixing reactants, letting them react, and then extracting products in a series of separate steps. In a continuous process, all the steps run simultaneously. A batch process can take place in a single container with one opening, but a continuous process will either need to flow through several containers, or at least require different places into which reactants can be introduced and from which products are removed.

**3.** (a) In salt domes, produced by bacteria from $CaSO_4$, $CH_4$, and $CO_2$
 (b) In coal, produced from plant matter via bacterial reduction, heat, and pressure
 (c) In phosphate rock, $Ca_5(PO_4)_3F$
 (d) In limestone (calcium carbonate, $CaCO_3$)

**5.** $Zn(s) + H_2SO_4(aq) \rightarrow Zn^{2+}(aq) + SO_4^{2-}(aq) + H_2(g)$    **7.** 960 K

**9.** (a) $3\ Fe(s) + 4\ H_2O(g) \rightarrow Fe_3O_4(s) + 4\ H_2(g)$
 (b) $\Delta H° = -151.1$ kJ. An increase in $T$ will *reduce* the thermodynamic yield

**11.** 99.9 L

**13.** $MnO_2(s) + \frac{4}{3}Al(s) \rightarrow Mn(s) + \frac{2}{3}Al_2O_3(s)$; $\Delta H° = -597.1$ kJ, $\Delta G° = -589.7$ kJ

**15.** 839 K = 566°C

**17.** $\Delta S°$ is positive, so $\Delta G$ will become negative at sufficiently high temperature (above 895 K).

**19.** 9.8 kg

**21.** (a) $Fe_2O_3(s) + 3\ CO(g) \rightarrow 2\ Fe(s) + 3\ CO_2(g)$
  $Fe_3O_4(s) + 4\ CO(g) \rightarrow 3\ Fe(s) + 4\ CO_2(g)$
  $FeCO_3(s) + CO(g) \rightarrow Fe(s) + 2\ CO_2(g)$
 (b) From $Fe_2O_3$, $-14.7$ kJ mol$^{-1}$ of Fe; from $Fe_3O_4$, $-4.4$ kJ mol$^{-1}$ of Fe; from $FeCO_3$, $+15.1$ kJ mol$^{-1}$ of Fe. $Fe_2O_3$ is thermodynamically easiest to reduce per mole of Fe produced.

**23.** $2\ Cl^- \rightarrow Cl_2(g) + 2\ e^-$    anode
 $Na^+ + e^- \rightarrow Na(\ell)$    cathode

**25.** $4.4 \times 10^4$ kg    **27.** $2\ Mg + TiCl_4 \rightarrow Ti + 2\ MgCl_2$; 102 kg    **29.** 42.4 min

## CHAPTER 21

**1.** Operate at high total pressures and low temperatures; continuously remove $SO_3$ as it is produced.

**3.** Yes. The reaction that would occur
 $H_2O_2(aq) + SO_2(g) \rightarrow H_2SO_4(aq)$
 has $\Delta G° = -321.7$ kJ < 0, assuming $H_2SO_4(aq)$ is entirely $HSO_4^-(aq)$ and $H^+(aq)$. Thus, the process is thermodynamically feasible.

**5.** $ZnS(s) + 2\ O_2(g) + H_2O(\ell) \rightarrow ZnO(s) + H_2SO_4(\ell)$

**7.** $H_2SO_4(\ell) + 2\ NaOH(s) \rightarrow Na_2SO_4(s) + 2\ H_2O(\ell)$
 $Na_2SO_4(s) + 2\ C(s) \rightarrow Na_2S(s) + 2\ CO_2(g)$

**9.** $5\ H_2SO_4(aq) + 2\ P(s) \rightarrow 5\ SO_2(g) + 2\ H_3PO_4(aq) + 2\ H_2O(\ell)$

**11.** $H_4P_2O_7$, diphosphoric acid. Two $PO_3$ groups are joined by an —O— linkage, with H atoms on four of the outer O atoms.

**13.** $3 \times 10^{26}$ s$^{-1}$

**15.** $CaCO_3(s) + 2\ C(s) + N_2(g) + 4\ H_2O(\ell) \rightarrow 2\ NH_3(g) + 2\ CO_2(g) + CO(g) + Ca(OH)_2(s)$; $\Delta H° = +374.4$ kJ. Dividing by 2 gives $+187.2$ kJ mol$^{-1}$ of $NH_3(g)$ produced.

**17.** Higher pressures mean higher yields, but also more expensive equipment. At fixed yield, a higher temperature can be used at high pressure, giving a faster reaction.

**19.** High yield will be favored by low temperature for all three steps.

**21.** $3\ HNO_3(aq) + Cr(s) + 3\ H_3O^+(aq) \rightarrow 3\ NO_2(g) + Cr^{3+}(aq) + 6\ H_2O(\ell)$
 $3\ HNO_3(aq) + Fe(s) + 3\ H_3O^+(aq) \rightarrow 3\ NO_2(g) + Fe^{3+}(aq) + 6\ H_2O(\ell)$

**25.** Unstable

**27.** $5\ N_2H_4(\ell) + 4\ HNO_3(\ell) \rightarrow 7\ N_2(g) + 12\ H_2O(\ell)$; $\Delta H° = -2986.71$ kJ

**29.** It appears that electronegative atoms reduce the basicity of the N atom to which they are attached: the lone electron pair is drawn closer to the N atom by the inductive effect of electronegative atoms. $NH_2OH$ should be less basic than $NH_3$ ($K_b = 1.07 \times 10^{-8}$ for hydroxylamine versus $1.76 \times 10^{-5}$ for $NH_3$).

**31.** $N_2(g) + 3\,H_2(g) \rightarrow 2\,NH_3(g)$ (Haber-Bosch)
$4\,NH_3(g) + 5\,O_2(g) \rightarrow 4\,NO(g) + 6\,H_2O(g)$ (Pt catalyst, first step of Ostwald process)
$2\,NO(g) + O_2(g) \rightarrow 2\,NO_2(g)$ (second step of Ostwald process)
$2\,NO_2(g) \rightarrow N_2O_4(s)$ (low temperature)

**33.** (a) Yes     (b) $P_4$

## Chapter 22

**1.** $3.2 \times 10^{16}$

**3.** $\Delta H° = +258.5$ kJ for the reaction
$$2\,NaCl(s) + H_2SO_4(\ell) + 2\,C(s) + CaCO_3(s) \rightarrow$$
$$Na_2CO_3(s) + 2\,CO_2(g) + CaS(s) + 2\,HCl(g)$$
Endothermic, so heat must be supplied to make it run.

**5.** Two moles of NaOH should form for each mole of $Cl_2$. The ratio of masses is 1.13, corresponding to 13% more caustic soda than chlorine. Presumably losses or side reactions give the different yield ratio obtained in practice.

**7.** $+3.87$ kJ (tabulated value $+3.93$ kJ)

**9.** $K = 1.0 \times 10^9$. Because $K$ is so large, the process is feasible. For example, suppose $[Br^-] = 10^{-6}$ M, $[Cl^-] = 0.1$ M. Then from $K$ we calculate $P_{Br_2} = 0.1$ atm.

**11.** $H_3PO_4(aq) + Br^-(aq) \rightarrow HBr(g) + H_2PO_4^-(aq)$     (b) 0.11 L

**13.** Unstable

**15.** (a) We find $I_2(s) + 2\,OH^-(aq) \rightarrow I^-(aq) + IO^-(aq) + H_2O(\ell)$
$\Delta\mathscr{E}° = +0.085$ V, so this is a feasible process. However,
$$3\,IO^-(aq) \rightarrow IO_3^-(aq) + 2\,I^-(aq)$$
has $\Delta\mathscr{E}° = +0.35$ V. This process is also feasible, and because it consumes $IO^-(aq)$ it is the ultimate reaction. The stable products are $IO_3^-(aq)$ and $I^-(aq)$.
(b) 0.26 V

**17.** (a) $2\,SrO(s) + 2\,F_2(g) \rightarrow 2\,SrF_2(s) + O_2(g)$
(b) $O_2(g) + 2\,F_2(g) \rightarrow 2\,OF_2(g)$     (c) $UF_4(s) + F_2(g) \rightarrow UF_6(g)$

**19.** $NaF\cdot2HF$     **21.** $ClF_3$

**23.** (a) Bent     (b) Trigonal planar     (c) Distorted T     (d) Square pyramidal
(e) Pentagonal bipyramid     (f) Octahedral

**25.** (a) Pyramidal     (b) Lewis base

**27.** $2\,BrF_3 \rightleftharpoons BrF_4^- + BrF_2^+$

Here, $BrF_2^+$ and $BrF_3$ are both Lewis acids (electron-pair acceptors), and $F^-$ is the Lewis base (electron-pair donor). Here, $BrF_3$ is amphoteric, acting both as a Lewis acid and a Lewis base. One $BrF_3$ is an electron-pair donor, as it donates an $F^-$ ion, which forms a covalent bond to the other $BrF_3$. The other $BrF_3$ is an electron-pair acceptor, as it accepts the $F^-$ ion.

**29.** $PtF_6 < PtF_4 < CaF_2$     **31.** 2740 kJ

**33.** (a) $-345$ kJ mol$^{-1}$     (b) 136 kJ mol$^{-1}$     **35.** $+190$ kJ mol$^{-1}$

**37.** $XeO_6^{4-}(aq) + 12\,H_3O^+(aq) + 8\,e^- \rightarrow Xe(g) + 18\,H_2O(\ell)$
$Mn^{2+}(aq) + 12\,H_2O(\ell) \rightarrow MnO_4^-(aq) + 8\,H_3O^+(aq) + 5\,e^-$
$5\,XeO_6^{4-}(aq) + 8\,Mn^{2+}(aq) + 6\,H_2O(\ell) \rightarrow 5\,Xe(g) + 8\,MnO_4^-(aq) + 4\,H_3O^+(aq)$

## CHAPTER 23

1. The structures of $P_2O_7^{4-}$ and $S_2O_7^{2-}$ have the same number of electrons, and are found by writing "P" or "S" in place of "Si" and adjusting the formal charges. The chlorine compound is $Cl_2O_7$, dichlorine heptaoxide.

3. (a) Tetrahedra; Ca: +2, Fe: +3, Si: +4, O: −2.
   (b) Infinite sheets; Na: +1, Zr: +2, Si: +4, O: −2
   (c) Pairs of tetrahedra; Ca: +2, Zn: +2, Si: +4, O: −2
   (d) Infinite sheets; Mg: +2, Si: +4, O: −2, H: +1

5. (a) Infinite network; Li: +1, Al: +3, Si: +4, O: −2
   (b) Infinite sheets; K: +1, Al: +3, Si: +4, O: −2, H: +1
   (c) Closed rings or infinite single chains; Al: +3, Mg: +2, Si: +4, O: −2

9. (a) $CaCO_3(s) + SiO_2(s) \rightarrow CaSiO_3(s) + CO_2(g)$
   (b) $\Delta H° = +89.41$ kJ; $\Delta S° = +160.81$ J K$^{-1}$    (c) 556 K    (d) 819 K

11. (a) $\Delta H° = -0.21$ kJ; $\Delta S° = -4.2$ J K$^{-1}$; $\Delta G° = +1.04$ kJ    (b) Calcite
    (c) Aragonite    (d) Aragonite    (e) $1.5 \times 10^6$ Pa K$^{-1}$ = 15 atm K$^{-1}$

13. $Mg_3Si_4O_{10}(OH)_2(s) \rightarrow 3\ MgSiO_3(s) + SiO_2(s) + H_2O(g)$    15. 234 L

17. 0.418 mol Si, 0.203 mol Na, 0.050 mol Ca, 0.0068 mol Al, 0.002 mol Mg, 0.0005 mol Ba

19. −113.0 kJ    21. $\frac{7}{3}$

23. (a) $SiO_2(s) + 3\ C(s) \rightarrow SiC(s) + 2\ CO(g)$    (b) +624.6 kJ mol$^{-1}$
    (c) Low conductivity, high melting point, very hard

25. −1188.2 kJ, unstable

## CHAPTER 24

1. $2.16 \times 10^{-19}$ J    3. $6.1 \times 10^{-27}$ electrons; that is, no electrons

5. (a) Electron movement ($n$-type)    (b) Hole movement ($p$-type)

7. 680 nm, red    9. Decrease    11. 8 molecules

## CHAPTER 25

1. $n\ CCl_2{=}CH_2 \rightarrow {+}CCl_2{-}CH_2{+}_n$    3. Formaldehyde, $CH_2{=}O$

5. (a) $H_2O$    (b) ${+}NH{-}CH_2{-}CO{+}$

7. 646 kg adipic acid and 513 kg hexamethylenediamine

9. $3.49 \times 10^{12}$ L = $3.49 \times 10^9$ m$^3$ = 3.49 km$^3$    11. 5 chiral centers

13. 27    (b)    15. In octane, because the side groups are all hydrocarbon

17. $C_9H_9NO$; 119 units

## APPENDIX A

1. (a) $5.82 \times 10^{-5}$    (b) $1.402 \times 10^3$    (c) 7.93
   (d) $-6.59300 \times 10^3$    (e) $2.530 \times 10^{-3}$    (f) 1.47

3. (a) 0.000537    (b) 9,390,000    (c) −0.00247    (d) 0.006020    (e) 20,000

5. 746,000,000 kg

7. (a) The value 135.6    (b) 111.34 g    (c) 0.22 g, 0.23 g

9. That of the mass in problem 7.

11. (a) 5    (b) 3    (c) Either two or three    (d) 3    (e) 4

13. (a) 14 L    (b) −0.0034°C    (c) 340 lb = $3.4 \times 10^2$ lb
    (d) $3.4 \times 10^2$ miles    (e) $6.2 \times 10^{-27}$ J

**15.** 2,997,215.55

**17.** (a) $-167.25$      (b) 76      (c) $3.1693 \times 10^{15}$      (d) $-7.59 \times 10^{-25}$

**19.** (a) $-8.40$      (b) 0.147      (c) $3.24 \times 10^{-12}$      (d) $4.5 \times 10^{13}$

**21.** $A = 337$ cm$^2$

## Appendix B

**1.** (a) $6.52 \times 10^{-11}$ kg      (b) $8.8 \times 10^{-11}$ s      (c) $5.4 \times 10^{12}$ kg m$^2$ s$^{-3}$
(d) $1.7 \times 10^4$ kg m$^2$ s$^{-3}$ A$^{-1}$

**3.** (a) 4983°C      (b) 37.0°C      (c) 111°C      (d) $-40$°C

**5.** (a) 5256 K      (b) 310.2 K      (c) 384 K      (d) 233 K

**7.** (a) 24.6 m s$^{-1}$      (b) $1.51 \times 10^3$ kg m$^{-3}$      (c) $1.6 \times 10^{-29}$ A s m
(d) $1.5 \times 10^2$ mol m$^{-3}$      (e) 6.7 kg m$^2$ s$^{-3}$ = 6.7 W

**9.** 1 kW-hr = $3.6 \times 10^6$ J; 15.3 kW-hr = $5.51 \times 10^7$ J

**11.** 6620 cm$^3$ or 6.62 L

## Appendix C

**1.** 50 miles per hour

**3.** (a) Slope = 4, intercept = $-7$
(b) Slope = $\frac{7}{2}$ = 3.5, intercept = $-\frac{5}{2}$ = $-2.5$
(c) Slope = $-2$, intercept = $\frac{4}{3}$ = 1.3333 . . .

**5.** The graph is not a straight line.

**7.** (a) $-\frac{5}{7}$ = $-0.7142857$ . . .      (b) $\frac{3}{4}$ = 0.75      (c) $\frac{2}{3}$ = 0.6666 . . .

**9.** (a) 0.5447 and $-2.295$      (b) $-0.6340$ and $-2.366$      (c) 2.366 and 0.6340

**11.** (a) $6.5 \times 10^{-7}$ (also two complex roots)      (b) $4.07 \times 10^{-2}$ (also 0.399 and $-1.011$)
(c) $-1.3732$ (also two complex roots)

**13.** (a) 4.551      (b) $1.53 \times 10^{-7}$      (c) $2.6 \times 10^8$      (d) $-48.7264$

**15.** 3.015      **17.** 121.477      **19.** 7.751 and $-2.249$      **21.** $x = 1.086$

**23.** (a) $8x$      (b) $3 \cos 3x - 8 \sin 2x$      (c) 3      (d) $\dfrac{1}{x}$

**25.** (a) 20      (b) $\frac{78125}{7}$      (c) 0.0675

# Index/Glossary

*Key:* c refers to a figure or photo caption; n refers to a footnote; p refers to a problem at the end of the chapter; t refers to a table.

nitrogen-fixing, 766
and photosynthesis, 612c, 619, 722c, 845–848
as sources of sulfur, 734
Bakelite, 856, 859
baking powder, 765–766
baking soda (sodium hydrogen carbonate), 354p, 359p, 393p, 786
Balmer, Johann Jakob (German chemist), 534
**Balmer series** A series of lines in the spectrum of atomic hydrogen that arises from the transitions of electrons between higher energy levels and the level $n = 2$, 534, 542
**band** A very closely spaced manifold of energy levels in a solid; also, a sequence of vibrational transitions in a molecule, 839
**band gap ($E_g$)** The energy difference between the top of one band and the bottom of the next higher band, 839
barium (Ba), 20, 816, A.42t
barium sulfate ("barium cocktail"), 398p
**barometer** A device to measure the pressure of the atmosphere, 101–102
Bartlett, Neil (American chemist), 94p, 800–801, 805p
**base [1]** A substance that, when dissolved in water increases the amount of hydroxide ion relative to that present in pure water. *See also* **Arrhenius base; Brønsted–Lowry base; Lewis base**
classification of, 315–318
indicator for, 172, 326–328
strength of, 322–334
strong, 320
titration of, 170–174
weak, 329–334
**base [2]** A number that is raised to a power, A.23
**base anhydride** A compound that reacts with water to form a base, 317
base ionization constant ($K_b$), 324
**base units** Physical units defined as standards of reference in the SI system, A.9, A.9t
baseball, 132p
**basic oxygen process** A process for the conversion of molten iron into steel that uses oxygen introduced at high pressure into the converter, 743
**batch process** A chemical process in which the reactants are mixed in a container and allowed to react, and the products are removed in sequence, 729
**battery** A galvanic cell or group of cells used to generate electrical work, 406–411, 424–428
bauxite, 745–746
**Bayer process** An industrial process for the purification of aluminum(III) oxide through its selective dissolution in a strongly basic solution, 745–746
Becquerel, Antoine Henri (French physicist), 499
**becquerel (Bq)** The SI unit of activity; equal to 1 $s^{-1}$, 502

Bednorz, J. Georg (German physicist), 831
Beer–Lambert law, 601
benzaldehyde, 310p
benzene, 637, 644–645
benzene-toluene-xylene, 645
benzyl acetate, 653
benzyl alcohol, 310p
berkelium (Bk), 8, A.49t
Berthollet, Claude (French chemist), 10, 783
**berthollides** Solids that exist over a small range of compositions, without the definite stoichiometry implied by a single molecular formula, 10
beryllia, 829
beryllium (Be), 20, A.42t
beta-alumina, 427
beta-carotene, 616–617
**beta decay ($\beta$-decay)** A mode of radioactive decay in which a nuclide emits a beta particle and an antineutrino as one of its constituent neutrons transforms into a proton, 496–497
**beta particle ($\beta$-particle) ($_{-1}^{0}\beta^-$ or $e^-$)** A high-energy electron when emitted in radioactive decay, 15, 496
Bethe, Hans (German-born American physicist), 512
Bhopal incident, 91p
bicarbonate of soda, 354p, 359p, 393p, 786
**bimolecular reaction** An elementary reaction that occurs through collision of two molecules, 458–459
**binding energy ($E_B$)** The energy required to separate a nucleus into its component protons and neutrons; also, the energy required to remove an electron from a specific orbital in an atom, 493–495, 536–538, 570–571
Binnig, Gerd (German physicist), 30
biosphere, 733, 848
bis(hydroxymethyl)aminomethane ("bis"), 356p
bismuth (Bi), 21–22, 156–157, A.43t
Black, Joseph (Scottish chemist), 100
**blackbody radiation** Electromagnetic radiation emitted from an idealized dark-colored solid with which it is in thermal equilibrium, 527–529
**blast furnace** A refractory-lined cylindrical reactor used for the smelting of ores, 741–742
bleach (sodium hypochlorite), 45, 784, 789
bleaching agents, 45, 159p, 416, 437p, 781
production of, 783–784, 789
bleaching powder, 784
blood
alcohol level in, 194p
and the bends, 197p
oxygen transport in, 485p
pH in, 334, 354p
red blood cells in, 198p
substitute for, 799
**body-centered unit cell** A nonprimitive unit cell having lattice points at the cell center as well as at each of the eight corners, 697, 701
Bohr, Niels (Danish physicist), 534, 538, 542, 543

**Bohr atom** A model for one-electron atoms in which the electron is pictured as moving in a circular orbit about the nucleus, 538–542
defects in, 545
**Bohr radius ($a_0$)** The radius of the first electron orbit in the Bohr model for the hydrogen atom; equal to $5.29 \times 10^{-11}$ m, 539
bohrium (Ns), A.49t
**boiling point ($T_b$)** The temperature at which the vapor pressure of a liquid equals the external pressure, 152–153
elevation of, 178–183, 179t
Boltzmann, Ludwig (Austrian physicist), 113, 116–117, 249
**Boltzmann distribution** An exponential distribution of population over the energy levels of a molecule, 602
**Boltzmann's constant ($k_B$)** The universal gas constant divided by Avogadro's number, $R/N_0 = 1.38066 \times 10^{-23}$ J $K^{-1}$, 117, 249–250
**bond** An attraction between atoms strong enough to maintain the atoms in a detectable unit, 55–95, 567
coordinate covalent, 316
covalent, 56, 65–76
double, 70
hydrogen, 147–148
ionic, 56, 59–65
$\pi$-, 587
polar covalent, 56, 77–80
as shared electron pair, 69
$\sigma$-, 583
triple, 70
**bond angle** The angle defined by two atoms bonded to a central atom in a molecule, 80–84
**bond energy** The amount of energy that must be absorbed to break a bond, 66–67, 67t, 68t
**bond enthalpy ($\Delta H_d$)** The enthalpy change $\Delta H$ in a reaction in which a chemical bond is broken in the gas phase, 227–228, 228t, 832t
**bond length** The distance between the nuclei of two atoms bonded in a molecule, 65–66, 67t, 68t
**bond order** A measure of the strength of a chemical bond; in the Lewis model defined as the number of shared electron pairs; in the molecular orbital description defined as one-half the number of electrons in bonding molecular orbitals minus one-half the number of electrons in antibonding ones, 67–68, 68t, 585, 586t, 589t
bonding
Lewis model of, 56, 59–60
in organic molecules, 633–638
quantum mechanical model of, 582–598
in transition metals and coordination complexes, 659–692
valence theory of, 600
**bonding orbital ($\sigma$)** A molecular orbital, the occupation of which by one or two electrons results in an increase in bond strength, 583

ceramic phase, 822

cerium (Ce), 8, A.48t

cesium (Cs), 8, 20, 576p, A.41t

**cesium-chloride structure** An ionic crystal structure that consists of two interpenetrating simple cubic lattices, one of cations, the other of anions, 707

CFC. *See* **chlorofluorocarbon**

Chadwick, James (British physicist), 507

**chain reaction** A reaction that proceeds through a series of linked elementary reactions; the general steps are initiation, propagation, and termination, 467

**chalcogen element** An element of Group VI of the periodic table, 20, A.44t

Charles, Jacques (French physicist), 106, 130p, 735

**Charles's law** At constant pressure, the volume of a sample of gas is a linear function of its temperature; $V = V_0 + aV_0 t$, 105–108

**chelate** A coordination complex in which a ligand coordinates through two or more donors to the same central atom, 666

**chemical amount (*n*)** The amount of a substance expressed in a unit that defines the number of chemically reactive entities present (the SI base unit is the mole), A.9t
    and concentration, 164

chemical bond. *See* **bond**

chemical energy, 32

**chemical equation** A representation of a chemical reaction that uses chemical formulas and other symbols and conforms to the laws of mass and charge conservation, 38–54
    balancing, 42–44

chemical equilibrium. *See* **equilibrium**

**chemical formula** A representation of the chemical composition of a substance; refers to either empirical formula or molecular formula, 11, 40–42

chemical industry, 729–735, 731t, 754–837

chemical properties, A.31t–A.49t

**chemical shift** In nuclear magnetic resonance (nmr) spectrometry, the position of a peak in the compound under examination relative to the position of a standard reference peak, 603

**chemical substance.** *See* **substance**

**chemical vapor deposition** A method used in the fabrication of ceramics, 834

chemistry, development of, 2–37

Chernobyl, 512, 519p

**chiral center** The central atom in a chiral structure; for tetrahedrally bonded carbon, the four attached groups must be different for the carbon atom to be a chiral center, 673

**chirality** The structural property in which a species is not superimposable on its mirror-image structure, 635, 673–675

chlor–alkali industry, 786–788

chlorine (Cl), 21, A.44t
    chemistry of, 788–791
    production of, 130p, 783–784

**chlorofluorocarbon** A compound in which chlorine and fluorine appear as side-groups on a chain of carbon atoms; sometimes used to refer to compounds that contain hydrogen as well, 228, 239p, 474, 622–623, 799–800

chloroform, 648

chlorophyll, 612c, 617, 845–846

**chromatography** A method of separating substances that exploits the difference in the partition coefficients of solute species between two phases, such as a gas and a liquid absorbed on a porous support, 366–368, 366t

chrome plating, 749

chromium (Cr), 20, A.46t

chromium(IV) oxide, 580p

cinnabar, 844–845

circular motion, A.17

***cis-trans* isomer** One of two or more forms of a species that has the same pattern of connections of chemical bonds but differs in geometrical structure, 634

cisplatin, 673, 688

citric acid, 53p, 354p

**Clapeyron equation** An equation that relates the pressure and temperature changes at the coexistence of two phases to the enthalpy and volume change; $(dp/dT)_{coex} = \Delta S/\Delta V = \Delta H/T\Delta V$, 819

classical mechanics, 523, 547. *See also* **quantum mechanics**
    paradoxes in, 528

**Claus process** The reaction of $SO_2$ with $H_2S$ to give sulfur and water, 760

Clausius, Rudolf (German mathematical physicist), 113, 252, 255–256

**clay** A hydrated aluminosilicate produced by the weathering of primary minerals, 812–813, 824

cleaning agents, 90p, 764–765

close packing, 701–704

**closed-shell species** An atom or ion the valence shell of which contains the maximum possible number of electrons, 567

**closed system** A system whose boundaries are impermeable to the flow of matter, 204

cobalt (Co), 816, A.47t
    radioactive uses of, 576p

Cockcroft, John (British physicist), 518p

codeine, 400p

**coefficient** A number in a chemical equation giving the relative quantity of a species involved in the reaction, 42

**coefficient of thermal expansion (α)** A value characterizing the rate at which a substance changes its volume with temperature at constant pressure, 139

coinage, 664

coke, 738

**colligative property** A physical property of a solution that depends only on the concentration of a solute species and not on its nature, 176–186

collisions
    frequency of molecular, 121–122

**colloid** A mixture of two or more substances in which one is suspended in the second as tiny particles that nonetheless exceed molecular size, 191–193

color
    in acid-base indicators, 327–328
    and band gaps, 844–845
    of blood, 686–687
    and conjugated π-electron systems, 615
    and coordination complexes, 680–681
    in minerals, 659c, 665, 824–825

**color center.** *See* **F-center**

column chromatography, 367

combination, laws of chemical, 8–15

**combining volumes, law of** The volumes of reacting gases at constant temperature and pressure stand in the ratio of simple integers, 13

**common-ion effect** The observation that if a solution and a solid salt to be dissolved in it have an ion in common, the solubility of the salt is depressed, 378–380

**common logarithm** A logarithm for which the base is 10, A.23–A.24

**complementary color** The color opposite a given color on the color wheel, 615

**completed octet.** *See* **octet, completed**

**complex ion** An ion in which a central metal ion is bound to one or more ligand molecules or ions that are capable of independent existence in solution, 384–391. *See also* **coordination complex**

**composite ceramic** A ceramic in which one ceramic is reinforced by the admixture before firing of another ceramic, 834

**compound** A substance containing two or more elements, 7. *See also* **covalent compound; ionic compound**

**compressibility** The measure of the rate of change of the volume of a substance with pressure, 138

**compressibility factor (*z*)** The ratio $PV/nRT$, which differs from 1 for a real gas, 123

**concentration (*c*)** The amount of a solute present in a given amount of solvent or solution, 164
    and Le Châtelier's Principle, 300–302
    and Nernst equation, 418–424
    time dependence of, 300–302

**concrete** A mixture of portland cement, sand, and aggregate in the proportions $1:3.75:5$ by volume, 828

**condensation** The formation of a liquid or solid from a gas, 150

**condensation polymerization** Polymerization via condensation reactions, 855

**condensation reaction** The joining together of two molecules as a third, small molecule is split out, 650–651 condensed phase. *See* **liquid; solid**

**conduction band**   A partially filled band of energy levels in a crystal through which the electrons can move easily, 839

confidence limit, A.4

configuration interaction, 600

**conjugate acid–base pair**   A Brønsted–Lowry acid and the base formed when it donates a hydrogen ion, or a Brønsted–Lowry base and the acid formed when it accepts a hydrogen ion, 315

**conjugated $\pi$-electron system**   A molecule or portion of a molecule in which double or triple bonds alternate with carbon–carbon single bonds, and electrons are delocalized across several atoms, 614–619, 616t, 636–638

**conservation of energy, law of**   In a chemical change, an equal amount of energy exists before and after the change, 4, 32–33. *See also* **first law of thermodynamics**
modification of, 490

**conservation of mass, law of**   In a chemical change, an equal quantity of matter exists before and after the change, 4, 9
modification of, 490

**constraint**   a portion of the apparatus in a thermodynamic experiment; it holds one of the properties of the system at a constant value throughout the experiment, 204

constructive interference, 544, 698–700

**contact process**   A means for the production of sulfuric acid in which sulfur dioxide is oxidized to sulfur trioxide by oxygen on a vanadium oxide catalyst, 757–759

**continuous process**   A chemical process in which reactants are fed in and products removed continuously, 729

**convection**   the net flow of one region of a fluid with respect to another region, 140

conversion of units, A.11–A.12

**coordinate covalent bond**   A covalent bond in which the shared electrons are both supplied by one of the bonded atoms; results from the interaction of a Lewis base and a Lewis acid, 316

**coordination complex**   A compound in which metal atoms or ions are bonded via coordinate covalent bonds to anions or neutral molecules that supply electron pairs, 384, 659–692. *See also* **complex ion**
in biology, 686–688
color of, 680–681
formation of, 665–670
naming of, 667–669
structure of, 670–675

**coordination number**   The number of coordinate bonds formed by a central metal atom to donor atoms in a coordination complex, 666

**coordination sphere**   The ligands coordinated to a metal ion or atom, 665

copolymer, 855

copper (Cu), 7–8, 737, A.47t
electrorefining of, 748–749
smelting of, 739–741

core [of the earth], 809

**core electron**   An electron that lies lower in energy than the valence electrons and plays only a minor role in chemical bonding, 59, 567

**correlation diagram**   A diagram that shows the relative energy of the molecular orbitals in a molecule and their derivation from the atomic orbitals of the constituent atoms, 585

**corrosion**   A destructive chemical process in which metals are spontaneously oxidized, 430–432, 773

corundum, 36p–37p

cotton, 857

Coulomb, Charles (French physicist), A.16

**coulomb (C)**   The SI unit of electric charge; equal to 1 A s, A.10t, A.16

**Coulomb energy**   Potential energy arising from electrical repulsions and attractions; $V(r) = q_1q_2/4\pi\epsilon_0 r$; 62–65, A.16–A.17

**Coulomb force**   The electrostatic attraction between particles of dissimilar charge or the repulsion between particles of similar charge, 142

**Coulomb's law**   The dependence of the force between two charges $q_1$ and $q_2$ on their separation $r$; $F = q_1q_2/4\pi\epsilon_0 r^2$, A.16

**covalent bond**   A bond formed by the sharing of two or more electrons between two atoms, 65–68. *See also* **coordinate covalent bond**
electron pair model of, 68–76
and molecular orbitals, 582–585

**covalent compound**   A compound formed from electron sharing; tends to be low-melting, low-boiling, and nonconducting, 69
naming of binary, 88, 88t

**covalent crystal**   A crystal in which structure is maintained by covalent bonds, 710–711

covalent fluoride, 797–798

**cracking**   The degradation of longer-chain alkanes to shorter-chain alkenes by means of heat or catalysts, 644

Crick, Francis (British biologist), 873

**critical point**   The point in the phase diagram at which the gas–liquid coexistence curve terminates, 154

**critical pressure ($P_c$)**   The pressure above which the distinction between gas and liquid disappears, 154

**critical temperature ($T_c$)**   The temperature above which a gas cannot be liquefied, 154

Crookes, William (British chemist and physicist), 767

**cross-linked polymer**   A polymer in which individual chains are linked together by lateral bonds that increase the rigidity of the structure, 855–857

crude oil. *See* petroleum

crust [of the earth], 809

cryolite, 746, 792

crystal
covalent, 710–711
ionic, 706–708, 714–715
liquid, 719–720
metallic, 708–710
molecular, 705–706, 713–714, 720c
symmetry of, 694–700

**crystal-field splitting energy ($\Delta_0$)**   The amount by which the (otherwise equal) energy levels for the $d$-electrons of a metal ion are split by the electrostatic field of the surrounding ligands in a coordination complex, 676

**crystal-field stabilization energy (CFSE)**   The energy-lowering in a coordination complex relative to that in a spherical crystal field, 677, 677t

**crystal-field theory**   A model for coordination complexes in which the central metal is treated as ionically bonded to surrounding ligands, 675–680

**crystal lattice**   A three-dimensional array of points that embodies the pattern of repetition in a crystalline solid, 696

crystal structure, 701–705

**crystal system**   A category (one of seven) into which a crystalline solid can be classified on the basis of the symmetry of its diffraction pattern, 696t

**crystalline solid**   A solid that possesses long-range order at the atomic or molecular level in its structure, 694–715

crystallography, 693

cubane, 633c

**cubic close packing**   A scheme of packing of equal spheres in which each sphere has 12 immediate neighbors and the symmetry of the lattice generated from the positions of the spheres is cubic, 701–702, 703t

cubic system, 696

Curie, Marie (French chemist), 8, 18

Curie, Pierre (French chemist), 18

**curie (Ci)**   A unit of activity; equal to $3.7 \times 10^{10}$ s$^{-1}$, 502

curium (Cm), 8, A.49t

Curl, Robert (American chemist), 618

current. *See* **electric current**

**cyanamide process**   A method of fixing nitrogen in the compound calcium cyanamide ($CaCN_2$), 768

cyanidin, 314c, 327

**cyclic process**   A process in which a system is taken from some initial state through a series of other states and then back to the original state, 253–256

**cycloalkane**   A saturated hydrocarbon with one or more closed rings of carbon atoms, 640–643

## Locations of Some Important Tables of Data